W9-BHV-676

PLANAR LIPID BILAYERS (BLMs)
AND THEIR APPLICATIONS

Membrane Science and Technology Series

Membrane Science and Technology Series, 7

PLANAR LIPID BILAYERS (BLMs) AND THEIR APPLICATIONS

Edited by

H.T. Tien

Department of Physiology, Michigan State University
East Lansing, MI 48824, USA

A. Ottova-Leitmannova

Department of Physiology, Michigan State University, East Lansing, MI 48824, USA

and

Center for Interface Sciences, Microelectronics Department, Faculty of Electrical Engineering & Information Technology, Slovak Technical University, 812 19 Bratislava, Slovak Republic

2003

ELSEVIER
Amsterdam - Boston - London - New York - Oxford - Paris
San Diego - San Francisco - Singapore - Sydney - Tokyo

ELSEVIER SCIENCE B.V.
Sara Burgerhartstraat 25
P.O. Box 211, 1000 AE Amsterdam, The Netherlands

First edition 2003

Library of Congress Cataloging in Publication Data
A catalog record from the Library of Congress has been applied for.

British Library Cataloguing in Publication Data
A catalogue record from the British Library has been applied for.

ISBN: 0-444-50940-2
ISSN: 0927-5193

♾The paper used in this publication meets the requirements of ANSI/NISO Z39.48-1992 (Permanence of Paper).
Printed in The Netherlands.

PREFACE

About four decades ago at a symposium on the Plasma Membrane, organized jointly by the American and New York Heart Association, Donald O. Rudin, Director of Basic Research at Eastern Pennsylvania Psychiatric Institute, in Philadelphia, and his associates reported in a paper entitled *'Reconstitution of Cell Membrane Structure in vitro and its Transformation into an Excitable System'*. As evidenced by the present volume, the rest as they say was history. The reconstituted system has been known under various names (black, bimolecular, bilayer lipid membrane, BLM for short, or simply planar lipid bilayer). Call whatever name you prefer, a conventional BLM about 5 nm thick is interposed between two aqueous solutions. It is, together with lipid vesicles (liposomes), the most widely used experimental model of biomembranes. This liquid-crystalline BLM, embodied in the lipid bilayer principle of biomembranes, and upon numerous years of development, above and beyond performing as a physical boundary, has been evolved, to serve:

- as a discriminatory barrier
- as a conduit for transport
- as a reactor for energy conversion
- as a transducer for signal processing
- as a bipolar electrode for redox reactions
- as a site for molecular recognition and/or
- other diverse functions such as apoptosis, signal transduction, etc.

To mark BLM's 40th anniversary, we have invited some of our colleagues to contribute a chapter describing their unique approach. It was impossible to invite all of the active researchers of planar lipid bilayers, past and present, to put in their work to this volume. The Editors offer their apologies and will endeavor to include other investigators' pioneering and outstanding work on planar BLMs and liposomes, not represented here, in future volumes.

The Editors

PLANAR LIPID BILAYERS (BLMs) AND THEIR APPLICATIONS

H.T. Tien and A. Ottova (Editors)

IV.   Light-induced phenomena and spectroscopy

Planar Lipid Bilayers (BLMs) and their Applications
H.T. Tien and A. Ottova-Leitmannova (Editors)

Chapter 1

The lipid bilayer concept:
Experimental realization and current applications

H. T. Tien and A. Ottova

Physiology Department, Biomedical and Physical Sciences Building
Michigan State University, East Lansing, MI 48824 (USA)

1. INTRODUCTION

It may sound far fetched but it is nonetheless true that the concept of the lipid bilayer of cell or biological membranes began with the observations of R. Hooke of Hooke's law fame, who in 1672 coined the word 'cell' to describe the array of a cork slice under a microscope he constructed. Using the microscope, Hooke discovered 'black holes' in soap bubbles and films. Years later, Isaac Newton estimated the blackest soap film to be about $3/8 \times 10^{-6}$ inch thick. Modern measurements give thickness between 5 and 9 nm, depending on the soap solution used.

Question: How does one embark from black soap films to black lipid membranes (BLMs) as models of biological membranes? Here we must go back in time to the early 1960s. A good starting point, perhaps, is the conference held in New York City.

1.1. Symposium on the Plasma Membrane

In December 1961, at the meeting sponsored by the American and New York Heart Association, when a group of unknown researchers reported the reconstitution of a bimolecular lipid membrane *in vitro,* the account was met with skepticism. Those present included some of the foremost proponents of the lipid bilayer concept, such as Danielli, Davson, Stoeckenius, Adrian, Mauro, Finean, and many others [1]. The research group led by Donald Rudin began the report with a description of mundane soap bubbles, followed by 'black holes' in soap films, etc. ending with an invisible 'black' lipid membrane, made from extracts of cow's brains. The reconstituted structure (60-90 Å thick) was created just like a cell membrane separating two aqueous solutions. The speaker then said:

"... *upon adding one, as yet unidentified, heat-stable compound... from fermented egg white.......to one side of the bathing solutions.... lowers the resistance... by 5 orders of magnitude to a new steady state.... which changes with applied potential... Recovery is prompt... the phenomenon is indistinguishable... from the excitable alga Valonia... , and similar to the frog nerve action potential...*"

As one member of the amused audience remarked, "... the report sounded like... cooking in the kitchen, rather than a scientific experiment!" That was in 1961, and the first report was published a year later [2]. In reaction to that report, Bangham, the originator of liposomes [3], wrote in a 1995 article entitled 'Surrogate cells or Trojan horses':

"... *a preprint of a paper was lent to me by Richard Keynes, then Head of the Department of Physiology* (Cambridge University), *and my boss. This paper was a bombshell... They* (Mueller, Rudin, Tien and Wescott) *described methods for preparing a membrane .... . not too dissimilar to that of a node of Ranvier... The physiologists went mad over the model, referred to as a 'BLM', an acronym for Bilayer or by some for Black Lipid Membrane. They were as irresistible to play with as soap bubbles.*"

Indeed, the Rudin group, then working on the 9th floor of the Eastern Pennsylvania Psychiatric Institute (now defunct) in Philadelphia, Pa., was playing with soap bubbles with the 'equipment' purchased from the local toyshop [4]. While nothing unusual for the researchers at work, it must have been a curious and mysterious sight for the occasional visitors who happened to pass through the corridor of laboratories there! However, playing with soap bubbles scientifically has a long, respectable antiquity, as already mentioned in the introductory paragraph.

1.2.   Origins of the Lipid Bilayer Concept

Nowadays it is taken for granted that the lipid bilayer comprises the fundamental structure of all biomembranes. The recognition of the lipid bilayer as a model for biomembranes dates back to the work of Gorter and Grendel published in 1925 [5]. However, the origin of the lipid bilayer concept is much older, and is traceable to black soap bubbles more than three centuries ago!

The early observation of 'black holes' in soap films by Hooke and Newton had a profound influence on the development of the lipid bilayer concept of biomembranes and its subsequent experimental realization in planar lipid bilayers and spherical liposomes. In this connection, there has been much discussion lately on self-assemblies of molecules, meaning the aggregation of molecular moieties into thermodynamically stable and more ordered structures. Without question, the inspiration for these exciting developments comes from the biological world, where, for example, Nature uses self-assembly, as a strategy to create complex, functional structures, such as viral protein coatings, and DNA, besides the above-mentioned *lipid bilayer* of cell membranes. Many researchers have reported self-assembling systems such as Langmuir-Blodgett monolayers and multi-layers [6]. A broad list of man-made, self-assembling systems involving amphiphilic molecules is given in Table 1.

Table 1
Experimental self-assembling interfacial amphiphilic systems

System	Interfaces
1. Soap films	air | soap solution | air
2. Monolayers	air | molecular layer | water
3. Micelles (water-in-oil)	water | monolayer | oil
Micelles (oil-in-water)	oil | monolayer | water
4. Multilayers	air | molecular layers | water
5. Bilayers (BLMs and Liposomes)	water | BLM | water
6. Nuclepore supported BLMs	water | BLM | water
7. Gold supported Monolayers	air | molecular layers | gold
8. Metal supported BLMs	water | lipid bilayer | metal
(s-BLMs)	
9. Salt-bridge supported BLMs	water | lipid bilayer | hydrogel
(sb-BLMs)	
10. Tethered BLMs (t-BLMs)	gold | SH-BLM | water

Where the vertical line | denotes an interface.

1.3.   Early experimental evidence for the lipid bilayer

Prior to the seminal work of Gorter and Grendel on the red blood cells (RBC), several pertinent questions had been raised:

- *Was there a plasma membrane (Pfeffer, 1877)?*
- *What was the nature of the plasma membrane (Overton, 1890)?*
- *What is the meaning of membrane potential (Bernstein, 1900)?*
- *How thick is the plasma membrane (Fricke, 1925)?*
- *What is the molecular organization of lipids in the plasma membrane (Gorter and Grendel, 1925)?*

Before answering the last posed question, let us digress for the moment to a different topic, namely, interfacial and colloid chemistry. It is no exaggeration that modern investigation of interfacial and colloid chemistry began in the kitchen sink! Over a span of 4 years (1891-1894), Agnes Pockels (a house wife working over the kitchen sink) reported in *Nature* (see Ref. [7]) how surface films could be enclosed by means of physical barriers to about 20 A^2/molecule. To be more accurate historically, the first landmark is generally bestowed to Benjamin Franklin, who in 1774 demonstrated that a teaspoonful of oil (olive ?) is able to calm a half-acre (~ 2000 m^2) surface of a pond. Pockels' observation was followed about a quarter of century later by the quantitative investigation of Langmuir, using a setup now known as Langmuir's film balance. Langmuir deduced the dimensions and structures of fatty acid molecules at the air-water interface. Our current understanding of the structure and function of biomembranes can be traced to the studies of experimental interfacial systems such as soap bubbles and Langmuir monolayers, which have evolved as a direct consequence of applications of classical principles of colloid and interfacial chemistry [6]. Accordingly, interfaces play a vial role in molecular biology, since most ligand-receptor interactions occurring at surfaces and interfaces. In the 19th century, the first observations were made that surfaces control biological reactions. The challenge is to fully develop the biological model for surface/interface science in the highly complex and interactive *in vivo* biological setting. Concerning these, surface and colloid scientists have known for decades that amphiphilic compounds such as phospholipids can self-organize or self-assemble themselves into supramolecular structures of emulsions, micelles, and bilayer lipid membranes (planar BLMs and spherical liposomes). Thermodynamics favors self-assembly, if the molecules of an amphipathic phospholipid (e.g. phosphatidylcholine, PC) are in water, the hydrocarbon chains will want to be 'away' from the environment. They could all go to the top (like oil on water), or they could have the hydrocarbon chains point toward each other. With the hydrocarbon chains pointing toward each other, amphiphilic molecules of phospholipids could adapt at least three different configurations (see Fig. 1). The three ways in which they do so are to form small spherical micelles, liposomes or to form a lamellar BLM. A micelle is a small structure, generally less than two μM in diameter, while a free-standing BLM typically has a thickness of about 5 nm and can have an area of several square millimeters or more. Mechanical treatment or sonication of lipids can produce spherical BLMs (i.e. liposomes or lipid vesicles), which contain aqueous solution enclosed within a lipid bilayer. Liposomes are considerably larger than micelles, ranging from tenth of μM to cm in diameter [3].

Studies of such BLMs have given us much insight into the functions and operation of biomembranes in living cells. Unmodified BLMs are highly permeable to water molecules, while ions such as Na^+ and K^+ can traverse them hardly at all (by nine orders of magnitude lower!). As shown in Fig. 1, a micelle may be depicted as a ball with the lipid polar groups on the outside and the hydrocarbon chains pointing together, as in w/o (water-in-oil) micelles, and vice versa as in o/w (oil-in-water) micelles. Interestingly, micelles are generally not formed by phospholipids in aqueous solutions, since the fatty acyl hydrocarbon chains in phosphoglycerides (PG), sphingomyelins (SM), and all glycolipids (GL), are too large to fit into the interior of a micelle. Thus, naturally occurring phospholipids have a rather low cmc (critical micelle concentration); it is on the order of 5×10^{-4} µM. The arrangement of atoms and molecules at the surface or interface of a substance differs from that in the bulk, owing to their surroundings.

Gibbs, Laplace and Young formulated the basic principles of interfacial chemistry. Specifically Gibbs articulated the thermodynamics of interfaces, whereas Young and Laplace introduced the concept of surface tension. It is thus useful in many fundamental and practical applications such as molecular films, biomembranes, and lipid-based biosensors to have some knowledge concerning the behavior of substances at the interface. The nature of interfacial phenomena is inherently interdisciplinary involving physics, chemistry, biology, and engineering [6].

Now, answering the question posed earlier: *"What is the molecular organization of lipids in the plasma membrane?"* Shortly after Langmuir's 1917 paper showed that his simple apparatus could provide the data to estimate the dimensions of a molecule, Gorter, a pediatrician and Grendel, a chemist, determined the area occupied by lipids extracted from red blood cell (RBC) ghosts. They found that there were enough lipids to form a layer two molecules thick over the whole cell surface, that is

$$\frac{\textit{Surface area occupied (from monolayer experiment)}}{\textit{Surface area of RBC (from human, pig or rat source)}} \cong 2 \qquad (1)$$

On this basis, Gorter and Grendel suggested that the plasma membrane of RBCs might be thought of as a bimolecular lipid leaflet [5]; they wrote

*"... of which the polar groups are directed to the inside and to the outside, in much the same way as Bragg supposes the molecules to be oriented in a 'crystal' of fatty acid, and as the molecules of a **soap bubble** are according to Perrin."*

Hence, the concept of a lipid bilayer as the fundamental structure of cell membranes was born, and has ever since dominated our thinking about the molecular organization of all biomembranes!

From the 1930s onward, research on biomembranes diverged into two major directions: (i) the further elaboration of the lipid bilayer concept using increasingly sophisticated physical chemical techniques including x-ray diffraction and electron microscopy, and (ii) the invention of model membrane systems so that physical, biochemical, and physiological processes may be isolated and analyzed in molecular terms. The former approach resulted in many insightful cartoons, prominently among these are the classic model of Harvey, Danielli and Davson, the spherical micelle model of Sjostrand and of Lucy, the various other models proposed by Benson, Lenard, Singer and Nicolson, and by Green, as well as the Chop-Suey (Smorgasbord) model [6,7]. Some of these models are in essence an elaboration of the 'unit membrane' hypothesis of Robertson, which led eventually to the so-called fluid mosaic model of biomembranes. The other direction of research culminated in the reconstitution of lipid bilayers *in vitro* [2,3], as described in the following section.

2.   EXPERIMENTAL REALIZATION OF THE BILAYER LIPID MEMBRANES

How did the pioneering researchers get the idea in 1960 to form the first experimental bilayer lipid membrane? In short, the early investigators conceived the idea by playing with soap bubbles and films whose molecular organization provided the key! What follows is a personal account that might be of some historic interest [1-7].

In the late 1950s, while a research group headed by Donald O. Rudin, was investigating the ion specificity of lipid monolayers and multilayers (the Langmuir-Blodgett type), two refreshing publications appeared, which exerted great influence on the thinking of Rudin and his associates, and altered the course of their research. The first was a reprint of C.V. Boys' classic book on soap bubbles. The second was a volume dedicated to N.K. Adam, in which A.S.C. Lawrence recounted succinctly the highlights in the development of monolayers, soap films, and colloid chemistry. Lawrence's account in particular brought into focus the relationship among the various topics. Most significant was the mention of Newton's observation of the black soap films (described in detail in Boys' book). The confirmation of the thickness of black soap films measured by several investigators, including Perrin and Dewar, since Hooke and Newton's time, and the bimolecular leaflet model for the plasma membrane suggested independently by Gorter and Grendel, and by

Fricke in the 1920's, which was then being corroborated by numerous investigations. The structure of black soap films led to the realization by Rudin and his co-workers in 1960 that a soap film in its final stages of thinning has a structure comprised of two fatty acid monolayers sandwiching an aqueous solution. With the above background in mind, Rudin and his associates simply proceeded to make a 'black lipid bubble' under water; their effort was successful [2-4]. As far as forming a 'black', bimolecular or bilayer lipid membrane (BLM) is concerned, there is not a lot to it. It is worth stating, however, that this experimental realization drew on three centuries of annotations! The preceding account, in essence, is the historical origin of the bilayer lipid membrane (dubbed BLM).[*]

Rudin and his colleagues showed that, an under water 'lipid film', or a BLM formed from brain extracts was self-sealing to puncture with many physical and chemical properties similar to those of biomembranes. Upon modification with a certain compound (called EIM, Excitability-Inducing-Molecule), this otherwise electrically 'inert' structure became excitable, displaying characteristic features similar to those of action potentials of the nerve membrane. An unmodified lipid bilayer separating two similar aqueous solutions, on the order of 5 nm in thickness, is in a *liquid-crystalline* state (viscosity $\sim$ 1 centipoise), and possesses the following electrical properties: membrane potential ($E_m \simeq 0$), membrane reisitance ($R_m \simeq 10^9$ Ω cm^2), membrane capacitance ($C_m \simeq 0.5\text{-}1\mu F$ cm^{-2}), and dielectric breakdown (V_b > 250,000 volts/cm). In spite of its very low dielectric constant ($\varepsilon \simeq 2\text{-}7$), this liquid-crystallline BLM is surprisingly permeable to water (8-24 μm/sec) [6].

Functionally speaking, in terms of the five traditional senses (i.e. seeing, smell, taste, touch, and hearing), almost everything we know about the world comes to us via the cell membrane, which is made of a lipid bilayer with receptors embedded in it. Indeed, it was through the investigation of ion selectivity and specificity of the nerve membrane that the pioneering researchers came upon the idea of forming the bimolecular (bilayer or black) lipid membrane *in vitro*. The BLM system, as evidenced by the unique chapters included in this book, has since been widely used for investigations into a variety of physical, chemical, and biological phenomena, including membrane reconstitution, molecular biology, bio-medical research, solar energy transduction, and biosensors development. The use of experimental bilayer lipid membranes (planar lipid bilayers and spherical liposomes), particularly planar BLMs, will be discussed in more detail in this book.

For those who are interested in the spherical BLMs (i.e., the liposome system), an autobiographical account by its originator is available [3].

---

[*]A short movie showing the formation of BLMs made in the 1960s is available on the Internet. URL: http://www.msu.edu/user/ottova/soap_bubble.html

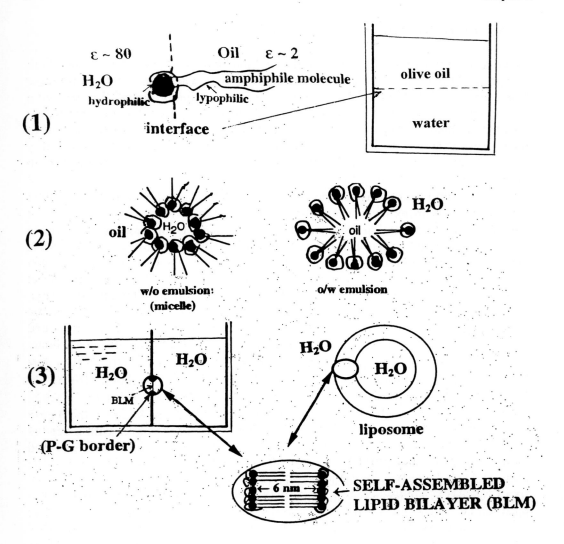

Figure 1 Amphipathic molecular structures at interfaces. (1) Oil-water interface (2) Emulsions: water-in-oil and oil-in-water (3) Bilayer lipid membrane (planar BLM and liposome); also shown is an enlarged view of a self-assembled bilayer lipid membrane structure separating two aqueous solutions.

3.   THE LIPID BILAYER PRINCIPLE

In spite of the variable compositions, the fundamental structural element of all biomembranes is a liquid-crystallline phospholipid bilayer. Thus, the lipid bilayer principle of cell or biological membranes may be summarily stated that all living organisms are made of cells, and every cell is enclosed by a plasma membrane, the indispensable component of which is a lipid bilayer. The most pivotal function of the lipid bilayer membrane is that it separates the environment by a permeability barricade that allows the cell to preserve its identity, take up nutrients and to remove waste. This 5 nm thick liquid-crystallline lipid bilayer serves not just as a physical barrier but also as a conduit for transposrt, a reactor for energy conversion, a transducer for signal processing, a bipolar electrode for redox reactions or as a site for molecular recognition.

In eukaryotic cells, there are also organelles (e.g. mitochondria, chloroplasts, endoplasmic reticulum, etc.) that are confined by their own bilayer lipid membranes. Some of these membranes are the plasma membrane of cells, the thylakoid membrane of the chloroplast, the cristae membrane of the mitochondrion, the visual receptor membrane of the eye, and the nerve membrane of the axon. Depending on cells, biomembranes are made up of different combinations and quantities of phospholipids and glycoproteins. Succinctly, the membrane architecture is that of a lipid bilayer; it is an assembly of lipid, protein, and carbohydrate molecules in various proportions, and held together by noncovalent interactions. The lipid bilayer, as its name implies, consists of two molecules of lipids, in particular phospholipids. Phospholipids belong to a class of amphiphilic (amphipathic) compounds that have a hydrophilic polar moiety at one end of the molecule and a hydrophobic hydrocarbon chain at the other. The polar group has a variable region that differs between the various phospholipids. Typically, phospholipids have two fatty acid chains esterified to a glycerol molecule and on the third glycerol hydroxyl there is a phosphate group that is associated with a hydrophilic moiety such as choline, (i.e. phosphatidylcholine, PC), phosphatidylethanolamine (PE), phosphatidylserine (PS), or phosphatidylinositol (PI). The acyl chains of fatty acids can vary both in chain length (usually 16, 18, or 20), and saturation, of which one is saturated (without double bonds) and the other unsaturated, having one or more double bonds. In general, all such double bonds are of the *cis* configuration. A *cis* double bond introduces a rigid kink in the otherwise flexible straight chain of a fatty acid. When a suspension of liposomes or a planar BLM composed of a single type of phospholipid is heated, it undergoes an abrupt change in physical properties over a very narrow temperature range. This phase transition is due to increased motion about the C-C bonds of the fatty hydrocarbon chains, which pass from a highly ordered, gel-like state to a more mobile liquid state. During the gel to liquid transition, a relatively large amount of heat is absorbed over a narrow temperature range, which is the melting temperature of the liquid-crystallline bilayer [6].

For mammalian cells, cholesterol, a small molecule compared to phospholipids, is usually present. Cholesterol gives the membrane constancy, by preventing crystallization of hydrocarbons and phase shifts, and its proportion to other molecules determines rigidity of the membrane. For bacteria containing no cholesterol, such cells need a cell wall for maintaining structural integrity. It is important to note that the membrane molecular make-up of a lipid bilayer is usually *not* symmetric. For example, the plasma membrane of red blood cells (RBC or erythrocytes) consists of neutral phospholipids (except the glycolipids) facing the surrounding (e.g. PC), whereas PS and PI are on the cytoplasmic side of the lipid bilayer (*see* Ref. [6]). This lipid bilayer asymmetry is vital in cell functions, as will be considered later (Section 3.2.1. see also Fig. 2).

3.1.   Membrane-solution interfaces

To appreciate the lipid bilayer principle of biomembranes, it is informative to consider the earliest model membrane experiments of Overton, who used an oil-water interface for penetration studies. At the interface between hydrocarbon and water (e.g. extra virgin olive oil and water, its interfacial tension, $\gamma \sim 50$ ergs/cm^2), there is a higher energy state for the water; it is energetically favorable for the hydrocarbon (HC) to associate with HC and to minimize the surface area of contact with water, because hydrogen bonds are forfeited. In terms of free energy change, the mixing of HC and water causes a decrease in entropy (i.e., the entropy of water is decreased at the interface between HC and water). Of particular interest to note here is that natural phospholipids, through trial-and-error, have nearly equal areas of hydrophilic and hydrophobic moieties (~ 0.5 nm^2/lipid molecule). This deliberate design of nature is of far-reaching consequence; it is in fact the physical basis of the lipid bilayer formation, for the fatty acids of phospholipids can stabilize HC in water by covering the hydrocarbon surface with their hydrophilic regions, while their hydrocarbon regions associate with HC (Fig. 1). The end result is an extremely low interfacial tension ($\gamma \sim 0.1$ ergs/cm^2) between phospholipid molecules and aqueous solution, and can be explained by the so-called hydrophobic effect, i.e. the inability of hydrocarbons to hydrogen bond with water. Thus, the major driving force for the formation of lipid bilayers is hydrophobic interaction between the fatty acyl chains of phospholipid and/or glycolipid molecules. Van der Waals interactions among the HC chains favor close packing of these hydrophobic moieties. Hydrogen bonding and electrostatic interactions between the polar head groups and water molecules also stabilize the bilayer. Perhaps, this state of affairs is best depicted as follows:

$$\text{aq. soln} \mid \text{polar-moiety-(HC} \sim \text{lipid bilayer} \sim \text{HC)-polar moiety} \mid \text{aq. soln} \qquad (2)$$

where aq. soln. stands for aqueous solution, HC~lipid bilayer~HC = hydrocarbon chains of the lipid bilayer, and | denotes an interface.

3.1.1. *Surfactants and lipid rafts (see Chapter 22)*

In concluding this section on membrane-solution interfaces, mention should be made on surfactants (also known as detergents) in BLM research. As far back as in 1967, detergents, together with cholesterol, were used to form BLMs with interesting properties [8]. Evidence is accumulating that microdomains exist in cell membranes (also referred to as rafts or detergent resistant membranes).   For example, Rinia et al [9] recently used atomic force microscopy to study supported lipid bilayers, consisting of a liquid-crystallline phosphatidylcholine (PC), sphingomyelin and cholesterol. The presence of cholesterol was found to induce bilayer coupling. In addition, Rinia and associates were able to directly visualize the resistance of domains (rafts) against non-ionic detergent. Today, lipid rafts (domains) are currently an intensely investigated topic of cell biology, owing to its demonstrated role in signal transduction of the host cell. As reviewed by London and Brown [10], the insolubility of lipids in detergents is an useful method for probing the structure of biomembranes. Insolubility in detergents like Triton X-100 is observed in lipid bilayers that exist in physical states in which lipid packing is tight. The Triton X-100-insoluble lipid fraction obtained after detergent extraction of eukaryotic cells is composed of detergent-insoluble membranes rich in sphingolipids and cholesterol. These insoluble membranes appear to arise from sphingolipid- and cholesterol-rich membrane rafts in the tightly packed liquid ordered state. Because the degree of lipid insolubility depends on the stability of lipid-lipid interactions relative to lipid-detergent interactions, the quantitative relationship between rafts and detergent-insoluble membranes is complex, and can depend on lipid composition, detergent and temperature. Nevertheless, when used conservatively, detergent insolubility is an invaluable tool for studying cellular rafts and characterizing their composition. Regarding these, Campbell, Crowe and Mak [11] reviewed lipid rafts that serve as entry and exit sites for microbial pathogens and toxins (such as FimH-expressing enterobacteria, influenza virus, measles virus and cholera toxin). Furthermore, caveolae, a specialised form of lipid raft, are required for the conversion of the non-pathogenic prion protein to the pathogenic scrapie isoform. Concerning Triton X-100, Gincel, Silberberg and Shoshan-

Barmatz [12] reported purification of voltage dependent anion channel (VDAC) from sheep brain synaptosomes or rat liver mitochondria, using a reactive red-agarose column, in addition to the hydroxyapatitate column. The red-agarose column allowed further purification (over 98%), concentration of the protein over ten-fold, decreasing Triton X-100 concentration, and/or replacing Triton X-100 with other detergents, such as Nonidet P-40 or octylglucoside. This purified VDAC, reconstituted into planar lipid bilayer, had a unitary maximal conductance of 3.7 ± 0.1 nS in 1 M NaCl, at 10 mV and was permeable to both large cations and anions. In the maximal conducting state, the permeability ratios for Na^+, acetylcholine (+), dopamine (+) and glutamate (-), relative to Cl^-, were estimated to be 0.73, 0.6, 0.44, and 0.4, respectively. In contrast, in the subconducting state, glutamate (-) was impermeable, while the relative permeability to acetylcholine (+) increased and to dopamine (+) remained unchanged. At the high concentrations (0.1- 0.5 M) used in the permeability experiments, glutamate eliminated the bell shape of the voltage dependence of VDAC conductance. Glutamate at concentrations of 1 to 20 mM, in the presence of 1 M NaCl, was found to modulate the VDAC activity. In single-channel experiments, at low voltages ($\pm$ 10 mV), glutamate induced rapid fluctuations of the channel between the fully open state and long-lived low-conducting states or short-lived closed state. Glutamate modification of the channel activity, at low voltages, is dependent on voltage, requiring short-time (20-60 sec) exposure of the channel to high membrane potentials. The effect of glutamate is specific, since it was observed in the presence of 1 M NaCl and it was not obtained with aspartate or GABA. These results suggest that VDAC possesses a specific glutamate-binding site that modulates its activity, according to Gincel et al [12].

In the article of Campbell et al [11] mentioned above, the authors provide a brief overview of the role of membrane-associated lipid rafts in cell biology, and to evaluate how HIV-1 has taken over this cellular component to support HIV-1 replication. Like a number of other pathogens, HIV-1 has evolved to rely on the host cell lipid rafts to support its propagation during multiple stages of the HIV-1 replication cycle. A number of reports have shown, directly or indirectly, that lipid rafts are important at various stages of the human immunodeficiency virus type-1 (HIV-1) replication cycle.

3.2.   An idealized cell

The liquid-crystallline lipid bilayer of biomembranes is most unique in that it not only provides the physical barrier for separating the cytoplasm from its extracellular surroundings; it also separates organelles inside the cell to protect important processes and events. More specifically, the lipid bilayer of cell membrane must keep its molecules of life (genetic materials and its variety of proteins) from scattering away. At the same time, the lipid bilayer must keep out foreign molecules that are harmful to the cells. To be viable, the cell must also communicate with the environment to continuously monitor the external conditions and adapt to them. Further, the cell needs to pump in nutrients and release toxic products of its metabolism. How does the cell carry out all of these multi-faceted activities? A brief answer is that the cell depends on its lipid-proteins-carbohydrate complexes (i.e., glycoproteins, proteolipids, glycolipids, etc) embedded in the lipid bilayer to gather information about the environment in various ways. Some examples are as follows: communication with hundreds of other cells concerning a variety of vital tasks such as growth, differentiation, and death (apoptosis). Glycoproteins are responsible for regulating the traffic of material to and from the cytoplasmic space. Paradoxically, the intrinsic structure of cell membranes creates a bumpy obstacle to these vital processes of intercellular communication. The cell shields itself behind its lipid bilayer, which is virtually impermeable to all ions (e.g. Na^+, K^+, Cl^-) and most polar molecules (except H_2O). This barricade of the lipid bilayer prevents the free interchange of materials between inside and outside of the cell. This barrier must be overcome, however, one way or another in order for a cell to inform itself on what is happening in the world outside, as well as to carry out vital functions. Thus, upon countless years of evolution, the liquid-crystalline lipid bilayer, besides acting as a physical restraint, has been modified to serve:

- as a discriminatory physical barrier (e.g. for polar and non-polar species)
- as a conduit for transport (e.g. ion channels and carriers)
- as a reactor for energy conversion (e.g. light-induced charge generation)
- as a transducer for signal processing (e.g. chemical to electrical or vice versa)
- as a bipolar electrode for redox reactions (e.g. one side serves as cathode and the other as anode), or
- as a site for molecular recognition (e.g. ligand-receptor interactions), and
- other diverse functions such as apoptosis, signal transduction, etc.

A good way to consider the lipid bilayer principle of biomembranes, perhaps, is to begin with an overview of a highly idealized cell shown in Fig. 2. The figure illustrates schematically the plasma membrane of an idealized eukaroytic cell, together with its organelle membrane systems. It may be seen that, even in this greatly simplified drawing, the plasma membrane is an extremely complex and elaborate supramolecular structure. The most important feature is that the plasma membrane of cells as consisting of a two-dimensional liquid-crystalline lipid bilayer in which various functional entities (e.g., enzymes, receptors, channels) are embedded. In order to comprehend and to explain the structure and function of

Captions for Fig. 2 (facing page)

(1)   **The Lipid Bilayer**: A universal element of all cell membranes. The asymmetry of membrane lipids is established during their biosynthesis and maintained throughout the lipids lifetime. In both natural and artificial lipid bilayers, a typical lipid molecule exchanges places with its neighbors in a leaflet about 10^7 times per second and diffuses several micrometers per second at 37°C. As a result of thermal motion, phospholipid and gycolipid molecules are known to rotate freely around their long axes and to diffuse laterally within the bilayer leaflet. Because such movements are rotational or lateral, the fatty acyl chains remain in the hydrophobic interior of the bilayer.
      In general, lipids with short or unsaturated fatty acyl chains undergo phase transition at lower temperatures than lipids with long or saturate chains.

(2)   **Membrane Proteins**: Manners by which they are associated with the lipid bilayer: (a) span the bilayer, (b) partially immerse in the bilayer, (c) held by non-covalent interactions with other proteins, and (d) attach to fatty acid chains by anchoring the protein in one surface of the bilayer. *All Integral Proteins Bind Asymmetrically to the Lipid Bilayer.* Each type of integral membrane protein has a single, specific orientation with respect to the cytosolic and exoplasmic faces of a cellular membrane. For example, an integral membrane protein, such as glycphorin, lie in the same direction. This absolute asymmetry in protein orientation give the two membrane faces their different characteristics. Membrane asymmetry is most obvious in the case of membrane glycoproteins and glycolipids. In the plasma membrane, all the O- and N-lined oligosacharides in glycoproteins and all of the oligosaccharides in glycolipids are on the exoplasmic surface. In the endoplasmic reticulum, they are found on the interior, or lumenal, membrane surface.

(3)   **Permeability and Transport**: Passive transport occurs either by simple diffusion or by facilitated diffusion. Active transport needs an input of the metabolic energy.

(4)   **The Electrochemical Potential**: μ, The driving force for membrane transport, oxidative- and photo-phosphorylation.

(5)   **The Na^+ - K^+ Pump**: Virtually all animal cells contain a Na^+/K^+ pump that operates as an antiport, actively pumping K^+ into the cell and Na^+ out against their respective electrochemical potential gradients.

(6)   **Ion Channels**: There are ligand-gated and voltage-gated channels made of proteins, which open in response to ligand interaction and voltage change. For example, a Ca^{2+} channel may be depicted as a port with negatively charged sites, to act as a 'selectivity filter' to distinguish among different cations, with voltage-sensors conferring voltage dependence on channel opening and closing.

(7)   **The Nucleus**: Where genetic material (DNA) is located. Its nuclear membrane has numerous pores (up to ~ 80 nm in diameter) which probably transport large particles selectively.

(8)   **Energy-transducing Organelle Membranes**: The mitochondrion converts foods into usable energy (ATP), whereas the chloroplast, in plants and other photosynthesizing organisms, transduces electromagnetic radiation (solar energy) into other forms of energy.

(9)   **Experimental Bilayer Lipid Membranes**: Since their inception in the early 1960s, such lipid bilayer systems, either in the form of a planar BLM or of a vesicular liposome, have been used extensively as models of biomembranes. Liposomes are spherical vesicles up to 1 cm or more in diameter consisting of a phospholipid bilayer that encloses a central aqueous compartment. Planar bilayer are formed across a hole in a partition that separates two aqueous solutions. The advantage of the planar lipid bilayer (BLM) system is that both sides of the membrane can be easily altered and probed by electrodes. For long-term investigations and practical applications, planar BLMs can also be formed on either gel or metallic supports.

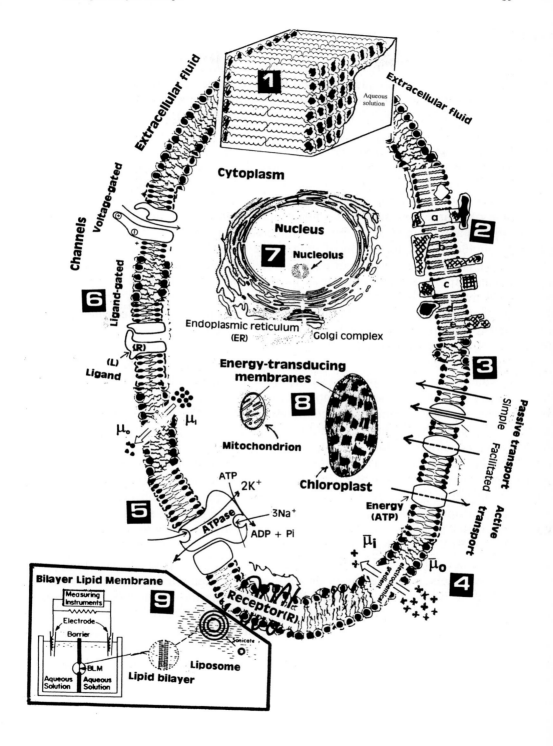

Extracellular fluid

Extracellular fluid

Channels

Voltage-gated

Ligand-gated

(R)

(L)
Ligand

μ_o μ_i

Cytoplasm

Nucleus

Nucleolus

Endoplasmic reticulum (ER)

Golgi complex

Energy-transducing membranes

Mitochondrion

Chloroplast

Aqueous solution

a

c

Passive transport
Simple
Facilitated

Active transport

Energy (ATP)

$\overline{\mu}_i$

μ_o

electrochemical gradient

ATP
2K$^+$

3Na$^+$

ATPase

ADP + Pi

Receptor (R)

Bilayer Lipid Membrane

Measuring Instruments

Electrode

Barrier

Aqueous Solution   Aqueous Solution

BLM

Lipid bilayer

Liposome

Sonicate

natural cell membranes in physical and biochemical terms, resorting to model systems has a long tradition in biosciences. Thus, insofar as experimental model membranes are concerned, there are planar bilayer lipid membranes and spherical liposomes, both of which are also depicted in Fig. 2. It is beyond the scope of this chapter to discuss what is shown in the figure other than pointing out the highlights as given in the caption of Fig. 2 (*see* Ref. [6] for more details).

3.2.1. Membrane potentials and transport

Since almost all interfaces are electrified, the key concept to describe these electrified interfaces is the electrochemical potential, an extension of which leads to the Nernst-Planck equation -- the crux of lipid bilayer electrochemistry [6,13]. Separation of charges is usually as a result of: (a) preferential adsorption of ions, (b) adsorption and orientation of dipolar molecules, (c) transfer of charge from one phase to another, (d) deformation of polarizable atoms or molecules (anisotropy, asymmetry), and (e) the presence of electrical field. Hence, these separated charges may be ions, dipolar molecules such as phospholipids; they may be free charges such as electrons and 'holes'.

At neutral pH, some phosphoglycerides (PC, PE) have no net electric charge; others (PS, phosphatidylglycerol, cardiolipin) have a net negative charge. Nonetheless, the polar head groups in all phospholipids can pack together into the characteristic bilayer structure. At other pHs, there is a very high density of negative charges on the cytoplasmic surface (0.5-0.8 charge/nm^2). Thus, it appears that the lipid bilayer asymmetry is important to the cell for transport and signaling. For example, energy-dependent concentration of PS has been observed and is a critical component in keeping cell viability, i.e. cell death (apoptosis) occurs when lipid bilayer asymmetry is lost.

Insofar as membrane transport is concerned, cells make use of three approaches: simple diffusion, facilitated diffusion, and active transport. Though simple diffusion is an effective transport mechanism for some substances such as water, the cell must make use of other mechanisms for moving substances in and out of the cell. Facilitated diffusion utilizes membrane channels to allow charged molecules, which otherwise could not diffuse across the lipid bilayer. These channels are especially useful with small ions such as K^+, Na^+, and Cl^-. The number of protein channels available limits the rate of facilitated transport, whereas the speed of simple diffusion is controlled by the concentration gradient. Concerning active transport, the expenditure of energy is necessary to translocate the molecule from one side of the lipid bilayer to the other, in opposition to the concentration gradient. Similar to facilitated diffusion, active transport is limited by either the capacity of membrane channels or the number of carriers present.

Further, active transport may be divided into two major categories: primary active transport involves using energy through ATP hydrolysis (e.g. the Na^+-K^+ pump) and secondary active transport involves using energy to establish a gradient across the cell membrane, and then utilizing that gradient to transport a molecule of interest up its concentration gradient (e.g. *E. coli*). In this case, *E. coli* establishes a pH gradient of H^+ ions across the cell membrane by using energy to pump protons out of the cell, which is then coupled to the lactose permease that uses the energy of the proton moving down its concentration gradient to transport lactose into the cell. A similar mechanism is used in the Na^+-glucose secondary transport. It is worth noting here that Romano-Fontes and associates [14] reported the transport of palmitic acid (PA) across planar lipid bilayers using a high specific activity ^{14}C palmitate as tracer for PA. An all-glass trans chamber was employed in order to minimize adsorbance of PA onto the surface. Electrically neutral (diphytanoyl PC) and charged (azolectin) planar bilayers were maintained at open electric circuit. The authors found a permeability to PA of 8.8 ± 1.9 µm/s in neutral and of 10.3 ± 2.2 µm/s in charged BLMs. These values fall within the order of magnitude of those calculated from desorption constants of PA in different vesicular systems. Differences between data obtained from planar and vesicular systems are discussed in terms of the role of solvent, radius of curvature, and pH changes [14].

3.2.2. Membrane Channels

The liquid-crystalline lipid bilayer plays a vital role in a number of life's processes and serves the functional environment for other membrane constituents such as proteins that include receptors and ion channels. Concerning the latter, ion channels are of crucial importance in organisms; they are vital for living and defense as well as for attack, and have been investigated for decades in reconstituted bilayer lipid membranes (planar BLMs and spherical liposomes). Evidence for single ion-channel activities in biomembranes were, in fact, first reported in BLMs in 1969. The structure and function of ion channels (pores) are well established in all sorts of biomembranes. For example, through these channels ions flow across the lipid bilayer and depolarize or hyperpolarize the membrane potential. There are three main types of ion channels: (a) non-gated, (b) voltage gated such as Na^+, K^+, Ca^{2+}, and Cl^-, and (c) ligand gated channels including ACh, GABA, glutamate, and glycine channels. Ion channels are macromolecular protein complexes embedded in the lipid bilayer. They are divided into distinct protein entities called subunits, each of which has a specific function and is encoded by a different gene [6,7].

Today, ion-channels are found ubiquitously. To name a few, they are in the plasma membrane of sperms, bacteria and higher plants, the sarcoplasmic reticulum

(SR) of skeletal muscle, synaptic vesicle membranes of rat cerebral cortex, and in the skin mucus of carps. As a weapon of attack, many toxins released by living organisms are polypeptide-based ion-channel formers such as dermonecrotic toxin, hemolysin, brevetoxin, and bee venom. Functioning of membrane proteins, in particular ionic channels can be modulated by alteration of their arrangement in membranes. For example, Krylov and colleagues [15] earlier addressed this issue by studying the effect of different chain length polylysines on the kinetics of ionic channels formed in a bilayer lipid membrane, using O-pyromellitylgramicidin carrying three negative charges at the C-terminus. The method of sensitized photoinactivation was applied to the analysis of the channel association-dissociation kinetics (characterized by the exponential factor of the curve describing the time course of the flash-induced decrease in the transmembrane current, τ). Addition of polylysine to the bathing solutions of BLM led to the deceleration of the photoinactivation kinetics, i.e. to the increase in τ. It was shown that for a series of polylysines differing in their chain lengths, the value of τ grew as their concentration increased above a threshold level until at a certain concentration of each polylysine, τ reached maximum. The increase in the ionic strength of the medium shifted the concentration dependence of τ to higher polylysine concentrations and decreased the maximum value of τ. Krylov et al [15] concluded that the increase in τ was caused by the formation of domains of O-pyromellitylgramicidin molecules induced by binding of polylysines.

With the availability of the patch-clamp technique, the presences of ion-channels have been established in all sorts of biomembranes, some of which have already been mentioned above. It has been assumed from the earlier years of BLM research that membrane-bound channels are polypeptides of proteinaous materials [6]. Typically, a simple picture concerning ion-channels in BLMs is that of an array of cylindrical structures bearing fixed charges and filled with aqueous solution. If so, electrokinetic phenomena based on the electrical double layer theory should be expected and have been actually reported [7,13]. In this regard is the work of Egorova [16] who has made critical remarks concerning the disagreement between experimental data and the Gouy-Chapman theory predictions. Experimental dependences of electrophoretic mobilities and zeta-potentials are on (i) the electrolyte concentration for latexes, lipid and cell membranes, (ii) zeta-potentials of lipid membranes on the fixed surface charge density at a given ionic strength, (iii) the surface potential ($\Delta\psi$) jump of lipid monolayers on the ionic strength of aqueous subphase, and (iv) the surface potential difference of $\Delta\psi$'s of planar lipid bilayer on the ionic strength of water solution at one side of a membrane. Many comprehensive reviews on membrane channnels and closely related BLM systems are available [6,7, *see* also Chapters 13,21,22]. Below, we will consider the effect of electrical field across the lipid bilayer, which is of major interest to both basic and practical applications of the BLM.

3.2.3 Electroporation

Transmembrane voltage pulses between 200 mV to 1 V are known to create transient 'pores' in the lipid bilayer, with a diameter about 2 nm. This phenomenon, termed electroporation, has been extensively investigated by Neumann, Tsong, Zimmermann, Weaver, and others [7,17-19]. Electroporation occurs following electric pulses up to 10^6 V/cm with duration between μsec and msec to membranes in close contact and is believed to initiate primarily in the lipid bilayer [17]. One of practical applications is cell transfection for gene expression. Other applications include encapsulation of drugs in controlled-release and insertion of proteins in living cells. For instance, Tung, Troiano et al [18] reported the effects of the channel-forming peptide gramicidin D (gD) on the conductance and electroporation thresholds of BLMs. BLM conductance before electroporation was measurable only after the addition of gD and increased monotonically as the peptide/lipid ratio increased. It seems likely that the presence of membrane proteins affect the electroporation of the lipid bilayer by changing its mechanical properties. Transport of ions such as Na^+, K^+, Cl^- through membrane pores discharge the membrane potential, and at times an external pulse of sufficient amplitude and duration tends to cause dielctric breakdown of the lipid bilayer. Molecular transport through primary pores and pores enlarged by secondary processes provides the basis for transporting molecules into and out of cells. Tissue electroporation, by longer, large pulses, is involved in electrocution injury. Tissue electroporation by shorter, smaller pulses is under investigation for biomedical engineering applications of medical therapy aimed at cancer treatment, gene therapy, and transdermal drug delivery.

To better understand the kinetics of membrane permeabilization, Bier et al. [19] reported that the plasma membrane by electroporation can be either transient or stable. Although the exact molecular mechanics have not yet been described, Bier and associates sought to determine the time constants for spontaneous transient pore sealing. By using isolated rat skeletal muscle cells and a two-compartment diffusion model, Bier and colleagues found that pore sealing times (τ) after transient electroporation were approximately 9 min., τ was not significantly dependent on the imposed transmembrane potential (ΔV). They also determined the ΔV thresholds necessary for transient and stable electroporation in the skeletal muscle cells (ΔV's ranging between 340 mV and 480 mV caused a transient influx of magnesium, indicating the existence of spontaneously sealing pores). An imposed ΔV of 540 mV or greater led to complete equilibration of the intracellular and extracellular magnesium concentrations. This finding suggests that stable pores are created by the larger imposed transmembrane potentials. These results may be useful for understanding nerve and skeletal muscle injury after an electrical shock and for developing optimal strategies for accomplishing transient electroporation, particularly for gene transfection and cell transformation. In this regard, nuclear ionic channels (NICs) represent ubiquitous structures of living cells, although little is known about their functional properties and encoding genes. To characterize NICs, liver nuclear membrane vesicles were reconstituted by Guihard et al [20] into either planar lipid bilayers or proteoliposomes. Reconstitution of nuclear envelope (NE) vesicles into planar lipid bilayer proceeded with low efficiency. NE vesicle reconstitution into proteoliposomes led to NIC observations by the patch-clamp technique. Large conductance. voltage-gated, K^+-permeant and Cl-permeant NICs were characterized. The data of Guihard and colleagues establish that NICs can be characterized upon reconstitution into giant proteoliposomes and retain biophysical properties consistent with those described for native NICs.

For further discussion on electroporation, interested readers may want to consult Chapter 23 of this book).

3.2.4. Membrane Receptors

At the membrane level, most cellular activities involve some kind of lipid bilayer-based receptor-ligand contact interactions. Outstanding examples among these are ion-sensing, molecular recognition (e.g. antigen-antibody binding and enzyme-substrate interaction), light conversion and detection, gated channels, and active transport. The development of self-assembled bilayer lipid membranes (BLMs and liposomes) have made it possible to investigate directly the electrical properties and transport phenomena across a 5 nm thick biomembrane element separating two aqueous phases. A modified or reconstituted BLM is viewed as a dynamic structure that changes in response to environmental stimuli as a function of time, as described by the so-called dynamic membrane hypothesis [6]. For example, each type of receptor (R) interacts specifically to its own ligand (L). That is, the so-called G-receptor is usually coupled to a guanosine nucleotide-binding protein that in turn stimulates or inhibits an intracellular, lipid bilayer-bound enzyme. G-protein-linked receptors mediate the cellular responses to a vast variety of signaling molecules, including local mediators, hormones, and neurotransmitters, which are as varied in structure as they are in function.

There are at least several hundred distinct ligand-receptor pairs in our body; each is devoted to the binding of a specific extracellular ligand. Animal cells like our own have wide and diverse types of transmembrane G-protein receptors, for a number of vital signaling functions, including vision, smell, and taste. For vision, a carotenoid molecule related to vitamin A is bound in the ligand position of rhodopsin in the rods and cones of our eyes. Rhodopsin, embedded in the lipid bilayer of the visual receptor membrane, serves to pick up photons, alters its conformation, and causes the photo-receptor to which it is bound to release signals into the rod/cone cytoplasm that results in our perception of light. For smell, hundreds of distinct receptor species in the cells of our olfactory bulbs in the nose convey information about the presence of odorant ligands (*see* Ref. [6]). G-protein-linked receptors usually consist of a single polypeptide chain which threads back and forth across the lipid bilayer up to seven times. The members of this receptor family have similar amino acid sequence and functional relationship. The binding sites for G proteins have been reported to be the second and third intracellular loops and the carboxy-terminal tail. As will be considered later (Section 4), the endogenous ligands such as hormones, neurotransmitters, and exogenous stimulants such as odorants, belonging to this class that are important target analytes for biosensor technology [7].

4.   BILAYER LIPID MEMBRANES: MODELS OF BIOMEMBRANES

Generally speaking, biomembranes may be divided into two types: plant and animal. Plant cell membranes generally contain little cholesterol. In terms of cells and organelles, the following categories are usually accepted according to their particular functions:
- the plasma membrane (e.g. red blood cells)
- the chloroplast (e.g. thylakoid membranes)
- the mitochondrion (e.g. cristae membranes)
- the nerve (e.g. axon membranes)
- the eye (e.g. the visual receptor membranes)
- other membrane systems (muscles, microtubules, cytoskeletal elements, etc.)

From the viewpoint of the lipid bilayer principle, most of these membranes have been reconstituted and characterized via experimental BLMs (planar lipid bilayer and spherical liposomes [6,7]). Below a brief description of each membrane system under separate headings will be delineated together with their studies using experimental lipid bilayers.

4.1.   The plasma membrane
Of interest to note, membrane preparations are often contaminated with cells from other organelles. One exception is the plasma membranes from erythrocytes

(RBCs) that can be isolated in high purity because these cells contain no internal membranes. As discussed recently by Van Dort et al. [21], three major hypotheses have been proposed to explain the role of membrane-spanning proteins in maintaining membrane stability. These hypotheses attribute the essential contribution of integral membrane proteins to (i) their ability to anchor the membrane skeleton to the lipid bilayer, (ii) their capacity to bind and stabilize membrane lipids, and (iii) their propensity to influence and regulate local membrane curvature. According to the authors, they have modified both the membrane skeletal and lipid binding interactions of band 3 (the major membrane-spanning and skeletal binding protein of the human RBC membrane) and have examined the impact of these modifications on erythrocyte membrane morphology. deformability, and stability. Samoilenko and colleagues [22] have examined the effects of proteolysis on the lipid bilayer of RBC membranes, and concluded that in erythrocyte ghosts, no products of lipid peroxidation were observed under proteolytic treatment. The selective splitting of proteins located at the external surface of resealed pink ghosts (5-7% of total erythrocyte Hb) was accompanied by a decrease of microviscosity in the hydrophobic core of the lipid bilayer.

Suwalsky *et al* [23] studied the effects of $AlCl_3$ on toad skin, human erythrocytes, and on BLMs. They found that aluminum, a very abundant metal, could play a toxic role in several pathological processes, including neurodegeneration. Although the effects of Al^{3+} on biological membranes have been extensively described, direct information concerning the molecular basis of its biological activity is rather scanty. To examine aluminum challenges on cell membranes, various concentrations of $AlCl_3$ in aqueous solutions were incubated with human erythrocytes, isolated toad skin, and molecular models of biomembranes. The latter consisted of multilayers of dimyristoyl PC and dimyristoyl PE, representing phospholipid classes located in the outer and inner monolayers of the human erythrocyte membrane. These specimens were studied by scanning electron microscopy, electrophysiological measurements, and x-ray diffraction [23]. The results indicate that Al^{3+} in the concentration range of 10-100 µM induced the following structural and functional effects: (i) change in the normal discoid shape of human eryhrocytes to echinocytes due to the accumulation of Al^{3+} ions in the outer moiety of the red cell membrane; (ii) perturbation of dimyristoyl-PC, and to a lesser extent of dimyristoyl-PE bilayers, and (iii) decrease in the short-circuit current and in the potential difference of the isolated toad skin, effects that are in accordance with a time-dependent modulation of ion transport in response to changes in the molecular structure of the lipid bilayer. The effect of fenitrothion (FS, an organophosphorous insecticide) on similar types of planar lipid bilayers was reported by Gonzalez-Baro, Garda, and Pollero [24]. The effect of FS is different from most perturbing agents for which an increased order is accompanied by a higher lifetime. The effect of FS on dipole relaxation at the hydrophilic-hydrophobic interface of 1-palmitoyl-2-oleoyl PC bilayers was also reported.

About 2 decades ago, Babunashvili and Nenashev reported the inclusion of erythrocyte (RBCs) directly into BLMs [6,7]. More recently Melikyan et al [25] described the fusion between voltage-clamped planar lipid bilayers and influenza virus infected cells, adhered to one side of the BLM, using measurements of electrical admittance and fluorescence. The changes in currents in-phase and 90-degrees out-of-phase with respect to the applied sinusoidal voltage were used to monitor the addition of the cell membrane capacitance to that of the lipid bilayer through a fusion pore connecting the two membranes. Fusion required acidic pH on the cell-containing side and depended on temperature. The incorporation of gangliosides into the planar bilayers greatly augmented fusion. According to the authors, the essential features of fusion pores are produced with proteins. In connection with membrane pores (channels), Agner and colleagues [26] have reported the effect of temperature on the formation and inactivation of syringomycin E (SRE) pores in human RBCs and in BLMs. The kinetics of ^{86}Rb ions and hemoglobin effluxes were measured at different temperatures (6-37 ^{0}C) and pore formation was found to be only slightly affected, while inactivation was strongly influenced by temperature. With BLMs, SRE induced a large current that remained stable at 14 ^{0}C, but at 23 ^{0}C it decreased over time while the single channel conductance and dwell time did not change. The results show that the temperature dependent inactivation of SRE channels is due to a decrease in the number of open pores, according to the authors. Pertaining to this, Basu and colleagues [27] investigated the effect of temperature on the reversible nonlinearity as manifested in the DC I-V (current and voltage) characteristic of BLMs. The values of at the onset of nonlinearity have been found to be a function of temperature, whereas the nonlinearity was found to decrease with increasing temperature.

Pore-forming properties of staphylococcal alpha-toxin was investigated in planar bilayer lipid membranes by Yuldasheva, Krasilnikov and colleagues [28] who found that single-channel conductance in planar bilayers was decreased about threefold. The anion selectivity of the channel was replaced with cation selectivity, and the asymmetry in the current-voltage relationship of the channel became more pronounced. At the same time the nicked toxin kept its full ability to form ion channels in lipid bilayers, although it lost a considerable part of its hemolytic activity. In planar lipid bilayers and in erythrocyte membranes, the proteolytically nicked toxin actually formed channels with a slightly smaller diameter ($\sim$ 1.2 times) than that formed by the native toxin. This decrease was not marked enough to explain changes in the biological effects of the nicked toxin. The change in channel selectivity induced by the cleavage is considered to be the major determinant of the changes in the biological effects of the nicked toxin. In a related study [28], the same group reported replacement of an amino acid residue at position 130 -Gly by Cys- in the primary structure of *Staphylococcus aureus* alpha-toxin decreases the single-channel conductance induced by the toxin in planar lipid bilayers. Concomitantly, the pH value is also decreased at which the channel becomes unable to discriminate between Cl^- and K^+ ions. By contrast, the pH dependence of the efficiency of the mutant toxin to form ion channels in lipid bilayers was unchanged (maximum efficiency at pH 5.5-6.0). The asymmetry and nonlinearity of the current-voltage characteristics of the channel were increased by the point mutation but the diameter of the water pore induced by the mutant toxin, evaluated in lipid bilayers and in erythrocyte membranes, was found to be indistinguishable from that formed by wild-type toxin and equal to 2.4-2.6 nm, according to Yuldasheva and colleagues [28].

Regarding toxins, blepharismins are polycyclic quinones found in the pigment granules of the ciliated protozoan, *Blepharisma*. Exposure to purified blepharismins results in lethal damage to several other ciliates. Muto and colleagues [29] reported that, at cytotoxic concentrations, blepharismins formed cation-selective channels in planar phospholipid bilayers. The channels formed in a diphytanoyl PC bilayer had a K^+/Cl^- permeability ratio of 6.6:1. Single channel recordings revealed the conductance to be quite heterogeneous, ranging from 0.2 to 2.8 nS in solutions containing 0.1 M KCl, possibly reflecting different states of aggregation of blepharismin. The observations of Muto et al suggest that channel formation is a cytotoxic mechanism of blepharismin's action against predatory protozoa. In this connection, the apical brush border membrane, the main target site of *Bacillus thuringiensis* toxins, was isolated from gypsy moth (*Lymantria dispar*) larval midguts and fused to planar lipid bilayers, as reported by Peyronnet et al [30]. Under asymmetrical N-methyl-D-glucamine-HCl conditions (450 mM cis/150 mM trans, pH 9.0), which significantly reduce endogenous channel activity, trypsin-activated Cry1Aa, a *B. thuringiensis* insecticidal protein active against the gypsy moth in vivo, induced a large increase in BLM conductance at much lower concentrations (1.1-2.15 nM) than in receptor-free bilayer membranes. At least 5 main single-channel transitions with conductances ranging from 85 to 420 pS were resolved. Peyronnet and associates showed no evidence of current rectification. Analysis of the macroscopic current flowing through the composite lipid bilayer suggested voltage-dependence of several channels. In comparison, the conductance of the pores formed by 100-500 nM Cry1Aa in receptor-free BLMs was significantly smaller (about 8-fold). This study provides a detailed demonstration that the target insect midgut brush border membrane material promotes considerably pore formation by a *B. thuringiensis* according to Peyronnet and colleagues [30].

In fish, a layer of mucus covers the external body surface contributing therefore, among other important biological functions, to the defense system of fish. The prevention of colonization by aquatic parasites, bacteria and fungi is mediated both by immune system compounds (e.g. IgM, lysozyme) and by antibacterial peptides and polypeptides. Ebran et al [31] have recently shown that only the hydrophobic components of crude epidermal mucus of fresh water and sea water fish exhibit strong pore-forming properties, which were well correlated with antibacterial activity The authors have isolated novel glycosylated proteins from the hydrophobic supernatant of tench eel and rainbow trout mucus. The study of their secondary structure was performed by circular dichroism and revealed structures in random coil and α-helix in the same proportions. When reconstituted in planar lipid bilayers, they induced the formation of ion channels. This pore-forming activity was well correlated with a strong antibacterial activity (minimal inhibitory concentration less than 1 μM for the three proteins) against both Gram-negative and Gram-positive bacteria. The results of Ebran and associates [31] suggest that fish secrete antibacterial glycoproteins able to kill bacteria by forming large channels (several hundreds to thousands of pS) in the target membrane. Regarding this, *Helicobacter pylori*, a gram-negative bacterium associated with gastritis, peptic ulceration, and gastric adenocarcinoma in humans, secretes a protein toxin, VacA, that causes vacuolar degeneration of epithelial cells. Several different families of *H. pylori* vacA alleles can be distinguished based on sequence diversity in the "middle" region (i.e., m1 and m2) and in the 5' end of the gene (i.e., s1 and s2). Type s2 VacA toxins contain a 12-amino-acid amino-terminal hydrophilic segment, which is absent from type sl toxins. To examine the functional properties of VacA toxins containing this 12-amino-acid segment, McClain and colleagues [32] analyzed a wild-type VacA and a chimeric VacA protein. Purified VacA from *H. pylori* strain 60190 induced vacuolation in HeLa and Vero cells, whereas the chimeric toxin lacked detectable cytotoxic activity. Type VacA from strain 60190 formed membrane channels in a planar lipid bilayer assay at a significantly higher rate than did s2/m1 VacA. However, membrane channels formed by type sl VacA and type s2 VacA proteins exhibited similar anion selectivities (permeability ratio, $P_{Cl}/P_{Na} = 5$). When an equimolar mixture of the chimeric s2/m1 toxin and the wild-type s1/m1 toxin was added to HeLa cells, the chimeric toxin completely inhibited the activity of the s1/m1 toxin [32].

Strichartz et al. [33] reported the potencies of synthetic saxitoxin (+/-STX) and six of its synthetic analogues, including the enantioselectively synthesized unnatural (-)enantiomer of decarbamoyl saxitoxin (dcSTX). These compounds were measured and compared to those of natural saxitoxin [(+)STX]. The analogues, all of which were racemic (+/-) mixtures except for dcSTX, varied in the substituents at the C6 position, the carbamoyl 'moiety', and the C12 position, the hydrated ketone. The ability of the toxins to inhibit the compound action potential (AP) and to displace radiolabeled natural saxitoxin (H-3-STX) from nerve membranes at equilibrium were both used as potency assays. Biological activity of both (+) and (-dcSTX was analyzed by the kinetics of block of single Na^+ channels reconstituted in planar lipid bilayers, where it was demonstrated that only (+)dcSTX had biological activity. Replacement of the ketone at the C12 position by a methylene group was accomplished in two derivatives, although both also had substituents at the C6 position [33].

To understand the physics of polymer equilibrium and dynamics in the confines of ion channels, partitioning of poly(ethylene glycol)s (PEGs) of different molecular weights into the bacterial porin, OmpF has been recently reported by Rostovtseva, Nestorovich, and Bezrukov [34]. Thermodynamic and kinetic parameters of partitioning are deduced from the effects of polymer addition on ion currents through single OmpF channels reconstituted into planar lipid bilayers. The equilibrium partition coefficient is inferred from the average reduction of channel conductance in the presence of PEG; rates of polymer exchange between the pore and the bulk are estimated from PEG-induced conductance noise. The PEG-induced conductance noise is compatible with the polymer mobility reduced inside the OmpF pore by an order of magnitude relatively to its value in bulk solution.

In search of protective antigens which can be used in a vaccine to prevent *Helicobacter pylori* infection, Peck et al [35] report on the identification of four genes, HopV, HopW, HopX and HopY, and the characterization of the corresponding proteins which belong to the *H. pylori* outer membrane protein (Hop) family containing 32 homologous members, some of which were shown to function as porins. Sequence analysis of 16 different *H. pylori* strains revealed that the proteins are highly conserved. Further, the localization of these proteins at the surface of the bacteria was investigated by immunofluorescence. Using a planar lipid bilayer system, the proteins HopV and HopX were shown to form pores with single-channel conductances of 1.4 and 3.0 nS, respectively [35].

4.2. The thylakoid membrane of the chloroplast (*see Chapter 35*)

The chloroplast membranes are highly regulated and biological active regions of the plant cell, which carry numerous essential proteinaceous components, For example, in the thylakoid membrane the photosynthetic apparatus, one of the most life-relevant biological machineries, is located [6]. Schleiff and

Klosgen [36] reported the mode of insertion of a class of proteins into the outer envelope of the thylakoid membranes, which share a unique feature. They inserted apparently directly into the lipid bilayer, i.e. without the help of a proteinaceous translocation pore. Interesting to note that more than 3 dacades ago light-induced effects and electronic processes were described in planar lipid bilayers, involving only simple photosynthetic pigments [6,36]. About a decade ago, Schonknecht, Althoff and Junge [37] discussed the dimerization constant and single-channel conductance of gramicidin in the lipid bilayer of thylakoid membrane. One important aspect of this work is electron transfer reactions in these membranes as described below.

4.2.1. Photoinduced charge generation and separation

During photosynthesis, pigment molecules organized into reaction centers by membrane-spanning proteins carry out photoinduced electron transport across membranes. The resulting transmembrane electrochemical potential is then coupled to the movement of protons across the membrane. Photoinduced electron transport followed by thermal electron transfer, leading to charge separation over distances of 8 nm, has been demonstrated in artificial mimics of the photosynthetic reaction center comprising covalently linked electron donors and acceptors [6,38]. It is worth noting that Steinberg, Moore and associates [39] reported the assembly of an artificial mimic of the photosynthetic apparatus, which transports protons across a lipid bilayer when illuminated. Their model reaction center is a molecular 'triad', consisting of an electron donor and acceptor linked to a photosensitive porphyrin group. When excited, it establishes a reduction potential near the outer surface of the bilayer and an oxidation potential near its inner surface, In response to this redox potential gradient, a freely diffusing quinone molecule alternates between its oxidized and reduced forms to ferry protons across the bilayer with an overall quantum yield of 0.004, creating a pH gradient between the inside and outside of the liposomal bilayer. In this regard, Shingles, Roh, and McCarty [40] studied proton-linked transport. They found that acidification is dependent upon ΔpH, with the inside of vesicles being alkaline with respect to the outside. In contrast, added nitrate had no effect on vesicle acidification. Nitrite, however, caused acidification of asolectin vesicles. Their results indicate that nitrite movement occurs by rapid diffusion across membranes as nitrous acid, and this movement is dependent on a proton gradient across the lipid bilayer. It should be noted that photoelectric effects in lipid bilayers are of practical interest (e.g. mimicking photosynthetic solar energy transduction and molecular electronics) [41]. In this conection, most interestingly is a report by Drain [57] who has found that linear porphyrin arrays can be self-assembled into planar lipid bilayers. The length of the transmembrane assemblies is determined both by the thermodynamics of the intermolecular interactions in the supermolecule and by the dimension and physical chemical properties of the bilayer. Thus, the size of the porphyrin assembly can self-adjust to the thickness of the bilayer. Consistent with prior findings [6,38,41], when an electron acceptor (e.g. Fe^{3+}) is placed on one side of the pigmented BLM and an electron donor (e.g. ascorbic acid) on the opposite side, substantial photocurrents are observed, when illuminated with white light. Only the assembled structures give rise to the photocurrent, as no current is observed from any of the component molecules. The fabrication of this photogated molecular electronic conductor from simple molecular components exploits several levels of self-assembly and self-organization (see Chapters 33 and 35). In this connection, Jiang et al [42] discussed the relvance of electron transfer in photosynthesis, and in mitochondrial respiration, as well as in the design of molecular systems for solar energy conversion [38,41]. In photosynthesis, solar energy harvested by antenna pigments is funneled to special electron-hole pairs by highly efficient energy migration and energy transfer in a photosynthetic reaction center; there multistep electron-transfer reactions proceed successively to electron-hole pairs separated in the lipid bilayer of thylakoid membrane. The well-organized molecular arrangement in the antenna systems and the reaction centers play an important role in this unidirectional energy transfer and charge separation [6]. Jiang and associates [42] further reported that gold supported octadecanethiol/PC hybrid lipid bilayer mediated by fullenene C_{60} (abbreviated C_{60}-BLM) can act as both a photosensitizer for electron transfer from a donor molecule and a mediator for electron transport across a hybrid bilayer membrane. The steady-state photocurrent behaviors in different concentration of ascorbate, $Co(bpy)_3^{2+/3+}$ or $Fe(CN)_6^{4-/3-}$ solution have been studied. The rate-limiting step of the whole photoinduced electron transfer depends on the applied potential and the redox concentration in solution (see Chapter 29 of this book).

In plant physiology, Davenport and Tester [43] in a comprehensive study reported the transporters responsible for toxic Na^+ influxes in wheat (Triticum aestivum) root plasma membrane preparations were screened using the planar lipid bilayer technique as an assay for Na^+-permeable ion channel activity. The predominant channel in the bilayer was a 44-pS channel that they called the nonselective cation (NSC) channel, which was nonselective for monovalent cations and weakly voltage dependent. Single channel characteristics of the NSC channel were compared with $^{22}Na^+$ influx into excised root segments. Na^+ influx through the NSC channel resembled $^{22}Na^+$ influx in its partial sensitivity to inhibition by Ca^{2+}, Mg^{2+}, and Gd^{3+}, and its insensitivity to all other inhibitors tested (tetraethylammonium, quinine, Cs^+, tetrodotoxin, verapamil, amiloride, and flufenamate). Na^+ influx through the NSC channel also closely resembled an instantaneous current in wheat root protoplasts in its permeability sequence, selectivity for K^+ over Na^+ ($\sim$ 1.25), insensitivity to tetraethylammonium, voltage independence, and partial sensitivity to Ca^{2+}. Comparison of tissue, protoplast, and single-channel data indicate that toxic Na^+ influx is catalyzed by a single transporter, and this is likely to be the NSC channel identified in planar lipid bilayers, according to the authors [43]. In another paper, White, Pineros, Tester, and Ridout [59] reported calcium channels in the plasma membrane of root cells that fulfill both nutritional and signaling roles. The permeability of these channels to different cations determines the magnitude of their cation conductances, their effects on cell membrane potential and their contribution to cation toxicities. The selectivity of a Ca^{2+}-permeable channel from the plasma membrane of wheat roots was studied following its incorporation into planar lipid bilayers. It was assumed that cations permeated in single file through a channel with three energy barriers and two ion-binding sites. Differences in permeation between divalent and monovalent cations were attributed largely to the affinity of the ion binding sites. The authors' model suggested that significant negative surface charge was present in the vestibules to the pore and that the pore could accommodate two cations simultaneously.

In connection with ion selectivity, LeMasurier et al [44] reported ion conduction and selectivity properties of *KcsA*, a bacterial ion channel of known structure, in a planar lipid bilayer. Determination of reversal potentials with submillivolt accuracy shows that K^+ is over 150-fold more permeant than Na^+. Variation of conductance with concentration under symmetrical salt conditions is complex, with at least two ion-binding processes revealing themselves: a high affinity process below 20 mM and a low affinity process over the range 100-1,000 mM. These properties are analogous to those seen in many eukaryotic K^+ channels, and they establish *KcsA* as a faithful structural model for ion permeation in eukaryotic K^+ channels. In this connection Mi, Peters, and Berkowitz [45] characterized a K^+-conducting protein of the chloroplast inner envelope. Studies of this transport protein in the native membrane documented its sensitivity to K^+ channel blockers. Further studies of native membranes demonstrated a sensitivity of K^+ conductance to divalent cations such as Mg^{2+}, which modulate ion conduction through interaction with negative surface charges on the inner-envelope membrane. Purified chloroplast inner-envelope vesicles were fused into a planar lipid bilayer to facilitate recording of single-channel K^+ currents. These single-channel K^+ currents had a slope conductance of 160 picosiemens (pS). Antibodies generated against the conserved amino acid sequence that serves as a selectivity filter in the pore of K^+

channels immunoreacted with a 62-kD polypeptide derived from the chloroplast inner envelope [45].

Of interest to note is a paper by Wright and associates [46], who reported that a series of novel resorcin[4]arenes have been synthesised and developed as potassium-selective transporters. Resorcin[4]arenes that feature crown ether moieties function as efficient carriers of K^+ across bulk liquid membranes showing enhanced selectivity over the other alkali metal ions relative to a model system (benzo[15]crown-5). Incorporation of functionalities suitable for pore formation, in addition to an extra annulus of aromatic residues, gives molecules that have remarkable ion-channel-mimicking behavior in a lipid bilayer with outstanding K^+/Na^+ selectivity. Regarding this, Kobayashi et al [47] prepared lipophilic cyclodextrin (CD) derivatives, synthetic ionophores to transport alkali metal cations across a BLM. The purpose of their study was to develop a new class of an artificial transport system of alkali metal cations via BLMs, by using CD derivatives as a cation carrier. A lipophilic CD derivative incorporated into a BLM forms a complex with an alkali metal cation at one surface of the membrane. This charged complex migrates to the opposite side of the membrane and then releases the cation into the subphase. CD derivatives have various types of acyl groups as a complexing site and formed a 1:1 complex with the alkali metal cation. The complex formation was interpreted by an induced-fit mechanism. it is found that the ability of CD derivative for forming a complex and/or transporting cations across the BLM depends on the bulkiness of acyl groups. The conductivities of heptakis (2,6-di-0-propyl-3-0-propionyl)-beta-CD were higher than those of valinomycin regardless of sizes of cations. The order of the conductivity in all derivatives is $Li^+ < Na^+ < K^+ \sim Rb^+ \sim Cs^+$, regardless of the types of acyl groups in the derivatives [47]. By means of the two-compartment system, Zhang and Shi [77] reported the first Pinellia ternata lectin (PTL) channels in planar lipid bilayers. Their results show (i) PTL-channels are voltage-independent and have apparent subunits; (ii) in 50 mmol/L KCl and 25 mmol/L $BaCl_2$ solutions a single channel has unit conductance of 21 pS and 42 pS, respectively; (iii) the channel exhibits a slightly higher permeability to divalent than monovalent cation ($P_{Ba}/P_K = 4.\,1$), and (iv) the selectivity among divalent or monovalent cations is poor. The cation selectivity sequence for the channel follows P_{Ba} (7.0) is similar to P_{Sr} (6.4) is similar to P_{Mg} (6.4) > P_K (1.7)> P_{Na} 1.0) is similar to P_{Li} (1.0).

Using a method of sensitized photoinactivation, Antonenko et al [49] studied the effects of different anionic polymers on the kinetic properties of ionic channels formed by neutral gramicidin A (gA) and its positively charged analogs gramicidin-tris(2-aminoethyl)amine (gram-TAEA) and gramicidin-ethylenediamine (gram-

EDA) in a bilayer lipid membrane. The addition of Konig's polyanion caused substantial deceleration of the photoinactivation kinetics of gram-TAEA channels, which expose three positive charges to the aqueous phase at both sides of the membrane. In contrast, channels formed of gram-EDA, which exposes one positive charge, and neutral gA channels were insensitive to Konig's polyanion. The effect strongly depended on the nature of the polyanion added, namely: DNA, RNA, polyacrylic acid, and polyglutamic acid were inactive, whereas modified polyacrylic acid induced deceleration of the channel kinetics at high concentrations. In addition, DNA was able to prevent the action of Konig's polyanion. In single-channel experiments, the addition of Konig's polyanion resulted in the appearance of long-lived gram-TAEA channels. The deceleration of the gram-TAEA channel kinetics was ascribed to electrostatic interaction of the polyanion with gram-TAEA that reduces the mobility of gram-TAEA monomers and dinners in the membrane via clustering of channels [49].

4.3. The mitochondria (outer and cristae membranes)

Although all biomembranes contain phospholipids and proteins, the lipid/ protein: ratio varies greatly. For example, the inner cristae membrane of mitochondria is about 75% protein, whereas the myelin membrane is 80% lipid. There is a correlation between high protein content and membrane function. Subcellular fractionation techniques can partially separate and purify several important biomembranes such as mitochondrial membranes, from many kinds of cells.. Electron micrographs revealing mobility of protein particles in freeze-fractured vesicles prepared from the inner mitochondrial membrane. Integral protein particles are visible as protrusions that are randomly distributed in the surface of a free-fractured vesicle. It is worth noting that, when the vesicle was subjected to a strong electric field and then rapidly frozen, all the particles clustered to one end, showing that the particles can move laterally within the plane of lipid bilayer, in response to a voltage perturbation. Electric fields promote pore formation in both biological and model membranes. Thus, of particular significance is the work reported by Melikov et al [50] who observed fast transitions between different conductance levels reflecting opening and closing of metastable lipid pores, using clamped, unmodified planar bilayers at 150-550 mV to monitor transient single pores for a long period of time. Although mean lifetime of the pores was 3 ± 0.8 ms (250 mV), some pores remained open for up to 1 s. The mean amplitude of conductance fluctuations (similar to 500 pS) was independent of voltage and close for bilayers of different area, indicating the local nature of the conductive defects. The distribution of pore conductance was rather broad. Based on the conductance value and its dependence of the ion size, the radius of the average pore was estimated to be about 1 nm. Short bursts of conductance spikes (opening and closing of pores) were often separated by periods of background conductance. Within the same burst the conductance between spikes was indistinguishable from the background. The mean time interval between spikes in the burst was much smaller than that between adjacent bursts. These data indicate that opening and closing of lipidic pores proceed through some electrically invisible (silent) pre-pores. Similar pre-pore defects and metastable conductive pores might be involved in remodeling of cell membranes in different biologically relevant processes [50].

During apoptotic cell death, cells usually release apoptogenic proteins such as cytochrome c from the mitochondrial intermembrane space. If Bcl-2 family proteins induce such release by increasing outer mitochondrial membrane permeability, then the pro-apoptotic, but not anti-apoptotic activity of these proteins should correlate with their permeabilization of membranes to cytochrome c. With this in mind, Basanez and associates [58] tested the hypothesis using pro-survival full-length Bcl-x(L) and pro-death Bcl-x(L) cleavage products (ΔN612Bcl-x(L) and ΔN76Bcl-x(L)). Unlike Bcl-x(L), ΔN61Bcl-x(L) and ΔN76Bcl-x(L) caused the release of cytochrome c from mitochondria in vivo and in vitro. Recombinant ΔN61Bcl-x(L) and ΔN76Bcl-x(L), as well as Bcl-xL, cleaved *in situ*. Furthermore, only ΔN61Bcl-x(L) and ΔN76Bcl-x(L), but not Bcl-x(L), formed pores large enough to release cytochrome c and to destabilize planar lipid bilayers through reduction of pore line tension. Because Bcl-x(L) and its C-terminal cleavage products bound similarly to lipid membranes and formed oligomers of the same size, neither lipid affinity nor protein-protein interactions appear to be solely responsible for the increased membrane-perturbing activity elicited by Bcl-x(L) cleavage. Taken together, these data are consistent with the hypothesis that Bax-like proteins oligomerize to form lipid-containing channels in the outer mitochondrial membrane, thereby releasing intermembrane apoptogenic factors into the cytosol. It is worth noting that *Vibrio cholerae* cytolysin, a water-soluble protein with a molecular mass of 63 kDa, forms small channels in target cell membranes. Yuldasheva and co-workers [28] used planar lipid bilayers under voltage clamp conditions to investigate the geometric properties of the pores, and established that all cytolysin channels were inserted into membranes with the same orientation. Sharp asymmetry in the I-V curve of fully open cytolysin channels persisting at high electrolyte concentrations indicated asymmetry in the geometry of the channel lumen. Using the nonelectrolyte exclusion method, evidence was obtained that the cis opening of the channel had a larger diameter (less than or equal to 1.9 nm) than the trans opening (less than or equal to 1.6 nm), according to the authors. The channel lumen appeared constricted, with a diameter of less than or equal to 1.2 nm. Cup-shaped lumen geometry was deduced for both channel openings, which appeared to be connected to each other via a central narrow part. The latter contributed significantly to the total electrical resistance and determined the discontinuous character of channel filling with nonelectrolytes. Comparisons of the properties of pores formed by cytolysins of two *V. cholerae* biotypes indicated that the two ion channels possessed a similar geometry. Of particular interest to note, Fischer et al [64] synthesized the putative transmembrane segment of the ion channel forming peptide NE from influenza B. Insertion into the planar lipid bilayer revealed ion channel activity with conductance levels of 20, 61, 107, and 142 pS in a 0.5 M KCl buffer solution. In addition, levels at -100 mV show conductances of 251 and 413 pS. A linear current-voltage relation reveals a voltage-independent channel formation. In methanol and in vesicles the peptide appears to adopt an α-helical-like structure. Computational models of ct-helix bundles using N = 4, 5, and 6 NE peptides per bundle revealed water-filled pores after 1 ns of MD simulation in a solvated lipid bilayer. Calculated conductance values of about 20, 60, and 90 pS, respectively, suggested that the multiple conductance levels seen experimentally must correspond to different degrees of oligomerization of the peptide to form channels [64].

The pore-forming domain of colicin A (pfColA) fused to a prokaryotic signal peptide (sp-pfColA) is transported across and inserts into the inner membrane of *Escherichia coli* from the periplasmic side and forms a functional channel. The soluble structure of pfColA consists of a ten-helix bundle containing a hydrophobic helical hairpin. Nardi and associates [67] generated a series of mutants in which an increasing number of sp-pfColA α-helices was deleted. These peptides were tested for their ability to form ion channels *in vivo* and *in vitro*. Nardi et al found that the shortest sp-pfColA mutant protein that killed *E. coli* was composed of the five last α-helices of sp-pfColA, whereas the shortest peptide that formed a channel in planar lipid bilayers similar to that of intact pfColA. The peptide composed of the last five α-helices of pfColA generated a voltage-independent conductance in planar lipid bilayers with properties very different from that of intact pfColA. Thus, helices 1 to 4 are unnecessary for channel formation, while helix 5, or some part of it, is important but not absolutely necessary. Voltage-dependence of colicin is evidently controlled by the first four α-helices of pfColA [67].

Anti-apoptotic protein (Bcl-2) or pro-apoptotic (BAX) members can form ion channels when embedded into lipid bilayers. This contrasts with the observation that Bcl-2 stabilizes the mitochondrial membrane barrier function and inhibits the permeability transition pore complex (PTPC), Brenner et al. [68] provide experimental data which may explain this apparent paradox. BAX and adenine nucleotide translocator (ANT), the most abundant inner mitochondrial membrane protein, can interact in artificial lipid bilayers to yield an efficient composite channel whose electrophysiological properties differ quantitatively and qualitatively from the channels formed by BAX or ANT alone. The formation of this composite channel can be observed in conditions in which BAX protein alone has no detectable channel activity, Cooperative channel formation by BAX and ANT is stimulated by the ANT ligand atractyloside (Atr) but inhibited by ATP, indicating that it depends on the conformation of ANT, In contrast to the combination of BAX and ANT, ANT does not form active channels when incorporated into membranes with Bcl-2. Rather, ANT and Bcl-2 exhibit mutual inhibition of channel formation. Bcl-2 prevents channel formation by Atr-treated ANT and neutralizes the cooperation between BAX and ANT. their data are compatible with a menage a trois model of mitochondrial apoptosis regulation in which ANT, the likely pore forming protein within the PTPC, interacts with BAX or Bcl-2 which influence its pore forming potential in opposing manners [68].

4.4.   The nerve membranes

The lipid composition varies among different membranes. Owing to its high phospholipid content, myelin can electrically insulate the nerve cell from its environment. All membranes contain a substantial proportion of phospholipids, predominantly phosphoglycerides, which have a glycerol backbone. However, sphingomyelin, a phospholipid that lacks a glycerol backbone, is commonly found in plasma membranes of the nerve. Instead of a glycerol backbone it contains sphingosine, an amino alcohol with a long unsaturated hydrocarbon chain. A fatty acyl side chain is linked to the amino group of sphingosine by an amide bond to for a ceramide. The terminal hydroxyl group of sphingosine is esterified to phosphocholine, thus the hydrophilic head of sphingomyelin is similar to that of phosphatidylcholine (PC). In a comprehesive paper, Ptacek [69] appraised the electrical excitability of skeletal and cardiac muscle cells and neurons, as resulting from a balance of inhibitory and excitatory influences. Tonic concentration gradients established by adenosine 5'-triphosphate (ATP)-dependent pumps can be maintained because the lipid bilayer is an extremely good insulator. Once ionic concentrations are established, movement of one or more ions down their respective concentration gradients can establish voltage differences across a membrane. The Nernst equation allows prediction of membrane potentials based on the particular ion involved and the concentration gradient for that ion in the cell, as further discussed by Ptacek. A large number of voltage-gated ion channels, ligand-gated channels, and transporters are involved in maintaining this balance. The specific channels and transporters involved differ in various cell types. In any case, the balance of these opposing influences tightly regulates normal membrane excitability. It is not surprising that the disruption of the balance of excitability of various cells might lead to neurological phenotypes. However, large changes in excitability of muscle or nerve may well be lethal, according to the author. Therefore, nature may select against such major changes. A growing body of evidence suggests that subtle changes in some ion channels can lead to a slight increase in membrane excitability that results in a neurological phenotype. Interestingly, these phenotypes are frequently episodic. That is, under many circumstances, the nerve or muscle may be functioning properly; however, under certain circumstances, a precipitating event can lead to abnormal excitability resulting in one of any number of phenotypes. In Ptacek's article, the author further focuses on a number of monogenic disorders of the nervous system where episodic phenotypes are known to result from specific mutations of ion channels. The similarities between a large group of seemingly disparate disorders are emphasized [69].

Receptor-mediated endocytosis, caveolae internalization and certain trafficking events in the Golgi all require dynamin for vesiculation. Hinshaw and colleagues [56] reported that members of the dynamin family of GTPases have unique structural properties that might reveal a general mechanochemical basis for membrane constriction. The dynamin-related protein (Dlp1) has been implicated in mitochondria fission and a plant dynamin-like protein phragmoplastin is involved in the vesicular events leading to cell wall formation. A common theme among these proteins is their ability to self-assemble into spirals and their localization to areas of membrane fission. Here, the authors present the first three-dimensional structure of dynamin at a resolution of similar to 2 nm, determined from cryo-electron micrographs of tubular crystals in the constricted state. The map reveals a T-shaped assign consisting of three prominent densities: leg, stalk and head. The structure suggests that the dense stalk and head regions rearrange when GTP is added, a rearrangement that generates a force on the underlying lipid bilayer and thereby leads to membrane constriction.

Clay and Kuzirian [48] have localized the classical voltage-gated K^+ channel within squid giant axons by immunocytochemistry using the Kv1 antibody. Widely dispersed patches of intense immunofluorescence were observed in the axonal membrane. Punctate immunofluorescence was also observed in the axoplasm and was localized to similar to 25-50-mum-wide column down the length of the nerve (axon diameter similar to 500 μm). Immunoelectronmicroscopy of the axoplasm revealed a K^+ channel containing vesicles, 30-50 nm in diameter, within this column, These and other vesicles of similar size were isolated from axoplasm using a novel combination of high-speed ultracentrifugation and controlled-pore size, glass bead separation column techniques. Approximately 1% of all isolated vesicles were labeled by K+ channel immunogold reacted antibody. Incorporation of isolated vesicle fractions within a lipid bilayer revealed K^+ channel electrical activity similar to that recorded directly from the axonal membrane. These K^+ channel-containing vesicles may be involved in cycling of K^+ channel protein into the axonal membrane, Additionally, Clay and Kuzirian [48] have also isolated an axoplasmic fraction containing similar to 150-nm-diameter vesicles that may transport K^+ channels back to the cell body.

Sodium channels from human ventricular muscle membrane vesicles were incorporated into planar lipid bilayers by Wartenberg, Wartenberg and Urban [65], and the steady-state behavior of single sodium channels were examined in the presence of batrachotoxin. In symmetrical 500 mM NaCl the averaged single channel conductance was 24.7 ± 1.3 pS and the channel fractional open time was 0.85 ± 0.04. The activation midpoint potential was -99.5 ± 3.1 mV. Extracellular tetrodotoxin blocked the channel with a k(1/2) of 414 nM at 0 mV. In 7 out of 13 experiments subconductance states were observed (9.2 ± 1.2 pS). When sodium chloride concentration was lowered to 100 mM, single channel conductance decreased to 19.0 ± 0.9 pS, steady-state activation shifted by -17.3 ± 5.1 mV, tetrodotoxin sensitivity increased to 324 nM, and sub-conductance states were invariably observed in single channel records (7.9 ± 0.7 pS). In planar lipid bilayers, the properties of cardiac sodium channels from different species are not very different, but there are significant differences between sodium channels from human heart and from human CNS.

Fu and Singh [71] reported the mode of botulinum neurotoxin action, involving binding of its heavy chain for internalization into the presynaptic end of a nerve cell through endocytosis. The low-pH conditions of endosomes trigger translocation of the light chain across the endosomal membrane to the cytosol, where the light chain cleaves specific target proteins involved in the docking and fusion of synaptic vesicles for acetylcholine release. In an effort to model the interaction of botulinum neurotoxin and its subunit chains with lipid bilayer at low pH during the translocation process, Fu and Singh have examined type A botulinum neurotoxin-mediated calcein release from asolectin liposomes.

Schote and Seelig [61] have found that the fluorescent dye (FM1-43) labels nerve terminals in an activity-dependent fashion and increasingly useful in exploring the exo- and endocytosis of synaptic vesicles and other cells by fluorescence methods. The dye distributes between the aqueous phase and the lipid membrane but the physical-chemical parameters characterizing the adsorption/partition equilibrium have not yet been determined. Fluorescence spectroscopy alone is not sufficient for a detailed elucidation of the adsorption mechanism since the method can be applied only in a rather narrow low-concentration window. In addition to fluorescence spectroscopy, they have therefore employed high sensitivity isothermal titration calorimetry (ITC) and deuterium magnetic resonance (H-2-NMR). ITC allows the measurement of the adsorption isotherm up to 100 µM dye concentration whereas H-2-NMR provides information on the location of the dye with respect to the plane of the membrane. Dye adsorption/partition isotherms were measured for neutral and negatively charged phospholipid vesicles. A non-linear dependence between the extent of adsorption and the free dye concentration was observed. Though the adsorption was mainly driven by the insertion of the non-polar part of the dye into the hydrophobic membrane interior, the adsorption equilibrium was further modulated by an electrostatic attraction/repulsion interaction of the cationic dye ($z = +2$) with the membrane surface. The Gouy-Chapman theory was employed to separate electrostatic and hydrophobic effects (see Ref.[6]). After correcting for electrostatic effects, the dye-membrane interaction could be described by a simple partition equilibrium ($X_b=Kc(dye)$) with a partition constant of 10^3-10^4 M^{-1}, a partition enthalpy of $\Delta H = -2.0$ kcal/mol and a free energy of binding of $\Delta G = -7.8$ kcal/mol. The insertion of FM1-43 (fluorescent dye) into lipid membranes at room temperature is thus an entropy-driven reaction following the classical hydrophobic effect. Deuterium nuclear magnetic resonance provided insight into the structural changes of the lipid bilayer induced by the insertion of FM1-43. The dye disturbed the packing of the fatty acyl chains and decreased the fatty acyl chain order. FM1-43 also induced a conformational change in the phosphocholine headgroup. The P—N$^+$ dipole was parallel to the membrane surface in the absence of dye and was rotated with its positive end towards the water phase upon dye insertion. The extent of rotation was, however, much smaller than that induced by other cationic molecules of similar charge, suggesting an alignment of FM1-43 such that the POPC phosphate group is sandwiched by the two quaternary FM1-43 ammonium groups. In such an arrangement the two cationic charges counteract each other in a rotation of the P—N+ dipole, according to Schote and Seelig [61]. It should be mentioned that Pappayee and Mishra [51] have tried to evaluate the usefulness of 1-naphthol as an excited state proton transfer fluorescent probe for studying the ethanol-induced interdigitation in lipid bilayer membranes. When ethanol concentration in liposomes is progressively increased, the neutral form fluorescence of 1-naphthol is found to decrease with corresponding increase in the anionic form intensity. This behavior is in contrast to that observed in the absence of lipid where a reverse effect is noticed. Modification of lipid bilayer is known to occur in the presence of ethanol, which increases the packing density of the membrane. Due to this induction of interdigitated gel phase, redistribution of naphthol between the inner core and interfacial region of the lipid bilayer takes places, accounting for the reduction in neutral form fluorescence intensity.

Radke, Frenkel and Urban [52] investigated the molecular effects of droperidol ($C_{22}H_{22}FN_3O_2$) on single sodium channels from the human brain, using the planar lipid bilayer technique. Droperidol (0.05-0.8 mM) induced a concentration dependent and voltage independent reduction in the time averaged single channel conductance by two mechanisms: a reduction in the fractional channel open time (major effect, approximately 90%) and a decrease in the channel amplitude (minor effect). These blocking effects of droperidol on central nervous system (CNS) sodium channels occurred at a concentration range comparable with other specific anaesthetic compounds but far beyond clinical serum levels (up to 0.002 mM). Therefore in contrast with animal preparations (frog peripheral nerve, sodium channel) the human brain sodium channel is not a major target site for droperidol during clinical anaesthesia. In this regard, Bosio, Binczek and Stoffel [53] reported that the lipid bilayer of the myelin membrane of the CNS and the peripheral nervous system (PNS) contains the oligodendrocyte- and Schwann cell-specific glycosphingolipids galactocerebrosides (GalC) and GalC-derived sulfatides (sGalC). Oligodendrocytes and Schwann cells are unable to restore the structure and function of these galactosphingolipids to maintain the insulator function of the membrane bilayer. The velocity of nerve conduction of homozygous cgt(-/-) mice is reduced to that of unmyelinated axons. This indicates a severely altered ion permeability of the lipid bilayer. GalC and sGalC are essential for the unperturbed lipid bilayer of the myelin membrane of CNS and PNS. The severe dysmyelinosis leads to death of the cgt(-/-) mouse at the end of the myelination period.

Clostridium perfringens enterotoxin (CPE) is an important cause of food poisoning with no significant homology to other enterotoxins and its mechanism of action remains uncertain. Although CPE has recently been shown to complex with tight junction proteins, Bai and Shi [54] have previously demonstrated that CPE increases ionic permeability in single Caco-2 cells using the whole-cell patch-clamp technique, thereby excluding any paracellular permeability. The membrane proteins extracted from the human sperm were reassembled into liposomes, and the liposomes were fused into giant liposomes with a diameter more than 10 μm by dehydrationrehydration procedure. The giant liposomes were used to study the Cl⁻ channel activities by patch-clamp technique. By patch clamping the giant liposome in an asymmetric NMDG (N-methyl-D-glucamine)-Cl solution system, three kinds of single-channel events with unit conductances of (74.1 ± 8.3), (117.0 ± 5.7) and (144.7 ± 4.5.) pS, respectively, were detected. Their activities were voltage-dependent and all were blocked by SITS (4-acetamido-4'-isothiocyanato-stilbene-2',2'-disulfonic acid) in a concentration-dependent manner. By constructing the open and close dwell time distribution histograms and then fitting them with exponential function, two time constants were obtained in both the open and the close states. In this connection, mention should be made that Hardy et al [55] have demonstrated that CPE forms channels in lipid bilayers in the absence of receptor proteins. The properties of the channels are consistent with CPE-induced permeability changes in Caco-2 cells, suggesting that CPE has innate pore-forming ability. In connection with cultured cells, it is worth mentioning that these cells are often used, for example, for permeability testing in drug screening. This owes to the fact that an injested pharmaceutical compound must permeate lipid bilayers of cell membranes that block the passage of polar or charged species. Instead using delicate conventional BLMs, phospholipid-coated filters (e.g. Nucleopore or Millipore filters) have been proven suitable many years ago [6].

Using patch clamp planar lipid bilayer reconstitution techniques, Jovov and colleagues [72] reported actin interactions that account for the signature biophysical properties of cloned epithelial Na^+ channels (ENaC) (conductance, ion selectivity, and long mean open and closed times) They found the following. (a) in bilayers, actin produced a more than 2-fold decrease in single channel conductance, a 5-fold increase in Na^+ versus K^+ permselectivity, and a substantial increase in mean open and closed times of wild-type alpha beta gamma-rENaC but hardly any effect on a mutant form of rENaC, and (b) when α(R613X), β γ-rENaC was heterologously expressed in oocytes and single channels examined by patch clamp, 12.5-pS channels of relatively low cation permeability were recorded, These characteristics were identical to those recorded in bilayers for either α (R613X) β, γ-rENaC or wild type α β γ-rENaC in the absence of actin, Moreover, the authors have shown that rENaC subunits tightly associate, forming either homo- or heteromeric complexes when prepared by in vitro translation or when expressed in oocytes. Jovov and associates [72] then concluded that actin serves an important regulatory function for ENaC and that planar bilayers are an appropriate system in which to study the biophysical and regulatory properties of these cloned channels.

The ubiquity of mechanosensitive (MS) channels triggered a search for their functional homologues in Archaea, the third domain of the phylogenetic tree. Two types of MS channels have been identified in the cell membranes of Haloferax volcanii using the patch clamp technique, as reported by Kloda and Martinac [75]. MS channels were identified and cloned from two archaeal species occupying different environmental habitats. These studies demonstrate that archaeal MS channels share structural and functional homology with bacterial MS channels. The mechanical force transmitted via the lipid bilayer alone activates all known prokaryotic MS channels. This implies the existence of a common gating mechanism for bacterial as well as archaeal MS channels according to the liquid-crystalline bilayer model. Based on recent evidence that the bilayer model also applies to eukaryotic MS channels, mechanosensory transduction probably originated along with the appearance of the first life forms according to simple biophysical principles [75]. In support of this hypothesis the phylogenetic analysis revealed that prokaryotic MS channels of large and small conductance originated from a common ancestral molecule resembling the bacterial MscL channel protein. Furthemore, bacterial and archaeal MS channels share common structural motifs with eukaryotic channels of diverse function indicating the importance of identified structures to the gating mechanism of this family of channels. The comparative approach used throughout their review should contribute towards understanding of the evolution and molecular basis of mechanosensory transduction in general [75].

From both patch-clamp and fluorescence microscopy experiments, Ryttsen and co-workers [73] determined the threshold transmembrane potentials for dielectric breakdown of NG108-15 cells, using 1-ms rectangular waveform pulses. The electric field strengths necessary for ion-permeable pore formation and investigated the kinetics of pore opening and closing as well as pore open times. The electroporation pulse preceded pore formation, and analyte entry into the cells was dictated by concentration, and membrane resting potential driving forces. In order to minimize adverse capacitive charging effects, the patch-clamp pipette was sealed on the cell at a 90 degrees angle with respect to the microelectrodes where the applied potential reaches a minimum, according to the authors [73]. Besides electric field, Ozeki, Kurashima and Abe [74] reported deformation in membranes of dipalmitoyl-PC (DPPC) led to the fusion of its liposome and large changes in the membrane potential of its BLM under high magnetic fields of up to 28 gauss. The magnetofusion of DPPC liposomes significantly depended on the particle size and aromatic compounds doped. Although a theory for the magnetic deformation of liposomes predicts magnetofusion and magnetodivision, only magnetofusion was experimentally observed. There seem to be discrete liposome sizes stabilized at 10 gauss. The changes in liposome size due to magnetofusion give an estimation of the local curvature of the membrane. The membrane potential of bilayer DPPC membranes markedly increased with high magnetic fields acid doped molecules, corresponding to magnetofusion, Undulation of a membrane due to high magnetic fields may relax any orientational defects in a BLM, leading to a ripple-like structure, which may cause the magnetoresponse in the membrane potential [74].

Cinnamomin, a new type ribosome-inactivating protein, purified from the seeds of Cinnamonum camphor, has been reconstituted by Zhang and associates [?77] into planar lipid bilayers and giant liposomes. The channel-forming activity of the cinnamomin and cation permeability of the channel was characterized by patch clamp. In an asymmetric solution system, bath 150/pipette 100 mM KCl, the unit conductance is 140 ± 7 pS and the reversal potential is 10.4 $\pm$ 0.6 mV, very close to the theoretical value of the K^+ electrode.

As models of ion channel proteins and naturally occurring pore-forming peptides, Hara and co-workers [63] designed a series of Aib rich peptides [Ac-(Aib-XXx-Aib-Ala)(5)-NH2 (Xxx = Lys, Glu, Ser, and Gly: BXBA-20)] to investigate the effects of the side chains of the amino acid residues Lys, Glu, Ser, and Gly on the conformation and electrophysiological properties of ion channels. The conformation of peptides and their affinity for phospholipid membranes were evaluated by CD spectroscopy. Patch-clamp experiments revealed that all BXBA-20 peptides form ion channels in DPhPC bilayers exhibiting clearly resolved transitions between the open and closed states. The channel forming frequency was in the order BKBA-20 > BEBA-20 > BSBA-20 > BGBA-20. In the case of BIKBA-20 and BEBA-20, the self-assembled conductive oligomers expressed homogeneous and voltage-independent single channel conductances. In contrast, heterogeneous conductance was observed in BSBA-20 and BGBA-20 ion channels under similar experimental conditions. From these results, Hara et al [63] then concluded that peptides with a high degree of helical conformation, high amphipathicity, high affinity for lipid membranes, and self-associating characters in vesicles are most suitable for inducing ion channels with a high frequency of occurrence. Moreover, BEBA-20, BSBA-20, and BGBA-20 channels were cation-selective, whereas the BEBA-20 channel was non-selective.

4.5. The Eye (Visual Receptors and Light Transduction)

First, let us have a brief background overview. At the back of the eye is the photosensitive retina; the visual organ that is sensitive to a small portion of the electromagnetic spectrum (wavelengths 400-800 nm). Photosensitive pigment rhodopsin, made of opsin and retinal, is bleached by light. The retinal of rhodopsin exists as two isomers: 11-*cis* retinal and *all*-trans retinal in the retina. A single photon of light is capable of converting one molecule of retinal from 11-cis to all-trans. In the dark, photoreceptor neurons continuously release an excitatory neurotransmitter (glutamate). This is so-called the 'dark current'. In the light, the photoreceptor turns off, releasing less glutamate. Visual sensitivity curves are different in high vs. low light. The two visual systems are scotopic (dark) and photopic (light), which are supported by two kinds of receptor: rods and cones, with the cone opsins providing the basis of color vision (*see* Ref. [6]).

Visual receptors are under continuous battering from light, which can initiate deleterious reactions, and a high level of oxidative catabolism. At the same time, visual membranes are constantly being renewed, by highly reactive polyunsaturated fatty acids and vitamin A metabolites. Thus, the question of visual receptor maintenance has been pivotal in vision research for about three decades. Visual receptors have, therefore, remained of intense interest even to the present era as the differing rhodopsins have been cloned and studied with molecular techniques. For example, Petrache, Salmon, and Brown [90] recently reported structural properties of docosahexaenoyl phospholipid bilayers investigated by solid-state H-2 NMR spectroscopy. One hypothesis is that polyunsaturated lipids are involved in second messenger functions in biological signaling. Another current hypothesis affirms that the functional role of polyunsaturated lipids relies on their ability to modulate physical properties of the lipid bilayer. These authors have shown that how introduction of DHA (docosahexaenoic acid) chains translates the order profile along the saturated chains, making more disordered states accessible within the bilayer central region. As a result, the area per lipid head group is increased as compared to disaturated bilayers. According to Petrache et al [90] the systematic analysis of the H-2 NMR data provides a basis for studies of lipid interactions with integral membrane proteins, for instance in relation to characteristic biological functions of highly unsaturated lipid membranes. Of speciat interest to note, Brown [90] pointed out about a decade ago the roles of the phospholipid head groups and the lipid acyl chains. A prevalent model then for the function of rhodopsin centered on the metarhodopsin I (MI) to metarhodopsin II (MII) conformational transition as the triggering event for the visual process. In bilayers of phosphatidylcholines (PC), the MI-MII equilibrium is shifted to the left; whereas in the native rod outer segment membranes it is shifted to the right, i.e., at neutral pH near physiological temperature. The lipid mixtures sufficient to yield full photochemical function of rhodopsin include a native-like head group composition, namely, comprising PC, PE, and PS, in combination with polyunsaturated docosahexaenoic acid (DHA) chains. The MI-MII transition is favored by relatively small head groups which produce a condensed bilayer surface, namely, a comparatively small interfacial area as in the case of PE, together with bulky acyl chains such as DHA which prefer a relatively large cross sectional area. The resulting force imbalance across the bilayer gives rise to a curvature elastic stress of the lipid-water interface, such that the lipid mixtures yielding native-like behavior form reverse hexagonal (H-II) phases at slightly higher temperatures. These findings of Petrache and co-workers [90] reveal that the membrane lipid bilayer has a direct influence on the energetics of the conformational states of rhodopsin in visual excitation.

In connection with metarhodopsin, Niu, Mitchell, and Litman [91] investigated the effects of membrane lipid composition on receptor-G protein coupling. Rhodopsin was reconstituted into large, unilamellar phospholipid liposomes with varying acyl chain unsaturation, with and without cholesterol. The association constant (K_α) for metarhodopsin II (MII) and transducin (G(t)) binding was determined by monitoring MII-G(t) complex formation spectrophoto-metrically. Increasing acyl chain unsaturation from 18:0,18:1PC to 18:0,22:6PC resulted in a 3-fold increase in K_α. The inclusion of 30 mol% cholesterol in the membrane reduced K_α in both 18:0,22:6PC and 18:0,18:1PC. These findings demonstrate that membrane compositions can alter the signaling cascade by changing protein-protein interactions occurring predominantly in the hydrophilic region of the proteins, external to the lipid bilayer.

Landin, Katragadda, and Albert [92] investigated the thermal stability of rhodopsin and its bleached form, opsin, using differential scanning calorimetry (DSC) [6]. The G-protein coupled receptor, rhodopsin, consists of seven transmembrane helices which are buried in the lipid bilayer and are connected by loop domains extending out of the hydrophobic core. The thermal stability of rhodopsin and its bleached form, opsin, was investigated using DSC. The thermal transitions were asymmetric, and the temperatures of the thermal transitions were scan rate dependent. This dependence exhibited characteristics of a two-state irreversible denaturation in which intermediate states rapidly proceed to the final irreversible state. These studies suggest that the denaturation of both rhodopsin and opsin is kinetically controlled. The denaturation of the intact protein was compared to three proteolytically cleaved forms of the protein. Trypsin removed nine residues of the carboxyl terminus, papain removed 28 residues of the carboxyl terminus and a portion of the third cytoplasmic loop, and chymotrypsin cleaved cytoplasmic loops 2 and 3. In each of these cases the fragments remained associated as a complex in the membrane. DSC studies were carried out on each of the fragmented proteins. Trypsin-proteolyzed protein differed little from the intact protein. However, the activation energy for denaturation was decreased when cytoplasmic loop 3 was cleaved by papain or chymotrypsin. This was observed for both bleached and unbleached samples. In the presence of the chromophore, 11-cis-retinal, the noncovalent interactions among the proteolytic fragments produced by papain and chymotrypsin cleavage were sufficiently strong such that each of the complexes denatured as a unit. Upon bleaching, the papain fragments exhibited a single thermal transition. However, after bleaching, the chymotrypsin fragments exhibited two calorimetric transitions. These data suggest that the loops of rhodopsin exert a stabilizing effect on the protein [92].

4.5.1. Rhodopsin (Rho) and bacteriorhodopsin (bR)

Intrinsic membrane proteins represent a large fraction of the proteins produced by living organisms and perform many crucial functions. Structural and functional characterization of membrane proteins generally requires that they be extracted from the native lipid bilayer and solubilized with a small synthetic amphiphile, for example, a detergent. Yu et al. [120] described the development of a small molecule with distinctive amphiphilic architecture, a 'tripod amphiphile', which solubilizes both bacteriorhodopsin (BR) and bovine rhodopsin (Rho). The polar portion of this amphiphile contains an amide and an amine-oxide: small variations in this polar segment are found to have profound effects on protein solubilization properties. The optimal tripod amphiphile extracts both BR and Rho from the native membrane environments and maintains each protein in a monomeric native-like form for several weeks after delipidation. Tripod amphiphiles are designed to display greater conformational rigidity than conventional detergents, with the long-range goal of promoting membrane protein crystallization. The results reported represent an important step toward that ultimate goal [120].

In connection with bacteriorhodopsin (bR), Curran, Templer, and Booth [78] have developed three different lipid systems. They have investigated the effect of physicochemical forces within the lipid bilayer on the folding of the integral membrane protein bacteriorhodopsin bR. Each system consists of lipid vesicles containing two lipid species, one with PC and the other with PE headgroups, but the same hydrocarbon chains: either L-alpha-1,2-dioleoyl, L-alpha-1,2-dipalmitoleoyl, or L-alpha-1,2-dimyristoyl. Increasing the mole fraction of the PE increases the desire of each monolayer leaflet in the bilayer to curve toward water. This increases the torque tension of such monolayers, when they are constrained to remain flat in the vesicle bilayer. Consequently, the lateral pressure in the hydrocarbon chain region increases, and the authors have used excimer fluorescence from pyrene-labeled PC to probe these pressure changes. Further, they showed that bR regenerates to about 95% yield in vesicles of 100% PC. The regeneration yield decreases as the mole fraction of the corresponding PE component is increased. The decrease in yield correlates with the increase in lateral pressure, which the lipid chains exert on the refolding protein. Curran et al [78] suggest that the increase in lipid chain pressure either hinders insertion of the denatured state of bR into the bilayer or slows a folding step within the bilayer, to the extent that an intermediate involved in bR regeneration is effectively trapped. Pertaining to these, Schmies and colleagues [88] investigated the light-activated proton transfer reactions of sensory rhodopsin II from Natronobacterium pharaonis (pSRII) and those of the channel-mutants D75N-pSRII and F86D-pSRII, using flash photolysis and BLM techniques. The electrical measurements prove that pSRII and F86D-pSRII can function as outwardly directed proton pumps, whereas the mutation in the extracellular channel (D75N-pSRII) leads to an inwardly directed transient current. The authors discussed that in a two-photon process a late intermediate (N- and/or O-like species) is photoconverted back to the original resting state; thereby the long photocycle is cut short, giving rise to the large increase of the photostationary current. The results presented in this work indicate that the function to generate ion gradients across membranes is a general property of archaeal rhodopsins (*see* Refs. [6, 108]).

Isele, Sakmar, and Siebert [93] used Fourier-transform infrared (FTIR) difference spectroscopy to study the interaction between lipid bilayer and rhodopsin, the latter being a member of a family of G-protein-coupled receptors that transduce signals across membranes. Upon receptor activation. a difference band was identified in the FTIR-difference spectrum of rhodopsin mutant. Rhodopsin and the mutant were reconstituted into various C-13-labeled 1-palmitoyl-2-oleoyl-sn-glycero-3-PC liposomes and probed. The band intensity scaled with the amount of rhodopsin but not with the amount of lipid, excluding the possibility that it was due to the bulk lipid phase, according to the authors. Isele et al [93] also excluded the possibility that the lipid band represents a change in the number of boundary lipids or a general alteration in the boundary lipid environment upon formation of metarhodopsin II. Instead, the data suggest that the lipid band represents the change of a specific lipid-receptor interaction i.e. coupled to protein conformational changes. Compared to ion-channel receptors, G-protein-coupled receptors-based sensors could, in principle be more sensitive, since the activation of one receptor can lead to cascade activation of many G-proteins, thereby providing an amplification of the signal.

Many eukaryotic cellular and viral proteins have a covalently attached myristoyl group at the amino terminus. One such protein is *recoverin*, a calcium sensor in retinal rod cells, which controls the lifetime of photoexited rhodopsin by inhibiting rhodopsin kinase. *Recoverin*, a recently discovered member of the EF-hand protein superfamily, serves as a Ca^{2+} sensor in vision has a relative molecular mass of 23,000, and contains an amino-terminal myristoyl group (or related acyl group). The binding of two Ca^{2+} ions to rccoverin leads to its translocation from the cytosol to the disc membrane. Ames et al [95] have used nuclear magnetic resonance (NMR) to show that Ca^{2+} induces the unclamping and extrusion of the myristoyl group, enabling it to interact with a bilayer lipid membrane. Their NMR data demonstrate that the binding of Ca^{2+} to recoverin induces the extrusion of its myristoyl group into the solvent, which would enable it to interact with a lipid bilayer or a hydrophobic site of a target protein. The conservation of the myristoyl binding site and two swivels in recoverin homologues from yeast to humans indicates that calcium-myristoyl switches are ancient devices for controlling calcium-sensitive processes [95].

More recently, Wakabayashi and colleagues [96] described a single-channel method for evaluating agonist selectivity in terms of the very number of Ca^{2+} ions passed through the epsilon4/zeta1 N-methyl-D-aspartate (NMDA) receptor ion channel in BLMs. The number of Ca^{2+} passed through the single-channel was obtained from single-channel recordings in a medium where the primary permeant ion is Ca^{2+}. The recombinant NMDA channel was partially purified from Chinese hamster ovary cells expressing the channel and incorporated in BLMs formed by the tip-dip method. It was found that the channel in BLMs is permeable to Ca^{2+} and Na^+, but the number of Ca^{2+} passed through the channel is much fewer than that of Na^+. The integrated Ca^{2+} currents induced by three typical agonists NMDA, L-glutamate and L-CCG-IV were obtained at concentration of 50 μM, where the integrated currents for all the agonists reached their saturated values. The integrated Ca^{2+} currents obtained for NMDA, and for L-glutamate suggesting that the three kinds of agonists have different efficacies to induce permeation of Ca^{2+}. The range of the agonist selectivity thus obtained is much narrower than that of binding affinities for the NMDA receptors from rat brain. The authors claimed that their method is able to detect Ca^{2+} permeation with a detection limit of approximate to $10^5 Ca^{2+}$ ions/sec [96].

4.6.   Miscellaneous Studies and Other Structures (brain tissues, muscles, microtubules, cytoskeletal elements, DNA, etc.)

4.6.1 Molecular dynamics simulations

Saiz and Klein [83] reported electrostatic interactions in a neutral phospholipid bilayer by molecular dynamics simulations, and obtained the results for a fully hydrated 1-stearoyl-2-docosahexaenoyl-sn-glycero-3-PC lipid bilayer at room temperature. The orientational distribution of the lipid dipole moments with respect to the membrane normal presents a maximum at 70 degrees (20 degrees above the plane of the interface, pointing toward the water region). The statistics of the main lipid-lipid interactions, the charge density profiles, the authors also analyzed the electrostatic potential along the bilayer normal, and the polarization of water molecules at the interfacial plane. Pertaining to this, Pasenkiewicz-Gierula and co-workers [99] applied a molecular dynamics simulation method to study membrane systems at various levels of compositional complexity. The studies were started from simple lipid bilayers containing a single type PC and water molecules (PC bilayers). As a next step, cholesterol (Chol) molecules were introduced to the PC bilayers (PC-Chol bilayers). These studies provided detailed information about the structure and dynamics of the membrane/water interface and the hydrocarbon chain region in bilayers built of various types of PCs and Chol. This enabled studies of membrane systems of higher complexity. The authors included the investigation of an integral membrane protein in its natural environment of a PC bilayer, and the antibacterial activity of magainin-2. The latter study required the construction of a model bacterial membrane, which consisted of two types of phospholipids and counter ions. Additionally, the results of the simulations were compared with published experimental data [99].

Yang et al [81] have investigated the configuration and the stability of a single membrane pore bound by four melittin molecules and embedded in a fully hydrated bilayer lipid membrane. They used molecular dynamics simulations up to 5.8 ns. It is found that the initial tetrameric configuration decays with increasing time into a stable trimer and one monomer. This continuous transformation is accompanied by a lateral expansion of the aqueous pore exhibiting a final size comparable to experimental findings. The expansion-induced formation of an interface between the pore-lining acyl chains of the lipids and the pore water ("hydrophobic pore") is transformed into an energetically more favorable toroidal pore structure where some lipid heads are translocated from the rim to the central part of the interface ("hydrophilic pore"). The expansion of the pore is supported by the electrostatic repulsion among the α-helices. It is hypothesized that pore growth, and hence cell lysis, is induced by a melittin-mediated line tension of the pore [86]. In connection with melittin, Sakai N, Gerard D, Matile [86] reported the design, synthesis, and structure of rigid-rod ionophores of different axial electrostatic asymmetry. Studies in isoelectric, anionic, and polarized bilayer membranes confirmed a general increase in activity of uncharged rigid push-pull rods in polarized bilayers. The similarly increased activity of cationic rigid push-pull rods with an electrostatic asymmetry comparable to that of bee toxin melittin (positive charge near negative axial dipole terminus) is shown by fluorescence-depth quenching experiments to originate from the stabilization of, transmembrane rod orientation by the membrane potential. The authors suggest that their structural evidence for cell membrane recognition by asymmetric rods is unprecedented and of possible practical importance with regard to antibiotic resistance

4.6.2 *Artificial ion channels*

In a unique approach, Kobuke and Nagatani [76] reported that cholic acid was modified so as to represent the simplest expression of an artificial supramolecular ion channel by converting three hydroxyls to methyl ethers and a carboxyl to a methylene(trimethyl)ammonium grouping. It gave stable single channel currents showing several conductance values in a planar BLM. Later, Goto et al [85] reported single ion channel properties using a planar lipid bilayer method under symmetrical 500 mM KCl at pH 8.2 or 7.2. Voltage-dependent artificial ion channels were synthesized. Two cholic acid derivatives were connected through a m-xylylene dicarbamate unit at 3-hydroxyl groups. Terminal hydrophilic groups, carboxylic acid and phosphoric acid introduced asymmetries for 3 and hydroxyl and carboxylic acid for 4. Under basic conditions, these head groups in 3 and 4 are expected to be dissociating into -1/-2 (pH 8.2) and 0/-1 (pH 7.2), respectively. When 3 and 4 were introduced into the bilayer membrane under application of positive voltage (a positive-shift method), the current values at positive applied voltage were larger than the corresponding ones at the negative applied voltage. The current-voltage plots were fitted by curves through a zero point to show clear rectification properties. The direction of rectification could be controlled by positive- or negative-shift methods. Vectorial alignment of terminal head group charges by the voltage-shift incorporation is essential for giving voltage-dependent rectified ion channels [85].

4.6.3   *Muscle cells and Ryanodine Receptors (RyRs)*

Neyton and Pelleschi [94] reported single-channel recordings of high-conductance Ca^{2+}-activated K^+ channels from rat skeletal muscle inserted into planar lipid bilayer to analyze the effects of two ionic blockers, Ba^{2+} and Na^+, on the channel's gating reactions. The gating equilibrium of the Ba^{2+}-blocked channel was investigated through the kinetics of the discrete blockade induced by Ba^{2+} ions. Gating properties of Na^+-blocked channels could be directly characterized due to the very high rates of Na^+ blocking/unblocking reactions. While in the presence of K^+ (5 mM) in the external solution Ba^{2+} is known to stabilize the open state of the blocked channel. The authors showed that the divalent blocker stabilizes the closed-blocked state, if permeant ions are removed from the external solution (K^+ < 10 μM). Ionic substitutions in the outer solution induce changes in the gating equilibrium of the Ba^{2+}-blocked channel that are tightly correlated to the inhibition of Ba^{2+} dissociation by external monovalent cations. In permeant ion-free external solutions, blockade of the channel by internal Na^+ induces a shift (around 15 mV) in the open probability-voltage curve toward more depolarized potentials, indicating that Na^+ induces a stabilization of the closed-blocked state, as does Ba^{2+} under the same conditions. A kinetic analysis of the Na^+-blocked channel indicates that the closed-blocked state is favored mainly by a decrease in opening rate. Addition of 1 mM external K^+ completely inhibits the shift in the activation curve without affecting the Na^+-induced reduction in the apparent single-channel amplitude. The results suggest that in the absence of external permeant ions, internal blockers regulate the permeant ion occupancy of a site near the outer end of the channel. Occupancy of this site appears to modulate gating primarily by speeding the rate of channel opening [94].

Of particular interest to note is the paper by Shtifman and colleagues [89], who have investigated the effects of imperatoxin A {IpTx(a)} on local calcium release events in permeabilized frog skeletal muscle fibers, using laser scanning confocal microscopy in linescan mode. IpTx(a) induced the appearance of Ca^{2+}

release events from the sarcoplasmic reticulum that are similar to 2 s and have a smaller amplitude ($31 \pm 2\%$) than the "Ca^{2+} sparks" normally seen in the absence of toxin, The frequency of occurrence of long-duration imperatoxin-induced Ca^{2+} release events increased in proportion to IpTx(a) concentrations ranging from 10 nM to 50 nM. The mean duration of imperatoxin-induced events in muscle fibers was independent of toxin concentration and agreed closely with the channel open time in experiments on isolated frog ryanodine receptors (RyRs) reconstituted in planar lipid bilayer, where IpTx(a) induced opening of single Ca^{2+} release channels to prolonged subconductance states. These results suggest involvement of a single molecule of IpTx(a) in the activation of a single Ca^{2+} release channel to produce a long-duration event. Assuming the ratio of full conductance to subconductance to be the same in the fibers as in bilayer, the amplitude of a spark relative to the long event indicates involvement of at most four RyR Ca^{2+} release channels in the production of short-duration Ca^{2+} sparks [89]. Recently, Tang and associates [109] designed an experiment to determine whether FK-506 binding protein 12.6 (FKBP12.6), an accessory protein of the RyRs, plays a role in cADPR-induced activation of the RyRs. A 12.6-kDa protein was detected in bovine coronary arterial smooth muscle (CASM) and cultured CASM cells by being immunoblotted with an antibody against FKBP12, which also reacted with FKBP12.6. With the use of planar lipid bilayer clamping techniques, FK-506 (0.01-10 μM) significantly increased the open probability (NPO) of reconstituted RyR/Ca^{2+} release channels from the SR of CASM. This FK-506-induced activation of RyR/Ca^{2+} release channels was abolished by pretreatment with anti-FKBP12 antibody. The RyRs activator cADPR (0.1-0 μM) markedly increased the activity of RyR/Ca^{2+} release channels. In the presence of FK-506, cADPR did not further increase the NPO of RyR/Ca^{2+} release channels. Addition of anti-FKBP12 antibody also completely blocked cADPR-induced activation of these channels, and removal of FKBP12.6 by preincubation with FK-506 and subsequent gradient centrifugation abolished cADPR-induced increase in the NPO of RyR/Ca^{2+} release channels. Tang et al. [109] conclude that FKBP12.6 plays a critical role in mediating cADPR-induced activation of RyR/Ca^{2+} release channels from the SR of bovine CASM.

Of interest is the paper by Kominkova and associates [62] who examined the effect of ethanol on single potassium channels derived from plasma membranes of bovine tracheal smooth muscles. The observed potassium channels had a conductance of 296 ± 31 pS in symmetrical 250 mmol/l KCl solutions, and exhibited a voltage- and Ca^{2+}-dependence similar to BKCa channels. Ethanol at 50, 100 and 200 mM concentrations increased the probability of open potassium channels to 112 ± 5, 127 ± 7 and $121 \pm 13\%$, respectively. Kominkova and associates suggested that increased activity of the BKCa channels by ethanol hyperpolarizes the plasma membrane and thus may contribute to relaxation of tracheal smooth muscle.

Maksaev, Mikhaliov, and Frolov [58] studied the low pH-induced fusion of single influenza viral particles with a small patch of bilayer lipid membrane, and showed that fusion is accompanied by short-term reversal conductance changes of the BLM patch. The mean amplitude of conductance changes depends on the virus strain, and the electrical activity observed was abolished by amantadine known as an inhibitor of the proton channels formed by influenza virus M2 membrane protein, In order to clarify the origin of observed electrical activity the special experiments were carried out. In these experiments the pH of the medium on the trans-side of the BLM, that had no contact with viral particles initially, was varied. Therefore, it was possible to change pH of the medium inside the virion through the fusion pore, bypassing the M2 proton channels. Maksaev et al. found that upon blockage of M2-channels the fusion was accompanied by electrical activity, only when the medium on the trans-side of the BLM (and therefore, inside the virus) was acidified (pH 5.0). Hence, the electrical activity is initiated by the lowering of the pH inside the virion. It is known that at low pH the inner viral core lining the viral membrane and formed by M1-protein dissociates into monomers. The authors assume that this may cause the formation of defects in the lipid viral membrane, and which are registered as conductance changes [58].

The properties of ryanodine receptors (RyRs) have been extensively investigated. Lokuta et al [95] studied RyRs from rat dorsal root ganglia (DRGs) The density of RyRs (B-max) determined by $[H_3]$ ryanodine binding was 63 fmol/mg protein with a dissociation constant (K_d) of 1.5 nM. $[H_3]$ ryanodine binding increased with caffeine, decreased with ruthenium red and tetracaine, and was insensitive to millimolar concentrations of Mg^{2+} or Ca^{2+}. Lokuta and associates reconstituted in planar lipid bilayers were Ca^{2+}-dependent and displayed the classical long-lived subconductance state in response to ryanodine; however, unlike cardiac and skeletal RyRs, they lacked Ca^{2+}-dependent inactivation. Antibodies against RyR3. but not against RyR1 or RyR2, detected DRG RyRs. Thus, DRG RyRs are immunologically related to RyR3, but their lack of divalent cation inhibition is unique among RyR subtypes. Of special significance is the work of Jona et al [70], who studied Mg^{2+}-induced inhibition of the skeletal ryanodine receptor/calcium-release channel (RyR), in the presence and absence of ATP under isolated conditions, and *in situ*, by examining the RyR incorporated into a planar lipid bilayer and the calcium release flux (R-rel) in isolated single fibres mounted in the double Vaseline gap system. When the incorporated RyR had been activated by calcium (50 μM) in the absence of ATP, the magnesium-induced inhibition showed co-operativity with a Hill coefficient (N) of 1.83 and a half-inhibitory concentration

(IC50) of 635 µM When the open probability was measured in the presence of 5 mM ATP and at a low calcium concentration, the magnesium-induced inhibition was non-cooperative (N=1.1, IC50= 860 muM) In isolated muscle fibres, in the presence of ATP, lowering the intracellular magnesium concentration $[Mg^{2+}]_i$ increased the maximal R-rel and shifted its voltage dependence to more negative membrane potentials. Increasing $[Mg^{2+}]_i$ had the opposite effect. At the concentration required for the measurements from isolated fibres, ATP had a full activatory effect on the isolated channel. At a low calcium concentration, the RyR had two ATP-binding sites with half-activatory concentrations of 19 and 350 µM, respectively [70].

Regarding RyRs, it is known that cADP ribose (cADPR) serves as second messenger to activate the RyRs of the sarcoplasmic reticulum (SR) and mobilize intracellular Ca^{2+} in vascular smooth muscle cells. However, the mechanisms mediating the effect of cADPR remain unknown. In this connection, Lee, Meissner, and Kim [97] examined the effect of quercetin on the behavior of single skeletal CRC (Ca^{2+} release channel) in planar lipid bilayer. Quercetin, a bioflavonoid, is known to affect Ca^{2+} fluxes in sarcoplasmic reticulum, although its direct effect on CRC in sarcoplasmic reticulum has remained to be elucidated, until now. The effect of caffeine was also studied for comparison. Caffeine showed a similar tendency. Analysis of lifetimes for single CRC showed that quercetin and caffeine led to different mean open-time and closed-time constants and their proportions. Addition of 10 µM ryanodine to CRC activated by quercetin or caffeine led to the typical subconductance state (similar to 54%) and a subsequent addition of 5 µM ruthenium red completely blocked CRC activity. Quercetin affected only the ascending phase of the bell-shaped Ca^{2+} activation/inactivation curve, whereas caffeine affected both ascending and descending phases. These characteristic differences in the modes of activation of CRC by quercetin and caffeine suggest that the channel activation mechanisms and presumably the binding sites on CRC are different for the two drugs [97].

4.6.1. Planar lipid bilayers in DNA and in disease research
DNA. Transferred DNA (T-DNA) transfer from *Agrobacterium tumefaciens* into eukaryotic cells is the only known example of interkingdom DNA transfer. T-DNA is a single-stranded segment of Agrobacterium's tumor-inducing plasmid that enters the plant cell as a complex with the bacterial virulence proteins VirD2 and VirE2, The VirE2 protein is highly induced on contact of *A. tumefaciens,* a plant host and has been reported to act in late steps of transfer. One of its previously demonstrated functions is binding to the single-stranded (ss) T-DNA and protecting it from degradation. Recent experiments suggest other functions of the protein. Dumas and

colleagues [98] used a combination of planar lipid bilayer experiments, liposome swelling assays, and DNA transport experiments demonstrated that VirE2 can insert itself into lipid bilayer and form channels. These channels are voltage gated, anion selective, and single-stranded DNA-specific and can facilitate the efficient transport of single-stranded DNA through membranes. These experiments demonstrate a VirE2 function as a transmembrane DNA transporter, which could have applications in gene delivery systems, according to the authors [98].

Diseases associated with ion channel malfunction. It is believed that more than one gene may control the functioning in an ion channel, therefore different genetic mutations may manifest with the same disorder. However, the complex nature of the genetic, and molecular structures of channels is still obscure. There is evidence that Na^+ channel may be involved in familial generalized epilepsy with febrile seizures, and hypokalemic periodic paralysis, whereas K^+ channel may be associated with benign infantile epilepsy, and episodic ataxia. Familial hemiplegic migraine, central core disease, malignant hyperthermia syndrome, and congenital stationary night blindness are associated with Ca^{2+} channels. Chloride ion channels may play a role in myotonia congenitas [85]. For instance, Kim et al [107] reported that the C-terminal 105 amino acid fragment of beta-amyloid precursor protein (CT105) is highly neurotoxic. To obtain insights into its cytotoxic effect, Kim and colleagues further examined the ionophoric effects of CT105 (10-1000 nM) on planar lipid bilayers. Macroscopic membrane conductance increased with CT105 concentration and its ionophoric effect. The mean unitary conductance of CT105-induced channels was 120 pS and open-state probability was close to 1 at voltages from -80 to +80 mV. CT105-induced channels were selective to cations (P_K/P_{Cl} = 10.2), being most selective to Ca^{2+}. These findings suggest that CT105, a C-terminal fragment of beta-amyloid precursor protein (AβP), can cause direct neurotoxic effects by forming Ca^{2+} permeable cation channels on neuronal membranes (*see* the following section).

Alzheimer's disease (AD). Amyloid beta protein/peptide (AβP) is the primary constituent of senile plaques, a defining feature of Alzheimer's disease (AD). Aggregated AβP is toxic to neurons, but the mechanism of toxicity is uncertain. One hypothesis is that interactions between AβP aggregates and cell membranes mediate AβP toxicity. More specifically, the presence of extracellular amyloid plaques that are composed predominantly of the AβP. Diffuse plaques associated with AD are composed predominantly of AβP42, whereas senile plaques contain both AβP40 and AβP42. However, their mechanistic roles in AD pathogenesis are poorly understood. Globular and nonfibrillar AβPs are continuously released during normal metabolism. Using techniques of atomic force microscopy (AFM), laser confocal microscopy, electrical recording, and biochemical assays, Lin, Bhatia, and Lal [100] have examined the molecular conformations of reconstituted globular AβPs as well as their real-time and acute effects on neuritic degeneration. AFM of AβP(1-42) shows globular structures that do not form fibers in physiological-buffered solution for up to 8 h of continuous imaging. AFM of AβP(1-42) reconstituted in a planar lipid bilayer reveals multimeric channel-like structures. Consistent with these AFM resolved channel-like structures, biochemical analysis demonstrates that predominantly monomeric AβPs in solution form stable tetramers and hexamers after incorporation into lipid membranes. Electrophysiological recordings demonstrate the presence of multiple single channel currents of different sizes. At the cellular level, AβP(1-42) allows calcium uptake and induces neuritic abnormality in a dose- and time-dependent fashion. At physiological nanomolar concentrations, rapid neuritic degeneration was observed within minutes; at micromolar concentrations, neuronal death was observed within 3-4 h. These effects are prevented by zinc (an AβP channel blocker) and by the removal of extracellular Ca^{2+}, but are not prevented by antagonists of putative AβP cell surface receptors. Thus, AβP channels may provide a direct pathway for Ca^{2+}-dependent AβP toxicity in AD, according to Lin et al [100]. In this connection, the recent study of Yip, Darabie, and McLaurin [101] demonstrated the association of AβP42 with planar bilayers composed of total brain lipids, which results initially in peptide aggregation and then fibre formation. Modulation of the cholesterol content is correlated with the extent of AβP42-assembly on the bilayer surface. Although AβP42 was not visualized directly on cholesterol-depleted bilayers, fluorescence anisotropy and fluorimetry demonstrate AβP42-induced membrane changes. The results demonstrate that the composition of the lipid bilayer governs the outcome of AβP interactions, according to the authors [101]. Regarding the above, Ji, Wu, and Sui [102] have shown that AβP is able to be inserted into lipid bilayer and cholesterol is an important factor affecting the membrane insertion of AβP40, which may potentially inhibit the fibril formation. Interactions between AβP and neuronal membranes have been postulated to play an important role in the neuropathology of AD. The membrane insertion ability of AβP is critically controlled by the ratio of cholesterol to phospholipids. At a low concentration of cholesterol AβP prefers to stay in membrane surface region mainly in an AβP structure. In contrast, as the ratio of cholesterol to phospholipids rises above 30 mol%, AβP could be inserted spontaneously into lipid bilayer by its C terminus. During lipid bilayer insertion, AβP generates about 60% α-helix and removes almost all β-sheet structure. Fibril formation experiments show that such membrane insertion can reduce fibril formation. These findings reveal a possible pathway by which AβP prevents itself from aggregation and fibril formation by membrane insertion, according to the authors [102]. In this regard, it is known that a variety of AβP fragments formed channels, as reported by Kourie, Henry, and Farrelly [103] in a detailed study. They used the lipid bilayer technique to characterize the biophysical and pharmacological properties of several ion channels formed by embedding AβP 1-40 into lipid membranes. Based on the conductance, kinetics, selectivity, and pharmacological properties, several AβP 1-40-formed ion channels have been identified. For example, the 1-40 formed "bursting" fast cation channel by (a) a single channel conductance of 63 pS (250/50 mM KCl cis/trans) at +140 mV, 17 pS (250/50 mM KC/ cis/trans) at -160 mV, and the nonlinear current-voltage relationship drawn to a third-order polynomial, (b) selectivity sequence $P_K > P_{Na} > P_{Li} = 1.0:0.60:0.47$, respectively, (c) P_o of 0.22 at 0 mV and 0.55 at +120 mV, and (d) Zn^{2+}-induced reduction in current amplitude, a typical property of a slow block mechanism. The authors have concluded that the formation of AβP based oligomers could be an important common step in the formation of cytotoxic amyloid channels [103].

Of speciat interest to note, Harroun, Bradshaw, and Ashley [87] reported human islet amyloid polypeptide (hIAPP), cosecreted with insulin from pancreatic β cells, misfolds to form amyloid deposits in non-insulin-dependent diabetes mellitus (NIDDM). Like many amyloidogenic proteins, hIAPP is membrane-active: this may be significant in the pathogenesis of NIDDM. Non-fibrillar hIAPP induces electrical and physical breakdown in planar lipid bilayers, and IAPP inserts spontaneously into lipid monolayers, markedly increasing their surface area and producing Brewster angle microscopy reflectance changes. Congo red inhibits these activities, and they are completely arrested by rifampicin, despite continued amyloid formation. The results of Harroun, Bradshaw, and Ashley support the idea that non-fibrillar IAPP is membrane-active, and may have implications for therapy and for structural studies of membrane-active amyloid. Using in situ AFM, Last, Waggoner, and Sasaki [105] have imaged nanoscale structural reorganization of a lipid bilayer membrane induced by a chemical recognition event. Supported lipid bilayers, composed of distearylphosphatidylcholine (a synthetic lipid), functionalized with a Cu^{2+} receptor, phase-separate into nanoscale domains that are distinguishable by the 0.9 nm height difference between the two molecules. Upon binding of Cu^{2+} the electrostatic nature of the receptor changes, causing a dispersion of the receptor molecules and subsequent shrinking of the structural features defined by the receptors in the membrane. Complete reversibility of the process was demonstrated through the removal of metal ions with EDTA.

In an interesting study, Kremer, Sklansky, and Murphy [104] reported that a positive correlation between the AβP aggregation state and surface hydrophobicity, and the ability of the peptide to decrease fluidity in the center of the membrane bilayer. Further, AβP complexes increased the steady-state anisotropy of 1,6-diphenyl-1,3,5-hexatriene (DPH) embedded in the lipid bilayer of phospholipids with anionic, cationic, and zwitterionic headgroups, suggesting that specific charge-charge interactions are not required for AβP-membrane interactions. AβP did not affect the fluorescence lifetime of DPH, indicating that the increase in anisotropy is due to increased ordering of the phospholipid acyl chains rather than changes in water penetration into the bilayer interior. AβP aggregates affected membrane fluidity above, but not below, the lipid phase-transition temperature; it did not alter the temperature or enthalpy of the phospholipid phase transition. AβP induced little to no change in membrane structure or water penetration near the bilayer surface. Overall, these results suggest that exposed hydrophobic patches on the AβP aggregates interact with the hydrophobic core of the lipid bilayer, leading to a reduction in membrane fluidity. Decreases in membrane fluidity could hamper functioning of cell surface receptors and ion channel proteins [104].

Concerning the membrane lipid bilayer, Mason, Olmstead, and Jacob [105] reported free radical-induced damage to lipid and protein constituents of neuronal membranes that may contribute to the pathophysiology of neurodegenerative diseases such as AD. The development of an effective inhibitor of oxidative stress represents an important goal for the treatment of AD. In the study of Mason and associates, the intrinsic antioxidant activity of lazabemide, a potent and reversible inhibitor of monoamine oxidase B (MAO-B), was tested in a membrane-based model of oxidative stress. Under physiologic-like conditions, lazabemide inhibited lipid peroxidation in a highly concentration-dependent manner. At low, pharmacologic levels of lazabemide (100 nM), there was a significant and catalytic reduction in lipid peroxide formation, as compared with control samples. The antioxidant activity of lazabemide was significantly more effective than that of either vitamin E or the MAO-B inhibitor, selegiline. The ability of lazabemide to inhibit oxidative damage is attributed to physico-chemical interactions with the membrane lipid bilayer, as determined by small angle x-ray diffraction methods. By partitioning into the membrane hydrocarbon core, lazabemide can inhibit the propagation of free radicals by electron-donating and resonance-stabilization mechanisms. These findings indicate that lazabemide is a potent and concentration-dependent inhibitor of membrane oxy-radical damage as a result of inhibiting membrane lipid peroxidation, independent of MAO-B interactions [105]. In this connection, Wang and Lee [109] reported that choline plus cytidine stimulate phospholipid production, and the expression and secretion of amyloid precursor protein in rat PC12 cells. The amyloid precursor protein (APP) is a transmembrane protein anchored in the membrane lipid bilayer. Choline and cytidine are major precursors of cell membranes, and are regulatory elements in membrane biosynthesis. Wang and Lee further examined the levels of cellular APP holoprotein and secreted APPs when rat PC12 cells are stimulated to undergo increase in membrane phospholipids by choline + cytidine (at various µM concentratons) treatment. They have shown that as phospholipids levels are increased by supplemental choline and cytidine treatment, the levels of cell-associated APP also rise stoichiometrically. Their results indicate that choline plus cytidine increase both phospholipid levels, and the expression and secretion in PC12 cells. The authors then concluded that agents that stimulate cellular membrane biosynthesis may be used in stimulating the secretion of neurotrophic APPs and neurite formation in Alzheimer's disease.

Cystic fibrosis. Using a planar lipid bilayer method for studying the self-assembly of rationally designed channel-forming peptides, Salay and colleagues [66] reported that a synthetic, channel-forming peptide, derived from the α-subunit of the glycine receptor (M2GlyR), has been synthesized and modified. In Ussing chamber experiments, apical N-K-4-M2GlyR (250 µM) increased transepithelial short-circuit current (I_{sc}) by 7.7 ± 1.7 and 10.6 ± 0.9 mA cm^{-2} in Madin-Darby canine kidney and T84 cell monolayers, respectively. These values are significantly greater than those previously reported for the same peptide modified by adding the lysines at the -COOH terminus. N-K-4-M2GlyR-mediated increases in I_{sc} were insensitive to changes in apical cation species. Pharmacological inhibitors of endogenous Cl$^-$ conductance had little effect on N-K-4-M2GlyR-mediated I_{sc}. Whole cell membrane patch voltage-clamp studies revealed an N-K-4-M2GlyR-induced anion conductance that exhibited modest outward rectification and modest time- and voltage-dependent activation. Planar lipid bilayer studies yielded results indicating that N-K-4-M2GlyR forms a 50-pS anion conductance, and shows therapeutic potential for the treatment of hyposecretory disorders such as cystic fibrosis. In this regard, it is of particular interest to note the work of Berdiev and co-workers [111], who in their search of the structural basis for gating of amiloride-sensitive Na$^+$ channels. These were studied in a cell-free planar lipid bilayer. Their results identify the N-terminus of the α-subunit as a major determinant of kinetic behavior of both homooligomeric and heterooligomeric ENaCs, although the carboxy-terminal domains of β- and χ-ENaC subunits play important role(s) in modulation of the kinetics of heterooligomeric channels. The authors also found that the cystic fibrosis transmembrane conductance regulator (CFTR) inhibits amiloride-sensitive channels, at least in part, by modulating their gating. The proposed mechanisms, however, do not completely override the gating mechanism of the α-channel [111].

Peptide. The horse apomyoglobin 56-131 peptide is a convenient object for studies on the recently discovered antimicrobial activities of haem-binding protein fragments called haemocidins. Mak and colleagues [112] reported the effect of this peptide on planar lipid bilayers and on liposomes of different lipid compositions. Micromolar concentrations of the apomyoglobin 56-131 fragment disrupt PS/PE planar lipid bilayers without discrete conductance changes. The observed detergent-like action is dependent on peptide concentration: the lower amount of peptide resulted in longer bilayer lifetime. The cholesterol has an inhibitory effect on peptide-induced liposome lysis from liposomes. Additionally, there was considerable lytic activity on liposomes formed from anionic lipids of the sort found in bacterial membranes. Circular dichroism (CD) experiments showed that the peptide had a disordered structure in aqueous solutions and folds gradually to form helices in both membrane-mimetic trifluoroethanol solutions as well as in liposome suspensions. Related to these is the work reported by Broughman et al [113], who have synthesized a channel-forming peptide, derived from the α-subunit of the glycine receptor (M2GlyR), and modified by adding four lysine residues to the NH_2 terminus (N-K-4-M2GlyR). Planar lipid bilayer studies yielded results indicating that N-K-4-M2GlyR forms a 50-pS anion conductance with a $k_{1/2}$ for Cl⁻ of 290 meq. These results indicate that N-K-4-M2GlyR forms an anion-selective channel in epithelial monolayers and shows therapeutic potential for the treatment of hyposecretory disorders such as cystic fibrosis.

4.6.2. Microtubules

The microtubules consisting of tubulin subunits are ubiquitous cytoskeletal elements implicated in a variety of cellular functions. It is probable that they play an important role in the integration of the membrane processes and in the structural-functional organization inside the cells by mediating long-range interactions between membranes and by functioning as an intermembrane linking system. To conclude this section, therefore, mention should be made concerning a two-BLM system made from brain lipids, formed in the presence of the brain tubulin, and having their electrical potentials measured simultaneously. Two different techniques have been developed for studying the interrelationships between two BLMs. In both systems, evidence for microtubule-assisted interconnections between the membranes was obtained. When electric pulses with amplitude 60-70 mV and duration 30 ms were applied across one of the BLMs, a displacement in the potential across the other BLM was observed, even when the membranes were separated. This effect was found only in the presence of microtubules. It was not observed in the presence of depolymerized microtubule protein, colchicine-treated tubulin, or albumin. On the basis of the data, Vassilev et al. have suggested that defined configurations of the microtubule fiber networks and bundles may form structural bridges and coupling between membranes, and that microtubules may play a similar role in the processes of information transfer in neurons and other cells. The findings suggest that the microtubule fiber networks may serve as an interconnecting system between membranes (see Ref. [6]).

5.   BLMs, MOLECULAR ELECTRONICS AND BIOTECHNOLOGY

In 1987 at the symposium on molecular electronics and biocomputers held in Budapest, Hungary, one of the participants suggested that '*Life is molecular electronics*!'. This is actually an extension of a remark made by Albert Szent-Gyorgyi who said a quarter of a century earlier that '*life, as we know it, is nothing but a movement of electrons*'. Indeed, in the 1970s the U. S. Air Force had a program called "Molecular Electronics", because the scientists thought they could find something in the basic structure of the molecule that would serve the function of traditional resistors, capacitors, diodes, etc. This was owing to a relentless decrease in the size of silicon-based microelectronics devices. The most important among these are limitations imposed by quantum-size effects and instabilities introduced by the effects of thermal fluctuations. As a result, these inherent problems of present-day systems have prompted scientists to look for alternative choices. Advancement in the understanding of biosystems such as photosynthetic apparatus and genetic engineering has allowed attention to be focused on the use of biomolecules (pigments, enzymes, receptors, etc.) Biomolecules have the advantages of specificity and functionality. The invention of scanning tunneling microscopy and atomic force microscopy has opened up the possibility of addressing and manipulating individual atoms and molecules [107,108]. Realization of the power of self-assembly principles has opened a novel approach for designing and assembling molecular structures with desired intricate architecture. The utility of molecules such as DNA as a three-dimensional, high-density memory element and its capability for molecular computing have been fully recognized but not yet realized. By the 1990s, symposiua were held, and monographies published, as well as an international journal devoted to molecular electronics was available (see Ref.[6] for details). Today, three overlapping areas of molecular electronics research are materials science, fabrication technology, and device architecture. These are based on the fusion of ideas and disciplines between membrane biophysics and molecular cell biology on one hand, and advances in microelectronics on the other. For instance, the visual receptor of the eye is one of basic structures of Nature's sensors and devices [6,82,114]. Thus, an experimental bilayer lipid membrane separating two phases (liquid/liquid or liquid/solid) is an unique approach in fabricating molecular junctions in which an ultrathin (< 5 nm) layer of oriented lipid/organic molecules is interposed between two electric conductors. The liquid-crystallline lipid bilayer becomes a component in an electronic circuit and exhibits properties that depend strongly on molecular structure without pinholes. It is worth noting that bonding between the substrate (e.g. metal or hydrogel) and the BLM is physical and electrostatic, and thus differs fundamentally from that of the monolayers of alkane thiols on metal surfaces. Self-assembled planar lipid bilayer-based molecular junctions represent a new paradigm for molecular electronics, which shows promising electronic behavior and is potentially low cost [6]. A few selected examples are given below.

5.1 Molecular Electronics

This discipline involves expertises from several branches of science, namely, various materials science, fabrication technology, and device archtecture. While passive biomaterials such as supported lipid bilayers are involved in anchoring the active biomolecules, the latter are involved in switching and/or signal transduction. Self-assembly provides a simple route to organize suitable organic molecules on metals (e.g. stainless steel, Pt, Au) and selected nanocluster surfaces by using monolayers of long chain organic molecules with various functionalities such as -COOH, -SH, -NH_2, silanes, etc. These surfaces can be effectively used to build-up interesting nano-level architectures. Flexibility with respect to the terminal functionalities of the organic molecules allows the control of the hydrophilicity or hydrophobicity of metal surface, while the selection of length scale can be used to tune the distant-dependent electron transfer behavior. Organo-inorganic materials tailored in this fashion are extremely important in nanotechnology to construct nanoeletronic devices, sensor arrays, supercapacitors, catalysts, rechargeable power sources etc. by virtue of their size and shape-dependent electrical, optical or magnetic properties (see Chapters 27,30,33).

Presented here are several convergent synthetic routes to conjugated oligo(phenylene ethynylene)s. Some of these oligomers are free of functional groups, while others possess donor groups, acceptor groups, porphyrin interiors, and other heterocyclic interiors for various potential transmission and digital device applications. The synthesis of oligo(phenylene ethynylene)s with a variety of end groups for attachment to numerous metal probes and surfaces are presented. Some of the functionalized molecular systems showed linear, wirelike, current versus voltage (I/V) responses, while others exhibited nonlinear I/V curves for negative differential resistance (NDR) and molecular random access memory effects. Finally, the syntheses of functionalized oligomers are described that can form self-assembled monolayers on metallic electrodes that reduce the Schottky barriers. Information from the Schottky barrier studies can provide useful insight into molecular alligator clip optimizations for molecular electronics. For example, Purrucker et al [115] deposited highly resistive supported planar lipid membranes onto highly doped p-type silicon-silicon dioxide electrodes. Physical parameters of the substrates (e.g. dopant, doping ratio and oxide layer thickness) were optimized by a combined study using the ellipsometry and ac impedance spectroscopy. Lipid bilayer was deposited by fusion of small unilamellar vesicles [116,117,119], and the self-assembling of the homogeneous bilayer could be monitored as a function of time. Impedance spectroscopy over a wide frequency range (from 20 kHz to 10 mHz) enables separating membrane resistance and capacitance from the background signals. Membrane resistance amounted to $1 \times 10^8 \ \Omega \ cm^2$, and the capacitance was 0.7 $\mu F \ cm^{-2}$. The resistance obtained here is comparable to that of the freestanding black lipid membrane (c-BLM). Although the area of the supported membrane (0.5 cm^2) is much larger than that of the BLM (similar to 0.002 cm^2), the electrical properties were stable for more than a week. Gramicidin D was inserted into the membrane from trifluoroethanol solution, and activity of the channels was checked in terms of membrane conductance and ion selectivity. Functional incorporation of ion channels into the supported membrane suggested that the gramicidin monomers could diffuse over the membranes to form transmembrane pores (see Chapters 23, 36).

Preparation and characterization of ordered ultrathin organic films (from a few to several hundred nanometers) has recently attracted considerable attention because of the possibility of controlling order and interactions at the molecular level and has triggered several innovative applications ranging from molecular electronics to tribology. For example, Fromherz [125] has suggested recently the electrical interfacing of individual nerve cells and semiconductor chips, as well as the assembly of elementary hybrid systems made of neuronal networks and semiconductor microelectronics. The rationale is based on that, without electrochemical processes, coupling of the electron-conducting semiconductor and the ion-conducting neurons relies on a close contact of cell membrane and oxidised silicon with a high resistance of the junction and a high conductance of the attached membrane. Neuronal excitation can be elicited and recorded from the chip by capacitive contacts and by field-effect transistors with an open gate (see also Ref. [6]). Alternatively, monomolecular films prepared by self-assembly are attractive for several exciting applications because of the unique possibility of making the selection of different types of terminal functional groups as well as length scales more flexible. Chapter 33 of this book further discusses various applications of self-assembled lipid bilayers (s-BLMs) in molecular electronics ranging from biosensors to optoelectronic devices with specific examples (see Chapters 27,30).

Ivanov and associates [116] created a phospholipid-containing biochip by covalently immobilizing phospholipids on the optical biosensor's aminosilane cuvette and employed to monitor the interactions of the membrane and water-soluble proteins in cytochrome P450-containing monooxygenase systems with planary layers of dilauroyl-PE (DLPE) and distearoyl-PE (DSPE), differing in acyl chain length. They showed that the full-length membrane proteins-cytochromes (d-2B4 and d-b5), and NADPH-cytochrome P450 reductase (d-Fp)-readily incorporated into the phospholipids. The incorporation was largely due to hydrophobic interactions of membranous protein fragments with the lipid bilayer. However, electrostatic forces were also but not always involved in the incorporation process. They promoted d-Fp incorporation but had no effect on d-b5 incorporation. In low ionic strength buffer, no incorporation of these two proteins into the DSPE lipid layer was observable. Incorporation of d-b5 into the DLPE layer was abruptly increased at temperatures exceeding phospholipid phase transition point. Incorporation of d-2B4 was dependent on its aggregation state and decreased with increasing protein aggregability.

To conclude this section on molecular electronics, it is worth pointing out that, since the First symposium in Budapest, Hungary in 1987, we have come a long way toward the discipline, as evidenced by the recent school held in Pisa, Italy [127], the proceedings of which may be soon available.

5.2   Biotechnology

Tominaga [118] incorporated ferritin, a large protein, into a cast film of an ammonium lipid, $2C(16)N(+)Br(-)$ via ion exchange. The incorporation of ferritin into $2C(16)N(+)Br(-)$ film was enhanced at temperatures above the phase transition temperature of the bilayer, Tc. In contrast, at temperatures below Tc, incorporation was inhibited. The ferritin incorporated into $2C(16)N(+)Br(-)$ film on a graphite electrode showed well-defined redox waves. The results obtained indicate that lipid films are useful for electrochemical interface for a large protein such as ferritin. In this regard, Cherry, Bjornsen, and Zapien [121] investigated the electron transfer reactions of horse spleen ferritin at tin-doped indium oxide electrodes. Cyclic voltammetry reveals that ferritin adsorbs from phosphate solution into a layer composed of two adsorbed states, the relative packing density of the states depending strongly on ionic strength. Upon reduction, the initial layer reconstructs into a new one with faster electron-transfer kinetics. Reduction of the adsorbed layer in the presence of ethylenediaminetetraacetic acid (EDTA) results in the disappearance of any further voltammetric response indicating that transport of iron from the protein shell had been induced electrochemically. In this connection, mention should be made concerning scanning electrochemical microscopy (SECM) that has been used in addition to potentiometric measurements, as reported by Tsionsky and associates [122] and by Gyurcsanyi et al [123]. These researchers have used the technique to follow the time-dependent buildup of a steady-state diffusion layer at the aqueous-phase boundary of lead ion-selective electrodes (ISEs). Differential pulse voltammetry is adapted to SECM for probing the local concentration profiles at the sample side of solvent polymeric membranes. Major factors affecting the membrane transport-related surface concentrations were identified from SECM data and the potentiometric transients obtained under different experimental conditions (inner filling solution composition, membrane thickness, surface pretreatment). The amperometrically determined surface concentrations correlated well with the lower detection limits of the ion-selective electrodes [122,123].

Cheng, Bushby, Evans, et al [124] have reported that the charge selectivity of staphylococcal alpha-hemolysin (alpha HL), a bacterial pore-forming toxin, is manipulated by using cyclodextrins as noncovalent molecular adapters. Anion-selective versions of alpha HL, including the wild-type pore and various mutants, become more anion selective when beta-cyclodextrin (beta CD) is lodged within the channel lumen. By contrast, the negatively charged adapter, hepta-6-sulfato-beta-cyclodextrin (s(7)beta CD), produces cation selectivity. The cyclodextrin adapters have similar effects when placed in cation-selective mutant alpha HL pores. Most probably, hydrated Cl- ions partition into the central cavity of beta CD more readily than K+ ions, whereas s(7)beta CD introduces a charged ring near the midpoint of the channel lumen and confers cation selectivity through electrostatic interactions. The molecular adapters generate permeability ratios (PK+/PCl-) over a 200-fold range and should be useful in the de novo design of membrane channels both for basic studies of ion permeation and for applications in biotechnology.

5.3.1 Supported BLMs as Devices and Sensors

To study BLMs in detail over a period of time, the task has been a daunting one until about a decade or so ago, since a 5 nm BLM separating two aqueous solutions is an extremely labile structure with limited lifetime. Planar BLMs can now be formed in conventional manners with long-term stability (*see* Section III. Biotechnology of this book). Further, planar lipid bilayers (BLMs) can also be formed on various substrates (e.g. metal, hydrogels, etc), thereby opening the way for basic research and development work in the domains such as biotechnology, catalysis, electrochemistry, microelectronics, and membrane biophysics. These supported BLMs are referred to, respectively, as s-BLMs (metal-supported), as sb-BLMs (hydrogel-supported), or as t-BLMs (tethered to Au via alkalthiol). They are of great scientific interest and of practical merit due to their ability to mimicking biomembranes [6,7]. For example, as a prerequisite for the development of neuroelectronic devices, Braun and Fromherz [80] reported the capacitive stimulation of nerve cells from semiconductor chips. They found that on the primary response of a cell membrane to a voltage step applied to oxidized silicon and observed with a luminescent voltage-sensitive dye; exponential voltage transients with a time constant of 1-5 µs. The authors have assigned the short response to an electrical decoupling by a thin film of electrolyte between oxide and lipid bilayer. The high-pass filtering of stimulation is a crucial constraint for the development of silicon-to-neuron interfaces. Similar to conventional BLMs, supported s- and sb-BLMs provide a natural environment for embedding proteins or other compounds under non-denaturating conditions. In respect to practical applications they permit the preparation of ultrathin, high-resistance film with a well-defined orientation. on metals or semiconductors, and the incorporation of receptor proteins into these insulating structures for the design of biosensors and bioelectronic devices using electrical methods of detection [6,7,42]. It is worth pointing out that, with all other systems (e.g. Langmuir-Blodgett films), the compound of interest is immobilized in a rigid, solid-like structure, whereas in the BLM it is embedded. By embedding is meant that the compound(s) (i.e. membrane modifiers such as electron acceptors, donors, mediators, polypeptides, proteins, etc) of interest in the lipid bilayer is relatively free to adapt to its surroundings. The functions of lipid bilayers are mediated via specific modifiers, which assume their active conformations only in a liquid-crystalline environment. Further, the presence of the lipid bilayer greatly reduces the background noise (interference) and effectively excludes unwanted hydrophilic electroactive compounds from reaching the detecting surface causing undesired reactions. From specificity, selectivity, and design points of view, a supported planar lipid bilayer (s-BLM or sb-BLM) is an ideal natural environment for embedding a host of materials of interest. Hence, the s-BLM system offers a wider opportunity for biosensor development. Interested readers may find more details in Section III of this book.

6.   CONCLUDING REMARKS

In this essay on the lipid bilayer principle of biomembranes, at the outset, we trace the origins of the notion of lipid bilayer starting with 'black holes' in soap bubbles and films; the very first ultrathin films quantitatively characterized. Next, using the monolayer technique of Langmuir, Gorter and Grendel in 1925 put forth the bimolecular leaflet model of the plasma membrane, which has dominated our thinking ever since. In the 1950s, the 'unit membrane hypothesis' of Robertson, and in 1961, the experimental realization of bilayer lipid membranes *in vitro* led eventually to the fluid mosaic model of biomembranes. Shortly afterwards, electrical activities in terms of 'discrete ion channel conductance' of the nerve and light-induced redox reactions in photosynthesis have been elucidated with the aid of experimental planar BLMs and liposomes. In nerves, translocation of ions across protein channels embedded in the lipid bilayer play the pivotal role. In photosynthesis, light absorption by pigments confined in the lipid bilayer, initiates electronic charge generation and separation, leading eventually to redox reactions on opposite sides of the membrane. Presently, electron and charge transfer processes through the lipid bilayer are being actively investigated, from both theoretical and biotechnological viewpoints. The crucial role played by the lipid bilayer may be summarized as follows. Living organisms are made of cells bound by their bilayer lipid membranes. Lipid bilayers are self-assembling entities; each is organizing a particular combination of phospholipids in the form of a liquid-crystalline matrix with other constituents (e.g. proteins) embedded in it. This lipid bilayer, existing in all biomembranes, is most unique; it serves not merely as a physical barrier but also functions as a two-dimensional medium for rejoinder. Also, the liquid-crystalline lipid bilayer acts as a conduit for ion transport, as a framework for antigen-antibody binding, as a bipolar electrode for redox reactions, as a reactor for energy conversion (e.g. light to electric to chemical). Additionally, a modified lipid bilayer serves as an outfit for signal transduction (e.g. as a component of a biosensor), and numerous other functions as well. All these myriad activities require the ultrathin, liquid-crystalline lipid bilayer [6,7].

The advancements in biomembranes that have occurred in the last half of the 20th century have significantly increased our ability to characterize the composition and molecular structure of biomaterials. Today, after 4 decades of research and development, bilayer lipid membranes (planar lipid bilayers), along with liposomes [3,119], are well-established disciplines in certain areas of membrane biophysics and cell biology, and in biotechnology [6,7]. For example, a modified lipid bilayer performs as a transducer for signal transduction (i.e. sensing). The development of conventional BLMs (black lipid membranes, or planar lipid bilayers) and later supported lipid bilayers (s-BLMs, sb-BLMs, and t-BLMs), have made it possible for the first time to study, directly, electrical properties and transport phenomena across a 5 nm lamella separating two phases (see Table 1). Supported BLMs, formed on the tip of metallic wires, conducting glasses, and gel substrates, as well as on microchips, possess properties resembling biomembranes. These self-assembled, supported BLMs have opened research opportunities in studying hitherto unapproachable phenomena at interfaces. Some recent findings demonstrate potentials for investigating processes at solid-liquid interfaces as well. As a result of these studies, biomembranes are now acknowledged as the basic structure of Nature's sensors and molecular devices. For example, the plasma membrane of cells provides sites for a host of ligand-receptor interactions such as the antigen-antibody binding. To impart relevant functions in BLMs, a variety of compounds such as ionophores, enzymes, receptors, pigments, and so on have been embedded. Some of these embedded compounds cause the BLMs to exhibit non-linear phenomena, or to display photoelectric effects. A modified or reconstituted BLM is viewed as a dynamic system that changes both in time and in response to environmental stimuli. The self-assembled lipid bilayer, the crucial component of most, if not all biomembranes, is in a liquid-crystalline and dynamic state. A functional cell membrane system, based on self-assembled lipid bilayers, proteins, carbohydrates and their complexes, should be considered in molecular and electronic terms; it is capable of supporting ion or/and electron transport, and is the site of cellular activities in that it functions as a 'device' for either energy conversion or signal transduction. Such a system, as we know intuitively, must act as some sort of a transducer capable of gathering information, processing it, and then delivering a response based on this information. In the past, we were limited by our lack of sophistication in manipulating and monitoring such a lipid bilayer system. Today, membrane biophysics is a matured field of research, as a result of applications of many disciplines and techniques including interfacial chemistry, electrochemistry, voltage- and patch-clamp techniques, and spectroscopy. Concurrently, much advancement has taken place in microelectronics and biomedical sciences. The combination of these advances have allowed the development of the membrane model for interface science, where the ultimate goal is to gain a detailed understanding of how the surface and interfacial properties of a lipid bilayer regulate the bioreactivity of a cell interacting with that interface [7,117]. Ample evidence shows that the interfacial properties of a biomembrane are directly related to biological arrangement such as protein adsorption and cell growth . We now know a great deal about the structure of biomembranes, 'ion pumps', electroporation, membrane channels, etc. In membrane reconstitution experiments, for example, the evidence is that intracellular signal transduction begins at membrane receptors. It should be reiterated that the research area covered in this chapter is highly interdisciplinary. Lipid bilayer research has been benefited by a cross-fertilization of ideas among various branches of sciences (see Table 2 at the end of this section that presents a chronology of the lipid bilayer research, since its inception).

It seems likely that the devices based on 'smart' materials may be constructed in the form of a hybrid structure, for example, utilizing both inorganic semiconducting nanoparticles and synthetic lipid bilayers. More time and effort are necessary before devices that can transcend existing ones will become readily available. The biomimetic approach to practical applications is unique and full of exciting possibilities. By gleaning the design principles from Nature's successful products and applying them to our research and development from which we believe advanced sensors and molecular devices will ultimately be based. Future directions and opportunities for research scientists and development engineers working in biomedical and environmental research include exploiting biological knowledge, precision immobilization, self-assembly, nanofabrication, control of non-specific reactions, and biomimetics. Finally, concerning the lipid bilayer principle of biomembranes, the significance of experimental bilayer lipid membranes for modern research will be increased incessantly in the years to come.

Table 2

Origins of the Lipid Bilayer Concept and its Experimental Realization:
A Chronology

Year	Major findings and insight
1672	Robert Hooke coined the word 'cell' used today, and his observation of 'black' holes in soap bubbles and films [7].
1704	Isaac Newton estimated the thickness of the 'blackest' soap film to be $3/8 \times 10^{-6}$ inch [7] Naegeli and C. Cramer described the cell membrane as barrier essential to explain osmosis in plant cells (see Ref. [6])
1877	Pfeffer accounted the osmotic behavior of plant cells and recognized the boundary between the protoplasma and its environment must constitute an osmotically semipermeable membrane, postulated the existence of an invisible (under light microscope) plasma membrane of the cell.
1888	W. Nernst developed a theory of electrical potentials based on diffusion of ions in solution. This theory (called Nernst potential) became and still is fundamental in modeling ion flux across biomembranes in electrophysiology
1890	Overton, over a ten-year period, carried out some 10,000 experiments with more than 500 different chemical compounds. He measured the rate of entrance of compounds into cells and compared this with the partition coefficients of those compounds between olive oil and aqueous solution. He found that fatty compounds such as diethyl ether with larger partition coefficients readily entered the cell. That is, following the chemists' rule of 'like-dissolves-like', the cell membrane must be oily or lipid-like. Overton concluded that there must be a lipoid film of lecithin and cholesterol in the cell membrane separating the cytoplasm from its surroundings. Overton's findings were later confirmed and extended by Collander who also found that compounds such as urea, glycerol, and ethylene glycol with small olive oil/water partition coefficients yet readily penetrated the cell. To explain this 'anomaly', water-filled pores in the plasma membrane were later proposed.
1900s	Hober, in 1910, found that suspensions of intact red blood cells (RBC) have a high electrical resistance, while the cytoplasm has conductivity similar to that of physiological saline. From this fact, Hober concluded that the cell membrane has a high electrical resistance. At that time the dominant theory for nerve cells was due to Bernstein, whose hypothesis postulated that a cell consisted of a semipermable membrane capable of electrical activities and recordable as an electric potential difference across the membrane. Changes in membrane permeability would ensure a change in this potential difference. At rest, the membrane was permeable to K^+ only. Although later shown to be incomplete, the Bernstein hypothesis stimulated much discussion on the action potential among neuroscientists. 1917 I. Langmuir developed the monolayer technique; and reported orientation of "soapy" (amphipathic) molecules at interfaces could be measured with a 'yard stick'.
1925	Fricke investigated the conductivity and capacity of the RBC suspension that was measured as a function of frequency. At low frequencies the impedance of the suspension of RBC is very high, whereas at high frequencies the impedance decreases to a low value. To explain his findings, Fricke proposed a model that the RBC was

surrounded by a thin layer of low dielectric material electrically equivalent to a resistor (R_m) and a capacitor (C_m) in parallel. Thus, the lines of current flow around the RBC at low frequencies. At very high frequencies, the resistance becomes very low because all the current is shunted through the capacitor. Using the formula for a parallel plate, Fricke determined the capacitance (C_m) of the RBC to be 0.81 μF cm^{-2}.

1925   Gorter and Grendel proposed lipid bilayer structure for cell membranes; surface area covered by lipids extracted from red blood cells on water surface is twice as large as original surface of red blood cells. Further, they concluded in their paper "*.. of which the polar groups are directed to the inside and to the outside, in much the same way as Bragg supposes the molecules to be oriented in a 'crystal' of fatty acid, and as the molecules of a* soap bubble *are according to Perrin.*" [5].

1935   Danielli and Davson's membrane model of globular proteins on surface of lipid bilayer; this model specifically excludes transmembrane proteins based on the previously shown hydrophilic surface of globular proteins

Voltage clamp technique was developed to study macroscopic ion currents in neurons; Hodgkin, Huxley and Katz published the resting and the action potential recordings form single nerve fibers

1949   Mitochondria are shown to be organelles responsible for oxidative phosphorylation and containing enzymes of Krebs cycle.

1950s   Unit membrane hypothesis was proposed by J. D. Robertson (*J. Cell Biol.*, 91 (1981) 189s-204s.

1961   Chemiosmotic theory by Peter Mitchell postulated that a proton gradient as energy source for ATP synthesis in mitochondria

1962   The BLM technique-- Black lipid membrane (BLM); successful formation of the first planar lipid bilayer (reported by a group led by D. O. Rudin). Reconstitution of a nerve membrane *in vitro* [2]

1964   Application of electrochemical impedance spectroscopy (EIS) to BLM research; Formation of BLMs with simple composition (see [6])

1965   Liposomes – Bangham reported vesicles with bilayer lipid structures [3]

BLM formation by the Langmuir-Blodgett technique [1974]

1966   Formation of 'black' lipid membranes by oxidation products of cholesterol", *Nature*, 212, 718 (1966); Antigen-antibody and enzyme-substrate interactions [1974]

1967   Ion selectivity – valinomycin for K$^+$ ions; formation of BLMs from surfactants (detergents); "Some physical properties of bimolecular lipid membranes produced from new lipid solutions", *Nature*, 215, 1199 (1967)

1968   "Light-induced phenomena in black lipid membranes constituted from photosynthetic pigments", *Nature*, 219, 272 (1968).

1969   "Carotenoid BLM: An experimental model for the visual receptor membrane", *Nature*, 224, 1107 (1969); Discrete ion channels in BLMs [7]; unit channel structure for Gramicidin A peptide confirmed by Hladky and Haydon using BLMs by demonstrating "the discreteness of conduction change in planar lipid bilayers in the presence of certain antibiotics"; Unilamellar liposomes used for membrane protein reconstitution to study ion flux (Huang); this single membrane liposomes revolutionize transport studies because they can be prepared in large quantities.

1970   "Electronic processes in bilayer lipid membranes", *Nature*, 227, 1232 (1970). Light-induced redox reactions in pigmented BLMs, water photolysis via pigmented BLM was proposed; redox reactions and electronic processes in BLMs.

1972   First monograph on BLMs published (M. K. Jian, The Bimolecular Lipid Membrane: A System, van Nostrand-Reinhold, NY); ATP-mediated active transport in BLMs demonstrated [1972].

1974   Second monograph on BLMs was published {Bilayer lipid membranes (BLM): Theory and Practice, Dekker, NY}; membrane reconstitution; kinetics of ion channels, pumps and transporters were demonstrated using these membrane systems.

       First electron microscopy derived structure of a membrane protein, bacteriorhodopsin at 0.7 nm resolution (Henderson and Unwin); structure shows seven transmembrane α-helices supporting the idea that α-helices but not β-sheets are the secondary structure of choice for membrane proteins; this idea strongly influences research on structure-function relationship of membrane protein; β-barrel models for bacterial porin; glycophorin, first integral membrane protein sequenced.

1978   Patch-clamp technique was developed by Neher and Sackmann; this modification of the voltage clamp setup was developed in the 1940s allows for the first time the measurement of single channel activity in BLMs and in living cells instead of the usual macroscopic currents; the existence of separate ion channels selective for either Na^+, K^+ or Ca^{2+} ions can be demonstrated at the single channel level.

1979   V.F. Antonov, Y.G. Rovin, L.T. Trifimov, A Bibliography of Bilayer Lipid Membranes: 1962-1975, All Union Institute for Scientific Information, Moscow.

1984   Electronic processes in BLMs in the absence of light, using cyclic voltammetry [6]

1989   BLMs on metal supports for practical applications [6]

1990   Electroporation, defined as a transiently increase.in membrane permeability due to a transient increase in the transmembrane voltage. The phenomenon is of great interest in understanding of basic molecular mechanism (S.Kakorin, E. Neumann, Bioelectrochem,, 56 (2002): 163, Tsong [17], Zimmermann [7], A Barnett, JC Weaver, Bioelectroch Bioener 25 (1991): 163)

1999   Application of Scanning Electrochemical Microscopy to BLMs [7,122,123]

2000   A textbook on BLMs and liposomes published (see Ref. [6])

2001   For recent references regarding planar lipid bilayer research in physiology and biotechnology, see Ref. [7]

2002   To mark the 40th anniversary of bilayer lipid membrane research, a short article has been prepared for the occasion: *see* A. Ottova and HT Tien, Bioelectrochemistry, 56: (2002) 171-173.

Note: Where no specific references are given, detailed citations may be found as cited by the year given above (e.g., 1972, 1974, 1979, 2000, 2001, 2002).

REFERENCES

[1]   P.Fishman (Ed.), Symposium on the Plasma Membrane, American Heart Association and
       New York Heart Association, New York City, December 8-9, 1961. Published as a
       supplement in Circulation, 26 (1962) 1167.
[2]   P. Mueller, D.O. Rudin, H.T. Tien and W.C. Wescott, Nature, 194 (1962) 979
[3]   A.D. Bangham, Surrogate cells or Trojan horses, BioEssays, 17 (1995) 1081
[4]   R. M. Burton (ed.), Sym. on lipid monolayers, bilayer models and cellular membranes,
       May 7-10, 1967, Am Oil Chemists' Soc., 45 (1968) 201
[5]   E.Gorter, and F. Grendel, J. Expt. Med., 41 (1925) 439
[6]   H. T Tien and A.L. Ottova, Membrane Biophysics: As viewed from experimental bilayer
       lipid membranes (planar lipid bilayers and spherical liposomes), Elsevier, Amsterdam,
       2000, 648 pp.
[7]   H.T. Tien and A.L. Ottova, J. Memb. Sci., 189 (2001) 83
[8]   H.T. Tien and A.L. Diana, Nature, 215 (1967) 1199; J. Theoret.Biol.16 (1967) 97
[9]   H.A. Rinia, Snel MME, van der Eerden JPJM, de Kruijff B., FEBS Lett. 501 (2001): 92
[10]   E. London and D.A. Brown, BBA-*Biomembranes*, 1508 (2000) 182
[11]   S. M.Campbell, S.M. Crowe and J. Mak, J. Clinical Virology, 22 (2001) 217
[12]   D.Gincel, Silberberg, SD, Shoshan-Barmatz V., J. Bioenerg. Biomemb.32 (2000) 571
[13]   G. S. Wilson (ed) Bioelectrochemistry, Vol. 9, Encyclopedia of Electrochemistry, Wiley-
       VCH, New York, 2002
[14]   L. G. Romano-Fontes, Curi R, Peres CM, Nishiyama-Naruke A, Brunaldi K, Abdulkader
       F, Procopio J.Lipids 35 (2000) 31
[15]   A.V. Krylov, Kotova EA, Yaroslavov AA, Antonenko YN., BBA-Biomembranes 1509
       (2000): 373
[16]   E. M. Egorova, COLLOID J.,. 62 (2000): 284-296
[17]   T.Y. Tsong, Bioelectrochem. Bioenerg., 24 (1990) 271; Biophys. J., 60 (1991) 297
[18]   L. Tung, Troiano GC, Sharma V, Raphael RM, Stebe KJ., Occupational Electrical Injury:
       New York Acad. Sci.,888 (1999): 249
[19]   M. Bier, Hammer SM, Canaday DJ, Lee RC., Bioelectromagnetics 20 (1999) 194
[20]   G. Guihard, Proteau S, Payet MD, Escande D, Rousseau E., FEBS Lett. 476 (2000) 234
[21]   H. M. Van Dort, Knowles DW, Chasis, JA, Lee G, Mohandas N, Low PS
       J.. Biol.. Chem. 276 (2001) 46968
[22]   S.G. Samoilenko, Aksentsev SL, Konev SV, Biol. Memb.18 (2001) 299-305
[23]   M. Suwalsky, Ungerer B, Villena F, Norris B, Cardenas H, Zatta P.,J. Inorg. Biochem. 75
       (1999): 263
[24]   M. R. Gonzalez-Baro, Garda H, Pollero R, BBA-Biomembranes 1468 (2000) 304
[25]   G. B. Melikyan, Niles WD, Peeples ME, Cohen FS, J.General Physiol. 102 (1993) 1131
[26]   G. Agner, Kaulin YA, Schagina LV, Takemoto JY, Blasko K. BBA-Biomembranes, 1466
       (2000): 79; ___ ___, Gurnev PA, Szabo Z, Schagina LV, Takemoto JY, Blasko K.,
       Bioelectrochem. 52 (2000) 161
[27]   R Basu, De S, Ghosh D, Nandy P., Physica A 292 (2001)146
[28]   L. N.. Yuldasheva, Merzlyak PG, Zitzer AO, Rodrigues CG, Bhakdi S, Krasilnikov OV.,
       BBA--Biomembranes 1512 (2001) 53
[29]   Y. Muto, Matsuoka T, Kida A, Okano Y, Kirino Y., FEBS Lett. 508 (2001) 423
[30]   O. Peyronnet, Vachon V, Schwartz JL, Laprade R., J. Membrane Biol. 184 (2001) 45

[31]   N.. Ebran, Julien S, Orange N, Auperin B, Molle G., BBA—Biomemb.1467 (2000) 271
[32]   M..S.. McClain, Cao P, Iwamoto H, Vinion-Dubiel AD, Szabo G, Shao ZF, Cover TL J. Bacteriol. 183 (2001): 6499-6508
[33]   G. R.Strichartz, Hall S, Magnani B, Hong Cy, Kishi Y, Debin Ja., Toxicon 33 (1995): 723
[34]   T. K. Rostovtseva, Nestorovich EM, Bezrukov SM., Biophys. J., 82 (2002) 160
[35]   B Peck, Ortkamp M, Nau U, Niederweis M, Hundt E, Knapp B.Microbes Infection 3 (2001)171
[36]   E. Schleiff, Klosgen RB,BBA -Molecular Cell Research, 1541 (2001)) 22
[37]   G.. Schonknecht, Althoff G, Junge W J Membrane Biol 126 (1992): 265
[38]   G. Volkov, D.W. Deamer, D.L. Tanelian and V.S. Markin, Liquid Interfaces in Chemistry and Biology, Wiley, Inc., New York, 1998
[39]   Y. G. Steinberg, Liddell PA, Hung SC, Moore AL, Gust D, Moore TA. Nature, 385 (1997): 239
[40]   R. Shingles, Roh MH, McCarty RE Plant Physiol 112 (1996): 1375
[41]   H. T. Tien, A. L. Ottova, Current Topics in Biophysics, 25 (2001) 39-60
[42]   D..L Jiang, Li JX, Diao P, Jia ZB, Tong RT, Tien HT, Ottova AL., J. Photochem. Photobiol. A-Chemistry, 132 (2000) 219-224
[43]   R. J.. Davenport, Tester M., Plant Physiol. 122 (2000) 823
[44]   M. LeMasurier, Heginbotham L, Miller C J.General Physiol. 118 (2001) 303
[45]   F. Mi, Peters Js, Berkowitz Ga., Plant Physiology, 105 (1994): 955
[46]   A..J. Wright, Matthews SE, Fischer WB, Beer PD. Chem.-A European J. 7 (2001) 3474
[47]   K. Kobayashi, Mittler-Neher S, Spinke J, Wenz G, Knoll W., BBA-Biomemb.1368 (1998): 35
[48]   J. R. Clay, Kuzirian AM., J. NeurobioL 45 (2000): 172-184
[49]   Y.N.Antonenko, Borisenko V, Melik-Nubarov NS, Kotova EA, Woolley GA., Biophys. J. 82 (2002): 1308
[50]   K.C Melikov, Frolov VA, Shcherbakov A, Samsonov AV, Chizmadzhev YA, Chernomordik LV., Biophys.ical J. 80 (2001): 1829-1836
[51]   N Pappayee, Mishra AK., Photochem. Photobiol. 73 (2001): 573-578
[52]   P.W. Radke, Frenkel C, Urban BW., EUR J Anaesth 15 (1998) 89-95
[53]   A. Bosio, Binczek E, Stoffel W, P Natl Acad Sci US, 93 (1996) 13280-13285
[54]   G. P. Zhang, Bai JP, Shi YL., Chinese Sci Bull 46 (2001): 1085-1088; Zhang H, Shi Yl., Science China Series B-Chemistry, 37 (1994) 547;
[55]   S.P. Hardy, Ritchie C, Allen MC, Ashley RH, Granum PE BBA--Biomembranes 1515 (2001) 38
[56]   Hinshaw JE, Zhang PJ., MOL BIOL CELL 12: (2001) 450 Suppl.
[57]   C.M. Drain, Proc.National Academy Sciences US 99 (2002): 5178-5182
[58]   G. Basanez, Zhang J, Chau BN, Maksaev GI, Frolov VA, Brandt TA, Burch J, Hardwick JM, Zimmerberg J., J. Biol. Chem. 276 (2001) 31083; Maksaev GI, Mikhaliov II, Frolov VA Biol. Memb.18 (2001) 489
[59]   P. J. White, Pineros M, Tester M, Ridout MS., J. Membrane Biol. 174 (2000) 71
[60]   G. P. Zhang, Shi YL, Wang WP, Liu WY., Toxicon 37 (1999) 1313; Zhang GP, Bai JP, Shi YL., Chinese Sci Bull 46 (2001): 1085-1088
[61]   U. Schote, Seelig J., BBA-Biomembranes 1415 (1998): 135
[62]   V. Kominkova, Magova M, Mojzisova A, Malekova E, Ondrias K Physiol.Res. 50 (2001) 507

[63] T. Hara, Kodama H, Higashimoto Y, Yamaguchi H, Jelokhani-Niaraki M, Ehara T, Kondo M., J.Biochem. 130 (2001) 749

[64] W.B. Fischer, Pitkeathly M, Wallace BA, Forrest LR, Smith GR, Sansom MSP., Biochem., 39 (2000)12708

[65] H. C. Wartenberg, Wartenberg JP, Urban BW.,Basic Res. Cardiol. 96 (2001) 645

[66] L.C. Salay, Aggeli A, Boden N, Knowles PF, Hunter M., Biophys. J. 80 (2001) 595

[67] A. Nardi, Slatin SL, Baty D, Duche D. J. Mol. Biol. 307 (2001) 1293

[68] C. Brenner, Cadiou H, Vieira HLA, Zamzami N, Marzo I, Xie ZH, Leber B, Andrews D, Duclohier H, Reed JC, Kroemer G., Oncogene 19 (2000): 329-336

[69] L. J. Ptacek., Seminars Neurology, 19 (1999) 363

[70] I. Jona, Szegedi C, Sarkozi S, Szentesi P, Csernoch L, Kovacs L Pflugers Archiv-Euro. J. Physiol. 441 (2001) 729

[71] F. N. Fu, Singh BR., J Protein Chem 18 (1999): 701-707

[72] B. Jovov, Tousson A, Ji HL, Keeton D, Shlyonsky V, Ripoll PJ, Fuller CM, Benos DJ J. Biol. Chem. 274 (1999) 37845

[73] F. Ryttsen, Farre C, Brennan C, et al. Biophys J 79 (2000): 1993-2001

[74] S. Ozeki, Kurashima H, Abe H., J. Phys. Chem.B 104 (2000): 5657-5660

[75] A. Kloda, Martinac B., Cell Biochem Biophys 34 (2001): 349-381

[76] Y. Kobuke and Nagatani T., Chem. Lett (2000) 298

[77] J. P. Bai, Shi YL Asian J Androl 3 (2001): 185

[78] A.R. Curran, Templer RH, Booth PJ., Biochemistry 38 (1999): 9328

[79] K. G. Klemic, Klemic JF, Reed MA, Sigworth FJ., Biosens. Bioeloelect 17 (2002): 597

[80] D. Braun and P. Fromherz, PHYS REV LETT 86 (2001): 2905-2908

[81] L Yang, Harroun TA, Weiss TM, Ding L, Huang HW Biophys J. 81 (2001): 1475-1485

[82] H. Nakanishi, Prog. Surf. Sci., 49(2) (1995) 197-205

[83] L. Saiz, Klein ML, J Chem Physics, 116 (2002) 3052

[84] A. J. Lokuta, Komai H, McDowell TS, Valdivia HH., FEBS Lett., 511 (2002): 90

[85] C. Goto, Yamamura M, Satake A, Kobuke Y., JACS, 123 (2001): 12152

[86] N. Sakai, Gerard D, Matile S., *JACS* 123 (2001): 2517-2524

[87] T.A. Harroun, Bradshaw JP, Ashley RH., FEBS Lett., 507 (2001) 200; Biophys J (2000) 967-76

[88] G. Schmies, Luttenberg, B. Chizhov I., M. Engelhard , A. Becker, and E. Bamberg

[89] A Shtifman, Ward CW, Wang JL, Valdivia HH, Schneider MF., Biophys J 79 (2000): 814

[90] M. F.. Brown Chem. Phys. Lipids 73 (1994) 159; H. I. Petrache, Salmon A, Brown MF., JACS, 123 (2001) 12611

[91] S. L. Niu, Mitchell DC, Litman BJ. J Biol. Chem. 276 (2001): 42807

[92] J..S. Landin, Katragadda M, Albert AD., Biochemistry-*US* 40 (2001) 11176

[93] J. Isele, Sakmar TP, Siebert F., Biophys J. 79 (2000): 3063-3071

[94] J. Neyton, Pelleschi M., J Gen. Physiol.97 (1991): 641

[95] J. B. Ames, Ishima R, Tanaka T, Gordon JI, Stryer L, Ikura M., Nature 389 (1997) 198

[96] M. Wakabayashi, Hirano A, Sugawara M, Uchino S, Nakajima-Iijima S J. Pharmaceut. Biomed.Analysis, 24 (2001): 453-460

[97] E. H.Lee, Meissner G, Kim DH, Biophy. J. 82 (2002): 1266

[98] F. Dumas, Duckely M, Pelczar P, Van Gelder P, Hohn B. Proc. Natl Acad. Sci., *US* 98 (2001) 485

[99]   M. Pasenkiewicz-Gierula, Murzyn K, Rog T, Czaplewski C. A. Biochim Polonica, 47 (2000) 601
[100]   H. Lin, Bhatia R, Lal R., FASEB J.15 (2001) 2433
[101]   C. M. Yip, Darabie AA, McLaurin J., J. Mol.Biol. 318 (2002): 97
[102]   S. R. Ji, Wu Y, Sui SF., J. Biol.Chem. 277 (2002): 6273
[103]   J. I. Kourie, Henry CL, Farrelly P., Cell.Mol.Neurobiol. 21 (2001): 255
[104]   JJ Kremer, Sklansky DJ, Murphy RM., Biochemistry40 (2001): 8563;
[105]   R. P. Mason, Olmstead EG, Jacob RF., Biochemical Pharmacology 60 (2000):709
[106]   C. S. Wang, Lee RKK., Neuroscience Letters, 283 (2000) 25
[107]   H. J. Kim, Suh YH, Lee MH, Ryu PD., Neuroreport 10 (1999) 1427
[108]   F. Hong, Prog. Surf. Sci., 62 (1999) 1-237
[109]   W. X. Tang, Chen YF, Zou AP, Campbell WB, Li PL., Am. J. Physiol.-Heart Circulatory Physiol. 282 (2002): H1304-H1310
[110]   J. A.Last, Waggoner TA, Sasaki DY., Biophys. J. 81 (2001): 2737-2742
[111]   P. Mak, Szewczyk A, Mickowska B, Kicinska A, Dubin A Int'l J. Antimicrobial Agents 17 (2001): 137
[112]   J. R. Broughman, Mitchell KE, Sedlacek RL, Iwamoto T, Tomich JM, Schultz BD., Am J.Physiol.-Cell Physiol. 280 (2001): C451-C458
[113]   B. K. Berdiev, Shlyonsky VG, Karlson KH, Stanton BA, Ismailov II., Biophys. J. 78 (2000) 1881
[114]   M. Ikematsu, Iseki M, Sugiyama Y, Mizukami A., J.Electroanal.Chem. 403 (1996): 61
[115]   O. Purrucker, Hillebrandt H, Adlkofer K, Tanaka M., Electrochim.A.47 (2001) 791-798
[116]   Y. D. Ivanov, Kanaeva IP, Gnedenko OV, Pozdnev VF, Shumyantseva VV, Samenkova NF, Kuznetsova GP, Tereza AM, Schmid RD, Archakov AI., J. Mol. Recognition,14 (2001): 185-196
[117]   T. Gropp, N. Brustovetsky, M. Klingenberg, V. Muller, K. Fendler and E. Bamberg, Biophys. J., 77 (1999) 714-726
[118]   M. Tominaga., Electrochemi. 69 (2001): 937-939
[119]   M. Seitz, E. Ter-Ovanesyan, M. Hausch, C.K. Park, J.A. Zasadzinski, R. Zentel, J.N. Israelachvili, Langmuir, 16 (2000) 6067
[120]   S.M. Yu, McQuade DT, Quinn MA, Hackenberger CPR, Krebs MP, Polans AS, Gellman SH., Protein Science, 9 (2000): 2518
[121]   R. J. Cherry, Bjornsen AJ, Zapien DC., Langmuir 14 (1998): 1971-1973
[122]   M. Tsionsky, Zhou JF, Amemiya S, Fan FRF, Bard AJ, Dryfe RAW Analytical Chem., 71(1999): 4300-4305
[123]   R. E. Gyurcsanyi, Pergel E, Nagy R, Kapui I, Lan BTT, Toth K, Bitter I, Lindner E. Analyt. Chem. 73 (2001): 2104-2111
[124]   Y. L. Cheng, R.J. Bushby, S.D. Evans, P.F. Knowles, R.E. Miles, S.D. Ogier, Langmuir, 17 (2001) 1240
[125]   P.Fromherz, Chemphyschem, 3 (2002): 276-284
[126]   R. Guidelli, G. Aloisi, L. Becucci, A. Dolfi, M.R. Moncelli, F.T. Buoninsegni, J. Electroanal. Chem. 504 (2001) 1.
[127]   P.Gualtieri and I. Willner, Molecular Electronics: Bio-sensor and Bio-computer, NATO Advanced Study Institute, Pisa, Italy, June 24 – July 4, 2002

Planar Lipid Bilayers (BLMs) and their Applications
H.T. Tien and A. Ottova-Leitmannova (Editors)

Chapter 2

Dielectric and Electrical Properties of Lipid Bilayers in Relation to their Structure.

Hans G. L. Coster[+]

[+]UNESCO Centre for Membrane Science and Technology, Department of Biophysics, School of Physics, University of New South Wales, Sydney 2052, Australia.

1. INTRODUCTION

Membranes play a crucial role in living cells and organisms. For instance, the plasma membranes enveloping living cells separate the exterior environment from the internal cytoplasm and provides a selective diffusion barrier to molecules moving into, and out of, the cell.

Early insights into the basic composition of cell membranes began with the observation, by Overton [1] in 1899, that the rates of intracellular accumulation of substances were proportional to their solubility in lipids. From this he concluded that the cell membrane, at least in part, must be composed of lipoidal substances. Danielli and Davson [2] in 1935 produced a model of the cell membrane in which a lipoidal layer is sandwiched between two proteinaceous layers. The first direct measurement of the thickness obtained from electronmicrographs by Robertson [3] in 1959 suggested that the lipoidal layer was probably only as thick as two molecules of common biological lipids. He introduced the idea of the "unit membrane". Interestingly that conclusion had already been reached much earlier by Gorter and Grendel [4] in 1925 who had spread the lipids extracted from red blood cells as a monolayer and determined that the area occupied by the lipids was equal to twice the total surface areas of the erythrocytes. Using electrical impedance measurements Fricke and Morse [5] in 1925 had provided the first estimate (2-3 nm) of the thickness of cell membranes from measurements of the electrical impedance of cell suspensions.

The concept of the "unit membrane" was clinched when in 1962 Mueller, Rudin, Tien and Wescott [6] produced the first *in vitro* planar lipid bilayer membrane from lipids extracted from bovine brain. This stimulated many other studies and lipid bilayers were formed from a wide variety of lipids extracted from biological tissue as well as synthetic amphiphiles (eg. see [7-11,12] and a recent review by Tien [13]).

2. THERMODYNAMICS AND ELECTRICAL PROPERTIES OF LIPID BILAYERS

2.1 Self Assembly
There are numerous different kinds of lipid molecules but they all share the feature that they are amphipathic; one region of the molecule is polar and hydrophilic whilst the other is non-polar and hydrophobic. When dispersed in water these molecules tend to form aggregates.

When lipids such as lecithin are dispersed in water, they can form[*] spherical and cylindrical micelles, vesicles and bi-molecular membranes. These aggregates form spontaneously. Indeed, on dispersion of the substance in water, the concentration of the monomer initially simply rises with increases in the material added. However, at some stage, micelles begin to form and additional lipid added to the mix then does not lead to an increase in the concentration of the monomer molecules; instead the concentration of micelles rises. When the concentration of micelles is sufficiently high, other aggregates, such as bi-molecular membrane structures, form.

The statistical mechanical condition for equilibrium between the molecules in the aggregates and the monomers [14] is that the chemical potential for the molecules in these two states must be the same. The chemical potential will contain terms that are dependent on the concentration or mole-fraction of the molecules, as well as energy terms that are *independent* of their concentration but dependent on the state of aggregation and therefore different for the monomers and the aggregate structures.

For equilibrium between monomer dispersions of the molecules and say, spherical aggregates, A, containing N lipids, we must therefore have:

$$\mu_1 = \mu_A$$

or $\qquad \mu_{1,0} + RT \ln X_1 = \mu_{A,0} + \dfrac{RT}{N} \ln \dfrac{X_A}{N}$ $\qquad\qquad$ (1)

---

[*] The possible aggregates for any given lipid molecule is determined to a large degree by the packing volume and head-group area of the lipids [23].

where:

$\mu_{1,0}$ and $\mu_{A,0}$ are the standard chemical potentials for the monomer and aggregates, A, containing N lipids (N-mers) respectively. The standard chemical potentials include all the contributions that are independent of the concentration.

X_1 and X_A are the mole fractions of the monomers and the mole fraction of total lipids tied up in aggregates, respectively. The aggregates each contain N lipid molecules and the mole fraction of aggregates, as a combined molecular entity is therefore X_A/N.

In Eq. (1), the terms in X represent the entropy, expressed per molecule; for the aggregate (N-mer) this is the "communal" entropy per molecule.

Consider now the special case of a bi-molecular layer of lipids (bilayer) in equilibrium with the monomers. For the lipid bilayer, we wish to explicitly separate out a term in the standard chemical potential that describes the interfacial free energy[*] of the membrane. For a macroscopic area of a lipid bilayer, N is very large and the entropy term (per lipid molecule) therefore becomes negligible (cf. Eq. (1)). The condition for equilibrium between the lipid bilayer membrane (subscript m) and lipid monomers is then:

$$\mu_{1,0} + RT \ln X_1 = \mu_{m,0} + \gamma_m a \tag{2}$$

where: γ_m is the interfacial free energy of the membrane per unit area,
a is the surface area occupied by the lipid molecules and
$\mu_{m,0}$ is the standard chemical potential for the lipids in the membrane.

For artificially constructed lipid bilayers [15], γ_m is of the order of 1-2 mJ/m^2. This compares with a value of ~ 50 mJ/m^2 for an oil-water interface.

A consequence of these considerations is that the surface free energy, or surface tension, γ_m, is dependent on the concentration of monomers of the lipid with which it is in equilibrium.

2.2 Electrical Properties

The lipid bilayers have a central, non-polar, layer that is about 2-3 nm in thickness. The partitioning of ions from the aqueous solution into this layer is

---

[*] This is an important parameter in determining the stability and electrical properties of the membrane as we will see later.

determined by the (Born) energy required to move an ion from the aqueous environment where the dielectric constant is $\varepsilon_w \sim 78$ to the bilayer interior where the dielectric constant is very low [9,16-18,64] (around 2.1). The Born energy can be calculated from the image forces, which arise as ions approach the membrane, which has a hydrophobic region with a very low dielectric constant from the aqueous medium. The Born energy for partitioning into the lipid bilayer interior from an aqueous medium is given by [see 28]:

$$W_B = \frac{z^2 e^2}{8\pi\varepsilon_0 R}\left[\frac{1}{\varepsilon_m} - \frac{1}{\varepsilon_w}\right] \qquad (3)$$

where: R is the radius of the ion,
         e is the electronic charge,
         z is the valency of the ions,
         ε_m is the dielectric constant of the membrane interior,
         ε_w is the dielectric constant for water ($\varepsilon_w \sim 78$) and
         ε_0 is the permittivity of the space.

For a potassium ion, for instance, this Born energy for partitioning [14] is of the order of 3eV. The electrical conductance of the lipid bilayers therefore would be expected to be very low since the ion concentrations in the membrane will be extremely low, even if the mobilities of the ions in the membrane were high. The experimentally measured conductance for artificial lipid bilayer membranes are indeed very low; of the order of 10^{-3} S/m^2, lower than that of an equivalent thickness of glass!

The low conductivity of this very flexible, fluid, membrane structure is of significance to the functional properties of the cell membrane. The low conductance of the lipid membranes are, however, several tens of orders of magnitude larger than that expected from the concentration of carriers in the lipid membrane! The explanation for this lies in the formation of pore "defects". We will return to that later.

3. THE DIELECTRIC STRUCTURE OF LIPID BILAYERS

In principle, knowledge of the response of lipid bilayers to alternating electric fields can furnish useful information relating to film thickness and dielectric constant as well as the organisation both of the molecules in the bilayers and of the ions in the electrical double layers. This prompted many early investigations into the impedance of a variety of bilayer systems as a function of frequency [8,16,18-20,22]. The early measurements indicated that the capacitance of the bilayer systems was independent of frequency at least up to frequencies of 20 kHz.

The early measurements of bilayer impedance as a function of frequency all involved "two terminal" measurements in which the impedance of the bilayer system is measured via two electrodes inserted into the electrolyte bathing the two sides of the bilayer.   The problem with such measurements is that the electrode-electrolyte system without a bilayer has an impedance that also disperses strongly with frequency, over a similar frequency range as the system including the bilayer.   Thus, it is difficult to deduce the bilayer impedance per sé from such measurements and early reports provided results only for the combined system (electrode-electrolyte-bilayer-electrolyte electrode). As the conductances of the bilayers, however, were very low, it was possible to estimate an approximate value of the bilayer capacitance. These pioneering studies provided estimates of the bilayer capacitance, which ranged from 3.8 mF/m^2 (eg [18]) to 6 to 7 mF/m^2 (eg [22]). It was not possible to obtain good estimates of the bilayer conductance, although it was possible to deduce that the conductances were extremely low. The wide range of values of the capacitance reported in the literature reflects not only shortcomings of the measurement technique available, but was further complicated by the fact that there are intrinsic variations in the lipid bilayer arising from different levels of alkane solvent retention. The latter depend on various factors, including the nature of the solvent from which the bilayers are formed – see section 4 below).

In principle, of course, it is possible to separate the real and imaginary components of the impedance of the bilayer from the electrodes-electrolyte-bilayer system by measurement of the real and imaginary components of the impedance with and without the bilayer. To do this successfully requires extremely precise measurements of the capacitance and conductance of the system with and without the bilayer. It also requires bilayers of large and precisely known, area.   The first attempt to do this [29] gave a value of the bilayer capacitance of 5.7 mF/m^2.

The problem associated with the frequency dependent impedance of the electrode-electrolyte interface can be eliminated by the use of "4 terminal" impedance measurements techniques[*]. Such techniques for low frequency measurements were not developed until much later [17,30].

3.1 Detecting the internal dielectric structure of lipid bilayer membranes

The presence within the BLM of substructural layers with distinctly different conductivities and dielectric constants will give rise to interfacial polarisations that lead to a Maxwell-Wagner dispersion in the membrane impedance with frequency. In such a dispersion the overall bilayer capacitance will decrease with frequency and the conductance increases with frequency until, at sufficiently high frequencies, both parameters reach new, frequency independent, values.

As a first approximation the bilayer substructure can be conceptualised to contain a central acyl layer sandwiched between two more polar layers containing the polar heads of the lipids. The polar head regions are likely to be more conductive and have a higher dielectric constant than the central acyl chain region. The latter is likely to be similar to that of a hydrocarbon fluid; with a dielectric constant [9,16-18,31,64] around 2. Based on ion partitioning (see section 2.2), the conductance of this region is likely to be extremely low. The equivalent circuit [17,32] of such a trilaminar structure is depicted in Fig. 1.

---

[*] *Four terminal Measurements*

In such techniques the current is injected into the system via two electrodes and the potential developed across membrane is measured with two additional electrodes. If the measurements of the AC potential difference developed across the bilayer are made with amplifiers with sufficiently high input impedances, and since there is no AC current flowing through the potential measuring electrodes, the potential drop across the electrode-electrolyte interfaces will be negligible for the potential-measuring electrodes. The independent measurement of potential together with measurement of the current injected into the system then allows the capacitance and conductance of the bilayer to be determined. The bilayer impedance so determined includes the impedance of the thin layers of the electrolyte in series with the bilayer (on both sides), between the bilayer surface and the plane containing the tips of the potential measuring electrodes on each side of the membrane. The impedance of these electrolyte layer is very small and can be readily separated from the total.

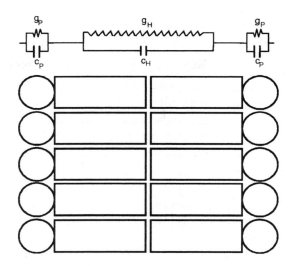

Fig. 1.   The lipid bilayer as a sandwich of three dielectric layers; the hydrocarbon region and two polar head regions.   The equivalent circuit for such a system is also shown (based on Coster and Smith [17]).   Typical parameters for this circuit are given in Table 1.

Using plausible estimates of the parameters in this equivalent circuit Coster and Smith estimated the magnitude of the dispersion in capacitance and conductance expected [17]. The precision required in phase and impedance magnitude to detect this small dispersion was about 0.02 degrees in phase angle and 0.3% in magnitude of the bilayer impedance (assuming a 4-terminal measuring technique is used[+]). Further, the dispersion was predicted to occur over the frequency range 0.1 Hz to 1kHz and was expected to similar to that shown in Fig. 2.

It should be pointed out that the detection and quantification of the dispersion in capacitance and conductance requires precise and independent measurement of *both* the phase and magnitude of the impedance. This allows the capacitance and conductance to be calculated at each frequency.

In principle, of course, the capacitance and conductance can be deduced also from the variation of the impedance with frequency. In practice, however, that would require extraordinary precision in the measurements. This is best illustrated by plotting the effect of adding additional layer (with an equivalent parallel combination of capacitance and conductance) in series with a layer which has a

---

[+] So that the measurements yield the capacitance and conductance of the bilayer itself and does not include the frequency dependent impedance of the double layers at the electrode-electrolyte interface.

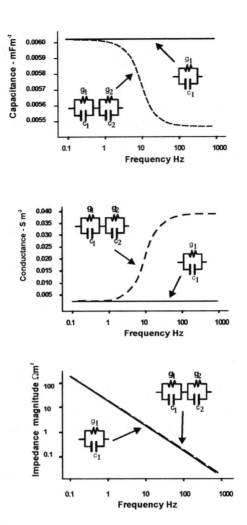

Fig. 2.   Plots of the capacitance, conductance and impedance magnitude for a single dielectric layer (subscripts "1") and a 2 layer Maxwell-Wagner system with parameters $C_1=6$ mF/m², $g_1=3$mS/m² and $C_2=59$ mF/m² and $g_2=4350$ mS/m². These values are typical of the acyl chain and polar head regions of bilayer. Note that the difference between the single layer and the 2 layer system is obvious in the C and G dispersions but can hardly be discerned in the impedance magnitude plot (based on Coster, Chilcott and Coster 1996 [33]).

distinctly different time constant. Examples of the impedance and capacitance and conductance of the single and double layered system are shown in Fig. 2.

The parameters of the two layers in this case were chosen to be representative of bilayer membranes. It is immediately apparent that the additional layer (which for a bilayer would be the polar-head layers) is almost undiscernible in the impedance dispersion plots but has a very dramatic impact on the capacitance and conductance dispersion with frequency [33,34]. To determine the latter, of course, require independent measurement of the phase angle of the impedance. As the total dispersion in capacitance is quite small, the phase angle needs to be determined with high precision.

A digital, 4-terminal, ultra low frequency impedance spectrometer was subsequently developed [17,30] which had the resolution required to detect this dispersion in lipid bilayers. This provided the first direct estimates of the dielectric properties of the polar head regions separate from the bilayer *in toto*; see Table 1.

Table 1.
Estimates of the polar head and acyl-chain dielectric layers of a lecithin bilayer in 1mM and 100 mM KCl solutions (data from Ref. [17]).

		1 mM KCL		100 mM KCl	
Hydrocarbon region	Capacitance	5.0 ± 1	mF/m^2	5.0 ± 1	mF/m^2
	Conductance	0.24	mS/m^2	9.0	mS/m^2
Polar group region	Capacitance	300 ± 60	mF/m^2	300 ± 60	mF/m^2
	Conductance	$7{,}000 \pm 2{,}000$	mS/m^2	$20{,}000 \pm 2{,}000$	mS/m^2

3.2 The effects of Cholesterol on the dielectric structure

X-ray diffraction [35,36] and ESR [37] studies of liposomes generated from mixtures of lecithin and cholesterol have suggested that cholesterol leads to an increase in the thickness of the central hydrophobic region of bilayers. Using an improved version of the ultra low frequency spectrometer [33,38,40], it was possible to determine how the dielectric structures of the bilayers are modified by the inclusion of steroids such as cholesterol [38,41]. The bilayers for these experiments were formed from solutions of lecithin or lecithin-cholesterol (2:1 mole ratio) in *n*-hexadecane. These experiments allowed the dielectric properties of the polar head region to be further subdivided into three dielectrically distinct layers. This is illustrated in Fig. 3.

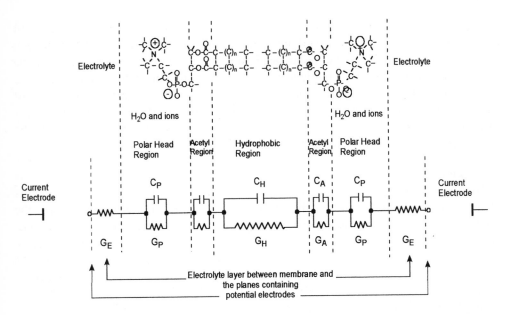

Fig. 3.   A finer subdivision of the internal dielectric structure of a lipid bilayer and the equivalent, Maxwell-Wagner, circuit.   To resolve the additional individual components also requires an increased precision in measuring the phase and magnitude of the impedance over the frequency range 0.1 Hz to 3000 Hz (based on reference [32]).

Experiments were performed using bilayers generated from solutions of the lipid using purified* (unoxidised) cholesterol. The experiments were repeated with bilayers generated from cholesterol that was deliberately oxidised by bubbling oxygen through a dispersion of the cholesterol in chloroform solvent for 4 hours. It had been shown earlier that it is also possible to produce planar membranes from oxidised cholesterol alone [42]. The dispersion with frequency of both the capacitance and conductance of the lecithin-cholesterol bilayers is presented [38] in Fig. 4.   The size of the data points indicates the error bars for repeated measurements on the same membrane. In terms of the equivalent circuit for the substructure, it is immediately clear that the system has more than two time constants. Indeed the solid line fitted to the data is that of a Maxwell-Wagner equivalent circuit containing elements representing, separately, the acyl-chain, carbonyl oxygen, glycerol-bridge and polar head regions of the lipid bilayer.

---

* The cholesterol was produced by converting it to the dibromide/acetic acid complex and then reacting it with zinc dust in ether solution. The cholesterol was recrystallised several times from methanol which yielded colourless crystals.

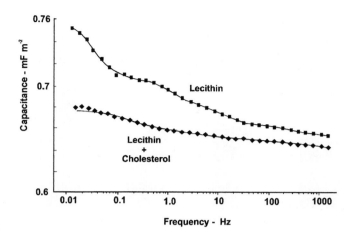

Fig. 4. The dispersion with frequency of the capacitance of lecithin and lecithin-cholesterol (2:1, ratio) bilayers formed from *n*-hexadecane solutions of the lipids. The points show the experimental data (size of the points represent standard errors for 5 runs). The curves are the theoretical Maxwell –Wagner dispersions derived from the putative dielectric structure depicted in figure 3 and fitted to the data. A corresponding dispersion with frequency occurred for the bilayer conductance. The data was obtained using a 4-terminal measuring technique. The membranes were formed under 100 mM KCl electrolyte at 25 °C (data from reference [38]).

Interestingly, for those membranes that lasted more than 10 hours, the impedance dispersion for the bilayers generated from unoxidised cholesterol, gradually approached that of those generated from the oxidised sterol. It was concluded that the cholesterol in the bilayers underwent *in situ* oxidation.

The values of these substructural parameters[*] determined from the impedance dispersion are shown in Fig. 5. Note the major effect of cholesterol on the carbonyl oxygen layer, the glycerol bridge layer and the phosphatidylcholine region [38,41]. It was concluded that the cholesterol molecules were located in the hydrocarbon interior of the membranes with their hydroxy group spanning the carbonyl interface.

---

[*] It should be pointed out that in Ref. 38, the scales for the carbonyl-oxygen, glycerol bridge and phosphatidyl choline regions are in error; the scale should read 10x the values shown. The scale for the acyl-chain region presented in Ref. 38 was correct.

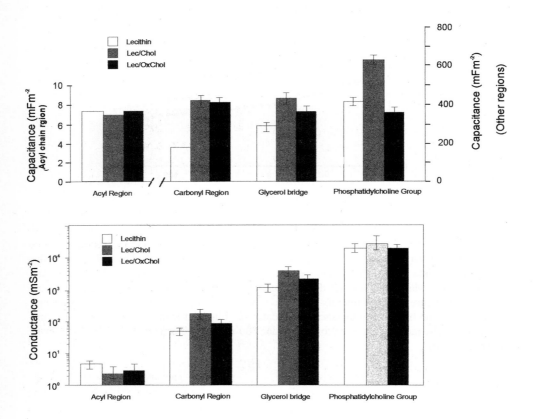

Fig. 5. The capacitances and conductances for the detailed dielectric substructure of lecithin, lecithin-cholesterol and lecithin-oxidised-cholesterol bilayers. Note the separate scales for the capacitance of the acyl-chain region and the other regions of the bilayer.

There is a strong effect of cholesterol on the carbonyl-oxygen region and polar-head regions of the membrane and a smaller effect on the glycerol-bridge region. There is also a slight increase in the thickness of the acyl-chain region (decrease in capacitance) and decrease in the conductance of the acyl-chain region. Oxidised cholesterol had much less effect on the polar-head region although the changes in the carbonyl oxygen region remained.

It should be noted also that the addition of cholesterol greatly enhanced the stability (lifetime) of the bilayers. The data is taken from [38] but it should be noted that in that reference the scales for the capacitance data for the carbonyl-oxygen, glycerol-bridge and phosphatidyl choline regions was in error (too small) by a factor of 10x.

Oxidised cholesterol[+] molecules were located just slightly out of the hydrocarbon interior with their enlarged polar regions spanning the glycerol bridge region of the bilayer. The latter was inferred from the change in the capacitance and conductance of the carbonyl and glycerol regions of the bilayer on substituting oxidised for unoxidised cholesterol. This effect could be seen to occur in time during *in situ* oxidation of the cholesterol.

The addition of cholesterol to the lecithin bilayer resulted in an increase in the polar nature of the hydrophobic-hydrophilic interface possibly due to an increase in water penetration into this region on addition of cholesterol. Its insertion into the membrane forces the lecithin molecules further apart but the small polar hydroxy group does not fill the space vacated by the bulkier polar group of the lecithin molecules. This is presumably also related to the small increase in the thickness of the acyl-chain region (decrease in capacitance) in lecithin-cholesterol bilayers. This is most likely due to an ordering effect [37,43,44] of the more rigid cholesterol molecules, which is also reflected in changes in the viscosity and permeability [39] properties.

4. HYDROCARBONS IN LIPID BILAYERS

The most common methods of generating planar lipid bilayers are to disperse the lipids in a hydrophobic solvent such as a liquid *n*-alkane [eg see 8, 13]. A film of this solution is then established across an aperture in a polymer septum dividing two aqueous electrolyte solutions. The bilayers spontaneously form as the alkane solvent drains (upwards) from the film and into the surrounding torus between the bilayer and the perimeter of the aperture in the septum. This method has indeed not changed since the generation of the first artificial bilayers were reported by Mueller, Rudin, Tien and Wescott [6] [1962].

The bilayers so formed contain varying amounts of alkane solvent that in the steady state is presumably in equilibrium with the bulk lipid-alkane solution in the torus as well as the lipid in the aqueous solution. The only major alternative method developed for generating freestanding, planar, bilayers is based on apposition of two lipid monolayers. This technique [45-49], however, also requires the use of a short alkane (usually pentane) to assist in the formation of the bilayer and maintain its stability.

[+] Oxidised cholesterol was produced by bubbling oxygen through a chloroform solution of the steroid for 4 h (cf. Tien and Dawidowicz [42]). The autooxidation of cholesterol yields a range of products including 7a and 7b hydroxperoxides.

It is known that the incorporation of hydrocarbon solvents significantly alters membrane function and can act as indiscriminate local anaesthetics [50].

The presence of *n*-alkanes has also been reported to modulate the conduction properties of artificial membranes containing passive ion channels such as those formed by gramicidin [51]. The presence of alkane solvents in bilayers also modulates their Young's modulus of elasticity (normal to the membrane surface) as determined by AC electrostriction techniques; the modulus being some 16 times larger for solvent free bilayers (of oxidised cholesterol) compared to bilayers formed from *n*-octane solvent [52]. There is very little data on the amount of hydrocarbons present in *cell* membranes.

The wide range of capacitances reported in the literature for lipid bilayers is probably in large part due to varying degrees of hydrocarbon solvent present in the membranes. The adsorption of alkanes into the interior of the bilayer is known to increase their thickness [16,31,32,53-56]. The alkane in a bilayer decreases over time after the bilayer is formed as the hydrocarbons leave the bilayer and approach their equilibrium concentration. This process can be relatively slow; of the order of several hours. This is illustrated in the results shown in Fig. 6, showing the capacitance, measured using a 4-terminal method at 1.05 Hz, of a totally "black" bilayer formed from *n*-tetradecane dispersions of egg phosphatidylcholine. At a frequency of 1 Hz, the measured capacitance was essentially that due to the central hydrophobic layer of the bilayer. This film had become entire "black" at about 15 minutes after the film had been formed. The capacitance, however, continued to increase for a considerable time as the *n*-tetradecane solvent was squeezed out from the bilayer interior (after Coster and Smith, 1974 [17]).

The amount of hydrocarbon solvent remaining in the membrane is also strongly dependent on the alkane chain length. White [57] found that bilayers could be formed with extremely small equilibrium concentrations of the solvent by using solvents such as squalene, which are large molecules. The solubility of alkanes in bilayers is also dependent on temperature [47,58]; decreasing the temperature decreases the amount of solvent retained.

The solvent retention in the bilayers can be readily monitored using low frequency measurements of the bilayer capacitances. As pointed out above, four terminal measurements at a frequency of 1 Hz yield capacitances that are essentially that of the central hydrophobic region of the bilayer. An example of measurements on "aged" membranes (that had reached steady-state values of the membrane capacitance) for bilayers generated from different *n*-alkane solvents, as a function of temperature is shown in Fig. 7.

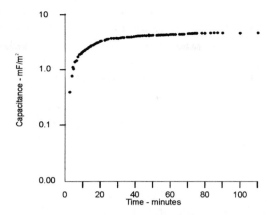

Fig. 6. The capacitance, as a function of time, of a lecithin bilayer formed under 10 mM KCl from *n*-tetradecane dispersions of egg-lecithin. The measurements were made at a frequency of 1.05 Hz. At this frequency the capacitance, measured using a 4-terminal method (which eliminates the electrode-electrolyte impedances) is essentially that of the hydrocarbon region of the bilayer. At approximately 15 minutes, the film had become completely black over the entire area of the circular aperture which had an area of $\sim$ 4.5 mm^2.

It is clear that after the film had become much thinner than the wavelength of light (hence its black appearance), the membrane continued to become thinner as the hydrocarbon solvent was expelled from the bilayer interior into the torus (after Coster and Smith 1974 [17]).

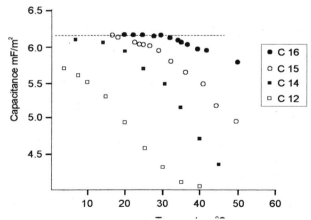

Fig. 7. The capacitance of lipid bilayers of lecithin (egg-phosphatidylcholine) generated from solutions of the lipid in various *n*-alkane solvents. The measurements were made at 1 Hz using a 4-terminal measuring technique and after the capacitance had become stabilised (increasing by less than 1% per hour). The electrolyte solution was 1 mM KCl. For the longer chained alkanes such as *n*-hexadecane, the capacitance at temperatures below 30 °C reached values that were independent of temperature and were essentially solvent free (after Coster and Laver [58]).

The results show that for the longer chain alkanes such as *n*-hexadecane, the bilayers retain very little solvent at equilibrium, although that point may only be reached after 2 hours. For shorter chained solvents such as decane or dodecane considerable amounts of the solvent are retained in the bilayer. Indeed it is possible to determine the equilibrium constant for such solvent partitioning by adding the shorter chain alkane to hexadecane and generating the bilayers from this mixture of solvents. For instance, a decrease in the mole fraction of dodecane in hexadecane in the torus caused an increase in the (temperature dependent) bilayer capacitance [58].

The measurements of capacitance of the bilayers, as a function of temperature, for solvents of different chain lengths allows some inferences to be made concerning the lipid chain order and its effect on *n*-alkane partitioning. The capacitance measurements[*] provide a means of quantifying the amount of alkane solvent in the bilayer and this allows a determination of the differences in the standard chemical potentials, for the different alkanes, in the torus and bilayer interior [56]. The results of these calculations are shown in Figs. 8 and 9.

----

[*] Provided that the increase in thickness of the solvent-free bilayer and the solvent retaining bilayer arises entirely from the partial molar volume of the solvent in the bilayer (that is, the area density of the lipids remains the same), one can calculate the molar concentration, c_a, of the alkane per unit area of the membrane from $c_a = \dfrac{\Delta d\rho}{M}$,

where ρ and M are the mass density and molecular weight of the alkane solvent. Δd is the change in thickness of the bilayer The volume averaged mole fraction of the alkane (relative to the acyl chains) is then given by:

$$X_a = \frac{c_a}{c_a + c_1} ,$$

where c_1 is the number of moles of acyl chains per unit area of membrane. Δd can be obtained from the difference in capacitance of the solvent free bilayer and solvent containing bilayer. Thus;

$$\Delta d = \varepsilon_o \varepsilon_r \left[\frac{1}{C} - \frac{1}{C'} \right],$$

where C and C' are the capacitances of the alkane-lipid bilayer and the same bilayer when it is free of alkane solvent. It should be noted that the capacitance at 1 Hz largely reflects that of the acyl chain region of the bilayer, but in any case any contribution of the carboxyl-ester oxygens on either side of the bilayer and any Gouy-Chapman double layers at the bilayer-water interface will cancel out when calculating the alkane concentration within the bilayer.

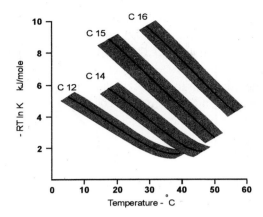

Fig. 8. The difference in the value of the standard chemical potential, $\Delta\mu_o$ = -RT ln K, between the bilayer and the torus for alkanes of various chain length. The lines represent the typical experimental results for $\Delta\mu_o$ for a given bilayer and the shaded regions represent the experimental scatter for 5-10 different bilayers. $\Delta\mu_o$ was calculated from the mole fraction of alkane remaining in the bilayer (which was determined from the capacitance as a function of temperature). Based on Coster and Laver [58].

The acyl chains in the lipid bilayer (above its phase transition temperature) are in a semi-ordered state. This order is imparted by the fact that the chains are anchored at one end to the polar heads which sit at the hydrophobic-hydrophilic interface. The alkane solvents are chemically and structurally similar to the acyl chains of the lipids. However, the alkanes have no polar group and consequently are not anchored at the bilayer-water interface. In those regions of the bilayer where the acyl chains are highly ordered, the alkanes are constrained to lie parallel to the acyl chains. Thus, the entropy of the alkane molecules will be much lower in these ordered regions than in bilayer regions where the acyl chains are less ordered. Further, if the alkanes were intercalated in the acyl chains close to the polar head regions, the acyl chains themselves would need to become more ordered in order to minimise the interfacial free energy that would arise from increased lipid-lipid spacing. A consequence is that shorter chain alkanes would partition deeper into the hydrophobic interior of the bilayer whilst the longer chain alkanes would be constrained to lie parallel to the acyl chains, particularly if the chain length is similar to that of the acyl chains of the lipid. In that case, these longer chained alkanes would have a portion of their chain located in the more ordered portions of the acyl chains. From both entropic and interfacial energy considerations, therefore, the longer chain alkanes would not partition as readily into the bilayer.

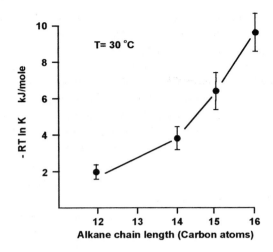

Fig. 9. $\Delta\mu_o$ = -RT ln K at 30 °C as a function of chain length of the alkane solvent from which the lecithin bilayer was formed. The difference in chemical potential increases sharply as the chain length increases. This is related to the fact that the longer chain alkanes partition more sparingly into bilayer from the torus. Data from Coster and Laver [58].

This is consistent with the results shown in Fig. 8 that show that the differences in standard chemical potential for the alkanes (between the torus and the bilayer interior) increase with decreasing temperature. This expresses quantitatively the free-energy cost of inserting alkanes into bilayer interior. The free-energy cost increases also strongly with increasing chain length (Fig. 9).

5. INTERACTIONS OF MOLECULES WITH LIPID BILAYERS

Many of the functional properties of cell membranes arise from embedded protein modules. Nonetheless, the insulating properties of the supporting lipid bilayer, that derive from the hydrophobic properties of the lipid molecules, play a very important role in the organisation of these modules and will strongly influence the electrical characteristics of the cell membrane. Many of the protein modules are composed of a number of subunits that span the membrane with transmembrane subunits connected by flexible strands that remain external to the membrane. The positioning and axial orientation of the fully functional modules results from a balance between opposing hydrophobic and hydrophilic forces between the aqueous environment, the non-polar lipid interior of the membrane and the polar and non-polar portions of the protein subunits [eg. 14,21,60,62]. Molecules that perturb the lipid bilayer might be expected therefore to modulate the functionality of cell membranes. Studies on the effects of various molecules on the basic structure of artificial lipid bilayers may thus provide some insights into the mode of action of such materials in cell membranes.

A good example of how studies on lipid bilayer membranes can shed light on the origin of physiological effects in cellular membranes is the mode of action of anaesthetics. Many, chemically very different molecules, including the relatively inert "noble" gases such as xenon, not only exert an anaesthetic effect but their effects also are additive when used in combination. This suggests that these molecules act not by binding to a receptor on the cell membrane, but more likely through some physical effect on the membranes. One suggestion would be that these molecules affect the substructure, such as the thickness of the hydrophobic region, or the degree of order in the lipids of the bilayer matrix of cell membranes. Indeed local anaesthetics such as benzyl alcohol, procaine, benzocaine and tetracaine lidocaine, etc. do affect the structure of lipids bilayers [40,52,53,59,60,63,65]. The picture, however, is complicated by the fact that the various experimental results reported depend on factors such as the alkane solvent used to generate the bilayers, the temperature and the concentration of the aqueous electrolyte. Molecules such as benzyl alcohol and lidocaine etc. are located at the hydrophobic-polar interface of the bilayer [40,51,52,54, 64,65,]. The hydrophilic as well as the hydrophobic part of these molecules are small compared to that of the lipids. Their location at the hydrophobic-polar interface would have the consequence of reducing the area density of the lipids without a proportional contribution to the hydrophobic volume of the bilayer. The acyl chains of the lipid then would become more disordered as they fold around the benzyl alcohol to fill the space created by the insertion of this molecule. This effect has also been detected in NMR studies [16,66]. Other local anaesthetics have been reported to have similar effects in planar lipid bilayers [51, 54, 65]. Some examples of these are shown in Figs. 10 and 11.

The increase in disorder of the acyl chains [40,49,65] will tend to encourage the partitioning of alkanes into the bilayer [27], thus increasing its thickness. An example is shown in Fig. 11. These effects also appear to dependent on the electrolyte concentration under which the bilayers are formed. This is illustrated by the results shown in Fig. 12.

This effect of the anaesthetic can be counteracted by a decrease in temperature and can be partly compensated also by the presence of cholesterol which has an ordering effect on the acyl chains [37,38,43,44,67]. These consequential effects of the anaesthetic might explain the sometimes apparently contradictory results reported in the literature.

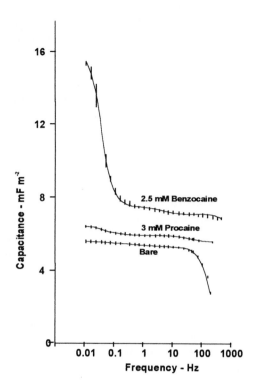

Fig. 10.   Effect of some common local anaesthetics on the dispersion of the bilayer capacitance with frequency. The bilayers were lecithin-cholesterol bilayers formed from *n*-tetradecane solutions of the lipid.   The electrolyte was 1 mM KCl.   The vertical bars represent the standard errors of the data at each frequency for 5 separate runs on the same bilayer.   The full curves are the theoretical dispersions of a Maxwell-Wagner system containing 2 polar head regions, 2 carbonyl-oxygen regions and a central acyl-chain region. The dispersions due to the internal substructure (below 100 Hz) for the "bare" membrane is hardly visible on the scale used to plot the capacitance (cf figure 4.).   Data from reference [40].

Acyl-chain region

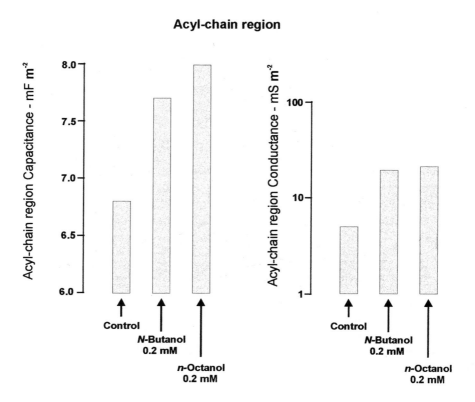

Fig. 11. Effect of *n*-butanol and *n*-octanol on the acyl-chain region capacitance and conductance of lecithin-cholesterol bilayers in 100 mM KCl. The bilayers were formed from *n*-tetradecane solutions of the lipids. The capacitance and conductance of the acyl-chain region were obtained from the dispersion of the bilayer capacitance and conductance with frequency. Data from reference [40].

It is well known that reduced temperatures reduce the potency of many anaesthetics. It is tempting to correlate this with the counteracting effect of reduced temperatures on the effects of local anaesthetics on the acyl chain order (deduced from their effects on alkane partitioning into the bilayer). In cell membranes therefore, the anaesthetic effects may arise not so much from an increase in the thickness of the hydrophobic portion of the lipid bilayer matrix, but rather because of the increased disorder of the acyl chains. The latter could play an important role in determining the equilibrium between various lipid components present and the local fluidity, particularly in boundary regions around ion channels.

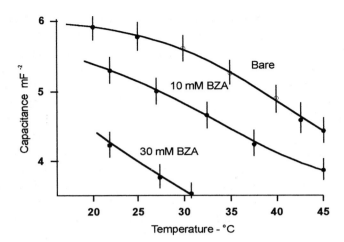

Fig. 12. Effect of benzyl alcohol (BZA) on the capacitance as a function of temperature of lecithin bilayers. The capacitances were measured using a 4-terminal technique and were made at a frequency of 1 Hz and reflect largely the capacitance of the acyl-chain region of the bilayers. The measurements were made after the capacitance had become essential time-independent (usually after at least 1 hour). The error bars arise from the uncertainty in estimating the size of the torus (or membrane area); usually <2%. The benzyl alcohol gives rise to increased disorder in the acyl chain region that enhances the partitioning of the alkane into the bilayer from the torus. Data from reference [67].

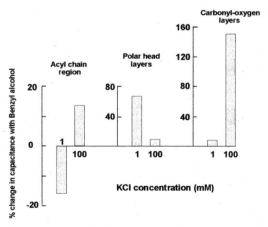

Fig. 13. Relative effects of benzyl alcohol on the capacitance of the acyl-chain, polar-heads and carbonyl-oxygen regions of lecithin-cholesterol bilayers formed from *n*-hexadecane solutions of the lipids. These parameters were obtained by fitting a Maxwell-Wagner model to the dispersion with frequency of the bilayer capacitance and conductance. Note the very different results obtained for 1 mM KCL and 100 mM KCL electrolytes (data from Coster, Laver and Smith [40]).

X-ray diffraction studies on the increase in the disorder of the acyl chains in lecithin-cholesterol liposomes, on addition of procaine [68], are consistent with this notion. In this context, it is noted that benzyl alcohol also is known to fluidise bilayer membranes [16].

6. STABILITY OF LIPID BILAYERS

6.1 Pore Defects

So far, we have not examined possible "defects" in the lipid bilayer structures that might provide aqueous transmembrane connections between the two aqueous phases that the bilayer membrane separates. It has been suggested [69-71] that pores like that shown in Fig. 14. can form in the lipid bilayer.

When such pores form, the curved surface of the pore would be lined with lipid molecules but the packing constraints on the lipid molecules would dictate that the number of lipid molecules per unit surface area in these curved regions would be less than in the planar surface of the bilayer itself. The consequence of this is that water molecules near the lipid lined curved surface in the pore would be able to penetrate further into the non-polar constituents of the lipid structure in this region. This would produce a greater hydrophobic interaction energy. Whilst the exact calculation of the increased hydrophobic energy is complicated, a good first approximation can be obtained by considering the increased surface area available to each lipid molecule and assuming the *additional* area has a surface free energy per unit area similar to that of an oil-water interface (~ 50 mJ/m^2).

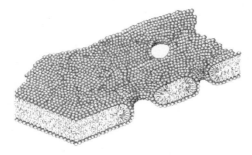

Fig. 14. A stylised view of a lipid bilayer membrane showing putative pores or "defects" in the bilayer structure. Lipid molecules line the pores that provide an aqueous bridge between the electrolyte solutions on the two sides of the bilayer. Image forces on the ions in the pore are smaller than the image forces operating for the partitioning of ions into the bilayer interior itself. Whilst the ion concentration in the pores therefore will be less than that in the bulk electrolyte, the conductance of the pores will nevertheless dominate the total conductance of the membrane.

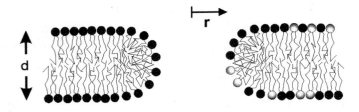

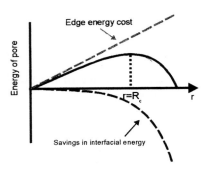

Fig. 15. Cross-section of a pore in the bilayer and the contributions to the energy of formation of such a pore as a function of the radius. The latter contains two contributions: an energy cost in creating the curved portion of the surface that constitutes the pore (which increases linearly with pore radius) and an energy saving which is due to the "missing" circular flat bilayer which previously occupied the area of the pore. The latter varies as r^2. At the critical radius R_c the total energy cost of creating a pore has a maximum value and thereafter decreases with increasing radius (from Coster [14]).

With reference to Fig. 15, for a pore of radius r, the energy required to create the curved surface is:

$$W_P = \left(2\pi \, r \frac{\pi d}{2} \right) \gamma_P = \pi^2 r \, d\gamma_P \;, \tag{4}$$

where γ_p is the surface free energy per unit area in the curved regions of the pore.

The energy cost of creating the curved surface of the pore is offset by the energy saved from not forming the disc of planar bilayer where the pore is formed. This energy is simply $2\pi r^2 \gamma_m$, where γ_m is the surface free energy, or surface tension, for the planar bilayer. The net cost of creating a pore of radius r is therefore given by:

$$E_P = \pi^2 r d\gamma_P - 2\pi r^2 \gamma_m \tag{5}$$

Now that we have the energy required to form a pore we can determine, at least the relative, probability of formation of pores of different radii using the Boltzmann distribution function.

These pores will contribute significantly to the electrical conductance of the membrane. Indeed the conductance of the bilayer itself (in regions where there are no pores) is so small that the conductance of a bilayer with pores would be entirely that due to the pores.

6.2 The energy of formation of bilayers

The interfacial free energy, γ_m, (or surface tension) of the planar bilayer affects the energy of formation of a pore. The interfacial free energy of the planar bilayer is therefore also a determinant of the electrical conductance of the bilayers containing the pores.

The interfacial free energy of a bilayer can be determined by measuring the pressure difference required to bow the membrane formed across a circular aperture in a septum dividing two aqueous compartments. This requires the determination of the radius of curvature of the bowed membrane. As the bilayers are "black", it is difficult to do that with optical measurements [25,26]. The area of the bowed membrane, however, can be directly monitored by measurement of the capacitance of the bilayer when it is bowed and when it is flat (where it will have a minimum capacitance and area [24]). Very small pressure differences can be applied to a bilayer by adjusting the levels of the electrolyte solutions in the two compartments that the bilayer separates. The latter can be done by lowering a very small glass rod into one compartment with a micromanipulator. The distance the rod is lowered into the compartment then allows the pressure difference to be calculated, once the volume displaced by the bowed membrane is allowed for. The latter can be obtained once the area of the bilayer and hence the radius of curvature of the bilayer has been determined.

The slope of the plot of the pressure difference as a function of the inverse radius of curvature than yields the interfacial tension (or free energy)- see Fig. 16. Using this method [15] the interfacial free energy for lecithin-cholesterol bilayers generated from *n*-tetradecane solutions of the lipids across a 7mm aperture was found to be ~1.7 mJ/m^2 (of each bilayer surface).

A   B

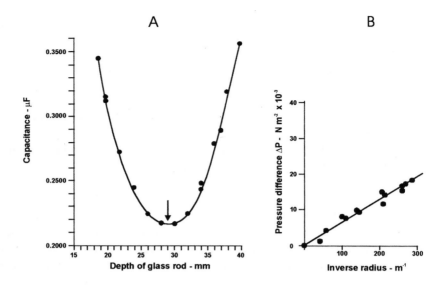

Fig.16.

A. The capacitance of a lecithin/cholesterol bilayer as a function of the depth of immersion of a glass rod immersed into one of the compartments. The minimum in capacitance (arrow) corresponds to the position of the rod when the membrane was flat.

B. The inverse radius of curvature plotted as a function of the pressure difference across the bilayer. (Data from Coster and Simons [15]).

The interfacial free energy did not appear dependent on the electrolyte concentration (1 mM to 2 M) or electrolyte type (NaCl, KCl, LiCl and CaCl$_2$). This value is, as expected, considerably smaller than the interfacial free energy for an oil-water interface. Nonetheless, γ_m remains an important term in the condition for equilibrium for the bilayer (Eq. (2)).

6.3 Electrical conductance of pores

The conductance of a pore filled with water will depend on the radius of the pore as well as the concentration of ions. The latter will continue to be influenced by image forces due to the surrounding lipid bilayer and will depend on the shape of the pore. For simple geometries such as a cylindrical pore, the Born energy for partitioning can be readily determined [72] and for a long and narrow pore and is given by:

$$W_B = \frac{z^2 e^2 \alpha}{4\pi \varepsilon_0 \varepsilon_w},$$ (6)

where: z is the valency of the charge carriers (ions),

e is the electronic charge,

α is a form factor and for a cylindrical pore has a value ~0.175 [72],

r is the radius of the pore,

ε_w is the dielectric constant of water and

ε_0 is the permittivity of free space.

The total conductance of the membrane also, of course, will depend on the total number of pores. These two factors can be separated experimentally, by measurement of the activation energies for electrical conduction [73]. An example of this is shown in Fig. 17 for lipid bilayers (2:1 Lecithin: Cholesterol) formed from *n*-hexadecane solutions of the lipids in 1 mM KCl electrolyte. The activation energies for conduction deduced from these results was ~ 38 kJ/mole[*].

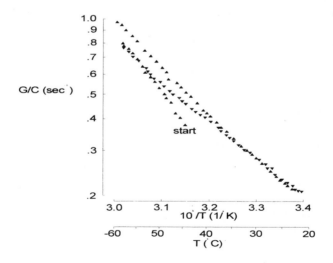

Fig. 17. An Arrhenius plot of the conductance, normalised with respect to the capacitance (and hence membrane area), of a lecithin-cholesterol lipid membrane in a 1 mM KCl solution. The symbols ? and ? refer to measurements whilst the temperature was rising and falling respectively. From the final slope, the activation energy for conduction gave a value of 38 kJ/mole. Data from Smith, Laver and Coster [73].

---

[*] This activation energy for electrical conduction is lower than the value of 28 kJ/ mole reported for the permeation of water through lipid bilayers [74].

This activation energy for conduction contains a contribution due to the activation energy for diffusion through the aqueous medium in the pore (~ 18 kJ/mole). Therefore, the diffusion through our putative pores relates to the effect of the pores themselves. The remaining energy of activation of 20 kJ/mole associated with that process has two contributions:

(a)   The Born energy of partitioning of the ions into the pore (image forces),
(b)   The temperature dependence of the average pore size (and perhaps the number of pores).

These two contributions cannot be readily separated. A lower limit on the *average* pore size may be obtained by assuming the activation energy to derive solely from the Born image forces which arise from partitioning[+] an ion into the narrow (albeit aqueous) pore. On the further assumption that the diffusion of ions in the pore is the same as that in the free solution (mobility ~ 2. 10^{-9} m^2/s) and using the value of the partition coefficient for the ions calculated from the image forces in a cylindrical pore [72], the radius of an equivalent cylindrical pore and its conductance then can be calculated. From this we can also calculate the total number of pores. This analysis [73] suggests that the pore radii are ~ 1.1 nm and the number of such pores is ~ 10^{10} per m^2.

Although the pores completely dominate the electrical conductance of the bilayer, they occupy only 2×10^{-7} of the total area and will contribute negligibly to the hydraulic permeability and capacitance of the membrane. The conductance of lipid bilayers is only weakly dependent on the nature of the electrolyte [75,76] although iodide complexes may represent an exception to this [eg see 77]. This is also consistent with the notion that the electrical conductance does not arise from conductance through the bilayer interior but rather through aqueous pores in the bilayer.

6.4 Membrane Stability

From Eq. (5) (see also Fig. 15), the maximum energy for the formation of a pore (when $\frac{\partial E_P}{\partial r} = 0$) occurs at a radius R_c given by:

[+] Part of the electric field of an ion located in the pore will penetrate the hydrophobic region of the bilayer (which has a low dielectric constant). This will give rise to image forces that produce a partitioning energy. This energy will decrease with increasing pores radius. The calculation of these image forces (Born energies) is very complicated except for simple pore geometries such as a long cylindrical pore [72].

$$R_c = \frac{\pi d.\gamma_P}{4\gamma_m} \, ,$$
(7)

and the energy of such a pore is given by:

$$E_c = \frac{\pi^3 (d.\gamma_P)^2}{8\gamma_m}$$
(8)

The radius R_c is a *critical* radius because a pore of that size will spontaneously grow and this would rupture the membrane. In a cell, this would cause lysis of the cell if it were not for the constraint on the growth of the pore imposed by the presence of intrinsic proteins in the bilayer matrix that are anchored also to extrinsic proteins (for instance intrinsic proteins linked to spectrin via anchorin in the erythrocyte membrane).

Lipid molecules which have a smaller volume of non-polar components or relatively larger polar heads would be expected to pack more favourably into the curved region of the pore. This would decrease the surface free energy in the curved surface of the pore, and hence decrease the critical pore energy and critical pore radius. It is interesting to note that a form of lecithin exists that has only one hydrocarbon-chain tail but the same polar head (trimethyl amonium-phosphate group). This lipid would favour the formation of pores and it is interesting to note that it is known that this lipid promotes the lysis of cells and is referred to as *lysolecithin*. Conversely, molecules such as cholesterol, which have a very small polar head (a single OH group) and a bulky hydrophobic part, would inhibit the formation of pores and lower the conductance of the bilayers. Similar effects would be expected and have been reported [78], on addition of molecules such as dolichyl phosphate to lipid bilayers. Indeed, in experimental studies on lipid bilayers the addition of cholesterol to bi-molecular membranes of lecithin greatly increases the stability of such membranes. Cholesterol is found in all mammalian cell membranes. Other organisms such as fungi contain steroids such as ergosterol that have a similar structure, although small differences may account for differential effects of certain antibiotics in mammalian cells and fungi.

6.5 The effect of trans-membrane potentials on pore defects

Living cells maintain an electric potential difference between the cell interior (cytoplasm) and the external medium. The potential differences vary greatly from cell type to cell type; ranging from ~ 10 mV in erythrocytes (red blood cells) to >250 mV in some fresh water aquatic plant cells such as *Chara corallina*. The interior of the cell is always negative with respect to the external medium. This potential difference appears across the plasma membrane. If we take a value of ~ 60 mV as a typical (e.g. neurones) this potential difference across a membrane which is ~ 6nm thick represents an average field strength in the membrane of ~ 10^7 V/m. This is an extremely large field strength. The local field within the membrane and in particular the central acyl-chain region, is likely to be even larger than this average value.

With such intense electric fields it becomes important to take into account the electrical energy stored in the membrane when considering its molecular organisation and stability.

The membrane capacitance is typically around 10^{-2} F/m^2, so the electrical energy stored at a membrane potential of 60 mV is ~ 1.8 x 10^{-5} J/m^2. That figure is considerably smaller than the 1-2 x 10^{-3} J/m^2 for the interfacial free energy. However, the electrical energy stored may be of significance in the energy of formation of the pores.

To gain some insights into the effect of the electric field in the formation of pores, we will begin by assuming that the pores are small compared to the Debye length for the surrounding aqueous medium. In that case, the potential difference appearing across the membrane will not substantially collapse in the vicinity of the pore. The pore, however, is filled with an aqueous medium with a dielectric constant of ~78 whilst the interior of the membrane has a dielectric constant of ~ 2.1. The formation of a pore therefore will lead to a change (increase) in the charge stored in the region occupied by the pores. As a consequence, the *Free Energy* of the system, including the source of emf. that maintains the potential difference, would then *decrease*[*]. Another way of looking at the process is that water is drawn into the pore by dielectrophoretic forces into the high field region in the pore (since its dielectric constant is higher than that of the membrane). This process is opposed only by the net energy costs of creating the pore in the membrane due to the interfacial free energy terms.

_____

[*] The charging of a capacitor from a battery decreases the overall energy in the system that comprises the battery plus the capacitor, although the energy stored in the capacitor increases. In a cell this energy is derived from the membrane potential which is maintained by differences in the concentration of various ions across the membrane. These concentration and composition differences are maintained via energy-driven ion "pumps" located elsewhere in the membrane. In experiments with artificial lipid bilayers, the membrane potential is maintained by an external source of emf. applied to electrodes immersed in the electrolytes on the two sides of the bilayer and the charging of the bilayer capacitance decreases the energy of this source of emf..

The energy of a pore of radius r (assumed to be much less than a Debye length), in the presence of a membrane potential, V_m, is therefore given by:

$$E_P = \pi^2 rd.\gamma_P - 2\pi r^2 \left(\gamma_m + \frac{C_m V_m^2}{2} \right)$$

(9)

where: C_m is the membrane capacitance (per unit area) and V_m is the transmembrane potential.

The critical radius is reached when $\dfrac{\partial E_P}{\partial r} = 0$ and is given by:

$$R_c = \frac{\pi d.\gamma_P}{4 \left(\gamma_m + \dfrac{C_m V_m^2}{2} \right)}$$

(10)

The energy to form such a critical pore is then given by:

$$E_c = \frac{\pi^3 \left(d.\gamma_P \right)^2}{8 \left(\gamma_m + \dfrac{C_m V_m^2}{2} \right)}$$

(11)

The critical radius and the energy to form such a critical pore, which would lead to rupture, is decreased with increasing transmembrane potential (Fig. 18). The probability of such a pore forming (via the Boltzmann probability function) thus will rapidly increase with increasing membrane potential

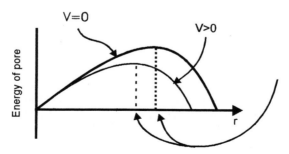

Fig. 18. Effect of transmembrane potential on the energy of creating a pore. The effect of the applied potential is to decrease the critical pore radius (beyond which the pore will grow uncontrollably). Based on reference [14].

For large pores (with a radius much larger than the Debye length), the membrane potential difference in the central portion of the pore will collapse and there will not be a contribution to the electrical energy of the pore from these regions. However, the peripheral region of the pore (within a Debye length of the perimeter) will continue to contribute to the Free Energy as before, although the functional dependence on the radius will then vary as r and not as r^2.

SUMMARY

The many studies of the structure and properties of lipid bilayer membranes have contributed a great deal to our understanding of the properties, function and structure of cellular membranes. Experimental and theoretical studies of the dielectric and other electrical properties of lipid bilayer membranes have featured prominently in these studies of lipid bilayers and continue to yield new insights into these fascinating systems.

Acknowledgements

The author is privileged to have had the opportunity to work with many learned scholars, mentors and students, most of whom are referred to in the references. He would also like to thank the Australian Research Council and the University of New South Wales (Sydney, Australia) for support to conduct much of the research described.

REFERENCES

[1]   E. Overton, Vjochr. Naturf. Ges. Zurich, 44 (1899) 88.
[2]   H. A. Danielli and J. F. Davson, J. Cell. Comp. Physiol. 5 (1935) 495
[3]   J. D. Robertson, Biochemical Society Symp. 16 (1959) 3.
[4]   E. Gorter and F. Grendel, J. Exp. Med. 41 (1925) 439.
[5]   H. Fricke and S. Morse, J. Gen. Physiol. 9 (1925) 153.
[6]   Mueller, Rudin, H. T. Tien and W. C. Wescott (1962). Nature. 194; 979.
[7]   P. Mueller, D. O. Rudin, H. T. Tien and W. C. Wescott. J.J. Danielli, K.C.A. Pankhurst and A.C. Riddeford (eds.) Recent Progress in Surface Science Vol I, Academic Press, 1964, p379.
[8]   H. T. Tien, A. L. Dianna and A. Louise, Chem. Phys. Lipids 2 (1968) 55.
[9]   H. T. Tien, A. L. Dianna and A. Louise, Nature. 215 (1967) 1199.
[10] H. T. Tien, Bilayer lipid membranes, theory and practice , Marcel Dekker Inc. New York. 1975.
[11] H. T. Tien , J. Mol. Biol. 16 (1966) 577. 1964, p379.
[12] A. Goldup, Ohki, S. and J. F. Danielli, Rec. Prog. Surf. Sci. 3 (1970) 193.
[13] H. T. Tien and A. L. Ottova, J. Membrane Sci. 189 (2001) 83.
[14] H. G. L. Coster, Aust. J. Phys. 52 (1999) 117.

[15] H.G. L. Coster and R. Simons, Biochim. Biophys. Acta 163 (1968) 234.

[16] R. Fettiplace, D. M. Andrews and D.A. Haydon, J. Membrane Biol. 5 (1971) 277.

[17] H. G. L. Coster and J. R. Smith, Biochim. Biophys. Acta 373 (1974) 151.

[18] T. Hanai, D. A. Haydon and J. L. Taylor, Proc. Roy. Soc. London, Ser A. 281 (1964) 377.

[19] J. Taylor, and D. A. Haydon, Discussions Faraday Soc. 42 (1966) 51.

[20] P.Läuger, W. Lesslaver, E. Marti and J. Richter, Biochim. Biophys. Acta 135 (1967) 20.

[21] H. Bloom, M. E. Evans and O. G. Mouritsen, Quart. Rev. Biophys. 24 (1991) 293.

[22] H. P. Schwan, C. Huang and T. E. Thompson, Proc. Ann. Meeting Biophys. Soc. 10th (1966) 51.

[23] J. N. Israelachvilli, D. J. Mitchell and B. W. Ninham, Biochim. Biophys. Acta. 470 (1977) 185.

[24] T. Hanai, D. A. Haydon and J. Taylor, J. Theoret. Biol. 9 (1965) 433.

[25] H. T. Tien, J. Theoret. Biol. 16 (1967) 97.

[26] H. T. Tien, J. Phys. Chem. 71 (1967) 3395.

[27] L. Ebihara, J. E., Hall, R. C., MacDonald, T. J. McIntosh and S. A. Simon, Biophys. J. 28 (1979) 185.

[28] B. Neumcke and P. Läuger, Biophys. J. 9 (1969) 1160.

[29] H. G. L. Coster and R. Simons, Biochim. Biophys. Acta. 203 (1970)17.

[30] D. J. Bell, H. G. L. Coster and J. R. Smith, J. Phys. E. 8 (1975) 66.

[31] R. Benz, O. Frohlich, P. Lager and M. Montal, Biochim. Biophys. Acta. 394 (1975) 323.

[32] R. G Ashcroft, H. G. L. Coster and J. R. Smith, Biochim. Biophys. Acta. 643 (1981) 191.

[33] H. G. L Coster, T. C. Chilcott and A. C. F. Coster, Bioelectrochem. Bioenergetics. 40 (1996) 79.

[34] H. G. L. Coster and T. C. Chilcott. T. S. Sørensen (ed.), Surface Chemistry and Electrochemistry of Membranes, Marcel Dekker, NY., 1999, p749.

[35] R. P. Rand and V. Luzzati, Biophys. J. 8 (1968)125.

[36] Y. K. Lecuyer and D. G. Dervichian, J. Mol. Biol. 45 (1969) 39.

[37] J. C. Hsia, H. Schneider and I. C. P. Smith, Can. J. Biochem. 49 (1971) 614.

[38] C. Karolis, H. G. L. Coster, T. C. Chilcott and K. D. Barrow, Biochim. Biophys. Acta. 1368 (1998) 247.

[39] A. Finkelstein and A. Cass, Nature 216 (1967) 717.

[40] H. G. L. Coster, D. R. Laver and J.R. Smith. H. Keyzer, F. Gutmann (Eds.), Bioelectrochemistry, Plenum Press. New York, (1980) p331.

[41] R. G., Ashcroft, H. G. L. Coster and J. R. Smith, Biochim. Biophys. Acta. 730 (1983) 23.

[42] H. T. Tien and E. A. Dawidowicz, J. Colloid Interface Sci. 22 (1966) 438.

[43] E. Oldfield and D. Chapman, Biochim. Biophys. Acta. 43 (1971) 610.

[44] K. W. Butler, H. Schneider and I. C. P. Smith, Biochim. Biophys. Acta. 219 (1970) 514.

[45] S. H. White, D. C. Peterson, S. Simon and M. Yafuso, Biophys. J. 16 (1976) 481

[46] R. Fettiplace, L. G. M. Gordon, S. B. Hladky, J. Requena, H. P. Zingsheim and D. A. Haydon. E. D. Korm, (ed.) Methods of Membrane Biology, Vol. 4, Plenum Press, N.Y. 1975, p1.

[47] R. Benz, O. Fröhlich, P. Läuger and M. Montal, Biochim. Biophys. Acta. 394 (1975) 323.

[48] M. Montal and P. Mueller, Proc. Nat. Acad. Sci. USA 69 (1972) 3561.

[49]   M. Takagi, K. Azuma and U. Kishimoto, Ann. Rep. Biol. Works. Fac. Sci. Osaka University. 13 (1965) 107.

[50]   D. A. Haydon, B. M. Hendry, S. R. Levinson and J. Requena, Biochim. Biophys. Acta. 470 (1977) 17.

[51]   B. M. Hendry, B.W. Urban and D. A. Haydon, Biochim. Biophys. Acta. 513 (1978) 106.

[52]   T. Hianik, J. Miklovicova, A. Bajci, D. Chorvat and V. Sajter, Gen. Physiol. Biophys. 3 (1984) 79.

[53]   S. H. White, NY Proc. Acad. Sci. 3 (1977) 243.

[54]   R.G. Ashcroft, H. G. L. Coster and J.R. Smith, Biochim. Biophys. Acta. 469 (1977) 13.

[55]   R.G. Ashcroft, K. R. Thulborn, J. R. Smith, H. G. L. Coster and W. H. Sawyer, Biochim. Biophys. Acta. 602 (1980) 299.

[56]   D. R. Laver, J. R. Smith and H. G. L. Coster, Biochim. Biophys Acta. 772 (1984) 1.

[57]   S. H. White, Biophys. J. 23 (1978) 337.

[58]   H. G. L. Coster and D.R. Laver, Biochim Biophys Acta. 857 (1986) 95.

[59]   J. Reys and Latorre, R. Biophys. J. 28 (1979) 259.

[60]   R.G. Ashcroft, H.G. L. Coster and J. R. Smith, Nature. 269 (1977) 819.

[61]   H. G. L. Coster and T. C. Chilcott, Bioelectrochem. 56 (2002) 141.

[62]   J. N. Israelachvilli, D. J. Mitchell and B. W. Ninham, J. Chem. Soc. Far. Trans. II 72 (1976) 1525.

[63]   J.C. Metcalfe, P. Seeman and A. S. V. Burgen, Mol. Pharmacol. 4 (1968) 87.

[64]   W.T. Huang and D.G. Levitt, Biophys. J. 17 (1977) 111.

[65]   T. Hianik, M. Fajkus, B. Tarus, P. T. Frangopol, V. S. Markin and D. F. Landers, Bioelectrochem. Bioenergetics. 46 (1998) 1.

[66]   G. L. Turner and E. Oldfield, Nature. 277 (1979) 669.

[67]   H. G. L. Coster and D. R. Laver, Biochim. Biophys. Acta. 861 (1986) 406.

[68]   H. G. L. Coster, V. J. James, C. Berthet and A. Miller, Biochim. Biophys. Acta. 641 (1981) 281.

[69]   C. Taupin, M. Dvolaitzky and C. Sauterey, Biochemistry. 14 (21) (1975) 4771.

[70]   J. G. Abidor, V. B. Arakelyan, L. V. Chernomordik, Yu. A. Chizmadzhev, V. F. Patushenko and M. R. Tarasevich, Bioelectrochem. Bioenergetics. 6 (1979) 37.

[71]   H. G. L. Coster. G. Eckert, F. Gutmann and H. Keyzer (eds.) Electropharmacology CRC Press, Boca Raton, 1990, p139.

[72]   V. A. Parsigian, Ann. N.Y. Acad. Sci. 264 (1975) 161.

[73]   J. R. Smith, D.R. Laver and H. G. L. Coster, Chem. Phys. Lipids. 34 (1984) 227.

[74]   H.T. Tien and Hie Ping Ting, J. Colloid Interface Sci. 27 (1968) 702.

[75]   R. C. MacDonald, Biochim. Biophys. Acta. 448 (1976) 161.

[76]   J. R. Smith, H. G. L. Coster and D. R. Laver, Biochim. Biophys. Acta. 812 (1985) 181.

[77]   C. J. Bender and H. T. Tien, Anal. Chem. Acta. 201 (1987) 51.

[78]   T. Janas and H. T. Tien, Biochim. Biophys. Acta. 939 (1988) 624.

Planar Lipid Bilayers (BLMs) and their Applications
H.T. Tien and A. Ottova-Leitmannova (Editors)
© 2003 Elsevier Science B.V. All rights reserved.

Chapter 3

Boundary potentials of bilayer lipid membranes: methods and interpretations

Yu.A.Ermakov and V.S.Sokolov

Institute of Electrochemistry RAS, Moscow, Russia

1. INTRODUCTION

Many physiologically important processes are accompanied by charge transfer across the membrane and adjacent layers, and the related electrostatic phenomena are commonly recognized as a fundamental aspect of membrane biophysics. Both, charge transfer and binding to the surface depend on the electric field distribution at the membrane boundary, which in most cases is determined by the presence of charged lipid species. These circumstances have stimulated extensive use of planar lipid bilayers (BLM) and liposomes as model systems for studies of electrostatic phenomena at the membrane boundaries induced by inorganic ions and substances of biological interest. It is impossible to present here an exhaustive list of publications related to this subject therefore we simply refer the reader to previous reviews and monographs [1, 2]. This paper focuses on principles and applications of a method of measurement of boundary potentials based on the principle of intramembrane field compensation (IFC). The method was developed and used extensively in our studies. When used in combination with other techniques, such as electrophoresis, or ionophore-induced membrane conductance, IFC reveals the 'electrostatic structure' of the membrane in a greater detail than any of these techniques would achieve separately. Below we present observations of electrostatic effects induced by adsorption of inorganic ions or amphiphilic molecules bearing ionized groups or dipole moment and generally characterized by high affinity to phospholipids. We discuss the problem of positioning the bound substances and probes within the membrane and correlation of the adsorption plane with their structure and hydrophobicity. Finally, we discuss IFC application to studies of the distribution of membrane-permeable substances and the possibility of detection of electrically neutral transport of weak acids or bases across the membrane using this technique.

The distribution of electric field across the membrane interface in a microscopic scale is extremely complex and generally may be only approximated by smooth changes of potential in both directions perpendicular and parallel to membrane. The goal of electrochemical methods is to evaluate the difference between the potential in a reference point, usually taken as zero in the bulk of solution, and the potential averaged over a certain plane parallel to the membrane interface. One important imaginary plane (membrane surface) separates the membrane from the aqueous solution. The position of other planes inside or outside of the membrane will be specified below for each of the experimental techniques considered here. A simplified approach to the 'electrostatic' structure of the interface is illustrated in Fig.1. The total potential drop across the interface, referred to as boundary potential, ϕ_b, can be defined as the potential difference between the points in two phases, one of which is in the bulk of electrolyte, another is inside the membrane, near its hydrophobic core. This potential ϕ_b is presented in Fig.1 as the sum of two parts (ϕ_s and ϕ_d). The first is the potential drop in the diffuse part of the electrical double layer and defined here as the surface potential, ϕ_s. It is determined by the processes of surface ionization and screening by ions of the electrolyte. The Gouy-Chapman-Stern (GCS) model provides an adequate description of these phenomena on the surface of biological membranes [2]. In the sections that follow we will use the GCS model to determine ϕ_s and ion binding parameters from electrokinetic data [3].

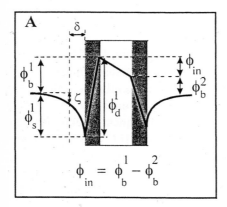
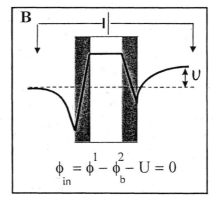

Fig.1. Electric potential distribution across the membrane under conditions of short-circuit (A) and intramembrane field compensation by an external voltage source (B). Hydrophobic core is indicated as light and polar regions as dark parts of the membrane. The vertical dashed line on the left shows the position of shear plane in the diffuse part of electrical double layer at distance δ from the surface of membrane.

The other component of the boundary potential corresponds to the voltage drop across the interface and can be defined as the potential difference between two imaginary planes placed in different phases: in the aqueous solution immediately adjacent to the membrane surface and in the hydrophobic core inside the membrane. In contrast to the surface potential formed by the double layer, this interfacial potential drop is not accessible by any experimental approach because the energy of charge transfer between different phases includes not only the electrostatic but also an unmeasurable chemical component. This important point had been discussed in many monographs and textbooks (see Refs. [4-6] for details). Only the change of this component and therefore the variation of the total boundary potential can be monitored by external devices. The exact potential distribution in this region of the membrane is unknown and is presented in Fig. 1 as linear. As follows from many experiments [2, 7, 8] and numerous estimations [9], the interior of the membrane is more positive then that the surrounding aqueous phase by about 200-300 mV. The physical nature of this potential difference reflects the molecular structure of the interface and may be ascribed to the orientation of dipole moments of lipid and water molecules or other moieties adsorbed or incorporated into lipid bilayer. Here we refer to it as dipole component of boundary potential and denote as ϕ_d bearing in mind that this term is rather conventional. It will be shown in section 4, that the origin of this potential may substantially differ from the layer of elementary dipoles.

2. INTRAMEMBRANE FIELD COMPENSATION

2.1. Principles of the measurements of boundary potentials

A typical experimental chamber consists of two compartments separated by a membrane. If the electrometer is connected to both compartments directly (open circuit conditions) it can monitor only the transient changes of the boundary potentials. Under steady state it will measure the electric potential difference between the two bulk water solutions, referred to as membrane potential, ϕ_m. This potential is determined by the membrane permeability to ions and their bulk concentrations on both sides of the membrane (see, e.g. [10]). In a completely symmetrical system ϕ_m is zero. The boundary potential at one BLM side can be changed, for example, by adding a charged substance to one compartment. If the substance binds to the surface but does not permeate through the membrane, it will alter the surface charge and the boundary potential, $\phi_b{}^1$, at the appropriate side of the membrane. If this change is fast enough and is not accompanied by charge redistribution across the membrane, then an electrometer would show an instantaneous shift of the membrane potential $\Delta\phi_m$ equal to $\Delta\phi_b{}^1$. The potential would then relax to steady-state value due to transfer of ions through the membrane [11]. It has to create the electric field (potential drop ϕ_{in}) inside of the membrane equal to the difference of boundary potentials: $\phi_{in}=\phi_b^1-\phi_b^2$. Therefore this method is suitable only for detection of the boundary potential changes, which are faster in comparison to the intrinsic time constant of the membrane, The latter is determined by the product of its resistance and capacitance, typically estimated as less than 100s [11]. To monitor electrostatic phenomena in a longer timeframe one may need alternative techniques. Particularly, the changes in boundary potentials can be detected and measured as the external voltage necessary to keep the electric field inside the membrane core, ϕ_{in}, close to zero. These methods are based on the intramembrane field compensation (IFC) principle.

Under short circuit conditions, when the potentials in the two bulk solutions are equal (Fig. 1a), a potential drop inside the membrane appears as the result of membrane asymmetry regarding the boundary potentials, $\phi_{in} = \Delta\phi_b$. The voltage across the membrane core can be detected by utilizing the fact that the electric capacitance of the lipid bilayer depends on voltage. The early experiments with planar BLM [12] showed that the membrane capacitance increases with voltage irrespective of its sign, i.e. is minimal when the voltage drop across the hydrocarbon core is zero. By measuring the position of this minimum on the potential scale, one may measure the difference of boundary potentials, as was proposed in [13]. Later this idea was implemented as a compensation technique [14], where the value of $\Delta\phi_b$ was determined by the value of externally applied voltage that is necessary to minimize the capacitance of the BLM. This state with a completely compensated intramembrane field is diagramed in Fig.1b. The IFC method is based on this idea.

The simplest explanation of dependence of membrane capacitance on applied voltage, C(U), is electrostriction [15]. The membrane is presented as an elastic capacitor, whose plates are attracted towards each other by electric field and pushed apart by elastic "spring" between them, and the capacitance increase is due to the thinning of the membrane. The capacitance minimum evidently corresponds to zero electric field between the plates., The increase of the capacitance, however, may also be associated with the increase of the area of the bilayer part due to the reduction of the meniscus and microlenses [16-18]. For the IFC method it is important that the two effects have significantly different characteristic time. As it follows from direct optical observations, the thinning of membrane under applied voltage is complete within a hundred microseconds, whereas the increase of its area takes minutes [17] The review of the experimental data and models of elastic compression is presented in a monograph [15]. An alternative to the viscoelastic model was proposed in [19], where the author explained the same phenomena by the electric field acting on the spectrum and amplitude of thermal fluctuation of the membrane. The increase of capacitance was presented as a function of surface tension and membrane bending elasticity. Note, that all these models were aimed initially to explain the effect of voltage on the membrane capacitance. The ideas to apply this phenomenon to measure the boundary potentials were developed later.

The potential drop inside the hydrophobic region of the membrane, φ_{in}, can be conventionally presented as the sum of external voltage, U, and the difference of boundary potentials:

$$\varphi_{in} = U + \Delta\varphi_b \tag{1}$$

The dependence of the membrane capacitance on voltage near its minimum may be approximated as:

$$C = C_0\left[1 + \alpha(U - \Delta\varphi_b)^2\right], \tag{2}$$

where C_0 is the minimum capacitance at zero potential inside the core, α is known as electrostriction coefficient. It should be noted that either of the physical models mentioned above predicts such parabolic dependence. The apparent agreement to the models of eq. (2) is not surprising, because a parabolic relation immediately follows from the experimental observation of the existence of capacity minimum. Mathematically, a quadratic term would be the main term in the serial expansion of any function around its minimum.

The coefficient α depends on the structure of the lipid bilayer and can be directly linked to the mechanical properties of the membrane. Its value is in the order of 0.1 V^{-2} for planar bilayers made from natural lipids dissolved in n-decane. In the model of «elastic capacitor» it particularly corresponds to the transversal elasticity of the membrane of about 10^7 Pa [15].

The position of the capacitance minimum on the scale of transmembrane potentials may be found by a direct measurement of BLM capacity with an AC generator [20, 21], or by comparing the shapes of current pulses in response to the "up" and "down" phases of triangular voltage ramps [22]. The most convenient method utilizes the analysis of higher harmonics of the capacitive current generated due to the non-linearity of the membrane capacitive response to a sine-wave voltage. The harmonic analysis of the capacitive current has been used by several groups [23-25], but the technique in which the amplitude of the second harmonic could be utilized to monitor the BLM boundary potential in an automatic regime was developed first in our laboratory in 1980 [26]. In this method, the external voltage applied to the membrane is the sum of a sine wave function and DC bias, $U = U_0 + V\cos(\omega t)$. Substitution of this sum into Eq.2 gives the three harmonics of the capacitive current in the following form:

$$I = -V\omega C_0 \left[1 + 3\alpha(U_0 - \Delta\varphi_b)^2 + \frac{\alpha V^2}{2} \right] \sin(\omega t) - 3\alpha C_0 \omega V^2 (U_0 - \Delta\varphi_b)\sin(2\omega t) - \frac{3}{4}\alpha C_0 \omega V^3 \sin(3\omega t), \qquad (3)$$

Note that the amplitude of the second harmonic, I_2, is proportional to the DC component of $\varphi_{in} = U_0 - \Delta\varphi_b$. Thus, this amplitude becomes zero at the point where the capacity reaches the minimum. Moreover, its sign changes to the opposite one with the sign of φ_{in}. This means that the external DC voltage applied to the BLM may suppress the signal of the second harmonic or change its phase by 180°. This property can be used to find the capacitance minimum and thereby to measure the difference of boundary potentials in an automatic regime as proposed in [26]. It can be implemented as either a digital or completely analog design.

2.2. Experimental setup

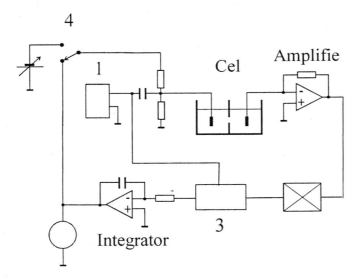

Fig.2. Block-diagram of the setup used to measure the difference between the boundary potentials on a BLM by the inner field compensation (IFC) method. The numbers designate: (1) sine wave generator; (2) notch filter rejecting the first harmonic; (3) lock-in amplifier; (4) regulated DC voltage source with a switch for manual compensation of the second harmonic

signal. The system permitted tuning of the phase of the signal at the reference input of the lock-in amplifier to obtain a negative feedback.

The setup is shown as a block-diagram in Fig.2 and the principal elements are listed in the legend. In a manual regime one can apply a DC voltage to the membrane from the battery and to follow the amplitude of I_2. It is enough to find the position of the capacitance minimum and roughly estimate the value of $\Delta\phi_{.b}$. In practice this procedure helps to adjust the phase shift of the signal corresponding to the maximal amplitude of I_2 that is necessary to set the negative feedback loop. The negative feedback is implemented via an integrator such that under a stable registration regime it zeroes the amplitude of I_2. The DC output from the integrator, which feeds back to the chamber eliminating the second harmonic signal, directly corresponds to the difference of boundary potentials.

The accuracy of compensation in the automatic regime depends on the ratio of amplitudes the first and second harmonics of the capacitive current (I_2/I_1). In addition to common noise, there might be sources of high harmonics the circuit, as the sine wave generator and amplifiers can be sources of nonlinear distortions and may introduce a second harmonic signal with the amplitude I_2^x. This point has particular importance for IFC applications for the accuracy of $\Delta\phi_b$ measurements which can be evaluated as $\delta\phi_b \sim I_2^x/(3\alpha V^2)$. The error decreases as the ac component of the input voltage, V, and the electrostriction coefficient α increase. Typically, the amplitude of the stimulus does not exceed 50 mV because at higher voltages BLM become unstable. The coefficient α depends on the lipid composition and the type and content of the solvent [15, 27, 28] in the membrane. The value of α and, hence, the accuracy of compensation are significantly lower in "solvent-free" bilayers, e.g. those formed by the method of [29]. The best conditions for IFC application were found with decane-containing painted BLM prepared according to [30]. The typical accuracy for these BLM is $\delta\phi_b \sim 1$ mV at V=30 mV at ω=300 Hz.

The IFC technique permits one to control the difference in the total boundary potentials between the two sides of the membrane, which could be a result of binding of a substance of interest. The procedure of measurements usually begins with the BLM formation in a symmetrical background electrolyte. Initially, the second harmonic signal is absent, corresponding to equal (but unknown) boundary potentials at both BLM surfaces. If the molecules of interest added to one (cis) compartment are membrane-impermeable, then the opposite (trans) side of BLM can be treated as reference. Therefore, the IFC method measures $\Delta\phi_b$, which is equal to the difference of boundary potentials between the cis and trans sides of BLM. Note that the electrodes (Ag/AgCl) are usually separated from the bulk by salt bridges filled by agarose, which slightly decreases the 2nd harmonic signal but precludes any influence of additions to the chamber on the potentials of Ag/AgCl electrodes.

A simple experiment that can be used to test the setup is to introduce a concentrated background (indifferent) electrolyte into the compartment at cis-side of BLM. In this particular case the absolute value of the surface potential, ϕ_s, decreases as the result of screening effect. The sign of $\Delta\phi_b$ is opposite for negatively or positively charged surfaces. The simplest theory of diffuse double layer (Gouy-Chapman model) allows one to interpret the increments in surface potential in terms of the surface charge density (see Eq.5 below). The IFC method thereby allows detection of even small surface charge of the BLM [20, 31]. Unfortunately, this straightforward result is rately realized in practice for several reasons. The experimental data usually deviate from the predictions of the theory at small electrolyte concentration (below 10 mM). It may be probably caused by the presence of minor traces of foreign ions and especially multivalent ones, which limit the highest value of surface potential at low concentration of background electrolyte. At high concentration the assumption of indifferent electrolyte is not supported by the experiment, and the adsorption of ions has to be taken into account. We'll discuss these and related problems of IFC applications in sections 3 and 4 in more details.

2.3. Advantages and limitations of IFC

Two principal assumptions are used in the IFC method design. i). The applied voltage drops mainly inside the hydrocarbon core of membrane and this part determines its capacitance; therefore the 2nd harmonic signal can be used to detect the compensation of the intramembrane field by an external voltage. Ii). The boundary potentials on both sides of the membrane are not affected by this voltage. Indeed, the hydrophobic core of the membrane is relatively thick and has lower dielectric permeability then the outer polar regions [1, 32]. Therefore, the external voltage should drop mainly within the membrane, where the electrical capacitance is minimal among the series of capacitances belonging to the hydrophobic interior, polar regions of the membrane and diffuse parts of the electric double layers. This assumption is correct for diffuse double layers at ionic strength > 1 mM. Apparently, these considerations are not absolutely rigorous and can be confirmed only by testing the data of IFC method with other methods sensitive to electric field distribution in the membranes.

In practice, the IFC method has a number of limitations related to permeability and conductivity of BLM. As mentioned above, the boundary potential on one side of the membrane is always presumed to be invariable and used as reference for the other surface that undergoes modification. An exception from this rule will be discussed in the last section, which describes the experiments with membrane-permeable remantadine. This example shows that the IFC technique under special circumstances may give interpretable results even if the substance penetrates through the membrane and changes boundary potentials on both sides of it.

It is important to note that the electrostatic potentials under investigation by IFC method (surface and dipole components of the boundary potential in Fig.1) are assumed to drop in the region, which does not overlape with the hydrophobic core of the membrane, responsible for electrostriction effect. Otherwise the conditions discussed above are not fulfilled and the correct measurement of these potentials becomes impossible. The adsorption of styryl dyes with the ionized groups located in hydrophobic region of the membrane at significant depth can be mentioned here as an example [33]. Fortunately, in the most practical cases the boundary potentials change either in water phase outside the membrane (surface potential), or in the polar regions of the membrane (evidently located outside its hydrophobic core). These changes are measured by IFC method correctly, and it will be illustrated below by comparison of the results obtained by this method and other ones.

Ideally, for IFC measurements the conductance of the membrane should be low. However even a significant conductance does not preclude the implementation of the technique as long as remains independent of voltage. Unfortunately, lipid membranes modified by hydrophobic ions or ionophores usually have nonlinear current- voltage characteristics, which produce higher harmonics undesirable for IFC applications. The only way to prevent the effect of nonlinear conductivity on the compensation procedure is to increase the capacitive component of the current by increasing the frequency. Sometimes the effect of the conductivity becomes negligible at a certain frequency, where the current-to-voltage characteristics of the membrane is close to linear. The automatic IFC procedure was tested at frequencies up to 40 kHz without noticeable limitations. The IFC method is restricted to frequencies above 100 Hz. At lower frequencies the contribution of changes in thickness and area of the membrane to its capacitance may be comparable and the assumptions mentioned above become invalid.

In comparison to other techniques designed for $\Delta\phi_b$ measurements, the automated IFC method has a number of specific features that make it especially convenient for some applications.

1. The stability of planar BLM increases under automatic compensation procedure because the electric field inside the membrane is kept close to zero at any time. This is important for ions or polyelectrolytes of extremely high affinity to phospholipids, which may induce changes of boundary potentials of several hundred mV.

2. The boundary potential changes are recorded continuously in real time. This allows recording of the kinetics of adsorption or desorption of charged substances, charge transfer in photoreactions, activity of phospholipases, etc. To test the reversibility of adsorption (essential for polyelectrolytes) it is reasonable to monitor the kinetics following a quick perfusion of the chamber [34].

3. The IFC method provides an opportunity to follow the changes of the boundary potential during phase transitions induced either by temperature or by adsorption of inorganic ions. These conditions are not suitable for obtaining information on boundary potentials from the membrane conductivity induced, for example, by hydrophobic ions or ionophores: the mobility of hydrophobic ions within the membrane core may change dramatically with the phase transition in a manner independent of the boundary potentials.

4. In parallel to the electrostatic phenomena, IFC method can be modified easily to monitor the elasticity of the membrane. This capability is based on the measurements of the 3-d harmonics of the capacitive current. Its amplitude, as follows from Eq.3, is independent on the DC-component of the applied voltage, and is proportional to the parameter α, which represents bilayer compressibility. We refer the reader to monograph [15] for a more detailed description of this approach.

3. DIFFUSE COMPONENT OF BOUNDARY POTENTIAL

Generally, the alterations of the lipid bilayer induced by membrane-active substances are accompanied by charge and electric field redistribution at the interfaces. According to the approach illustrated by Fig.1, an important step in studies of electrostatic phenomena is discrimination of the effects related to the changes in diffuse and dipole components of the boundary potential. The electrokinetic and IFC methods complement each other by monitoring potentials at different planes. The first technique measures ζ-potentials at the shear plane in the electrolyte at a distance δ from the surface of liposomes. IFC is sensitive to changes of total boundary potential at an imaginary plane within the membrane. Both potentials are defined relative to the potential in the bulk electrolyte. Below we discuss the procedure of how to compare the data of these two methods. The correspondence is expected when the changes of the boundary potential take place in the diffuse part only.

3.1. Charge and potential of diffuse layer

The theory of the electrical double layer has been presented in physical chemistry textbooks and monographs that specifically discuss electrokinetic methods [1, 3]. Numerous reviews include examples of its application to cell membranes and lipid bilayers; the earliest and most comprehensive analysis of the membrane electrostatics is presented in reference [2]. The description of ion equilibria at membrane surfaces by a combination of the Gouy-Chapman model and Langmuir-type isotherm, known as Gouy-Chapman-Stern (GCS) model, has been commonly accepted. The experimental and theoretical arguments in favor of the GCS model are summarized in Ref. [35]. An extensive analysis of the experimental data and models concerning the electric field distribution at the membrane surface is reviewed in [36], and their application to electrokinetic studies of liposomes is presented in [37]. The models alternative to GCS are reviewed in [38].

The GCS formalism is used below to distinguish the events restricted to the diffuse part of the boundary potential as opposed to the "dipole" phenomena. The essential point in this respect is to define a reasonable number of adjustable parameters and to find their values, thus making the model agree with experimental data. This practical aspect was discussed in [39] with the conclusion that the accuracy of measurements and other methodological limitations provide no experimental basis for theories more sophisticated then the classical version of the GCS model. The ability of this simplified model to explain a broad spectrum of phenomena with a minimal number of free parameters may be illustrated by electrokinetic studies of inorganic cation adsorption of different affinity to phospholipids. The same set of parameters, naturally, has to be applied to IFC data. In this section we discuss experiments that show good agreement between the methods and the GCS model.

In its initial form, the GCS theory considered a symmetrical electrolyte of valency z, and the surface charge density, σ, presumed to be a fixed parameter [40, 41]. The theory used the Poisson equation together with Boltzmann relation, which implies an exponential relationship between the concentration of ions c_i, their bulk concentration, c_{bulk}, and the potential in the plane positioned at the distance x from the surface [3]:

$$c_i(x) = c_{i,\,bulk}\ \exp(-ez_i\ \varphi(x)/kT), \tag{4}$$

The potential at the surface, $\varphi(0)$, is determined by well known Gouy-Chapman formula:

$$\sigma = \sqrt{8kT\,\varepsilon\varepsilon_0 c_{\text{bulk}}}\ \sinh(\frac{ez\,\varphi(0)}{2kT}) \tag{5}$$

The decay of the potential with the distance from the surface can be described by the relation:

$$\tanh\left(\frac{ez\,\varphi(x)}{4kT}\right) = \exp\left(-\kappa x\right)\tanh\left(\frac{ez\,\varphi(0)}{4kT}\right). \tag{6}$$

Here $\kappa = \sqrt{\dfrac{2e^2 c}{\varepsilon\varepsilon_0 kT}}$ is the inverse Debye length. Later, the theory was extended to electrolytes of arbitrary composition [42] and Eq.5 could be rewritten in a more generalized form:

$$\sigma^2 = 2kT\,\varepsilon\varepsilon_0 \sum_i c_{i,\text{bulk}}\left[\exp\left(-\frac{ez_i\varphi(0)}{kT}\right)-1\right]. \tag{7}$$

For complex electrolytes with ions of different valence, Eq. (6) may be used as an approximation. In these instances, the valence z has to be substituted with the valency of a dominant counterion, and the concentration, c, - with the ionic strength, $I = 0.5\sum_i c_{i,bulk} z_i^2$. No significant deviation was found between the approximation given by Eqn. (6) and the exact numerical solutions of initial differential equations for complex electrolytes in most practical situations [43].

In the electrokinetic experiment, the measurable quantity is the electrophoretic mobility of colloid particles (liposomes), μ. The Smoluchowsky theory links the mobility with the ζ-potential at shear plane, viscosity, η, and dielectric permeability ε of the medium [3]:

$$\mu = \frac{\zeta\varepsilon\varepsilon_0}{\eta}. \tag{8}$$

The application of eq.(8) to liposome suspensions has been discussed previously (see Ref. [37]). So far, the mobility of multilayer liposomes of about 1 μm diameter in a 10-300 mM electrolyte obeys Eq.(8) well under typical experimental conditions [39].

3.2. Shear plane position

ζ-potential in Eq.(8) is the potential at the shear plane, the distance to which from the charged surface, δ, is generally unknown. In order to find the surface potential one need to evaluate the distance between the shear plane and the physical surface by comparing the electrokinetic data with data of other methods sensitive to surface potential. The value of $\delta=0.2$ nm was initially estimated by the comparison of electrostatic potentials measured in liposome suspensions and in BLM modified by nonactin [44]. In the latter case the BLM conductivity depends on the distribution of the ions at the boundaries of the membrane and therefore can be used to detect the changes of boundary potential with electrolyte concentration [2]. These changes were compared with changes of ζ-potentials using Eq.(6) and their difference reported about the shear plane position. Later, the value of δ was estimated by ionized molecular probes [45]. The application of IFC method to planar lipid membranes gave us another independent way to verify these results.

The difference between surface and ζ-potentials is demonstrated in Fig.3 by experiments with membranes made from the zwitterionic lipid phosphatidylcholine (PC). The electrophoretic mobility of liposomes from PC became positive in the presence of divalent cations due to their adsorption. After reaching a maximum, the mobility begins to decline with salt concentration due to a screening effect. A similar non-monotonic dependence of ζ-potential on concentration was observed with different divalent cations [46]. The changes of ζ-potentials were much more pronounced specifically with Be^{2+}, which we had a chance to study in greater detail using several methods [43, 47, 48]. Measurements performed with high concentrations of this cation revealed a significant difference between the ζ-potential and changes of the boundary potential measured by IFC method (Fig.3). The discrepancy between the two methods could be minimized if the IFC data was compared quantitatively with surface potential found from ζ-potential, taking into account the position of the shear plane. Substitution of $\delta=0.18$ nm into Eq.(6) gave the best fit of the data such that the changes of surface and boundary potentials coincide. This result supports the assumption that both methods detect the changes of the potential drop in the diffuse double layer, although measured at two different planes.

It is important to mention that no assumption was made up to now about the nature of surface charge and ion adsorption. To obtain the value of surface potential only the effect of screening of the charged surface by the electrolyte was taken into account and described quantitatively by the single parameter $\delta=0.2$ nm. This value appears too small in the molecular scale to have real physical meaning. Nevertheless, this formal parameter is essential for comparing the electrostatic effects observed with liposomes and planar BLM. Experimental data in Fig.3 show clearly that the position of the maximal electrophoretic mobility does not correspond to the maximum surface potential measured by IFC method. This difference can be explained simply by screening effect at different planes and is independent of the nature of ions present in the electrolyte.

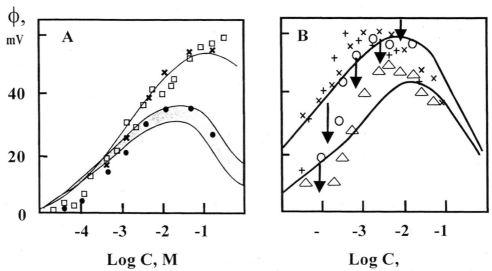

Fig.3 Boundary and ζ-potentials of membranes from phosphatidylcholine at different concentrations of Be^{2+} in the background electrolyte containing 0.1 M KCl, 20 mM imidazole, pH 6.4. A. Boundary potentials measured by IFC (open symbols) and surface potentials (crosses) obtained from ζ-potentials (filled circles) according to Eq.(6) with the parameter δ=0.18 nm.. Theoretical curves are calculated for surface and ζ-potentials by the GCS model with binding parameters S=0.2 C/m², K₂=400 M⁻¹ in the isotherm defined by Eq.(8). The shaded area illustrates the variation in the isotherm position obtained by varying the values of S and K by two orders of magnitude while keeping their product constant and equal to 80 C m M⁻¹.
B. Electrokinetic and IFC data for DPPC membranes. ζ-potential of liposomes from DPPC measured at 22 C in heavy (+) and regular (x) water solutions, and at 45 C in regular (Δ) or heavy (o) water. Arrows indicate the changes of boundary potential measured by IFC in planar BLM during the temperature ramp from 50 to 39 C. Theoretical curves correspond to S=0.2 C/m² and K=400 (45 C) and 10000 M⁻¹ (22 C). Data reproduced from [47, 49] [43]

3.3. Adsorption of inorganic cations

Equations (5) or (7) link surface potentials with the surface charge density, σ. The charge can be then attributed to the adsorption/desorption equilibrium of ions at the surface. The problem of ion adsorption is interesting here in several respects. In the framework of the GCS model, the ion equilibrium at the surface is usually described by a Langmuir-type isotherm with assumptions that i) every lipid molecule is a binding site for inorganic ions, and ii) ion adsorption does not change the packing of lipid molecules in the bilayer. Both assumptions had to be proved by experiments. The data discussed below demonstrate that the electrokinetic method is unlikely to detect deviations from the GCS model. The IFC method, in contrast, measures changes of the entire boundary potential, including the components apart from the diffuse double layer and this may be the source of significant deviations from the model (see Fig.1).

To describe ion adsorption, the GCS model introduces at least two additional parameters: the surface density of binding sites, S, and binding constants, K_i. For uncharged lipids (e.g. phosphatidylcholine, PC), the contribution of monovalent cations to adsorption is negligibly small. Adsorption of multivalent cations of valence z_2 may be described in these cases by the isotherm [44]

$$\frac{\sigma}{S} = \frac{z_2 K_2 c_2(0)}{1 + K_2 c_2(0)}. \qquad (8)$$

The maximal density of binding sites, S, is expressed here in units of charge density. The theoretical curves (solid lines in Fig.3) were calculated with the common value of S=0.2 C/m^2 corresponding to the area of approximately 0.6 nm^2 per lipid.

At first glance, both parameters in Eq.(8), S and K, may be evaluated taking into account the position of the maximum of the experimental curves in Fig.3 [50]. However, electrokinetic experiments show this position nearly the same for electrolytes of different nature [46]. This result can be explained again by screening effect, which masked the specificity of cation adsorption. We should note here that the parameters K and S in Eq.(8) are not completely independent in the fitting procedure and the efficiency of this procedure depends mostly on their product. The shaded area in Fig.3A shows the approximate position of theoretical curves, which correspond to the values of S and K varied by two orders of magnitude while keeping their product constant. This product is proportional to the occupancy of the surface by adsorbed ions. Therefore, the agreement between the model and the experiment neither automatically defines the density of binding sites not does indicate the real stoichiometry of adsorption.

The data presented in Fig.3b show electrokinetic measurements similar to that in Fig.3a, but with liposomes made from DPPC. These measurements were carried out below and above the temperature of the main phase transition for this lipid or in electrolyte solutions prepared with normal and heavy water. The data show that the isotope effect is quite similar to the effect of a decrease of temperature: in both cases the electrophoretic mobility increases. The binding constant (the product of S and K) calculated according to the GCS model for the gel state of the bilayer became 20 times higher than that for liposomes in the liquid state [43, 48]. However, the application of the GCS model under these circumstances is questionable because the cation adsorption induces phase transition.

Most biological membranes are negatively charged due to the presence of acidic phospholipids. Their negatively charged headgroups represent binding sites for cations and it would be natural to assume an universal binding stoichiometry of one ion per headgroup. In terms of GCS-model, this means that the maximal density of binding sites, S, used in isotherm equations must be generally equal to the surface density of phospholipid headgroups. According to the general form of Langmuir isotherm, the relative part of the surface occupied by cations or the net density of surface charge depends on the product of their concentration in the solution adjacent to the surface, $c(0)$, and binding constants [2]:

$$\frac{\sigma}{S} = \frac{(z_2 - 1)K_2 c_2(0) - 1}{1 + K_1 c_1(0) + K_2 c_2(0)}. \tag{9}$$

Parameters K_1 and c_1 refer to protons or monovalent cations of the background electrolyte, whereas terms that include K_2 and c_2 describe the cations of valence z_2. The values of proton binding constants (pK) found in the literature [32] are scattered in a broad range, showing no systematic dependence on the membrane organization. The electrokinetic measurements in suspensions of liposome made from phosphatidylserine (PS), when corrected according to the shear plane position, show the value of pK = 3.0, which correlates well with IFC data [51].

If *a priori* the number of ionized groups per lipid molecule is unknown, electrophoretic measurements become the only way to obtain that information by determining the density of binding sites, S. On the other hand, there is evidence that lipid packing in PS monolayers may depend on the type of cations present in the solution and that their adsorption may induce more complex effects then simple surface charge neutralization according to Eq.(9) [52]. This means that K and S may depend on the electrolyte composition and possibly could be determined by fitting different models to the data. An appropriate fitting procedure was proposed in [53] and applied to electrophoretic mobility of liposomes composed from the mixtures of PS and PC. Measurements made in the presence of alkali metals (K^+, Na^+) or organic cations (TMA^+) indicated different values of S (-0.15 and −0.21 C/m^2, respectively), with binding constants corresponding well with literature values [32, 54]. This result suggests that the lipid bilayer may have different packing in different electrolytes, because the ions may influence the proton equilibrium, lipid hydration, or hydrogen bonding in different ways. Therefore, the nature of ion interaction with the lipid bilayer may not be restricted to surface ionization and screening effects in the diffuse double layer.

According to Eq.(9), in the presence of multivalent cations with $z_2 > 1$ at a concentration equal to K^{-1}, the charge of the surface is zero. This concentration is therefore defined as a zero charge point. In the vicinity of the zero charge point, the GCS model predicts the slope of the surface potential dependence on cation concentration of about 30 mV and 20 mV per decade for di- and trivalent cations, respectively. This prediction was verified in experiments with membranes made from phosphatidylserine in the presence of several kinds of divalent cations such as Mg^{2+}, Ca^{2+} and Ni^{2+} [43, 48, 49]. As previously, the position of the shear plane at $\delta=2$ nm was used to correct the electrokinetic data and to correlate them with the boundary potentials measured by IFC. No deviation was observed between the two experimental methods, and both types of data were described quantitatively by the GCS model with the same set of parameters (but different K for these cations) [55]. We concluded that these electrostatic phenomena reflect the changes in surface charge density due to the adsorption of divalent cations, but without any detectable reorganization of the lipid. This conclusion, however, does not hold for cations with an extremely high affinity to phospholipids. We have found that Be^{2+} and Gd^{3+} demonstrate significant deviations between the electrokinetic and IFC methods. Neither of these techniques yielded results consistent with the predictions of the GCS model [48, 49]. The reasons for these deviations are discussed in the next section.

The use of all equations above is straightforward when the bulk concentrations of ions are well defined. It is more complicated, however, for cations of high affinity to the surface, which demonstrate a dependence of ζ-potential on their concentration that is considerably steeper than the GCS model predicts. Indeed, if the volume of the experimental chamber is too small, and the charged surface is large, c_{bulk} may significantly deviate from c_{tot}, which is equal to the ratio of total amount of ions introduced to the chamber volume. The effect of bulk depletion in these instances should be accounted for with the mass balance condition:

$$c_{tot} = c_{bulk} + c_{dif} + c_{ads}, \qquad (10)$$

where c_{dif} and c_{ads} are the amounts of ions in the diffuse double layer and ions bound to the surface, respectively, each divided by the total volume of the chamber, V. The number of adsorbed ions is proportional to the occupancy of binding sites and to the total amount of lipid in the chamber. The surface excess of ions in the diffuse double layer near the charged surface of total area A can be defined as:

$$\frac{c_{dif}}{c_{bulk}} = \Omega \frac{A}{V} \qquad (11)$$

Here Ω is the excess of ions in the double layer per unit area. Ω can be found using the Boltzmann relationship Eq. (4) and the profile of electric potential along the x-axis normal to the surface Eq. (6):

$$\Omega = \int \left(\frac{c(x)}{c_{bulk}} - 1 \right) dx \qquad (12)$$

On the assumption that the total lipid concentration in the chamber is c_{lip}, and that the every molecule contributes to the surface and is accessible to ions, the A/V ratio in Eq. (11) can be estimated as $N_A A_{lip} c_{lip} \cdot 1000 = 6.4 \cdot 10^8 c_{lip}$ (M^{-1}). In a typical liposome suspension used in electrokinetic experiments, c_{lip} is approximately 1 mM.

The procedure of calculation is as follows. The theoretical value of surface potential is calculated at a cation concentration c_{bulk} by GCS equations listed above, also taking into account the isotherm of adsorption Eq.(9). Then the total amount of cations at the surface and in the diffuse double layer is calculated using the area-to-volume ratio taken from the experiment. Eq.(10) gives the theoretical value of c_{tot} and the curve $\phi_s(c_{tot})$ can be fit to the experimental data. The GCS theory modified with the mass balance condition was shown to be in good agreement with the electrokinetic data in the case of adsorption of trivalent cation Gd^{3+} on membranes made from mixtures with different PS/PC ratio [49, 56] (Fig.4). For uncharged membranes made of pure PC, the data of the electrokinetic and IFC methods were highly consistent and showed little deviation from the modified theory. This implied no changes of the dipole component of boundary potential and no structural reorganization of phosphatidylcholine bilayers in the presence of Gd^{3+}.

3.4. Conclusions

The theoretical arguments and experimental data summarized above illustrate that the knowledge of the position of the shear plane is necessary for a comparison of the data obtained by IFC and electrokinetic methods. The latter has to take into account the shape of potential decay near the charged surface. The parameter $\delta=0.2$ nm in Eq.(6) provides good agreement between the methods. This conclusion is independent of the type of isotherm or binding parameters. On the other hand, the good correspondence of the two methods indicates that the observed effects take place exclusively in the diffuse part of the electrical double layer and are determined by the ionization of surface groups and their screening by the electrolyte.

The agreement of the GCS model with the experimental data does not prove the model itself. It has an excessive number of parameters, some of which are not completely independent. The problem of unambiguous determination of the density of binding sites or the binding stoichiometry for some ions remains unsolved. In order to see the possible alterations of the lipid structure induced by adsorbed ions one has to measure the changes of the entire boundary potential, which may deviate from the changes of the surface potential. The IFC method has helped us to detect electrostatic phenomena far more complex then simple charging of the surface.

4. ELECTRICAL 'STRUCTURE' OF MEMBRANE BOUNDARIES

4.1 Electrostatic phenomena correlate with phase transitions

Phase transitions of lipids were monitored by electrokinetic measurements in suspensions of liposomes made from phosphatidylserine [57] and different phosphatidylcholines [48, 58, 59]. At the temperature corresponding to the phase transition between the liquid and gel states of the bilayer, a shift of ζ-potential in the positive direction was observed in electrolytes containing ions of different valence. This effect was similar to the shifts in ζ-potential measured in liposome suspensions prepared in regular or heavy water (see Fig.3b). The effects were especially pronounced in membranes made of DPPC in the presence of Be^{2+}, which has the largest binding constant among the divalent cations [47]. These observations agreed with the data of microcalorimetry, which showed that Be^{2+} induces larger shifts of the main transition temperature in DPPC than Mg^{2+} or Ca^{2+}, even when the latter ions are present at 100 times higher concentration. The magnitude of the effect on phase transitions correlated well with the ion's binding constant. Similar experiments conducted in heavy water demonstrated observable effects when the concentration of Be^{2+} was further lowered by one order of magnitude [43], suggesting that the dynamics of the solvent and hydration of headgroups are affected by ion binding and vice versa.

The nature of the dipole component of the boundary potential has its direct relevance to the structure of the interface, encompassing parts of the non-polar and polar layers of the membrane. Dipole effects during a thermotropic lipid phase transition can be detected by IFC. It has been observed repeatedly that the stability of a bilayer decreases dramatically near the temperature of the main phase transition [60]. Therefore, by using IFC simultaneously with ramps of decreasing temperature we could follow the changes of the boundary potential for as long as the bilayer remained in the liquid state. These changes are shown in Fig.3b by arrows. The measurements were implemented under asymmetrical conditions with Be^{2+} added only in one compartment of the chamber. Note that the signs of the changes of boundary and surface potentials are opposite, and their absolute value are different ($\Delta\phi_b > \Delta\phi_s$). It is not possible yet to distinguish the role of orientation of lipid headgroups or associated water molecules in these observations. We may postulate only that the average dipole moment and/or their surface density changes upon phase transition, which may be possible due to a change of the phosphatidylcholine headgroup hydration. Recent experimental data support the idea that ordered water molecules at the membrane surface provide the main contribution to the dipole potential [61] and the hydration-related repulsion [8].

The effect of inorganic ions on the dipole potential was demonstrated with negatively charged membranes made from phosphatidylserine. Again, Be^{2+} was the most effective divalent cation and it clearly revealed the difference between the electrokinetic and IFC data. But the most significant difference between surface and boundary potentials was observed with trivalent lanthatides (Gd^{3+} and others). Gd^{3+} is known as a nonspecific blocker of different kinds of mechanosensitive channels irrespective of their origin and selectivity [62, 63]. It is difficult to imagine that all mechanosensitive channels possess a specific binding site for Gd^{3+}, however a high-affinity binding of lanthanides to the phospholipid component of membranes has been previously documented [64-66]. Therefore, a more plausible explanation of the Gd^{3+}-induced block of mechanosensitive channels would be binding of Gd^{3+} to the lipid matrix, its rearrangement and changes of physical properties, which in turn of may alter the sensitivity of the channels to membrane tension [67]. Our experiments with lipid membranes supported this hypothesis and show a direct correlation between the changes of dipole potential and membrane tension induced by adsorption of Gd^{3+} at the membrane surface [49]. As previously, the information about the dipole potential was extracted by comparing the effects of Gd^{3+} on the boundary and surface potentials, i.e. from the data of IFC and electrokinetic measurements.

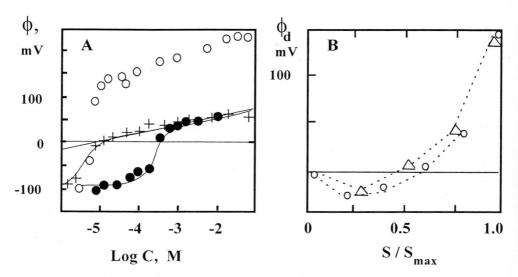

Fig. 4. Binding of Gd^{3+} to membranes made of PS or PS + PC mixtures
A. Boundary potentials (open circles), and surface potentials (filled circles) measured with membranes from PS or from the mixture PS:PC=3:2 (crosses) (10 mM KCl, pH7.1). The IFC data are shifted by −114 mV for liposomes made from pure PS and by −101 mV for the mixture. The surface potentials are obtained from electrokinetic data by correcting for the shear plane position (δ=0.2). Theoretical curves are calculated in the framework of GCS model taking into account competitive binding of K^+ and Gd^{3+} to PS in Eq.(9) with constants K_1=4, K_2=5 10^4 M^{-1}, and binding of Gd^{3+} to PC Eq.(8) with K_2= 10^3 M^{-1} respectively. The parameter c_{lip} in Eq.10 introducing the condition of mass balance was 0.02, and 0.3 mM, for experiments with BLM and liposomes, respectively.
B. Mean values of dipole potentials induced by Gd^{3+} adsorption on membranes with different content of negatively charged PS molecules. The density of negative charges, S, is normalized to the maximal value S_{max}=0.2 C/m^2 which corresponds to the charge density on liposomes made from pure PS at pH 7.0. Open circles are for different PS/PC mixtures, the triangles are for liposomes from pure PS measured at different pH. Reproduced from [49].

The data of several experiments with liposomes and planar BLM composed from PS and PC/PS mixtures are presented in Fig.4A. The changes of the boundary potential in the presence of Gd^{3+} exceed the changes of surface potential (calculated from the ζ-potential and the position of shear plane) by about 200 mV. The quantitative analysis of data was performed in the framework of GCS model expanded by the condition of mass balance, which was necessary in view of the very high affinity of this cation to phospholipids. Its binding constants to PC and PS used as parameters in Eqs.(8) and (9) were

10^3 and $5*10^4$ M^{-1} , respectively [56]. This resulted in a very steep dependence of the potentials observed in PS membranes around the zero charge point. Comparison of the methods required the introduction of one extra parameter in Eq.(10), the total amount of the lipids in the cell c_{lip}. Pretreatment of the chamber septum with lipids and the BLM formation procedure itself does not allow controlling the exact amount of lipid in the chamber. Therefore, c_{lip} could be either roughly estimated or used as an adjustable parameter (see ref. [49] for details).

The differences between the boundary and surface potentials, which report on the changes of the dipole component, were extracted from the two data sets. As shown in Fig.4B, the magnitude of the dipole component change is determined solely by the presence of the ionized form of PS. The change of the dipole potential was independent of the way the concentration of charged PS groups was achieved; by adding more PS to the mixture at neutral pH, or by varying pH, and thereby changing the state of ionization of the membrane made of pure PS. The increase of the dipole component also correlated with the changes of membrane tension [49]. This result and our preliminary observations of effects that Gd^{3+} exerts in lipid monolayers made from PS and PC convince us that the adsorption of this cation induces a dramatic condensation of films prepared from negatively charged lipids. In the condensed state of monolayer the dipole potential varied with the surface density of lipid headgroups such a way as it was described for elementary dipoles in [68]. It should be mentioned that the mechanism of lipid condensation by Gd^{3+} is not fully consistent with the GCS model. The model is based on the assumption of independent ion binding to non-interacting sites, which is not true in the case of cooperative phase transition in lipid bilayers. Lipid condensation induced by multivalent ions of high affinity to phospholipids and related electrostatic phenomena require certainly further and more detailed investigation. According to our observations the affinity of such cations (Gd^{3+}, Be^{2+}) increases to a more compact lipid layers.

The specific features of IFC method have been crucial for the above studies. A continuous intramembrane field compensation in automatic regime prevented BLM breakdown at high asymmetry of boundary potentials, which could be as high as 350 mV (Fig.4). The technique worked reliably even when the membranes undergo dramatic structural rearrangements induced by Gd^{3+}. The change in visco-elastic properties and general rigidification of the membrane in the presence of the high-affinity ions made it impossible to implement measurements of dipole potentials using the alternative ionophore permeation technique in these studies.

4.2. Adsorption of amphiphilic ions

The information on electric fields at the membrane interface can be obtained using methods, which are sensitive to potentials at different planes inside or outside the membrane. The combination of electrokinetic and IFC methods described above was used to detect changes of the dipole component of the boundary potential. The same approach was used in studies of the adsorption of amphiphilic ions characterized by the presence of hydrophobic groups, which help them to intercalate into the membrane and to incorporate their ionized groups into the bilayer. This is an intrinsic property of many membrane-active compounds and drugs, fluorescent dyes or other probes used in studies of the membrane structure. Therefore, the knowledge of their location in the membrane is essential. We have studied the adsorption of several amphiphilic ions and dyes [69-73], and below we discuss the data concerning 8-aniline-1-naphtalene-sulfonate (ANS), the fluorescent dye widely used in studies of lipid dynamics in membranes [74].

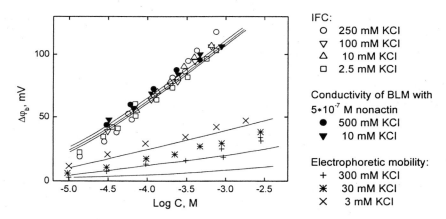

Fig.5. Increments of boundary potential $\Delta\varphi$ and ζ-potential $\Delta\zeta$ induced by adsorption of ANS on membranes of dioleoyl lecithin at different ANS concentrations in the bulk and several ionic strengths. Theoretical curves are calculated by the GCS model and Eqs.14-17 with capacitance $C_a=0.1$ F m^{-2} and the coefficient $K = 10^3$ C m^{-2} M^{-1}. Data reproduced from [74].

The measurements were carried out using three independent techniques applied to planar BLM and liposomes. Data of measurements with initially uncharged phosphatidylcholine membranes in the presence of ANS are shown in Fig.5. The electrokinetic measurements in liposome suspensions show that the ζ-potential gradually increases with ANS concentration; the slope of curves decreases with ionic strength of the background electrolyte. In contrast to the ζ-potential, the IFC method detects much larger changes of boundary potential without any visible effect of ionic strength. This result was confirmed by

conductivity measurements in planar membranes modified by the ionophore
nonactin. In the later case, ANS was added to both compartments of the
chamber and the changes of boundary potential were calculated from BLM
conductivity by the formula [2]:

$$\Delta\varphi_b = \frac{RT}{zF} \ln\left(\frac{g}{g_0}\right), \tag{13}$$

where z is the valence of the charged form of ionophore conferring the
conductivity, g_0 and g stand for the initial BLM conductivity and its value in the
presence of ANS; R, T and F have conventional meaning. The data of IFC
method coincide perfectly with conductivity measurements (Fig.5). Thus both
methods are sensitive to the changes of potential in the plane of adsorption of
ANS. On the other hand, this potential significantly exceeds the surface
potential determined by electrokinetic measurements. This observation indicates
that ANS may adsorb at a certain depth beneath the surface of the membrane,
which makes it inaccessible to the screening by the electrolyte. The total
boundary potential was essentially independent of the ionic strength. In
principle this result could be attributed to some sort of compensation effect:
higher ionic strength facilitates the adsorption of ANS and thus increases the
negative surface charge. In the framework of the GCS model it can be shown,
however, that this effect decreases the slope of the dependence of surface
potential on ionic strength to about 20 mV per decade. The experiments
demonstrated a much weaker dependence.

The origin of a potential inaccessible to electrolyte screening was explained in
the literature by several models. It has been proposed that amphiphilic ions
possess (or may induce) a dipole moment normal to the membrane surface [75].
Another model has been suggested for dipicrylamine hydrophobic ions, which
assumed a dielectric saturation in the vicinity of the membrane/solution
interface due to the water polarization. This could decrease the dielectric
constant and increase the potential drop compared to that predicted by the GCS
model. Penetration of charged groups of hydrophobic ions tetraphenylborate
into the interface was proposed in [2]. The potential drop between the
adsorption plane and the membrane surface was described as the potential of a
charged capacitor.

The latter model was used for the quantitative description of our data [74]. The
total boundary potential φ_b can be represented as the sum of the surface
potential φ_s, and the potential drop embedded in the boundary layer inside the

membrane, φ_δ :

$$\varphi_b = \varphi_s + \varphi_\delta \tag{14}$$

The second electrolyte inaccessible term can be found by the capacitor formula:

$$\varphi_\delta = \sigma / c_\delta \tag{15}$$

Note that the same expression formally applies to models in which the "capacitor" originates from the adsorption of molecules carrying intrinsic dipoles. The value of φ_s is determined according to the theory of the diffuse double layer:

$$\frac{\sigma}{\sqrt{8\varepsilon\varepsilon_0 RTC_e}} = sh\left(\frac{zF\varphi_s}{2RT}\right) \tag{16}$$

where C_e is the ionic strength of background electrolyte, z is the charge of adsorbing ions (z=-1 for ANS). This equation is similar to Eq.(5) or Eq.(7), except the origin of the surface charge in the present system is different. If BLM consist of neutral lipids, its initial surface charge is zero, and the charge density σ is determined by amphiphilic molecules incorporated into the bilayer. In a typical experimental situation, the density of absorbed molecules is far from saturation and can be described by a linear relationship between their concentration in the adjacent electrolyte and on the surface, which determines the surface charge (isotherm of Henry). Taking into account the Boltzmann relation Eq.(4), the surface charge density can be presented as:

$$\sigma = KC_a e^{-\frac{zF\varphi_b}{RT}} \tag{17}$$

where C_a is bulk concentration of ANS. These equations describe the curves presented in Fig.5 well with capacitance $C_a=0.1$ F m^{-2} and coefficient K $= 10^3$ C m^{-2} M^{-1}.

The capacitance c_δ characterizes the thickness and dielectric permeability of the layer of the membrane between the plane of adsorption of the amphiphilic ions and the surface. Therefore, its value depends on the depth of adsorption and may change with hydrophobicity of the molecules. For instance, in the case of adsorption of haematoporphyrins of different structures, there was a direct correlation of c_δ with the number of hydrophobic groups in the molecule. This parameter was shown to be important for their effectiveness in photodynamic damage of membranes [69].

The shortcomings of this simple model are evident from the asymptotic behavior of the experimental curves. For instance, the model predicts that the dependence of the potential on the ANS concentration may have a maximal slope of about 58 mV per decade, whereas in the experiments with negatively charged membranes this slope was shown to be twice larger [73]. To provide an adequate description to these data, more elaborate models were developed. They took into account the discreetness of charge distribution in the adsorption plane and nonelectrostatic interaction of adsorbed ions in the membrane [76].

4.3. Adsorption of dipole modifiers

An important class of membrane-active compounds is characterized by large dipole moments, having at the same time zero net charge of the molecule. These dipole moments, when oriented normally to the membrane plane, may create an electric field similar to one described above. Phloretin and phloridzin are well known examples of such dipole modifiers. They were studied extensively by the IFC method and by measuring the conductance of BLM in the presence of nonactin [77].

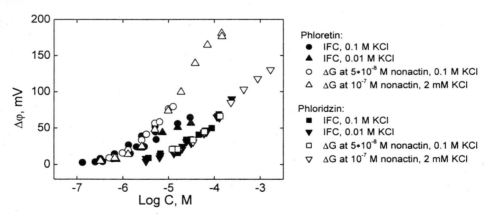

Fig.6. Boundary potentials induced by adsorption of dipolar molecules of phloretin and phloridzin as a function of their concentrations in the solution. BLM are formed from azolectin (50 mg/ml of decan). The boundary potentials are determined by IFC method (filled symbols) or from the conductivity of BLM modified by nonactin BLM (open symbols). Data reproduced from [77].

Phloretin and phloridzin are neutral under normal conditions and had no effect on the electrophoretic mobility of liposomes. This observation is consistent with the data presented in Fig.6: the experiments do not show any dependence of the boundary potentials on ionic strength even when the measurements were conducted in membranes formed from charged lipids. Adsorption of each of the substances added either asymmetrically (for IFC) or symmetrically (for conductance measurements) changed the boundary potential in a concentration-dependent manner. In the case of phloretin, potentials measured using nonactin were much larger then those measured by IFC (Fig.6). This discrepancy was explained by the phloretin transfer through the membrane and the appearance of these molecules at the opposite side of the membrane [77]. Such a transfer was studied in more details in [78]. The same measurements were done with phloridzin, which is structurally different from phloretin by one additional glucose moeity. The more polar molecule was expected to be less permeable. Indeed, in experiments with phloridzin no difference between the two methods was observed (Fig.6).

4.4. Transfer of amphiphilic ions across the membrane

As mentioned above, IFC cannot be implemented directly in studies of molecules permeable through the membrane. IFC measurements rely on strictly asymmetrical conditions when the substance is added to one compartment and $\Delta\phi_b$ reflects the equilibrium between the surface of the membrane and the bulk, while the opposite (unmodified) side of the membrane serves as reference. If the molecules exchange between the two sides of the membrane, we can no longer measure the true value of $\Delta\phi_b$. We confronted such a situation studying the membrane effects of the antiviral drug remantadine (α-methyl-1-adamantanemethylamine, REM) [79].

In spite of the difficulties posed by REM permeation, after several preliminary trials we found conditions, in which the IFC method could detect true changes of boundary potential caused by its adsorption. REM is a weak base, which permeates the membrane in neutral form [80]. The electrically neutral transport of any weak acid or base through a BLM is coupled to protonation/deprotonation reactions on the membrane boundaries [81]. The important consequence of this mechanism is that the distribution of charged forms of these molecules between the two boundaries of the membrane depends on the concentrations of hydrogen ions in the solutions and their transmembrane gradient [82]. For this reason ΔpH between the two sides of the membrane becomes critical in determining the value and even the sign of $\Delta\phi_b$ measured by the IFC method.

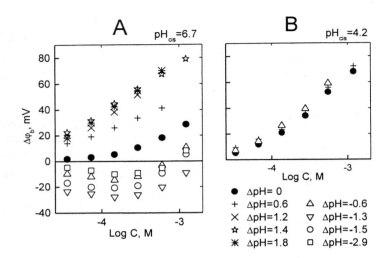

Fig.7. The dependence of $\Delta\varphi_b$ on concentration of REM in the cis solution at different transmembrane gradients of pH (the values of $\Delta pH=pH_{trans}-pH_{cis}$ are indicated). The values of pH varied only in the trans solution, whereas in the cis solution they were kept constant. BLM was formed from dioleoyl PC in the buffers containing: (left) 50mM KCl+2.5mM multicomponent buffer (phosphate, citrate, borate), pH 6.7, or (right) 50mM KCl+2.5mM citrate, pH 4.2. Data reproduced from [80]

Fig. 7 represents $\Delta\varphi_b$ as a function of REM concentration on the *cis* side of BLM measured by IFC method under different pH gradients ($\Delta pH=pH_{trans}-pH_{cis}$, created by adding an acid or a base to the *trans* compartment of the cell). If the membrane was totally impermeable to REM, no effects of pH at the *trans* side of the BLM would be observed on the REM adsorption at the *cis* side. However, the data in Fig.7A clearly show these effects. The positive values of $\Delta\varphi_b$ reflect a larger presence of the charged form of REM at the *cis* side of the membrane. This distribution of REM is enhanced by a positive ΔpH. Note that in the case of unbuffered solutions in both compartments, a pH gradient of the same sign appeared in the unstirred layers at both sides of BLM due to the transmembrane flux of REM. It generated higher $\Delta\varphi_b$ values compared to those measured in buffered solutions [80]. Negative ΔpH resulted in negative $\Delta\varphi_b$ reflecting the opposite distribution of the charged form of REM between the two sides of BLM.

One of the important predictions of the model [80] is that the rate of REM transport decreases with ΔpH (i. e. as the medium becomes more alkaline at the *trans* side). In the extreme situation (very high ΔpH), the positively charged form of REM is concentrated entirely at the *cis* side and approaches equilibrium with the bulk. In this particular case the IFC method measures an exact value of $\Delta\phi_b$ due to a negligibly small flux of REM across the membrane. The potentials recorded at high pH gradients (ΔpH$>$1.5) are maximal as illustrated by the data in Fig.7A.

The transport of REM across the membrane is mediated by the diffusion of the neutral form. Because REM is a weak base, the fraction of neutral REM should decrease at low pH. Conceivably, the flux of this form will decrease and its effect on $\Delta\phi_b$ should diminish in acidic solutions. Indeed, IFC measurements detected substantial changes of $\Delta\phi_b$ with ΔpH when the pH in the cis-compartment was 6.7 (Fig.7A), and there was no effect of the pH gradient when the pH in the *cis* side was 4.2 (Fig.7B).

Above we described an unusual application of the IFC method in studies of the electrically neutral transport of remantadine. Apparently, this version of the technique applies to studies of any weak acid or base that may interact with the membrane and traverse to the trans-compartment. By varying pH on both sides of the membrane it is possible to find conditions at which the flux of permeable molecules is minimal and the distribution between the bulk and the *cis* side of the membrane approaches the equilibrium. Under such conditions $\Delta\phi_b$ measured by IFC is close to the value that would be observed for an impermeable molecule.

5. SUMMARY

The IFC method proved to be crucial in studies of interactions of several diverse groups of substances with planar bilayers. These were inorganic and organic ions, as well as molecules with intrinsic dipole moments. This method was especially informative when it was used in combination with the electrokinetic technique sensitive to the potential at a plane in the aqueous phase near the membrane interface. This allowed us to separate the changes of potential in the outer diffuse layer from those within the membrane. This approach revealed significant changes of the dipole component of the boundary potential induced by multivalent cations characterized with a high affinity to phospholipids. The changes in the layer of the boundary (or membrane-embedded) dipoles were found to be of the same or even higher magnitude then the surface effects in the diffuse part of the electrical double layer. The changes of dipole potential could be attributed to lipid condensation likely as a result of liquid–gel phase transition initiated by ion adsorption. These conclusions were supported by the data of microcalorimetry and studies in lipid monolayers [83, 84]. The shifts of potentials and transition temperatures observed in heavy water also implicate changes of lipid hydration. Thus, the electrostatic measurements provide information on three types of closely related phenomena – ion adsorption itself, its effect on the phase state of the lipids and on lipid hydration. The Gouy-Chapman-Stern model provides an adequate quantitative treatment of data outside of the concentration range corresponding to the phase transition.

The IFC method was applied successfully in studies of adsorption of amphiphilic ions. The existence of an electrolyte-inaccessible component of the boundary potential was demonstrated; this component characterizes the depth of the charge penetration into the membrane. The distance of the absorption plane from the surface was found to correlate with the hydrophobicity of the particular molecule.

The IFC method was helpful in studies of membrane permeation of remantadine, a potent antiviral agent. It has been shown to traverse the membrane in the neutral form that leads to its deprotonation and protonation at the boundaries of the membrane. It exemplified a common transport mechanism of various hydrophobic molecules of weak basic or acids. By adjusting the external conditions permits to minimize the permeation of a substance one can then apply IFC in its original way to characterize the equilibrium between the free and membrane-bound form.

It is important that in every system that permitted simultaneous application of IFC and the more traditional ionophore-based technique for boundary potential measurement, the two methods gave identical results. Only complications caused either by permeation of molecules or by changes of membrane structure led to discrepancies. The method opens many possibilities for studies of electrostatically-driven membrane interactions with charged polymers, peptides and macromolecules [34]. IFC is capable of detecting enzymatic reactions at the membrane surface such as cleavage of lipids by phospholipase A [85]. It makes IFC belonging in armamentarium of modern analytical techniques.

ACKNOWLEDGMENTS.
The authors thank Prof. S. Sukharev and Prof. T.E.DeCoursey for critical reading of the manuscript. The work was supported by INTAS-2001-0224 and Russian Foundation for Basic Reseach (00-04-48920 and 01-04-49246) .

REFERENCES

[1]   G.Cevc and D.Marsh. Phospholipid Bilayers. Physical Principles and Models. Vol.5. 1987. New York, Willey-Interscience Publication. Cell Biology: A Series of Monographs. Bittar, E. E. (ed.)

[2]   S.McLaughlin. Electrostatic Potentials at Membrane-Solution Interfaces. Vol.9, 71-144. 1977. New York. Current Topics Membranes and Transport. Bronnen, F. and Kleinzeller, A. (eds.)

[3]   R.J.Hunter. Zeta Potential in Colloid Science. Principles and Applications. 1981. London, Academic Press. Colloid Science. Ottewill, R. H. and Rowell, R. L. (eds.)

[4]   E.A.Guggenheim. Thermodynamics. Wiley, New York. 1950.

[5]   A.W.Adamson. Physical Chemistry of Surfaces. Wiley&Sons, 1967.

[6]   J.T.Davies and E.K.Rideal. Interfacial Phenomena. Academic Press, 2nd ed., New York. 1963.

[7]   A.D.Pickar and R.Benz. J. Membrane Biol. 44 (1978) 353.

[8]   K.Gawrisch, D.Ruston, J.Zimmerberg, A.Parsegian, R.P.Rand, and N.Fuller. Biophys. J. 61 (1992) 1213.

[9]   R.F.Flewelling and W.L.Hubbell. Biophys. J. 49 (1986) 541.

[10] A.Kotyk and K.Janachek. Membrane transport. An Interdisciplinary approach. Plenum Press, New York, London. 1977.

[11] R.C.MacDonald and A.D.Bangham. J. Membrane Biol. 7 (1972) 29.

[12] A.V.Babakov, L.N.Ermishkin, and E.A.Liberman. Nature 210 (1966) 953.

[13] P.Schoch and D.F.Sargent. Experimentia 32 (1976) 811.

[14] O.Alvarez and R.Latorre. Biophys. J. 21 (1978) 1.

[15] T.Hianik and V.I.Passechnik. Bilayer Lipid Membranes: Structure and Mechanical Properties. Ister Science, Bratislava. 1995.

[16] J.Requena, D.A.Haydon, and S.B.Hladky. Biophys. J. 15 (1975) 77.

[17] G.N.Berestovskii. Biofizika 26 (1981) 474.

[18] S.H.White and W.Chang. Biophys. J. 36 (1981) 449.

[19] I.G.Abidor, Y.A.Chizmadzhev, I.Y.Glazunov, and S.L.Leikin. Biol. Membrany 3 (1986) 627.

[20] P.Schoch, D.F.Sargent, and R.Schwyzer. J. Membr. Biol. 46 (1979) 71.

[21] C.Usai, M.Robello, F.Gambale, and C.Marchetti. J. Membr. Biol. 82 (1984) 15.

[22] I.G.Abidor, S.H.Aityan, V.V.Chernyi, L.V.Chernomordik, and Y.A.Chizmadzhev. Dokl. Akad. Nauk SSSR 245 (1979) 977.

[23] V.I.Passechnik and T.Hianik. Kolloidn. Zh. 38 (1977) 1180.

[24] W.Carius. J. Colloid Interface Sci. 57 (1976) 301.

[25] C.Shimane, V.I.Passechnik, S.El-Karadagi, V.A.Tverdislov, N.S.Kovaleva, N.V.Petrova, I.G.Kharitonenkov, and O.V.Martzenyuk. Biofozika (Rus). 29 (1984) 419.

[26] V.S.Sokolov and S.G.Kuzmin. Biofizika (rus.) 25 (1980) 170.

[27] R.Benz and K.Janko. Biochim. Biophys. Acta 455 (1976) 721.

[28] R.Benz and K.Janko. Biochim. Biophys. Acta 455 (1976) 721.

[29] M.Montal and P.Mueller. PNAS USA 69 (1972) 3561.

[30] P.Mueller, D.O.Rudin, H.T.Tien, and W.C.Wescott. J. Phys. Chem. 67 (1963) 534.

[31] V.V.Cherny, V.S.Sokolov, and I.G.Abidor. Bioelectrochemistry and Bioenergetics 7 (1980) 413.

[32] J.F.Tocanne and J.Teissie. Biochim. Biophys. Acta 1031 (1990) 111.

[33] V.I.Passechnik and V.S.Sokolov. Bioelectrochemistry 55 (2002) 47.

[34] Yu.A.Ermakov, I.S.Fevraleva, and R.I.Ataulakhanov. Biol. Membr. 2 (1985) 1094.

[35] S.McLaughlin. Annu. Rev. Biophys Biophys Chem 18 (1989) 113.

[36] G.Cevc. Biochim. Biophys. Acta 1031 (1990) 311.

[37] G.Cevc. Chem. Phys. Lipids 64 (1993) 163.

[38] M.Blank. Electrical Double Layers in Biology. Plenum Press, New York, London. 1985.

[39] Yu.A.Ermakov. Kolloidn. Zh. 62 (2000) 389.

[40] D.L.Chapman. Philos. Mag. 25 (1913) 475.

[41] M.Gouy. J. Phys. (Paris) 9 (1910) 457.

[42] D.C.Grahame. Chem. Rev. 41 (1947) 441.

[43] Yu.A.Ermakov, S.S.Makhmudova, E.V.Shevchenko, and V.I.Lobyshev. Biol. Membr. 10 (1993) 212.

[44] A.McLaughlin, C.Grathwohl, and S.G.A.McLaughlin. Biochim. Biophys. Acta 513 (1978) 338.

[45] S.McLaughlin. Experimental Test of the Assumptions Inherent in the Gouy-Chapman-Stern Theory of the Aqueous Diffuse Double Layer. *In* Physical Chemistry of Transmembrane Ion Motions. *Ed. by* G.Spach. Amsterdam. 1983. p. 69.

[46] S.A.Tatulian. Ionization and ion binding. *In* Phospholipid Handbook. *Ed. by* G.Cevc. New York. 1993. p. 511.

[47] Yu.A.Ermakov, V.V.Cherny, and V.S.Sokolov. Biol. Membr. 9 (1992) 201.

[48] Yu.A.Ermakov, S.S.Makhmudova, and A.Z.Averbakh. Colloids and Surfaces. A. :Physicochemical and engineering aspects 140 (1998) 13.

[49] Yu.A.Ermakov, A.Z.Averbakh, A.I.Yusipovich, and S.I.Sukharev. Biophys. J. 80 (2001) 1851.

[50] S.A.Tatulian. J. Phys. Chem. 98 (1994) 4963.

[51] Yu.A.Ermakov, A.Z.Averbakh, A.B.Arbuzova, and S.I.Sukharev. Membr. Cell Biol. 12 (1998) 411.

[52] H.Hauser. Chem. Phys. Lipids 57 (1991) 309.

[53] Yu.A.Ermakov. Biochim. Biophys. Acta 1023 (1990) 91.

[54] M.Eisenberg, T.Gresalfi, T.Riccio, and S.McLaughlin. Biochemistry 18 (1979) 5213.

[55] S.McLaughlin, N.Mulrine, T.Gresalfi, G.Vaio, and A.McLaughlin. J. Gen. Physiol. 77 (1981) 445.

[56] Yu.A.Ermakov, A.Z.Averbakh, and S.I.Sukharev. Membr. Cell Biol. 11 (1997) 539.

[57] R.C.MacDonald, S.A.Simon, and E.Baer. Biochemistry 15 (1976) 885.

[58] S.A.Tatulian. Biochim. Biophys. Acta 736 (1983) 189.

[59] S.A.Tatulian. Surface Electrostatics of Biological Membranes an Ion Binding. *In* Surface Chemistry and Electrochemistry of Membranes. *Ed. by* T.S.Soersen. N.-Y., Basel. 1999. p. 871.

[60] V.F.Antonov, E.V.Shevchenko, E.Yu.Smirnova, E.V.Yakovenko, and A.V.Frolov. Chem. Phys. Lipids 61 (1992) 219.

[61] S.A.Simon, C.A.Fink, A.K.Kenworthy, and T.J.McIntosh. Biophys. J. 59 (1991) 538.

[62] O.P.Hamill and D.W.McBride, Jr. Pharmacol. Rev. 48 (1996) 231.

[63] S.I.Sukharev, W.J.Sigurdson, C.Kung, and F.Sachs. J. Gen. Physiol 113 (1999) 525.

[64] J.Bentz, D.Alford, J.Cohen, and N.Duzgunes. Biophys. J. 53 (1988) 593.

[65] B.Z.Chowdhry, G.Lipka, A.W.Dalziel, and J.M.Sturtevant. Biophys. J. 45 (1984) 633.

[66] V.V.Kumar and W.J.Baumann. Biophys. J. 59 (1991) 103.

[67] R.S.Cantor. Biophys. J. 77 (1999) 2643.

[68] H.Brockman. Chem. Phys. Lipids 73 (1994) 57.

[69] I.N.Stozhkova, V.V.Cherny, V.S.Sokolov, and Yu.A.Ermakov. Membr. Cell Biol. 11 (1997) 381.

[70] V.S.Sokolov, V.V.Chernyi, M.V.Simonova, and V.S.Markin. Biologicheskie Membrany 7 (1990) 872.

[71] M.V.Simonova, V.V.Chernyi, E.Donat, V.S.Sokolov, and V.S.Markin. Biologicheskie Membrany 3 (1986) 846.

[72] A.V.Cherny, V.S.Sokolov, and V.V.Cherny. Russ. J. Electrochem. 29 (1993) 321.

[73] V.V.Chernyi, M.M.Kozlov, V.S.Sokolov, I.A.Ermakov, and V.S.Markin. Biofizika 27 (1982) 812.

[74] Yu.A.Ermakov, V.V.Chernyi, S.A.Tatulian, and V.S.Sokolov. Biofizika 28 (1983) 1010.

[75] D.Cros, P.Seta, C.Gavach, and R.Benz. J. Electroanal. Chem. Interfacial Electrochem. 134 (1982) 147.

[76] M.M.Kozlov, V.V.Chernyi, V.S.Sokolov, Yu.A.Ermakov, and V.S.Markin. Biofizika 28 (1983) 61.

[77] V.S.Sokolov, V.V.Chernyi, and V.S.Markin. Biofizika 29 (1984) 424.

[78] P.Pohl, T.I.Rokitskaya, E.E.Pohl, and S.M.Saparov. Biochim. Biophys. Acta 1323 (1997) 163.

[79] V.V.Cherny, M.Paulitschke, M.V.Simonova, E.Hessel, Y.Ermakov, V.S.Sokolov, D.Lerche, and V.S.Markin. Gen. Physiol Biophys. 8 (1989) 23.

[80] V.V.Cherny, M.V.Simonova, V.S.Sokolov, and V.S.Markin. Bioelectrochemistry and Bioenergetics 23 (1990) 17.

[81] Y.N.Antonenko and L.S.Yaguzhinskii. Biofizika 29 (1984) 232.

[82] V.S.Markin, V.I.Portnov, M.V.Simonova, V.S.Sokolov, and V.V.Cherny. Biologicheskie Membrany 4 (1987) 502.

[83] A.Averbakh, D.Pavlov, and V.I.Lobyshev. Journal of Thermal Analysis and Calorimetry 62 (2000) 101.

[84] Yu.A.Ermakov and S.I.Sukharev. Condensation of phosphatidylserine monolayers by pressure and lanthanides. manuscript in preparation. 2002.

[85] V.M.Mirsky, V.V.Cherny, V.S.Sokolov, and V.S.Markin. J. Biochem. Biophys. Methods 21 (1990) 277.

Planar Lipid Bilayers (BLMs) and their Applications
H.T. Tien and A. Ottova-Leitmannova (Editors)
© 2003 Elsevier Science B.V. All rights reserved.

Chapter 4

Effect of anisotropic properties of membrane constituents on stable shapes of membrane bilayer structure

Ales Iglic[a] and Veronika Kralj-Iglic[b]

[a]Laboratory of Applied Physics, Faculty of Electrical Engineering, University of Ljubljana, Trzaska 25, SI-1000 Ljubljana, Slovenia

[b]Institute of Biophysics, Faculty of Medicine, University of Ljubljana, Lipiceva 2, SI-1000 Ljubljana, Slovenia

1. INTRODUCTION

From a physical point of view, we can outline two properties of the lipid bilayer. The first is that the dimension of the lipid bilayer in at least one of the lateral directions is much larger than its thickness, while the second is that the shape attained by the bilayer reflects the shape of the molecules that constitute the bilayer and the interactions between them. Due to the first property, the bilayer resembles a two dimensional surface, while the second property contributes to the particularities of the shape that this "surface" attains in three-dimensional space (Fig. 1).

A two-dimensional surface in three-dimensional space can be described in an elegant way by using the equations of differential geometry based on a local 2×2 curvature tensor. The equilibrium shape attained by the membrane corresponds to the minimum of the free energy of the system at relevant geometrical constraints. The link between the membrane shape and its free energy is provided by expressing the free energy of each part of the membrane by the invariants of the local curvature tensor.

The lipid bilayer is of scientific interest especially due to its relation to the membrane of cells and organelles [1]. Following many years of thorough study, the fluid mosaic model has been proposed [2] where it is assumed that the cell membrane, regardless of its specific function, can be described as a lipid bilayer with intercalated proteins or other large molecules. The lipid bilayer exhibits the properties of a two dimensional liquid that is laterally isotropic,

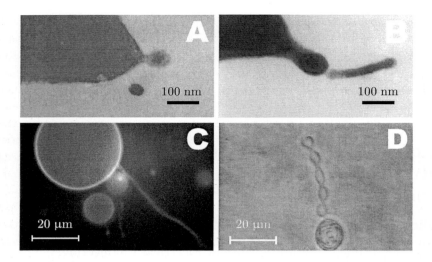

Figure 1. Examples of bilayer membrane shapes: A: transmission electron microscope (TEM) image of a spherical bud at the top of the echinocyte spicule induced by adding dodecylzwittergent to an erythrocyte suspension (from [10]), B: TEM image of a cylindrical bud at the top of the echinocyte spicule induced by adding the dimeric detergent dioctyldiQAS to an erythrocyte suspension (from [10]), C: fluorescence microscope image of the vesicle made of POPC and 1.5 % NBD-PC probe in sugar solution; the length of the myelin-like protrusion was several diameters of the spherical part (from [32]); D: phase contrast microscope image of a vesicle made of POPC in pure water; the shape exhibits an undulated protrusion while there is a multilamellar structure within the globular part (from [12]).

within which the proteins and other large molecules are more or less laterally mobile. In agreement with the fluid mosaic model [2], the membrane was considered as a laterally isotropic continuum so that its energy was expressed by the invariants of the local curvature tensor: the mean curvature and the Gaussian curvature [3].

The properties of the cell membrane can also be studied by introducing exogenously added molecules into the membrane. A convenient system for such study is represented by mammalian erythrocytes. These cells have no internal structure, so that the properties of their membrane (composed of a lipid bilayer with intercalated proteins and a membrane skeleton), are reflected in their shape. It was observed that intercalation of various substances induces observable shape changes in mammalian erythrocytes [4, 5, 6].

Study of the effects of detergents on the erythrocyte membrane showed that continuous intercalation of detergent molecules into the membrane leads to microvesiculation of the membrane [7, 8, 9]. The released microexovesicles are so small that they cannot be observed by an optical microscope. However, the vesicles could be isolated and observed by an electron microscope [8]. It was found that the microexovesicles are spherical and cylindrical [8, 10] while the endovesicles are spherical, cylindrical and torocytic [11], depending on the species of exogenously added detergent. The difference between the main curvatures of the membrane in these structures is rather large. It can be expected that in the regions where there is a large difference between the main curvatures the membrane can no longer be described as a laterally isotropic continuum. A new physical description should be created taking into account the specific structure of the membrane constituents.

In this contribution, evidence is given that thin strongly anisotropic structures that are attached to the mother vesicle/cell can be found in one-component phospholipid bilayers and in erythrocytes under certain conditions. Starting from a microscopic description of the membrane constituents, we describe the thin anisotropic structures of the bilayer membrane by a mechanism that is based on the orientational ordering of the anisotropic membrane constituents. It is derived that the relevant invariants of the curvature tensor for description of such systems are the mean curvature and the curvature deviator [10, 12, 13]. The shapes of the membraneous structures are studied within the frame of these invariants and in correspondence with the experimentally observed shapes.

2. INTERDEPENDENCE BETWEEN THE CONFIGURATION OF THE INCLUSIONS AND THE SHAPE OF THE MEMBRANE

The membrane inclusion can be any membrane constituting molecule or any assembly of molecules that can be distinguished from the surrounding membrane constituents. The surrounding membrane constituents are then treated as a curvature field, the membrane inclusion being subject to this field. The inclusion may be located in a single monolayer or it may protrude through both layers.

We imagine that there exists a shape that would completely fit the inclusion. This shape is referred to as the shape intrinsic to the inclusion. The corresponding main curvatures are denoted by C_{1m} and C_{2m} [14, 15]. Fig. 2 gives a schematic presentation of four different intrinsic shapes. The inclusion is called isotropic if $C_{1m} = C_{2m}$ while it is called anisotropic if $C_{1m} \neq C_{2m}$.

It would be energetically most favourable if the membrane had the intrinsic shape over all its area. However, if we consider a closed shape subject to geometrical constraints, the membrane cannot have such a curvature at

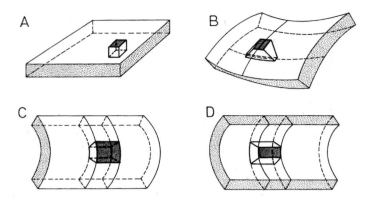

Figure 2. A schematic presentation of four different intrinsic shapes: A: flat shape $(C_{1m} = C_{2m} = 0)$, B: saddle shape $(C_{1m} > 0, C_{2m} < 0)$, C: cylinder $(C_{1m} > 0, C_{2m} = 0)$, D: inverted cylinder $(C_{1m} < 0, C_{2m} = 0)$. The shape A is isotropic while the shapes B,C and D are anisotropic (from [12]).

all its points. In general, the local membrane shape differs from the intrinsic shape. This means that the principal curvatures and the principal directions of the actual shape differ from the principal curvatures and the principal directions, respectively, of the intrinsic shape. The mutual orientation of the two principal systems describes the orientation of the inclusion if $C_{1m} \neq C_{2m}$ *i.e.* if the inclusion is anisotropic.

In a one-component bilayer membrane every membrane constituent can be treated as an inclusion that is confined to the corresponding monolayer. In a membrane containing proteins or some other molecules the inclusion is formed by the seed molecule and adjacent membrane constituent molecules that are significantly distorted due to its presence. The difference in the treatment of the respective systems lies in the statistical mechanical description that yields consistently related expressions for the membrane free energy and positional and orientational distribution functions. A lattice with an adjustable lattice constant is introduced. In the case when we consider every constituent as an inclusion embedded in the continuum formed by the other constituents, all of the lattice sites are occupied [12]. In the case when the number of inclusions is much smaller than the number of membrane constituents (*e.g.* when the inclusions are formed by the membrane proteins or induced by intercalated drugs), most of the lattice sites are empty [14, 15]. As the inclusions are laterally mobile over the membrane area, they would accumulate in the regions of favourable curvature, while they would be depleted from the regions of unfavourable curvature. In both statistical mechanical approaches there would be a different degree of orientational ordering of the inclusions in different regions. Correspondingly, the whole membrane would attain a shape exhibiting large regions of favourable local shape and small regions of unfavourable shape, as such a configuration would yield the minimal free energy of the whole membrane.

The free energy of the membrane is subject to the local thermodynamic equilibrium (which is considered by using canonical ensemble statistics) and the global thermodynamic equilibrium with respect to the positional and orientational distribution functions of the inclusions and with respect to the membrane shape, *i.e.* the principal membrane curvatures at each point of the membrane. The relevant geometrical constraints such as the requirement for a fixed membrane area and enclosed volume are taken into account. In this work we present some approximate solutions of this variational problem that seem to be relevant for the particular experimentally observed features.

3. SINGLE-INCLUSION FREE ENERGY

The origin of the coordinate system is chosen at the site of the inclusion. The membrane shape at this site is described by the diagonalized curvature tensor $\underline{C}$,

$$\underline{C} = \begin{bmatrix} C_1 & 0 \\ 0 & C_2 \end{bmatrix}, \tag{1}$$

while the intrinsic shape is described by the diagonalized curvature tensor $\underline{C}_m$

$$\underline{C}_m = \begin{bmatrix} C_{1m} & 0 \\ 0 & C_{2m} \end{bmatrix}. \tag{2}$$

The principal directions of the tensor $\underline{C}$ are in general different from the principal directions of the tensor $\underline{C}_m$, the systems being mutually rotated by an angle ω.

We introduce the mismatch tensor $\underline{M}$ [13, 16],

$$\underline{M} = \underline{R}\,\underline{C}_m\,\underline{R}^{-1} - \underline{C} \tag{3}$$

where $\underline{R}$ is the rotation matrix,

$$\underline{R} = \begin{bmatrix} \cos\omega & -\sin\omega \\ \sin\omega & \cos\omega \end{bmatrix}. \tag{4}$$

The single-inclusion energy is defined as the energy that is spent in adjusting the inclusion into the membrane and is determined by terms composed of two invariants of the mismatch tensor $\underline{M}$. Terms up to the second order in the tensor elements are taken into account. The trace and the determinant are considered as the fundamental invariants [13, 16],

$$E = \frac{K}{2} \left(\text{Tr}(\underline{M}) \right)^2 + \bar{K} \, \text{Det}(\underline{M}), \tag{5}$$

where K and $\bar{K}$ are constants. Performing the necessary operations and using the expressions (1) - (5) yields the expression for the single-inclusion energy [14, 15]

$$E = \frac{\xi}{2}(H - H_{\text{m}})^2 + \frac{\xi + \xi^*}{4}(\hat{C}^2 - \hat{C}_{\text{m}}\,\hat{C}\cos 2\,\omega + \hat{C}_{\text{m}}^2), \tag{6}$$

where

$$H = \frac{1}{2}(C_1 + C_2), \quad H_{\text{m}} = \frac{1}{2}(C_{1\text{m}} + C_{2\text{m}}) \tag{7}$$

are the respective mean curvatures while

$$\hat{C} = \frac{1}{2}(C_1 - C_2), \quad \hat{C}_{\text{m}} = \frac{1}{2}(C_{1\text{m}} - C_{2\text{m}}). \tag{8}$$

The constants used in Eq. (6) are $\xi = 2\bar{K} + 4K$ and $\xi^* = -6\bar{K} - 4K$.

The partition function of a single inclusion q is [17]

$$q = \frac{1}{\omega_0} \int_0^{2\pi} \exp\left(-\frac{E(\omega)}{k\,T}\right) d\omega, \tag{9}$$

with ω_0 an arbitrary angle quantum and k the Boltzmann constant. In the partition function of the inclusion the contribution of the orientational states q_{orient} is distinguished from the contribution of the other states q_{c}, $q = q_{\text{c}}\, q_{\text{orient}}$ [15],

$$q_{\text{c}} = \exp\left(-\frac{\xi}{2k\,T}(H - H_{\text{m}})^2 - \frac{\xi + \xi^*}{4k\,T}(\hat{C}^2 + \hat{C}_{\text{m}}^2)\right), \tag{10}$$

$$q_{\text{orient}} = \frac{1}{\omega_0}\int_0^{2\pi} \exp\left(\frac{(\xi + \xi^*)\hat{C}_{\text{m}}\,\hat{C}\cos(2\omega)}{2k\,T}\right) d\omega. \tag{11}$$

Integration in Eq. (11) over ω yields

$$q_{\text{orient}} = \frac{1}{\omega_0}I_0\left(\frac{(\xi + \xi^*)\hat{C}_{\text{m}}\,\hat{C}}{2k\,T}\right), \tag{12}$$

where I_0 is the modified Bessel function. The free energy of the inclusion is then obtained by the expression $F_i = -kT \ln q$,

$$F_i = \frac{\xi}{2}(H - H_m)^2 + \frac{\xi + \xi^\star}{4}(\hat{C}^2 + \hat{C}_m^2) - kT \ln \left(I_0 \left(\frac{(\xi + \xi^\star)\hat{C}_m \hat{C}}{2kT} \right) \right). \quad (13)$$

We introduce the curvature deviator

$$D = |\hat{C}| \quad (14)$$

that is an invariant of the curvature tensor as it can be expressed by its trace and determinant,

$$D = \sqrt{(\text{Tr}(\underline{C})/2)^2 - \text{Det}(\underline{C})} = \sqrt{H^2 - C_1 C_2}. \quad (15)$$

Here, it was considered that $\text{Tr}(\underline{C}) = 2H$ and $\text{Det}(\underline{C}) = C_1 C_2$. Since the modified Bessel function and the quadratic function are even functions of the difference $\hat{C}$, the quantity $\hat{C}$ in (Eq. (13)) can be replaced by the curvature deviator D (Eq. (14)). Thereby the single inclusion free energy is expressed in a simple and transparent way by two independent invariants of the curvature tensor: the trace and the absolute value of the difference of the main curvatures *i.e.* by the mean curvature H and the curvature deviator D,

$$F_i = \frac{\xi}{2}(H - H_m)^2 + \frac{\xi + \xi^\star}{4}(D^2 + D_m^2) - kT \ln \left(I_0 \left(\frac{(\xi + \xi^\star) D_m D}{2kT} \right) \right), \quad (16)$$

where $D_m = |\hat{C}_m|$.

The average orientation of the inclusion may be given by $\langle \cos(2\omega) \rangle$ [18],

$$\langle \cos(2\omega) \rangle = \frac{I_1 \left(\frac{(\xi + \xi^\star) D_m D}{2kT} \right)}{I_0 \left(\frac{(\xi + \xi^\star) D_m D}{2kT} \right)}. \quad (17)$$

where I_1 is the modified Bessel function.

Fig. 3 shows the average orientation of the inclusion as a function of the curvature deviator D. For small D, *i.e.* in nearly isotropic regions, the inclusions are randomly oriented. The orientational ordering increases with increasing D and approaches the state where all the inclusions are aligned at large D, *i.e.* in strongly anisotropic regions.

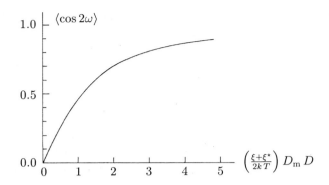

Figure 3. Average orientation of the inclusion $\langle\cos(2\omega)\rangle$ as a function of the curvature deviator D (from [18]).

4. SHAPES OF EXTREME AVERAGES OF CURVATURE TENSOR INVARIANTS

The shapes of the extreme average invariants of the curvature tensor are distinct shapes in the set of possible shapes. In order to obtain the shapes of the membrane of an extreme average mean curvature

$$\langle H\rangle = \frac{1}{A}\int H\,\mathrm{d}A \tag{18}$$

and of an extreme average curvature deviator

$$\langle D\rangle = \frac{1}{A}\int D\,\mathrm{d}A \tag{19}$$

at a given area of the membrane surface A and a given volume enclosed by the membrane V, the variational problems are stated by constructing the respective functionals [18, 16]

$$\mathcal{G}_H = \langle H\rangle - \lambda_A \cdot \left(\int \mathrm{d}A - A\right) + \lambda_V \cdot \left(\int \mathrm{d}V - V\right), \tag{20}$$

$$\mathcal{G}_D = \langle D\rangle + \lambda_A \cdot \left(\int \mathrm{d}A - A\right) - \lambda_V \cdot \left(\int \mathrm{d}V - V\right), \tag{21}$$

where λ_A and λ_V are the Lagrange multipliers. The analysis is restricted to axisymmetric shapes. The shape is given by the rotation of the function $y(x)$ around the x axis. In this case the principal curvatures are expressed by $y(x)$ and its derivatives with respect to x as $C_1 = \pm 1/y\sqrt{1+y'^2}$ and

$C_2 = \mp y'' / \sqrt{1+y'^2}^3$, where $(y' = \partial y/\partial x$ and $y'' = \partial^2 y/\partial x^2)$. The area element is $dA = 2\pi \sqrt{1+y'^2}\, y\, dx$, and the volume element is $dV = \pm \pi\, y^2\, dx$. By $\pm$ it is taken into account that the function $y(x)$ may be multiple valued. The sign may change at the points where $y' \to \infty$. The variations

$$\delta \mathcal{G}_H = \delta \int g_H(x,\, y,\, y',\, y'')\, dx = 0 \tag{22}$$

and

$$\delta \mathcal{G}_D = \delta \int g_D(x,\, y,\, y',\, y'')\, dx = 0 \tag{23}$$

are performed by solving the corresponding Poisson - Euler equations

$$\frac{\partial g_i}{\partial y} - \frac{d}{dx}\left(\frac{\partial g_i}{\partial y'}\right) + \frac{d^2}{dx^2}\left(\frac{\partial g_i}{\partial y''}\right) = 0, \quad i = H,\, D. \tag{24}$$

By inserting g_H and g_D into Eq. (24) we can express both variational problems by a Poisson-Euler equation of single form. After obtaining the necessary differentiations, this Poisson - Euler equation is [18, 16]

$$\delta_1 \frac{2y''}{(1+y'^2)^2} + \lambda_A \left(\frac{1}{\sqrt{1+y'^2}} - \frac{yy''}{(\sqrt{1+y'^2})^3}\right) - \delta_2 y\, \lambda_V = 0, \tag{25}$$

where δ_1 and δ_2 may be $+$ or $-$, depending on the actual situation. It follows from the above that the solutions for the extremes of the average invariants of the curvature tensor are equal. The nature of the obtained extreme may, however, be different. So it is possible that some solution corresponds to a maximal average mean curvature and maximal average curvature deviator. Some other solution may correspond to the minimum average mean curvature and maximum average curvature deviator *etc.*

Some simple analytic solutions of Eq. (25) were found: the cylinder $y = const$ [19, 10] and the circle of the radius r_{cir}, $y = y_0 \pm \sqrt{r_{cir}^2 - (x - x_0)^2}$ where $(x_0,\, y_0)$ is the centre of the circle. If $x_0 \neq 0$ and $y_0 = 0$ the ansatz fulfills equation (25) for two different radii [19], representing spheres with two different radii. If $x_0 = 0$ and $y_0 \neq 0$, the circle is the solution of the equation (25) only when the Lagrange multipliers are interdependent; for $r_{cir} < y_0$, the solution represents a torus and a torocyte [18].

As the sum of the solutions of the differential equation within each of the above categories is also a solution of the same equation at the chosen constraints, different combinations of shapes within the corresponding category are possible, provided that the combined shape fulfills the constraints [20]. In these cases, the Lagrange multipliers may be interdependent [19, 18, 10].

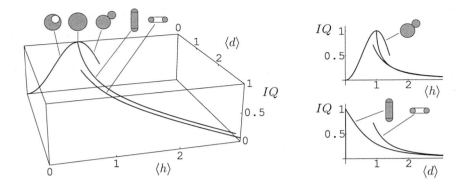

Figure 4. The $(\langle h \rangle, \langle d \rangle, IQ)$ phase diagram. The lines pertaining to three sets of limiting shapes are depicted: the set of shapes composed of two spheres, the set of shapes composed of a cylinder ended by two hemispheres and the set of tori. The corresponding projections on the $\langle d \rangle = 0$ plane and on the $\langle h \rangle = 0$ plane are also shown (from [16]).

The equilibrium shapes can be characterized by the volume to area ratio defined as the isoperimetric quotient $IQ = 36\pi\, V^2/A^3$ and both average invariants of the curvature tensor. Dimensionless quantities are used to represent the average invariants: the dimensionless average mean curvature is $\langle h \rangle = R \langle H \rangle$ and the dimensionless average curvature deviator is $\langle d \rangle = R \langle D \rangle$, where $R = \sqrt{A/4\pi}$. The possible equilibrium shapes can be represented in a $(\langle h \rangle, \langle d \rangle, IQ)$ phase diagram. The shapes of extreme average invariants of the curvature tensor form curves in this phase diagram (Fig. 4). These lines in turn form limits of the trajectories that correspond to the processes with changing average curvature invariants.

Fig. 5 shows the budding of a spherical vesicle from a planar lipid bilayer with increasing average mean curvature. All the shapes in the sequence but the first and the last are obtained by minimization of the membrane bending energy [22]. The first and the last shape are obtained by the solution of the

Figure 5. The sequence of shapes simulating the budding of a spherical vesicle from a planar lipid bilayer. The average mean curvature increases from left to right (from [22]).

variational problem of the extreme average mean curvature of a segment with fixed area ($\lambda_V = 0$). It can be seen that the limiting shape corresponding to the minimal average mean curvature consists of a section of a sphere while the limiting shape corresponding to the maximal average mean curvature consists of a segment of a flat surface and a spherical vesicle [21, 22, 23].

5. STATISTICAL MECHANICAL DESCRIPTION OF THE MEMBRANE

5.1. Statistical mechanical description of a bilayer membrane composed of only one kind of constituents

The contribution to the membrane free energy due to local interaction between the molecules and the mean curvature field is in the first approximation obtained by summing the contributions of the individual molecules of both layers [12],

$$F = \int m_{\text{out}} \, F_{\text{i}}(C_1, \, C_2) \, \mathrm{d}A + \int m_{\text{in}} \, F_{\text{i}}(-C_1, \, -C_2) \, \mathrm{d}A, \tag{26}$$

where m_{out} and m_{in} are the area densities of the molecules in the outer and the inner membrane layer, respectively, while F_{i} is given by Eq. (16). The integration is performed over the membrane area A. Note that the principal curvatures in the inner layer have signs opposite to the signs of the principal curvatures of the outer layer due to the specific configuration of the phospholipid molecules within the layers - touching by the tails.

If we assume for simplicity that the area densities are constant over the respective layers and also equal, $m_{\text{out}} = m_{\text{in}} = m_0$, and insert the expression for the single-molecule energy (Eq. (16)) into Eq. (26), we obtain [12]

$$F = m_0 \, \xi \int H^2 \, \mathrm{d}A + m_0 \frac{\xi + \xi^\star}{2} \int D^2 \, \mathrm{d}A -$$
$$- \; 2 m_0 \, k \, T \int \ln \left(I_0 \left(\frac{\xi + \xi^\star}{2k \, T} D_{\text{m}} \, D \right) \right) \mathrm{d}A. \tag{27}$$

In integrating, the differences in the areas of the inner and the outer layer were disregarded, so that the contributions proportional to the intrinsic mean curvature H_{m} of the inner and the outer layer cancel and there is no spontaneous curvature for bilayer vesicles composed of a single species of molecules. Also, the constant terms were omitted from Eq. (27).

The above procedure leading to Eq. (27) is a statistical mechanical derivation of the expression of continuum elastomechanics. It can be seen from Eq. (27) that the free energy of the membrane inclusions is expressed in a simple and transparent way by the two invariants of the local curvature tensor: by the mean curvature and the curvature deviator.

The first and the second term of Eq. (27) can be combined by using Eq. (15) to yield [12]

$$F = m_0 \frac{3\xi + \xi^\star}{8} \int (2H)^2 \, \mathrm{d}A - m_0 \frac{\xi + \xi^\star}{2} \int C_1 C_2 \, \mathrm{d}A$$
$$-2k\,T\,m_0 \int \ln\left(I_0\left(\frac{\xi + \xi^\star}{2k\,T} D_\mathrm{m} D\right)\right) \mathrm{d}A, \tag{28}$$

as to compare the expression (28) to the bending energy of an almost flat thin membrane [3] with zero spontaneous curvature

$$W_\mathrm{b} = \frac{k_c}{2} \int (2H)^2 \, \mathrm{d}A + k_\mathrm{G} \int C_1 C_2 \, \mathrm{d}A, \tag{29}$$

where k_c and k_G are the membrane local and Gaussian bending constants, respectively. We can see that the statistical mechanical derivation recovers the expression (29), where

$$m_0(3\xi + \xi^\star)/4 = k_c \tag{30}$$

and

$$-m_0(\xi + \xi^\star)/2 = k_\mathrm{G}, \tag{31}$$

and yields also an additional contribution (third term in Eq. (28)) due to the orientational ordering of the phospholipid molecules. This contribution, which is always negative, is called the deviatoric elastic energy of the membrane (originating in the curvature deviator D (Eq. (14))). Also, it follows from Eqs. (28) and (29) that the saddle splay modulus is negative for a one-component bilayer membrane.

5.1.1. *Estimation of the strength of the deviatoric effect in a phospholipid bilayer membrane*

Introducing the dimensionless quantities, the free energy F (Eq. (28)) is normalized by $2\pi\,m_0(3\xi + \xi^\star)$,

$$f = \frac{1}{4} \int (2h)^2 \, \mathrm{d}a + \kappa_G \int c_1 c_2 \, \mathrm{d}a - \kappa \int \ln\left(I_0(\vartheta\, d_\mathrm{m}\, d)\right) \mathrm{d}a, \tag{32}$$

where $\mathrm{d}a = \mathrm{d}A/4\pi\,R^2$, $R = (A/4\pi)^{1/2}$, $\kappa_G = -(\xi + \xi^\star)/(3\xi + \xi^\star)$, $\kappa = 4k\,T\,R^2/(3\xi + \xi^\star)$, $\vartheta = (\xi + \xi^\star)/2k\,T\,R^2$, $c_1 = R\,C_1$, $c_2 = R\,C_2$, $h = R\,H$, $d = R\,D$ and $d_\mathrm{m} = R\,D_\mathrm{m}$. We also obtain the dimensionless bending energy w_b if we normalize the expression (29) by $8\pi\,k_c$. Thereby, $\kappa_G = k_\mathrm{G}/2k_c$.

To estimate the interaction constants, we assume that the conformation of the phospholipid molecules is equal all over the membrane and for simplicity take that $\xi = \xi^*$. In this case, $\kappa_G = -1/2$, so that $k_G = -k_c$. By comparing the constants before the first terms of Eqs. (28) and (29) we can express the interaction constant ξ by measured quantities: the local bending constant k_c and the area density of the number of phospholipid molecules m_0, so that [12]

$$\xi = k_c/m_0, \tag{33}$$

$$\kappa = 1/\vartheta = k\,T\,R^2\,m_0/k_c. \tag{34}$$

We consider that $k_c \simeq 20k\,T$ [24, 25] and that $m_0 = 1/a_0$ where a_0 is the area per molecule, $a_0 \simeq 0.6$ nm^2 [26], $T = 300$ K and $R = 10^{-5}$ m. This gives $\kappa = 1/\vartheta \simeq 8.3 \cdot 10^6$. We estimate that the upper bound of D_m is the inverse of the molecular dimension ($\simeq 10^8$ m^{-1}) so that in our case $d_m = R\,D_m$ would be of the order of 10^3.

It follows from the above estimation that the argument of the Bessel function in Eq. (32) is very small unless $1/D$ is smaller than about a micrometre. In effect, the deviatoric contribution to the area density of the free energy of the phospholipid bilayer membrane is important only in those regions of the vesicle shape where there is a large absolute value of the difference between the two principal curvatures.

5.2. Statistical mechanical description of anisotropic inclusions within the approximation of a two-dimensional ideal gas

It is imagined that the membrane layer is divided into patches that are so small that the curvature is constant over the patch; however, they are large enough to contain a large number of inclusions that can be treated by statistical methods. It is taken that the inclusions protrude through both membrane layers. A chosen patch is a system with a well defined curvature field $\underline{C}$, given area A^P, number of inclusions M and temperature T and can therefore be subject to a local thermodynamic equilibrium. To describe the local thermodynamic equilibrium we chose canonical statistics [27] where we treat the inclusions as a two-dimensional ideal gas confined to the membrane surface. Within this approximation, the inclusions are treated as dimensionless and explicitly independent. The inclusions are also considered as indistinguishable. The canonical partition function of the inclusions in the small patch of the membrane is $Q = q^M/M!$, where q is the partition function of the inclusion (Eq. (16)) and M is the number of inclusions in the patch. Knowing the canonical partition function of the patch Q, we obtain the Helmholtz free energy of the patch, $F^P = -k\,T \ln Q$. The Stirling approximation is used and the area density of the number of molecules $m = M/A^{\mathrm{p}}$ is introduced. This gives for the area density of the free energy [15]

$$\frac{F^{\mathrm{p}}}{A^{\mathrm{p}}} = -k\,T\,m\,\ln\Big(q_{\mathrm{c}}\,\mathrm{I}_0\Big(\frac{\xi+\xi^\star}{2k\,T}\,D_{\mathrm{m}}\,D\Big)\Big) + k\,T(m\ln m - m). \qquad (35)$$

To obtain the free energy of the whole membrane F_{m} the contributions of all the patches are summed, *i.e.*, integration over the membrane area A is performed

$$F_{\mathrm{m}} = \int \frac{F^{\mathrm{p}}}{A^{\mathrm{p}}}\,\mathrm{d}A. \qquad (36)$$

The explicit dependence of the area density m on position can be determined by the condition for the free energy of all the membrane inclusions to be at its minimum in the thermodynamic equilibrium $\delta F_{\mathrm{m}} = 0$. It is taken into account that the total number of inclusions M_{T} in the membrane is fixed,

$$\int_A m\,\mathrm{d}A = M_{\mathrm{T}} \qquad (37)$$

and that the area of the membrane A is fixed. The above isoperimetric problem is reduced to the ordinary variational problem by constructing a functional

$$F_{\mathrm{m}} + \lambda_m \int_A m\,\mathrm{d}A = \int_A \mathcal{L}(m)\,\mathrm{d}A, \qquad (38)$$

where

$$\mathcal{L}(m) = -k\,T\,m\,\ln\Big(q_{\mathrm{c}}\,\mathrm{I}_0\Big(\frac{\xi+\xi^\star}{2k\,T}\,D_{\mathrm{m}}\,D\Big)\Big) + k\,T(m\ln m - m) + \lambda_m\,m \qquad (39)$$

and λ_m is the Lagrange multiplier. The variation is performed by solving the Euler equation $\partial\mathcal{L}/\partial m = 0$. Deriving (39) with respect to m and taking into account Eq. (37) gives the Boltzmann distribution function modulated by the modified Bessel function I_0 [15]

$$\frac{m}{m_{\mathrm{u}}} = \frac{q_{\mathrm{c}}\,\mathrm{I}_0\big(\frac{\xi+\xi^\star}{2k\,T}\,D_{\mathrm{m}}\,D\big)}{\frac{1}{A}\int q_{\mathrm{c}}\,\mathrm{I}_0\big(\frac{\xi+\xi^\star}{2k\,T}\,D_{\mathrm{m}}\,D\big)\mathrm{d}A}, \qquad (40)$$

where q_{c} is given by Eq. (10) and m_{u} is defined by $m_{\mathrm{u}}\,A = M_{\mathrm{T}}$.

To obtain the equilibrium free energy of the inclusions the expression for the equilibrium area density (Eq. (40)) is inserted into Eq. (35) and integrated over the area A. Rearranging the terms yields [15]

$$F_{\mathrm{m}} = -k\,T\,M_{\mathrm{T}}\,\ln\Big(\frac{1}{A}\int q_{\mathrm{c}}\,\mathrm{I}_0\Big(\frac{\xi+\xi^\star}{2k\,T}\,D_{\mathrm{m}}\,D\Big)\,\mathrm{d}A\Big). \qquad (41)$$

The equilibrium free energy of the inclusions cannot in general be expressed as an integral of the area density of the free energy. We say that the contribution of the inclusions is a nonlocal one. A change of the local conditions affects the cell shape and the distribution of the inclusions through the minimization of the free energy of the whole membrane. Although the inclusions are explicitly treated as independent, their mutual influence is taken into account through the mean curvature field which in turn depends on the lateral and orientational distribution of the inclusions. It can also be seen from Eq. (41) that the energy of the membrane with inclusions is not scale invariant. The inclusions favour a certain packing arrangement that depends on the values of the principal membrane curvatures.

To estimate the strength of the deviatoric effect, the free energy $F_{\rm m}$ (Eq. (41)) is normalized by $8\pi\,k_{\rm c}$

$$f_{\rm m} = -\kappa_{\rm m} \ln \int \big(q_{\rm c}\, {\rm I}_0(\vartheta\, d_{\rm m}\, d) \big)\, {\rm d}a,, \tag{42}$$

where $\kappa_{\rm m} = M_{\rm T}\, k\, T/8\pi\, k_{\rm c}$, $\vartheta = (\xi + \xi^\star)/2k\,T\,R^2$, while h, d, $d_{\rm m}$, R and ${\rm d}a$ are defined as below Eq. (32). We took $\xi = \xi^\star$, $\kappa_{\rm m} = 1$, $H_{\rm m} = h_{\rm m}/R = 0$, $D_{\rm m} = 1/300\ \text{Å}^{-1}$ and $R = 6\mu\,{\rm m}$ [15]. The interaction constant ξ was estimated by assuming that the energy cost of distorting the tail of a phospholipid molecule within the inclusion is approximately equal to the energy difference corresponding to the tail packing in different aggregation geometries. Such a difference is of the order of $(0.1 - 0.5)\ k\,T$ per tail of a phospholipid molecule [28]. If we assume that there are 10 molecules (20 tails) involved in the inclusion, the energy of the inclusion can reach several $k\,T$. For the mother cell $h \simeq 1$ while $d \simeq 0$. Then, from the above choice of $h_{\rm m}$ and $d_{\rm m}$ and Eq. (6), we estimated that ϑ is of the order of 10^{-3}.

Fig. 6 shows the relative free energy of the inclusions $f_{\rm m}$ and the relative membrane bending energy $w_{\rm b}$ as a function of increasing average mean curvature $\langle h \rangle$ for a sequence in which the shape with one spherical vesicle is formed from a pear shape. The shapes in the sequence correspond to the minimum of the membrane bending energy. It can be seen that the bending energy monotonously increases along the sequence. The free energy of the inclusions only slightly decreases as long as the neck is wide. When the neck shrinks, the free energy of the inclusions sharply decreases, reaches a minimum and then increases towards the initial value. When the neck becomes infinitesimal, the curvature deviator becomes very large, however, the area of the neck becomes very small. The deep minimum of the $f_{\rm m}(\langle h \rangle)$ curve provides a possible explanation for the stability of thin necks connecting the daughter vesicle and the mother cell [29].

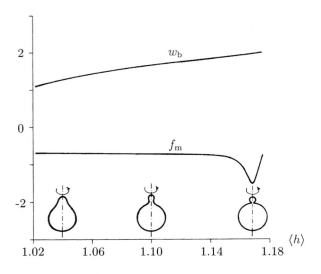

Figure 6. The normalized free energy of the inclusions f_m (Eq. (42)) and the normalized membrane bending energy w_b (first term of Eq. (32)) as a function of increasing normalized average mean curvature $\langle h \rangle$; $\vartheta = 10^{-3}$, $\kappa_m = 1$, $h_m = 0$, $d_m = 100$ and $IQ = 0.9$. The energy f_m is determined up to a constant (adapted from [15]).

6. MEMBRANE SHAPES EXHIBITING DEVIATORIC ELASTICITY

6.1. Myelin-like protrusions of phospholipid bilayer vesicles

An experiment showing fast recovery of fluorescence in photobleached giant phospholipid vesicles [30] indicated that the giant phospholipid vesicles obtained in the process of electroformation [31] are connected by thin tubular structures [30]. Later, it was observed [32] that giant palmitoyloleylphosphatidylcholine (POPC) bilayer vesicles, which immediately after formation appear spherical, spontaneously transform into flaccid fluctuating vesicles in a process where the remnants of thin tubular structures that are attached to the vesicles become thicker and shorter and eventually integrate into the membrane of the mother vesicle. The thin tubular network acts as a reservoir for the membrane area and importantly influences the future shape and dynamics of the globular phospholipid vesicles.

The POPC vesicles were prepared and observed in sugar solution [32] and in pure water [12]. Also, the vesicles were labelled with the fluorescent probe NBD-PC [32]. The observed features are the same in all cases. Immediately after being placed into the observation chamber the vesicles are spherical; the protrusions are not visible under phase contrast and the long

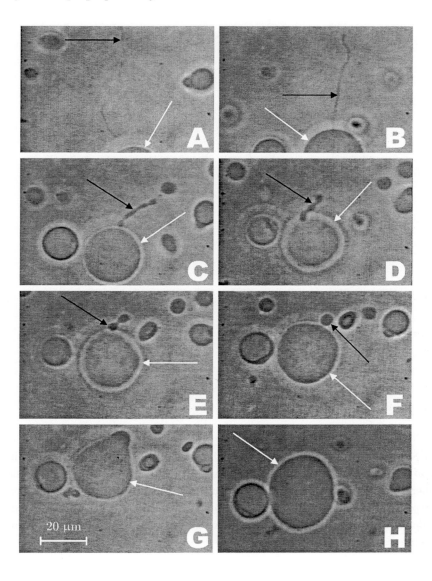

Figure 7. Shape transformation of a giant phospholipid vesicle (made of POPC and 1.5 % NBD-PC in sugar solution) with time. The times after the preparation of the vesicles are A: 3 h, B: 3 h 20 min, C: 4 h, D: 4 h 2 min, E: 4 h 4 min 30 s, F: 4 h 8 min 15 s, G: 4 h 14 min 25 s, H: 4 h 14 min 30 s. The white arrows indicate the protrusion while the black arrows indicate the mother vesicle. The vesicle was observed under an inverted Zeiss IM 35 microscope with phase contrast optics (from [32]).

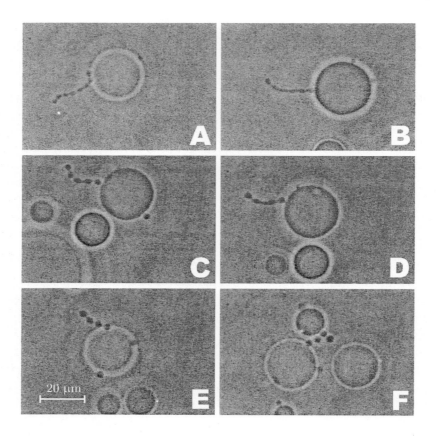

Figure 8. Shape transformation of a giant phospholipid vesicle (made of POPC in sugar solution as in [32]) with time. A small void was left in the grease closing the observation chamber to allow water evaporation. Undulations of the protrusion were observed at longer protrusions. When the "beads" became sphere-like (F), the shape stayed stable for about two hours. The vesicle was observed under an inverted Zeiss IM 35 microscope with phase contrast optics.

wavelength fluctuations of the spherical part are not observed. After some time, long thin protrusions become visible (Fig. 1C); the protrusions appear as very long thin tubes that are connected to the mother vesicle at one end while the other end is free. With time, the protrusion becomes shorter and thicker; however the tubular character of the protrusion is still preserved (Fig. 7A-C) while the fluctuations of the mother vesicle increase in strength. Later, undulations of the protrusion appear and become increasingly apparent. Shortened protrusions look like beads connected by thin necks (Fig. 7D-F). Eventually, the protrusion is completely integrated into the vesicle membrane to yield a fluctuating globular vesicle (Fig. 7H). The transformation of the protrusion is usually very slow, indicating that all the observed shapes may be considered as quasiequilibrium shapes. In the sample, the tubular protrusions are still observed several hours after the formation of the vesicles. The timing of the transformation may vary from minutes to hours, as the protrusions are initially of very different lengths.

The observed shape transformation may be driven by the inequality of the chemical potential of the phospholipid molecules in the outer solution and in the outer membrane layer which causes a decrease of the difference between the outer and the inner membrane layer areas. Possible mechanisms that were suggested to contribute to this are the drag of the lipid from the outer solution by the glass walls of the chamber, chemical modification of the phospholipid and phospholipid flip-flop [32]. A decrease of the volume to area ratio (isoperimetric quotient) of the vesicle occurs due to slight evaporation of water from the chamber. The vesicle then looses water in order to equalize the respective chemical potentials inside and outside the vesicle. The tubular/beadlike character of the protrusion seems to depend on the speed of the loss of lipid molecules from the outer membrane layer relative to the speed of the decrease of the enclosed volume. The undulations of the protrusion are more evident when a small void is left in the grease thereby enhancing the evaporation of water from the outer solution (Fig. 8).

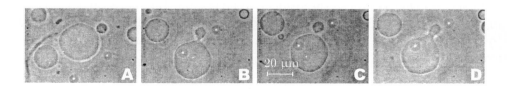

Figure 9. Oscillations of the neck width before opening of the neck of giant phospholipid (POPC) vesicle.

Another interesting observed feature is oscillation of the neck width. This is especially notable before the opening of the neck that connects the last bead with the mother vesicle (Fig. 7F-H). The neck shrinks and widens several times before the last bead integrates with the mother vesicle (Fig. 9) indicating increased stability of the neck. A similar effect - the persistence of the neck connecting a spherical daughter vesicle and a mother vesicle - was also observed in the opening of the neck induced by cooling, while the formation of the neck by heating was quick and took place at higher temperature, indicating hysteresis [33]. In the case when a small void is left in the observation chamber longer protrusions already have a beadlike form (Fig. 8). When the protrusion undergoes transformation into a shape with one bead less, the necks open to yield an almost tubular shape (Fig. 8B,D). If the protrusion develops into spherical beads connected by very thin necks (Fig. 8F), the shape stays stable for hours.

A complete picture of the dynamics of shape transformation seems at this point beyond our understanding. However, we can point to some facts that can be elicited with a certain confidence. In comparing the protrusions at an early time and at a later time, the protrusions at the early time appear considerably more tubular. Therefore we think that the protrusions also have tubular character even at earlier times when they are too thin to be seen by the phase contrast microscope. The possibility should be considered that the radius of the tubular protrusion immediately after its formation is very small - as small as the membrane thickness.

To describe these features theoretically, the equilibrium shape is determined by the minimum of the membrane free energy (Eq. (28)) under given constraints. It is considered that the membrane area A, the enclosed volume V and the average mean curvature are fixed. The average mean curvature of a thin membrane is proportional to the difference between the two membrane layer areas, $\langle H \rangle = \Delta A / 2A \delta$, where δ is the distance between the two layer neutral areas which is considered to be small with respect to $1/H$ and $\Delta A = \delta \int (C_1 + C_2) \mathrm{d}A$. The constraint for $\langle H \rangle$ therefore reflects the number of molecules that compose the respective layers and therefore the conditions in which the vesicle formation took place. If the area difference is normalized by $8\pi \delta R$ ($\Delta a = \Delta A / 8\pi \delta R$), for a thin bilayer, it is equal to the normalized average mean curvature,

$$\langle h \rangle = \Delta a = \frac{1}{2} \int (c_1 + c_2) \mathrm{d}a. \tag{43}$$

For the sake of simplicity, we compare two shapes that represent the limits of the class of shapes with a long thin protrusion. In the first case the protrusion consists of equal small spheres (Fig. 10A), while in the second

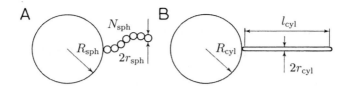

Figure 10. A: schematic presentation of a shape composed of the mother sphere and protrusion composed of small spheres connected by infinitesimal necks, B: schematic presentation of a shape composed of the mother sphere and a thin cylinder closed by hemispherical caps (from [12]).

case the protrusion consists of a cylinder closed by hemispherical caps (Fig. 10B). It is expected that these two shapes are continuously connected by a sequence of shapes with decreasingly exhibited undulations of the protrusion.

Each of these two limiting cases involves three geometrical model parameters (Fig. 3) that can be determined from geometrical constraints for the area, volume and average mean curvature. In the limit $IQ \simeq 1$, the geometrical parameters and therefore also the energies can be expressed analytically [12]. It has been shown that the relative membrane bending energy of the shape with the cylindrical protrusion $w_{b,\,cyl}$ is always higher than the membrane bending energy of the shape with the spherical beads $w_{b,\,sph}$ [12],

$$w_{b,\,cyl} = w_{b,\,sph} + 1. \tag{44}$$

As within the theory of elasticity of an isotropic bilayer membrane [3] the shape with a protrusion composed of small spheres that are connected by infinitesimal necks would always be favoured over the shape with a tubular protrusion (Eq. (44)), this theory is unable to explain stable tubular protrusions.

A possible mechanism that can explain the stability of the long thin tubular protrusions is that of a deviatoric elasticity which is a consequence of the orientational ordering of the membrane constituent molecules in the thin tubular protrusion. If we chose high $\langle h \rangle$ the shape has a long protrusion. As the membrane area and the enclosed volume are fixed, this protrusion is very thin and consequently its mean curvature is large. For tubular protrusions the deviatoric contribution to the normalized free energy ($f_{\rm d}$) (third term in Eq. (32))

$$f_{\rm d} = -\kappa \int \ln\big(I_0(\vartheta \, d_{\rm m} \, d)\big) \, {\rm d}a \tag{45}$$

is large enough to compensate for the less favourable bending energy of the cylinder Eq. (44). On the other hand, for lower $\langle h \rangle$, a protrusion of the same

membrane area and enclosed volume is shorter and broader, and therefore its
mean curvature is lower. The corresponding deviatoric term of the cylinder
is therefore too small to be of importance and the shape with the beadlike
protrusion has lower free energy. At a chosen intrinsic anisotropy d_m, the
shapes with small spheres are energetically more favourable below a certain
$\langle h \rangle$ while above this threshold the shapes with cylinders are favoured.

Fig. 11 shows a $(\langle h \rangle, d_m)$ phase diagram exhibiting the regions corresponding to the calculated stable shapes composed of a spherical mother
vesicle and a tubular protrusion and to a stable shapes composed of a spherical mother vesicle and a protrusion consisting of small spheres connected by
infinitesimal necks.

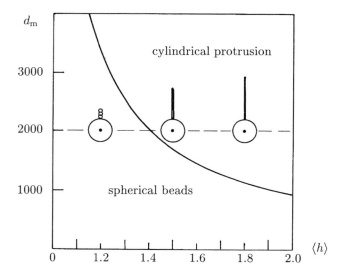

Figure 11. The $(\langle h \rangle, d_m)$ phase diagram of calculated equilibrium shapes with protrusions. The regions where shapes with a particular kind of protrusion are energetically
more favourable are indicated. The sequence of shapes shown in the figure indicates
the process of diminishing $\langle h \rangle$ at constant IQ that could be observed experimentally.
We chose $a_0 = 0.6$ nm^2, $R = 10^{-5}$ m, $k_c = 20\,k\,T$, so that $\kappa = 1/\vartheta = 8.3 \cdot 10^6$ while
$IQ = 0.90$. The shapes corresponding to different $\langle h \rangle$ are depicted with the centre of
the spherical part at the corresponding $\langle h \rangle$ values (adapted from [12]).

The radii of the stable tubular protrusion are $200 - 400$ nm while the
corresponding deviatoric energies are larger than the estimated energy of
thermal fluctuations [12]. The sequence of shapes shown in the figure roughly
simulates the transformation observed experimentally (Fig. 7). Initially,

$\langle h \rangle$ is large and the shape is composed of a mother sphere and a long thin nanotube. Assuming that the volume of the vesicle remains constant with time, the number of phospholipid molecules in the outer layer diminishes, so that $\langle h \rangle$ decreases and the tubular protrusion becomes thicker and shorter. In the experiment [32], the undulations of the protrusion became increasingly notable during the process. Our theoretical results shown in Fig. 11 exhibit a discontinuous transition from a tubular protrusion to a protrusion composed of small spheres connected by infinitesimal necks, as we consider only the limits of the given class of shapes. Therefore, the phase diagram and the sequence (Fig. 11) should be viewed only as an indication of the trend of shape transition and not as to the details of the shape.

6.2. Detergent-induced anisotropic structures of the erythrocyte membrane

Continuous intercalation of detergent molecules into the outer layer of the erythrocyte membrane eventually leads to microexovesiculation [7, 8, 9]. Upon intercalation of the detergent molecules into the membrane the mother-cell becomes spherical while a few percent of the erythrocyte membrane area is released in the form of microexovesicles. Analysis of the protein composition of the isolated microexovesicles [9, 21] showed that the microexovesicles are depleted in the membrane skeletal components spectrin and actin, suggesting that a local disruption of the interactions between the membrane skeleton and the membrane bilayer occurrs prior to microexovesiculation [7, 9, 34, 21] and indicating that the shape of the microexovesicles is determined by the properties of the membrane.

It was observed [8] that some species of added amphiphiles induce predominantly spherical microexovesicles while other species induce predominantly tubular microexovesicles. Among these, the cationic dimeric amphiphile dioctyldiQAS, in which two head - tail entities are connected by a spacer at the headgroup level, induces stable tubular microexovesicles [10] (Fig. 13). An elongated tubular shape was exhibited even in the buds (Fig. 1B). For comparison, Fig. 1A shows budding of a spherical vesicle.

Fig. 13 shows a sequence of prolate shapes connecting two limiting shapes: the shape composed of a cylinder and two hemispherical caps and the shape composed of three spheres connected by infinitesimal necks. The corresponding normalized average mean curvatures and normalized average curvature deviators are also depicted. It can be seen that the shape composed of spheres corresponds to the maximal average mean curvature and minimal average curvature deviator while the shape composed of the cylinder and two hemispherical caps corresponds to the maximal average curvature deviator and minimal average mean curvature.

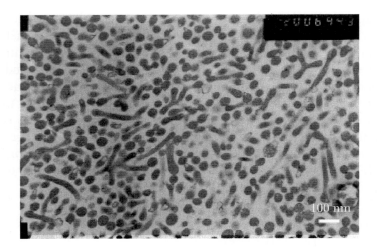

Figure 12. TEM image of the isolated tubular daughter microexovesicles induced by adding dioctyldiQAS to an erythrocyte suspension (from [10]).

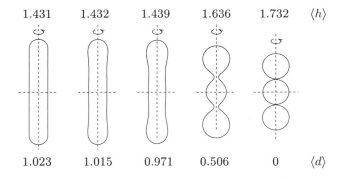

Figure 13. A sequence of axisymmetric vesicle shapes with $IQ = 1/3$. The corresponding values of the average mean curvature $\langle h \rangle$ and of the average curvature deviator $\langle d \rangle$ are given. All the shapes but the first from the left were obtained by minimizing the membrane bending energy. The first shape from the left was obtained by combining the solutions of Eq. (25) representing the sphere and the cylinder (adapted from [10]).

A B

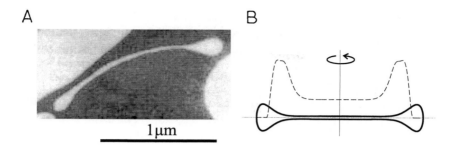

Figure 14. A: TEM micrograph of the torocyte endovesicle induced by adding $C_{12}E_8$ to an erythrocyte suspension (from [11]), B: calculated equilibrium shape of the torocyte (full line) and the corresponding distribution of the anisotropic inclusions (broken line); $\kappa_m = 100$, $\vartheta = 1$, $h_m = -1.5$, $d_m = 1.5$ (adapted from [16, 40]).

After adding both, dodecylzwittergent or dioctyldiQAS, to the erythrocyte suspension, the erythrocytes first underwent a discocyte-echinocyte-spheroechynocyte transformation [10, 35], so it is evident that the average mean curvature $\langle h \rangle$ continuously increased in the process [35]. According to the bilayer couple model [5, 36, 37, 38, 39, 35] expressed by the bending of the laterally isotropic membrane [3], the process of increasing the area difference $\triangle A$ *i.e.* - $\langle h \rangle$ due to the intercalation of detergent molecules into the outer membrane layer would lead to the shape of the maximal average mean curvature *i.e.* to spherical microexovesicles (Fig. 13). While the bilayer couple model explains the spherical shape of the microexovesicles that are induced by intercalating dodecylzwittergent into the erythrocyte membrane, it cannot explain the tubular shape of the microexovesicles induced by dioctyldiQAS.

A possible mechanism that can explain the observed stable tubular microexovesicle shape is that of a deviatoric elasticity which is a consequence of the orientational ordering of the detergent-induced inclusions on the tubular buds/vesicles [17, 14, 15, 10]. Besides increasing the average mean curvature, the intercalation of anisotropic dioctyldiQAS-induced inclusions also increase the average curvature deviator [10]. If the second effect prevails, the process may continue until the limiting shape composed of cylinder and two hemispherical caps (Fig. 12), that corresponds to the maximal average curvature deviator, is reached.

On the other hand, continuous intercalation of detergent molecules into the inner layer of the erythrocyte membrane eventually leads to endovesiculation. It was reported [11] that octaethyleneglycol dodecylether ($C_{12}E_8$)

may induce (usually one) stable endovesicle having a torocyte shape (Fig. 14A). It was observed that the torocyte endovesicle originates from a primarily large stomatocytic invagination which upon continuous intercalation of $C_{12}E_8$ molecules loses volume. The invagination may finally close, forming an inside-out endovesicle of small isoperimetric quotient. Three partly complementary mechanisms were suggested in order to explain the formation and stability of the observed torocytes [11, 18, 40]. The first is preferential intercalation of the $C_{12}E_8$ molecules into the inner membrane layer, the second is preference of the $C_{12}E_8$ induced inclusions for zero or slightly negative local mean curvature, and the third is the lateral and orientational ordering of $C_{12}E_8$ induced inclusions. It was suggested [40] that lipid molecules and membrane proteins may be involved in $C_{12}E_8$ induced inclusions. To determine the equilibrium shape of the torocyte endovesicle, the membrane free energy (including the free energy of the inclusions) is minimized at fixed membrane area and fixed enclosed volume [40]. Eq. (42) is used to calculate the free energy of the inclusions. Fig. 14 shows the calculated equilibrium shape of the torocyte vesicle and the corresponding distribution of the $C_{12}E_8$ induced inclusions. It can be seen that the inclusions favour the region of highly different main curvatures while the contour of the calculated shape agrees with the observed TEM image. In order to calculate the characteristic torocyte shape (Fig. 14B) it was necessary to include the deviatoric effect [16, 40]. Within the standard bending elasticity model of the bilayer membrane [3, 36, 37, 38, 39] the calculated torocyte vesicle shapes, corresponding to the minimal bending energy, have a thin central region where the membranes on the both sides of the vesicle are in close contact, *i.e.* the resultant forces on both membranes in contact are balanced [18]. However, as it can be seen in Fig. 14A the adjacent membranes in the flat central region are separated by a certain distance indicating that the stability of the observed torocyte shape can not be explained by the standard bending elasticity model.

The same substance, *i.e.* the detergent $C_{12}E_8$, was shown to stabilize transient pores in cell membrane that are created by electroporation [41]. By using a simple geometrical model of the pore where the pore was described as the inner part of the torus, and by minimizing the free energy of the inclusions (Eq. (42)), the equilibrium configuration of the system was predicted at a finite size of the pore - if the inclusions were anisotropic [42, 16].

The examples presented indicate that the model in which the membrane is treated as a laterally isotropic two dimensional liquid should be upgraded in order to describe the observed features. The deviatoric elasticity provides an explanation for some of the observed features; however, further refinement of the model such as inclusion of the role of the membrane skeleton [43, 44,

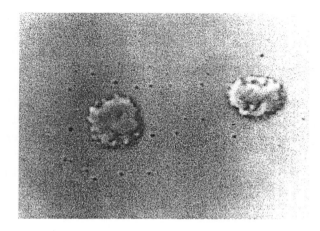

Figure 15. Vesiculation of human erythrocytes at high pH with exogeneously added dibucaine. The vesicles around the mother cell move synchronously with the mother cell indicating that they are connected to the mother cell by thin tethers (from [29]).

45, 46, 35, 47, 48] should be taken into account in order to obtain a more realistic description.

7. CONCLUSION

Considerable knowledge has been gathered on lipid bilayer membranes. The interdependence between the membrane elastic properties, the conditions in solution and the vesicle and cell shape has been thoroughly investigated [49, 3, 39, 50, 43, 51, 35, 52, 53]. Theoretical approaches based on statistical mechanical methods mostly describe the configuration of the hydrocarbon tails [54, 28, 55], while the link between the detailed description of the tails and the membrane shape has not been made. On the other hand, this link was achieved in describing the effect of membrane inclusions [56] on the vesicle shape [57, 14, 17, 15, 40]. Most of the work was devoted to globular shaped vesicles and cells where the membrane could be considered as an almost flat and laterally isotropic two-dimensional liquid [2].

Recently, the experiments have drawn attention to thin anisotropic structures attached to the globular part of the vesicle [30, 10, 32, 12]. Further, such structures seem to be common also in cells [29] (Fig. 15). It was indicated that these structures form an important auxiliary pool of the membraneous material that hitherto remained obscure. We have found that the features involving the auxiliary pool cannot be explained if the membrane is treated as a two dimensional liquid. We propose a simple mechanism which considers that the membrane constituents are intrinsically anisotropic. While the collective effect on almost flat regions yields the state of a laterally isotropic two dimensional liquid, the anisotropic properties become expressed if the membrane for some reason develops regions of highly different main curvatures. The proposed mechanism provides an explanation for the stability of the phospholipid micro and nanotubes attached to the giant phospholipid vesicle, for the stability of thin tethers connecting the mother cell and the daughter vesicle in erythrocytes, for the stability of tubular daughter microexovesicles and torocytic daughter endovesicles of the erythrocyte membrane and for the stability of detergent-induced pores in the cell membrane. However, the auxiliary pool is yet to be explored.

REFERENCES

[1] H. Ti Tien and A.L. Ottova, J. Membr. Sci., 189 (2001) 83.

[2] S.J. Singer and G.L. Nicholson, Science, 175 (1972) 720.

[3] W. Helfrich, Z. Naturforsch., 28c (1973) 693.

[4] B. Deuticke, Biochim. Biophys. Acta, 163 (1968) 494.

[5] M.P. Sheetz and S.J. Singer, Proc. Natl. Acad. Sci. USA, 72 (1974) 4457.

[6] S. Schreier, S.V.P. Malherios and E. de Paula, Biochim. Biophys. Acta, 1508 (2000) 210.

[7] S.C. Liu, L.H. Derick, M.A. Duquette and J. Palek, Eur.J.Cell Biol., 49 (1989) 358.

[8] H. Hagerstrand and B. Isomaa, Biochim. Biophys. Acta, 1109 (1992) 117.

[9] H. Hagerstrand and B. Isomaa, Biochim. Biophys. Acta, 1190 (1994) 409.

[10] V. Kralj-Iglic, A. Iglic, H. Hagerstrand and P. Peterlin, Phys. Rev. E, 61 (2000) 4230.

[11] M. Bobrowska-Hagerstrand, V. Kralj-Iglic, A. Iglic, K. Bialkowska, B. Isomaa and H. Hagerstrand, Biophys. J., 77 (1999) 3356.

[12] V. Kralj-Iglic, A. Iglic, G. Gomiscek, F. Sevsek, V. Arrigler and H. Hagerstrand, J. Phys. A.: Math. Gen., 35 (2002) 1533.

[13] V. Kralj-Iglic, M. Remskar, M. Fosnaric, G. Vidmar and A. Iglic, Phys. Lett. A, 296 (2002) 151.

[14] V. Kralj-Iglic, S. Svetina and B. Zeks, Eur. Biophys. J., 24 (1996) 311.

[15] V. Kralj-Iglic, V. Heinrich, S. Svetina and B. Zeks, Eur. Phys. J. B, 10 (1999) 5.

[16] M. Fosnaric, A. Iglic and V. Kralj-Iglic, in: I.M. Mladenov and G.L. Naber (eds.), Geometry, Integrability and Quantization, Coral Press,

Sofia 2002, p.224.

[17] J.B. Fournier, Phys. Rev. Lett., 76 (1996) 4436.

[18] A. Iglic, V. Kralj-Iglic, B. Bozic, M. Bobrowska-Hagerstrand, B. Isomaa and H. Hagerstrand, Bioelectrochemistry, 52 (2000) 203.

[19] A. Iglic, V. Kralj-Iglic and J. Majhenc, J. Biomech., 32 (1999) 1343.

[20] L.E. Elsgolc, Calculus of Variations, Pergamon Press, Oxford, 1961.

[21] H. Hagerstrand, V. Kralj-Iglic, M. Bobrowska-Hagerstrand and A. Iglic, Bull. Math. Biol., 61 (1999) 1019.

[22] A. Iglic and H. Hagerstrand, Med. Biol. Eng. Comp., 37 (1999) 125.

[23] H. Hagerstrand, M. Danieluk, M. Bobrowska-Hagerstrand, V. Pector, J. M. Ruysschaert, V. Kralj-Iglic and A. Iglic, Biochim. Biophys. Acta, 1421 (1999) 125.

[24] H.P. Duwe, J. Kas and E. Sackmann, J. Phys. (France), 51 (1990) 945.

[25] U. Seifert, Adv. Phys., 46 (1997) 13.

[26] G. Cevc and D. Marsh, Phospholipid Bilayers, Wiley - Interscience, New York, 1987.

[27] T.R. Hill, An Introduction to Statistical Thermodynamics, Dover Publications, New York, 1986.

[28] A. Ben-Shaul and W.M. Gelbart, in: W.M.Gelbart, A. Ben-Shaul and D. Roux (eds.), Micelles, Membranes and Monolayers, Springer Verlag, New York, 1994, 47.

[29] V. Kralj-Iglic, A. Iglic, M. Bobrowska-Hagerstrand and H. Hagerstrand, Coll. Surf. A, 179 (2001) 57.

[30] L. Mathivet, S. Cribier and P.F. Devaux, Biophys. J., 70 (1996) 1112.

[31] M.I. Angelova, S. Soleau, Ph. Meleard, J.F. Faucon and P. Bothorel, Prog. Colloid Polym. Sci., 89 (1992) 127.

[32] V. Kralj-Iglic, G. Gomiscek, J. Majhenc, V. Arrigler and S. Svetina, Coll. Surf. A, 181 (2001) 315.

[33] J. Kas and E. Sackmann, Biophys.J., 60 (1991) 825.

[34] A. Iglic, S. Svetina and B. Zeks, Biophys. J., 69 (1995) 274.

[35] A. Iglic, V. Kralj-Iglic and H. Hagerstrand, Eur. Biophys. J., 27 (1998) 335.

[36] E.A. Evans, Biophys. J., 14 (1974) 923.

[37] W. Helfrich, Z. Naturforsch., 29c (1974) 510.

[38] E.A. Evans, Biophys. J., 30 (1980) 265.

[39] S. Svetina, A. Ottova-Leitmannova and R. Glaser, J. Theor. Biol., 94 (1982) 13.

[40] M. Fosnaric, M. Nemec, V. Kralj-Iglic, H. Hagerstrand, M. Schara and A. Iglic, Coll. Surf. B, 26 (2002) 243.

[41] G. Troiano, K. Stebe, V. Sharma and L. Tung, Biophys. J., 74 (1998) 880.

[42] M. Fosnaric, H. Hagerstrand, V. Kralj-Iglic and A. Iglic, Cell. Mol. Biol. Lett., 6 (2001) 167.

[43] A. Elgsaeter and A. Mikkelsen, Biochim. Biophys. Acta, 1071 (1991) 273.

[44] A. Iglic, J. Biomech., 30 (1997) 35.

[45] M. Bobrowska - Hagerstrand, H. Hagerstrand and A. Iglic, Biochim. Biophys. Acta, 1371 (1998) 123.

[46] R.E. Waugh, Biophys. J., 70 (1996) 1027.

[47] H. Hagerstrand, M. Danieluk, M. Bobrowska-Hagerstrand, A. Iglic, A. Wrobel, B. Isomaa and M. Nikinmaa, Biochim. Biophys. Acta, 1466 (2000) 125.

[48] H. Hagerstrand, M. Danieluk, M. Bobrowska-Hagerstrand, T. Holm-strom, V. Kralj-Iglic, C. Lindqvist and M. Nikinmaa, Mol. Membr. Biol., 16 (1999) 195.

[49] P.B. Canham, J. Theor. Biol., 26 (1970) 61.

[50] R. Lipowsky, Nature, 349 (1991) 475.

[51] L. Miao, U. Seifert, M. Wortis and H.G. Dobereiner, Phys. Rev. E, 49 (1994) 5389.

[52] M. Bobrowska-Hagerstrand, H. Hagerstrand, and A. Iglic, Biochim. Bio-phys. Acta, 1371 (1998) 123.

[53] A. Iglic, P. Veranic, U. Batista and V. Kralj-Iglic, J. Biomech., 34 (2001) 765.

[54] S. Marcelja, Biochim. Biophys. Acta, 367 (1974) 165.

[55] I. Szleifer, D. Kramer, A. Ben-Shaul, W.M. Gelbart and S.A. Safran, J. Chem. Phys., 92 (1990) 6800.

[56] S. Marcelja, Biochim. Biophys. Acta, 455 (1976) 1.

[57] S. Leibler and D. Andelman, J. Phys., 48 (1987) 2013.

Chapter 5

Elastic properties of bilayer lipid membranes and pore formation

Dumitru Popescu[a,b], Stelian Ion[b], Aurel Popescu[c@], and Liviu Movileanu[d]

[a]Laboratory of Biophysics, University of Bucharest, Faculty of Biology, Splaiul Independentei 91-95, Bucharest R-76201, Romania

[b]Institute of Applied Mathematics, Calea 13 Septembrie 13, P.O.Box 1-24, Bucharest, Romania

[c]Department of Electricity and Biophysics, University of Bucharest, Faculty of Physics, Bucharest-Magurele, P.O.Box MG-11, R-76900, Romania

[d]Department of Medical Biochemistry and Genetics, The Texas A&M University System Health Science Center, 440 Reynolds Medical Building, College Station, Texas 77843-1114, USA

1. INTRODUCTION

Phospholipid bilayer membranes (BLM's) represent a useful model system to examine fundamental aspects of the lipid bilayer components of biological cell membranes and, particularly, to investigate their elastic properties. They are self-assembled structures of amphipatic molecules with physical features closely similar to those of smectic liquid crystals [1]. Lipid bilayer matrix is capable of incorporating both hydrophobic and amphipatic molecules like proteins, other lipids, peptides, steroids and cosurfactants. The elastic properties of lipid membranes regarded as continuous media have been used in a variety of studies ranging from local phenomena, such as lipid-lipid [1-11], lipid-protein [12-25] and protein-protein interactions [26-27], to shape fluctuations of the whole cells [28-32]. In addition, the liquid hydrocarbon nature of the bilayer is maintained by inter-molecular interactions between phospholipids at nanoscopic scale: electrostatic and dipole-dipole interactions between the polar headgroups [33, 34], interactions mediated by water molecules [35], and van der Waals dispersion interactions between hydrocarbon chains [36-38].

[@]The correspondence author. Fax: +(331) 69 075 327; +(401) 4 208 625;
E-mails: Aurel.Popescu@curie.u-psud.fr; p.aurel@mailbox.ro

The surfaces of a BLM are neither perfectly planar nor rigid [39, 40]. The BLM system is a quasi-two-dimensional flexible structure that undergoes continuously a variety of conformational and dynamic transitions [41-44]. Furthermore, the artificial and natural BLMs are not insulating systems, but permeable for water and electrolytes that diffuse across by a diversity of transmembrane pores.

Stochastic transmembrane pores are generated by either of the following mechanisms: random and biased thermal fluctuations (thermoporation), and electrical triggering (electroporation). Lipid molecules inside BLM follow three distinct categories of random thermal movements: lateral translations, parallel to the bilayer surface, with the lateral diffusion coefficient in the order of 10^{-7} m^{2}s^{-1} (D_l) [45], oscillations and rotations around the lipid axes perpendicularly to the bilayer surface [46].

Lateral translations with random directions induce local fluctuations of the density of the lipid polar headgroups at bilayer surfaces. Therefore, a snapshot of the bilayer surface reveals local domains of nanoscopic dimension with a higher density of polar headgroups (i.e. clusters) as well as zones with a lower density. For certain physical conditions of BLM (pH, temperature, lipid components, electrochemical potential, etc.), the latter zones represent small local defects (i.e. vacancies) of the membrane. In these domains, the water molecules can penetrate the hydrophobic matrix of the bilayer. Let us consider the case of two independent defects from each monolayer that are aligned in a perpendicular direction on the membrane surface. They can generate a cylindrical hydrophobic pore with the inner surface flanked by the hydrophobic chains of the lipids. Therefore these types of transient pores are of hydrophobic nature. It is also possible that the polar headgroups, situated in the proximity of a hydrophobic pore to obey rotations towards its interior. In this case, the internal hydrophobic surface of the pores will become coated with polar headgroups. Thus, these pores have a hydrophilic nature, have no more a cylindrical geometry and have more stability than the hydrophobic ones [47]. In other words, the random thermal fluctuations of the polar headgroup density in the two monolayers of BLM are able to generate stochastic transmembrane pores.

The presence of hydrophobic thickness fluctuations inside BLM was demonstrated both by theory [3] and experiment. This was achieved by determination of the values of bilayer thickness (h) from three independent procedures: electrical capacitance measurements (h_c) [48], optical reflectance measurements (h_r) [49, 50], and direct computation (h_{av}). Tanford (1980) [44] calculated the bilayer thickness using the following formula $h_{av} = N_l\,M/\rho$, where N_l, M and ρ are the number of lipids per unit area, the molecular weight of the hydrophobic chains and the density of the hydrophobic zone, respectively. Because of the "thickness fluctuations" of the hydrophobic regions, h_c should be equal to h_{av}, while, in this case, both of them should be smaller than h_r by the thickness of the polar layer (h_{tp}): $h_c \cong h_{av} = h_r - h_{tp}$. If the lipid bilayer would have a uniform thickness, then h_c should be equal to h_{av}. In the case of BLM composed by a binary mixture of lipids, a selective association between phospholipids takes place following appearance of phospholipid domains. Their thickness is dependent on the length of the hydrocarbon chain of lipid components [51-55]. Popescu et al. (1991) [56] demonstrated the appearance of stochastic pores in BLMs due to fluctuations in bilayer thickness. The height of the energy barrier for membrane perforation following such a mechanism is large though (about 91 *kT*, where *k* and *T* are the Boltzmann constant and absolute temperature, respectively). In this case, the geometrical profile of the pore is of an elliptical toroidal form. It was also shown that such a transmebrane pore could evolve to a stable state [56]. The results obtained by this model were pretty surprising, because of the rapid time scale for closure of statistical pores in membranes. Two years latter on, Zhelev and Needham (1993) [57] have created large, quasistable pores in lipid bilayer vesicles, thus keeping with the previous model prediction [56]. The membrane's resistance to rupture [58, 59] in terms of a line tension for a large pore in bilayer vesicles was calculated by Moroz and Nelson (1997) [60].

Stochastic transmembrane pores can also be formed by biased thermal motion of lipids [42]. This mechanism is sometimes called thermoporation. The pores appear in the membrane via a thermally induced activation process. Alternatively, the activation process for pore formation can be induced via an external electrical field (also, called electroporation) [61, 62]. The pore generated by electroporation is larger and more stable [63]. The electroporation mechanism was proposed for delivery of drugs and genes to cells and tissues [61, 62, 64].

Transmembrane protein pores are formed by proteinaceous systems covering a wide range from small peptide channels (e.g. gramicidin, alamethicin, melittin etc.) to large protein multimeric assembled channels. Since these pores are large and water-filled, the hydrophilic substances, including ions, can diffuse across them, thus dissipating the membrane electrical potential. Transmembrane protein pores are consisted of integral proteins from two major structural classes: (1) selective channels formed by bundled transmembrane α-helical structures, and (2) selective channels, pores and porins formed by monomeric (e.g. OmpG), dimeric (e.g. selective Cl⁻ channels), trimeric (e.g. OmpF) or multimeric transmembrane β-barrel structures (e.g. α-hemolysin, leukocidins, cytolysins) [65]. The lipid bilayer may be used as an *in vitro* system to study these protein channels when they are reconstituted into a functional membrane [66]. In addition, the BLMs can be used as a tool for membrane protein engineering and its applications in either single-molecule biophysics [67-70] or biotechnological area [68].

In another example, colicin Ia, a protein secreted by *Escherichia coli*, forms voltage-gated ion channels both in the inner membrane of target bacteria and in planar BLMs [71-73]. Colicin Ia is a membrane transporter belonging to the class of bacterial toxins that share the same strategy: they are inserted into the membrane of the other nutrient competing bacteria, thus generating pores of large dimensions. Accordingly, these pores will damage the electrochemical membrane potential and, finally, will provoke the death of these competing bacteria. As compared with the stochastic pores mentioned above, the proteinaceous pores have a different mechanism of formation and also different properties. While a stochastic pore "forgets" its mechanism of generation, some of the transmembrane protein pores (e.g. colicin Ia) seem to exhibit "memory" effects, at least under the influence of a specific sequence of pulses used for BLM electrical stimulation [74].

Genetic pores were encountered in the wall of sinusoid vessels from the mammalian liver. The endothelial cells of these vessels have numerous sieve plate pores [75]. These pores of about 0.1 μm diameter enable a part of blood plasma and chylomicrons to pass from sinusoidal space into the space of Disse. Therefore, the endothelial pores control the exchange of fluids, solutes and particles between the sinusoidal blood and the space of Disse [75, 76].

In this work, we used the elasticity theory of continuous media to describe the appearance of stochastic pores across planar BLMs.

2. BLMs – SMECTIC LIQUID CRYSTALS

Phospholipids possess a very exciting feature – the amphiphilic nature, that is, they have a hydrophobic and hydrophilic character due to their hydrocarbon chains and polar headgroups, respectively. Therefore, if the phospholipids are dissolved in a strong polar medium (e.g. water), they spontaneously aggregate into ordered structures: planar lipid bilayers or spherical vesicles. In these ordered structures the polar headgroups have the orientation towards the exterior of the bilayer (i.e. towards the polar medium, in this case, water), while the hydrocarbon chains tend to be hidden into hydrophobic interior, thus avoiding a direct contact with the polar medium [44, 77, 78].

If the planar bilayer membranes are analysed only from the point of view of their physical properties, one can find out:

1. their component molecules are oriented along a preferential direction, almost perpendicularly to the BLM surface;

2. the two monolayers have closely similar thickness and can glide one along to another;

3. there is no exchange of molecules between the two monolayers. Translocation events from one monolayer to another (e.g. flip-flop transitions) are very improbable processes (see the section 8 of this chapter). The flip-flop diffusion coefficient (D_{ff}) is in the order of 10^{-23} m^{2}s^{-1}, which is much smaller than D_l [45];

4. in each monolayer there is a short range order, which is characteristic to the liquids. This is because both the interactions between the polar headgroups and those of van der Waals-London nature between the hydrophobic chains are known to be of short range;

5. the bilayer possesses a unique optical axis, *Oz*, perpendicularly to the bilayer plane;

6. the two orientations, *Oz* and *–Oz*, perpendicularly to the bilayer surface are equivalent;

These properties are characteristic to a smectic liquid crystal of type A [79].

The X-rays diffraction diagrams [80] and the analysis of the thermal phase transitions [81, 82] suggest a special BLM architecture in the case of the bilayers composed of lipids whose chains differ by more than four methylene groups. In this BLM structural organization, the phospholipid hydrocarbon chains are positioned so that the shorter molecules of one monolayer are aligned along with the longer molecules of the other one, and vice versa. Both of these molecules are tilted over the bilayer surfaces. These peculiar structures seem to be unstable. They appear only in the case of an adequate tilting of the support walls used to form the bilayer. Even in this case the lipid bilayer is a liquid crystal, but of the type C.

The mesomorph state of BLM, considered as a smectic liquid crystal of the type A, allows the application of both experimental and theoretical methods that are specific for the study of liquid crystals.

3. DEFORMATION FREE ENERGY OF BLM

In 1974, Pierre G. de Gennes was first that applied the theory of elasticity of continuous media to liquid crystals [79]. Taking into account a single axis liquid crystal, the average orientation of the long molecular axis is described in the point *r* by the direction vector *n(r)*. The direction of *n(r)* coincides with the optical axis of the liquid crystal. In the first part of this approach, we shall consider that the elastic deformations do not induce local modifications of the liquid crystal density.

Let us consider a rectangular system of coordinates *Oxyz*, in which *n(r)* is parallel to the *Oz* axis. Also, let us consider the case of small deformations. The components of the direction vector *n(r)* can be equated by the following Taylor series:

$$n_x = n_{x0} + e_1 x + e_2 y + e_3 z + O(r^2)$$
$$n_y = n_{y0} + e_4 x + e_5 y + e_6 z + O(r^2)$$
$$n_z = 1 + O(r^2)$$

(1)

where deformation coefficients e_i (i=1÷6) as well as the deformation types from the Eqs. (1) are the following:

$$e_1 = \frac{\partial n_x}{\partial x} \quad ; \quad e_5 = \frac{\partial n_y}{\partial y} \qquad (the\ splay\ distorsion)$$

$$e_2 = \frac{\partial n_x}{\partial y} \quad ; \quad e_4 = -\frac{\partial n_y}{\partial x} \qquad (the\ twist\ distorsion)$$

$$e_3 = \frac{\partial n_x}{\partial z} \quad ; \quad e_6 = \frac{\partial n_y}{\partial z} \qquad (the\ bend\ distorsion)$$

(2)

and $O(r^2)$ are terms of the order of magnitude of r^2.

The free energy density is a quadratic form of the deformation coefficients:

$$F(x,y,z) = F_0 + \sum_{i=1}^{6} k_i e_i + \frac{1}{2} \sum_{i=1}^{6} \sum_{j=1}^{6} k_{ij} e_i e_j \qquad (3)$$

where F_0 denotes the free energy of the initial non-deformed state. k_i and k_{ij} are coefficients that will be discussed later on.

Due to special properties of the single axis liquid crystals, the expression of the free energy density is invariant to any rotation around the Oz axis and, also to the inversion $z \rightarrow -z$ (Oz is equivalent to $-Oz$, as we already stated above). If the surface energy is considered to be negligible too, then only three independent coefficients k_{ij} will remain in Eq. (3). They will be renamed as follows: k_{11}=K_1, k_{22}=K_2, k_{33}=K_3. In this way, the variation of the free energy, ΔF, derived from the expression (3), will have a simpler form:

$$\Delta F(x,y,z) = \frac{1}{2} K_1 (\text{div } \mathbf{n})^2 + \frac{1}{2} K_2 (\mathbf{n} \cdot \text{curl } \mathbf{n})^2 + \frac{1}{2} K_3 (\mathbf{n} \times \text{curl } \mathbf{n})^2 \qquad (4)$$

This is the fundamental expression of the theory of the continuous elastic media applied to liquid crystals. The crystal deformation is characterised by the $n(r)$. Its modulus is unity. However, it has a variable direction unlike a unit vector. We also assume that n varies slowly and smoothly with r. The coefficients K_i (i = 1, 2, 3) are positive elastic constants that correspond to each type of deformations. In Eq. (4), the first, second and third terms indicate the splay free energy, the torsion free energy (twist deformation) and the bending free energy, respectively.

Let us take into account a smectic liquid crystal of the type A, which is composed of many parallel and equidistant layers. The thickness of a layer is denoted by h. The component molecules with an elongated conformation have their long axes perpendicularly oriented to the layer surfaces. The system of ordinates has its origin in the mass centre of the liquid crystal with the axis Oz oriented perpendicularly on the layer surfaces. Let us assume that the liquid crystal undergoes a small deformation, which is characterised by a long wavelength. The shift of the *n*-th layer, in the direction *Oz*, as a consequence of this small deformation, is denoted by w_n *(x, y)*. If the crystal is considered a continuous medium, then the discrete variable *n* can be converted in a continuous variable $z = nh$. Accordingly, w_n *(x, y)* will be replaced by $w(x, y, z)$. Following this deformation, the liquid crystal molecules have their long axis no longer parallel to *Oz*. In the particular case of a large wavelength deformation, one can consider that the molecular deviations versus *Oz* could be small, and therefore neglected. To take into consideration the BLM thickness fluctuations, we implemented the compression deformations as well. Therefore, a local change in the liquid crystal density, ρ, will be generated:

$$\rho = \rho_0(1 - \psi(\mathbf{r})) \tag{5}$$

where ρ and ρ_0 are the layer densities in the presence and absence of compression deformation, respectively. Here, $\psi(\mathbf{r})$ is a function expressing the volume compression.

To find out an explicit form of the compression deformation, we consider the function $\psi(\mathbf{r})$ split in two terms:

$$\psi(\mathbf{r}) = \theta(\mathbf{r}) + \beta(\mathbf{r}) \tag{6}$$

The first and second term accounts for the compression parallel and perpendicularly oriented to the liquid crystal layer surfaces, respectively. Thus, one can express the volume compression as a function depending on the transversal compression by means of the parameter $\beta(\mathbf{r})$. This is also described as a function depending on $w(\mathbf{r})$: $\beta(\mathbf{r}) = \partial w / \partial z$.

In the case of smectic liquid crystal of the type A, both torsion and bending deformations are excluded. Therefore, the last two terms of the Eq. (4) are cancelled. However, if the compression deformation is taken into consideration, then the Eq. (4) is replaced by the following:

$$\Delta F(x, y, z) = \frac{1}{2} K_1 (\text{div } \mathbf{n})^2 + A\theta^2 + B\beta^2 + 2C\beta\theta \tag{7}$$

where *A* and *B* are the compression coefficients for parallel and perpendicular directions to the liquid crystal layers, respectively. *C* is an interaction constant between the two types of compressions.

From the stability condition $(A \times B > C^2)$ and static equilibrium $(\Delta F = 0)$, we can obtain an expression for parameter $\theta(\mathbf{r})$:

$$\theta(\mathbf{r}) = -\frac{C}{A} \frac{\partial w}{\partial z} \tag{8}$$

   If we take into account the compression deformation, then the free energy of deformation per unit volume of a smectic liquid crystal is the following:

$$\Delta F(x, y, z) = \frac{1}{2}\overline{B}\left(\frac{\partial w}{\partial z}\right)^2 + \frac{1}{2}K_1(\text{div }\mathbf{n})^2 \qquad (9)$$

where

$$\overline{B} = B - \frac{C^2}{A} \qquad (10)$$

   Let us consider a reference domain $G = \{(x, y, z) | -h \le z \le 0, (x, y) \in S$ occupied by the unperturbed liquid crystal (S is an arbitrary 2D domain). We suppose that the upper boundary undergoes a deformation given by $u(x, y)$. We also consider that the bottom boundary rests unperturbed. We assume that inside the two boundaries the deformation is given by $w(x, y, z) = b(z)u(x, y)$. It is conceivable that the components of the direction vector $\mathbf{n}$ do not depend on z, so that its projections on the Ox and Oy axes are given by:

$$n_x = -\frac{\partial u}{\partial x} \ll 1$$
$$n_y = -\frac{\partial u}{\partial y} \ll 1 \qquad (11)$$

   To find out the unknown function $b(z)$, we apply the boundary conditions as well as a condition on the minimum of the free energy. Therefore, the function $b(z)$ must satisfy:

$$\begin{cases} -\dfrac{d^2 b}{d z^2} = 0, \\ b(0) = 1, \quad b(-h) = 0. \end{cases} \qquad (12)$$

   The solution of the problem stated above is given by:

$$b(z) = z/h + 1 \qquad (13)$$

   Therefore, in this case the deformation free energy per unit area is:

$$\Delta F(x, y) = h\overline{B}\left(\frac{u}{h}\right)^2 + hK_1\left(\frac{\partial^2 u}{\partial x^2} + \frac{\partial^2 u}{\partial y^2}\right)^2 \qquad (14)$$

If we take into account the compression deformation, then the free energy of deformation per unit volume of a smectic liquid crystal is the following:

$$\Delta F(x, y, z) = \frac{1}{2}\overline{B}\left(\frac{\partial w}{\partial z}\right)^2 + \frac{1}{2}K_1(\operatorname{div}\mathbf{n})^2 \tag{9}$$

where

$$\overline{B} = B - \frac{C^2}{A} \tag{10}$$

Let us consider a reference domain $G = \{(x, y, z) | -h \le z \le 0, (x, y) \in S\}$ occupied by the unperturbed liquid crystal (S is an arbitrary 2D domain). We suppose that the upper boundary undergoes a deformation given by $u(x, y)$. We also consider that the bottom boundary rests unperturbed. We assume that inside the two boundaries the deformation is given by $w(x, y, z) = b(z)u(x, y)$. It is conceivable that the components of the direction vector $\mathbf{n}$ do not depend on z, so that its projections on the Ox and Oy axes are given by:

$$n_x = -\frac{\partial u}{\partial x} \ll 1$$
$$n_y = -\frac{\partial u}{\partial y} \ll 1 \tag{11}$$

To find out the unknown function $b(z)$, we apply the boundary conditions as well as a condition on the minimum of the free energy. Therefore, the function $b(z)$ must satisfy:

$$\begin{cases} -\dfrac{d^2 b}{d z^2} = 0, \\ b(0) = 1, \quad b(-h) = 0. \end{cases} \tag{12}$$

The solution of the problem stated above is given by:

$$b(z) = z/h + 1 \tag{13}$$

Therefore, in this case the deformation free energy per unit area is:

$$\Delta F(x, y) = h\overline{B}\left(\frac{u}{h}\right)^2 + hK_1\left(\frac{\partial^2 u}{\partial x^2} + \frac{\partial^2 u}{\partial y^2}\right)^2 \tag{14}$$

The variation of the unit area Δs, due to the BLM deformation, is given by:

$$\Delta s = \left[1 + \left(\frac{\partial u}{\partial x}\right)^2 + \left(\frac{\partial u}{\partial y}\right)^2\right]^{\frac{1}{2}} - 1 \cong \frac{1}{2}\left[\left(\frac{\partial u}{\partial x}\right)^2 + \left(\frac{\partial u}{\partial y}\right)^2\right] \tag{15}$$

Let us consider the surface tension coefficient denoted by γ. Taking into account the energetic contribution due to the change in the cross-sectional area of the BLM surface, one can obtain a complete expression for the deformation free energy per unity area:

$$\Delta F(x,y) = h\overline{B}\left(\frac{u}{h}\right)^2 + hK_1\left(\frac{\partial^2 u}{\partial x^2} + \frac{\partial^2 u}{\partial y^2}\right)^2 + \gamma\left[\left(\frac{\partial u}{\partial x}\right)^2 + \left(\frac{\partial u}{\partial y}\right)^2\right] \tag{16}$$

In this formula, the first term represents the elastic compression energy of the BLM, which is characterized by the compression elastic constant, $\overline{B}$. The second term indicates the elastic energy of splay distortion, which is characterized by the elastic coefficient, K_1. The third term is the free energy due to the surface tension, which is characterised by the surface tension coefficient, γ.

In our case, the deformation of a monolayer obey to a cylindrical symmetry, so that Eq. (14) can simply be written in a system of polar ordinates:

$$\Delta F(r) = \overline{B}\frac{u^2}{h} + hK_1\left(\frac{\partial u}{r\partial r} + \frac{\partial^2 u}{\partial r^2}\right)^2 + \gamma\left(\frac{\partial u}{\partial r}\right)^2 \tag{17}$$

Therefore, the total free energy of BLM deformation on a planar surface perturbation of radius, R, is given by:

$$\Delta\overline{F} = 2\pi\int_0^R\left[\overline{B}\frac{u^2}{h} + hK_1\left(\frac{\partial u}{r\partial r} + \frac{\partial^2 u}{\partial r^2}\right)^2 + \gamma\left(\frac{\partial u}{\partial r}\right)^2\right]rdr \tag{18}$$

4. PORE FORMATION DUE TO BLM THICKNESS FLUCTUATIONS

In the next section of this chapter, we consider only the oscillations that are perpendicularly oriented to the bilayer surface. There are the following interactions between phospholipids: (a) between the polar headgroups, (b) between the polar headgroups and the hydrophobic chains, and (c) between hydrophobic chains. Individual motions of the phospholipids are transformed in a collective thermal motion normally to the BLM surface. This phenomenon can trigger local changes in the BLM thickness. Here, we would like to address the following question: can these collective motions themselves induce the BLM perforations?

We consider only a local surface perturbation, $u(r)$, with a wavelength, λ, and its amplitude equal to half of bilayer thickness, h:

$$u(r) = -h\cos\frac{2\pi r}{\lambda} \tag{19}$$

The elasticity theory, applied to the lipid bilayer regarded as continuous media, has a great advantage because it takes into account the intrinsic properties of bilayer membrane (e.g. the constants $\overline{B}$, K_1 and γ). This was the main reason why we have used the elasticity theory to calculate the BLM free energy, ΔF, induced by a surface perturbation. In fact, we want to demonstrate that the thermal movements of the phospholipids, normally oriented to the BLM surfaces, can produce such thickness fluctuations that trigger the pore formation.

The initial state of the BLM was considered that in which its molecules do not move at all (i.e. we consider the initial temperature 0 K). At this temperature, both surfaces of the BLM are planar. By warming the BLM up to the temperature T > 0 K, the phospholipid molecules will get kinetic energies and their thermal motions will start deforming the lipid bilayer. This phenomenon favours the appearance of thickness fluctuations. Thus, this thermal energy must be equal or greater than the free energy of deformation:

$$\frac{a}{\pi R^2}\Delta\overline{F} \le (3N - N_g)\frac{kT}{2} \tag{20}$$

where N is the number of atoms of a single molecule, N_g is the number of chemical intra-molecular bonds, a represents the cross-sectional area of the polar headgroups [83] and k is the Boltzmann constant. For the sake of simplicity, we have chosen the case for which $3N - N_g = 6$.

We are interested to find the parameters λ and R that satisfy the following relation:

$$\frac{2a}{R^2}\int_0^R rF(r,\lambda)dr = (3N - N_g)\frac{kT}{2} \tag{21}$$

After somehow tedious calculus, the last equation can be written as:

$$a_0(y)x^4 + b_0(y)x^2 + c_0(y) = 0 \tag{22}$$

where the variables x and y and the functions a_0, b_0, c_0 are given by:

$$x = h\frac{2\pi}{\lambda}, \; y = R\frac{2\pi}{\lambda}$$

$$a_0(y) = \frac{K_1}{Bh^2}\left(1 + \frac{\sin 2y}{y} - 3\frac{\cos 2y - 1}{2y^2} + \frac{4}{y^2}\int_0^1 \frac{\sin^2 ty}{t}dt\right)$$

$$b_0(y) = \frac{\gamma}{Bh}\left(1 - \frac{\sin 2y}{y} - \frac{\cos 2y - 1}{2y^2}\right)$$

$$c_0(y) = 1 + \frac{\sin 2y}{y} + \frac{\cos 2y - 1}{2y^2} - (3N - N_g)\frac{kT}{Bha} \qquad (23)$$

Unfortunately the Eq. (22) is too complicated to easily obtain the explicit formulae for λ and R, but one can obtain their parametrical representation. We shall consider x as a unknown and y as a parameter in such a way that the equations are reduced to an algebraic equation for x. Since the functions a_0 and b_0 are always positive, the equation has real solutions only if the function c_0 is negative. If these conditions are fulfilled, we get only one positive solution of Eq. (22):

$$\lambda(y) = 2\pi h\left(\sqrt{\frac{-b_0(y) + \sqrt{b_0^2(y) - 4a_0(y)c_0(y)}}{2a_0(y)}}\right)^{-1}$$

$$R(y) = \frac{y\lambda(y)}{2\pi} \qquad (24)$$

It is worth mentioning that the wavelength depends on the interactions between the polar headgroups and, also, on the relative motions of the neighboured molecules. In other words, the wavelength depends on the dephasing that appears between the neighboured molecules along the radius of the BLM surface occupied by the molecules involved in the collective thermal motion. Obviously, this dephasing can be influenced by the motion of the molecules from the adjacent medium (e.g. the motion of ions and waters).

5. CALCULATION OF THE PORE RADIUS

After perforation of the BLM, a rearrangement of molecules situated in the deformation zones may take place, so that the surface of the sinusoidal torus is modifying its concavity becoming a surface of a revolution elliptic torus. By equating the volumes delimited by the two revolution surfaces (i.e. sinusoidal and ellipsoidal), before the BLM perforation and after molecular rearrangement, the pore radius can be obtained. In the next steps, we calculate these two volumes.

We consider the space domain, D, of an ellipsoidal torus, defined by the relations:

$$z \in (0, h), \ r \le \lambda / 4$$

$$\frac{z^2}{h^2} + \frac{(r - \lambda / 4)^2}{b^2} \le 1 \tag{25}$$

where b is a parameter that fixes the space configuration of the pore.

Using a system of polar coordinates, the volume, D, of an ellipsoidal torus, is given by:

$$V = 2\pi \iint_D r \, dr \, dz \tag{26}$$

With some standard calculations, one can obtain the following expression for V:

$$V_a(\lambda, b) = 2\pi h \begin{cases} b\left(\dfrac{\lambda\pi}{16} - \dfrac{b}{3}\right), & b \le \dfrac{\lambda}{4} \\[2ex] \dfrac{\lambda^2}{32}\sqrt{1 - \left(\dfrac{\lambda}{4b}\right)^2} + \dfrac{\lambda b}{8}\arcsin\dfrac{\lambda}{4b} + \dfrac{b^2}{3}\left[\left(\sqrt{1 - \left(\dfrac{\lambda}{4b}\right)^2}\right)^3 - 1\right], & b \ge \dfrac{\lambda}{4} \end{cases} \tag{27}$$

For any $R \in (\lambda / 4, \ \lambda / 2)$, let us define the function $f(r)$ by:

$$f(r) = \begin{cases} h\left(1 - \cos\dfrac{2\pi}{\lambda} r\right), & r \in (0, R) \\[2ex] h\left(1 - \cos\dfrac{2\pi}{\lambda}(2R - r)\right), & r \in (R, 2R - \lambda / 4) \end{cases} \tag{28}$$

Using this function we define the following two domains:

$$D^\lambda = \left\{ (z, r) \mid r \in (0, \lambda / 4), \ z \in (0, f(r)) \right\} \tag{29}$$

and

$$D_{ext}^{\lambda, R} = \left\{ (z, r) \mid r \in (\lambda / 4, 2R - \lambda / 4), \ z \in (h, f(r)) \right\} \tag{30}$$

For $R \in (0, \lambda / 4)$ we define also other two domains:

$$D_{int}^{\lambda, R} = \left\{ (z, r) \mid r \in (0, R), \ z \in (0, f(r)) \right\} \tag{31}$$

and

$$G_{int}^{\lambda, R} = \left\{ (z, r) \mid r \in (R, \lambda / 4), \ z \in (0, h) \right\} \tag{32}$$

Finally, we define the domain $\Omega^{\lambda, R}$ by:

$$\Omega^{\lambda, R} = \begin{cases} D^\lambda \cup D_{ext}^{\lambda, R}, & R \ge \lambda / 4 \\[1ex] G_{int}^{\lambda, R} \cup D_{int}^{\lambda, R}, & R \le \lambda / 4 \end{cases} \tag{33}$$

$z \in (0, h), \; r \le \lambda / 4$

$$\frac{z^2}{h^2} + \frac{(r - \lambda / 4)^2}{b^2} \le 1 \tag{25}$$

where b is a parameter that fixes the space configuration of the pore.

Using a system of polar coordinates, the volume, D, of an ellipsoidal torus, is given by:

$$V = 2\pi \iint_D r \, dr \, dz \tag{26}$$

With some standard calculations, one can obtain the following expression for V:

$$V_a(\lambda, b) = 2\pi h \begin{cases} b\left(\dfrac{\lambda \pi}{16} - \dfrac{b}{3} \right), & b \le \dfrac{\lambda}{4} \\[2ex] \dfrac{\lambda^2}{32} \sqrt{1 - \left(\dfrac{\lambda}{4b} \right)^2} + \dfrac{\lambda b}{8} \arcsin \dfrac{\lambda}{4b} + \dfrac{b^2}{3}\left[\left(\sqrt{1 - \left(\dfrac{\lambda}{4b} \right)^2} \right)^3 - 1 \right], & b \ge \dfrac{\lambda}{4} \end{cases} \tag{27}$$

For any $R \in (\lambda / 4, \; \lambda / 2)$, let us define the function $f(r)$ by:

$$f(r) = \begin{cases} h\left(1 - \cos \dfrac{2\pi}{\lambda} r \right), & r \in (0, R) \\[2ex] h\left(1 - \cos \dfrac{2\pi}{\lambda} (2R - r) \right), & r \in (R, 2R - \lambda / 4) \end{cases} \tag{28}$$

Using this function we define the following two domains:

$$D^\lambda = \left\{ (z, r) \middle| r \in (0, \lambda / 4), \; z \in (0, f(r)) \right\} \tag{29}$$

and

$$D_{ext}^{\lambda, R} = \left\{ (z, r) \middle| r \in (\lambda / 4, 2R - \lambda / 4), \; z \in (h, f(r)) \right\} \tag{30}$$

For $R \in (0, \lambda / 4)$ we define also other two domains:

$$D_{int}^{\lambda, R} = \left\{ (z, r) \middle| r \in (0, R), \; z \in (0, f(r)) \right\} \tag{31}$$

and

$$G_{int}^{\lambda, R} = \left\{ (z, r) \middle| r \in (R, \lambda / 4), \; z \in (0, h) \right\} \tag{32}$$

Finally, we define the domain $\Omega^{\lambda, R}$ by:

$$\Omega^{\lambda, R} = \begin{cases} D^\lambda \cup D_{ext}^{\lambda, R}, & R \ge \lambda / 4 \\[1ex] G_{int}^{\lambda, R} \cup D_{int}^{\lambda, R}, & R \le \lambda / 4 \end{cases} \tag{33}$$

The volume of the domain $\Omega^{\lambda,R}$, characterising a sinusoidal torus, is given by:

$$V_t(\lambda,R)=2\pi h\left(\frac{\lambda}{2\pi}\right)^2\begin{cases}\dfrac{\pi^2}{8}+1-\cos\dfrac{2\pi R}{\lambda}-\dfrac{2\pi R}{\lambda}\sin\dfrac{2\pi R}{\lambda}, & R\le\dfrac{\lambda}{4}\\[2ex]\dfrac{\pi^2}{8}+1-\dfrac{\pi}{2}+\dfrac{4\pi R}{\lambda}\left(1-\sin\dfrac{2\pi R}{\lambda}\right), & R\ge\dfrac{\lambda}{4}\end{cases}\quad(34)$$

Equating the two volumes given by the Eqs. (27) and (34), namely $V_a(\lambda,b)=V_t(\lambda,R)$, one can calculate the magnitude of the pore radius r.

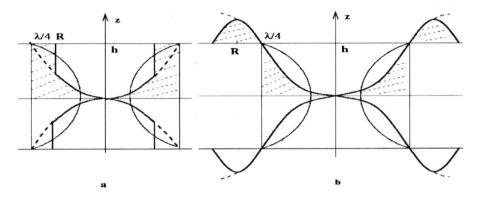

Fig. 1. Cross-sectional view through the BLM deformed in the case in which the radius, R, of the perturbation zone is: a) smaller than $\lambda/4$, and b) greater than $\lambda/4$. The forms of the pores are sketched, just immediately after the BLM perforation. h is the half of the BLM thickness. As a consequence of the lipid rearrangement (situated in the hatched zones), the molecules from the sinusoidal torus ($\Omega^{\lambda,R}$ domain) will be found in the volume limited by the elliptical surface of the ellipsoidal torus (D domain).

6. PORE FORMATION BY LATERAL THERMAL TRANSLATIONS

As stated above, when two defects, created due to lateral thermal translations, are aligned on the same perpendicular axis on both bilayer surfaces, they can generate cylindrical hydrophobic pores. Furthermore, these defects can vanish or extend towards the interior of the hydrophobic zones. Simultaneously, rotations of the phospholipids may appear, so that their polar groups cover the hydrophobic surface of the newly formed defect. In this way, the hydrophilic pores are generated as a consequence of random, but permanent thermal motions.

One can simply consider that the variation of the free energy, after a single-pore formation, contains two terms. The first term represents an energy decrease due to disappearance of a BLM surface (occupied by the pore). The second term represents an energy increase due to the pore contour generation (so-called edge energy). Therefore, the BLM free energy associated with a single-pore formation of radius r is the following [84, 85]:

$$\Delta F = 2\pi r \sigma - \pi \gamma r^2 \tag{35}$$

where γ represents a surface tension coefficient. σ is the strain energy per unit length of the pore edge.

The two coefficients are related by the formula: $\gamma = 2h\sigma$, where $2h$ indicates the hydrophobic width of the BLM.

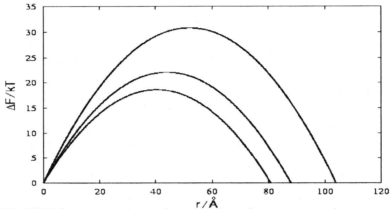

Fig. 2. The BLM free energy due to the generation of a pore versus the pore radius r. The hydrophobic width of the BLM is $2h$: 40.4 Å *(lower curve)*, 44 Å *(middle curve)*, and 52 Å *(upper curve)*.

According to this model, there is a critical energetic barrier $\Delta F_c = 4\pi\gamma h^2$, whether the radius r of the pore attains a critical value $r_c = 2h$ (Fig. 2). As noticed, the energy barrier value is dependent on the thickness of the hydrophobic domain and the surface tension as well, while the pore radius is equal to the bilayer thickness. If the pore radius is smaller than the critical value ($r < r_c$), then the force on the pore edge is inward and the pore tends to reseal itself. By contrast, if the pore radius is greater than this value, then the pore becomes unstable. Its diameter may increase indefinitely, thus triggering the BLM rupture.

In Fig. 2, the BLM free energy associated with the appearance of a single pore, is represented as a function of pore radius for the following values of the hydrophobic zone thickness: $2h = 40.4$ Å, $2h = 44$ Å, and $2h = 52$ Å. In all the cases, $\gamma = 15 \times 10^{-4}$ N m^{-1}.

7. THICKNESS FLUCTUATIONS AND PORE APPEARANCE

In this section, we analyse the conditions under which the thermal fluctuations of the hydrophobic domain of the BLM, composed from single-chain lipids, can generate transbilayer pores. In this respect, it is necessary to have information regarding the parameters from Eq. (22). Because the phospholipid membranes were extensively studied, there are sufficient data within the specific literature for required elastic constants. The thickness of the hydrophobic zone, in the case of the membrane composed of double-chain lipids is $2h = 40.4$ Å [86]. The compressibility coefficient obtained by Hladky and Gruen [3] from the experimental data of White [87] is equal to 5.36×10^7 Nm^{-2}.

The BLM surface area occupied by a single phospholipid is 38.6 Å^2, while the surface tension coefficient, γ, is 8×10^{-4}Nm^{-1}[88]. The splay coefficient K_1 was obtained from the experimental data regarding the modulus of the curvature elastic coefficient, $K_c = (2.8 \div 6.5) \times 10^{-20}$J [89], obtained in the case of lecithin vesicles. Thus, from $K_1 = K_c/2h$ [1, 90] one can easily obtain the splay coefficient K_1, which is needed in Eq. (22). Because the vesicle thickness used for the experimental measurement of K_c was 30 Å, then it results that $K_1 \in (0.93 \div 2.17) \times 10^{-11}$N.

The results presented in this section deal with the case when the deformation energy is equal to the total thermal energy of the molecules participating to the collective motion. More precisely, the results were obtained by solving the Eq. (22) in each particular case. The BLM formed by dehydrated phosphatydilcholine, with the thickness of the hydrophobic zone $2h = 40.4$ Å, has $K_1 = 0.93 \times 10^{-11}$N.

One can say that the collective thermal motion of the molecules can generate three types of transbilayer pores: open stable pores, closed stable pores (zippered pores) and unstable open pores, which determine the membrane rupture. Because this approach does not imply a condition to distinguish between open, stable and unstable pores, we have used the condition obtained in the case of pores generated by surface defects induced by lateral thermal motion. Therefore, we have considered that the pores are stable, if their radius is smaller than the hydrophobic thickness of a lipid bilayer ($r < r_c = 2h$).

Generally, the wavelength bilayer deformation is a two-value function on the perturbed zone radius. There is an interval in such a way so that for each value of the R belonging to it, the solution of Eq. (22) gives two values for λ. On the other hand, the pore radius is a monotonously decreasing function on the bilayer deformation wavelength. Due to these properties of the functions $\lambda = \lambda(R)$ and $r = r(\lambda)$, for each value of R, belonging to the non-univocity domain, it is possible the formation of two pores of different sizes. The selection between these pores is made by the magnitude of deformation: that one with a smaller wavelength will cause the appearance of a smaller pore radius and vice versa.

The Tables 1-3 reflect the three characteristic features of the BLM pore formation. Their upper part comprises the parameters R, λ and r_p for the case of two pore formation, both of them open and stable; the middle part describes the case of two stable pore formation: one open pore and the other closed; the lower part contains the values of R, λ and r_p in the case of appearance of two open pores, one stable and the other unstable.

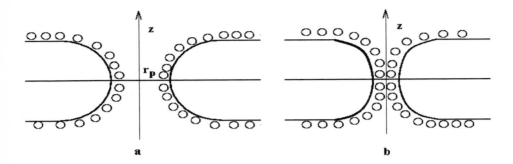

Fig. 3 The two types of pores generated by the BLM thickness fluctuations: a) open pore; b) closed pore. The circles represent the polar headgroups of the lipids. r_p is the pore radius.

In the last column of the upper part are also indicated the figures and the corresponding number associated to the curves from the figures. For an easier understanding of the tables and figures presented bellow, we describe in details the obtained data in the case of the reference BLM (Table 1 and curves 1 of the Fig.4). For each value of perturbed zone belonging to the non-univocity interval $[R_m \div R_1] = [36.0 \div 36.5]$ Å one can appear either a pore with a small radius or a pore with a large radius. If the deformation wavelength has its value in the smaller value interval $\lambda \in [\lambda_0 \div \lambda_{1m}] = [117.0 \div 112.4]$ Å, then smaller pore will appear. Its radius will be embedded into the interval: $[r_{p0} \div r_{pm}] = [10.3 \div 0.8]$ Å.

Table 1. The radia of the collective thermal motion zones of the BLM perturbation, the deformation wavelengths and the radia of the pores (or of the closing distance) in the case of reference BLM ($h = 20.2$ Å) and of other two BLMs differing by this one by their thickness ($h = 22$ Å) and their polar headgroup magnitude ($a_0 = 44$ Å^2). R, λ and r are measured in Å. N_m represents the minimal number of molecules which are participating to the collective thermal motion. d/h is the close ratio of the closed pores. Cv is the curve number within the Fig. 4.

	N_m	$R_m \div R_1$	$\lambda_0 \div \lambda_{1m}$	$r_{p0} \div r_{pm}$	$\lambda_0 \div \lambda_{1M}$	$r_{p0} \div r_{pM}$	Fig/Cv
$h = 20.2$	105	36.0-36.5	117.0-112.4	10.3-0.8	117.0-125.2	10.3-15.0	4/1
$h = 22.0$	137	41.0-41.6	131.4-127.7	9.6-0.3	131.4-138.4	9.6-14.6	4/3
$a_0 = 44.0$	124	41.7-42.1	132.4-129.2	8.5-0.3	132.4-138.4	8.5-13.5	4/2
	N_m	$R_1 \div R_2$	$\lambda_{1m} \div \lambda_m$	d/h	$\lambda_{1M} \div \lambda_2$	$r_{pM} \div r_{p2}$	
$h = 20.2$	105	36.5-37.1	112.4-111.4	0.10-0.94	125.2-130.9	15.0-16.9	
$h = 22.0$	137	41.6-42.3	127.7-126.8	0.21-0.95	138.4-144.8	14.6-16.8	
$a_0 = 44.0$	124	42.1-42.9	129.2-128.7	0.21-0.95	138.4-144.0	13.5-15.5	
	N_m	$R_2 \div R_c$	$\lambda_2 \div \lambda_c$	$r_{p2} \div r_{pc}$	$R_c \div R_r$	$\lambda_c \div \lambda_r$	$r_{pc} \div r_r$
$h = 20.2$	105	37.1-69.9	130.9-275.7	16.9-40.4	69.9- 96.4	275.7-384.1	40.4-53.1
$h = 22.0$	137	42.3-80.4	144.8-305.3	16.8-44.0	80.5-115.7	305.3-441.9	44.0-63.8
$a_0 = 44.0$	124	42.9-77.3	144.0-286.7	15.5-40.4	77.3-121.3	286.7-452.7	40.4-56.1

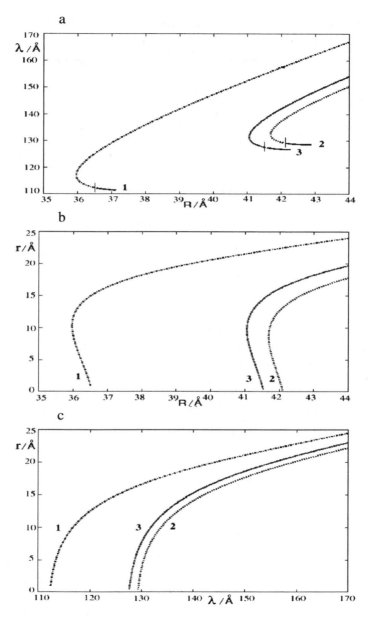

Fig. 4 The graphs, for: the deformation wavelength dependence of the perturbation zone radius (a); the pore radius dependence of the perturbation zone radius (b); the pore radius dependence of the BLM deformation wavelength (c). Curves 1: reference BLM. Curves 2: BLM with the hydrophobic thickness of 22 Å. Curves 3: BLM similar with the reference BLM, but with a polar headgroup area a_0 of 44 Å^2. The information in the Fig. 4 are supplemented by those from the Table 1.

If the deformation wavelength has its value in the greater value interval, $\lambda \in [\lambda_0 \div \lambda_{1M}] = [117.0 \div 125.2]$ Å, then a larger pore will be generated. Its radius will be in the following interval: $[r_{p0} \div r_{pM}] = [10.3 \div 15.0]$ Å. If the radius magnitude of the deformation zone of the collective thermal motion will be embedded in the following non-bijection interval: $[R_1 \div R_2] = [36.5 \div 37.1]$ Å, then it is possible to appear an open pore or a closed one. The magnitude of deformation will determine which of the two pores will be generated.

Following the Table 1, one can notice that the deformation wavelength has its value in the smaller value interval, $\lambda \in [\lambda_{1m} \div \lambda_m] = [112.4 \div 111.4]$ Å, then the closed pore is characterized by ratio d/h. Here, d is the magnitude of the distance on which the pore's surfaces (i.e. envelopes) are in a direct contact, and h is the half of the thickness of the hydrophobic zone. In our case, $d/h \in [0.10 \div 0.94]$. If the deformation wavelength has its value in the greater value interval, $\lambda \in [\lambda_{1M} \div \lambda_2] = [125.2 \div 130.9]$ Å, the open pore will appear. Its radius pertains to the interval: $[r_{pM} \div r_{p2}] = [15.0 \div 16.9]$ Å. The one-to-one correspondence interval between λ and R is divided in two sub-intervals. The first one is $R \in [R_2 \div R_c] = [37.1 \div 69.9]$ Å and it is in correspondence with the following wavelength interval: $\lambda \in [\lambda_2 \div \lambda_c] = [130.9 \div 275.7]$ Å. In this case, a single open stable pore will appear. The second interval: $R \in [R_c \div R_r] = [69.9 \div 96.5]$ Å is correspondent to another wavelength interval: $\lambda \in [\lambda_c \div \lambda_r] = [275.7 \div 384.1]$ Å. In this case, an open unstable pore will appear that will evolve towards the BLM rupture.

On the curves describing the dependence of the wavelength of deformation of the perturbing zone radius, vertical bars are figured. The parts of the curves situated at the right of these bars describe the values of λ and R for which the closed pores (zipper pores) are formed.

7.1 The effect of lipid molecule nature

Planar BLM with a high temporal stability are generally obtained by a single species of lipid molecules. We shall analyse such a situation. Therefore, the BLM will be characterised by: the thickness of the hydrophobic core, the thickness of the polar group zone, the electrical state and dipolar orientation of the polar groups, the BLM area occupied by a single lipid molecule equal to that of the hydrophobic chains, and the cross sectional area of a polar headgroup.

Taking into account BLMs differing by the reference BLM by the hydrophobic zone ($h = 22$ Å), one can study the effect of the hydrophobic zone thickness both on the conditions of pore formation and on the pore magnitude (Fig. 4, curves 3 and Table 1, row 2).

The BLM thickness increase has a slight effect on the magnitude of pore radius. This effect can be observed only in the case of the small pores, whose radii are decreasing. Instead, this influence of the thickness increase is stronger on the conditions of pore generation. As it can be easily seen from the Table 1, the pore formation, in the case $h = 22$ Å, is effective when the perturbed zone radius and the corresponding BLM deformation wavelength have great values.

In the case of a critical radius equal to 44 Å ($r_c = 2h$) the small pore radius is decreasing, the pores with larger radius are not affected, while the closing distance of the closed pores is increasing.

The cross-sectional area of the polar headgroup determines indirectly the deformation energy by means of the number of lipid molecules, which must participate to the collective thermal motion, in order to assure the energy required by BLM deformation. In order to appreciate the effect of the area magnitude of polar headgroups on the pore formation, we have approached a BLM, similar in its thickness with the reference BLM, but differing by this one in the cross section area of lipid molecules which, in this case, was of 44 Å^2.

The increasing of the polar headgroup area determines an increase both of the zone magnitude of the collective thermal motion (implicitly, of the number of the lipid molecules involved) and of the magnitude of BLM deformation. Nevertheless, the open pore radius is decreasing, but the closing distance of the closed pore is increasing (Fig. 4, curve 2 and Table 1, row 3).

7.2 Temperature effect

It is evident, that by the temperature raising, the average thermal energy of lipid molecules is increasing. Because, in our case, the thermal energy is the single source of energy involved in BLM deformation, it results that the number of the lipid molecules, which simultaneously participate to the BLM perturbation, must also decrease as the temperature is increasing. Consequently, the radius of the zone occupied by these lipid molecules will decrease, too. This

Table 2. The radia of the collective thermal motion perturbation, the wavelengths of BLM deformation and the radia of pores, in the case of the reference BLM ($T = 300$ K) and of other two different BLMs for $T = 290$ K and $T = 310$ K. R, λ and r are measured in Å. N_m is the minimum number of molecules participating to the collective thermal motion. d/h is the close ratio of the closed pores. In the last column, Cv indicates the curve number within the Fig. 5.

$T(K)$	N_m	$R_m \div R_1$	$\lambda_0 \div \lambda_{1m}$	$r_{p0} \div r_{pm}$	$\lambda_0 \div \lambda_{1M}$	$r_{p0} \div r_{pM}$	Fig/Cv
290	113	37.2-37.7	120.1-115.9	9.8-0.3	120.1-128.1	9.8-14.5	5/8
300	105	36.0-36.5	117.0-112.5	10.3-0.8	117.0-125.2	10.3-15.0	5/1
310	99	34.9-35.5	114.2-109.3	10.5-0.8	114.2-123.2	10.5-15.3	5/9

$T(K)$	N_m	$R_1 \div R_2$	$\lambda_{1m} \div \lambda_m$	d/h	$\lambda_{1M} \div \lambda_2$	$r_{pM} \div r_{p2}$	
290	113	37.7-38.3	115.9-115.0	0.22-0.94	128.1-133.5	14.5-16.4	
300	105	36.5-37.1	112.4-111.4	0.10-0.94	125.2-130.9	15.0-16.9	
310	99	35.5-36.1	109.3-108.3	0.09-0.94	123.3-128.3	15.3-16.9	

$T(K)$	N_m	$R_2 \div R_c$	$\lambda_2 \div \lambda_c$	$r_{p2} \div r_{pc}$	$R_c \div R_r$	$\lambda_c \div \lambda_r$	$r_{pc} \div r_r$
290	113	38.3-71.3	133.4-315.6	16.4-40.4	71.3-101.5	315.6-397.9	40.4-58.0
300	105	37.1-69.9	130.9-275.7	16.9-40.4	69.9- 96.4	275.7-384.1	40.4-53.1
310	99	36.1-55.1	128.2-219.5	16.9-32.1	–	–	–

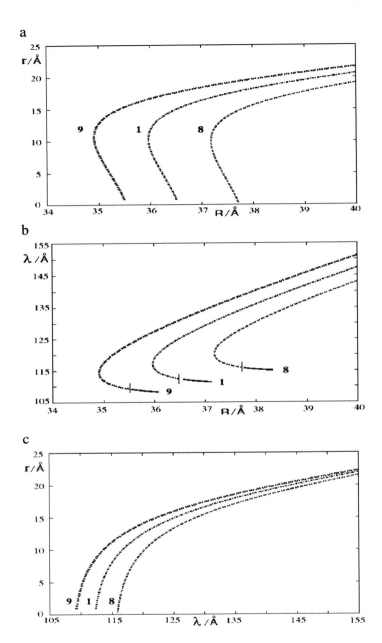

Fig. 5 The graphs, in the case of reference BLM, for: the deformation wavelength dependence of the perturbation zone radius (a); the pore radius dependence of the perturbation zone radius (b); the pore radius dependence of the BLM deformation wavelength (c). Curves 8: $T = 290$ K. Curves 1: $T = 300$ K. Curves 9: $T = 310$ K. The information in the Fig. 5 are supplemented by those from the Table 2.

conclusion is confirmed by the solutions of the Eq. (22) for the reference BLM, in the case of two temperatures 290 K and 310 K, under the phase transition temperature (Fig. 5 and Table 2). In exchange, the open pore radius is increasing with temperature, while the closed pore closing distance is decreasing.

7.3 Effect of the intermolecular forces

In this section we shall approach the effect of: a) the variation of the splay elastic properties, b) the elastic compression properties, and c) the surface tension.

a) The effect of the variation of the splay elastic properties. Molecular rotation energy against the normal direction to the BLM is known as the splay energy, being characterised by the splay elastic constant, K_1. This energy depends on the interaction between BLM lipid molecules. The splay constant values, experimentally determined for the glycerolmonooleat BLM, are belonging to the following interval: $[0.93 \div 2.17] \times 10^{-11} \text{N}$. The values of the splay constant for phosphatidilcholine, are belonging to the same interval [16].

In order to appreciate the role of the splay properties on the pore formation, this elastic constant was modified, namely: the minimal value of the splay elastic constant, $K_1 = 0.93 \times 10^{-11} \text{N}$, of the reference BLM, was replaced by the maximal value, $K_1 = 2.17 \times 10^{-11} \text{N}$. It can be observed that all calculated values (R, λ, r_p) are shifted to greater values, while the closing distance of the closed pore is shifted towards smaller values (Fig. 6, curves 6 and Table 3, row 2). The influence of splay constant increase is stronger on the open pore radius.

b) The effect of the elastic compression properties. The experimental value of the compression constant, in the case of BLM composed of double chain lipids without solvent, is $\bar{B} = 5.36 \times 10^7 \text{ Nm}^{-2}$, while in the case of the same BLM, but containing solvent, is $\bar{B} = 5.75 \times 10^4 \text{ Nm}^{-2}$.

Compressibility properties have the strongest effect on the transbilayer pore formation. If the Eq. (22) is solved, in the case of $\bar{B} = 1.5 \times 10^7 \text{ Nm}^{-2}$, it results that the dependence between λ and R becomes monotonically decreasing, which eliminates the possibility of pore appearance with two geometric states (forms). Both the radius of the perturbed zone and the wavelength of the deformation are strongly decreasing. The pores that appear are either open pores, with larger radii, or closed ones, with smaller closing distances (Fig. 6, curves 5 and Table 3, row 3). The decreasing of compression constant eliminates the possibility of the membrane rupture, as a consequence of the pore formation. On the other hand, the modification of the compression elastic constant, from the value, $\bar{B} = 5.36 \times 10^7 \text{ Nm}^{-2}$ to the value, $\bar{B} = 1.5 \times 10^7 \text{ Nm}^{-2}$, determines a modification of the curves: $\lambda = \lambda (R)$, $r_p = r_p (R)$, $r_p = r_p (\lambda)$. There is a critical value, $B_c = 14.5 \times 10^7 \text{ Nm}^{-2}$, beyond which these changes are taking place. For the values of $\bar{B}$ greater than this critical value, the wavelength dependence of the perturbed zone radius has no one-to-one correspondence. From the Fig.6, it results that only the curve "branch" of the greater values of the wavelength are influenced by the increasing of the compression elastic constant. The width of the non-univocity interval is decreasing with the increasing of the $\bar{B}$ constant, so that, for $\bar{B} = B_c = 14.5 \times 10^7$ Nm^{-2}, the superior branch of the curve (representing the wavelength dependence of the perturbed zone radius), becomes parallel with $O\lambda$ axis. This is equivalent with an uncertainty of the radius dependence of the wavelength.

For the values of $\bar{B}$ smaller than the critical value ($\bar{B} < B_c$) the dependence of wavelength of the radius, R, becomes of a univocity type, and the interval of the wavelength spectrum is narrowing with the decreasing of $\bar{B}$. For values of $\bar{B}$ around the value, $\bar{B} = 12.5 \times 10^6$ Nm^{-2}, the sensitivity of wavelength dependence of R is very weak. The compression elastic constant can be also modified in the case when the double chain lipids of BLM are replaced by lisophospholipids.

c) The effect of surface tension. The coefficient of surface tension, characterizing the molecular interactions at the separation surface of two non-miscible media, is depending of the composition of these two media. In the case of BLM, the surface tension coefficient is dependent of the pH of the BLM adja-

Table 3. The values of the zone radia of collective thermal motion perturbation and of the wavelength BLM deformation, in the case of the reference BLM and of other three bilayers which differ from this one by: the elastic splay constant, $K_1 = 2.17 \times 10^{-11}$N, the compression constant, $\bar{B} = 1.5 \times 10^7$ Nm^{-2}, and by surface tension coefficient, $\gamma = 2.4 \times 10^{-4}$Nm^{-1}, respectively. The parameters R, λ and r are measured in Å. N_m is the minimum number of molecules participating simultaneously to the collective thermal motion. In the last column, Cv indicates the curve number within the Fig. 6. d/h is the close ratio of the closed pores. The information from the Table 3 are supplemented by those in the Fig. 6.

N_m	$R_m \div R_1$	$\lambda_0 \div \lambda_{1m}$	$r_{p0} \div r_{pm}$	$\lambda_0 \div \lambda_{1M}$	$r_{p0} \div r_{pM}$	Fig/Cv
105	36.0-36.5	117.0-112.5	10.3-0.8	117.0-125.2	10.3-15.0	6/1
156	43.7-44.5	142.4-136.7	12.5-1.0	142.4-152.3	12.5-18.2	6/6
98	34.8-35.3	113.1-108.5	9.9-0.8	113.1-120.9	9.9-14.5	6/7
51	25.0-28.2	99.6- 86.7	4.6-0.6	–	–	6/5
N_m	$R_1 \div R_2$	$\lambda_{1m} \div \lambda_m$	d/h	$\lambda_{1M} \div \lambda_2$	$r_{pM} \div r_{p2}$	
105	36.5-37.1	112.4-111.4	0.10-0.94	125.2-130.9	15.0-16.9	
156	44.5-45.2	136.7-135.5	0.10-0.94	152.3-158.2	18.2-20.2	
98	35.3-35.8	108.5-107.7	0.10-0.94	120.9-125.3	14.5-16.0	
51	28.2-28.5	86.7- 85.7	0.09-0.94	–	–	
N_m	$R_2 \div R_c$	$\lambda_2 \div \lambda_c$	$r_{p2} \div r_{pc}$	$R_c \div R_r$	$\lambda_c \div \lambda_r$	$r_{pc} \div r_r$
105	37.1-69.9	130.9-275.7	16.9-40.4	69.9- 96.4	275.7-384.1	40.4-53.1
156	45.2-70.4	158.2-277.0	20.2-40.4	70.4-111.4	277.0-444.5	40.4-14.9
98	35.8-68.4	125.3-271.7	16.0-40.4	68.4- 81.8	271.7-326.1	40.4-47.6
51	–	–	–	–	–	–

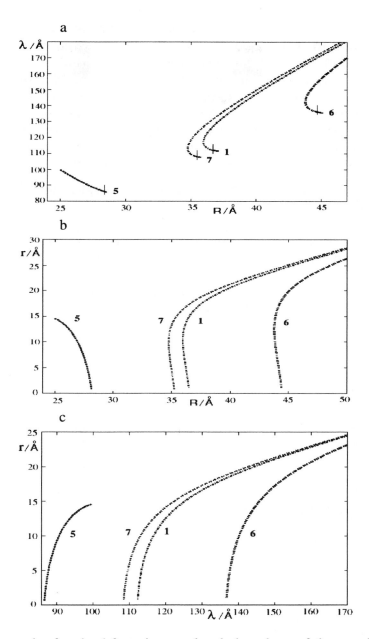

Fig. 6 The graphs, for: the deformation wavelength dependence of the perturbation zone radius (a); the pore radius dependence of the perturbation zone radius (b); the pore radius dependence of the BLM deformation wavelength (c). Curves 1: reference BLM. Curves 5: BLM with the elastic compression coefficient, $\overline{B} = 1.5 \times 10^7$ Nm^{-2}. Curves 6: BLM with the elastic splay coefficient, $K_1 = 2.17 \times 10^{-11}$N. Curves 7: BLM with the surface tension coefficient, $\gamma = 2.4 \times 10^{-4}$ Nm^{-1}.

cent medium. The experimental values of surface tension coefficient, γ, are situated between $10^{-8}\,\mathrm{Nm^{-1}}$ and $5\times10^{-2}\,\mathrm{Nm^{-1}}$.

In this work, we have chosen two values for this coefficient: $\gamma = 15\times10^{-4}\,\mathrm{Nm^{-1}}$, for the reference BLM, and $\gamma = 2.5\times10^{-4}\,\mathrm{Nm^{-1}}$. As can be seen from Table 3 and Fig. 6 (curves 7), the surface tension influences quite slightly the conditions of transmembrane pore apparition. If one would take into account that by BLM deformation, the space among polar headgroups is increasing, favouring thus the contact area between water and hydrophobic zone, it would be possible that the surface tension effect to be more important.

8. STOCHASTIC PORES AND TRANSVERSAL DIFFUSION COEFFICIENTS

Even if there is a very low probability, however the phospholipids translocate from one monolayer to the other. This transversal diffusion is caused by their perpendicular oscillations on the bilayer surface.

In this section, we demonstrate that the pores following self-oscillations of the bilayer membrane might represent a possibility for the passive flip-flop diffusion of phospholipid molecules.

The appearance of statistical pores through lipid membranes due to the self-oscillations of the lipid bilayer involves the existence of three energetic barriers [56] for:

1) perforation of the membrane, with the energy height, ΔW_1;
2) recovery of the membrane, with the energy height, ΔW_2;
3) breaking of the membrane, with the energy height, ΔW_3.

The membrane can be found in one of the following states [56]:

-state 1, without pores, namely in the phase preceding the pore formation;
-state 2, with a single pore on it, regardless the diameter of the pore;
-state 3, the broken membrane.

The scheme for the kinetics of state transitions of the membrane is the following:

$$1 \underset{k_2}{\overset{k_1}{\rightleftharpoons}} 2 \xrightarrow{k_3} 3 \qquad (35)$$

with the rate constants:

$$k_1 \approx \exp(-\beta\Delta W_1) \qquad (36)$$
$$k_2 \approx \exp(-\beta\Delta W_2) \qquad (37)$$
$$k_3 \approx \exp(-\beta\Delta W_3) \qquad (38)$$

where $\beta = 1/kT$.

If we denote the probabilities that at a given time the membrane is found to be in the states 1, 2 and 3, by $P_1(t)$, $P_2(t)$ and $P_3(t)$, respectively, then the kinetic description of this system is given by the following system of equations:

$$dP_1 / dt = k_2 P_2(t) - k_1 P_1(t) \tag{39}$$
$$dP_3 / dt = k_3 P_2(t) \tag{40}$$
$$P_1(t) + P_2(t) + P_3(t) = 1 \tag{41}$$

with the initial conditions: $P_1(0) = 1$, $P_2(0) = 0$, $P_3(0) = 0$.

Let us introduce a dimensionless rate variable defined by the following expression:

$$\tau = (k_1 + k_2 + k_3)t \tag{42}$$

Then, the probabilities to find each of the three states are:

$$P_1 = [(x_1 + v)\exp(x_2\tau) - (x_2 + v)\exp(x_1\tau)] / y \tag{43}$$
$$P_2 = v[\exp(x_1\tau) - \exp(x_2\tau)] / y \tag{44}$$
$$P_3 = 1 - [x_1 \exp(x_2\tau) - x_2 \exp(x_1\tau)] / y \tag{45}$$

with the auxiliary relations:

$$v = k_1(k_1 + k_2 + k_3)^{-1} \tag{46}$$
$$u = k_1 k_2 (k_1 + k_2 + k_3)^{-2} \tag{47}$$
$$y = \sqrt{1 - 4u} \tag{48}$$
$$x_1 = (y - 1)/2 \tag{49}$$
$$x_2 = -(y + 1)/2 \tag{50}$$

We can say that the distribution function of the lifetime of the pore is:

$$F(\tau) = P_3(\tau) \tag{51}$$

Thus, the probability density has the following expression:

$$f(\tau) = dF(\tau) / d\tau = u \exp(x_1\tau)[1 - \exp(-y\tau)] / y \tag{52}$$

Finally, the mean lifetime, $\overline{\tau}$, of the pore is equal to:

$$\bar{\tau} = \int_0^{\infty} \tau f(\tau) d\tau = 1/u \tag{53}$$

The mean number, $\bar{n}$, of pores is equal to the probability of a pore appeared on a stable (unbroken) membrane [58]:

$$\bar{n} = P_2 /(P_1 + P_2) = v[\exp(y\tau) - 1]/x_1 \tag{54}$$

The number of phospholipid molecules that cross the median plane of the bilayer through statistical pores is equal to:

$$dN_{ff} = ndN_t \tag{55}$$

where dN_t is the number of molecules that perform a translocation through the pore. Therefore, the flip-flop diffusion coefficient is $D_{ff} = nD_l$.
    Taking into consideration that:

$$n = \lim_{a \to \infty} \frac{1}{a} \int_0^a n(\tau) d\tau = v /|x_2| \tag{56}$$

we obtain:

$$n \approx k_1 /(k_1 + k_2 + k_3) \tag{57}$$

From the results of Ref. [56], the energy barriers for ΔW_1, ΔW_2 and ΔW_3 are 91.2, 65.6 and 2,090.1 kT, respectively. Then it was obtained $n = exp(-25)$. Because the lateral diffusion coefficient, D_l, of the phospholipids at the surface of the planar bilayer is in the order 10^{-12} m^{2}s^{-1}, this gives the value $D_{ff} = 10^{-12} exp(-25) = 10^{-23}$ m^{2}s^{-1}. If we consider that formula for D_{ff}, is similar to that for the translation diffusion:

$$D_{ff} = \overline{r^2} /4\tau$$

$$(58)$$

and adopt the model of jumps in an ordered 2-dimensional network, then

$$d = \sqrt{\overline{r^2}} \tag{59}$$

will be the thickness of the hydrophobic domain of the lipid bilayer. Here and henceforth, τ is the duration of the jump from one monolayer into the other.
    If we consider a lipid bilayer with a thickness of 30 Å, then the translocation time, τ, is:

$$\tau = \overline{r^2} / D_{ff} = 2.25 \times 10^5 \text{ s} \tag{60}$$

It is interesting to observe that this is result in the same order of magnitude as the lifetime of phospholipids in a lipid bilayer. This calculation demonstrates the passive flip-flop translocation of phospholipids across a statistical pore formed in a lipid membrane is a very rare but detectable event [91].

This mechanism for the translocation of phospholipids from one monolayer to another one is possible because: (1) there are oscillations of the lipid bilayer surface due to thickness fluctuations [56]; (2) there are perpendicular oscillations of the phospholipids on the monolayer plane [92].

In addition, the oscillations into the neighbour water medium are forestalled by the hydrophobic effect within the hydrocarbon domain, which appear via image forces due to polar headgroups of phospholipids.

9. CONCLUSIONS

Since their use for the first time as the reconstitution *in vitro* of cell membrane structure [93], BLMs were and continue to be the most suited artificial membrane models in numerous experiments that supplied the extremely important information necessary to a better understanding of the very intricate life processes occurring in living cells [94].

Among the membrane related phenomena, the membrane transport occupies a central role. It is important to emphasize that in this transport mechanism, among other complex structures (e.g. ionic pumps, ionic channels, etc.) the different types of pores are involved, too.

We tried to demonstrate that the general theory of continuous elastic media can predict the generation of all types of pores due to the BLM thermal thickness fluctuations.

Our results are explaining both the pore formation and their properties and, besides are confirming the existence of a new mechanism of transbilayer pore formation.

The values of the open pore radii, obtained by us, are in agreement with those encountered in the literature, while the closed pore formation confirms an old hypothesis concerning the existence of water "little threads" into the hydrophobic region of the lipid bilayers [95].

Taking into account that the thermal energy necessary to provoke BLM deformation is proportional to R^2, it results that the closer to R_m is the radius of the collective thermal motion zone, the greater is the probability of a pore appearance.

The closed pores can be very stable due to the coupling between the polar headgroup dipoles, the space between the polar headgroups being filled with a row of water molecules. Practically, the closed pore radius is not zero.

The mechanism of pore formation is based on the variation of BLM thickness due to the collective thermal motion, but it doesn't exclude the surface defects induced by lateral thermal motions. However, the mechanism of pore formation by thermal fluctuations still continue to be a matter of challenge and interesting future developments.

REFERENCES

[1] W. Helfrich, Z. Naturforsch., 28 (1973) 693.
[2] E. A. Evans and R. Skalak, Mechanics and Thermodynamics of Biomembranes, CRC Press, Boca Raton, 1980.
[3] S. B. Hladky and D. W. R. Gruen, Biophys. J., 38 (1982) 251.
[4] I. R. Miller, Biophys. J., 45 (1984) 643.
[5] S. B. Hladky, Biophys. J. 45 (1984) 645.
[6] E. Evans and W. Rawicz, Phys. Rev. Letters 64 (1990) 2094.
[7] J. N. Israelachvili, Intermolecular and Surface Forces, 2nd edn, Academic Press, New York, 1992.
[8] A. Ben-Shaul, In: R. Lipowsky and E. Sackmann (eds.), Structure and Dynamics of Membranes, 2nd edn, vol 1, Elsevier, Amsterdam, 1995.
[9] S. May, J. Chem. Phys., 103 (1995) 3839.
[10] S. May, J. Chem. Phys., 105 (1996) 8314.
[11] W. Rawicz, K. C. Olbrich, T. McIntosh, D. Needham and E. Evans, Biophys. J. 79 (2000) 328.
[12] E. H. DeLacey and J. Wolfe, Biochim. Biophys. Acta 692 (1982) 425.
[13] I. R. Elliott, D. Needham, J. P. Dilger and D. A. Haydon, Biochim. Biophys. Acta 735 (1983) 95.
[14] O. G. Mouritsen and M. Bloom, Biophys. J. 36 (1984) 141.
[15] J. R. Abney and J. C. Owicki, In: A. Watts and J. De Pont (eds.), Progress in Protein-Lipid Interactions, Elsevier, New York, 1985.
[16] H. W. Huang, Biophys. J., 50 (1986) 1061.
[17] P. Helfrich and E. Jakobsson, Biophys. J., 57 (1990) 1075.
[18] S. M. Gruner, In: L. Pelity (ed.), Biologically Inspired Physics, Plenum Press, New York, 1991.
[19] N. Dan, P. Pincus and S. A. Safran, Langmuir 9 (1993) 2768.
[20] M. F. Brown, Chem. Phys. Lipids 73 (1994) 159.
[21] N. Dan, A. Berman, P. Pincus and S. A. Safran, J. Phys. II France 4 (1994) 1713.
[22] A. Ring, Biochim. Biophys. Acta, 1278 (1996) 147.
[23] C. Nielsen, M. Goulian and O. S. Andersen, Biophys. J., 74 (1998) 1966.
[24] S. May and A. Ben-Shaul, Biophys. J., 76 (1999) 751.
[25] S. May, Eur. Biophys. J., 29 (2000) 17.
[26] M. M. Sperotto, Eur. Biophys. J., 26 (1997) 405.
[27] K. S. Kim, J. Neu and G. Oster, Biophys. J., 75 (1998) 2274.
[28] F. Brochard and J. F. Lennon, J. Physique, 36 (1975) 1035.
[29] R. Waugh and E. A. Evans, Biophys. J., 26 (1979) 115.
[30] T. M. Fischer, Biophys. J., 63 (1992) 1328.
[31] L. Miao, U. Seifert, M. Wortis and H.-G. Döbereiner, Phys. Rev., E 49 (1994) 5389.
[32] R. T. Tranquillo and W. Alt, J. Math. Biol., 34 (1996) 361.
[33] L. Movileanu and D. Popescu, BioSystems, 36 (1995) 43.
[34] L. Movileanu and D. Popescu, J. Biol. Systems, 4 (1996) 425.

[35] D. W. R. Gruen, S. Marcelja and V. A. Parsegian, In: A. S. Perelson, C. Delisi and F. W. Wiegel (eds.), Cell Surface Dynamics, Dekker, New York, 1984.

[36] J. N. Israelachvili, S. Marcelia and R. G. Horn, Quarterly Rev. Biophys., 13 (1980) 121.

[37] D. Popescu, Biochim. Biophys. Acta, 1152 (1993) 35.

[38] D. Popescu, Biophys. Chem., 48 (1994) 369.

[39] M. H. Saier, Jr. and C. D. Stiles, Biological Membranes, Springer Verlag, New York, 1975.

[40] D. Popescu and C. Rucareanu, Mol. Cryst. Liq. Cryst., 25 (1992) 339.

[41] E. Sackmann, In: R. Lipowsky and E. Sackmann (eds.), Structure and Dynamics of Membranes, Elsevier/North-Holland, Amsterdam, 1995.

[42] J. C. Shillcock and U. Seifert, Biophys. J., 74 (1998) 1754.

[43] J. F. Nagle and S. Tristram-Nagle, Curr. Op. Struct. Biol., 10 (2000) 474.

[44] C. Tanford, The Hydrophobic Effect: Formation of Micelles and Biological Membranes, John Wiley, New York, 1980.

[45] D. Popescu and G. Victor, Bioelectrochem. Bioenerg., 25 (1991) 105.

[46] M. Shinitzki and P. Hencart, Int. Rev. Cytol., 60 (1979) 121.

[47] I. G. Abidor, V. B. Arakelyan, I. V. Chernomondik, Yu. A. Chizmadzhev, V. F. Pastushenko and M. R. Tarasevich, Bioelectrochem. Bioenerg., 6 (1979) 37.

[48] R. Benz, O. Frőhlich, P. Lauger and M. Montal, Biochim. Biophys. Acta, 394 (1975) 323.

[49] R. J. Cherry and D. Chapman, J. Mol. Biol., 40 (1969) 19.

[50] J. P. Dilger, Biochim. Biophys. Acta, 645 (1981) 357.

[51] D. Popescu, and G. Victor, Biochim. Biophys. Acta, 1030 (1990) 238.

[52] L. Movileanu, D. Popescu and M. L. Flonta, J. Mol. Struct., 434 (1998) 213.

[53] L. Movileanu, D. Popescu, G. Victor and G. Turcu, Bull. Math. Biol., 50 (1997) 60.

[54] K. Jorgensen and O. G. Mouritsen, Biophys. J., 95 (1995) 942.

[55] K. Jorgensen, M. M. Sperotto, O. G. Mouritsen, J. H. Ipsen and M. J. Zuckermann, Biochim. Biophys. Acta, 1152 (1993) 135.

[56] D. Popescu, C. Rucareanu and G. Victor, Bioelectrochem. Bioenergetics, 25 (1991) 91.

[57] D. Zhelev and D. Needham, Biochim. Biophys. Acta, 1147 (1993) 89.

[58] D. Popescu and D. G. Margineanu, Bioelectrochem. Bioenerg., 8 (1981) 581.

[59] J. C. Shillcock and D. H. Boal, Biophys. J., 71 (1996) 317.

[60] J. D. Moroz and P. Nelson, Biophys. J., 72 (1997) 2211.

[61] J. C. Weaver and Yu. A. Chizmadzhev, Bioelectrochem. Bioenerg., 41 (1996) 135.

[62] G. Saulis, Biophys. J., 73 (1997) 1299.

[63] W. Sung and P. J. Park, Biophys. J., 73 (1997) 1797.

[64] E. Neumann, S. Kakorin and K. Toensing, Bioelectrochem. Bioenerg., 48 (1999) 3.

[65] L. K. Tamm, A. Arora and J. H. Kleinschmidt, J. Biol. Chem., 276 (2001) 32399.

[66] L. Movileanu, S. Cheley, S. Howorka, O. Braha and H. Bayley, J. Gen. Physiol., 117 (2001) 239.

[67] S. Howorka, L. Movileanu, X. Lu, M. Magnon, S. Cheley, O. Braha and H. Bayley, J. Am. Chem. Soc., 122 (2000) 2411.

[68] L. Movileanu, S. Howorka, O. Braha and H. Bayley, Nature Biotechnol., 18 (2000) 1091.

[69] L. Movileanu and H. Bayley, Proc. Natl. Acad. Sci. USA, 98 (2001) 10137.

[70] S. Howorka, L. Movileanu, O. Braha and H. Bayley, Proc. Natl. Acad. Sci. USA, 98 (2001) 12996.

[71] J. Konisky, Ann. Rev. Microbiol., 36 (1982) 125.

[72] R. Cassia-Moura, Bioelectrochem. Bioenerg. 32 (1993) 175.

[73] P. K. Kienker, X. Qiu, S. L. Slatin, A. Finkelstein, and K. S. Jakes, J. Memb. Biol., 157 (1997) 27.

[74] R. Cassia-Moura, A. Popescu, J. R. A. Lima, C. S. Andrade, L. S. Ventura, K. S. A. Lima and J. Rinzel, J. Theor. Biol., 206 (2000) 235.

[75] E. Wisse and D. L. Knook, In: Progress in Liver Diseases, vol. VI. H. Popper and F. Schaffner (eds.), Straton, Grune, 1979.

[76] D. Popescu, L. Movileanu, S. Ion and M. L. Flonta, Phys. Med. Biol., 45 (2000) N157.

[77] H. T. Tien and A. L. Ottova, J. Membr. Sci., 4886 (2001) 1.

[78] S. J. Singer and G. L. Nicholson, Science, 175 (1972) 720.

[79] P. G. de Gennes, The Physics of Liquid Crystals. Clarendon Press, Oxford, 1974.

[80] I. Pascher, M. Lundmark, P. G. Nyholm and S. Sundell, Biochim. Biophys. Acta, 1113 (1992) 339.

[81] H. Hauser, I. Pascher, R. H. Pearson and S. Sundell, Biochim. Biophys. Acta, 650 (1981) 21.

[82] C. H. Huang and J. T. Mason, Biochim. Biophys. Acta, 864 (1986) 423.

[83] D. Popescu and G. Victor, Biophys. Chem., 39 (1991) 283.

[84] J. D. Litster, Phys. Lett., 53A (1975) 711.

[85] J. C. Weaver and R. A. Mintzer, Phys. Lett., 86A (1981) 57.

[86] B. L. Silver, The Physical Chemistry of Membranes, Allen & Unwin, London, 1985.

[87] S. H. White, Biophys. J., 23 (1978) 337.

[88] E. Neher and H. Eibl, Biochim. Biophys. Acta, 464 (1977) 37.

[89] H. Engelhardt, H. P. Duwe and E. Sackman, J. Physique Lett., 46 (1985) L395.

[90] M. B. Schneider, J. T. Jenkins and W. W. Webb, J. Physique, 45 (1984) 457.

[91] L. Movileanu, D. Popescu, G. Victor and G. Turcu, BioSystems, 40 (1997) 263.

[92] M. Shinitzky, Biomembranes. Physical Aspects, Balaban Publishers VCH, Weinheim, 1993.

[93] P. Mueller, D. O. Rudin, H. T. Tien, W. C. Wescott, Nature 194 (1962).

[94] H. T. Tien and A. Ottova-Leitmannova, Membrane Biophysics as Viewed from Experimental Bilayer Lipid Membranes. Elsevier, Amsterdam and New York, 2000.

[95] H. Trauble, J. Membr. Biol., 4 (1971) 193.

Planar Lipid Bilayers (BLMs) and their Applications
H.T. Tien and A. Ottova-Leitmannova (Editors)

Chapter 6

Mechanoelectric properties of BLM

Alexander G. Petrov

Biomolecular Layers Department, Institute of Solid State Physics, Bulgarian Academy of Sciences, 72 Tzarigradsko chaussee, 1784 Sofia, Bulgaria

1. INTRODUCTION

This Chapter presents an account on some basic mechanoelectric properties of membranes: flexoelectricity, mechanocapacitance and mechanoconductivity. Their investigation became possible thanks to the BLM model, namely to the fact that a BLM represent a freestanding bilayer that is supported along its periphery only. Experimentally, the way to probe BLM mechanoelectricity is to subject a membrane to a pressure differential, either static or dynamic, or both. BLM models used by us comprise the classic, celebrated Mueller-Rudin-Tien-Wescott discovery of BLM painting a over a hydrophobic hole [1], as well as the tip-dip method where a microsize BLM is patch-clamped inside a hydrophilic borosilicate glass micropipette [2]. In the present review we concentrate on the first system.

Flexoelectricity [6] stands for curvature-induced membrane polarization (direct flexoeffect) or voltage-induced membrane curvature (converse flexoeffect), mechanocapacitance is lateral stress-induced capacitance variation, while mechanoconductivity means lateral stress-sensitive ion conduction of a membrane. Stress sensitivity is usually mediated by the presence of ion channels in the lipid bilayer, synthetic or native ones. On the other hand, flexoelectricity and mechanocapacitance can be observed even in unmodified lipid bilayers. All these properties are closely related to some vital function of native membranes, mechanosensitivity and electromotility. They involve two generalized degrees of the membrane, mechanical and electrical ones. Strictly speaking, mechanical degrees of freedom involved are two: curvature change and lateral stretching.

Furthermore, in photoactive BLM flexoelectricity becomes light-dependent, i.e. a photoflexoelectricity phenomenon can be demonstrated. This effect involves a further degree of freedom of the membrane, the optical one.

In the experiments it is easy to apply controlled amounts of pressure difference, either static or dynamic ones, and measure the current or voltage response of BLM. Quite often, though, these properties are superimposed in the resulting response. In the experimental arrangement of a BLM (and especially in a patch clamped membrane) the two mechanical degrees of freedom are coupled: inducing a curvature of the membrane by its blowing is inevitably related to a lateral stress variation. Nevertheless, since both degrees of freedom are of different symmetry (curvature variation is polar, while lateral stretch is not), usually by applying a phase-sensitive analyser it is possible to discriminate between the two contributions to the electric response (see below).

1.1. Theoretical remarks

The first experimental hints about the generation of a.c. currents by BLMs subjected to oscillating gradients of pressure were obtained in the early 1970s. When a BLM was clamped to certain voltage, these currents were related to mechanoconductivity [4] or mechanocapacitance [5], and were found to be 2nd harmonics of the driving pressure's frequency. Subsequently, we pointed out that the membrane c u r v a t u r e is also a variable in these experiments. Consequently, we related the accidental appearance of "subharmonic" current under zero voltage clamp to the oscillating flexoelectric polarization of the curved membrane [6]:

$$P_S = f(c_1 + c_2), \tag{1}$$

where $(c_1 + c_2)$ is the total membrane curvature, f is flexoelectric coefficient, measured in Coulombs [C] and P_S is flexopolarization per unit area [6-8].

Phenomenologically, the generation of a "subharmonic noise", i.e. a 1st harmonic a.c. current by a zero voltage-clamped BLM, follows from the appearance of flexopolarization. Assuming (for simplicity) oscillating spherical curvature and representing it in the form

$$c_1 + c_2 = c(t) = 2c_m \sin \omega t, \tag{2}$$

where ω is the angular frequency of oscillations and $c_m = 1/R_m$ is the maximal curvature, we will obtain time-dependent flexopolarization as well. Let us assume that polarization variations instantaneously follow curvature variations [3]:

$$P_S(t) = f c(t) = f 2 c_m \sin \omega t \ ,\tag{3}$$

Surface polarization leads to a transmembrane voltage difference (Helmholtz equation, Eq. (4)) that can be measured in an external circuit of two electrodes immersed in the bathing electrolyte on each side of the membrane, connected to a very high impedance electrometer (open circuit conditions, zero current clamp):

$$U_f(t) = \frac{P_s}{\varepsilon_0} = \frac{f}{\varepsilon_0} 2 c_m \sin \omega t = U_\omega \sin \omega t \ .\tag{4}$$

The flexoelectric voltage is a first harmonic with respect to curvature oscillation and its amplitude is $U_\omega = (f/\varepsilon_0) 2 c_m$.

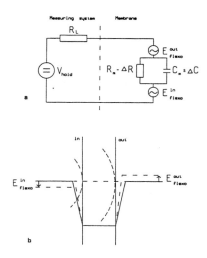

Fig. 1. a) Equivalent circuit of an oscillating membrane connected to a measuring system; b) Graph of the potential distribution across a planar and a curved membrane. E^{out} and E^{in} are two e.m.f. generators that modulate the surface potentials of the two surfaces of a curved membrane for the case of a positive flexoelectric coefficient, negative surface charge and zero intramembrane field, i.e., zero current clamp. Since the two surfaces' generators operate in counter-phase they can be combined in one flexoelectric generator U_f, as in text (from Ref. 3, with permission from the Publisher).

When the two electrodes are effectively shorted out *via* low impedance ammeter (shorted circuit conditions, zero voltage clamp), the displacement current through the meter can be calculated by adopting an equivalent circuit (Fig. 1), containing an a.c. voltage generator U_f (describing the oscillating flexoelectric voltage according to Eq. (4)) and a capacitor C (where C is the membrane capacitance):

$$I_f(t) = \frac{d}{dt}(CU_f) = \frac{dC}{dt}U_f + C\frac{dU_f}{dt}. \tag{5}$$

Since P_S is by definition the t o t a l membrane polarization due to both permanent and induced dipoles, the relevant membrane capacitance is simply $C_0 = \varepsilon_0 S_0/d$, where S_0 is the flat membrane area and d is the c a p a c i t i v e thickness of the membrane. When the membrane curvature oscillates, the membrane area will oscillate as well [7]:

$$S = S_0 + \Delta S\sin^2\omega t, \quad \text{where} \quad \Delta S = S_0 r^2 c_m^2/4 \tag{6}$$

for spherical curvature, with r being the membrane radius. The membrane capacitance will then also oscillate like

$$C = C_0 + \Delta C\sin^2\omega t. \tag{7}$$

The relationship between $\Delta C/C_0$ and $\Delta S/S_0$ is, in general, frequency-dependent (see below). With flexoelectricity we are mainly interested in the first harmonic of the response. Capacitance oscillations will contribute to higher harmonics. Therefore, from (4) and (5) we have for the 1st harmonic amplitude of the membrane flexoelectric current $I_\omega = f(C_0/\varepsilon_0)2c_m\omega$.

In this way, measuring U_ω and c_m, or I_ω, C_0 and c_m, we can determine experimentally the value of the phenomenologically introduced flexocoefficient f in Eq. (1). Measuring membrane curvature c_m or $(c_1 + c_2)$ is a separate problem to be discussed below.

We have shown in Ref. 3 that several electric multipoles (charges, dipoles, quadrupoles) of membrane components (lipids and proteins) contribute to the flexoelectric coefficient. Experimental elucidation of the importance of different multipoles by their independent variation (especially the partial electric charge per lipid head, β, expressed in percent of proton charge) is of substantial interest. Another important subject is the polarization relaxation as it relates to the lipid exchange rate. Such information can be obtained from the frequency dependence of the flexoelectric response.

Lipid exchange can be free or blocked. We have proven [3] that lateral lipid exchange within each monolayer with the torus of a BLM is equivalent to transbilayer lipid exchange (flip-flop). Otherwise, flip-flop is a slow process (half time of hours). When lateral exchange is blocked, polarization-related processes (head group conformational changes, adsorption/desorption of counterions, proton equilibrium shift) are much faster than the minimum oscillation period of 1 ms used in our studies. Much faster (*ca.* 1 MHz) also is the relaxation rate of the double electric layer. That relaxation rate governs the establishment of a displacement current in the external circuit. Therefore, we can assume that polarization does indeed follow curvature variations instantaneously and Eq. (2) does hold true. This means that Eqs. (3) and (9) hold for the high frequency case (blocked exchange), the flexoelectric coefficient f in them being actually f^B.

In the low frequency limit, $\omega\tau < 1$, where τ is the relaxation time of the lateral exchange, Eq. (2) no longer holds. Time-dependent analysis has been performed [9, 10], showing that now both U_f and I_f feature an in-phase and a quadrature component [3]:

$$U_f(t) = \frac{1}{\varepsilon_0}\sqrt{\frac{\left(f^F\right)^2}{1+\omega^2\tau^2} + \frac{\left(f^B\right)^2\omega^2\tau^2}{1+\omega^2\tau^2}}\, 2c_m \sin\left(\omega t + \phi\right) = \frac{f(\omega)}{\varepsilon_0}c(t), \qquad (8)$$

$$I_f(t) = \frac{C_0}{\varepsilon_0}\omega\sqrt{\frac{\left(f^F\right)^2}{1+\omega^2\tau^2} + \frac{\left(f^B\right)^2\omega^2\tau^2}{1+\omega^2\tau^2}}\, 2c_m \sin\left(\omega t + \phi + \frac{\pi}{2}\right), \qquad (9)$$

where $\tan\phi = \dfrac{\left(f^B - f^F\right)\omega\tau}{f^B\omega^2\tau^2 + f^F}$, f^F and f^B are free and blocked flexocoefficient, respectively, and $f^B > f^F$.

2. FLEXOELECTRIC CURRENT RESPONSE OF A BLM

Our first experiments, aimed specifically at registering the flexoelectricity of
BLMs from biological lipids (egg yolk lecithin (EYPC) [11], bacterial
phosphatidylethanolamine (PE) [9]) were performed in the current registration
regime by using a low impedance current-to-voltage converter. The influence of
some modifiers of the surface charge and surface dipole, as well as of the
membrane conductivity, upon the value of the effect was also studied.

2.1. Lipids. Surface charge and dipole modifiers

BLMs were formed by the classical method [1] on a 1 mm conical hole in
a Teflon cup, immersed in a spectrophotometric glass cuvette, at room
temperature. Bacterial phosphatidylethanolamine (PE) *ex. E.coli* (Koch Light)
dissolved in *n*-decane (20 mg/ml) was used as a membrane-forming solution. As
a bathing electrolyte either unbuffered KCl, 10 mM to 1 M (pH 6.2) or buffered
KCl with higher or lower pH values was used (KH_2PO_4 pH 3.0; MES pH 6.0;
TRIS pH 8.5; CHES pH 9.3; KH_2PO_4 pH 10.4). The surface charge of the BLM
was varied either by pH variation or by adsorption of the ionic detergent
cetyltrimethylammonium bromide (CTAB, Chemapol) added to the electrolyte
as a 1 mM water solution. The surface dipole was varied by adsorption of the
dipole modifier phloretin (Serva), added as a 10 mM solution in ethyl alcohol.

2.2. Method for excitation of membrane curvature

Membrane curvature oscillations were excited by the method first
employed in Ref. 4. The experimental set-up is shown in Fig. 2. The Teflon cup
with a very small internal volume (0.25 cm^2) was closed by a Teflon cap. Two
reversible Ag/AgCl electrodes were used for current measurements. One of
them was mounted in the cup; another one was immersed in the glass cuvette.
Through an opening in the cap connected to a flexible pneumatic pipeline,
oscillating air pressure was applied to the electrolyte surface and was transferred
in this way to the membrane. An electrically shielded earphone generated air
pressure. A function generator fed the earphone. The applied frequency was in
the range of 10 to 1000 Hz. Using a T-pipe, an open branch filled with cotton
wool was introduced into the pipeline in order to equilibrate any static pressure
difference between the two membrane sides while transferring the dynamic air
pressure.

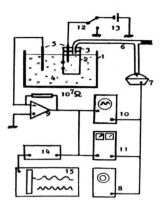

Fig 2. Scheme of the experimental set-up for current registration: (1) Glass cuvette; (2) Teflon cup; (3) Teflon cap; (4) Electrolyte: 10 mM to 1 M KCl, pH 6.2; (5) Reversible electrodes, Ag/AgCl; (6) Plastic pipeline; (7) Ear-phone; (8) Function generator, Textronics FG504; (9) Current-to-voltage converter, Keithley 427; (10) Oscilloscope, C1-78; (11) Phase sensitive analyzer, PAR 5204; (12) Switch; (13) D.c. voltage source; (14) Boxcar averager, PAR Model 162; (15) X-Y recorder (from Ref. 9, with permission from the Publisher).

2.3. Evaluation of membrane curvature by mechanocapacitance

After spreading the membrane, its planarity was adjusted by varying the solution level in the cuvette, and membrane capacitance was measured. The first important point in the study of curvature-electric effects is to evaluate membrane curvature and to assure its constancy at each frequency applied, in order to study the frequency dependence of the effects. To this aim, we made use of the "capacitance microphone effect" [5]. Briefly, a small voltage difference $\Delta\varphi$ (100 mV) was applied to the membrane and the displacement current due to the oscillations of the membrane capacitance was recorded. From the amplitude of the capacitance oscillation the curvature was evaluated.

Area oscillations of an initially flat membrane obey Eq. (7), i.e., for o n e period of curvature oscillation membrane area reaches a maximum t w i c e. Membrane capacitance oscillations follow the same type of equation, Eq. (8). However, the relationship between $\Delta C/C_0$ and $\Delta S/S_0$ is frequency-dependent [11, 12]. In general,

$$\frac{\Delta C}{C_0} = \frac{\Delta S}{S_0} Q(\omega) \ , \tag{10}$$

where the shape of the factor $Q(\omega)$ depends on the interrelation of the elastic and viscosity parameters of the torus and the BLM. Three particular regimes of membrane bulging could be recognized in this respect: slow, $Q(\omega)=1$, medium, $Q(\omega)=2\alpha$, and fast, $Q(\omega)=2$.

Medium bulging is mostly relevant for flexoexperiments. It occurs when the frequency is too high to expect thinning down to BLM thickness of the stretched portion of the torus, ΔS, but not high enough for torus viscosity to be effective. When an incompressible membrane is stretched, an equal relative decrease of its thickness must take place (volume conservation). Respective capacitance change increases by a factor of two. The percentage of BLM stretching is governed by the ratio, α, of the torus elasticity to the sum of both BLM and torus elasticity. Under ambient temperature it holds that $\alpha \ll 1$. Usually, values of $\alpha \approx 0.05$ are quoted. The medium bulging regime displays a mid-frequency plateau in the capacitance response. The low frequency limit is about 0.1 Hz. The high frequency limit is in general higher than 100 Hz, and (depending on the torus viscosity) can reach 1000 Hz.

Now, with $\Delta\varphi$ applied the charge on the membrane capacitance is $q = \Delta\varphi C$ and the current is (in view of Eq. 7)

$$I = \Delta\varphi\frac{dC}{dt} = \Delta\varphi\Delta C\omega\sin 2\omega t = I_{2\omega}\sin 2\omega t . \qquad (11)$$

The current is the second harmonic with respect to the vibration and its amplitude is $I_{2\omega} = \Delta\varphi\Delta C\omega$. Consequently, a calibration of $I_{2\omega}$ following a linear growth with the frequency will assure a constant value of ΔC. The desired value of ΔC was adjusted by increasing the amplitude of the driving signal, i.e., by increasing the peak volume enclosed in the bulged membrane segment. The actual bulging regime was assumed to be a medium one in the whole range from 10 Hz to at least 400-500 Hz. From Eqs. (6) and (10) it follows that the constant curvature maintained by our procedure is

$$c_m = \sqrt{\frac{\Delta C}{2\alpha C_0}}\frac{2}{r} , \quad \text{where} \quad r = \sqrt{\frac{C_0 d}{\pi\varepsilon\varepsilon_0}} . \qquad (12)$$

The flat membrane radius r is accurately determined from the measured membrane capacitance if membrane thickness d is known from previous area-capacitance measurements.

With typical values of the membrane capacitance of 2.5 nF, calibrated value of capacitance amplitude of 2.25 pF, membrane radii of about 0.4 mm and membrane thickness of PE/n-decane membrane of 4.2 nm, by assuming $2\alpha \approx$ 0.1, we can estimate the radii of curvature, $R_m = 1/c_m$, as about 2 mm. This is a quite substantial curvature, if compared to the flat membrane radius.

2.4. Flexoelectric coefficient of bacterial PE membranes

Boxcar-averaged traces of the curvature-electric currents and "capacitance microphone" currents of oscillating membranes demonstrate the doubled frequency of the capacitance response (recorded with 100 mV membrane voltage), the fundamental frequency of the flexoelectric response (recorded with zero membrane voltage) and the phase shift between them.

The frequency characteristics of the flexoelectric effect for several membranes in 50 mM KCl are shown in Fig. 3. Experimentally, the existence of two frequency regions (corresponding to free and blocked lipid exchange) is evident. The median frequency between them is about 150 Hz.

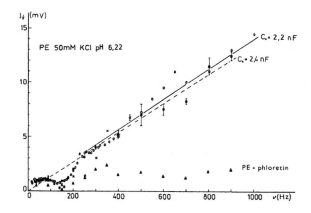

Fig. 3. Frequency dependence of the amplitude of flexoelectric effect. I_v represents the RMS value of the flexoelectric current, voltage-converted with 10^7 V/A amplification factor. BLM is formed from *E.coli* PE/n-decane. The electrolyte is unbuffered KCl, 50 mM, pH 6.22. Experimental points: ($\bullet$) flat BLM capacitance C_0 = 2.4 nF; (O) C_0 = 2.2 nF; $\times$, C_0 = 2.0 nF; (Δ) phloretin-modified BLM: 40 nM phloretin in the bathing electrolyte (from Ref. 9, with permission from the Publisher).

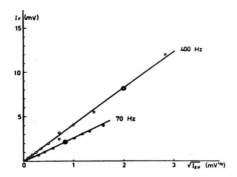

Fig. 4. Amplitude dependence of the flexoelectric effect. Membrane curvature is proportional to $\sqrt{I_{2V}}$ (Eqs. 12 and 11). (•) oscillation frequency 70 Hz; (O) oscillation frequency 400 Hz. (◎) calibrated curvatures employed for the frequency characteristics in Fig. 3 (from Ref. 9, with permission from the Publisher).

In the low frequency region a slight increase of the response from 10 to 50 Hz is followed by a plateau up to 120 Hz and by a marked drop of the response with some membranes around 150 Hz.

In the high frequency region a more or less linear growth of the response is observed. Extrapolation of the high-frequency straight line goes approximately to zero. Qualitatively, the shapes of the characteristics correspond to the theory, Eq. (9), although the deviations from the straight line in the low frequency range are somewhat stronger and the initial plateau is unexpected. Calculation of the flexocoefficient is given below.

Fig. 4 shows the amplitude characteristic of the flexoelectric effect at 70 Hz (below the median frequency) and at 400 Hz (above the median frequency). The flexoelectric response I_V (with $\Delta\varphi = 0$) as a function of the square root of the second harmonic $\sqrt{I_{2V}}$ (with $\Delta\varphi = 100$mV) is plotted. $\sqrt{I_{2V}}$ is taken as a variable, because the amplitude of the membrane curvature, c_m, is proportional to it, according to Eqs. (12) and (11). Both characteristics are linear down to very small curvatures, as the nature of flexoelectricity requires. Deviations from linearity can be seen at higher curvature amplitudes only. We noted, e.g., that at the last three points of 70 Hz line overtones were excited. Double circles show the calibrated values for the frequency characteristic in Fig. 4. These are well within the linear region.

Phase measurements of capacitance currents from pre-curved membranes were performed with the aim of determining the sign of the flexoelectric coefficient in the two frequency regions (at 70 and 400 Hz). Both capacitance and flexoelectric currents should be in phase when curvature-induced polarization is directed the same way as the transmembrane electric field. For both frequencies this took place when the electric field pointed towards the centre of the membrane curvature. This means that the flexoelectric coefficient was negative in both frequency regions [9].

In the high frequency limit (blocked lateral exchange) the flexoelectric coefficient f^B can be calculated from the peak value of the flexoelectric current at a given frequency v (or from the slope $I_v/v = 2\pi f(C_0/\varepsilon_0) 2c_m$) by means of Eq. (12):

$$f^B = \frac{(I_v/v)}{8\pi} \sqrt{\frac{2\alpha\varepsilon\varepsilon_0 d}{\pi\Delta C}} .$$
(13)

With a slope of 2 pA/Hz (Fig. 3), $2\alpha \approx 0.1$, $\varepsilon = 2$, $d = 4.2$ nm and $\Delta C = 2.25$ pF we obtained $f^B = -25.5 \cdot 10^{-19}$ C. The relative error was estimated to be 72.5%, since an error of the adopted value of α as high as 100% is quite possible.

The low frequency limit is not fully understood; therefore, from the shape of the frequency curve it can only be estimated that $f^F < f^B$.

In order to study the influence of the ionic strength, electrolyte solutions of non-buffered KCl 10 mM, 50 mM and 1 M were employed. In the low frequency range the only noticeable influence was upon the median frequency, which seemed to increase with decreasing ionic strength. In the high frequency region a marked correlation between the slope of the frequency characteristic (i.e., the value of the flexocoefficient) and the inverse ionic strength was found. The frequency dependence with 10 mM electrolyte was actually nonlinear, with gradually increasing slope. For comparison, the slope of 1 M curve was the lowest and the generated responses were strictly sinusoidal.

The dipolar modifier phloretin is known to decrease the surface potential of lipid head dipoles by about 100 mV because of its oppositely directed dipole moment. With the symmetric addition of 40 μl of 10 mM ethanol solution to 10 ml bathing membrane electrolyte before BLM formation, its amplitude-frequency characteristic demonstrated a drastic drop of the flexoresponse in the high frequency region and reduction of the overall slope by more than 7 times (Fig. 3). Unilateral addition to a pre-formed membrane did not produce any definite effect on the first harmonic, but led to the appearance of the second harmonic; this was clearly a capacitance microphone effect from the surface potential difference of the asymmetric membrane.

3. FLEXOELECTRIC VOLTAGE RESPONSE OF A BLM

Further experiments on membrane flexoelectricity were performed in the regime of voltage registration (open external circuit) by using a high impedance selective nanovoltmeter. These experiments provided a check of surface charge contribution in the presence of univalent or divalent ions [13].

3.1. Lipids and modifiers

BLMs were formed from chromatographically pure lecithin extracted from egg yolk (EYPC). The membrane-forming solution was made by dissolving lecithin in *n*-decane, in concentration 40 mg/ml. A drop of this solution was applied by a glass pipette to a 1 mm pretreated orifice in a Teflon plate, partitioning a Teflon chamber into two compartments. In some experiments mixed BLMs of lecithin with negatively charged phosphatidyl serine (PS, Serva, 20 mg/ml chloroform solution) were formed. The necessary amount of PS solution was dried on a piece of glass coverslip and then dipped in the membrane-forming solution EYPC/*n*-decane until completely dissolved.

EYPC BLMs modified by uranyl acetate (UA, Merck) were prepared by adding different amounts of concentrated UA solution in water to the bathing electrolyte. As a bathing solution either unbuffered saline (0.1 M NaCl in distilled water) or a saline with phosphate buffer (pH 6.0) was used. Strong adsorption of UO_2^{2+} ions on the bilayer surface is expected, thus resulting in a large positive surface charge depending on the UA concentration.

The purity of EYPC was controlled by thin layer chromatography on Merck plates. Surface charge was estimated by measurements of the electrophoretic mobility of large multilamellar liposomes prepared after Bangham from EYPC in 0.1 M NaCl. Measurements were performed using Mark II (Rank Brothers) electrophoretic equipment.

3.2. Experimental set-up

The scheme of the experimental set-up is shown in Fig. 5. A two-compartment Teflon chamber (after Ref. 5, with some modifications) was used. The front compartment was open to the air and had a glass window on its front face, enabling the observation of membrane formation and thinning in reflected

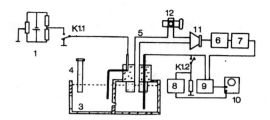

Fig. 5. Scheme of the experimental set-up for voltage registration: (1) D.c. voltage source; (2) coupled switches (K1.1 and K1.2); (3) Teflon measuring cell; (4) Levelling piston (5) Plastic pipeline; (6) Amplifier; (7) Function generator; (8) Selective nanovoltmeter Unitra 237; (9) Lock-in nanovoltmeter Unitra 232; (10) Oscilloscope, Textronics 2246A; (11) Compression loudspeaker; (12) Valve (from Ref. 13, with permission from the Publisher).

light. The rear compartment was closed by a Teflon cap carrying the two reversible Ag/AgCl electrodes and a pipeline transmitting oscillating air pressure generated by a compression loudspeaker. A Teflon plate with a 1 mm cone-shaped orifice was pressed between the two compartments and used as a BLM support. Membrane planarity was controlled by Teflon leveling piston immersed in the front compartment and attached to a micrometer screw. It was also used to deliberately impose a membrane curvature for sign determination of the flexocoefficient.

Membrane thinning was followed electrically by the membrane capacitance increase over time. Since the dynamics of membrane oscillations depends on α, the ratio of the torus elasticity to the sum of both BLM and torus elasticities the aim was to study BLMs with possible smaller tori. Such membranes displayed capacitances of more than 2.5 nF; therefore, BLMs of lesser capacitance were discarded. Flexoelectric potentials generated by the oscillating BLMs were registered by a lock-in nanovoltmeter (Fig. 5).

3.3. Measuring procedure

After thinning of the BLM, measuring its capacitance C_0 and ensuring its planarity visually and electrically (planarity corresponded also to a capacitance minimum), the loudspeaker was put into operation at a particular frequency, carefully increasing its driving voltage from zero. The input of the selective nanovoltmeter, shorted out by a 1 MΩ resistor, was connected to the rear electrode by a K 1.2 switch; the selective nanovoltmeter was adjusted to the second harmonic of the driving frequency and a small d.c. bias (100 mV) was applied across the BLM by a K 1.1 switch. In this manner the potential drop of the "capacitance microphone" current was measured. We used the same principle of calibration as above, i.e., adjusting second harmonic amplitude to obey a relationship $U_{2v}^{rms}[V] = v[Hz]/10^7$. Practically, this meant 10 μV at 100 Hz, 50 μV at 500 Hz, etc. In this manner we assured a frequency-independent amplitude of the capacitance oscillation, 10 times lower than before, i.e., $\Delta C = 0.225$ pF. Assuming again a medium bulging regime for an unmodified BLM in the whole frequency range with $\alpha \approx 0.05$, we evaluated from Eq. (12) $R_m = 9$ mm for a BLM with C_0 of 3 nF. The UA-modified BLM probably entered the fast bulging regime above 300 Hz. In this regime the calculation gave a definite value of $R_m = 40$ mm for the same ΔC.

Having adjusted the second harmonic amplitude, the transmembrane voltage bias was set at zero and the lock-in was connected in order to measure the frequency dependence of the flexoelectric voltage. From the plateau value U_{rms}^B of this voltage, the blocked flexocoefficient can be calculated from Eq. (4)

$$f^B = \varepsilon_0\sqrt{2}U_{rms}^B\frac{R_m}{2} = \varepsilon_0\sqrt{2}U_{rms}^B\sqrt{\frac{2\alpha C_0}{\Delta C}}\frac{r}{4}. \tag{14}$$

Determination of the sign of the flexoelectric coefficient was made with pre-curved BLM. Since in that case the capacitance current I_C and flexocurrent I_P have an identical time-dependent part, by simple comparison of the amplitude of I_P at $\Delta\varphi = 0$ and the amplitude of $I_P + I_C$ at a definite sign of $\Delta\varphi$ and ΔC we can judge the sign of ΔP, i.e., determine the sign of f^B. In the free exchange case an additional phase difference arises between curvature and polarization, precluding a definite conclusion about the sign of f^F from such a simple method.

3.4. Flexoelectric coefficient of egg yolk PC membranes

From electrophoretic measurements of the surface charge of EYPC bilayers yield a negative sign of the surface charge and a rather small value of the electrophoretic charge density. We could estimate a value of partial charge per lipid head $\beta = -0.4\%$ for a freshly opened lecithin ampoule. A partial charge $\beta = -4.6\%$ was found for an older sample, three months post opening.

The frequency dependence of the amplitude of the flexoelectric response of pure EYPC membranes is shown in Fig. 6. The graph represents the data

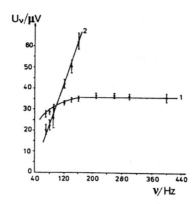

Fig. 6. Frequency dependence of the flexoelectric voltage of BLMs. Curve 1: egg yolk lecithin/n-decane BLM (averaged data from 21 BLMs). Curve 2: egg yolk lecithin + 2% mol phosphatidyl serine/n-decane BLM (averaged data from 4 BLMs). Electrolyte: unbuffered 0.1 M NaCl (from Ref. 13, with permission from the Publisher).

average over 21 membranes. One can see that above 160 Hz the response is essentially frequency-independent. This is in accordance with Eq. (8) and makes it possible to draw the conclusion that blocked lipid exchange prevails above 160 Hz, quite the same frequency as in the current registration regime. The low-frequency plateau corresponding to the free exchange case could not be revealed because of experimental problems below 70 Hz. In any case, the curve shape from 70 to 160 Hz again demonstrates that $f^F < f^B$, as predicted by the theory. From the plateau value of the response of each individual membrane we calculated the blocked flexocoefficient, Eq. (14), assuming $\alpha \approx 0.05$. The mean value from the 21 measurements is $f^B = (26.5 \pm 5.5) \cdot 10^{-19}$ C. The sign of this coefficient turned out to be positive (see below). The uncertainty of the final result is actually higher. It depends on the provisional choice of α. Since a relative error of 100% is quite admissible, then only this source of error would increase the relative error of f^B up to 50%.

The data from mixed EYPC + 2 mol% PS BLMs also are given in Fig. 6. The points are averaged over 4 measurements. These measurements were complicated by the shorter lifetime of the BLM in the presence of PS, so a high frequency plateau was not revealed. However, the increased response at higher frequencies is evident.

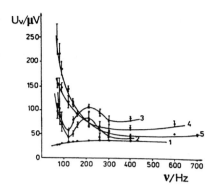

Fig. 7. Frequency dependence of the flexoelectric voltage of BLMs modified by various concentrations of uranyl acetate (UA) in the electrolyte (0.1 M NaCl). Curve 1: no UA (same as curve 1, Fig. 8). Curve 2: 1 mM UA. Curve 3: 3 mM UA. Curve 4: 5 mM UA. Curve 5: 10 mM (from Ref. 13, with permission from the Publisher).

The data for UA-modified EYPC membranes are shown in Fig. 7 for various UA concentrations in the bathing electrolyte, ranging from 1 to 10 mM. In the upper part of this range BLMs were markedly stabilized and measurements could be extended up to higher frequencies. The results with 1 and 3 mM UA differ from those with 5 and 10 mM. In any case, modified BLMs give a higher response than unmodified ones for all frequencies. From the plateau values above 300 Hz under the condition of fast bulging ($\alpha = 2$ in Eq. (14)) flexocoefficients f^B in the range of 120 to $200 \cdot 10^{-19}$ C were calculated, with a negative sign (see below); the minimum value was at 1 mM UA, the maximum at 3 mM. Since with fast bulging the error from $Q(\omega)$ is eliminated, the relative error of these results is not higher than 10%. Hence: $f^B = 120 \cdot 10^{-19}$ ($\pm 10\%$) C.

Finally, we describe the results for the sign of the flexocoefficient. Measurements were performed at 300 Hz and $\Delta\varphi = 40$ mV (rear electrode positive). With EYPC membranes the comparison of the flexoelectric amplitude to that of the sum of the flexoelectric and capacitance amplitudes revealed an increase at switching $\Delta\varphi > 0$ on and a positive (outward) initial curvature, or a decrease at switching $\Delta\varphi > 0$ on and a negative (inward) initial curvature. This means that at positive static curvature both currents have the same phase, while at negative static curvature the phases are opposite. Since in the first case

$\Delta\varphi\Delta C > 0$, while in the second case $\Delta\varphi\Delta C < 0$, from both experiments it follows that $\Delta P > 0$, i.e., $f^B > 0$. This positive sign is opposite to the sign of the surface charge established electrophoretically.

Repeating the experiment with the UA-modified BLM where recharging of the surface takes place, we see just the opposite behavior, showing that $f^B < 0$ in the whole UA concentration range studied. This negative sign is also opposite the positive surface charge due to the strong UO_2^{2+} ion adsorption.

4. STROBOSCOPIC INTERFEROMETRY

So far, BLM curvatures were evaluated indirectly from the second harmonic capacitance responses by assuming, necessarily, completely spherical deformation of an initially flat BLM. A new development was later concentrated on the actual membrane shape under oscillating pressure.

4.1. Membrane-forming lipids

Glyceryl monooleate (Sigma), egg yolk phosphatidylcholine (Sigma), bovine brain phosphatidylserine (Sigma), n-decane (purity 99.99%, Aldrich), and potassium chloride (KCl, purity 99.9%, Fisher), were used as received. Water was distilled and then purified by a Millipore (Milli-Ro, Milli-Q) system with final filtration by a Milli-pak filter (0.22 μm pore size). Final water resistivity was 18 MΩ.cm. Ag/AgCl electrodes (Ag wire, 1 mm, purity 99.9%, Sigma) were freshly prepared each week. The pH of the electrolyte solution was measured before each experiment. Electrolyte solutions were buffered to the desired pH values (Fisher Scientific SB101-500 pH 4.0±0.01, SB107-500 pH 0±0.01, and SB115-500 pH 10.0±0.01).

The BLM-forming solution had a concentration of 40 mg GMO (or PC) per ml of decane or 25 mg PS per ml of decane. BLMs were formed in either a 0.30±0.1mm or a 1.0±0.1mm hole, punched into a Teflon (Crucible, NY) plate (thickness 0.75 mm). The Teflon plate was clamped diagonally in a black Teflon (Crucible, NY) chamber designed especially for interferometric experiments. BLMs were formed using a modification of the "brush" method deposition of the BLM-forming solution across the Teflon aperture was delivered by the piston of a Hamilton syringe that had been bent to an appropriate angle. Membrane formation and thinning were monitored by interferometry and capacitance. Capacitance was measured by an RLC Digibridge (1689 M, Genrad).

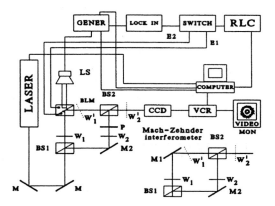

Fig. 8. Schematics of the experimental set-up used for simultaneous electrical and real-time stroboscopic interferometric measurements. A Quantronics 4116 Q-switched, mode-locked and pulse-picked Nd:YAG laser provided second harmonic 532 nm, 120 ps, 40 µJ pulses at 50 - 5000 Hz repetition rate. A computer-driven Hewlett-Packard 8116 A function generator (GENER) was used to excite membrane oscillations. Real-time BLM movements were observed by stroboscopic effect (see text). The lock-in was a Dynatrac 391 A Ithaco lock-in amplifier. Key: RLC, 1689 M GENRAD Digibridge; LS, a piezotransducer (Radio Shack, Cat. No. 273-073); BLM, the cell compartment for the bilayer lipid membrane; BS1 and BS2, beam splitters; M1 and M2, mirrors; CCD, an NEC colour camera; VCR, a commercial VHS video recorder; VIDEO MON, NEC colour monitor; and COMPUTER, a Hewlett-Packard IBM-compatible PC. For comparison, the scheme of a simple Mach-Zehnder interferometer is included (from Ref. 16, with permission from the author).

4.2. Experimental set-up

Schematics of the experimental set-up used for synchronized electrical and real-time stroboscopic interferometric measurements of BLMs are shown in Fig. 8. Curvature oscillations were excited by a piezotransducer generating oscillating air pressure in the range of 100 to 800 Hz with constant amplitude when fed by the function generator. The transducer was attached to the chamber by a short pipeline (3 cm) and the air pressure was applied to a thin flexible wall of the chamber (a rubber membrane), thus being transformed into oscillating hydrodynamic pressure that then acted to excite BLM oscillations. In this way, we avoided the standing wave and vibration problems that were encountered with longer pipelines and obtained vibration-free stroboscopic images.

The second harmonic pulses (532 nm, 120 ps, 40 µJ, 50 - 5000 Hz repetition rate) of a Quantronics 4116 Q-switched, mode-locked and pulsed-picked NdYAG laser were used for optical interferometry. The laser beam was divided by a beam splitter (BS1). The reference beam was reflected by a mirror (M2) through a polarizer (P) toward a second beam splitter (BS2), as shown in Fig. 8. The intensities of the object and reference beams were balanced by P during thinning and subsequent measurements of the optical interference fringes. The object beam was directed at the BLM, which acted as a black mirror and reflected some 0.01% of the incident light to BS2. The superimposed reference and object beams were imaged onto the CCD camera target.

This arrangement is analogous to that used in the Mach-Zehnder interferometer, with the exception that mirror 1 was replaced by the BLM (Fig. 8, inset). Changes in the optical path of the object beam after reflection from the deformed BLM were manifested in the appearance of an interference pattern. Perfect parallel alignment of the reference and object beams was found to be absolutely essential for the observation of interference fringes. This requirement is difficult to meet since BLMs never position themselves exactly in the same place in the aperture, and consequently, they reflect the incident beam at random. Strategically, it was simpler to align the BLM in the optical path. This was achieved with a horizontally positioned rotator (RSA-1, Newport) that rotated the BLM along the Z axis (normal to the optical table top). The rotator was mounted on a translator assembly that moved the BLM in the X-Y-Z direction. The rotator and translators were bolted together and placed under a Gimball mount (7330, Inrad), which allowed tilting and a small rotation along the Z axis. The entire BLM set-up was mounted on a vibration isolation table and covered by a Faraday cage.

For static images, synchronization was accomplished by keeping the frequency of the oscillating pressure the same as that of the laser pulses. Adding a delay to the synchronization signal (coming from the generator) allowed the observation of the oscillating BLM at different positions.

Real-time movements of the BLM were most conveniently observed as follows: the mode-locked frequency of the laser (41.0007 MHz) was fed into the computer and then divided appropriately to the frequency range of membrane oscillations (100-1000 Hz). This new frequency was returned to the laser to serve as the Q-switch and pulse-picking system frequency. On the other hand, the computer fed the function generator with a slightly different frequency in order to obtain the stroboscopic effect. With a piezotransducer driven by the function generator, a slowly oscillating fringe pattern was produced.

The apparent rate of oscillations was slowed by stroboscopy so that optical interferograms could be continuously videorecorded at the normal speed of a commercial VHS recorder (every 33 ms, a BLM interferogram was recorded). A time sequence of images was read from the videotape (played in slow motion) into the computer memory using a PC-Vision Plus frame-grabber (Imaging Technology, Inc.). Playing the videotape in slow motion and acquiring a new image each time a fringe was created at, or moved past, the center accomplished height correlation among the images. Thus, the absolute order of the most central fringe increased (or decreased, depending on the direction of movement) by one from each image to the next, indicating its absolute height (the procedure of computer analysis of the interferograms is described in detail in Ref. 14).

ΔZ_x and ΔZ_y were defined as the height differences between BLM center and edge along the x and y axes, respectively. The x axis was chosen as horizontal and the y axis as vertical (as seen by the video camera). In fact, curvatures along any two orthogonal axes could be measured, as $c_x + c_y$ is an invariant. Curvatures at any BLM position are calculated from

$$c_{mj} = \frac{1}{R_{mj}} = \frac{2\Delta Z_{mj}}{\Delta Z_{mj}^2 + r^2} \, , \quad m = x, y; \; j = 1, 2 \, . \tag{15}$$

When calculating the flexoelectric coefficient, ΔZ_x and ΔZ_y values are needed for two extreme BLM positions in order to calculate the curvature amplitude. The two extremes of a BLM oscillation were taken as ΔZ_{x1}, ΔZ_{y1} and $\Delta Z_{x2}, \Delta Z_{y2}$. Now, the flexoelectric coefficient can be calculated by the equation

$$f = \varepsilon_0 \frac{2\sqrt{2} U_f^{rms}}{\left[\left(c_{x1} + c_{y1} \right) - \left(c_{x2} + c_{y2} \right) \right]} \, . \tag{16}$$

For comparison, along with stroboscopic measurements of curvature, capacitance measurements by the second harmonic were also performed, as described above.

Analysis of the errors of this new method was given in [15]. The final error of the flexoelectric coefficient was determined to be 25%. The reproducibility between different BLM preparations was of the same order.

4.3. Flexoelectric coefficients of GMO, egg yolk PC and bovine brain PS BLM

It is important to reemphasize the experimental limitations of the two different methods (as described above) to measure flexoelectric coefficients. Capacitance method is indirect, contains a poorly accessible parameter α and requires the assumption of completely spherical membrane deformation. It permits, however, a much greater flexing amplitude of the BLM (i.e., the application of much higher oscillating pressure) than that possible in the optical method. Greater flexing is likely to produce a more homogeneously spherical deformation of the BLM, which would be an advantage in the stroboscopic method, too (see below). However, flexing a BLM to a curvature greater than 330 m^{-1} (i.e., ΔZ_x and $\Delta Z_y > 15$ µm) results in the disappearance of the concentric optical interference fringes and, therefore, renders the optical method unusable. An inevitable consequence of smaller BLM flexing is a decrease of the capacitance second harmonic U_{2v} below a detectable level. Limitations of these two methods have precluded the determination of curvature, and hence the flexocoefficient, by simultaneous capacitance and stroboscopic interferometric measurements on the same BLM in the present set of investigations. Under the conditions of capacitance measurements, interference fringes could not be observed, and under the conditions of interferometric measurements U_{2v} could not be measured. Fortunately, curvature-induced transmembrane potentials (U_v) were within the range of adequate sensitivity, in both the capacitance and interferometric method. Taking these limitations, we believe that the interference method provides more precise values for the flexoelectric coefficients than those obtained by the indirect capacitance measurements, especially in the low frequency range.

The surface charge of GMO and EYPC BLM were determined by voltage-dependent capacitance measurements under an ionic strength gradient. We obtained $\beta = -18.8\%$ for EYPC. For GMO membranes measurements demonstrated complete absence of any surface charge with this dipolar lipid.

Selected interferograms were obtained by capturing distinct videorecorded images from those replayed in slow motion. Height differences between the center of the BLM and its edge along the x and y-axis were calculated (Fig. 9). As seen, the middle position of this membrane was not flat, but rather, saddle-shaped. This was a very typical situation with most of the membrane preparations. Probably, it is due to imperfections of the orifice and small wetting irregularities of the Teflon film.

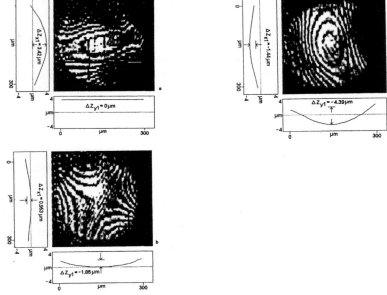

Fig. 9. Typical computer-enhanced interferograms of a GMO BLM at three delay times t: (a) $t = 0$ ms: first extreme position (forward curved). (b) $t = 0.91$ ms: middle position (saddle shaped). (c) $t = 1.67$ ms: second extreme position (backward curved). Notice the difference in scales between x (0 to 300 μm) and z (−4 to +4 μm) (from Ref. 16, with permission from the author).

In Fig. 10 we show a plot of the results for the flexoelectric coefficients of GMO as a function of frequency, with curvatures being obtained by stroboscopic interferometry and capacitance. Capacitance data were initially calculated supposing $\alpha = 0.05$, as before [14]. Here we also show normalized capacitance data, divided by a factor of 5, in order to achieve a coincidence with interferometric values in the low frequency end. According to Eq. (12) this means to take for α a 25 times lower value, i.e., $\alpha = 0.002$.

Comparing interferometric values and normalized capacitance values, we see that capacitance data reflect well the expected frequency dependence of $f(\omega)$, Eq. (8): the higher frequency values (blocked exchange) are about 2 times higher than the lower frequency ones (free exchange). The median frequency v_m between the two regimes is about 250 Hz. On the contrary, interference data

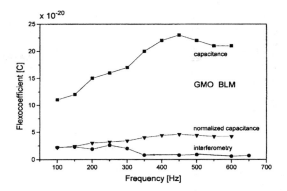

Fig. 10. Flexoelectric coefficients of a GMO/n-decane BLM as a function of frequency, where curvatures were obtained by either by stroboscopic interferometry or capacitance. Capacitance data were initially calculated supposing $\alpha = 0.05$. Normalized capacitance data are divided by a factor of 5 (i.e., $\alpha = 0.002$), in order to achieve a coincidence with interferometric values in the low frequency end (from Ref. 3, with permission from the Publisher).

display the opposite frequency dependence (higher frequency values lower than the lower frequency ones). This is probably due to the appearance, at higher frequencies, of a few curvature maxima over the membrane surface, so that its surface deviates from a spherical one. This situation was especially well seen on the interferograms of PS BLM (see below). Here the median frequency v_m is about 350 Hz. Having measured the displacement of the BLM in the central point only, we have probably overestimated the a v e r a g e curvature and, hence, underestimated the high frequency values of the flexocoefficient.

Anyway, combining the two methods, we could claim with a fair degree of confidence $f^B = (4.3 \pm 0.2) \cdot 10^{-20}$ C, while f^F is about a factor of 2 lower.

Same type of data processing was performed regarding the flexoelectric coefficients of EYPC as a function of frequency, with curvatures being obtained by stroboscopic interferometry and capacitance. Proceeding as above, capacitance data were initially calculated supposing $\alpha = 0.05$, as in Sect. 3.3. The normalized capacitance data are then divided by a factor of 4.57, in order to coincide with interferometric values in the low frequency end. According to Eq. (12) this means to take for α a 20.88-times lower value, i.e., $\alpha = 0.0024$.

Comparing interferometric and normalized capacitance values, we again found that capacitance data better reflect the expected frequency dependence, with higher frequency value also 2 times higher than the lower frequency ones. The median frequency v_m was about 275 Hz. Combining again the two methods, we could claim with a reasonable confidence: $f^B = (13.0 \pm 0.3) \cdot 10^{-19}$ C for EYPC. This value is 30 times higher than that of the GMO, demonstrating the surface charge effect ($\beta = -18.8\%$ for EYPC *vs.* $\beta = 0$ for GMO).

Computer-enhanced interferograms and cross sections of the membrane surface of a bovine brain PS BLM at the two extreme positions of its curvature oscillations were further obtained [15, 16]. The images in [16] emphasized the nonhomogeneous curvature at higher frequencies (590 Hz), while at lower ones (310 Hz) the corresponding images showed that curvature distribution is rather homogeneous [15]. BLM curvatures were determined from such interferograms by the described method (which clearly overestimates a v e r a g e curvature, when more than one curvature maximum is present). Experimental data on flexocoefficients for PS BLMs at pH 4.0, 7.0, and 10.0 are plotted in Fig. 11. Flexoelectric coefficients were found to be, within experimental error, pH-independent in the frequency ranges below 200 Hz and above 450 Hz (Fig. 11). Therefore, a pH-independent high frequency value of PS flexocoefficient could be accepted as follows: $f^B = 20 \ (\pm 25\%) \cdot 10^{-19}$ C, with the provision that this value might be underestimated (see above).

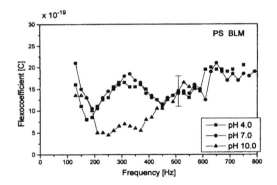

Fig. 11. Flexoelectric coefficients for bovine brain phosphatidyl serine BLMs bathed in aqueous solutions at pH 4.0, 7.0 and 10.0, respectively, as functions of curvature oscillation frequency. Conditions: ionic strength 0.1 M KCl, membrane radius 487 μm (from Ref. 3, with permission from the Publisher).

5. CONVERSE FLEXOELECTRIC EFFECT. MAXWELL RELATION

According to the theory [3] both the direct and the converse flexoeffect must exist and both flexoelectric coefficients must be equal (Maxwell relation). The converse flexoeffect stands for voltage-induced bending stress in a BLM, which should be able to produce substantial membrane curvature in the absence of other constraints. However, no indication of any voltage-induced BLM curvatures could be found in the literature. Indeed, as the theoretical analysis demonstrated, a flexoelectric bending of BLMs is strongly damped by another mechanical constraint – the lateral BLM tension that is due to the presence of the Plateau-Gibbs border. Also, BLMs are electrically fragile structures and are unable to withstand higher voltages because of electric breakdown. Therefore, our strategy to register the converse flexoeffect was concentrated on using the most sensitive and up-to-date techniques for BLM curvature detection (interferometry and real-time stroboscopic interferometry) and on the stabilization of BLMs by uranyl acetate (UA) against electric breakdown.

5.1. Theoretical remarks

We shall describe curvature deformations of a membrane by its mean curvature $c_m = \frac{1}{2}(c_1 + c_2)$, the mean value of its two principal curvatures. Describing electric field (E) induced deformations in BLMs, we must note that they are limited not only by the curvature elastic energy $\frac{1}{2}K(c_1 + c_2)^2$ with its curvature elasticity K, but also by additional mechanical forces, most notably to lateral membrane tension σ. Curvature deformations are coupled to the volume changes of the bathing electrolyte between the two sides of the curved membrane, which involves gravity effects as well. However, complete theoretical analysis [17] shows that all these limiting factors are negligible compared to membrane tension and (as soon as transmembrane pressure difference p is identically zero) with a fair degree of accuracy it holds:

$$c_m = \frac{2fE}{4K + \sigma r^2/2} \ , \qquad \text{i.e.,} \qquad f = \frac{\sigma r^2}{4(U/d)} c_m, \qquad (17)$$

after neglecting K in the denominator.

5.2. Experimental investigation of converse flexoelectricity of BLM

L-α-phosphatidyl-L-serine from a bovine brain (PS; purity approximately

98%; Sigma), n-decane (purity 99.99%; Aldrich), potassium chloride (KCl; purity 99.9%; Fisher), and uranyl acetate (UA; purity AG; General Chemical Division, NY) were used as received. The other experimental conditions were the same as above. The pH of the electrolyte solution was measured before each experiment and an average value of 6.8 to 7.0 was observed. Dissolving PS in n-decane at a concentration of 25 mg/ml made a BLM-forming solution. BLMs were formed in a (0.3 ± 0.1) mm hole. Schematic diagram of the experimental set-up was very similar to that shown in Fig. 9, except that now curvature was electrically induced (and not pressure-induced like with direct flexoeffect). Transmembrane potential was applied by Ag/AgCl electrodes using the bias option of the Digibridge from a function generator (Hewlett-Packard 8116A) controlled digitally by an IBM PC computer.

Computer-controlled stroboscopic illumination permitted the detection of the motions of the membrane surface in real time by introducing a small frequency difference (~1Hz) between the BLM oscillation and the laser pulse frequency. The real-time images were recorded, frame-grabbed (PC-Vision Plus, Imaging Technology, Inc.), and analyzed by using the software developed by us. Displacements to the maximum forward (ΔZ_{x1}, ΔZ_{y1}) and backward (ΔZ_{x2}, ΔZ_{y2}) excursions of the membrane surface were determined from the captured interferometric images as in the case of direct flexoeffect. The curvature of the BLM along the x and y axes of the two extreme deviations of the BLM surface, (c_{x1}, c_{y1}) and (c_{x2}, c_{y2}), were calculated by Eq. (15). Zero-to-peak amplitude of the mean curvature was then calculated as

$$2c_m = \frac{1}{2}[(c_{x1} + c_{y1}) - (c_{x2} + c_{y2})]. \qquad (18)$$

Prior to the taking of measurements, the planarity of the BLMs was adjusted using a micrometer burette (0.2 ml; RG Laboratory Supplies Co., Inc., NY). The pressure corresponding to the addition of 1 μl of water to one side of the membrane was on the order of 0.125 Pa. As in our previous experiments on the direct flexoeffect, the equilibrium membrane geometry was never flat, but rather saddle-shaped, its two principal curvatures being opposite in sign. Efforts were then made in each experiment to adjust the m e a n BLM curvature to as close to zero as possible in order to minimize the effects related to transmembrane pressure difference.

5.3. Converse flexoelectric coefficient of an UA-stabilized bovine PS BLM

The basic experimental finding was that, under an a.c. electric potential, curvature oscillations were elicited in the BLM at the same frequency as the applied a.c. voltage. This was confirmed by our method of stroboscopic interferometry at each of the applied frequencies, from 100 to 450 Hz. The excitation of the fundamental frequency was in accordance with the flexoelectric nature of the effect. A voltage-dependent effect on the BLM tension coupled to some residual pressure differences would have generated a second harmonic (because the tension would have reached two minima within one period of the electric field). This was not observed.

The mean curvature of the saddle-shaped BLM was slightly changed in opposite directions during one period of the a.c. field (see Fig. 12 for typical interferograms). Stroboscopic imaging permitted us to record BLM oscillations in real time and to identify the maximum forward and backward excursions of the membrane surface. Having calculated c_m by Eq. (18), flexocoefficients were then calculated according to Eq. (17).

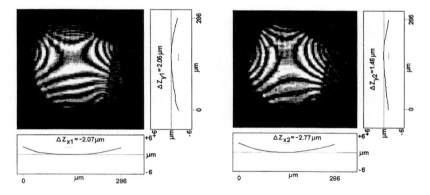

Fig. 12. Typical interferograms obtained at the two extreme positions of a bovine brain PS BLM bathed in 0.1 M KCl, 1 mM uranyl acetate, exposed to 600 mV (peak-to-peak) transmembrane potential at 300 Hz. Notice that if the two central hyperbolic fringes along the x axis are assigned to the zeroth order, then the two corresponding fringes along the y axis must be assigned to the next, first order. Membrane displacements (ΔZ_{ij}) were measured with respect to the line connecting the ends of the arc representing the section of the membrane surface (bars). Note the difference in scales between x, y (0-286 μm) and ΔZ_{ij} (−6 to +6 μm) (from Ref. 18, with permission from the Publisher).

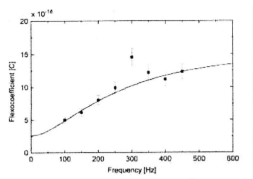

Fig.13. Converse flexocoefficient of a bovine brain PS (UA^{2+}) BLM *vs.* oscillation frequency of a.c. transmembrane potential. Conditions: 25 mg PS in 1 ml n-decane at 25° C; BLM modified by 1 mM UA in bathing electrolyte (0.1 M KCl); membrane radius was 143 μm; the amplitude (peak-to-peak) of the applied a.c. potential was 600 mV. Each point was averaged over at least six successive extreme deviations of the BLM and the relative error in f was typically 9%. The solid line is a fit according to Eq. 8 for $f(\omega)$: fitting parameters are $f^F = (3.0 \pm 1.5) \cdot 10^{-18} \mathrm{C}$, $f^B = (15.1 \pm 2.9) \cdot 10^{-18} \mathrm{C}$ and $\tau = 440 \mathrm{ms}$, i.e. $v_m = 362 \mathrm{Hz}$ (from Ref. 3, with permission from the Publisher).

PS bilayer thickness d is 5.4 nm [3]. The same value of d was used for UA-modified PS membranes, as BLM thickness changes upon UA adsorption were found to be negligible [17]. The measured capacitance of the BLM was 212 pF; therefore, with the above value of d, a membrane radius (r) of 143 μm was calculated. This value was also confirmed by the optical images. Membrane tension at 1 mM UA was taken to be 5 mN/m (according to Ref.17). The relative error of f is estimated as 9% [18].

The frequency dependence of the flexocoefficient values (Fig. 13) reveals two domains: a low frequency one below 300 Hz and a high frequency one above 300 Hz. Such behaviour is commonly observed in direct flexoeffect as a transition from free to blocked lipid exchange regime. The transition frequency varies from one BLM system to another, but is usually in the range of 150 to 300 Hz. From the fit of the data we obtain $f^B = 151 \ (\pm 19\%) \cdot 10^{-19}$ C.

This value is in very good agreement with that evaluated earlier at 300 and 400 Hz by the direct flexoeffect in a (1 mM) UA-modified, egg-lecithin BLM: $120(\pm 10\%) \cdot 10^{-19}$C. Egg lecithin had a lower negative surface charge than PS, which was used in the present case. However, once UA^{2+} ion adsorption on the membrane surface was completed (usually at micromolar concentrations), recharging to a high positive surface charge value had taken place [17]. Then, evidently, electric and flexoelectric BLM properties should be dominated by the adsorbed UA ion layer that is known to have a substantial dipole moment [17]. For comparison, the direct flexocoefficient of a n o n - m o d i f i e d lecithin BLM in the same frequency range is $(13.0\pm0.3)\cdot10^{-19}$C and that of a n o n - m o d i f i e d PS BLM (34) is $20(\pm25\%)\cdot10^{-19}$C. We can conclude that the flexoelectricity of UA-modified membranes is, indeed, dominated by the adsorbed UA^{2+} ion layers.

Furthermore, our finding demonstrates the validity of the Maxwell relation for the flexocoefficients of the direct and converse flexoelectric effect.

The first observation of the converse flexoeffect in BLMs permits the potential use of stabilized BLM systems as microtransducers, micro sound wave generators, and microactuators in molecular electronics. Indeed, flexoelectrically induced displacements of a membrane surface that is only nanometers thick represent (on a molecular scale) a huge, coordinated motion in space of a whole molecular assembly. This effect may, then, find some interesting applications.

Table 1

Flexocoefficients of BLMs made from different lipids under various ionic conditions. 1 - 3: electrical estimation of curvature; 4 - 7: interferometric measurement of curvature, sign determination of f not attempted. All experiments except 7 concern the direct flexoeffect. All data refer to the high frequency range above 200-300 Hz, i.e., to the blocked flexocoefficient.

Lipid, electrolyte	Partial charge of lipid head β (%)	Flexocoefficient $f^B \times 10^{19}$ C
1. Phosphatidyl ethanolamine from *E.coli*; 50 mM KCl	−3.7%	−25.5(±72.5%)
2. Phosphatidyl choline from egg yolk; 0.1M NaCl	− 0.4%	+26.5 (±50%)
3. Phosphatidyl choline from egg yolk; 0.1M NaCl + 1mM UA	*ca.* 100%	−120(±10%)
4. Phosphatidyl choline from egg yolk; 0.1M KCl	−18.8%	13 (± 0.2%)
5. Glyceryl monooleate (synthetic); 0.1 M KCl	0	0.43 (± 0.5%)
6. Phosphatidyl serine from bovine brain; 0.1 M KCl	*ca.* −100%	20 (± 25%)
7. Phosphatidyl serine from bovine brain; 0.1M KCl + 1mM UA	*ca.* 100%	151 (± 19%)

6. SUMMARY OF THE RESULTS

The experimental results reported so far are summarized in Table 1. All data in this table refer to the high frequency range of curvature oscillations, above 200 - 300 Hz, i.e., to the blocked flexocoefficient f^B.

   The pure dipolar contribution, order of magnitude $1 \cdot 10^{-20}$ C, is observed only in the case of fully synthetic GMO (5), with strictly zero surface charge. Still, the value is a few times higher than the theory predicts. In all other cases the charge contribution, order of magnitude $1 \cdot 10^{-18}$ C predominates, as already mentioned. The difference in signs between PE and PC flexocoefficients, 1 and 2, is puzzling and needs further attention.

   The equality between direct and converse flexocoefficients of UA-modified membranes (3 and 7) demonstrates the validity of the Maxwell relation, as we noted before.

7. PHOTOFLEXOELECTRICITY

We have seen that flexoelectricity provides an intricate interrelation between mechanical and electrical degree of freedom of a BLM. A third, optical degree of freedom of membrane systems called for further exploration in this context. An opto-mechano-electric model membrane has a number of possible combinations between its three degrees of freedom (serving either as inputs or outputs). Above all, due to the ultimate existence of these degrees of freedom in some native membranes, new hints about the structure-function relationship in photosynthetic membranes, retinal rods and discs, and other photoactive membranes could be obtained.

   Black lipid membranes decorated with nanoparticles of photo-semiconductors, "nanomembranes" [19], represent a model system having the three degrees of freedom into consideration. Like some biomembranes, they are then capable of the transformation of the three types of energy one into another.

   Recently, an effect named "photoflexoelectricity" was demonstrated in nanomembranes [20]. Both optical and mechanical degrees of freedom acting as inputs, the electric degree served as output.

7.1. Preparation of nanomembranes

   The model system chosen for this investigation was a GMO BLM that was covered by *in situ*-generated nanosized semiconductor cadmium sulphide (CdS) particles. Development of this system has been prompted by its mimetic relevance to biological photosystems and by the availability of information on semiconductor-coated BLMs [19, 21].

Semiconductor particles of this size exhibit intermediate properties between the bulk solid state and a single macromolecule. Electronic excitation of semiconductors leads to loosely bound electron-hole pairs (i.e., to excitons). The existence of a vast interface between the particle and the surrounding medium can have a profound effect on the particle's properties. Most of the particles have imperfect surfaces, which provides electron and hole traps upon optical excitation. These trapped electron and hole levels on the surface can cause a reduction of the overlap of the electron and hole wave function and particle polarization. The band-gap excitation and generation of electron-hole pairs is shown to result in the reduction of O_2- molecules and the formation of superoxide radicals ($O_2^{\cdot-}$) at the *cis* (CdS-covered) side of the BLM together with the oxidation of H_2S and HS^- and the formation of H^+ at the *trans* BLM side [19]. These photoprocesses increase charge separation across the membrane and, thus, can result in changes of the flexoelectric response. Enhancement of the flexoelectric response of the oscillating nano-BLM upon illumination was, indeed, observed in the present investigation.

Cadmium perchlorate ($Cd(ClO_4)_2.6H_2O$,99.9%, Alfa Products), potassium chloride (KCl, purity 99.9%, Fisher) and H_2S (Matheson Gas Products) were used as received. Water was purified using a Millipore Milli-Q system provided with a Millistak filter at the outlet (0.22 μm pore size). The final water resistivity was 18 MΩ.cm. GMO (1-monooleoyl-*rac*-glycerol (c18;1,[*cis*]-9); Sigma, purity 99%) was used as received. BLM-forming solutions, 50 mg/ml GMO in *n*-decane (Decane, purity 99.9%, Aldrich), were freshly prepared just prior to their use.

Experimental set-up was very similar to the one used before (Sect. 4). A 150 W xenon lamp with an optical fibre was used for membrane illumination. Membrane curvature oscillations were excited by a piezotransducer, as described above.

Semiconductor particles were formed on the BLM by introducing a freshly prepared solution of $Cd(ClO_4)_2$ (typically 50 μl of 0.1 M solution) to the *cis* side of the BLM (containing a 2 ml solution of 0.1 M KCl). After a 10 min incubation period, 20-25 μl of H_2S gas was slowly injected into the *trans* compartment. Within approximately 5 min, the semiconductor particles became visible on the BLM when observed through the microscope, and the amount and sizes of the particles grew for the next 1-2 h. With smaller concentration of the precursors (30 μl of 0.1 M $Cd(ClO_4)_2$ and 15 μl H_2S) the semiconductor growth

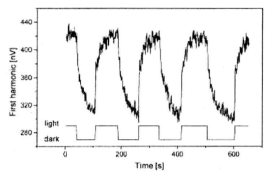

Fig. 14. Flexoelectric response of a GMO BLM coated with CdS nanoparticles, oscillating at 700 Hz under zero current clamp. The first harmonic of the membrane potential is measured. The increase of the flexoelectric signal after light flashing shows exponential rise kinetics and reversible decay (from Ref. 20, with permission from the Publisher).

was initiated when oscillating pressure was applied to the membrane. Switching off the oscillation later stopped the process of growth.

7.2. Photoflexoelectricity of nanomembranes

The formation and development of CdS nanocrystalline particulate films on the GMO BLM surface were followed by optical microscopy. Small isolated CdS clusters became visible on the *cis* side of the BLM within a few minutes of H_2S introduction. With time, the number of clusters increased, the clusters coalesced into islands and covered an increasingly greater part of the BLM surface. Concomitant with increasing CdS coverage, the BLM became more rigid until it could not be flexed under oscillating pressure. Photoflexoelectric effects were, therefore, examined on GMO BLMs with only up to 20% CdS coverage.

Significant changes in both the amplitude and phase of the response were found between the light and dark state of the membrane. The increase of the flexoelectric response accompanying illumination, referred to as photoflexoelectricity, was followed by a return to the initial dark value when the light was switched off. For a given GMO BLM, this photoflexoelectric response was reproducible for many cycles. The curvature-induced membrane potential (Fig. 14) measured under zero current clamp (open external circuit) showed a reversible increase of its amplitude following light-dark cycles. Photokinetic curves of the flexoelectric voltage were fitted with single exponentials demonstrating a faster rise after illumination ($\tau^L = 13$ s) and a slower decay to the dark level ($\tau^D = 20$ s).

The flexoelectric amplitude increments and the phase shifts caused by illumination of BLMs with increasing amount of particles generated on the membrane were studied. The dark response decreased as the amount of CdS particles increased (as judged from the visual appearance), but the difference (light-dark) became more and more pronounced. At the same time, a tendency towards larger phase shift at higher photoflexoelectric response was observed.

A nonlinear dependence of the photo-induced flexoelectricity *vs.* generator amplitude (and membrane curvature amplitude, respectively) was found, in contrast to the fairly linear increase of the flexoresponse *vs.* generator amplitude for the dark state. A possible explanation for nonlinearity can be a disintegration of some particle clusters at higher curvature amplitude and a concomitant increase of the total amount of CdS particles participating in the photo-flexoresponse. The frequency dependence of the effect was, however, found to be nearly linear in both the light and dark state.

In conclusion, a black lipid membrane modified with photosensitive nano-sized CdS particles serves as a model of a photosensitive native cell membrane, where CdS crystallites mimic membrane proteins [19]. First observation of photoflexoelectricity in such a system permits us to speculate upon the possible existence of this phenomenon in native membranes as well. The existence of a converse effect (i.e., photopotential-induced membrane curvature upon illumination) also seems highly probable. Experimentally, though, the converse flexoeffect is rather difficult to observe in BLMs because it is strongly suppressed by non-zero lateral tension in these membrane models. On the other hand, in native photosensitive membranes with vanishing lateral tension the photopotential-induced membrane curvature could be very strong. The connectivity between the membrane shape (curvature), the light intensity and the membrane potential could be related to the mechanism responsible for the phototaxis and other photoadaptive phenomena on the cell membrane level.

The present findings could also stimulate the development of cell-like photo-mechano-electric micromachines based on lipid vesicles and featuring cyclic energy transformation expressed in cyclic variations of the vesicle shape under constant illumination.

REFERENCES

[1] P.Mueller, D.O.Rudin, H.Ti Tien and W.C.Wescott, Nature, 194 (1962) 979.
[2] R.Coronado and R.Latorre, Biophys. J., 43 (1983) 231.
[3] A.G.Petrov, The Lyotropic State of Matter, Gordon and Breach Publs., NJ, 1999.
[4] V.I. Passechnik and V.S.Sokolov, Biofizika, 18 (1973) 655.
[5] A.L.Ochs and R.M.Burton, Biophys.J., 14 (1974) 473.
[6] A.G. Petrov, in J.Vassileva (Ed.), Physical and Chemical Bases of Biological Information Transfer, Plenum Press, N.Y.-L. (1975) p.111.
[7] A.G.Petrov and A.Derzhanski, J.Physique suppl., 37 (1976) C3-155.
[8] A.Derzhanski, A.G.Petrov and Y.V.Pavloff, J. Physique Lett., 42 (1981) L-119.
[9] A.G.Petrov and V.S.Sokolov, Eur.Biophys.J., 13 (1986) 139.
[10] K.Hristova, I.Bivas and A.Derzhanski, Mol.Cryst.Liq.Cryst., 215 (1992) 237.
[11] J.G. Szekely and B.D.Morash, Biochim.Biophys.Acta, 599 (1980) 73.
[12] D. Wobschall, J.Colloid Interface Sci., 36(1971) 385.
[13] A.Derzhanski, A.G.Petrov, A.T.Todorov and K.Hristova, Liquid Crystals,7 (1990) 439.
[14] A.T. Todorov, A.G.Petrov, M.O.Brandt and J.H.Fendler, Langmuir, 7(1991) 3127.
[15] A.T.Todorov, A.G.Petrov and J.H.Fendler, Langmuir, 10 (1994) 2344.
[16] A.T. Todorov, Ph.D. Thesis, Syracuse University, N.Y. (1993)
[17] S.I..Sukharev, L.V.Chernomordik, I.G.Abidor and Yu.A.Chizmadzhev, Elektrokhimiya, 17 (1981)1638.
[18] A.T. Todorov, A.G.Petrov and J.H.Fendler J.Phys.Chem. 98 (1994) 3076.
[19] J.H. Fendler, Membrane Mimetic Chemistry, Willey-Interscience, N.Y., 1982.
[20] M.Spassova, A.G.Petrov and J.H.Fendler, J.Phys.Chem., 99 (1995) 9485.
[21] H.T Tien,. Prog.Surf.Sci., 30 (1989) 1.

Planar Lipid Bilayers (BLMs) and their Applications
H.T. Tien and A. Ottova-Leitmannova (Editors)

Chapter 7

Chain-ordering phase transition in bilayer: Kinetic mechanism and its physicochemical and physiological implications

D.P. Kharakoz
Institute of Theoretical and Experimental Biophysics, RAS
142290 Pushchino, Moscow region, Russia

1. INTRODUCTION

Current paradigm of the physiology of cellular membranes is based on the understanding of two major functions of lipid bilayers [1]. First, the bilayers serve as inert self-assembling insulators preventing the uncontrolled transfer across the biological membranes and thus separating the cells and intracellular compartments from their environment. Second, they serve as matrix for membrane proteins that are considered to be the only active participants in the membrane physiological processes. Even implementation of these passive functions requires the lipid bilayers to have certain structural and dynamic properties dependent on their phase state. Indeed, the ground state of biological membranes is known to be lamellar fluid [2] providing sufficient freedom for the lateral, rotational and internal motions of embedded proteins.

On the other hand, there are numerous indirect evidences for the idea that the lipid phase transitions themselves may take active part in the cellular processes (reviewed in Refs. 3-9). Noteworthy is, e.g., a tendency observed in studying the temperature adaptation of some microorganisms: the transition temperature in their membranes being beneath the growth temperature is often close to the latter [7, 8]. It gives an impression that, although the basic physiological state of the membranes is fluid, the cell needs the solid state to be attainable. Clearly, the insight into what is the active physiological role of the fluid to solid transition is impossible without information on the mechanism of this process and—what is of particular importance from the biological point of view—its kinetics.

The chain-ordering/melting transition (main transition) in biologically relevant model lipid membranes is characterized by pronounced discontinuities in the first derivatives of free energy [10, 11] thus demonstrating the first-order character of the transition. At the same time, in the vicinity of transition temperature, lipid systems manifest dynamic features characteristic of second-order transitions: enhanced fluctuations of in-plane density [12], drop in bending rigidity [13], and anomalous increase in interbilayer spacing [14, 15]. There are two alternative approaches to the explanation of the combination of the first- and second-order character of the transition. A commonly accepted one is based on the assumption that the main transition in lipid bilayers is close to a critical state [14-16]. An alternative idea is that this is a weak first-order transition occurring far from any critical state and that the anomalies result from the early steps of new phase nucleation—formation of the nuclei of sub-critical size that can occur in a wide range of temperatures around the main phase transition temperature [17-19]. This idea presents consistent quantitative explanation for the properties associated with the in-plane packing of lipid and, as a consequence, some phenomena related to the out-of-plane motions.

In this chapter, we first consider a phenomenological kinetic theory of the early steps of the solid-phase nucleation and compare the theoretical predictions with relevant experimental data. Then, we analyze the dynamic consequences of the theory as applied to the long-range processes, such as formation of the nucleus of critical size and kinetics of the entire phase transition. Finally, some physiological implications of the mechanism of the solid phase creation and propagation are discussed.

2. THE MECHANISM OF SOLID-STATE NUCLEATION AND HETEROPHASE FLUCTUATIONS

In a system near first-order transition, small clusters resembling the structure of the new phase state the system approaches to can create and dissipate within the parent stable phase [20-22]. A theory of this phenomenon, called heterophase fluctuations, has been first suggested by Frenkel [20] and recently applied to the chain-ordering transition in lipid bilayers [17-19]. Formation of the small (sub-critical) solid nuclei within the parent fluid phase is a matter of consideration in this section.

2.1. Phenomenological theory

2.1.1. Basic concepts

The process of creation, growth, and dissipation of small nuclei is considered *an intrinsic property of monolayers*, independent of the intermonolayer or interbilayer interactions. It can be represented by a chain of reversible chemical reactions corresponding to the consecutive attachment of lipid molecules to a nucleus according to the scheme

$$
\begin{array}{ccccccccccc}
& f_1 & & f_2 & & f_3 & & & f_{N-1} & & f_N & \\
L_0 & \leftrightarrow & L_1 & \leftrightarrow & L_2 & \leftrightarrow & \dots & \leftrightarrow & L_{N-1} & \leftrightarrow & L_N & \dots \\
& b_1 & & b_2 & & b_3 & & & b_{N-1} & & b_N &
\end{array}
\qquad (1)
$$

where each step represents the attachment of a single molecule with the forward and backward rate constants f_i and b_i, respectively; L_0 designates the fluid state; and L_1 is the solid nucleus of minimum size, containing n_{min} molecules. Practically, the chain length can be restricted by $N = 30$ without a detectable error because of negligible probability of larger nuclei. Therefore, the indexes $i = 1, 2, 3, \dots, N$ correspond to the following numbers of molecules in the nuclei: $n_i = n_{min} + i - 1$.

The equations presented below are valid until the *total fraction of lipid embedded in solid nuclei is small* and, thus, the interaction between the nuclei is negligible. In the case of a biologically relevant lipid studied in Ref. 18 this simplification is reasonable. *If the system is flat and the nucleus shape circular* the chemical potential of *i*th nucleus, μ_i, defined relative to the ground state L_0, is written as follows:

$$
\mu_i = \mu_{bulk} + \mu_{interface} = \Delta G \times n_i + 2\gamma N_A \times (\pi \sigma n_i)^{1/2}
\qquad (2)
$$

where the first term, μ_{bulk}, proportional to the number of molecules in the nucleus, describes the bulk contribution to the potential; the second term, $\mu_{interface}$, proportional to the nucleus perimeter, describes the interfacial contribution; ΔG is the molar free energy of fluid-to-solid transition in an infinite system; γ the interfacial line tension coefficient; N_A the Avogadro's number; and σ the surface area per molecule in the monolayer plane in solid state. The free energy of chain-ordering transition is an almost linear function of temperature [18]. Hence,

$$
\Delta G = -\Delta S (T - T_0)
$$

where the apparent transition entropy, ΔS, and the transition temperature, T_0, are considered to be constant. In the equilibrium state, the relative number of nuclei, x_i, defined relative to the total number of molecules in the system, is given by the relation [20, 21]:

$$
x_i = e^{-\mu_i / N_A k_B T}
\qquad (3)
$$

where k_B is the Boltzmann constant. This relation, being strongly valid in the case of infinitesimally small fraction of nuclei, is employed here as an approximation.

Kinetics of heterophase fluctuations is determined by the rate constant of an elementary chemical step of nucleation, b_0, which is defined as the probability of a given boundary molecule to leave the nucleus during a unit of time. *If the probability is assumed to be independent of the curvature of boundary*, then the backward rate constants are proportional to the number of boundary molecules and, thus, are constrained by the approximate relation

$$b_i = 2\pi\, b_0 \times (n_i^{1/2}/\pi^{1/2} - \tfrac{1}{2}) \tag{4}$$

where the term ½ takes into account that the radius of the circumference where the centers of external molecules are located is less than the radius of the nucleus by half the linear dimension of a molecule in membrane plane. The forward rate constants are then defined by the equilibrium condition

$$f_i = b_i\, x_i\, /\, x_{i-1} \tag{5}$$

Eqs. (4) and (5) allow reducing the set of rate constants to a single kinetic parameter, b_0, whose physical meaning was defined above. Therefore, the kinetics of the system of reactions (1) is explicitly described by the following differential equations neglecting the changes in the mole fraction of fluid lipid and, hence, assuming $x_0 = 1$:

$$
\left.
\begin{aligned}
\dot{x}_0 &= 0, \\
\dot{x}_1 &= f_1 x_0 - (b_1 + f_2) x_1 + b_2 x_2, \\
&\cdots \\
\dot{x}_i &= f_i x_{i-1} - (b_i + f_{i+2}) x_{i+1} + b_{i+1} x_{i+1}, \\
&\cdots
\end{aligned}
\right\} \tag{6}
$$

If the function $\Delta G(T)$ or the values ΔS and T_0 are known, then, for a given set of the nucleation parameters (γ, n_{min}, and b_0), a variety of experimentally available thermodynamic and kinetic properties of heterophase fluctuations can be calculated. Thus, the parameters of nucleation can be determined by fitting to experimental data [18], as shown below.

2.1.2. Extensive thermodynamic properties and static susceptibilities
    The total mole fraction of lipids, X, embedded into solid nuclei is given by the sum

$$X_{\text{nucl}} = \sum_{i=1}^{N} n_i x_i \tag{7}$$

The nuclei contribute to thermodynamic properties resulting in pretransition phenomena. Of our interest will be the nucleation-induced changes in molar enthalpy, H_{nucl}, volume, V_{nucl}, surface area in membrane plane, $N_A \sigma_{\text{nucl}}$, and the static relaxation parts (measured at infinite time) of mechanical susceptibilities: isothermal bulk compressibility of membrane, $K_{\text{rel}} \equiv -(\partial V_{\text{nucl}} / \partial P)_T$, and isothermal lateral compressibility of a lipid monolayer, $\Lambda_{\text{rel}} \equiv -(N_A \partial \sigma_{\text{nucl}} / \partial \Pi)_T$, where P is the bulk pressure, Π the lateral pressure, and σ_{nucl} the area per a molecule in nucleus. These changes are given by the equations

$$H_{\text{nucl}} = X_{\text{nucl}} \Delta H = \sum_{i=1}^{N} n_i x_i \Delta H \tag{8}$$

$$V_{\text{nucl}} = \sum_{i=1}^{N} n_i x_i \Delta V \tag{9}$$

$$K_{\text{rel}} = (N_A k_B T)^{-1} \sum_{i=1}^{N} x_i (n_i \Delta V)^2 \tag{10}$$

$$\Lambda_{\text{rel}} = (N_A k_B T)^{-1} \sum_{i=1}^{N} x_i (n_i N_A \Delta \sigma)^2 \tag{11}$$

Note that the nucleation-induced lateral compressibility, given by Eq. (11), refers to a lipid monolayer. The values ΔG, ΔS, ΔH, ΔV, $N_A \Delta \sigma$ are the corresponding molar effects of the transition from parent fluid phase to solid nuclei. It should be noted that, in general case, the thermodynamic characteristics of the short-range nucleation process (i.e., the above-listed molar effects and also T_0) might differ from those observed in the macroscopic experiment.

2.1.3. Heterophase fluctuations and acoustic relaxation
    The solution of differential equations (6) can be found in the general form
of a sum of exponents

$$\delta x_i(t) = \sum_{i=1}^{N} a_{i,j} \exp(-t/\tau_j)$$

where t is the current time; $a_{i,j}$ is the particular amplitude of the jth exponential
component of relaxation of ith nucleus; and τ_j is the characteristic relaxation
time of the component. The amplitudes $a_{i,j}$ are defined by initial boundary
conditions. This problem has been solved for a particular case of acoustic
relaxation [17, 18] as the acoustic spectroscopy is of primary importance in
studying the kinetics of heterophase fluctuations. Skipping details, we present
here only the final expressions.
    The molar relaxation strength of jth exponential component is determined
by expression

$$R_j = -\Delta V_S (2\beta_S M)^{-1} \sum_{i=1}^{N} n_i (\partial a_{i,j}/\partial P)_S \ \sim (\Delta V_S)^2 \tag{12}$$

where β_S is the adiabatic compressibility coefficient of the system (lipid
dispersion, e.g.); M is the molecular mass of lipid; and V_S is the adiabatic volume
effect of the elementary step of nucleation. The latter value may substantially
differ from the isothermal one, ΔV, due to the heat production during the process
(the difference depends on the characteristic size of lipid assembly and the heat
wave decay length; e.g., in the case of multilamellar lipid vesicles in water at
40 °C, $\Delta V_S \approx \Delta V/1.65$ [18]). The relaxation part of either sound absorption or
velocity at a given circular frequency ω is the sum of particular relaxation
components:

$$[A]_{\text{rel}} = 2\pi \sum_{i=1}^{N} R_j \omega \tau_j (1 + \omega^2 \tau_j^2)^{-1} \tag{13}$$

$$[U]_{\text{rel}} = -\sum_{i=1}^{N} R_j (1 + \omega^2 \tau_j^2)^{-1} \tag{14}$$

The acoustic quantities are expressed here in terms of the absorption number,
defined as $[A] \equiv (\alpha - \alpha_0)/c$, and velocity number, defined as $[U] \equiv (u - u_0)/u_0 c$,
where c is the molar concentration of lipid; α and u are the absorption per

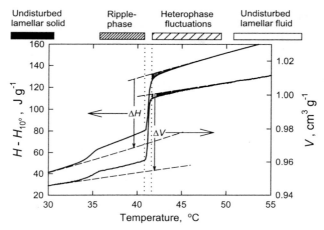

Fig. 1. Specific partial enthalpy (left axis) and volume (right axis) as functions of temperature for DPPC multilamellar vesicles in aqueous dispersion [18]. The enthalpy scale is presented as difference between the current value and that measured at 10 °C. Dashed lines were extrapolated from 10-25 °C for the undisturbed solid state and from 55-70 °C for the undisturbed fluid state. The whole temperature range was restricted in this picture to 30-55 °C in order to magnify the deviations (shown by vertical solid lines) caused by heterophase fluctuations. Vertical dotted lines show the quadruple width of liquid-crystalline to ripple phase transition. The horizontal bars at the top of figure show the specified ranges of temperature.

wavelength and velocity, respectively; and the subscript "0" refers to pure solvent. Eqs. (13) and (14) determine the ultrasonic relaxation spectra represented, e.g., by the absorption spectrum, $[A(\omega)]_{\text{rel}}$. The spectrum of the heterophase fluctuations falls within a rather narrow range of circular frequencies about $10^{8}\,\text{s}^{-1}$ and, thus, it can be approximated by a single effective quantity. The one commonly used is the apparent relaxation time, τ_{app}, defined as the reciprocal frequency of maximum absorption: $\tau_{\text{app}} = 1/\omega_{\text{max}}$. Another one is the effective relaxation time, τ_{eff}, defined by the expression

$$\tau_{\text{eff}} = - (2\pi\omega)^{-1}\,[A]_{\text{rel}}\,/\,[U]_{\text{rel}}$$

which is convenient when the relaxation is studied at a single frequency ω [17, 18, 23].

2.2. Experimental quantification of the nucleation parameters

There are many experimental observations of the heterophase fluctuations near the chain-ordering transition in lipid membranes. They display themselves

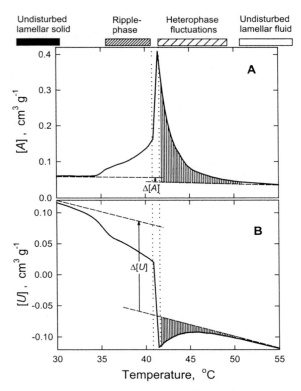

Fig. 2. Specific sound absorption (A) and velocity (B) numbers as functions of temperature for DPPC multilamellar vesicles in aqueous dispersion [18]. The ultrasound frequency was 7.4 MHz. The designations are as in Fig. 1.

as the deviations of temperature dependencies of measured quantities from simple sigmoid curves [18]. Particularly pronounced deviations are observed in the ultrasonic characteristics [16-18, 24-26] due to extensive fluctuations of density associated with the nucleation. These features are shown in Figs. 1 and 2 for dipalmitoylphosphatidycholine (DPPC) vesicles where the heterophase fluctuations have been most comprehensively studied [18].

The structure of the solid nuclei in DPPC is assumed to be like that of undisturbed gel phase (see Ref. 18 for argumentation), and the molar effects (ΔY) of nucleation are thus assumed to be equal to those of the transition from undisturbed liquid-crystalline (fluid) to undisturbed gel (solid)—a hypothetical "ideal" chain-ordering transition. Therefore, these quantities can be evaluated by means of extrapolation of the temperature dependences from a range above

Table 1

Characteristics of the undisturbed-fluid to undisturbed-solid phase transition ("ideal" chain-ordering transition) in vesicular membranes of DPPC at 40-50 °C

Parameter		Value	Comments and references
Transition temperature	T_0	40.1 + 273.15 K	[17]; $T_0 \equiv \Delta H/\Delta S$
(that for fluid to ripple phase	(T_m)	(41.2 + 273.15 K)	[18]; presented for
transition)			comparison
Enthalpy change	ΔH	−44.6 kJ mol^{-1}	[18]
Entropy change	ΔS	−0.143 kJ K^{-1} mol^{-1}	[18]
Gibbs free energy change	ΔG	0.143 $(T - T_0)$ kJ mol^{-1}	[18]
Volume change, isothermal	ΔV	−34.5 cm^3 mol^{-1}	[18]
Volume change, adiabatic	ΔV_S	−20.9 cm^3 mol^{-1}	[18]
Area occupied by a molecule			
in monolayer plane:			[12]
fluid	σ_{fluid}	0.63 nm^2	
solid	σ_{solid}	0.50 nm^2	
change	$\Delta\sigma$	−0.13 nm^2	
Thickness of monolayer:			[12]
fluid	h_{fluid}	2.0 nm	
solid	h_{solid}	2.4 nm	
change	Δh	0.4 nm	
Lateral compressibility of			[12]; assumed to be twice
monolayer:			that of bilayer
fluid	λ_{fluid}	14 m N^{-1}	
solid	λ_{solid}	2 m N^{-1}	
change	$\Delta\lambda$	−11 m N^{-1}	

The values have been obtained by extrapolation from the temperatures where the undisturbed gel and undisturbed liquid-crystalline states are assumed to exist (below 25 °C and above 55 °C, respectively).

55 °C, for solid, and from a range below 25 °C, for fluid (illustrated in Figs. 1 and 2). A number of the thermodynamic, acoustic, and structural characteristics of the "ideal" transition are collected in Table 1. As follows from Eqs. (8) and (9), the nucleation-induced deviations of enthalpy and volume are proportional to the mole fraction of lipid embedded in solid nuclei. Therefore, the mole fraction of solid at any temperature can be determined directly from the deviations of the experimental curves, as shown in Fig. 1, assuming that $Y_{nucl} = Y_{measured} - Y_{undisturbed\ fluid}$.

In the case of acoustic properties, the deviations contain substantial contributions from the relaxation of heterophase fluctuations:

$$[A]_{nucl} = [A]_{measured} - [A]_{undisturbed\ fluid} = X_{nucl} \Delta[A] + [A]_{rel}$$

$$[U]_{nucl} = [U]_{measured} - [U]_{undisturbed\ fluid} = X_{nucl} \Delta[U] + [U]_{rel}$$

Table 2
Minimum nucleus size, n_{min}, fluid-solid interphase line tension per monolayer, γ, and the elementary rate constant of nucleation, b_0, for vesicular DPPC membrane near the main transition temperature

Experimental method	n_{min}	$\gamma, 10^{-12}$ J m^{-1}	$b_0, 10^7$ s^{-1}
Calorimetric	7 ± 3	3.7 ± 0.5	
Volumetric	4 ± 1	4.4 ± 0.3	
Acoustic	9 ± 1	3.6 ± 0.2	5.2 ± 0.6
Recommended values	7	4	5

Adjustable parameters obtained by fitting of the phenomenological theory of nucleation to experimental data [18].

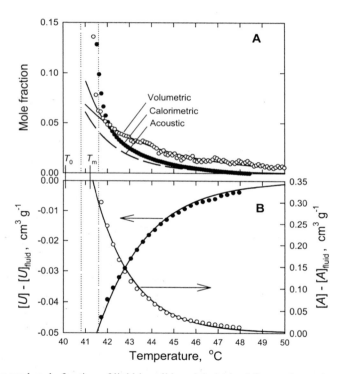

Fig. 3. The total mole fraction of lipid in solid nuclei obtained from volumetric, calorimetric, and acoustic data, A, and the nucleation-induced deviations in acoustic quantities, B, as functions of temperature [18]. Theoretical curves correspond to the best-fit values of adjusting parameters (Table 2). Vertical dotted lines show the quadruple width of the liquid-crystalline to ripple-phase transition.

where the relaxation terms $[A]_{rel}$ and $[U]_{rel}$ are calculated with Eqs. (13) and (14).

Therefore, all parameters of nucleation but three—the minimum nucleus size n_{min}, the solid-fluid interphase line tension γ, and the elementary rate constant b_0—are determined explicitly from experimental data by extrapolation of temperature dependencies from the ranges of undisturbed states, as discussed above (Table 1). The remaining three parameters are considered adjustable and evaluated by fitting of theoretical equations to the temperature dependencies of the nucleation-induced deviations [18]. It should be noted that the adjusting parameters are essentially independent of each other in the fitting procedure because they determine different features of the temperature dependences of the deviations: the value γ mainly influences the magnitude of the deviations, n_{min} influences the curvature of the temperature dependences, and the parameter b_0 defines the ratio of the relaxation part of sound velocity to that of sound absorption.

The results of fitting to calorimetric, volumetric, and ultrasonic data are presented in Table 2 and Fig. 3. High quality of fitting, illustrated in the figure, and a good agreement between the values obtained from the three independent experimental approaches demonstrate that the phenomenological theory of nucleation is self-consistent. The results are also confirmed by the fact that the value for solid-fluid line tension per monolayer obtained from the heterophase fluctuations is the same as that determined by independent methods in Langmuir monolayers of stearic acid (4 ± 1.5 pJ/m [27]).

3. PHYSICOCHEMICAL ISSUES

3.1. Solid-fluid interphase tension and the phase transition nature

The interfacial tension coefficient is an essential parameter determining the physical nature of phase transition. At a high tension, the interface becomes energetically expensive thus making the transient states highly unfavorable. Then, the system can only "jump" from one state to another, the transition being thus first-order with the discontinuities in first derivatives of free energy. Otherwise the transition is second-order or critical-like. Is the solid-fluid line tension in lipid membrane high or low in this respect?

A lipid molecule occupies about 7×10^{-10} m in the interface line. Therefore, the interfacial energy per molecule is close to the heat motion energy, $0.7\ k_B T$. Evidently, this value is large enough to make any extended boundary highly unfavorable and, hence, to make the transition first-order. On the other hand, it is small enough to allow detectable heterophase fluctuations. Hence, the transition should be qualified as a weak first-order transition. Furthermore, a quantitative analysis of the fluctuations performed in Ref. 18 under the assumption that γ is a function of temperature shows that, although the interfacial tension is as low as

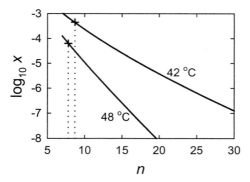

Fig. 4. Size distribution of solid nuclei at two selected temperatures [18]. Crosses with vertical dotted lines show the position of mean size ($\Sigma_i n_i x_i / \Sigma_i x_i$) for each temperature.

might be expected for a sub-critical state, the apparent critical temperature (where γ vanishes) is far below the main transition temperature. Therefore, the approach to the anomalous fluctuations from the viewpoint of a "pseudocritical" behavior appears to be not applicable to the in-plane chain-ordering process. However, such approach might be valid for the out-of-plane motions related, e.g., to the anomalous swelling of multilamellar lipid systems.

3.2. Heterophase fluctuations and the equilibrium dynamics

3.2.1. Size distribution of nuclei

The minimum nucleus size $n_{min} = 7$, obtained by fitting, is a reasonable value. Indeed, seven is the minimum number of molecules in a cluster when at least one of them finds itself surrounded by the like molecules (given that the packing is nearly hexagonal in plane). It is also supported by a consideration of the magnitude of density fluctuations [18]: it has been shown that this is the minimum size of a subsystem within which the squared magnitude of the heterophase fluctuations greatly exceeds the total mean-square magnitude of fluctuations in the subsystem (otherwise, they could not be considered heterophase).

The probability of nuclei decreases with their size (Fig. 4). The decay is so fast that the mean size (shown in figure by crosses) turns to be close to the minimum value. The nuclei with $n > 30$ are several orders of magnitude less probable than the nuclei of mean size. The size distribution has a tendency to shift towards larger nuclei when temperature approaches transition point. This is seen from the temperature dependence of both the mean size and the slope of the

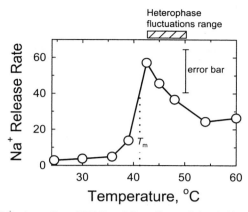

Fig. 5. Rate of Na⁺ release from DPPC multilamellar vesicles (arbitrary units) as function of temperature (data from Ref. 28). The main transition temperature is shown by vertical dotted line. Error bar represents the mean value for the points above the transition temperature (for lower temperatures, the data are more accurate).

curves in Fig. 4. This tendency is an important factor underlying the mechanism of slowing the relaxation of heterophase fluctuations (Subsection 3.3.3).

3.2.2. Anomalous permeability of bilayers

An anomalous peak in the permeability of some ions and small neutral molecules as function of temperature is observed in lipid bilayers at main transition [28, 29]. Fig. 5 shows the temperature dependence of the permeability of sodium ions across DPPC membrane. Despite the transition itself proceeds within a narrow range of temperatures (less than a degree, as discussed above) the sodium permeability peak is spread over 10 degrees above the transition point—just in the range where heterophase fluctuations occur. Comparing Fig. 5 with Fig. 2 one finds a qualitative similarity between the peak of permeability and the peak of heterophase fluctuations. This correlation may indicate that the anomalous permeability is due to packing defects at the boundaries between the solid-state nuclei and fluid phase. Another explanation of this phenomenon is also not excluded—that based on considering the anomalous increase of lateral compressibility near transition point as the cause of enhanced permeability [30]. Although the two possibilities cannot be strictly discriminated on phenomenological level, because the heterophase fluctuations and the lateral compressibility are closely related to each other through Eq. (11), one should take note of the following fact favoring the idea of heterophase fluctuation as an immediate factor determining the anomalous permeability. If the lateral

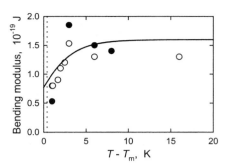

Fig. 6. Anomalous drop in the temperature dependence of bending rigidity of bilayers. Comparison of the theoretical curve (*solid line*) for DPPC with experimental data [13] on DPPC (*bold circles*) and dimyristoylphosphatidylcholine (*open circles*). Calculations were performed with Eqs. (15) and (16) assuming the nucleation parameters from Table 2 ("recommended" values) instantaneous λ from Table 1. Temperature scale is presented as difference between the current and main transition temperatures. Vertical dotted line shows the upper limit of quadruple width of transition. (Figure from Ref. 18.)

compressibility were the primary reason of enhanced permeability then one might expect the peak to be exactly at the transition point where the highest compressibility is attained. The peak is however not at T_m but shifted towards higher temperatures (see Fig. 5), similarly to what is observed for heterophase fluctuations (Fig. 2).

3.2.3. Binding rigidity, out-of-plane motions, and anomalous swelling

Mechanical properties of bilayers experience anomalous changes near transition point [12, 13]. These changes are perfectly accounted for within the concept of heterophase fluctuations. If constituting monolayers are bound, the bending modulus of the bilayer, B, will be related to the lateral compressibility coefficient ($\lambda \equiv -\partial \ln \sigma / \partial \Pi$) of a monolayer as follows [31]:

$$B = h^2/2\lambda \tag{15}$$

where h is the distance between the centers of monolayers. This relation, although approximate, is nevertheless in sufficient agreement with experimental data on synthetic lipid membranes [12]. The lateral compressibility coefficient of a monolayer can be expressed as a sum of the instantaneous part, λ_∞, measured in the absence of heterophase fluctuations, and the nucleation-induced relaxation part, λ_{rel}, as follows:

$$\lambda = \lambda_\infty + \lambda_{rel} = \lambda_\infty + \Lambda_{rel} / \sigma N_A \tag{16}$$

where Λ_{rel} is given by Eq. (11). The instantaneous term is also a function of solid fraction as the two states are different in compressibility (Table 2). Eq. (16) is similar to that presented in Ref. 18 differing only in that the compressibility coefficients are attributed here to the monolayer instead of bilayer. Predictions of the heterophase fluctuation theory well agree with experimental data on bending rigidity of bilayers [13], as shown in Fig. 6.

In conclusion, we touch upon the phenomenon of anomalous swelling in lipid multilayers near main transition studied by many workers [14, 15]. The anomalous swelling is observed in the same range of temperature above main transition where the heterophase fluctuations occur. The highly reduced bending rigidity and, therefore, enhanced undulation of bilayers causes an entropy-driven inter-bilayer repulsion force [32, 33]. Probably, this is a significant if not the major factor contributing to the anomalous swelling of multilayers. If so, the heterophase fluctuation theory gives ground for quantitative description of this factor of swelling. Indeed, our calculations (unpublished) show that the experimentally observed increase in the interbilayer distance is perfectly accounted for within the concept of heterophase fluctuations.

3.3. Long-range processes

3.3.1. The speed of phase propagation
The speed of the new-phase-front propagation, u, given that the front is flat, follows the expression

$$u = u_{max} [1 - \exp(\Delta G / N_A k_B T)] \tag{17}$$

where ΔG is the molar free energy of the transition from the parent unstable phase to the new stable one [21]. The maximum speed, u_{max}, corresponds to the case the system is far from the equilibrium, when the backward process (dissipation of new phase) is negligible. In terms of the kinetic theory of nucleation, the maximum speed is defined as $u_{max} = b_0 \times \sigma^{1/2}$ [18]. Using the data from Tables 1 and 2, one finds $u_{max} = 3.5$ cm/s. The curve for $u(T)$ is shown in Fig. 7. Within 5 K near transition temperature, the speed of propagation is an almost linear function of temperature

$$u_{\Delta T = 0-5 \, K} \approx 2 \times |T - T_0| \text{ mm/s} \tag{18}$$

Biological significance of this estimate will be discussed in Section 4.

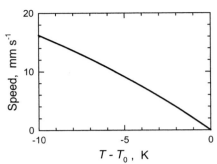

Fig. 7. Speed of solid phase front propagation in supercooled DPPC membrane as function of temperature. Calculated with Eq. (17). See text for details.

Unfortunately, no direct experimental data on the speed of phase propagation along a bilayer are available. The speed cannot be derived directly from the time course of structural changes monitored, for instance, by time-resolved X-ray diffraction [34] because the integral time course is determined by a combination of two processes: permanent creation of new nuclei (discussed below) and expansion of the nuclei having already been created.

3.3.2. The nucleus of critical size

In ideal infinite system at a temperature above the chain-ordering transition point, the new-phase nuclei of any size are thermodynamically unfavorable and, thus, solidification is impossible. The probability of the transition is non-zero only if a nonequilibrium metastable state is achieved (supercooled in the case of fluid-to-solid transition). The transition proceeds through the formation of a nucleus of critical size containing $n*$ molecules. Upon achieving this size, the nucleus starts to spread over the system irreversibly. The existence of the critical size is a result of interplay between the bulk and interfacial terms of the chemical potential in Eq. (2). The interfacial term of the potential is always positive and proportional to $n^{1/2}$ while the bulk term is negative in the supercooled state and proportional to n thus resulting in an extremum where $\mu*$ is a maximum. The values of $n*$ and $\mu*$ are therefore defined by the condition $(\partial\mu/\partial n)_{n=n*} = 0$. In conjunction with this condition, Eq. (2) yields for an infinite system

$$n* = \pi\sigma\gamma^2 N_A^2 / (\Delta S)^2 (T - T_0)^2 \tag{19a}$$

$$\mu* = \pi\sigma \gamma^2 N_A^2 / \Delta S(T - T_0) \tag{20a}$$

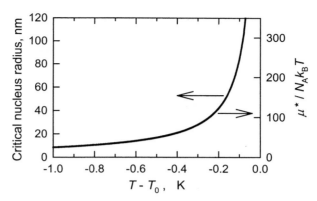

Fig. 8. Critical nucleus parameters (its size, left scale, and normalized chemical potential, right scale) as functions of the depth of supercooling in an infinite system. Calculated with Eqs. (19a) and (20a). The critical size is presented in terms of radius for further convenience when discussing the physiological aspects of phase transition. The ordinates are scaled in such a way that the two curves are superimposed one on another, that has been possible due to the

Both n^* and μ^* tend to infinity when T approaches T_0 as shown in Fig. 8.

Note that the membrane is assumed to be flat in this consideration (Subsection 2.1.1). Due regard should be given to this simplification when the equations are applied to vesicular membranes. In this case, Eqs. (19) and (20) are valid only if the radius of the critical nucleus is much less than that of the vesicle. Otherwise, due correction for the membrane curvature should be included into the interfacial term of the chemical potential expressed by Eq. (2), and this correction will change also the equations for the critical nucleus parameters. This factor however will be disregarded in our further qualitative analysis.

For a finite system comprised of N_d molecules, n^* does not exceed N_d, and the whole range of temperatures in this case splits onto two ranges, where

$$\left.\begin{array}{ll} n^* = \pi\sigma\gamma^2 N_A^2 / (\Delta S)^2 (T - T_0)^2 & \text{if } T < T_0 + (\pi\sigma)^{1/2}\gamma N_A/\Delta S \\[2mm] n^* = N_d & \text{if } T \geq T_0 + (\pi\sigma)^{1/2}\gamma N_A/\Delta S \end{array}\right\} \qquad (19b)$$

(note that ΔS is a negative value for fluid to solid transition). Then, the chemical potential of critical nucleus follows Eq. (2)

$$\mu^* = -\Delta S(T - T_0)\, n^* + 2\gamma N_A (\pi\sigma n^*)^{1/2} \qquad (20b)$$

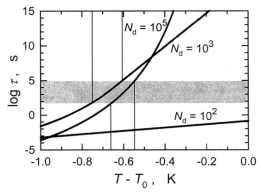

Fig. 9. Characteristic time, $\tau = k^{-1}$, of solidification of a population of DPPC membranes as a function of the depth of supercooling in a finite system. Calculated with Eqs. (19b), (20b), and (21) for the virtual domains of different sizes, N_d. The typical time range of laboratory experiment (from 1 min to 1 day) is shown in gray and the corresponding temperature intervals are shown with vertical lines.

and is a finite value at $T = T_0$.

Consider a population of virtually independent domains within a lipid membrane (the term domain is used here as an effective description of the structure correlation range, i.e., the domain is an area within which the structure is highly correlated being at the same time virtually independent of the state of neighboring areas). In the simplest case when the formation of critical nucleus but not the propagation of new phase front is the rate limiting step of transition, the critical nucleus formed in a domain can be considered the activated state of the domain in the sense of the theory of the rates of chemical reactions [35]. Then, the solidification is described by an approximate kinetic formula [19, 36]

$$k = (1/2)N_d\, b_0\, (4\pi n^*)^{1/2} \exp(-\mu^*/N_A k_B T) \tag{21}$$

where k is the probability for a given domain to be converted into solid state during unit time; N_d is the average number of molecules per domain; $(4\pi n^*)^{1/2}$ is an estimate for the number of boundary molecules in the critical nucleus; the factor ½ takes into account the equal probabilities for the critical nucleus to decompose or to grow further. The quantity k has the meaning of the rate constant of the reaction "fluid domains—solid domains".

Fig. 9 shows the temperature dependencies of the decay time $\tau = k^{-1}$ calculated for three domain sizes: 10^5, 10^3, and 10^2. The reasons for this choice and the conclusions from the results will be discussed in Subsection 3.3.4. The

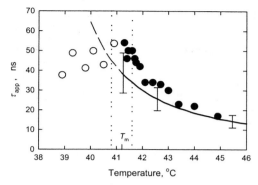

Fig. 10. Apparent acoustic relaxation time of heterophase fluctuations in multilamellar vesicles of DPPC. The quantity τ_{app} is defined as the reciprocal frequency of the peak of the sound absorption spectrum measured at a given temperature. Experimental data [25] for the stable fluid phase are shown with bold circles and for stable solid (ripple) phase in open circles. Line represents the theoretical results for the relaxation of solid nuclei in the stable fluid phase. The dashed part of the line is the extension of the small-scale solid-state nucleation process to the range of the metastable supercooled state. The error bars correspond to uncertainty in adjustable parameters of the phenomenological theory of nucleation (Table 2). Vertical dotted lines show the quadruple width of the phase transition. (Redesigned figure from Ref. 19.)

only what has to be mentioned here is the fact that the closer the temperature to T_0 the slower the process.

3.3.3. Relaxation slowing-down near transition point

There is a wide spectrum of relaxation times associated with the main-transition dynamics. Some of them display anomalous temperature dependence: when temperature (or pressure) approaches that of transition point, the relaxation time progressively increases. This behavior is characteristic of processes occurring on the time scales of about 50 ns (as observed with ultrasonic spectroscopy [24, 25]) and about 50 ms and several seconds (as observed by means of time-resolved x-ray diffractionin the pressure-jump experiments [34], nephelometry in the temperature-jump experiments [37], and also, by means of modulation calorimetry [10, 38, 39]). The slowing-down phenomenon is often considered an indication of proximity to a critical state. However, the same behavior can be a direct consequence of first-order transition mechanism. The most clear quantitative evidence of this statement is provided by an analysis of ultrasonic relaxation spectra reflecting the relaxation of heterophase fluctuations. The temperature dependence of apparent relaxation time is shown in Fig. 10. The points above transition temperature correspond to solid nuclei formation.

The closer is the temperature to transition point the slower is the relaxation. This feature is perfectly explained by the heterophase fluctuations, as revealed from the theoretical curve presented in the figure. For a qualitative idea of the reasons underlying the slowing of relaxation, one should consider that when temperature approaches the transition point, the size distribution of nuclei shifts towards larger sizes (Fig. 4) thus contributing to the slower pole of the acoustic relaxation spectrum.

The same anomaly displayed by the longer relaxation times (occurring on the scales of seconds and tens of millisecond) has not yet been explained quantitatively. I only point out that two factors may be responsible for such behavior: the probability of the critical nucleus formation and the speed of new phase propagation. Both quantities tend to zero at the transition point, as discussed in preceding subsections. This is a general feature of first-order transitions [20, 21, 40] also valid for the chain ordering/melting transition in membranes [18, 41, 42]. Therefore, any process depending on the new phase creation and growth should inevitably be retarded when the lipid bilayers approach this first-order transition—no assumption on the proximity of the system to a critical state is necessary for explanation of this phenomenon. The relaxation times may be of any conceivable time-scale, depending on the characteristic size and structure of particular subsystems responsible for the process.

3.3.4. Transition width, kinetic hysteresis, and virtual domains

The main transition in lipid membranes proceeds within a finite temperature interval called transition width. This fact is interpreted as a sign of existence of the cooperative domains of finite size (N_d) whose phase state is virtually independent of the state of surrounding areas. The thermodynamic basis for a phenomenological consideration of the domains is given by van't Hoff equation

$$\Delta H_{VH} = 4N_A k_B T / \delta T$$

where δT is the transition width and ΔH_{VH} the apparent transition enthalpy considered to be a sum of the contributions of N_d molecules embedded into the cooperative unit: $\Delta H_{VH} = N_d \Delta H$ (ΔH being the molar transition enthalpy directly measured with calorimetry). This equation is valid only if the transition occurs under truly equilibrium conditions.

One of the most important criteria of the equilibrium condition is the reversibility of the process—the absence of a hysteresis in the heating-cooling cycles. However, most calorimetric studies of lipid vesicles use a technique that allows measurements in the heating but not cooling courses. Due to this technical restriction a more simple criterion is often applied for checking if the

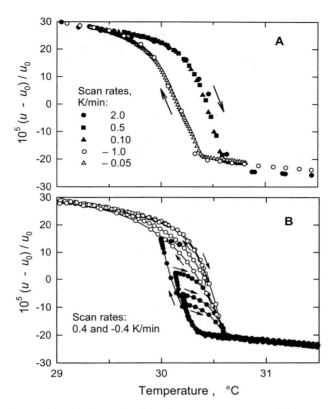

Fig. 11. Ultrasound velocity as function of temperature measured in aqueous dispersion of dilauroylphosphatidylethanolamine multilamellar vesicles during heating-cooling cycles around main transition [45]. Lipid weight fraction is 7×10^{-4}%. Ordinate is the relative lipid-induced increment of sound velocity. A: The entire transition hysteresis loop measured at different temperature-scan rates. B: The incomplete cycles started from the gel state to a point in transient zone and back (open circles) and those started from the liquid-crystalline state to a point in transient zone and back (bold circles); measured at scan rates 0.4 and −0.4 K/min.

equilibrium condition is fulfilled: it is assumed that if the melting curve for a given sample is many times reproduced displaying no detectable dependence of the curve shape on the temperature scan-rate, then the equilibrium condition holds and van't Hoff equation is applicable. This criterion however may be an oversimplification as judged by both theory and experiment.

Indeed, Fig. 9 shows that, if the cooperative domain is large, $N_d \sim 10^5$ (as if the whole monolayer in a vesicle of 100 nm in diameter is considered a domain) or $N_d \sim 10^3$ (a commonly accepted estimate for the virtual domain size in large multilamellar vesicles; see, e.g., Ref. 43), then a wide time scale of typical laboratory experiments (1 min to 1 day) correspond to a narrow interval of temperatures (~0.1 K) near a supercooling depth of 0.6-0.7 K. This should necessarily lead to a hysteresis character of phase transition in such membranes, and, furthermore, the width of the hysteresis loop should only slightly depend on the temperature scan-rate. Therefore, performing the temperature-scanning measurements in only one direction and observing the transition curve shape to be independent of scan rate one can falsely conclude that the process is equilibrium.

Only if the virtual domain consists of about 10^2 molecules (as it takes place in unilamellar vesicles), the measurements performed within the usual laboratory time-scale are expected to be truly equilibrium, because in this case the critical nucleus is small and the characteristic time is extremely short (see Fig. 9). In unilamellar vesicles, the transition indeed does not display a hysteresis [44].

This consideration is supported by many experimental works [10, 37, 38, 43, 45]. For instance, the hysteresis in dilauroylphosphatidylethanolamine (DLPE) multilamellar vesicles was thoroughly studied by means of sound velocity measurements [45] and the following results were obtained. The transition curve shape in either cooling or heating courses and the width of the hysteresis loop are almost independent of scan rate varied within 0.05-2.0 K/min (Fig. 11A). At the same time, the hysteresis loop shape displays a considerable dependence on the thermal pre-history of the sample as shown in Fig. 11B: the incomplete cycles started from solid are quite different in the shape from those started from fluid. Therefore, the entire transition curves in these experiments, being independent of the scan rate and perfectly reproducible, are nevertheless certainly irreversible, and hence, van't Hoff equation does not hold.

3.3.5. On the problems of macroscopic transition

It has to be stressed that, although the kinetic theory of the early steps of nucleation being extrapolated to the long-range processes appears to be in accord with experimental data at least on the qualitative level, there are some experimental observations that are inconsistent with the theory in its current state.

One of the most serious problems is encountered in attempts to rationalize the difference in the behavior of multi- and unilamellar vesicles. As follows from basic concepts of the theory the kinetic hysteresis is an intrinsic property of monolayers resulting from a high kinetic barrier associated with the critical nucleus formation. However, some delicate features of the hysteresis loop in multilamellar DLPE vesicles [45] indicate that it is unlikely that the nucleus formation is indeed a rate-limiting step of the transition at least in this particular system. These features (whose detailed description can be found in the original paper) suggest that a different factor, probably related to mechanical stresses accumulated between the concentric bilayer spheres, plays a major role. The stresses may arise, for example, because of violation of the geometric self-similarity in the multilayer vesicle during solidification of the constituting bilayers. The surface area of each bilayer considerably decreases, while its thickness increases [12]. As a result, the distance between the adjacent bilayers deviates from its equilibrium value, and the system becomes strained. A relaxation to the equilibrium distance can only occur via redistribution of lipid molecules between the bilayers. In the absence of a high destructive stress the process of the redistribution would be extremely slow—it might take many days [46] as it follows from the rate of interlayer lipid exchange measured under equilibrium conditions. Therefore, the role of steric interactions between the bilayers cannot be completely neglected when the entire phase transition is considered.

Another problem is related to the in-plane packing defects in bilayers. In the basic concepts of the theory, a lipid layer is considered to be a truly two-dimensional system with a high order of lipid packing in solid state. However, as any molecular membrane, the lipid bilayer is a *quasi*-two-dimensional system in which the out-of-plane collective motions are not forbidden and, hence, the energy of dislocations is largely compensated by membrane buckling: the less the bending rigidity of a membrane the higher the probability of "buckled" dislocations [47]. Thus, the long-range order is destroyed and the apparent cooperativity of phase transition decreased. This factor, important in the long-range transition processes, has not been explicitly considered in our theoretical background. It has been taken into account only by involving an "outside" phenomenological concept of the virtual cooperative domains (Subsection 3.3.4). A more developed kinetic theory of the long-range nucleation and macroscopic transition must consider inherently the accumulation of packing defects.

4. PHYSIOLOGICAL IMPLICATIONS

4.1. General remarks on the physiological role of chain-ordering transition

From the viewpoint of possible physiological role of the lipid chain-ordering phase transition, its kinetic features discussed in this chapter are of indubitable importance. Indeed, the heterophase fluctuations near the transition correlate with the permeability of bilayers, as discussed above. The solidification of a bilayer, associated with a considerable surface contraction [12], may cause formation of large water pores and even the membrane rupture [48]. It is also important that, if the solid state is provoked in a bilayer asymmetrically, i.e., in only one monolayer the opposite one being unchanged, a mechanical stress will develop resulting in ether a change of membrane curvature or even local rupture, provided that the solidification of the monolayer proceeds with a sufficiently high rate. The curvature and rupture may, in turn, stimulate membrane fusion [49]. Therefore, the transition may be an important factor regulating the membrane plasticity and transfer processes and, hence, its kinetic properties become of primary importance for understanding the molecular mechanisms of these phenomena.

As mentioned in Introduction, there are numerous reports on the observations favoring the idea that chain-ordering transition participates in physiological processes. Related literature has been the most comprehensively reviewed in recent works [5, 6]. Only the most important aspects will be outlined in this section with a focus on the points related to the phase transition kinetics.

Two general points have to be stressed first.

(i) A universal property of biological membranes is the negative net charge of their surface and, on the other hand, Ca^{2+} is a universal regulator of cellular processes [1]. As the interaction of Ca^{2+} with negatively charged membranes stabilizes the solid state, its concentration can be considered a factor equivalent to cooling of the system. In the biological membranes, as estimated from the data on model phospholipid membranes, the Ca^{2+}-induced increase of phase transition temperature can amount to 3-10 K per 1 mM of free Ca^{2+} [5, 6]. This is a sufficiently strong effect as the inner concentration of these ions in the vicinity of calcium channels in excited cells can achieve even higher values [50-52].

(ii) An entire biological membrane is a multicomponent and heterogeneous system where the phase transition is spread over a wide range of temperature. However, structurally distinct self-assembling domains are known to exist in the membranes [53]. They are cooperative structures and, hence, the functional phase transitions in such domains are, in principle, possible.

4.2. Phase transition and synaptic transmission (discussion of a hypothesis)

Except for the data on the temperature adaptation of the cellular membranes in microorganisms, mentioned in Introduction, there are numerous data indicating that, first, the chain-ordering phase transition plays an important physiological role in animals, and second, this is the brain where a phase-transition-dependent process is vital for higher animals [3, 5, 6].

4.2.1. The hypothesis of phase-transition-dependent synaptic transmission

A hypothesis has been proposed [4-6] suggesting the Ca^{2+}-induced solidification of self-assembling functional lipid domains in synaptic terminals to be the driving force of the fast neurotransmitter release in the central synapses in brain. The phase transition, resulting in surface contraction, causes a mechanical force that leads to the membrane rupture and fusion of a presynaptic vesicle with the plasma membrane resulting in a quick ejection of the vesicle

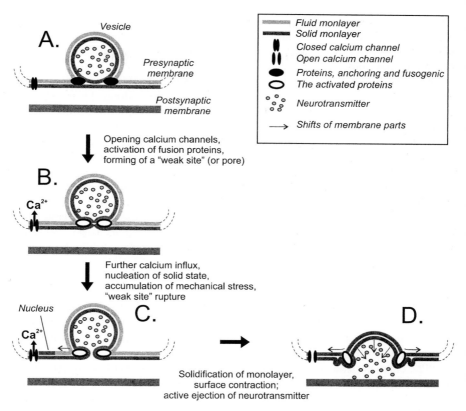

Fig. 12. Schematic diagram illustrating the phase-transition-dependent synaptic exocytosis (see Ref. 6 for details). The scheme is generally similar to that commonly accepted with one modification: the Ca^{2+}-induced phase transition is suggested to be the driving force of neurotransmitter release at the last step of the process (C → D). (Figure from Ref. 6.)

content into the synaptic gap between the pre- and postsynaptic membranes. The consideration of the release mechanism is, thus, in line with a general physical idea that, in the molecular "machines", the fastest large-scale structural rearrangements are only possible when they are essentially mechanical processes [54].

A molecular model based on this hypothesis and, also, accumulated the current state of knowledge about the molecular mechanisms of synaptic transmission is schematically illustrated in Fig. 12. It has been postulated that the presynaptic plasma membrane is composed of highly cooperative domains whose phase transition temperature is close to but less then the body temperature. Then, the Ca^{2+} influx induced by a nerve impulse will cause a metastable (supercooled) state at which the process of solidification of the membrane may start.

Within the context of this chapter, the most interesting events are the creation of the critical solid nucleus and its propagation along the membrane. Both take a finite time thus contributing to a synaptic transmission delay, which is a crucial characteristic for the brain excitability (and, hence, for evolutionary selection). These events develop within the synaptic active zone—a morphologically distinct part of the presynaptic plasma membrane about 200 nm in length, tightly conjunct with the postsynaptic membrane. It is bordered with dense structures and can be considered as a structural domain within which the presynaptic vesicles have been docked and activated awaiting for a nerve impulse to eject their content into the synaptic gap. Both the dense structures at the borders and the docked vesicles could be considered as large "foreign" inclusions that might serve as pre-existing "seeds" favoring solidification—this is the most essential point in view of the following consideration.

4.2.2. Kinetics of nucleation within the synaptic active zone

Let us assume that the phase transition kinetics in the active zones is, within an order of magnitude, the same as in the model DPPC membrane. Then, we can estimate an essential part of the delay time of synaptic transmission—the part associated with the creation and propagation of solid phase.

The characteristic size of the active zone (200 nm in length) corresponds to about 10^5 lipid molecules. Considering the whole active zone to be a cooperative domain ($N_d = 10^5$) and assuming the depth of the Ca^{2+}-induced supercooling to be about 1 K (which is a reasonable value [5, 6]) one finds that the time for the nucleus creation is in the order 0.1 ms (see Fig. 9). The speed of the solid phase propagation at this depth of supercooling is 2 mm/s (Eq. (18) and Fig. 7). Hence, the solid state, having been created, will cover the whole active zone in 0.1 ms or less (depending on where exactly in the zone the nucleus has been created). The two consecutive steps thus takes about 0.2 ms. This time is compatible with the known synaptic delay time in fast synapses [55].

What if the depth of supercooling in synaptic terminals does not reach 1 K even at the highest possible concentration of Ca^{2+}? In this case, the time of phase propagation will increase only in a linear proportion with the inverse depth of supercooling and, hence, the slow-down of this step of the process will be not as dramatic as the increase in the time of the critical nucleus formation. Indeed, the latter time is changing with temperature extremely fast (Fig. 9): twofold decrease in the depth of supercooling causes 10^8-fold increase (!) in the time of the nucleus creation, the time thus getting an absolutely unreasonable value from the biological point of view.

However, these estimates were based on the consideration of a homogeneous lipid system, which is not the case in the biological membrane. As mentioned above, the large inclusions embedded into the membrane could serve as preexisting "seeds" promoting the new stable phase state. To serve as the "seed", the inclusion must have the radius of curvature to be comparable with the radius of critical nucleus at the given temperature; then the interfacial part of chemical potential becomes negligible or even negative, the kinetic barrier thus being vanished. One can refer to Fig. 8 to get an idea of how large these inclusions must be at any depth of supercooling. An inclusions of 20 nm in radius, comparable with the size of presynaptic vesicles, might abolish the barrier at the depths of supercooling 0.4 K and deeper. Moreover, if the dense morphological structures surrounding the active zones (mentioned above) might serve as the "seeds", then the situation could be even more profitable because the radius of curvature of the inner side of such a ring-like "seed" is negative—the line tension force will thus be directed towards the center of the active zone and favor the phase front propagation at any temperature and any size of the growing solid phase.

Therefore, kinetic limitations associated with the critical nucleus formation might be got over by a special molecular arrangement of the active zones.

4.2.1. Phase transition and the stochastic character of synaptic exocytosis

Predictions of the phase-transition-dependent mechanism of neurotransmitter release are consistent with experimental data in different fields of physiology of animals [5, 6]. Here, only one example directly related to the kinetics of nucleation is presented.

Two unique features surprising investigators characterize the exocytosis in central synapses [56]. (i) Despite an active zone contains a dozen of the vesicles having been fully prepared to release their content, most of the time a nerve impulse does not elicit exocytosis. In some parts of brain fewer than a third of the impulses result in neurotransmitter release indicating that the exocytosis is essentially stochastic process. (ii) Moreover, when the release does occur, fusion is limited to the exocytosis of only a single vesicle indicating that there is a mechanism of lateral inhibition: one fusion event becomes immediately "known" to other vesicles within the same active zone.

Both features find a simple explanation within the phase-transition-dependent mechanism. The stochastic character of fusion is a direct consequence of the stochastic nature of solid-state nucleation: if a critical nucleus does not form during the calcium peak, fusion does not occur. If the nucleus has formed and starts to propagate over the membrane, the stress is accumulated in plasma membrane until the very first rupture and fusion event—then incorporation of the vesicle's lipid into the plasma membrane takes immediately the stress away, and thus, no more vesicles in the same zone can fuse until the initial state of the plasma membrane is recovered.

5. CONCLUDING REMARKS

Heterophase fluctuations at the main transition are observed not only in dipalmitoylphosphatidylcholine vesicles but also in other lipids. The magnitude of the fluctuations is virtually independent of whether the lipid system is multi- or unilamellar indicating that this is indeed a short-range process inherent in lipid bilayers (Kharakoz and Shlyapnikova, unpublished). The fluctuations exist also beneath the transition point reflecting the fluid-state nucleation. Their study is however aggravated by the existence of the irreversible ripple-phase in this range of temperatures.

Although the kinetic theory of nucleation in its current state cannot be directly applied to the entire phase transition in membranes, its extrapolation to the long-range scale can be considered a first approximation that is consistent with experimental data at least on qualitative level. The theory presents a unitary approach to the analysis of various "critical-like" anomalies in terms of the first-order transition mechanism. Further development of the theory requires that the interlayer interactions, the accumulation of packing defects, and other long-range effects be taken into account explicitly.

Predictive power of the theory appears to extend also on the physiological processes dependent on the main phase transition in biological membranes. In this connection, the most interesting directions of further physicochemical experimental studies might be aimed to elucidate the role of the structurally distinct self-assembling multicomponent domains and the transition-promoting large "foreign" inclusions in the kinetics of chain-ordering phase transition in lipid bilayers.

REFERENCES

[1]   B. Alberts, D. Bray, J. Lewis, M. Raff, K. Roberts, and J.D. Watson, Molecular Biology of the Cell. Garland Publishing, Inc., New York, 1994.
[2]   M. Bloom, E. Evans, and O.G. Mouritsen, Q. Rev. Biophys., 24 (1991) 293.
[3]   S.E. Shnol, Physicochemical Factors of Biological Evolution. Hardwood Academic Publishers GmbH, Amsterdam, 1981. Chapter 11.
[4]   D.P. Kharakoz, Biophysics 45 (2000) 554.
[5]   D.P. Kharakoz, Uspekhi Biol. Khim. [Progr. Biol. Chem.] 41 (2001) 333.
[6]   D.P. Kharakoz, Biosci. Rep., 21 (2001) Issue 5 (in press).
[7]   D.L. Melchior and J.M. Steim, Annu. Rev. Biophys. Bioeng. 5 (1976) 205.
[8]   J.R. Hazel, Annu. Rev. Physiol. 57 (1995) 19.
[9]   J.R. Hazel, S.J. McKinley, and M.F.Gerrits, Am. J. Physiol. 275 (1998) R861.
[10] R.L. Biltonen, J. Chem. Thermodynamics 22 (1990) 1.
[11] M. Porsch, U. Rakusch, Ch. Mollay, and P. Laggner, J. Biol. Chem., 10 (1983) 10.

[12] T. Heimburg, Biochim. Biophys. Acta. 1415 (1998) 147.
[13] L. Fernandez-Puente, I. Bivas, M.D. Mitov, and P. Meleard, Europhys. Lett. 28 (1994) 181.
[14] J. Lemmich, K. Mortensen, J.H. Ipsen, T. Hønger, R. Bauer, and O.G. Mouritsen, Phys. Rev. Lett. 75 (1995) 3958.
[15] J.F. Nagle, H.J. Petrache, N. Gouliaev, S. Tristram-Nagle, Y. Liu, R.M. Suter, and K. Gawrisch, Phys. Rev. E. 58 (1998) 7769.
[16] S. Mitaku, A. Ikegami, and A. Sakanishi, Biophys. Chem. 8 (1978) 295.
[17] D.P. Kharakoz, A. Colotto, C. Lohner and P. Laggner, J. Phys. Chem. 97 (1993) 9844.
[18] D.P. Kharakoz and E.A. Shlyapnikova, J. Phys. Chem. B, 104 (2000) 10368.
[19] D.P. Kharakoz, Russian J. Phys. Chem., 74, Suppl. 1 (2000) S177.
[20] J. Frenkel, Kinetic Theory of Liquids. Dover, New York, 1946. Page 366.
[21] A.B. Ubbelohde, Melting and Crystal Structure. Clarendon Press, Oxford, 1965. Chapters 11 and 14.
[22] V.I. Yukalov, Physics Reports—Review Sect. Phys. Lett. 208 (1991) 208, 395.
[23] D.P. Kharakoz, J. Acoust. Soc. Amer., 91 (1992) 287.
[24] S. Mitaku and T. Date, Biochim. Biophys. Acta. 68 (1982) 411.
[25] S. Mitaku, T. Jipo, and R. Kataoka, Biophys. J. 42 (1983) 137.
[26] D.B. Tata and F. Dunn, J. Phys. Chem. 96 (1992) 3548.
[27] R. Muller and F. Gallet, Phys. Rev. Lett. 67, 1991 1106.
[28] D. Papahadjopoulos, K. Jacobson, S. Nir, and T. Isac, Biochim. Biophys. Acta 311 (1973) 330.
[29] V.F. Antonov, E.Yu. Smirnova, and E.V. Shevchenko. The Lipid Membranes by Phase Transition. Nauka, Moscow, 1992. Chapter 4.
[30] J.F. Nagle and H.L.Scott, Biochim. Biophys. Acta, 513 (1978) 236.
[31] E.A. Evans and R. Skalak, Mechanics and Thermodynamics of Biomembranes. CRC Press, Boca Raton, 1980. Chapter 4.
[32] W. Helfrich, Z. Naturforsch. 33A (1978) 305.
[33] J.F. Nagle and S. Tristram-Nagle, Biochim. Biophys. Acta, 1469 (2000) 159.
[34] A. Cheng, B. Hummel, A. Mencke, and M. Caffrey, Biophys. J., 67 (1994) 293.
[35] H. Eyring, S.H. Lin, and S.M. Lin, Basic Chemical Kinetics, Wiley, New York, 1980.
[36] D.P. Kharakoz, Biophysics, 40 (1995) 1379.
[37] Y. Tsong and I. Kanehisa, Biochemistry, 16 (1977) 2674.
[38] B. Tenchov, H. Yao, and I. Hatta, Biophys. J., 56 (1989) 757.
[39] W.W. Van Osdol, M.L. Johnson, Q. Ye, and R.L. Biltonen, Biophys. J., 59 (1991) 775.
[40] K. Huang, Statistical Mechanics. Wiley, New York, 1987. Chapter 2.
[41] W.W. Van Osdol, R.L. Biltonen, and M.L. Johnson, J. Biochem. Biophys. Methods, 20 (1989) 1.
[42] B. Tenchov, Chem. Phys. Lipids, 57 (1991) 165.
[43] M. Posch, U. Rakusch, C. Mollay, and P. Laggner, J. Biol. Chem. 258 (1983) 1761.
[44] H. Nagano, T. Nakanishi, H. Yao, and K. Ema, Phys. Rev. E. 52 (1995) 4244.
[45] D.P. Kharakoz and G.V. Shilnikov, Biophysics, 42 (1997) 371.
[46] V.G. Ivkov and G.N. Berestovsky Dinamicheskaya Struktura Lipidnogo Bisloya [Dynamic Structure of Lipid Bilayer], Nauka, Moscow, 1981. Chapters 5 and 7.
[47] D.R. Nelson and B.I. Halperin, Phys. Rev. B 19 (1979) 2457.
[48] V.F. Antonov, E.T. Kozhomkulov, E.V. Shevchenko, A.N. Vasserman, E.Yu. Smirnova, S.A. Vosnesensky, and Yu.V. Morozov, Biofizika 31 (1986) 252.
[49] G. Cevc and H. Richardsen, Adv. Drug Deliv. Rev. 38 (1999) 207.

[50] R. Heidelberger, C. Heinemann, E. Neher, and G. Matthews, Nature 371 (1994) 513.

[51] R. Llinas, M. Sugimori, and R.B. Silver, Science 256 (1992) 677.

[52] G. Matthews, G. Annu. Rev. Neurosci. 19 (1996) 219.

[53] K. Simons and E. Ikonen, Nature, 387 (1997) 568.

[54] L.A. Blumenfeld, Problems in Biological Physics. Heidelberg: Springer-Verlag, 1981.

[55] B.L. Sabatini and W.G. Regehr, Nature 384 (1996) 170.

[56] Y. Goda and T.C. Südhof, Curr. Opin. Cell Biol. 9 (1997) 513.

Planar Lipid Bilayers (BLMs) and their Applications
H.T. Tien and A. Ottova-Leitmannova (Editors)

Chapter 8

Coupling of chain melting and bilayer structure: domains, rafts, elasticity and fusion

T.Heimburg

Membrane Biophysics and Thermodynamics Group,
Max-Planck-Institute for Biophysical Chemistry, 37077 Göttingen, Germany

1. INTRODUCTION

Lipid membranes in biology have a complex composition, consisting of hundreds of different lipids and proteins, plus various steroids like cholesterol. Therefore extensive research on the physical and physico-chemical properties of lipids took place in the 1970s and 1980s [1, 2]. However, the interest in lipids somewhat declined among biochemists, because at the time there was no obvious connection between the physical properties of the model systems and biological membranes. At the same time, the emphasis shifted to the development of single molecule recording methods (e.g. patch clamp [3]), which strengthened the belief in the predominant relevance of single proteins for biological function [4]. Textbooks

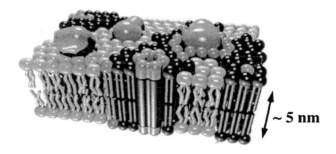

Figure 1: Schematic drawing of a biological membrane, showing lipids of different nature and state distributed inhomogeneously within the membrane plane. Proteins penetrate through the membrane or are bound to its surface.

of Biochemistry and Physiology [5] focus on reactions on the single molecular level, for example on ion conductance of single channel proteins as the potassium channel. The lipid membrane is considered by many scientists as the supporting matrix for the proteins, and no independent role is attributed to it. In this contribution we challenge this view. Many properties of surfaces cannot be understood on a single molecule level, for example the the lateral distribution of molecules, and physical features such as elastic constants, curvature, and fusion and fission events of vesicles.

The lipid composition is known to vary between different organelles within one cell. Mitochondria display a high fraction of charged lipids (mainly cardiolipin) and hardly any sphingolipids, whereas plasma membranes are rich in cholesterol and sphingolipids [6]. Nerve membranes, on the other hand, are rich in lipids with polyunsaturated fatty acid chains. The lipid composition of bacterial membranes depends on their growth temperature [7]. The reason for the large diversity in membrane composition is largely unknown.

Bilayers composed of a single lipid species display an order-disorder transition (the so-called melting transition) in a temperature regime, which is of biological relevance in a broader sense. This means that membranes consisting of extracted biological lipids have melting points close or not too far away from physiological temperature (-20° to +60° C). The low temperature lipid state has hydrocarbon chains predominantly ordered in an all-trans configuration and is for historical reasons called the 'gel' state. The high temperature state with unordered chains is called the 'fluid' state. Unsaturated lipids (containing double bonds in the hydrocarbon chains) display significantly lower melting temperatures as saturated lipids. Distearoyl phosphatidylcholine (DSPC) is a saturated lipid with 18 carbons per hydrocarbon chain, displaying a melting transition at about 55°C. Dioleoyl phosphatidylcholine (DOPC) is identical to DSPC except for one double bond in the center of each hydrocarbon chain. This small change in chemical structure leads to a lowering of the melting point to about -20°C. Since a large fraction of the lipids of biological membranes are unsaturated, there is a common but unjustified belief that melting transitions in biological membranes occur below the physiological temperature regime.

We will now outline why this is incorrect in our opinion. Fig. 2 shows heat capacity profile of different artificial and biological membranes. Fig. 2a is the melting of dipalmitoyl phosphatidylcholine (DPPC) multilamellar membranes, a lipid system that forms spontaneously upon dissolving the dry lipid in water. This system displays a highly cooperative melting peak at 41°C with a half width of about 0.05 K. A fine detail in this melting profile is the so called pretransition (see insert) which will play a role in section 5, but will not be considered here in more detail. Fig. 2b displays an equimolar mixture of two lipids, dimyristoyl phosphatidylcholine (DMPC) and distearoyl phosphatidylcholine (DSPC).

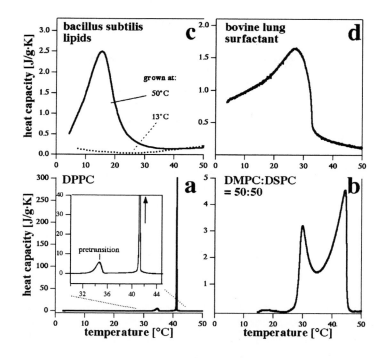

Figure 2: Melting profiles of different artificial and biological samples: a. Dipalmitoyl phosphatidylcholine (DPPC) multilamellar vesicles. The insert is a magnification which shows the pretransition in more detail. b. Dimyristoyl phosphatidylcholine(DMPC):distearoyl phosphatidylcholine(DSPC) equimolar mixture. c. Lipid extract from bacillus subtilis cells grown at 50°C and at 13°C (adapted from [7]). d. bovine lung surfactant [8].

DMPC alone melts at 24°C, whereas DSPC melts at 55°C. The melting profile of the mixture, however, displays a continuous melting event between 27°C and 46°C. Thus, the melting transition of a mixture is not just a cooperative melting at a temperature which represents the arithmetic mean of these temperatures (which for this mixture is about 39°C), but is rather represented by an extended temperature regime. This behavior can be understood by simple theoretical concepts such as regular solution theory, which is based on the macroscopic separation of gel and fluid domains in the melting regime [9]. Now regular solution theory is very useful for the understanding of the principles of phase diagrams, but is not accurate enough to explain details about the distribution of molecules. Mixtures of many components - as in biological membranes - can result in very broad melting profiles. The absence of a pronounced melting peak, found in many biological systems in no way means that there are no melting events. The chain melting, for example, may be so spread out over a large temperature regime that it becomes difficult to distinguish it from the base line.

The melting profile given in Fig. 2b was analyzed In more detail by Sugar and coworkers [10] using Monte Carlo simulations. In these simulations one can obtain snapshots of the distribution of lipids in the two-dimensional plane as a function of temperature. Seven representative snapshots, calculated at different temperatures, are displayed in Fig. 3. They show that the cause for the continuous nature of the melting is the lateral separation of lipids of different state and nature into nanoscopic, mesoscopic and macroscopic domains. Domains have also been found recently in biological membranes, where they are often called 'rafts' [11, 12, 13, 14, 15]. Rafts are domains, usually rich in sphingolipids and cholesterol, but also in certain proteins. The finding of these structures has refueled the interest in the lipids of biological membranes. In human erythrocytes sphingomyelin has to a very high percentage saturated chain and a high number of carbons (C24) [6]. Cholesterol is known to even further immobilize chain mobility [16] in the quantities found in these membranes (about 20%). Thus, rafts consist of lipids that have high melting points. The finding of rafts is itself a proof for the heterogeneous nature of biological membranes.

We will now discuss the melting profiles of biological samples. Fig. 2c displays

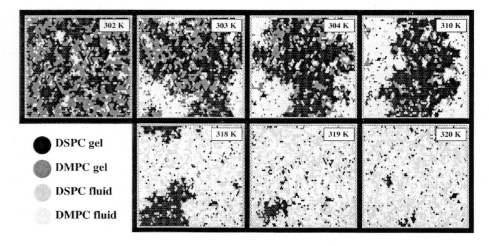

Figure 3: Series of snapshots from a computer simulation of the melting process of DMPC:DSPC=50:50 mixtures shown in Fig. 2b at various temperatures. Given in four different grey shades are gel state lipids of DSPC and DMPC, and fluid state lipids of DSPC and DMPC, respectively. The progress of domain formation on different length scales and the macroscopic demixing into fluid and gel domains at 310 K can clearly be seen.

the melting of a lipid extract from Bacillus subtilis [7], grown at two different temperatures. The lipids of the population grown at 50°C show a pronounced but broad melting peak at about 15°C which extends to much higher temperatures. The the lipids of the same cells grown at 13°C display no obvious melting anomaly, indicating that the lipid composition in this population is different and the cells felt a need to adjust their lipid composition differently from the population grown at higher temperatures. Van de Vossenberg and coworkers pointed out that this change in physical properties is mainly due to the change in iso-branched and antiso-branched fatty acid chains within the lipids.

Lung surfactant (Fig. 2d) forms a film on the lung surface and prevents it from collapsing. It contains several proteins, known as the surfactant proteins A, B and C (the sample in Fig. 1d had been washed to remove soluble proteins). Lung surfactant displays a pronounced but broad melting peak at 26°C. The upper end of this transition extends to physiological temperature. There are therefore cases where melting events in biological membranes can clearly be shown. However, as pointed out above, mixtures of many components can result in very broad melting profiles, which may be difficult to distinguish from the base line.

We can now pose the following question: Why may nature bother to adjust its lipid composition to environmental conditions?

2. CHAIN MELTING AND FLUCTUATIONS

2.1 Fluctuations in the state of the system

During the melting transition several membrane properties change. The enthalpy increases by about 20-40 kJ/mol, depending on lipid chain length. The volume increases by about 4% and the area by about 25%. Fig. 4 shows the fraction of fluid lipids for dipalmitoyl phosphatidylcholine (DPPC) unilamellar vesicles at the melting point (41°C, obtained in a computer simulation [17, 18]. Although the mean fraction of fluid lipids is, as expected, 50%, at any given time this fraction deviates from the mean. The deviations from the mean value are called fluctuations. During the melting process the fluctuations are strong. At the melting temperature the Gibbs free energies of the fluid and the gel state of a lipid are equal. This implies that it costs no free energy to shift the system from gel to fluid. Therefore, thermal fluctuations are sufficient to induce large alterations in the state of the membrane. The number of gel and fluid state domains will vary in time and between different vesicles, meaning that there are space and time dependent fluctuations in the enthalpy, the volume and the area of the membrane.

The 'fluctuations-dissipation' theorem relates the fluctuations in enthalpy (closely related to the fluctuations in the number of fluid lipids) to the heat capacity:

$$c_P = \frac{\overline{H^2} - \overline{H}^2}{RT^2} \tag{1}$$

Thus, the mean square deviation of the distribution shown in Fig.4 is proportional

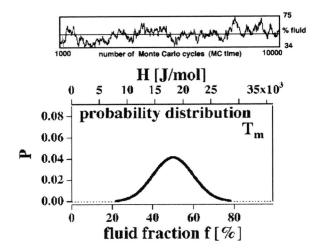

Figure 4: Top: At the melting point of a single lipid membrane, the number of fluid lipids (and its enthalpy) fluctuates around a mean value (50% fluid). Bottom: The distribution of states derived from these fluctuations is roughly given by a Gaussian profile (fat line). The half width of this distribution is related to the heat capacity. Data are taken from computer simulations of the melting profile of DPPC unilamellar vesicles at the melting point [18].

to the heat capacity. This is used in Monte Carlo simulations to derive the heat capacity from the noise produced in the simulation of the enthalpy of a system.

2.2 Fluctuations in composition: Domains and rafts

During a Monte Carlo simulation both the the percentage of fluid lipids and the domain sizes fluctuate around a mean value. Since with time lipids undergo a Brownian motion laterally within the membrane, the composition is also subject to fluctuations. The Monte Carlo snapshots shown in Fig. 3 clearly show that depending on temperature domains of different size and composition form. The lateral distribution of molecules has mainly been analyzed by computer simulations making use of experimental heat capacity profiles. Mouritsen and his group working in Denmark have contributed considerably to this understanding of the thermodynamics of lipid mixing [19, 20]. Domain formation is shown to be a function of the physico-chemical properties of the components, and it is strongly influenced by the presence of proteins. Another biomolecule of large importance is cholesterol which is very abundant in biological membranes (up to 30% of the lipid). The general influence of cholesterol on heat capacity profiles is a broadening and shift to higher temperatures. This has been used to construct phase diagrams of lipid bilayers containing cholesterol [16, 21, 20, 20, 22], leading to a new terminology for lipid phases.

The terms 'gel' and 'fluid' phase are somewhat misleading, since they do not specify what kind of order is changing. In physics the loss of lateral order (from crystalline to random) is called the solid-liquid transition. Lipids, however, also possess internal degrees of freedom, which may change from an all-trans chain conformation (ordered) to a random chain arrangement (disordered) via trans-gauche isomerizations. Thus, the phases in the lipid-cholesterol diagram can be referred to as solid-ordered (all-trans chains arranged on a crystalline lattice), liquid-ordered (all-trans chains in an unordered or glass-like lateral arrangement) or liquid-disordered (random chains with random lateral arrangement). The first phase resembles the 'gel'-phase, whereas the latter phase represents the 'fluid' phase [22].

The terminology of lipid-cholesterol phase diagrams is now also used to charac-

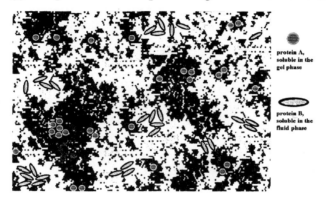

Figure 5: Fig.5. Schematic drawing of a membrane consisting of two lipids in the gel-fluid coexistence regime. Two different kinds of proteins (A and B) are designed to be better soluble in either gel or fluid phase. Thus, these two proteins do not interact in the gel-fluid coexistence regime.

terize certain domain types called rafts which are found in biological membranes. Originally rafts were identified by washing cell membranes in detergent. A certain detergent resistant fraction of the membranes ('rafts'), was found to be rich in sphingolipids, cholesterol and certain proteins. This was characterized in terms of islands of ordered lipid surrounded by a fluid matrix [11, 12, 14, 13, 15]. There is an ongoing debate between biochemists and physical chemists on whether these rafts are rigid structures or whether they are domains subject to fluctuations in physical state and composition. The author of this article favors the latter view, as he believes that nature in general tries to avoid stable structures. Fluctuating systems change as a response to changes in temperature, pH, and ionic strength, as well to the changes in membrane composition as induced by the binding of proteins or the action of phospholipases. This provides control mechanisms for nature to adjust to changes in environmental conditions.

It is quite clear that the lateral segregation of membrane components into domains will have a major impact on biological function, because it can influence reaction cascades. A two component lipid matrix in the gel-fluid coexistence regime is shown in Fig. 5. Two proteins are imbedded into the lipid matrix, one with a preference for the gel and the other one with a preference for the fluid phase. Assume furthermore that for biological activity these two proteins have to interact with each other. It is obvious that in the setting of Fig. 5 proteins A and B would not interact with each other because they are located in different domains. Thus, biological activity would be low. Everything that alters the domain arrangement thus changes the function of the membrane. This represents a major control mechanism of a membrane, which is based on the physics of the membrane ensemble rather than on single molecular properties. This feature of a membrane has been overlooked in recent decades and it is exciting to see the recent development of an understanding for the macroscopic control mechanisms of membranes.

2.3. Fluctuations in volume

An important observation is that for most lipid systems the change in volume and enthalpy during the melting transition is exactly proportional [23, 8]. The volume expansion coefficient (dV/dT) and the heat capacity (dH/dT) of a lipid sample are shown in Fig. 6. These two functions are exactly superimposable within experimental error. This has also been demonstrated for other artificial and biological lipid samples [8]. Thus

$$\frac{d\Delta V}{dT} = \gamma \frac{d\Delta H}{dT} \qquad \longrightarrow \qquad \Delta V(T) = \gamma \, \Delta H(T) \tag{2}$$

with $\gamma = 7.8 \cdot 10^{-4} cm^3/J$. Therefore, the fluctuations in enthalpy must be related to the fluctuations in volume in a proportional manner.

3. THE ELASTIC CONSTANTS AND RELAXATION TIMES

3.1. Volume compressibility

The fluctuation theorem can also be used to derive an expression for the isothermal volume compressibility:

$$\kappa_T = \frac{\overline{V^2} - \overline{V}^2}{V \cdot RT} \tag{3}$$

Similarly, the area compressibility of a membrane is given by

$$\kappa_T^{area} = \frac{\overline{A^2} - \overline{A}^2}{A \cdot RT} \tag{4}$$

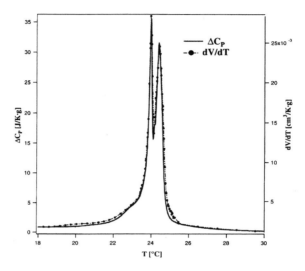

Figure 6: Heat capacity (solid line) and volume expansion coefficient (symbols) of DMPC large unilamellar vesicles (from extrusion). Both functions display identical temperature dependence.

Since enthalpy and volume are proportional to each other, it follows from Eqs. 2 and 3 that

$$\Delta\kappa_T = \frac{\gamma^2 \cdot T}{V} \Delta c_P \quad , \tag{5}$$

meaning that the change of the isothermal volume compressibility in the melting regime is proportional to the heat capacity. In other words, the compressibility is high in the melting regime. If the heat capacity is given in molar units, the volume V is the volume per mol of lipid. Interestingly, the constant γ was found to be the same for all lipids, even in a biological sample such as lung surfactant ($\gamma = 7.8 \cdot 10^{-4} cm^3/J$.). Eq.5 was shown to be correct from ultrasonic experiments [24, 25].

3.2. Area compressibility and bending elasticity

A proportional relation can also be assumed between area changes and enthalpy of the membrane. This relation is much more difficult to measure and therefore it was first used as a postulate [23]. If this were the case, one would also obtain a proportional relation between area compressibility changes and excess heat capacity:

$$\Delta \kappa_T^{area} = \frac{\gamma_{area}^2 \cdot T}{A} \Delta c_P \qquad . \tag{6}$$

From crystallographic and NMR data it has been deduced that the total area change in the transition is about 25%. It can also be concluded from these data that the γ_{area} is approximately $9 \cdot 10^3 cm^2/J$. If the heat capacity is given in molar units, the area A is the area per mol of lipid. Eq. 6 is important for the determination of the curvature elasticity of membranes. Evans [26] derived an expression for the bending rigidity of membranes, based on the simplifying assumption that bending requires lateral expansion of the outer monolayer and compression of the inner monolayer. Using his derivations, simple relations can also be obtained for the bending elasticity (or the bending modulus, respectively). For a symmetric homogeneous membrane the curvature Gibbs free energy is given by

$$G_{curv} = K_{bend} \left(\frac{1}{R}\right)^2 \qquad , \tag{7}$$

where the bending modulus, K_{bend}, is a function of temperature. It is related to the heat capacity as follows:

$$K_{bend}^{-1} = f \cdot \frac{1}{K_{bend}^{fluid}} + (1-f) \cdot \frac{1}{K_{bend}^{gel}} + \frac{16\gamma_{area}^2 \, T}{D^2 \, A} \Delta c_P \qquad , \tag{8}$$

where f is the fraction of fluid lipid, K_{gel} and K_{fluid} are the bending moduli of the pure gel or fluid phases, respectively, A is the membrane area and D is the membrane thickness [23]. The bending modulus, as predicted from Eq.8 and the experimental heat capacity profile of DPPC unilamellar vesicles are shown in Fig. 7. The symbols are measurements of the bending modulus as obtained from an optical trapping method [28, 27]. The good agreement between prediction and experiment justifies the above assumption that $\Delta A(T) \propto \Delta H(T)$.

3.3 Relaxation times

When a lipid membrane is perturbed by an external change in pressure or temperature, it relaxes into the new equilibrium state within a period of time called the relaxation time. The system is driven back to equilibrium by the thermodynamic forces, which represent a concept from non-equilibrium thermodynamics. It can be demonstrated that the fluctuations shown in Fig.7 can be used to calculate these forces [31]. Essentially it can be concluded that the relaxation times close to the

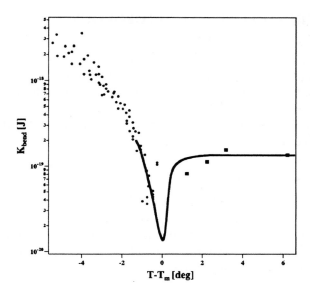

Figure 7: Temperature dependence of the bending rigidity as estimated from the heat capacity of unilamellar DPPC vesicles (solid line) [23] and as measured for unilamellar DMPC vesicles (symbols) [27, 28, 29, 30]. The membranes are more flexible by about one order of magnitude in the melting transition regime.

melting transition are proportional to the heat capacity

$$\tau = \frac{R\,T^3}{L} \cdot \Delta c_p \qquad (9)$$

where $L \approx 2 \cdot 10^{12}\,J^2 K / mol^2 s$.

3.4 Summary response functions

In summary it must be stated that membranes become very flexible close to the melting transition in a simple relation with the heat capacity changes. Thus, for all lipid systems (single lipids, lipid mixtures, biological membranes) showing melting events, the changes in the elastic constants can be predicted from heat capacities. If structural changes of membrane assemblies are possible - e.g. changes in vesicular shape or in membrane topology, they will be most likely to occur in the chain melting regime. Since large fluctuations are also equivalent to the relaxation times, equilibration to a new state will be slow when the elastic constants (and the heat capacity) are large.

4. CHANGES IN STRUCTURE AND TOPOLOGY

The large change in the elastic constants close to the melting transition gives rise to structural changes, because the free energy of bending becomes small. However, the bending free energy is still positive and requires a driving force to favor different geometries. One possible factor in the determination of membrane structure is its interaction with the solvent. Another possibility is the interaction with other molecules or surfaces.

4.1 General considerations

Let us now consider a membrane patch which undergoes a transition between two different geometries. It has been shown both, theoretically and experimentally [32, 18] that curvature broadens the melting transition, meaning that the heat capacity profiles of curved membranes are different from those with a flat geometry. That heat capacities should change can be understood intuitively by looking at

Figure 8: Equilibrium between a flat and a curved membrane patch. The curvature leads to a rearrangement of gel and fluid lipids on the two monolayers.

Fig.8. This figure shows the equilibrium between a flat and a curved geometry at the melting point (equal number of gel and fluid lipids). However, for the curved geometry the number of fluid lipids on the outer monolayer must be larger than on the inside, because gel and fluid lipids possess different areas. The flat geometry has equal numbers of gel and fluid lipids on both sides. Thus, the number of ways in which gel and fluid lipids can be arranged on both monolayers is different in the two cases, and the entropy of the two configurations is different, even though the total number of fluid lipids is identical for flat and curved geometry. Using statistical thermodynamics models the broadening of the heat capacity profile induced by a well defined curvature change can be calculated [18]. The free energy of a membrane with respect to the gel state is given by

$$G(T) = G_0 + \underbrace{\int_{T_0}^{T} \Delta c_P dT}_{H} - T \underbrace{\int_{T_0}^{T} \frac{\Delta c_P}{T} dT}_{S} \tag{10}$$

with G_0 being a term describing the interaction of the membrane with the environment. From the changes in heat capacity the elastic free energy change can be

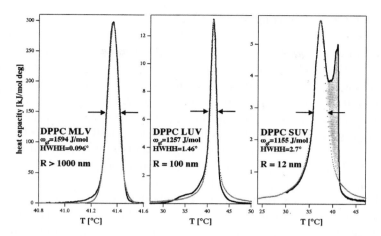

Figure 9: The effect of curvature on the heat capacity profiles: three different vesicular preparations of DPPC with different mean curvature are shown. Left: Multilamellar vesicles with a nearly flat surface. Center: Larger unilamellar vesicles with $R \approx 100nm$. Right: Small unilamellar vesicles from ultrasonication with $R \approx 12nm$ (the shaded area represents a remaining fraction of large vesicles). Solid lines represent experimental profiles, dotted lines are simulated profiles from Monte Carlo simulations [18].

calculated via

$$\Delta G_{elast}(T) = \underbrace{\int_{T_0}^{T} (\Delta c_P^{curved} - \Delta c_P^{flat})dT}_{\Delta H_{elast}} - T \underbrace{\int_{T_0}^{T} \frac{(\Delta c_P^{curved} - \Delta c_P^{flat})}{T} dT}_{\Delta S_{elast}} \qquad (11)$$

The total free energy change between a flat and a curved membrane segment is the sum of the elastic component and the interaction with the environment, which includes interactions with solvent, membrane associated proteins or surfaces. None of these terms depend on the melting process and they are collectively described by ΔG_0.

$$\Delta G(T) = \Delta G_0 + \Delta G_{elast} \qquad (12)$$

Defining $K = \exp(-\Delta G(T)/RT)$, the free energy of the membrane patch is given by

$$G_{membrane}(T) = G_{flat}(T) \cdot \frac{1}{1+K} + G_{curved}(T) \cdot \frac{K}{1+K} \qquad , \qquad (13)$$

where G_{flat} and G_{curved} have been obtain from heat capacity profiles of flat and curved membranes using Fig. 9 and Eq. 10. The heat capacity profile of a system in which equilibrium between flat and curved membrane segments is allowed can be calculated from Eq.13 using the relation $c_P = -T(\partial^2 G/\partial T^2)_P$. The elastic free energy term in Eqs.11 and 12 is generally positive for a symmetric membrane (Fig.10, left upper panel). If G_0 is zero, the equilibrium constant, K, between the two geometries (flat and curved) is small at all temperatures and the heat capacity of the membrane system is nearly identical to the of a flat membrane (Fig.10, left bottom panel). If, however, G_0 is sufficiently negative, the free energy $\Delta G(T)$ can become negative close to the melting transition (Fig.10, right upper panel). Thus, the equilibrium constant, K, may become larger than unity close to the melting transition and the membrane can then undergo a transition from a flat to a curved geometry. The heat capacity profile, calculated from Eq.13 now displays three peaks (Fig.10, right bottom panel). The left peak represents the structural transition from flat to curved, the right peak represents the transition from curved to flat and the center peak is the melting peak of the curved membrane. This splitting of the heat capacity profile only occurs if G_0 is negative, meaning that the interaction of the curved membrane segment with the environment is more favorable than the interaction of the flat segment with the environment.

Outside of the transition regime the flat geometry will prevail, because the bending elasticity is low and the free energy required to bend the membrane is high.

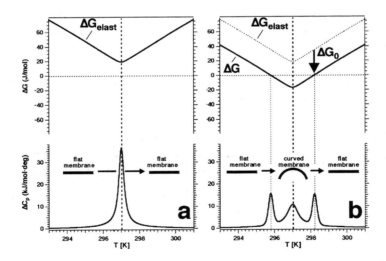

Figure 10: Left: Assuming two states of a membrane patch, one flat and the other curved, one can calculate the bending free energy difference ΔG_{elast} from the heat capacity profiles (top). If $\Delta G_{elast} > 0$, the flat geometry is thermodynamically stable. Right: If the free energy difference between flat and curved is shifted to lower values due to a different interaction of both conformations with the aqueous medium, ΔG can change its sign (top). Under these conditions conformational changes and the heat capacity profile splits into a pattern with three maxima, corresponding to the geometrical changes and the melting of the curved phase (bottom).

Within the transition regime, however, the membrane is very soft and can adapt to the most favorable interaction with the environment.

The theoretical considerations are of course based on some very simplifying thoughts. However, there are several experimental systems that exactly behave in this way. The splitting of heat capacity profiles into several peaks indicates the existence of a coupling of the melting transition with changes in membrane geometry. As examples, we will show in the next sections the transitions of anionic system from vesicles to extended membrane networks, and ripple phase formation. Both examples fit into the above pattern.

4.2 Membrane networks

The anionic lipid dimyristoyl phosphatylglycerol (DMPG) is a negatively charged lipid. It has been observed that this lipid at low ionic strength displays a rather peculiar behavior [33, 34, 35, 36, 37]. The basic features of the melting process are shown in Fig.11a. DMPG dispersions display a heat capacity profile which extends over a wide temperature regime. The integrated heat yields the heat of melting expected for this lipid. Thus, the profile represents a melting process which extends over a broad range of temperatures [33]. The profile shows three distinct peaks, one at the onset of the profile, on in the center and one at the upper end of the transition. Simultaneously, the viscosity of the lipid dispersion increases considerably between the two outer peaks (Fig.11a, center trace), and the dynamic light scattering intensity is reduced [33, 34], indicating a change in the geometry of the membrane assembly [36]. A detailed analysis (using negative stain electron microscopy, freeze fracture electron microscopy and cryo transmission electron microscopy) reveals a microscopic picture of the chain of events [33, 36] (Fig.11a, top): Below the heat capacity events the lipid dispersion exists as a dispersion of large vesicles, which undergo a structural change to a long range membrane network upon shifting the temperature to within the melting regime. Above the melting regime a dispersion with vesicles was found, with vesicle sizes much smaller than in the gel phase. From this it can be concluded that the vesicles in the gel phase have fused into a long range membrane network. Further heating to temperatures above the melting events resulted in spontaneous formation of vesicles. All of these events are fully reversible upon heating and cooling.
The splitting of the heat capacity profile is dependent on the interaction with the solvent. Increasing the lipid concentration [33], addition of polyethyleneglycol and increase of the ionic strength [36] reduce the temperature range of the calorimetric events. Since at low ionic strength electrostatic interactions are long ranged, we take these findings as a proof for that the free energy contribution, ΔG, which describes the interaction with solvent (or environment), favors the long range network. Reducing the water content or increasing the ionic strength

reduces the favorable interaction with solvent.

Obviously the formation of long range networks is a consequence of the long

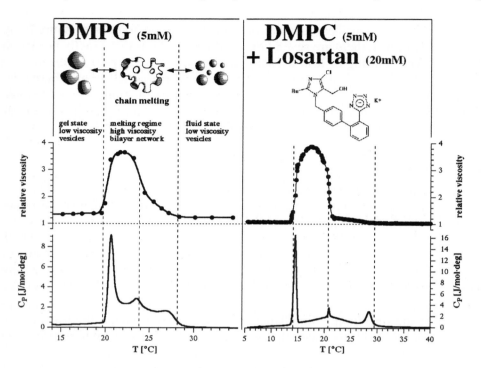

Figure 11: Left: Experimental heat capacity profile of a dispersion of the anionic lipid DMPG at 20mM ionic strength (bottom) and the increase in viscosity in the phase transition regime (center). The heat capacity profile is very similar to that calculated in Fig.10. The increase in viscosity indicates changes in the long range geometry of the lipid membranes. The chain of events is a transition from large unilamellar vesicles to bilayer networks back to smaller unilamellar vesicles upon temperature increase (top). Right: In zwitterionic membranes on can induce similar heat capacity profiles when adding Losartan to the membranes. The viscosity again increases in the melting regime. Thus, external charging of neutral membranes can lead to a coupling of chain melting and membrane structural changes.

range electrostatic contribution of the charged membranes. Interestingly, one can charge membranes with small drugs. The pharmaceutical company Merck distributes a drug called Losartan^{TM} against high blood pressure, which acts as an angiotensin receptor antagonist. Since this drug is rich in aromatic groups it binds to the head group region of zwitterionic membranes. Losartan carries one net negative charge and can thus be considered as an agent that charges the surface. The heat capacity profiles of DMPC in the presence of Losartan looks surprisingly similar to the heat capacity profiles of DMPG dispersions [38]. Fig. 11 (right) similar to the heat capacity profiles of DMPG dispersions [38]. Fig. 11 (right) shows the heat capacity trace of a 5mM dispersion of extruded unilamellar vesicles in the presence of 20mM Losartan and the corresponding viscosity profiles. Furthermore, in the melting regime the dynamic light scattering amplitude is reduced and the viscosity is increased, indicating structural changes very similar to those occurring in DMPG dispersions.

It seems likely the the charging of membranes generally promotes structural changes close to the chain melting transition. In our group we have further investigated DMPC at low pH (protonation of the phosphates), and phoshpatidylglycerols with different chain lengths and modified head groups (unpublished data). All of these lipids displayed structural changes close to the melting points. Thus, structural changes close to melting events seem to be a general phenomenon.

4.3 The pretransition

The pretransition is a low enthalpy transition found in many lipid systems below the main chain melting transition. Above the pretransition and below the main chain melting transition the membrane surface adopts a pattern of periodic ripples. This phase has therefore been called the ripple phase (or P'_β-phase, see Fig.12). In the literature the pretransition is commonly considered as an event indepen-

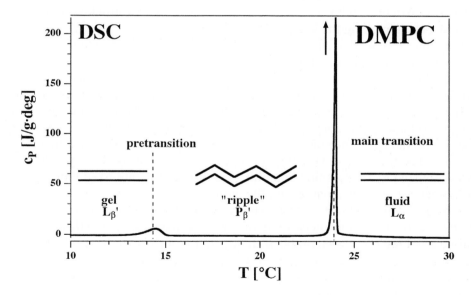

Figure 12: Heat capacity profile of DMPC multilamellar membranes, showing pre- and main transition. In between the two transitions the membrane adopts a different geometry, the so-called ripple phase, which displays periodic surface undulations with periodicities of about 10-20 nm. Above the main transition and below the pretransition the membrane is flat.

Figure 13: Ripple phase formation by antisymmetric, periodic fluid domain formation (black: gel lipids, white: fluid lipids).

dent from chain melting, caused by tilting of lipid chains and undulations in lipid position, and various models exist in the literature based on purely geometrical arguments. This line of thought, however, seems to be unplausible considering the following experimental findings: During the chain melting transition both enthalpy and volume change, and so do surface area and the chain order parameter (as determined from electron spin resonance [39, 31]). All these system properties also change in the pretransition. Furthermore, the ratio between enthalpy and volume change, as well as the ratio between change in order parameter and enthalpy, display very similar values in both, the main and the pretransition [31]. Since all physical events in the pre- and the main transition are very similar, it is likely that both transitions are caused by a similar physical phenomenon, namely chain melting. In fact, the pretransition may be a phenomenon very similar to the formation of extended bilayer networks in so far as it is also a structural transition linked to a chain melting transition.

One may thus assume that ripple formation takes place via the formation of a periodic pattern of fluid chains arranged along the principle axes of the hexagonal array of lipids (Fig.13, see also Fig.8). Such a behavior would automatically explain the enthalpy, volume, area and order changes in the transition regime, but also the observation that membrane ripple displays typically 120°-angles. Such assumptions have been implemented in a Monte-Carlo simulation [31], assuming that fluids domains on one side and gel domains on the other side result in a local curvature. The periodic arrangement of fluid and gel domains in such a calculation is stabilized by the fact that in multilayers or vesicles with fixed ratios of the areas of outer and inner monolayers, the number of fluid domains on both sides must be equal. Furthermore, it is known that the ripple phase is stabilized by hydration (as is the membrane network in the previous paragraph), and therefore the locally curved regions are favorable if the elasticity of the membranes is large as is the case close to the melting transition. Fig.14 shows the result of such a simulation. The heat capacity profile splits into two peaks due to the coupling with the change in geometry caused by the ripple formation. On the right hand side of Fig.14 some representative Monte Carlo snapshots show how the periodic ripples occur in between the pre and the main transition, whereas the membrane is nearly flat outside of this temperature regime.

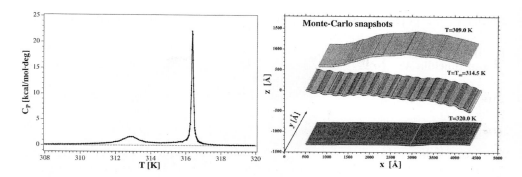

Figure 14: Left: Heat capacity profile of a simulated DPPC bilayer showing pre- and main transition. Right: Monte Carlo snapshots of membranes below the pretransition (top), in the ripple phase regime (center), and above the main transition (bottom). Adapted from [31].

4.4 Summary

Summarizing it must be stated that in general geometric transitions in vesicular structure become more likely close to the chain melting transition. If structural transitions occur in this temperature regime, they lead to a splitting of the heat capacity profile into several maxima which can be attributed to the structural changes and the melting of the intermediate phase. Thus, chain melting and structural transitions are coupled events which are likely to be quite common phenomena. If structural transitions are not taken into account, the interpretation of heat capacity profiles can be quite misleading.

The next chapter will focus on how proteins may influence the fluctuations and thus the elastic constants.

5. HEAT CAPACITY CHANGES INDUCED BY PROTEINS

The interaction of peptides with lipids can also be described using simple statistical thermodynamics models [17, 40, 18]. As in the lipid mixtures, the models are based on relatively few parameters: The melting enthalpy and entropy of the lipid system, and three nearest neighbor interaction parameters, describing the interfacial energy between gel and fluid lipids, between gel lipid and peptide, and between fluid lipid and peptide. These parameters can be obtained from experiment by fitting experimental heat capacity profiles. Some simple cases of peptide mixtures with lipid membranes were discussed by Ivanova et al. [18]. It can be shown that the influence of peptides on the c_P-profiles is mainly due to the miscibility of the peptide with the two lipid phases, gel and fluid, respectively. If,

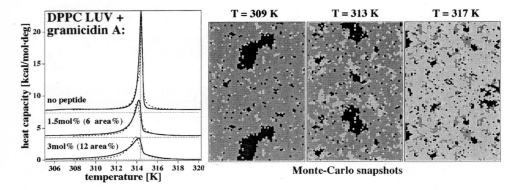

Figure 15: Interaction of gramicidin A with DPPC membranes. Left: heat capacity profiles of DPPC membranes containing 0, 1.5 and 3 mol% of gramicidin A. Solid lines: calorimetric experiments, dashed lines: Monte-Carlo simulations. Right: Monte-Carlo snapshots of the lipid/peptide-matrix from simulations at three different temperatures. The peptide (black dots) tends to aggregate in both, gel (dark grey dots) and fluid (light gray dots) phase. The aggregates in the fluid phase are smaller.

for instance, the peptide mixes well with the fluid phase (low nearest neighbor interaction energy) and does not mix well with the gel phase (high nearest neighbor interaction energy), the peptide would homogeneously distribute in the fluid phase but aggregate in the gel state. The corresponding heat capacity profile will be shifted to lower temperatures and display an asymmetric broadening at the low temperature side of the transition [18]. As an example we can take the interaction of gramicidin A with DPPC large unilamellar vesicles (LUV).

Fig.15 (left hand panel) shows the heat capacity profiles of DPPC LUVs with various concentrations of gramicidin A. It can be seen that the peptide leads to a slight shift of the c_p-profile to lower temperatures, and to an asymmetric broadening at the low temperature end. However, the influence on the heat capacity profiles is not very pronounced. One may conclude from this that the peptide interacts better with the fluid phase than with the gel phase, but tends to form aggregates or clusters in both phases. The Monte Carlo snapshots obtained during the simulation of the heat capacity profiles are shown on the right hand side of Fig.15. Peptides are shown as black dots, and the tendency to aggregate into clusters can well be seen in this simulation. This finding predicts the behavior of gramicidin A in real membranes and we have shown by atomic force microscopy that gramicidin A indeed aggregates in both lipid phases (Ivanova et al., submitted 2002).

In order to fully understand the physical behavior of lipid/peptide systems, it is important to investigate the effect of fluctuations in state (see above). Since the lipid/peptide matrix is not homogeneous, the fluctuations are also different at different locations of the matrix. To calculate such local fluctuations, we proceeded as follows: The lipid/peptide mixture was simulated with the parameters necessary to describe the experimental heat capacity profile. Usually the Monte-Carlo simulation was performed over several thousand cycles to equilibrate the system, including steps to switch the lipid state between gel and fluid state, and steps to let the peptides diffuse. Now we switch the diffusion off, and average the lipid state and derive the fluctuation in state for each lipid site. The fluctuations of the system already shown as snapshots in Fig.15 can be seen in Fig.16 for three temperatures, two of which are below the melting point of the pure lipid (314.3K) and one above the melting point. The height profiles correspond to the fluctuation strength. It can be seen that the fluctuations in the pure lipid matrix at all three temperatures are small, whereas in the peptide containing system the fluctuations are significantly altered. They are especially high close to the peptides.

Taking into account that high fluctuations are equivalent to flexible membranes

DPPC, no peptide   DPPC, 10% peptide

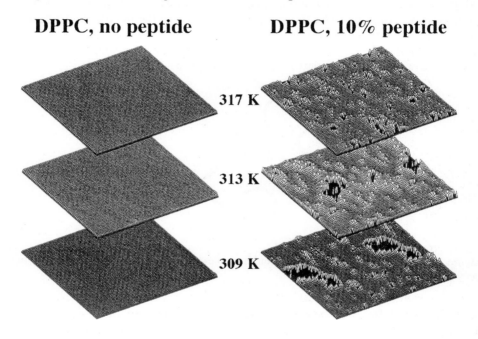

Figure 16: Local lipid fluctuations in a membrane without gramicidin A (left) and containing 10 mol% gramicidin A (right) at three different temperatures below and above the melting point of the lipid (314.3 K). The fluctuations of the lipid matrix in the absence of peptides are small and homogeneous at all given temperatures, but depending on temperature, the fluctuations in state are altered by peptides and can be significantly enhanced close to the gramicidin aggregates (black regions).

with slow relaxation rates, one has to conclude that proteins are able to alter the mechanical properties and the response times in the direct environment. It is very likely that nature will make use of this mechanism to control the elastic constants of the environment and its relaxation behavior. It has been estimated by Grabitz et al. [41] that in biological membrane relaxation times in the range from 1-10ms can be expected, which roughly corresponds to many enzymatic turnover rates and to opening times of membrane ion channels. Furthermore, since fusion of bilayers requires local curvature, proteins may locally induce fluctuations that facilitate membrane fusion.

6. STRUCTURAL CHANGES IN BIOLOGY

Structural changes in biology are quite important, for example in secretion or in synaptic fusion and fission, or endo- and exocytosis, respectively [42, 43] (Fig.17). Endocytosis is thought to be the key process by which the nerve pulse gets across

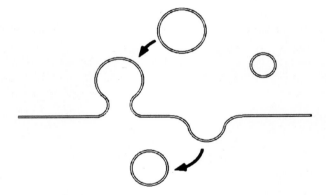

Figure 17: Schematic drawing of fusion and fission events in membranes, as they occur in endocytosis and exocytosis.

the synaptic cleft. The synapse contains a large number of vesicles of about 100nm in diameter. They contain various neurotransmitters, which are mostly small aromatic molecules that partition quite well into the head group region of lipid membranes. Endocytosis involves the fusion of synaptic vesicles with the presynaptic membrane. After the fusion initiation, neurotransmitters are released into the synaptic cleft, where they trigger an excitation of the postsynaptic membrane. In detail, the endocytosis is assumed to be initiated by calcium influx into the synapse, where the calcium interacts with a group of proteins called the snare proteins. These proteins are thought to undergo a conformational change, which then causes the fusion of synaptic vesicles with the presynaptic membrane. In these models, the main event in the fusion is the formation and a subsequent conformational change of the snare complex [42, 43]. Thus, it represents a model on a single molecular basis.

In contrast, the view of synaptic fusion by a certain theoretical physicists school is purely based on macroscopic elasticity theory. The theoretical concepts of membrane elasticity, introduced by Helfrich [44], have been used by various authors to describe fusion intermediates on the basis of local curvature [45, 46]. In these models proteins or individual molecules don't play a defined role. The elastic constants are assumed to be constant throughout the whole membrane of both, synaptic vesicle and presynaptic membrane. Therefore, the molecular biology vision and the theoretical physicists calculations are based on very different assumptions and concepts. None of these models make any use of heterogeneities in the membrane or fluctuations.

It seems to be a fact that snare proteins are involved in the fusion process. Furthermore, synaptic membranes are not homogeneous but contain cholesterol rich domains, where snare proteins are colocalized. When cholesterol is removed, synaptic fusion is inhibited [47]. Thus, it is also very probable that the concepts of fluctuations, closely linked to domain formation and protein clustering, will contribute to the physics of membrane fusion in synapses. Fluctuations are the physical way to make a system adaptable to changes in the environment, may they be changes in calcium concentration, protein binding, pH changes or sudden changes in temperature.

7. SUMMARY

In this contribution we have outlined how chain melting events are linked to changes of the elastic constants, based on a theory for the fluctuations of lipid membranes. Using these concepts we were able to explain several known examples of structural changes of membranes systems, including to reversible fission and fusion of anionic lipid vesicles and the formation of the ripple phase. Fluctuations were also shown to be the basis for domain formation. They are strongly influenced by temperature, but also by pH changes and calcium. The presence of proteins can locally change the membrane state and thus influence elastic constants, relaxation times and domain formation. Additional to the high chemical specificity of individual proteins, these physical and unspecific mechanisms provide a powerful mechanism for the control of biological function.

Acknowledgments

I am grateful to Thomas Schlötzer, Heiko Seeger and Denis Pollakowski from the 'Biophysics and Thermodynamics of Membranes Group' to let me use some of their data prior to publication. Lung surfactant was a kind donation of Fred Possmeyer (University of Western Ontario, London, Canada). Martin Zuckermann helped proof-reading the manuscript. TH was supported by the Deutsche Forschungsgemeinschaft (DFG, Grants He1829/6 and He1829/8).

References

[1] G. Cevc and D. Marsh. Phospholipid Bilayers: Physical Principles and Models. Wiley, New York (1989).

[2] R. B. Gennis. Biomembranes. Molecular structure and function. Springer, New York (1989).

[3] E. Neher and B. Sakmann. Nature, 260 (1976) 779.

[4] B. Hille. Ionic channels of excitable membranes. Cambridge University Press, Cambridge (1957).

[5] C. K. Mathews and K. E. van Holde. Biochemistry. The Bejamin/Cummings Publishing Company, Redwood City, Ca (1990).

[6] E. Sackmann. Physical basis of self-organization and function of membranes: Physics of vesicles. in: Structure and Dynamics of Membranes: From Cells to Vesicles. R.Lipowski and E. Sackmann, eds., Elsvier, Amsterdam (1995) 213–304.

[7] J. L. C. M. van de Vossenberg, A. J. M. Driessen, M. S. da Costa and W. N. Konings. Biochim. Biophys. Acta, 1419 (1999) 97.

[8] H. Ebel, P. Grabitz and T. Heimburg. J. Phys. Chem. B, 105 (2001) 7353.

[9] A. G. Lee. Biochim. Biophys. Acta, 472 (1977) 285.

[10] I. P. Sugar, T. E. Thompson and R. L. Biltonen. Biophys. J., 76 (1999) 2099.

[11] K. Simons and E. Ikonen. Nature, 387 (1997) 569.

[12] D. A. Brown and E. London. Ann. Rev. Cell Develop. Biol., 14 (1998) 111.

[13] T. Harder, P. Scheiffele, P. Verkade and K. Simons. J. Cell Biol., 141 (1998) 929.

[14] A. Rietveld and K. Simons. Biochim. Biophys. Acta, 1376 (1998) 467.

[15] M. Bagnat, S. Keranen, A. Shevchenko and K. Simons. Proc. Natl. Acad. Sci. USA, 97 (2000) 3254.

[16] J. H. Ipsen, O. G. Mouritsen and M. J. Zuckermann. Biophys. J., 56 (1989) 661.

[17] T. Heimburg and R. L. Biltonen. Biophys. J., 70 (1996) 84.

[18] V. P. Ivanova and T. Heimburg. Phys. Rev. E, 63 (2001) 1914.

[19] O. G. Mouritsen, M. M. Sperotto, J. Risbo, Z. Zhang and M. J. Zuckermann. Computational approach to lipid-protein interactions in membranes. in: Advances in Computational Biology. H.O. Villar, ed., JAI Press, Greenwich, Conneticut (1995) 15–64.

[20] O. G. Mouritsen and K. Jorgensen. Mol. Membr. Biol., 12 (1995) 15.

[21] L. Cruzeiro-Hansson, J. H.Ipsen and O. G. Mouritsen. Biochim. Biophys. Acta, 979 (1990) 166.

[22] M. Nielsen, L. Miao, J. H. Ipsen, M. J. Zuckermann and O. G. Mouritsen. Phys. Rev. E, 59 (1999) 5790.

[23] T. Heimburg. Biochim. Biophys. Acta, 1415 (1998) 147.

[24] S. Halstenberg, T. Heimburg, T. Hianik, U. Kaatze and R. Krivanek. Biophys. J., 75 (1998) 264.

[25] W. Schrader, H. Ebel, P. Grabitz, E. Hanke, T. Heimburg, M. Hoeckel, M. Kahle, F. Wente and U. Kaatze. J. Phys. Chem. B, (2002). In press.

[26] E. A. Evans. Biophys. J., 14 (1974) 923.

[27] T. Heimburg. Curr. Opin. Colloid Interface Sci., 5 (2000) 224.

[28] R. Dimova, B. Pouligny and C. Dietrich. Biophys. J., 79 (2000) 340.

[29] L. Fernandez-Puente, I. Bivas, M. D. Mitov and P.Meleard. Europhys. Lett., 28 (1994) 181.

[30] P. Meleard, C. Gerbeaud, T. Pott, L. Fernandez-Puente, I. Bivas, M. D. Mitov, J. Dufourcq and P. Bothorel. Biophys. J., 72 (1997) 2616.

[31] T. Heimburg. Biophys. J., 78 (2000) 1154.

[32] T. Brumm, K. Jorgensen, O. G. Mouritsen and T. M. Bayerl. Biophys. J., 70 (1996) 1373.

[33] T. Heimburg and R. L. Biltonen. Biochemistry, 33 (1994) 9477.

[34] K. A. Riske, M. J. Politi, W. F. Reed and M. T. Lamy-Freund. Chem. Phys. Lett., 89 (1997) 31.

[35] K. A. Riske, O. R. Nascimento, M. Peric, B. L. Bales and M. T. Lamy-Freund. Biochim. Biophys. Acta, 1418 (1997) 133.

[36] M. F. Schneider, D. Marsh, W. Jahn, B. Kloesgen and T. Heimburg. Proc. Natl. Acad. Sci. USA, 96 (1999) 14312.

[37] K. A. Riske, H.-G. Döbereiner and M. T. Lamy-Freund. J. Phys. Chem. B, 106 (2002) 239.

[38] E. Theodoropoulou and D. Marsh. Biochim. Biophys. Acta, 1461 (1999) 135.

[39] T. Heimburg, U. Würz and D. Marsh. Biophys. J., 63 (1992) 1369.

[40] T. Heimburg and D. Marsh. Thermodynamics of the interaction of proteins with lipid membranes, Birkhäuser, Boston (1996) 405–462.

[41] P. Grabitz, V. P. Ivanova and T. Heimburg. Biophys. J., 82 (2002) 299.

[42] R. Jahn and T. C. Südhof. Ann. Rev. Neurosci., 17 (1994) 219.

[43] R. Jahn and T. C. Südhof. Ann. Rev. Biochem., 68 (1999) 863.

[44] W. Helfrich. Z. Naturforsch., 28c (1973) 693.

[45] Y. Kozlovsky and M. M. Kozlov. Biophys. J., 82 (2002) 882.

[46] V. S. Markin and J. P. Albanesi. Biophys. J., 82 (2002) 693.

[47] T. Lang, D. Bruns, D. Wenzel, D. Riedel, P. Holroyd, C. Thiele and R. Jahn. EMBO J., 20 (2001) 2202.

Planar Lipid Bilayers (BLMs) and their Applications
H.T. Tien and A. Ottova-Leitmannova (Editors)

Chapter 9

Water transport

Peter Pohl

Forschungsinstitut fuer Molekulare Pharmakologie, Robert-Roessle-Str. 10, 13125 Berlin, Germany

1. INTRODUCTION

Membrane water transport is of central importance for any living cell. A variety of regulatory mechanisms enables the cell to maintain electrolyte and water homeostasis. For example, special signaling cascades allow adaptation to changes in external osmotic pressure [1, 2]. Mechanosensitive channels may transform the osmotic signal into an electric signal [3]. Despite its fundamental significance, the mechanism of water transport across lipid bilayers and epithelia as well as its coupling to ion transport is not clarified yet.

Two alternative mechanisms (Fig. 1A) have been proposed to account for solute permeation of lipid bilayers, (i) the solubility-diffusion mechanism and (ii) the pore mechanism:

(i)   Water partitions into the hydrophobic phase and then diffuses across [4-6]. The process of water diffusion across the hydrophobic barrier is related to membrane fluidity [7] and membrane structure [8].

(ii)   Water permeates through defects arising in the lipid bilayer due to thermal fluctuations [9, 10]. This "hopping" mechanism, which proposes that small solutes fit into holes available in the acyl-chain region of the bilayer, is supported by molecular dynamic simulation of solute diffusion carried out at an atomic level [11].

After a century of investigation, the mechanism of solute-solvent coupling and the pathways taken by water across an epithelium are still not entirely resolved [12]. Commonly, it is considered that osmosis best accounts for water movement [13]. Alternatively, molecular water pumps (Fig. 1C) are described which are able to transport water in the absence of an osmotic gradient or against it [14-17]. The uphill water transport requires its direct linkage to solute transport by cotransport proteins such as the brush border Na^+/glucose cotransporter [18].

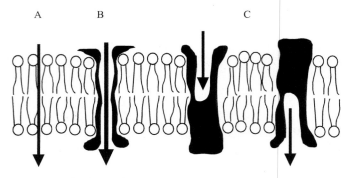

Fig. 1. Mechanisms of membrane water transport are (A) diffusion across the lipid bilayer or across transient defects, (B) facilitated diffusion across pores and (C) carrier induced cotransport with other substrates (e.g. sodium and glucose)

However, secondary active water transport by protein co-transporters is not the only explanation for the water flux associated with Na^+/glucose cotransport in oocytes. Instead, it may be driven by local osmotic gradients [19]. The role of molecular water pumps was further questioned by the finding that the water channel protein (Fig. 1B), aquaporin-1 (AQP1), accounts for 80 % of the water permeability in proximal kidney tubules [20]. AQP1 is also required for NaCl-driven water transport across descending thin limbs of kidney [21].

Members of the family of membrane channel proteins termed "aquaporins" are found in all life forms, including archaea, eubacteria, fungi, plants, and all phyla of animals [22]. Aquaporins accommodate the requirement posed by many physiological processes for rapid transmembrane water diffusion. Several studies have suggested that water ion cotransport may occur in some aquaporins [23-28], raising the question of whether ion conductivity is a general property of aquaporins. For example, before molecular water channels were discovered, major intrinsic protein of lens (now known as AQP0) was reported to conduct anions in a voltage dependent manner when added to planar lipid bilayers [29].

One of the most significant impediments to the accurate determination of the water transport mechanism across lipid bilayers, epithelia and aquaporins is posed by near-membrane stagnant water layers, the so called unstirred layers (USLs). As has been pointed out repeatedly over a period of four decades [13, 30, 31], USLs result in solute polarization during transmembrane fluid flow and act to oppose any imposed osmotic gradient and thereby diminish the magnitude of the driving force for fluid movement. Methods that fail to correct for osmolyte dilution within the USL, e.g. space averaged measurements of dye dilution in the vicinity of airway epithelia, give water permeability values, P_f, that are subject to an error of 100 % and more [32, 33]. With an error smaller than 10 %, the accuracy of water flux measurements based on scanning ion sensitive microelectrodes is much better. They allow the determination of the exact near-membrane solute concentration distribution [34, 35] and thus the separation of the resistances to water flow that are originated by the membrane itself and by the adjacent unstirred (stagnant) water layers (USL). The microelectrode technique enabled an improvement of the USL theory [36, 37], the investigation of the reconstituted lipid barrier of epithelial cells [38], the visualization of ion and water flux coupling by solvent drag [39, 40], monitoring of membrane dehydration (decrease of membrane water permeability) upon protein adsorption onto the membrane [41] and accurate measurements of AQP water selectivity [42, 43].

2. UNSTIRRED LAYERS

2.1 Nernst's model

The USL concept was introduced by Nernst [44]. It implies the existence of a completely unstirred water layer in the immediate membrane vicinity that yields abruptly to a perfectly stirred region [for review see 31]. Transport across the USL is assumed to occur solely by diffusion. It is well known that even in vigorously stirred systems there is a USL that leads to concentration differences, i. e. water that passes through the membrane dilutes the solution it enters and concentrates the solution it leaves [45]. The size and importance of the solute concentration gradients depends on their rate of dissipation through backdiffusion of the solute.

Within the USL, the solute concentration is a function of the distance to the membrane. The USL thickness, δ, is defined in terms of the concentration gradient at the membrane water interface [see e.g. 30]:

$$\frac{|c_s - c_b|}{\delta} = \frac{\partial c}{\partial x}\bigg|_{x=0},$$

(1)

where x is the distance from the membrane. c_b and c_s denote the solute concentrations in the bulk and at the interface respectively.

Assuming that the only motion in the USL is the osmotic flux itself and that c depends only on the distance x from the membrane [46], the steady state flux J of a membrane impermeable solute can be expressed as:

$$J = D\,dc/dx + vc = 0$$

(2)

where v is the linear velocity of osmotic volume flow. Integrating between the edge of the USL ($x = \pm \delta$) where $c(\pm\delta) = c_b$ and the membrane surface ($x = 0$) one obtains (compare Fig. 2) for the hypotonic USL [31]:

$$c(x)=c_b e^{(\delta-x)v/D} \tag{3}$$

The volume flow is directed toward the other side of the membrane, where the concentration distribution of the solute is given by:

$$c(x)=c_b e^{(x-\delta)v/D} \tag{4}$$

after replacing v by $-v$ in Eq. (2).

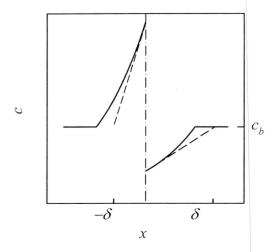

Fig. 2. Theoretical concentration changes as a function of the distance x to the membrane (dash-dotted line). The profiles were calculated with Eqs. (3) and (4). From the concentration gradient at the membrane water interface, the thickness of the USL was derived according to Eq. (1).

This approach reveals at least two shortages:

(i)   According to Eqs. (3) and (4), the greatest gradient of solute concentration is located at the outer edge of the USL at the hypertonic side of the membrane [47]. This highly unrealistic result (Fig. 2) is a consequence of the inconsistency of the commonly accepted model of the USL, assuming the existence of a completely unstirred region that yields abruptly to a perfectly stirred region.

(ii)   In contrast to the assumption $c(\pm\delta) = c_b$ that has lead to Eqs. (3) and (4), the concentration at the distance - δ is not equal to c_b (Fig. 2). This inadequacy has long been known but in the absence of an alternative treatment, the assumption was not taken literally [48]. δ determined in terms of the concentration gradient at the wall (Fig. 2) was commonly regarded as the effective unstirred layer thickness that if inserted into the Eqs. (3) and (4) allows a rough estimation of the near-membrane concentration shift [49].

2.2   Microelectrode measurements of near-membrane solute concentration gradients

Microelectrode measurements of solute polarization in the vicinity of lipid bilayers provide an excellent opportunity to improve the USL model. An ion sensitive electrode can easily be made of a glass capillary by filling its tip with a commercially available ionophore cocktail according to the procedure described by Amman [50]. The tip should have a diameter of about 1 - 2 μm and the 90 % rise time should be below 1 s. In this case, artifacts due to a very slow electrode movement (1 – 4 μm/s) are unlikely. Possible effects of time resolution or distortion of the unstirred layer can be tested by making measurements while moving the microelectrode toward and away from the bilayer. If no hysteresis is found, it can be assumed that an electrode of appropriate time resolution is driven at rate that is slow relative to the rate at which any electrode-induced disturbance of the USL reaches a "stationary" state [36]. As an example, the polarization of sodium ions is shown in Fig. 3. The size of the interfacial sodium concentration gradients depends on the osmotic gradient and on the lipid used for membrane formation. In the vicinity of a membrane made from fully saturated lipids (diphytanoyl-phosphatidylcholine, DPhPC), the concentration shift is smaller than in the vicinity of an asolectin membrane that contains unsaturated lipids (Fig. 3). This result agrees well with the observation that P_f increases with the degree of lipid unsaturation [51].

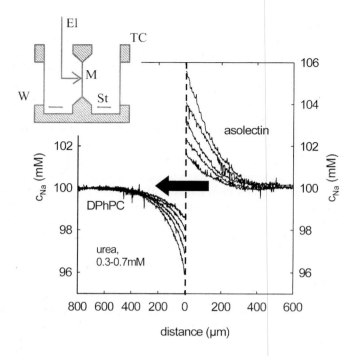

Fig. 3 Experimental solute concentration changes as a function of the distance x to the membrane (dotted line). The profiles were obtained with a scanning Na^i-sensitive microelectrode. Increases of the osmolyte concentration in steps of 0.1 M resulted in a stepwise polarization increase. Changing the lipid composition from DPhPC (left) to asolectin (right) increased the interfacial concentration gradient as well (adopted from [36]). The abbrevations are: El – microelectrode, M – planar membrane, St – stirrer, TC – Teflon chamber, W – observation window

The experiment (Fig. 3) clearly shows that the greatest gradient of solute concentration is not located at the outer edge of the USL (compare Fig. 2) but at the membrane/water interface.

Calculation of P_f from the measured concentration profile according to Eqs. (3), (4) and:

$$P_f = \frac{v}{c_{osm}V_w} \tag{5}$$

results in an overestimated value [36]. c_{osm} and V_w are, respectively, the effective transmembrane osmolyte concentration gradient and the molar volume of water.

In the case of Na$^+$ dilution (Fig. 3), calculation of δ (Eq. 1) reveals a dependency from c_{osm}. Intuitively this result is expected because in a gentle stirred system, the osmotic volume flow is the only origin of convection in the membrane vicinity. However, its effect on δ is neglected by the conventional USL model where δ is a fixed parameter.

A unique δ is assumed for all diffusing substances [34, 52-54]. Once determined from experiments with an indicator, δ was used to assess diffusive restrictions of other compounds [31, 55-59]. However, the validity of this approach is questionable. The distance from the membrane, from where a particle migrates by diffusion depends on the competition between diffusion and convection. The distance is shortened due to vigorous stirring or, on contrary, lengthened by an increase in the mobility of the solute.

Inadequacies of the conventional USL model have been known for almost as long as the approximation has been used. Introducing the correction for USL effects (see Eqs. 2 - 4) Dainty [46] has emphasized that it is approximate and probably an overcorrection. An improved model has to drop the assumption about the existence of a completely unstirred region. A complete description of the transmembrane convective flow requires instead that fluid motions are taken into account.

2.3   Hydrodynamic model

Proper solutions of the equations for simultaneous convection and diffusion have only proved possible for a few special geometries. According to Levich's theory [60], δ is a function of the diffusion coefficient of the solute, the velocity and viscosity of the solution. Pedley [61] has developed a model for the interaction between stirring and osmosis, where he proposed that the stirring motions in the bulk solution, which counter the osmotic advection, can be represented as a stagnation-point flow. His examination of the hydrodynamic description of the two dimensional flow reveals that it is possible for the solute concentration to be independent of the coordinate parallel to the membrane. Both Levich's and Pedley's models assume that in a region very close to the membrane, the stirring velocity increases with the square distance to the membrane. Assuming that v is small in comparison to the convection velocity in the bulk, Eq. (2) may be transformed into:

$$J = D\,dc/dx + (v - ax^2)c = 0 \tag{6}$$

where a is a stirring parameter. Integrating in a region, where $-\delta < x < \delta$ and assuming that the solute concentration at the membrane surface $c(0) = c_s$, one obtains for both the hypotonic and the hypertonic USLs:

$$c(x) = c_s e^{\frac{-vx}{D} + \frac{ax^3}{D}} \tag{7}$$

Fitting of Eq. (7) to experimental concentration profiles allows to obtain v and thus provide a potentially useful way of determining P_f with the help of Eq. (5). Although the hydrodynamic description of the transmembrane volume flow given in Eq. (7) is derived due to an oversimplification of the system, the concentration distribution within the USL is described accurately. Experimental proof has been obtained by microelectrode aided measurements of the flux of ions that are dragged by water across gramicidin channels (compare §5 of this chapter). In this case, the ion flux determined simultaneously by current measurements was identical [39].

The hydrodynamic model, Eq. (7), predicts δ to dependent on the diffusion coefficient of the solute. Herein it differs from the conventional model (compare Eqs. 3 and 4). Indeed, a detailed theoretical analysis of the interaction between stirring and osmosis reveals that δ depends on the diffusion coefficient of the solute [61]. The same conclusion is drawn in the case of transmembrane solute diffusion [60]. By means of the microelectrode technique, experimental evidence for these theoretical predictions was obtained [37]. The experiments have shown that, under given experimental conditions, there is a δ corresponding to each molecule diffusing in the solution [37]:

$$\frac{\delta_1}{\delta_2} = \frac{\sqrt[3]{D_1}}{\sqrt[3]{D_2}} \tag{8}$$

Because δ depends on which species is diffusing, several USL thicknesses exist simultaneously.

Table 1. Comparison of the USL models

Nernst's model	Hydrodynamic model
A completely unstirred region adjacent to the membrane yields abruptly to a perfectly stirred region.	With increasing distance from the membrane, the convective flow velocity increases continuously.
δ is unique for all molecules.	δ is a function of D.
The greatest concentration gradient is located on the outer edge of the USL.	The greatest concentration gradient is located on the membrane water interface.
δ does not depend on v.	δ is a function of v.

Taking into account a gradual change of the convective flow velocity, the hydrodynamic model of USL overcomes the main shortages of Nernst's model (Tabl. 1). Applying the new model to concentration profiles obtained with the scanning microelectrode technique, the "nuisance" concentration polarization within the USL that previously hampered water flux measurements, transforms into a useful phenomenon enabling accurate water flux measurements across bilayers [38] model [39, 40] and water [42, 43] channels.

3. WATER DIFFUSION ACROSS LIPID BILAYERS

In artificial vesicular and planar lipid bilayers, P_f tends to be comparatively high. Usually, it varies between 20 and 50 µm/s [45]. Based on this observation, the necessity of water channel proteins has long been questioned. For example, being equipped with AQP3 and AQP4 on the basolateral membrane and AQP2 on the apical membrane, epithelial cells from the kidney inner medulla have a transcellular water permeability of only 26 µm/s [62]. Thus their P_f seems to be comparable to the P_f of a "normal" lipid matrix. It should however be noted that cells which transport water have evolved specialized membranes. These are characterized by:

1) the inclusion of cholesterol which condenses membranes and elicits tighter packing [63],
2) an increase in the degree of acyl chain saturation and

3)   utilization of sphingolipids which by virtue of the unconventional amide linkage are able to form hydrogen bonds with each other and with cholesterol [64, 65].

In addition, these cells are able to maintain membrane lipid asymmetry by incorporating different lipids into the outer and inner leaflets of the bilayer. Thus, the outer leaflet of epithelial cells is enriched in glycosphingolipids and the inner leaflet in phosphatidylethanolamine and phosphatidylserine [66]. In order to understand the functional consequences of lipid composition and the role it might play in creating a barrier to water permeation, the inner and outer leaflet of the MDCK Type 1 apical membrane was modeled by liposomes which had the same composition as either the inner or outer leaflet [65]. Liposomes, however, do not allow to test whether each leaflet of the bilayer would resist solute permeation independently.

To form asymmetric planar bilayers, which mimic the MDCK apical membrane, the Montal technique [67] was exploited. Microlelctrode aided measurements of the interfacial Na^+ concentration gradient enabled determination P_f (at 36 °C). From Eqs. (5) and (7), P_f was derived to be equal to 34.4 ± 3.5 μm/s for symmetrical inner leaflet membranes and to 3.40 ± 0.34 μm/s for symmetrical exofacial membranes (Fig. 4). If P_{AB} is the permeability of a bilayer composed of leaflets A and B, and if P_A is the permeability of leaflet A and P_B is the permeability of leaflet B, the permeability of an asymmetric membrane can be *estimated* from [38]:

$$\frac{1}{P_{AB}} = \frac{1}{P_A} + \frac{1}{P_B}$$ (9)

Eq. (9) returned 6.2 μm/s. Water permeability *measured* for the asymmetric planar bilayer was 6.7 ± 0.7 μm/s (Fig. 4), which is within 10 % of the calculated value. Thus, direct experimental measurement of P_f for an asymmetric planar membrane validates the use of Eq. (9).

Fig. 4 indicates that most of the resistance offered by the MDCK apical membrane to water permeation is contributed by the outer leaflet which is enriched in cholesterol and sphingolipids. Thus, microelectrode aided measurements of the concentration polarization confirmed that each leaflet independently resists water permeation and that the resistance is additive [38] in line with previous predictions [65, 68, 69]. It is concluded that cells specialized in water transport achieve barrier function due to special lipid composition. This is the background required for rapid water flux adaptation by regulation of water channel protein abundance in the plasma membrane.

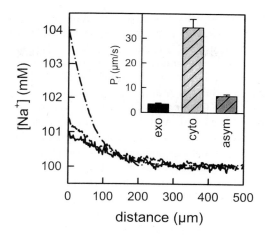

Fig. 4. Water permeability across symmetric and asymmetric planar lipid membranes. The polarization of Na⁺ was induced by a transmembrane urea gradient of 1 M. Symmetrical membranes were formed from exofacial (spline line) or cytoplasmic lipids (dashed-dotted line). P_f for the asymmetric planar membrane (dotted line) is mainly determined by the exofacial leaflet (adopted from [38])

4.   WATER CHANNELS

Transport of water across cell membranes must be sufficiently selective to prevent the movement of other solutes, ions, and even protons [49]. At the same time, water absorption, release or redistribution occurs with remarkable speed and precision. A molecular explanation for these cellular processes was provided by the discovery of the protein water channel, AQP1 [70]. With limited exceptions, every organism seems to have some members of AQP family including vertebrates, invertebrates, microbials, and plants [71]. The human AQP family contains now ten homologous proteins [72]. A much more expanding number of AQP is identified in plants - at least 25 only in Arabidopsis thaliana [73].

Water selectivity of AQP1 was tested after reconstitution into liposomes. P_f of AQP1 proteoliposomes was up to 50-fold above that of control liposomes, but permeability to urea and protons was not increased [70]. Contrasting evidence was obtained by method that is much more sensitive to ion flow – by current measurements of planar bilayers reconstituted with AQP. The plant AQP homologue, nodulin-26 exhibited multiple, discreet conductance states ranging from 0.5 to 3.1 nS in 1 M KCl. Nodulin-26 channels were voltage-sensitive [26]. Similar results were obtained with AQP0. The addition of the vesicles containing the major intrinsic protein of lens to both sides of preformed lipid films induced voltage-dependent channels with a conductance of 200 pS in 0.1 M salt solutions [29, 74]. The ion channel activity was absent when AQP0 was expressed in *Xenopus* oocytes [75]. It is difficult to establish whether AQP reconstitution into planar lipid bilayers alters the properties of the channel or whether expression in oocytes leads to regulatory events that also may alter the transport properties. To clarify these discrepancies, a defined system was employed that allows to measure water and ion flux simultaneously, on one single bilayer.

With respect to the high background P_f of the lipid bilayer, the membrane was reconstituted with a large number of AQP copies. Because it is impossible to insert a million protein molecules by proteoliposome fusion with the planar membrane [76, 77], we have used Schindlers technique [78]. The technique is based on the finding that monolayers spontaneously form at the air - water interface of any vesicle suspension [79, 80]. Two such AQP-containing monolayers were combined in an aperture of a Teflon septum separating the two aqueous phases of the chamber. First the shortest and simplest member of the aquaporin family, *E. coli* AQPZ, was studied since it can be expressed in the homologous system and purified in high concentrations [81]. After insertion of about 10^7 AqpZ molecules, P_f of the planar membrane was about 3-fold above that of control bilayers but the electrical conductivity was that of the control membrane (Fig. 5). To obtain an upper estimate for ion/water permeability ratio, it is assumed that the measured membrane conductance, G, is solely due to ion movement through AQPZ. G allows to calculate the ion flux density, j_{ion} [82, 83]:

$$j_{ion} = \frac{RT}{z^2 F^2} G \tag{10}$$

where R, T, F and z are the gas constant, the absolute temperature, Faraday's constant and the valence of the ion, respectively. According to Eq. (10), the maximum flux of ions in Fig. 5 is about $7 \cdot 10^{-16}$ mol $s^{-1} cm^{-2}$.

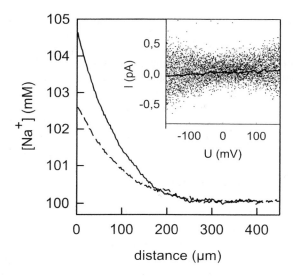

distance (µm)

Fig. 5. Representative sodium concentration profiles at the hypertonic side of planar membranes. P_f was equal to 20 µm/s in the absence of the protein (dashed line) and reached 54 µm/s after the reconstitution of AqpZ at a mass ratio of 1:100 (solid line). Membrane conductivity both in the presence and in the absence of the protein was (2.6   0.6) nS/cm^2 (inset). The osmotic flux was driven by 1 M urea. pH was 7.5 (adopted from [42]).

The simultaneously measured incremental water flux density due to AQPZ was equal to 4.3 µmol s^{-1} cm^{-2} (Fig. 5). Consequently, less than one ion was transported per 10^9 water molecules [42]. Reconstitution of AQP1 revealed essentially the same result [43].

The atomic structures of AQP1 [84-86] and the homologue protein GlpF [87] provide the explanation for the uniquely selective mechanism for free permeation by water and the exclusion of ions. The main barrier to cation permeation is a positive charge (Arg-195) located at the narrowest constriction of the aqueous pathway through AQP1, approximately 8 Å above the middle of the bilayer. This positive charge inside the channel is expected to inhibit also the movement of proton vehicles, for example H$_3$O$^+$. Nevertheless, structural proton diffusion according to the Grotthuss mechanism may still be possible, due to charge delocalization along the hydrogen-bonded water chain. However, planar membranes reconstituted with AQPZ revealed no proton permeability – even in an acidic milieu in the presence of a large transmembrane potential [42]. This experimental result is in agreement with the lack of continuous hydrogen-bonded water chains through the channels in molecular dynamics simulations [88]. The reason is the electrostatic potential generated by two half membrane spanning loops that dictates opposite orientations of water molecules in the two halves of the channel [89]. Thus, studies of reconstituted AQP revealed that there is no ion cotransport along the main water pathway [42, 43].

WATER AND ION COTRANSPORT

Passive water movement across permanent transmembrane aqueous pores [90] should be coupled to the flux of a solute that is also permeant through the channels. During osmosis, for example, the solute is pushed through pores by the solvent, i.e. the solute flux is increased in the direction of water flow and is retarded in the opposite direction. This phenomenon resulting in an incremental solvent flux, $J_{m,t}$, is called true solvent drag. Solvent drag studies have been widely used to establish a porous transport pathway [91]. For example, solvent drag experiments suggested that:

(i) the passage for nonelectrolytes may be a water transport pathway in salivary epithelium [92],

(ii) the hamster low-density lipoprotein is sieved by the solvent through pores of the endothelial barrier in perfused mesentery microvessels [93].

Potassium reabsorption in rat kidney proximal tubule is mediated both by solvent drag and K^+ diffusion along an existing concentration gradient. It is, however, impossible to resolve their contributions [94]. Even in such a well-defined system as represented by planar bilayers it is very difficult to demonstrate a true solvent drag effect [49]. The reason is the transmembrane solute concentration gradient, Δc, that is settled up in both USLs by the osmotic flow (compare Fig. 3). Δc causes a solute flux, $J_{m,p}$, in the same direction as the water flux. Because $J_{m,p}$ masquerades as solvent drag, it is named pseudo solvent drag [31]. Thus, an experimentally observed increase of transmembrane solute flux, J_m, that accompanies water flow does not necessarily mean that true solvent drag is involved.

With respect to these difficulties, visualization of true solvent drag awaited the introduction of the scanning microelectrode technique [39]. The only planar bilayer study undertaken before [95] remained unconvincing because the evidence presented in favor of the predicted solvent drag effect across amphotericin channels was rather indirect [49].

Visualization of coupled ion and water movements across gramicidin channels (Fig. 6) was achieved by simultaneous measurements of the near-membrane concentration distributions of an impermeable (calcium ions) and a permeable (sodium ions) solute [39]. In a first step, v was obtained from the concentration distribution of the impermeable solute according to Eq. (7). In a second step, the sodium flux densities, J_m, were derived from the concentration profiles of the permeable solute shown in Fig. 6. They equaled to (i) 26 pmol cm^{-2} s^{-1} and (ii) 16 pmol cm^{-2} s^{-1}, corresponding to (A) $J_{m,p} + J_{m,t}$ and (B) $J_{m,t}$. J_m was determined by fitting the parametric Eq. (11):

$$c_p(x) = \frac{J_m}{v} + \left(c_{p,s} - \frac{J_m}{v} \right) e^{\frac{-vx}{D}}$$

(11)

to the experimental data set in the interval $0 < x < 50$ μm. In this region, $ax^2 \ll v$ as revealed by the analysis of the Ca^{2+}-concentration distribution. p in Eq. (11) indicates the permeable solute (Na^+). The only difference between (A) and (B) was the addition of 0.1 mM NaCl to the hypertonic compartment. It was sufficient to establish a situation where $\Delta c = 0$ (Fig. 6).

In parallel to the concentration profiles, the current density, I, was measured. I is related to J_m by:

$$I = zFJ_m$$

(12)

According to Eq. (12), a flux of 13 pmol cm^{-2} s^{-1} was found for case (B) in Fig. 6. Considering that the current measurements returned the average flux density whereas the microelectrode measurements reflected the maximum flux density in the center of the membrane, it had to be acknowledged that both methods are in reasonable agreement with each other. Thus, simultaneous current and microelectrode measurements confirmed the validity of the hydrodynamic model (compare § 2.3 of this chapter).

If the membrane conductance, G, is monitored along with v, it is possible to calculate the single channel water permeability coefficient, p_f, and the number of water molecules, N, that are required to transport one ion across the channel. In general, there exist three different experimental strategies:

(i) The absence of an osmotic pressure gradient refers to electroosmotic experiments where the number of water molecules transported per ion is simply the ratio of water and ion fluxes ($N = J_W/J_m$) [96, 97].

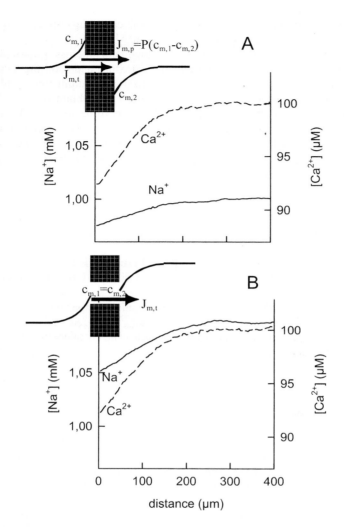

Fig. 6. Visualization of solvent drag. (A) Under short-circuited conditions potassium is dragged by the osmotic water flow across gramicidin channels ($J_{m,t}$) and diffuses along its concentration gradient ($J_{m,p}$). Here, the total Na^+ flux is equal to 26 pmol cm^{-2} s^{-1}. (B) The Na^+ bulk concentration in the hypertonic compartment had been augmented to 1.1 mM. $J_{m,p}$ was equal to 0. In this situation, $J_m = J_{m,t} = 16$ pmol cm^{-2} s^{-1}. With respect to the conductivity of 0.74 mS cm^{-2}, the number of water molecules moving along with one cation is calculated to be 4.8 ± 0.7 (Figure adopted from [39]).

(ii)   Under open circuited conditions, i.e. in the absence of a transmembrane current, a streaming potential develops that also can be used to calculate N [96, 97]. The accuracy of streaming potential measurements is increased if measured by ion sensitive microelectrodes. For streaming potential measurements, the microelectrodes are placed at a fixed distance from the membrane [98].

(iii)   Solvent drag experiments, in which the transmembrane potential $\varphi = 0$ [39]:

$$J_{m,t} = RT\chi c_{osm} V_w NG / z^2 F^2 \qquad (13)$$

where χ is the osmotic coefficient.

The number of water molecules that are moving along with one sodium ion was calculated after correction of the osmolyte concentration for volume flow dilution. In the particular experiment described above (Fig. 6), N was equal to 4.8 ± 0.7 [39]. Thus, the value obtained from solvent drag measurements is consistent with that determined via electroosmosis [96, 97] and streaming potential [97-99] measurements.

The microelectrode technique enabled the first direct visualization of the solvent drag effect; i. e. in the absence of a transmembrane ion concentration difference, the osmotic water flow is shown to push cations across gramicidin channels.

REFERENCES

1. J.L.Brewster, T.de Valoir, N.D.Dwyer, E.Winter, and M.C.Gustin, Science., 259 (1993) 1760.
2. M.J.Droillard, S.Thibivilliers, A.C.Cazale, H.Barbier-Brygoo, and C.Lauriere, FEBS Lett., 474 (2000) 217.
3. S.H.Oliet and C.W.Bourque, Nature, 364 (1993) 341.
4. T.Hanai and D.A.Haydon, J. Theor. Biol., 11 (1966) 370.
5. S.Paula, A.G.Volkov, A.N.Vanhoek, T.H.Haines, and D.W.Deamer, Biophys. J., 70 (1996) 339.
6. S.Paula, A.G.Volkov, and D.W.Deamer, Biophys. J., 74 (1998) 319.
7. M.B.Lande, J.M.Donovan, and M.L.Zeidel, J. Gen. Physiol., 106 (1995) 67.
8. W.K.Subczynski, A.Wisniewska, J.J.Yin, J.S.Hyde, and A.Kusumi, Biochemistry, 33 (1994) 7670.
9. D.W.Deamer and J.Bramhall, Chem. Phys. Lipids, 40 (1986) 167.
10. M.Jansen and A.Blume, Biophys. J., 68 (1995) 997.

11. D.Bassolino-Klimas, H.E.Alper, and T.R.Stouch, Biochemistry, 32 (1993) 12624.
12. K.R.Spring, News in Physiological Sciences, 14 (1999) 92.
13. K.R.Spring, Annu. Rev. Physiol., 60 (1998) 105.
14. A.K.Meinild, D.D.Loo, A.M.Pajor, T.Zeuthen, and E.M.Wright, Am. J. Physiol. Renal. Physiol., 278 (2000) F777-F783.
15. A.Meinild, D.A.Klaerke, D.D.Loo, E.M.Wright, and T.Zeuthen, J. Physiol. (Lond.), 508 (1998) 15.
16. T.Zeuthen, Int. Rev. Cytol., 160 (1995) 99.
17. T.Zeuthen, Rev. Physiol. Biochem. Pharmacol., 141 (2000) 97.
18. D.D.Loo, T.Zeuthen, G.Chandy, and E.M.Wright, Proc. Natl. Acad. Sci. U. S. A., 93 (1996) 13367.
19. P.P.Duquette, P.Bissonnette, and J.Y.Lapointe, Proc. Natl. Acad. Sci. U. S. A., 98 (2001) 3796.
20. J.Schnermann, C.L.Chou, T.H.Ma, T.Traynor, M.A.Knepper, and A.S.Verkman, Proc. Natl. Acad. Sci. U. S. A., 95 (1998) 9660.
21. L.S.King, M.Choi, P.C.Fernandez, J.P.Cartron, and P.Agre, N. Engl. J. Med., 345 (2001) 175.
22. D.Kozono, M.Yasui, L.S.King, and P.Agre, J. Clin. Invest., 109 (2002) 1395.
23. A.J.Yool, W.D.Stamer, and J.W.Regan, Science, 273 (1996) 1216.
24. T.L.Anthony, H.L.Brooks, D.Boassa, S.Leonov, G.M.Yanochko, J.W.Regan, and A.J.Yool, Mol. Pharmacol., 57 (2000) 576.
25. M.Yasui, A.Hazama, T.H.Kwon, S.Nielsen, W.B.Guggino, and P.Agre, Nature, 402 (1999) 184.
26. C.D.Weaver, N.H.Shomer, C.F.Louis, and D.M.Roberts, J. Biol. Chem., 269 (1994) 17858.
27. J.W.Lee, Y.Zhang, C.D.Weaver, N.H.Shomer, C.F.Louis, and D.M.Roberts, J. Biol. Chem., 270 (1995) 27051.
28. R.L.Rivers, R.M.Dean, G.Chandy, J.E.Hall, D.M.Roberts, and M.L.Zeidel, J. Biol. Chem., 272 (1997) 16256.
29. G.R.Ehring, G.Zampighi, J.Horwitz, D.Bok, and J.E.Hall, J. Gen. Physiol., 96 (1990) 631.
30. J.Dainty and C.R.House, J. Physiol. (Lond.), 182 (1966) 66.
31. P.H.Barry and J.M.Diamond, Physiol. Rev., 64 (1984) 763.
32. H.Matsui, C.W.Davis, R.Tarran, and R.C.Boucher, J. Clin. Invest., 105 (2000) 1419.
33. Y.Song, S.Jayaraman, B.Yang, M.A.Matthay, and A.S.Verkman, J. Gen. Physiol., 117 (2001) 573.
34. Y.N.Antonenko, G.A.Denisov, and P.Pohl, Biophys. J., 64 (1993) 1701.
35. P.Pohl, Y.N.Antonenko, and E.H.Rosenfeld, Biochim. Biophys. Acta, 1152 (1993) 155.
36. P.Pohl, S.M.Saparov, and Y.N.Antonenko, Biophys. J., 72 (1997) 1711.
37. P.Pohl, S.M.Saparov, and Y.N.Antonenko, Biophys. J., 75 (1998) 1403.
38. A.V.Krylov, P.Pohl, M.L.Zeidel, and W.G.Hill, J. Gen. Physiol., 118 (2001) 333.
39. P.Pohl and S.M.Saparov, Biophys. J., 78 (2000) 2426.
40. S.M.Saparov, Y.N.Antonenko, R.E.Koeppe, and P.Pohl, Biophys. J., 79 (2000) 2526.
41. P.Pohl, S.M.Saparov, E.E.Pohl, V.Y.Evtodienko, I.I.Agapov, and A.G.Tonevitsky, Biophys. J., 75 (1998) 2868.

42. P.Pohl, S.M.Saparov, M.J.Borgnia, and P.Agre, Proc. Natl. Acad. Sci. U. S. A., 98 (2001) 9624.
43. S.M.Saparov, D.Kozono, U.Rothe, P.Agre, and P.Pohl, J. Biol. Chem., 276 (2001) 31515.
44. W.Nernst, Z. Phys. Chem., 47 (1904) 52.
45. R.Fettiplace and D.A.Haydon, Physiol. Rev., 60 (1980) 510.
46. J.Dainty, Adv. Botan. Res., 1 (1963) 279.
47. T.J.Pedley and J.Fischbarg, J. Theor. Biol., 70 (1978) 427.
48. T.J.Pedley, Q. Rev. Biophys., 16 (1983) 115.
49. A. Finkelstein Water Movement Through Lipid Bilayers, Pores, and Plasma Membranes, Wiley & Sons, New York. (1987)
50. D. Amman Ion-Selective Microelectrodes. Principles, Design and Application, Springer, Berlin. (1986)
51. D.Huster, A.J.Jin, K.Arnold, and K.Gawrisch, Biophys. J., 73 (1997) 855.
52. M.D.Levitt, T.Aufderheide, C.A.Fetzer, J.H.Bond, and D.G.Levitt, J. Clin. Invest., 74 (1984) 2056.
53. A.Strocchi, G.Corazza, J.Furne, C.Fine, A.Disario, G.Gasbarrini, and M.D.Levitt, Am. J. Physiol., 33 (1996) G487-G491.
54. Y.N.Antonenko, P.Pohl, and G.A.Denisov, Biophys. J., 72 (1997) 2187.
55. R.Holz and A.Finkelstein, J. Gen. Physiol., 56 (1970) 125.
56. P.A.Rosenberg and A.Finkelstein, J. Gen. Physiol., 72 (1978) 341.
57. E.Orbach and A.Finkelstein, J. Gen. Physiol., 75 (1980) 427.
58. T.X.Xiang and B.D.Anderson, J. Membr. Biol., 140 (1994) 111.
59. F.J.Burczynski, Z.S.Cai, J.B.Moran, T.Geisbuhler, and M.Rovetto, Am. J. Physiol., 268 (1995) H1659-H1666.
60. V. G. Levich Physicochemical Hydrodynamics., Englewood Cliffs, Prentice-Hall. (1962)
61. T.J.Pedley, J. Fluid. Mech., 101 (1980) 843.
62. K.Maric, B.Wiesner, D.Lorenz, E.Klussmann, T.Betz, and W.Rosenthal, Biophys. J., 80 (2001) 1783.
63. J.M.Smaby, H.L.Brockman, and R.E.Brown, Biochemistry, 33 (1994) 9135.
64. J.M.Smaby, M.Momsen, V.S.Kulkarni, and R.E.Brown, Biochemistry, 35 (1996) 5696.
65. W.G.Hill and M.L.Zeidel, J. Biol. Chem., 275 (2000) 30176.
66. K.Simons and G.van Meer, Biochemistry, 27 (1988) 6197.
67. M.Montal and P.Mueller, Proc. Natl. Acad. Sci. U. S. A., 69 (1972) 3561.
68. H.O.Negrete, R.L.Rivers, A.H.Gough, M.Colombini, and M.L.Zeidel, J. Biol. Chem., 271 (1996) 11627.
69. W.G.Hill, R.L.Rivers, and M.L.Zeidel, J. Gen. Physiol., 114 (1999) 405.
70. G.M.Preston, T.P.Carroll, W.B.Guggino, and P.Agre, Science., 256 (1992) 385.
71. M.J.Borgnia, S.Nielsen, A.Engel, and P.Agre, Annual Rev. Biochem., 68 (1999) 425.
72. L.S.King, M.Yasui, and P.Agre, Molecular Medicine Today, 6 (2000) 60.
73. K.Ishibashi, M.Kuwahara, and S.Sasaki, Rev. Physiol. Biochem. Pharmacol., 141 (2000) 1.
74. G.A.Zampighi, J.E.Hall, and M.Kreman, Proc. Natl. Acad. Sci. U. S. A., 82 (1985) 8468.

75. S.M.Mulders, G.M.Preston, P.M.T.Deen, W.B.Guggino, C.H.van Os, and P.Agre, J. Biol. Chem., 1995) 9010.
76. D.J.Woodbury and J.E.Hall, Biophys. J., 54 (1988) 1053.
77. D.J.Woodbury and C.Miller, Biophys. J., 58 (1990) 833.
78. H.Schindler, Methods Enzymol., 171 (1989) 225.
79. F.Pattus, P.Desnuelle, and R.Verger, Biochim. Biophys. Acta., 507 (1978) 62.
80. H.Schindler, Biochim. Biophys. Acta., 555 (1979) 316.
81. M.J.Borgnia, D.Kozono, G.Calamita, P.C.Maloney, and P.Agre, J. Mol. Biol., 291 (1999) 1169.
82. A.L.Hodgkin, Biol. Rev., 26 (1951) 339.
83. A.Walter, D.Hastings, and J.Gutknecht, J. Gen. Physiol., 79 (1982) 917.
84. K.Murata, K.Mitsuoka, T.Hirai, T.Walz, P.Agre, J.B.Heymann, A.Engel, and Y.Fujiyoshi, Nature, 407 (2000) 599.
85. B.L.de Groot, A.Engel, and H.Grubmuller, FEBS Lett., 504 (2001) 206.
86. H.Sui, B.G.Han, J.K.Lee, P.Walian, and B.K.Jap, Nature, 414 (2001) 872.
87. D.Fu, A.Libson, L.J.Miercke, C.Weitzman, P.Nollert, J.Krucinski, and R.M.Stroud, Science., 290 (2000) 481.
88. B.L.de Groot and H.Grubmuller, Science, 294 (2001) 2353.
89. E.Tajkhorshid, P.Nollert, M.O.Jensen, L.J.W.Miercke, J.O'Connell, R.M.Stroud, and K.Schulten, Science, 296 (2002) 525.
90. A.E.Hill, Int. Rev. Cytol., 163 (1995) 1.
91. B.Rippe and B.Haraldsson, Physiol. Rev., 74 (1994) 163.
92. T.Nakahari, H.Yoshida, and Y.Imai, Exp. Physiol., 81 (1996) 767.
93. J.C.Rutledge, F.E.Curry, P.Blanche, and R.M.Krauss, Am. J. Physiol., 268 (1995) H1982-H1991.
94. R.W.Wilson, M.Wareing, and R.Green, J. Physiol. (Lond.), 500 (1997) 155.
95. T.E.Andreoli, J.A.Schafer, and S.L.Troutman, J. Gen. Physiol., 57 (1971) 479.
96. D.G.Levitt, S.R.Elias, and J.M.Hautman, Biochim. Biophys. Acta, 512 (1978) 436.
97. P.A.Rosenberg and A.Finkelstein, J. Gen. Physiol., 72 (1978) 327.
98. S.Tripathi and S.B.Hladky, Biophys. J., 74 (1998) 2912.
99. D.G.Levitt, Curr. top. membr. transp., 21 (1984) 181.

Planar Lipid Bilayers (BLMs) and their Applications
H.T. Tien and A. Ottova-Leitmannova (Editors)

Chapter 10

Membrane-Macromolecule Interactions and their Structural Consequences

S. May[a] **and A. Ben-Shaul**[b]

[a]Institut für Molekularbiologie, Friedrich-Schiller-Universität Jena, Winzerlaer Str. 10, Jena 07745, Germany

[b]Department of Physical Chemistry and the Fritz Haber Research Center, The Hebrew University, Jerusalem 91904, Israel

1. INTRODUCTION

The lipid bilayer, constituting the central structural component of biological membranes, is a *flexible, self-assembled, two-dimensional fluid mixture*. Each of these bilayer characteristics, separately and more often jointly with others, plays a crucial role in the interaction between biomembranes and biological macromolecules, such as integral or peripheral proteins and DNA. Being *flexible* (or "elastic") with respect to bending deformations, the bilayer can respond to interactions with both peripheral and intrinsic macromolecules through local or even global variation of its curvature. Changes in membrane area (and hence thickness) are also possible, but usually involve a higher energetic cost. A local change in membrane structure occurs, for example, when a hydrophobic α-helical peptide is inserted into a lipid bilayer whose thickness does not match its length. By locally changing its curvature (and thickness) around the incorporated peptide the lipid bilayer minimizes the exposure of hydrophobic (protein or lipid) moieties to water [1].

A more drastic, global, change in membrane curvature may occur, for instance, when a double stranded DNA molecule adsorbs onto a lipid bilayer composed of cationic and neutral lipids [2]. Driven by the electrostatic attraction between the negatively charged DNA and the positively charged lipids, the bilayer tends to wrap around the DNA, paying the energetic toll of membrane bending deformation. This bilayer-coated DNA complex (later referred to as the "spaghetti"-like complex [3]) is usually metastable with respect to other

DNA-lipid complexes. In general, mixing DNA and cationic (liposomal) bilayers results in the spontaneous formation of periodic composite phases of lipid and DNA, characterized by either lamellar or hexagonal symmetry [4,5]; (see Fig. 2 in Sec. 3). As in the ordinary lamellar lipid phase (L_α), the lamellar complex phase (L_α^c) consists of a stack of bilayers, but with monolayers of parallel DNA strands sandwiched between them. Other lipid mixtures end up forming hexagonal complexes (H_{II}^c), consisting of double stranded DNA's intercalated within the aqueous tubes of the inverse hexagonal lipid phase (H_{II}). The formation of such lipid-DNA complexes exhibits both the *fluidity* and *self-assembly* properties of lipid mixtures.

The cationic lipids in the L_α^c complex tend to concentrate in the vicinity of the DNA. Similar phenomena occur when, say, basic proteins adsorb onto lipid membranes containing a small fraction of acidic lipids. Driven by electrostatic attraction to the macromolecule, the charged lipids diffuse into the interaction zone, now paying the (tolerable) toll of lateral *demixing* entropy. Clearly, this ability of the membrane to "optimize" its interaction with a neighboring macromolecule, is possible because the bilayer is a two-dimensional *fluid mixture*.

In forming the H_{II}^c complex with DNA, the lipid bilayer undergoes a much more dramatic change than a "simple" bending or stretching deformation. This process is a real, first order, thermodynamic phase transition. An analogous, lamellar-to-inverse hexagonal, lipid phase transition can be induced by the incorporation of short hydrophobic peptides into lipid bilayers [6] (see below). Again, the occurrence of these, and various other morphological transformations and phase transitions in lipid systems, are direct consequences of the fact that lipid molecules in water spontaneously organize in *self-assembling* aggregates.

Belonging to the wider class of amphiphilic molecules, lipids in aqueous solution exhibit a rich variety of self-assembled aggregates and ordered phases [7]. The aggregation geometry of ultimate biological importance is, of course, the planar bilayer. Other familiar lipid assemblies include the inverse-hexagonal (H_{II}) phase mentioned above, as well as a multitude of cubic and micellar phases. The spontaneous formation of H_{II}^c complexes by adding DNA into a solution of vesicles, i.e., *planar lipid bilayers*, demonstrates how interacting macromolecules can modify the self-assembly characteristics of the lipid species. Still, since only certain mixed lipid bilayers end up forming hexagonal complexes while others remain lamellar, there must be some intrinsic preference of the lipid constituents for one packing geometry over the other. Indeed, the propensity of lipids and other amphiphiles to self-assemble into one of many possible packing geometries, may be understood, at least qualitatively, by considering their "molecular shape" [8,9]. Most ionic surfactants, as well as certain lipids, like the single-tailed lyso-PC (phosphatidylcholine), comprise a single hydrophobic tail and a relatively large polar head group. (The head group "size" is dictated by the range of the repulsive interaction with neighboring molecules. The interaction may be steric, electrostatic or both.) The overall shape of such molecules may be depicted as a cone, suggesting that their optimal packing geometry is that of a highly curved aggregate, e.g., a spherical or cylindrical micelle, a notion corroborated by many experiments. Other lipids may be depicted as "inverted cones", with their head group positioned at the cone vertex. DOPE (dioleoylphosphatidylethanolamine) for instance. involves a relatively small head group and, like most phospholipids, a bulky double-chain tail. As expected, the preferred aggregation geometry of such molecules is the inverse hexagonal, H_{II}, or some other inverted (that is, "hydrocarbon-continuous" rather than "water-continuous") phase.

The molecular packing characteristics of most lipid species, DOPC (dioleoylphosphatidylcholine) to mention one example, are intermediate between the above two extremes. Namely, their head groups and tails are of comparable size. These molecules spontaneously assemble into planar bilayers. Recall, however, that two monolayers combine to form a bilayer only because a single monolayer in solution is unstable, owing to the high interfacial energy associated with exposing its hydrophobic surface to water. Yet, based only on molecular packing considerations we expect that most monolayers should possess a nonzero *spontaneous curvature* [10]. For instance, if inter-head group repulsion is stronger than tail-tail repulsion (imagine a "truncated cone" with the head group at its wider base) the monolayer will tend to curve "positively", i.e., with the head groups on the concave side. In forming a *planar* bilayer the two monolayers rid off the highly unfavorable hydrocarbon-water interfacial energy, paying a smaller yet nonzero curvature elastic energy associated with flattening the (spontaneously curved) monolayers. As we shall see below, this *elastic frustration* energy, plays a subtle yet crucial role in the phase behavior of membrane-macromolecule assemblies.

In our discussion so far we have emphasized the various modes in which a lipid bilayer can respond to interactions with biopolymers, implicitly assuming that the latter are rigid macromolecules. While not invariably so, in most interacting lipid-macromolecule systems the soft-mixed-fluid membrane is, indeed, considerably more amenable to structural modifications as compared to most proteins and certainly in comparison to DNA. Consistent with this notion we shall treat the interacting biopolymer as a rigid macromolecule of a given size and shape, as well as of given charge distribution in the case of DNA and charged proteins. Our goal is to present a simple, yet powerful, theoretical framework for treating the roles of membrane elasticity, its two-dimensional fluidity, and its ability to re-assemble in lipid-macromolecule interactions. As specific examples we shall consider lipid-DNA complexes and two types of lipid-protein systems, exhibiting both local and global structural re-organization of the lipid matrix.

The theoretical description of membrane-macromolecule interaction involves two complementary aspects. First, on a 'molecular' (or 'local') level one has to calculate the interaction free energy between the macromolecule, e.g., an adsorbed or trans-membrane protein, and the lipid bilayer. This calculation should yield the interaction free energy as a function of the molecular and thermodynamic characteristics of the system such as membrane composition, protein size and charge and the ionic strength of the solution. Within this molecular-level theory, and using appropriate (typically mean-field like) approximations one may also account for macromolecule concentration effects, e.g., the variation of protein adsorption energy on the density of already adsorbed proteins. The second stage of the theoretical treatment involves the inclusion of the calculated interaction potential in a statistical-thermodynamic treatment of the membrane-macromolecule system. The phase behavior of this multi-component system is then derived by analyzing its (usually rather complicated) thermodynamic free energy.

The thickness of lipid bilayers, as well as typical diameters of proteins and DNA are all in the nanometer-range. This range defines the relevant length scale for lipid-macromolecule interactions, e.g., the range of elastic perturbations of lipid hydrocarbon chains by a trans-membrane protein, or the range of modulations in lipid composition induced by protein adsorption. Atomistic-level theories or computer simulations have proved useful in studying specific (e.g., protein-lipid) systems over a limited range of time and length scales [11]. At present they do not yet provide an efficient means for analyzing global, nanometer-range phenomena such as membrane curvature modulations or phase transitions following protein adsorption or insertion. Many interesting general questions can be answered using "mesoscopic-level" theories; e.g., how does the adsorption isotherm of basic proteins on acidic membrane vary with protein charge and size and the mole fraction of acidic membrane lipids, or, what is required from a lipid mixture in order to form hexagonal rather than lamellar DNA-lipid complexes. By mesoscopic-level theories we refer here to continuum models, based on expressing the free energy of the interacting system as an integral over locally varying contributions. In the present context we mainly refer to two rather simple and most useful theoretical tools. The phenomenological, Helfrich's, formalism of membrane elasticity, and the Poisson-Boltzmann (PB) theory for the electrostatic interaction between charged surfaces in ionic solutions. We find it appropriate to briefly recapitulate – in Sec. 2 – these tools before proceeding to discuss lipid-macromolecule interactions. Then, in Sec. 3 we shall discuss the major structural and thermodynamic characteristics of cationic lipid-DNA complexes, focusing on the coupling between electrostatic, elastic and compositional degrees of freedom of the lipid mixture. In Sec. 4 we shall discuss the electrostatic adsorption of charged proteins on oppositely charged membranes, emphasizing the role of lipid mobility and lateral adsorbate interactions. Sec. 5 is devoted to the interaction of membranes with integral, hydrophobic, proteins. Our goal here is to demonstrate how, on one hand, membrane lipids can mediate the interaction between embedded proteins, possibly leading to two-dimensional phase transitions and, on the other hand, how proteins can drive morphological phase transitions of lipid bilayers. A few comments pertaining to partially membrane-penetrating biopolymers are given in the final section.

2. ELASTIC AND ELECTROSTATIC FREE ENERGIES

2.1. Curvature elasticity

Lipid membranes are quite "soft" with respect to bending deformations. In contrast to that, changes in membrane thickness generally involve large deformation energies. Frequently, membrane deformations cannot be simply classified as pure bending or stretching deformations. This is the case, for instance, when a lipid bilayer is perturbed by the incorporation of either short or long trans-membrane proteins, as will be discussed in Sec. 5. Nevertheless, the principles of membrane curvature elasticity are essential for understanding membrane-macromolecule interaction phenomena. To this end let us consider a lipid bilayer of total, equilibrium, area A; ignoring area deformations we treat A as a constant. The equilibrium curvature of the symmetric bilayer, i.e., a bilayer composed of two identical monolayers is, of course, zero. However, as mentioned in the previous section, depending on the relative magnitude of lipid head group and tail interactions, the monolayers tend to adopt some nonzero equilibrium curvature, c_{eq}. Similarly a non-symmetric bilayer will also possess a nonzero equilibrium curvature.

A most useful expression for the elastic energy of lipid films (i.e., monolayers, bilayers or, in fact, any thin membrane) has been proposed by Helfrich [10]. Let R_1 and R_2 denote the local principal radii of curvature at some given point of the membrane, (hence $c_1 = 1/R_1, c_2 = 1/R_2$ are the corresponding principal curvatures), and da a small area element around this point. The curvature elastic free energy of the film can now be expressed in the form

$$F = \int_A da \left[\frac{k}{2}(c_1 + c_2 - c_0)^2 + \bar{k}c_1c_2 \right] \tag{1}$$

where $k, \bar{k}$ and c_0 are material constants depending on the lipid composition of the film; k and $\bar{k}$ are elastic moduli known as splay modulus (or bending rigidity), and saddle-splay modulus, respectively. The bending rigidity of common lipid layers is typically of order $k = 10\, k_B T$; with k_B denoting Boltzmann's constant, and T is the absolute temperature. It is much harder to measure or calculate the saddle-splay modulus $\bar{k}$. However, the Gauss-Bonnet theorem states that the second term in Eq. (1) represents a topological invariant, and is thus irrelevant whenever the topology is the lipid aggregate remains unaffected.

The constant c_0 in Eq. (1) is known as the *spontaneous curvature* and is closely related to the equilibrium curvature of the film, i.e., to the optimal packing curvature of the constituent lipids. More specifically, minimizing F with respect to c_1 and c_2 it is easy to show that at the equilibrium configuration of the membrane $c_1 = c_2 \equiv c_{eq} = kc_0/(2k + \bar{k})$. Another result which follows directly from Eq. (1) involves the curvature frustration energy associated with joining two monolayers to form a planar bilayer. This energy is always positive and given by $\Delta F_{frus} = 2kc_0 c_{eq} A$. It may be compared with the surface energy associated with exposing the two hydrophobic surfaces of the separated monolayers to water; $\Delta F_{surface} = 2\gamma A$, with $\gamma \approx 10 k_B T/\text{nm}^2$ denoting the hydrocarbon-water surface energy. For typical monolayer parameters, $k \approx 10 k_B T$ and $c_0 \approx c_{eq} \leq 1/(10\text{ nm})$ we find that $\Delta F_{surface}$ is at least two orders of magnitude larger than ΔF_{frus}, explaining why the bending frustration energy is a small penalty compared to the large energetic gain resulting from the association of two monolayers into a bilayer. Yet, the transition from the bilayer to the inverse hexagonal lipid geometry does not involve exposure of hydrocarbon tails to water, and the curvature frustration energy, though nominally small, becomes a crucial factor. This and other issues pertaining to membrane elasticity are further discussed in the following sections.

2.2. Electrostatic free energy

Given the electrostatic potential in a system containing macroions (e.g., charged membranes, proteins or DNA) in solution, one can calculate the electrostatic ("charging") energy for arbitrary macroion configurations. Interaction potentials between macroions, e.g., those depicted in Fig. 1, can then be derived by comparing the energies (actually free energies, see below) corresponding to different distances. In principle, from classical electrostatics we know that for any given distribution of charges, the electrostatic potential, $\Phi = \Phi(\mathbf{r})$, is determined by the solution of Poisson's equation. An exact treatment of a system such as in Fig. 1, assuming the macroions to be fixed in space, requires solving Poisson's equation for all possible distributions of the mobile ions. Then, after assigning each distribution its proper statistical weight one could derive the interaction free energy of this system. This can only be achieved through computer simulations. However, because of the long range nature of the Coulomb potential, and the high complexity of most systems of biological relevance, computer simulations are so far available only for a small number of model systems. A most useful alternative for studying electrostatically interacting macroions in solution is provided by an approximate, "mean-field" approach, known as Poisson-Boltzmann (PB) theory [12]. All the electrostatic treatments in the following sections are in the spirit of PB theory.

Driven by Coulomb attractions, the mobile ions in solution tend to concentrate near oppositely charged (macroion) surfaces, thus lowering the electrostatic *energy* of the system. This tendency is opposed by the *entropic* penalty associated with any deviation of the local ion densities, $n_i(\mathbf{r})$, from their bulk (i.e., far from any charge surface) values, n_i^0. The average, equilibrium, distribution of mobile ions represents a compromise between these opposing tendencies; corresponding, as usual, to the minimum of the system free energy. Ignoring spatial correlations between the mobile ions (hence the "mean-field" nature of the theory), one could solve the Poisson equation corresponding to the thermally averaged distribution of mobile ions, thus deriving a "thermally averaged" electrostatic potential for the system. Clearly, however, the average ion densities depend on the electrostatic potential through the Boltzmann distribution; namely $n_i(\mathbf{r}) = n_i^0 \exp(-z_i e \Phi(\mathbf{r})/k_B T)$ with e denoting the elementary charge, and z_i the valency of ion species i ($\Phi \equiv 0$ far away from charged surfaces). This "circular" dependence of the electric potential on the local ion densities through Poisson's equation, and the dependence of the latter on the former through Boltzmann's formula, leads to a *self-consistent* relationship which $\Phi(\mathbf{r})$ must fulfill. This relationship is the Poisson-Boltzmann (PB) equation. For a symmetric 1:1 electrolyte solution ($i = \{-1, 1\}$) it is $z_1 = -z_{-1} = 1$, and $n_1^0 = n_{-1}^0 = n_0$. In this case the PB equation reads

$$\frac{\partial^2 \Psi}{\partial x^2} + \frac{\partial^2 \Psi}{\partial y^2} + \frac{\partial^2 \Psi}{\partial z^2} = \frac{1}{l_D^2} \sinh \Psi \qquad (2)$$

with $\Psi = e\Phi/k_B T$ denoting the reduced (dimensionless) electrostatic potential and l_D is a (temperature and concentration dependent) constant, known as Debye's *screening length*. This important constant sets the length scale beyond which the Coulombic interaction between two charged particles (or surfaces) is effectively screened by the intervening salt ions. For 1:1 electrolyte solutions at

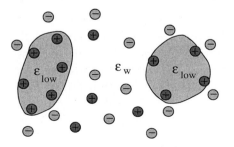

Figure 1. *A common scenario in cellular biology: two macroions of low dielectric constant ε_{low} interact in an aqueous solution (where the dielectric constant is $\varepsilon_w \approx 80$), containing mobile salt ions. Electrostatic interactions between the macroions occur if their respective diffuse counterion layers overlap.*

room temperature, the dependence of the Debye length on the molar concentration of salt is $l_D = 0.304/\sqrt{M}$ nm. Under "physiological" conditions $M \approx 0.1$ and hence $l_D \approx 1$ nm.

Writing down the PB equation for multivalent, nonsymmetric or multicomponent salt solutions is straightforward. However, being a nonlinear partial differential equation, analytical solutions of PB equation (even in its symmetric form, Eq. (2)) are available for just a few special cases; see e.g., Ref. [13]. Even numerical solutions usually require advanced computer methods [14]. Yet, it should be noted that once the PB equation is solved one can calculate any desired characteristic of the interacting system. In particular, given the electrostatic potential $\Psi(\mathbf{r})$ is known, we can calculate the local concentrations of mobile ions, $n_{\pm 1} = n_0 \exp(\mp \Psi)$, around and between the interacting macromolecules. Using these ion densities (or equivalently Ψ) we can also calculate the electrostatic energy, the (counterion) entropy and hence the electrostatic free energy. Following the variation of the free energy as a function of, say, the distance between two macroions, see e.g., Fig. 1, one can derive the solvent-mediated electrostatic interaction between such macromolecules.

The approximate nature of the PB approach is mainly reflected in the neglect of charge-charge correlations [15]. Computer simulations reveal (see e.g., Ref. [16]) that despite its approximate nature, PB theory works well for biopolymers immersed in monovalent salt solutions. Thus, in the following sections all calculations based on PB theory refer to a symmetric, 1:1 electrolyte solution.

3. CATIONIC LIPID-DNA COMPLEXES

In Sec. 1 the complexes formed between DNA and cationic-neutral lipid mixtures were brought up as model system, featuring many of the unique properties characterizing the interaction between lipid membranes and biopolymers. In this section we further elaborate on these systems, emphasizing the ability of the DNA molecules to promote pronounced structural changes of the lipid matrix through strong electrostatic and elastic interactions.

Cationic lipids (CL) are not abundant in living cells. However, recently they have attracted much attention because of their ability to compact DNA by forming the above-mentioned CL-DNA complexes [2], often called *lipoplexes* [17]. Lipoplexes play a central role in current approaches to gene therapy [18], serving as potent transfection vectors, i.e., as carriers of genomic material across the plasma membrane into the cell, on its way to the nucleus. This offers a promising alternative to viral transport vector methods which often suffer from antibody responses. Not yet well understood, however, is how these aggregates enter into living cells, how the DNA molecules dismantle from the cationic lipids, and how do they finally enter the cell nucleus [19,20].

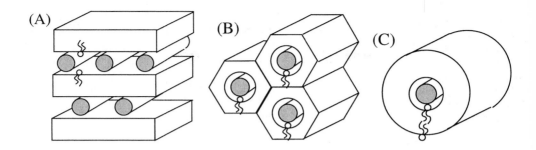

Figure 2. Structural models of CL-DNA complexes. The figure shows, schematically, a two-dimensional cross-section through a plane perpendicular to the DNA (shaded circles). The sandwich-like (L_α^c) structure (A) consists of smectic-like lamellar stacks of charged bilayer in which one-dimensional arrays of DNA strands are intercalated. The honeycomb-like (H_{II}^c) complex (B) is an hexagonally packed bundle of monolayer-coated DNA units. Bilayer-coating of single (or double) stranded DNA leads to the formation of spaghetti-like complexes.

At the very basis of lipoplex formation is simple electrostatics [21]: the negatively charged phosphate groups of DNA attract cationic lipids vesicles. This interaction is strong enough to totally disrupt the vesicles and transform them into compact, often micron-sized, aggregates [22]. Most lipoplex-based transfer vectors use a *mixture* of cationic and neutral lipids [23]. The reason is that by judicious choice of the neutral colipid – also called *helper* lipid – one can significantly enhance the efficiency of DNA transfection as well as reduces the lipoplex toxicity. From a physico-chemical point of view, the helper lipid is helpful in tuning both the charge density of the lipid layers and their curvature elasticity. Commonly used colipids are double-tailed phospholipids, like DOPC or DOPE. While both these lipids are uncharged they exhibit very different structural phase behavior. As already mentioned in Sec. 1, DOPC prefers the lamellar, L_α, phase, whereas DOPE shows a strong tendency for the inverse hexagonal, H_{II}, phase [7]. Cationic lipids, like DOTAP (dioleoyltrimethylammoniumpropane), are generally characterized by very small spontaneous curvatures, and thus self-assemble into planar bilayers.

The choice of different colipids is reflected in the appearance of different lipoplex morphologies, and through this affects the efficiency of gene delivery [24]. With DOPC as the helper lipid the lipoplexes are usually lamellar, (L_α^c). Their "sandwich" structure, as illustrated in Fig. 2, has been determined by X-ray diffraction measurements [4,25] and a variety of complementary methods [22,26–28]. The monolayers of DNA in each gallery of the sandwich complex form a one-dimensional array, characterized by a well defined repeat distance d. For reasons explained below, d depends on lipid composition.

Unlike the lamellar complexes, the structures shown in Fig. 2 B and Fig. 2 C involve highly-bent lipid layers. Fig. 2 B depicts the hexagonal (H_{II}^c), or "honeycomb"-like, complex mentioned in Sec. 1. Its inverse-hexagonal symmetry has unambiguously been confirmed by X-ray diffraction [5,29]. Noting the similarity to the H_{II} phase it is no surprise that the colipid DOPE promotes the formation of the H_{II}^c complex. Note that the repeat unit of the honeycomb-like complex consists of a DNA "rod" wrapped by a lipid monolayer. Like the H_{II} lipid phase, the H_{II}^c complex is stabilized by attractive (hydrophobic) interactions between the lipid chains of neighboring (cylindrically bent) monolayers.

Another possible complex geometry, the "spaghetti"-like structure shown in Fig. 2 C, has been observed in electron microscopy studies [3,30]. As in the H_{II}^c phase this complex consists of a DNA rod wrapped around by a lipid monolayer. However, instead of being surrounded by identical units, the monolayer coated DNA is further surrounded by an oppositely curved, monolayer, giving rise to a bilayer-enveloped DNA complex. Actually, it appears very likely that these complexes are formed upon "direct encounter" between DNA molecules and lipid bilayers, possibly as intermediate structure prior to the formation of H_{II}^c complexes.

The experimental observation and subsequent characterization of the structures shown in Fig. 2 have motivated a number of computer simulations [31,32] and theoretical studies of the structure, stability, and phase behavior of CL-DNA complexes [33–35]. Overall, these studies confirm that the structural and thermodynamic characteristics of these systems indeed reflect a delicate interplay between electrostatic interactions and membrane elasticity. In the following we elaborate on this interplay utilizing the theoretical tools reviewed in Sec. 2.

3.1. Stability of spaghetti-like and honeycomb-like lipoplexes

In both the spaghetti-like and honeycomb-like complexes the negatively charged DNA rods are symmetrically and tightly enveloped by oppositely charged, cationic, lipid monolayers. The concentric geometry of these structures allows, in principle, for exact electrostatic neutralization of the charged phosphate groups on the DNA surface by the cationic lipid charges. When this optimal charge matching condition is fulfilled, the counterions originally bound to the DNA and cationic surfaces are no longer needed for charge neutrality and can thus be released into the bulk solution, thereby increasing their translational entropy. Note, however, that for any given complex radius membrane-DNA charge neutralization requires a very specific cationic/neutral lipid mixture. Although one can experimentally adjust the original lipid mixture, the actual composition in the lipoplex, as well as its radius will be determined by the tendency of the system to minimize its total free energy (see below).

While the concentric DNA-lipid geometry is most favorable on electrostatic grounds, it generally involves a nonzero elastic deformation energy penalty, resulting from the high ("negative") curvature of the lipid monolayer. This energy penalty may be substantial if the spontaneous curvature of the lipid layer, c_0, is markedly different from its actual curvature in the complex; see Eq. (1). High bending rigidity will also increase the elastic energy penalty. Clearly then, the energetics as well as the detailed structure of the hexagonal-like and spaghetti-like complexes is governed by a delicate balance between the electrostatic and elastic energies.

The interplay between these forces can be understood in terms of a simple model [36], which treats the DNA-monolayer unit as a concentric cylindrical *capacitor*. Its inner surface is that of the DNA, modeled as a rod of fixed radius R^D and surface charge density σ^D. The outer cylinder, of *adjustable* radius R and surface charge density $\sigma = \sigma^D R^D / R$, represents the lipid surface, whose total charge exactly matches the negative DNA charge. The total free energy of the system is then written as a sum of the elastic monolayer energy, as given by Eq. (1) and the charging energy of the capacitor, both of which depend on R. The optimal outer radius is obtained by minimizing the total free energy with respect to R. For lipid mixtures characterized by (the rather typical) spontaneous curvature $c_0 = 0$ one finds the optimal monolayer radius

$$R = \frac{\pi k}{k_B T l_B} l^2 \tag{3}$$

Here $l_B = 7.14 \,\text{Å}$ is the Bjerrum length, and $l = 1.7 \,\text{Å}$ is the average separation between the charged phosphate groups, measured along the DNA axis. With a typical bending rigidity of $k = 10 \, k_B T$ we obtain $R = 12.7 \,\text{Å}$, just slightly larger than the DNA radius $R^D \approx 10 \,\text{Å}$. Clearly, a lower R is impossible, even if k gets much lower. A larger value of k would lead of course to a larger R.

The simple capacitor model embodies several assumptions whose validity deserves further explanation. These assumptions pertain to the following questions:

1. Why should the monolayer seek to adjust its charge density according to $\sigma = \sigma^D R^D / R$, as required by the capacitor model?

2. How is the tendency of certain helper lipids (like DOPE) to form inversely bent structures related to the complex stability?

3. What is the energetic cost associated in protecting the hydrophobic side (of the monolayer enveloping the DNA) from exposure to the aqueous environment?

The first question is related to the tendency of the monolayer-coated DNA to be *"isoelectric"*; namely to ensure that all the negative phosphate charges on the

DNA are exactly neutralized by cationic lipid charges. Charge neutrality of the complex implies that any deviation from isoelectricity must be compensated by transferring mobile counterions from the bulk solution into the confines of the water gap separating the DNA and monolayer surfaces. This process implies a substantial entropy loss of the counterions involved or, equivalently, a highly unfavorable increase of the free energy. In other words, the lipid-DNA complex is of maximal stability at the isoelectric point. Solutions of the PB equation discussed in Sec. 2.2 reveal these conclusions both qualitatively and quantitatively. As noted above the lipid charge density, σ, is regulated by the presence of the colipid. The surface charge density of a monolayer composed of a 1:1 mixture of cationic and colipid is roughly equal to that on the DNA surface.

The answer to the second question is rather clear: a lipid mixture with an intrinsic tendency to form the inverse hexagonal phase (i.e., $c_0 < 0$) will naturally favor the honeycomb and spaghetti-like complexes. Since the spontaneous curvature of cationic lipids is generally small, a simple way to achieve negative spontaneous curvatures is by using helper lipids with this propensity; e.g., DOPE whose most stable aggregation geometry is the inverse hexagonal phase. Tuning the spontaneous curvature of the lipid mixture is, indeed, an important role of the colipid. Detailed phase diagrams demonstrating the richness of lipoplex structures and their variation with lipid composition will be described and analyzed in Sec. 3.3.

Finally, regarding the third question there are two options for the monolayer-coated DNA. That is, the coating of the DNA may involve a whole bilayer instead of only a single monolayer, leading to the spaghetti-like structure shown in Fig. 2 C. Or, alternatively, the monolayer-coated DNA units aggregate into an hexagonally packed array, namely the H_{II}^c complex. Our theoretical investigation suggests the H_{II}^c complex to be generally more stable than the spaghetti-like structure. Still, the latter represents a *metastable* state; it is stable with respect to a spontaneous unwrapping of the DNA-coating bilayer. The formation of the H_{II}^c complex is characterized by a free energy gain of order 1 $k_B T$ per Å length of the DNA [36].

3.2. The sandwich-like, L_α^c-structure

The lamellar (L_α^c) phase of CL-DNA complexes appears if bilayer-forming helper lipids are used. Indeed, lamellar lipoplex structures were detected for mixtures of cationic lipids and neutral lipids with the phosphatidylcholine headgroup [4,25,37]. The DNA strands between the lipid bilayers are parallel to each other, forming a one-dimensional lattice with a correlation length extending over a few DNA strands. The average DNA-to-DNA distance, d, is not fixed but can adjust according to the experimental conditions, namely the lipid-to-DNA charge ratio, ρ, and the bilayer composition, m (ratio of neutral to

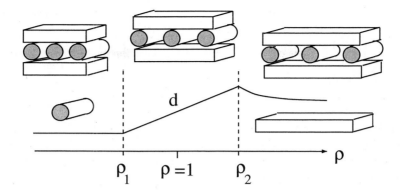

Figure 3. *Schematic illustration of the phase behavior of lamellar CL-DNA self-assembled complexes. Shown is the DNA-to-DNA distance, d, as a function of the lipid-to-DNA charge ratio, ρ. For $\rho < \rho_1$ uncomplexed DNA coexists with L_α^c complexes. For $\rho_1 \leq \rho \leq \rho_2$ the sytem proceeds through a one-phase region where d is proportional to ρ. Note that this region includes the isoelectric point $\rho = 1$ where the fixed charges on the DNA and on the cationic bilayer balance each other. For $\rho > \rho_2$ the L_α^c complexes coexist with excess cationic bilayers. Generally, the composition of the excess bilayer can be different from that in the L_α^c complex.*

cationic lipids). The variations in d with ρ reflect the phase state in which the system resides, as is schematically illustrated in Fig. 3. In particular, for $\rho < \rho_1$ there appears coexistence of sandwich-like complexes with excess DNA. The corresponding DNA-to-DNA distance is small and constant. For intermediate $\rho_1 < \rho < \rho_2$, all lipids and DNA strands are accommodated into the sandwich-like complex. In this, single phase, region d varies linearly with ρ, as implied by the conservation of material. Finally, in the high ρ regime (that is for $\rho > \rho_2$) the sandwich complexes coexist with excess membranes, and the spacing d remains nearly constant. It turns out that the phase boundaries, ρ_1 and ρ_2, depend somewhat on m. But the single phase region, $\rho_1 < \rho < \rho_2$, generally includes the "isoelectric" point, $\rho = 1$, where the fixed negative charges on the DNA balance the same number of positive charges on the cationic lipids. The fact that the single phase region, $\rho_1 < \rho < \rho_2$, is not confined to the point $\rho = 1$ has its origin in the adjustability of the DNA-to-DNA distance d. This additional degree of freedom allows a charge regulation mechanism to take place: for $\rho_1 < \rho < 1$ there are more DNA strands in the complex than would be necessary to neutralize all the lipid charges. The sandwich-like structure then carries a negative excess charge that must be balanced by an appropriate number of positive counterions. Analogously for $1 < \rho < \rho_2$, the L_α^c structure incorporates all lipids, even though this leads to an "overcharging" of the complex.

The experimental observation of the above-described phase behavior raises

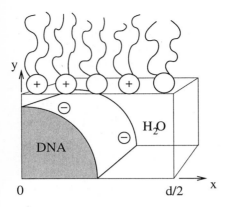

Figure 4. *Schematic representation of the* L_α^c *complex. Shown is one-quarter of the unit cell. It is sufficient to solve the PB equation in the aqueous interior. The (dimensionless) electrostatic potential* $\Psi = \Psi(x, y)$ *depends only on the directions,* x *and* y*, parallel and normal to the bilayer. The figure also illustrates the lateral mobility of the lipids. That is, the distribution of the cationic lipids can adjust along the* x*-axis such as to minimize the free energy of the lipoplex.*

a principal question: why do DNA molecules attract each other when being sandwiched between two cationic bilayers? After all, naked DNA does not show attractive forces in monovalent salt solutions. Hence, the formation of one-dimensional DNA arrays between the galleries of the L_α^c structure must somehow be mediated by the cationic lipid layers. Indeed, lipid-mediated attractive forces can arise from elastic perturbations induced by membrane-associated macromolecules. In case of DNA, such indirect attractions could compete with the direct electrostatic repulsion between the DNA strands, thus giving rise to an equilibrium distance, d. This mechanism was suggested by Dan [38].

Another, entirely different, scenario is based solely on the electrostatic interactions between the cationic lipid layers and the DNA [39,40]. Here, the L_α^c complex adopts its most favorable free energy at the isoelectric point. A corresponding theoretical model that explains the observed phase evolution of the L_α^c complex (see Fig. 3) was recently suggested on the basis of PB theory [40] (see also Ref. [39]). To this end, the corresponding nonlinear PB equation was solved within one-quarter of the unit cell of the L_α^c structure, as shown in Fig. 4. The individual lipids within the fluid-like bilayers are able to diffuse in lateral direction. Hence, the presence of the DNA can induce a spatial modulation ("polarization") of the lipids along the x-axis, as schematically illustrated in Fig. 4. In fact, the most efficient polarization would lead to a constant electrostatic surface potential on the cationic membranes. However, the tendency of the membrane to become polarized is partially opposed by the loss of entropy, associated with the in-plane demixing of the cationic and neutral lipids. A compromize results where the bilayer does neither keep its surface charge density nor the electrostatic surface potential constant. What is kept constant along the x-axis is the electro-chemical potential of the lipid molecules. This constancy can be taken into account through a special boundary condition of the PB equation. It appears that especially for small compositions, ϕ, of the L_α^c

complex (where the number of cationic lipids in the bilayer is small) there is a pronounced accumulation of cationic lipids in the vicinity of the DNA. This leads to a more favorable interaction beween the DNA strands and the bilayers, and thus stabilizes the L_α^c complex.

From the numerical solutions of the nonlinear PB equation [40] it appears that the optimal electrostatic free energy is adopted for an isoelectric complex where a maximal number of counterions is released into solution. Notable, maximal release of counterions at the isoelectric point is also found experimentally [41]. At the isoelectric point, the DNA-to-DNA distance is $d^\star = a\phi/l$ where $a = 65$ Å^2 is the cross-sectional area per lipid (which is roughly the same for both lipid species). The existence of a finite d is thus a consequence of isoelectricity, i. e. the tendency to release as many mobile ions as possible into solution. Electrostatics also explains the stability of the L_α^c complexes for deviations from isoelectricity, that is for $\rho_1 < \rho < \rho_2$ and $\rho \neq 1$ where $d \neq d^\star$. Here, the complexes take up either excess DNA (for $\rho_1 < \rho < 1$) or cationic membranes (for $1 < \rho < \rho_2$). Both cases lead to an "overcharging" of the L_α^c complex that is accompanied by an energetic penalty. Yet this is still more favorable than to form an isoelectric complex where excess DNA or excess cationic bilayer would be present that both must shield their charges by immobilizing a diffuse counterion layer. This is the reason why the most stable L_α^c complex, namely the isoelectric one, is in fact unstable with respect to the absorption of either DNA or additional lipids. At $\rho = \rho_1$ or $\rho = \rho_2$ the L_α^c complex becomes unstable. In the former case (at $\rho = \rho_1$) the DNA uptake stops because of the direct DNA-DNA repulsion. In the latter case (at $\rho = \rho_2$) the electrostatic repulsion between neighboring bilayers in the L_α^c complex becomes too large.

Interestingly, the correleations of the DNA strands may extend through the bilayers, giving rise to an intermembrane coupling between the DNA arrays in different galleries of the L_α^c complex. Indeed, for sufficiently low temperatures, where the lipids reside in the gel phase, there is experimental evidence [42] for the formation of a rectangular superlattice of DNA similar to that shown in Fig. 2 A. Even for fluid-like membranes, fingerprint-like patterns in cryo-electron micrographs are indicative of inter-membrane correlations between the DNA arrays [27]. The most reasonable explanation for the appearance of such correlations seems to be a DNA-induced, elastic perturbation of lipid bilayers. This mechanism was analyzed by Schiessel and Aranda-Espinoza [43] on the basis of the linearized PB equation.

3.3. The phase behavior of CL-DNA mixtures

Our discussion of the H_{II}^c and L_α^c structures so far in this section has revealed distict energetic features of each complex type: The H_{II}^c structure benefits from the close proximity between opposite charges of the DNA and cationic

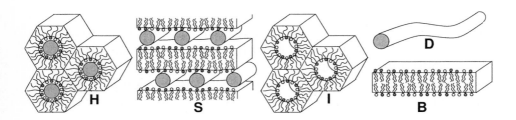

Figure 5. *Schematic illustration of the five macroscopic phases (from Ref. [48], with permission), commonly observed in CL-DNA systems. The phases denoted by H and S are the H_{II}^c and L_α^c complex structures, respectively. The symbols I and B mark the (DNA-free) H_{II} and L_α phases, respectively. D represents uncomplexed DNA. The shaded regions correspond to the DNA cross sectional area. The lipid layers are mixed, consisting of cationic and uncharged (helper) lipids.*

lipids. Yet this electrostatic preference is partially opposed by the elastic energy needed to bend the lipid layers around the DNA strands. On the other hand, the L_α^c complex involves flat membranes that comply with the typical tendency of common cationic lipids to form lamellar phases. In addition to that, the L_α^c structure can adust the DNA-to-DNA distance, d; this extra degree of freedom offers more viability. But here, the geometry of the L_α^c structure implies less perfect electrostatic matching between the DNA and the lipid bilayers. Already these qualitative considerations suggest neither the H_{II}^c complex nor the L_α^c structure to be the ultimate stable complex type. Indeed, recent experiments [5,24,44] evidence the possibility of a transition between the L_α^c and H_{II}^c structure. In fact, there appear two different mechanisms to induce the $L_\alpha^c \to H_{II}^c$ transition that both lower the energy needed to coat the DNA by a highly bent monolayer. One mechanism employs the colipid as a means to shift the spontaneous curvature, c_0, of the cationic layer [5]. This applies for the common helper lipid DOPE which itself has a preference to assemble into the inverse hexagonal (H_{II}) phase. When used as colipid, DOPE introduces a tendency into the cationic layer to adopt negative curvature, similar to that in the H_{II} phase. The second mechanism is based on a reduction of the elastic energy needed to wrap the cationic layer around DNA. This was achieved experimentaly by the addition of DOPC plus a membrane-soluble cosurfactant (like hexanol) [5]. Here the preference of the cationic layer for the bilayer structure is not affected but the cosurfactant drastically softens the bilayer [45]. This can be understood as a reduction of the bilayer bending rigidity, k (see Eq. (1)), nearly eliminating the energetic penalty needed to bend the cationic lipid layer [46,47].

We shall give a short account of a recent phase calculation that included the

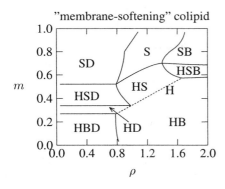
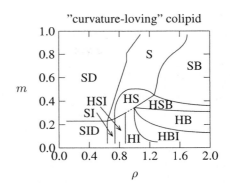

Figure 6. *Calculated phase diagrams of CL-DNA mixtures (from Ref. [48], with permission), as a function of the lipid-to-DNA charge ratio, ρ, and the bilayer composition, m (ratio of neutral to cationic lipids). The left diagram was derived for very soft lipid layers (the actual calculation was carried out for $k = 0.2\,k_BT$ and $c_0 = 0$). This illustrates the effects of adding to the commonly used colipid DOPC a membrane-softening cosurfactant like hexanol. The right diagram models the usage of "curvature-loving" colipids like DOPE. The calculation was performed for a constant bending rigidity $k = 10\,k_BT$ and a spontaneous curvature that changes linearly with the content of colipid from $c_0 = 0$ to $c_0 = -1/25\,\text{Å}^{-1}$. In both diagrams, the straight dashed line marks the single H_{II}^c phase region. The symbols S, B, H and D, denote, respectively, the L_α^c, L_α, H_{II}^c and uncomplexed DNA phases, see Fig. 5.*

appearance of the five structures shown in Fig. 5 [48]. The theoretical tools, sufficient for a detailed model of the phase behavior of CL-DNA mixtures are again PB theory and the quadratic bending energy given in Eq. (1). The spaghetti-like structure was not considered because this complex type does not appear in expriments as a macroscopic phase. Each of the structures in Fig. 5 involves charged surfaces (either the DNA or the cationic lipid layers); the corresponding free energies were calculated using PB theory. The bending energies of the curved lipid layers (concerning structures H and I in Fig. 5) were extracted on the basis of Eq. (1), with linearily composition-dependent spontaneous curvature c_0. The phase calculations were carried out as a function of the lipid-to-DNA charge ratio, ρ, and the bilayer composition, m (ratio of neutral to cationic lipids). For each given values of ρ and m, the actual calculation of the appearing phases involves the minimization of the overall free energy which accounts for all possible combinations of existing phases. The resulting phase diagrams appear rather complex, exhibiting a multitude of phase transitions and coexistences. Yet, the two above-mentioned mechanisms to induce the $L_\alpha^c \rightarrow H_{II}^c$ can be reproduced as shown in Fig. 6. The left diagram in Fig. 6 is derived in the limit of a nearly vanishing bending rigidity of the cationic membrane (in the calculations $k = 0.2\,k_BT$ was used) where the bending of the lipid layers does not cost an appreciable amount of elastic energy. We have argued above that in this case the H_{II}^c complex is energetically most favorable. This however does not exclude the presence of the L_α^c structure. In fact, Fig. 6 (left diagram) displays regions where the L_α^c complex either coexists with other structures or is even present as a single phase. This is a result of the additional degree of freedom pertaining to the L_α^c complex which allows it to adjust to combinations of ρ and m that do not match the actual need of the H_{II}^c geometry. The pure H_{II}^c phase exists only on a single line in the phase diagram (the broken line in Fig. 6) because it has only one single degree of freedom, namely to adjust its composition. The right diagram in Fig. 6 models the influence of a curvature-loving helper lipid (like DOPE). The calculation was carried out for a constant bending rigidity $k = 10\,k_B T$ and a spontaneous curvature that changes linearly with the content of colipid from $c_0 = 0$ to $c_0 = -1/25\,\text{Å}^{-1}$. This is reflected in the phase diagram by the preferential appearance of the H_{II}^c phase only in the low-m region.

4. ELECTROSTATIC ADSORPTION OF PROTEINS ONTO LIPID MEMBRANES

The adsorption of peripheral proteins onto lipid membranes is frequently driven by electrostatic interactions, and can be characterized by a binding constant that describes the equilibrium between the membrane-bound and free proteins. Determination of the binding constant is one of the main objectives in modeling protein adsorption. There are a number of recent approaches to calculate the adsorption energy of charged proteins onto oppositely charged surfaces, ranging from simple generic models up to atomic-level representations of the protein structures [49–53]. Yet so far, the predictive power of the existing models is still limited. We see three reasons for the current deficiencies: First, the long-range nature of the Coulombic interaction makes exact calculations computationally expansive. A common alternative is the utilization of PB theory which, however, fails if the aqueous solution contains multivalent salt [16,54]. Second, the binding is often influenced by non-electrostatic interactions like the hydrophobic effect [55,56] or conformational changes of the proteins [57]. These effects can be rather specific, depending on the structural details and microscopic interactions of the adsorbing protein. Third, multicomponent lipid bilayers are flexible surfaces and have internal degrees of freedom that affect the adsorption of proteins [58]. That is, unlike a solid surface the lipid membrane actively participates in the adsorption process; the lipid membrane also seems to be a crucial factor for the lateral organization of adsorbing proteins [1,59]. Fig. 7 illustrates two membrane-mediated mechanisms that affect the adsorption energy and interactions between adsorbed proteins. One mechanism involves local curvature modulations of the lipid membrane that enhance the strength of interaction be-

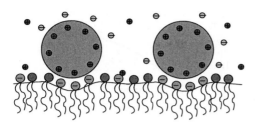

Figure 7. *Mixed lipid membranes are flexible materials with compositional degrees of freedom. They can adapt to the shape and charge density of peripherally adsorbed proteins (illustrated as two large circles). Both the ensuing elastic membrane perturbations and lipid "polarization" effects increase the adsorption strength and give rise to membrane-mediated protein-protein interactions.*

tween the membrane and associated proteins [60,61]. The other one is based on local, protein-induced, demixing of individual membrane components. That is, those membrane lipids that interact more favorably with the adsorbed proteins may migrate towards the protein adsorption sites, replacing the less favorably interacting lipids [62]. In the following we shall focus on the second mechanism.

4.1. Protein-induced lipid demixing

The lateral mobility of the individual lipids in a mixed membrane affects the adsorption energy of proteins. Consider for example a flat, two-component membrane where one lipid species is negatively charged and the other one is electrically neutral. The adsorption of a positively charged protein generally leads to lateral reorganization of the membrane lipids. Negatively charged lipids may accumulate in the vicinity of the protein where they can interact more favorably with the fixed positive charges at the protein surface. This change in the local lipid composition is counteracted by a demixing free energy. One may ask what is the resulting compromise between the two opposing tendencies, and how important is the lateral lipid mobility for the protein adsorption energy?

Let us focus on a recent theoretical investigation that addressed the extent of protein-induced demixing upon the adsorption of a charged protein [63]. The protein was simply modeled as a uniformly charged sphere, and the membrane was kept flat as schematically shown in Fig. 8 (left). PB theory was used to calculate the adsorption free energy of the protein-sphere as function of the adsorption height h. The protein-induced lateral lipid demixing was taken into account by using a special boundary condition for the PB equation which leads to a constant electro-chemical potential of all membrane lipids. In fact, this boundary condition, which reflects the mobility of the charged lipids, is the same as for the calculations of the lamellar CL-DNA complexes (see Sec. 3). The case of constant electro-chemical potential is intermediate between two familiar, limiting, boundary conditions that correspond to either fixed membrane

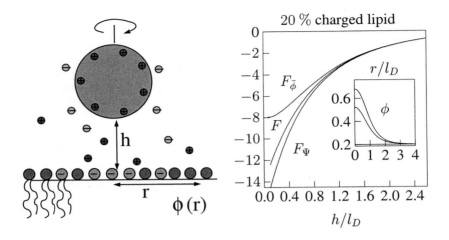

Figure 8. *Left: a uniformly charged sphere adsorbs onto a flat two-component membrane. The individual membrane lipids are mobile; the charged lipid species can optimize its radial composition $\phi(r)$ for any given protein-to-membrane distance h. Right: Predictions from PB theory for the adsorption of a single sphere (of radius $R = 10$ Å with a uniform charge density corresponding to 7 positive charges) onto a mixed membrane that contains 20% negatively charged lipids (from Ref. [63], with permission). The three adsorption free energies (in units of k_BT) correspond to fixed surface charge density ($F_{\bar\phi}$), constant electro-chemical potential (F), and constant membrane surface potential (F_Ψ). The inset shows the local membrane composition, ϕ, for the three cases at $h/l_D = 0.3$. The Debye length is $l_D = 10$ Å.*

surface charge density or constant membrane potential. They correspond to two hypothetical limits: infinitely large or vanishingly small in-plane demixing penalty of the lipids. The results of a representative calculation are shown in Fig. 8 (right), corresponding to 20% charged lipids in the membrane, a radius $R = 10$ Å of the protein sphere with 7 positive charges attached to its surface. The three adsorption free energies correspond to fixed surface charge density ($F_{\bar\phi}$), constant electro-chemical potential (F), and constant membrane potential (F_Ψ). The inset shows the local membrane composition, ϕ, for the three cases at $h/l_D = 0.3$. Clearly, the binding energy is significantly enhanced if the lipid mobility is taken into account. It turns out that the lipid demixing is particularly important for the biologically most relevant case where highly charged proteins adsorb to weakly charged membranes. The described method was also used to investigate the electrostatic interactions between membrane-adsorbed proteins [63]. These (always repulsive) interactions become relevant if the counterion atmospheres of two proteins start to overlap. Inter-protein repulsion leads to pronounced suppression of the adsorption compared to the simple Langmuir-like adsorption where protein-protein repulsion is not taken into account.

The situation becomes more involved if additional short-range attractions are present between individual lipids of the same type. This generally introduces a tendency of the lipids for macroscopic phase separation. However, this tendency can easily be counterbalanced by the electrostatic repulsion acting between charged lipids. This opens an interesting possibility: positively charged proteins that adsorb onto the membrane would effectively neutralize the negative lipid charges. The repulsive part of the interactions between the charged lipids would disappear and, consequently, the tendency for macroscopic phase separation reappears. Indeed, simple electrostatic models that are based on PB theory predict the ability of protein adsorption-induced phase separation [64,65]. The existence of membrane domains has recently received much attention [64,66–68]. Hence, it would be interesting to further develop microscopic models that describe biopolymer-induces phase separations in lipid membranes.

5. MEMBRANE REORGANIZATION INDUCED BY INTEGRAL PROTEINS

A considerable fraction of all proteins is sealed into the lipid bilayer by hydrophobic interactions. Nearly all integral proteins span entirely the lipid bilayer. What varies is the number and secondary structure of the membrane-spanning segments. Yet, despite the structural and functional variety of transmembrane proteins there is a distinct difference to the host bilayer: transmembrane protein segments are much more rigid than the surrounding lipid phase [69]. Thus, the strong hydrophobic forces that anchor the protein into the membrane create an interface between a rigid membrane inclusion and the fluid-like membranous environment. Note that this is an entirely non-specific effect which is present for all protein-containing membranes; its consequences are evidenced experimentally for many different systems. The focus in this section is on our own recent efforts to model non-specific interactions between integral proteins and lipid membranes.

There is an immediate consequence of the different rigidities between integral proteins and membranes: The membrane must adjust to the prescribed shape and size of the protein (and not vice versa). For example, if the length, $2h_P$, of the hydrophobic span of an integral protein differs from the hydrophobic thickness, $2h_0$, of the host bilayer, then there will be an elastic deformation of the bilayer so as to *locally* match the protein size. The mismatch between h_P and h_0 is generally referred to as *hydrophobic mismatch* which can be positive or negative as schematically shown in Fig. 9. Even though the hydrophobic mismatch is usually quite small (a few Å) its consequences can be observed by appropriate experiments (for a recent summary see Killian [6]). With regard to the lipids there is an increase in chain order close to integral proteins [70], there are shifts in the main phase transition temperature, and there is a lipid sorting

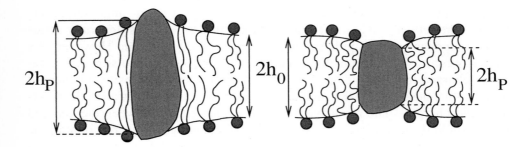

Figure 9. *Schematic illustration of positive (left) and negative (right) hydrophobic mismatch. The length of the hydrophobic span of the protein is denoted by* $2h_P$. *The hydrophobic thickness of the membrane far away from the protein in* $2h_0$.

mechanism for mixed membranes [71]. Perhaps most exiting however, integral proteins can mediate structural phase transitions of bilayers as will be discussed below. Also the protein itself can be affected through hydrophobic mismatch in various ways . Experimental evidence exists for modifications in protein activity and conformation, for mismatch-induced protein aggregation, for a tilt of the membrane spanning segments, and for protein delocalization on the membrane surface [6,72,73].

Most of the theoretical efforts in the past have focused on the lateral organization of membrane proteins [1]. The key here is to understand how the membrane mediates interactions between integral proteins in membranes. In the following we shall discuss our own theoretical investigations to model lipid-protein interactions using a microscopic molecular-level model and a phenomenological approach.

5.1. Predictions for lipid-protein interaction from microscopic models

On a microscopic scale, the presence of an integral protein perturbs the packing of the neighboring lipids. Such a perturbation is present even if there is no hydrophobic mismatch between the protein and membrane. Generally, membrane perturbations may arise from specific interactions between the lipid chains and the amino acid residues of the transmembrane protein segments. Another, non-specific, mechanism is the loss of conformational freedom experienced by the lipid chains in the vicinity of a rigid protein surface. This effect was recently investigated on the basis of molecular-level, mean-field, chain packing theory for a single (and sufficiently large) protein with given hydrophobic mismatch [74]. Our objective was to investigate the implications of the conformational chain restrictions on the interactions between integral membrane proteins [75]. To this end, two interacting (and sufficiently large) integral pro-

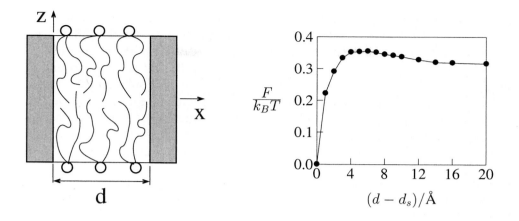

Figure 10. *Left: schematic illustration of a lipid membrane bounded by two rigid, impenetrable, walls. The distance between the walls is d. According to chain packing theory, the probability distributions for all accessible lipid chain conformations inside the bilayer are calculated so as to achieve uniform density of chain segments everywhere within the hydrocarbon core. Right: The wall-induced perturbation free energy (per monolayer and per unit length $L = 1$ Å of the walls) as a function of the wall-to-wall separation $d - d_s$ (where d_s is a constant on the order of the cross-sectional extension of a single lipid chain).*

teins were represented by two rigid walls that are perpendicularly embedded in a symmetric lipid bilayer as shown schematically in Fig. 10 (left). The conformational chain properties were calculated on the basis of the above-mentioned chain packing theory. The key point in this approach [76,77] is the assumption of uniform density of lipid chain segments everywhere within the hydrocarbon core. This is achieved by an appropriate statistical behavior of the flexible lipid chains. In an unperturbed membrane, the average statistical behavior is the same for all lipids. Not so in the presence of the rigid protein-walls. Here, different lipid chains have different conformational space available. The most restricted lipid chains are those in immediate neighborhood to either one of the walls. More generally, the average conformational behavior of the lipid chains depends on the distance to the walls. If the walls are sufficiently close to each other, a given chain can be influenced by both walls. This gives rise to short-range interactions between the walls. The prediction of the chain-packing theory for the interaction potential between walls of distance d is shown in Fig. 10 (right). Most notably, the potential is non-monotonic, exhibiting a maximum at some intermediate wall-to-wall distance. The appearance of a maximum can be understood in terms of a simple *director model* in which each lipid chain is represented as a fluctuating director [75].

Non-monotonic interaction potentials are also found using other microscopic models to study membrane-mediated protein-protein interactions. Recent Monte-Carlo simulations of a coarse-grained membrane model [78,79] revealed a rather complex interaction potential between rigid membrane inclusions: short and long-range attraction was separated by a repulsive barrier for distances between the inclusions somewhat larger than the lipid diameter. Similarly, a recent approach based on hypernetted chain integral equation formalism [80] (where the lateral density-density response function of the hydrophobic core was extracted from a molecular dynamics simulation [81]) also resulted in an energetic barrier upon the approach of two integral proteins.

5.2. Membrane elasticity theory

Membrane elasticity theory treats the bilayer as an elastic continuum. The membrane free energy is expressed with respect to a certain (and usually small) number of *order parameters* where each order parameter represents a degree of freedom of the system. The interaction of a rigid biopolymer with the membrane imposes a localized perturbation of some lipid molecules. This perturbation induces an elastic response of the bilayer; the response and the corresponding membrane free energy can be calculated in terms of elastic moduli through which the material properties of the membrane are characterized. In the past, membrane elasticity theory was frequently used to model interactions between bilayers and integral proteins. Two different cases appear depending on whether an integral protein is symmetric or asymmetric [82]. Symmetric ones leave the midplane of a bilayer planar (unless, of course, the bilayer itself is asymmetric) and cause exponentially decaying perturbations and interactions. In contrast, interactions between asymmetric proteins are long-ranged [83,84].

The most significant experimental motivation to apply membrane elasticity theory comes from studies of interactions between model membranes and gramicidin A [85]. This miniprotein consists of two monomers that dimerize in a bilayer to form a (symmetric) cation-selective channel. The average channel lifetime depends on the properties of the host membrane [86–88]; via monitoring its lifetime, the channel can be used as a force transducer that reflects the energetics of a lipid bilayer. Let us give a basic introduction into the theory for symmetric proteins (like gramicidin A). To this end, it is sufficient to describe, say, the upper monolayer by the local (effective) thickness, h, of the lipid chain region or, equivalently, the relative thickness dilation $u = h/h_0 - 1$. In the most simple case, two modes of deformation are taken into account in the free energy of a perturbed bilayer. Then, F can be written as

$$F = \int_A da \left[\frac{K}{2} u^2 + \frac{k}{2} (c_1 + c_2 - c_0)^2 \right] \tag{4}$$

The first term in Eq. (4) is due to the stretching (or compression) of the lipid chains; K is the corresponding stretching modulus of a lipid monolayer. The

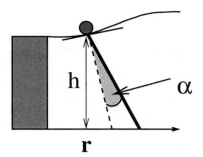

Figure 11. *Schematic illustration of the lipid tilt degree of freedom. Generally, lipid tilt refers to the angle, α, of the average lipid chain director (thick line) at position $\mathbf{r} = \{x, y\}$ of the head group (circle), measured with respect to the normal direction (broken line) of the hydrocarbon-water interface. Another, commonly used, order parameter is the local hydrophobic thickness dilation, $u = h/h_0 - 1$.*

second term in Eq. (4) is equivalent to the first term in Eq. (1), and accounts for the splay energy of the lipid molecules. To a good approximation the principal curvatures c_1 and c_2 can be measured at the hydrocarbon chain-water interface. Realizing that $c_1 + c_2 \approx h_0(\partial^2 u/\partial x^2 + \partial^2 u/\partial y^2)$ we find F to be a function of a single order parameter, namely $u(x, y)$ where the x, y-plane coincides with the bilayer midplane. Minimization of F then requires to solve an appropriate Euler equation for which the proteins enter as boundary conditions. For example, in case of a given hydrophobic mismatch the Euler equation must be solved such that at the protein boundaries we find $u(x, y) = (h_p/h_0 - 1)$.

Already the simple approach according to Eq. (4) seems to well describe real membranes. In fact, the predictions of membrane elasticity theory according to Eq. (4) agree with experimental results. This concerns the so-called spring constant [89] that describes the energetic response of a bilayer membrane to a given, gramicidin A-imposed, hydrophobic mismatch [90]. It also appears to reproduce well the experimental findings of gramicidin A-induced membrane thinning [91]. Note that a number of previous theoretical investigations have basically used Eq. (4) to calculate lipid-protein [92,93] and membrane-mediated protein-protein interactions [94,95].

Recently, it was suggested that another elastic mode of deformation should be taken into account when applying membrane elasticity theory to structurally perturbed membranes, namely the ability of the lipid chains to tilt with respect to the hydrocarbon chain-water interface [96–102] as schematically illustrated in Fig. 11. Lipid tilt is possible despite the fact that the membrane is in the fluid-like state. In fact, here the tilt refers to the *average* orientation of the flexible chains. The ability to tilt gives rise to a second order parameter, the tilt angle α (besides u), which generally lowers the free energy, F, and facilitates the spatial relaxation of the membrane. With the possibility of a local tilt angle α the free energy becomes

$$F = \int_A da \left[\frac{K}{2} u^2 + \frac{k}{2} (\tilde{c}_1 + \tilde{c}_2 - c_0)^2 + \frac{k_t}{2} \alpha^2 \right] \tag{5}$$

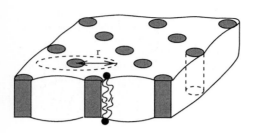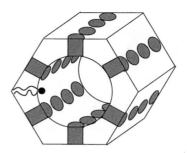

Figure 12. The structures of the inclusion-containing L_α (left) and H_{II} phases (right), the latter one as suggested by Killian [104]. The shaded regions represent the peptide locations (from Ref. [101], with permission).

where k_t is the tilt modulus which was very roughly estimated to be on the order of $k_t \approx 0.1\ k_B T/\text{Å}^2$ [99]. The quantities $\tilde{c}_1$ and $\tilde{c}_2$ account for the splay of the monolayer. For $\alpha = 0$ the splay $\tilde{c}_1 + \tilde{c}_2$ becomes equal to the sum of the principal curvatures $c_1 + c_2$ of the chain-water interface. But generally, lipid splay can also arise through changes in the tilt angle α as was recently analyzed by Hamm and Kozlov [100]. The lipid tilt degree of freedom gives rise to new interesting phenomena, like a "ripple" instability of a pure membrane [97], or the formation of sharp membrane edges [103]. It also opens new ways to model structurally perturbed lipid phases like the inverse-hexagonal (H_{II}) phase [101,99].

5.3. Protein-induced structural phase transitions of lipid membranes

Aside from the effects on the lateral organization of planar lipid bilayers, integral proteins are able to affect the structural phase behavior of membranes. This was shown for gramicidin A and a number of synthetic α-helical trans-membrane peptides residing in common lipid membranes [104–106]. In fact, these peptides are able to induce the $L_\alpha \to H_{II}$ transition in membranes that without the peptide have a strong tendency to self-assemble into planar bilayers. It was found that the $L_\alpha \to H_{II}$ transition could be induced above a critical concentration of the peptide in the membrane. Moreover, the transition only takes place for a sufficiently large *negative* hydrophobic mismatch, and only if the peptides contain interface-anchored tryptophans (probably to prevent peptide aggregation in the membrane). A structural model suggested by Killian et al [104] for the arrangement of the peptides in both the L_α and H_{II} phases is shown in Fig. 12. According to this model, the peptides in the H_{II} phase are arranged in such a way that they span the entire hydrophobic region between neighboring tubes.

We have recently presented a theoretical investigation of the peptide-

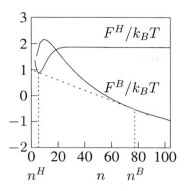

Figure 13. The free energy per inclusion in the inverse hexagonal phase (F^H), and in the bilayer phase (F^B), as a function of the number of lipids per inclusion, n. The dotted lines characterize the 'common tangent', and the coexistence values of n^B and n^H. The results shown correspond to lipid layers with $c_0 = 0$, containing cylindrical inclusions with 35% negative hydrophobic mismatch (adapted from Ref. [101]).

induced $L_\alpha \rightarrow H_{II}$ transition [101]. It was based on membrane elasticity theory applied to the peptide-containing L_α and H_{II} phases. As is common in membrane elasticity theory, the peptides were represented as rigid inclusions of given shape and size. A significant feature of our investigation was the usage of a molecular-level model that accounts for the principal forces governing the structural phase behavior of lipid molecules, namely head group repulsion (f_{hg}), tail repulsion (f_c), and the interfacial tension (f_i) between the hydrocarbon-water interface. In fact, the resulting free energy per lipid, $f = f_{hg} + f_c + f_i$, can be used to calculate the elastic properties of a lipid layer. The lipid tilt degree of freedom appears naturally in this approach, and all the elastic constants defined in Eq. (5) can be calculated in terms of the molecular model. Hence, our model of the inclusion-containing L_α and H_{II} phases is based on Eq. (5) and an appropriate minimization of the free energies, F^B and F^H, for the bilayer and the inverse hexagonal phase, respectively. Once F^B and F^H are calculated as a function of the peptide concentration, the coexisting phases can by derived from common thermodynamic equilibrium conditions. In the present case, this is equivalent to finding the common tangent of F^B and F^H as shown in Fig. 13. In agreement with experimental results we found that only sufficiently short peptides are able to drive the $L_\alpha \rightarrow H_{II}$ transition, once the peptide concentration exceeds a critical value. In the coexisting bilayer the peptide concentration is much smaller than in the H_{II} phase (see Fig. 13).

6. CONCLUSIONS AND OUTLOOK

Lipid membranes are complex fluids that adapt their structure and conformation to various kinds of associated biopolymers. This ability gives rise to exciting phenomena, ranging from lateral modulations of membrane structure to morphological phase transitions. Our understanding is challenged because membrane-biopolymer interactions are ultimately relevant in biology and biotechnology. The present chapter focuses entirely on elastic and electrostatic interactions which is only a fraction of what is present in real systems. However, it appears that already these two types of interaction are sufficient to principally understand a number of experimental observations, like the formation of CL-DNA complexes or protein induced structural phase transitions of lipid membranes.

Note that this chapter does not deal with partially membrane penetrating biopolymers, like amphipathic peptides. Even though there are intensive experimental activities to understand the action of amphipathic, membrane-active, peptides on lipid bilayers [107–109], there are only a few theoretical investigations [110,111] and computer simulations performed so far. One reason for our current deficiencies in modeling the association between amphipathic peptides and lipid bilayers is perhaps the profound coupling between electrostatic and elastic interactions: amphipathic peptides insert partially into membranes and cause substantial perturbations of the hydrophobic core. At the same time, most of these peptides are highly charged which influences their mutual interactions. Hence, cooperative behavior, like peptide-induced formation of membrane pores can be expected to sensitively depend on the interplay between electrostatic and elastic interactions. Further work is required to better understand these important systems.

ACKNOWLEDGMENTS

We would like to thank our collaborators and co-workers, in particular D. Harries, W. M. Gelbart, M. M. Kozlov, and A. Ulrich. We apologize to all colleagues who work on lipid-biopolymer interaction but are not directly referenced. SM is generously supported by TMWFK.

REFERENCES

[1]   T. Gil, J. H. Ipsen, O. G. Mouritsen, M. C. Sabra, M. M. Sperotto and M. J. Zuckermann, Biophys. Biochim. Acta., 1376 (1998), 245.
[2]   C. R. Safinya, Curr. Opin. Struct. Biol., 11 (2001), 440.
[3]   B. Sternberg, F. L. Sorgi and L. Huang, FEBS Letters, 356 (1994), 361.
[4]   J. O. Rädler, I. Koltover, T. Salditt and C. R. Safinya, Science, 275 (1997), 810.
[5]   I. Koltover, T. Salditt, J. O. Rädler and C. R. Safinya, Science, 281 (1998), 78.
[6]   J. A. Killian, Biophys. Biochim. Acta., 1376 (1998), 401.
[7]   J. M. Seddon and R. H. Templer, in Structure and Dynamics of Membranes, (eds.) R. Lipowsky and E. Sackmann, vol. 1, Elsevier, Amsterdam, 1995, second edn. 98–160.

[8] J. N. Israelachvili, Intermolecular and Surface Forces, Academic Press, 1992, second edn.

[9] D. F. Evans and H. Wennerström, The colloidal domain, where physics, chemistry, and biology meet, VCH publishers, 1994, second edn.

[10] W. Helfrich, Z. Naturforsch., 28 (1973), 693.

[11] D. P. Tieleman, S. J. Marrink and H. J. C. Berendsen, Biophys. Biochim. Acta., 1331 (1997), 235.

[12] E. J. W. Verwey and J. T. G. Overbeek, Theory of the stability of lyophobic colloids, Elsevier, Elsevier, New York, 1948.

[13] D. Andelman, in Structure and Dynamics of Membranes, (eds.) R. Lipowsky and E. Sackmann, vol. 1, Elsevier, Amsterdam, 1995, second edn. 603–642.

[14] W. Rocchia, E. Alexov and B. Honig, J. Phys. Chem. B, 105 (2001), 6507.

[15] V. Vlachy, Annu. Rev. Phys. Chem, 50 (1999), 145.

[16] H. Wennerström, B. Jönsson and P. Linse, J. Chem. Phys, 76 (1982), 4665.

[17] P. L. Felgner, T. R. Gadek, M. Holm, R. Roman, H. W. Chan, M. Wenz, J. P. Northrop, G. M. Ringold and M. Danielsen, Proc. Natl. Acad. Sci. U.S.A., 84 (1987), 7413.

[18] P. L. Felgner, Scientific American, 276 (1997), 102.

[19] M. J. Hope, B. Mui, S. Ansell and Q. F. Ahkong, Molecular Membrane Biology, 15 (1998), 1.

[20] Y. Xu and F. C. Szoka Jr., Biochemistry, 35 (1996), 5616.

[21] W. M. Gelbart, R. Bruinsma, P. A. Pincus and V. A. Parsegian, Physics today, 53 (2000), 38.

[22] S. Hübner, B. J. Battersby, R. Grimm and G. Cevc, Biophys. J., 76 (1999), 3158.

[23] S. W. Hui, M. Langner, Y.-L. Zhao, R. Patrick, E. Hurley and K. Chan, Biophys. J., 71 (1996), 590.

[24] J. Smisterova, A. Wagenaar, M. C. Stuart, E. Polushkin, G. ten Brinke, R. Hulst and J. B. E. D. Hoekstra, J. Biol. Chem, 276 (2001), 47615.

[25] D. D. Lasic, H. Strey, M. C. A. Stuart, R. Podgornik and P. M. Frederik, J. Am. Chem. Soc., 119 (1997), 832.

[26] Y. S. Tarahovsky, R. S. Khusainova, A. V. Gorelov, T. I. Nicolaeva, A. A. Deev, A. K. Dawson and G. R. Ivanitsky, FEBS Letters, 390 (1996), 133.

[27] B. J. Battersby, R. Grimm, S. Hübner and G. Cevc, Biophys. Biochim. Acta., 1372 (1998), 379.

[28] B. Pitard, O. Aguerre, M. Airiau, A. M. Lachages, T. Boukhnikachvili, G. Byk, C. Dubertret, C. Herviou, D. Scherman, J. F. Mayaux and J. Crouzet, Proc. Natl. Acaf. Sci. USA, 94 (1997), 14412.

[29] T. Boukhnikachvili, O. AguerreChariol, M. Airiau, S. Lesieur, M. Ollivon

and J. Vacus, FEBS Letters, 409 (1997), 188.

[30] B. Sternberg, J. Liposome Res., 6 (1996), 515.

[31] D. A. Pink, B. Quinn, J. Moeller and R. Merkel, Phys. Chem. Chem. Phys., 2 (2000), 4529.

[32] S. Bandyopadhyay, M. Tarek and M. L. Klein, J. Phys. Chem. B, 103 (1999), 10075.

[33] L. Golubovic and M. Golubovic, Phys. Rev. Lett., 80 (1998), 4341.

[34] C. S. O'Hern and T. C. Lubensky, Phys. Rev. Lett., 80 (1998), 4345.

[35] C. Fleck, R. R. Netz and H. H. von Grünberg, Biophys. J., 82 (2002), 76.

[36] S. May and A. Ben-Shaul, Biophys. J., 73 (1997), 2427.

[37] N. S. Templeton, D. D. Lasic, P. M. Frederik, H. H. Strey, D. D. Roberts and G. N. Pavlakis, Nature Biotechnology, 15 (1997), 647.

[38] N. Dan, Biophys. J., 73 (1997), 1842.

[39] R. Bruinsma, Eur. Phys. J. B, 4 (1998), 75.

[40] D. Harries, S. May, W. M. Gelbart and A. Ben-Shaul, Biophys. J., 75 (1998), 159.

[41] K. Wagner, D. Harries, S. May, V. Kahl, J. O. Rädler and A. Ben-Shaul, Langmuir, 16 (2000), 303.

[42] F. Artzner, R. Zantl, G. Rapp and J. Rädler, Phys. Rev. Lett., 81 (1998), 5015.

[43] H. Schiessel and H. Aranda-Espinoza, Eur. Phys. J. E, 5 (2001), 499.

[44] D. Simberg, D. Danino, Y. Talmon, A. Minsky, M. E. Ferrari, C. J. Wheeler and Y. Barenholz, J. Biol. Chem., 276 (2001), 47453.

[45] C. R. Safinya, E. B. Sirota, D. Roux and G. S. Smith, Phys. Rev. Lett., 62 (1989), 1134.

[46] I. Szleifer, D. Kramer, A. Ben-Shaul, W. M. Gelbart and S. A. Safran, J. Chem. Phys., 92 (1990), 6800.

[47] S. May and A. Ben-Shaul, J. Chem. Phys., 103 (1995), 3839.

[48] S. May, D. Harries and A. Ben-Shaul, Biophys. J., 78 (2000), 1681.

[49] P. Warszyńsky and Z. Adamczyk, J. Colloid Interface Sci, 187 (1997), 283.

[50] S. A. Palkar and A. M. Lenhoff, J. Colloid Interface Sci, 165 (1994), 177.

[51] D. McCormack, S. L. Carnie and D. Y. C. Chan, J. Colloid Interface Sci, 169 (1995), 177.

[52] N. Ben-Tal, B. Honig, R. M. Peitzsch, G. Denisov and S. McLaughlin, Biophys. J., 71 (1996), 561.

[53] N. Ben-Tal, B. Honig, C. Miller and S. McLaughlin, Biophys. J., 73 (1997), 1717.

[54] N. Gronbech-Jensen, R. J. Mashl, R. F. Bruinsma and W. M. Gelbart, Phys. Rev. Lett., 78 (1999), 2477.

[55] S. McLaughlin and A. Aderem, Trends in Biochem. Sci., 20 (1995), 272.

[56] S. H. White and W. C. Wimley, Biophys. Biochim. Acta., 1376 (1998), 339.

[57] S. H. White and W. C. Wimley, Annu. Rev. Biophys. Biomol. Struct., 28 (1999), 319.

[58] A. Ben-Shaul, N. Ben-Tal and B. Honig, Biophys. J., 71 (1996), 130.

[59] P. Schiller and H. J. Mögel, Phys. Chem. Chem. Phys., 2 (2000), 4563.

[60] H. Schiessel, Eur. Phys. J. B, 6 (1998), 373.

[61] P. Schiller, Mol. Phys., 98 (2000), 493.

[62] T. Heimburg, B. Angerstein and D. Marsh, Biophys. J., 76 (1999), 2575.

[63] S. May, D. Harries and A. Ben-Shaul, Biophys. J., 79 (2000), 1747.

[64] G. Denisov, S. Wanaski, P. Luan, M. Glaser and S. McLaughlin, Biophys. J., 74 (1998), 731.

[65] D. Harries, S. May and A. Ben-Shaul 2001, Colloids and Surfaces, in press.

[66] D. A. Brown and E. London, J. Mem. Biol., 164 (1998), 103.

[67] A. J. Bradley, E. Maurer-Spurej, D. E. Brooks and D. V. Devine, Biochemistry, 38 (1999), 8112.

[68] A. K. Hinderliter, P. F. F. Almeida, C. E. Creutz and R. L. Biltonen, Biochemistry, 40 (2001), 4181.

[69] K. Gekko and H. Noguchi, J. Phys. Chem., 83 (1979), 2706.

[70] M. R. R. de Planque, D. V. Greathouse, R. E. Koeppe, H. Schäfer, D. Marsh and J. A. Killian, Biochemistry, 37 (1998), 9333.

[71] F. Dumas, M. M. Sperotto, M. C. Lebrun, J. F. Tocanne and O. G. Mouritsen, Biophys. J., 73 (1997), 1940.

[72] M. R. R. de Planque, E. Goormaghtigh, D. V. Greathouse, R. E. Koeppe, J. A. W. Kruijtzer, R. M. J. Liskamp, B. de Kruijff and J. A. Killian, Biochemistry, 40 (2001), 5000.

[73] B. Bechinger, FEBS Letters, 504 (2001), 161.

[74] D. R. Fattal and A. Ben-Shaul, Biophys. J., 65 (1993), 1795.

[75] S. May and A. Ben-Shaul, Phys. Chem. Chem. Phys., 2 (2000), 4494.

[76] A. Ben-Shaul and W. M. Gelbart, in Micelles, Membranes, Microemulsions, and Monolayers, (eds.) W. M. Gelbart, A. Ben-Shaul and D. Roux, Springer, New York, 1994, first edn. 359–402.

[77] A. Ben-Shaul, in Structure and Dynamics of Membranes, (eds.) R. Lipowsky and E. Sackmann, vol. 1, Elsevier, Amsterdam, 1995 359–402.

[78] T. Sintes and A. Baumgärtner, J. Phys. Chem. B, 102 (1998), 7050.

[79] T. Sintes and A. Baumgärtner, Physica A, 249 (1998), 571.

[80] P. Lagüe, M. J. Zuckermann and B. Roux, Biophys. J., 79 (2000), 2867.

[81] S. E. Feller, R. M. Venable and R. W. Pastor, Langmuir, 13 (1997), 6555.

[82] S. May, Curr. Op. Coll. Int. Sci., 5 (2000), 244.

[83] M. Goulian, R. Bruinsma and P. Pincus, Europhys. Lett., 22 (1993), 145.

[84] K. S. Kim, J. Neu and G. Oster, Biophys. J., 75 (1998), 2274.

[85] T. A. Harroun, W. T. Heller, T. M. Weiss, L. Yang and H. W. Huang, Biophys. J., 76 (1999), 937.

[86] J. A. Lundbæk, A. M. Maer and O. S. Andersen, Biochemistry, 36 (1997), 5695.

[87] J. A. Lundbæk, P. Birn, J. Girshman, A. J. Hansen and O. S. Andersen, Biochemistry, 35 (1996), 3825.

[88] J. A. Lundbæk and O. S. Andersen, J. Gen. Physiol., 104 (1994), 645.

[89] C. Nielsen, M. Goulian and O. S. Andersen, Biophys. J., 74 (1998), 1966.

[90] J. A. Lundbæk and O. S. Andersen, Biophys.J, 76 (1999), 889.

[91] T. A. Harroun, W. T. Heller, T. M. Weiss, L. Yang and H. W. Huang, Biophys. J., 76 (1999), 3176.

[92] N. Dan and S. A. Safran, Biophys. J., 75 (1998), 1410.

[93] C. Nielsen and O. S. Andersen, Biophys. J., 79 (2000), 2583.

[94] N. Dan, P. Pincus and S. A. Safran, Langmuir, 9 (1993), 2768.

[95] H. Aranda-Espinoza, A. Berman, N. Dan, P. Pincus and S. A. Safran, Biophys. J., 71 (1996), 648.

[96] F. C. MacKintosh and T. C. Lubensky, Phys. Rev. Lett., 67 (1991), 1169.

[97] J. B. Fournier, Europhys. Lett., 43 (1998), 725.

[98] J. B. Fournier, Eur. Phys. J.E, 11 (1999), 261.

[99] M. Hamm and M. M. Kozlov, Eur. Phys. J. B, 6 (1998), 519.

[100] M. Hamm and M. M. Kozlov, Eur. Phys. J. E, 3 (2000), 323.

[101] S. May and A. Ben-Shaul, Biophys. J., 76 (1999), 751.

[102] S. May, Eur. Biophys. J., 29 (2000), 17.

[103] Y. Kozlovsky and M. M. Kozlov, Biophys. J., 82 (2002), 882.

[104] J. A. Killian, I. Salemink, M. R. R. de Planque, G. Lindblom, R. E. Koeppe and D. V. Greathouse, Biochemistry, 35 (1996), 1037.

[105] S. Morein, E. Strandberg, J. A. Killian, S. Persson, G. Arvidson, R. E. Koeppe and G. Lindblom, Biophys. J., 73 (1997), 3078.

[106] P. C. A. van der Wel, T. Pott, S. Morein, D. V. Greathouse, R. E. Koeppe and J. A. Killian, Biochemistry, 39 (2000), 3124.

[107] R. M. Epand, Y. Shai, J. P. Segrest and G. M. Anantharamaiah, Biopolymers (Peptide Science), 37 (1995), 319.

[108] B. Bechinger, Phys. Chem. Chem. Phys., 2 (2000), 4569.

[109] M. Dathe and T. Wieprecht, Biophys. Biochim. Acta., 1462 (1998), 71.

[110] H. W. Huang, J. Phys. II France, 5 (1995), 1427.

[111] M. J. Zuckermann and T. Heimburg, Biophys. J., 81 (2001), 2458.

Chapter 11

Investigation of substrate-specific porin channels in lipid bilayer membranes

Roland Benz

Lehrstuhl für Biotechnologie, Biozentrum der Universität Würzburg, Am Hubland, D-97074 Würzburg, Germany

1. INTRODUCTION

The cell envelope of gram-negative bacteria such as *Escherichia coli*, *Salmonella typhimurium*, and *Pseudomonas aeruginosa* consists of three distinct layers, the outer membrane, the peptidoglycan (murein) layer, and the inner membrane. The inner membrane acts as a real diffusion barrier and contains, in addition to the respiration chain and the H^+-ATPase, a large number of uptake systems for hydrophilic substrates [1]. All these hydrophilic solutes have to pass the outer membrane on their way into the cell. This membrane must have very special permeability properties because substrates may have various structures and should all fit through the same transport pathways. The outer membrane protects the bacterial cells from the action of bile acid detergents and the degradation by digestive enzymes [2-3]. Furthermore, the outer membrane of enteric and some other gram-negative bacteria represents a strong permeability barrier to many antibiotics that are highly effective against gram-positive bacteria that do not contain a permeability barrier in the cell wall [4-5].

The outer membrane of gram-negative bacteria acts as a molecular filter for hydrophilic compounds. The active components of these molecular sieving properties are a major class of proteins called porins [6]. The porins have molecular masses between 30,000 and 50,000 Daltons and normally form oligomers in the outer membrane that are, in many cases, highly resistant against detergents [7-8]. The outer membrane of any gram-negative bacteria contains at least one porin in high copy number. Certain bacteria may contain several different constitutive porins. Others may be induced under special growth conditions. Dependent on their permeability properties for hydrophobic solutes the porins can be devided into two classes. Most porins form in the outer membrane and reconstituted systems large water-filled channels and are general diffusion porins because they sort solutes mostly according to their molecular mass [7-9]. This means that they form wide water-filled channels in the outer membrane. Other porins possess solute specificity and contain binding sites for carbohydrates [10-12], phosphate [13-14], and nucleosides [15] inside the pores. This binding site represents a considerable advantage for the diffusion of the solutes through the outer membrane. Prominent examples for these specific porins are LamB of *Escherichia coli* [16] and *Salmonella typhimurium* [17] and OprP of *Pseudomonas aeruginosa* [13]. Porins from a large number of gram-negative bacteria have been investigated (see Ref. [18] for a list of physical properties and channel characteristics of porins).

2. ISOLATION AND RECONSTITUTION OF PORINS

2.1. Isolation and purification of bacterial porins

In the following a short description of porin isolation and purification is provided. The porins of many gram-negative bacteria are tightly associated with the peptidoglycan layer [6, 19]. This allows the rapid isolation of the porins through the preparation of the murein-protein complex [19]. The cells are harvested in the late logarithmic phase. They are washed and resuspended in a small volume before they are passed several times through a French pressure cell to disrupt them. The pellet of a subsequent centrifugation step contains the cell envelope. Most components of the cell envelope are soluble in detergents. The insoluble material is composed of murein and a few proteins either covalently bound to or tightly associated with the peptidoglycan. The porins can be released by standard methods either by digestion of the murein or by the salt extraction method [6, 19-20]. When the porins are not tightly associated with the murein first outer membrane and inner membrane have to be separated. This can be achieved by sucrose step density gradient centrifugation [21]. Other methods use the treatment of the cell envelope with mild detergents to solubilize most components of the inner membrane and to separate thus the outer membrane. After release from the murein or the outer membrane, pure porin may be obtained by column chromatography using gel filtration or affinity chromatography. An elegant way to purify carbohydrate-specific porins is affinity chromatography across a starch column (starch coupled to Sepharose 6B as described by Ref. [22])

2.2. Reconstitution of porins in lipid bilayer membranes

It is possible to study the *in vitro* properties of porins in the liposome swelling assay [11, 23]. However, many more *in vitro* studies have been performed using the lipid bilayer technique [24-27]. Two basically different methods have successfully been used for formation of lipid bilayer membranes and for the study of membrane channels. The first method was proposed by Mueller et al. [28], and the reconstitution of membrane channels using this method has been described in a number of different publications [24, 26, 29-31]. The cell used for membrane formation consists of a Teflon chamber with a thin wall separating two aqueous compartments. The Teflon divider has small circular holes with areas between 0.4 and 2 mm^2. For membrane formation, the lipid is dissolved mainly in n-decane in a concentration of 1 of to 2% (wt./vol.). However, also other organic solvents, such as n-hexadecane, squalene or triolein have successfully been used for bilayer studies. The basic difference between them is the amount of residual solvent in the membrane after it is in the "black" state, i.e. the solvents influence the membrane thickness [32-33]. On the other hand, the type of solvent has no influence on the properties of reconstituted channels. The lipid solution is painted over the holes to form a lamella. The membrane experiments start after the lamella thins out and turns optically black in reflected light, which suggests that the membrane is much thinner than the wavelength of the light [33]. The addition of small amounts of n-butanol 10% (vol./vol.) to the membrane forming solutions stabilizes the membranes.

The second method is the formation of "solvent-free" or "solvent-depleted" lipid bilayer membranes according to Montal and Mueller [34]. The cell for membrane formation is basically the same as described above for the formation of solvent-containing membranes with the exception that the holes in the Teflon divider are extremely small (diameter 10 to 50 μm). They have to be pretreated with small amounts of vaseline, hexadecane or other organic solvents for successful formation of membranes [32]. The surfaces of the aqueous phases on both sides of the hole are adjusted below it. Lipid dissolved in hexane is now added to both surfaces to form monolayers (in fact, the lipid has to be sufficient for many monolayers). Then the aqueous phases are raised over the hole and a lipid bilayer membrane may be formed. Solvent-depleted membranes cannot be controlled in reflected light because of their small surface area. The control of successful membrane formation (as opposed to the possibility that the hole is simply plugged with lipid) is only possible on the basis of the measurement of the membrane capacitance [32, 34].

The lipid bilayer technique allows the very sensitive detection of current through the membrane. It is, however, not very well suited for the study of fluxes of uncharged solutes. For sensitive electrical measurements, the membrane cell has to be surrounded by a Faraday cage to avoid the 50 or 60 cycles noise of the line and other pertubations caused by electric fields. It is also necessary to insulate the membrane cell against mechanical oscillations. Normally, Ag/AgCl or calomel electrodes are inserted into the aqueous phases on both sides of the membrane. Electrodes with salt bridges have to be used in case of salt gradients across the membrane or if the aqueous phase does not contain a sufficient concentration of chloride. The electrodes are switched in series with a voltage source and a highly sensitive current amplifier. The amplified signal is monitored with a storage oscilloscope and recorded with a tape or a strip chart recorder or is directly transferred via an A/D card in a personal computer for further analysis. The lipid bilayer technique allows good access from both sides of the membrane. This means that with this technique the ionic composition on both sides of the membrane can be controlled. It is possible to establish salt gradients across the membrane by the addition of concentrated salt solution to one side of the membrane or by exchange of the aqueous solution. Zero-current membrane potential measurements allow the measurement of the ionic selectivity of channels if the membrane contains a sufficient number of channels [26].

3. THEORETICAL DESCRIPTION OF SUBSTRATE TRANSPORT THROUGH SPECIFIC PORIN CHANNELS

3.1. Analysis of substrate transport through specific porins

The theoretical treatment of substrate transport through channels has been given in full in the literature [35-36]. Here only the simple two-barrier one site-model is discussed in some detail, which provides a good description of substrate transport through specific porins [12, 14, 37-38]. This model assumes a binding site in the centre of the channel. The rate constant k_1 describes the jump of the substrates from the aqueous phase (concentration c) across the barrier to the central binding site, whereas the inverse movement is described by the rate constant k_{-1}. There is only little evidence for channel asymmetry; we therefore assume symmetrical barriers with respect to the binding site.

The stability constant of the binding between a substrate molecule and the binding site is $K = k_1/k_{-1}$. Furthermore, it is assumed that the channel is a single file channel; only one substrate can bind to the binding site at a given time and that no substrate or other molecule [12, 14, 35] can pass the channel when the binding site is occupied. This means that a substrate can enter the channel only when the binding site is free. The probability, p, that the binding site is occupied (identical concentrations on both sides) is given by:

$$p = \frac{K \cdot c}{1 + K \cdot c} \tag{1}$$

and that it is free by:

$$1 - p = \frac{1}{1 + K \cdot c} \tag{2}$$

3.2. Ion transport

When a voltage is applied to the two-barrier one site-channel the barrier levels change (i.e. k_1 and k_{-1} have to be replaced by k_1', k_1'' and k_{-1}', k_{-1}'', respectively, because of the applied voltage) and a net flux of ions is observed. The net flux of ions, ϕ, through the channel under stationary conditions is equal to the net movement of ions across one barrier (for instance, across the left-hand side barrier):

$$\phi = k_1' \cdot c \frac{1}{1 + K \cdot c} - k_{-1}'' \frac{K \cdot c}{1 + K \cdot c} \tag{3}$$

The rate constants k_1' and k_{-1}'' are multiplied by the probabilities that the binding site is free or occupied, respectively. These probabilities are independent of voltage when the top of the barriers is halfway between the binding site and the membrane surfaces. The current, I, through the channel is given by [14]:

$$I = e_0 \cdot \phi = e_0 \frac{K \cdot c}{1 + K \cdot c} (k_1' / K - k_2'') \tag{4}$$

Eq. (4) can be used for the calculation of the single-channel conductance, $G^0(c)$, in the limits of small voltages ($V_m \ll 25$ mV):

$$G^0(c) = G^0_{max} \frac{K \cdot c}{1 + K \cdot c} \qquad (5)$$

where G^0_{max} is the maximum single-channel conductance at very high ion concentration:

$$G^0_{max} = \frac{e_0 \cdot F}{2 \cdot R \cdot T} k_{-1} \qquad (6)$$

3.3. Evaluation of the stability constant for binding of neutral solutes to the binding site inside specific porins

The transport of neutral solutes through substrate-specific porins has been studied in detail in recent years [12, 15, 37-40]. As pointed out above the minimum requirements for the description of the solute transport are given by a simple two-barrier one site-model. The stability constant of the binding between a substrate molecule and the binding site is $K = k_1/k_{-1}$. Furthermore, its assumed that the channel is a single file channel. This means that a solute can enter the channel only when the binding site is free. The channel (given by P) is open when no substrate S is bound, and closed when it is occupied by a substrate, S, to form the non-conducting substrate-channel complex PS:

$$P + S \underset{k_{-1}}{\overset{k_1 c}{\rightleftharpoons}} PS \qquad (7)$$

The probability, p, that the binding site is occupied by a substrate molecule is given by Eq. (1) and that it is free and the channel conducts ions is given by Eq. (2). The conductance, $G(c)$ ($=I(c)/V_m$), of a membrane containing many specific porins in the presence of a solute with the stability constant, K, and a solute concentration, c, is given by the probability that the binding site is free:

$$G(c) = \frac{G_{max}}{1 + K \cdot c} \qquad (8)$$

where G_{max} is the membrane conductance before the start of the solute addition to the aqueous phase when all channels are in the open configuration (I_0 is the initial current and $I(c)$ is the current at the solute concentration c). Eq. (8) may also be written as:

$$\frac{G_{\text{max}} - G(c)}{G_{\text{max}}} = \frac{I_0 - I(c)}{I_0} = \frac{K \cdot c}{1 + K \cdot c} \tag{9}$$

which means that the titration curves can be analysed using Lineweaver-Burke plots as has been shown in previous publications [12, 15, 37]. The half saturation constant, K_S is given by the inverse stability constant $1/K$.

3.4. Analysis of substrate binding kinetics using analysis of current noise

The numbers of open and closed channels are coupled through a chemical reaction. Assuming that we have 1,000 channels in a membrane 500 channels are blocked on average when the solute is added in a concentration equal to the half saturation concentration K_S. However, small perturbations of the number of closed channels still exist due to microscopic variations of the number of bound substrate molecules. These variations are controlled by the reaction rate $1/\tau$ of the second order reaction given in Eq. (7) [41-42]:

$$\frac{1}{\tau} = 2\pi \cdot f_c = k_1 \cdot c + k_{-1} \tag{10}$$

Where f_c is the corner frequency of the power density spectrum, $S(f)$, which is given by a "Lorentzian" function [42-43]:

$$S(f) = S_0 / (1 + (f / f_c)^{\alpha}) \tag{11}$$

a is the slope of the decay of the Lorentzian function (usually close to 2) and S_0 is the plateau value of the power density spectrum at small frequencies. It is given by [41-43]:

$$S_0 = 4 \cdot N \cdot i^2 \cdot p \cdot (1 - p) \cdot \tau \tag{12}$$

N is the total number of channels (blocked and unblocked) within the membrane and i is the current through one single open channel. The membrane current, $I(c)$, is given by the number of open channels times the current through one single-channel and the probability that the channel is open:

$$I(c) = i \cdot N \cdot (1 - p) \tag{13}$$

Eqs. (12) and (13) can be used to calculate the single-channel conductance, g, of the channels from the applied membrane potential V_m and the current through one single channel:

$$g = \frac{i}{V_m} = \frac{S_0}{4 \cdot V_m \cdot I(c) \cdot p \cdot \tau} \tag{14}$$

4. STUDY OF SPECIFIC PORINS IN LIPID BILAYER MEMBRANES

4.1. OprP of *Pseudomonas aeruginosa* outer membrane

The outer membrane of *Pseudomonas aeruginosa* has very special sieving properties, which make this organism quite resistant against most antibiotics [44]. It contains a variety of different outer membrane channels, which are mostly specific porins and are involved in the uptake of different classes of substrates. Protein D1 is part of the sugar uptake system and is induced in the presence of glucose [45]. Protein D2 plays probably an important role in the permeation of amino acids and peptides since it has been demonstrated that it contains a binding site for basic amino acids and peptides [46]. OprP with a molecular mass of 48 kDa [13] is another outer membrane porin of *P. aeruginosa*. It is induced in the outer membrane when the organisms grow in media with low phosphate concentrations together with an alkaline phosphatase, a periplasmic binding protein and the inner membrane transport system [13,47]. In this respect OprP shares some similarities with PhoE of *E. coli*, which is also induced under the conditions of phosphate limitation. However, in contrast to the general diffusion pore PhoE [9], OprP and the closely related OprO (which is probably a channel for polyphosphate uptake [48]) form highly anion-selective channels in lipid bilayer membranes which are at least 100-times more permeable for chloride than for potassium [14, 48].

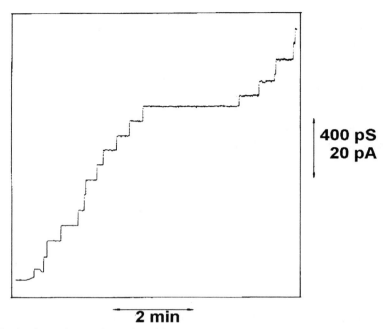

400 pS
20 pA

2 min

Fig. 1. Single-channel recording of a diphytanoyl phosphatidylcholine/n-decane membrane in the presence of 10 ng/ml OprP and 30 mM KCl in the aqueous phase. V_m = 50 mV; T = 25°C. The average single-channel conductance of the current steps was about 100 pS.

Fig. 1 shows a single-channel recording of a diphytanoyl phosphatidyl-choline/n-decane membrane in the presence of 30 mM KCl and 10 ng/ml OprP. Typical as for most porins the conductance increase was in a stepwise fashion. Closing events were very rarely observed with OprP and other gram-negative bacterial porins [12-13, 24]. Interestingly, the single-channel events were found to be fairly homogeneous as has been observed for most specific porins in contrast to general diffusion pores that show very often broad histograms [24, 49].

The single-channel conductance was not a linear function of the bulk aqueous salt concentration and saturated above 30 mM KCl. Furthermore, The single-channel conductance was independent from the type of cation in the aqueous phase. Fig. 2 shows the conductance of the OprP channel as a function of the KCl-concentration [14, 50]. The curve could be fitted to Eq. (5) with the following parameters: G_{max}^0 = 280 pS and K = 20 1/M. This result indicates that a one-site, two-barrier model provides a good fit of the experimental data. It means also that Eqs. (5) and (6) can be used to calculate the

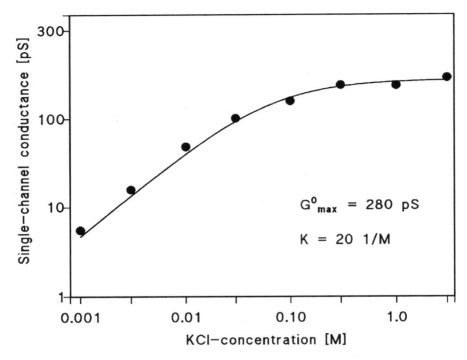

Fig. 2. Single-channel conductance of OprP as a function of the KCl concentration in the aqueous phase. Membranes were formed from diphytanoyl phosphatidylcholine/n-decane. The data were derived as a mean from at least 100 single events. $T = 25°C$; $V_m = 50$ mV. The data were taken from Ref. [50].

rate constants of anion transport through the OprP channel. Table 1 contains the results for a variety of different anions. The stability constants for anion binding was found to be highest for chloride, whereas G_{max}^0 was maximal for fluoride, which means that the latter followed the Eisenman sequence AVI.

The OprP channel has the highest single channel conductance in 0.1 M chloride solution [14]. A variety of other anions also permeate through the channel, whereas in the presence of salts with large organic anions, such as HEPES (N-2-hydroxyethylipiperazine-N'-2-ethanesulfonic acid) no conductance fluctuations could be observed at all [14]. Interestingly, the single channel conductance was extremely small in phosphate (pH 6) although OprP should be a channel for phosphate. This argues in principle against a phos-

Table 1

Stability constants, maximum single-channel conductance and rate constants of anion transport through the OprP channel of *Pseudomonas aeruginosa* outer membrane. The anion radii, r, are given for comparison. The cation was in all cases potassium. The data were taken from Refs. [14] and [38]. Note that the data were derived for OprP trimers, which contain presumably three individual channels.

Anion	K [1/M]	G^0_{max} [pS]	k_1 [10^7 1/(Ms)]	k_{-1} [10^7 1/s]	r [nm]
Fluoride	3.5	515	60	17	0.133
Chloride	20	280	180	9.0	0.181
Bromide	4.7	265	40	8.5	0.195
Iodide	1.3	110	4.6	3.5	0.216
Nitrite	4.6	263	39	8.4	-
Nitrate	2.0	140	9	4.5	0.198
Phosphate (pH 6)	11,000	9	3,500	0.28	-

phate-specific channel. However, phosphate limitation *in vivo* means that its concentration is below 1 mM in the growth media. At this concentration the channel offers a considerable advantage for the transport of phosphate as the following consideration demonstrates. The net flux of substrates, ϕ, through the two-barrier one-site channel in the case of a gradient (c" = 0; c' = c) and zero voltage is given by [37]:

$$\phi = \phi_{max} \cdot K \cdot c / (2 + K \cdot c) \tag{15}$$

where $\phi_{max} = k_{-1}$ is the maximum flux of substrates at very high concentration through the channel (similar to the turnover number of an enzyme). The maximum permeability is given by $k_1 / 2$ at very low substrate concentration. This means that the permeability of OprP at low substrate concentration is about 20-times higher for phosphate as compared to chloride (see Table 1). This is consistent with the *in vivo* situation of phosphate starvation. The channel should have a high permeability at low external phosphate, which together with the sink provided by the high-affinity phosphate binding protein in the periplasmic space should provide maximum presentation to the cytoplasmic membrane transport system for phosphate. It is noteworthy that the high on-rates for phosphate binding are close to those of diffusion controlled reaction processes, which means that the individual channels in an OprP trimer have large capture radii that are presumably positively charged.

4.2. Carbohydrate binding to LamB of *Escherichia coli*

The LamB-proteins of *E. coli* [16] and *S. typhimurium* [17] are inducible specific porins. The properties of LamB of *E. coli* and *S. typhimurium* have been studied in detail *in vivo* [16-17, 22, 51] and *in vitro* [11-12, 37, 39-40, 43, 52]. The sequences and also the 3D-structures of both proteins are known [53-54]. Their structures are different from those of the general diffusion porins because they have 18 instead of 16 β-strands and do not show any immunologic cross reaction with them [55]. The results of liposome swelling experiments in the presence of maltoporin showed that the rate of penetration of maltose was much larger than that of sucrose or lactose despite similar molecular masses [11]. This facilitated diffusion has been explained by the assumption of a specific binding site for maltose and maltooligosaccharides inside the channel [56]. This possibility was checked in lipid bilayer experiments. LamB of *Escherichia coli* has a single channel conductance of 155 pS in 1 M KCl. The addition of maltotriose reduces the single-channel conductance, and large concentrations completely block the channel [12, 37]. In fact, it is possible to titrate the LamB conductance with maltose, maltooligosaccharides and other carbohydrates and to calculate the binding constant of these solutes to the channel interior.

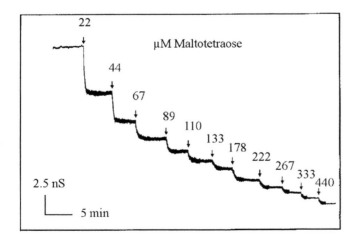

Fig. 3. Titration of LamB-induced membrane conductance with maltotetraose. The membrane was formed from diphytanoyl phosphatidylcholine/n-decane. The aqueous phase contained 50 ng/ml LamB, 1 M KCl, and maltotetraose at the concentrations shown at the top of the figure. The temperature was 25°C and the applied voltage was 20 mV.

Figure 3 shows an experiment of this type. LamB was added in a concentration of 50 ng/ml to a black lipid bilayer membrane. Thirty minutes after the addition, the membrane conductance was almost stationary. Different concentrations of maltotetraose were added to the aqueous phase with stirring (arrows). The membrane conductance decreased as a function of the maltotetraose concentration (Fig. 3). Using Eq. (9), a binding constant of about 15,500 1/M for maltotetraose to the channel interior can be calculated from the conductance decrease. Table 2 shows the results derived from similar titration experiments with a variety of carbohydrates. These results indicate that it is possible to study the specificity of channels for neutral solutes in lipid bilayer membranes. Table 2 demonstrates that the stability constant for the binding of carbohydrates increased with the number of residues in a carbohydrate chain. For example in the series glucose, maltose to maltopentaose, the binding constant increased about 1800-fold, whereas no further increase between occurred between 5 and 7 glucose residues in the maltooligosaccharides. This may be explained by assuming that the binding site has approximately the length of five glucose units, i.e. it is about 2.5 nm long. All disaccharides such as maltose, lactose, sucrose and others had stability constants between 18 and 250 1/M for the binding to LamB [37]. This means that sucrose and maltose had approximately the same affinity to LamB. It is noteworthy that the relative rates of permeation of these sugars as derived *in*

Table 2

Stability constants, K, for carbohydrate binding to the carbohydrate-specific LamB channel of *Escherichia coli*. The data were derived from titration experiments similar to that shown in Fig. 3 using Lineweaver-Burke plots (Eq. (9)). K_S is the half saturation constant. The aqueous phase contained 1 M KCl and about 50 ng/ml LamB. T = 25°C; V_m = 50 mV. The data were taken from Ref. [37].

Carbohydrate	K [1/M]	K_S [mM]
D-Glucose	9.5	110
Maltose	100	10
Maltotriose	2,500	0.40
Maltotetraose	10,000	0.10
Maltopentaose	17,000	0.059
Maltohexaose	15,000	0.067
Maltoheptaose	15,000	0.067
Trehalose	46	22
Lactose	18	56
Sucrose	67	15

vitro [11] and *in vivo* [57], differed substantially. This can be explained by the kinetics of the sugar transport and means that widely separated rate constants k_1 and k_{-1} may have the same ratio K (i.e. the same stability constant K; see below).

4.3. Analysis of carbohydrate-induced current noise of LamB-channels

The results of the liposome-swelling assay allow together with the binding data derived from titration experiments only a qualitative description of the kinetics of sugar movement through the LamB channel [37]. This means that the rate constants k_1 and k_{-1} cannot be derived from the analysis of the sugar-induced block of the LamB-channels in single-channel experiments because of the pure time resolution of this method. Therefore an alternative method was used, the analysis of the sugar-induced current noise of the LamB channel [39-40, 43, 58-59]. The experiments were performed in the following way: black lipid bilayer membranes were formed and LamB was added to both sides of the membrane. Instead of 20 to 30 min (as in the case

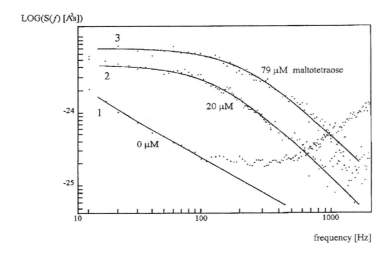

frequency [Hz]

Fig. 4. Power density spectra of maltotetraose-induced current noise of a diphytanoyl phosphatidylcholine/n-decane membrane containing 690 LamB-channels. Trace 1 shows the control (1 M KCl). Trace 2: the aqueous phase contained 20 μM maltotetraose and the power density spectrum of Trace 1 was subtracted ($\tau = 1.1$ ms; $S_o = 4.3 \cdot 10^{-24}$ A²s). Trace 3: the aqueous phase contained 79 μM maltotetraose and the power density spectrum of Trace 1 was subtracted ($\tau = 0.69$ ms; $S_o = 7.2 \cdot 10^{-24}$ A²s). $V_m = 25$ mV. The number, N, of channels was calculated by dividing the membrane conductance through the single-channel conductance of LamB.

of the titration experiments) about 60 min had to be waited to avoid any further conductance increase during the experiments since a change of the membrane conductance had a large influence on the noise measurements. Subsequently, the power density spectrum of the current noise was measured at a voltage of 25 mV (see Fig. 4, trace 1). Then maltotetraose was added in a concentration of 20 μM and the power density spectrum was measured again and the background noise was substracted (trace 2). At another concentration of maltotetraose (c = 79 μM) the power density spectrum (minus the background) corresponded to trace 3. The power density spectra of traces 2 and 3 are of the Lorentzian type expected for a random switch with unequal on and off probabilities [41, 43]. Both could be fitted to Eq. (11) with sufficient accuracy. The corner frequency obviously increased with increasing maltooligosaccharide concentration. The corner frequencies of the experiments shown in Fig. 4 and of other maltotetraose concentrations (not shown) could reasonably well be fitted to Eq. (10) as Fig. 5 indicates. The rate constants of the chemical reaction were $k_1 = 8.9 \cdot 10^6$ 1/(M·s) and $k_{-1} = 760$ 1/s. The stability constant was $K = 11,750$ 1/M, which agrees reasonably well with the results of the titration experiments (see Table 2). Similar experiments were performed with a variety of carbohydrates but not with maltose and glucose since the

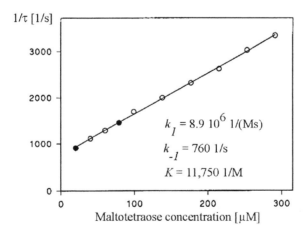

Fig. 5. Dependence of $1/\tau = 2 \cdot \pi \cdot f_c$ on the maltotetraose concentration in the aqueous phase. The data were derived from the fit of the power density spectra with Lorentzians similar to those given in Fig. 4 (closed circles; 20 and 79 μM) and for other maltotetraose concentrations ranging between 40 and 291 μM (open circles). The aqueous phase contained 1 M KCl. The applied membrane potential was 25 mV; $T = 25 ^\circ C$.

corner frequency, f_c, was in these cases outside the time resolution of the experimental set-up. The rate constants for the on- (k_1) and off-processes (k_{-1}) for the binding of these sugars are given in Table 3 [58]. The data for glucose and maltose are calculated on the basis of the relative rates of permeation of glucose and maltose taken from Luckey and Nikaido [11] and using the binding constants of Table 2. The results shown in Table 3 suggest that the rate constants for the on-process is more or less the same for the maltooligosaccharides. Substantial differences were observed for the movement of the sugars out of the channel, probably because the interaction between binding site and sugar increases with the number of glucose residues, which means that the maltooligosaccharides move many times forth and back until they can leave the channel.

The interesting result of the analysis of current noise is that the transport kinetics of maltose and sucrose can be compared. The data of Table 3 clearly indicate that sucrose transport through the LamB channel is extremely slow because the on- and the off-rates for its transport are considerably smaller than those for maltose (see Table 3). No wonder that enteric bacteria may contain a plasmid that codes for proteins of an uptake and degradation system for sucrose [60-61]. This plasmid (pUR400) contains also the gene *scrY* that codes for the outer membrane protein ScrY [62]. It exhibits a high homology to LamB but has an N-terminal extension of approximately 70 amino acids [62]. The results of *in vitro* experiments with ScrY have demonstrated that it is a specific porin and contains a binding site for sugars.

Table 3
Kinetics of sugar-transport through the LamB-channel of *Escherichia coli* as derived from noise-analysis of sugar-induced block of ionic current through the channel. The kinetic constants for glucose and maltose binding were derived from the data of the liposome swelling assay using the stability constants of glucose and maltose binding [58]. The data were taken from Ref. [58].

Carbohydrate	k_1 [10^6/(Ms)]	k_{-1} [1/s]	K [1/M]
Glucose	1.0	110,000	9.5
Maltose	0.8	8,000	100
Maltotriose	10	1,300	7,700
Maltotetraose	8.0	670	11,900
Maltopentaose	7.0	530	13,200
Maltohexaose	5.0	240	20,800
Maltoheptaose	4.0	120	33,000
Sucrose	0.004	50	80

Again the binding constants are very similar for sucrose and maltose [63] but the kinetics of the transport are different [64], which means that in the case of ScrY the kinetics of sucrose transport is much faster than in LamB. The reason for this is the larger cross section of the ScrY channel (see below).

4.4. 3D-Structure of LamB and ScrY

The LamB-proteins of *E. coli* and *S. typhimurium* and ScrY of the single copy plasmid pUR400 have been crystallized and their 3-D structures are known from X-ray crystallography [53-54, 65]. They are organized as trimers of three identical subunits and show a similar organization as the general diffusion porins with the exception that the monomers contain 18 instead of 16 antiparallel β-strands, tilted by an angle of about 45° to 60° regarding the membrane surface (see Fig. 6 for the 3D-structure of a LamB monomer). The periplasmic side (bottom) is very smooth and contains mainly short β-turns, whereas the external loops (top) are very long and form a complicated architecture, which tends to collapse when single loops are removed [66]. The size of the channel is restricted by external loop 3 from the N-terminal end in a similar way as it is the case in the 3-D structure of general

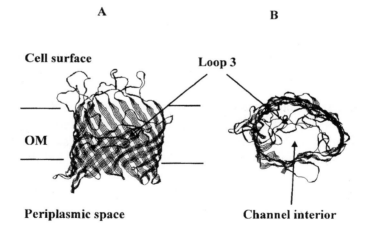

A   **B**

Cell surface   Loop 3

OM

Periplasmic space   Channel interior

Fig. 6. Schematic view of the 3-D structure of the *E. coli* LamB monomer. The coordinates were taken from the crystallographic data of Ref. [53].
A: Side view of the LamB (maltoporin) monomer.
B: View of the monomer from the cell surface exposed side. The other two monomers are attached from the bottom side of the monomer, opposite to the location of loop 3.

diffusion porins [67-68]. The size of the channel itself is much smaller as compared to that of the general diffusion pores, which means that its permeability for non carbohydrate-solutes is reduced. Moreover its single-channel conductance is much smaller than that of a general diffusion pore such as OmpF of *E. coli*. LamB mutants have shown that the binding of the carbohydrates inside the channel is mediated by H^+-bridges. In particular arginine R8 localized on the first β-strand from the N-terminal end has a strong influence on the binding affinity and its mutation in a histidine lowers the stability constant for maltopentaose substantially (from 10,000 1/M to 114 1/M [59]). It seems that aromatic residues play also an important role in carbohydrate transport and form some sort of slide (greasy slide [69]) along which the carbohydrates move along on the way through the channel from the cell surface to the periplasmic space with the non-reducing end in advance [66]. The sucrose specific ScrY (sucroseporin) combines features of a general diffusion pore and of carbohydrate-specific porins because it has a much higher permeability for ions [63-64]. This is consistent with the 3D-structure, which demonstrates that LamB forms definitely a much smaller channel, which is obviously less favorable for the permeation of sucrose than ScrY [53, 65].

4. 5. Nucleoside binding to Tsx of *Escherichia coli*

Besides such called major proteins the outer membrane of gram-negative bacteria contains also minor proteins that have an important function in transport of solutes. The *tsx* gene of *E. coli* encodes for such a protein (Tsx). Expression of Tsx is coregulated with the systems for nucleoside uptake through the inner membrane and nucleoside metabolism, suggesting that it plays an important role in the permeation of nucleosides across the outer membrane [70-71]. In fact, *E. coli* strains lacking Tsx are impaired in the uptake of all nucleosides, with the exception of cytidine and deoxycytidine [70] when the outer membrane is the limiting factor for the overall transport process. At high substrate concentrations of substrate, the Tsx protein becomes dispensable, and the nucleosides diffuse across the outer membrane primarily through OmpF. Investigation of the Tsx-mediated translocation of nucleosides across the outer membrane has revealed a number of remarkable properties of Tsx. Interestingly, the Tsx-channel apparently discriminates between the closely structurally related pyrimidine nucleosides, cytidine and thymidine [70]. Furthermore, the uptake of deoxynucleosides is more strongly dependent on Tsx than that of the corresponding nucleosides [70-71].

Table 4

Stability constants, K, for the binding of nucleobases, nucleosides, and deoxynucleosides to the Tsx-channel (half saturation constant $K_S = 1/K$). K was calculated from titration experiments similar to that shown for LamB (Fig. 3). n.d. means not detectable. The data were taken from Ref. [15].

Compound	K [1/M]	K_S [mM]
Adenine	500	2.0
Adenosine	2,000	0.50
Deoxyadenosine	7,100	0.14
Guanine	n.d.	n.d.
Guanosine	1,000	1.0
Deoxyguanosine	3,100	0.32
Thymine	170	5.8
Thymidine	5,000	0.20
5'-Deoxythymidine	20,000	0.050
Cytosine	n.d.	n.d.
Cytidine	46	22
Deoxycytidine	100	10
Uracil	50	20
Uridin	1,900	0.54
Deoxyuridin	19,000	0.053

Tsx forms ion-permeabel channels upon reconstitution in lipid bilayer membranes. These channels have an extremely small single channel conductance of about 10 pS in 1 M KCl [15, 72]. This is more than 100-times less than the conductance of the general diffusion pore OmpF (1,500 pS) and more than 10-times less than that of the carbohydrate-specific LamB channel (155 pS) under otherwise identical conditions. Similarly as described above for LamB and carbohydrates it is possible to block the flux of ions through Tsx with nucleosides. This means that it is possible to measure the binding of different nucleosides, deoxynucleosides and free bases to the binding site by using titration experiments with lipid bilayer membranes. Table 4 shows the results of these titration measurements. From the different pyrimidines, thymine had the largest stability constant for the binding to the Tsx-channel, whereas the half saturation constant for the binding of cytidine could not be given and was possibly much larger than 20 mM. A comparison of the structures of the different pyrimidines shows that thymine contains in the 5-position an additional methyl-group as compared to uracil. This group obviously results in a larger stability constant for the binding. Cytosine has in the 4-position an amino group instead of the carbonyl group as compared to uracil. This results in a considerable decrease of the binding affinity of this compound, probably because one hydrogen bound between an OH-group of the protein and the carbonyl is missing or because of the increased hydrophilic nature of the whole molecule.

Similar considerations apply also to a comparison of the binding affinities of adenine and guanine. However, the relationship between binding affinity and the structure of the purines and the pyrimidines cannot be understood from of the data of Table 3 alone. The binding between Tsx and purines and pyrimidines is stabilised if a ribofuranose is bound to the 9-position in the case of the purines or to the 1-position in the case of the pyrimidines. This increase of the stability constant could in principle be caused by the formation of additional hydrogen bounds. However, the removal of the hydroxyl-group in the 2'-position (deoxynucleosides) resulted in even larger stability constant, which makes it rather unlikely that hydrogen bonds are the reason for the larger stability constants. It seems moreover that the binding is also dependent on some kind of hydrophobic interaction between binding site and the different molecules. This hypothesis may be supported by the rather small single-channel conductance (10 pS) of the Tsx-channel in 1 M KCl [15, 72].

5. CONCLUSIONS

The lipid bilayer technique represents a powerful tool for the study of membrane active material. It is particularly well suited for the study of membrane channels. The high sensitivity of the method allows the detection of single events. Experiments with gram-negative bacterial porins suggest that lipid bilayer membranes have an outstanding sensitivity and only little material is needed to study the porin properties. In this review structural and functional studies of specific porin channels are discussed. These channels are highly structured and contain binding sites for one class of specific solutes. The presence of a binding site leads to the saturation of solute flux above the half saturation constant $1/K$. However, the presence of a binding site offers a considerable advantage for the flux of solutes across the outer membrane at small external solute concentrations, the *in vivo* situation, at which the general diffusion pores are rate-limiting [73].

Fig. 7 shows the net flux of maltose and maltotetraose across LamB calculated on the basis of Eq. (15), i.e. the concentration of the carbohydrates on one side (the periplasmic side) is zero. The data was calculated using the rate constants given in Table 3. The flux of maltose through a general diffusion pore (straight line) was estimated according to the dimensions of OmpF [68]. It was corrected by using the Renkin correction factor [74] for the hit of the maltose to the rim of the OmpF channel, which leads to the reflection of the sugar and a decrease of the channel permeability. Specific porins have their maximum permeability in the linear range of Fig. 7 and it is proportional to the rate constant for the movement from the external solution to the binding site inside the channels (i.e. by $k_1/2$). The saturation of the flux at very high substrate concentrations is given by the rate constant k_{-1}, which is the maximum turnover number of ions through a one-site two-barrier channel. The comparison of the different channels of Fig. 7 demonstrates again the advantage of a binding site for the maximum scavenging of substrates at small substrate concentration. It also shows that the flux through a general diffusion pore can exceed that through a specific porin at high substrate concentration.

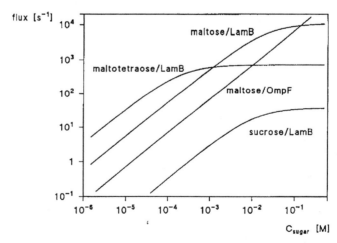

Fig. 7. Flux of maltose, sucrose and maltotetraose through one single LamB channel as a function of the corresponding sugar concentration on one side of the channel. The concentration on the other side was set to zero. The flux was calculated using eqn. (15) and the rate constants given in Table 3. The flux of maltose through OmpF was calculated on the basis on its dimensions [68] and using the Renkin correction factor [74].

REFERENCES

[1] G. Winkelmann (ed.), Microbial Transport Systems, WILEY-VCH Verlag GmbH, Weinheim/Germany, 2001.
[2] J. Patterson-Delafield, R.J. Martinez and R.I. Lehrer, Infect. Immun. 30 (1980) 180.
[3] R.F. Rest, M.H. Cooney and J.K. Spitznagel, Infect. Immun. 16 (1977) 145.
[4] H. Nikaido and T. Nakae, Adv. Microb. Physiol. 20 (1979) 163.
[5] H.-G. Sahl, M. Kordel and R. Benz, Arch. Microbiol. 149 (1987) 120.
[6] T. Nakae, J. Biol. Chem. 251 (1976) 2176.
[7] R. Benz, Ann. Rev. Microbiol. 42 (1988) 359.
[8] H. Nikaido, Mol. Microbiol. 6 (1992) 435.
[9] R. Benz, A. Schmid and R.E.W. Hancock, J. Bacteriol. 162 (1985) 722.
[10] T. Ferenci, M. Schwentorat, S. Ullrich and J. Vilmart, J. Bacteriol. 142 (1980) 521.
[11] M. Luckey and H. Nikaido, Proc. Natl. Acad. Sci. US. 77 (1980) 167.
[12] R. Benz, A. Schmid, T. Nakae and G. Vos-Scheperkeuter, J. Bacteriol. 165 (1986) 978.
[13] R.E.W. Hancock, K. Poole, and R. Benz, J. Bacteriol. 150 (1982) 730.
[14] R. Benz and R.E.W. Hancock, J. Gen. Physiol. 89 (1987) 275.
[15] R. Benz, A. Schmid, C. Maier and E. Bremer, Eur. J. Biochem. 176 (1988) 699.
[16] S. Szmelcman and M. Hofnung, J. Bacteriol. 124 (1975) 112.
[17] E.T. Palva, J. Bacteriol. 136 (1978) 286.
[18] R. Benz and K. Bauer, Eur. J. Biochem. 176 (1988) 1.
[19] H. Nikaido, Methods Enzymol. 97 (1983) 85.
[20] K. Nakamura and S. Mizushima, J. Biochem. 8O, (1976) 1411.
[21] T. Miura and S. Mizushima, Biochim. Biophys. Acta 150 (1968) 159.
[22] T. Ferenci and K.-S. Lee, J. Bacteriol. 171 (1989) 855.
[23] H. Nikaido and E.Y. Rosenberg, J. Gen. Physiol. 77 (1981) 121.
[24] R. Benz, K. Janko, W. Boos and P. Läuger, Biochim. Biophys. Acta 511 (1978) 305.
[25] H. Schindler and J.P. Rosenbusch, Proc. Natl. Acad. Sci. U.S.A. 75, (1978) 3751.
[26] R. Benz, K. Janko and P, Läuger, Biochim. Biophys. Acta 551 (1979) 238.
[27] H. Schindler and J.P. Rosenbusch, Proc. Natl. Acad. Sci. U.S.A. 78 (1981) 2302.
[28] P. Mueller, D.O. Rudin, H.T. Tien and W.C. Wescott, Nature 194 (1962) 979.
[29] S.J. Schein, M. Colombini and A. Finkelstein, J. Membr. Biol. 30 (1976) 99.
[30] C. Miller and M.M. White, Ann. N. Y. Acad. Sci. 341 (1980) 534.
[31] V. Vachon, D.I. Lyew and J.W. Coulton, J. Bacteriol. 162 (1985) 918.
[32] R. Benz, O. Fröhlich, P. Läuger and M. Montal, Biochim. Biophys. Acta 394 (1975) 323.
[33] J.P. Dilger and R. Benz, J. Membrane Biol. 85 (1985) 181.
[34] M. Montal and P. Mueller, Proc. Natl. Acad. Sci. U.S.A. 69 (1972) 3561.
[35] P. Läuger, Biochim. Biophys. Acta 300 (1973) 423.
[36] W.R. Lieb and W.D. Stein, Biochim. Biophys. Acta 373 (1974) 165.
[37] R. Benz, A. Schmid and G.H. Vos-Scheperkeuter, J. Membrane Biol. 100 (1987) 21.
[38] R. Benz, C. Egli and R.E.W. Hancock, Biochim. Biophys. Acta, 1149 (1993) 224.

[39] P. Van Gelder, F. Dumas, J.P. Rosenbusch and M. Winterhalter, Eur. J. Biochem. 267 (2000) 79.
[40] C. Hilty and M. Winterhalter, Phys. Rev. Lett. 86 (2001) 5624.
[41] A.A. Verveen and L.J. De Felice, Prog. Biophys. Mol. Biol. 28 (1974) 189.
[42] F. Conti and I. Wanke. Quarter. Rev. Biophys. 8 (1975) 451.
[43] S. Nekolla, C. Andersen and R. Benz, Biophys. J. 86 (1994) 1388.
[44] R.E.W. Hancock and D.P. Speert, Drug Resist. Updat. 3 (2000) 247.
[45] J.Trias, E.Y. Rosenberg and H. Nikaido, Biochim. Biophys. Acta 938 (1988) 493.
[46] J. Trias and H. Nikaido, J. Biol. Chem. 265 (1990) 15680.
[47] R.J. Siehnel, N.L. Martin and R.E.W. Hancock, Mol. Microbiol. 4 (1990) 831.
[48] R.E.W. Hancock, C. Egli, R. Benz and R.J. Siehnel, J. Bacteriol. 174 (1992) 471.
[49] R. Benz, J. Ishii and T. Nakae, J. Membrane Biol. 56 (1980) 19.
[50] R. Benz, M. Gimple, K. Poole and R.E.W. Hancock, Biochim. Biophys. Acta 730 (1982) 387.
[51] H.-G. Heine, J. Kyngdon and T.Ferenci, Gene 53 (1987) 287.
[52] K. Schülein, and R. Benz, Mol. Microbiol. 4 (1990) 625.
[53] T. Schirmer, T.A. Keller, Y.F. Wang and J.P. Rosenbusch, Science 267 (1995) 512.
[54] J.E. Meyer, M. Hofnung and G.E. Schulz, J. Mol. Biol. 266 (1997) 761.
[55] M.-A. Bloch and C. Desaymard, J. Bacteriol. 163 (1985) 106.
[56] M. Luckey and H. Nikaido, Biochem. Biophys. Res. Commun. 93 (1980) 166.
[57] M. Schwartz, Methods Enzymol. 97 (1983) 100.
[58] C. Andersen, M. Jordy and R. Benz, J. Gen. Physiol. 105 (1995) 385.
[59] M. Jordy, C. Andersen, K. Schülein, T. Ferenci, and R. Benz, J. Mol. Biol. , 259 (1996) 666.
[60] K. Schmid, R. Ebner, J. Altenbuchner, R. Schmitt and J.W.Lengeler, Molec. Microbiol. 2 (1988) 1.
[61] C. Hardesty, G. Colon, C. Ferran, and J.M. DiRienzo, Plasmid 18, (1987) 142.
[62] K. Schmid, R. Ebner, K. Jahreis, J.W. Lengeler and F.Titgemeier, Mol. Microbiol. 5 (1991) 941.
[63] K. Schülein, K. Schmid and R. Benz, Mol. Microbiol. 5 (1991) 2233.
[64] C. Andersen, R. Cseh, K. Schülein and R. Benz, J. Membrane Biol. 164 (1998) 274.
[65] D. Forst, W. Welte, T. Wacker and K. Diederichs, Nat. Struct. Biol. 5 (1998) 37.
[66] C. Andersen, C. Bachmeyer, H. Täuber, R. Benz, J. Wang, V. Michel, S.M.C. Newton, M. Hofnung, and A. Charbit, Mol. Microbiol. 32 (1999) 851.
[67] M.S. Weiss, U. Abele, J. Weckesser, W. Welte, E. Schiltz and G.E. Schultz, Science 254 (1991) 1627.
[68] S.W. Cowan, T. Schirmer, G. Rummel, M. Steiert, R. Gosh, R.A. Pauptit, J.N. Jansonius and   J.P. Rosenbusch, Nature 356 (1992) 727.
[69] R. Dutzler, Y.F. Wang, P. Rizkallah, J.P. Rosenbusch and T. Schirmer, Structure 4 (1996) 127.
[70] H.J. Krieger-Brauer and V. Braun, Arch. Microbiol. 124, (1980) 233.
[71] E. Bremer, P. Gerlach, and A. Middendorf, J. Bacteriol. 170 (1988) 108.
[72] C. Maier, E. Bremer, A. Schmid and R. Benz, J. Biol. Chem. 263 (1988) 2493.
[73] I.C. West and M.G.P. Page, J. Theor. Biol. 110 (1984) 11.
[74] E.M. Renkin, J. Gen. Physiol. 38 (1954) 225.

Planar Lipid Bilayers (BLMs) and their Applications
H.T. Tien and A. Ottova-Leitmannova (Editors)

Chapter 12

Planar lipid bilayer analyses of bacterial porins; the role of structure in defining function

M. A. Arbing and J. W. Coulton

Department of Microbiology and Immunology, McGill University, 3775 University Street, Montreal, Quebec, Canada, H3A 2B4

---

1. **Introduction**
2. **The outer membrane**
    2.1 Lipid components
    2.2. Proteinaceous components
    2.3. β-barrel structure
    2.4. Classes of outer membrane proteins
    2.5. Porin structure
3. **Assessment of porin function**
    3.1 Porin purification
    3.2 Liposome reconstitution
    3.3 Electrophysiology
    3.3.1 Comparison of planar lipid bilayers and the patch clamp method
    3.3.2 Bilayer formation
    3.3.3 Porin reconstitution
4. **Measurements**
    4.1 Single channel conductance
    4.2 Ionic selectivity
    4.3 Voltage gating
5. **Conclusion**
6. **Acknowledgements**
References

1.   INTRODUCTION

All cells require the presence of membranous structures to contain and protect their cytoplasmic contents. The Gram-negative bacterial cell envelope has three structural components: the cytoplasmic membrane (CM), the outer membrane (OM) and the cell wall.   The cell wall which confers structural rigidity is situated within the periplasm, a compartment of the cell located between the CM and the OM, the most external component of the Gram-negative cell envelope.   The result is that Gram-negative bacteria are protected from the environment by two distinct membranous barriers.

 The OM limits the ability of hydrophobic compounds to penetrate the cell but allows for the passage of small hydrophilic molecules. The cytoplasmic membrane is the true osmotic barrier of the cell and molecules that cross this barrier must be selectively imported or exported. Most hydrophilic molecules enter the periplasm of the cell by traversing the OM through water-filled proteinaceous channels.   These channels have been termed porins and they allow the diffusion of molecules between the periplasm and the external milieu along concentration gradients.

2.   THE OUTER MEMBRANE

The OM differs substantially from typical lipid bilayers in that it has an asymmetrical structure. Typical lipid bilayers such as the CM are composed of two symmetrical leaflets of phospholipids. The Gram-negative bacterial OM contains an inner leaflet of phospholipids facing the periplasm and an outer leaflet composed of a unique glycolipid termed lipopolysaccharide (LPS) that is found only in Gram-negative bacteria.   The asymmetrical structure of the OM has evolved to protect Gram-negative bacteria from the array of noxious compounds that are found in the diverse environments encountered by bacteria.   The result is an OM with a highly impermeable barrier that requires specialized transport systems to facilitate the exchange of metabolites between its internal and external environments.   This exchange of molecules is mediated by a diverse collection of proteins involved in both the active and passive transport of macromolecules into and out of the cell.

2.1   Lipid components

 Many barrier properties of the OM can be attributed to the LPS component of the membrane. LPS serves as a formidable barrier for two main reasons. The first is that a hydrophobic barrier is generated by efficient packing of LPS acyl chains. Deep rough mutants that lack the repeating O-antigen chain and the outer core sugars and many of the inner core sugars show increased sensitivity to hydrophobic agents compared to smooth or rough phenotype bacteria. This increase in susceptibility has been attributed to increased amounts of phospholipids in the OM [1,2] and demonstrates the contribution of acyl chain packing in the outer leaflet to OM permeability. Secondly, ionic bridges between adjacent LPS molecules are formed through association with the negatively charged sugar moieties of LPS and divalent cations. Likewise, LPS molecules are associated with OM proteins (OMPs) by divalent cations that bridge negatively charged groups on LPS and negatively charged amino acids in the loops of OMPs. The result is a rigid barrier impermeable to hydrophobic compounds. The impermeability of the OM to hydrophobic compounds has been demonstrated by experiments that used hydrophobic probes to evaluate membrane permeability. The probes penetrated through the OM at one-fiftieth to one-hundredth the rate at which they penetrated the CM [3].

2.2 Proteinaceous components

The proteinaceous components of the OM play an important role in microbial physiology. They are located at the interface of the bacterial cell and the extracellular milieu and thus mediate interactions between the bacterial cell and its environment. Uptake of extracellular nutrients and protection against noxious environmental compounds involve both the hydrophobic barrier of the lipid component of the bacteria and the proteinaceous components of the OM. The proteinaceous components of the OM are present in either exceedingly large amounts and are considered major OMPs by virtue of their numbers (10^5 copies per cell), or are present in lesser numbers and are considered minor OMPs (10^3 copies per cell).

2.3 β-barrel structure

The unique asymmetrical structure of the OM, with an outer leaflet of LPS and inner leaflet of phospholipids, is in turn complemented by a unique protein architecture found only in the OM of Gram-negative bacteria, mitochondria, and chloroplasts [4]. Eubacterial and mitochondrial OMs contain exclusively proteins with the β-barrel architecture, in contrast to proteins in eukaryotic cytoplasmic and organellar membranes and eubacterial and archael CMs that have α-helical structures [5]. To date, the structures of seventeen outer membrane protein structures with the β-barrel fold have been determined by X-ray crystallography (Table 1).

The β-barrel structure is composed of anti-parallel membrane-spanning β-strands that display an amphipathic character. Approximately every other residue is hydrophobic and faces the interior of the membrane, while the alternating residues generally have a hydrophilic character and face the interior of the barrel. The β-strands are connected by long extracellular loops of variable length while short turns of one to four amino acids connect the β-strands on the periplasmic face of the

Table 1
Bacterial outer membrane proteins for which crystal structures have been determined.

Protein	Source	Function	β-strands	Forms channels (y/n)	Reference
Porin (1.8 Å)	*Rhodobacter capsulatus*	Porin	16	Y	[6]
OmpF (2.4 Å)	*Escherichia coli*	Porin	16	Y	[7]
PhoE (3.0 Å)	*Escherichia coli*	Porin	16	Y	[7]
Porin (1.96 Å)	*Rhodopseudomonas blastica*	Porin	16	Y	[8]
Porin (3.1 Å)	*Paracoccus denitrificans*	Porin	16	Y	[9]
OmpK36 (3.2 Å)	*Klebsiella pneumoniae*	Porin	16	Y	[10]
Omp32 (2.1 Å)	*Comamonas acidovorans*	Porin	16	Y	[11]
LamB (3.1 Å)	*Escherichia coli*	Specific porin	18	Y	[12]
ScrY (2.4 Å)	*Salmonella typhimurium*	Specific porin	18	Y	[13]
FhuA (2.5 Å)	*Escherichia coli*	High-affinity receptor	22	Y*	[14,15]
FepA (2.4 Å)	*Escherichia coli*	High-affinity receptor	22	N	[16]
FecA (2.0 Å)	*Escherichia coli*	High-affinity receptor	22	N	[17]
OmpA (2.5 Å)	*Escherichia coli*	OM stability	8	Y*	[18]
OmpX (1.9 Å)	*Escherichia coli*	Undetermined	8	N	[19]
OmpLA (2.1 Å)	*Escherichia coli*	Phospholipase	12	N	[20]
TolC (2.1 Å)	*Escherichia coli*	Efflux pump	12	Y	[21]
OmpT (2.6 Å)	*Escherichia coli*	OM protease	10	N	[22]

* Channel formation observed only under specific conditions.
Resolution (Å) of the structure is included in brackets after the name of the protein.

membrane. The barrel structure is stabilized by hydrogen bonding between adjacent β-strands [23]. The position of the barrel within the membrane is stabilized by two "girdles" of aromatic amino acids located at the level of the phospholipid head groups within both leaflets of the membrane, a feature that was demonstrated by the crystallization of an LPS molecule with the high-affinity receptor FhuA [14,24]. The geometry of the barrel is determined by the number of β-strands and by the tilt of the strands relative to the barrel axis [25]. All structures solved to date have revealed barrels with an even number of β-strands, between 8 and 22 strands per protein (Table 1). The degree of tilt of the β-strands with respect to the barrel axis varies between 30° and 60°. Both the number of β-strands and the tilt angle determine the radius of the barrel [25].

2.4 Classes of outer membrane proteins

Bacterial OMPs have been traditionally grouped into three classes: non-specific porins, specific porins, and high-affinity receptors [26]. Atomic resolution structures are known for all three classes and for other OMPs that do not fall into these categories (Table 1).

Non-specific porins have 16 transmembrane β-strands connected by eight long extracellular loops and allow the passive diffusion of solutes between the periplasm and the external environment along concentration gradients. Porins form large channels and thus are highly suitable to planar lipid bilayer (PLB) analysis. Porins have properties that include a molecular mass exclusion limit, single channel conductance, ionic selectivity, and voltage gating [27]. The latter three physical qualities can be measured in PLBs and a number of porins have been characterized extensively in PLBs. Non-specific porins will be the primary focus of this chapter and we will refer to our studies on the model porin from *Haemophilus influenzae* type b (Hib) and other bacterial porins including OmpF of *Escherichia coli* (Fig. 1B), the best characterized bacterial porin.

The second class of OMPs has been termed specific porins and they are involved in the uptake of a specific sugar such as maltose or sucrose. Structures are known for two specific porins: LamB of *E. coli* [12] which facilitates the uptake of maltodextrins, and ScrY of *Salmonella typhimurium* [13] which promotes the acquisition of sucrose. These channels are composed of 18 transmembrane β-strands connected by nine extracellular loops, three of which fold into the pore to restrict access to the channel. They form channels with smaller conductances than non-specific porins and their binding of carbohydrate ligands can be studied using PLB analyses [28-30].

The third major class of OMPs is the high-affinity receptors that are involved in the active transport and intracellular accumulation of essential molecules against concentration gradients. Each of these proteins is specific for the uptake of a single ligand for which the protein has extremely high affinity. The primary function of this class of proteins is to obtain iron, although at least one member is involved in vitamin B12 uptake [31]. High-resolution structures have been obtained for three examples of this class of OMP. The structure of FhuA (Fig. 1C), responsible for the uptake of the iron-containing siderophore ferrichrome, was determined by two independent

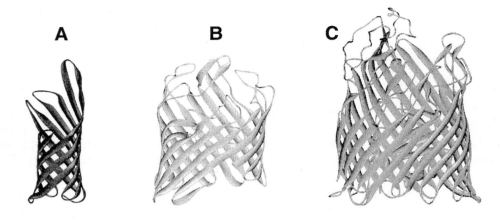

Figure 1. Examples of outer membrane proteins with β-barrel architecture. Structures of 8- (panel A, OmpA), 16- (panel B, OmpF) and 22- (panel C, FhuA) stranded β-barrels. The top of the figure represents the extracellular environment.

groups [14,15]; the structures of FepA [16] for uptake of ferric enterobactin, and FecA [17] for transport of ferric citrate have been solved at high resolution.

    Under specific conditions, high-affinity receptors are capable of channel-forming behavior in PLBs as are other OMPs that fall outside of the three major classes. In general, these proteins can be considered ligand-gated channels because they contain a channel that is normally in the closed state. Opening of the gate can be induced by either the ligand or by applying molecular biology to delete portions of the gate that occlude the channel. The ligand that gates the channel may be either the cognate ligand, which may require an additional element to provide the energy to open the channel, or the channel may be opened by another mechanism. Examples of channel-forming activity by these proteins have been demonstrated for proteins involved in the secretion of proteins and chemical agents into the external environment [32-35]. The channel-forming activity of the high-affinity receptor FhuA has been demonstrated in PLBs by deleting parts of the protein which physically occlude the channel [36,37] and by interaction with a secondary ligand, the bacteriophage T5 which uses FhuA as an OM receptor [38].

    There are other OMPs that do not belong to the three main classes and for which high-resolution structures are known. They apparently do not form channels. These proteins demonstrate the common β-barrel fold of the channel-forming OMPs and include OmpLA, a phospholipase [20]; OmpA, a protein involved in structural integrity (Fig. 1A), [18]; OmpX, a protein postulated to be involved in bacterial virulence[19]; and OmpT, an outer membrane protease [22].

2.5   Porin structure

The genes for porins have been isolated from a number of bacterial species and their DNA sequences determined. All analyses of porin sequences have revealed the presence of a characteristic signal peptide that is responsible for targeting of the porin to the OM. This signal peptide is cleaved during the translocation of the porin from the cytoplasm to its final destination in the OM. A mature primary amino acid sequence of 250-450 amino acids is generated with resulting molecular masses of 30 to 50 kDa [39].

Porins are organized as homotrimers of 16 stranded monomers (Fig. 1B). The β-strands are tilted from 30° to 60° with respect to the membrane and have lengths between 7 and 17 amino acids. The strands are connected by short periplasmic turns (1 to 4 amino acids) and long extracellular loops of various lengths (3 to 50 amino acids). One of the loops (loop 3) is not surface-exposed but rather is folded back into the lumen of the barrel. Each monomer contains approximately 50 to 60% β-sheet and a small percentage, approximately 5%, of α-helix [23]. The monomers are slightly elliptical in shape with axes of approximately 30 Å by 40 Å.

Other structural features of porins differ between species. At the trimer interface the height of the barrel wall can range from 20 Å (*R. capsulatus*) to 35 Å (*E. coli* porins OmpF and PhoE). The membrane-exposed surface has a uniform height of 25 Å in all solved structures and comprises a hydrophobic belt around the protein with hydrophobic residues from the β-sheets protruding into the membrane interior. Flanking the hydrophobic belt are girdles of aromatic amino acids [40]. These aromatic girdles have been seen in all OMP structures to date and their interaction with the membrane environment was highlighted by the first co-crystallization of an LPS molecule with the high-affinity receptor FhuA [14]. The non-polar moieties of the aromatic sidechains are postulated to interact with the membrane interior while their polar moieties interact with the polar head groups of phospholipids at the periplasmic face or the extracellular leaflet of the membrane. This structural organization anchors the trimer within the membrane [41]. The trimer itself is stabilized by hydrophobic and hydrophilic interactions between monomers at the subunit interface [7]. Further stabilization of the trimer may occur via salt bridges between the N- and C- termini of different monomers that are located within the trimer interface.

The long loops that connect the anti-parallel β-strands on the extracellular side of the membrane are responsible for some properties of the channel. The β-barrel itself can almost be regarded as a structural scaffold upon which to place different loops that establish specific properties [39]. The loops have a predominantly hydrophilic character and many charged amino acids that contribute to the transport properties of the porin. The association of monomers within the trimer is stabilized through the interaction of loop 2 of a porin monomer with the barrel of an adjacent monomer [42]. This loop, also known as the latching loop, has polar and ionic interactions with the barrel rim between loop 2 and loop 4 of the adjacent subunit and thereby serves to fill the gap left by the folding of loop 3 into the barrel lumen. Mutation of residues involved in ionic interactions between the two subunits significantly reduces the thermal stability of the trimer [43].

Loop 3 is of interest for a number of reasons. As mentioned, loop 3 folds into the barrel lumen to constrict the size of the pore. This constriction divides the channel into three regions: the external mouth or vestibule, the constriction zone, and the periplasmic mouth or vestibule. The length of the constricting loop varies from 30 to 50 amino acids depending upon the porin species [44]. Data from the crystal structures of bacterial porins reveal that in contrast to the extracellular loops, loop 3 has a low temperature factor indicating low mobility. The structural basis for the reduced mobility of loop 3 is a network of hydrogen bonds between the charged and polar residues located between loop 3 and the barrel wall [7]. A sequence (PEFGG) that is well conserved among a number of porins [44] is located at the turn in the tip of loop 3. Proline initiates the turn; glutamate contributes to the electrical field at the constriction zone and the phenylalanine anchors the tip to the barrel wall through hydrophobic interaction; two glycine residues allow for completion of the turn due to unique torsion angles.

Initially, the extracellular loops screen solutes that diffuse through porins. The nature of the charged amino acids in these loops determines in part the selectivity of the channel. In cation selective channels, for example, negative charges increase the local concentration of cationic molecules and thus promote passage of these compounds. The critical determinant for passage through the channel is determined by the nature of the constriction zone, a feature attributed to loop 3 [45]. Size and shape constraints are imposed on permeant compounds by the constriction zone because the geometry of this area is oval. The dimensions of the constriction zone vary from 5 by 7 Å (Omp32 of *C. acidovorans*) to 10 by 11 Å for *R. capsulatus* porin [11]. Observed in all structures solved to date is a cluster of positively charged residues on the barrel wall opposing a cluster of negatively charged residues on the constricting loop. Because these charged regions are in close proximity, they result in a strong transversal electrical field [46] that has two consequences. First, the sidechains of both groups are fully extended, thus diminishing the size of the constriction zone. Second, the charged constriction zone allows only the passage of those molecules which can orient themselves within the electrical field; thus, nonpolar molecules are excluded from entering the periplasm. Once the extracellular loops and the constriction zone have accomplished the screening process, solutes enter the internal vestibule, a relatively open and uncharged area restricted only by the β-strands of the barrel wall. The absence of protruding structures and charged amino acids allows for the easy passage of solutes into the periplasm.

3. ASSESSMENT OF PORIN FUNCTION
3.1 Porin purification

Porins may be purified from bacterial OMs by a variety of techniques, all of which aim to produce a homogeneous protein preparation that allows for structural or functional characterization. Most techniques involve isolation of the OM followed by a detergent extraction step to isolate the membrane proteins. One or several chromatographic procedures follow the detergent extraction, with the objective of separating porins from other OM proteins that are often non-specifically associated with porins. In recent years, chromatographic processes have been simplified with the introduction of hexahistidine tags into surface-exposed determinants of the protein allowing for rapid and efficient purification by affinity chromatography techniques. In other cases, proteins are expressed in large amounts without signal sequences in a heterologous system, usually as insoluble inclusion bodies, and then the protein is renatured to its native conformation. Refolding of bacterial OMPs into their native conformations has been demonstrated for porins [47,48], for a high-affinity receptor [49], and for OmpA [50].

3.2 Liposome reconstitution

Initial characterization of porin properties focused on determining the molecular mass exclusion limits of the pore and hence the OM of a bacterium. Most porins can be easily reconstituted into vesicles or liposomes of artificial lipids allowing for two types of experiments that measure pore size: vesicle permeability assays and liposome swelling assays [51]. Vesicle permeability assays measure the diffusion of radioactive compounds out of liposomes formed in the presence of the radioactive compound and porin. By using radiolabeled compounds of various sizes, one can quantitate the apparent pore size of the porin. The liposome-swelling assay provides data about diffusion rates of solutes and an accurate assessment of pore size. In this assay, proteoliposomes are formed in the presence of a large, impermeable dextran thereby trapping the dextran within the liposome. The newly formed liposomes are transferred to an isotonic solution containing a test solute. If the test solute can penetrate through the pore, its influx is followed by an influx of water and results in swelling of the liposome. The liposome swelling changes the refractive index of the liposome solution and can be measured with a spectrophotometer.

Moreover, the ability of the pore to discriminate between differently charged species may be crudely assessed through varying the net charge of the test solutes.

The Renkin equation predicts pore size by virtue of the logarithm of the rate of permeation being a linear function of the molecular mass of the solute. By applying the Renkin equation, liposome swelling assays performed with solutes of different sizes can provide an estimate of pore size. Values generated from vesicle permeability assays and liposome swelling assays [52,53] indicated that OmpF of *E. coli* had a molecular mass exclusion limit of 600 Da and a pore diameter of 11 Å, the latter value being similar to the actual size revealed by the X-ray structure. When these techniques were applied to Hib porin [54,55] a molecular mass exclusion limit of 1400 Da was determined and the Renkin equation predicted a channel diameter of 18 Å, a value consistent with the higher molecular mass exclusion limit of Hib porin.

3.3   Electrophysiology

Liposome experiments can elucidate some details of porin function, such as molecular mass exclusion limits and, in a limited way, the ability of the pore to discriminate between differently charged species. However, to provide a more accurate assessment of porin properties, more sophisticated and sensitive approaches must be used. Electrophysiological measurements through the use of PLBs or the patch clamp method allow for a rigorous and detailed analysis of pore properties. Single channel conductance, ionic selectivity, voltage gating properties, the effects of buffer solutions (pH), and the function of specific amino acids (through site-directed mutagenesis) can all yield detailed information about porin properties.

3.3.1 Comparison of planar lipid bilayers and the patch clamp method

Two major experimental techniques exist for the electrophysiological study of ion channel function: PLBs and the patch clamp method. Whereas PLBs involve the reconstitution of detergent-solubilized protein into artificial bilayers, the patch clamp method is applied to ion channels within their native membranous environment. When compared to PLBs, the patch clamp method benefits from reduced electrical noise and allows for higher time resolution. In addition, patch-clamp eliminates artifacts that may arise through the purification of proteins from their native membranes [56]. Conversely, there are also major disadvantages to applying the patch clamp method to bacterial channels. The first limitation is that no heterologous system exists for the expression of channels in an ion channel-free environment. The second disadvantage of patch clamp is that it cannot be directly applied to bacterial cells due to their small size. Therefore, OM fragments must be isolated and fused with liposomes to generate vesicles large enough for patch clamp analyses. During the aforementioned procedure the OM preparations may be contaminated with fragments of the CM. Finally, the third drawback is that proteins in bacterial OM preparations may contain some endogenous channel-forming activity [56]. For these reasons PLBs have tended to be the system of choice for the electrophysiological characterization of bacterial OM proteins.

3.3.2 Bilayer formation

In the study of porins, two basic PLB techniques are used: the painted (Mueller-Rudin) technique and the folded (Montal-Mueller) technique [57]. A variation of the Montal-Mueller technique known as the Schindler technique [58] has also been used. Regardless of which technique is used, the principle of all PLB techniques remains the same. A bilayer is formed over a small aperture that connects a two-sided chamber. The volume of the chamber varies depending on the experimental setup but is usually 1-5 ml. It is possible to acquire most bilayer supplies including cup and chamber systems from commercial sources. In this arrangement, a cup fits into a chamber and divides it into two chambers of equal volume. The aperture is pre-drilled in the cup and the cup itself is available in a number of materials and aperture sizes. For our experiments we use a chamber with a volume of 5 ml and a 250 μm aperture. We find that a cup manufactured from an opaque material is ideal as it presents more visual contrast than other available materials and allows for easier application of lipids and for optical monitoring of the bilayer.

The choice of lipids and salt solution is a matter of preference. In general, commercially available phospholipids are used. The lipids that are most often selected are phosphatidylcholine, phosphatidylethanolamine, and monoolein. In order to solubilize the lipids they are dissolved in an organic solvent such as *n*-decane or *n*-hexadecane to a final concentration of 10-30 mg/ml. The salt or electrolyte solutions used in most PLB experiments with porins are often 1 M KCl or 1 M NaCl. The salt solution may be buffered with Tris or another biological buffer to adjust the pH.

Folded bilayers of the Montal-Mueller type are formed by spreading lipids on the surface of the aqueous phase of each chamber while the buffer is below the level of the aperture. The buffer levels are then raised above the aperture resulting in the apposition of a lipid monolayer from each side of the chamber forming a bilayer across the aperture. The Schindler technique is essentially the same except that lipids are added to the chamber in the form of liposomes. The liposomes form monolayers on the surface of the aqueous phase and the buffer levels are then raised above the aperture. Furthermore, protein may be previously incorporated into the liposomes that are added to one or both sides of the chamber resulting in the formation of bilayers with incorporated protein. The advantage of the Schindler technique is that the liposome preparation and the resulting bilayer contain minimal amounts of organic solvent.

Painted (Mueller-Rudin) bilayers are formed by expelling a small amount of lipid over the aperture while visually monitoring the aperture with a stereoscope. The lipid can be expelled from a small volume capillary (volume of 1 to 2 μl) directly at the level of the aperture or slightly below. This is the method that we use and we have found that 0.3 to 1.0 μl of lipids dissolved at 25 mg/ml in *n*-decane is adequate to form a bilayer. After expelling the lipids it takes 5 to 10 minutes for the excess solvent in the lipids covering the aperture to drain into the aqueous solution, thereby forming a stable bilayer.

Upon achieving bilayer formation, stability should be assessed before beginning an experiment; premature membrane breakage must be avoided. Criteria for assessing bilayer stability include visual inspection, capacitance measurements, and measurement of current in the absence of channels. A bilayer that has not completely formed or that is overly thick will appear to be multi-colored when observed with white light. When excessive amounts of solvent have drained into the aqueous medium and a true bilayer has formed, the bilayer will appear black; this serves as the first indication that the bilayer may be properly formed. The capacitance of the bilayer is the second criteria for proceeding with an experiment. Bilayer capacitance varies according to the diameter of the bilayer and should be calculated for the selected aperture size. High capacitances can result in large background current noises that can potentially obscure ion channel currents. The capacitance of the bilayer can be reduced by minimizing the size of the annulus and the amount of solvent in the bilayer and by careful selection of experimental equipment. A more detailed discussion of capacitance and practical considerations is given in Refs. 58 and 59. The final bilayer stability measurement is the amount of current flowing across a protein-free bilayer in response to test potentials in the voltage range that is to be used in the experiment. If test potentials generate substantial currents across the membrane in the absence of protein, the bilayer is deemed unstable and should be broken and reformed before beginning an experiment.

3.3.3 Porin reconstitution

Bacterial porins must be extracted from membranes with detergents to maintain them in a soluble state. However, the presence of detergents in PLB experiments can have deleterious effects on bilayer stability. Porins are abundant OM proteins and can be isolated in large amounts and concentrated. Prior to use in a PLB experiment, the protein sample can be much diluted in an appropriate buffer such as 10 mM Tris pH 8.0 to reduce the amount of detergent introduced into the bilayer chamber but still allow for a sufficient amount of protein to be introduced into the chamber. In general, porins may be added to the chamber to a final concentration between 1 to 20 ng/ml. After addition of porin to the aqueous phase it requires up to 15 minutes for the protein to diffuse to the bilayer and insert.

4.   MEASUREMENTS

Three properties of porins can be measured in PLBs: single channel conductance, ionic selectivity, and voltage gating properties.

4.1   Single Channel Conductance

The conductance of a single porin channel is measured by adding small amounts of porin to the bilayer chamber so that the insertion of individual porin molecules occurs on a time-resolvable scale. Figure 2 shows the insertion of Hib porin molecules into a PLB. The current mediated by each channel is proportional to the applied voltage; dividing the current flowing through each channel by the applied voltage gives the conductance of the channel. The conductances of hundreds of individual channels are measured to determine the average conductance of the species of channel. The conductances of OmpF and OmpC of *E. coli* in 1 M NaCl were determined [60] to be 0.84 ± 0.06 nanoSiemens (nS) and 0.47 ± 0.04 nS respectively. The experimentally determined value for Hib porin was 0.85 nS using 1 M KCl as the salt solution [61]. There is an apparent discrepancy in the single channel conductance values for OmpF in comparison to Hib porin; the molecular mass exclusion limit is approximately three fold lower for OmpF yet the two channels have the same single channel conductance. This discrepancy may explained by the observation that Hib porin has a lower trimeric stability than OmpF and dissociates into monomers upon purification [55]. Thus, the observed conductance increments for Hib porin are those of individual monomers inserting into the membrane while those of *E. coli* OmpF represent trimeric conductance. Therefore, the single channel conductance of a single OmpF molecule would represent one-third of this value, or approximately 0.28 nS.

Depending on whether positive or negative polarity of voltage is applied, the increase in conductance will be seen as all upward or all downward deflections, respectively. Deflections of the opposite nature, representing the closure of a channel, are rarely seen at lower voltages (> ±40 mV). Closures that are seen at potentials lower than these values often reflect the lateral movement of the channel into an area of the membrane in which the channel is incapable of passing current across the membrane, such as the annulus that is formed at the interface of the lipid bilayer and the aperture.

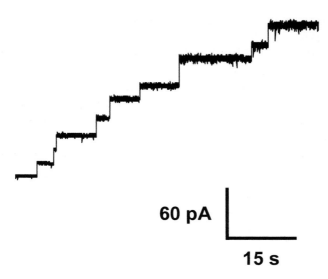

60 pA

15 s

Figure 2. Current traces showing incorporation of porin molecules into PLBs. The bilayer was formed from 1 μl of monoolein dissolved at a concentration of 25 mg/ml in *n*-decane. Hib porin was added to the *cis* chamber at a final concentration of 1 ng/ml. The electrolyte solution was 1 M KCl and a potential of +10 mV was applied. Each upward deflection indicates the insertion of one porin monomer.

4.2   Ionic selectivity

Ionic selectivity of porins reflects the ability of a porin to allow preferential passage of one species of ion over an oppositely charged species. Measuring ionic selectivity involves creating a stable membrane, adding porin to one side of the bilayer chamber and then waiting for the insertion of porin molecules to stabilize. Once the membrane has stabilized with respect to porin insertion, concentrated salt solution is added to one side of the chamber (termed the *cis* chamber) to increase the salt concentration by a predetermined amount; generally an increase in salt concentration between two- to ten-fold is used.

In the absence of an applied voltage, the ions for which the channel is selective will be propelled from the *cis* chamber through the channel into the other (termed *trans*) chamber. If the channel is cation-selective the movement of cations into the *trans* chamber creates a positive current across the membrane and results in a positive potential in the *trans* chamber. The membrane potential is stabilized when the concentration gradient is balanced by the membrane potential. The generated membrane potential, known as the zero-current membrane potential or reversal potential (V_{rev}), with the known salt concentrations is used to solve the Goldman-Hodgkin-Katz equation [62] and to determine the ratio of the permeabilities of cations to anions (P_c/P_a). As a result of the diverse environmental niches of bacterial species, the ionic selectivity of porins varies considerably between bacterial species. *E. coli* OmpF is a weakly cation selective porin with a P_{Na}/P_{Cl} of 4.5 ± 0.8 [63] while the porin Omp32 of *C. acidovorans* is highly anion selective with a P_{Cl}/P_K of 17 [64]. We investigated [65] selectivity of Hib porin and determined that it is weakly cation selective with a P_K/P_{Cl} of 1.6.

PLB analyses of porins with amino acid mutations, either naturally-occurring or introduced by site-directed mutagenesis, have established the molecular basis of ion selectivity as being highly dependent on the charged constriction zone. For *E. coli* OmpF, the mutation of one of the three arginines in the constriction zone to an uncharged amino acid resulted in a three-fold higher selectivity for cations; mutation of a negatively charged amino acid to an uncharged amino acid reduced cation selectivity by four-fold [63]. Likewise, the mutation of two positively charged amino acids in the constriction zone of porin from *Paracoccus denitrificans* [66] to negatively charged amino acids resulted in a major change in selectivity from slightly anion selective (P_K/P_{Cl} of 0.35) to strongly cation selective (P_K/P_{Cl} of 14.0). Mutagenesis experiments with PhoE [67], the anion selective (P_K/P_{Cl} of 0.30) porin of *E. coli*, showed that the substitution Lys125Glu at the pore's constriction zone resulted in a cation-selective porin (P_K/P_{Cl} of 3.8). The substitution Lys64Glu, located in the extracellular vestibule, also decreased the anion selectivity (P_K/P_{Cl} of 0.70) but not as dramatically as the mutation at the constriction zone. Thus, charged amino acids at the constriction zone determine the molecular mass exclusion limit and also play an important role in determining the ionic selectivity of the pore.

4.3   Voltage gating

The third porin property that can be measured in PLBs is voltage gating. Voltage gating has yet to be observed *in vivo* but may be an effective method of modulating OM permeability at the post-translational level. Voltage gating has been measured for porins from numerous bacterial species including *E. coli* [63], *H. influenzae* [61], *P. denitrificans* [66], *Neisseria gonorrhoeae* [68], and *Enterobacter aerogenes* [69]. Voltage gating is observed in PLBs when the applied potential exceeds a critical voltage (V_c). Above V_c the pore closes and the current mediated by the pore is reduced, presumably in response to conformational changes of the channel. Because the critical voltage required for gating of some porins is above that found across the bacterial OM, the significance of voltage gating is controversial.

Microscopic current measurements of single channels incorporated into PLBs reveal that the channels within a trimer close independently, although cooperativity between channel closures may be seen.   Macroscopic current measurements on membranes containing many channels show that the each channel closes independently and current reduction therefore shows an exponential decay.   Current traces from an experiment measuring macroscopic currents mediated by Hib porin with the mutation Lys165Glu [70] are shown in Figure 3.   Fitting macroscopic current relaxation curves with exponential functions has provided details on the time constants for pore closure and the gating kinetics of Hib porin [27,70] and porin Omp34 of *Acidovorax delafeldii* [71].   As long as the applied potential exceeds V_c, the channels remain closed for both macroscopic and microscopic current measurements.   However, channel gating is reversible and when the voltage is stepped below V_c, the channels re-open.   The V_c for pore closure varies between species, with OmpF and PhoE of *E. coli* having a V_c of + 145 mV [63] and +/- 135 mV [72], respectively, while V_c of Hib porin has been determined to be + 50 mV [61].

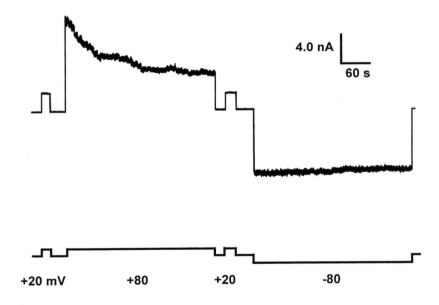

Figure 3.   Current traces demonstrating the response of Hib porin mutant Lys165Glu [70] to holding potentials of +20 mV and +/-80 mV.   The bilayer conditions are the same as in the legend to Figure 2.   The voltage was stepped from 0 mV to various test potentials as indicated by the lines below the current trace.   The scale bar (upper right) indicates current and time values.

Wild-type Hib porin is essentially a voltage-independent pore. However, PLB experiments have revealed that naturally occurring mutations in Hib porin which involve loss of a positive charge in extracellular loop 4 result in asymmetric voltage sensitivity. The Hib porin subtypes close in response to voltages in excess of + 50 mV but remain open in the presence of negative potentials of up to − 80 mV [61]. Further studies on Hib porin with chemical modification [27] and site-directed mutagenesis [27,70] revealed that the voltage sensitivity of Hib porin is increased by the introduction of negative charges into two distinct domains of the protein, extracellular loops 4 and 6 and the barrel lumen at the level of the constriction zone. Likewise, mutagenesis experiments that substituted charged residues in the constriction zone of OmpF [63] confirm that mutations introducing negative charge into cation selective pores results in increased voltage sensitivity. The introduction of positive charge into anion selective pores increases their voltage sensitivity [66,73]. This demonstrates the ability of PLB analyses, in conjunction with genetic engineering and chemical modification, to elucidate details of porin function. It is unlikely, however, that the mechanism by which voltage gating occurs can be fully resolved with PLB analyses and mutagenesis studies. Ultimately, additional biophysical approaches will be required to understand the conformational changes that underlie voltage gating.

5.   CONCLUSION

The Gram-negative bacterial OM contains many types of proteins, of which porin are the most abundant and principally responsible for OM permeability. Bilayer studies of porins have revealed the molecular details of single channel conductance and ionic selectivity and they have shed light on the voltage gating properties of the pore. A complete understanding of porin function may be elucidated only in the context of the complex environment of the intact bacterial cell. Further studies are required to examine the effects of molecules that may act as modulators of porin function and that are found either in the external or the internal environment of the bacterium.

6.   ACKNOWLEDGEMENTS

Research in the laboratory of the authors is supported by the Canadian Institutes of Health Research. We thank C.M. Khursigara and P.D. Pawelek for insightful discussions and their critical reviews of this chapter. We also gratefully acknowledge the editorial support of R.H.A. Arbing and J.A. Kashul.

REFERENCES

[1]   Y. Kamio and H. Nikaido, Biochemistry 15 (1976) 2561.
[2]   D.S. Snyder and T.J. McIntosh, Biochemistry 39 (2000) 11777.
[3]   P. Plesiat and H. Nikaido, Mol. Microbiol. 6 (1992) 1323.
[4]   B. Bolter and J. Soll, EMBO J. 20 (2001) 935.
[5]   M.H. Saier, Jr., J. Membr. Biol. 175 (2000) 165.
[6]   M.S. Weiss, U. Abele, J. Weckesser, W. Welte, E. Schiltz and G.E. Schulz, Science 254 (1991) 1627.
[7]   S.W. Cowan, T. Schirmer, G. Rummel, M. Steiert, R. Ghosh, R.A. Pauptit, J.N. Jansonius and J.P. Rosenbusch, Nature 358 (1992) 727.
[8]   A. Kreusch and G.E. Schulz, J. Mol. Biol. 243 (1994) 891.
[9]   A. Hirsch, J. Breed, K. Saxena, O.M. Richter, B. Ludwig, K. Diederichs and W. Welte, FEBS Lett. 404 (1997) 208.
[10] R. Dutzler, G. Rummel, S. Alberti, S. Hernandez-Alles, P. Phale, J. Rosenbusch, V. Benedi and T. Schirmer, Structure Fold. Des. 7 (1999) 425.
[11] K. Zeth, K. Diederichs, W. Welte and H. Engelhardt, Structure Fold. Des. 8 (2000) 981.
[12] T. Schirmer, T.A. Keller, Y.F. Wang and J.P. Rosenbusch, Science 267 (1995) 512.
[13] D. Forst, W. Welte, T. Wacker and K. Diederichs, Nat. Struct. Biol. 5 (1998) 37.
[14] A.D. Ferguson, E. Hofmann, J.W. Coulton, K. Diederichs and W. Welte, Science 282 (1998) 2215.
[15] K.P. Locher, B. Rees, R. Koebnik, A. Mitschler, L. Moulinier, J.P. Rosenbusch and D. Moras, Cell 95 (1998) 771.
[16] S.K. Buchanan, B.S. Smith, L. Venkatramani, D. Xia, L. Esser, M. Palnitkar, R. Chakraborty, H.D. van der Helm and J. Deisenhofer, Nat. Struct. Biol. 6 (1999) 56.
[17] A.D. Ferguson, R. Chakraborty, B.S. Smith, L. Esser, H.D. van der Helm and J. Deisenhofer, Science 295 (2002) 1715.
[18] A. Pautsch and G.E. Schulz, Nat. Struct. Biol. 5 (1998) 1013.
[19] J. Vogt and G.E. Schulz, Structure Fold. Des. 7 (1999) 1301.
[20] H.J. Snijder, I. Ubarretxena-Belandia, M. Blaauw, K.H. Kalk, H.M. Verheij, M.R. Egmond, N. Dekker and B.W. Dijkstra, Nature 401 (1999) 717.
[21] V. Koronakis, A. Sharff, E. Koronakis, B. Luisi and C. Hughes, Nature 405 (2000) 914.
[22] L. Vandeputte-Rutten, R.A. Kramer, J. Kroon, N. Dekker, M.R. Egmond and P. Gros, EMBO J. 20 (2001) 5033.
[23] W. Welte, U. Nestel, T. Wacker and K. Diederichs, Kidney Int. 48 (1995) 930.
[24] A.D. Ferguson, W. Welte, E. Hofmann, B. Lindner, O. Holst, J.W. Coulton and K. Diederichs, Structure Fold. Des. 8 (2000) 585.
[25] R. Koebnik, K.P. Locher and P. Van Gelder, Mol. Microbiol. 37 (2000) 239.
[26] H. Nikaido, Mol. Microbiol. 6 (1992) 435.
[27] M.A. Arbing, D. Dahan, D. Boismenu, O.A. Mamer, J.W. Hanrahan and J.W. Coulton, J. Membr. Biol. 178 (2000) 185.
[28] C. Andersen, R. Cseh, K. Schulein and R. Benz, J. Membr. Biol. 164 (1998) 263.
[29] P. Van Gelder, F. Dumas, J.P. Rosenbusch and M. Winterhalter, Eur. J. Biochem. 267 (2000) 79.
[30] R. Benz, G. Francis, T. Nakae and T. Ferenci, Biochim. Biophys. Acta 1104 (1992) 299.

[31] G.S. Moeck and J.W. Coulton, Mol. Microbiol. 28 (1998) 675.
[32] R. Benz, E. Maier and I. Gentschev, Zentralbl. Bakteriol. 278 (1993) 187.
[33] F. Tardy, F. Homble, C. Neyt, R. Wattiez, G.R. Cornelis, J.M. Ruysschaert and V. Cabiaux, EMBO J. 18 (1999) 6793.
[34] N. Nouwen, N. Ranson, H. Saibil, B. Wolpensinger, A. Engel, A. Ghazi and A.P. Pugsley, Proc. Natl. Acad. Sci. U. S. A. 96 (1999) 8173.
[35] U.W. Konninger, S. Hobbie, R. Benz and V. Braun, Mol. Microbiol. 32 (1999) 1212.
[36] H. Killmann, R. Benz and V. Braun, EMBO J. 12 (1993) 3007.
[37] V. Braun, H. Killmann and R. Benz, FEBS Lett. 346 (1994) 59.
[38] M. Bonhivers, A. Ghazi, P. Boulanger and L. Letellier, EMBO J. 15 (1996) 1850.
[39] T. Schirmer, J. Struct. Biol. 121 (1998) 101.
[40] G.E. Schulz, Curr. Opin. Struct. Biol. 6 (1996) 485.
[41] B.A. Wallace and R.W. Janes, Adv. Exp. Med. Biol. 467 (1999) 789.
[42] G.E. Schulz, Curr. Opin. Cell Biol. 5 (1993) 701.
[43] P.S. Phale, A. Philippsen, T. Kiefhaber, R. Koebnik, V.P. Phale, T. Schirmer and J.P. Rosenbusch, Biochemistry 37 (1998) 15663.
[44] D. Jeanteur, J.H. Lakey and F. Pattus, Mol. Microbiol. 5 (1991) 2153.
[45] T. Schirmer and P.S. Phale, J. Mol. Biol. 294 (1999) 1159.
[46] A. Karshikoff, V. Spassov, S.W. Cowan, R. Ladenstein and T. Schirmer, J. Mol. Biol. 240 (1994) 372.
[47] D. Dahan, R. Srikumar, R. Laprade and J.W. Coulton, FEBS Lett. 392 (1996) 304.
[48] T. Surrey, A. Schmid and F. Jahnig, Biochemistry 35 (1996) 2283.
[49] S.K. Buchanan, Biochem. Soc. Trans. 27 (1999) 903.
[50] A. Pautsch, J. Vogt, K. Model, C. Siebold and G.E. Schulz, Proteins 34 (1999) 167.
[51] R. Benz and K. Bauer, Eur. J. Biochem. 176 (1988) 1.
[52] T. Nakae, Biochem. Biophys. Res. Commun. 71 (1976) 877.
[53] H. Nikaido and E.Y. Rosenberg, J. Bacteriol. 153 (1983) 241.
[54] V. Vachon, D.J. Lyew and J.W. Coulton, J. Bacteriol. 162 (1985) 918.
[55] V. Vachon, D.N. Kristjanson and J.W. Coulton, Can. J. Microbiol. 34 (1988) 134.
[56] A.H. Delcour, FEMS Microbiol. Lett. 151 (1997) 115.
[57] R. Benz, Annual review of microbiology 42 (1988) 359.
[58] W. Hanke and W.-R. Schlue, Planar lipid bilayers: methods and applications, Academic Press, San Diego, CA, 1993.
[59] R. Sherman-Gold, (ed.), The Axon Guide for electrophysiology & biophysics: laboratory techniques, Axon Instruments, Inc., Foster City, CA, 1993.
[60] A. Prilipov, P.S. Phale, R. Koebnik, C. Widmer and J.P. Rosenbusch, J. Bacteriol. 180 (1998) 3388.
[61] D. Dahan, V. Vachon, R. Laprade and J.W. Coulton, Biochim. Biophys. Acta 1189 (1994) 204.
[62] B. Hille, Ionic channels of excitable membranes, Sinauer Associates, Sunderland, MA, 1992.
[63] P.S. Phale, A. Philippsen, C. Widmer, V.P. Phale, J.P. Rosenbusch and T. Schirmer, Biochemistry 40 (2001) 6319.
[64] A. Mathes and H. Engelhardt, Biophys. J. 75 (1998) 1255.
[65] V. Vachon, R. Laprade and J.W. Coulton, Biochim. Biophys. Acta 861 (1986) 74.

[66] K. Saxena, V. Drosou, E. Maier, R. Benz and B. Ludwig, Biochemistry 38 (1999) 2206.
[67] K. Bauer, M. Struyve, D. Bosch, R. Benz and J. Tommassen, J. Biol. Chem. 264 (1989) 16393.
[68] T. Rudel, A. Schmid, R. Benz, H.A. Kolb, F. Lang and T.F. Meyer, Cell 85 (1996) 391.
[69] E. De, A. Basle, M. Jaquinod, N. Saint, M. Mallea, G. Molle and J.M. Pages, Mol. Microbiol. 41 (2001) 189.
[70] M.A. Arbing, J.W. Hanrahan and J.W. Coulton, Biochemistry 40 (2001) 14621.
[71] A. Mathes and H. Engelhardt, J. Membr. Biol. 165 (1998) 11.
[72] E.F. Eppens, N. Saint, P. Van Gelder, R. van Boxtel and J. Tommassen, FEBS Lett. 415 (1997) 317.
[73] P. Van Gelder, N. Saint, P. Phale, E.F. Eppens, A. Prilipov, R. van Boxtel, J.P. Rosenbusch and J. Tommassen, J. Mol. Biol. 269 (1997) 468.

Planar Lipid Bilayers (BLMs) and their Applications
H.T. Tien and A. Ottova-Leitmannova (Editors)

Chapter 13

Reconstitution in planar lipid bilayers of ion channels synthesized *in ovo* and *in vitro*

L.K. Lyford and R.L. Rosenberg

Departments of Pharmacology and Cell and Molecular Physiology,
CB# 7365, University of North Carolina at Chapel Hill, Chapel Hill, NC 27599
USA

1. INTRODUCTION

Ion channels are essential for many cellular functions, including electrical excitability, synaptic transmission, hormone release, intracellular Ca^{2+} signaling, salt/water balance, fluid secretion, and cell volume regulation. In a broad sense, ion channels can be classified as voltage-gated, ligand-gated, intracellular messenger-gated, or constitutively active [1]. Ion channels, in addition to being gated by their primary activator (or inhibitor), are also substantially modulated by phosphorylation, dephosphorylation, accessory proteins, ion concentrations, pH, interactions with scaffolding and/or cytoskeletal proteins, and other "secondary" effectors. Thus, cellular regulation of ion channel activity can be extremely complex, with multiple modulatory processes that can interact and compete to set the overall level of channel activity.

To characterize the fundamental properties of ion channels and how their activity is regulated it is necessary to obtain experimental control over their environment. A major advance in experimental control was taken when Hodgkin and Huxley used voltage-clamp technology to control the voltage of squid giant axons and thereby determine the voltage-dependence of the Na^+ and K^+ channels responsible for the nerve action potential [1]. Another significant advance was the development of patch-clamp techniques, which allowed the voltage clamp of a wide variety of cell types including very small cells, increased experimental access to the intracellular aqueous milieu, and allowed the measurement of single-channel activity in membrane patches [2,3].

While patch-clamp technology has increased the ability to experimentally control the environment of ion channels, there are limitations. Protein-protein interactions can remain intact in both whole-cell and excised-patch configurations, making them difficult to control. Kinases and phosphatases in close proximity to ion channels can remain active in patches [4]. It is difficult to control simultaneously the ionic milieu in both the intracellular and extracellular compartments. It is impossible to control the hydrophobic lipid environment surrounding ion channels.

Reconstitution of ion channels into planar lipid bilayers [5] offers improved experimental control. Even without solubilization and protein purification, the biochemical preparation of membrane fragments from tissues that express the ion channels of interest usually removes cytoplasmic and extrinsic membrane proteins that are not tightly associated with the membranes. Simultaneous control of both intracellular and extracellular ionic environments is easy. Modification of the lipid environment is also possible. Tight protein-protein interactions, however, are difficult to control, and it is impossible to manipulate the isoforms or subunit composition of the channels.

The most significant advance in the experimental control over subunit composition of ion channels was the identification and isolation of genes coding for many different types of ion channels. This, combined with heterologous expression in target cells and patch-clamp approaches, have allowed the functional characterization of a large number of cloned ion channels. In addition to having the ability to control the isoforms and subunit composition of the channel under study, it is possible to make specific mutations in the channel protein.

It has also been possible to reconstitute into planar lipid bilayers cloned ion channels that are expressed heterologously. These approaches combine the advantages of heterologous expression (uniformity of receptor populations, mutagenesis, etc.) with the high level of experimental control available in planar lipid bilayer recording. A wide variety of heterologous expression systems have been used to synthesize cloned ion channels that are reconstituted into planar lipid bilayers. For example, $\alpha 7$ nicotinic receptors (nAChRs), voltage-gated K^+ channels, epithelial Na^+ channels (ENaC), cystic fibrosis transmembrane regulator (CFTR) Cl^- channels, and ClC Cl^- channels have been expressed in *Xenopus* oocytes, and then reconstituted (see Section 2.2, below). In addition, infection of insect cells with recombinant baculovirus has been used to express cloned ion channel protein in high abundance, followed by protein purification and bilayer reconstitution. This combination of methods has been used to characterize functional CFTR channels [6,7], rat $\beta 2$ connexons [8], and ENaC [9]. CFTR destined for bilayer reconstitution has also been expressed in other heterologous systems, such as Chinese hamster ovary (CHO) cells [7,10], NIH-3T3 cells [10], yeast [11], and retrovirally transduced mouse L cells [12]. Other examples of ion channels studied via of cellular expression and bilayer reconstitution include Ca^{2+}-dependent K^+ channels expressed in COS cells [13], voltage-gated K^+ channels expressed in vaccinia-virus transfected African green monkey kidney cells [14], and rat inositol (1,4,5)-triphosphate receptors expressed in HEK293 cells [15].

An alternative to the cellular expression of ion channels is the synthesis of cloned channels in cell-free expression systems. In this approach, cell lysates for *in vitro* protein expression are supplemented with endoplasmic reticulum (ER) microsomal membranes. Components of these microsomes recognize membrane proteins as they emerge from ribosomes, form a docking structure between the ribosome and microsomal membrane, and catalyze the transmembrane folding and oligomerization of new-born proteins [16]. Reconstitution of the microsomes into planar bilayers then allows the functional characterization of the channels. The primary advantage of this approach is that the ion channels are never inserted in a true target membrane and are not exposed to an intact cytoplasm, so auxiliary and regulatory proteins that normally associate with the ion channels in the target membrane are not likely to be present.

This chapter will focus on the reconstitution of ion channels expressed either in *Xenopus* oocytes (*in ovo*) or in cell-free expression systems (*in vitro*). We will discuss the advantages and disadvantages of reconstitution compared to other functional approaches such as patch-clamp, the rationale for expression *in ovo* and *in vitro*, describe the techniques used in these reconstitution studies, and briefly review the knowledge of ion channel function that these reconstitution approaches have provided.

1.1. Advantages and disadvantages of traditional reconstitution approaches

Reconstitution of ion channels into planar lipid bilayers allows detailed measurements of the ion permeation and gating properties of the channels under conditions that are often impossible to achieve in intact cells. For ion permeation especially, it is informative to make measurements in extremely non-physiological conditions and to control the intracellular milieu. For example, reconstitution experiments showed that Ca^{2+} permeation through L-type Ca^{2+} channels was functionally symmetrical; when Ca^{2+} concentrations are equal on both sides of the channels, the current-voltage relationship is linear [17].

Another power of the reconstitution approach is the ability to control the lipid composition surrounding the ion channels. The use of charged and neutral lipids has shown that the outer mouth of sarcoplasmic reticulum K^+ channels is likely to be separated from the membrane by approximately 10-20 Å [18]. Control of membrane composition has also shown that the functional properties of nAChRs are modulated by fatty acids, alcohols, cholesterol, and phospholipids [19]. Modification of membrane lipids with eicosanoid sidechains has shown that the transmembrane domain is probably the target for lipid metabolite regulation of cardiac Ca^{2+} channels [20].

Ion channel reconstitution is at its best when solubilized, purified channel proteins are reconstituted into membranes for functional characterization. This has been accomplished for several ion channels, including nAChR [21], Na^+ channels [22,23], and CFTR [6,7]. The solubilization and isolation of purified protein permits the determination of both the subunit composition of the channels and their functional properties in the presumed absence of associated proteins. However, because reconstitution approaches usually study only a few ion channels, and biochemical approaches characterize large populations of proteins, it is impossible to be certain that the functional properties of the few reconstituted channels are not influenced by associated proteins that are present at concentrations too low to be detected by biochemical means.

For most channels, it has not been possible to solubilize and purify the channel protein from native tissue or from cells that heterologously express the channels. Typically, the density of channel protein is not high enough to permit purification by biochemical approaches. Thus, the reconstitution of most channels is from semi-purified membranes rather than from solubilized and purified protein. In membrane preparations, auxiliary proteins are likely to remain associated with ion channels throughout membrane fractionation and reconstitution. For example, protein kinases and phosphatases are associated with reconstituted large-conductance Ca^{2+}-activated K^+ channels [24,25]. G proteins and phosphatases are associated with L-type Ca^{2+} channels [26,27]. Native membranes may also contain multiple channel isoforms due splice variations, heterogeneous gene products, and variations in subunit composition in addition to auxiliary proteins. For example, reconstituted large-conductance Ca^{2+}-activated K^+ channels were classified as Type 1 or Type 2 based on gating kinetics [24,25]. Reconstituted L-type Ca^{2+} channels show wide channel-to-channel variability in the rates of voltage-gated inactivation [28], mimicking results obtained in cell-attached patches [29]. Thus, when the goal of a reconstitution experiment is to determine the functional properties of an ion channel in isolation from its binding partners, the reconstitution of channels from intact membrane cannot usually meet the necessary criteria of channel isolation or uniformity.

Whereas the number of ion channel proteins that have been solubilized and purified is small, the number of ion channels whose genes have been cloned is very large. Thus, the heterologous expression of cloned ion channels has become the most common means to characterize specific isoforms of ion channels. Although the channels are expressed in a cell type that does not normally express the channel of interest, regulatory pathways and membrane proteins that modify channel properties are still present and may be co-incorporated into bilayers.

1.2. Rationale for reconstitution of ion channels expressed *in ovo*

Expression in *Xenopus laevis* oocytes is probably the most common means to study the functional properties of cloned ion channels. This approach has been combined with ion channel reconstitution to characterize the functional properties of the channels at the single channel level and in relative isolation from an intracellular environment. The overall experimental strategy is to express the channel in *Xenopus* oocytes, isolate plasma membranes, fuse the membranes with planar lipid bilayers to incorporate the channels, and characterize the semi-isolated channels.

The advantages of oocyte expression generally stem from the enormous size of the oocytes (0.5-1 mm in diameter). Thus, injection of cRNA or cDNA is relatively easy and high levels of surface expression are routinely obtained. In addition, it is also relatively simple to use two-microelectrode voltage clamp to obtain high-quality voltage-clamp data from the expressed channels. The functional characteristics of ion channels expressed *in ovo* are generally very similar to those obtained in cells that normally express the channel or in other types of heterologous expression systems. The main disadvantages arise from limitations in rapid voltage-clamp of the large membrane surface area and confounding currents from endogenous Ca^{2+}-activated Cl^- channels [30,31].

For reconstitution studies, oocyte expression offers both advantages and disadvantages. The high levels of expression suggest that surface membrane preparations will have an acceptable ion channel density. The expressed channels are derived from cDNA, so a single isoform of the channel may be characterized in isolation from other isoforms. Heteromeric channels that incorporate multiple different subunits can also be expressed and characterized. Specific mutations can be introduced and their functional effects characterized. Disadvantages stem from the fact that each oocyte must be individually injected, making the expression of channels in the large number of oocytes required for membrane preparations labor-intensive. The isolation of surface membranes from yolk and other contaminants can be challenging. In addition, because oocytes are cells, obviously, it is difficult to ascertain to what extent the expressed ion channels are coupled to endogenous proteins. One would like to assume that the oocyte leaves the expressed ion channels in a relatively naïve state and does not couple the expressed channels to endogenous membrane proteins or signal transduction pathways, but auxiliary proteins derived from the oocytes may still be present in the isolated surface membranes.

1.3. Rationale for reconstitution of ion channels expressed *in vitro*

The expression of ion channels *in vitro* solves many of the problems associated with cellular expression strategies. *In vitro* systems are used for transcription and translation, so the expressed protein is never in contact with an intact cell. Translation is performed in the presence of endoplasmic reticulum microsomes to achieve transmembrane folding, assembly, and glycosylation. Functional reconstitution is achieved by incorporating the ER microsomes into planar bilayers.

Although the expressed channel protein comes in contact with cell lysates and ER membranes during synthesis, it is not incorporated in a surface membrane where channels and regulatory protein may be clustered. Even if regulatory proteins are present in the lysates or the ER membranes, they are unlikely to be localized with the nascent channel protein. Scaffolding that may be necessary for formation of regulatory complexes is absent in cell lysates. In addition, the cell types used for the cell-free lysates (e.g. reticulocytes) and ER microsomes (secretory cells from mammalian pancreas or avian oviduct) are generally quite different from the cells that normally express the channel of interest, so regulatory pathways are not likely to be well-developed. Ion channels expressed *in vitro* are thus likely to be relatively free of interactions with endogenous regulatory molecules. Ion channels that are endogenous to the ER microsomes are generally few in number and easy to identify (see Section 3.1.6, below).

The main disadvantage to expression *in vitro* stems from the low level of protein synthesis and difficulties with efficient folding and oligomerization. Thus, it can be difficult to synthesize enough functional channels so that reconstitution is reliable and rapid enough to characterize the expressed channels in a reasonable time.

2. *IN OVO* EXPRESSION

This section describes techniques used in the reconstitution of ion channels expressed in *Xenopus* oocytes, provides examples of ion channels that have been studied following this approach, and highlights those results that could not easily be obtained from traditional approaches.

2.1. General methods for *in ovo* expression and reconstitution

2.1.1. Oocytes

The harvesting, defolliculation, selection and maintenance of oocytes from the South African frog *Xenopus laevis*, as well as the preparation and injection of RNA, are well-established and commonly-used protocols. Standard references for these techniques can be found in Methods in Enzymology volumes 207 [33,34] and 293 [35,36].

2.1.2. Oocyte membrane preparations

One method for preparing oocyte membranes for ion channel reconstitution utilizes a one-step sucrose gradient to isolate plasma membrane [36,37]. Most studies of ion channels reconstituted from oocytes employ this procedure, as this technique is relatively easy to perform and provides a reasonable yield of membranes containing functional channels.

Two other methods for preparing membranes from oocytes have been described. One protocol employs differential centrifugation to prepare plasma membranes [38] and was used to study CFTR Cl⁻ channels [39]. A second protocol involves the detachment of cortical vesicles [40] and the subsequent isolation of the oocyte plasma membranes [41] by differential centrifugation and a multi-layer sucrose gradient. This approach was used to study reconstituted inward rectifier K^+ channels [42].

2.1.3. Reconstitution methods

Several experimental approaches have been taken to incorporate functional channels into lipid bilayers and characterize their functional activity. Solvent-free bilayers [43] can be formed on various partitions [44], including horizontal apertures [45], and the end of patch pipettes [46]. Solvent-containing bilayers [47] can also be formed on various partitions. Ion channels are reconstituted into bilayers via fusion of membrane vesicles [48,49], direct application of vesicles to bilayers [37], or incorporation into monolayers prior to the formation of solvent-free bilayers [50]. In our experience, the approach most successful for the incorporation of channels from *Xenopus* oocytes has been via direct application [37]. In this approach, membranes are suspended in a high-sucrose solution and a smooth, fire-polished glass probe is used to transfer the suspension directly to the bilayer surface. The sucrose aids in this transfer by increasing the viscosity of the solution and increasing the rate of vesicle fusion to the bilayer. The diameter of the glass at the end of the probe is usually slightly greater than that of the bilayer aperture. Bilayers are momentarily disturbed by the contact with the probe, but they typically re-form spontaneously when the probe is removed.

There does not appear to be any strong requirement for a particular phospholipid composition of the bilayers, although lipid composition has not been explored comprehensively. We typically use synthetic palmitoyl-oleoyl phosphatidyl-ethanolamine and phosphatidyl-serine, (75% PE, 25% PS). For reconstitution of nicotinic receptors expressed in oocytes, bilayers usually include cholesterol (~40 mol%) to increase receptor function and bilayer stability [51]. Perez *et al.* [37] employ bilayers comprised of phosphatidyl-ethanolamine (50%), phosphatidyl-choline (30%), and phosphatidyl-serine (20%).

2.1.4. Endogenous channels

Surface membrane microsomes from *Xenopus* oocytes contain ion channels that can complicate attempts to record activity from heterologously-expressed channels. When we reconstituted surface membrane preparations from uninjected oocytes into planar lipid bilayers, a wide variety of Cl⁻ selective channels, with conductances ranging between 50 and 150 pS, were recorded (unpublished observations). The open probability of these channels was often very high, and did not show a strong dependence on Ca^{2+} concentration or voltage. The well-known Ca^{2+}-activated Cl^- channel in oocyte surface membranes [30,52] has a conductance of only 3 pS [53], so is unlikely to be resolved at moderate membrane potentials because the single-channel current amplitude and signal-to-noise ratio are small.

Cation selective channels were also recorded after reconstitution of oocyte membranes into bilayers, although the frequency of recording these was much lower than that of the Cl^- selective channels. For example, in a single experiment we recorded a large-conductance (600 pS) non-selective cation channel.

2.2. Results from *in ovo* expressed ion channels

2.2.1. α7 nicotinic acetylcholine receptors

The α7 nicotinic acetylcholine receptors (nAChR) are ligand-gated ion channels activated by the neurotransmitter acetylcholine. These receptors play a role in the presynaptic modulation of neurotransmitter release [54] and mediate excitatory neurotransmission at select synapses [55]. Nicotinic receptors are pentameric structures comprised of homologous subunits arranged around a central, ion-conducting pore [56,57]. Each subunit has four transmembrane helices, with the second transmembrane domain lining the ion-conducting pore [1]. Heterologously expressed α7 nAChRs can assemble as homopentamers, although the native subunit composition of receptors containing the α7 subunit in some species may contain other nAChR subunits as well [58].

The Ca^{2+} permeability of α7 nAChRs is higher than other nicotinic receptors [59] and is important in their functional roles. In order to characterize the interaction of Ca^{2+} ions with the pore, we used reconstitution approaches to measure the effects of intracellular and/or extracellular Ca^{2+} on single-channel amplitudes. To avoid the fast desensitization properties of the wild-type α7 nAChRs, we used a non-desensitizing mutant ($S^{240}T/L^{247}T$). This allowed us to record channel activity in the continual presence of agonist [51].

Membrane vesicles from oocytes expressing the $S^{240}T/L^{247}T$ α7 receptors were prepared and applied directly onto planar lipid bilayers [37]. Channel openings were evoked by the continuous presence of acetylcholine in both chambers. The orientation of the channels in the bilayer (and the nicotinic identity of the channels) was determined by observing the sidedness of block with nicotinic antagonists that interact with extracellular binding sites. Only when blockers were present in the *cis* chamber (the side of the bilayer to which the membranes were applied) was there substantial channel blockade, indicating that the extracellular side of the channel faced the *cis* chamber. The channels had a unitary conductance of 52 pS and were not selective between Na^+ and K^+, as expected [60]. The unitary conductance was reduced in a dose-dependent fashion as the concentration of Ca^{2+} was increased stepwise from 0.5 mM to 20 mM. From this information, the K_i for the block by Ca^{2+} of monovalent current was estimated to be 1.5 mM Ca^{2+}. Thus, the $\alpha7$ nicotinic receptors have one or more low-affinity Ca^{2+}-binding sites. There was no evidence of high-affinity Ca^{2+} binding sites. Thus, despite their high Ca^{2+} permeability, $\alpha7$ nAChRs do not promote the dramatic ion-ion interactions observed in highly Ca^{2+}-selective, voltage-gated Ca^{2+} channels [61].

2.2.2. *Large-conductance Ca^{2+}-activated K^+ channels*

Ca^{2+}-activated K^+ channels (K_{Ca}) are found in many cell types [62]. They are comprised of a tetramer of α subunits, each with seven transmembrane domains and a re-entrant pore-lining loop. Native K_{Ca} channels also contain β subunits that modulate Ca^{2+}- and voltage-dependence. Expression of α subunits in oocytes and other cells suggests that β subunits are not required for channel function.

One of the first demonstrations that a cloned, oocyte-expressed ion channel could be functionally reconstituted into planar lipid bilayers was demonstrated by the reconstitution of *Drosophila Slowpoke* (*dSlo*) K_{Ca} channels [37]. The reconstituted channels had properties that were similar to those in native tissue, including a single-channel conductance of 260 pS, Ca^{2+}-sensitivity, voltage-dependence, intracellular block by a peptide that produces inactivation of voltage-gated K^+ channels, and extracellular block by tetraethylammonium ions. The *dSlo* K_{Ca} channels were insensitive to charybdotoxin, a specific blocker of most mammalian large-conductance, Ca^{2+}-activated K^+ channels, suggesting a difference between *Drosophila* and mammalian isoforms of *Slowpoke*.

There was a wide variability between individual channels in open probability, despite identical voltages and internal Ca^{2+} concentrations. This variability has been observed in native K_{Ca} channels in patch clamp and reconstitution experiments [63], and has been ascribed to heterogeneous channel populations [64]. However, when cloned channels are reconstituted, the primary sequence of the channels is uniform, suggesting that the functional variability derives from an intrinsic modal gating property, differences in posttranslational modification, or variable association with β subunits or other proteins endogenous to the oocytes. Thus, the reason for divergent channel properties within one preparation may be more complex than originally thought.

Similar results were obtained by Wallner *et al.* [65] with cloned human K_{Ca} channels. Channel activity in excised patches from oocytes was carefully compared to that after reconstitution in bilayers. Single-channel conductance, voltage-dependence, Ca^{2+}-dependence, and pharmacological blockade were all very similar in the two sets of experiments, indicating that the reconstituted channels were not altered during membrane isolation and reconstitution. In addition, channel activity was characterized with and without addition of β subunits during channel synthesis *in ovo*. In bilayers, as in excised patches, the voltage-dependence of α/β-containing channels was shifted relative to those containing α subunits alone, indicting that the β subunits remained associated during membrane isolation and fusion to bilayers. As with *dSlo* K_{Ca} channels, there was some heterogeneity of channel behavior at the single-channel level following reconstitution of the cloned α subunit of *hSlo*, possibly indicating involvement of post-translational modifications or unknown protein-protein interactions that remain intact during reconstitution. The heterogeneity was not fully characterized for *hSlo* channels synthesized as α/β hetero-oligomers, leaving open the possibility that some of the heterogeneity in channel behavior of α subunit homo-oligomers was due to low-level synthesis of an oocyte β-subunit or related protein.

2.2.3. Inward rectifier K^+ channels

Inward rectifier K^+ channels (IRK) are important for maintaining the resting membrane potential of most cells. They are a tetramer of subunits, each with two transmembrane domains and a pore-lining loop [66]. The mechanisms of inward rectification derive at least in part from blockade by intracellular Mg^{2+} [67] and intracellular polyamines [68].

The properties of these channels were characterized further by expression in oocytes and reconstitution into planar lipid bilayers [42]. Highly purified plasma membranes were isolated as described in Refs. 41 and 40. The functional properties of the reconstituted channels were similar to those in isolated patches from oocytes, showing distinctive inward rectification and high K^+ selectivity. Interestingly, reconstituted IRK channels showed clear inward rectification even when they were recorded in the complete absence of intracellular Mg^{2+} and polyamines. Thus, these reconstitution experiments showed that there is an intrinsic element of inward rectification of IRK channels in addition to voltage-dependent Mg^{2+} and polyamine block.

2.2.4. Epithelial sodium channels

Amiloride-sensitive epithelial Na^+ channels (ENaC) are important in Na^+ reabsorption and fluid balance in kidney, lung, and other epithelia [69]. They are formed as heteromultimers of three homologous subunits (α, β, and γ), although functional channels can be formed following the expression of α subunits alone. They are believed to be formed as tetramers with an $\alpha_2\beta\gamma$ stoichiometry. Each subunit has two transmembrane domains, and the pore-forming region is lined by the second hydrophobic domain.

Several studies have used *in ovo* expression and bilayer reconstitution to examine the properties of cloned ENaC channels [70]. The single channel properties of α,β,γ-ENaC were examined, and the functional consequences of intracellular N-terminal deletions and point mutations in the extracellular and intracellular domains were characterized [71,72]. The bilayer approach permitted studies of the effects of amiloride, pH, and Ca^{2+} at the single-channel level on cloned wild-type and mutant channels. Detailed characterizations of open probability, unitary conductance, amiloride dose-responses, and ion selectivity were also performed.

2.2.5. Cystic fibrosis transmembrane regulator Cl⁻ channels

The cystic fibrosis transmembrane regulator (CFTR) is a multifunctional membrane protein that can act as a cyclic AMP-regulated Cl⁻ channel. Its function is important in airway fluid secretion, and mutations generate the human disease cystic fibrosis [73]. The transmembrane architecture and pore-lining domains are not completely known.

The reconstitution of oocyte-expressed CFTR (both normal and the deletion mutant ΔF508 most commonly found in cystic fibrosis patients) showed that both forms have the same single-channel conductance of 10 pS [39]. The primary difference between normal and ΔF508 CFTR was that the mutant CFTR had a lower open probability. The reconstitution approach showed that both forms had a reduced open probability at acidic intracellular pH.

2.2.6. ClC-2G Cl⁻ channels

ClC chloride channels are double-barreled, anion selective channels that regulate electrical activity in muscle and the flow of salt and water in epithelia. The channels have eighteen alpha helices and are formed as a dimer of two identical subunits, each with its own ion-conducting pore [74]. The ClC-2G Cl⁻ channel subtype is a voltage- and pH-activated anion channel that is involved in acid secretion in the stomach and is expressed in several diverse tissues [75].

The properties of cloned rabbit and human ClC-2G channels were described following oocyte expression and planar lipid bilayer reconstitution. The ClC-2G channels have a linear current-voltage relationship with a unitary conductance of 22 pS. The reconstitution approach allowed the channels to be characterized under conditions of extremely low extracellular pH, with the simultaneous control of intracellular PKA activity. Both experimental manipulations increased the open probability of the channel in an additive fashion [75].

Because the experiments focused on cloned channels, specific mutations could be made to test the mechanism for pH activation [76]. A potential pH sensor region, EELE, was identified in an extracellular loop region. Mutation of the motif to QQLE had no effect on pH activation, indicating that the first two glutamate residues were not important in sensing extracellular pH. Channels with the EELQ mutation, however, had no pH sensitivity and a reduced open probability. Thus, the combination of specific mutagenesis and bilayer reconstitution made it possible to identify E^{419} as the predominant amino acid in pH sensing and/or pH-dependent gating.

3. *IN VITRO* EXPRESSION

As with channels expressed *in ovo*, *in vitro* expression uses cDNA encoding ion channels, so that specific isoforms and mutations can be studied. As described in the Section 1.3 above, *in vitro* expression of ion channels uses cell lysates for protein transcription and translation, and endoplasmic reticulum microsomes for transmembrane folding, glycosylation, and oligomerization. The main advantages of expression *in vitro* over other cellular expression approaches (including expression *in ovo*) are that the expressed channels *in vitro* are likely to be free from interactions with endogenous subunits, auxiliary proteins, and regulatory molecules.

This section describes techniques used for *in vitro* expression of ion channels and highlights results obtained following functional reconstitution in bilayers.

3.1. General methods for *in vitro* expression and reconstitution

3.1.1. In vitro transcription

RNA transcripts can be synthesized *in vitro* from cDNA templates using one of three bacteriophage RNA polymerases (SP6, T3, or T7) driven by its corresponding promoter. DNA templates should be linearized to produce optimal yields of cRNA transcripts. Restriction enzymes that produce blunt ends or 5′ overhangs are preferred, as DNA templates with 3′ overhangs will result in the production of double-stranded RNA (Ref. 77; pp. 9.29-9.37). The presence of a 5′ cap on the RNA is critical for efficient *in vitro* translation and other *in vitro* reactions, and improves the stability of the RNA in oocytes (Ref. 77; pp. 9.87-9.88). Care must be taken to avoid RNA degradation by ubiquitous RNases.

3.1.2. In vitro translation

In vitro translation systems are commercially available from several sources, including Promega, Novagen, Ambion, and Roche. The three most commonly used cell extract preparations for *in vitro* translation are rabbit reticulocyte lysates (RRL), wheat germ extracts, and S30 extracts from *E. coli*. Some of these preparations are treated with a Ca^{2+}-dependent micrococcal nuclease to remove endogenous RNA.

Rabbit reticulocyte lysates [78] are optimal for translating animal proteins [79]. In addition, RRL-based translation systems can be used in conjunction with microsomal membranes for co-processing membrane proteins. Wheat germ extracts are optimal for translating plant proteins and eukaryotic transcription factors. Wheat germ extracts are generally not recommended for studies that include microsomal membranes for processing, because the signal recognition particle is removed during preparation. The S30 extract of *E. coli* is the optimal system for translating prokaryotic proteins, but it is not recommended for non-bacterial proteins.

Commercial systems are available that couple the transcription and translation reactions in one step. For these systems, cDNA is used as the template rather than transcribed cRNA. Extracts from rabbit reticulocytes or wheat germ are used in the presence of an RNA polymerase. Depending on the system, the templates can be linear, circular, or PCR-derived DNA.

3.1.3. Microsomal membrane sources

Microsomes from canine pancreas are the most widely used preparation of endoplasmic reticulum membranes used for co-translational processing of membrane proteins [80]. These preparations can be obtained commercially from the sources listed above. ER microsomes derived from avian oviduct are also efficient at co-translational processing of nascent proteins [81]. Because this membrane preparation is not as widely used as the commercially available microsomal membranes from dog pancreas, details of this preparation are provided here.

The magnum portion of the oviduct of fully mature, laying hens or quail is minced finely and suspended in 5 volumes of 50 mM Tris-HCl, pH 7.7, 25 mM KCl, 2.5 mM MgCl$_2$, and 3 mM dithiothreitol (TMKD buffer) plus 0.88 M sucrose. All procedures are carried out at 4°C. The tissue is homogenized with 4-6 strokes in a motorized teflon-glass homogenizer and 4-6 strokes with a tight-fitting dounce homogenizer. The homogenate is centrifuged at 4000 x *g* for 10 minutes. The supernatant is diluted with TMKD to a final sucrose concentration of 0.62 M and 20-25 ml is layered onto a step gradient of 1.5 M (10 ml) and 2.0 M (5 ml) sucrose in TMKD. The gradient is centrifuged for 16 hours at 100,000 x *g* in an SW28 rotor. The membranes at the interface between the 1.5 and 2 M sucrose layers are collected, diluted with 10 volumes of Mg^{2+}-free (TKD) buffer and centrifuged for 1 hour at 55,000 x *g* (Ti45 rotor). The pelleted microsomes are resuspended by gentle trituration with a Pasteur pipette and gentle homogenization with a dounce homogenizer in 20 mM HEPES, pH 7.5. The resuspended microsomes are diluted with 20 mM HEPES to an A$_{260}$ of 60, and then 0.5 volumes of 750 mM sucrose, 20 mM HEPES, pH 7.5 is added. Membranes are flash-frozen in liquid nitrogen and stored at –80°C.

3.1.4. Co-translational membrane processing

Membrane processing in eukaryotes occurs co-translationally [16]. ER microsomal membranes contain signal recognition particles that bind the signal sequence, temporarily stop translation, and dock the ribosome to the ER membranes. Glycosylation, signal sequence cleavage, and transmembrane processing also occur during translation. For *in vitro* translation of membrane proteins, ER microsomal membranes are included in the *in vitro* translation or coupled transcription/translation reaction during translation.

The presence of oxidizing conditions in the lumen of ER-derived microsomes may be critical for the formation of disulfide bonds and protein maturation during the *in vitro* translation of some membrane proteins [82]. For example, the proper folding of connexon proteins for subsequent functional studies in bilayers required oxidized glutathione during *in vitro* translation [83]. The presence of glutathione was also shown to be important for the maturation of nAChR subunits *in vitro* [84] and was used during cell-free synthesis of nAChRs for functional reconstitution [85].

After translation is terminated, it is essential to separate the ER microsomes from the lysate because use of the entire lysate mixture during reconstitution causes bilayer instability and disruption. The microsomes can be pelleted from the translation mixture by gentle centrifugation (15 min, 9,000 rpm) in a microfuge [85] or by centrifugation through a 250-500 mM sucrose cushion in an appropriate buffer [83]. The pelleted microsomes can then be resuspended in relatively high concentrations of permeable salt (i.e. KCl for K^+ channels) to help stabilize the ion channel [36]. To aid in the direct application of the ER microsomes to planar bilayers, we resuspend ER microsomes in a high concentration of sucrose buffer (~250-300 mM) [85]. The sedimented membranes were allowed to stand in the resuspension buffer for two hours at 4°C prior to gentle resuspension by trituration with a yellow-tip pipette [32].

Alternatively, Awayda *et al.* [86] report that the nascent proteins can be purified from the microsomes by a variety of techniques, reconstituted into proteoliposomes, and then fused with lipid bilayers. Miles *et al.* [87] report that leukocidins from *Staphlococcus aureus* can be eluted from rehydrated SDS-PAGE gel slices and incorporated into planar lipid bilayers.

3.1.5. Reconstitution of microsomal membranes into planar lipid bilayers

Several methods have been used to reconstitute *in vitro* expressed proteins into planar lipid bilayers. Rosenberg and East [32] used the traditional approach of adding microsomal vesicles to the one of the bilayer chambers, stirring occasionally, and waiting for spontaneous fusion to occur between the microsomes and the planar lipid bilayer. The presence of ~0.5 mM divalent ions is often included to promote membrane fusion. Another approach is to apply the vesicles directly onto one face of the bilayer as described above (Section 2.1.3). This method has the advantage of requiring a much smaller volume of vesicles that can be used repeatedly with several bilayers, rather than committing the entire translation reaction to one bilayer experiment. Microsomes can also be incorporated into lipid bilayers formed at the tip of a microelectrode (the dip-tip technique; Ref. 46) either by filling the electrode with solutions containing the microsomes or by dipping the electrode into a microsome-containing solution [83].

3.1.6. Endogenous ion channels in ER microsomal membranes

Occasionally, ion channels endogenous to ER microsomal membranes from canine pancreas have been observed in planar lipid bilayer reconstitution experiments. Channels with unitary conductances of 11.5 pS, 28 pS, and 105 pS (in 200 mM KCl) and undetermined ion selectivity [83] have been reported. We have observed cation channels from canine microsomes with a wide variety of unitary conductances, ranging from 13-250 pS (unpublished data). One chloride channel (140 pS) and one nonselective channel (32-49 pS) were also observed.

In quail oviduct microsomes, Cl⁻ channels were frequently observed in reconstitution experiments. Both Ca^{2+}-dependent and Ca^{2+}-independent Cl⁻ channels were observed, with mean conductances of ~110 pS and ~80 pS, respectively, in 300 mM KCl (*cis*) and 50 mM KCl (*trans*) (unpublished data). These channels displayed high open probability and many subconductance states. Similar Cl⁻ channel activity was observed in one of 52 trials studying the cell-free expression of the *Shaker* K⁺ channel [32], and in many experiments with *in vitro* expressed α7 nAChRs [85]. A few cation channels have also been observed in quail oviduct microsomes, with conductances ranging from 20-38 pS and >100 pS [85].

3.2. Results from *in vitro* expressed ion channels

3.2.1. Shaker K⁺ channel

Delayed rectifier K⁺ channels of the *Shaker* subtype (Kv1.1) are voltage gated, K⁺-selective ion channels that are important in setting action potential frequency [1]. The wild-type channels inactivate rapidly after activation at depolarized potentials. The channels are formed as a tetramer of subunits [88], each with six transmembrane domains and a pore-lining re-entrant loop [1,66].

Rosenberg and East [32] were the first to combine *in vitro* expression and bilayer reconstitution to study the functional properties of an ion channel. *Shaker* channels were chosen for this first effort because of their biological significance, distinctive voltage-dependent activation, K⁺ selectivity, homomeric structure (requiring the translation of a single cDNA), and modestly complex transmembrane architecture and oligomerization requirements. To overcome the difficulties associated with rapid inactivation, Rosenberg and East [32] used a *Shaker* mutant in which amino acids in the N-terminal inactivation domain were removed (ShIR, similar to ShΔ6-46; Ref. 89). These channels activate normally but have very slow inactivation kinetics.

To determine if the full-length protein could be synthesized and glycosylated *in vitro*, ShIR cRNA was translated with rabbit reticulocyte lysates in the presence and absence of avian oviduct microsomes. In the absence of the oviduct microsomes, a protein of approximately 70 kDa was translated [32]. This was the expected molecular mass, because approximately 5 kDa was deleted from the N-terminus of the 75 kDa *Shaker* sequence. When translation mixtures included oviduct microsomes, a 77 kDa protein was produced in addition to the 70 kDa polypeptide. Membranes pelleted by centrifugation from translation mixtures contained most of the 77 kDa protein. This larger polypeptide was not found in the supernatants. Treatment of the 77 kDa protein with endoglycosidase H shifted the molecular mass back to 70 kDa, indicating that the difference was due to the glycosylation.

To determine if the products of protein translation were functional, the oviduct microsomes containing the 77 kDa protein were incorporated into planar lipid bilayers [32]. Channel activation required depolarizations from a holding potential of −80 or −100 mV to >0 mV, as expected for this voltage-gated channel [89]. The voltages of the bilayers were assigned as *cis* relative to *trans*, and the voltage-dependence of channel activation indicated that the *cis* chamber represented the intracellular solution. This was the expected orientation of channels incorporated by fusion from the inside-out ER microsomes. Channel activation was rapid, and the first opening event usually occurred within 10-20 ms. In some of the recordings at +20 mV the channel remained open during the entire depolarization, but closed immediately upon repolarization, as evidenced by the absence of detectable "tail" currents after repolarization. This indicated that channel deactivation was also faster than 10-20 ms. At +40 and +60 mV, the channel showed a tendency to slowly enter an inactivated state before the end of the depolarization, similar to "C-type" inactivation of the channels [90,91]. The time constant for inactivation was approximately 600 ms, a value similar to those seen when ShBΔ6-46 channels were expressed in *Xenopus* oocytes.

The channels had a single-channel conductance of 14 pS in 50 mM (external) // 350 mM (internal) K^+. This was very close to the 13 pS conductance of ShIR K^+ channels recorded in patches from *Xenopus* oocytes [92]. The reversal potential of current-voltage plots was -50 mV, a value identical to the equilibrium potential for K^+ in these experiments, indicating ideal selectivity for K^+ over Cl^-.

There were some difference between the reconstituted channels expressed *in vitro* and those recorded from patches of oocytes. The apparent mean open time of reconstituted channels was >100 ms, a value much greater than the ~4 ms recorded in oocytes [90]. This could be due to the 200 Hz low-pass filtering that made it impossible to resolve brief closing events (<1 ms), effects of bilayer lipids, incomplete carbohydrate processing, or other differences between oviduct microsomes and *Xenopus* oocytes. The reconstituted channels appeared to have only a 10-fold selectivity for K^+ over Na^+, a value lower than that determined for *Shaker* channels (P_K/P_{Na}~65). However, the selectivity ratio determined from bilayer experiments was likely to be an underestimate because the reversal potential was from an extrapolation of data at positive voltages.

Thus, these results showed that oviduct ER microsomes can promote the transmembrane folding and oligomerization of complicated membrane proteins to produce functional ion channels. The results also indicated that simple high-mannose carbohydrate is sufficient for most voltage-gated K^+ channel behavior, and that β subunits and auxiliary proteins are probably not required.

3.2.2. $\alpha7$ nAChRs

The $\alpha7$ nicotinic receptor (described in section 2.2.1) is another homomeric ion channel, formed from five subunits each with four transmembrane domains. To study the $\alpha7$ nAChR expressed *in vitro*, we used the $S^{240}T/L^{247}T$ mutant because of its non-desensitizing property [85].

The $S^{240}T/L^{247}T$ $\alpha7$ nAChRs were synthesized *in vitro* using a coupled transcription/translation system that is derived from rabbit reticulocyte lysates. Reactions were supplemented with DTT and oxidized glutathione to promote an oxidizing lumenal environment necessary for proper folding and assembly [84,93]. Microsomes from quail oviducts or canine pancreas were used for biochemical analysis, but channel activity was observed only when quail oviduct microsomes were used for functional analysis. The apparent requirement for quail ER microsomes may be due to the enhanced glycosylation function of these membranes [81], or the possible presence of foldases or chaperonins that are necessary for nicotinic receptor maturation [94]. The difference was not due to the level of membrane-processed protein, because the amount of protein produced was actually lower in quail membranes than in canine pancreatic microsomes.

The $S^{240}T/L^{247}T$ $\alpha7$ nAChRs synthesized in the absence of microsomes had a molecular weight of 41 kDa. When microsomes were included in the *in vitro* transcription/translation reaction, proteins of 41 and 51 kDa were observed. The 41 kDa product was found the supernatant and most likely represents an unprocessed form of the protein. The 51 kDa form was associated with the microsomal membranes and contained high-mannose oligosaccharides, as determined by endoglycosidase H sensitivity. Proteinase K digestion of intact microsomes revealed a 36 kDa membrane-protected fragment. This size was consistent with the molecular weight of the glycosylated, extracellular N-terminal domain predicted by most topological models [1]. Thus, the data showed that the extracellular domain of the receptors was on the lumenal side of the ER microsomes, as expected. The membrane-associated protein was integrally processed through the membrane, not associated extrinsically [85].

Sucrose gradient analysis of translated protein revealed a 3.6 S peak in the absence of microsomes and an ~11 S peak in their presence. The 3.6 S peak corresponded to the expected size of the nAChR monomer [95]. The 11 S peak was similar to the sedimentation values of *Torpedo* and $\alpha 7$ nAChR pentamers [96,97]. Thus, sedimentation analysis indicated that the $\alpha 7$ subunits were oligomerized in the presence of ER microsomes.

The functional properties of the $S^{240}T/L^{247}T$ $\alpha 7$ nAChRs expressed *in vitro* were assessed in planar lipid bilayers [85]. Quail oviduct microsomes containing the nascent protein were applied directly to bilayers. The reconstituted channels had a unitary conductance of ~50 pS, were highly selective for cations over anions, and were not selective between K^+ and Na^+ (P_K/P_{Na} of 1.2). These are the functional properties that are common among many nAChR subtypes. In addition, channel activity was blocked by α-bungarotoxin, a peptide neurotoxin that blocks $\alpha 7$ nAChRs [98]. These experiments showed that functional ligand-gated ion channels, in addition to voltage-gated ion channels, could be synthesized *in vitro*.

3.2.3. ENaC

Numerous studies by the Benos group have used the *in vitro* translation/reconstitution approach to characterize epithelial sodium channels. Cell-free expression with rabbit reticulocyte lysates in the presence of canine pancreatic microsomes was used to study cloned ENaC α subunits alone [86] and α, β, and γ subunits in combination [99]. A wide variety of modulators have been tested, including hydrostatic pressure, protein kinases, intracellular Ca^{2+}, and actin [70].

3.2.4. Connexons

Gap junctions are intercellular channels that permit ionic and small molecule flow between two cells. They are formed from a pair of connexons, one from each cell membrane. The connexons are comprised of six subunits, called connexins. Each connexin subunit has four transmembrane-spanning segments, with intracellular N- and C-termini [100].

The cell-free expression system was used to synthesize functional connexon hexamers for reconstitution in planar lipid bilayers [83]. Connexins were translated in rabbit reticulocyte lysates from transcribed cRNA in the presence of canine pancreatic microsomes. Oxidized glutathione was included to promote appropriate folding. Wild-type connexins are aberrantly cleaved by canine pancreatic microsomes at a site following the first transmembrane domain. To prevent this proteolytic cleavage and produce full-length protein, the cleavage site was mutated to an N-linked glycosylation site. Carbohydrate added during membrane processing was then removed with endoglycosidase H.

The parts of the connexins responsible for co-assembly were explored using the *in vitro* translation technique [83]. When full-length $\alpha 1$ and $\beta 1$ were co-translated in the same reaction, co-immunoprecipitation analysis showed that they did not co-assemble into heteromeric complexes, but remained segregated in separate connexon hexamers. On the other hand, wild-type connexin subunits lacking the N-terminus and first transmembrane domains due to proteolytic cleavage assembled indiscriminately. Thus, the N-terminal and/or the first transmembrane domain were critical for subunit segregation during assembly. Deletion of the C-terminal domain had no effect.

Connexons synthesized *in vitro* were reconstituted into bilayers formed on the tip of a patch pipette [83]. Microsomes were added to the intra-pipette solution. Reconstituted wild-type connexons had unitary conductances of 52 pS (β2) and 170 pS (β1), similar to the conductances reported for baculovirus-expressed β2 connexons [8] and reconstituted β1 connexons purified from rat liver [101].

These results showed that functional, gap junction connexons can be synthesized *in vitro* and can assemble into hexameric structures in ER membranes. The results also suggest that the N-terminus and first transmembrane domains are not essential for ion channel properties.

4. CONCLUSIONS

This chapter has discussed techniques used to express and reconstitute ion channels using *in ovo* and *in vitro* methods. The advantages and disadvantages of these approaches were described. In addition, we provided several examples of channels expressed *in ovo* and *in vitro*, and highlighted some of the results obtained.

Future efforts will likely benefit from an improved understanding of the requirements for efficient membrane processing, folding, assembly, and post-translational modifications of ion channels. In addition, advances in electronic design and single-channel analysis could improve the resolution of small and/or rapid channel events. The detailed characterization of ion channel function under highly controlled experimental conditions, combined with the site-directed mutagenesis and an explosion of structural information from x-ray crystallography, will continue to increase our understanding of how ion channels work.

REFERENCES

[1]   B. Hille (ed.), Ion Channels of Excitable Membranes, Sinauer Associates, Inc., Sunderland, MA, 2001.
[2]   B. Sakmann, Neuron, 8 (1992) 613.
[3]   E. Neher, Neuron, 8 (1992) 605.
[4]   I. Favre, Y.M. Sun, and E. Moczydlowski, Methods Enzymol., 294 (1999) 287.
[5]   C.Miller (ed.), Ion channel reconstitution, Plenus Press, New York, N.Y., 1986.
[6]   C.E. Bear, C.H. Li, N. Kartner, R.J. Bridges, T.J. Jensen, M. Ramjeesingh, and J.R. Riordan, Cell, 68 (1992) 809.

[7]   C.R. O'Riordan, A. Erickson, C. Bear, C. Li, P. Manavalan, K.X. Wang, J. Marshall, R.K. Scheule, J.M. McPherson, and S.H. Cheng, J.Biol.Chem., 270 (1995) 17033.

[8]   L.K. Buehler, K.A. Stauffer, N.B. Gilula, and N.M. Kumar, Biophys.J., 68 (1995) 1767.

[9]   U.S. Rao, R.E. Steimle, and P. Balachandran, J.Biol.Chem., 277 (2002) 4900.

[10] B.C. Tilly, M.C. Winter, L.S. Ostedgaard, C. O'Riordan, A.E. Smith, and M.J. Welsh, J.Biol.Chem., 267 (1992) 9470.

[11] P. Huang, K. Stroffekova, J. Cuppoletti, S.K. Mahanty, and G.A. Scarborough, Biochim.Biophys.Acta, 1281 (1996) 80.

[12] B. Jovov, I.I. Ismailov, B.K. Berdiev, C.M. Fuller, E.J. Sorscher, J.R. Dedman, M.A. Kaetzel, and D.J. Benos, J.Biol.Chem., 270 (1995) 29194.

[13] D.A. Sullivan, M.H. Holmqvist, and I.B. Levitan, J.Neurophysiol., 78 (1997) 2937.

[14] R.H. Spencer, Y. Sokolov, H. Li, B. Takenaka, A.J. Milici, J. Aiyar, A. Nguyen, H. Park, B.K. Jap, J.E. Hall, G.A. Gutman, and K.G. Chandy, J.Biol.Chem., 272 (1997) 2389.

[15] E. Kaznacheyeva, V.D. Lupu, and I. Bezprozvanny, J.Gen.Physiol, 111 (1998) 847.

[16] R.J. Keenan, D.M. Freymann, R.M. Stroud, and P. Walter, Annu.Rev.Biochem., 70 (2001) 755.

[17] R.L. Rosenberg, P. Hess, J.P. Reeves, H. Smilowitz, and R.W. Tsien, Science, 231 (1986) 1564.

[18] J.E. Bell and C. Miller, Biophys. J., 45 (1984) 279.

[19] F.J. Barrantes, FASEB J., 7 (1993) 1460.

[20] J. Chen, J.H. Capdevila, D.C. Zeldin, and R.L. Rosenberg, Mol.Pharmacol., 55 (1999) 288.

[21] M.Montal, R.Anholt, and P.Labarca, The reconstituted acetylcholine receptor, in C. Miller (ed.) Ion Channel Reconstitution, Plenum Press, New York, N.Y., 1986, pp. 157-204.

[22] R.L. Rosenberg, S.A. Tomiko, and W.S. Agnew, Proc.Natl.Acad.Sci.U.S.A, 81 (1984) 5594.

[23] R.P. Hartshorne, B.U. Keller, J.A. Talvenheimo, W.A. Catterall, and M. Montal, Proc.Natl.Acad.Sci.U.S.A, 82 (1985) 240.

[24] S. Chung, P.H. Reinhart, B.L. Martin, D. Brautigan, and I.B. Levitan, Science, 253 (1991) 560.

[25] P.H. Reinhart and I.B. Levitan, J.Neurosci., 15 (1995) 4572.

[26] Y. Wang, C. Townsend, and R.L. Rosenberg, Am.J.Physiol, 264 (1993) C1473.

[27] Y. Imoto, A. Yatani, J.P. Reeves, J. Codina, L. Birnbaumer, and A.M. Brown, Am.J.Physiol, 255 (1988) H722.

[28] J.A. Haack and R.L. Rosenberg, Biophys J., 66 (1994) 1051.

[29] A. Cavalie, D. Pelzer, and W. Trautwein, Pflugers Arch., 406 (1986) 241.

[30] M.E. Barish, J.Physiol., 342 (1983) 309.

[31] Z. Qu and H.C. Hartzell, J.Gen.Physiol, 116 (2000) 825.

[32] R.L. Rosenberg and J.E. East, Nature, 360 (1992) 166.

[33] A.L. Goldin, Methods Enzymol., 207 (1992) 266.

[34] W. Stühmer, Methods Enzymol., 207 (1992) 319.

[35] W. Stühmer, Methods Enzymol., 293 (1998) 280.

[36] T.M. Shih, R.D. Smith, L. Toro, and A.L. Goldin, Methods Enzymol., 293 (1998) 529.

[37] G. Perez, A. Lagrutta, J.P. Adelman, and L. Toro, Biophys.J., 66 (1994) 1022.

[38] K. Geering, I. Theulaz, F. Verrey, M.T. Häuptle, and B.C. Rossier, Am.J.Physiol.Cell Physiol., 257 (1989) C851.

[39] A.M. Sherry, J. Cuppoletti, and D.H. Malinowska, Am.J.Physiol.Cell Physiol., 266 (1994) C870.
[40] W.H. Kinsey, G.L. Decker, and W.J. Lennarz, J.Cell Biol., 87 (1980) 248.
[41] G. Bretzel, J. Janeczek, J. Born, M. John, and H. Tiedemann, Roux's Arch.Devel.Biol., 195 (1986) 117.
[42] A. Aleksandrov, B. Velimirovic, and D.E. Clapham, Biophys.J., 70 (1996) 2680.
[43] M. Montal and P. Mueller, Proc.Natl.Acad.Sci.U.S.A, 69 (1972) 3561.
[44] O.Alvarez How to set up a bilayer system, in C. Miller (ed.) Ion Channel Reconstitution, Plenum, New York, NY, 1986, pp 115-130.
[45] R.A. Brutyan, C. DeMaria, and A.L. Harris, Biochim Biophys Acta, 1236 (1995) 339.
[46] R. Coronado and R. Latorre, Biophys J., 43 (1983) 231.
[47] H.T. Tien and A.L. Diana, Chem.Phys.Lipids, 2 (1968) 55.
[48] F.S.Cohen Fusion of liposomes to planar bilayers, in C. Miller (ed.) Ion Channel Reconstitution, Plenum, New York, NY, 1986, pp 131-140.
[49] W.Hanke Incorporation of ion channels by fusion, in C. Miller (ed.) Ion Channel Reconstitution, Plenum, New York, NY, 1986, pp 141-153.
[50] H. Schindler and U. Quast, Proc.Natl.Acad.Sci.U.S.A, 77 (1980) 3052.
[51] L.K. Lyford, J.W. Lee, and R.L. Rosenberg, Biochim.Biophys.Acta, 1559 (2002) 69.
[52] K. Machaca and H.C. Hartzell, Biophys.J., 74 (1998) 1286.
[53] T. Takahashi, E. Neher, and B. Sakmann, Proc.Natl.Acad.Sci.U.S.A, 84 (1987) 5063.
[54] L.W. Role and D.K. Berg, Neuron, 16 (1996) 1077.
[55] S. Jones, S. Sudweeks, and J. Yakel, Trends Neurosci., 22 (1999) 555.
[56] K. Brejc, W.J. van Dijk, R.V. Klaassen, M. Schuurmans, J. van der Oost, A.B. Smit, and T.K. Sixma, Nature, 411 (2001) 269.
[57] N. Unwin, J.Struct.Biol., 121 (1998) 181.
[58] C.R. Yu and L.W. Role, J.Physiol., 509 (1998) 651.
[59] S. Vernino, M. Amador, C.W. Luetje, J.W. Patrick, and J.A. Dani, Neuron, 8 (1992) 127.
[60] Z. Shao and J. Yakel, J.Physiol., 527.3 (2000) 507.
[61] R.W. Tsien, P. Hess, E.W. McCleskey, and R.L. Rosenberg, Annu.Rev.Biophys Biophys Chem., 16 (1987) 265.
[62] C. Vergara, R. Latorre, N.V. Marrion, and J.P. Adelman, Curr.Opin.Neurobiol., 8 (1998) 321.
[63] O.B. McManus and K.L. Magleby, J.Physiol, 443 (1991) 739.
[64] L. Toro, L. Vaca, and E. Stefani, Am.J.Physiol, 260 (1991) H1779.
[65] M. Wallner, P. Meera, M. Ottolia, G.J. Kaczorowski, R. Latorre, M.L. Garcia, E. Stefani, and L. Toro, Receptors.Channels, 3 (1995) 185.
[66] D.A. Doyle, C.J. Morais, R.A. Pfuetzner, A. Kuo, J.M. Gulbis, S.L. Cohen, B.T. Chait, and R. MacKinnon, Science, 280 (1998) 69.
[67] C.A. Vandenberg, Proc.Natl.Acad.Sci.U.S.A, 84 (1987) 2560.
[68] E. Ficker, M. Taglialatela, B.A. Wible, C.M. Henley, and A.M. Brown, Science, 266 (1994) 1068.
[69] d.l.R. Alvarez, C.M. Canessa, G.K. Fyfe, and P. Zhang, Annu.Rev.Physiol, 62 (2000) 573.
[70] D.J. Benos and B.A. Stanton, J.Physiol, 520 Pt 3 (1999) 631.
[71] I.I. Ismailov, T. Kieber-Emmons, C. Lin, B.K. Berdiev, V.G. Shlyonsky, H.K. Patton, C.M. Fuller, R. Worrell, J.B. Zuckerman, W. Sun, D.C. Eaton, D.J. Benos, and T.R. Kleyman, J.Biol.Chem., 272 (1997) 21075.

[72] B.K. Berdiev, T.B. Mapstone, J.M. Markert, G.Y. Gillespie, J. Lockhart, C.M. Fuller, and D.J. Benos, J.Biol.Chem., 276 (2001) 38755.

[73] E.M. Schwiebert, D.J. Benos, M.E. Egan, M.J. Stutts, and W.B. Guggino, Physiol Rev., 79 (1999) S145-S166.

[74] R. Dutzler, E.B. Campbell, M. Cadene, B.T. Chait, and R. MacKinnon, Nature, 415 (2002) 287.

[75] A.M. Sherry, K. Stroffekova, L.M. Knapp, E.Y. Kupert, J. Cuppoletti, and D.H. Malinowska, Am.J.Physiol, 273 (1997) C384.

[76] K. Stroffekova, E.Y. Kupert, D.H. Malinowska, and J. Cuppoletti, Am.J.Physiol, 275 (1998) C1113.

[77] J. Sambrook (ed.), Molecular Cloning: A Laboratory Manual, Cold Spring Harbor Laboratory Press, Cold Spring Harbor, NY, 2001.

[78] H.R. Pelham and R.J. Jackson, Eur.J.Biochem., 67 (1976) 247.

[79] Promega protocol and applications guide, Madison WI, 1996.

[80] P. Walter and G. Blobel, Methods Enzymol., 96 (1983) 84.

[81] R.C. Das, S.A. Brinkley, and E.C. Heath, J.Biol.Chem., 255 (1980) 7933.

[82] C. Hwang, A.J. Sinskey, and H.F. Lodish, Science, 257 (1992) 1496.

[83] M.M. Falk, L.K. Buehler, N.M. Kumar, and N.B. Gilula, EMBO J., 16 (1997) 2703.

[84] S.S. Shtrom and Z.W. Hall, J.Biol.Chem., 271 (1996) 25506.

[85] L.K. Lyford and R.L. Rosenberg, J.Biol.Chem., 274 (1999) 25675.

[86] M.S. Awayda, I.I. Ismailov, B.K. Berdiev, and D.J. Benos, Am.J.Physiol.(Cell Physiol.), 268 (1995) C1450.

[87] G. Miles, S. Cheley, O. Braha, and H. Bayley, Biochemistry, 40 (2001) 8514.

[88] R. MacKinnon, Nature, 350 (1991) 232.

[89] T. Hoshi, W.N. Zagotta, and R.W. Aldrich, Science, 250 (1990) 533.

[90] T. Hoshi, W.N. Zagotta, and R.W. Aldrich, Neuron, 7 (1991) 547.

[91] K.L. Choi, R.W. Aldrich, and G. Yellen, Proc.Natl.Acad.Sci.U.S.A, 88 (1991) 5092.

[92] G. Yellen, M.E. Jurman, T. Abramson, and R. MacKinnon, Science, 251 (1991) 939.

[93] M.S. Gelman and J.M. Prives, J.Biol.Chem., 271 (1996) 10709.

[94] S.A. Helekar, D. Char, S. Neff, and J.W. Patrick, Neuron, 12 (1994) 179.

[95] H.L. Paulson, A.F. Ross, W.N. Green, and T. Claudio, J.Cell Biol., 113 (1991) 1371.

[96] J.A. Reynolds and A. Karlin, Biochemistry, 17 (1978) 2035.

[97] R. Anand, X. Peng, and J. Lindstrom, FEBS Lett., 327 (1993) 241.

[98] S. Couturier, D. Bertrand, J.-M. Matter, M.-C. Hernandez, S. Bertrand, N. Millar, S. Valera, T. Barkas, and M. Ballivet, Neuron, 5 (1990) 847.

[99] M.S. Awayda, I.I. Ismailov, B.K. Berdiev, C.M. Fuller, and D.J. Benos, J.Gen.Physiol, 108 (1996) 49.

[100] D.A. Goodenough, J.A. Goliger, and D.L. Paul, Annu.Rev.Biochem., 65 (1996) 475.

[101] S.K. Rhee, C.G. Bevans, and A.L. Harris, Biochemistry, 35 (1996) 9212.

Planar Lipid Bilayers (BLMs) and their Applications
H.T. Tien and A. Ottova-Leitmannova (Editors)

Chapter 14

MULTI-CHANNEL AND SINGLE-CHANNEL INVESTIGATION OF PROTEIN AND PEPTIDE INCORPORATION INTO BLM

E. Gallucci, S. Micelli and V. Picciarelli *

Dept Farmaco-biologico and * Dept. Interateneo di Fisica, Università degli Studi di Bari, I- 70125 Bari, Italy

1. INTRODUCTION

Investigating the mechanisms of how proteins/peptides are incorporated into lipid membranes may be of relevance in understanding how proteins/peptides and lipids interact and how the lipid composition modulates the physicochemical properties of the membranes. The information gathered can be transferred :

• to other biological phenomena such as signal transduction or membrane protein traffic [1], the receptor targeting mitochondria [2-3], or membrane dysfunction due to improper protein incorporation [4,5]

• to pharmacological/clinical applications as in the case for example of calcitonin /magainin peptides [6,7]

• to the technological development of biosensors [8].

One way to perform such a study is to use reconstitution systems such as lipid vesicles, planar lipid membranes (PLM's) or black lipid membranes (BLM's). The former have been used mainly for physico-chemical studies and for the process of insertion into bilayers, whereas PLM's or BLM's have been used to investigate phenomena associated with channel incorporation and their relation with various parameters such as voltage, ionic strength, protein/peptide concentration, pH, temperature, medium composition and membrane composition.

In particular, as is well known, BLM's used as model membranes have been characterized by using two electrical parameters, conductance and capacitance, connected in parallel. The first is related to the ionic current passing through the bilayer, while the second is associated with the gating current due to charge rearrangements in the membranes when voltage-dependent ion channels have been incorporated [9,10].

These two electrical parameters have usually been studied independently.

There have been various experiments, based on different measurement techniques, using conductance variation to study channel insertion and electrical properties under various conditions. The principal methods used are as follows:

i) the relaxation type. For example, gramicidin insertion into lipid bilayer membranes was investigated by means of a relaxation experiment in which the rise-time due to a voltage jump was measured, showing that the initial and stationary current ratio was related to the increase in the number of conducting channels inserted after the voltage jump [11]

ii) the I-V type . In this case the voltage (the current) is clamped to a given value and the current (the voltage) reflects the change in membrane conductance. For example, recently there has been discussion regarding data obtained via the voltage clamp method of conductivity during porin insertion into phospholipid membranes under urea perfusion [12]. With a similar voltage clamp method, porin insertion into an asymmetric membrane (made up of glycospingolipid or lipopolysaccharide on one side and phospholipids on the other) was studied [13];

iii) the noise analysis type [14-18] based on Fourier analysis and autocorrelation function investigations.

While conductance, especially when related to incorporation of different proteins, has been widely investigated due to its evident relation to channel formation, there are less data on capacitance, even though this would provide important new information on functional properties at a molecular level.

Capacitance measurements, based on the discharge through a known resistor, have been mainly limited to black lipid membranes and used to gather information on thickness, dielectric properties, capacitance voltage dependence and capacitance dependence on the radius of the BLM [19-23].

New techniques for capacitance measurements have been developed and used to collect data on capacitance variation during membrane bilayer formation [24].

Interest in the simultaneous measurements of conductance/capacitance has been growing recently, and data have been collected on various characteristics of gating mechanisms as a function of the applied voltage in natural and artificial membranes such as auditory cells [25], the Shaker K^+ channel in Xenopus oocytes [26;27], vesicle fusion events occurring during exocytosis [28] and Alzheimer's disease's amyloid β-peptides [5]. Chanturya [29] and Chanturya and

Nikoloshina [30] separate the capacitive and the reactive component of the current by using a compensating capacitor. Such an approach was used for investigating the correlation between change in the membrane capacitance induced by changes in ionic environment and the conductance of channels incorporated into the membranes due to toxins.

From the previous discussion it could be of interest to investigate the two electric parameters characterizing the BLM with two different approaches :

• macroscopic investigations. In this case the apparatus for the measurements and the experimental conditions under which the data are collected (i.e. BLM dimensions, protein/peptide concentrations) prevent the identification of a "single act of incorporation" (i.e. rapid and significant variation in the electrical parameter measured). For this reason, the term macroscopic incorporation is used to describe an effect due to many "single acts of incorporation"

• single channel investigation . In this case, unlike the previous one, the apparatus' sensitivity and the experimental conditions are such that a rapid and significant variation in the variable under study is detectable as an effect of a single process of incorporation.

In the last few years we have been involved in the development and testing of two reliable and low cost apparatus:

• for simultaneously measuring conductance and capacitance during macroscopic protein/peptide incorporation,

• for investigating the single channel incorporation into BLM,

Using these apparatus, which will be discussed in the next section (Sect. 2), we have been able to collect data on, amongst others, the following topics :

• macroscopic porin incorporation as measured by the total current flowing through the BLM made up of phosphotydinositol charged membranes (PI) and oxidized cholesterol neutral membranes (OxCh). A mathematical model has been developed to interpret the kinetics of porin incorporation. Having identified two concurrent processes ascribed to positive/negative cooperativity, we investigated the dependence of the main parameters associated with these processes on the external applied potential (Sect. 3)

• simultaneous measurements of ionic and gating currents during the dynamic phase and at the steady state of porin incorporation into PI and OxCh BLM's. In particular we investigated (Sect. 4) the dependence of conductance and capacitance on the membrane applied potential at the steady state

• indirect evidence of porin's assembly properties once it is inserted into PI and OxCh BLM's. To obtain this information, a systematic investigation of the steady state conductance under different conditions (i.e. ionic strength and porin concentration) has been successfully undertaken (Sect. 5)

• channel formation and macroscopic incorporation of gramicidin A, human and salmon calcitonin and magainin peptides into different BLM . The macroscopic properties of the incorporation of these peptides, in particular gramicidin, have been investigated together with the evidence of single channel formation in the case of calcitonin and magainin. In particular we have studied the single channel conductance dependence on the membrane applied voltages. Moreover in the case of magainin we have also investigated the role played by the lipid bilayer's polar head in the channel formation process (Sect. 6)

During our systematic investigations, we were confronted with some "problems" which, far from being well understood, in our opinion, merit attention due to their potential interest in understanding the mechanism involved in the incorporation and channel formation into BLM's and biotechnological application. Sect. 7 will be devoted to some of these "problems".

The final section (Sect. 8) will summarize the main results obtained and briefly discuss their relation to the current literature in the field of protein/peptide interaction with model systems and their possible application.

2. THE SET-UP USED AND THE HOME-MADE APPARATUS FOR INVESTIGATING MACROSCOPIC AND SINGLE-CHANNEL INCORPORATION

Let us first briefly describe the cell used during the experiments and the method by which we were able to obtain BLM.

All the experiments which we will deal with were performed in a classical teflon chamber similar to that described by Läuger [31]. The volume of each compartment was 4 ml, and the circular aperture between the two compartments had a variable diameter ranging from 100 μm to 1300 μm depending on the experiment.

The compartments were filled with a salt solution of different composition and concentration and mechanically stirred.

The BLM were formed by brushing various lipid solutions across the circular aperture.

In our investigations into macroscopic incorporation, we developed an ac home-made apparatus which would simultaneously measure the two electrical parameters characterizing the BLM: conductance and capacitance

The advantages of using an ac technique are twofold: 1) any complications arising from polarization effects are avoided, 2) it allows the signals of two appropriate frequencies to be mixed, thus enabling both of the electrical parameters (i.e. resistance and capacitance) to be measured simultaneously

In Fig. 1A we report a schematic set-up of the electronic apparatus used for the measurements.

To measure the two quantities simultaneously, we mixed two frequencies, 1 Hz variable V_{pp} and 1 kHz ($2mV_{pp}$). The signal V_s was used as input voltage to a Pt

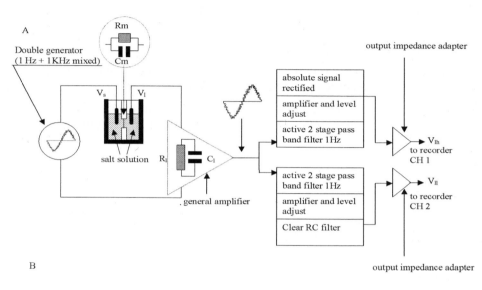

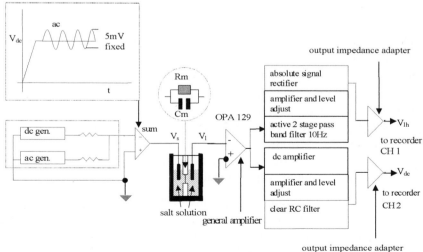

Fig. 1 Experimental setup for ac (A) and dc (B) measurements

electrode on one side of the cell : the output signal V_1 was acquired through a second Pt electrode placed on the other side of the cell.

V_1 was then used as input to the measuring instruments, which consist of a variable impedance input circuit and a passband amplifier, followed by a frequency splitter able to separate the 1 Hz output signal from the 1 kHz component. Once split, the 1 Hz component (V_{ll}) and the 1 kHz (V_{lh}) component were collected via a PC computer interfaced with two Voltage-frequency converters and stored for further analysis. The signal V_{ll} (V_{lh}) is mainly associated with the membrane conductance (capacitance).

A detailed discussion of the relation between the two measured quantities V_{ll} and V_{lh} and the significant electrical parameters is reported elsewhere [9].

The home-made device used for the single channel measurements, shown in Fig. 1B, consists of five stages, which are easily identifiable in the circuit diagram (not detailed) :

• a generator for dc and ac (10 kHz) voltages

• a sum amplifier

• a general amplifier

• an amplifier for measuring current

• circuitry for measuring capacitance

To be able to measure very weak currents, much attention has been paid to avoiding/minimizing electric fields and electrostatic charges. Therefore, to reduce noise the power was supplied by batteries, double screening was applied and the connections to the ground were all relative to the same junction mass.

Four low-pass filters for each frequency cut-off were used in order to optimize the cut-off frequency in relation to a given signal-to-noise ratio. Special attention was paid to stabilizing the voltages used for the off-set, by building the circuit with high stability and very low noise components, such as IC at "reference voltage". Also as a first amplifier a "ultralow bias current" type OPA 129, well screened and separated from the rest of the circuit, was used as an impedance adapter. Two further amplifiers, one for the current and the other for the capacity, measured the corresponding output signals.

In order to avoid the effect of parasitic capacitance, an off-set regulator is introduced. The capacitance which can be measured (resulting from the calibration curve obtained by using known capacitance) is in the range from 0 to 120 pF. It is clear that, as the parasitic capacitance is parallel to the membrane capacitance, the higher the parasitic capacitance the lower the sensitivity of the device to the effective membrane capacitance variation which can even get so low that it is no longer detectable.

3. MACROSCOPIC PORIN INCORPORATION

In this section we report on macroscopic porin incorporation as a function of the external applied voltage, and concentrate our experimental investigations on the kinetics part (i.e. the time dependence). As a probe of the incorporation process, we consider the total current passing through the membranes. We must point out that apart from a constant, the V_{II} signal essentially measures the total current passing through the membrane. This total current is made up of a resistive and a capacitive contribution of the membrane model. The first is related to the usual conductance measured by the dc technique , while the second one is peculiar to an ac device.

The above-mentioned investigations were performed with two kinds of BLM, namely PI and Ox Ch, which were chosen for their different physico-chemical characteristics; i) Ox Ch has a rigid hydrophobic structure with a small neutral polar head, while PI has a negative polar head; ii) moreover, from a functional point of view it must be considered that PI is present in the outer mitochondrial membrane [32] but at low concentrations, whereas sterols are fundamental components with which porin is associated [33-35].

Typical on-line data for the output V_{II} acquired with the bilayers having a hole diameter of 1300 μm and a porin concentration of 250 ng/ml for the PI membranes and 5 ng/ml for the OxCh membranes are reported in Fig. 2. These

a
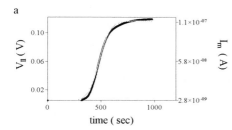

b
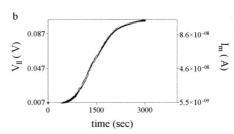

Fig.2 Example of on-line data for the output signal V_{II} acquired with the two kinds of bilayers: a) PI membrane, porin concentration 250 ng/ml; b) Ox Ch membrane, porin concentration 5 ng/ml. Experimental conditions were: $V_s = 0.12$ V, $R_l = 1$ MΩ and $C_l = 10$ nF, hole diameter φ = 1.3 mm, KCl = 1M, temperature 24 ± 0.5 °C. The curve superimposed on the data is the result of the fit with the model: $V_{vv}(t) = A + B/(1 - Ce^{-K_1 t})$ containing the four parameters (A, B, C, K_1). On the secondary ordinate axis the total current is indicated. The initial and the steady state conductance values for PI and (Ox Ch) membranes were: $1.1*10^{-8}$S , $2.6*10^{-6}$S and ($1.8*10^{-8}$S, $5*10^{-6}$S) respectively.

highly discrepant concentrations used are due to the different affinity of porin for the two membranes.

The behavior of the curve suggests some simple underlying incorporation mechanism. In fact:

1) after porin has been dissolved in both cell compartments, the total current increases continually. This means that porin is being incorporated and that channels are opening within the membrane. One can realistically hypothesize that this process depends (for fixed porin concentration, temperature and pH) on the nature of the protein and the bilayer, and is independent of the applied voltage V_s.

2) the growth rate (i.e. total current variation with respect to time), taking into account the cooperative nature (positive cooperativity) of the incorporation, should be proportional to the actual amount of porin already present within the membrane, which can be measured by the total current through the membrane (i.e. to the voltage V_{ll}) . This phenomenon has been already proposed by Zizi et al. [36] and Gallucci et al. [37]

3) a saturation value is reached for V_{ll}. In other words, porin incorporation proceeds until a sort of equilibrium condition is reached between the two competitive mechanisms (positive/negative cooperativity, where the latter means porin disappearing from the membrane). We wish to point out that this condition can also be observed by allowing free diffusion into the membrane and monitoring the process after a sufficiently long time. So we suppose that a sort of equilibrium is established between the three components of the system considered - solution, torus and membrane. But as reported by Brunen and Engelhardt [38], the exchange between membrane and aqueous phase at the steady state can be excluded. However, channel disappearance in micro lenses or in the membrane torus has been proposed by Nekolla et al. [39].

4) one simple assumption regarding the V_{ll} dependence of the negative cooperativity, which is compatible with the observed stationary state (i.e. allowing to reach a steady state) might be of the quadratic form.

Mathematically, the above-mentioned hypothesis can be expressed by the formula:

$$v(t) = dV_{ll} / dt = K_1 * V_{ll} - K_2 * V_{ll}^2 \tag{1}$$

where

K_1 = growth rate parameter of the total current related to the positive cooperativity

K_2 = growth rate parameter of the total current related to the negative cooperativity

By integrating this expression and imposing the boundary conditions $V_{ll}(-\infty) = A$ (i.e. the initial value) and $V_{ll}(+\infty) = A + B$ (i.e. the steady state V_{ll} value) we obtain the following formula for $V_{ll}(t)$:

$$V_{vv}(t) = A + B/\left(1 - Ce^{-K_1 t}\right) \qquad (2)$$

which contains four parameters (A, B, C, K_1) that are easily obtainable from the experimental data by a fitting procedure.

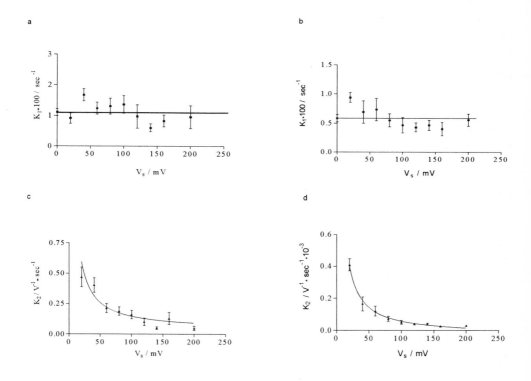

Fig.3 The most significant parameters obtained by the kinetics fitting procedure as a function of the applied external potential V_s for the PI (a,c) and OxCh (b,d) membrane. $-K_1$ is the growth rate parameter of the total current describing porin insertion (or positive cooperativity), the straight line corresponds to the mean values of K_1 at the different voltages applied. K_2 is the growth rate parameter related to negative cooperativity. The curve superimposed on the data is the result of the fit with the model: $K_2 = A/V_s + B$, where for PI membrane $A = 11.17 \pm 1.9$ (sec^{-1});$B = 0.036 \pm 0.0.34$ (V^{-1} · sec^{-1}); $R^2 = 0.97$, and for OxCh membrane $A = 8.58 \pm 0.27$ (sec^{-1}); $B = 0.028 \pm 0.0051$ (V^{-1} · sec^{-1}); $R^2 = 0.999$. Experimental conditions: porin was 250 ng/ml (5 ng /ml) for Pi (OxCh), $R_l = 1$ MΩ and $C_l = 10$ nF, hole diameter $\phi = 1.3$ mm, KCl = 1M, temperature 24 ± 0.5 °C

Moreover the derivative of (2) becomes zero (i.e. the system is in its steady state) for K_2 given by:

$$K_2 = K_1 / (A + B) \tag{3}$$

In other words, formula (2) allows the direct evaluation of K_1, while K_2 can be deduced from the three parameters K_1, A and B. Assuming that, as discussed before, K_1 could be independent of V_s and as A is usually quite small, the K_2 dependence could reflect the saturation value B's dependence on V_s (i.e. on the membrane potential V_m). For the latter, in the hypothesis of a voltage-gating mechanism, a decrease in K_2 due to pore closing is expected as V_s increases.

We must point out that this model in reality is a way to mathematically parameterize our data in order to allow a quantitative comparison of the presumed underlying mechanisms.

Data were collected for both the PI and Ox Ch bilayer membranes at different V_s values. The data obtained were fitted by using the Graphpad Prism 2 software with model formula (2). In general, good fits were obtained (see Fig. 2 for an example) in spite of the fact that we were attempting to reproduce by a single formula data relative to the whole growth process until the steady state was reached (i.e. not less than 250 data points).

From the best fit, the most significant parameters were extracted.

The results on the dependence of the two incorporation parameters K_1 and K_2 reported in Fig. 3 show that:

i) K_1 values are practically independent of the V_s applied, in spite of the fact that we investigated a broad range of values; from 20 to 200 mV for PI and Ox Ch respectively.

ii) on the other hand, K_2 values are, as expected, strongly dependent on V_s.

The data were fitted with a hyperbolic function of the type :

$$K_2 = a / V_s + b \tag{4}$$

Both these results are in agreement with the hypothesized underlying mechanisms of porin incorporation.

Evidence of the dynamic events between torus- and membrane-inserted porin have been reported elsewhere [9]

4. SIMULTANEOUS MEASUREMENTS OF IONIC AND GATING CURRENT DURING INCORPORATION AND AT THE STEADY STATE

Our apparatus is able to simultaneously investigate the ionic and gating current behavior during incorporation and at the steady state.

4.1 Ionic and gating currents during incorporation

The kinetics of porin incorporation were investigated with both bilayers (hole diameter of 1300 μm) and, owing to its relatively different affinity, a porin concentration ranging from 0.67 to 600 ng/ml (PI membranes) and from 0.3 to 62.7 ng/ml (Ox Ch membranes).

In Fig. 4, we report an example of the data obtained during the process of VDAC incorporation for two different porin concentrations.

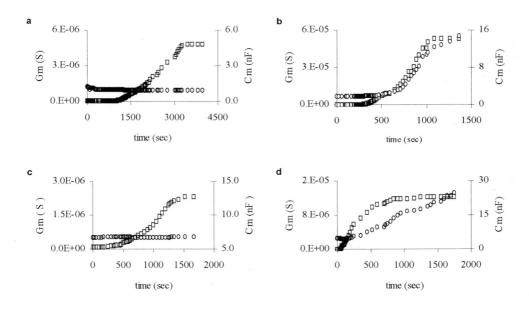

Fig.4 Time course of the membrane conductance (G_m , □) and capacitance (C_m , ○) during porin incorporation into bilayer lipid membranes. PI membranes, porin concentration 62.7 ng/ml (a) and 600 ng/ml (b). OxCh membranes, porin concentration 1.25 ng/ml (c) and 5 ng/ml (d). Electrical resistance and capacitance of the measuring circuit were R_l= 1MΩ and C_l =20 nF, hole diameter φ = 1300 μm, KCl = 1M, temperature 24± 0.5 °C . Source voltage (V_s) 80 mV.

In all cases, the time-course of the conductance presents an increase due to channel formation, and this increase shows a co-operative behavior with a clear evidence of steady-state level, which, as previously discussed, was explained through an equilibrium between channel incorporation and channel migration in the torus. These steady-state characteristics seem to be independent of the porin concentration and the bilayer used. Moreover the Ox Ch membrane shows a larger affinity for the porin, as the process of incorporation is more rapid even at lower porin concentrations.

During porin incorporation, when the porin concentration is greater than a threshold value, both capacitance and conductance increase; however when the conductance has reached a steady state, a slight increase in capacitance can still be observed (Fig.4 b,d), probably due to a delay in the conformational change of the protein.

From a detailed investigation into capacitance variations as a function of the porin concentration, it results that porin concentrations of about 250 ng/ml (about 12.5 ng/ml) are required to detect a moving charge process during porin incorporation into PI (Ox Ch) bilayers.

As the same behavior is shown by porin incorporation in the "solvent free" membranes of PI, we can assume that solvent leakage can be excluded as a primary cause of the capacitance increase observed. In 5 experiments with such membranes, an increase of $C_m = 114.4 \pm 40.1$ % is already observed at 250 ng/ml of porin concentration.

In order to evaluate if, in the above-mentioned conditions (i.e. high porin concentration), the capacitance variation is due to protein charge movements we performed experiments on PI membranes at a threshold concentration of porin (250 ng/ml) and at different external applied voltages.

It was observed that the main capacitance variation (C_{mf}), at the steady state of total current flowing through the membrane, is obtained at applied voltages of 20,40,60,80 and 100 mV . We believe that at higher applied voltages (V_s>100 mV), the potential on the membrane (V_m) is also high and will impede conformational variation of VDAC, responsible for the capacitance variation. However, further experiments at the current steady state of porin incorporation on voltage-dependent capacitance are reported below.

4.2 Conductance and capacitance dependence on the membrane applied voltage

As regards the dependence of conductance and capacitance on membrane applied voltage at the steady state we observe that :

• to study the gating mechanism, the number of inserted pores should be reduced so that a detailed analysis of the conductance versus membrane voltage (V_m) can be performed. In our voltage-gating investigations, we found a useful compromise by using a hole diameter of 900μm and a porin concentration of 0.67 (0.3) ng/ml for PI (Ox Ch).

In our approach to the measurements we first of all required that the insertion process had been completed (i.e. the steady state conditions for conductance had been reached); then V_s was switched to a maximum value (220 mV). A series of decreasing voltages V_s, each lasting 5 minutes, was applied to reach a very low voltage on the membrane (V_m). This time was considered to be long enough to allow the system to reach a new stable steady state. For each applied voltage V_s, we collected a series of stationary values V_{ll}, V_{lh}.

The data measured were finally used in the formula discussed in [9] to obtain the conductance (G_m) and membrane potential (V_m).

We are confident that with such an approach we were able to distinguish between channel insertion and channel opening/closing. The presence of steady states at each stage of the procedure can be considered indicative of a stable mean number of inserted channels, while the variations in conductance versus V_m can almost certainly be ascribed to a voltage-gating mechanism for the inserted channels.

We observe that for each V_s applied, there is a "new" steady state different from the previous one only in that the opening/closing state of the inserted channel has changed. Moreover if the above-mentioned procedure were applied before complete channel insertion had taken place (i.e. before reaching the steady state), the two phenomena (i.e. insertion and voltage-gating mechanism) probably overlap, thus making it difficult to interpret the voltage-gating mechanism. In fact in such a case it can be seen that the steady state is reached a long time after each voltage step, and only when the insertion process has been completed.

The relevance of reaching a steady state has been stressed by AA [39; 49].

In Fig. 5a-b we report examples of the data relative to the dependence of the conductance ratio ($G - G_c$) / ($G_o - G$), where G_o (G_c) correspond as usual to open (closed) states of conductance, on the V_m obtained, using the above method, with the two different bilayer membranes. . We observe that the data points are in both cases indicative of two gating processes with high (low) dependence in the range of low (high) membrane applied voltages.

A simple way to take into account the aforementioned gating behavior is to parameterize the conductance dependence by a dual-exponential law in agreement with the hypothesis that the pores switch from open to closed through an intermediate subsystem [34;41-44]. We point out that this approach merely represents a way to parameterize the data, since no direct evidence of the underlying processes are obtained in our experiments. The formula used is:

$$(G - G_c)/(G_o - G) = A * e^{-\alpha_1(V_m - V_0)} + (1 - A) * e^{-\alpha_2(V_m - V_0)}$$
(5)

where α_1 and α_2 are related to the gating charge n_1, n_2 where $n_{1,2} = \alpha_{1,2} RT/F$, R= 8.314 Joule / mole$_*$°K, F = 96320 Coulomb/mole and T = 298 ° K and V_0 indicates the potential at which the conductance ratio is equal to 1.

Parameter A allows us to estimate the contribution made by the two mechanisms.

In 7 (8) experiments for PI (Ox Ch) bilayer membranes using this fitting procedure, we obtained the following mean values: α_1 = 46.0 ± 3.1 (V^{-1}) and α_2 = 427.3 ± 25.4 (V^{-1}) ; n_1 = 1.36 ± 0.30; n_2 = 11.0 ± 0.6 (α_1 = 57.7 ± 8.5 (V^{-1}); α_2 = 560.5 ± 87.0 (V^{-1}); n_1 = 1.48 ± 0.22; n_2 = 14.38 ± 2.23). We observe that between the n_1 and n_2 values there is one order of magnitude of difference, independent of the bilayer membranes used.

In order to study the voltage-dependent capacitance of VDAC, the experimental conditions used and the measurement approach are the same as those previously reported for the conductance .

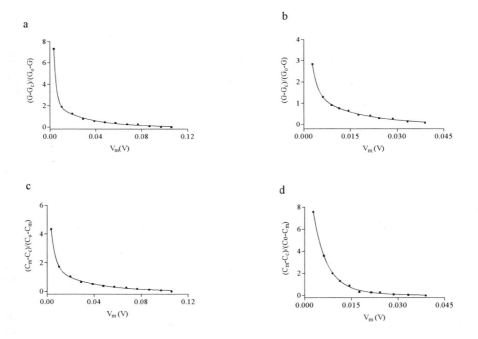

Fig.5) A typical example of the conductance/capacitance-dependence ratio on the effective potential V_m acting on the membranes. PI (a, c) and Ox Ch (b, d) membranes. Data were parameterized with two-exponential equations (5). Experimental conditions were: porin concentration 0.67 ng/ml (0.3 ng/ml) for PI (Ox Ch), KCl = 1M, R_1 = 1 MΩ, C_1 = 20 nF, temperature = 24 ± 0.5 °C.

The results relative to the capacitance ratio: $(C - C_c) / (C_o - C)$ (where C_o = the capacitance for the lower voltage value and C_c = the capacitance for the highest voltage value) are reported in Fig. 5 c-d and show a pattern similar to those obtained for the conductance measurements [8].

The data have been parameterized with two exponential forms (5).

The value obtained for the two exponential coefficients are $\alpha_1 = 64.7 \pm 17.5$ V^{-1} and $\alpha_2 = 753.3 \pm 174.3$ V^{-1} ($\alpha_1 = 118.1 \pm 37.3$ V^{-1} and $\alpha_2 = 549.4 \pm 350.8$ V^{-1}) for PI (Ox Ch) for a total of 5 (4) experiments respectively. We suspect that when VDAC is inserted it is subjected to a polarization, induced by membrane applied potential V_m, which in turn is responsible for the capacitance variation. It is worth mentioning that applying tension or potential to the outer hair cell (OHC) from the mammalian cochlea has been found to be responsible for the observed exponential capacitance variation and explained as a charge movement across the membrane[25].

We point out that in the same voltage range (< 100 mV for PI and < 40 mV for OxCh membranes respectively) of Fig.5 the variation of the bare BLM is less than 1% for PI and less than 2.3 % for OxCh membranes respectively.

Further, we also attempted to describe both data on voltage-dependent conductance and capacitance, by a simple exponential law which is perhaps included in formula (8) (i.e. A =1). In order to choose between the two models, we performed an appropriate statistical test of the results obtained by the two best-fits (F-test Graphpad Prism 2).

The results obtained indicate that in all the analyses the single exponential fails to give a statistically significant better description ($P < 0.05$) in any of the comparisons.

5. INDIRECT EVIDENCE OF PORIN'S ASSEMBLY PROPERTIES

In our investigations performed with the apparatus for macroscopic measurements, we also aimed to answer the following question : are the assembly properties of porin dependent on the ionic strength?
As regards the assembly properties, we point out that this information can be obtained by studying the steady state conductance as a function of porin concentration at different ionic strength [9;37].
We report in Fig. 6a (fig.6b) the data obtained for PI (Ox Ch) bilayer at different ionic strengths.
A sigmoidal pattern was found for PI bilayer, a result confirming a previous observation [37]. From the best-fit to data point with a four-parameter logistic equation:

$$G_m = G_{m0} + (G_{mss} - G_{m0})/1 + 10^{D\{\log EC50 - \log[porin]\}} \tag{6}$$

where G_{m0}=conductance at time zero, G_{mss}= steady state conductance, D is the Hill coefficient or slope factor and EC50 = porin concentration that yields half-maximum of the G_{mss} ; we obtain a Hill coefficient of 5.41 ± 0.94 , 3.03 ± 0.05 and 1.94 ± 0.65 at 1M, 0.5 M and 0.1 M KCl respectively.

Unlike the Gm_{ss} of surface charged membranes, the steady state conductance for Ox Ch obeys a Michelis-Menten kinetics:

$$(G_{mss} - G_{m0}) = [porin]^n G_{max} / (K_m)^n + [porin]^n \tag{7}$$

where G_{max} is the plateau value of conductance , K_m is the dissociation constant, and n is the order of the reaction, with K_m = 14.6 ± 2.4 (ng/ml) and G_{max} = 2720.4 ± 168.4 ($\mu S * cm^{-2}$) and a power of 0.94 ± 0.13 (1.03 ± 0.03) at 1 M (0.1 M) KCl.

These results could indicate that besides lipids the ionic strength can influence the assembly of porin in the membranes.

a

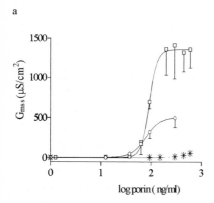

b

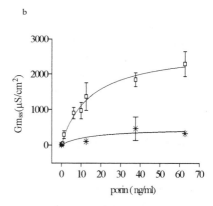

Fig.6 Dependence of the steady state conductance G_{mss} for PI (a) and OxCh (b) membrane as a function of porin concentration and ionic strength . Bathing solution: KCl 1 M (□) ,0.5M (○) and 0.1M (∗).The curve superimposed on the data in (a) is the result of the fit with equation (6). The curve superimposed on the data in (b) is the result of the fit with equation (7).

6. MACROSCOPIC INCORPORATION OF PEPTIDES AND CHANNEL FORMATION

Systematic investigations of gramicidin, calcitonin and magainin peptide incorporation have been undertaken in order to provide an answer to various research questions. Namely:

• are the incorporation and the assembly properties of the gramicidin A peptide (GA) dependent on the ionic strength of the bathing solution? It is well known that GA forms channels and its dependence on the membrane applied potential has been extensively investigated . However, very little information is available on its assembly properties, which have been limited to high GA concentrations and detected with quite sophisticated techniques [45;46].

• human and salmon calcitonin (indicated as hCt and sCt respectively) show channel forming activity in BLM. Are these activities related to different peptide concentrations and membrane applied voltages?

• does the Magainin 2 peptide form channels when incorporated into BLM? Is the channel conductance affected by the BLM surface charge?

6.1 Investigations on Gramicidin macroscopic incorporation and assembly properties

The investigations into the macroscopic incorporation of the different peptides into different BLM's is characterized, in all the cases studied, by an initial rapid increase in the incorporation rate which, depending on the nature of

the BLM and peptide, may or may not end up with a steady state. Unfortunately not all the peptides reach the steady state. In fact, the BLM breaks before such a state is reached, and this prevents systematic investigations of the type described in the previous section for porin.

An exception among the peptides under study is GA incorporation into OxCh BLM. In this case, we were able to perform investigations into the dependence on KCl ionic strength:

• of the two main incorporation parameters K_1 and K_2 describing the kinetics processes

• of the assembly configuration at steady state

The results are summarized in Fig. 7-8. It is evident that the affinity, as measured by K_1, increases almost linearly with KCl concentration, and that higher values of

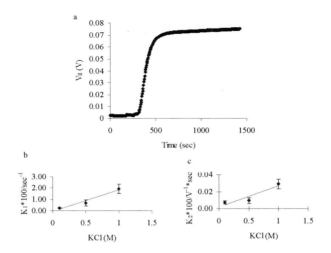

Fig.7 Kinetics of Gramicidin A (GA) incorporation (a) into Ox Ch membranes, and K_1 (b) and K_2 (c) parameters obtained by the kinetics fitting procedure as a function of the ionic strength. Typical on-line data for the output signal V_{ll} acquired with a PI membrane, GA concentration $5*10^{-9}$M Experimental conditions: $V_s = 0.12$ V, $R_l = 1$ MΩ and $C_l=10$ nF, hole diameter $\phi = 1300$ μm, KCl = 1M, temperature 24 ± 0.5 °C. The curve superimposed on the data is the result of the fit with equation (2). The initial and steady state conductance values were: $G_{m0}= 1.58 \cdot 10^{-9}$ S and $G_{mf}= 2.1 \cdot 10^{-5}$ S .

K_2 are found at higher ionic strengths. Moreover from the Hill coefficients $4.1\pm$

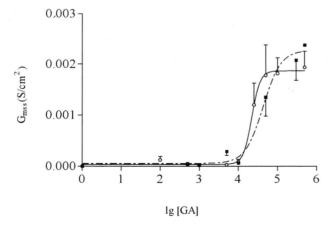

Fig.8 Dependence of the steady state conductance (G_{mss}) as a function of GA concentration and ionic strength in OxCh membrane. Bathing solution KCl 1 M (O) and 0.1M (■). The curve superimposed on the data is the result of the fit with equation (6).

0.8 (1.9± 0.9) at 0.1 M (1M) obtained with the fitting procedure described in Sect. 5 formula (6), it is clear that at low (0.1M) ionic strengths the GA channels assembled as tetramers, while at higher ionic strengths (1M) the GA channels are assembled as dimers.

6.2 Investigations into hCt and sCt incorporation

Two different techniques were used to study the interaction of calcitonins with lipid bilayer membranes: alternating current measurements (AC method) and single channel measurements. In both cases, Teflon chambers were used that had two aqueous compartments connected by small circular holes. The hole diameter was 1300 μm for the AC method or 300 μm for the single-channel experiments. Membranes were formed with a 1% (w/v) solution of a mixture of DOPC and DOPG (molar ratio 85:15) in n-decane or from pure DOPC dissolved in n-decane. The salts used in the experiments were of analytical grade. The aqueous solutions were used unbuffered and had a pH of 7.

6.2.1 Macroscopic conductance measurements

In a first set of experimental conditions, we studied the effect of sCt on the conductance of lipid bilayer membranes using the AC-method. After addition of 85nM sCt to DOPC/DOPG the specific conductance (capacitance) increased on average from $4.7*10^{-7} \pm 8.4*10^{-8}$ S/cm^2 to $1.53*10^{-5} \pm 3.24*10^{-6}$ S/cm^2 (0.51 ± 0.04 and 1.14 ± 0.3 μF/cm^2) (average ± SE of four experiments). Fig. 9 shows a representative experiment of the current/capacitance increase of membrane after addition of sCt to the aqueous phase. This indicates that sCt molecules insert and aggregate in a cooperative manner into the membrane to form channels. It is noteworthy, however, that the membranes were found to be fairly stable under these conditions (85nM). Higher sCt concentrations tended to reduce membrane stability (data not shown). It is possible that the instability of the membrane at higher sCt concentrations reflected some sort of detergent effect mediated by the peptide. The effect of hCt on membranes made from the same lipids was less reproducible than that shown for sCt. However, a higher reproducibility was obtained at a lower hCt concentration (24.5 nM), where the conductance also increased about two orders of magnitude within 200 min (data not shown).

It is possible that the conductance increase described above was caused by an unspecific artifact. To rule that possibility out, we performed control experiments where trypsin was added to the aqueous phase prior to the addition of Ct. Under these conditions we observed only an insignificant increase in membrane specific conductance of about a factor of two within 200 min. Similarly, we found also an insignificant effect on membrane specific conductance when the hormones were pretreated with trypsin before they were added to the salt solution on one or both

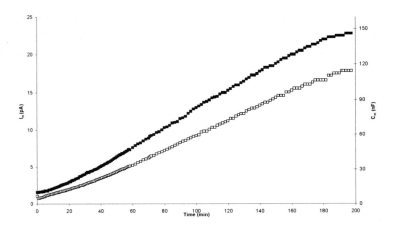

Fig.9 Time-course of the membrane current, I_m (■), and of the capacitance, C_m (□) of a lipid bilayer membrane made from DOPC/DOPG after addition at zero time of 85nM sCt . The aqueous phase contained 1 M KCl; pH 7. The temperature was 20°C; $R_l = 1$ MΩ; $C_l = 20$ nF; f = 0.5 Hz; V_s =80mV. Basal conductance $G_{mb} = 1,58*10^{-8}$ S/cm², final conductance $G_{mf} = 8.32$ *10^{-6} S/cm².

sides of the membrane. These results indicated that the conductance increase reported in Fig. 9 and similar experiments were caused by sCt and did not represent an artifact.

6.2.2 Single channel recording and conductance dependence on the membrane voltage

To study the interaction between Ct and lipid bilayer membranes in more detail, we performed single-channel conductance measurements. These experiments are considered to be a sensitive tool for the study of channel-forming substances. For these measurements, it is important that the substances form defined channels and do not act as detergents inducing lipid perturbation. The observation of non-random discrete step-like increases in conductance could be considered to conform to these criteria. Preliminary experiments indicate that all calcitonins tested (eel, porcine, carbocalcitonin and salmon) are able to form channels when added to one or both sides of the black lipid bilayer membranes kept under voltage-clamp conditions (Table 1). Here we focused our attention on two Ct's, namely sCt and hCt. Shortly after addition of the Ct's, we observed a stepwise increase in membrane conductance, which indicated the formation of ion-permeable channels in the membranes. Fig. 10 shows a single-channel recording with associated histogram of channel conductance distribution derived from such an experiment with sCt (125 nM), where two different applied voltages and/or two different bathing media were used. As a comparison the single-channel recording with an associated histogram of hCt channel conductance

Table 1. Average single-channel conductance Λ of the CTs (125 nM) for 1M KCl solution, at +50 mV voltages. The membranes were formed from DOPC/DOPG. The aqueous solutions were at pH 7; T=20°

Calcitonin	Λ ± SE (nS)	N° exp.
Eel	0.42 ± 0.0014	2
Porcine	0.56 ± 0.005	2
Salmon	0.090 ± 0.007	7
Carbocalcitonin	0.042 ± 0.0006	2

distribution at the same concentration (125 nM) and at + 150 mV was also reported (see legend of figure).

The conductance fluctuations were not uniform in size and most of the steps were directed upwards, while downward steps (closing channel) were more rarely observed under these conditions. The current fluctuations were distributed over a conductance range of 0.0066-0.125 nS for sCt (125 nM, +150 mV) (Fig.10c). The central values (Λ_c) of single-channel conductance for sCt at concentrations of 125nM (+150mV) was 0.0135 ± 0.0003 nS .

At +50 mV, it was not possible to observe channels in membranes when hCt was present in a concentration of 125 nM. Interestingly, the channel-forming activity of hCt increased when the voltage across the membrane was increased. The current fluctuations were distributed over a conductance range of 0.0066-0.085 nS (125 nM, +150 mV)(Fig.10 d), with a central value Λ_c of 0.0138 ± 0.0005 nS (+150 mV). It is noteworthy that we observed a similarly broad distribution of the single-channel conductance in experiments where the membranes were bathed in 1M NaCl, 1M $CaCl_2$, and 1M KCH_3COO (data not shown). The results suggest that considerable variations in single-channel conductance were observed under different conditions. One interesting result is that the single-channel conductance decreased as a function of the applied voltage. The preliminary investigation was performed by running the experiment at different voltages (data not reported) or by changing the applied potential during the experiment (Fig. 11). The hCt channel showed the same

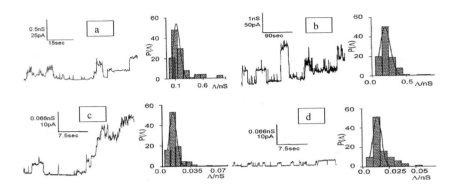

Fig.10 Chart recorder tracing with associated distribution histogram of sCt and hCt channel conductance events. Experiments on DOPC/DOPG (molar ratio 85:15) membrane in the presence of 125 nM sCt (a,b,c) and hCt (D) added to both sides. T = 20°C. (a) The voltage was set to + 50 mV and the aqueous phase contained 1 M KCl (pH 7), (166 events) . (b) The voltage was set to + 50 mV and the aqueous phase contained 1 M NaCl (pH 7) (194 events). (c) The voltage was set to + 150 mV and the aqueous phase contained 1 M KCl (pH 7) (253 events) (d) The voltage was set to + 150 mV and the aqueous phase contained 1 M KCl (pH 7) (175 events). Histogram of the probability P(Λ) for the occurrence of a given conductivity unit. Gaussian fit is shown as a solid curve.

pattern (data not shown). The dependence on membrane voltage was fitted by means of two exponential equations (5) whose parameters α_1 and α_2 can give an indication of the number of charges involved in the formation of the sCt channel. It is noteworthy that many other small peptides that form α-helical structures in lipid membranes such as mellitin [47], gramicidin A [48],or alamethicin [49] tend to show increased channel size and increased channel-forming activity when the voltage increases. This probably means that sCt (and probably hCt as well) has no dipole moment.

6.2.3 Ion selectivity

The ion channel selectivity of the channels formed by sCt was investigated by measuring the membrane potential under zero current conditions. DOPC/DOPG membranes were formed in 0.3 M solution of KCl or 0.3M NaCl. After blackening the membranes, sCt was added to both sides in a final concentration of 125 nM. About 20 min after the addition, the membrane conductance reached a virtually stable value; then the salt concentration on one side of the membranes was raised up to 0.5 M by addition of concentrated salt solution. In all cases, the more diluted side (trans to salt addition) became

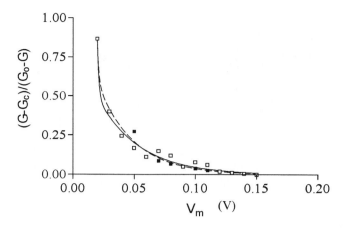

Fig.11 The conductance ratio $(G - G_c) / (G_o - G)$ where G_o (G_c) correspond to open (closed) states of conductance of the sCt (125 nM) as a function of applied positive/negative (■)/(□) voltage. Data were parameterized with the two-exponential equation (5) where $\alpha_1 = 34.5$ (V^{-1}) and $\alpha_2 = 812$ (V^{-1}) (■) , and $\alpha_1 = 29.7$ (V^{-1}) and $\alpha_2 = 806$ (□) (V^{-1}). The membranes were formed from DOPC/DOPG (molar ratio 85:15). The aqueous solutions were at pH 7; T=20°C

slightly positive (+ 1 mV for 0.5 M at the trans side) indicating that channels are almost equally permeable to anions and cations. Analysis of the zero current membrane potentials using the Goldman-Hodgkin-Katz equation [50] gives a P_c/P_a of 1.03.

6.3 Investigations on Magainin 2 incorporation

Magainins belong to a family of a linear 23 residues peptides produced by the skin of the African frog, Xenopus leavis . Magainin 2 , which is a member of this family, has been extensively studied for its killing action on various microorganisms as well as on tumor cells, mitochondrial membranes and liposomes [7;51;52] by permeabilizing their membranes, in a detergent-like manner or making well structured channels [7;54]. This aspect has aroused great interest among researchers due to its potential relevance in overcoming the resistance to most antimicrobials in use.

Our studies on the incorporation of magainin 2 in BLM was aimed to provide insight into the interaction modality as probed by channel formation.

Membranes were formed with a mixture of palmitoyl-oleoyl-phosphatidylcholine (POPC) and dioleoyl phosphatidylglycerol (DOPG) (85:15; w:w)

6.3.1 Macroscopic conductance measurement

The conductance measurements were performed at the same magainin 2 concentration used for single-channel experiments ($4.05 * 10^{-8}$ M) by means of the AC-method, on POPC:DOPG membranes. After magainin 2 addition to the solutions bathing the black membrane, an increase in the membrane current and in the capacitance(Fig.12) occurred at variable times. This means that magainin 2 molecules inserted and aggregated into the membrane to form pores in a cooperative manner. In a few experiments, the current reached a steady state. This behavior cannot be generalized because rupture of the membrane occurs after a certain time. The relative % of variation (i.e. (final value – basal value)/basal value*100) of the value of currents and capacitance between the basal (before drug addition) and the final values , depending on the experimental conditions, were 286.4 ± 36.7 (169.3 ± 87.2) respectively.

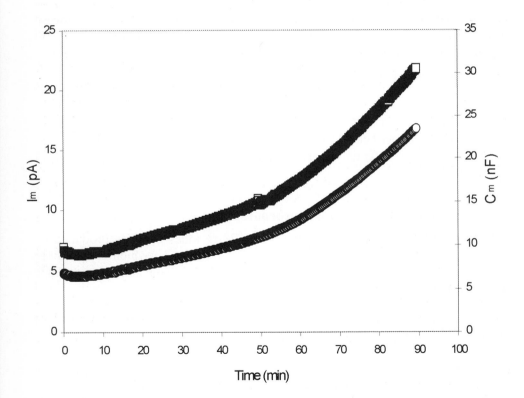

Fig.12 Time-course of membrane current, I_m (□) and capacitance, C_m (○) of a lipid bilayer membrane made up of DOPC/DOPG after addition at zero time of magainin 2 ($1*10^{-7}$ M). The aqueous phase contained 0.5 M KCl; pH 7. The temperature was 20°C; $R_l = 1$ MΩ; $C_l = 20$ nF; f = 1 Hz; V_s =160mV. The initial and final conductance of black lipid membranes was $4.67*10^{-7}$S/cm² and $1.56*10^{-6}$S/cm²

6.3.2 Single channel recording and conductance dependence on the membrane voltage

Planar lipid membranes were used to see whether and in which experimental conditions magainin 2 at a concentration of 4.05*10-8M is able to form channels. With this aim, POPC , POPC:DOPG (85:15, w:w) and PS membranes were used as zwitterionic, mixed, and negatively-charged membranes, respectively. In a large number of experiments with POPC or PS membranes, carried out in different periods throughout the year, we were not able to observe current fluctuations for many hours and with a wide range of applied voltage after magainin 2 addition. In other experiments in PS membranes, the magainin 2 concentration was increased stepwise up to a final value of 1.94*10-6M. Again in this case, no channels were observed. To test whether membrane thickness causes a lack of magainin 2 channel formation, Gramicidin A ($1*10^{-10}$ M) was added. Shortly after gramicidin addition stepwise fluctuations of the current, indices of channel formation, were observed (results not shown).

On the other hand, non-random discrete current fluctuations in mixed membranes of POPC:DOPG, after magainin 2 addition, compatible with channel-type openings and closures with many different conductance levels and lifetime, were evident. To verify that these current fluctuations were due to magainin2, control experiments were performed, in which protease was added to the medium before magainin 2 addition. In this case, no channels were observed.

As is shown in Fig.13, the occurrence of single-channel activity sometimes occurred in steps that were twice the size of the basal unit step. Furthermore, we observed alternating periods of paroxystic channel activity followed by quiescent periods and the open time interrupted by brief closure. All the histograms show single-peaked conductance distributions.

These results indicate that a small amount of negatively-charged lipids play an important role for the electrostatic interaction between cationic magainin 2 and the membrane, as proposed by AA [7;55;56] .

Channel formation by magainin 2 has been found to be potential dependent. A threshold potential of about 160 mV is required to induce channel activity. This result is in accordance with AA's findings [7].

6.3.3 Ion selectivity

The ion channel selectivity of magainin 2 was calculated by means of two methods: reversal potential and I-V relationship.

A twofold concentration gradient was used, with 200 mM of KCl on one side (cis) and 100 mM of KCl on the other side (trans), because the reversal potential is not linearly correlated with the salt gradient at elevated salt gradients.

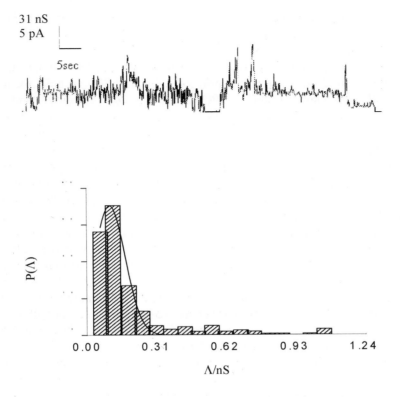

Fig.13 Chart recorder tracing with associated distribution histogram of magainin 2 channel conductance. Experiment on DOPC/DOPG (molar ratio 85:15) membrane in the presence of magainin 2 (4.05∗10-8M). The voltage was set to + 160 mV and the aqueous phase contained 0.5 M KCl (pH 7), T= 20°C.Gaussian fit is shown as a solid curve (Λ_c= 0.103 ± 0.049nS ; 560 channels)

About 60 min after magainin 2 addition on the cis side, the membrane conductance reached a virtually stable value; then the salt concentration on the cis side of the membrane was raised up to 200 mM by addition of concentrated salt solution. Single-channel current were observed when the holding potential ± 120 mV in membrane was applied. Most probably the low ionic strength and the chemical potential reduce the threshold potential. The reversal potentials were determined by changing the holding potential of ± 2 mV step by step, and the potential at which the current was zero was taken as the reversal potential for the open channel. The mean reversal potential was -16.34 mV .

The permeability ratio P_K^+/P_{Cl}^- was 3.4.

Approximately the same result was obtained by means of I-V curve. In fact the reversal potential was -16.6 mV.

7. UNSOLVED QUESTIONS TO BE INVESTIGATED

In our investigations into protein/peptide interactions with BLM's, we were challenged with some problems which far from being understood, from our point of view, merit further investigations in the attempt to clarify them.

Here we intend to mention briefly just a few of them :

7.1 ac signal distortions

During our systematic investigations into the macroscopic incorporation of protein/peptides, we observed no distortion of the ac output signal, except for

alamethicin. Fig.14a-c shows an example of the output signals during the incorporation of porin , gramicidin and alamethicin. Similar behavior was observed at the steady state conditions.

A distortion (in the form of a well) can only be observed for the alamethicin peptide. The analogy between the relevant output signals of many physical/chemical systems satisfying non-linear equations for the dependence of the relevant variables (on time) and the ac output signal for alamethicin suggested that some non-linear phenomenon could govern alamethicin insertion, and in such a case be the source of ultraharmonic generation during signal transmission. A Fast-Fourier-Transform analysis performed in our preliminary investigations evidences (see Fig.14d-f) the presence of such an effect only in the case of alamethicin.

What is its origin? Do any other peptides show the same behavior? Is the harmonic composition related to the assembly characteristics?

We are presently developing a home-made apparatus with a variable frequency as input ac signal, that could allow us to investigate the above-mentioned observation, as well as a possible dependence of the macroscopic incorporation parameters on the frequency.

7.2 Analysis of the kinetics growth during incorporation : is there a phase transition?

Incorporation of porin/peptides and the related conductance variations cannot be a continuously increasing process (provided that the BLM does not break) : a steady state must be reached. In fact, we observe such an effect in many cases.

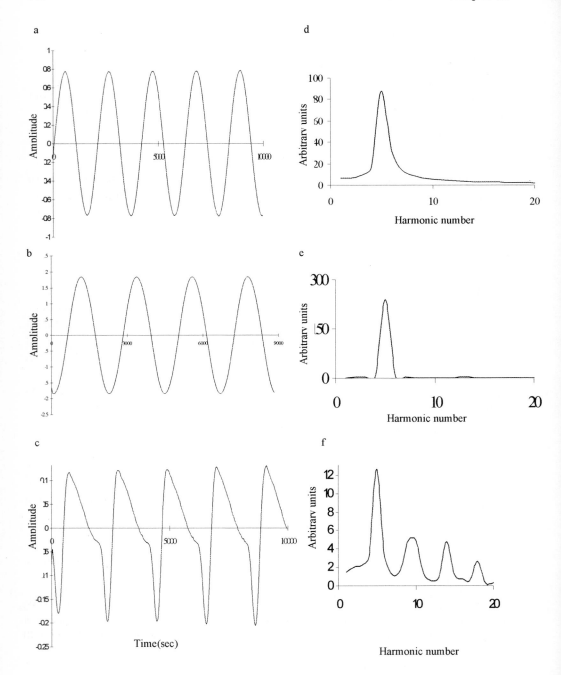

Fig.14 Amplitude and Fourier analysis of the output signal V$_{ll}$ of PI membrane in the presence of porin (a,d), gramicidin A (b,e) and alamethicin (c,f).

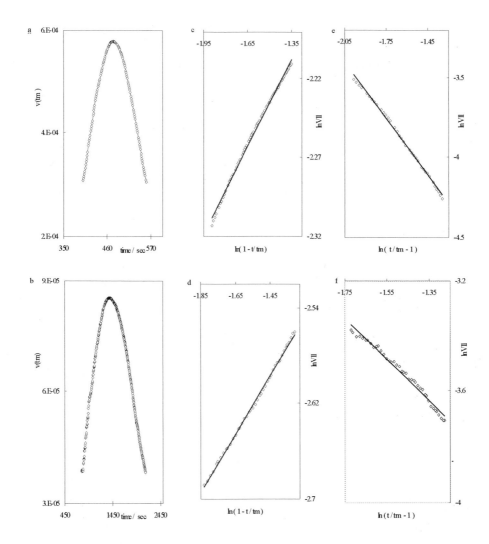

Fig.15 Detailed study of the kinetics around the maximum depolarization instant t_m. An example of the data $v(t)$ around the instant t_m of maximum depolarization rate for PI (a) and for Ox Ch (b) respectively. To investigate how the incorporation mechanism proceeds when approaching t_m from below and above, the time scale is reported as $\ln(1- t /t_m)$ and as $\ln(t/t_m - 1)$ (c and e, d and f for PI and Ox Ch respectively); on the vertical axis the $\ln V_{II}$ is reported. The t_m value is 486 sec and 1394 sec for PI and OxCh membranes respectively. The straight lines are the results of the linear best fit with: $r^2 = 0.997$ and 0.996 for (c) and (e) respectively, and $r^2 = 0.998$ and 0.982 for (d) and (f) respectively. For experimental conditions see Fig. 2.

Table 2. α_a and α_b as a function of applied voltage in PI and OxCh membranes. The slope values α_a and α_b obtained by using the formula: $V_{II} \approx (1 - t/t_m)^{\alpha_b}$ for $t < t_m$ and $V_{II} \approx (t/t_m - 1)^{\alpha_a}$ for $t > t_m$ in the study of the kinetics around the t_m instant.

	PI membranes		OxCh membranes	
V_s / mV	α_b	α_a	α_b	α_a
20	-1.09 ± 0.12	0.20 ± 0.04	-0.99 ± 0.06	0.20 ± 0.02
40	-1.01 ± 0.08	0.17 ± 0.01	-0.88 ± 0.09	0.19 ± 0.02
60	-1.07 ± 0.27	0.29 ± 0.03	-1.14 ± 0.12	0.18 ± 0.01
80	-1.19 ± 0.25	0.21 ± 0.03	-0.95 ± 0.07	0.17 ± 0.02
100	-1.30 ± 0.31	0.19 ± 0.01	-0.91 ± 0.14	0.20 ± 0.01
120	-0.95 ± 0.18	0.19 ± 0.02	-0.84 ± 0.11	0.24 ± 0.01
140	-1.08 ± 0.17	0.21 ± 0.01	-1.02 ± 0.19	0.21 ± 0.03
160	-1.01 ± 0.07	0.24 ± 0.02	-1.00 ± 0.08	0.22 ± 0.02
200	-1.10 ± 0.15	0.24 ± 0.01	-0.99 ± 0.07	0.18 ± 0.01
Mean $\pm$ S.E	-1.09 ± 0.04	0.22 ± 0.01	-0.97 ± 0.03	0.20 ± 0.01

Particular attention should be paid to the way in which the steady state is approached, since, as observed in other "growing" processes, interesting effects could be present. In particular the incorporation rate as measured by the total current flowing through the BLM could show some "discontinuity" in its time dependence.

We report an example of the data $v(t)$ (i.e. the derivative of V_{II}) around t_m – the instant of maximum depolarization (fig.15). An appropriate change in the time scale was performed (i.e. we used the quantities (t/t_m -1) or (1- t/t_m) for t values higher or lower than t_m respectively) since we intended to investigate how the incorporation mechanisms proceed when approaching t_m from below and above. The data were parameterized by the formula

$$V_{ll} \approx \begin{cases} \left(1 - t/t_m\right)^{\alpha_b} & \text{for } t < t_m \\ \\ \left(t/t_m - 1\right)^{\alpha_a} & \text{for } t > t_m \end{cases} \qquad (7)$$

Where α_b and α_a are the two parameters describing the approach from below and above t_m.

Following the usual procedure adopted in this kind of analysis (i.e. neglecting the region near t_m), we obtained the results reported in Figs. 18 c-f. A summary of the results obtained in all experiments is reported in Tab 2 .

For both bilayers, the system approaching t_m is adequately described by (7), with completely different values for the two parameters α_b and α_a. We observe that:

i) both parameters are independent of V_s, a fact which can be considered as indicative of the kinetic nature of the phenomenon, which is typical of the phase transition phenomenon.

ii) the mean value of α_a is: 0.21 ± 0.01 and 0.2 ± 0.01, the mean value of α_b is: -1.09 ± 0.04 and -0.97 ± 0.03 for PI and Ox Ch respectively – the phenomenon therefore seems to be independent of the nature of the bilayer membrane. These values are compatible with what is typically expected for a phase-change process in which there is an abrupt variation in the order parameter [57-61].

The same analysis performed with GA incorporation in OxCh membrane at KCl 1M gives similar α_a (0.214 ± 0.042) and α_b (-1.00 ± 0.22) values .

Is this characteristic common to other proteins/peptides reaching steady state conditions? How can the phenomenon be interpreted?

7.3 capacitance/conductance single channel variations : is there a correlation?

During macroscopic incorporation we have found evidence, at least for porin, of similarities in the dependence of capacitance and conductance on membrane applied voltages. Moreover, in our investigations into macroscopic incorporations, both ionic and gating currents were variable.

Are these phenomena strictly related?

Obviously it would be very useful to have an apparatus available for use in single channel investigations that were able to detect capacitance variation related to channel incorporation (i.e. simultaneous jumps in the currents associated with the two parameters). Using our home-made apparatus, we have actually only collected very preliminary results on such an issue, which is undoubtedly very interesting from both the theoretical and experimental points of view, and will therefore deserve much more attention in our future investigations.

8. CONCLUSIONS

Since the BLM were developed it was realized that, owing to the central role of bilayer as component of biological membrane, many phenomena regarding the functions of membrane could be clarified: such as transport, bioelectrical phenomenon, interaction between lipid-protein and protein assembly.

However, despite the many studies on biological membrane structures, there is still a relatively poor understanding of lipid-protein/peptide interactions, although a multidisciplinary approach has been applied to the biophysics of BLM 's.

A method to study the kinetics of incorporation of proteins/peptides into BLM's by monitoring simultaneously both capacitance and conductance has been optimized and a mathematical parameterization of kinetic growth developed. A cooperative behavior has been found for both porin and peptide incorporation, especially gramicidin A, and in lipid membrane structures whose intrinsic properties seem to play a pivotal role in determining protein/peptide assembly in the membrane.

Furthermore, the capacitance parameter, beside probing the stability of lipid bilayer membranes, has assumed an important role in the mechanism of voltage-gating. Due to the fact that proteins/peptides during the process of incorporation or, when already incorporated, are subjected to potential-induced conformational variation, membrane capacitance assumes a further role. In this context, it can be considered as a useful probe for monitoring dynamic phenomena such as conformational variation in channel gating, or protein insertion into the bilayer [62-65].

The dependence of membrane conductance on protein/peptide concentration at the steady state of incorporation into different lipid membranes or different ionic strengths evidences a different assembly of the protein/peptide in the membrane, indicating that the nature of lipid-protein/peptide interaction originates from the intrinsic properties of lipids. It is worth mentioning that the number of channels that are assembled in the membrane has been found, at least for high ionic strengths, with other techniques [66].

These studies, which aimed to understand the biological phenomenon, may turn out to be useful for developing biosensors for biodiagnostic and technological applications.

In this context, the implementation of an easy and stable procedure for forming phospholipids on metals is of great relevance. An appropriate strategy could be required to understand:

•the nature of lipid/protein or protein/protein interactions

•the role of lipids in protein binding and folding and of polypeptide toxins in cell-killing actions

•how the peptides are assembled

• the actions of some peptides at the molecular level

•the design of new peptides for a specific pharmacological action and for selectively binding to other organic molecules or metals

ACKNOWLEDGMENTS The work described in this chapter was supported by 60% MURST,2001. The authors wish to tank Dr. Daniela Meleleo for her invaluable assistance.

REFERENCES

[1] W. J. Els and K. Y. Chou, J. Physiol. , 462 (1993) 447

[2] M. Kiebler, K. Becker, N. Pfanner and W. Neupert. J. Membr. Biol., 135 (1993) 191

[3] M. Kiebler, P., I. Keil van der Klei, N. Pfannen and W. Neupert, Cell, 74 (1993) 483

[4] A. Brandner, Nature 379 (1996) 339

[5] J. Vargas, J. M. Alarcon and E. Rojas, Biophys. J. 79 (2000) 934

[6] V. Stipani, E. Gallucci, S. Micelli, V. Picciarelli and R. Benz. Biophys. J., 81 (2001) 3332

[7]R.A. Cruciani J.L. Barker, M. Zasloff H.C. and O. Colamonici. Proc. Natl.Acad.Sci. U.S.A. 88 (1991) 3792

[8] H. T. Tien and A. Ottova-Leitmannova (eds.),Membrane Biophysics- As viewed from experimental bilayer lipid membranes, Elsevier,2000

[9] S. Micelli, E. Gallucci and V. Picciarelli, Bioelectrochem., 52 (2000) 63

[10] S. Micelli, E. Gallucci, D. Meleleo, V. Stipani and V. Picciarelli, Bioelectrochem., (2002) in press

[11] E. Bamberg and P. Läuger, J. Membrane Biol., 11 (1973) 177

[12] X. Xu and M. Colombini, J. Biol. Chem., 271 (1996) 23675

[13] A. Wiese, J. O. Reiners, K. Brandenburg, K. Kawahara, U. Zahringer and U. Seydel, Biophys. J., 70 (1996) 321

[14] E. Neher and H. P. Zingsheim, Pflugers Arch., 351(1974) 61

[15] H. A.. Kolb, P. Läuger and E. Bamberg, J. Membrane Biol., 20 (1975) 133

[16] L. E. Moore and E. Neher,. J. Membrane Biol., 27 (1976) 347

[17] H. A. Kolb and E. Bamberg, Biochim. Biophys. Acta, 464 (1977) 127

[18] H. A Kolb, and G. Boheim, J. Membrane Biol. 38 (1977) 151

[19] H.T. Tien (ed.), Bilayer Lipid Membranes : theory and practice, Marcel Dekker, inc, N.Y. chapter 4 and 5 (1974)

[20] H. T.Tien and J. D. Mountz , Martinosi A. N. ed., Protein-Lipid Interaction in Bilayer Lipid Membranes (BLM), In: The enzyme of biological membranes. N Y Plenum, 1 (1977) 139-170.

[21] S. H. White, Ann. N. Y. Acad. Sci. 303 (1977) 243

[22] S. H. White. Christopher Miller (ed.), Ion channel reconstitution .Plenium Press, N.Y., chapter 1(1986)

[23] U. Seydel,G. Schröder and K. Brandenburg, J. Membr. Biol. 109 (1989) 95

[24] Kalinowski S. and Z. Figaszewski, Biochem. Biophys. Acta 1112 (1992) 57

[25] K. H. Iwasa, Biophys. J. 65 (1993) 492

[26] F. Bezanilla, E. Peroso and E. Stefani, Biophys. J. 66 (1994) 1011

[27] E. Stefani, L. Toro, E. Peroro and F. Bezanilla, , Biophys. J. 66 (1994) 996

[28] D. F. Donnelly , Biophys. J. 66 (1994)   873

[29] Chanturiya A. N., Biochem. Biophys. Acta 1026 (1990) 248

[30] A. N. Chanturiya and H. V. Nikoloshina, J. Membrane Biol. 137 (1994) 71

[31]P. Lauger, W. Lesslauer, F. Marti and J. Richter. Biochem. Biophys. Acta 135 (1967)20

[32] R. Benz, Biochem. Biophys. Acta 1197 (1994) 167

[33] H. Freitag, W. Neupert and R. Benz, Eur. J. Biochem. 123 (1982) 629

[34] V. De Pinto, R. Benz. and F. Palmieri, Eur. J. Bioch. 183 (1989) 179

[35] B. Popp, A. Schmid and R. Benz, Biochemistry 34 (1995) 3352

[36] M. Zizi , L.Thomas, E. Blachly-Dyson, M. Forte and M. Colombini, J. Membrane Biol., 144 (1995) 121

[37] E. Gallucci , S. Micelli and G. Monticelli, Biophys. J., 71 (1996) 824

[38] M. Brunen and A. Engelhardt, Eur. J. Biochem. 212 (1993) 129

[39] S. Nekolla, C. Andersen and R. Benz, Biophys. J., 66 (1994) 1388

[40] F. Wohnsland and R. BenzJ. Membrane Biol., 158 (1997) 77

[41] S. J. Schein, M. Colombini and A. Finkelstein, J. Memb. Biol., 30 (1976) 99

[42] M. Colombini, Nature, 279 (1979) 643

[43] N. Roos, R. Benz and D. Brdiczka, Biochim. Biophys. Acta, 686 (1982) 204

[44] R. Benz, CRC Crit. Rev. Biochem., 19 (1985) 145

[45] H. Tanaka and J.H. Freed, J. Phys. Chem., 89 (1985) 350

[46] Ge Hingtao, J.H. Freed, Biophys. J. 76 (1999) 264

[47] M.T.Tosteson, O. Alvarez, W. Hubbel, R.M. Bieganski, C. Attenbach, L.H.,Caporales, J.J. Levy, R.F. Nutt, M. Rosenblatt, and D.C. Tosteson, Biophys. J. 58 (1990) 1367

[48] E. Bamberg, H. J. Apell, and H. Alpes, ,Proc. Natl. Acad. Sci. USA 74 (1977) 2402

[49] J.E. Hall, I. Vodyanoy, T.M. Balasubramanian, and G.R. Marshall, Biophys. J. 45 (1984) 233

[50] R. Benz, K. Janko, and P. Läuger, ,Biochim. Biophys. Acta 551 (1979) 238

[51]H.V. Westerhoff, D. Juretic, R.W. Hendler and M. Zasloff Proc.Natl.Acad.Sci. U.S.A., 86 (1989)6597

[52]H.V. Westerhoff, R.W.Hendler, M.Zasloff, D.Juretic, Biochim. Biophys. Acta,975 (1989) 371

[53] E. Grant, T.J. Beeler, K.M.P. Taylor, K. Gable, M.A. Roseman, Biochemistry,31 (1992)9912

[54] B. Bechinger J. Memb. Biol. , 156 (1997)197

[55]K. Matsuzaky, K. Sugishita, N. Ishibe, M. Ueha, S. Nakata, K. Miyajima and R. Epand, Biochem. 37 (1998)11856

[56] T. Wieprecht, M. Beyermann and J. Seelig, Biochemistry 38 (1999) 10377

[57] E. B. Zambrowicz and M. Colombini, Biophys. J., 65 (1993) 1093

[58] T. Hodge and M. Colombini, J. Membr. Biol., 157 (1997) 271

[59] F. Jahnig,I. Theoretical description. Biophys. J., 36 (1981) 329

[60] F. Jahnig, Biophys. J., 36 (1981) 347

[61] A. C. Mowery and D. T. Jacobs, Am. J. Phys., 51(1983) 542

[62] H. J. Hermann, Physics Report, 136 (1986) 153

[63] M. Basta, V. Picciarelli and R. Stella, Eur. J. Phys.12 (1991) 210

[64] F. Bezanilla, E. Peroso and E. Stefani, Biophys. J. 66 (1994) 1011

[65] E. Stefani, L. Toro, E. Peroro and F. Bezanilla, Biophys. J. 66 (1994) 996

[66] J.Vargas, J. M. Alarcon and E. Rojas, Biophys. J. 79 (2000) 934-944

[67] O. Alvarez and R. Latorre, Biophys. J. 21 (1978) 1-17

Planar Lipid Bilayers (BLMs) and their Applications
H.T. Tien and A. Ottova-Leitmannova (Editors)

Chapter 15

Structure and function of plant membrane ion channels reconstituted in planar lipid bilayers

M. Smeyers, M. Léonetti, E. Goormaghtigh and F. Homblé

Laboratoire de Physiologie végétale, SFMB, Université Libre de Bruxelles, Campus Plaine (CP206/2), Bd du Triomphe, B – 1050 Brussels, Belgium

Channels are transmembrane proteins that allow the passive transport of solutes, viz. down their electrochemical potential gradient, at a high rate (10^7-10^8 ions/s). The understanding of the basic molecular processes that control the transport of solutes through channels requires detail knowledge of their function and structure and how they change during regulation. Several techniques such as the planar lipid bilayer and the patch-clamp permit the study of the function of ion channels at the single-protein level. X-ray crystallography can provide information about the protein structure at the atomic level. However, the crystallization of membrane proteins remains a very tedious task and it requires an amount of proteins (few mg/ml) often difficult to obtain. Nonetheless, a general knowledge of the structure of membrane proteins can be acquired using spectroscopic and microscopic techniques. In particular, circular dichroism and infrared spectroscopy are widely used to investigate the structure of membrane proteins. The attenuated total reflection Fourier transform infrared spectroscopy (ATR-FTIR) has become one of the most powerful methods to study the static and the dynamic structure of proteins embedded in planar lipid bilayers [1-3].

Several authors have exhaustively reviewed the various channels found in plant membranes [4-12]. In 1976, Aleksandrov et al. [13] have reconstituted in planar lipid bilayers a calcium-selective channel from Nitellopsis obtusa. Since, ion channels from chloroplasts, mitochondria, plasmamembrane, tonoplast and peroxysomes have been characterized in planar lipid bilayers [11, 14-16]. In this chapter, we will mainly be concerned with methodological considerations. A particular attention will be directed to the mitochondrial VDAC purified from plant tissues for which both functional and structural data have been published.

The <u>v</u>oltage-<u>d</u>ependent <u>a</u>nion-selective <u>c</u>hannel (VDAC) is the main pathway for solute transport across the mitochondrial outer envelope membrane. A variety of metabolites and proteins are known to interact with the VDAC of yeast and mammals. Some of them regulate the function of the channel. The literature has already been reviewed [17] and will not be further discussed here, for the sake of brevity. VDAC has been purified from corn [18], pea [19], wheat [20], potato [21] and kidney beans [22]. They all share similar functional properties. Though little is known about the regulation of plant VDACs, it has recently been speculated that plant VDAC could play a role in programmed cell death. Cytochrome c is located in the intermembranar space of the mitochondria. The release of cytochrome c from the mitochondria is an early event in many forms of mammal apoptosis. *In vitro*, it has been shown that pro-apoptotic factors promote cytochrome c flux through VDAC whereas anti-apoptotic factors close the VDAC [23]. Whether a similar process occurs in plants remains an open question. In plants, programmed cell death is thought to play a role in xylogenesis, reproduction, senescence, and in defence against pathogen attack as part of the hypersensitive response: a fast and spatially confined cell death that impedes the spread of the pathogen. By contrast to what is known in animal biology, the key regulators of apoptosis have not been identified in plants. Nonetheless, transgenic plants expressing animal anti-apoptotic genes like bcl-2 are not able to activate their programmed cell death pathway [24-25]. Thus, this suggests that there could be a functional similarity between plant and animal programmed cell death process.

1. TECHNIQUES USED TO STUDY THE FUNCTION OF PLANT ION CHANNELS IN PLANAR LIPID BILAYERS

Planar lipid bilayers are currently used to investigate the function and the regulation of plant ion channels at a molecular level. Either native membrane fragments or purified proteins can be used to reconstitute ion channels in planar lipid bilayers. They are particularly useful to monitor the activity of channels during the course of their purification and to characterize ion channels of small-size organelles (e.g., mitochondria, peroxisomes, chloroplasts) that are difficult to patch. Phospholipids and glycolipids are the major lipids found in plant membranes. These lipids are amphiphilic with a polar head and a hydrophobic hydrocarbon tail. In solution the formation of a bilayer structure is a self-assembly process during which the lipid are organized so that the polar head faces the solvent and the hydrophobic tail forms the core of the membrane. Several methods are used to reconstitute ion channels in planar lipid bilayers [26-28]. We will describe here those that have been used to incorporate plant ion channels in planar lipid bilayers.

1.1. Painted planar lipid bilayers

The painted bilayer was the first method devised to form a planar lipid bilayer [29-31]. Though it has evolved with time the general procedure has not changed. Basically, the method consists in spreading a small amount of a lipid solution over a hole drilled in a hydrophobic material and wait until the lipids

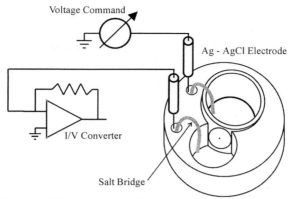

Fig. 1. Schematic diagram of the experimental chamber used to form painted planar lipid bilayers.

self-organizes into a bilayer. The experimental chamber used in our laboratory is shown in Fig. 1. It consists of two parts milled in a cross-linked styrene copolymer (Goodfellow, U.K.): the main block that contains two circular compartments and a vessel that fits perfectly to one of them. A hole of 300 μm diameter is drilled in the wall of the vessel, which separates the two compartments, and this wall is thinned to about 150 μm at the point where the hole is drilled.

Prior to planar lipid bilayer formation, about 2 μl of the lipid solution (1% lipid (w/v) dissolved in n-decane) is applied over the hole. After drying, the vessel is put in the main block and each compartment is filled with the experimental solution. Painting the lipid solution over the hole forms planar lipid bilayers. Usually, the planar lipid bilayer is formed in the few seconds following the application of the lipids over the hole. This method was used to reconstitute ion channels from the chloroplast, the mitochondrion, the plasma membrane, the tonoplast and the peroxisomes isolated from plant cells, e.g., [32-36]

1.2. Folded planar lipid bilayers

The method of folded planar lipid bilayers was originally developed by Montal and Mueller [37]. They have shown that folding two lipid monolayers over the hole made in a hydrophobic partition separating two compartments can form a planar lipid bilayer. The partition consists of a Teflon sheet 5 to 25 μm thick. There are several ways to make a hole in the partition [27, 38]. We used a laser beam to locally melt the Teflon. The size of the hole depends on the intensity and duration of exposure to the laser beam. These parameters must be determined empirically. Their value will vary according to the thickness of the Teflon film and the laser power. We usualllly made a hole of 200 μm

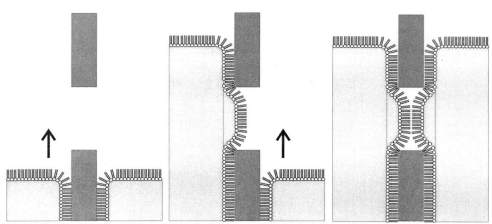

Fig. 2. Formation of a folded planar lipid bilayer from two monolayers.

diameter. Prior to the planar lipid bilayer formation, 2 µl of squalene (0.5 % (w/v) in hexane) was spread over the hole to increase the oleophilicity of the Teflon. The experimental chamber consists of the Teflon film tightly sealed with silicone grease between two compartments containing the aqueous solutions. Initially, the level of solutions must be below the hole and an aliquot of lipids (0.5% (w/v) in hexane) is spread over the surface to form a lipid monolayer. At least five minutes are required to evaporate the hexane. To form a planar lipid bilayer, the two monolayers are folded together by raising successively the solution level of each compartment above the hole (Fig. 2). This method was used to study the ion permeability through thylakoid lipids e.g., [39-40] and the VDAC from the outer membrane of mitochondria e.g.,.[41].

1.3. Dip-tip technique

Taking advantage of the patch-clamp technique [42] Coronado and Latorre [43] have shown that a planar lipid bilayer can be formed at the tip of a pulled pipet. A borosilicate pipet with a tip of a micron-size and filled with an aqueous solution is dipped just below the surface of a solution containing proteoliposomes. After a few minutes, a monolayer of proteolipids is formed at the water-air interface [44-45]. Lifting the pipet into the air results in the formation of a monolayer around the tip of the pipet (Fig. 3). When the tip is again dipped into the bath a planar lipid bilayer is formed around the tip. This method has been used to screen ion channels during the preparation of plasma membrane vesicles from corn roots [45].

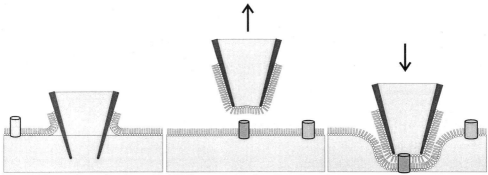

Fig. 3. Formation of a planar lipid bilayer at the tip of a micropipet.

1.4. Tricks and tips

The success of the reconstitution depends on the fusogenic propensity of the membrane vesicles and proteins. Factors that increase the adsorption of membranes or proteins and their fusion to planar lipid bilayers will raise the chances of success. This can be achieved by decreasing the electrostatic repulsion forces, by increasing the membrane mechanical tension of vesicles or by using fusogenic molecules. Several experimental conditions have been used to improve the fusion between native membrane vesicles or proteoliposomes and planar lipid bilayers:

- Addition of calcium (1-10 mM) in the presence of negatively charged lipids [46-47]
- The presence of cationic lipids [48]
- Performing the fusion below the phase transition temperature of the lipids [49]
- An osmotic gradient through the planar lipid bilayer, the side to which the vesicles are added being hyperosmotic with respect to the other side [50]
- An osmotic gradient leading to the swelling of membrane vesicles [51-52]
- Proteoliposomes doped with polyene antibiotics (e.g., nystatin) and cholesterol fuse readily to planar lipid bilayers. The fusion process can be observed because a transient peak of current related to the inactivation of nystatin channels occurs after the fusion. This method only works with cholesterol-free planar lipid bilayers [53]

The mechanical stability of the planar lipid bilayer can be improved after addition of cholesterol to its lipid composition.

Channel-like current fluctuations can occur in the presence of a high density of lipid vesicles [54] and high detergent concentration [55]. Unfortunately, there is no way to get rid of these problems during the course of an experiment. As a rule, any source of artifact should be checked when a new experimental protocol is used.

1.5. Experimental solutions

1.5.1. Aqueous solutions
Planar lipid bilayers can be formed in solution containing millimolar to molar salt concentration. They are stable in a broad range of pH (4-10). Excessive acidification of the aqueous solutions (pH<3) can lead to channel-like artifacts [56]. Aqueous solutions are made with milliQ or tridistilled water and filter through 0.22 μm Millipore filters.

1.5.2. Lipid solutions
To prepare a stock solution (2.5–10 mg/ml), the lipids are dissolved in chloroform. For bilayer experiments an aliquot of the stock solution is evaporated under a stream of nitrogen and the dried lipid are dispersed in 50 μl of n-hexane or n-decane to have the required concentration (0.5-1%, w/v). Both synthetic and purified lipids can be used. Lipids purified from biological tissues may contain channels as impurity. The stock of lipids must be protected from oxidation and it must be kept at less than –20°C (preferably at – 80°C). The oxidation of lipids may be the cause of leaky planar lipid bilayers. There is no restriction on the type of lipid composition applicable for a planar lipid bilayer formation, unless the lipids are known to form non-bilayer structures. For instance, the monogalactosyldiacylglyceride (the most abundant chloroplast lipid) alone do not form lipid bilayers [39-40]. The composition of the lipid solution will generally depend on the constraint of a particular experiment.

1.6. Electrical recording
The two compartments of the experimental chamber are called cis and trans. There is a convention whereby the trans side (or trans compartment) is always defined as zero voltage. Thus, the voltage is applied to the cis side while the trans side is held at ground. The experimental chamber and the head stage of the amplifier (the current-voltage converter) are enclosed in an aluminum box that serves as a shield against atmospheric electrical noise. Mechanical vibrations are damped by placing the experimental chamber on a heavy metal plate that rests upon a partially inflated rubber inner tube of a wheel. Each compartment is connected to the electronic equipment through an Ag-AgCl electrode and a salt bridge (1M KCl, 2% agar) (Fig. 1). The electrical equipment consists of a current-to-voltage converter, a low-pass filter, a waveform generator and a data storage device (tape recorder, CD writer and/or computer). The bilayer formation is monitored on an oscilloscope by following the increase in capacitance between each compartment. To measure the capacitance, a triangular voltage wave is used (10 mV, 50 Hz). Since the capacitive current is directly proportional to the time derivative of the applied voltage, the result is a square wave with amplitude proportional to the

membrane capacitance during the thinning. To measure the bilayer conductance, a voltage pulse is applied (±100 mV) and the steady-state current is recorded. Typically, the membrane conductance should be about 10^{-9} S cm^{-2} and the membrane capacitance should be about 0.4 μF cm^{-2} or 0.8 μF cm^{-2} for painted and folded bilayers, respectively. In the presence of chloroplast lipids the conductance can increase ten times [39].

2. TECHNIQUES USED TO STUDY THE STRUCTURE OF PLANT ION CHANNELS IN PLANAR LIPID BILAYERS

2.1. Attenuated total reflection Fourier transform infrared spectroscopy

Attenuated total reflection Fourier-transform infrared (ATR-FTIR) spectroscopy is used to investigate the structure of the reconstituted proteins. The internal reflection element is a germanium (Ge) ATR plate (50 $\times$ 20 $\times$ 2 mm, Harrick EJ2121) with an aperture angle of 45°, yielding 25 internal reflections (Fig.4). A sample of proteoliposomes containing 15-100 μg proteins are spread on one side of the ATR plate and slow evaporation of the solvent under a continuous stream of nitrogen results in the formation of aligned lipid multibilayers oriented parallel to the ATR surface [1]. The lipid/protein weight ratio of the proteoliposomes must be around 10.

ATR-FTIR spectra are obtained on a Bruker IFS-55 spectrophotometer equipped with liquid nitrogen cooled MCT detector and a polarizer mount assembly (KRS-5 polarizer). The spectra are recorded at a nominal resolution of 4 cm^{-1} and 128 scans were averaged for each measurement. The spectrophotometers were continuously purged with dry air.

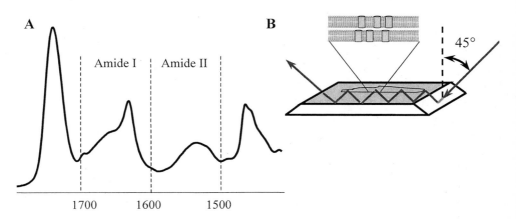

Fig. 4. The ATR-FTIR technique. A) Typical infrared spectra of a membrane protein reconstituted in planar lipid bilayers. B) Schematic diagram of the internal reflection element.

A considerable absorbance of the lipids in the amide I and amide II regions that overlapped the protein spectra was observed during the IR measurements. To obtain the pure protein spectra, the spectrum of a liposome blank is recorded under identical conditions and subtracted from the sample spectrum. To account for varying absorbance intensities in sample and lipid spectra, the integrated area of the lipid $v(C=O)$ band between 1705 and 1770 cm^{-1} is used as a reference to obtain the subtraction coefficient.

2.1.1. Secondary structure determination

The secondary structure composition is determined by the analysis of the amide I band shape between 1700 and 1600 cm^{-1}, which arises mainly from the C=O stretching vibration of the peptide group [57-58]. The analysis is performed on the deuterated sample. Only deuteration of the sample allows the differentiation of α-helical and irregular structures, because the absorption band of irregular structures shifts from about 1655 cm^{-1} to near 1642 cm^{-1} upon deuteration while the helix contribution remains around 1655 cm^{-1}. Since the different secondary structure components of a protein (α-helix, β-sheet, turn and other) are simultaneously recorded, their absorption bands overlap in the amide I region. A method for the analysis of the fractional composition of a secondary structure type in a protein has been developed in our laboratory by K. Oberg (K. Oberg, J.-M. Ruysschaert, E. Goormaghtigh, manuscript in preparation). The method is based on the multivariate statistical analysis method for band shape recognition as reported previously [59]. For a reference set, 50 commercially available soluble proteins with known high-resolution crystal structure were chosen. The selected proteins represent a wide distribution of α-helix and β-sheet compositions with 60 different domain folds. The reference set (RaSP50)

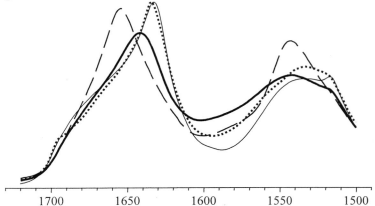

Fig. 5. FTIR spectra of proteins with different secondary structure. (---) hemoglobin, 69% α-helix; (——) trypsin inhibitor, 22% α-helix and 19% β-strand; (——) xylanase, 6% α-helix ans 61% parallel β-strand; (•••) concanavanlin 45% anti-parallel β-strand.

is generated by collecting the ATR-FTIR spectra (a few examples are depicted in Fig. 5) as well as the circular dichroism (CD) spectra (see below) of the 50 proteins. The RaSP50 represents currently the most complete reference set available for the secondary structure analysis of proteins. For the analysis of the IR spectra a combination of the amide I (between 1720 and 1600 cm^{-1}) and the amide II band (between 1600 and 1500 cm^{-1}) of the non-deuterated protein is used.

2.1.2. Secondary structure Orientation

ATR-FTIR spectroscopy provides information about the orientation of the peptide secondary structure embedded in the hydrated lipid film where the lipid acyl chains are oriented perpendicular to the surface of the ATR plate [1]. The method is based on the fact that the IR light absorption is maximal if the transition dipole moment of a molecular vibration is parallel to the electric field component of the incident light. The absorption is minimal if the transition dipole moment is perpendicular to the electric field component of the incident light. Thus, information about the orientation of the dipoles can be obtained by recording IR spectra with parallel and perpendicular polarized incident light, because the absorption intensities of a band will be different when measured with either light (Fig. 6). This permits the computation of the dichroic ratio, R^{ATR}, that contains all the information about the orientation of the protein secondary structure in an oriented lipid bilayer.

The protein spectra are recorded with parallel and perpendicular polarized light with respect to the plane of incidence, averaging 512 scans at a resolution of 2 cm^{-1}. To determine the dichroic ratio of the different secondary structure components, the amide I band (1700 to 1600 cm^{-1}) is decomposed by a least-square curve-fitting procedure using a Cauchy (Lorenzian/Gaussian) function as

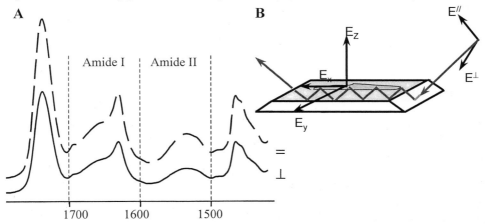

Fig. 6. ATR-FTIR experiments with polarized infrared light. A typical spectra of a membrane protein recorded with an IR beam oriented parallel and perpendicular to the incident plane.

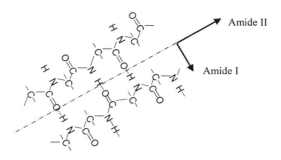

Fig. 7. Orientation of the transition dipole moments in a β-strand.

described previously [60]. The integrated areas corresponding to the β-sheet components (near 1630 cm^{-1}) were determined, and the ratio of the integrated areas from the two polarized spectra then yielded the dichroic ratio of the β-sheet, R_β^{ATR}. The dichroic ratio of the amide II band is determined from the ratio of the integrated areas between 1590 and 1505 cm^{-1} from the two polarized spectra. The dichroic ratio of the lipid ν(C=O), called below R^{ATRiso}, is obtained from the parallel and perpendicular polarized spectra. R^{ATRiso} was used to compute the film thickness and the values of the electric field components E_X, E_Y and E_Z at 1631 cm^{-1} as described previously [3]. For these calculations refractive indexes of 4.0 and 1.44 were used for the Ge-plate and the sample film, respectively.

In an anti-parallel β-sheet the amide I transition dipole moment that gives rise to the major component of the amide I band (at ~1630 cm^{-1}) has an orientation perpendicular to the β-strand axis. The transition dipole moment that contributes mainly to the amide II band is oriented parallel to the β-strand axis [61-62] (Fig. 7). This allows calculating the orientation of the β-strands relative to the barrel axis from the dichroic ratio of the amide I band and of the amide II band [63-64]

2.2. CIRCULAR DICHROISM

2.2.1. Circular dichroism measurements

Circular dichroism (CD) spectroscopy is uniquely sensitive to the molecular conformation of proteins. The far-UV CD bands of proteins (between 180 to 260 nm) derive primarily from the peptide bonds and reflect the secondary structure of the protein (see Fig. 8.) [65]. Circular dichroism (CD) spectra are recorded from proteins in detergent solution (1 mM Hepes, 0.1% Genapol X-080, pH 7.2). The samples are filtered (0.2 μm) before use. The

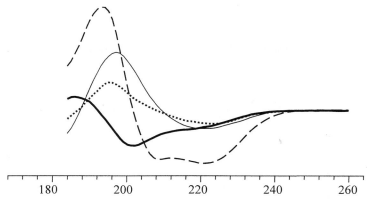

Fig. 8. CD spectra of proteins with different secondary structure. (---) hemoglobin, 69% α-helix; (   ) trypsin inhibitor, 22% α-helix and 19% β-strand; (   ) xylanase, 6% α-helix ans 61% parallel β-strand; (•••) concanavanlin 45% anti-parallel β-strand.

samples are placed in a quartz cuvette of 0.1 cm path length (Hellma) and measurements are performed using a JASCO-710 spectra polarizer. Spectra are recorded at room temperature from 260 to 185 nm at a scan speed of 50 nm/min and a resolution of 0.5 nm. 8 scans are accumulated and averaged. A spectrum of the buffer in absence of protein is recorded in the same manner.

2.2.2. Secondary structure determination

The determination of protein secondary structure from the CD spectra uses the same multivariate statistical analysis method for band shape recognition as described above. Here, the CD spectra of the 50 proteins (Fig. 8) were collected and used to generate the reference set (RaSP50). Spectra of the buffer are subtracted from the CD-spectra of the proteins.

3. CASE STUDY: THE MITOCHONDRIAL VDAC OF PLANT SEEDS

The seeds of Fabaceae, like kidney bean (*Phaseolus vulgaris*) contain a little embryo and two large cotyledons surrounded by the seed coat. The cotyledons of the Fabaceae are the organs of storage reserves (starch (58%), protein (20%), and lipid). During germination the storage materials are mobilized to support the growth and the development of the embryo. During imbibition rapid water uptake results in an approximately two-fold increase in fresh weight of the seed. Imbibition is accompanied by a strong increase in oxygen consumption due to the onset of mitochondrial respiration. Depending on the specific plant, increase in mitochondrial activity lasts for 5 to 10 days and then decays. At this stage, the reserves of the storage tissue have been consumed. Senescence of the cotyledons commences, marked by a decrease in respiration activity [66]. Thus, though most plants are autotrophic, they are heterotrophic at the early stages of their development during which the mitochondria play a key role in the metabolism.

3.1. The VDAC

The exchange of solutes between the mitochondrial matrix and the cytoplasm proceeds through the two mitochondrial membranes. The voltage-dependent anion-selective channel, VDAC, is considered to constitute the general pathway for hydrophilic solutes across the mitochondrial outer membrane. VDAC was first identified in crude extracts of mitochondria of *Paramecium aurelia* upon reconstitution into planar lipid bilayers [67] and has then been localized to the mitochondrial outer membrane where it constitutes a major component [68]. Since then, VDAC has been purified from a great variety of organisms, e.g., yeast [69], *N. crassa* [70], mammals [71-72], and plants [19, 20, 22, 73].

Several *vdac* genes have been found in plants [21, 74-75]. The presence of a multigene family raises the question of how many of the *vdac* genes are actually expressed and translated into functional proteins. The detection of three different VDAC messenger RNAs (mRNA) in different wheat tissues indicated that transcription of three *vdac* genes occurs [74] but purification of VDAC from the plant yielded only one of the three wheat VDAC isoforms [20]. The physiological role of the different isoforms has yet to be elucidated.

3.2. Biochemical methods

Separation of membrane fractions and channels purification vary with plant species and organs. The reader is referred to the specific literature to consult the details concerning specific protocols. Here, we will describe the procedure used in our laboratory to purify the mitochondrial VDAC isoforms from seeds.

3.2.1. Isolation of Mitochondria

Mitochondria are isolated from plant seeds of kidney bean (*P. vulgaris*), following the protocol of Douce et al. [76] with some minor modifications (Fig. 9A). Usually, 200 g (dry weight) of seeds are soaked for 24 h in aerated tap water at room temperature. The seed coat and the embryo are removed. The cotyledons are homogenized in 500 ml extraction media (0.3 M mannitol, 0.6% PVP (25 kDa), 4 mM L-cysteine, 1 mM EDTA, 0.1% BSA, 25 mM MOPS, pH 7.5) at low speed for 2 x 3 s in a 1 l Warring blender. The homogenate is squeezed through nylon net (60 µm pores) and the homogenization step is repeated once. Crude mitochondria are obtained after four differential centrifugation steps (Fig. 9A). The crude mitochondria are further purified on a 28% Percoll step gradient. The final mitochondrial pellet is resuspended in approximately 15 ml 0.3 M Mannitol, 10 mM Hepes and 1 mM EDTA, pH 7.2

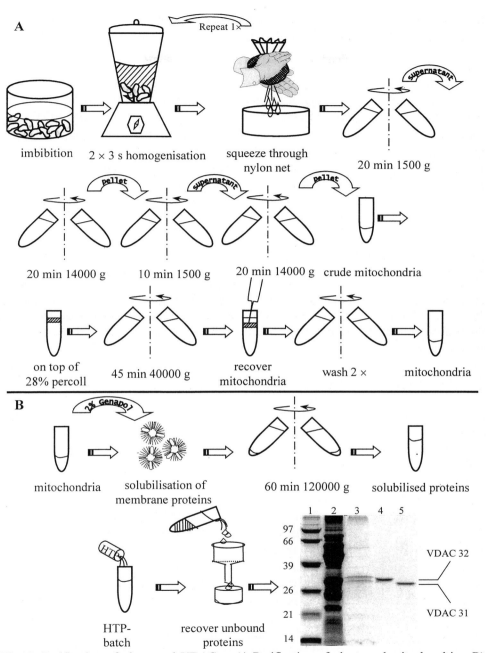

Fig. 9. Purification of plant seed VDACs. A) Purification of plant seed mitochondria. B) Purification and separation of VDAC isoforms. The final step is a 12% SDS-PAGE gel. Lane 1: MW standard, lane 2: mitochondrial proteins solubilized in 1% Genapol X-080, lane 3: VDACs after HTP, lanes 4 and 5: separation of two isoforms after a chromatofocusing.

at a protein concentration of 25 to 30 mg/ml. The Protein concentration was determined with the BCA-kit from Pierce using BSA (Sigma) as the standard.

3.2.2. VDAC purification

VDAC was purified from whole mitochondria according to De Pinto et al., [72] with the modification of Heins et al., [21] (Fig. 9B). Mitochondria (300 to 400 mg protein) are solubilized in 2% (v/v) Genapol X-080, 1 mM EDTA, 10 mM Hepes (pH 7,2) at a protein concentration of 5 mg/ml (final concentrations). The suspension is then frozen overnight (-20°C), thawed at room temperature and incubated for 30 min on ice. Non-solubilized proteins are pelleted at 35,000 rpm for 60 min (Beckman L7, rotor Ti70). Dry hydroxyapatite (1g HTP/10 mg protein) is added in a batch procedure to the detergent-solubilized proteins (~70 ml containing 2 to 2.5 mg protein/ml), incubated for 30 min at 4°C, and the unbound protein fraction, which was highly enriched in the two VDAC isoforms, is recovered by filtration.

3.2.3. Separation of the two kidney bean VDAC isoforms

In contrast to most purification procedures reported in the literature, the HTP purification step resulted in the co-purification of two VDAC isoforms with mitochondria from Fabaceae seeds. For the investigation of the functional and structural properties of each individual kidney bean VDAC isoform, it is essential to develop a method for the separation of the two proteins. Chromatofocusing chromatography in the presence of 4 M urea is employed to separate the two VDAC isoforms [22].

The recovered fraction from the HTP-purification step is concentrated five to six times (Centriprep-30, Amicon) and dialyzed overnight against the chromatofocusing start buffer (see below). Low-pressure chromatofocusing chromatography with polybuffer exchanger PBE 94 is performed according to the instruction manual from Pharmacia with a pH-gradient from pH 9.4 to pH 7.0 at room temperature. The 10 ml column (1.6 cm × 5 cm) is equilibrated with the start buffer containing 25 mM ethanolamine, 0.1% (v/v) Genapol X-080, 4 M urea, pH 9.4. 2 ml aliquots of the sample (in start buffer) are loaded onto the chromatofocusing column and proteins are eluted with 10% Polybuffer 96, 0.1% Genapol X-080 and 4 M urea, pH 7.0. In the absence of urea the two isoforms are co-eluted. The collected fractions are checked by SDS-PAGE for proteins and those containing the same proteins are pooled and concentrated (Centriprep-30, Amicon). To remove the ampholytes and the urea present in the elution buffer, the proteins are applied to a gel filtration column (Sephadex G-75, 1.6 × 18 cm) equilibrated with 0.1% Genapol, 10 mM Hepes, pH 7.2. The column is run at room temperature at a low flow-rate of 7.5 cm/h in order to avoid precipitation of proteins during the elimination of urea.

3.2.4. VDAC purification for ATR-FTIR spectroscopy

For ATR-FTIR measurements purification of the two VDAC proteins is achieved in a single chromatofocusing chromatography purification step i.e. the HTP-batch procedure described above was omitted. This modification resulted in a small but sufficient increase in the yield of VDAC proteins, albeit slightly compromising the purity. Mitochondria are isolated as described above. The final mitochondrial pellet (300 to 400 mg protein, 20 to 25 mg/ml) is mixed with an equal volume of two-fold concentrated chromatofocusing start buffer (see above) supplemented with 4% (v/v) Genapol-X-080 to solubilize the membrane proteins. This suspension was frozen at −20°C in 2 ml aliquots until use. An aliquot of the mitochondria suspension was thawed at room temperature and incubated for 30 min at 4°C. About 20 mg proteins are loaded onto the chromatofocusing column and proteins are eluted as described above. The fractions containing the individual VDAC isoforms from 3 chromatofocusing columns (ca. 25 ml) are pooled and concentrated about 20 fold (Centriprep-30, Amicon). The elution buffer is exchanged against 1 mM Hepes 0.1% (v/v) Genapol-X-080, pH 7.2 by gel filtration (Sephadex G75; 1.6 × 18 cm) at room temperature at a flow rate of 7.5 cm/h. The proteins elute at a concentration of about 0.07 to 0.12 mg/ml.

3.2.5. Reconstitution of the VDAC in liposomes

Asolectin (soybean phosphtidylcholine type II, Sigma) is washed by the procedure of Kagawa and Racker [77]. For the infrared measurements, the proteins are reconstituted into liposomes composed of purified asolectin and 5% stigmasterol. 950 µg asolectin and 50 µg stigmasterol are dissolved in 0.5 ml chloroform and the solvent is evaporated under a stream of N_2 to have a thin lipid film, which was dried further overnight under vacuum. The dry lipids are resuspended in 2 ml protein solution containing about 0.1 mg protein/ml. The lipid-protein-detergent mixture is incubated 30 min at room temperature under constant agitation followed by 3 freeze/thawing cycles. The removal of detergent is achieved by 4 subsequent additions of 40 mg Bio-Beads (previously washed with methanol and water). Incubation times were 45, 45, 30, and 30 min under constant agitation at 4°C. After removal of the final Bio-Beads, the proteoliposome suspension is diluted with 1 mM Hepes, pH 7.2 to a total volume of 4.5 ml. The proteins associated with the lipid vesicles are pelleted at 37,000 rpm for 2 h (Beckman L7, SW-60 rotor). This washing step is repeated once. Finally, the proteoliposomes are suspended in 20 to 30 µl of 1 mM Hepes, pH 7.2. Protein recovery was in the range of 25 to 35%.

3.3. Two mitochondrial VDAC isoforms are synthesised in plant seeds

The experimental procedure described here above permits to purify and separate two VDAC isoforms from plant seeds. In some seeds (e.g., beans) each

pH   3   4   5   6   7   8   9   10

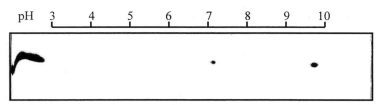

Fig. 10. Isoelectric point of two VDAC isoforms purified from lentil seeds.

isoform has a different molecular mass but in others (e.g., lentils) the isoforms have an identical apparent molecular mass [34]. However, all the isoforms have a specific isoelectric point (pI) ranging between 7 and 10 when 2D-gel electrophoresis is performed. For instance, the single 31.5 kDa band of lentil VDACs (on the left side of the 2D-gel) separate into two polypeptides with isoelectric points of 7.1 and 9.8, respectively (Fig. 10). Previous purifications of a plant VDAC protein from maize and wheat [20, 73] revealed that VDAC formed a single band on one-dimensional SDS-PAGEs. Our results suggest that this does not mean that mitochondria of these plants contain a single VDAC protein. 2D-gel electrophoresis would be required to assess that different VDAC isoforms exist in mitochondria of these plants. To demonstrate that the purification procedure did not alter the function of the proteins, each purified VDAC isoform were reconstituted in planar lipid bilayers. The two isoforms purified from kidney beans were called VDAC31 and VDAC32 according to their apparent molecular weight on a SDS-PAGE gel [34].

3.4. Functionnal analysis

A minimal description of a channel should include measurement of the conductance, selectivity and voltage-dependence [78]. The results shown here were obtained using the painted planar bilayer technique.

3.4.1. VDAC reconstitution

Purified VDACs were added to the cis side of a planar lipid bilayer at a concentration of 50 to 100 ng/ml. A constant electrical potential difference of +10 mV was applied to the membrane. At this voltage, VDAC is mainly in its fully open state and each insertion of a channel into the membrane yields to a discrete current step, giving rise to a typical "staircase" time course of the current as illustrated in Fig. 11A. The initial current increments occurred with rather uniform amplitudes. The histogram shows an amplitude distribution of the observed current steps (Fig. 11B). Each of the major peaks accounts for 85% of all observed channel incorporations with a mean conductance of 3.7 ± 0.1 nS (n = 79) and 4.0 ± 0.1 nS (n = 83) for VDAC 31 and VDAC 32, respectively. These conductances correspond to the fully open state of a single channel and are consistent with those reported for VDACs from other sources, ranging from

A   B

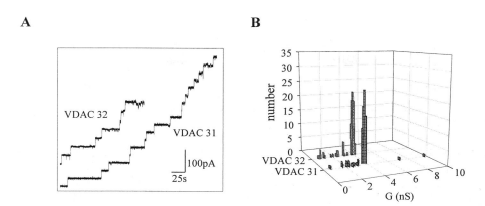

Fig. 11. Reconstitution of kidney bean VDAC isoforms into planar lipid bilayers. A) Time course of successive reconstitution (V = +10mV). B) Amplitude histogram of the current steps.

3.6 nS to 4.5 nS in 1 M KCl [79-81]. The two isoforms exhibited slight differences in their single channel conductances similar to that reported for the wheat VDAC isoforms [41]. The lower conductances that were observed during these experiments (Fig. 11B) likely represent the insertion of channels in low conducting states, whereas the higher ones probably result from the simultaneous incorporations of several channel proteins.

3.4.2. Single channel records

For single channel recordings the proteins (about 10 ng/ml) were added to the cis compartment and a voltage of +10 mV was applied to the planar lipid bilayer, in order to detect channel incorporations. After the first channel insertion into the bilayer was observed, stirring of the compartment solution was stopped to prevent the insertion of further channels during the course of an experiment. Voltages from ±10 mV to ±50 mV were then applied to the planar lipid bilayer for 20 s and the current flow through a single VDAC channel was recorded. Fig. 12 shows the time course of typical current fluctuations recorded in symmetrical conditions (1M KCl) at different voltages. The single-channel records indicate that the VDAC has several subconducting states. At the lower voltages, the channel spends most of the time in the fully open state and transitions to low-conducting states occurred only rarely. At the higher voltages transitions to subconducting states became more frequent. At least four different subconducting states ranging from 0.5 to 3 nS were recorded. Transitions between the fully open state and the subconducting states were reversible.

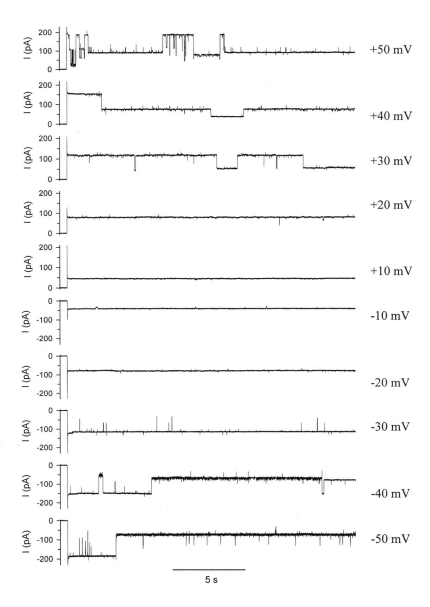

Fig. 12. Current fluctuations of a single VDAC31 recorded at various voltages.

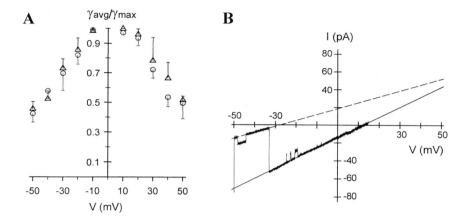

Fig. 13. A) Voltage-dependence of VDAC31 (o) and VDAC32 (Δ). B) Selectivity of VDAC32.

All VDACs examined so far are voltage-dependent *in vitro*. The voltage-dependence of a multichannel-containing membrane is usually obtained from the relationship between the probability of being in the fully open state and the voltage applied to the membrane [67, 82].   Practically, the normalized conductance G/G_0 is plotted versus the voltage, where G is the steady state membrane conductance at a given voltage and G_0 is the membrane conductance when all channels are open.   To obtain the voltage-dependence of a single-channel, the current flowing through a single-channel were recorded during 20 seconds at various voltages. The magnitude of the current flowing through the channel in its fully open state was measured (I_{max}) and the average current (I_{avg}) flowing through the channel during 20 seconds was calculated.   These two currents were converted into a conductance (γ and γ_{avg}, respectively) using the Ohm law.   The ratio γ_{avg}/γ was plotted against the voltage and a typical bell-shape voltage-dependence was obtained (Fig. 13A).

To gain information about the ion selectivity of a channel, experiments in asymmetrical conditions, i.e. in the presence of a salt gradient, have to be performed. The voltage difference at zero current is the reversal potential, V_r. The selectivity of the channel can be quantified in terms of a permeability ratio, $P_K{}^+/P_{Cl}{}^-$, from the reversal potential, using the Goldman-Hodgkin-Katz equation:

$$V_r = \frac{RT}{F} \ln \frac{P_{K^+}\left[K^+\right]^{trans} + P_{Cl^-}\left[Cl^-\right]^{cis}}{P_{K^+}\left[K^+\right]^{cis} + P_{Cl^-}\left[Cl^-\right]^{trans}} \tag{1}$$

where P_j is the permeability coefficient of the ion j, R is the gas constant, F is the Faraday's constant and T is the absolute temperature. To get information about the ion selectivity of the VDAC, a symmetrical 10 mHz triangular voltage wave was applied to the lipid bilayer membrane containing a single channel and the current was recorded. In a first step, experiments were performed in asymmetric 500/100 mM KCl (cis/trans) conditions. As a control, the concentration of the trans compartment was then raised to 500 mM KCl, to have symmetrical conditions. Both voltage and current were digitized and used to draw a current-voltage curve as illustrated in Fig. 13B. The current-voltage relationship obtained with a single VDAC 32 channel in asymmetric conditions shows different conducting states of the channel (Fig. 13B). The large conducting state, corresponding to the fully open state of the channel, had a reversal potential of 12.5 mV ± 0.4 (n = 7) which, according to the Goldman-Hodgkin-Katz equation, corresponds to a permeability ratio P_K^+/P_{Cl}^- equal to 0.47. Thus, being in the fully open state the channel is slightly anion selective. The low conducting state (dashed regression line) had a permeability ratio $P_K^+/P_{Cl}^- = 6.6$. This means the channel is cation selective when being in a low conducting state. The change of the ion selectivity upon transition between the fully open state and a low conducting substate is a typical feature of VDAC [83-84].

3.5. Structural analysis

3.5.1. Theoretical prediction

Theoretical prediction of the structure of a membrane protein can be achieved using bioinformatic tools. Comparison between theoretical and experimental data can be particularly useful to suggest new hypothesis that could be check experimentally. The VDAC secondary structure prediction was

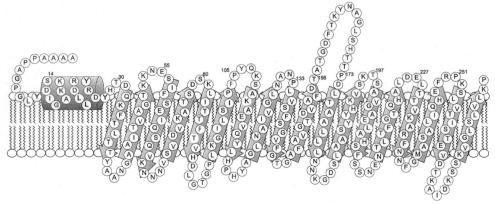

Fig. 14. Predicted topology for the rice VDAC (AJ251562)

first done with the Prof method [85] on rice (Y18104, AJ251562, AJ 251563) and then compared to the result obtained from yeast (P04840, P40478), Neurospora (P07144), human (P21796, P45880, Q94277), pea (CAA80988) and Arabidopsis (CAA10363, CAC01828). A secondary structure was considered as acceptable if it was detected in all VDAC sequences. This prediction was then compared to that obtained with the PHD method [86-87] and the neural network-based predictor method [88] to check for consistency. In the VDACs, all the β-strands should be transmembrane segments. Therefore, only the β-strand prediction consisting of at least seven residues were considered to be correctly assigned. The results suggest that there is 18 membranes spanning β-strands and an α-helix of about 14 residues at the amino terminal end in the sequence tested. When plotted on a helical wheel, the α-helix displays an amphipatic character, suggesting that it could be partly embedded in the membrane laying parallel to the surface. The predicted topology of one of the rice VDAC sequenced in our laboratory is shown in Fig. 14.

3.5.2. FTIR spectra

For the analysis of proteins, the most important bands are the amide I band between 1700 and 1600 cm^{-1} and the amide II band between 1595 and 1505 cm^{-1}. Amide I results to 75 to 85% from the stretching vibrations of the peptide C=O group (ν(C=O)). Amide II is more complex and is therefore less frequently used for secondary structure determination. It derives to 60% from the in-plane bending vibration of the peptide N-H group (δ(N-H)) with some contribution from the C-N stretching [58, 61]. The exact frequencies of the absorption bands depend on the nature of H-bonding involving the peptide groups which, in turn, is determined by the particular secondary structure (see Fig. 5). Fig. 15 shows the ATR-FTIR spectra of the plant seed VDAC in the amide I and amide II region. The amide I bands are characterized by a major component located around 1631 cm^{-1} which is assigned to β-sheet. The position of the major component at 1631 cm^{-1} and a weak component at about 1694 cm^{-1} in the amide I region, together with the major amide II band at 1535 cm^{-1}, indicate that the β-strands are in an anti-parallel configuration [58, 61]. The shape of the amide I band of the VDAC is very similar to that of the OmpF porin, suggesting a similar secondary structure composition.

3.5.3. Secondary structure: band shape recognition

The secondary structure analysis of both plant seed VDAC and OmpF are shown in Table 1. The good agreement between the secondary structure of OmpF determined by FTIR analysis and that from the crystal structure strongly supports the validity of the method uses to find the structure of the VDAC.

Table 1
Secondary structure of the reconstituted plant seed VDAC and bacterial OmpF porin determined by ATR-FTIR.

	% α-helix	% β-sheet	% turn	% other
VDAC 31	14	50	12	27
VDAC 32	9	53	11	35
OmpF	0 (3.5)	58 (56)	13 (13)	42

A small shoulder in the VDAC IR-spectra at about 1660 cm^{-1} (Fig. 15) is an indication for a higher α-helix content.

3.5.4. Orientation of the β-strands in a β-barrel

The ATR-FTIR spectroscopy technique provides information about the orientation of proteins inserted in a lipid membrane when IR spectra are recorded with parallel and perpendicular polarized incident light. The dichroic ratio R^{ATR} is obtained by dividing the absorbance of a band measured with parallel incident light by the absorbance of the band measured with perpendicular incident light (R^{ATR} = $A^{//}/A^{\perp}$, [3]). R^{ATR} contains all the information about the orientation of the protein secondary structures in an oriented lipid bilayer.

Protein spectra were recorded with parallel and perpendicular polarized

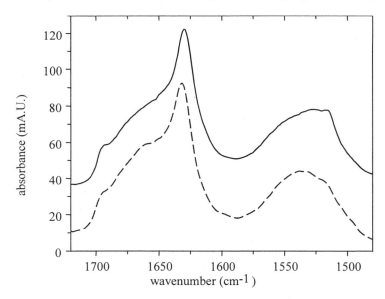

Fig. 15. ATR-FTIR spectra of the kidney bean VDAC31 (—) and of the bacterial OmpF porin (---).

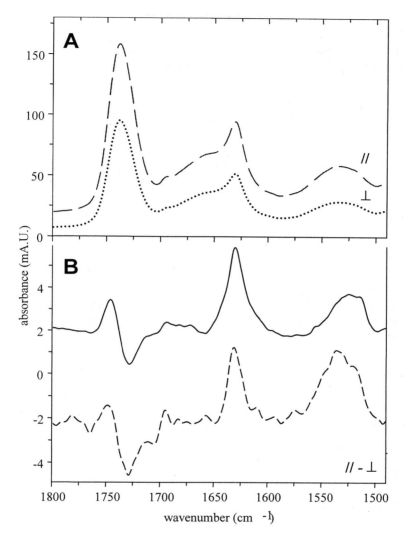

Fig. 16. Dichroic spectra. A) ATR-FTIR spectra of the kidney bean VDAC31 recorded with parallel and perpendicular polarized incident light. B) Dichroic spectra of VDAC31 (——) and bacterial OmpF porin (---) obtained after substraction of the perpendicular spectra from the parallel spectra.

incident light (Fig. 16A). The dichroic spectra (// - ⊥) of the VDAC and of the OmpF yield qualitative information about a preferred orientation of a secondary structure (Fig. 16B). The three dichroic spectra show a positive dichroism signal for the main component of the amide I band at ~1630 cm^{-1} and a positive deviation of the amide II band. This indicates that in all three proteins the β-sheet structure has a preferred orientation.

Table 2

The dichroic ratio of the β-sheet component, R_β^{ATR}, of the amide I band and the dichroic ratio of the amide II band of plant seed VDAC32 and bacterial OmpF porin. The data are the mean of three experiments (± SEM).

	R_β^{ATR}	$R_{amideII}^{ATR}$
VDAC 32	1.94 (± 0.07)	1.83 (± 0.03)
OmpF	1.72 (± 0.1)	1.81 (± 0.09)

For a quantitative determination of the β-strand orientation, the direction of the transition dipole moments must be known. For the special case of an anti-parallel β-sheet, it has been shown that the amide transition dipole moment that gives rise to the most intense component of the amide I band at around 1630 cm^{-1} is oriented perpendicular to the β-strand direction and parallel to the plane of the β-sheet. The amide transition dipole moment that gives rise to the component of the amide II band near 1530 cm^{-1} is oriented parallel to the strand direction and parallel to the plane of the β-sheet (Fig.) [2, 61-62]. To determine the orientation of the β-sheet secondary structure, the amide I bands of the polarized spectra were decomposed into their different components by a least-square curve-fitting procedure. The integrated areas corresponding to the β-sheet component (near 1630 cm^{-1}) were determined, and the ratio of the integrated areas from the two polarized spectra then yielded the dichroic ratio of the β-sheet component, R_β^{ATR}. The dichroic ratio of the amide II band, $R_{amideII}^{ATR}$, was determined from the ratio of the integrated areas between 1590 and 1505 cm^{-1}. The dichroic ratios are given in Table 2. From the knowledge to the dichroic ratio of the secondary structure components of amide I and amide II bands we can calculate the tilt angle of the β-strands with respect to the barrel axis and the orientation of the barrel with respect to the normal to the membrane surface, see [2, 63, 64, 89, 90] and references therein. We have shown that the β-strands of plant seed VDACs are tilted by 46°-47° with respect to the barrel axis and that the whole barrel should be slightly tilted in the membrane [90].

3.5.5. Circular dichroism

Previously, CD measurements have been performed with VDAC from yeast and *N. crassa*. The analysis of these CD spectra indicated 40 to 60% β-sheet and they were the only experimental data about the secondary structure of VDAC available in the literature [91-92]. The CD spectra of VDAC 32 and OmpF in 0.1% Genapol are compared in Fig. 17. The CD spectrum of VDAC 32 in non-denaturing detergent solution has a negative band at 216 to 218 nm and a maximum at 196 nm. This CD spectrum is typical for proteins rich in β-sheet structure. The CD spectrum of the plant VDAC closely resembles the CD

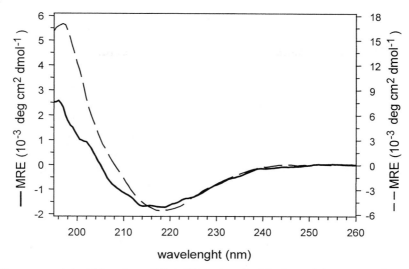

Fig. 17. CD spectra of the kidney bean VDAC32 (——) and of the bacterial OmpF porin (---).

spectrum of VDAC from N. crassa [92]. The OmpF CD spectrum corresponds to that reported previously for the native OmpF [93]. The similarity between the VDAC and the OmpF spectra indicate a similar secondary structure composition. An estimate of the fractional secondary structure content of VDAC and OmpF from their CD spectra was obtained using the method of band shape recognition Table 3).

Comparing the values obtained here with the values deduced from the crystal structure of OmpF, indicates that the CD measurements underestimate the β-sheet content. It is well known that CD measurements generally provide more accurate estimations of the protein α-helix content since α-helices show a strong and characteristic CD spectrum (see Fig. 8) with a double minimum at 222 nm and 208 to 210 nm and a maximum at 191 to 193 nm. On the contrary, the CD spectrum of β sheets is weak and depends on the length and twist of the β-strands and can therefore be highly variable for different proteins [94]. Thus, the determination of the β-sheet content is less reliable when estimated from CD spectra.

Table 3
Secondary structure of VDAC 32 and OmpF in 0.1 % Genapol determined by CD spectroscopy. The values in parenthesis are deduced from the crystal structure of OmpF.

	% α-helix	% β-sheet	% turn	% other
VDAC 32	0	38	13	45
OmpF	9 (3.5)	41 (56)	13 (13)	39

3.5.6. Topological model for plant mitochondrial VDAC

According to our results, plant VDACs are made up of 50 to 57% β-sheet in an anti-parallel conformation. Although the exact amino acid sequences of the kidney bean VDAC is not known, a length of ~280 may be assumed corresponding to the length of the known VDAC sequences from fungi, human and plants [21, 75, 95-96]. This means that approximately 140 to 150 residues are in a β-sheet conformation. The study of the function showed that each isoform can form a large pore. Taken together, this suggests that the pore of VDAC is most likely formed by a β-barrel. The structural geometry of a regular β-barrel is mainly determined by the number, n, of β-strands and the tilt angle, β, of the β-strands [97]. The radius, R, of a cylindrical barrel is related to the strand tilt by

$$R = \frac{d}{2\sin(\pi/n)\cos\beta} \tag{2},$$

where d = 0.472 nm is the distance between two residues in a β-strand. From the knowledge of the barrel diameter and the mean orientation of the β-strands determined here by polarized ATR-FTIR spectroscopy, the number of β-strands in VDAC can be calculated. A diameter of VDAC of 3.7 nm has been determined by electron microscopy studies of yeast VDAC 2D-crystals and from low-resolution crystals of human VDAC [98-99]. Thus, using equation 5 and a mean β-strand tilt of 46° to 47° results in a barrel that is formed from 17 β-strands. Because β-barrels have an even number of β-strands, VDAC could have

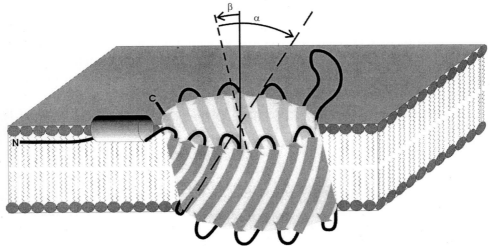

Fig. 18. Putative model for the mitochondrial plant VDAC reconstituted in a planar lipid bilayer.

either 16 or 18 β-strands (a tilt of 48° would give 16 β-strands and a tilt of 44° would result in 18 β-strands). Knowing the approximate number of β-strands of the VDAC barrel and the number of amino acids involved (see above) the average length of the strands can be calculated. The 16 (or 18) β-strands would have a mean length of approximately 9 (or 8) amino acid residues per strand. Thus, according to our structural analysis, VDAC could form a somewhat tilted β-barrel (Figure 18) composed of 16 (or 18) β-strands with a mean length of 9 (or 8) residues per strand. The strands are tilted at about 46 to 47° relative to the barrel axis to form a barrel with a diameter of 3.7 nm.

Acknowledgments. F.H. and E.G. are Research Directors and M.L. is a Foreign Postdoctoral Fellow from the National Fund for Scientific Research (Belgium). We thank F. Van Eycken for her technical assistance. This work was supported by a grant of the "Communnauté Française de Belgique- Actions de Recherches Concertées"

REFERENCES

[1] U.P. Fringeli and H.H. Gunthard, Mol. Biol. Biochem. Biophys., 31 (1981) 270.
[2] L.K. Tamm and S.A. Tatulian, Q. Rev. Biophys., 30 (1997) 365.
[3] E. Goormaghtigh, V. Raussens and J.-M. Ruysschaert, Biochim. Biophys. Acta 1422 (1999) 105.
[4] R. Hedrich and J.I. Schroeder, Annu. Rev. Plant Physiol., 40 (1989) 539.
[5] M. Tester, New Phytol., 14 (1990) 305.
[6] M. R. Blatt, J. Membrane Biol., 124 (1991) 95.
[7] S.D. Tyerman, Annu. Rev. Plant Physiol. Mol. Biol., 43 (1992) 351.
[8] D.P. Schachtman, Biochem. Biophys. Acta, 1465 (2000) 127.
[9] P.J. White, Biochem. Biophys. Acta, 1465 (2000) 171.
[10] H. Barbier-Brygoo, M. Vinauger, J. Colombet, G. Ephritikhine, J.M. Frachise and C. Maurel, Biochem. Biophys. Acta, 1465 (2000) 199.
[11] H.E. Neuhaus and R. Wagner, Biochem. Biophys. Acta, 1465 (2000) 307.
[12] F.J. Maathuis and D. Sanders, Curr. Opinion Plant Biol., 2 (1999) 236.
[13] A.A. Aleksandrov, G.N. Berestovsky, S.P. Vokova, I.Y. Vostikov, O.M. Zherelova, S. Kravchik and V.Z. Lunevsky, Dokl. Acad. Nauk S.S.S.R., 227 (1976) 723.
[14] P.J. White and M. Tester, Physiol. Plant., 91 (1994) 770.
[15] B. Bölter and J. Soll, EMBO J., 20 (2001) 935.
[16] S. Reumann, Biol. Chem., 381 (2000) 639.
[17] M. Colombini, *In* Ion Channel Reconstitution, C. Miller (ed.), Plenum, New York, (1986) 533.
[18] D.P. Smack and M. Colombini, plant Physiol., 79 (1985) 1094.
[19] A. Schmid, S. Krömer, H.W. Heldt and R. Benz, Biochim. Biophys. Acta, 1112 (1992) 174.
[20] A. Blumenthal, K. Kahn, O. Beja, A. Galum, M. Colombini and A. Breiman Plant Physiol., 101 (1993) 579.

[21] L. Heins, H. Mentzel, A. Schmid, R. Benz and U.K. Schmitz, J. Biol. Chem., 269 (1994) 26402.

[22] H. Abrecht, R. Wattiez, J.-M. Ruysschaert and F. Homblé, Plant Physiol., 124 (2000) 1181.

[23] S. Shimizu, M. Narita and Y. Tsujimoto, Nature, 399 (1999) 483.

[24] I. Mitsuhara, K.A. Malik, M. Miura and Y. Ohashi, Curr. Biol., 9 (1999) 775.

[25] M.B. Dickman, Y.K. Park, T. Oltersdorf, W. Li, T. Clemente and R. French, PNAS, 98 (2001) 6957.

[26] T.H. Tien, Bilayer Lipid Membranes, Marcel Dekker, New York, 1974.

[27] W. Hanke and W.-R. Schlue, Planar Lipid Bilayers, Academic Press, London, 1993.

[28] C. Miller (ed.), Ion Channel Reconstitution, Plenum, New York, 1986.

[29] P. Mueller, D.O. Rudin, H.T. Tien and W.C. Wescott, Nature 194 (1962) 979.

[30] P. Mueller, D.O. Rudin, H.T. Tien and W.C. Wescott Circulation 26 (1962) 1167.

[31] P. Mueller, D.O. Rudin, H.T. Tien and W.C. Wescott J. Phys. Chem., 67 (1963) 534.

[32] B. Fuks and F. Homblé, J. Biol. Chem., 270 (1995) 9947.

[33] P.J. White and M. Tester, Planta 186 (1992) 188.

[34] H. Abrecht, R. Wattiez, J.M. Ruysschaert and F. Homblé, Plant Physiol., 124 (2000) 1181.

[35] B. Klughammer, M. Betz, R. Benz and K.-J. Dietz, J. Membrane Biol., 128 (1992) 17.

[36] S. Reuman, E. Maier, H.W. Heldt and R. Benz, Eur. J. Biochem., 251 (1998) 359.

[37] M. Montal and P. Mueller, Proc. Natl. Acad. Sci. USA (1972) 69: 3561.

[38] M. Colombini, Methods in Enzymol., 148 (1987) 465.

[39] B. Fuks and F. Homblé, Biophys. J., 66 (1994) 1404.

[40] B. Fuks and F. Homblé, Plant Physiol., 112 (1996) 759.

[41] A. Elkeles, A. Breiman and M. Zizi, J. Biol. Chem., 272 (1997) 6252.

[42] B. Sakmann and E. Neher Plenum Press, New York, 1995.

[43] R. Coronado and R. Latorre, Biophys. J., 43 (1983) 231.

[44] B.A. Suarez-Isla, K. Wan, J. Lindstrom and M. Montal, Biochemistry, 22 (1983) 2319.

[45] N. Galtier, F. Simon-Plas, G. Miquel, F. Homblé, C. Grignon and P. Seta, Plant Cell Physiol., 33 (1992) 585.

[46] C. Miller, J. Membrane Biol., 40 (1978) 1.

[47] C. Miller and E. Racker, J. Membrane Biol., 30 (1976) 283.

[48] K. Anzai, M, Masumi, k; Kawasaki and Y. Kirino, J. Biochem., 114 (1993) 487.

[49] W. Hanke, H. Eibl and G. Boheim, Biophys. Struct. Mech., 7 (1981) 131.

[50] C. Miller, P. Arvan, J.N. Telford and E. Racker, J. Membrane Biol., 30 (1976) 271.

[51] F. Cohen, J. Zimmemberg, A. Finkelstein, J. Gen. Physiol., 75 (1980) 251.

[52] F. Cohen, M. Abakas, A. Finkelstein, Science, 217 (1982) 458.

[53] D.J. Woodbury and C. Miller, Biophys. J., 58 (1990) 833.

[54] D.J. Woodbury, J. Membrane Biol., 109 (1989) 145.

[55] T.K. Rostovtseva, C.L. Bashford, A.A. Lev and C.A. Pasternak, J. Membrane Biol. 141 (1994) 83.

[56] K. Kaufmann and I. Silman, Biophys. Chem., 18 (1983) 89.

[57] D.M. Byler and H. Susi, Biopolymer, 25 (1986) 469.

[58] E. Goormaghtigh, V. Cabiaux and J.M. Ruysschaert, Subcell. Biochem., 23 (1994) 329.

[59] V. Cabiaux, K.A. Oberg, P. Pancoska, T. Waltz, P. Agre and A. Engel, Biophys. J., 73 (1997) 406.

[60] E. Goormaghtigh, V. Cabiaux and J.M. Ruysschaert, Eur. J. Biochem., 190 (1990) 409.

[61] T. Miyazawa, J. Chem. Phys., 32 (1960) 1647.

[62] D. Marsh, Biophys. J., 72 (1997) 2710.

[63] D. Marsh, Biophys. J., 75 (1998) 354.
[64] D. Marsh, J. Mol. Biol., 297 (2000) 803.
[65] J.T. Yang, C.S. Wu and A.D. Martinez, Methods Enzymol., 130 (1986) 208.
[66] H. Öpik, J. Exp. Bot., 16 (1965) 667.
[67] S.J. Schein, M. Colombini and A. Finkelstein, J. Membrane Biol., 30 (1976) 99.
[68] M. Colombini, Nature, 279 (1979) 643.
[69] O. Ludwig, J. Krause, R. Hay and R. Benz, Eur. Biophys. J., 15 (1988) 269.
[70] H. Freitag, G. Genchi, R. Benz, F. Palmieri and W. Neupert, FEBS Lett., 145 (1982) 72.
[71] M. Linden, P. Gellefors and B.D. Nelson, FEBS Lett., 141 (1982) 189.
[72] V. De Pinto, G. Prezioso and F. Palmieri, Biochim. Biophys. Acta, 905 (1987) 499.
[73] J.A. Aljamal, G. Genshi, V, De Pinto, L. Stefanizzi, A. De Santis, R. Benz and F. Palmieri, Plant Physiol, 102 (1993) 615.
[74] A. Elkeles, K.M. Devos, D. Graur, M. Zizi and A. Breiman, Plant Mol. Biol., 29 (1995) 109.
[75] N. Roosens, F. Al Bitar, M. Jacobs and F. Homblé, Biochim. Biophys. Acta, 1463 (2000) 470.
[76] R. Douce, J. Bouguignon, R. Brouquisse and M. Neuberger, Methods Enzymol., 148 (1987) 403.
[77] Y. Kagawa and E. Racker, J. Biol. Chem., 246 (1971) 5477.
[78] F. Homblé, V. Cabiaux and J.M. Ruysschaert, Mol. Microbiol., 27 (1998) 1261.
[79] M. Colombini, J. Membrane Biol; 111 (1989) 103.
[80] M.C. Sorgato and O. Moran, Crit. Rev.Biochem. Mol. Biol., 28 (1993) 127.
[81] R. Benz, Biochim. Biophys. acta., 1197 (1994) 167.
[82] M. Colombini, E. Blachy-Dyson and M. Forte, *In*: Ions Channels, T. Narahashi (ed.), 4 (1996) 169.
[83] R. Benz, M. Kottke and D. Brdiczka, Biochim. Biophys. Acta, 1022 (1990) 311.
[84] D.W. Zhang and M. Colombini, Biochim. Biophys. Acta, 1025 (1990) 127.
[85] M. Ouali, and R.D. King, Protein Sci., 9 (2000) 1162.
[86] B. Rost, and C. Sander, J. Mol. Biol., 232 (1993) 584.
[87] B. Rost, and C. Sander, Proteins, 19 (1994) 55.
[88] I. Jacoboni, P.L. Martelli, P. Fariselli, V. De Pinto and R. Casadio, Prot. Sci., 10 (2001) 779.
[89] N.A. Rodionova, S.A. Tatulian, T. Surrey, F. Jahnig and L.K. Tamm, Biochemistry, 34 (1995) 1921.
[90] H. Abrecht, E. Goormaghtigh, J.M. Ruysschaert and F. Homblé, J. Biol. Chem., 275 (2000) 40992.
[91] L. Shao, K.W. Kinnally and C.A. Mannella, Biophys. J., 71 (1996) 778.
[92] D.A. Koppel, K.W. Kinnally, P. Masters, M. Forte, E. Blachy-Dyson and C.A. Mannella, J. Biol. Chem., 273 (1998) 13794.
[93] J.L. Eisele and J.P. Rosenbusch, J. Biol. Chem., 265 (1990) 10217.
[94] R.W.Woody, *In* G.D. Fasman (ed.), Circular dichroism and the conformational analysis of biomolecules, Plenum Press, New York, (1996) 25.
[95] K. Mihara and R. Sato, EMBO J., 4 (1985) 769.
[96] E. Blachly-Dyson, E.B. Zambronicz, W.H. Yu, V. Adams, E.R. McCabe, J. Adelman, M. Colombini and M. Forte, J. Biol. Chem, 268 (1993) 1835.
[97] A.G. Murzin, A.M. Lesk and C. Chothia, J. Mol. Biol., 236 (1994) 1369.
[98] X.W. Guo, P.R. Smith, B. Cognon, D. D'Arcangelis, E. Dolginova and C.A. Mannella, J. Struct. Biol., 114 (1995) 41.

[99] M. Dolder, K. Zeth, P Tittmann, H. Gross, W. Welte and T. Wallimann, J. Struct. Biol., 127 (1999) 64.

Planar Lipid Bilayers (BLMs) and their Applications
H.T. Tien and A. Ottova-Leitmannova (Editors)

Chapter 16

Reconstituting SNARE proteins into BLMs

K. T. Rognlien[a] and D. J. Woodbury[b]

[a]Department of Physiology, Wayne State University School of Medicine, 540 E. Canfield Ave., Detroit, Michigan 48201, USA.

[b]Department of Physiology and Developmental Biology, Brigham Young University, 574 Widtsoe Building, Provo, Utah 84602, USA.

1. INTRODUCTION

SNARE proteins are molecular motors that drive the biological fusion of two membranes [1]. Part of the motor assembly is in the vesicle membrane (v-SNAREs) and part is in the target membrane (t-SNAREs) [2,3]. During fusion many matched pairs of v- and t-SNAREs intertwine to pull opposing membranes close so that they fuse together. This intertwining conjures up an image suggested by the proteins' acronym, SNARE. It is this image however, not the meaning of an acronym that is likely to be remembered. SNARE stands for SNAP receptor, SNAP stands for soluble NSF attachment protein, and NSF stands for N-ethylmaleimide sensitive factor! Recently a less informative but more direct use of the SNARE acronym has been used: soluble N-ethylmale-imide-sensitive factor-attachment protein receptors [1, 4-5]. Before fusion occurs, the vesicle is delivered to and docks with the target membrane. Once the vesicle is docked, SNARE proteins can be activated to fuse the vesicle and target membranes together. Although the delivery and docking steps are interesting and under intense investigation, it is the fusion step that is the focus of this chapter. The fusion of two membranes is a process that lends itself to investigation using principles of electrophysiology and the black lipid membrane (BLM) technique.

2. BACKGROUND

During fusion, SNARE proteins windup into a highly stable ternary configuration typically comprised of the v-SNARE, synaptobrevin (VAMP) and the t-SNAREs, syntaxin and SNAP-25 (or SNAP-23). Formation of this complex leads to fusion of the vesicular membrane and plasma membrane, driving the release of secretatogogue or the insertion of membrane proteins from the vesicle [6-9]. Recycling of the SNARE machinery begins when α-SNAP and NSF bind syntaxin and SNAP-25. Hydrolysis of ATP by NSF leads to the unwinding of the SNARE complex [10-11]. As a result, the SNARE machinery becomes primed for further rounds of fusion.

The process of SNARE fusion, appears rather generic at first glance, however cells often regulate the process carefully. For example, epithelial cells produce different types of vesicles that use different SNARE proteins to fuse at specific locations within the cell. Also, many endocrine and neuronal cells regulate timing of hormone or neurotransmitter release by activating SNARE proteins with an appropriate signal. These examples are worth exploring in greater detail.

Experiments examining the SNARE machinery in kidney epithelial cells (Madin-Darby canine kidney, MCDK) have demonstrated that the SNARE machinery, although ubiquitous, employs different mechanisms on the baso-lateral and apical surfaces [12-13]. Transport from the trans-Golgi network to the basolateral membrane is NSF-dependent. However, fusion of vesicles with the apical membrane is SNARE-dependent but NSF is not required, suggesting a different, as yet unidentified NSF-like protein may be involved on the apical side. This difference may be due to differences in the type of t-SNARE, syntaxin, found in the membranes. Syntaxins 2 and 3 are localized to the apical membrane, while syntaxin 4 is specific to the basolateral membrane. SNAP-23 interacts with all three of these syntaxin isoforms and resides in both the apical and basolateral surface [14]. Thereby suggesting that vesicular cycling in polarized cells not only differs at each pole, but that it may also be different from cycling in non-polarized cells.

In 1998, Inoue *et al.* [15], demonstrated that the SNARE complex was responsible for the targeting of aquaporin-2 to the apical membrane of the collecting duct epithelia, following vasopressin stimulation. Aquaporin-2 is an integral membrane protein found in both the intracellular vesicle membrane and the apical membrane and not simply a soluble protein destined for secretion. The collecting duct SNARE utilizes the syntaxin-4 and SNAP-25 isoforms of the t-SNAREs in the apical plasma membrane and the neuronal VAMP-2 isoform in the vesicle to deliver aquaporin-2. Noteworthy, since the syntaxin-4 isoform only resides in the basolateral membrane of MDCK cells, yet resides in the apical membrane of collecting duct epithelia, while SNAP-23 does not interact with aquaporin-2 containing vesicles at all. Also interesting is the finding that VAMP-2 is the v-SNARE isoform, rather than the VAMP-7 or VAMP-8 isoforms of MDCK cells. This provides evidence to suggest that varying combinations of v and t-SNAREs may lead to a range of product delivery to cellular surfaces.

Banerjee, *et al.* [8] provides another example of SNARE complex diversity found in rat ICMD (inner medullary collecting duct) cells. In the ICMD cells the SNAREs are responsible for the regulated trafficking of vesicles containing the H^+-ATPase, which regulate cellular pH levels. IMCD cells in culture express the neuronal v- and t-SNARE isoforms, VAMP-2 and syntaxin-1, respectively. These cells express the non-neuronal SNAP-23 isoform, yet utilize the neuronal soluble factors α- and γ-SNAP, as well as NSF. Translocation of these proton pumps to the apical membrane results from an acute lowering of cellular pH. The exocytotic SNAREs form a fusion complex, which shares homology to the SNARE protein complement necessary for the formation of the neuronal fusion complex. Yet, these vesicles contain the H^+-ATPase, rather than neurotransmitter.

Experimental evidence reveals a requirement of the t-SNARE, SNAP-23 in the transcytotic recycling of transferrin. Leung, *et al.*, reconstituted transferrin recycling in streptolysin-O-permeabilized MDCK cells to elucidate the role of SNAP-23 in the endocytotic trafficking of this protein. They found that at the basolateral membrane, transferrin recycling is NSF-dependent and transferrin recycling could be blocked. Blockage was achieved using exogenous SNAP-23 or anti-SNAP-23 antibodies. The SNAP-23 also demonstrates susceptibility to cleavage by botulinum neurotoxin serotype E, indicating that it shares further homology with the neuronal SNAP-25, which is also cleaved by this neurotoxin.

The best understood and most extensively studied SNARE pathway to date, involve neurosecretion and vesicular cycling. Neuroexocytosis is initiated by the formation of the SNARE complex [2-4]. Formation of this complex leads to fusion of the vesicular membrane and presynaptic membrane, driving the release of neurotransmitter into the synaptic cleft [9,16-17]. Completion of the fusion event and recycling of the vesicle requires α-SNAP and NSF. NSF binds to SNAP-25 and hydrolysis ATP to disassemble the SNARE complex [3,10,17,18-21]. At the synapse, the vesicular proteins synaptotagmin [5,22-23] and rabphilin [24-25] act as Ca^{2+} sensors that help regulate the neurosecretion process. Syntaxin and SNAP-25 may also be responsible to some degree in Ca^{2+} sensing and, along with the vesicular proteins, may participate in the recycling of vesicles [16,26].

Although much has been discovered, the complete understanding of the SNARE motor in mediating cell trafficking still remains elusive. It is readily apparent however, that this mystery needs to be fully solved. The diversity by which the SNARE protein families are combined and intermingled suggests a vast variety of possibilities. Current evidence gathered from *in vitro* and *in vivo* observations in all cell types leads us to their key role in physiological function and management of homeostasis in key organ systems. Pathogenic studies indicate that malfunction in these SNARE proteins can lead to human disease.

Thus, elucidation of the mechanism of SNARE proteins should prove indispensable to the basic understanding of cell functions.

3. MEASURING SNARE-INDUCED FUSION WITH BLMS

3.1 Biological Membranes and BLMs

All cells are surrounded by a lipid bilayer. The bilayer is comprised of an inner and outer monolayer of phospholipids. Phospholipids are amphipathic molecules, consisting of both a polar and non-polar region. These lipids align such that the polar head groups face outward. This lipid bilayer forms an ideal barrier for the cell. This barrier is modified by proteins that span the cell membrane.

The black lipid membrane, or BLM, is a model of the cell membrane. It is made of lipids, but proteins can also be reconstituted into it. The BLM was first demonstrated in 1962 by Mueller *et al.* [27]. Because the membrane is formed in a controlled environment, it provides the ideal stage to study the properties of membrane proteins, especially properties that change the electrical conductance of the membrane. Although it is not yet clear if SNARE proteins can directly alter membrane conductance, a method is described later that allows SNARE-directed fusion event to be detected electrically. With such a method in hand, it is possible to measure directly if a certain set of t- and v-SNAREs are sufficient to induce fusion of the vesicles to the BLM. Reconstitution of SNAREs into membranes is described below.

3.2 Reconstitution of SNARE Proteins into a BLM

As illustrated in the background section, many researchers have been examining the role of SNARE proteins within cell cultures. However, large amounts of SNARE proteins can now be made easily in *E. Coli* using recom-binant technology. The purified SNARE proteins can be reconstituted into liposomes or lipid bilayers. This allows individual SNAREs, as well as combinations of SNAREs, to be intensely studied under the totally controlled experimental conditions of the BLM.

We have successfully reconstituted the t-SNARE, syntaxin, into planar lipid bilayers using two different methods. The first is simply to "brush" the aqueous solution containing purified syntaxin protein onto a preformed BLM, as we have previously reported [28]. More recently, we have developed a second method that more consistently reconstitutes larger amounts of syntaxin into a BLM. For this method, purified syntaxin is added directly to the lipid solution prior to bilayer formation. The aqueous solution containing syntaxin forms an emulsion with the organic solution containing lipids. The emulsion is formed by adding 20 μl of SNARE protein (0.1-0.2 mg/ml aqueous buffer) to 50 μl of lipid solution (20 mg lipid/ml decane). Upon protein addition the mixture is vortexed gently for 5 minutes. This emulsion is used to precoat the hole and to brush over the hole to form a BLM. Typically, we use a lipid composition of 70% brain phosphatidylethanolamine (PE) and 30% brain phosphatidylcholine (PE), although mixtures containing 10% phosphatidyl-serine (PS) have also been used. BLMs formed using the syntaxin-containing emulsion have classic values for membrane conductance and capacitance, although small (10-30 pS) fast flickery conductances are often observed after membrane formation. These conductances may be due to clustering of many syntaxin molecules to briefly form a channel, or due to another contaminant(s) in the syntaxin preparation.

Both of these bilayer reconstitution techniques allow the investigator to incorporate either a single type of SNARE or combinations of SNAREs into the membrane. Before reconstitution, the lipid composition of the membrane can be selected to suit the needs of the SNAREs under investigation. Also, as noted above, reconstitution of SNARE protein into bilayers may prove useful in detecting channel activity of individual or combinations of SNAREs.

3.3 Formation of SNARE Proteoliposomes

SNARE-containing vesicles can be made in two ways. One method is to directly add purified SNAREs to artificial vesicles. This is described at the end

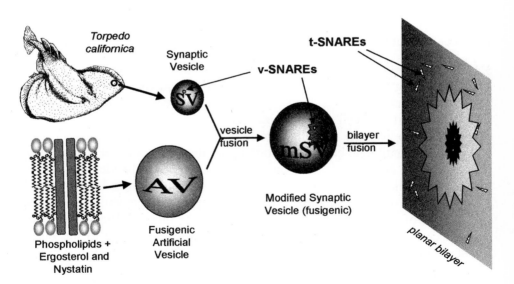

Fig. 1. Formation of mSV and their fusion with a BLM. Synaptic vesicles (SV) are isolated from the electric fish, *Torpedo californica*, and mixed with artificial vesicles (AV) containing nystatin channels. After three freeze-thaw-sonicate cycles, membranes from the two populations of vesicles intermingle to form a hybrid vesicle called an mSV. These hybrid vesicles have unique fusion properties as described in the text.

of this section.   We have previously reported a second method to make
SNARE-containing vesicles [29-30].   In brief, the procedure is to join artificial
vesicles with native secretory vesicles (see Fig. 1).   The vesicles are fused
together to form modified secretory vesicles (mSV) that have important
properties obtained from each original population of vesicles.   The artificial
vesicles provide components that make it possible to induce and measure fusion
of vesicles to a BLM.   The secretory vesicles contribute their full compliment
of v-SNAREs, such as VAMP and membrane-bound accessory proteins.

Vesicles containing a single type of SNARE or a predefined combination
of SNAREs can be formed as well.   In this way it is possible to examine any
possible protein-protein interactions that may occur between SNAREs.

Proteoliposomes are basically prepared as follows.   250 μl of lipids
containing PE, PC, PS and ergosterol are combined in mol% of 50%, 10%, 20%
and 20%, respectively in a glass test tube.   Stock lipids, obtained from Avanti
Polar Lipids (AL, USA), are all stored as solutions of 10 mg lipid per ml
chloroform.   To the lipids, 5-6 μl nystatin stock solution (2.5mg/ml methanol)
are added so as to give a final aqueous solution concentration of 50-60 μg/ml.
The organic solution is dried under a stream of nitrogen.   The resulting dried
plaque is then re-hydrated with 250 μl of 150 mM KCl solution, buffered with 8
mM HEPES at pH 7.2.   To the liposome solution an aliquot of SNARE protein
is added.   This solution is then vortexed and treated with three freeze-thaw-
sonicate cycles.   Each sonicate cycle is for 60 seconds using a bath-type
sonicator, except for the last sonication, which is for 5-15 seconds.

3.4 Electrophysiological Detection of SNARE-induced Fusion

Assuming that the fusion event can be detected easily, reconstitution of
SNARE proteins into BLMs allows researchers to directly determine the role of
SNAREs as part of a cell's fusion machinery.   One simple method for detection
of fusion is to use the nystatin/ergosterol (Nys/Erg) fusion system [29,31].
Nystatin is an antibiotic capable of forming an ion channel in membranes when
combined with ergosterol.   The ion channel can be used to identify fusion of
individual vesicles with a planar lipid bilayer.   Each fusion can be identified by
the appearance of a current spike.   As shown in Fig. 2, the upstroke of the spike
is the result of an abrupt, momentary upsurge in the bilayer conductance due to
nystatin channels in the vesicular membrane that are relocated to the bilayer.
Membrane conductance slowly returns to baseline following each fusion event
as the nystatin channels close.   The nystatin channels are unable to remain open
once transferred to the bilayer because, the bilayer, unlike the vesicle, contains
no ergosterol and ergosterol is essential to formation of the small nystatin pore.
Thus, the channel closes as ergosterol, originally in the vesicle, diffuses away
from the site of fusion.

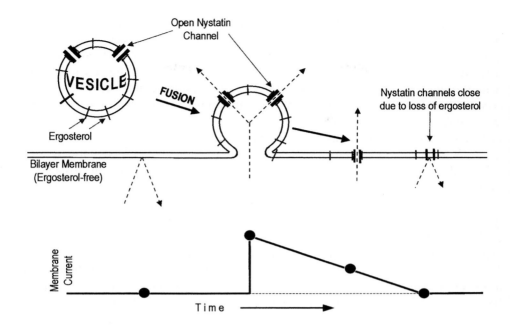

Fig. 2. Fusion of a channel-containing vesicle with a BLM. In the upper left a vesicle containing open nystatin channels approaches the BLM. Middle top: the vesicle fuses transferring its nystatin channels to the BLM and increasing membrane current (bottom trace). Right: After fusion into an ergosterol-free membrane, the nystatin channels close, returning membrane current to zero.

This fusion detecting system has an additional benefit. It provides a way to induce fusion independent of SNARE proteins. Introduction of an osmotic gradient, elicits vesicle swelling, which pulls membrane lipids apart and exposes hydrophobic domains, allowing fusion to occur [32-33]. This provides for a built in control for each population of vesicles that are tested. Since all vesicles are capable of fusing with the bilayer in the presence of an osmotic gradient, the general competency of individual vesicle populations, as well as vesicle delivery to the membrane is confirmed in each experiment. In experiments where no fusion occurs following addition of an osmotic gradient, the vesicles are deemed incompetent and the experiment does not infer a false negative result. A similar lack of fusigenisity is also observed occasionally in experiments when using unaltered Nys/Erg vesicles (AV). This is presumed indicative of one or more of the following: compromised vesicle formation, low delivery to the membrane, or an inability to record (electrically) the fusion events.

In contrast to osmotic-induced fusion, the SNARE hypothesis implies that interactions between t- and v-SNAREs possibly expose hydrophobic domains and facilitate the fusion event. We have found that mSV are capable of fusion with bilayers containing the t-SNARE syntaxin-1, in the absence of an osmotic gradient. In other words, mSV fuse spontaneously with syntaxin-containing bilayers [28]. Therefore, reconstituting various t and/or v-SNARES into bilayers and/or into Nys/Erg vesicles, will allow fusion events to be measured. Using this method various SNARE protein-protein interactions absolute for fusion of secretory vesicle and target membrane can be mapped.

We have started the mapping process by using syntaxin truncation mutants to determine which domains of syntaxin are essential for inducing spontaneous fusion of mSV. Table 1 lists the results for four mutants tested to date. All the mutants contain the membrane-spanning region, amino acids 268-288. The first entry, syntaxin 1A-1, is the closest we have to full-length syntaxin. It is missing just the first seven of 288 amino acids. When syntaxin 1A-1 is reconstituted into a bilayer it induces fusion (spontaneous fusion) of mSV at a rate similar to the rate induced by an osmotic gradient (gradient fusion). For the experiments in Table 1, a gradient of 500 mOsM was used (150mM KCl in the *trans* side and 400 mM KCl in the *cis* or vesicle side of the bilayer). As presented in the table, the spontaneous rate is 78% of the gradient rate. Without syntaxin, the spontaneous rate is essentially zero. Two other truncation mutants, 1A-7 and 1A-9, have a suppressed fusion rate about 1/7 of the rate observed with 1A-1. The shortest mutant, 1A-12, contains less than 50 amino acids of the soluble portion of the protein. We did not observe any

Table 1
Fusion rates for mSV fusion to bilayers containing different syntaxin truncation mutants.

Syntaxin 1A Mutant (amino acids)	Spontaneous Fusion Rate (per min.)	Gradient Fusion Rate (per min)	Spontaneous Fusions / 100 Gradient Fusions	Total Number of Gradient Fusions	N
1A-1 (a.a. 8-288)	2.38	3.04	78	182	6
1A-7 (a.a. 77-288)	0.26	2.66	10	133	5
1A-9 (a.a. 194-288)	0.24	1.95	12	156	8
1A-12 (a.a. 221-288)	0.00	1.91	0	134	7

Note that the fastest spontaneous fusion rate is with the longest mutant, syntaxin 1A-1. The shortest syntaxin mutant, 1A-12, did not induce spontaneous fusion. For each truncation mutant, the range of amino acids expressed, is given in column 1.

spontaneous fusion with this mutant. These data suggest that fusion is induced by specific interactions between syntaxin and SNARE proteins in the mSV, and that although amino acids 194-288 are sufficient to induce some fusion, full length syntaxin works much better.

4. ADDITIONAL APPLICATIONS

Incorporation of SNAREs in bilayer membranes provides multiple uses. The bilayer system will allow experimenter's the freedom to adjust or introduce new variables into the experimental environment. One example is that the bath solution and or proteoliposomal makeup can be altered to reflect either an extracellular or intercellular milieu. The experimenter is also free to add toxins or newly developed pharmaceuticals to the bath or the proteoliposome, or even introduce an agent into the bilayer itself. Plausible co-factors or enzymes can be added and the subsequent effect studied.

SNAREs can be fluorescently labeled and then incorporated into liposomes. Thereby allowing visualization of protein interactions. Utilizing total internal reflection fluorescence microscopy (TIRFM) individual SNAREs or combinations are reconstituted into planar lipid bilayers. Wagner and Tamm [34], used this method to visualize the t-SNAREs, Syntaxin 1A/SNAP-25 as they interacted with lipids. This technique allows them to study t-SNARE interactions with not only various combinations of lipids but also interactions with v-SNAREs.

It is clear that reconstitution of SNARE proteins into BLMs offers an experimenter a wide range of possibilities. Much as of yet remains unclear as to the exact roles that each of the SNARE protein superfamily members are capable of playing. The reconstitution of each member into BLMs lends to investigators an innovative and proven approach to study characteristics of each. In the years to come, each SNARE will ultimately be characterized and the function of each no longer a mystery which needs to be solved.

REFERENCES

[1] S.J. Scales, M.F.A. Finley, and R.H. Scheller, Science, 294 (2001) 1015.
[2] T. Weber, B.V. Zemelman, J.A. McNew, B. Westermann, M. Gmachl, F. Parlati, T.H. Söllner, and J.E. Rothman, Cell, 92 (1998) 759.
[3] B.J. Nichols, C. Ungermann, H.R.B. Pelham, W.T. Wickner, and A. Haas, Nature, 387 (1997) 199.
[4] S. Schoch, F. Deák, A. Kőnigstorfer, M. Mozhayeva, Y. Sara, T.C. Sűdhof, and E.T. Kavalali, Science, 294 (2001) 1117.
[5] C-T. Wang, R. Grishanin, C.A. Earles, P.Y. Chang, T.F.J. Martin, E.R. Chapman, and M.B. Jackson, Science, 294 (2001) 1111.
[6] J.B. Bock and R.H. Scheller, Proc Natl Acad Sci USA, 96 (1999) 12227.
[7] P.I. Hanson, J.E. Heuser, and R. Jahn, Curr Opin Neurobiol, 7 (1997) 310.

[8] A. Banerjee, G. Li, E.A. Alexander, and J.H. Schwartz, Am J Cell Physiol, 280 (2001) C775.

[9] J. Pevsner, S.C. Hsu, J.E.A. Braun, N. Calakos, A.E. Ting, M.K. Bennett, and R.H. Scheller, Neuron, 13 (1994) 353.

[10] R.B. Sutton, D. Fasshauer, R. Jahn, and A.T. Brunger, Nature, 395 (1998) 347.

[11] P.I. Hanson, H. Otto, N. Barton, and R. Jahn, J Biol Chem, 270 (1995) 16955.

[12] T. Galli, A. Zahraoui, V.V. Vaidyanathan, G. Raposo, J.M. Tian, M. Karin, H. Niemann, and D. Louvard, Mol Biol Cell, 9 (1998) 1437.

[13] S.H. Low, P.A. Roche, H.A. Anderson, S.C. Vanljzendoorn, M. Zhang, K.E. Mostov, and T. Weimbs, J Biol Chem, 273 (1998) 3422.

[14] B. Quiñones, K. Riento, V.M. Olkkonen, S. Hardy, and M.K. Bennett, J Cell Sci, 112 (1999) 4291.

[15] T. Inoue, S. Nielsen, B. Mandon, J. Terris B.K. Kishore, and M.A. Knepper, Am J Physiol, 275 (1998) F752.

[16] Y.A. Chen, S.J. Scales, S.M. Patel, Y-C. Doung, and R.H. Scheller, Cell, 97 (1999) 165.

[17] P.I. Hanson, J.E. Heuser, and R. Jahn, Curr Opin Neurobiol, 7 (1997) 310.

[18] H. Otto, P.I. Hanson, and R. Jahn, Proc Natl Acad Sci USA, 94 (1997) 6197.

[19] T. Söllner, M.K. Bennett, S.W. Whiteheart, R.H. Scheller, and J.E. Rothman, Cell, 75 (1993) 409.

[20] T. Hayashi, S. Yamasaki, S. Nauenburg, T. Binz, and H. Niemann, EMBO Journal, 14 (1995) 2317.

[21] M.R. Block, B.S. Glick, C.A. Wilcox, F.T. Wieland, and J.E. Rothman, Proc Natl Acad Sci USA, 85 (1988) 7852.

[22] I. Fernandez, J. Ubach, I. Dulubova, X. Zhang, T.C. Südhof, and J.Rizo, Cell, 94 (1998) 841.

[23] T.C. Südhof and J. Rizo, Neuron, 17 (1996) 379.

[24] T. Sasaki, H. Shirataki, H. Nakanishi, and Y. Takai, Adv Second Messenger Phosphoprotein Res, 31 (1997) 279.

[25] E. Fykse, M.C. Li, and T.C. Südhof, J Neurosci, 15 (1995) 2385.

[26] V.E. Degtiar, R.H. Scheller, and R.W. Tsien, J Neurosci, 20 (2000) 4355.

[27] P. Mueller, D. Rudin, H.T. Tien, and W.C. Wescott, Nature, 194 (1962) 979.

[28] D.J. Woodbury and K. Rognlien, Cell Biol Int, 24 (2000) 809.

[29] D.J. Woodbury, Methods in Enzymology, 294 (1999) 319.

[30] M.L. Kelly and D.J. Woodbury, Biophys J, 70 (1996) 2593.

[31] D.J. Woodbury and C. Miller, Biophys J, 58 (1990) 833.

[32] W.D. Niles, F.S. Cohen, and A. Finkelstein, J Gen Physiol, 93 (1989) 211.

[33] D.J. Woodbury and J.E. Hall, Biophys J, 54 (1988) 1053.

[34] M.L. Wagner and L.K. Tamm, Biophys J, 81 (2001) 266.

Planar Lipid Bilayers (BLMs) and their Applications
H.T. Tien and A. Ottova-Leitmannova (Editors)

Chapter 17

Mitochondrial ion channels, their isolation and study in planar BLMs. A comparison of the properties of the ion channels with the properties of ion transport systems in intact mitochondria

G. D. Mironova

Institute of Theoretical and Experimental Biophysics, Russian Academy of Sciences, Pushchino, 142 290 Russia

1. INTRODUCTION

Considerable progress that has been made in studies of single channels of native cell membranes is due to the development of the patch-clamp technique. This method involves placing a glass micropipette in contact with a cell membrane, which makes it possible to isolate one or several ion channels of the same type. With the advent of this method, membrane biophysics was brought to the molecular level.

At the same time, the standard patch-clamp technique is not quite suitable for investigating the channel activity in intracellular membranes. The use of the patch-clamp technique for isolated cell organelles (mitochondria and sarcoplasmic reticulum) is difficult due to a small size of these organelles. In this particular case the best technique for characterizing the single-channel activities remains the planar bilayer lipid membrane (BLM). There are two approaches to studying channels reconstituted into BLM. The first technique involves the formation of liposomes into which some patches from cell organelle membranes are introduced. These liposomes are then fused with the BLM, and the electrical characteristics of the BLM are measured. In the second approach, channel proteins isolated from itracellular membranes are incorporated into the BLM, and the electrical characteristics of this membrane are recorded. These approaches make it possible to study the properties of single channels and the molecular mechanisms of their functioning. The latter technique, however, requires the solubilization of channel proteins in the active and rather homogeneous state.

2. METHODS OF SOLUBILIZATION OF MEMBRANE PROTEINS

Our laboratory has been engaged in studies of the solubilization of membrane proteins and their reconstitution into BLM for more than 25 years [1-3]. At the time when we started to isolate proteins from mitochondrial membranes, many membrane proteins were solubilized by detergents [4-6]. Although detergents are successfully used for the solubilization of a number of membrane proteins, this approach has some drawbacks. The most noticeable problem is that these detergents bind tightly to membrane proteins and influence significantly their structure and functions [7]. Since detergents can induce the nonspecific permeability of BLM [8], their use for the solubilization of membrane proteins with the consequent reconstitution of these proteins into BLM cannot be recommended. Detergents, however, can be successfully used if proteins are reconstituted into liposomes, since in this case they can be removed by long-term dialysis of proteoliposomes [9] or by using a Bio-beads CM-2 column [10] and other columns.

Another approach to solubilizing membrane proteins is the use of organic solvents [11-14]. Erythrocyte membrane proteins, for example, are readily solubilized and released from the most of membrane lipids if the water suspension of hemoglobin-free ghosts is extracted with n –butanol [13-14]. The technique provides a mean recovery of 85% of the membrane protein in the water phase. The mean amount of lipids present in the water phase after a single extraction is 5% of protein weight. There is substantial evidence that the dissolved protein is not extensively denatured by butanol [13-14]. Since most of the lipids are retained in the butanol phase, the one-step procedure has the advantage that separation from lipids takes place simultaneously with solubilization. Our experiments demonstrated that butanol can also be used for solubilization of ion-selective channels from the mitochondrial membranes [1]. In this case, however, the channel protein is isolated into the water phase together with other proteins, which creates some difficulties in its purification.

We have developed a method in which membrane proteins are extracted by ethanol or acetone. By using ethanol, we have isolated a number of membrane proteins, and some of them were found to possess the ion-channel activity. This method allows the proteins to be released from the membrane into water, since ethanol has the dielectric constant intermediate between those of water and carbohydrates. This facilitates the penetration of ethanol into the area of protein-lipid interactions, decreases the binding of lipids to protein and releases the protein into water. Protein conformation can change significantly in a water environment or even a multimolecular complex can be formed. Since the ratio of hydrophobic and hydrophilic amino acids in the membrane proteins is usually about 1:1 [15], upon contact of the protein with water the hydrophobic parts of the protein become buried within protein interior or within intermolecular complexes. In both cases, the hydrophilic parts of the protein molecule are localized on the protein surface in contact with water or become exposed to water. Upon reconstruction of these molecules or intermolecular complexes into the membrane they can acquire in a lipid environment the same conformation as in the native membrane, thus ensuring the restoration of activity [14].

We have applied this approach to the isolation and reconstitution in lipid bilayer of a few mitochondrial ion channels [2-3] and the large subunit of the (Na^+,K^+)-ATPase from microsomes [16-17]. The properties of these channels closely resemble those in native membranes.

3. ATP-DEPENDENT POTASSIUM CHANNEL FROM RAT LIVER AND BEEF HEART MITOCHONDRIA

The potassium selective channel was isolated in our laboratory from beef heart mitochondria by solubilizing proteins from the membrane by ethanol [18]. Later, using Triton X-100 as a solubilizing agent, we have isolated in Garlid's lab a protein with similar properties and molecular weight [10]. The reconstitution of the Triton-isolated protein into liposomes showed that potassium transport stimulated by the protein was inhibited by ATP and glybenclamide with high affinity. The ATP- and glybenclamide-dependent potassium channel has also been identified by direct patch clamping by fusing giant mitoplasts from rat liver mitochondria [19].

3.1. Ethanol extraction and purification of the mitochondrial K_{ATP} channel

The mitochondrial K_{ATP} channel (mitoK $_{ATP}$) was solubilized from the inner mitochondrial membrane by 60% ethanol. Mitochondria were diluted with 10 mM Tris-HCl buffer (pH 7.4) up to a concentration of 44 mg/ml and then extracted with 66% (w/v) ethanol. During extraction, one part of a 44 mg/ml mitochondrial suspension was added to ten parts of 66% ethanol solution in water (-20°C). The final protein and ethanol concentrations were 4 mg/ml and 60%, respectively. After stirring for 30 min at 4°C, the suspension was centrifuged at 7000 g for 15 min. The pellet was dissolved in an equal volume of 50% (w/v) ethanol (-20°C), and the above-described extraction procedure was repeated. Both supernatants were combined and evaporated to 10-30 ml in vacuum at 30°C. Since this fraction contained lipids, the extract was defatted by the addition of a chloroform-methanol (2:1, v/v) mixture (1 ml of the extract per 3-4 ml of mixture). The water phase was evaporated to 2-3 ml.

For purification, we used the well known methods of protein chemistry. In accordance with the protocol, the homogeneous protein was obtained from a lipid-free ethanol extract by gel filtration on a Sephadex G-15 column with subsequent purification of the channel-forming fraction on a Sephadex G-50 column and by preparative gel electrophoresis [18].

In another protocol, we excluded the chloroform-methanol treatment from the procedure of purification of the mitochondrial potassium channel since it was found that it decreases the channel activity of the isolated protein [20]. After extraction with 60% ethanol, the supernatant was dialyzed at 4°C overnight in buffer A containing 5 mM Tris-HCI (pH 7.4) and 0.02 % β-mercaptoethanol, and centrifuged at 105 000 g for 30 min. The supernatant from a 300-mg mitoplast was loaded onto a DEAE cellulose column (1 ml volume) at a rate of 40 ml/h with the buffer B containing 50 mM Tris-HCI (pH 7.4), 0.05 % β-mercaptoethanol, and 1 mM EDTA, and the column was washed with 2 ml of the buffer B. The bound proteins were eluted with a KCI step gradient (50, 100, 150, 200, 250, 300 mM) in the buffer B at a rate of 40-60 ml/h. The eluted fractions were dialyzed overnight in the buffer A and then during 2 h in the same buffer without β-mercaptoethanol. All the procedures were carried out at 4°C.

3.2. Reconstitution of the protein into BLM and the properties of the channel formed by the protein

The ion-transporting properties of the proteins isolated as described previously were measured using the BLM technique [21]. A teflon cell with a circular hole in the plastic sheet separating the cell into two compartments was used. The cell contained 4 ml of 20 mM Tris-HCl buffer, pH 7.4, and 100 mM KCI in the *cis* and *trans* compartments. For the formation of BLM, 1.2 mg of total brain lipids and 100 μg of cardiolipin were mixed, dried, and dissolved in 60 μl of *n*-decane. The conductivity was estimated by the voltage clamp method. Current was measured by an MAX406 operational amplifier connected to an IBM-compatible computer. The conductivity of the nonmodified membrane was in the range of 1-3 pS. The protein was added to one of the cell compartments. The experiments were carried out at 20-22 °C.

The fraction eluted from the DEAE column with a buffer containing 250 mM KCI had a potassium channel activity. The major band of this fraction corresponded to a 55-kDa protein in SDS-PAGE [20]. The same protein we detected by using preparative gel electrophoresis [18]. The amino acid composition of this protein was determined and presented in Ref. 20.

The purified 55-kDa protein, added on one (*cis*) side of the membrane, exhibited the ion-channel properties in the presence of 100 mM KCI. The discrete character of changes in the current of the protein-modified membrane indicates that the protein forms the channels in BLM (Fig. 1).The conductivity of the open state of a single channel was 10 pS under these conditions. This is in a good agreement with the value of 9.7 pS observed in intact mitoplasts by the patch-clamp technique [19]. The channels were selective for K^+, the reversal

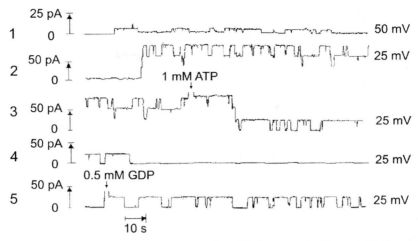

Fig. 1. Clusterization of the potassium channel and influence of ATP and GDP on channel clusters. The protein (3 µg/ml) was added into the *cis*-compartment, ATP and GDP were added into the *trans*-compartment. The bath solutions in both bompatment contained 20 mM Tris-HCI buffer (pH 7.4) and 100 mM KCl. Curves 1-5 are the records taken for one membrane at different times. The sensitivity on record 1 is fourfold higher than on records 2-5, due to a higher potential applied to this channel and a higher sensitivity of the recorder. All these factors enable us to distinguish on record 1 a single channel with the open state conductivity of 10 pS. After decreasing the membrane potential (records 2-5), step-like changes in transmembrane current were observed, indicating the formation of three channel clusters each with a conductivity of 100 pS. The standard values of conductivity and potentials applied are presented in the figure.

potential in a KCl gradient 1:3 was 29 mV, which is close to the theoretical value of the Nernst potential under the conditions described above. Ca^{2+} and Mg^{2+} could not pass through this channel [18, 20]. The high potassium selectivity characteristic of the channels was preserved as the total conductivity of membrane was increased. This increase in conductivity was probably due to the incorporation of a large number of channels into membrane.

In a protein-modified membrane, several elementary channels can be switched simultaneously as a large cluster. The record shown in Fig. 1 (1-5) illustrates different current levels and different types of channel behavior on the modified membrane. Figure 1.1 shows an elementary channel with a conductivity of 10 pS. It is believed that the opening and closing of channels are activated by changing the conformation of the channel protein complex. When the applied membrane potential was decreased (Fig. 1.2), step-like changes in transmembrane current were observed, indicating the formation of three channel clusters each with a conductivity of 100 pS. The 100-pS channel was

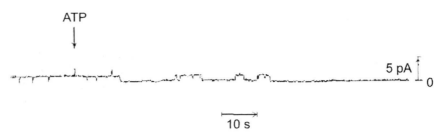

Fig. 2. Influence of ATP (1 mM) on a single channel. The protein (1 µg/ml) and experimental conditions were as in Fig. 1. A voltage applied to the membrane was 100 mV.

characterized as a cluster of ten single channels [22-23]. The ion selectivity of elementary channels was the same as that of clusters.

It was found that some channels are closed while being active in BLM. Spontaneous loss of activity over time, "channel rundown", is commonly observed in patch-clamp measurements of cellK$_{ATP}$ [24]. We found that small concentrations of ATP and ADP prevent the channel rundown of the mitochondrial K$_{ATP}$ channel (mitoK$_{ATP}$), and the selectivity of the channels for potassium [23, 25]. The SH-reagent DTT at a 1 mM concentration reactivates the mitoK$_{ATP}$ which lost their activity at some time but decreases the selectivity of the channels for potassium. Our results are consistent with the data on the effects of small thiol-containing molecules on gating and selectivity of the bacterial potassium channel [26].

3.3. Regulation of the mitochondrial potassium channel reconstituted into BLM

The mitoK$_{ATP}$ channel reconstituted into BLM is inhibited by physiological concentrations of ATP and belongs, probably, to the family of cell K$_{ATP}$ channels [24]. ATP has to be added to the *trans* compartment. The inhibition is reversible and is eliminated by the removal of ATP from the compartment. ATP inhibits both a single channel (Fig. 2) and a cluster of channels (Fig. 1) and Mg^{2+} is not needed for this inhibition. As it is shown on Fig. 1.3, 1 mM ATP inhibits two clusters at first and then the rest of channel cluster (Fig. 1.4). K$_{1/2}$ for the inhibition of the channels by ATP is 522 µM.

This channel is not inhibited either by 5-hydroxydecanoate (50-100 µM) or glybenclamide (2-10 µM) and is not activated by 10-50 µM cromakalim and diazoxide (10-30 µM), which are used as the modulators of mitoK$_{ATP}$ channel [27-28].

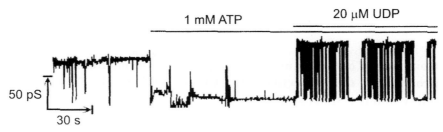

Fig. 3. UDP activation of the channel formed in BLM by the 55-kD protein (3 µg/ml). The inhibition of the channel by ATP (1 mM) and its reactivation by UDP (20 µM); a potential applied across the membrane was 75 mV. The experimental conditions were the same as in Fig. 1.

3.4. Activation of mitochondrial ATP-dependent potassium channel

In accordance with the $K_{1/2}$ value for ATP inhibition of the reconstituted channel (39 µM [10]) and the high (2-3 mM) concentration of ATP in the cell, this channel should be closed under physiological conditions. It is known, however, that activators of mitoK$_{ATP}$ have preconditioning-like effects on heart tissue and this channel plays a key role in ischemic cardioprotection [29-30]. Moreover, we found that the activation of this channel can be important for the regulation of nonshivering thermogenesis during hibernation [31]. Therefore, it is of extreme importance to find natural and pharmacological activators of mitoK$_{ATP}$ which can act as cardioprotectors and thermoproduction regulators.

It was found in Garlid's laboratory that GTP can activate mitoK$_{ATP}$ [32]. Our experiments on the reconstruction of the 55-kDa protein showed that another phosphonucleotide, GDP, taken at a concentration of 0.5 mM reactivates the channels (Fig. 1.5). It does not influence the reconstituted potassium channel in the absence of ATP. We also found another effective inhibitor of this channel, UDP. As it is shown in Fig. 3, the ATP inhibition of the 100 pS channel formed in BLM by the 55-kDa mitochondrial protein is reversed by UDP. A half maximal effect was observed with 9.6 µM UDP and complete reversal of the closing effect of ATP occurred after the addition of 20 µM UDP. The channel selectivity for K^+ was retained after reactivation. Further studies showed that UTP is a less effective channel activator than UDP, while UMP, CTP and CDP have no effect on the channel.

The uncharged local anesthetic p-diethylaminoethylbenzoate (DEB) also reactivates the potassium channel after its inhibition by ATP [25]. At a concentration of 4-8 µM it restores the ion-transporting activity to the initial level. Presumably, the activating effect of DEB is due to its electron donor properties, since the electron acceptor pilargonidine abolishes the activating effect of DEB.

We suggest that the thiol groups on mitoK$_{ATP}$ are the targets for redox ligands [25].

3.5. Comparison of the channel properties of the reconstituted 55 kDa protein with the properties of the ATP-dependent mitochondrial potassium transport system

The ATP-dependent potassium transport in intact mitochondria was discovered and studied in detail in the Garlid's laboratory [27-29, 33]. It was found that in mitochondria, K$_{1/2}$ for ATP inhibition is 2-3 µM, and Mg^{2+} is needed for inhibition. The channel is inhibited by 5-hydroxydecanoate (50-100 µM) and glybenclamide (2-3 µM) and is activated by potential activators of the cellK$_{ATP}$ channel cromakalim (10-50 µM) and diazoxide (10-30 µM). For the inhibition of channels reconstituted into BLM, a substantially higher concentration of ATP is needed. This fact and the absence of sensitivity to channel modulators in the case of the 55-kDa protein reconstituted into BLM suggest the subunit structure of mitoK$_{ATP}$. Most likely, the 55-kDa protein is a channel subunit of the rectifying K$^+$ mitochondrial channel (mitoKIR), whereas in mitochondria both the channel subunit and the sulfonylurea receptor (mitoSUR) are present, as was previously hypothesized [30]. The suggestion that the 55-kDa protein plays the role of mitoKIR is also confirmed by our immunochemical and electrophysiological studies. We clearly demonstrated that the antibody to the 55-kDa protein specifically inhibits the ATP-dependent transport of K$^+$ in mitochondria but does not influence other mitochondrial functions [34], and the poential affects this channel [3].

A similar subunit structure was observed on cellK$_{ATP}$ [35]. It the case of cellK$_{ATP}$, the expression of the inwardly rectifying K$^+$ channel (cellKIR) in the absence of cellSUR increases the blocking concentration of the nucleotide from 10 to 100 µM and leads to a loss of the sensitivity of cellKIR to glybenclamide and the openers of channels.

We observed that, as in the case of the reconstituted 55-kDa protein (mitoKIR), UDP reestablishes K$^+$ uptake in intact mitochondria whose mitoK$_{ATP}$ was preliminarily inhibited by ATP-Mg^{2+} (unpublished data). DEB also eliminates the ATP inhibition of the reconstituted channel and reactivates the ATP-inhibited potassium transport in intact mitochondria. This activation is eliminated, however, by 5-hydroxydecanoate, which being used as a specific inhibitor of mitoK$_{ATP}$ [27-29].

Thus, the organic solvent solubilizes the channel protein and retains therewith the properties of channel subunits of the mitochondrial potassium transport system. The use of BLM technique together with the study of intact mitochondria allowed us to clarify the subunit structure and the mechanism of regulation of this physiologically important channel.

4. CA²⁺ UNIPORTER FROM RAT LIVER AND BEEF HEART MITOCHONDRIA

Mitochondria have a very effective Ca^{2+} uniporter, which provides the influx of Ca^{2+} inside mitochondria by using the electrochemical potential created on the inner mitochondrial membrane by the respiratory chain. Ruthenium Red at a 1-5 µM concentration strongly inhibits this ion transport. During the last 30 years, attempts to isolate the selective calcium uniporter from mitochondria have been made in several laboratories [36]. However, the structure of the uniporter has not yet been identified. In our laboratory we isolated the Ca^{2+} uniporter by using the ethanol extraction of mitochondria.

4.1. Ethanol extraction and purification of mitochondrial Ruthenium Red-sensitive Ca^{2+} channel

The Ca^{2+} channel was solubilized from the rat liver and beef heart mitochondria by 88% ethanol. Mitochondria (30-50 mg protein/ml) were treated at 4°C for 20-40 min with 96% (w/v) ethanol (-20°C) at pH 7.0 under continuous stirring. The final concentration of ethanol was 88%. The suspension was centrifuged at 5000 g for 15 min at 4°C. The pellet was dissolved in an equal volume of 50% (w/v) ethanol (-20°C) and the above procedures of extraction and centrifugation were repeated. Both supernatants were combined, evaporated, and delipidized (see subsection 3.1). The chloroform phase obtained upon delipidization was evaporated to dryness, and the residue was dissolved in a mixture of chloroform and methanol (2 : 1) with consequent purification of the Ca^{2+}-binding lipid (see below). Then the water phase was evaporated to 2-3 ml and used to purify the calcium channel.

In accord with the protocol [3,37], the protein was obtained from a lipid-free ethanol extract by gel filtration on a Sephadex G-15 column with the subsequent purification of the high-molecular weight channel-forming fraction by preparative gel electrophoresis. The glycoprotein (GP) of a 40 kDa molecular weight was obtained using this procedure. The protein contains SH-groups essential for its activity and the trace amounts of protein-bound phospholipids [3, 37]. Earlier a glycoprotein with similar properties was isolated from mitochondria by using detergents[36], this protein, however, formed nonselective channels. We suggest that the detergent changes the structure of the glycoprotein and influences its activity. This glycoprotein had about 50% of phospholipids bound to the molecule, which confirms our suggestion that ethanol penetrates into the area of protein-lipid interaction and thus decreases the binding of lipids to protein.

Besides GP, a low-molecular weight active component was detected upon gel-filtration of the ethanol fraction on a Sephadex G-15 column [3, 37]. Since the pronase treatment of the component resulted in the loss of Ca^{2+}-transporting activity, we suggested that this is a peptide (P). The peptide was purified by chromatography on a Thiol-Sepharose 4B column, a Sephadex G-15 column and by HPLC [38].

As it was found, the GP is a GP-P complex [39]. After loading it on a Thiol-Sepharose 4B column, we eluted the proper GP, which does not bind to the column, and a peptide resembling the above-mentioned P, which binds to the column. Treating this P with pronase leads to a loss of activity. The channel activity was exhibited only by P, whereas the proper GP had no acivity. Hence, in both fractions eluted from the first Sephadex G-15 column, the active component was presumably the same peptide that can be dissociated from the GP-P complex during isolation. It was found that the proper GP has high-affinity sites for Ca^{2+} [39]. We assumed that its function is related to the creation of higher calcium concentration in the vicinity of the gate of the channel formed by the peptide. The amino acid sequence analysis of GP showed that it is highly homologous to human plasma orosomuoid [38]. Nevertheless, the polyclonal monospecific antibodies to the ethanol-solubilized glycoprotein-peptide complex inhibit the uniporter-mediated transport of Ca^{2+} in mitoplasts from the rat liver mitochondria and specifically bind to mitochondria in human fibroblasts [40], while the antibodies to orosomuoid did not possess these effects [38].

4.2. Reconstitution of the GP-P complex and the peptide into BLM

The GP-P complex and P have similar Ca^{2+}-channel properties and upon reconstitution into BLM have to be added to both compartments of the membrane [37, 38]. The necessity of adding P on two sides of the membrane can be explained if one assumes that P, as gramicidine, is assembled in the membrane from two equal subunits added on different sides of membranes. More frequently channels with conductivities of 20 and 40 pS were observed [3, 38] in the presence of 10 mM Ca^{2+}. However, the channels with a conductivity of 20 pS combine to form channel with a high conductivity (Fig. 4A), which probably are a cluster, judging from the kinetics of its inhibition (Fig. 4B-D). The Ca^{2+}-uniporter channels are most of the time in the open state, and their kinetics are different from that of the mitoK$_{ATP}$ channel. The potential obtained on the protein-modified membrane at a twofold gradient of Ca^{2+} was 8 mV, which is close to the Nernst potential for bivalent cations, indicating the selectivity of the channel for Ca^{2+}. The inhibitor of the mitochondrial calcium uniporter Ruthenium Red at a 1–4 µM concentration closed the channels formed by both the peptide and the GP-P complex.

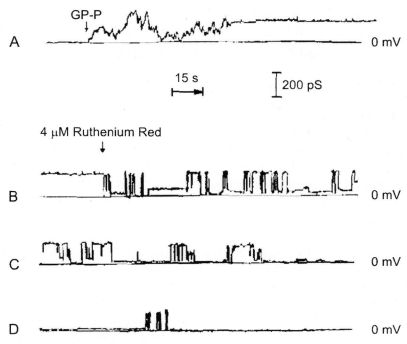

Fig. 4. Inhibition by Ruthenium Red of the current fluctuations on the BLM modified by the GP-P complex. The compartment solution contained 20 mM Tris-HCl, pH 7.4, and 10 mM CaCl$_2$. The protein (GP-P, 6 µg/ml) was introduced to the compartments on each side of the membrane. Curves A-D are the records taken for one membrane at different times.

5.   UNCOUPLING PROTEIN FROM BROWN ADIPOSE TISSUE MITOCHONDRIA

The mitochondrial uncoupling protein (UCP1) is a key and rate-limiting component of the thermogenic machinery of brown adipose tissue (BAT). It belongs to the family of mitochondrial carriers, such as adenine nucleotides transporter, 2-oxoglutarate carrier and phosphate carrier, which probably are the derivatives of a common ancestor [41]. It was found that partial uncoupling is physiologically important not only for thermogenesis but for prophylaxis against obesity, prevention of the formation of reactive oxygen species, etc. [42-43].

5.1. Mechanism of H$^+$ transport and the role of fatty acids

UCP1 has a molecular weight of 32 kDa and functions as a dimmer [44]. It mediates the reentry of protons pumped across the inner mitochondrial membrane by respiratory chain enzymes, thus converting the energy of proton gradient into heat. Its transport properties and regulatory mechanism well fit to fulfill the role and respond to the signals received by the adipocyte to initiate thermogenesis [41]. It was assumed that there are two pathways inside the protein- one for protons and the other for anions, such as chloride [44]. These two ion-conducting pathways are not identical and characterized by different sensitivities to inhibition by purine nucleotides and activation by free fatty acids [45]. It was found that H$^+$ transport is much more active, and a strong competition is observed with Cl$^-$ transport by alkyl sulfonates in BAT mitochondria and in reconstituted vesicles [46].

There are two models of fatty acid-facilitated proton conductivity. The first model proposes that UCP1 catalyzes the transport of fatty acid anions and the protonophoric action is accomplished by the flip-flop of their protonated forms in the membrane [47]. The "cycling" model provides a mechanistic explanation to the fatty acid uncoupling, in this model, UCP1 works as a "carrier" [48]. In accord with the second "channel" model, the binding of fatty acid to UCP1 induces conformational changes in the protein molecule and the formation of a channel by the carboxylic groups of fatty acids within the protein molecule, which facilitates H$^+$ traffic [49]. Since the transport of Cl$^-$ is not regulated by fatty acids, the question remains open whether H$^+$ and Cl$^-$ transport proceeds by a common or by different pathways [50].

Experiments with the reconstitution of detergent-solubilized UCP1 in liposomes indicated that the activation of UCP1 is accompanied by the transport of either proton [51] or chloride [52]. One explanation for the discrepancies in the results is that these authors used different detergents, which might alter the functional and structural properties of UCP1. Moreover, by using the reconstitution of proteins into liposomes it is impossible to chose between the "carrier" and "channel" models. To solve these questions, we used nondetergent methods for isolation of UCP1 with subsequent reconstitution of the protein into BLM.

5.2. Solubilization of the uncoupling protein by 66% ethanol

To obtain a membrane fraction, BAT mitochondria (80 mg/ml) of adult golden hamsters were diluted ten times in bidistilled water and subjected three times to freezing/thawing using liquid nitrogen and a water bath (20°C). The membrane fraction was centrifuged for 30 min (4°C) at 100 000 g and resuspended in 2 mM Tris-HCl, pH 7.4, at a concentration of 80-90 mg protein/ml [53]. Two approaches were used for extraction of UCP1.

The mitochondrial membrane fraction (20 mg protein/ml) was treated at 4°C with 66% (w/v) ethanol. One milliliter of the membrane fraction was homogenized in a glass/teflon homogenizer with 15 ml of 70% ethanol (-20°C), and the suspension was stirred for 20-30 min and then centrifuged at 5000 g for 15 min at 4°C. The pellet was dissolved in 15 ml of 40% (w/v) ethanol (-20°C), and the above procedures of extraction and centrifugation were repeated. Both supernatants were combined and evaporated in vacuum at 30°C to 5 ml in the case of using 1 ml of the membrane fraction. Lipids were removed from the membrane fraction by extraction with chloroform/methanol [54]. The yield of total proteins was about 0.6 mg and 0.3 mg before and after delipidization, respectively. The water phase was evaporated to 2-3 ml. The gel electrophoresis and immunoblotting using antiserum against hamster UCP1 showed that the major constituent of the fraction obtained was a protein with an apparent molecular weight of 32 –34 kDa [53]. Immunoblotting indicated only a faint UCP1 protein band in this case. These data indicate that ethanol releases mitochondrial membrane proteins, however, in addition to UCP1, this fraction also contains other proteins. These proteins might be phosphate carries or other carriers that also have a molecular mass of 32 kDa [41].

5.3. Acetone extraction as a method for solubilization of membrane proteins

The procedure of acetone extraction of mitochondria and other membrane systems was developed in order to understand the contribution of phospholipids and neutral lipids to the structure and function of membrane proteins [55-56]. After lipid extraction, proteins usually lose their enzymatic activity; however, their rebinding to the lipids leads to the restoration of their activity. Some enzymes are damaged by extraction using such solvents. In this case the conditions for optimal stability of proteins have to be taken into account in the methodology applied for removal of lipids and resolving of proteins.

We solubilized significantly larger amount of total proteins from the BAT mitochondrial membrane by using acetone compared to ethanol, i.e. 10-15 mg and 0.3 mg of total proteins were extracted from 80-90 mg of the mitochondrial fraction in these cases, respectively. According to the immunobloting analysis, the amount of UCP1 in the case of acetone extraction was 30-fold larger [53]. Our preliminary data demonstrated that acetone extraction can be more effective also in the isolation of other ion-transporting proteins in the physiologically active state.

5.4. Solubilization of UCP1 from brown adipose tissue mitochondria by acetone

The mitochondrial membrane fraction (80 mg/ml, 1 ml) was homogenized in a glass/teflon homogenizer in 6 ml of acetone (-20°C) for 1 min. Then 17 ml of acetone (-20°C) was added, and the suspension was stirred at -20°C for 30 min. After filtration through a glass filter No. 4; (Kavalier, Czechoslovakia), the residue was washed twice using 24 ml of acetone (-20°C), dried under vacuum at 30°C and then over NaOH for 3 days at 4°C. The obtained "acetone powder" was ~ 60-70 mg dry weight. It was homogenized in a glass/glass homogenizer in 25 ml of the solution containing 5 mM Tris-HCI, 1 mM EDTA, 0.02% mer-captoethanol, pH 7.4 and stirred for 45 min at 4°C. The suspension was centri-fuged at 10 000 g for 30 min at 4°C. Then the pellet was re-extracted as de-scribed above. The supernatants were concentrated under vacuum to 3-5 ml, dialyzed against 5 mM Tris-HCI, 0.02% mercaptoethanol, pH 7.4, for 3 days at 4°C, then they were dried by lyophilization and dissolved in 1 ml of water.

The amount of proteins extracted into the water phase was ~30% of the original membrane protein. The SDS gel electrophoresis and immunoblotting using antiserum against hamster UCP1 were applied. They demonstrated that the major constituent of the protein fraction was a protein with an apparent mo-lecular weight of 32 –34 kDa, similar to what was observed upon ethanol ex-traction. Immunoblotting indicated a strong protein band, which was much more intensive than in the case of ethanol extraction. The partially purified UCP1 ac-counted for 10% of all extracted proteins, which coincided well with the amount of this protein present in mitochondria [44]. The development of the next purifi-cation steps such as DEAE and immunoaffinity chromatography and preparative gel electrophoresis is in progress.

It was noted that prolonged acetone treatment of the membrane (30 min, -20°C) is essential for subsequent optimal extraction of UCP1 into the water phase. The volume of acetone used and the presence of 1 mM EDTA during extraction in the water phase were also important for the maximum release of UCP1.

5.5. Reconstitution in BLM of UCP1 solubilized from the membrane by acetone

The ion-transport activity of solubilized proteins was measured upon their incorporation into BLM as described above for the reconstitution of the mito-K_{ATP} channel. It is well known that UCP1 has a remarkable anion-transporting capacity [57]. In order to estimate H^+ transport, the buffer in the cell compart-ments contained 10 mM Tris-SO_4, pH 7.4. This salt was used to eliminate the possibility of Cl^- transport. It was found that the protein obtained can be recon-stituted into BLM with the retention of the properties of UCP1. It is shown on

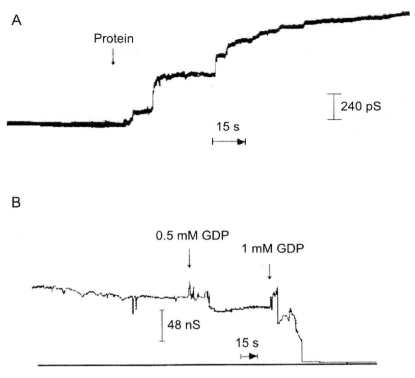

Fig. 5. Current changes induced in BLM by a protein fraction extracted by Tris-EDTA from acetone-treated BAT mitochondrial membrane. A, ionic current across BLM was measured in the absence and in the presence (arrow) of extracted proteins (30 µg/ml); B, inhibition of protein-induced current in BLM by GDP. The bath solution contained 3 ml of 10 mM Tris-SO$_4$ buffer, pH 7.4. A potential 50 mV was applied to the membrane. The protein (70 µg/ml) was introduced on both sides of the membrane.

Fig. 5A that this protein increases membrane permeability, which is entirely inhibited by the specific inhibitor of UCP1 GDP (Fig. 5B). The character of the induced current indicates that the protein behaves as a carrier, which supports the idea that UCP1 acts as a carrier rather than as a channel. The potential of the protein-modified membrane at the pH gradient was close to the Nernst potential for H$^+$. The transport proceeded in Tris-SO$_4$, indicating that H$^+$ movement is independent on CI$^-$. These data confirm the hypothesis that two ion-transporting ways can exist independently.

In accordance to this data, the H$^+$ transport occurs without the addition of fatty acids. The activating effect of fatty acids on the H$^+$ transport cannot be excluded, however, since free fatty acids are present in natural membrane [58] used for BLM formation. It was also found earlier that a low level of endoge-

nous free fatty acids is sufficient for GDP-sensitive uncoupling in BAT mitochondria [49].

Further purification of the protein and the reconstitution of the pure protein
into BLM will give insight into the role of fatty acids in the regulation of UCP1,
the effect of pH on the transport, the role of long-chain acyl-CoA, etc. The reconstitution of pure UCP1 into BLM will provide a definite answer to the question whether H^+ and CI^- transport occurs by a common or different pathways
within UCP1.

6. THE SELF-OSCILLATING CHANNEL FORMED BY THE HYDROPHOBIC CA^{2+}-BINDING MITOCHONDRIAL COMPONENT

Recently, a new channel-forming component of the inner mitochondrial
membrane was isolated in our laboratory [59]. It is a hydrophobic low-
molecular-weight component, which can bind Ca^{2+} and is, probably, the subunit
c of H^+ ATPase .

6.1. Isolation and identification of the hydrophobic Ca^{2+}-binding component from mitoplasts

A Ca^{2+}-binding component (CaBC) was extracted from mitoplasts or purified fractions of mitochondria by 60% ethanol as it was described above for the
isolation of mitoK$_{ATP}$ channel. The obtained fraction was separated into hydrophilic and hydrophobic components in accordance with the Folch procedure
[54]. As it has been described previously, all mitochondrial ion channels were
purified from the hydrophilic fraction, whereas the Ca^{2+}-binding components
were isolated from the hydrophobic fraction [59]. After performing the Folch
procedure, the chloroform phase was evaporated to dryness and dissolved in a
mixture of chloroform : methanol (2:1).

The hydrophobic fraction on SDS-PAGE in 14% polyacrylamide gel
showed two bands, both bands bound Ca^{2+} and were stained by Coomassie
Briliant Blue [59]. In order to estimate the Ca^{2+}-binding capacity, SDS-PAGE
gels were washed by 20 mM Tris-HCI, pH 7.5, for 30 min and incubated overnight in the washing buffer containing 10 μM $^{45}Ca^{2+}$ (Amersham Radiochemical Co.). An excess of the radioactive calcium was removed by washing in a
calcium-free buffer for 30 min, and the gel was dried and allowed to expose to a
Kodak film for 2 h. One of the bands, localized on the gel in the area of the
front of line, contained the lipids. This band was purified to identify the nature
of Ca^{2+}-binding lipid [60-61]. The other band was, probably, a Ca^{2+}-binding
peptide with a molecular weight of ~8 kDa compared to the protein standard

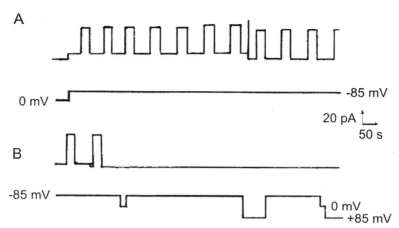

Fig. 6. Self-oscillations of channel activity. Ca^{2+} concentrations were 10 mM on the side of the *cis*-compartment and 40 mM from the side of the *trans*-compartment of the membrane. A, is a record of the channel in the beginning of the experiment, and B, is a record of the channel after 2 h. The lower trace shows the applied voltage. The chamber solution contained 20 mM Tris-HCI buffer, pH 7.4.

[59]. No more protein bands were observed in the mitochondrial hydrophobic fraction.

The presence of the peptide in the hydrophobic fraction upon solubilization of the brain plasma membrane by chloroform (a method similar to that used in our experiments) was confirmed by McGeoch and Guidotti [62]. They showed that, in accordance with the sequence of this 7.6 kDa peptide, it should be the subunit *c* of the mitochondrial F_0 ATPase. Earlier it was shown that this subunit is the most hydrophobic peptide of the mitochondrial membrane and it can be isolated from the membrane by butanol and ether [63].

6.2. Reconstitution of the mitochondrial hydrophobic Ca^{2+} -binding component into BLM

The channel activity was estimated using an operation amplifier (OPA 101, Burr-Brown Corp. Tucson. Arizona) connected to a Tectronix 2211 digital storage oscilloscope to record the *trans*-membrane current at different clamping voltages. BLM was formed from 2.2 mg of total brain lipids + 100 μg of cardiolipin and an extract from 5 mg of mitoplasts after all the components were dried and diluted in 60 μl of *n*-decane. The conductivity of BLM without extract was 3-5 pS/mm^2 in the presence of 10 mM CaCI$_2$.

Being added to brain lipids and cardiolipin (10%), the hydrophobic component forms a Ca^{2+}-activated channel with a conductivity of 100 pS or its multiples (500 pS) in the presence of 10 mM Ca^{2+}. Generating the ion gradient (4:1)

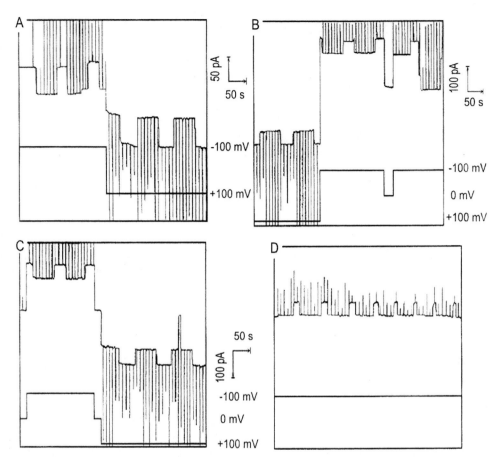

Fig. 7. Oscillations and spiking of channel activity. The experiment conditions were as in Fig. 6. Potentials applied to the membrane are given. A, the channel activity in the beginning of the experiment; B, the same membrane after 0.5 h; C, after 1 h; D, after 3.5 h.

in the membrane induces a self-oscillating channel activity, which is retained for several hours. Fig. 6 illustrates a channel that functions during 2.5 h at a membrane potential of -85 mV. It can be seen that the channel suddenly closed after 2.5 h when the potential was eliminated from the membrane. This was an irreversible process and a subsequent application of the potential did not open this channel.

In another experiment, if the 500 pS channel also oscillated during 2-3 h, the time of the open state of the channel gradually decreased without a signifi-cant change in the time of the closed state of the channel (Fig. 7). In this case, a regular spike-like channel was observed against the background of the regular oscillating channel. The amplitude of this spiking-like channel was usually high

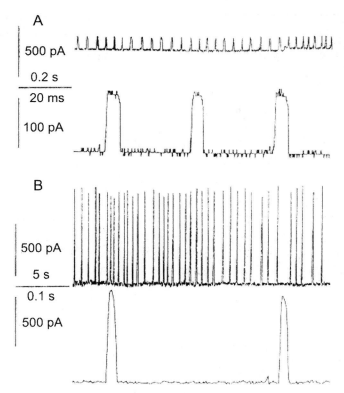

Fig. 8. Variations of current oscillations in the reconstituted subunit *c* of ATP synthase [62]. In each part of the figure, the zero of the upper trace coincides with the horizontal time bar. The lower trace is expanded from the highlighted part of the upper trace. A- bovine brain crystallized subunit *c* clamped at 70 mV and in pipette was 50 mM NaCI; B- rat hippocampal extract clamped at 60 mV and in the pipette was 150 mM NaCI.

if the 500 pS channel was in the close state and decreased if the channel was open. At a zero potential no spiking was observed. (Fig. 7 B and C). The time of open and close states of this spiking-like channel remained unchanged throughout the experiment.

Reversing the potential did not change the general behavior of the system (Fig. 7 A-C). However, the time of the open state of the channel at a potential +100 mV was longer throughout the experiment than at a potential of –100 mV. This can be explained by the presence of Ca^{2+} gradient on the membrane.

It is important to note that during this experiment only one channel was oscillating, even if three channels functioned on the membrane. (Fig. 7 B and C). During the experiment, a gradual decrease in spike amplitude was observed, coinciding with a decrease in the duration of the open state of the main channel

(Fig. 7 D). The interval between spikes did not change significantly and after 3.5 h of work, the main channel was opened during the interval between three spikes. It is also remarkable that, in some cases, all three channels were closed during the period between two spikes (Fig. 7 C). This suggests that there is a functional relationship between spikes and the channel state.

6.3. Reconstitution of the subunit *c* and patch-clamp assay

Similar self-oscillating channels were found by using patch-clamp assay of plant lipid vesicles. The vesicles were modified by the pure subunit *c* of H^+ATPase obtained from brain membranes [62]. Typical records of the channels formed by the brain subunit *c* are presented in Fig. 8. A similar oscillating channel was found upon reconstitution of the liver subunit *c* into BLM. The current step was typically 50-200 pA, and the oscillation frequency is usually in the range from 0.5 to 200 Hz. In order to generate the oscillations, calcium must be present in the pipette solution. The antibody to the subunit *c* closed the channels, while antibodies to the α_1 subunit of Na,K-ATPase did not produce any effect on the channel activity [62]. The authors suggested that the opening and closing transitions can be related to cooperative binding of calcium to the peptide. This suggestion is in agreement with our data that the small peptide in the presence of which self-oscillating channels are formed on the membrane has a Ca^{2+}-binding capacity [59]. Considering the channel properties of the ethanol-isolated peptide, it can be assumed that it is the subunit *c* of ATP synthase.

Oscillating ion fluxes could be observed in the mitochondrial membranes [64] and in artificial bilayers in the presence of some lipids [65]. One plausible model suggests that cations passing through the channel become bound to charged groups of the membrane, thus causing the phase transition and closing of the channel. If the cations are lost from the sites, the phase transition reverts and the channel is reopened.

7. CA^{2+}-BINDING AND ION-TRANSPORTING PROPERTIES OF PALMITIC AND STEARIC ACIDS

The hydrophobic fraction on SDS-PAGE in 14% polyacrylamide gel shows two Ca^{2+}-binding and Coomassie Briliant Blue staining bands (*see* subsection 6.1 and Ref. [59]) during the isolation of a self-oscillating channel from mitochondria. In subsequent procedures we found that the Ca^{2+}-binding capacity of the low-molecular-weight fraction is due mainly to palmitic and stearic acids [61].

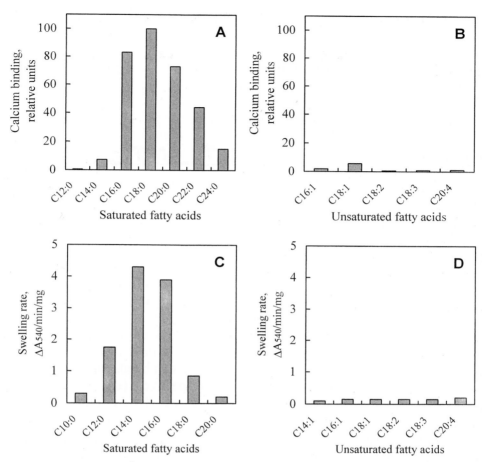

Fig. 9. The affinity to Ca^{2+} and the capacity of saturated and unsaturated fatty acids to form pores in mitochondrial membranes varies with the length of the aliphatic chain. A, B- affinity to Ca^{2+} (our data); C, D- cyclosporin-insensitive pore-inducing capacity [71]. A, C –saturated fatty acids; B, D- unsaturated fatty acids.

7.1. Ca^{2+}-binding capacity of various lipids

The Ca^{2+}-binding capacity of various lipids was measured with ^{45}Ca^{2+} (Amersham Radiochemical Co., U.K.) using the following protocol- 5 μl of a 2 mM solution of each lipid was applied on a polyvinylidene difluoride membrane (PVDF Immobilon-P Millipore) and incubated in 20 mM Tris-HCI buffer, pH 8.5, for 5 min. Calcium was removed from PVDF by incubation during 10 min in the same buffer in the presence 0.5 mM EDTA, and PVDF was then washed three times, 15 min each time, in the buffer in the absence of EDTA. In order to estimate the Ca^{2+} binding, the membrane was incubated for 1 h in the

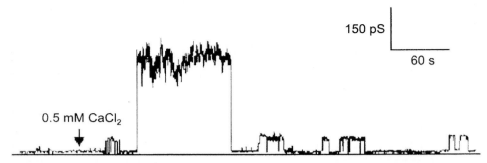

Fig. 10. Channel activity in a bilayer lipid membrane reconstituted with stearic acid. The chamber solution contained 20 mM Tris-HCl buffer, pH 8.5, and $CaCl_2$. The applied voltage was 40 mV. BLM was formed using a mixture of 1.2 mg of total brain lipids, 100 µg of cardiolipin and 6 µg of stearic acid dissolved in 60 µl of *n*-decane (see Materials and Methods).

same buffer in the presence of 5 µM $^{45}CaCl_2$ and washed with the buffer (three times for 15 min). Radioactivity was measured by liquid scintillation.

It was found that saturated fatty acids (SFA) have a higher Ca^{2+}-binding capacity than unsaturated acids, phospholipids, lysophospholipids, and other lipids [61]. The affinity of SFA to Ca^{2+} varies with the length of the aliphatic chain, the highest Ca^{2+}-binding capacity being exhibited with palmitic acid (PA), stearic acid (SA) and eicosanoic acids, respectively (Fig. 9 A and B). Total mitochondrial lipids were separated into classes (cholesterol, neutral lipids, free fatty acids, phospholipids and others) by using an aminopropyl-bound silica column and by elution in different organic solvents [66]. The high-affinity Ca^{2+}-binding activity was found, however, only in the mitochondrial lipid fraction containing free fatty acids [61].

7.2. Ca^{2+}-Binding properties of palmitic and stearic acids

Binding of Ca^{2+} with PA and SA was found to be pH-dependent. At pH 8.5 the affinity was maximum, and at pH 6.8 Ca^{2+}-binding was not observed. The K_d and B_{max} values for both fatty acids were estimated to be 5 µM at pH 8.5 and 15 µM at pH 7.5 for K_d and 0.48 ± 0.08 mmol of Ca^{2+} per g of PA at both pH for B_{max} [61]. Considering the molecular weight of PA (256 Da), a maximal PA/Ca^{2+} ratio was estimated to be 8 : 1; therefore, it is possible to bind up to eight molecules of PA to a Ca^{2+} ion.

Apparently, the bonds between PA molecules and a Ca^{2+} ion are coordinate. It is known that in coordinate complexes Ca^{2+} has 6-8 bonds [67]. Evidence that Ca^{2+}-PA complexes are of coordinate nature was demonstrated also

Table 1
Influence of free fatty acids on BLM conductance

Additions	BLM conductance, 10^{-9} S/cm^2			
	Palmitic acid		Stearic acid	Palmitoleic acid
	Total brain lipids	Mitochondrial lipids	Total brain lipids	Total brain lipids
None	1÷1.5	4.08÷6.1	1.02÷2.04	1÷1.6
0.1 mM CaCl$_2$	1÷1.5	10.2÷14.3	1÷1.5	–
0.5 mM CaCl$_2$	10÷50	40.8÷48.9	28.5÷53.1	–
1 mM CaCl$_2$	30÷150	581÷618	53.14÷114.6	1.1÷3.04
0.1 mM CaCl$_2$ +30 mM KCl	10÷12	51÷81	12÷15	–
1 mM CaCl$_2$ +30 mM KCl	–	15600÷30000	122.14÷341.06	1.05÷1.9
1 mM CaCl$_2$ +100 mM KCl	300÷500	40700÷40900	–	–
100 mM KCl	1÷1.5	9÷6.5	1.02÷2.04	–

The experimental cell contained 20 mM Tris/HCl buffer, pH 8.5. BLM was formed from the mixture of 1.2 mg of total brain lipids, 100 μg of cardiolipin and 6 μg of palmitic or stearic acid dissolved in 60 μl of *n*-decane, or from the mixture of 1.2 mg of mitochondrial lipids and 3.9 μg of palmitic acid dissolved in 60 μl of *n*-decane (see Materials and Methods).

by IR-spectroscopy. The IR spectrum of PA measured under anhydrous conditions changes in the presence of Ca^{2+} [61]. The observed changes indicate the formation of PA-Ca^{2+} complexes (the appearance of a more narrow and intensive band at 3403 cm^{-1}). There are no indications, however, of salt bonds between PA and Ca^{2+} as the band of the carbonyl group does not shift from 1705 to 1558 cm^{-1}.

7.3. Influence of palmitic and stearic acids on BLM permeability

As it follows from the data in Table 1, the incorporation of PA or SA into BLM did not change the level of membrane conductivity. The addition of Ca^{2+} to BLM, however, resulted in a 30-fold increase in membrane permeability. This BLM was formed from a mixture of total brain lipids and 0.5% (w/w) PA or SA. It was not merely a rupture of the membrane, as it was able to work for more than an hour, however, the channels can be closed during some period of time (Fig. 10). Subsequent addition of 100 mM KCl caused further growth of conductivity. In the presence of the threefold Ca^{2+} gradient, the membrane potential was only 3 mV, indicating that the permeability is nonspecific. If mito-chondrial lipids were used for BLM formation, the same effects could be achieved using smaller content of PA or SA in the membrane and using a lower concentration of Ca^{2+} in the medium (Table 1). In contrast to PA and SA, pal-mitoleic acid, which had an unsaturated bond and bound Ca^{2+} with low affinity, did not alter the membrane permeability in the presence of Ca^{2+} and K^+. We ob-served similar results with liposomes loaded by sulforhodamine B. It was found that PA in the presence of Ca^{2+} increased the permeability of liposomes in a PA and Ca^{2+} concentration-dependent way (unpublished data).

7.4. Comparison of the effects of palmitic acid on ion permeability of BLM and intact mitochondria

Recently it has been found that PA is a physiological activator of pro-grammed cell death and that mitochondria play a central role in PA-activated apoptosis [68-69]. It has been shown that PA promoted an increase in perme-ability of the inner mitochondrial membrane with subsequent release of cyto-chrome *c*, which is a well known trigger of apoptosis. In some pathological states, the PA content in tissues increases substantially, causing significant changes in membrane permeability and function (from Ref. [68]). The nature of this permeability, however, has not been finally established.

The mechanism of the PA-induced permeability of mitochondrial mem-brane can be explained by the capacity of PA to bind Ca^{2+} and increase nonspe-cific permeability in BLM, which has been described above. This was recently confirmed by an observation that PA can promote a nonclassical cyclosporin-insensitive permeability transition in mitochondria [70]. It is interesting to note that all results described above, including the ability to bind Ca^{2+}, to induce nonspecific conductivity in BLM and liposemes, as well as nonclassical transi-tions in mitochondria, correlate well with the length of the aliphatic chain of fatty acids (Fig. 9 and Table 1). These allow us to assume that the mechanism of PA-activated apoptosis is related to the formation in the membrane of a PA/Ca^{2+} complex that creates non-specific pores. The data obtained in the Sokolove's laboratory showed that this pore is opened only during some time and then it closes [70, 71], probably, due to the ejection of this complex from the mem-brane, which prevents the destruction of mitochondria. In this case, the integrity of mitochondria should be preserved, which is necessary for the final stage of apoptosis [72].

8. CONCLUSION

This review demonstrates a new approach to the solubilization of mem-brane proteins by organic solvents. We used this approach for the isolation of ion channels of biological membranes, primarily mitochondrial. The isolation of channel proteins in a water-soluble state without the use of detergents makes it possible to apply a wider range of biochemical methods for their purification. In addition, this approach eliminates the effect of detergents on membranes into which proteins are reconstituted. It was found that the properties of solubilized ion channel proteins reconstituted into BLM are close to the properties of the corresponding ion-transport systems of intact mitochondrial membranes. Such a complex approach enables one not only to study the mechanisms of functioning of these channel proteins but also find the pathways of their regulation in the cell.

9. ACKNOWLEDGMENTS

I thank Prof. Marie Kondrashova and my collaborators in ITEB RAS. This study was supported by the grant RN 2000 from the International Science Foundation, INSERM 189 (France), the Sigrid Juselius Foundation (Finland), the grants nos. 01-04-97005, 01-04-48551 of the Russian Foundation for Basic Research to G.D.M. and Fogarty Inernational Research Collaboration Award TW01116 from the National Institute of Health to K. D. G.

REFERENCES

[1]   G. Mironova, I. Dovgi and S. Salnikova. G.P.Curtis, (ed), Boiphysics of membrane, 454, Palanga, 1973.
[2]   G. Mironova, L. Pronevich and M. Kondrashova, Dokl. Acad. Nauk USSR, 226, (1976) 1203.
[3]   G. Mironova, T. Sirota, L. Pronevich, N. Trofimenko, G. Mironov and M. Kondrashova, Biofizika, 25, (1980) 276.
[4]   J. Heller, Biochemistry, 17, (1968) 2906.
[5]   P. Jorgensen, Biochim. biophys. Acta, 356 (1974) 36
[6]   C. Schnaitman and J. Greenwalt, J. Cell. Biol., 38, (1968) 158
[7]   J. Reynalds and C.Tanford, J. Biol. Chem., 245, (1970) 5161.
[8]   J. Bangham and E. Lee, Biochim. biophys. Acta, 511, (1978) 388.
[9]   J. Diwan, T. Haley and D. Sanadi, Bioch. Biophys. Res. Commun.,153, (1988) 224.
[10] P. Paucek, G. Mironova, F. Mahdi, A. Beavis, G. Woldegiorgis and K. Garlid, J.Biol. Chem., 267, (1992) 26062.
[11] P. Zahler, E. Weilbe, Biochim. biophys. Acta, 219, (1970) 320.
[12] R. Juliano, Biochim. biophys. Acta, 266, (1972) 301.
[13] A. Maddy, Biochim. biophys. Acta, 117, (1966) 193.
[14] A. Rega, R. Weed, C. Reed, G. Berg and A. Rothstein, Biochim. biophys. Acta, 147, (1967) 297.
[15] A. Maddy and M. Dann , A Maddy (ed.), Biochemical Analysis of Membranes, 160, London, New York, 1974.
[16] G. Mironova and N. Mirsalikhova, Biofizika, 26, (1981) 731.

[17] G. Mironova, N. Bocharnikova, N. Mirsalikhova and G. Mironov, , Biochim. biophys. Acta, 861,(1986) 224.

[18] G. Mironova, N. Fedotcheva, P. Makarov, L. Pronevich and G. Mironov, Biofizika, 26, (1981) 451.

[19] I. Inoue, H. Nagase, K. Kishi and T. Higuti, Nature, 352, (1991) 244.

[20] G. Mironova, Yu. Skarga, S. Grigoriev, V. Yarov-Yarovoi, A. Alexandrov and O. Kolomytkin, Membr. and Cell Biol., 10, (1996) 429.

[21] P. Mueller, D.O. Rudin, H. Tien, and W. Wescott , Nature 194, (1962) 979.

[22] G. Mironova, Yu. Skarga, S. Grigoriev, A. Negoda, O. Kolomytkin, B. Marinov, J. Bioener. Biomem., 31, (1999), 157.

[23] G. Mironova, Yu. Skarga, S. Grigoriev, V. Yarov-Yarovoi, A. Negoda and O. Kolomytkin, Membr. and Cell Biol., 10 (1997) 583.

[24] U.R. Maurer, E.L. Boulpaep and A.S. Segal, J. Gen. Physiol. 111, (1998) 139.

[25] S.M. Grigoriev, Yu.Yu. Skargs, G.D. Mironova, B.S. Marinov, Biochim. biophys. Acta, 1410, (1999) 91.

[26] J. Meury and A. Kepes, EMBO J., 1, (1982), 339.

[27] M. Jaburek, V. Yarov-Yarovoy, P. Paucek and K.D. Garlid, J. Biol. Chem., 273, (1998) 13578.

[28] K.D. Garlid, P. Paucek, V. Yarov-Yarovoy, X Sun and P.A. Schindler, J. Biol. Chem., 271, (1996) 8796.

[29] K.D. Garlid, P. Paucek, V. Yarov-Yarovoy, H.N. Murray, R.B. Darbenzio, A.J. D'Alonzo, N.J. Lodge, M.A. Smith and G.J. Grover, Circ. Res., 81, (1997) 1072.

[30] G.J. Grover, and K.D. Garlid, J. Mol. Cell, Cardiol. 32, (2000) 677.

[31] N.I. Fedotcheva, A.A. Sharyshev, G.G. Mironova and M.N. Kondrashova, Comp. Biochem. Physiol., 82B, (1985) 191.

[32] P. Paucek, V. Yarov-Yarovoy, X Sun and K.D. Garlid, J. Biol. Chem., 271, (1996) 32084.

[33] A.D. Beavis, Y. Lu and K.D. Garlid, J. Biol. Chem., 268, (1993) 997.

[34] Yu.Yu. Skarga, L.P. Dolgacheva, N.I. Fedotcheva and G.D. Mironova, Ukr. Biochem. J.,59, (1987) 54.

[35] S.J. Tucker, F.M. Gribble, C. Zhao, S. Trapp and F.M. Ashcroft, Nature, 387, (1997) 179.

[36] E. Carafoli, Molec.& Cell Biochem, 8, (1975) 133.

[37] G. Mironova, T. Sirota, L. Pronevich, N. Trofimenko, G. Mironov and M. Kondrashova, J. Bioenerg. Biomem., 14, (1982) 213.

[38] G. Mironova, M. Baumann, O. Kolomytkin, Z. Krasichkova, A. Berdimuratov, T. Sirota, I. Virtanen and N-E. Saris, J. Bioenerg. Biomem., 26, (1994) 231.

[39] G. Mironova and Zn. Utesheva, Ukr. Biochem. J.,61, (1989) 48.

[40] N-E. Saris, T. Sirota, I. Virtanen, K. Niva, T. Penttila, L. Dolgachova and G. Mironova, J. Bioener. Biomem., 25, (1993) 307.

[41] F. Boullaud, S. Raimbault. L. Cassard, and D. Ricquier, A. Azzi (ed.), Anion Carriers of Mitochondrial Membranes. Springer-Verlag Berlin Heidelberg, 1989, 251.

[42] L. Wojtczak, M. Wieckowski, J. Bioenerg. Biomem., 31, (1999) 447.

[43] V.P, Skulachov, J. Bioenerg. Biomem., 31, (1999) 431.

[44] D.G. Nicholls, E. Rial, J. Bioenerg. Biomem., 31, (1999) 399.

[45] J. Kopecky, F. Guerrieri, P. Jezek, Z. Dragota and J. Houstek, FEBS Lett., 170, (1984) 186.

[46] P. Jezeck and K.D. Garlid, J. Biol. Chem., 265,(1990) 19303.

[47] K.D. Garlid, D. Orosz, M. Modriansky, M. Vassanelli and P. Jezek, J. Biol . Chem., 271 (1996) 2615.

[48] V. P.Skulachev, Febs Lett., 294, (1991) 158.

[49] E. Winkler, and M. Klingenberg, J. Biol. Chem., 269, (1994) 2508.

[50] M. Klingenberg, J. Bioenerg. Biomem., 31, (1999) 419.

[51] M. Klingenberg, and E. Winkler, EMBO J., 4, (1985) 3087

[52] P.J. Strieleman, K.L. Schalinske and E. Shrago, Bioch. Biophys. Res. Commun., 127, (1985) 509.

[53] G. Mironova, N. Fedotcheva, Yu. Skarga, J. Kopetsky, and J. Houstek. S. Kolaeva, L. Wang, (eds.), Mechanisms of natural hypometabolic state., Pushchino, 1992, 255.

[54] J. Folch, M. Lees, and G. Stanley, 226, (1057) 497.

[55] S. Fleischer, H. Klouwen and G. Brierley, J. Biol. Chem., 236, (1961) 2936.

[56] S. Fleisher and B. Fleischer, S.Colowick ans N. Kaplan (eds.), Methods in Enzymology, Academic Press, New York and London, X, 1997, 406, 1967.

[57] D.S. Nichols and O. Landberg, Eur. J. Bioch., 37, (1973) 523.

[58] D.R. Pfeiffer, P.C. Schmid, M.C. Beatrice, and H.Y.O. Schmid, J. Biol. Chem., 254, (1979) 11485.

[59] G. Mironova, A.Lazareva, O. Gateau-Roesch, J. Tyynela, E. Pavlov, M.T. Vanier and Saris, N.-E.L. J. Bioenerg. Biomembr. 29, (1997) 561.

[60] O. Gateau-Roesch, E., Pavlov, A. Lazareva, E.Limarenko, C. Levrat, N.-E.L. Saris, P. Louisot and G. Mironova, J. Bioenerg. Biomembr. 32, (2000) 105.

[61] G. Mironova, O. Gateau-Roesch, C. Levrat, E. Gritsenko, E. Pavlov, A.Lazareva, E. Limarenko, C. Rey, P. Louisot and Saris, N.-E.L. J. Bioenerg. Biomembr., 33, (2001) 319.

[62] J.E. McGeoch and G. Guidotti, 766, (1997) 188.

[63] H. Sigrist, K. Sigrist-Nelson and C. Gitler, Bioch. Biophys. Res. Com., 74, (1977) 178.

[64] V. Gooch and L. Packer, Bioch. biophys. Acta, 346, (1074) 245.

[65] K. Toko, N. Ozaki, S. Iiyama, K. Yamafuji, Y. Matsui, K. Yamafuji, M. Saito and M. Kato, Biophys. Chem., 41, (1991) 143.

[66] A. Pietsch, and R.L. Lorenz, Lipids 28, (1993) 945.

[67] R. Williams, Symp. Soc. Exp. Biol 30, (1976) 1.

[68] J.Y. Kong and S.W. Rabkin, Biochim. Biophys. Acta ,1485, (2000) 45.

[69] G.C. Sparagna, D.L. Hickson-Bick, L.M. Buja and J.B. McMillinl, Am. J. Physiol. Heart Circ. Physiol., 279, (2000) H2123.

[70] A. Sutlan and P.M. Sokolove, Arch. Biochem. Biophys. 386, (2001) 31.

[71] A. Sutlan and P.M. Sokolove, Arch. Biochem. Biophys 386, (2001) 52.

[72] J.J. Lemasters, T.Qian, C.A. Bradham., W.E. Cascio, L.C. Trost, Y. Nishimura, A.L. Neiminen and B. Herman, J. Bioenerg. Biomembr., 31, (1999) 305

Planar Lipid Bilayers (BLMs) and their Applications
H.T. Tien and A. Ottova-Leitmannova (Editors)

Chapter 18

The use of liposomes to detect channel formation mediated by secreted bacterial proteins

V. Cabiaux, S. Vande Weyer and J.M. Ruysschaert

Structure et fonction des membranes biologiques, Université Libre de Bruxelles, CP 206/2, Boulevard du Triomphe 1050 Brussels, Belgium.

1. INTRODUCTION

Infectious diseases are the third leading cause of death in the United States and the leading cause worldwide [1]. In the last 10-15 years an amazing number of data have been accumulated about the molecular mechanisms of bacterial pathogenesis, both at the bacterial and the eucaryotic cell levels. In this review, we will focused on a particular step which is the insertion into the cell membrane of secreted bacterial proteins that leads to channel formation into the lipid membrane.

1.1 Toxins.

Roughly, two types of channel forming toxins can be described:

- pore forming toxins: these toxins act by forming a pore in a lipid membrane and the presence of such a structure in the cell membrane is directly correlated with toxicity either by disrupting the membrane permeability or by activating signal transductions pathways. For reviews on pore forming toxins see for example Refs. 2-4. These proteins are quite hydrophobic and in some cases, require high concentration of urea to be maintained in solution after purification (e.g. Ref. 5). In other cases, there is an activation step at the cell surface which leads to membrane insertion (e.g. aerolysin from *A. hydrophila* [2]). Often, purification of the activated form, which may be required for molecular characterization, results in quite unstable proteins or requires urea or the presence of a detergent.

- the translocation toxins : these toxins have been termed "A-B" type toxins because they are organized into two components: an A domain which carries the enzymatic activity, leading to cell death, and a B domain which binds to a cellular receptor and which plays a key role in translocation of the A fragment into the eukaryotic cell cytoplasm. Quite often, translocation of the A moieties is associated to a channel formation whose role in translocation has not been clearly understood yet for any toxin. Secreted as soluble proteins, these toxins undergo a conformational change at the surface or within the cells, allowing their insertion into the target lipid membrane (for a review see Ref. 6). The activation can be induced, for example, by a proteolytic cleavage (Protective antigen (PA) of *Bacillus anthracis*) or low pH (diphtheria toxin). In both cases, the activated form has a much lower solubility in aqueous buffer than the native protein. For example, DT precipitates at low pH and among several detergents tested only high concentration of Li-dodecylsulfate kept it soluble (V. Cabiaux personal data).

1.2 Type III secreted virulence factors.

Gram negative bacteria use several secretion mechanisms to allow the proteins to cross both the inner and the outer membranes. For example, *E. coli* uses a type I secretion system, composed of three proteins forming a channel across the inner and the outer membrane, types II and IV secretion machineries involving the sec machinery and a type III secretion machinery which has the originality to inject secreted bacterial proteins into the cytosol of the host cells (for a review, see Ref. 7).

A number of bacteria that are pathogenic for animals (Yersinia, Shigella, Salmonella, Escherichia) and plants (Pseudomonas syringae) use type III secretion systems to translocate virulence factors into host cells. These systems consist of a secretion apparatus and an array of proteins released by this apparatus. Some of the released proteins are effectors, injected into the cytosol of the target cell, the others are "translocators", which help the effectors to cross the membrane of the eukaryotic cell (for reviews, see Refs. 8-10).

Type III secretion systems produced by *S. typhimurium*, *S. flexneri* and *Y. enterocolitica* have been purified and visualized by electron microscopy [11-13]. These assemblages exhibit a basal body that spans the periplasm and a hollow projection that extents out from the bacterial surface like a needle. The basal body likely functions to allow protein transport from the bacterial cytoplasm across the inner and outer membranes. The "needle" is thought to allow protein translocation into the host cell cytoplasm. Depending upon the bacterial system, this translocation is associated to several phenomena, such as the induction of phagocytosis of the bacteria by the eucaryotic cells (Shigella, Salmonella) [8-10]. Translocation requires a very close contact between the bacterial and the target cell membranes [12,14]. For instance, addition of a supernatant of *S. flexneri* containing the secreted proteins responsible for the induction of phagocytosis do not cause the membrane rearrangements associated to the bacterium entry. This suggests that the proteins must be secreted at a very small distance from the membrane to efficiently interact with it. Three proteins (termed ipa : invasive plasmid antigen) IpaB, IpaC and IpaD are essential for entry into Hela cells. Ipa B and IpaC have hydrophobic regions that could form transmembrane helices. IpaB, IpaC and IpaD have been purified separately in buffers containing 8M urea (IpaB), 6M urea (IpaC) and 4M urea (IpaD) respectively [15,16]. In the absence of urea, the proteins aggregate, which is probably why a culture supernatant containing all secreted Ipa proteins is unable to induce the bacterium phagocytosis.

A common feature of all the systems described above (toxins and type III-secreted proteins) is the requirement of interaction with a lipid membrane that, in turn, requires the presence or the exposure at the protein surface of hydrophobic regions. Several methods of purification can be used to purify hydrophobic proteins but all of them might induce a partial unfolding of the proteins. To bypass the purification, we designed a set up in which the type III-secreted proteins or the toxins in their lipid-interacting conformation (activated form) can be inserted in one step. After purification, the proteoliposomes are used as such to determine the protein topology, its secondary structure or fused to a planar lipid bilayer to characterize the ability of the protein(s) inserted into the membrane to induce channel formation.

2. PROTEIN INSERTION AND MODULATION OF PROTEIN CONTENT OF THE LIPOSOMES

2.1 Insertion upon secretion.
Y. enterocolitica secretes a set of proteins, called Yops (Yersinia Outer Protein), that are exported by the type III secretion machinery. Some of the Yops are effectors (YopE, YopH, YopO, YopP/YopJ, YopM and YopT) that are delivered by extracellular bacteria into the cytosol of macrophages where they destroy the cytoskeleton, subvert the signalling network and induce apoptosis. Others such as YopB and YopD, have been shown by genetic analysis to be required for the translocation of the effectors across the eukaryotic cell membranes [10]. Similarly, *S. flexneri* secretes several proteins, among with four Ipas proteins, specifically secreted by the type III machinery. Secretion of these pre-synthezised proteins is induced by the contact of the bacteria with the eukaryotic cell [17]. Genetic analysis indicated that IpaB, IpaC and IpaD are essential for entry into epithelial cells [18]. Once secreted, IpaB and IpaC associate in a complex in the extracellular medium. This complex is necessary and sufficient to induce membrane ruffling and entry of latex beads into epithelial cells, suggesting that a IpaB/IpaC complex plays an essential role in the bacterial entry [19]. IpaB, IpaC and IpaD have been purified separately and their interaction with a lipid membrane has been characterized [15,16]. However, protein purification requires high urea concentration (from 8M to 4M) that could interfere with the protein structure, at least before the interaction with the lipid membrane. We therefore designed a protocol that allowed the insertion of the secreted Ipas in a lipid membrane, in a single step, avoiding

protein purification. In the absence of cells, secretion of Ipas proteins has been induced by addition of Congo Red [17]. The protocol is summarized Figure 1. Bacteria were grown under aerated conditions in 60 or 100 ml of TCS medium at 37°C until they reached the end of the exponential growth phase. 10 ml of culture were harvested by centrifugation at 3,300g for 12 min at 37°C, and resuspended in a 1/20 volume of PBS (1.47mM KH_2PO_4, 0.87mM Na_2HPO4, 10mM KCl, 137mM NaCl), pH 7.4, at 37°C. Asolectin large unilamellar vesicles (LUV [20]) prepared in PBS were added to the bacterial suspension (7 mg LUV/ml suspension) and Ipa secretion was induced by addition of Congo red at a final concentration of 150 µM. The concentration step of the bacterial culture was found essential to maximise the bacteria/liposomes interaction. In the absence of Congo Red, the liposomes have no effect on secretion, indicating they do not interfere with the secretion machinery. After 15 minutes of incubation at 37°C, bacteria were pelleted by centrifugation at 14,000g for 5 min. PMSF (final concentration 5.10^{-4} M) was added to the supernatant which was mixed with an equal volume of 80% sucrose in PBS, pH 7.4. A 30%-2% sucrose gradient was poured on top of the 40% sucrose layer and samples were centrifuged at 126,000g for 16 h at 4°C. Liposomes were collected, mixed with 3-4 vol of PBS containing 0.5 M NaCl (to remove adsorbed proteins), pH 7.4, and centrifuged at 126,000g for 35 min. The pellet was resuspended in PBS, and analyzed by gel electrophoresis on a 12% sodium dodecyl sulfate-polyacrylamide gel (Figure 1). The presence of liposomes does not modify the amount and the nature of the secreted proteins but slightly interfere with their mobility on the electrophoretic gel. Comparison of the total amount of secreted proteins in the absence of liposomes with the amount of proteins collected with the proteoliposomes indicates that about 5 to 10% of the secreted proteins were associated to the lipid membrane.

Protein bands were excised from the gel, transferred onto a polyvinylidene difluoride (PVDF) membrane. The protein N-terminal amino acid sequences were determined by automated Edman degradation using a Beckman LF 3400 D protein-peptide microsequencer. Four proteins were associated with the liposomes. From their N-terminal sequence (MHNVNNTQAPTFLYK, MHNVSTTTTGFPLAKI, MEIQNTKPTQTLY, MHITNLGLHQVSFQ), these proteins were identified as IpaA, IpaB, IpaC and IpgD, respectively. Among the 4 proteins, IpaB and IpaC are essential for induction of bacterial entry into the target cells and have been demonstrated to deeply insert into and to destabilize a lipid membrane [15,16].

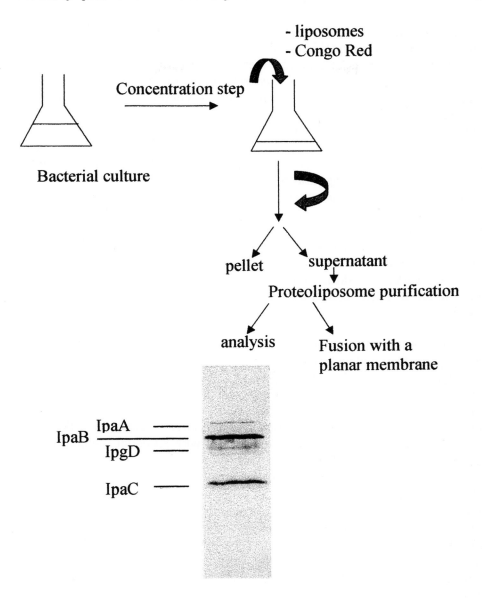

Figure 1 : Insertion of *S. flexneri* type III secreted proteins into asolectin large unilamellar vesicles. Protocol (top) and 12% SDS electrophoresis gel (silver staining) (bottom).

A similar experience was carried out with *Y. enterocolitica* but we had to adapt the above described protocol [21] : in *S. flexneri*, the Ipas proteins accumulate in the bacterial cytoplasm and can therefore be released immediately upon addition of Congo Red whereas in *Y. enterocolitica*, secretion is correlated with protein synthesis. The secretion is therefore a continuous process and the secreted proteins could not be released in one step upon induction. In the absence of eukaryotic cells, type III secretion of *Y. enterocolitica* can be induced by low calcium concentration and temperature shift to 37°C. This results in the secretion of 14 Yops (Yersinia outer protein) in the surrounding medium [22]. The following protocol was designed [21]:

Yersinia grown overnight at room temperature (22 to 26°C) were inoculated to an optical density at 600 nm of 0.2 in 10 ml BHI supplemented with 5 mM EGTA [Ethylene glycol-bis (b-aminoethyl ether)-N',N',N',N'-tetra acetic acid]. The culture was incubated with rotary shaking (150 rpm) for 1h30 at room temperature and then shifted to 37°C and grown for an additional 4 hours. Just before the temperature shift, large unilamellar vesicles of asolectin were added to the culture to a final concentration of 8 mg/ml. The incubation time at 37°C was determined by comparing the amount of proteins inserted in the liposomes after 1, 2, 3 and 4 hours by densitometric analyses of Coomassie blue stained gels. This amount reached a steady state maximal level after 4 hours. The bacteria were then pelleted by centrifugation for 6 min at 14000 g (Sigma 201 centrifuge, B. Braun Biotech International, Rotor Nr.12001). The 10 ml liposome-containing supernatant was concentrated by a 30 min centrifugation at 126.000 g at 4°C (Beckman L7 ultracentrifuge, SW 60 rotor) on a 0.5 ml 80 % sucrose layer. Concentrated liposomes formed a large turbid band at the top of that layer. This band was collected and mixed with an equal volume of 80% sucrose solution. A 25%-5% sucrose gradient was then poured on top of the liposome suspension. An overnight centrifugation at 126.000 g at 4°C allowed the separation of proteoliposomes from proteins that were not associated with liposomes. The band containing the proteoliposomes was collected and the proteoliposomes were washed in PBS NaCl 0.8 M and centrifuged 1 hour at 126.000 g to remove the sucrose from the samples and eliminate the proteins that may interact electrostatically with liposomes. The pellet was resuspended in PBS. When the wild type strain was used, in addition to YopB and YopD, several effectors proteins such as YopO, YopH, YopM, YopN and YopE were associated to the proteoliposomes (Figure 2 lane B). Since the genetic analysis suggested that the two major proteins involved in the interaction with the cell membrane were YopB and YopD, the experiment was repeated with a polymutant bacteria devoid of Yops H, O, P, E, M, N (HOPEMN strain). Figure 2 lane A shows that, in this case, only YopB and YopD were inserted in the lipid membrane. We went a step further and by using another mutant strain we succeeded in inserting only YopB in the proteoliposomes.

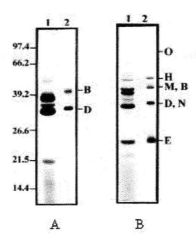

Figure 2 : 12% SDS-PAGE electrophoresis gel (silver staining). Lanes 1 culture supernatants. Lanes 2 : purified proteoliposomes. A: HOPEMN strain. B. Wild type strain.

The protocol we designed allowed not only to insert type III secreted proteins in a lipid membrane without any purification step that could interfere with the protein structure but the use of mutants bacteria secreting different sets of proteins allowed the modulation of the proteoliposome content. This has proven very useful in determining the role of each protein in channel formation (see below).

The above described protocol has been recently used to insert into the membrane HrpZ$_{Psph}$, a protein from the plant pathogen *Pseudomonas stringae* pv. *phaseolicola* secreted by a type III secretion mechanism. In that case, the proteoliposomes were precipitated in 80% (vol/vol) acetone and the protein was detected by immunoblotting [23].

2.2 Insertion of toxins in the lipid-interacting conformation (activated toxins)

Many secreted toxins require an activation step to interact with the lipid membrane and quite often, the activated form is difficult to handle in terms of water solubility. It is not the goal of this paper to review the procedures used to purify and keep in solution activated, hydrophobic toxins. However, since the concept we have developed could be an alternative approach to purification, we will give one example on how to use liposomes to insert in one step an activated toxin. PA from *B. anthracis* is a 83 kDa protein which binds to a specific cell surface receptor. Upon binding, PA must be activated by cleavage, leading to a 63 and a 20 kDa fragment. PA63 then binds either the edema or the lethal factor (EF and LF, two other proteins responsible for toxicity). The complexes are endocytosed and, in response to a cellular low pH, PA63 inserts into the membrane and allows the translocation of LF and EF across the membrane [24,25]. Although it is possible to purify PA63, the protein has a high tendency to oligomerize in solution. To bypass this problem, we incubated PA83 with large unilamellar vesicles of asolectin at neutral pH. Then trypsin was added (trypsin/PA, 1/100, w/w, 15 minutes at 30°C) and the pH was lowered to pH 5.0. The proteoliposomes were purified by centrifugation on a sucrose gradient (30-2%). Figure 3 shows that although PA63 and PA20 were present in the suspension, only PA63 is associated to the lipid membrane after liposomes purification. Several characterizations, such as topology or structure determination can then be undergone.

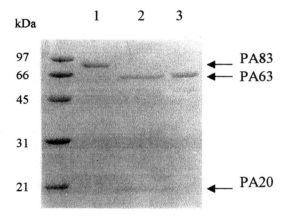

Figure 3: Insertion of PA63 into large unilamellar vesicles of asolection upon trypsin activation of PA83. 1. PA83 2. Trypsin-treated PA83 3. Purified proteoliposomes.

2.3 Insertion from a chaperone-protein complex.

As described in paragraph 2.1, mutated bacterial strains have been used to modulate the protein content of liposomes, a process essential to decipher the individual role of the inserted proteins in the biological process. However, there are some limits to the use of mutant strains. A good example is given by *S.*

flexneri: to evaluate the role of IpaB in the lipid interaction, it would be useful to have a mutant devoided of IpaB. Such a mutant has been constructed but it leads to a phenotype characterized by a non-regulated secretion. The proteins are secreted as they are synthesized and it is therefore quite difficult to synchronize secretion with the addition of liposomes. We are therefore designing another strategy based onto the properties of both IpaC and IpaB to be stored in the cell cytoplasm in association with a chaperone protein, IpgC. In the absence of IpgC, IpaB and IpaC associate in the cell cytoplasm. IpgC was found to be necessary for bacterial entry into epithelial cells, to stabilize the otherwise unstable IpaB protein, and to prevent the proteolytic degradation of IpaC that occurs through its association with unprotected IpaB [26]. The interaction between IpaC, IpaB and their chaperone has been recently characterized [27,28] by using a His-tagged IpgC protein to purify IpgC-containing complexes.

E. *coli* encoding a His-tagged IpgC/IpaC complex were grown at 37°C in 100 ml of LB medium until $OD_{600nm}=1$. IPTG was added to a final concentration of 1mM. After 4 hours, the culture was harvested by centrifugation (1,500 g for 10 min at 4°C) and resuspended in a 1/10 volume of Hepes buffer (Hepes 10 mM, NaCl 100 mM) pH 7. The bacteria were lyzed by sonication (60w, 3x30s). The cytoplasm was separated from membranes and insoluble fractions by centrifugation (1 hour at 300,000g, 4°C). The supernatant containing the IpgC/IpaC complex was immediately incubated with a cobalt resin (12.5µl/ml of culture, 1 hour at 4°C). The unspecifically bound proteins were washed in presence of a 10 mM Imidazole Hepes buffer, pH 7 at 4°C (2 x 10 volumes of bed resin, batch, 5 min) and centrifuged 2 minutes at 700g. 4 ml of asolectin LUV prepared in Hepes buffer containing imidazole 50mM were added to the washed resin (final concentration of 5 mg LUV/ml) and incubated for 15 min at 4°C. Liposomes were separated from the resin by centrifugation, washed with 1 M NaCl Hepes buffer, pH 7, centrifuged at 126,000 g for 45 min and analyzed by SDS-PAGE electrophoresis (Figure 4). In these conditions, IpaC spontaneously dissociates from its chaperone and inserts into the lipid vesicles (S. Vande Weyer and V. Cabiaux unpublished results). Several high molecular weight bands, which were not present in the complex sample (lane 2) are observed in the proteoliposomes sample (lane 3). Whether they are contaminants or partially denatured forms of IpaC remains to be investigated.

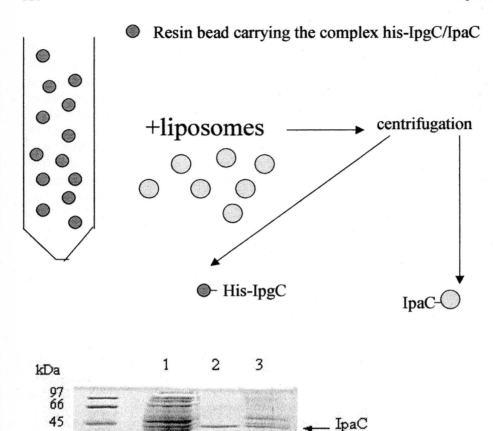

Figure 4 : Insertion of IpaC into large unilamellar vesicles of asolectin from the chaperon/protein complex. Protocol (top) and 12% SDS PAGE gel (bottom): 1. cytoplasmic extract 2. His-tagged IpgC/IpaC complex 3. Purified proteoliposomes.

3. FUSION BETWEEN THE PROTEOLIPOSOMES AND THE PLANAR MEMBRANE.

Induction of the fusion between proteoliposomes containing a membrane protein and a planar membrane to characterize the channel properties of the reconstituted protein is a quite commonly used technique. For examples, see Refs. [29] (MIP26 from bovine lens), [30] (K channels from the luminal membrane of rabbit proximal tubule) or [31] (liver gap-junction channels). Reconstitution of plant channels into a membrane and characterization of their properties is the topic of a chapter in this book (M. Smeyers, M. Leonetti, E. Goormaghtigh and F. Homblé: Structure and function of plant membrane ion channels reconstituted in planar lipid bilayers).

It is not the purpose of this paper to review all the literature about the interactions between liposomes and planar membrane. We will nevertheless review the aspects connected to the procedure we developed in our laboratory. Membrane fusion of phospholipid vesicles with a planar membrane is preceeded by an initial prefusion stage in which a region of the vesicle membrane adheres to the planar membrane. In the absence of an osmotic gradient, the vesicles stably adhere to the planar membrane. The total area of the contact region depends biphasically on the Ca^{2+} concentration but the distance between the bilayers in this zone decreases with increasing Ca^{2+} concentration. Membrane fusion may be initiated at these sites of closest membrane apposition [32].

Several parameters play a role in liposome-planar membrane fusion:

- <u>Ca^{2+} concentration</u>: Ca^{2+} concentration influences the adsorption step of the vesicles onto the planar membrane. This adsorption can lead to hemifusion (fusion of the outer monolayers) of the two membranes. For a characterization of the lipid organization during the absorption step, see [33] where the adhesion to planar lipid membrane of lipid vesicles containing calcein (a fluorescent probe) in the aqueous compartment or fluorescent phospholipid in the membrane has been examined by phase contrast, differential interference contrast and fluorescence microscopy. Vesicles can be "docked" onto the planar membranes in the absence of Ca^{2+} but they remain largely intact until a treshold concentration of Ca^{2+} induces the fusion. This threshold concentration is dependent on the charge of the vesicles and not on the planar membrane charge [34].
- <u>Osmotic swelling</u>: Osmotic swelling of phospholipid vesicles tightly adhered to planar bilayer membrane drives their fusion. This swelling can be induced either by replacing the bathing solution containing a solute impermeant through

the channel by a solute permeant through the channel (this will induce a movement of both the solute and water into the docked vesicles) or by making the bathing solution of the vesicles hyperosmotic compared to the inside of the vesicles [35]. Swelling of vesicles can also be achieved by creating an osmotic gradient across the planar membrane that drives water into the vesicles adhering to the planar membrane. This method of swelling is routinely used to reconstitute ion channels into planar membranes via fusion.
- planar membrane charge: the presence of cationic lipids in the planar membrane greatly enhances the fusion of negatively charged liposomes, in the absence of calcium ions [36].

We carried fusion experiments with proteoliposomes containing secreted proteins from either *Y. enterocolitica* [21] or *S. flexneri* (unpublished results) (see below for more details) and the two sets of proteoliposomes behaved quite differently. The proteoliposomes containing the yersinia proteins were formed in 137mM NaCl, 1.5 mM KH_2PO_4, 8.1 mM Na_2HPO_4, 10mM KCl buffer, pH 7.4. The two chambers of the set up were filled with a phosphate buffer (1.5mM KH_2PO_4, 8.1 mM Na_2HPO_4 pH 7.4) supplemented with various NaCl concentration from 50 mM to 1M. In the cis chamber (the chamber in where the proteoliposomes were added), the buffer also contained 1mM $CaCl_2$ to promote the fusion of the liposomes with the lipid bilayer. An electrical potential difference between 50 and 70 mV was applied across the planar lipid bilayer. In these conditions, in about 90% of our experiments, a fusion leading to a detectable jump of the bilayer conductance was observed within about 15 min. The salt concentration and therefore the osmolarity conditions had no detectable effect upon the fusion process, indicating that in this case, the presence of Ca^{2+} was sufficient.

The liposomes containing the secreted proteins of *S. flexneri* were formed in a NaCl 137mM, KCl 10mM, Na_2HPO_4 8.1 mM, KH_2PO_4 1.5mM pH 7.4 buffer. The proteoliposomes-containing chamber was made hyperosmotic (100 mM NaCl, 10 mM Hepes buffer, pH 7.4) compared to the "trans" compartment (5mM NaCl, 10 mM Hepes buffer, pH 7.4) and an electrical potential difference of –30mV was applied across the membrane. In these conditions, only 10% of fusion was obtained, indicating that one need to adapt the parameters to obtain a good and reproducible level of fusion between the proteoliposomes and the planar membrane.

4. CHANNEL FORMATION

Many bacterial toxins have the ability to induce channel formation into lipid bilayers. Channel formation is quite often characterized by measuring the

electric current fluctuations through the membrane. However, current fluctuations can also be induced by change in the lipid packing, yielding to a channel-like activity that could even be observed in the absence of any protein. This has been discussed in [37]. We will not go into the details of this discussion here but we will nevertheless describe the criteria that should be fulfilled before interpreting current fluctuations in terms of channel formation. One must define a series of features allowing channel and non-channel activities to be distinguished. Electrophysiological methods such as patch clamp and planar lipid bilayers techniques, allow the current flowing through the membrane to be measured. As ions carry a net charge, their net flux will produce an electrical current that can be amplified and recorded. Channels are gated: they undergo discrete random transitions between open and closed states. Because a channel opens and closes stochastically, the flow of ions through the channel will generate an electrical signal that fluctuates randomly between discrete current steps. This will be illustrated below with the Yersinia proteins YopB and YopD. Each discrete level of conductance has a specific and reproducible value. Characterization of a channel involves:
-   determination of the unitary conductance of the open channel defined as the slope of the I:V curve.
-   probability of being in a specific state, which can be calculated from the total amplitude histogram.
-   selectivity, which can be calculated from the reversal potential, namely the electrical potential difference at which current do not flow the channel.

It must be emphasized that channel activity is intrinsically a stochastic process. Therefore a statistical analysis of a sufficient large number of events must be done in order to draw a significant inference from the experimental data, and the simple observation of the square shape of a single current fluctuation through a membrane is not sufficient to conclude that the membrane contains an ion channel. We will show here in quite details the results obtained with proteoliposomes prepared as described paragraph 2.1 and containing the two proteins of *Y. enterocolitica*, YopB and YopD. Bilayers were formed from 1%, (w/v) asolectin dissolved in n-decane on a 200 μm hole drilled in a styrene-copolymer partition separating two chambers by using the Mueller-Rudin technique [38]. The two chambers were connected to the voltage generator through Ag/AgCl electrodes and a bridge of 3 M KCl, 2% agar. The cis chamber was connected to the headstage input of a Biologic RK 300 patch clamp amplifier (Claix, France). The trans chamber was held at ground. After formation of a stable asolectin membrane, proteoliposomes were added to the cis chamber under vigorous stirring and an electrical potential difference between 50 and 70 mV was applied across the planar lipid bilayer. Stirring was stopped immediately after the first fusion event. As shown in Figure 5 (A and

B), upon application of a voltage, random fluctuations between discrete levels of current amplitude occurred, which suggested that a channel had been reconstituted in the planar lipid bilayer. Gating was observed in symmetrical salt conditions (both sides of the membrane : phosphate buffer (PB) + NaCl from 25mM to 1M), as illustrated in Figure 5A for a single channel, and in asymmetrical salt conditions (cis side: PB + 100, 150, 180 mM NaCl, trans side: PB + 5mM NaCl) as shown in Figure 5B where at least three channels were inserted in the planar lipid bilayer and were closing and re-opening.

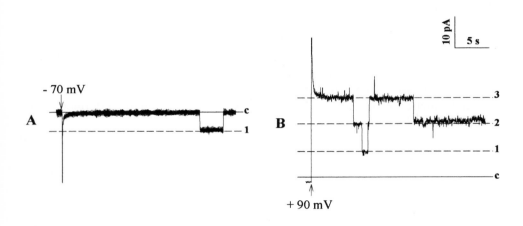

Figure 5: Current records obtained upon fusion of proteoliposomes containing YopB and YopD with a planar membrane. A. Current trace in symmetrical salt conditions : both cis and trans sides: PB + 180 mM NaCl. The arrow marks the time when the voltage is switched on to -70 mV. B. Current trace in asymmetrical salt conditions (cis side: PB + 100 mM NaCl, trans side : PB + 5mM NaCl). The arrow marks the time when the voltage is switched on to +90 mV. The letter "c" refers to the closed state of the channel and the different levels of single-channel current are identified by a number quoted on the right of the current traces.

4.1 Determination of the unitary conductance and selectivity.

The relationship between the magnitude of the current flowing through an open channel and the applied voltage (I/V curve) was linear in both symmetrical and asymmetrical salt conditions. The unitary conductance of the channel (105 ± 4 pS, n=8) was calculated from the slope of individual I/V curves (Figure 6). In asymmetrical conditions, the membrane potential under zero current conditions (the reversal potential) was −11.5 ± 0.3 mV (n=4) as determined from the intercept of the I/V curve with the x axis in asymmetrical salt conditions (arrow on the figure). Using the Goldman-Hodgkin-Katz equation, we calculated that it corresponds to a ratio of the permeability coefficient P_{Cl-}/P_{Na+} of 1.4 ± 0.3 (n=4). This value is very close to the ionic mobility ratio found in solution (u_{Cl-} =1.52 u_{Na+}) indicating that the observed channel has no ionic selectivity.

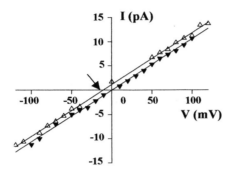

Figure 6 : Mean I/V curves. Values are the average of at least 3 independent (new batch of proteoliposomes and a new planar lipid bilayer) experiments realized in various symmetrical ((▼), PB + 180 mM NaCl, or 150 mM NaCl or 100 mM NaCl) and asymmetrical salt conditions ((Δ), trans = PB + 5mM NaCl, cis = PB +180 mM NaCl or 150 mM NaCl or 100 mM NaCl).

4.2 Amplitude histogram.

The amplitude histogram from which the probability of being in a specific state is calculated is shown Figure 7. Clearly, the YopB/YopD channel was characterized by two peaks corresponding to a single open state with a conductance of 105pS.

From the given example, it is clear that the signal produced by YopB and/or YopD fulfills all the criteria characteristics of channel formation. The discussion about the role of this channel in the effectors translocation across the cell membrane can be found in [21].

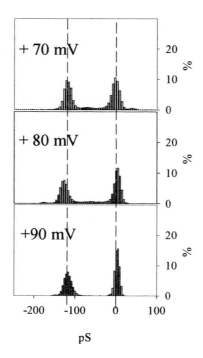

Figure 7: Amplitude histograms of the conductance levels for the channel formed by the YopB/YopD complex. The histogram was constructed from the frequency distributions of the current amplitude measured on representative current traces analyzed for 20 sec and showing at least 5 gating events. The current values were converted in conductance (pS, x-axis) using Ohm's law. The y-axis indicate the percentage of total time spent in one given conductance level

As stated before, we can modulate the proteoliposome composition to evaluate the respective role of proteins in channel formation. Proteoliposomes containing only YopD did not produce any electrical signal whereas those containing only YopB gave rise to channel-like activities characterized by a broad distribution of low conductive levels (Figure 8).

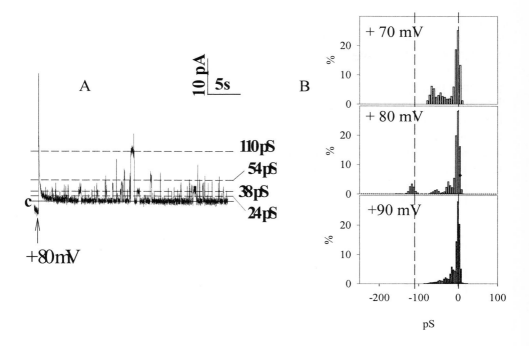

Figure 8: Current trace (A) and amplitude histogram (B see also legend figure 7) obtained after fusion of proteoliposomes containing YopB with planar membrane. The current trace has been selected in asymmetrical salt conditions : cis side : PB + 100mM NaCl, trans side : PB + 5 mM NaCl. The arrow marks the time when the voltage is switched on to +80 mV.

The letter "c" refers to the closed state (full line) of the channel and the multiple open states (dashed lines) are identified by their corresponding conductance value calculated using Ohm's law.

4.3 Contamination by porins.

Porins are pore proteins contained in the outer membrane of Gram-negative bacteria and they mediate the diffusion of small hydrophilic molecules. *Yersinia enterocolitica* contains a porin, OmpC, which has been purified and characterized by [39]. Since this protein is located in the bacterial outer membrane, one can not exclude that OmpC could be transferred spontaneously in the membrane of the proteoliposomes, in addition to the secreted proteins. Because this protein gives rise to a channel activity when incorporated in a planar membrane, it is essential to assess that the observed signal is due to type III secreted proteins and not to a porin contamination. If anti porin antibodies are available, one can always perform a western blot on a sample of the proteoliposomes. However, one should not forget that electrophysiology is a very sensitive method that allows the detection of a single channel and therefore proteoliposomes. However, one should not forget that electrophysiology is a very sensitive method that allows the detection of a single channel and therefore the insertion of a single molecule. Such a level of contamination could not be detected by immunoblotting. We therefore relied upon the comparison of the electrical properties of the channels induced respectively by YopB/YopD and the OmpC porin of *Y. enterocolitica*.

The YopB/YopD channel has a conductance of 105-137 pS for salt concentration between 100mM and 1M NaCl (Figure 9) and has almost no selectivity whereas the Yersinia porin has a conductance of 1.3nS in 1 M KCl and is cation selective (permeability coefficient of 14.4 for K^+) [39].

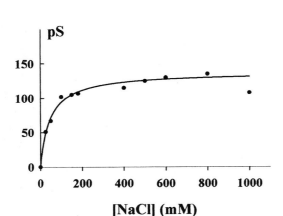

Figure 9 : Effect of salt concentration on the single channel conductance. Experiments were performed in symmetrical NaCl concentrations at room temperature.

5. OTHER APPLICATIONS.

The theme of this review was focused on the ability of bacterial secreted proteins reconstituted in liposomes to induce channel formation in planar membrane. However, when the proteins are associated to the proteoliposome's membrane, many other characterizations than channel formation can be envisaged:
- topology characterization by proteolysis: the proteoliposomes are incubated with several proteases, the protein domains protected from digestion because of their interaction with the membrane are separated by gel electrophoresis (1 or 2

dimensional electrophoresis) and identified by N-terminal sequencing or mass spectrometry [40]
- labeling: any membrane probe can be added to the liposomes before incubation, allowing, after purification of the proteoliposomes the identification of membrane domains [41].
- fluorescence spectroscopy on protein's Trp for example.
- determination of secondary structure by Fourier Transformed Infrared Spectroscopy. This technique allows the determination of the secondary structure of proteins in a lipid environment. It allows also to perform kinetics of deuteration which, in turn, give insights about the tertiary structure of the proteins [42].

Other types of proteins could also be inserted in lipid membrane using the same procedure. Indeed, this approach can be applied to all mechanisms that involve secretion of hydrophobic proteins immediately followed by insertion of these proteins in the target cell membrane, such as the type IV secretion mechanism [43] or the Tir-mediated adhesion of *E. coli* to eucaryotic cell surface [44]. It could also be applied to the characterization of eukaryotic membrane proteins such as receptors, ATPases etc. that could be expressed from a plasmid in a reticulocyte lysate. In this case, liposomes would be added to the reaction mixture before protein expression is induced. The new proteins should fold and insert into the lipid membrane as they are synthezised, avoiding hydrophobic aggregation denaturation.

6. CONCLUSION.

We developed a simple technique to reconstitute membrane proteins in a lipid membrane in one step, bypassing the need for denaturating purification steps. So far, we mainly focused on bacterial proteins that are secreted upon contact with the eucaryotic cell and which insert into the plasma membrane but we are convinced it could be applied to a broad range of membrane proteins. We therefore believe this technique is a straightforward protocol to bypass protein purification and membrane reconstitution, allowing characterization of biological parameters such as activity, structure or topology.

BIBLIOGRAPHY

[1] S. Binder, A.M. Levitt, J.J. Sacks and J.M. Hughes, Science 284 (1999) 1311.
[2] L. Abrami, M. Fivaz and G. van der Goot, Trends in Microbiology 8 (2000) 168.
[3] S.J. Billington, B.H. Jost and J.G. Songer, FEMS Microbiology 182 (2000) 197.
[4] G. Prevost, L. Mourey, D.A. Colin and G. Menestrina, Curr. Top. Microbiol. Immunol. 257 (2001) 53.

[5] A. Soloaga, M.P.Veiga, L.M. Garcia-Segura, H. Ostolaza, R. Brasseur and F.M. Goñi, Mol. Microbiology 31 (1999) 1013.

[6] V. Cabiaux, C. Wolff and J.M. Ruysschaert, Int. J. Biol. Macromolecules 21 (1997) 285.

[7] B. China and F. Goffaux, Vet. Res. 30 (1999) 181.

[8] S. Yang He, Ann. Rev. Phytopathol. 36 (1998) 363.

[9] A. Collmer, J.L. Badel, A.O. Charkowski, W.L. Deng, D.E. Fouts, A.R. Ramos, D.M. Anderson, O. Schneewind, K. van Dijk and J.R. Alfano, Proc. Natl. Acad. Sci. USA 97 (2000) 8770.

[10] G.R. Cornelis and F. Van Gijsegem, Annu. Rev. Microbiol. 54 (2000) 735.

[11] T. Kubori, A. Sukhan, S.I. Aizawa and G.E. Galan, Proc. Natl. Acad. Sci. USA 97 (2000) 10225.

[12] A. Blocker, N. Jouihri, E. Larquet, P. Gounon, F. Ebel, C. Parsot, P. Sansonetti and A. Allaoui, 39 (2001) 652.

[13] E. Hoiczyk and G. Blobel, Proc. Natl.Acad.Sci. USA 98 (2001) 4669.

[14] A. Blocker, P. Gounon, E. Larquet, K. Niebuhr, V. Cabiaux, C. Parsot and P. Sansonetti, J. Cell. Biol. 147 (1999) 683.

[15] C. De Geyter, R. Wattiez, P. Sansonetti, P. Falmagne, J.M. Ruysschaert, C. Parsot and V. Cabiaux, Eur. J. Biochem. 267 (2000) 5769.

[16] C. De Geyter, B. Vogt, Z. Benjelloun-Touimi, P.J. Sansonetti, J.M. Ruysschaert, C. Parsot and V. Cabiaux, FEBS Lett. 400 (1997) 149.

[17] F.K. Bahrani, P.J . Sansonetti and C. Parsot, Infect. Immun. 65 (1997) 4005.

[18] R. Ménard, P.J. Sansonetti and C. Parsot, J. Bacteriol. 175 (1993) 5899.

[19] R. Ménard, M.C. Prevost, P. Gounon, P.J. Sansonetti and C. Dehio, Proc. Natl. Acad. Sci. USA 93 (1996) 1254.

[20] M.J. Hope, M. Bally, G. Webb and P.R. Cullis, Biochim. Biophys. Acta 812 (1985) 55.

[21] F. Tardy, F. Homblé, C. Neyt, R. Wattiez, G.R. Cornelis, J.M. Ruysschaert and V. Cabiaux, EMBO J. 18 (1999) 6793.

[22] T. Michiels, P. Wattiau, R. Brasseur, J.M. Ruysschaert and G.R. Cornelis, Infect. Immun. 58 (1990) 2840.

[23] J. Lee, B. Klüsener, G. Tsiamis, C. Stevens, C. Neyt, A.P. Tampakaki, N.J. Panopoulos, J. Nöller, E.W. Weiler, G.R. Cornelis, J.W. Mansfield and T. Nürnberger, Proc. Natl. Acad. Sci. USA 98 (2001) 289.

[24] A.M. Friedlander J. Biol. Chem. 261 (1986) 7123.

[25] N.S. Duesbury, C.P. Webb, S.H. Leppla, V.M. Gordon, K.R. Klimpel, T.D. Copeland, N.G. Ahn, M.K. Oskarsson, K. Fukasawa, K.D. Paull and G.F. Vande Woude, Science 280 (1998) 734.

[26] R. Ménard, P. Sansonetti, C. Parsot and T. Vasselon, Cell 79 (1994) 515.

[27] A.L. Page, H. Ohayon, P.J. Sansonetti and C. Parsot, Cell. Microbiol. 1 (1999) 183.

[28] A.L. Page, M. Fromont-Racine , P. Sansonetti, P . Legrain and C. Parsot, Molecular Microbiology 42 (2001) 1133.

[29] L. Shen, P. Shrager, S.J. Girsch, P.J. Donaldson, C. Peracchia, J. Membr. Biol. 124 (1991) 21.

[30] F. Bellemare, N. Morier and R. Sauve, Biochim. Biophys. Acta 1105 (1992) 10.

[31] J.L. Mazet, T. Jarry, D. Gros and F. Mazet, Eur. J. Biochem. 210 (1992) 249.

[32] W.D. Niles, J.R. Silvius and F.S. Cohen, J. Gen. Physiol. 107 (1996) 329.

[33] M.S. Perin and R.C. MacDonald, J. Membrane Biol. 109 (1989) 221.

[34] A. Chanturiya, P. Scaria and M.C. Woodle, J. Membrane Biol. 176 (2000) 67.

[35] F.S. Cohen, W.D. Niles and M.H. Akabas, J. Gen. Physiol 93 (1989) 201.

[36] K. Anzai, M. Masumi, K. Kawasaki and Y. Kirino, J. Biochem. (Tokyo) 114 (1993) 487.

[37] F. Homblé, V. Cabiaux and J.M. Ruysschaert, Molecular Microbiology 27 (1998) 1261.

[38] P. Mueller, D.O. Rudin, H. Ti Tien and W.C. Wescott, Nature 194 (1962) 979.

[39] K. Brzostek, J. Hrebenda, R. Benz and W. Boos, Res. Microbiol. 140 (1989) 599.

[40] P. Quertenmont, R. Wattiez, P. Falmagne, J.M. Ruysschaert and V. Cabiaux, Molecular Microbiology 21 (1996) 1283.

[41] X-M. Wang, R. Wattiez, F. Brossier, M.Mock, P. Falmagne, J.M. Ruysschaert and V. Cabiaux, Eur. J. Biochem. 256 (1998) 179.

[42] Goormaghtigh, V. Cabiaux and J.M. Ruysschaert, Subcellular Biochemistry Volume 23, eds H.J. Hilderson and G.B. Ralston, Plenum Press, New York, (1994) 329.

[43] M. Stein, R. Rappuoli and A. Covacci, Proc. Natl. Acad. Sci. USA 97 (2000) 1263.

[44] R. DeVinney, A. Gauthier, A. Abe and B.B. Finlay, Cell. Mol. Life Sci. 55 (1999) 961.

Planar Lipid Bilayers (BLMs) and their Applications
H.T. Tien and A. Ottova-Leitmannova (Editors)

Chapter 19

The quest for ion channel memory using a planar BLM

R. Cassia-Moura

Instituto de Ciencias Biologicas, DCF - Biofisica, Universidade de Pernambuco, Caixa Postal 7817, Recife 50732-970 Brasil
E-mail: rita@npd.ufpe.br

1. INTRODUCTION

All living organisms are dependent on information from the past. Memory is the process of storing and retrieving this information, such that a lack of memory is incompatible with life. The aim of this chapter is to present the deployment of a membrane model in which it is possible to induce memory control. Many ion channels formed by colicin Ia incorporated into a planar bilayer lipid membrane (BLM) are investigated by the voltage clamp technique using different step voltage stimuli. Each artificial membrane demonstrates a critical resting interval, Δt_c, between two successive voltage pulses, in which a predictable current response is produced when the second pulse is applied within this Δt_c and an unpredictable current response is produced when the second pulse is applied after the Δt_c. The behaviour of the voltage-gated ion channels may be interpreted as a transient gain, loss or resetting of memory, as revealed by a specific sequence of electrical pulses used for stimulation. In this sense, it is possible to induce memory control of the artificial membrane in accordance with the experimental conditions imposed on it.

This chapter is organised as follows: Section 1 is the introduction; Section 2 describes the structure and functions of the plasma membrane and some transport processes across it; Section 3 describes the structure of the ion channel and some physiological implications of its functions and properties; Section 4 describes the structure and function of colicin Ia; Section 5 focuses on the artificial membrane as a plasma membrane model and the experimental procedure of the planar BLM incorporated by colicin Ia is described; Section 6 presents the ion channel kinetics as a process with memory and the manner in which the colicin Ia interacts with a planar BLM, revealing the memory control of the artificial membrane; Section 7 consists of the concluding remarks.

2. PLASMA MEMBRANE

The cell is the smallest unit of life that can exist independently and is the basic structure capable of fundamental vital processes, such as reproducing, taking in nutrients and expelling waste. All living systems are composed of cells. Some microscopic organisms consist of a single cell; the others are multicellular, that is, they are composed of many cells working in concert. Even the smallest cells are fantastically complicated at the physical and chemical level. Despite the apparent diversity of cell size, function, morphology and shape, several of them are amazingly similar in their basic structural characteristics. Moreover, one of these similarities is that all cells present a membrane. The membrane is the interface that isolates a system from its environment, delimiting processes that could be compromised by the surroundings; separately, the process achieves a greater efficiency than if it existed in a noncompartmentalised form and, at the same time, the membrane should selectively allow a constant exchange of information between its interfaces. A number of fundamental membrane functions such as conversion, transfer and storage of energy and information in living organisms make the membrane a component essential to life.

All living organisms may be classified into two major groups, the unicellular prokaryotes and the eukaryotes, which may be either unicellular or multicellular. Prokaryotes have no membrane-bound internal organelles. In the eukaryotes, the cell interior is separated into discrete compartments such as the nucleus and various organelles, and membranes play a central role in the functional and structural organisation of eukaryotic cells. However, membranes are not static inert boundaries of the cellular and intracellular compartments; they are dynamic structures intimately involved in many, if not most of the physical and chemical processes of all cells, including the ability of a cell to grow and to divide.

Every cell contains at least one membrane, known as plasma membrane, which defines its area, causing it to be self-contained and to a certain extent self-sufficient. The plasma membrane is a universal entity, which separates the cell from its surroundings and protects it from fluctuating or adverse conditions in the physicochemical environment. There are cells that exhibit a continuous change of shape, a fact that signifies a considerable plasticity of the edge of the plasma membrane, while in others the rigidity of the membrane is due to its interaction with cytoskeletal proteins. In an ordinary diagram of a cell or when it is examined under a microscope, the plasma membrane appears as a line, indicating the periphery of the cell. Surprisingly, however, we are dealing here with a structure organised in an extremely complex way and it should be emphasised that only a few cell activities could exist in its absence.

Membrane fusion is a very important phenomenon in all cells, where life may start with the sperm-egg fusion or death may be the result of a virus-cell fusion. In multicellular organisms, the activities of individual cells may be controlled both by their respective surroundings and by remotely located cells, a process that includes a communication between cells with the fundamental participation of the plasma membrane. The plasma membrane holds the cells together and contains sites that recognise other cells of the same kind, which can promote their association during the orderly development of tissue structure and limit cell growth in an organ. Cells in aggregates can carry out their designated functions only if the plasma membrane is polarised, i.e. organised into at least two regions, each of them specialised for very different tasks. The plasma membrane expresses the genetic individuality of the cell, which is used by the immune system to distinguish between self and non-self. All these manifold functions show the high level of organisation in the plasma membrane, which is only 7.5 $\pm$ 2 nm thick!

Although plasma membranes exhibit a great diversity of functions and a plasma membrane of a single cell may present regions with different functions and morphologies, a number of general features appear to be similar in all cells. All membranes have an invariant molecular general pattern, namely an arrangement of proteins distributed in a fluid mosaic comprising one bilayer of lipids. Such a structure is tightly packed and held together in an aqueous medium by a non-covalent hydrophobic interaction, a fact that permits its high degree of flexibility and fluidity, influencing even the lipid-protein interactions. In the water-excluding mosaic, the surfaces exposed to the aqueous phases must be predominantly polar, while the nonpolar region of proteins and lipids can interact in the core of the membrane. No other option is open if the orientation is to be stable and energetically feasible. In mammalian cells, small amounts of carbohydrates are associated either with proteins or with lipids, an association that has been implicated in important cellular events such as cell-cell recognition. Lipid and protein compositions are extremely variable from one membrane to another and the composition of one face of the membrane differs from that of the other. In addition, some proteins are immobile but most of them and all lipids are astonishingly mobile. Individual components are free to move about in the plane of the membrane, being continually renewed at different rates, and a transfer of components occurs from one membrane to another within a single eukaryotic cell. Lipids rotate on their axis, wiggle their fatty acid chains and change their lipid neighbours several million times a second [1]. Thus, we can describe only a time-averaged plasma membrane structure.

A large number of different kinds of lipids are found in membranes, in a composition that is constant and peculiar to each cell type. For structural purposes almost all of them may be treated as a class since they are amphipathic: one end of the rod is hydrophilic and the remainder is hydrophobic. The hydrophobic nonpolar tails tend to align themselves in such a way as to minimise their contacts with water, while the hydrophilic polar head interacts favourably with water, and they therefore spontaneously form bilayers in aqueous solution. The most abundant membrane lipids are phospholipids (e.g. phosphatidylcholine and phosphatidylethanolamine) and the phospholipid composition of the two leaflets may differ in a single membrane. Cholesterol is a major membrane component in animal cells. The lipid bilayer has two important properties: fluidity and impermeability to ions and to polar compounds [2]. Membrane fluidity plays an essential role in several life processes such as intercellular and intracellular communications, membrane elasticity, secretory processes and membrane-bound enzyme activities. Some disease states may be accompanied by altered membrane fluidity, which may be due to fusion with the virus membrane, the use of certain drugs or nutritional changes. Thus, a question of importance to molecular biology concerns the basic interaction of drugs with the cell membrane, enabling the drugs to exert their effects.

While lipids determine the basic structure of the plasma membrane, there does not appear to be any type of protein involved in the maintenance of its structure nor is there any evidence of the existence of a structural protein common to all membranes. On the other hand, proteins constitute the major component of most functional membranes and the specific activities of a given membrane vary in accordance with the quantity and type of proteins it contains, a fact that results in the radically different metabolic functions presented by the various membranes. Even a membrane patch owes much of its individuality to the properties of its proteins. Among their other attributions, proteins may function as a structural link between the cytoskeleton and the extracellular matrix; they may form assemblies which join neighbouring cells in a tissue; they may function as enzymes that catalyse reactions in specialised membranes; they may transport specific molecules into or out of the cell or organelle; and they may also function as receptors that receive and transduce information from the cell environment or organelle. The membrane lipid ordering influences the structure and function of the membrane proteins and an optimum level of membrane lipid ordering is necessary for the normal functioning of the cell. There are lipid-protein interactions that play a direct part in a variety of membrane functions. Some proteins require specific phospholipids for the expression of their enzymatic activity. In addition, the nature of fatty acid incorporated into the phospholipid in some cases also affects the protein function.

The protein molecules exhibit a more complex structure than the lipids and, in accordance with their amino acid sequence, come to be associated in several ways with the lipid bilayer. By the ease with which the membrane protein may be removed, despite the fact that this property varies continuously, some proteins are peripheral, but most are integral proteins. Peripheral proteins are weakly bound to the membrane surface, primarily if not exclusively, by protein-protein interaction. Integral proteins are amphipathic and interact directly with the phospholipid bilayer. All molecules of a single kind of integral protein lie in the same direction of the lipid matrix and are asymmetrically embedded in the structure.

A thorough knowledge of membranes is of crucial importance for our understanding of normal and abnormal physiology. Many of the major activities of living cells involve reactions associated with membranes. Among these functions we may mention the cell-cell interaction, nervous excitation, production of ATP and the conversion of energy into photosynthesis. In addition, the membranes are involved in many of the biochemical processes of the cell.

2.1 Transport through the plasma membrane

In a living organism, the different composition of its various liquid compartments is largely due to the nature of the barrier separating them, i.e. the structure of the membrane delimiting them. The plasma membrane is a selectively permeable continuous barrier around the cell, which ensures that essential molecules enter the cell and waste compounds leave it. Besides regulating the components that will cross the membrane, this property also regulates those that will be kept inside or outside the cell, thereby generating different compositions between the cytoplasm on the one hand and the extracellular environment on the other. Thus this selectivity is responsible for preserving the essential differences between the cell content and its surroundings, maintaining a steady state constancy of the cell's internal medium. In virtually all cells, the intracellular pH must be kept at around 7, the intracellular potassium ion concentration must be much higher than that of sodium ion and the extracellular sodium ion concentration must be much higher than that of potassium ion. Because of membrane selectivity and an asymmetrical ion distribution into both sides of the membrane, all cells maintain a significant transmembrane voltage. This difference in the electrical transmembrane potential between the cytoplasm and the external fluid is called resting membrane potential. It is essential for driving a variety of cellular functions such as signalling, movement, regulation, development and energy balance.

Living cells exchange many different kinds of components with their environment and two forces govern the movement of components across the plasma membrane: the membrane voltage and the chemical gradient. Chemical gradient is the result of the difference in the component concentration on the two sides of the membrane, and the direction of flow is from the higher concentration site to the lower. All charged molecules generate an electric field and if the component has a net charge like an ion, then both its chemical gradient and the membrane voltage, i.e. the electrochemical gradient, determine the direction of transport. These forces may act in opposite directions or in the same direction. The spontaneous movement of component down its gradient is a passive transport, in which no metabolic energy is expended. However, the maintenance of the unequal compositions of the cell and its environment signifies the existence of active transport, in which a component crosses the plasma membrane against its gradient. Active transport is divided into primary active transport, which utilises metabolic energy, and secondary active transport, which utilises energy provided by the gradient for another component.

Macromolecules such as proteins, or particles as large as a few micrometers such as broken cell parts or bacteria, may cross the plasma membrane by a membrane fusion process, such as endocytosis, in which the component is surrounded by a small plasma membrane region that internalises it to form intracellular vesicles, or exocytosis, in which a storage vesicle inside the cell fuses with the plasma membrane and expels its contents. Small uncharged molecules and gases such as urea, oxygen and carbon dioxide may cross the plasma membrane by direct passage through the lipid bilayer. On the other hand, the lipid bilayer of the membrane is an effective dielectric shield preventing the passage of most polar and charged molecules, and this transmembrane charge transport is the key to several processes that play an important role in the survival of the cell. Ion transport underlies many essential physiological processes, such as the regulation of cell volume, excitation and propagation of electrical signals in nerve and muscle cells, the secretion of fluids by organs such as the intestine and kidney, and the electrolyte levels in the blood. Defects in transport systems severely affect the metabolism of the cell, usually with pathological consequences.

Small polar and charged molecules, such as amino acids, sugars and ions, may cross the plasma membrane by means of transport proteins. Multiple forms of membrane transport proteins coexist in the different cells of the various organisms, so that each region of the membrane has its own mixture, differently regulated and with distinct functions. Some act as carriers and the others form ion channels. Both types are invariably present in all plasma membranes, each of which is responsible for transferring only a specific group of closely related molecules that do not come into direct contact with the hydrophobic core of the lipid bilayer. During the twentieth century, major cellular roles have been discovered for most ions of body fluids, such as Na^+, K^+, Ca^{2+}, H^+, Mg^{2+}, Cl^-, HCO_3^- and PO_4^{2-}, so the list of ions and their uses will continue to lengthen [1]. Each has been assigned at least one special regulatory, transport or metabolic task. None is just passively distributed across the cell membrane and each has at least one carrier-like device coupling its movement to the movement of another ion. The Na^+-K^+ pump, the Ca^{2+} pump, Na^+-Ca^{2+} exchange, Cl^--HCO_3^- exchange, glucose transport, the Na^+-coupled co- and countertransporters are some of the important carrier transports, in which small motions within the protein leave the macromolecule fixed in the membrane while exposing the transport binding site(s) alternately to the intracellular and extracellular media. Carrier proteins have the characteristics of membrane-bound enzymes, while proteins that form channels are macromolecular water-filled pores in cell membranes; not surprisingly, transport through ion channels can occur at a much faster rate than transport mediated by carriers [1, 3].

In the human body, the control of the flow of components through the plasma membrane is of fundamental importance for homeostasis as the plasma membrane influences the pH of the environments separated by it and participates effectively in several metabolic functions. An improved understanding of selective ion transport across the plasma membrane clearly informs a large variety of fundamental problems in biology and medicine, which will certainly have repercussions for the treatment and diagnosis of a variety of diseases.

3. ION CHANNELS

Ion transport through biological channels is a phenomenon that has attracted the attention of biophysicists and biologists for many years [1]. The main reason for this profound interest is that ion channels, abundant in most, if not all, eukaryotic cells, are involved in many physiological processes [1, 4]. Within our body, ion channels detect the sounds of chamber music or guide the artist's paintbrush, by producing the flickers of electrical activity that stir neurones and muscle cells; even cells not connected to the brain, such as those in the blood, immune system, liver and other organs, use ion channels for signalling processes; the transmission of signals within and between cells is also mediated by ion channels [1]. Ion channels contribute to gating the flow of messenger Ca^{2+} ions, establishing a resting membrane potential, controlling cell volume, regulating the net flow of ions across epithelial cells of secretory and resorptive tissues, and cell-to-cell coupling through gap junctions. Ion channels have also been identified on intracellular organelles but for most of them no role has yet been determined.

The electrical currents arising from the ion flows at a macroscopic level have been studied since the 1950s, when Alan Hodgkin and Andrew Huxley invoked [5] the concept of ion channels in their classic analysis of currents through nerve membranes. They used the giant nerve axon of the *Loligo* squid to demonstrate that action potentials are due to increases in membrane permeability to Na^+ and K^+ ions and recognised three different components of current, which they called leakage, sodium and potassium. Today, the names sodium channel and potassium channel are universally accepted for the corresponding ion channels in axons. The sodium and potassium channels were the first two channels to be recognised and described in detail: both have voltage-dependent kinetics, together accounting for the action potential of the axon, and are distinguished by their ionic selectivity and clearly separable kinetics [1]. In the Hodgkin-Huxley theory [5] voltage clamp data was used as a basis for a set of differential equations capable of predicting all the major excitation phenomena concerned with the initiation and propagation of the nerve action potentials; the membrane conductance changes are described in terms of time- and voltage-dependence of the channels that underlie the nerve impulse.

The above membrane ionic theory of excitation was transformed from untested hypotheses to established fact when the voltage clamp offered for the first time a quantitative measure of the ion currents flowing across an excitable membrane. Voltage clamp analysis [6] was introduced by Kenneth Cole in 1949, a technique which "tames the axon" by controlling the desired potential across the cell membrane, and the resulting membrane currents can then be measured and interpreted. By using the voltage clamp method, the dynamic behaviour of the membrane conductance is determined from the membrane current responses to a series of constant voltage steps. If current is applied as a stimulus, typically it flows locally across the membrane both as ion current and as capacity current, and also spreads laterally to distant patches of membrane; the ensuing changes in potential are measured. The voltage clamp may reverse this process [1]: simplifying conditions are used to minimise capacity currents and the spread of local circuit currents so that the observed current may be a direct measure of ion movements across a known membrane area at a known uniform membrane potential; by applying a voltage, the current is measured.

Only since the 1970s, however, have single ion channels on cell membranes been studied by the patch clamp technique [7], a very simple and powerful technique developed by Erwin Neher and Bert Sakmann, in which a thin glass pipette of suitable shape is tightly sealed against a cell membrane, thereby isolating a small patch of the membrane and ion channels it contains. These channels can then be manipulated and their properties deduced by applying voltage clamps to small cells; it is possible to study signalling mechanisms at a cellular level, and also to probe how ion channels affect membrane voltage and cell processes. Before the patch clamp technique, we knew more about the action potential in the famous squid giant axon than the nerve impulses of the human brain, because the voltage clamp technique requires that at least two microelectrodes be inserted into a cell, and mammalian cells can barely tolerate impalement with a single standard microelectrode. However, the refinement of the patch clamp technique took several years, making possible detailed observations of the behaviour of single molecules, in the form of ion currents in their natural environment and in real time.

One of the great advances since the 1980s has been the cloning and sequencing of DNA and hence the determination of the primary sequence of amino acids in ion channels by methods of molecular genetics. These have had an enormous impact on the way we view single ion channels and have led to an extraordinary corpus of information on their functional diversity, gene organisation, tissue-specific expression and structure-function relationship. The relationships between the primary, secondary, tertiary and quaternary structure of the protein and channel properties have been elucidated by several approaches, from site-selected mutagesis to modelling. By locating and changing critical sequences of amino acids in the channel protein, it is possible to study the effect of the alterations on the function of the channel. In addition, high-resolution electron microscopy and X-ray and electron diffraction make it possible to determine the physical conformation of the ion channel. The last 15 years have witnessed an amazing increase in our understanding of the cellular and molecular diversity of ion channels.

Ion channels are transmembrane proteins, some of them consisting of 1,800-4,000 amino acids arranged in one or several polypeptide chains with some hundreds of sugar residues covalently linked to amino acids on the outer face [1]. Analysis of the primary amino acid sequences of several channels has led to proposals concerning the folding of the channel proteins based on the number of multiple hydrophobic membrane spanning segments. Moreover, the pore wall is lined by hydrophilic amino acids and, if the channel opens, these form a water-filled pore extending fully across the plasma membrane when organised into the native three-dimensional structure. The predicted three-dimensional structure and function of amino acids that form one channel may reveal strong structural similarities among groups of channels. They all come in various isoforms coded by different genes that may be selectively expressed in certain cell types or in certain periods of development and growth of the organism [1]. Multiple forms of ion channels with different functions and regulation may coexist in each cell to suit its special purposes, and at least two classes of closely related channels are activated in the same membrane patch. In some cells a particular channel protein is synthesised, in others it is not, a regulation that occurs at many different levels, beginning with the control of gene expression. Like other membrane proteins, it will then be subject to turnover through degradation and replacement, a rate of turnover that can be adjusted and is highly variable. Typically, a channel protein survives for several days in the membrane before being internalised by endocytosis and degraded by lysosomal enzyme [3].

There is a great diversity of ion channels. Their structure and functions are only partly understood but in all of them solute movement across the channel is passive. Solutes of appropriate size and charge cross their continuous hydrophilic pathway by simple diffusion or by interacting with internal binding sites through the pore. In either case, movement through an open channel is influenced by the electric and chemical gradients across the plasma membrane. The rate of passage of ions through one open channel – often more than 10^6 ions per second – is far too high for any mechanism other than a pore [1]. For our purposes it is sufficient to note that the size of an outer shell of water molecules, with which the metal ion such as Li^+, Na^+ and K^+ forms a bound complex and which moves with the ion through the aqueous channel, determines ion mobility - the smaller the metal atomic diameter, the lower the ion velocity, because it attracts water molecules more strongly.

It is well established that multi-occupancy exists in some ion channels. The permeation of different ions across cell membranes is not independent: in excitable membranes, the flow of Na^+ ions in sodium channels influences the flow of K^+ ions in potassium channels and vice versa, a kind of interaction that may be removed by using the voltage-clamp technique to control the membrane potential. Ion channel permeability is the property of allowing ions to pass into or out of the cell. The conductance of an ion channel for a single ion species depends on the intrinsic ionic permeability of the channel and on the concentration of the ion in the region of the channel. The driving force for the movement of ions through the ion channels is the difference between the resting membrane potential and the equilibrium potential (which is the electrical potential that reduces to exactly zero the net flux of ions across ion channels due to a chemical gradient).

Ion channels vary considerably in their selectivity. The pore is much wider than an ion over most of its length and may narrow to atomic dimensions only in a short stretch, the selectivity filter, where ionic selectivity is established [1]. For electrical excitability of neurone and muscle, it is essential that different ion channels select ions that will pass easily while retarding or rejecting others. Individual channels are selective for either cations or anions. Some cation channels are selective for a single ion species – some allow only Na^+ almost exclusively, K^+ or Ca^{+2} to pass; still others are less selective. However, no channel is perfectly selective, such as the sodium channel of axons is highly permeable to Na^+ and Li^+, fairly permeable to NH_4^+ ions and even slightly permeable to K^+ ions. Voltage-gated channels have high ionic selectivity and non-voltage-sensitive channels have a low one. Anion channels are relatively nonselective for smaller anions and are referred to as chloride channels because Cl^- is the major permeant anion in biological solutions. Several types of chloride channels have been observed in a variety of tissues including spinal neurones, epithelial cells, Schwann cells and skeletal and cardiac muscle, whose roles are not determined.

Because of their thermal energy all large molecules are inherently dynamic (e.g., chemical bonds that stretch and relax, and twist and wave around their equilibrium positions). Molecular transitions in ion channels occur between open and closed states. During the resting membrane potential some ion channels open frequently and may be deactivated by an appropriate stimulus; the others are predominantly in the closed state and may be activated by an appropriate stimulus. Activation or deactivation of an ion channel means an increase or decrease in the probability of channel opening. The probability of channel opening is controlled in a variety of modes. Some of these modes of activation are physical, such as changes in membrane potential (voltage-gated channels) or membrane tension (stretch-gated channels); others are chemical, involving the binding of molecules to sites on the channel protein (ligand-gated channels). Ligand-gated channels are further divided into two subgroups, depending on whether the binding sites are extracellular or intracellular.

The mechanism whereby an ion channel is opened and closed is called gate. Gating requires a conformational change of the protein that moves a gate into and out of an occluding position. The probability of opening and closing of an ion channel is controlled by a sensor. Thus, the stimulus is detected by sensors, which in turn instruct the channels to open or to close. In some channels, the stimulus acts directly on the intrinsic sensor on the channel to affect the gating function; in others, the sensor is a physically separate molecule that communicates with the channel through intracellular second messenger molecules. The second messenger diffuses through the cytoplasm, carrying signals from the cell surface to its interior, and influences the ion channel and vice versa. There be a direct mutual interaction between the second messenger and the ion channel or the interaction may require other molecules, such as kinases (which are enzymes that add phosphate groups to molecules) and G proteins (which couple receptors to enzymes that catalyse the formation of many second messengers).

The nomenclature of the great diversity of ion channels has not been systematic and classifications are not rigid. Channels have been named on the basis of the most permeant ion, but if there is no major ion or if the ions involved are not adequately known, this situation has led to names such as qr, si and x_1 in cardiac Purkinje fibres; channels have also been named after anatomical regions, as in the endplate channel; after inhibitors, as in the amiloride-sensitive sodium channel; after neurotransmitters, as in glutamate channels; or on the basis of the amino acid sequence, such as brain-type-I (-II or –III) sodium channels [1]. Owing to space limitations, only few types of ion channels will be considered below. Many comprehensive reviews on ion channels are available (see Ref. 1 for detailed citations).

All ion channels produce electrical signals in living cells because ion flows are electric currents across the membrane and therefore the flows have an immediate effect on the membrane voltage. Voltage-sensitive ion permeability is found in virtually all eukaryotic cells. Voltage-gated channels are abundant in neurones and muscles and are present at a relatively low density in non-excitable cells. They open in response to, and then cause, changes in membrane voltage. Voltage-dependent gating requires intrachannel dipole reorientation or charge movement in response to changes in membrane voltage. If the voltage-gated channel enters a conformational state in which activation no longer occurs, even though the activating stimulus is still present, this condition is known inactivation. Inactivation of voltage-gated channels probably never involves their total closure and some ion flow persists; incomplete inactivation may be observed when large enough pulses are applied. Inactivated channels cannot be activated to the conducting state until their inactivation is removed. Neurones use inactivating potassium channels to modulate their firing frequency. Some toxins, phosphorylation, mutation, ions in the medium and chemical treatments may alter voltage sensor and gates. There are several subtypes for each type of voltage-gated channel, with different gating properties or different conductance that form a channel family. Voltage-gated potassium, sodium and calcium channels are cation selective channels and structurally homologous.

Potassium channels are the largest and most diverse known family of ion channels and the most amazing diversification has occurred among voltage-gated potassium channels. Potassium channels may exhibit complex kinetics, indicating the existence of many states and transition rates, some opening rapidly and some slowly; some which are voltage-dependent open only after the membrane is depolarised and some only after it is hyperpolarised; and some are only weakly coupled to the membrane voltage or completely indifferent to it. Potassium channels are ubiquitous in cellular membranes and play a vital role in the functioning of diverse cell types because the important part they play in stabilising the resting membrane potential. Each excitable membrane uses a different mix of the several potassium channels to fulfil its need, the heterogeneity of which extends to the single-cell level. The *Shaker* channel is an archetypal potassium channel and the gene encoded by the *Shaker* locus in the fruitfly *Drosophila melanogaster* was isolated using genetic techniques in combination with molecular strategies.

Less functional diversification has been noticed among sodium channels in excitable cells and not all excitable cells use sodium channels, but where they do exist (e.g. in axons, neurone cell bodies, vertebrate skeletal and cardiac muscles, and many endocrine glands), one is impressed more with the similarity of function than with the differences. Voltage-gated sodium channels have been classified on the basis of species and tissue source and may be distinguished by their sensitivities to various toxins and by certain functional differences. Based on sequence comparisons between the channels of rats and humans, there is a greater similarity among the channels from the same tissue than from the same species, suggesting that each sodium channel type may be optimised either for subcellular localisation in the tissue of origin or for specialised function. Voltage-gated sodium channels are responsible for the rapid membrane depolarisation that occurs during the initial upstroke phase of the action potential in neurones and muscles. Only the binding of ligands, some of which also activate the voltage-gated sodium channels, can activate some sodium channels. The most informative probes for the sodium channel have been tetrodotoxin and saxitoxin, which are known to block the pore of the channel.

The message that ion channels convey to cells very often involves a change in intracellular Ca^{2+} concentration as an internal messenger. Ca^{2+} ions are indispensable for membrane excitability, by acting both inside and outside the cell. The Ca^{2+} that flows into the cytoplasm is altered because the membrane voltage change opens or closes calcium channels, either on the surface membrane or on an internal membrane, causing a change in the concentration of the Ca^{2+} ions in the cytoplasm that may control the gating of some channels, gene expression and the activities of many enzymes. This may be how the nervous system controls the contraction of a muscle fibre or the neurotransmitters release, neurohormones, digestive enzymes, and so on [1]. Calcium channel types differ in voltage dependence, inactivation rate, ionic selectivity and pharmacology. They show extreme diversity of function, having major roles in triggering secretory vesicular release and muscle contraction. They have been recognised as ubiquitous and essential for a host of important biological responses.

Extracellular ligand-gated channels may have several extracellular binding sites for chemical transmitters. They receive extracellular chemical stimulus and spread electrical messages over the entire cytoplasm surface. The maintenance of the open state of the channel depends on the continuous interaction between the chemical transmitter and its receptor, thus giving the open state different durations. On the other hand, the ion current decreases in the maintained presence of the transmitter. When the ligand-gated channel enters a conformational state in which activation no longer occurs, even though the activating stimulus is still present, the condition is known as desensitisation. Most of extracellular ligand-gated ion channels have little voltage sensitivity.

The superfamily of extracellular ligand-gated ion channels includes nicotinic acetylcholine-gated receptor (AChR)-channels, extracellular 5-hydroxytryptamine (5-HT)-gated receptor channels, inhibitory receptor-channels gated by extracellular γ-aminobutyric acid (GABA), inhibitory receptor-channels gated by extracellular glycine, extracellular ATP-gated receptor-channels and ionotropic glutamate-gated receptor (iGluR)-channels. Glutamate is the native neurotransmitter for iGluR-channels, which have been classified into α-amino-3-hydroxy-5-methyl-4-isoxazole propionate (AMPA)-selective glutamate receptor-channels, N-methyl-D-aspartate (NMDA)-selective glutamate receptor-channels and high-affinity kainate glutamate receptor-channels. These three cation channel types have distinct functional properties and are distinguished experimentally by their different sensitivities to glutamate analogues. In the brain of vertebrates, glutamate is the major excitatory neurotransmitter, iGluR channels mediate most of the fast excitatory synaptic transmission, and nicotinic AChR channels are present in smaller amounts. In the peripheral nervous system and skeletal muscle, AChRs are the primary excitatory ligand-gated cation channel. The 5-HT$_3$ receptor appears to be distributed ubiquitously in the peripheral nervous system and its excitation may evoke neurotransmitter release, give rise to a sensation of pain or trigger reflex hypotension and bradycardia. GABA and glycine are the predominant inhibitory neurotransmitters in the brain, both acting by increasing the flow of Cl$^-$ ions into the postsynaptic neurones. GABA acts mainly via GABA$_A$ receptors, whose alterations may result in a number of neurological and psychiatric disorders. GABA$_A$ receptors ligands include a number of clinically useful neuroactive pharmacological agents such as benzodiazepines, barbiturates, anaesthetics and convulsants.

Ligand-gated channels responding to intracellular stimuli include channels that are sensitive to cyclic nucleotides or to local changes in the concentration of specific ions. Calcium channels sensitive to inositol 1,3,4,5-tetrakisphosphate, intracellular sodium-activated potassium channels, calcium channels gated by the arachidonic acid metabolite and cyclic nucleotide-gated channels are intracellular ligand-gated channels. Cyclic nucleotide-gated channels require cyclic nucleotide binding and the direct gating of ion channels by cyclic nucleotides invariably involves kinases and phosphorylation of effector proteins. Receptors in the retina and olfactory epithelium are activated by intracellular cyclic AMP or cyclic GMP. Certain extracellular ligand-gated channels and voltage-gated channels may be sensitive to intracellular ligands.

Gap junction couples most cells of most multicellular animal tissues. Gap junction is a region of contact between two cells and is formed by a collection of connexons. Connexons form aqueous channels that connect the cytoplasm of adjacent cells, allowing the movement of small ions and molecules from cell to cell, without leakage to the extracellular space. In response to physiological variations, gap junctions change the extent of coupling between cells, but few of them have a steep voltage-dependence. Transfer of electrical signal from one cell to the next requires gap junctions.

Transducers may serve as our sensory receptors, namely the senses of taste, sight, smell, touch and hearing; some may transduce temperature, position, heat radiation or electric sense; still others may serve in the regulation of osmotic balance, pH and circulating metabolites. Each transducer is responsible for a specific aspect of the environment or a particular type of event, such as light, temperature, a specific chemical, or a mechanical force or displacement [3]. A common feature of nearly all transduction processes is that stimulus absorption results in a change in the probability of a specific type of ion channel being open. However, the mechanisms employed to change this open probability are quite varied. In the simplest cases the stimulus appears to be absorbed directly by the ion channel structure. In some transducers, an appropriate stimulus causes local depolarisation of the plasma membrane, making the membrane potential less negative; in others, an appropriate stimulus causes local hyperpolarisation, making the membrane potential more negative. Many questions are open with respect to the cellular transduction mechanism.

During the past decade, mutations in genes encoding ion channels have been shown to cause inherited diseases. There are many diseases related to ion channels, and the channelopathies (inherited ion channel disorders) can affect any tissue, but the majority affect skeletal muscle or the central nervous system. These disorders include muscular dystrophy, linked to the nifedipine-binding calcium channels, while epilepsy, Alzheimer's disease, Parkinson's disease and schizophrenia may result from dysfunction of voltage-gated potassium, sodium and calcium channels, or acetylcholine- and glycine-gated channels. Cystic fibrosis, the most common lethal genetic disease in Caucasian people, is characterised by abnormal ion transport in the lung, an unusual saltiness of sweat and a deficit of pancreatic digestive juices, all due to a defective regulation of the chloride channels.

4. COLICIN Ia

Bacterial toxins are secreted as soluble proteins and have to interact with a plasma membrane either to permeabilise the cell (pore formation) or to enter the cytoplasm to express their enzymatic activity. Pore-forming colicins are produced by *Escherichia coli* and related bacteria and share a common strategy: they are inserted into the plasma membrane of the target cells, punching huge holes in them. *In vivo* these holes allow the entry of foreign particles and the exit of intracellular components. This dramatic exchange of charged particles results in the loss of the vital electrochemical potential stored in the plasma membrane and cellular death. Colicins are unusual bacterial toxins because they are directed against close relatives of the producing strain, i.e. colicins are capable of killing *E. coli* cells along with cells of closely related species such as Shigella and Salmonella. They are classified into groups based on the cell surface receptor to which they bind.

Colicins provide a useful means of studying questions such as toxin action, polypeptide translocation across and into membranes, voltage-gated channels and receptor function. Colicin Ia may be obtained as a purified protein in copious amounts. It forms voltage-gated ion channels both in the inner membrane of target bacteria and in BLMs [8-10]. It demonstrates rapid turn-on and turn-off kinetics, making it possible to clearly identify the incorporation of channels into the lipid matrix, and the opening and closing of channels already in the membrane. Colicin Ia is a valuable system for studying voltage-dependent conductance phenomena.

The action of colicin Ia is initiated by adsorption of each toxin to a specific outer membrane receptor, the presence or absence of which is a critical factor in defining the activity of a particular colicin against members of the *Enterobacteriacae* [10]. Its structure reveals several domains that primarily encode receptor attachment, translocation across the outer membrane, movement across the periplasmic space, and insertion into the bilayer membrane [11]. Colicin Ia is a protein of 626 amino acid rich in charged residues imparting hydrophilicity to it, except for a hydrophobic segment of 40 residues near the carboxyl terminus [9, 12]. The amino acid sequence rich in charged residues may be necessary in the earlier stage of attachment of the colicin Ia molecule to the membrane, whose external surfaces are also hydrophilic. Colicin Ia then binds with the hydrophobic segment parallel to the membrane and this portion is inserted into a transmembrane orientation [9]. Colicin Ia translocates a large hydrophilic part of itself completely across the lipid bilayer in conjunction with the formation of an ion-conducting channel [13]. Channel-forming colicin makes contact with the inner and outer membranes simultaneously during its functioning and at least 68 residues flip back and forth across the membrane in association with channel opening and closing, and several open channel structures can exist [14]. Thus, a remarkable series of conformations may be adopted by colicin Ia during translocation across the outer membrane and spanning of the periplasmic space.

5. ARTIFICIAL MEMBRANES

Many attempts have been made to get a clearer picture of biological membranes. Because of the high level of complexity involved in biological membranes, their study is a vast and complicated area. Today, membrane biophysics is a mature field of research, as a result of the applications of many disciplines and techniques, including interfacial chemistry, electrochemistry, voltage- and patch-clamp techniques, spectroscopy and microelectronics [2].

Membranes may be broadly classified into biological or natural and artificial or man-made. Artificial membranes may be required for performing the separation of the constituents of biological fluids using an artificial kidney or may form the basis for developing advanced materials that serve as chemical or biological sensors, having controlled interfacial properties such as adhesion and lubrication. In many cases, they may be used in technological areas as diverse as molecular electronics and optical switch applications. In biomedical applications, such as in tissue engineering, artificial membranes may provide a replacement for tissues lost as a result of disease, deformity or aging. The reconstitution of the functions of natural membranes is currently making amazing progress.

In the membrane phenomena investigation normally inaccessible to experiments on plasma membrane, whether for technical reasons or simply biological complexity, since the 1960s wide use has been made of artificial membranes, which are assembled from bilayer lipid membranes (BLMs) *in vitro*. Two types of BLMs are valuable models in experimental studies: bilayers made in the form of spherical vesicles called liposomes [15] or as a planar BLM [16]. The two models complement each other and each has its own advantages and shortcomings. They have been adopted to elucidate the molecular mechanisms of biomembrane function such as ion selectivity, signal transduction and membrane reconstitution. BLMs have been used as a model for the mitochondria membranes, light sensitive-membranes and visual receptor membranes [2].

Although the BLM is an extremely simple model, many of its physicochemical properties are strikingly similar to those of biological membranes, a fact that makes it an experimental model without much loss of generality. Its properties closely resembling the plasma membrane are flexibility, fluidity, membrane thickness, electrical capacitance, and its surface tension substantially matches that of the plasma membrane. All plasma membranes that have been studied are organised into bilayers and the BLM is an *in vitro* self-assembling system that constitutes this fundamental spatial structure. Lipids in water form a spontaneous bimolecular leaflet, which is a system of two monolayers. The lipid bilayer is probably the most stable arrangement for the kinds of phospholipids normally found in biological membranes, when present in an aqueous environment. It is of special interest that mixtures of phosphatidylcholine, phosphatidylethanolamine and different anionic lipids are able to form BLMs, since these lipids occur very often in the membrane of both eukaryotes and prokaryotes.

The functional properties of the individual components in BLMs can be examined without ambiguity. Membrane transport proteins can be extracted from biological membranes, purified and re-incorporated into a BLM, by making the artificial membrane composition quite similar to that of the plasma membrane, which is composed primarily of two types of molecules - lipids and proteins. Several transport proteins have been shown to aggregate and form ion channels or carriers in BLMs. Many compounds have been embedded into BLMs, such as pigments, electron acceptors, donors, mediators, redox proteins, metalloproteins, substances partaking in ligand-receptor contact interactions and a wide range of antibiotic substances (such as valinomycin, nigericin, nonactin, gramicidin A, amphotericin B, alamethicin, monazomycin and colicin).

The BLM has been extensively used and is particularly suitable for studying the plasma membrane's electrical properties and the nature of gated channel formation, since it is ideally suited to performing electrical measurements with the sensitivity and time resolution necessary to record the opening and closing of individual channels [17]. The effect of lipid composition on channel function may be studied by the incorporation of ion channels into a BLM. The method allows the asymmetric distribution of the proteins with respect to the bilayer. BLMs may also be useful for voltage clamp ion channels of tiny organelles.

When a solution of phospholipid is pipetted onto an aqueous solution, the lipid molecules spontaneously orient themselves at the interface between the air and water – the hydrophilic part is oriented in the aqueous medium and the hydrophobic part is lifted into the air. When phospholipid molecules are surrounded on all sides by an aqueous environment, they tend to aggregate to bury their hydrophobic tails and leave their hydrophilic heads exposed to water [3]. They do this in one of two ways: they can form spherical micelles, in which polar head groups are surrounded by water and hydrocarbon tails are sequestered inside, facing one another; or they can form liposomes, which are bilayers with the hydrophobic tails sandwiched between the hydrophilic head groups.

Liposomes can vary in size depending on how they are produced. The development of different types of unilamellar or multilamellar vesicles with controlled size may permit them to perform some functions such as binding, biocompatibility, controlled permeability and solute release, due to their similarities to real cells and their encapsulation properties for drugs and gene delivery vehicles. Liposomes may fuse with the plasma membrane of many kinds of cells for introducing a variety of substances into cells. Somatic gene therapy depends on the successful transfer and expression of extracellular DNA to the nucleus of eukaryotic cells, with the aim of replacing a defective gene or adding a missing gene at the corrective molecular level, and liposomes may be used as a carrier of recombinant DNA molecules for gene therapy. Such a synthetic membrane opens a new perspective of topical delivery for the treatment of skin diseases and some metabolic disorders.

Planar BLM is a planar bimolecular structure formed across a hole in a partition between two aqueous environments. The use of a planar BLM allows the preparation of a lipid matrix with higher electrical resistance than most plasma membranes, which makes it possible to measure small ion currents. The lipid composition of each monolayer may be controlled, making possible free access to the solutions bathing the matrix, and both sides of the membrane can be easily altered and probed by electrodes. By the addition of transporter proteins that are inserted into the planar BLM from the water phase and increase the ionic permeability by several orders of magnitude, the artificial membrane resistance can be brought into the physiological range. Although BLM is an electrical insulator, after transport protein incorporation it becomes an electrical conductor.

Planar BLM is an extremely delicate structure, rarely lasting more than 8 h. But for long-term stability, it may be supported (s-BLM) on metallic wires, conducting glasses or hydrogel substrates, as well as on microchips, permitting the preparation of an ultra-thin, high-resistance film with a well-defined orientation on metals or semiconductors, which is useful for practical applications and for mimicking biomembranes. The compound in question is immobilised in a rigid, solid-like structure, whereas in the BLM it is embedded.

BLMs have been widely used for investigations into a variety of physical, chemical and biological phenomena, including biotechnology, biomedical research, catalysis, solar energy transduction, electrochemistry, microelectronics and membrane biophysics studies, such as charge transfer experiment, redox reaction, antigen-antibody reaction, sensor development, evaluation of apoptosis and DNA investigation. The versatility of the BLMs is discussed and their uses are presented in an elegant paper published by Tien and Ottova [2].

5.1 Planar BLM preparation

According to the method developed by Montal and Mueller [18], bimolecular membranes are formed from two lipid monolayers at an air-water interface by the apposition of their hydrocarbon chains in a Teflon partition separating two aqueous phases. Formation of the planar BLM is monitored by an increase in electrical capacity, as measured with a voltage clamp. In a simplified version of this method, a thin Teflon film with one hole is clamped between two halves of a device and kept stationary. The artificial membrane is formed by filling the two compartments with saline to below the hole and then spreading a lipid monolayer on each side, raising first one then the other saline level slowly above the hole.

The lipid matrix is formed by opposing two monolayers across a small central hole (diameter: 180-200 μm) in a Teflon (tetrafluoroethylene, 25 μm thick) vertical partition separating two compartments containing an aqueous solution. An electrically heated platinum wire forms the hole in the Teflon. The partition is sealed with silicone grease to the walls of the device and insulates the two water compartments electrically. In order to increase the stability of the bilayer, the partition is precoated with a 2% Vaseline (ICN Pharmaceuticals Inc.) solution in pentene (Merck) before membrane formation. Experiments are performed at room temperature, i.e. at 25 ± 2 °C.

An equal amount of aqueous solution consisting of 500 mM KCl + 5 mM $CaCl_2$ + 5mM HEPES (N-2-hydroxyethylpiperazine-N'-2-ethane sulfonic acid) + 1 mM EDTA (ethylenediaminetetraacetate), final pH 7.00, is placed in each compartment of the device, to a level below the hole in the wall. Both solutions are connected to the equipment by Ag/AgCl electrodes via salt bridge (2.5% agar in the chamber medium, electrodes immersed in 3 M KCl). Deionised water, double distilled in glassware, is used for the preparation of all solutions. All chemicals are of analytical grade.

Monolayers are made using 1% phosphatidylcholine in hexane (L-α-phosphatidylcholine type II, Sigma Chemical Co.) with no additional preparation. After 10 µl of this lipid solution has been deposited on the surface of the aqueous solutions, it is necessary to wait approximately 20 minutes for the solvent evaporation and the spontaneous formation of the monolayers on the surface of the aqueous solution in each compartment of the device. With the aid of a syringe connected to the base of one compartment, the level of the aqueous solution is carefully raised to slightly above the hole to enable the monolayer to be deposited on the latter; with the aid of another syringe connected to the base of the other compartment, the aqueous solution is raised to exactly the same level as in the first compartment, thus enabling the second monolayer to be likewise deposited on the hole, thereby forming the planar BLM. The formation of the bilayer is monitored continuously by rapid and repeated measurements of the membrane capacity by the application of constant voltage pulses (amplitude: ±10 mV; duration: 2-4 ms; frequency: 500 Hz), and displays the capacitative current on a storage oscilloscope.

Colicin Ia is added directly to the aqueous solution in just one compartment of the device, attaining a final concentration of 1-5 µg/ml. The duration of the interaction between the colicin Ia and the planar BLM is 1 minute at the most for the artificial membrane to reach a steady state. Bearing in mind that the durability of each artificial membrane is approximately 3 hours, on the occasions when the membrane breaks following the addition of colicin, the experiment is terminated with all the reagents being discarded.

A pulse generator with DC voltage of ±200 mV is connected to the side of the membrane containing colicin Ia. The ion current flowing through the artificial membrane is measured under conventional voltage-clamp conditions using an operational amplifier (Burr-Brown model OPA111) in the current-to-voltage converter configuration. The converter input is connected to the colicin-free side of the membrane, and the output to a digital oscilloscope (Nicolet Instrument Corporation, model 201).

6. EVIDENCE OF ION CHANNEL MEMORY

All ion channels generate an electrical signal due to the ion flows across them and much of what we know about ion channels was deduced from electrical measurements. Electricity, life and the relation between them have long been the subject of intense curiosity. Investigations of living cells based on electrical concepts and using electrical techniques have been amazingly successful and significant advances in this field have been achieved over the last 50 years. Electrical pulses may also be used to gate some ion channels and even to demonstrate the ion channel memory.

The analysis of ion current recordings has been a tremendous challenge for many theoreticians [4] and over the past few years many attempts have been made to get a clear picture of ion channel kinetics. As applied to ion channels, kinetics is often used in order to find some specific mechanism channels undergo when changing from one state to another. Ion channel kinetics has been treated as either a deterministic process or a stochastic one. Here we shall not describe each process rigorously. Rather, we shall present some experimental and simulated data that indicate that ion channel kinetics may be a process with memory. Memory is usually considered a diffusely stored associative mechanism that puts together information from many different sources. It is the process of storing and retrieving information, reproducing what has been retained.

The molecular nature of ion channels may be revealed by unitary current steps seen with the patch-clamp method [7] or by using BLMs [15, 16]. The combination of these two endeavours has generated a wealth of phenomenological information on ion channel kinetics at the level of single molecular events. By recording the noise from each ion current through the ion channel, the sequence of its open and closed states may be measured. When the best available time resolution is used, individual channels appear to pop open and close suddenly without any evidence of gradualness in the transition [1]. Thus, noise measurements provide important information on the structure and function of the ion channel protein. The analysis of ion currents recorded through a single ion channel may provide the mechanism by which channels gate their pores and explain the phenomena of jumps, bursts, ligand interactions, etc. The transition kinetics between different conformational states of the ion channel may also be studied in the macroscopic noise from the ion current passing through many ion channels.

In cell membranes, the ion channels make repeated transitions between open and closed states during their activity. Ion channels have no threshold for opening and, at any membrane potential, there are several types of channel open [1]. The distribution of ion channels between the open and closed states determines the transmembrane voltage rather than the total number of channels. When a channel opens, the ion current appears abruptly, and when it closes, the current shuts off abruptly [1], but in almost all the ion channels a closer look reveals that the elementary event is not a single current but displays bursts of opening, i.e. a series of transient steps separated by brief gaps. The tiny, picoampere current through an individual ion channel clamped at constant voltage does not lead to a constant signal but presents a peak-valley landscape pattern of high and low current values, with millisecond time resolution [4]. The randomness of the length of time for which the ion channel stays open and closed reflects the probabilistic nature of the interaction between the stimulus and the gating. Both the jump magnitude and the number of discrete current levels of a single channel may vary among the several kinds of ion channels.

At the single-channel level, the gating transitions may be predicted only in terms of probability, since each trial with the same depolarising step shows a new pattern of openings of the individual ion channel. Information on the number of states and the transitional pathway between them can be obtained from the durations of open and closed intervals of different mean open time around which durations of opening fluctuate [19]. Identification of the minimum number of ways in which open and closed states communicate is a crucial step in defining the gating kinetics of multistate channel [17]. Widely varying durations of open and closed intervals suggest multiple open and closed states of the ion channel: if there is only one open conformation or state, then the duration of open intervals would be independent of the durations of adjacent closed intervals; but if a channel has two or more open states with different mean durations, and if each open state is entered directly from a different closed state with a different mean duration, then the open intervals should be related to the adjacent closed intervals [19].

The ion channel may behave as if it has forgotten all information prior to the last stimulus, so that the probability of interconversion at any given moment depends only on the state in which the channel is at that moment, and not on its history up to that moment, i.e. the probability of transition from one state to another is a constant, independent of time [20, 21]. On the other hand, the ion channel may behave as if it has [22, 23] memory due to different kinetic processes, some of which may be slow and some rapid, and which are not independent but interrelated; once the channel closes, the rate of re-opening is dependent on the time it remained closed, such that the rate of re-opening declines as the time of closure increases. Using the patch-clamp technique, single-channel currents through both a chloride channel and a large conductance calcium-activated potassium channel in skeletal muscle show that the durations of adjacent open and closed intervals are inversely related: shorter open intervals are adjacent to longer closed intervals [19]. Our previous studies [8, 24] have shown that colicin Ia channels incorporated into a planar BLM may also behave as if they have memory. It is possible to induce memory control of the artificial membrane in accordance with the experimental conditions imposed on it and on the basis of theoretical modelling [24]. The memory here must be interpreted in a general sense: if the current state of a system depends on its previous states, then this system is endowed with memory.

Short-term correlation may indicate that ion channel kinetics is a process with short-term memory [17, 19]. Autocorrelation analysis provides a simple and general strategy to extract information on channel gating kinetics, providing information concerning the identification of the open and closed states and on the pathways connecting them, considering transition dependencies between successive events at single channel currents [17]. Using a planar BLM containing a single channel from the electric organ of *Torpedo californica*, the autocorrelation analysis shows that for the acetylcholine receptor channel there are at least two entry/exit states through which the open and closed aggregates communicate; the chloride channel fluctuates between three conductance substates, identified as closed, intermediate and high conductance, and correlation analysis shows that although there are at least two distinct entry/exit states in the intermediate aggregates, there is no evidence for the existence of more than one entry/exit state in the closed or high conductance aggregates [17].

Ion current fluctuations occurring within open and closed states of a large-conductance locust potassium channel (BK channel), from cell-attached patches of adult locust (*Schistocerca gregaria*) extensor tibiae fibres, reveal [4] that the memory effect is present not only between successive conducting states of the channel, but also independently within the open and closed states themselves. Autocorrelation, Hurst (rescaled range) [25] and detrended fluctuation (DFA) [26] analyses showed the existence of correlation between successive duration of closed and open states and, thus, the existence of long-range memory effects in the BK channel [4, 27]. As the ion current fluctuations give information about the dynamics of the channel protein, this result points to the correlated character of the protein movement regardless of whether the channel is in its open or closed state [4].

Types II and IIA voltage-gated sodium channels are the most abundant voltage-gated channels in the mammalian brain. Sodium channels may be efficiently expressed in oocytes, allowing one to record macroscopic currents in a detached patch of membrane. The expression of proteins by messenger RNA in oocytes, as well as in other host cells, has been a valuable tool for examining the properties of ion channels. Using patch-clamp and voltage-clamp techniques, currents measured from the membrane of *Xenopus* oocytes, injected with cRNA coding for rat brain channel types NaII or NaIIA, have shown direct experimental evidence for the existence of memory in the functioning of the ion channels: the longer the voltage-gated sodium channel is held at a given voltage, the longer it takes to recover from inactivation, as a power law scaling between activity and availability [28]. Ion channels in the plasma membrane of the hippocampal neurones, fibroblasts and gramicidin channels in BLM have open and closed durations that can also be adjusted, all or in part, by power law distributions [23]. However, the scaling relationship exhibited by some ion channels is not a universal property of such channels. Power law scaling can act as a molecular memory mechanism that preserves traces of the previous activity, over a wide range of time scales, in the form of modulated reaction rates, owing to the fact that the scaling of recovery time course is intrinsic to the channel protein [28]. The channel exists in a large number of energy states and the transitions among these states are determined by rate constants that change continuously in time [23]. Analysis of patch clamp recordings from rabbit corneal endothelium also shows [22] that the ion channel behaves as if it has memory.

In a variety of brain locations, repetitive activity can produce change in synaptic efficacy. Glutamate-gated channels are essential elements in synaptic function and plasticity. There is a general agreement that long-term potentiation (LTP) is produced by an increase in calcium concentration in the postsynaptic cell. In the brain of vertebrates, NMDA-selective glutamate receptor-channels are involved in memory processing by mediating the induction of LTP, a long lasting and activity-dependent enhancement of synaptic efficacy. These channels are characterised by high single channel conductance and high calcium permeability, and are blocked by magnesium ions in a voltage-dependent manner. NMDA receptor-dependent synaptic plasticity at pyramidal cells in area CA1 of the hippocampus is required for both the acquisition of spatial memory and the formation of normal CA1 place fields [29]. This relationship suggests that robust place fields may be essential for spatial memory [29].

Hippocampal theta (4-10 Hz) oscillation represents a well-known brain rhythm implicated in spatial cognition and memory processes. Experimental data indicate that GABAergic cells in the medial septum play a pacemaker role for the theta rhythm and a biophysical model predicts that theta oscillations of septal GABAergic cells depend critically on a low-threshold, slowly inactivating potassium current [30]. Simulations show that theta oscillations are not coherent in an isolated population of pacemaker cells and robust synchronization emerges with the addition of a second GABAergic cell population [30].

Modulation of threshold potential by changes in the availability of conductance is a powerful memory mechanism that depends solely on the intrinsic properties of a neurone, since the neurone has a memory capacity that is embedded in the machinery of excitability and that is not delimited by particular time scales [28]. In mammals, the brain is one of the organs containing the highest proportion of membranes and, despite the significant biological role of brain membranes, knowledge of its protein and lipid content is incomplete but important in the understanding of the functioning of the nervous system. Moreover, little is known about the physiology of memory storage in the brain.

6.1 Ion channel memory control using a planar BLM

Using macroscopic measurements of the total ion current flow through a large number of voltage-gated channels, we demonstrated [8, 24] a critical resting interval (Δt_c) on the artificial membrane prepared as described in Section 5.1. The Δt_c value is shorter than 120 seconds and specific to each particular artificial membrane, by means of which it is possible to show the memory effect and its control by applying square voltage pulses of at least 20 seconds duration from a holding potential. The BLM's behaviour seems to be robust and general. Since the observed memory effect is mainly due to the applied field, it can be concluded that external voltage influences the ion channel action on an amazingly broad time scale [4].

We focus on the dynamic behaviour of the colicin Ia incorporated into a planar BLM in response to sequences of stimuli with various amplitudes and resting intervals. Before colicin incorporation into a planar BLM, the artificial membrane does not respond because there is no ion current across the BLM. After the incorporation of many colicin Ia channels into the BLM, the artificial membrane also does not respond to negative pulses. The artificial membrane response performs inconsistently to positive pulses smaller than 50 mV. In addition, positive pulses greater than 90 mV lead to a brief response and an irreversible membrane breakdown. When the artificial membrane is stimulated by positive P pulses, between 50 mV and 90 mV, the recording of the ion current versus time shows a sudden exponential rise that, on a first approximation, may be expressed as: $I(t) = (a + bt)[1 - \exp(ct)]$, where a and b are positive constants and c is negative [24]. The constants a, b and c might differ, since the ion current may vary in both amplitude and time course for a given P pulse (Fig. 1). In a trial, successive identical P pulses may generate a systematic variability. In addition, a long-time interval between pulses or between trials generates an unpredictable variability.

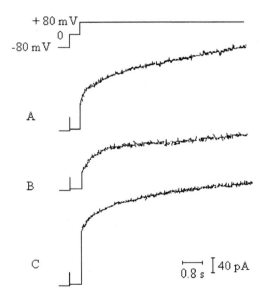

Fig. 1. If the artificial membrane is stimulated by −80 mV, the ion current is practically zero. If it is stimulated by +80 mV, the ion current response as a function of time may not be reliably predicted. Such a response may vary in amplitude and time course as shown by the curves A, B, C or many other responses. The time and vertical scales apply to all curves. (From Cassia-Moura, 1993.)

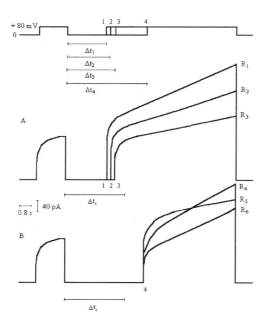

Fig. 2. Stimulation of the artificial membrane by two successive pulses of +80 mV, so that the second stimulus of each pair of stimuli is applied after four different resting intervals (Δt_1, Δt_2, Δt_3 and Δt_4). Three superimposed trials of the ion current response are shown as a function of time. (A) When the Δt_c value is greater than the resting interval between pulses, the second pulse always produces a predictable response, such that if $\Delta t_1 < \Delta t_2 < \Delta t_3 < \Delta t_c$, then responses decrease in the order R_1, R_2 and R_3. (B) When the resting interval between pulses is greater than Δt_c value, the response to the second pulse is not predictable, such that if $\Delta t_c < \Delta t_4$, then the response may be R_4, R_5, R_6 or any other. The time and vertical scales apply to all curves. (From Cassia-Moura *et al.*, 2000.)

By applying two successive P pulses interposed by a resting interval (Δt), if the second pulse is applied within the Δt_c value, it generates a deterministic ion current response across the artificial membrane, so that an ion current shows a smaller amplitude when the resting period between P pulses is greater or vice versa, as shown in Fig. 2a. With the growing increase in the rest intervals (Δt_1, Δt_2, Δt_3) between the application of the two consecutive pulses, the R_2 response is shorter than the R_1 response and the R_3 response is shorter than the R_2 response. Thus, when the length of time between the end of the first stimulus and the beginning of the second, interposed by a rest period, falls within the Δt_c value, the diminution or enhancement of the response to the second pulse can be attributed to the "memory" of the artificial membrane because the current state of the system depends on its previous state. On the other hand, it is clear from Fig. 2b that if this length of time between the end of the first stimulus and the beginning of the second, interposed by a rest period, increases and becomes greater than the Δt_c value, the ion current response across the artificial membrane can no longer be reliably predicted, so that R_4, R_5, R_6 or any other response may be triggered. This being the case, for certain parametric values the experimental system response is independent of the previous state, while for other parametric values the response depends on the previous state. Thus, the behaviour of the voltage-gated channels may be interpreted as a transient gain, loss or resetting of memory, as revealed by a specific sequence of electrical pulses used for stimulation [24].

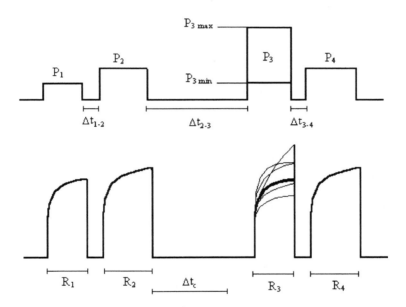

Fig. 3. Stimulation of the artificial membrane by four successive P pulses, interposed by resting intervals (Δt_{1-2}, Δt_{2-3} and Δt_{3-4}). P_3 may be any value within the $P_{3\,min}$ range, which is equal to +50 mV and $P_{3\,max}$, which is equal to +90 mV; P_1, P_2 and P_3 should have a fixed value within the same range. Since $\Delta t_{2-3} > \Delta t_c$, the artificial membrane responds stochastically to P_3 and R_3 may show different patterns. When by chance the bold R_3 is identical to R_1, the P_3 value for predicting the response R_4 does not matter, because if Δt_{1-2} is identical to Δt_{3-4}, and if P_4 is applied and is identical to P_2, the response R_4 can be predicted and will be identical to R_2. (From Cassia-Moura *et al.*, 2000.)

It is possible to have a memory proof of the artificial membrane by applying four successive P pulses, interposed by resting intervals. Moreover, in this sense, it is possible to induce memory control of the artificial membrane in accordance with the experimental conditions imposed on it. The first and the second pulses should be interposed by a resting period shorter than the Δt_c value, such that the response to the second pulse is predictable. The resting period between the end of the second pulse and the beginning of the third should be greater than the Δt_c value, such that the third pulse will thus generate an unpredictable response. The response to the third pulse may be a random one identical to that obtained from the first pulse, even if the third pulse is different from the first one; in this case, when a fourth pulse identical the second one is applied, provided that both occur after identical resting intervals, the fourth response will necessarily be predictable again and identical to the second response, as shown in Fig. 3. The artificial membrane is thus "remembering" the previous state installed after the P_1 action; moreover, different P pulses may induce this state. A phenomenological model devised using explicit mathematical formulation captures many dynamic features of this experimental result, suggesting that the system's intrinsic parameters and dynamic rules do not change after an unpredictable response. In this sense, memory is disrupted only transiently [24].

In spite of the simplicity of the above model, the phenomenon is, however, widespread and may have important implications for the functioning of voltage-gated channels in a variety of situations, since periodic rhythms underlie biological processes as diverse and fundamental as secretion, muscle contraction and neuronal firing. It is important to note that our biophysical model is proposed to explain specific observations in a simplified qualitative form and the pulses applied were continuous and optimised for characterization of ion channel kinetics rather than for physiological compatibility. Thus, our biophysical model cannot point directly to a physiological significance, despite showing a phase dependence of the effect of brief current pulses.

7. CONCLUDING REMARKS

Molecular research on ion channels is one of the essential topics in understanding the functions of living cells. Although a thorough knowledge of biological membranes is still a long way off, we hope that the deployment of a membrane model, in which it is possible to induce memory control, will be our contribution to the store of increasing information on the functioning of ion channels, helping to unlock cellular secrets. Despite the difference in interaction between the ion channels and a natural membrane or a planar BLM, we feel that the properties exhibited by the experimental model examined in Section 6.1 of this chapter may be applied to a greater or lesser extent to plasma membranes *in vivo.* We do not yet know how much our picture of just one kind of ion channel affects the understanding at the molecular level of current-voltage behaviour as a function of time of many types of ion channels, measured under many conditions, or whether our interpretation will have implications for multicellular organisms. Since time dominates life, we hope that the ideas introduced here may serve as a step towards a rigorous investigation of the memory effect phenomenon associated with the critical resting interval of a plasma membrane patch, but obtaining rigorous proof of this *in vivo* is no easy task. Knowledge gained in the past 15 years about the ways in which ion channels work calls for re-evaluation of the role played by intrinsic neural excitability properties in memory acquisition and learning. Undoubtedly, we can expect to witness major advances in knowledge of the memory effect on ion channels over the next few years, developments that will be of both fundamental biological and clinical significance.

Acknowledgements
No words are enough to express my sincere gratitude to Professor John Rinzel (Center for Neural Science of the New York University, USA) for many constructive comments and his valuable suggestions on this chapter. This chapter embodies the research work carried out by me as a scientist associate of The Abdus Salam International Centre for Theoretical Physics, in Italy. I am very grateful to Professor Hilda Cerdeira (from the Centre). I thank Mr. David Randall (University of Pernambuco, Brazil) for his technical assistance.

REFERENCES

[1] B. Hille, Ionic channels of excitable membranes, Sinauer Associates Inc, Sunderland, Mass., 1992.
[2] H. Ti Tien and A. L. Ottova, J. Membr. Sci. 189 (2001) 83.
[3] B. Alberts, D. Bray, J. Lewis, M. Raff, K. Roberts and J. D. Watson (eds.), Molecular Biology of the Cell, Garland Publishing, New York & London, 1983.
[4] Z. Siwy, M. Ausloos and K. Ivanova, Phys. Rev. E 65 (2002) 031907.
[5] A. L. Hodgkin and A. F. Huxley, J. Physiol. (London) 117 (1952) 500.
[6] K. S. Cole, Arch. Sci. Physiol. 3 (1949) 253.
[7] O. P. Hamil, A. Marty, E. Neher, B. Sakmann and F. J. Sigworth, Pflügers Arch. 391 (1981) 85.
[8] R. Cassia-Moura, Bioelectrochem. Bioenerg. 32 (1993) 175.
[9] P. K. Kienker, X. Qiu, S. L. Slatin, A. Finkelstein and K.S. Jakes, J. Memb. Biol.157 (1997) 27.
[10] J. Konisky, Annu. Rev. Microbiol. 36 (1982) 125.
[11] R. M. Stroud, Biophys. J. 74 (1998) A229.
[12] M. Wiener, D. Freymann, P. Ghosh and R. M. Stroud, Nature 385 (1997) 461.
[13] K. S. Jakes, P. K. Kienker, S. L. Slatin and A. Finkelstein, Proc. Natl. Acad. Sci. USA 95 (1998) 4321.

[14] X. Q. Qiu, K. S. Jakes, P. K. Kienker, A. Finkelstein and S. L. Slatin, J. Gen. Physiol. 107 (1996) 313.
[15] A. D. Bangham in Progress in Biophysics and Molecular Biology, J. A. V. Butler and D. Nobile (eds.), Pergamon Press, New York, 1968, pp.29-95.
[16] P. Mueller, D. O. Rudin, H. T. Tien and W. C. Wescött, Nature 194 (1962) 979.
[17] A. Labarca, J. A. Rice, D. R. Fredkin and M. Montal, Biophys. J. 47 (1985) 469.
[18] M. Montal and P. Mueller, Proc. Natl. Acad. Sci.USA 69 (1972) 3561.
[19] O. B. McManus, A. L. Blatz and K. L. Magleby, Nature 317 (1985) 625.
[20] M. S. P. Sansom, F. G. Ball, C. J. Kerry, R. McGree, R. L. Ramsey and P. N. R. Usherwood, Biophys. J. 56 (1989) 1229.
[21] D. Colquhoun and A. G. Hawkes, Proc. R. Soc. Lond. B. Biol. Sci. 211 (1981) 205.
[22] L. S. Liebovitch, Biophys. J. 55 (1989) 373.
[23] L. S. Liebovitch, Math. Biosci. 93 (1989) 97.
[24] R. Cassia-Moura, A. Popescu, J. R. S. A. Lima, C. A. S. Andrade, L. S. Ventura, K. S. A. Lima and J. Rinzel, J. Theor. Biol. 206 (2000) 235.
[25] J. Feder, Fractals, Plenum Press, NY, 1988.
[26] K. Hu, P.Ch. Ivanov, Z. Chen, P. Carpena and H. E. Stanley, Phys. Rev. E 64 (2001) 011114.
[27] Z. Siwy, Sz. Mercik, K. Weron and M. Ausloos, Physica A 297 (2001) 79.
[28] A. Toib, V. Lyakhov and S. Mahom, J. Neurosci. 18 (1998) 1893.
[29] M. A. Wilson and S. Tonegawa, Trends Neurosci. 20 (1997) 102.
[30] X. J. Wang, J. Neurophysiol. 87 (2002) 889.

Planar Lipid Bilayers (BLMs) and their Applications
H.T. Tien and A. Ottova-Leitmannova (Editors)

Chapter 20

Symmetric and asymmetric planar lipid bilayers of various lipid composition: a tool for studying mechanisms and lipid specificity of peptide/membrane interactions

A. Wiese, T. Gutsmann, and U. Seydel

Research Center Borstel, Center for Medicine and Biosciences, Division of Biophysics, Parkallee 10, D-23845 Borstel, Germany

1. INTRODUCTION

Membranes, in general, constitute the border between a cell or a cell compartment and its environment. They are composed of (glyco)lipids and proteins. They function as a permeability barrier, maintain constant ion gradients across the membrane, and guarantee a steady state of fluxes in the cell. Furthermore, the cell membranes carry recognition structures for the immune system and for the interaction with other cells.

These functions to work properly require a particular lipid composition and distribution on each side and between both sides of the lipid bilayer. Thus, a membrane is built up from a large variety of different lipids, varying in their charge and fatty acid substitution (length and degree of saturation), and these lipids are in a delicate equilibrium providing a suitable environment for protein function. By a complex interaction of passive - by diffusion through the lipid matrix or protein-aligned transmembrane channels- and active, energy-dependent transport processes - by ion pumps and transport proteins - an ion gradient is built up, which contributes together with the charge distribution of the lipids on the two leaflets of the membrane to a transmembrane potential.

The lipid matrix of eukaryotic membranes is composed of a mixture of mainly phospholipids on both leaflets. The content of a certain phospholipid may, however, vary between different cell types and, of course, also between the two leaflets of a given cell type, i.e., that the lipid bilayers are asymmetric with respect to the phospholipid composition. This asymmetry is even more pronounced in membranes containing glycolipids which are usually located only in the outer leaflet. An example for an extreme asymmetry between a glycolipid outer and a phospholipid inner leaflet is represented by the outer membrane of Gram-negative bacteria (Fig.1). Here, the outer leaflet is composed solely of lipopolysaccharides

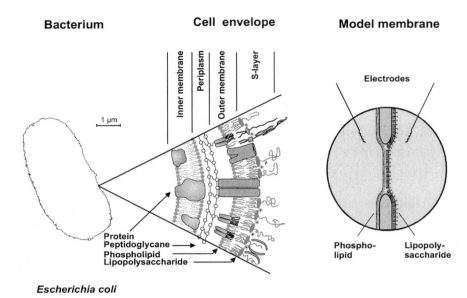

Bacterium   **Cell envelope**   **Model membrane**

Escherichia coli

Fig. 1. Gram-negative bacterium and cartoons of its cell envelope and of the reconstituted asymmetric planar bilayer.

(LPS) or, in a few cases, of glycosphingolipids (GSL) and the inner leaflet of a phospholipid mixture made up from phosphatidylethanolamine (PE), phosphatidylglycerol (PG) and diphosphatidylglycerol (DPG) [1].

Different techniques have been described for the preparation of reconstituted lipid bilayers. One of these is the formation of painted black lipid membranes (BLM) which has first been described in 1962 by Mueller et al. [2]. Briefly, BLM are prepared from a dispersion of phospholipids in a nonpolar solvent such as n-decane which is spread beneath an aqueous phase over an aperture of up to several millimeters diameter separating two plastic compartments. The lipid thins out in the center of the aperture until it forms a bilayer that is optically black when viewed in incident light. The membrane is essentially impermeable to ions, and thus, application of a voltage across the pure lipid bilayer does not result in any detectable electrical current. Induced membrane disturbances can, therefore, be monitored *via* current measurements (for review see [3] and chapter 1 of this book). This method is obviously not suitable for the reconstitution of asymmetric lipid matrices such as that of the outer membrane of Gram-negative bacteria, because the experimentator has no influence on the arrangement of the different lipids in regard to the specific - asymmetric - composition.

This difficulty is overcome by techniques introduced by Montal and Mueller [4] and by Schindler [5]. In both methods, bilayers are formed over a small aperture from two lipid monolayers on top of bathing solutions in two compartments separated by a septum containing the aperture. In the case of the Montal-Mueller technique, lipid solutions in a highly volatile solvent (e.g., chloroform) are spread on the air-water interface, whereas in the case of the Schindler technique the monolayers are built from vesicles added to the bathing solution. When monolayer formation is completed - either after solvent evaporation or after vesicle fusion at the air-water interface - the monolayers are successively raised over the aperture to form the bilayer membrane. Asymmetric membranes can thus be obtained, if different lipids or lipid mixtures are used on the two sides of the aperture (Fig. 1). For obvious reasons, the Montal-Mueller and Schindler techniques are most suitable for studying bacterial outer membrane function, the Schindler technique having the advantage of absolute solvent-freeness, but the disadvantage of the presence of lipids in the subphase. Furthermore, these techniques allow the use of a fairly wide variety of different lipids as long as these are soluble in the organic solvent and guarantee a sufficient fluidity of the acyl chains. This fact becomes extremely important when studying the lipid specificity of peptide-membrane interactions and lipid-mediated resistances against antimicrobial peptides. A detailed review on the preparation of planar lipid membranes as black lipid films or according to the Montal-Mueller- or Schindler techniques is given by S.H. White [6].

Reconstituted membranes are mainly utilized for studying peptide-membrane interactions and pore formation and function in membranes. Information on the underlying molecular processes is derived from time-resolved electrical measurements of membrane current, membrane capacitance, and innermembrane potential difference. This information may be complemented by the recruitment of other techniques such as film balance measurements on monolayers at the air-water interface or fluorescence resonance energy transfer spectroscopy with liposomes to study peptide incorporation, or infrared spectroscopy on lipid aggregates to study the influence of peptides on lipid fluidity and binding to defined functional groups or of lipids on peptide conformation.

In this review, we will focus on own work using the reconstitution system according to Montal and Mueller and the characterization of peptide-membrane interactions on the molecular level with regard to lipid and peptide specificity, especially for antimicrobial peptides targeting the outer membrane of Gram-negative bacteria and on the electrical measurements to characterize various interactions.

As already mentioned above, the outer leaflet of the outer membrane of Gram-negative bacteria (Fig. 1) is composed solely of glycolipid - usually LPS, in some cases GSL - and the inner leaflet of a phospholipid mixture (PL) of phosphatidylethanolamine (PE), phosphatidylglycerol (PG), and diphosphatidylglycerol (DPG) - in the case of *Salmonella enterica* serovar Thyphimurium in molar ratios of PE : PG : DPG = 81 : 17 : 2 [7].

Chemically, LPS consist of a hydrophilic heteropolysaccharide which is covalently linked to a hydrophobic lipid portion, termed lipid A, which anchors the

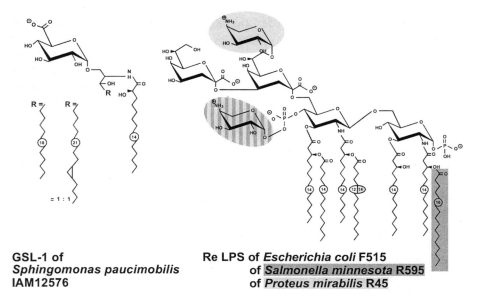

GSL-1 of
Sphingomonas paucimobilis
IAM12576

Re LPS of *Escherichia coli* F515
of *Salmonella minnesota* R595
of *Proteus mirabilis* R45

Fig. 2: Chemical structure of GSL-1 from *Sphingomonas paucimobilis* IAM 12576 (left) and of deep rough mutant lipopolysaccharide from different Gram-negative bacterial species (right).

molecule in the outer membrane. In wildtype strains, the polysaccharide portion consists of an O-specific chain and the core oligosaccharide (S-form LPS). Rough mutant strains do not express the O-side chain, but retained core oligosaccharides of varying length. The LPS of the various rough mutants are characterized by chemotypes in the sequence of decreasing length of the core sugar as Ra (complete core), Rb, Rc, Rd, and Re, the latter representing the minimal structure of LPS consisting only of lipid A and two 2-keto-3-deoxyoctonate (Kdo) monosaccharides. Lipid A is composed of a 8-glucosaminyl-1→6)-α-*D*-glucosamine disaccharide backbone which is phosphorylated in positions 1 and 4' of the disaccharide backbone and carries - for Enterobacteriaceae - six or seven (hydroxy) fatty acid residues which are either ester- or amide-linked (for review see [8]). The minimal LPS structure, the Re LPS of different Gram-negative species may not only differ in the acylation pattern but also in the number of negative charges. Thus, Re LPS of *Escherichia coli* strain F515 (F515 LPS) carries four negative charges (two from the phosphates and two from the carboxyl groups of the Kdo). In *Proteus mirabilis* strain R45, the first Kdo and the 4' phosphate are substituted to 50% by aminoarabinose leading to a reduction of the effective number of negative charges to three. In Re LPS from *Salmonella enterica* sv. Minnesoata strain R595 (R595 LPS) only the 4' phosphate is substituted by aminoarbinose to about 65%, there is, however, found a seventh fatty acid in non-stoichiometric substitution (Fig. 2, right).

The hydrophobic portion of GSL – as isolated from *Sphingomonas paucimobils* - was found to be heterogeneous with respect to the dihydrosphingosine residue but was, in any case, quantitatively substituted by a (S)-2-hydroxymyristic acid in amide linkage. The oligosaccharide portion of the two main fractions, GSL-4A and GSL-1, consists of a Man-Gal- GlcN-GlcA tetrasaccharide, and of a GlcA monosaccharide, respectively [9]. Only GSL-1 (Fig. 2, left) has been used in the experiments described later.

From this brief description of the complex architecture of the outer membrane, it becomes obvious that a detailed characterization of its various functions is feasible only with simpler reconstitution systems, in a first step of the unmodified lipid matrix. In such a model system, for example, the influence of the glycolipids on the function of transmembrane proteins may be studied by reconstitution of the proteins into the bilayer and, furthermore, the role of the glycolipid in its potential interaction with membrane active substances like drugs, detergents, and components of the immune system may be characterized.

2. SETUP FOR MEMBRANE RECONSTITUTION AND ELECTRICAL MEASUREMENTS

Here, only a relatively short description of the apparatus and the various steps of membrane preparation and electrical measurements is given, a very detailed procedure is given elsewhere [10]. The apparatus consists of a Faraday cage, a closed metal cylinder, which contains the measuring chamber in the bottom and the electrical amplifier (headstage) on top. This cage is placed on an electrically controlled heating plate on top of a magnetic stirrer. The measuring chamber consists of two conical semi-circled teflon compartments carrying various bore holes. The major bore, the actual preparation chamber with a cylindrical opening, facing the planar side of the compartment, is connected via vertical bores with the bore hole for the electrodes. Two smaller bores are used for the adjustment of the levels of the bathing solution. The two compartments are pressed tightly together by a conical steel ring with their planar faces only separated by a teflon septum containing the aperture for membrane formation.

The lower part of the Faraday cage contains two bores for the hose connection to the syringes for the adjustment of the levels of the bathing solution. Two further rectangular openings at the upper rim on the front and back sides are used for optical control of the membrane and its illumination by a flexible light guide. These openings can be closed after membrane formation by a moveable cylindrical shield around the Faraday cage. The headstage contains the amplifier unit of an L/M-PCA patch clamp amplifier. To reduce mechanical vibrations, the whole setup for the formation of planar bilayers is placed on a vibration isolation table.

The amplifier is connected to the controller of the L/M-PCA patch clamp amplifier. The clamp voltage is supplied either by the built-in power supply or by an external voltage source attached to the stimulus input connector. For this purpose, a digital analog converter (DAC) interface connected to a PC is used. Input voltage and output current are displayed on a double-trace oscilloscope and stored with a digital tape recorder with a sampling frequency of 48 kHz. Before being read into the computer *via* an analog-digital converter (ADC) interface, the stored signals can further be deep-pass filtered using a 4-pole bessel filter.

Membranes are formed over a small aperture (diameter between 100 and 200 μm) in the center of a three-layered foil septum. The inner PTFE foil has a thickness of 12 μm, whereas the outer PFE foils are 30 μm thick and serve for mechanical stability. The outer ring-shaped foils have an outer diameter of 16 mm and an inner diameter of 5 mm, the diameter of the PTFE foil is 11 mm. The three foils are sintered between two aluminum blocks under moderate pressure to one tight unit to avoid leakage at 600°C for about 2.5 min.

The aperture is shot into the center foil by a spark discharge. The septa are then cleaned in chloroform:methanol solution (2:1 v/v), air dried, and stored before use in the presence of silica gel to reduce air humidity.

The success of membrane formation is significantly increased, if the septum is pretreated with a mixture of hexan:hexadecan (1:20 v/v). For membrane formation, on each side, 2-5 μl lipid solution (2.5 mg/ml) are spread on top of the bathing solutions in the preparation chambers, e.g., LPS solution on the front side and the PL mixture on the back side. By removal of bathing solution (*via* syringes), the monolayers are adjusted to a position just below the lower rim of the aperture. Before the preparation of the membranes, the solvent has to be allowed to evaporate for 15 min. During this time, the bathing solution can adjust to the prefixed temperature.

After the evaporation of the solvent, the two monolayers are successively raised over the aperture by adding bathing solution from the syringes. The correct formation of the lipid bilayer is checked by determining the membrane capacitance as a measure for membrane thickness. For apertures with a diameter in the range of 100 to 150 μm, membrane capacitances in the range of 90 to 120 pF have to be expected. A smaller value would be indicative of too thick a membrane, because of, e.g., multilayer formation or because of silicon grease in the aperture.

The measurement starts after a further waiting period (5 min) for equilibration of the membrane. A low clamp voltage (10 - 20 mV) is applied to check, whether the incorporation of the external molecules takes place in the absence of any external forces possibly dragging the molecules into the membrane.

3. ELECTRICAL MEMBRANE PARAMETERS: DETERMINATION AND MEANING

The influence of externally applied proteins or drugs on the membrane can be manifold. Their adsorption to the membrane leaflet facing the side of their addition may influence the membrane potential profile due to changes in the electrostatic environment of the lipid bilayer. Furthermore, the interaction with the lipids may influence the state of order of the acyl chains of membrane lipids, resulting in their fluidization or rigidification depending, among other parameters, on the depth of intercalation of the molecules into the membrane and on the functional groups involved in the interaction. These interactions may, in turn, result in membrane permeabilization either by disturbance of the lamellar structure of the bilayer or by the formation of transmembrane pores.

Utilizing electrical measurements on planar bilayer membranes, the following membrane parameters can be monitored: i) applying a directed voltage $U_=$, an increase in membrane current I gives information on the formation of membrane pores/lesions. From the quotient $I/U_=$, the electrical conductance S of the membrane can be calculated. If the increase in membrane conductivity occurs in a stepwise manner, the height of the single steps gives the single channel conductivity Λ. The quotient Λ/σ, σ being the specific conductance of the bathing solution, is the so-called size parameter which determines the size of the formed membrane pores/lesions. Assuming a cylindrical shape of the membrane pores the pore diameter d can be calculated according to $d = 2 \, (\Lambda/\sigma \cdot l/\pi)^{1/2}$ [11], where l is the membrane thickness; ii) applying a rectangular voltage the membrane capacitance can be determined with the voltage-induced capacitance relaxation method [4,12,13] from the current following a voltage jump. Also other methods utilizing alternating voltages are used [14-16]. The membrane capacitance gives information on the thickness and the dielectric properties of the membrane and can be used to monitor the adsorption of molecules to the membrane surface [17]; iii) applying an alternating triangular voltage with a low sweep rate (about 4 mV/s) and an amplitude of about 180 mV or by stepwise switching to different voltages within this range the innermembrane potential profile (for detailed information see next section) can be derived from the transport characteristics of carrier molecules [14,18-21], and gating phenomena of transmembrane proteins can be investigated [21-26].

Electrostatic properties of the membrane are, besides their hydrophobic characteristics, of particular importance for the interaction of biomolecules with the membrane and may play a crucial role in many membrane-associated biological effects. In lipid bilayers, several distinct electrostatic potentials arising from different sources can be defined which superimpose to a characteristic intrinsic potential profile. These potentials are briefly summarized in the following.

Living cells actively maintain specific intracellular cation concentrations differing from the respective extracellular concentrations and resulting in a negative net charge inside the cells. The charge separation across the membrane causes a bulk-to-bulk potential difference of up to -100 mV. This transmembrane potential (resting membrane potential), which can be measured by microelectrodes, is important for the regulation of the activity of voltage sensitive ion channels [27]. Charged groups of lipid molecules (as phosphate-,carboxyl-, amino-groups) expressed at the membrane surface generate a surface potential which - according to Gouy and Chapman (for review see Cevc [28])- decreases exponentially within a distance of a few nanometers from the membrane surface. According to the lipid composition, surface potentials of biomembranes are typically negative in the order of a few tens of mV and play an important role in controlling biological processes in the immediate vicinity of bacterial and host cell membranes (e.g., repulsion of anionic or attraction and binding of polycationic compounds causing, in turn, alterations of the potential [29]). Changes of the surface potential of lipid aggregates (liposomes) can be followed by measuring the Zeta (ζ-) potential, the electrophoretically effective potential of the aggregates by laser-Doppler-anemometry [30] or by fluorescence spectroscopy using potential-sensitive dyes [31].

The largest contribution arises from the dipole potential (U_D) inside the lipid bilayer ($\leq$ +300 mV). Experiments as well as electrostatic calculations allow the conclusion that this potential is caused by oriented dipoles of bound water molecules [32]. Unlike the surface and the transmembrane potential, the internal dipole potential is independent on the ionic strength of the electrolyte solutions on both sides of the membrane. The dipole potential is responsible for observed differences in the passive permeability of cationic and anionic molecules through the membrane and for conformational changes in proteins, when they are incorporated into the membrane [33,34]. An additional energy barrier arises from the Born self-energy (Born-potential, U_B), which is the energy necessary to transfer a charge from a medium with a high dielectric constant ε (water) to one with a low ε (membrane) [14].

The transmembrane potential is the sum of the intrinsic membrane potential and the external potentials resulting from an ion gradient between both sides on the membrane and an externally applied clamp voltage (for review on membrane electrostatics see [28]).

Lipid asymmetry may provoke a potential difference between the two surfaces of the bilayer membrane. An extreme asymmetry in charge densities as well as in headgroup conformations occurs for asymmetric LPS/phospholipid-bilayers resembling the lipid matrix of the outer membrane of Gram-negative bacteria. At neutral pH, of the phospholipids of the inner leaflet of the outer membrane only the PG-molecule carries one negative charge, whereas each F515 LPS-molecule (outer leaflet) carries four negative charges. The resulting surface charge densities are - 0.52 As/m^2 (-3.25 e$_0$/nm^2) for F515 LPS and -0.05 As/m^2 (-0.31 e$_0$/nm^2) for the PL-mixture, respectively, since the molecular cross-section of an F515 LPS molecule is 1.23 nm^2 and that of a diacyl phospholipid 0.55 nm^2 (determined from monolayer isotherms with a film balance). From these surface charges, the surface potential can be calculated according to the Gouy equation [35].

Of the different contributions to the intrinsic membrane potential profile, only the difference in the innermembrane potential difference $\Delta\Phi$ [14] and changes in the height of the potential wall can be determined from the I/U-curves of carrier-doped bilayer membranes. For asymmetric phospholipid membranes, the innermembrane potential difference $\Delta\psi$ as determined by the method of inner-field compensation [36] is comparable to $\Delta\Phi$, however, in the case of asymmetric LPS/phospholipid membranes, the innermembrane regions for the determination of $\Delta\psi$ and $\Delta\Phi$ are different and, thus, lead to different results (own unpublished data).

4. ACTIVATION OF THE COMPLEMENT SYSTEM BY LPS-SURFACES

The complement system is an important part of the host defense against invading bacteria. Activation can occur *via* two distinct pathways, the classical and the alternative pathway, respectively. Both pathways finally result in the assembly of C5b-9 complexes (membrane attack complexes, MAC) that can insert into cell membranes and form pores (for review see [37-41]). Fully assembled complement lesions appear in the electron microscope as hollow cylinders with a diameter of approximately 10 nm and a length of 15 nm [42,43]. The complement pore is formed mainly from C9 molecules, with the other components occupying a peripheral position in the MAC [44].

Although the observation that fresh serum is capable of killing certain bacteria led to the discovery of the complement system, the mechanism of resistance against complement, which is an attribute of many Gram-negative strains, are poorly understood. Of particular importance is the role of LPS in complement activation. On the one hand it is known that the outer membrane of many Gram-negative bacteria as well as isolated lipopolysaccharides directly activate complement without the participation of antibodies [45,46], on the other hand LPS with a longer sugar chain [5,47] and capsular polysaccharides [48,49] are discussed as a physical barrier against terminal complement complex insertion. Furthermore, the questions, whether and *via* which pathway complement is activated by GSL have not yet been elucidated.

Concerning the complement activation by isolated LPS and lipid A in dependence on the length of the polysaccharide chain, fundamental works have been published by Vukajlovich et al. [50,51]. In their studies they found that LPS and lipid A lead to an antibody-independent activation of the complement cascade, but the pathway of activation depends on the length of the polysaccharide chain: lipid A and deep rough mutant LPS activate the complement system *via* the classical pathway, whereas lipopolysaccharides with a longer polysaccharide chain activate *via* the alternative pathway. Mey et al. [52] could show that the acylation of the lipid A of a *Klebsiella pneumoniae* LPS is important for the activation of the alternative pathway of the complement system and Clas et al. [53] pointed out the importance of the length of the sugar chains for the binding of C1q. The disadvantage of all these studies is the fact that LPS or lipid A were presented as aggregates in water not in their natural structural environment - as lipid bilayer.

With our reconstitution model of the lipid matrix of the outer membrane [13] we could on the one hand determine the activation pathway and on the other hand investigate the influence of the lipid matrix on the pore size [21,54]. Whereas no changes in conductance were detected when serum was added to the PL-side of the membrane, serum addition to the LPS Re-side led to an increase in membrane conductivity (Fig. 3). At the beginning of the conductance trace, the incorporation of

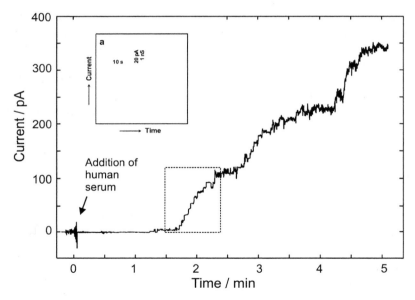

Fig. 3. Current trace after the addition of whole human serum to the lipopolysaccharide side of an asymmetric lipopolysaccharide/PL bilayer. The insert shows the stepwise formation of one complement pore from C9 monomers. Bathing solution: 100 mM KCl, 5 mM MgCl$_2$, 5 mM HEPES; pH7, T = 37°C. Clamp voltage: 20 mV, side of serum addition grounded. PL: mixture of phosphatidylethanolamine, phosphatidylglycerol, and diphosphatidylglycerol in molar ratios of 81:17:2. (Adopted from [54] with permission).

C9 monomers was reflected by a stepwise increase of the conductance - about nine steps - until the formation of one individual pore was completed and the assembly of the next pore begun. The small current steps either were of the same amplitude or increased in amplitude quadratically. From these current traces, a model for the complement pores could be derived which is based on the subsequent incorporation of about nine C9 monomers to form either a cylindrical pore (corresponding to the quadratic increase in conductance) or a so-called leaky patch which is only on one side surrounded by C9-monomers (corresponding to current steps of same amplitude). From the amplitude of the current increase related to the formation of one complement pore, a pore diameter of about 8 nm was determined [54]. Experiments with LPS differing in the polysaccharide moiety (from *S. enterica* sv. Minnesota strains R595, Rz and R5) and GSL-1 [21] led to same results.

In studies aiming at the elucidation of the pathways of the complement activation we found that neither the Ca^{2+}-chelator EGTA nor the depletion of C1q had an influence on the activation by GSL-1, whereas in the case of Re LPS a significant deactivation was observed. From these results we conclude that Re LPS activates complement predominantly *via* the classical pathway without the participation of antibodies, whereas GSL-1 activates the alternative pathway.

In recent investigations we could, furthermore, show that a higher fluidity of the lipid matrix increases pore formation by complement and the susceptibility of Gram-negative bacteria towards human serum [55]. This topic has been discussed controversially so far (e.g., [56,57]). Our investigations were, however, the first using a model membrane system closely resembling the lipid matrix of the natural outer membrane, and this model membrane activated the complement system directly.

5. EXAMPLES FOR THE INTERACTION OF ANTIBACTERIAL PEPTIDES WITH THE OUTER MEMBRANE

In this section we want to present some examples for the role of LPS in the interaction with the polycationic antibacterial peptides polymyxin B and the cathelicdin CAP18 (cationic antibacterial pepitide 18 kDa).

5.1. Polymyxin B
For the polycationic decapeptide polymyxin B (PMB), which possesses antibacterial activity against most genera of Gram-negative bacteria including *Escherichia*, *Salmonella*, or *Pseudomonas* at minimal inhibitory concentrations less than 2 µg/ml [58], the outer membrane is known to be the primary target [59]. A rapid increase of the membrane permeability with respect to charged or polar molecules of low molecular weight [60] and visible morphological alterations have been observed [61]. In the following decades, numerous studies have been reported

on the action of PMB including lipid monolayers [62,63], liposomal membranes [64-68], and black lipid membranes [41,69]. For these studies a variety of physical techniques such as electron spin resonance, nuclear magnetic resonance, fluorescence polarization, calorimetric methods, and X-ray diffraction were applied. These investigations provided mainly information on structural properties of PMB and on binding of PMB to lipids. As should be expected from the polycationic character of the antibiotic, strong interactions have been observed with negatively charged amphiphiles including phospholipids like phosphatidic acid [70] or PG [71] and, particularly, LPS [61,72]. From measurements of the electrical conductivity of planar lipid bilayers contradictory results were obtained. Antonov et al. [41] proposed a carrier mechanism for the action of PMB. In contrast, Miller et al. [41,69] found that the electrical resistance of BLM made from the negatively charged phosphatidylserine or an mixture of PE and PG only led to an unspecific destabilization of the membranes. In these studies, however, LPS has not been used as a membrane component.

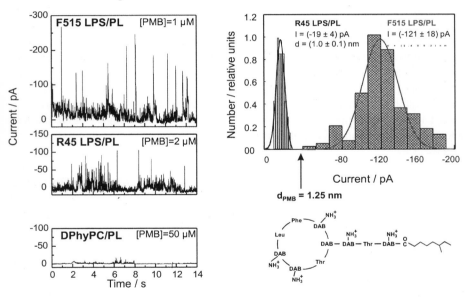

Fig. 4. Current fluctuations in the initial phase after the addition of Polymyxin B (PMB) to the lipopolysaccharide (LPS) side of membranes containing LPS from PBM sensitive (*E. coli* F515) and PMB-resistant (*P. mirabilis* R45) Gram-negative bacteria and to diphytanoylphosphatidylcholine/PL membranes (left); distributions of the amplitudes of the current fluctuations in LPS/PL membranes fitted by Gaussian distributions (solid curves) and chemical structure of PMB (right). Bathing solution: 100 mM KCl, 5 mM MgCl$_2$, 5 mM HEPES; pH7, T = 37°C. Clamp voltage: -80 mV, side of PMB addition grounded. PL: mixture of phosphatidylethanolamine, phosphatidylglycerol, and diphosphatidylglycerol in molar ratios of 81:17:2. (Adopted from [74] with permission).

Using asymmetric LPS/PL bilayers and adding PMB to the LPS side, single-channel current fluctuations occurred which then superimposed to a stationary current [73,74]. From a single-channel analysis, a correlation between the charge of the lipid molecules forming the membrane leaflet to which the PMB was added and the size of the membrane lesions was found (Fig. 4). For those membranes, which were formed from LPS from PMB-sensitive deep rough mutant strains (*E. coli* F515 and *S. enterica* sv. Minnesota R595), the lesions were large enough to allow the passage of the PMB molecules through the membrane. The lesions were, however, too small when LPS from the PMB-resistant *Proteus mirabilis* R45 was used [74]. These observations point out the importance of negative charges for PMB-sensitivity/resistance as discussed in previous publications [75-77]. The data clearly support the importance of the self-promoted uptake [78] of PMB (the drug molecules form the lesions for their own membrane passage) as a prerequisite for its activity. As could be seen using phopsholipid membranes (Fig. 4) PMB also is able to permeabilize the cytoplasmic membrane finally killing the bacteria (and, by the way, explaining its cytotoxicity). Furthermore, from the current traces, which were very similar to those found after the addition of detergents to the bilayer system, and the critical concentration dependence of the observed effects, a model for the PMB-induced pore formation was developed which is based on the detergent-like character of the antibiotic [73,74].

5.2. Cationic antibacterial peptide 18 kDa, CAP18

The host´s defense recruits antibacterial peptides and proteins against invading bacteria. They provide immediate protection by direct physico-chemical attack on the surface membrane of the microorganisms [79,80]. One group of these antibacterial proteins are the cathelicidins which have been identified in various mammals including humans [81]. They contain a highly conserved N-terminal domain called cathelin and a C-terminal domain that comprises an antimicrobial peptide. Human and rabbit cathelicidins are termed 18 kDa cationic antibacterial protein hCAP18 and rCAP18, respectively. The C-terminal domain of CAP18 exhibits LPS-binding, LPS-neutralizing, antibacterial, and anticoagulant activities [82]. A major difference in the interaction of human and rabbit CAP18 with cell membranes is observed in their effect on human red blood cells: the hCAP18-fragment FALL-39 (hCAP18$_{102-140}$) is hemolytic whereas rCAP18$_{106-142}$ is not [83]. Gram-negative bacteria such as *E. coli*, *S. enterica* sv. Minnesota and Typhimurium and Gram-positive bacteria such as *Streptococcus pneumoniae* and *Staphylococcus aureus* are known to be sensitive towards the antimicrobial action of rCAP18$_{106-142}$. There is, however, no activity of CAP18 against the Gram-negative strain *P. mirabilis*, fungi, and multiple-drug-resistant strains of *Mycobacterium avium* and *Mycobacterium tuberculosis* [84].

In the following, we discuss results obtained from electrical experiments on the molecular interaction between CAP18-derived peptides and reconstituted planar bilayers mimicking the symmetric phospholipid membrane of host's cells and the asymmetric outer membranes of Gram-negative bacteria.

A change of membrane capacitance could be observed only for membranes consisting of a negatively charged leaflet on the side of CAP18 addition, and no changes were observed for the neutral DPhyPC (Fig. 5, left). The decrease depended significantly on the LPS type used: in the case of F515LPS/PL and R595 LPS/PL membranes, minimal capacitance values were lower than that of the undisturbed bilayer, whereas in the case of R45 LPS/PL membranes the final value was higher. Furthermore, the effects were reduced in the case of R45 LPS/PL but not for F515 LPS/PL or R595 LPS/PL membranes (Fig. 5, left). It is known that the 4′-phosphate of the lipid A of the CAP18-sensitive *S. enterica* sv. Minnesota and the first Kdo of the LPS of the CAP18-resistant *P. mirabilis* are substituted by cationic L-Arap4N [85] (Fig. 2, right). The reduction of the molecular net charge from F515 LPS (-4 e_0), R595 LPS (-3.4 e_0) to R45 LPS (-3 e_0) and an increase of sterical hindrance could explain a lower binding stoichiometry between the R45 LPS and CAP18.

Changes induced in the transmembrane potential may influence the membrane permeability and with that the sensitivity of the organisms to antibiotics. Information on innermembrane potential changes was derived from I/U-curves of K^+-carrier-doped planar membranes in the absence and presence of CAP18. In the case of asymmetric F515 LPS/PL membranes, the addition of CAP18 causes a rise of the potential difference $\Delta\Phi$ from -35 mV to -30 mV. Similar results were obtained for R595 LPS/PL membranes. The addition of CAP18 to asymmetric bilayers made from R45 LPS from the resistant *P. mirabilis* led to a completely opposite effect, a decrease of $\Delta\Phi$ from -30 mV to -82 mV. This observation can be explained by the assumption that in the case of R45 LPS/PL membranes CAP18 only influence the negative charges at the surface of the R45 LPS leaflet, whereas the protein is inserted more deeply into the F515 LPS/PL and R595 LPS/PL bilayers, thus also acting on the dipole potential or even on the surface layer on the PL side [86].

CAP18 induced current fluctuations above a peptide- and lipid-specific clamp voltage V_{min} (Fig. 5, right). The transient lesions had lifetimes in the range of some ms to some min and diameters between 0.2 (detection limit) and 5 nm or, in some cases, even larger. Thus, current fluctuations are not caused by well-defined pores, but rather by irregular lesions or defects in the bilayer formed by CAP18 oligo- or multimers. A number of CAP18 molecules are forced into a transmembrane configuration by a positive clamp-voltage (cytoplasmic side is grounded). Lateral diffusion of the protein molecules in the membrane leads to their aggregation and formation of a so-called multipore clusters [87] with heterogeneous diameters. At low numbers, the lesions destabilize the planar membranes, and at higher numbers they disrupt their functional integrity. For the bacterial organism this implies that

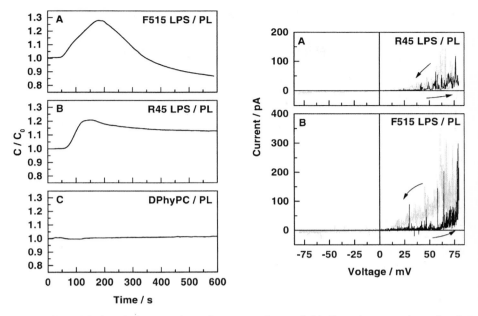

Fig. 5. CAP18-induced changes of membrane capacitance (left). Capacitance vs time after the addition of 520 nM CAP18 to asymmetric bilayers (the side of CAP18 addition is named first): (A) F515 LPS/PL bilayer; (B) R45 LPS/PL; (C) DPhyPC/PL. Voltage dependence of current fluctuations (right). Current/voltage curves for asymmetric bilayers after addition of 520 nM CAP18 to the lipopolysaccharide (LPS) side. (A) F515 LPS/PL bilayer (B) R45 LPS /PL. Bathing solution: 100 mM KCl, 5 mM MgCl$_2$, 5 mM HEPES, pH 7, T = 37 °C. F515 LPS: LPS from deep rough mutant strain *E. coli* F515; R595 LPS: LPS from deep rough mutant strain *S. enterica* sv. Minnesota R595; R45 LPS: LPS from deep rough mutant strain *P. mirabilis* R45; DPhyPC: diphytanoylphosphatidylcholine; PL: mixture of phosphatidylethanolamine, phosphatidylglycerol, and diphosphatidylglycerol in molar ratios of 81:17:2. (Adopted from [86] with permission).

CAP18, after having destroyed the permeation barrier of the outer membrane, could induce lethal leakages in the cytoplasmic membrane.

The V_{min} necessary for the induction of these lesions differ significantly for the peptides and also for membranes composed of different LPS. V_{min} is lowest for combinations of membranes and CAP18 peptides which allow the highest degree of peptide intercalation. In Gram-negative bacteria, the periplasmic space is strongly anionic as compared to the external medium, mainly due to the presence of anionic membrane-derived oligosaccharides (MDO) [88]. The MDO contribute to the Donnan potential across the outer membrane, and this contribution was determined for *E. coli* in the presence of an external cation concentration of 100 mM to be V_{OM} = 26 mV (inside negative) [89]. This value is in good agreement with two observations: (i) V_{min} obtained from experiments using rCAP18$_{106-137}$ is lower for asymmetric bilayers representing the lipid matrix of the CAP18-sensitive bacteria (F515 LPS/PL and R595 LPS/PL membranes) and that for R45 LPS/PL membranes, representing the lipid matrix of the resistant strain *P. mirabilis* R45, is higher than V_{OM}; (ii) All antibacterial peptides induce lesions in F515 LPS/PL membranes at clamp-voltages below V_{OM}, and all inactive have a $V_{min} > V_{OM}$. Moreover, in *in vitro* experiments on bacterial killing of *E. coli* J5, the 50 % inhibition concentration for the antibacterial peptides correlate linearly with V_{min} of the respective peptides. V_{min} discriminates between active and inactive antibacterial peptides and, moreover, between sensitive and resistant bacterial strains against rCAP18$_{106-137}$. Polycationic peptides kill Gram-negative bacteria in a two step mechanism: In a first step, they permeabilize the OM for their own "self-promoted" translocation (see above). In the second step, they permeate the IM, thus allowing ions to equilibrate across the cytoplasmic membrane and to reduce or destroy the membrane potential (depolarization). The observation that the active CAP18-derived peptides also induce lesions in DPhyPC/PL membranes at clamp-voltages > 60 mV (outside positive), which are significantly lower than the potential gradient of about 150 mV across the cytoplasmic membrane [90], clearly shows that the peptides can induce permeabilization of the IM which then results in bacterial killing.

The data provide clear evidence that the interaction between antibacterial peptides and membranes is determined by the physico-chemical properties of the peptides as well as by the chemical properties and the transmembrane potential of the target membranes.

6. CONCLUSIONS

The examples outlined above clearly reveal the superiority of the Montal-Mueller method, which allows the preparation of symmetric or asymmetric lipid bilayer of nearly any composition, when questions with respect to the lipid specificity or the innermembrane potential for molecular interactions with membranes are to be answered. This superiority becomes particularly evident when the membrane of interest is extremely asymmetric as in the case of glycolipid/phospholipid bilayers.

REFERENCES

[1] H. Nikaido and M. Vaara, Microbiol. Rev., 49 (1985) 1.
[2] P. Mueller, D. O. Rudin, and H. T. Tien, Nature, 194 (1962) 979.
[3] H. T. Tien, Bilayer Lipid Membranes (BLM): Theory and Practice, Marcel Dekker, New York, (1974).
[4] M. Montal and P. Mueller, Proc. Natl. Acad. Sci. USA, 69 (1972) 3561.
[5] K. A. Joiner, C. H. Hammer, E. J. Brown, R. J. Cole, and M. M. Frank, J. Exp. Med., 155 (1982) 797.

[6] S. H. White, The Physical Nature of Planar Bilayer Membranes C. Miller (ed.), pp.3-35, New York, (1986).
[7] M. J. Osborn, J. E. Gander, E. Parisi, and J. Carson, J. Biol. Chem., 247 (1972) 3962.
[8] U. Zähringer, B. Lindner, and E. T. Rietschel, Adv. Carbohydr. Chem. Biochem., 50 (1994) 211.
[9] K. Kawahara, U. Seydel, M. Matsuura, H. Danbara, E. Th. Rietschel, and U. Zähringer, FEBS Lett., 292 (1991) 107.
[10] A. Wiese and U. Seydel, Methods Mol. Biol., 145 (2000) 355.
[11] R. Benz, K. Janko, W. Boos, and P. Läuger, Biochim. Biophys. Acta, 511 (1978) 305.
[12] R. Benz and K. Janko, Biochim. Biophys. Acta, 455 (1976) 721.
[13] U. Seydel, G. Schröder, and K. Brandenburg, J. Membr. Biol., 109 (1989) 95.
[14] P. Schoch, D. F. Sargent, and R. Schwyzer, J. Membr. Biol., 46 (1979) 71.
[15] S. Toyama, A. Nakamura, and F. Toda, Biophys. J., 59 (1991) 939.
[16] C.-C. Lu, A. Kabakov, V. S. Markin, S. Mager, G. A. Frazier, and D. W. Hilgemann, Proc. Natl. Acad. Sci. USA, 92 (1995) 11220.
[17] I. N. Babunashvili, A. Y. Silberstein, and V. A. Nenashev, Studia Biophys., 83 (1981) 131.
[18] R. Latorre and J. E. Hall, Nature, 254 (1976) 361.
[19] J. E. Hall and R. Latorre, Biophys. J., 15 (1976) 99.
[20] U. Seydel, W. Eberstein, G. Schröder, and K. Brandenburg, Z. Naturforsch., 47c (1992) 757.
[21] A. Wiese, J. O. Reiners, K. Brandenburg, K. Kawahara, U. Zähringer, and U. Seydel, Biophys. J., 70 (1996) 321.
[22] D. A. Haydon, Ann. N. Y. Acad. Sci., 264 (1975) 2.
[23] L. K. Buehler and J. P. Rosenbusch, Biochem. Biophys. Res. Commun., 190 (1993) 624.
[24] M. Brunen and H. Engelhardt, Eur. J. Biochem., 212 (1993) 129.
[25] A. Wiese, G. Schröder, K. Brandenburg, A. Hirsch, W. Welte, and U. Seydel, Biochim. Biophys. Acta, 1190 (1994) 231.
[26] H. Kleivdal, R. Benz, and H. B. Jensen, Eur. J. Biochem., 233 (1995) 310.
[27] E. K. Gallin and L. C. McKinney, Monovalent Ion Transport and Membrane Potential Changes During Activation in Phagocytic Leukocytes S. Grinstein and O. D. Rotstein (eds.), Vol. 35, pp.127-152, New York, (1990).
[28] G. Cevc, Biochim. Biophys. Acta, 1031 (1990) 311.
[29] G. Beschiaschvili and J. Seelig, Biochemistry, 29 (1990) 10995.
[30] G. Cevc, Chem. Phys. Lipids, 64 (1993) 163.
[31] D. Cafiso, A. McLaughlin, S. McLaughlin, and A. Winiski, Meth. Enzymol., 171 (1989) 342.
[32] C. Zheng and G. Vanderkooi, Biophys. J., 63 (1992) 935.
[33] J. C. Franklin and D. S. Cafiso, Biophys. J., 65 (1993) 289.
[34] D. Cafiso, Curr. Opin. Struct. Biol., 1 (1991) 185.
[35] S. McLaughlin, Annu. Rev. Biophys. Biophys. Chem., 18 (1989) 113.
[36] V. S. Sokolov and V. G. Kuz'min, Biofizika, 25 (1980) 170.
[37] S. Bhakdi and J. Tranum-Jensen, Biochim. Biophys. Acta, 737 (1983) 343.
[38] H. J. Müller-Eberhard, The Membrane Attack Complex H. J. Müller-Eberhard (ed.), pp.227-278, New York, (1985).
[39] S. Bhakdi and J. Tranum-Jensen, Rev. Physiol. Biochem. Pharmacol., 107 (1987) 148.
[40] A. F. Esser, Toxicology, 87 (1994) 229.
[41] V. F. Antonov, E. A. Korepanova, and Y. A. Vladimirov, Studia Biophys., 58 (1976) 87.

[42] J. Tranum-Jensen, S. Bhakdi, B. Bhakdi-Lehnen, O. J. Bjerrum, and V. Speth, Scand. J. Immunol., 7 (1978) 45.

[43] J. Tranum-Jensen and S. Bhakdi, J. Cell Biol., 97 (1983) 618.

[44] J. Tschopp, J. Biol. Chem., 259 (1984) 7857.

[45] D. C. Morrison and L. F. Kline, J. Immunol., 118 (1977) 362.

[46] S. W. Vukajlovich, J. Hoffman, and D. C. Morrison, Mol. Immunol., 24 (1987) 319.

[47] K. A. Joiner, C. H. Hammer, E. J. Brown, and M. M. Frank, J. Exp. Med., 155 (1982) 809.

[48] S. Merino, S. Camprubí, S. Albertí, V.-J. Benedí, and J. M. Tomás, Infect. Immun., 60 (1992) 2529.

[49] C. K. Ward and T. J. Inzana, J. Immunol., 153 (1994) 2110.

[50] S. W. Vukajlovich, Infect. Immun., 53 (1986) 480.

[51] S. W. Vukajlovich, P. Sinoway, and D. C. Morrison, EOS J. Immunol. Immunopharmacol., 6 (Suppl.3) (1986) 73.

[52] A. Mey, D. Ponard, M. Colomb, G. Normier, H. Binz, and J.-P. Revillard, Mol. Immunol., 31 (1994) 1239.

[53] F. Clas, G. Schmidt, and M. Loos, Curr. Top. Microbiol. Immunol., 121 (1985) 19.

[54] G. Schröder, K. Brandenburg, L. Brade, and U. Seydel, J. Membr. Biol., 118 (1990) 161.

[55] A. Wiese, P Grünewald, K.-J. Schaper, and U. Seydel, J. Endotoxin Res., 7 (2001) 147.

[56] T. R. Hesketh, S. N. Payne, and J. H. Humphrey, Immunology, 23 (1972) 705.

[57] D. V. Devine, K. Wong, K. Serrano, A. Chonn, and P. R. Cullis, Biochim. Biophys. Acta, 1191 (1994) 43.

[58] D. R. Storm and K. Rosenthal, Ann. Rev. Biochem., 46 (1977) 723.

[59] M. Vaara and T. Vaara, Antimicrob. Agents Chemother., 24 (1983) 114.

[60] P. R. G. Schindler and M. Teuber, J. Bacteriol., 135 (1978) 198.

[61] P. R. G. Schindler and M. Teuber, Antimicrob. Agents Chemother., 8 (1975) 95.

[62] A. Theretz, J. Theissie, and J. F. Tocanne, Eur. J. Biochem., 142 (1984) 113.

[63] G. Beurer, F. Warncke, and H.-J. Galla, Chem. Phys. Lipids, 47 (1988) 155.

[64] M. Imai, K. Inoue, and S. Nojima, Biochim. Biophys. Acta, 375 (1975) 130.

[65] J. L. Ranck and J. F. Tocanne, FEBS Lett., 143 (1982) 175.

[66] E. Mushayakarara and I. W. Levin, Biochim. Biophys. Acta, 769 (1984) 585.

[67] F. Sixl and A. Watts, Biochemistry, 24 (1985) 7906.

[68] D. S. Feingold, C. C. HsuChen, and I. J. Sud, Ann. N. Y. Acad. Sci., 235 (1974) 480.

[69] I. R. Miller, D. Bach, and M. Teuber, J. Membr. Biol., 39 (1978) 49.

[70] P. Kubesch, J. Boggs, L. Luciano, G. Maass, and B. Tümmler, Biochemistry, 26 (1987) 2139.

[71] J. M. Boggs and G. Rangaraj, Biochim. Biophys. Acta, 816 (1985) 221.

[72] A. Peterson, R. E. W. Hancock, and E. J. McGroarty, J. Bacteriol., 164 (1985) 1256.

[73] G. Schröder, K. Brandenburg, and U. Seydel, Biochemistry, 31 (1992) 631.

[74] A. Wiese, M. Münstermann, T. Gutsmann, B. Lindner, K. Kawahara, U. Zähringer, and U. Seydel, J. Membr. Biol., 162 (1998) 127.

[75] K. Nummila, I. Kilpeläinen, U. Zähringer, M. Vaara, and I. M. Helander, Mol. Microbiol., 16 (1995) 271.

[76] I. M. Helander, Y. Kato, I. Kilpeläinen, R. Kostiainen, B. Lindner, K. Nummila, T. Sugiyama, and T. Yokochi, Eur. J. Biochem., 237 (1996) 272.

[77] G. Seltmann, B. Lindner, and O. Holst, J. Endotoxin Res., 3 (1996) 497.

[78] R. E. W. Hancock, Ann. Rev. Microbiol., 38 (1984) 237.

[79] R. E. W. Hancock, The Lancet, 349 (1997) 418.

[80] J. M. Schröder, Biochem. Pharmacol., 57 (1999) 121.

[81] M. Zanetti, R. Gennaro, and D. Romeo, FEBS Lett., 374 (1995) 1.

[82] M. Hirata, J. Zhong, S. C. Wright, and J. W. Larrick, Prog. Clin. Biol. Res., 392 (1995) 317.

[83] S. M. Travis, N. N. Anderson, W. R. Forsyth, C. Espiritu, B. D. Conway, E. P. Greenberg, P. B. McCray, Jr., R. I. Lehrer, M. J. Welsh, and B. F. Tack, Infect. Immun., 68 (2000) 2748.

[84] J. W. Larrick, M. Hirata, Y. Shimomoura, M. Yoshida, H. Zheng, J. Zhong, and S. C. Wright, Antimicrob. Agents Chemother., 37 (1993) 2534.

[85] W. Kaca, J. Radziejewska-Lebrecht, and U. R. Bhat, Microbios, 61 (1990) 23.

[86] T. Gutsmann, J. W. Larrick, U. Seydel, and A. Wiese, Biochemistry, 38 (1999) 13643.

[87] D. S. Cafiso, Annu. Rev. Biophys. Biomol. Struct., 23 (1994) 141.

[88] R. Benz, Uptake of Solutes Through Bacterial Outer Membranes J.-M. Ghuysen and R. Hakenbeck (eds.), pp.397-423, Amsterdam, (1994).

[89] K. Sen, J. Hellman, and H. Nikaido, J. Biol. Chem., 263 (1988) 1182.

[90] E. P. Bakker and W. E. Mangerich, J. Bacteriol., 147 (1981) 820.

Planar Lipid Bilayers (BLMs) and their Applications
H.T. Tien and A. Ottova-Leitmannova (Editors)

Chapter 21

Insights into ion channels from peptides in planar lipid bilayers

H. Duclohier

Interactions Cellulaires et Moléculaires, UMR 6026 CNRS –
Université de Rennes I, Campus de Beaulieu, 35042 Rennes, France*

1. INTRODUCTION AND BRIEF HISTORICAL ACCOUNT

A few years after Hodgkin & Huxley's classical work [1] and the emerging notion of ion channels whose protein nature was still a matter of conjecture, the advent of methods to form reliable artificial membranes and foremost planar lipid bilayers (PLBs or BLMs : black lipid membranes) in the early 1960s [2] was an important breakthrough. Without impinging on the special introductory chapter by one the pioneer, H. Tien, devoted to the historical aspects of BLMs development, the demonstration of membrane excitability by EIM (excitability inducing material) [3], then by the better defined alamethicin (a natural 20-residue long peptide) [4] had great impact for many biophysicists. The straightforward methods to form these bimolecular lipid membranes, especially the folded ones apposing two monolayers that was later proposed [5] certainly played a role in the final acceptance of the 'fluid mosaic model' [6] for biomembranes, incidentally published the same year (1972). In the seventies, work continued with pore- or channel-peptides, especially gramicidin and alamethicin (for reviews written at the end of this period, see [7, 8]).

Despite these two peptides being endowed with quite different structural and functional attributes and also different from the important physiological ion channels which were intensively studied through electrophysiology and whose aminoacid sequence will only be known in the eighties, the aim of these studies were three-fold. First, to investigate the physical chemistry of lipid-peptide/protein interactions [9, 10], secondly, to give a molecular basis for the action of antimicrobial peptides [11, 12]. Finally, a recurrent aim was to contribute to decipher either voltage-sensitivity or ion selectivity of ion channels and membrane receptors [13]. Inded, if the functional properties of these membrane proteins are relatively well understood, high-resolution structures are

* Tel./Fax : +33 299 28 61 39 ; e-mail : Herve.Duclohier@univ-rennes1.fr

still scanty despite some recent and remarkable achievements albeit mostly on prokaryotic ion channels and receptors (see [14-18]).

An alternative was -and still is- the so-called «peptide approach or strategy» which consists in functional and structural assays of synthetic peptides designed to reproduce or mimick some of the hypothetically-important transmembrane segments of these proteins. This approach can be considered as an off-shot of both the above-mentioned studies on antibiotic peptides and of reconstitution studies of purified ion channels and receptors. Indeed, PLBs or BLMs had already been extensively used for the characterization of functional properties of purified membrane transport proteins (for reviews, see e.g. [19, 20]. The easy access to both sides of the channels, the negligible 'series resistance', the control of lipid bilayer composition are among the factors that greatly contributed to the success of these reconstitution studies. After the elucidation of their aminoacid sequences and the first topological models in the eighties (see e.g . [21-23] as regards the voltage-dependent sodium channel), biophysicists took advantage of the availability of planar bilayers methods, sensitive electrophysiological methods as patch-clamp, solid-phase peptide synthesis, biochemical methods of purification and also spectroscopies (NMR, FTIR and Circular Dichroism, essentially) to assess their conformation.

Transmembrane segments and fragments (i.e. longer than 20-22 residues) issued from voltage-gated channels (sodium, potassium and calcium) as well as ligand-gated ones (acetycholine and glycine receptors) were thus studied, e.g. [24-33]. Alternatively, synthetic peptides designed from 'scratch' or along first chemical principles were shown to form e.g. proton channels [34]. These methods and results having been described in excellent previous reviews, e.g. [35-37], we shall now concentrate on our own contribution with the voltage-sensors and P-regions (forming the selectivity filter) of all the four homologous domains of the voltage-dependent sodium channel. Our previous reviews on the topic ought to be mentioned [38-40]. After summarizing the methods and main data, we shall criticaly assess the contributions of the peptide strategy in the light of recent functional and structural findings obtained on whole channels.

2. VOLTAGE-GATED ION CHANNELS : THE ACTION POTENTIAL SODIUM CHANNEL AS A CASE STUDY

Underlying action potential initiation and propagation, voltage-gated sodium channels are of paramount importance in the physio-pathology of the nervous and muscular systems (see e.g. [41, 42]). Sodium channels from excitable membranes are made of one main subunit (α) and several accessory and regulatory subunits (β_i) except in the electric eel sodium channel. The α-subunit underlying the main functional attributes, i.e. voltage-sensitivity and ion selectivity, is composed of four homologous domains made up of six transmembrane segments and a loop between S5 and S6 (for a recent review, see

[43]. These channels have so far resisted to structural elucidation by crystallographic means. On the other hand, their function has been extensively studied through electrophysiology: they are dynamic molecules characterized by dramatic rapid voltage-driven conformational changes that switch the molecule between activated, inactived and closed resting states [44].

A body of experimental evidence, essentially based on site-directed mutagenesis and chimeras constructs, support the notion that these channels can be considered as made of two main structural and functional modules [45]. The N-terminal part of the protein up to S4 and possibly the short cytoplasmic link between S4 and S5 is the voltage-sensing module whereas the remaining of the molecule, from S5 to the C-terminus, is the pore module. Within each of these modules, a greater contribution to i) voltage-sensing is brought about by S4 (four to eight positive charges), and to a lesser extent by S2 and S3 (three negative charges), and to ii) ion selectivity by the P-regions, i.e. the short intramembrane loops between S5 and S6.

The role of the highly positively-charged S4s (arginine residues every third position) as main voltage-sensors had been substantiated in site-directed mutagenesis on recombinant sodium channels heterologously expressed in oocytes. The slope of the activation curve vs. voltage and the gating charges moving upon depolarization were found to be correlated with the number of arginines, after their replacement by neutral residues [46]. Likewise, mutations in the P-regions, the once putative selectivity filter, strongly modulate ion selectivity, in particular the replacement of a single alanine by an aspartic acid residue can confer calcium selectivity to a 'sodium' channel [47].

The peptide strategy pushes further the apparent 'reductionist' point of view mentioned above, that is of individual functional modules coupled to structural units of the channel architecture, since it aims at a molecular dissection: not only determining the functional role of single transmembrane segments in lipid bilayers but also their structure. It assumes that the structures of these elements are roughly maintained in the whole channel, a dire assumption indeed but that is proving to be justified in the face of former as well as recent evidence (e.g. [48-50]) . However, such an approach needs to be as extensive as possible, i.e. in the case of the voltage-dependent sodium channel, the apparent homologous segments from the different domains have to be studied and compared, as well as longer segments or fragments prone to fold in the bilayer. Finally, there should be coherence between single-channel and macroscopic currents levels of analysis, and attempts to reconstruct or mimick the whole from interacting parts.

For our part, we endeavoured to study two peptide families :

- the four isolated main voltage sensors extended with the short cytoplasmic linker between transmembrane segments S4 and S5, hence the denomination iS4L45 used thereafter, i (in roman numerals) refering to one of the four homologous domains of the electric eel *Electrophorus electricus* voltage-gated sodium channel.

- the four P segments (formerly called SS1-SS2), intramembrane loops making up the main selectivity filter, and forming with the cytoplasmic ends of helical segments S5 and S6 the pore lumen.

   The sodium channel from *Electrophorus electricus* was selected since, not only it was the first one to be purified and functionally reconstituted in lipid vesicles and planar lipid bilayers [51-53], but also it was the first to be cloned and sequenced [21]. In addition, the electric eel sodium channel offers the advantage of being simpler than most of the other channels subsequently sequenced since it is only composed of the main α–subunit, i.e. without the smaller β_i subunits. On the other hand, and contrasting with most of the members of this family, this channel proved later that it could not be heterologously expressed from cDNA or mRNA such that comparisons between the effects of residues substitution in our peptides with electrophysiological data of mutated channels are not straightforward. Hence, a few other peptides derived from the human skeletal muscle sodium channel (hskm1 or $Na_v1.4$ [54-55]) were more recently studied.

3. METHODS AND TECHNIQUES OF THE PEPTIDE STRATEGY

As mentioned above, the peptide strategy to ion channel structure and function involves the following sequential steps : 1) selection of the aminoacid sequences of those channel segments that are thought to be important –from models and/or electrophysiology- in either gating or ion selectivity; 2) chemical synthesis, purification and verification of the peptides; 3) functional assays with electrical measurements in planar lipid bilayers interacting with these peptides; 4) finally, conformational or structural studies. This experimental approach will be detailed below with examples drawn from our studies on the voltage-dependent sodium channel.

3.1. Synthetic peptides, selection and preparation

   The sequences of the four P-peptides and the four S4L45 mentioned above, and issued from the whole sequence of electric eel sodium channel sequence [21], are shown in Fig. 1. The N- and C-termini limits of each segment were chosen as the best fit (or average) between several topological models which differ by only a few residues.

Selective pore-formers :

PI	GYTNYDNFAWTFLCLFRLML<u>QD</u>Y<u>W</u>ENLYQMT
PII	HMNDFFHSFLIVFRAL<u>CGEW</u>IETMWDCMEVG
PIII	VRWVNLKVNYDNAGMGYLSLLQVST<u>FKGW</u>MDIMYA
PIV	NFETFGNSMICLFEITTS<u>AGW</u>DGLLLPTLNTG
PR	DLAFFYSELNTWTVLDWGINLQMYRCFMGVG

Voltage-sensors :

IS4L45	SAL**R**TF**RV**L**R**ALKT I T I F*P*GLKT - IV**R**ALI ESM**K**Q
IIS4L45	SVL**R**SL**R**LL**R** I F**K**LA**K**SW*P*TLN - I LI **K** I I CNSVGA
IIIS4L45	GAI**K**NL**R**TI **R**AL**R** *P* L**R**ALS**R**FEG - M**K**VV**R**ALLGA
I9A,L15A	-------------- A ---------- A --
L15A	--------------------------- A --
IVS4L45	TLF**RV** I **R**LA**R** IA**RV**L**R**LI**R**AA**KG** – I**R**TLLFALMMS
IVhskm1	LF**RV** I **R**LA**R** IG **RV**L**R**LI**R**GA**KG** - I**R**TL
R1457A	LF**RV** I **R**LA**R** IG *P*VL**R**LI**R**GA**KG** – I**R**TL
VSR	LTIM**K**S*P***R**INLVLFC**R**LIAESAL**KL**VAFT**R**G**RR**A

Fig. 1. Aminoacid sequences of the peptides reviewed here (one-letter code) and taken from the electric eel sodium channel, except for IVhskm1 (voltage-sensor from domain IV of the human skeletal muscle sodium channel). In P-peptides, the presumed β-bends are underlined with the signature of the selectivity filter (DEKA) in bold. In S4L45s, positively-charged residues are in bold and proline is underlined bold. The approximate boundary between transmembrane segment S4 and L45, the linker between S4 and S5, is indicated by an hyphen. PR and VSR stands for scrambled or random peptides used as controls for both families.

Voltage-sensors peptides (S4 *stricto sensu*) were extended towards the C-termini with linkers L45 since, apart from a previous report [56] dealing with IVS4, these linkers present other positively-charged residues. In addition, at least for the *Shaker* K$^+$, there were preliminary evidence that these L45s might be involved in the inner exit of the pore [57] and that they may also form a 'receptor site' for the inactivation gate [58].

Peptides were chemically-synthesized by the Merrified solid-phase method [59] on an automatic apparatus (Applied Biosystems, model 433A) with initially the t-Boc strategy then with fmoc strategy, the latter allowing a more gentle cleavage of the final product from the resin (for review, see [60]). N- and C-termini were acetylated and amidated, respectively. Peptides were purified by high performance liquid chromatography (HPLC) using C18 columns and acetonitile/water gradient with 1% trifluoacetic acid. The molecular mass and identity of the final purified products was checked by MALDI-TOF and FAB-mass spectrometries.

3.2. BLMs or PLBs, the different configurations for electrical measurements

When the acronym BLM was proposed in the early sixties, it stood either for 'Bimolecular Lipid Membranes' or more commonly 'Black Lipid Membranes' [2] since the thinning to standard bilayer thickness of decane-painted lipids over the aperture was followed optically. With the advent of 'virtually solvent free bilayers' or folded bilayers [5], let alone with bilayers formed at the tip of patch pipettes [61, 62], it is more convenient to assess bilayer formation through capacitance monitoring. Moreover, the term 'membranes' can be misleading, hence at the risk of looking iconoclast, we prefer to use the term 'Planar Lipid Bilayers' (PLBs). An excellent practical handbook on bilayers techniques ought to be mentioned [63].

Supplementing flux studies in reconstituted proteolipisomes and extending the analysis towards molecular events, a thorough investigation of conductance properties induced by pore- or channel-forming peptides in PLBs should deal both with macroscopic and single-channel conductances. The first level of analysis yields more readily voltage-dependence or ohmic behavior, concentration-dependence and ion specificity (cations vs. anions) and selectivity (within the class of preferred ions). As for the single-channel configuration, not only the single-channel condutance amplitude – or amplitudes in case of open sub- or multi-states – is available, but also mean open and closed durations (*via* conductance and mean times histograms or distribution) as well as open probability P_o. There should be consistency between both levels of analysis, as regards for instance the voltage-dependence of the exponential branch of macroscopic Current-Voltage curves and P_o as a function of applied voltage.

- Methods for macroscopic conductance

An aperture of 100-250 μm diameter is made in a 10-25 μm thick PTFE film through an electric spark discharge. The film is then sandwiched between two half glass cells of 2 mL each clipped together with a minute amount of vacuum grease assuring etancheity. In our hands, we found glass a much more convenient material, easier to clean than the much more commonly-used teflon. Besides, due to the hydrophobicity of the latter, higher concentration of channel-forming material are needed in order to deliver a given conductance. The aperture in the film is made more hydrophobic by treating it with a few μL of an hexadecane/hexane (1/25) that are allowed to dry. Each half glass cell is then filled with ~ 2 mL of electrolytic solution (usually 0.5 M NaCl, 10 MM Hepes, pH 7.4, filtered through 25 μm) ensuring the levels both sides are equivalent and that the hole is submerged. The chamber is then transferred to the conductance set-up with Ag/AgCl electrodes dipping in each compartment (see Fig. 2).

Fig. 2. Glass chamber used by the author for 'folded' bilayers. The front or *cis*-side (on the right here) to which peptides will be later added is electrically connected *via* a headstage (inside the shielded *dural* box) to the patch-clamp amplifier -current-voltage converter and voltage generator). The back or *trans*-side (on the left here) is grounded.

Before any lipid is applied, it is wise to check that the 'way is clear' through the aperture, i.e. the electrical circuit is open. A triangular voltage waveform of amplitude ~ 20 mV and frequency ~ 100 Hz is applied to check the open or closed circuit and capacitance values after bilayer formation. 10-20 µL of a lipid (a mixture, whose proportions are indicated in the examples shown later, of 1-palmitoyl, 2-oleoyl-phosphatidylcholine -POPC- and 1,2 dioleoyl-phosphatidylethanolamine -DOPE-; from Avanti Polar Lipids, Alabaster, AL, USA) solution 10 mg/mL in hexane are spread on the liquid interfaces both sides, and allowed to dry to form monolayers. The electrolytic solution level is then lowered, past the hole, then raised back to its initial level, whilst monitoring capacitance on the oscilloscope. Planar bilayers can be readily obtained on first trial, but several 'pump and flush' manipulations may be necessary as well as raising the lipid concentration. After checking bilayer stability and electrical silence (under applied voltage), peptides are added from stock solutions to the *cis*-side, and the bilayer is submitted to slow (5 mV/s.) voltage ramps.

- Methods for single-channel conductance

The same protocole as above –designed to record the activity of thousands of channels– may be used but, in order to increase the signal/noise (S/N) ratio for a good resolution of single-channel events, it may be necessary to reduce the aperture diameter (still respecting a minimal aperture diameter/film thickness of about 3-4 to ensure bilayer formation) and to reduce the aqueous peptide concentration. We found more expedient to use the 'tip-dip' method, i.e. forming bilayers at the tip of patch-clamp pipettes [61, 62]. The much reduced bilayer area leads to a drastic reduction in capacitance-associated noise and thus a favorable S/N. A smaller glass cell (1 mL beaker) for the bath and a lipid solution at reduced concentration (1 mg/mL in hexane) are used. In the course of an experiment, we employed either the droplet technique applying a few microliters of the lipid solution to the micropipette shank just dipping at the air-salt solution interface, or the patch pipette is gently raised and dipped again at the interface by means an hydraulic micromanipulator.

In both configurations, a shielded Faraday box encloses chamber, micromanipultors, headstage or pre-amplifier. The remainder of the set-up is quite similar to standard modern electrophysiology equipment : voltage generator, oscilloscope, patch- and whole cell voltage-clamp amplifier, low-pass and high-pass Bessel filters, analog/digital converter interface, digital recorder on magnetic cartridges, and PC computer for acqusition and analysis.

Fig. 3. Electrical set-up, enclosed in a shielded rack, for planar bilayers experiments

3.3. Associated spectroscopies to assess secondary structures

Complementing the methods and techniques described above, various spectroscopies including Fourier-Transform InfraRed (FTIR) and Circular Dichroism (CD)might be implemented to evaluate the secondary structures, both in organic solvents and in lipid systems, of the peptides under investigation. As a rule, we employed CD which is more versatile for measuring helicities. Since these methods are not strictly relevant to the scope of this chapter, the interested reader is referrred to the excellent textbook edited by G.D. Fasman [64] and to our research papers, e.g. [65, 66]. Conformational contents are estimated within 5% from 51 mean residue ellipticity values (every nanometer, from 190 to 240 nm) and computed using published standards [67].

4. CONDUCTANCE ASSAYS OF TRANSMEMBRANE SEGMENTS INTO PLANAR LIPID BILAYERS

4.1. P-peptides and the selectivity filter

All four P-peptides, loops between transmembrane segments S5 and S6 that were assumed to form the main selectivity filter, have been assayed for conductances developed in planar lipid bilayers [66]. A control random peptide (whose aminoacid content was averaged from the composition of the four specific sequences but with a random order, Fig. 1) did not yield any appreciable conductance. As expected, no voltage-dependence was revealed apart for PIII

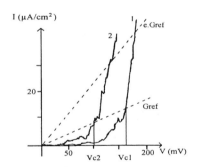

Fig. 4. Macroscopic current-voltage (I-V) curves for two aqueous PIII concentrations (*cis*-side) in the range of 1 mM, curve 2 being the response after doubling the concentration that yielded curve 1 (from [66]). The bilayer lipid mixture is POPC/DOPE (3/7, w/w) and the solution both sides is buffered 0.5 M NaCl. Dashed lines represent a reference conductance of 12.5 nS, allowing to define characteristic voltages Vc1 and Vc2, and e.Gref an e-fold change in conductance. If Ve is the voltage increment resulting in an e-fold conductance change on either exponential branch and Va the shift of characteristic voltages Vc for an e-fold change in peptide aqueous concentration, then an analysis drawn from alamethicin [68] allows to derive the apparent and mean number of monomers per conducting aggregate <N>=Va/Ve. Here <N> is 4.8, the averages value over the four P-peptides is close to 4.

for which both macroscopic Current-Voltage curves (Fig. 2) and single-channel analysis pointed to a significant voltage-sensitivity. Incidentally, PIII is the only member of this family to bear a net positive charge. It is also the one that displayed the larger selectivity for sodium over sodium although this remains quite modest ($PNa/PK = 1.5 - 3$) when compared to the whole channel.

Obviously, this argue for tighter filter in the native channel being due to geometric constraints from neighboring layers of transmembrane helical segments. It also now appears that the inner parts of S5 and S6 contribute to the whole pore, in series with the P-loops, as shown by the crystal structure of the KCSA channel [15]. Even negatively-charged residues located in the extracellular loops between S6s and the P-elements, notably in domain III, seem to play a role in permeation [69]. It is also likely that, assembling the four P-peptides, unto a template to give an heteroteramer, would enhanced sodium selectivity. Besides, the concentration-dependence of macrosocopic conductance yielded four monomers per transmembrane conducting aggregate whose single-channel conductance averaged 15 pS in 0.5 M NaCl [66], i.e. in good agreement with the whole channel [52, 70].

As regards secondary structure, the CD analysis (qualitatively confirmed by FTIR spectroscopy) for P-peptides both in organic solvents and in lipid vesicles pleads for a predominant β-sheet structure. Only when the peptides were interacting with lipid vesicles do they show a significant content (10%) of β-turn. Although this β-hairpin structure which we modelled [71], that sharply contrasts with S4L45s' conformation (see below), was already an hypothesis for these regions in the context of the whole channel (see e.g. molecular modelling studies, [72]), our study was one of the very first to substantiate this.

4.2. Voltage sensors : S4L45s from the four domains
- Implication of the linker L45 in the permeation pathway, and polarity-dependent conformational switching

A 34 residue-long peptide, encompassing S4 and L45 of domain IV (see Fig. 1), was first synthesized in order to test the potential implication of L45 in the gating or permeation pathway. The macroscopic conductance properties of neutral and negatively-charged Montal-Mueller planar lipid bilayers doped with IVS4L45 were compared to those of S4 [73]. From the concentration-dependence of I-V curves, the mean apparent number of monomers per conducting aggregate could be estimated to be 3-5. In single-channel experiments, the most probable events had amplitudes of 8 pS and 5 pS in neutral and negatively-charged bilayers respectively (Fig. 5). Ionic selectivity under salt gradients conditions, both at macroscopic and single-channel levels, was in favour of sodium ions ($PNa/PK = 3$). When compared to S4 alone [25], the reduced unit conductance and the increased selectivity for sodium support the implication of the L45 region in the inner lining of the open configuration of sodium channels [73, 74].

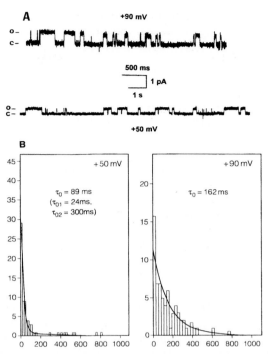

Fig. 5. Voltage-dependence of single-channel currents induced by IVS4L45 (25 nM in buffered 0.5 M NaCl both sides) in a POC/DOPE (7/3) bilayer formed at the tip of a patch pipette. A : single-channel traces at two voltages, and B : associated open lifetime histograms, τ_0 being the mean open lifetime. (From [73]).

The secondary structure of peptide IVS4L45 assessed by circular dichroism was found mainly helical, both in organic solvents and in lipid vesicles, especially negatively-charged ones. The ellipiticity of the L45 moiety previously shown to be involved in the permeation pathway was sharply reduced with a transition α-helix ⇨ β-sheet transition upon exposure to aqueous solvents [75], a situation that is likely to occur during activation and opening of the whole *in situ* channel. An independent and recent NMR and molecular modeling [76], confirms our own findings both from the CD analysis of small fragments of S4L45 and simulations studies on the differential insertion of N- and C-terminal parts [77], the helicity of the former being more stable whilst the latter tend to remain at the interface.

- *Voltage sensitivity and conformational change are tuned to proline*
In the following step, peptide fragments reproducing the sequences of S4 segments extended with L45 linkers from *all* four homologous domains of the electric eel sodium channel were prepared to allow circular dichroism studies in various solvents and conductance assays in planar lipid bilayers. Repeats III (with proline) and IV (lacking proline) present the lowest and highest helicities, respectively [78]. The conformational transition (from α-helix to β-strand) shown to occur on an increase of solvent dielectric constant is broader with

repeat III. Analytical ultracentrifugation (interference fringe pattern) was consistent with a monodispersion of the peptide. Macroscopic conductance experiments conducted with synthetic S4L45s in neutral lipid planar bilayers pointed to a moderate voltage-sensitivity for repeat IV which has no proline, whereas S4L45 of repeats I and II (Pro 19) and especially of repeat III (Pro 14) were much more voltage-sensitive [78]. The influence both of Pro and its position within the sequence was confirmed by comparing the human skeletal muscle channel isoform D4/S4 wild-type and the R4P analogue (compare the sequences in Fig. 1) The apparent and averaged number of monomers per intramembrane conducting aggregate was 4-5.

Thus, both conformational switching and voltage-sensitivity appear correlated to the presence and position of a single proline residue. Since voltage sensors are likely to experience different polarity environments in the channel open and closed states, our results suggest an alternative gating mechanism to ouward movement of voltage sensors during depolarization [79], i.e. a voltage-driven conformational change of S4L45s. The data also implies a plausible functional asymmetry, namely a "four-stroke" activation sequentially involving the four domains of the sodium channel [80].

5. SUBSEQUENT VALIDATION ON THE WHOLE CHANNEL

Such specializations of the four domains and within them of 'homologous' segments both in gating and selectivity are in agreement with the notion of structural-functional modules as propounded again recently [81]. In spite of the apparent reductionism of our approach, a number of subsequent studies on the whole voltage-gated sodium channel quite happily confirm some of our main findings or hypotheses. For instance, very recent coupled extrinsic (i.e. using probes) fluorescence and electrophysiology, do suggest the sequence of domains activation discussed above (DIII first, then DII + DI, and finally DIV) [82]. For a full description of this approach, see e.g. [83]. Also, there is now some evidence that the pore domain moves during gating [84], as might have been suggested by the voltage-dependence of PIII (see above). But perhaps, the most striking parallelism comes form an ongoing study making use of first the peptide strategy then electrophysiology on mutants. As it was suggested that the sodium channel might be considered as a ferrolectric material on one hand [85], and since branched sidechains of hydrophobic residues confer ferroelectricity to aminoacids [86], we set out to replace Ileu9 and Leu15, next to the third and fifth arginines respectively, by alanine in IIIS4L45. Voltage-dependence was abolished with the first substitution but preserved in the second mutated peptide [87]. This finding proved to be verified when these mutations in domain III were applied on the recombinant sodium channel of the human skeletal muscle expressed in a human cell line [88]. This study is now being extended to the other three domains and to other positions [89].

Acknowledgements. *This was very much a team work spanning the years 1994-present and involving my fellows : Drs S. Bendahhou, Marc Brullemans, Olivier Helluin, Pascal Cosette. I am indebted to collaborators abroad, especially in the UK : Drs. J. Breed, P. Biggin, I. Kerr, M. Sansom (Oxford University); Drs. N. Cronin and B. Wallace (University of London). Finally, I also give credit to A. O'Reilly for the photographs shown in this chapter.*

REFERENCES

[1] A.L. Hodgkin and A.F. Huxley, J. Physiol. (Lond.) 116 (1952) 449.

[2] P. Mueller, D.O. Rudin, H.T. Tien and W.C. Wescott, Nature 194 (1962) 979.

[3] P. Mueller and D.O. Rudin, J. Theor. Biol. 4 (1963) 268.

[4] P. Mueller and D.O. Rudin, Nature 217 (1968) 713.

[5] M. Montal and P. Mueller, Proc. Natl. Acad. Sci. USA 69 (1972) 3561.

[6] S.J. Singer and G.L. Nicolson, Science 175 (1972) 720.

[7] G. Ehrenstein and H. Lecar, Q. Rev. Biophys. 10 (1977) 1.

[8] R. Latorre and O. Alvarez, Physiol. Rev. 61 (1981) 77.

[9] E.T. Kaiser and F.J. Kézdy, Annu. Rev. Biophys. Biophys. Chem. 16 (1987) 561.

[10] D.S. Cafiso, Curr. Topics Membr. 48 (1999) 197.

[11] H. Duclohier and H. Wroblewski, J. Membrane Biol. 184 (2001) 1.

[12] H. Duclohier, Mini-Rev. Med. Chem. In the press (2002)

[13] B.A. Wallace, BioEssays 22 (2000) 264.

[14] S. Subramaniam and R. Henderson, J. Struct. Biol. 128 (1999) 19.

[15] D. Doyle, J.M. Cabral, R.A. Pfuetzner, A. Kuo, J.M. Gulbis, S.L. Cohen, B.T. Chait, R. MacKinnon, Science 28 (1998) 69.

[16] N. Unwin, Philos. Trans. R. Soc. Lond. B Biol. Sci. 355 (2000) 1813.

[17] P. Nollert, W.E.C. Harries, D. Fu, L.J.W. Miercke and R.M. Stroud, FEBS Lett. 504 (2001) 112.

[18] R. Dutzier, E.B. Campbell, M. Cadene, B.T. Chait and R. MacKinnon, Nature 415 (2002) 287.

[19] W. Dubinsky and O. Mayorga-Wark, in : Molecular Biology of Membrane Transport Disorders, S.G. Schultz et al. (eds.), Plenum Press, New York, 1996, pp. 73-86.

[20] P. Labarca and R. Latorre, Meth. Enzymol. 207 (1992) 447.

[21] M. Noda, S. Shimuzu, T. Tanabe, T. Takai, T. Kayano, T. Ikeda, H. Takahashi, H. Nakayama, Y. Kanaoka, N. Minamino, K. Kangawa, H. Matsuo, M.A. Raftery, T. Hirose, S. Inayama, H. Hayashida, T. Miyata, T. and S. Numa, Nature 312 (1984) 121.

[22] R.E. Greenblatt, Y. Blatt and M. Montal, FEBS Lett. 193 (1985) 125.

[23] H.R. Guy and P. Seetharamulu, Proc. Natl. Acad. Sci. USA 83 (1986) 508.

[24] S. Oiki, W. Danho and M. Montal, Proc. Natl. Acad. Sci. USA 85 (1988) 2393.

[25] M.T. Tosteson, D.S. Auld and D.C. Tosteson, Proc. Natl. Acad. Sci. USA (1989) 86, 707.

[26] D. Rapaport, M. Danin, E. Gazit and Y. Shai, Biochemistry 31 (1992) 8868.

[27] P.I. Haris, B. Ramesh, S. Brazier, FEBS Lett. 349 (1994) 371.

[28] P. I. Haris, B. Ramesh, M.S.P. Sansom, I.D. Kerr, K.S. Srai and D. Chapman, Protein Engng 7 (1994) 255.

[29] K. Shinozaki, K. Anzai, Y. Kirino, S. Lee and H. Aoyagi, Biochem. Biophys. Res. Commun. 198 (1994) 445.

[30] Y. Pouny and Y. Shai, Biochemistry 34 (1995) 7712.

[31] A. Grove, T. Iwamoto, J.M. Tomich and M. Montal, Protein Sci. 2 (1993) 1918.

[32] M. Oblatt Montal, T. Iwamoto, J.M. Tomich and M. Montal, FEBS Lett. 320 (1993) 261.

[33] D. Langosch, K. Hartung, E. Grell, E. Bamberg and H. Betz, Biochim. Biophys. Acta 1063 (1991) 36.

[34] W.F. DeGrado and J.D. Lear, Biopolymers 29 (1990) 205.

[35] A. Grove, T. Iwamoto, M.S. Montal, J.M. Tomich and M. Montal, Meth. Enzymol. 207 (1992) 510.

[36] M. Montal, Annu. Rev. Biophys. Biomol. Struct. 24 (1995) 31.

[37] D. Marsh, Biochem. J. 315 (1996) 345.

[38] P. Cosette, O. Helluin, J. Breed, Y. Pouny, Y. Shai, M.S.P. Sansom and H. Duclohier, in: Ion Channels, Neurons, and the Brain, V. Torre and F. Conti (eds). Plenum Press, New York, 1996, pp. 41-62.

[39] H. Duclohier, O. Helluin, P. Cosette, A.R. Schoofs, S. Bendahhou and H. Wroblewski, Chemtracts - Biochemistry and Molecular Biology. 10 (1997) 189.

[40] H. Duclohier, O. Helluin, P. Cosette and S. Bendahhou, Bioscience Reports 18 (1998) 279.

[41] P.B. Bennett, K. Yazawa, N. Makita and A.L. George Jr., Nature 376 (1995) 683.

[42] H. Lerche, K. Jurkat-Rott and F. Lehmann-Horn, Am. J. of Med. Genetics 106 (2001) 146.

[43] W.A. Catterall, Neuron 26 (2000) 13.

[44] E. Marban, T. Yamagishi and G.F. Tomaselli,. J. Physiol. 508 (1998) 647.

[45] C.D. Patten, M. Caprini, R. Planells-Cases and M. Montal, FEBS Lett. 463 (1999) 375.

[46] W. Stühmer, F. Conti, H. Suzuki, X. Wang, M. Noda, N. Yahagi, N. Kubo and S. Numa, Nature 339 (1989) 597.

[47] S.H. Heinemann, H. Terlau, W. Stühmer, K. Imoto and S. Numa, Nature 356 (1992) 441.

[48] I.L. Barsukov, D.E. Nolde, A.L. Lomize and A.S. Arseniev, Eur. J. Biochem. 206 (1992) 665.

[49] T.W. Kahn and D.M. Engelman, Biochemistry 31 (1992) 6144.

[50] J.K. Myers, C.N. Pace and J.M. Scholtz, Proc. Natl. Acad. Sci. USA 94 (1997) 2833.

[51] W.S. Agnew, S.R. Levinson, J.S. Brabson and M.A. Raftery, Proc. Natl. Acad. Sci. USA 75 (1978) 2606.

[52] E. Recio-Pinto, D.S. Duch, S.R. Levinson and B.W. Urban, J. Gen. Physiol. 90 (1987) 375.

[53] A.M. Correa, F. Bezanilla and W.S. Agnew, Biochemistry 29 (1990) 6230.

[54] A.L. George, J. Komisarof, R.G. Kallen and R.L. Barchi, Ann. Neurol. 31 (1992) 131.

[55] A.L. Goldin, R.L. Barchi, J.H. Caldwell, F. Hofman, J.R. Howe, J.C. Hunter, R.G. Kallen, G. Mandel, M.H. Meiser, Y. Berwald Netter, M. Noda, M. Tamkun, S.G. Waxman, J.N. Wood and W.A. Catterall, Neuron 28 (2000) 365.

[56] M.T. Tosteson, D.S. Auld and D.C. Tosteson, Proc. Natl. Acad. Sci. USA 86 (1989) 707.
[57] P.A. Slesinger, Y.N. Jan and L.Y. Jan, Biophys. J. 64 (1993) A114.
[58] E.Y. Isacoff, Y.N. Jan and L.Y. Jan, Nature 353 (1991) 86.
[59] R.B. Merrifield, J. Am. Chem. Soc. 89 (1963) 2149.
[60] E. Atherton and R.C. Sheppard, Solid phase peptide synthesis: a practical approach, IRL Press, Oxford, 1989.
[61] W. Hanke, C. Methfessel, U. Wilmsen, G. Boheim, Biochem. Bioenerg. J. 12 (1984) 329.
[62] R. Coronado and R. Latorre R, Biophys J. 43 (1983) 231.
[63] W. Hanke and W.R. Schlue, Planar Lipid Bilayers, Methods and Applications, Academic Press, 1983.
[64] G. D. Fasman (ed.), Circular dichroism and the conformational analysis of biomolecules, Plenum Press, New York and London, 1996.
[65] O. Helluin, J. Breed and H. Duclohier, Biochim. Biophys. Acta 1279 (1996) 1.
[66] P. Cosette, L. Brachais, E. Bernardi and H. Duclohier, Eur. Biophys. J. 25 (1997) 275.
[67] C.T. Chang, C.S.C. Wu and J.T. Yang, Anal. Biochem. 91 (1978) 13.
[68] J.E. Hall, I. Vodyanoy, T.M. Balasubramanian and G.R. Marshall, Biophys. J. 45 (1984) 233.
[69] R.A. Li, P. Vélez, N. Chiamvimonvat, G.F. Tomaselli and E. Marban, J. Gen. Physiol. 115 (2000) 81.
[70] A.M. Correa, F. Bezanilla and W.S. Agnew, Biochemistry 29 (1990) 6230.
[71] P. Cosette, I.D. Kerr, P. LaRocca, H. Duclohier and M.S.·P. Sansom, Biophys. Chem. 69 (1997) 221.
[72] S.R. Durell and H.R. Guy, Neuropharmacology 35 (1996) 761.
[73] M. Brullemans, O. Helluin, J.-Y. Dugast, G. Molle and H. Duclohier, Eur. Biophys. J. 23 (1994) 39.
[74] G.N. Filatov, T.P. Nguyen, S.D. Kraner and R.L. Barchi, J. Gen. Physiol. 111 (1998) 703.
[75] O. Helluin, J. Breed and H. Duclohier, Biochim. Biophys. Acta 1279 (1996) 1.
[76] K. Mattila, R. Kinder and B. Bechinger, Biophys. J. 77 (1999) 2102.
[77] O. Helluin, P. Cosette, P.C. Biggin, M.S.P. Sansom and H. Duclohier, Ferroelectrics 220 (1999) 329.
[78] O. Helluin, S. Bendahhou and H. Duclohier, Eur. Biophys. J. 27 (1998) 595.
[79] Yang, N., George, A.L. Jr. and Horn, R. Neuron 16 (1996) 113.
[80] H. Duclohier, O. Helluin, P. Cosette and S. Bendahhou, Biosci. Reports 18 (1998) 279.
[81] C.D. Patten, M. Caprini, R. Planells-Cases and M. Montal, FEBS Lett. 463 (1999) 375.
[82] B. Chanda and F. Bezanilla, Biophys. J. 82 (2002) 174a.
[83] Cha, A., Ruben, P.C., George, A.L. Jr., Fujimoto, E. and Bezanilla, F. (1999). Neuron 22, 73-87.
[84] J.P. Benitah, Z. Chen, J.R. Balzer, G.F. Tomaselli and E. Marban, J. Neurosci. 19 (1999) 1577.
[85] H.R. Leuchtag, J. Theor. Biol. 127 (1987) 321.
[86] K. Yoshino and T. Sakurai, *Ferroelectrics Liquid Crystals: Principles, Properties and Applications*, J.W. Goodby *et al.* (Edit.), Gordon & Breach (1991) pp. 317-363.

[87] O. Helluin, M. Beyermann, H.R. Leuchtag and H. Duclohier, IEEE Trans. Diel. Electr.
      Insul. 8 (2001) 637.
[88] S. Bendahhou, L.J. Ptacek, R.H. Leuchtag and H. Duclohier, Biophys. J. 80 (2001) 229a.
[89] S. Bendahhou, H. Duclohier, T.R. Cummins, R.H. Leuchtag, S.G. Waxman and L.J.
      Ptacek, Soc. for Neuroscience, Abst. (2001)

Planar Lipid Bilayers (BLMs) and their Applications
H.T. Tien and A. Ottova-Leitmannova (Editors)

Chapter 22

Permeation property and intramembrane environments of synthetic phytanyl-chained glyceroglycolipid membranes

T. Baba[a], H. Minamikawa[b], M. Hato[a] and T. Handa[c]

[a]Nanotechnology Research Institute, National Institute of Advanced Industrial Science and Technology, AIST Tsukuba Central 5, Tsukuba 305-8565, Japan

[b]Nanoarchitectonics Research Center, National Institute of Advanced Industrial Science and Technology, AIST Tsukuba Central 5, Tsukuba 305-8565, Japan

[c]Graduate School of Pharmaceutical Sciences, Kyoto University, Sakyo-ku, Kyoto 606-8501, Japan

1. INTRODUCTION

Naturally occurring lipid molecules are known to show a remarkable degree of structural diversity in both hydrophilic groups (headgroups) and hydrophobic groups. Phospholipids and glycolipids are two major categories in the large diversity of lipids found in biological membranes from the viewpoint of headgroups. Glycolipids can be subdivided into two groups: sphingoglycolipids (SGLs) and glyceroglycolipids (GGLs) [1-8]. While SGLs show considerable variance with respect to saccharide composition, GGLs show remarkable uniformity [1, 4, 6]. In most bacterial and eukaryal cell membranes as well as their intracellular membranes, phospholipids can be found as major structure-forming components. In those membranes, SGLs are considered to play important roles involved in a variety of physiological events such as inter-cellular adhesion, recognition phenomena, cell growth regulation, and signal transduction [1-5, 7, 8] rather than to be structure-forming components, because SGLs are present at the outer surface of cells and form large clusters [1, 8, 9] such as "rafts" [9]. In contrast to SGLs, GGLs occur abundantly as major structure-forming components in Gram-positive bacteria, cyanobacterial (blue-green algae) thylakoids, and higher plant chloroplasts [1-3, 6].

Diethers

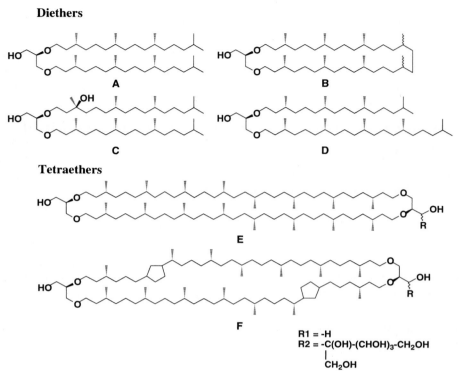

Fig. 1. Chemical structures of typical backbone of archaeal lipids: (A-D) diethers; (E, F) tetraethers. (A) 2,3-di-O-phytanyl-*sn*-glycerol (archaeol); (B) macrocyclic archaeol; (C) β-hydroxy archaeol; (D) α-sesterterpanyl archaeol; (E) tetraethers without cyclized chains, (F) tetraethers with monocyclized chains. For tetraethers, R = R₁: caldarchaeol (GDGT, glycerol-dialkyl-glycerol tetraethers), R = R₂: nonitolcaldarchaeol (GDNT, glycerol-dialkyl-nonitol tetraethers). Adapted from Ref. 15.

The utilization of lipid membranes, mainly phospholipid membranes, has been extensively examined to develop liposomal drug delivery systems as well as biotechnological devices such as biosensors, bioreactors, etc., as structural elements of such devices [10-13]. On the other hand, glycolipids have been employed not as structural elements but as functional ones [14]. In general, however, the problems of conventional lipids for these applications are their chemical, physical, and biological instability. To overcome these problems, several methods for stabilization of lipids have been developed and one of the methods is to utilize lipids from archaea.

Archaea have recently been identified as the third biological kingdom and inhabit under extreme conditions such as very high temperature, low pH or high salt concentrations [15, 16]. Although archaea have a cell wall structure lacking a muramic acid-peptidoglycan layer, their lipid membranes are considered to be chemically and physically stable [15, 16]. These lipids are characteristic in the molecular structure: ether linkages between glyceryl and highly branched hydrophobic chain (isopranyl chain) groups through an exceptional *sn*-2,3 configuration, and have been classified into two groups: diethers (monopolar lipids) and tetraethers (bipolar lipids) as shown in Fig. 1 [15, 16]. The former diethers are found in membranes of methanogenic archaea and extremely halophilic archaea; the latter are mainly found in methanogenic archaea and extremely thermoacidophilic archaea as shown in Fig. 2. Although the properties of archaeal membranes are less well understood than those of bacterial ones, archaeal membranes have been shown to be stable with low permeabilities to proton and other ionic or nonionic solutes [15-24]. Because the barrier function of membranes is critical for the functioning of cell and intracellular organelles, where a proton motive force is established [16], archaeal membranes are expected advantageous for avoiding short-circuiting of the proton gradient and therefore promising for reconstitution matrices of various functional proteins which generate and/or consume the proton motive force. Although it is of great interest to utilize archaeal lipids as biochemical or biotechnological materials [10-13], and accordingly several trials have been made by employing exclusively tetraethers [15, 18, 22, 23, 25], these lipids are not readily available in sufficient amounts and are often mixtures of heterogeneous lipids.

Recently, our group has developed and proposed novel synthetic phytanyl-chained glyceroglycolipids, a series of 1,3-di-*O*-phytanyl-2-*O*-(glycosyl) glycerols: 1,3-di-*O*-phytanyl-2-*O*-(β-D-glucosyl)glycerol (Glc(Phyt)$_2$) and 1,3-di-*O*-phytanyl-2-*O*-(β-D-maltooligoglycosyl)glycerols (Mal$_N$(Phyt)$_2$, $N = 2, 3, 5$; N is the number of glucose residue in the headgroup) as model archaeal diethers [26-28]. The stereochemically pure saccharide residue is linked

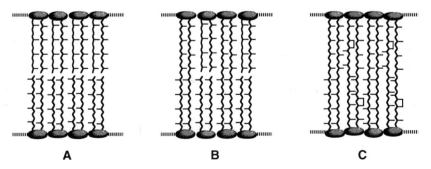

A   B   C

Fig. 2. Membrane models in several archaea. (A) extremely halophilic archaea (diethers only); (B) methanogenic archaea (mixture of diethers and tetraethers); (C) extremely thermoacidophilic archaea (tetraethers only). Adapted from Ref. 15.

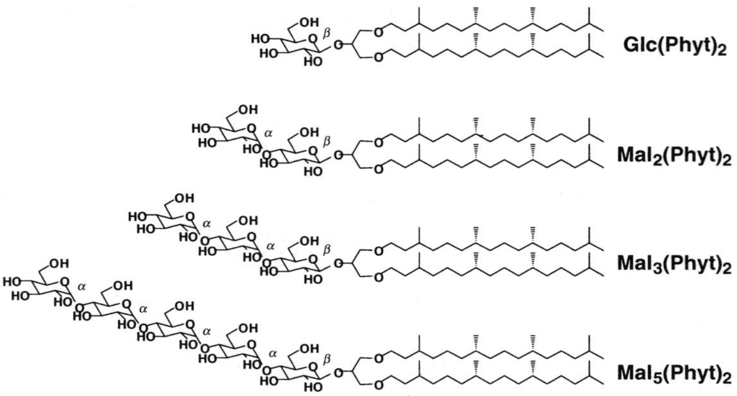

Fig. 3. Chemical structures of novel synthetic phytanyl-chained glyceroglycolipids 1,3-di-*O*-phytanyl-2-*O*-(β-D-maltooligoglycosyl)glycerols (Glc(Phyt)$_2$, Mal$_N$(Phyt)$_2$ (N = 2, 3, 5)). The maltooligosaccharide (Glcα1 → 4Glc type) headgroup is linked to the hydrophobic tail group through a β-*O*-glycosidic bond. The tail group is composed of highly branched phytanyl chains linked to glycerol backbone in C-1 and C-3 positions through ether bonds. Adapted from Refs. 26-28.

through a β-*O*-glycosidic bond of C-1 of glucose to C-2 of 1,3-di-*O*-phytanylglycerol as shown in Fig. 3. These glycolipids are chemically stable, i.e., insusceptible to hydrolysis under low or high pH condition due to lack of ester bonds, and also insusceptible to oxidation due to lack of double bonds in hydrophobic chains. These glycolipids form various kinds of self-assemblies in water depending on the headgroup size N, and exhibiting no transition from fluid liquid-crystalline states to hydrated solid states down to -120°C [26, 28]. The characteristics of the proposed glycolipids are expected to lead to a new class of artificial membrane matrices for basic science such as glycobiology as well as for biochemical and biotechnological applications.

In this chapter, we will describe firstly the membrane formability, the membrane stability, and the membrane permeation property of the novel synthetic phytanyl-chained glyceroglycolipids, studied by BLM conductance measurements and proton permeation measurements on vesicle membranes. Subsequently, we will describe the intramembrane environments, i.e., degrees of hydration and lipid molecular motions in the proposed glycolipid membranes, studied by nanosecond time-resolved fluorescence spectroscopy, and discuss the role of branched chains (phytanyl chains) in membrane permeation properties, in connection with archaeal lipid membranes.

2. PERMEATION PROPERTY OF SYNTHETIC PHYTANYL-CHAINED GLYCEROGLYCOLIPID MEMBRANES

2.1. Membrane formability

In general, the membrane formability is strongly governed by molecular properties of individual lipids, e.g., a geometric packing property [29, 30]. For the proposed phytanyl-chained glyceroglycolipids, only the volume and the occupied area of headgroup vary with the number of glucose residue N. Therefore, the membrane formability can be examined from the viewpoint of the headgroup size.

The phase diagram study on Glc(Phyt)$_2$/water and Mal$_N$(Phyt)$_2$/water systems revealed that Mal$_2$(Phyt)$_2$ and Glc(Phyt)$_2$ can not form liquid-crystalline lamellar (L$_\alpha$) phases but Mal$_2$(Phyt)$_2$ forms a reverse hexagonal (H$_{II}$) phase [26] and Glc(Phyt)$_2$ forms a reverse micellar cubic phase [27] in the presence of an excess water at ambient temperatures, respectively. On the other hand, Mal$_3$(Phyt)$_2$ and Mal$_5$(Phyt)$_2$ can form L$_\alpha$ phases as the stable phases [26]. These results suggest that the phytanyl-chained glyceroglycolipids bearing the smaller headgroups ($N \leq 2$) tend to pack into negative-curvature lipid assemblies, however, those bearing the larger headgroups ($N \geq 3$) prefer zero-curvature ones. Now then, we further attempted to clarify the most optimal headgroup size yielding a bilayer membrane conformation among the proposed glycolipids by means of BLM technique [31].

For BLM experiments, we employed two kinds of lipids other than the phytanyl-chained glycolipids, i.e., a branch-chained phospholipid diphytanoyl-phosphatidylcholine (DPhPC) and natural phospholipids bearing straight chains soybean phospholipids (SBPL), to compare the performances of the novel glycolipids with those of conventional lipids. BLM were prepared by the folding method [32, 33] in the aperture (100-250 μmϕ) located in a Teflon thin sheet (25 μm thick) as shown in Fig. 4. Two Teflon chambers (internal volume: ca. 2 ml) separated by the Teflon sheet were filled with 100 mM KCl unbuffered solution (pH $\approx$ 6). Before the formation of bilayer membranes, the aperture was treated with hexadecane. After placing a drop of a chloroformic lipid solution at the air-water interface and allowing to evaporate the solvent, all of the lipids formed fluid state (liquid expanded) monolayers under the examined conditions. The original monolayers were always set at their highest surface pressures in equilibrium with the collapse state. The setup for electrical measurements of BLMs is also shown in Fig. 4. Teflon chambers and a preamplifier were put in a grounded shield box on a vibration absorbing base. A couple of Ag/AgCl electrodes via saturated KCl salt bridges was used to apply external potentials and to monitor current changes. Electrical signals were preamplified with a

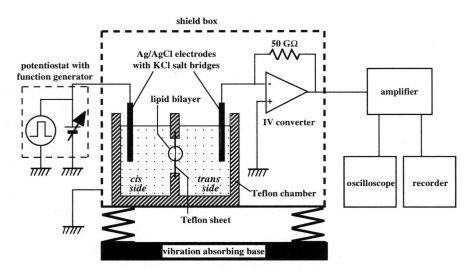

Fig. 4. Schematic diagram of the setup for electrical measurements of BLMs. A preamplifier in a shield box contains an IV converter and a 50 GΩ feed-back resistor. The *cis* side of a Teflon chamber is grounded. BLMs are formed in the aperture located in a Teflon thin sheet by the folding method [32, 33].

current to voltage converter equipped with a 50 GΩ feed-back resistor. The converted signals were amplified with a patch/whole cell clamp amplifier, displayed on a digital storage oscilloscope via a low-pass filter, and recorded on a video tape recorder or a chart recorder. Membrane capacitance C_m was measured by applying a triangular wave (usually 20 Hz and 4-25 mV peak to peak) with a function generator at ambient temperatures (22-25°C).

Among the glyceroglycolipids examined, it was found that $Mal_3(Phyt)_2$ easily forms BLMs. From measurements of the capacitive current passing through the $Mal_3(Phyt)_2$ bilayer, the membrane capacitance C_m was evaluated as 0.6-0.7 μF/cm^2, indicating the existence of a solvent-free single bilayer in the aperture. A geometric thickness of a hydrophobic region of a bilayer d_m was evaluated from Eq. (1),

$$d_m = \varepsilon_0 \varepsilon_m / C_m \qquad\qquad (1)$$

where ε_m is the relative dielectric coefficient of a hydrophobic region. By assuming the value of ε_m is the average value of a long chain hydrocarbon ε_m = 2.1 [34], the calculated value of d_m for $Mal_3(Phyt)_2$ BLMs is 2.9 ± 0.2 nm, in moderate accord with the estimated thickness (2.55 nm) of the hydrophobic region of Mal3(Phyt)2 lamellar structure [26]. In contrast to Mal3(Phyt)2, Glc(Phyt)2 and Mal2(Phyt)2 did not successfully form the stable single bilayers. These observations are consistent with the phase diagram data mentioned above, i.e., the glycolipids bearing the smaller headgroups ($N \leq 2$) prefer negative-curvature lipid assemblies to zero-curvature ones such as planar lipid bilayers. On the other hand, Mal5(Phyt)2 also failed to form stable planar lipid bilayer membranes although this glycolipid forms an L_a phase [26]. This unexpected result might be attributed to the possibility that the larger headgroup of Mal5(Phyt)2 causes lateral constraints between the headgroup and hydrophobic portions in a zero-curvature bilayer, and as a result, this glycolipid prefer gentle positive-curvature lipid assemblies (curved bilayers) to zero-curvature ones. From the above observations, it is concluded that the headgroup geometry is much important for the membrane formability, and Mal3(Phyt)2 is the most optimal glycolipid yielding BLMs among a series of novel synthetic phytanyl-chained glyceroglycolipids [31].

On the basis of these results, subsequent experiments were carried out by employing Mal3(Phyt)2 exclusively for membrane preparations.

2.2. Membrane stability

For practical uses of lipid membranes in various fields, stable membranes are much required, and therefore, an estimation of the membrane stability is also an important subject. To examine the membrane stability, we employed only BLMs showing the higher capacity (> 0.6 μF/cm^2) and stable current at the applied potential of $\pm$ 100 mV [31].

At first, the long-term stability of BLMs was checked at the applied potential of 0 mV, and as a result, it was found that Mal3(Phyt)2 BLMs are considerably stable, in the best preparation case, at least for 3 days, whereas SBPL BLMs are usually stable only for less than 8 h under the same conditions (100 mM KCl unbuffered solution, 22-25°C).

The membrane stability against the external electric field was also evaluated as the membrane rupture potential Ψ_r. Electrical rupture of BLMs (aperture diameter: 220-250 μm) was induced by applying the linearly elevating electric fields at the rate of 8.0 mV/s. Membrane rupture potential Ψ_r was defined as the applied potential at which the current abruptly increases and no longer recovers the original membrane conductance. The obtained values of Ψ_r for DPhPC (411 $\pm$ 42 mV) and Mal3(Phyt)2 (344 $\pm$ 63 mV) BLMs were higher than that for SBPL (293 $\pm$ 44 mV) [31]. This result implies that DPhPC and Mal3(Phyt)2 BLMs are more stable than SBPL BLMs against the external electric field change.

A significant increase in conductance of pure lipid BLMs is often observed due to reversible or irreversible electrical membrane breakdown at higher applied potentials [35, 36]. According to the data reported by Shchipunov and Drachev [37], the potential sweep rate of 8.0 mV/s as an external electric field change is slow enough that the membrane rupture is mainly caused by the transient pores or defects in a bilayer. At such low potential sweep rates, BLMs from various types of lipids have been reported to break down irreversibly at 100-300 mV [37-39]. Moreover, Robello and Gliozzi [40] reported the critical potential Ψ_c, at which conductance transition occurs due to the formation of aqueous pores under current-clamp conditions. The value of Ψ_c for DPhPC (390 ± 70 mV) BLMs [40] is very close to that of membrane rupture potential Ψ_r for this lipid (411 ± 42 mV) [31]. Similarly, the value of Ψ_c for the natural straight-chained phospholipid, egg yolk phosphatidylcholine (EPC, 280 ± 70 mV) BLMs [40] is very close to that of Ψ_r for the analogous straight-chained phospholipids SBPL (293 ± 44 mV) BLMs [31]. Therefore, the observed electric behavior of BLMs in our study is probably attributed to the aqueous pore formation in BLMs. In this context, the aqueous pore formation in BLMs seems to be much suppressed for branch-chained lipids $Mal_3(Phyt)_2$ and DPhPC, compared with a straight-chained lipid SBPL. According to the aqueous pore theory [36, 41, 42], the pores with an overcritical radius are responsible for the irreversible membrane breakdown and the energy barrier height of pore formation can be related to the pore edge line tension and the surface tension of a bilayer. Thus, BLMs with higher pore edge line tension may require the higher applied potentials to induce the membrane rupture [36, 41, 42]. Although the pore edge line tension and surface tension for the lipid bilayers can not be evaluated, it is likely that the higher stability of branch-chained lipid BLMs to the external electrical field change is ascribed to the higher pore edge tension, i.e., the higher lateral cohesion in this type of membrane. These findings are consistent with the results showing that liposome membranes composed of archaeal lipids are very stable over broad ranges of temperature and pH [15, 21, 22, 24, 43, 44].

2.3. Permeation property

The characteristics of the ionic conduction across bilayers are also easily examined by means of BLM technique. By using the setup shown in Fig. 4, the steady-state ionic currents were recorded. Membrane conductance G_m was determined by applying a rectangular wave and then eliminating the capacitive currents. The membrane conductance at 0 mV of the applied potential $G_m(0)$ was practically evaluated in the applied potential Ψ region of ± 40 mV where current-voltage relationship was linear [45], and the data points were fitted

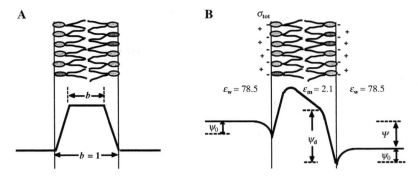

Fig. 5. Schematic diagrams of (A) the trapezoidal model of an ion potential energy barrier in a bilayer membrane and (B) the electrical potential profile for a positive test charge across a bilayer membrane. Ψ is the applied potential across the membrane by electrodes. ψ_0 is the surface potential from charged residues (the total surface charge density σ_{tot}) at the membrane-water interface. ψ_d is the dipole potential from the alignment of dipolar residues of the lipids and associated water molecules. In the trapezoidal model, b is the fraction of the membrane spanned by the minor base of the trapezoid at $\Psi = 0$. Adapted from Refs. 46, 48.

with simulation curves calculated from Eq. (2) for dielectrically symmetrical bilayer membranes [38, 45, 46],

$$\frac{G_m(\Psi)}{G_m(0)} = \frac{b\sinh(e_0\Psi/2kT)}{\sinh(be_0\Psi/2kT)} \tag{2}$$

where $G_m(\Psi)/G_m(0)$ is the ratio of the conductance at Ψ to that at 0 mV, b is the fraction of the membrane spanned by the minor base of the trapezoid as defined in Fig. 5A. Fig. 6 shows the current-voltage characteristics for three kinds of BLMs formed in 100 mM KCl unbuffered solution (pH $\approx$ 6) at 23°C. The current-voltage fitting curves were found to be symmetric and deviate from linearity at higher voltages. For Mal3(Phyt)2 and DPhPC BLMs, a value of $b = 0.6$-0.7 gives the best fit. For SBPL BLM, the fitting curve is almost linear ($b = 0.9$-1.0) over the examined voltage region. Obviously, Mal3(Phyt)2 (Fig. 6A) and DPhPC (Fig. 6B) BLMs showed similar conductances, which were much lower compared with those of SBPL BLMs (Fig. 6C). The observed lower conductances over branch-chained lipids imply the lower ionic permeability across these type of membranes and the ionic permeant is most probably proton. The values of $G_m(0)$ for Mal3(Phyt)2 and DPhPC are close to each other although their headgroups are totally different, suggesting that the rate-limiting step for small ion permeation is the translocation of the ions through a hydrophobic chain region rather than a headgroup region. The nonlinear current-voltage characteristics have been reported in many cases,

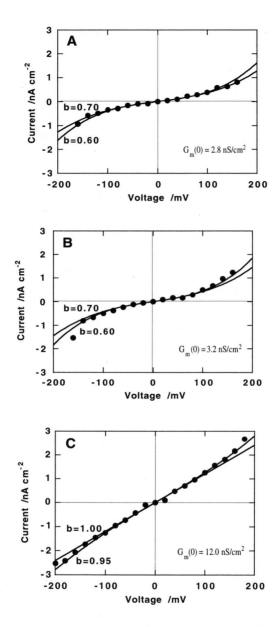

Fig. 6. Current-voltage characteristics for various BLMs: (A) Mal$_3$(Phyt)$_2$; (B) DPhPC; (C) SBPL. Membranes were formed in 100 mM KCl unbuffered solution (pH ≈ 6). Surface areas of membranes were 4.0-4.9 × 10^{-2} mm^2. Temperature was 23°C. The fitting curves were calculated from Eq. (2) changing the value of b. The membrane conductance at 0 mV $G_m(0)$ was practically evaluated from the conductance at ± 40 mV. From Ref. 31.

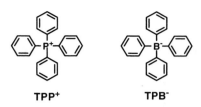

TPP⁺   **TPB⁻**

Fig. 7. Chemical structures of the lipophilic ions, tetraphenylphosphonium cation (TPP⁺) and tetraphenylborate anion (TPB⁻).

e.g., voltage-driven proton/hydroxyl ion (OH⁻) permeation [45], inorganic ion permeation [38], ion carrier-mediated transport [46, 47], and lipophilic ion transport [48] through BLMs, which were often ascribed to the trapezoidal potential energy barrier in a membrane against the charged solute permeation [38, 45, 46, 49] as shown in Fig. 5A. In this case, the current density strongly depends on the height of this potential energy barrier. Because the value of $(1 - b)/2$ is a measure of displacement of the barrier corner toward the bilayer center, as can be seen in Fig. 5A, due to the reduction of the Born self-energy of an ion by attraction of an ion to its image, the estimated value b may provide the information concerning the bilayer-water interface dielectric structure [46]. Judging from the b values for three kinds of BLMs, the dielectric structure of the bilayer-water interface for the Mal3(Phyt)2 BLM is analogous to that for the DPhPC BLM, regardless of the headgroup structure.

In addition to observations of small ion permeation behavior through the branch-chained lipid bilayers, measurements of current-voltage characteristics of large lipophilic ions were performed by employing tetraphenylphosphonium cation (TPP⁺; Fig. 7) and tetraphenylborate anion (TPB⁻; Fig. 7) [31]. They are often used as probes of membrane electrostatics because these ions are of the same size, the same hydrophobicity, and oppositely charged each other [49]. For TPP⁺ and TPB⁻ permeation experiments, 100 mM NaCl unbuffered solution was used instead of KCl solution and salt bridges were removed. In Fig. 8, variations of $G_m(0)$ of Mal3(Phyt)2 and DPhPC BLMs are shown as a function of the concentration of lipophilic ions. For TPP⁺, $G_m(0)$ did not significantly change upon the addition of a small amount of lipophilic ion and was practically constant in the examined concentration region in both BLMs. For TPB⁻, however, if only a slight amount of ion was added, greater than 10^3-fold increases in $G_m(0)$ were observed in both BLMs. These results are consistent with the previously reported observations that TPB⁻ has a much greater permeability than TPP⁺ [49-51]. By comparing these permeation rates, we can consider the contribution of the molecular structure of the lipid headgroup to a bilayer membrane dipole potential ψ_d shown in Fig. 5B [48-51].

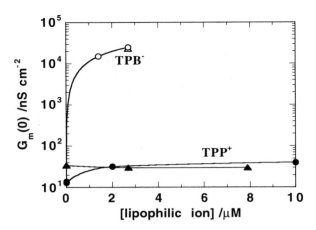

Fig. 8. Effects of lipophilic ions TPP+ and TPB- on the membrane conductances at 0 mV $G_m(0)$ of Mal3(Phyt)2 and DPhPC BLMs. (Δ, $\blacktriangle$) Mal3(Phyt)2; (O, $\bullet$) DPhPC. Closed symbols: TPP+; open symbols: TPB-. Membranes were formed in 100 mM NaCl unbuffered (pH $\approx$ 6). Temperature was 22-25°C. Surface areas of membranes were 3.0-4.0×10^{-2} mm^2.

On the basis of the bilayer membrane dipole potential model, the source of dipole potential is ascribed to the dipole organization of molecules at the bilayer -water interface, with the most significant contribution from the carbonyl groups in glycerol backbone, the headgroup, and surface water molecules [49-51]. The observation in our study showed that the permeation characteristics of lipophilic ions through BLMs of branch-chained lipids Mal3(Phyt)2 and DPhPC, appear to be analogous to each other, suggesting that there is no significant difference in the dipole potential barrier between two types of branch-chained lipids. Therefore, it can be presumed that the carbonyl groups in glycerol backbone of the lipid molecule are not necessarily required for the total dipole potential barrier against cations in Mal3(Phyt)2 BLM. The anion-selective permeation through Mal3(Phyt)2 BLM may be accounted for by the saccharide headgroup bearing a number of hydroxyl dipoles, making up the lack of the carbonyl groups and contributing to the total dipole potential barrier. Therefore, it is considered that the bilayer-water interface dielectric structure of Mal3(Phyt)2 BLM is analogous to that of the DPhPC BLM, regardless of the headgroup structure.

Our attention will hereinafter be focused on proton permeation property of Mal3(Phyt)2 membranes. For the estimation of proton permeation rates, large unilamellar vesicles (LUVs) entrapping a pH-sensitive fluorescent probe pyranine were prepared [52]. LUVs composed of a main lipid and an anionic sulfoglycolipid sulfoquinovosyldiacylglycerol (SQDG; Fig. 9) (9 : 1 mol/mol) were suspended in 10 mM phosphate/50 mM K_2SO_4 buffer solution (pH 7.0). Because the vesicles composed of $Mal_3(Phyt)_2$ alone exhibit a significant aggregation in the presence of inorganic salts [53], the addition of SQDG is required to support the suspension stably. The chemical structures of SQDG and several lipids mentioned below are shown in Fig. 9. For the LUV preparation from the main tetraether lipid (MPL) derived from the extremely thermoacidophilic archaeon *Thermoplasma acidophilum*, the addition of SQDG could be omitted. Rates of the proton permeation across membranes were estimated by measuring rates of changes in fluorescence intensity of pyranine on a spectrofluorometer at 25°C. After the removal of CO_2 by bubbling of N_2 gas, small pH gradients (0.14 pH unit) were applied across membranes by injecting a small amount of sulfuric acid under continuous stirring. To abolish a diffusion potential, valinomycin was added into vesicles prior to measurements and the molar ratio of valinomycin to lipid did not exceed 3×10^{-3}. Excitation light was alternately changed at a frequency of 1 Hz between 416 nm

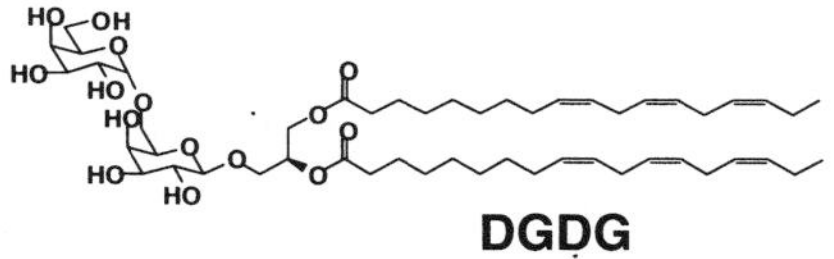

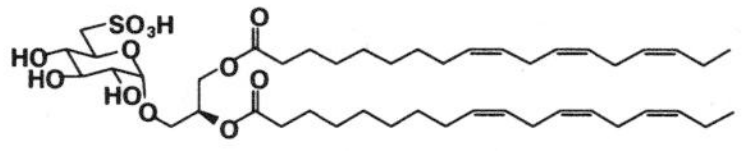

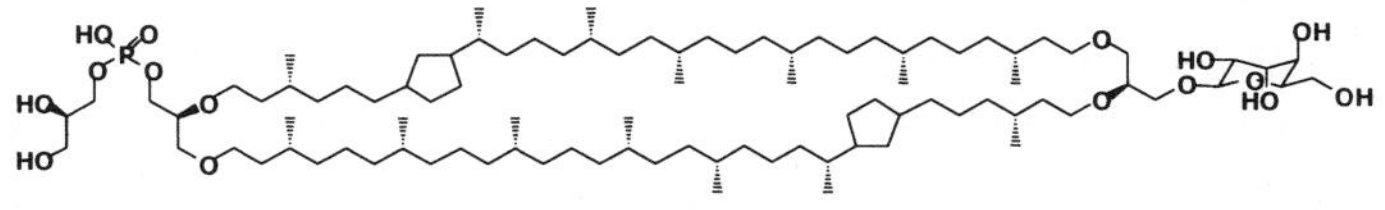

Fig. 9. Chemical structures of digalactosyldiacylglycerol (DGDG), sulfoquinovosyldiacyl-glycerol (SQDG), the main polar lipid from *Thermoplasma acidophilum* (MPL) and the main lipid of the polar lipid fraction E from *Sulfolobus acidocaldarius* (PLFE). For DGDG and SQDG, the hydrophobic chains are α-linolenates ($C_{18:3}$). For MPL, the hydrophobic back-bone is GDGT with cyclopentane rings and the two headgroups are a sn-3-glycerophosphate as one headgroup and a β-L-gulose as the other. Adapted from Ref. 54. For PLFE, the hydro-phobic backbone is GDNT with cyclopentane rings and the two headgroups are an inositol-phosphate as one headgroup and a β-D-glucose as the other. Adapted from Ref. 55.

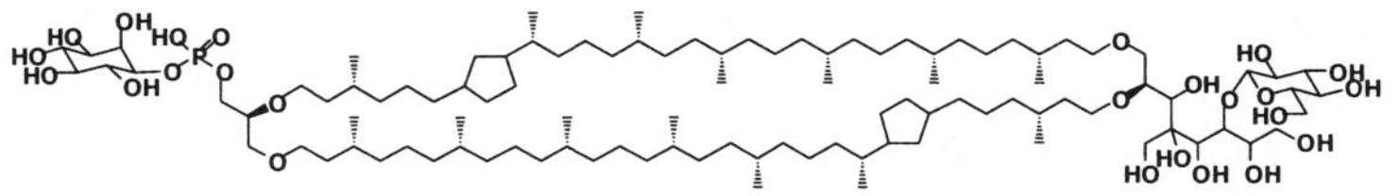

(isosbestic point) and 450 nm (pH-sensitive point). The emission was monitored at 510 nm. The internal pH was calculated from the ratio of these emission intensities. The net proton/hydroxyl ion permeability coefficients $P_{H/OH}$ were calculated from the decay time of pH gradient τ_H expressed by

$$\Delta pH(t) = \Delta pH_0 \exp(-t/\tau_H) \tag{3}$$

where $\Delta pH(t)$ and ΔpH_0 are pH differences across a membrane at the time t after the addition of an acid and at $t = 0$, respectively [56]. τ_H can be related with $P_{H/OH}$ as follows,

$$\tau_H = V_m B / A_m P_{H/OH}[H^+]_0 \ln 10 \tag{4}$$

where V_m, A_m and B are the volume and surface area of a vesicle and the buffer capacity, respectively. $[H^+]_0$ is the initial proton concentration of the external phase. As the observed average diameters of all LUVs were 100 ± 10 nm and the size distribution was narrow, the ratio V_m/A_m could be expressed in terms of the vesicle average radius r_m as $V_m/A_m = r_m/3$.

The observed decay times of pH gradient τ_H for LUVs of various lipids/SQDG (9 : 1 mol/mol) at pH 7.0 and 25°C are shown in Table 1. By considering the proton enrichment factor at the surface: $\exp(-e_0\psi_0/kT)$ due to the negative surface potential ψ_0 (-15 to -20 mV calculated from 10 mol% of SQDG), the proton concentration at the membrane surface is estimated to be about twice that in the bulk phase. The calculated values of $P_{H/OH}$ for various LUVs were also shown in Table 1. It should be noted that a small amount of SQDG is considered to have little effect on the permeability because it has barely influenced the membrane conductance. $P_{H/OH}$ values obtained in our study are in moderate accord with the previously reported data of 10^{-3}-10^{-4} cm/s for several lipid vesicle membranes. Thus proton permeation rates across membranes of branch-chained lipids are lower than those of straight-chained lipids such as EPC by a factor of about 4.

The thickness of the hydrophobic region of the membrane can be one of the important factors governing proton permeation rates, and this is true when acyl chains are straight and shorter than C_{20} [58]. Considering the chain lengths of the monopolar lipids examined in our study are almost the same (mainly saturated C_{16} or unsaturated C_{18}), one would expect the permeation rates to be almost the same. The results of membrane conductance measurements as well as pH-dependent fluorescence measurements, however, show that membranes of branch-chained lipids Mal3(Phyt)2 and DPhPC exhibit lower permeability than those of straight-chained lipids such as EPC and SBPL

Table 1
Net proton/hydroxyl ion permeability of various lipid LUV membranes

Lipid	τ_H (s) [a]	$P_{H/OH}$ (10^{-4} cm/s)	Condition	Ref.
Straight-chained				
EPC/SQDG (9:1)	20 ± 4	17 ± 3	pH 7.0, ΔpH: 0.14, 25°C	52
EPC [b]	—	9.6%min^{-1}[c]	pH 7.4, ΔpH: 2.4	57
	—	0.1-1	pH 7.2, $\Delta\psi$: 186 mV[d]	23
DC$_{16:1}$PC [e]	—	18.2	pH 7.0, ΔpH: 0.5, 30°C	58
DGDG/SQDG (9:1)	< 15	> 23	pH 7.0, ΔpH: 0.14, 25°C	52
DGDG [b]	—	15.6 ± 6.3 % min^{-1}[c]	pH 7.4, ΔpH: 2.4	57
SBPL	—	27	pH 7.2, ΔpH: 0.2, 25°C	56
Branch-chained				
DPhPC/SQDG (9:1)	101 ± 30	3.6 ± 1.1	pH 7.0, ΔpH: 0.14, 25°C	52
DPhPC	—	1.4	pH 7.2, ΔpH: 0.2, 25°C	56
Mal$_3$(Phyt)$_2$/SQDG (9:1)	81 ± 25	4.5 ± 1.4	pH 7.0, ΔpH: 0.14, 25°C	52
MPL	—	4.5×10^{-2}	pH 7.2, $\Delta\psi$: 186 mV[d]	23

[a] Values are means $\pm$ SD ($n = 3$).
[b] MLV membranes.
[c] Only an initial slope of the pH change curve has been described.
[d] Applied diffusion potential with K$^+$–valinomycin.
[e] Dipalmitoleoyl-PC.
 Adapted from Ref. 52.

by a factor of about 4-6. Further, membranes of tetraethers, e.g., MPL and the main lipid of the polar lipid fraction E from the extremely thermoacidophilic archaeon *Sulfolobus acidocaldarius* (PLFE [55]) shown in Fig. 9 were found to exhibit exceptionally low proton permeability which can not be accounted for solely by the thickness of hydrophobic region [21, 23, 24].

It is well known that proton permeability across membranes is considerably higher than the permeabilities to other monovalent ions, and is almost in the range of water permeability: 10^{-2}-10^{-3} cm/s [58, 59]. Although the coupling between proton and water permeation is less clear, two kinds of water-mediated proton transport mechanisms have been proposed: transient hydrogen -bonded chains of water molecules [58-61], and passive diffusion of hydrated protons [58]. The former mechanism assumes the transient formation of strands of water molecules called "proton wires"; that are connected through

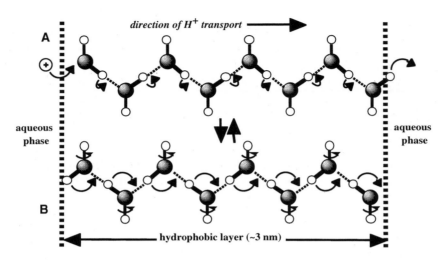

Fig. 10. Transient hydrogen-bonded chain (proton wire) model for the proton translocation across a membrane. When a strand of water molecules is connected through the membrane to the opposing water phases due to thermal fluctuations, a proton charge is transported along a water chain by way of a hopping mechanism (A), followed by the reorientation of the hydrogen-bonded chain (B). White and black balls represent hydrogen and oxygen atoms, respectively. Dashed lines between water molecules and vertical broken lines represent hydrogen-bonds and membrane–water interfaces, respectively. Adapted from Ref. 61.

the membrane to the opposing water phases and rapidly transport protons from one side to the other as shown in Fig. 10 [59-61]. According to molecular dynamics calculations [61, 62], the proton conduction along a proton wire in nonpolar medium includes two complementary steps: the rapid proton translocation along a water chain by way of a proton hopping mechanism, and the reorientation of water molecules in the chain. The latter step involves an activation energy of ca. 33 kJ/mol and is a rate-limiting step for the passage of several protons along the wire [62]. Based on this model, water penetration from membrane-water interface into membranes and molecular motions in membranes seem to govern the formation frequency of the proton wire spanning membranes. Actually, both proton permeability data for the thinner phospholipid membranes and nearly pH-independent conductance data can be explained not by passive diffusion mechanism but by the proton wire mechanism [58, 60]. Water penetration (hydration) and molecular motions, required for the proton wire formation in membranes, are considered to be possibly mediated by lipid headgroup fluctuations, lateral diffusive motions and the migration of *gauche-trans-gauche* conformations along lipid chains [63]. Because these conformational changes occur on the time scale of nanoseconds

[64], nanosecond time-resolved fluorescence spectroscopy is a suitable method to obtain information on molecular motions involved in permeation behavior. In the following section, we will describe the intramembrane environments of synthetic phytanyl-chained glyceroglycolipids, i.e., degrees of hydration and lipid molecular motions in membranes on the time scale of nanoseconds, and will suggest which more affects proton permeation across branch-chained lipid membranes, hydration or lipid molecular motions.

3. INTRAMEMBRANE ENVIRONMENTS OF SYNTHETIC PHYTANYL -CHAINED GLYCEROGLYCOLIPID MEMBRANES

3.1. Hydration in membranes

For probing membranes, several kinds of lipidic fluorescent probes were employed and their chemical structures are shown in Fig. 11. For the estimation of degrees of hydration in membranes, a series of n-(9-anthroyl-oxy)fatty acids (nAFs): n-(9-anthroyloxy)stearic acids (nAS, n = 2, 6, 9, 12) and 16-(9-anthroyloxy)palmitic acid (16AP) as well as N-dansyl-phosphatidyl-ethanolamine (DnsPE) were used. LUVs were labeled with nAFs or DnsPE at lipid to probe molar ratios of 500 : 1 or 200 : 1, respectively. Fluorescence lifetimes and anisotropy decays were measured with a single photon counting fluorometer equipped with a pulsed hydrogen lamp (FWHM: ~ 2 ns), suitable

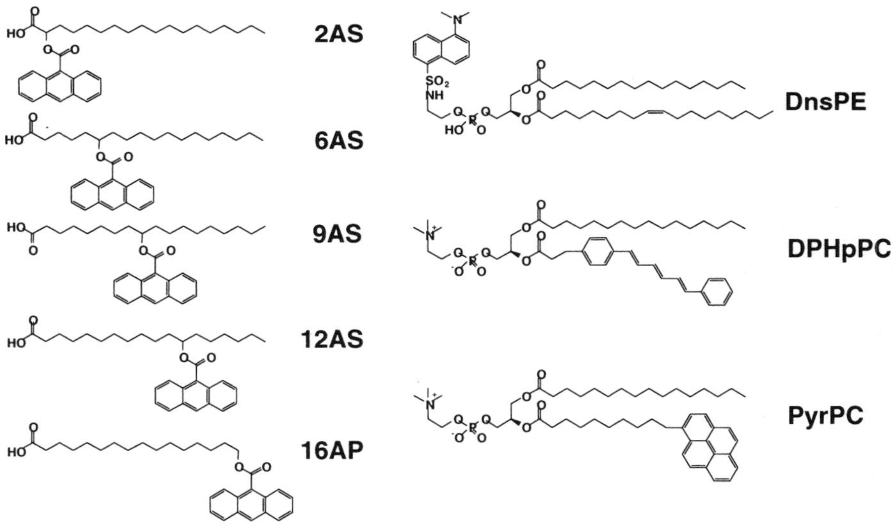

Fig. 11. Chemical structures of the fluorescent probes, a series of n-(9-anthroyloxy) fatty acids (nAFs), N-dansyl-phosphatidylethanolamine (DnsPE), 1-palmitoyl-2-(3-(diphenyl-hexatrienyl)propanoyl)PC (DPHpPC), and 1-palmitoyl-2-(10-pyrenedecanoyl)PC (PyrPC).

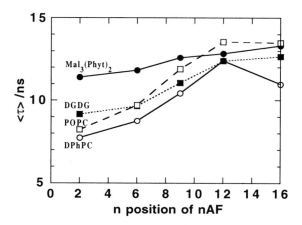

Fig. 12. Dependence of the average fluorescence lifetime $<\tau>$ for nAFs on the position of an anthroyloxy fluorophore in various lipid/SQDG (9 : 1 mol/mol) LUV membranes at 25°C. Data are the means for two experiments ($n = 2$). From Ref. 52.

filters, and polarizers. All samples were purged with Ar or N_2 gas to remove dissolved O_2 prior to measurements.

Fig. 12 shows the average fluorescence lifetimes $<\tau>$ of nAFs in various lipid/SQDG membranes at 25°C. Gradients of hydration in membranes can be estimated from the $<\tau>$ values of nAFs because the fluorophore of nAFs is located in membranes depending on the position number n; the location ranges from the membrane-water interface, most probably the glycerol backbone region for 2AS, to the center of the membranes for 16AP [64]. The $<\tau>$ value is more dependent on the local concentration of water surrounding the fluorophore than solvent viscosity or dielectric constant [65, 66]. For 1-palmitoyl-2-oleoyl-PC (POPC), EPC (data not shown but superimposable on the POPC data), DPhPC, and digalactosyldiacylglycerol (DGDG), the steeper increases in the $<\tau>$ value, suggesting decreases in hydration, were observed from the position $n = 2$ to 12. The $<\tau>$ values leveled off at $n = 12$ in most cases. Interestingly, nAFs in DPhPC/SQDG membrane showed the shortest lifetimes at all depths among the lipid membranes examined, suggesting that DPhPC/SQDG membrane has the highest water content. Mal3(Phyt)2/SQDG membrane, however, yielded the longest lifetimes of nAFs from the position $n = 2$ to 9. This result suggests that degrees of hydration and water penetration in Mal3(Phyt)2/SQDG membrane are the lowest in most parts of the membrane. Fig. 13 shows the relationship between the $<\tau>$ values for 2AS and those for DnsPE in various membranes at 25°C. As the fluorophore of DnsPE is located near the headgroup region or at the membrane-water interface [67],

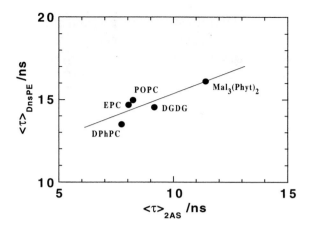

Fig. 13. Relationship between the average fluorescence lifetimes of DnsPE $<\tau>_{DnsPE}$ and 2AS $<\tau>_{2AS}$ in various lipid/SQDG (9 : 1 mol/mol) membranes at 25°C. DnsPE and 2AS were incorporated into multilamellar vesicle and LUV membranes, respectively. Data are the means for two experiments ($n = 2$). From Ref. 52.

its location may be close to that of 2AS. The $<\tau>_{2AS}$ value was positively correlated with the $<\tau>_{DnsPE}$ value, suggesting that both probes similarly sense the local concentration of water. Both probes showed the shortest lifetimes for DPhPC/SQDG membrane, and oppositely the longest lifetimes for Mal3(Phyt)2/SQDG membrane. The former result can be attributed to the highest hydration and the latter to the lowest hydration at the membrane-water interfaces. The measurements of the deuterium isotope effect [68, 69] on DnsPE fluorescence also revealed that DGDG and Mal3(Phyt)2 membranes show the lowest hydration and DPhPC membrane shows the highest hydration at the membrane-water interfaces, respectively [52].

According to Melo et al. [66], local concentrations of water surrounding the fluorophore of nAFs C_{H2O} can be directly estimated from the relation

$$C_{H2O} = (\tau_0 - \tau)/k_a \, \tau\tau_0 \qquad (5)$$

where τ, τ_0 (13.5 ns) are the lifetimes in the presence and the absence of water respectively, and k_a is the quenching rate constant. By adopting the k_a value of 5.5×10^6 M^{-1}s^{-1} in dioxane at 25°C [66], water concentration could be roughly evaluated as $C_{H2O} \approx 2.5$ M for Mal3(Phyt)2/SQDG membrane-water interface and $C_{H2O} \approx 10$ M for DPhPC/SQDG one. Straight-chained lipid membrane-water interfaces showed $C_{H2O} \approx 8.8$ M for POPC/SQDG and EPC/SQDG, and $C_{H2O} \approx 6.4$ M for DGDG/SQDG. Meanwhile, the centers of all membranes except for DPhPC/SQDG were found practically water-free. Why is there a striking contrast with respect to the hydration between Mal$_3$(Phyt)$_2$/SQDG and DPhPC/SQDG membranes? It is most likely that the difference in headgroup structure largely affects the extent of water penetration into membranes of Mal$_3$(Phyt)$_2$ and DPhPC because the chain structures of these lipids are similar. These observations are consistent with the structural views on DGDG/water and DPhPC/water systems previously reported: DGDG membrane surface, from which water tends to be excluded, yields a "dry surface" (relatively less hydrated surface) owing to the tightly packed saccharide moieties at the membrane-water interface [70]. On the contrary, DPhPC membrane surface requires more water molecules than those of straight-chained PCs such as POPC to adopt a "normal" liquid-crystalline lamellar structure, because of the relatively large cross-sectional area in DPhPC [71]. If the former is the case, the low hydration at Mal$_3$(Phyt)$_2$/SQDG membrane-water interface can be ascribed to the tight packing of saccharide moieties. The extent of water penetration into the deeper region likely reflects the higher density of the interchain free volume into which water molecules can be accommodated. If this is the case, such density in DPhPC/SQDG membrane is the highest among the monopolar lipids examined. Oppositely, the density seems the lowest in Mal$_3$(Phyt)$_2$/SQDG membrane; which is unlikely because the chains of both lipids are similar, and the molecular occupied area for DPhPC monolayer is smaller than that for Mal$_3$(Phyt)$_2$ under a constant surface pressure [31]. It is most likely that the bulkier saccharide moiety may act as a barrier and reduce the accessibility of water molecules to the interchain free volume. The order of decreasing hydration in membranes is obviously inconsistent with the order of decreasing proton permeability. Therefore, the lower proton permeability of branch-chained lipid membranes can not be ascribed solely to the lower hydration in membranes.

3.2. Molecular motions of lipids in membranes

For the estimation of molecular motions of lipids in membranes, 1-palmitoyl-2-(3-(diphenylhexatrienyl)propanoyl)-PC (DPHpPC) and 1-palmitoyl-2-(10-pyrenedecanoyl)-PC (PyrPC) shown in Fig. 11 were used as fluorescent probes. LUVs were labeled with DPHpPC or PyrPC at lipid to probe molar ratios of 500 : 1 or 9 : 1, respectively. In the case of DPHpPC, one can estimate the rotational diffusive motions of DPHpPC in membranes (Fig. 14A) by analyzing the fluorescence anisotropy decay data [72]. By comparing the values of the rotational diffusion coefficient D_W of DPHpPC in various membranes at 25°C, the rotational motions of DPHpPC in branch-chained lipid membranes: $D_W \approx 0.05$ GHz for Mal$_3$(Phyt)$_2$; 0.03 GHz for MPL, were found

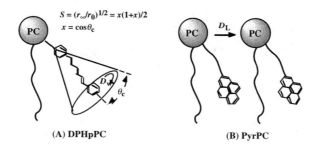

$$S = (r_\infty/r_0)^{1/2} = x(1+x)/2$$
$$x = \cos\theta_c$$

(A) DPHpPC   (B) PyrPC

Fig. 14. Dynamic parameters of (A) DPHpPC; S: order parameter of DPH moiety; r_0, r_∞: fluorescence anisotropies at $t = 0$ and at $t \to \infty$, respectively; θ_c: cone angle; D_w: rotational (wobbling-in-cone) diffusion coefficient, and (B) PyrPC; D_L: lateral diffusion coefficient.

lower than those in straight-chained ones: $D_w \approx 0.09$ GHz for DGDG; 0.1 GHz for EPC and POPC, however, intermediate motions of DPHpPC were observed in DPhPC membranes: $D_W \approx 0.08$ GHz [52].

In the case of PyrPC, one can estimate the lateral diffusive motions of PyrPC in membranes (Fig. 14B) by measuring the excimer complex formation efficiency. As all the membranes in our study are in fluid states at the temperatures examined, PyrPC are considered to be well miscible with the matrix lipids and the excimer complex formation due to lateral phase separation can be neglected. For PyrPC molecules which perform a random walk in a plane, by taking steps of length L corresponding to the average lipid-lipid spacing in membranes, and at PyrPC stepping frequency f, the lateral diffusion coefficient D_L of PyrPC is given by

$$D_{\rm L} = fL^2 / 4 \tag{6}$$

$$f = <N>(1/\kappa)(I_{\rm e}/I_{\rm m})(k_{\rm m}/k_{\rm e})(1/\tau_{\rm e}) \tag{7}$$

where κ is a constant characteristic for the probe used, (I_e/I_m) is the ratio of the excimer emission intensity I_e to the monomer one I_m, and (k_m/k_e) is the ratio of the radiative decay constant of the excited monomer k_m to that of the excimer k_e. The values of κ and (k_m/k_e) for PyrPC were assumed to be 0.8 and 0.1, respectively [73]. τ_e is the excimer lifetime. $<N>$ is the average step number between collisions of probe molecules and expressed in terms of the molar fraction of the probe X_{PyrPC} as follows [74],

$$<N> = (2/\pi X_{\rm PyrPC})\ln(2/X_{\rm PyrPC}). \tag{8}$$

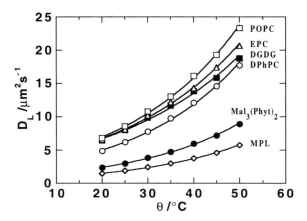

Fig. 15. Temperature dependence of the lateral diffusion coefficient D_L of PyrPC in various lipid/SQDG (9 : 1 mol/mol) LUV membranes. In the case of MPL, SQDG was not added. The fitting curves were obtained from Eq. (9) by using the best values for parameters. Data are the means of two experiments. From Ref. 52.

By evaluating several physical quantities appeared in Eqs. (6)-(8) experimentally and/or theoretically, lateral diffusion coefficients D_L of PyrPC in various membranes could be obtained as a function of temperature shown in Fig. 15. The L values were assumed to be independent of temperature, e.g., $L^2 = 1.0$ nm^2 at 25°C for Mal$_3$(Phyt)$_2$ [26]. The D_L values for branch-chained lipid membranes were lower than those for straight-chained lipid ones at the temperatures examined. The D_L values for DPhPC were, however, intermediate between those for straight-chained lipids and Mal$_3$(Phyt)$_2$, and became closer to those for straight-chained lipids at higher temperatures. In the Stokes-Einstein model, the temperature dependence of D_L can be described by the activation energy E_a associated with the effective viscosity as follows,

$$D_L = AT \exp(-E_a / RT) \tag{9}$$

where A is a temperature-independent constant and R is the gas constant. From the slope of the Arrhenius plots of D_L/T, the E_a values of PyrPC diffusion in various lipid membranes were determined to be 25-33 kJ/mol [52].

For straight-chained lipids, rotational diffusion coefficients of DPHpPC and lateral diffusion coefficients of PyrPC (Fig. 15) were relatively large. The molecular motions in DGDG membrane seem somewhat low compared to those of POPC and EPC, and this is possibly due to the laterally forming hydrogen-bond network among the headgroups. For branch-chained lipids, the tetraether MPL membrane showed the most restricted molecular motions among the lipids examined. The D_L values for MPL were 2.4, 3.8, and 5.8 $\mu m^2/s$ at 30, 40, and 50°C, respectively (Fig. 15), while those for total lipids from *Thermoplasma acidophilum* containing MPL have been reported to be 0.5-0.6, 0.99, and 2 $\mu m^2/s$ at 30, 40, and 50°C, respectively, on the basis of ^{31}P-NMR technique [75]. As the D_L values estimated by PyrPC excimer complex formation technique are usually about an order of magnitude larger than those estimated by other techniques [76], the present result for MPL is not irrelevant to the ^{31}P-NMR results. The very restricted motions of PyrPC were also reported in another tetraether PLFE membrane [77]. Mal$_3$(Phyt)$_2$ /SQDG membrane also showed highly restricted motions. These findings suggest that the higher intermolecular constraint is imposed on lipid molecules in the branch-chained membranes. ^2H-NMR technique has provided important information on dynamic properties of branch-chained lipid membranes and showed that methyl branches in a phytanyl chain reduce the segmental motion, e.g., *gauche-trans-gauche* kink formation at the tertiary carbons and lower the rate of the motion of the phytanyl chain itself [78, 79]. The E_a values for PyrPC in branch-chained lipid membranes are somewhat higher than those in straight-chained lipid ones [52]. This may be due to the steric hindrance of methyl branches against the lateral diffusive motions of PyrPC. These restricted motions in branch-chained membranes should lead to the reduction of translocational and reorientational motions of water molecules responsible for proton permeation. It should be noted that DPhPC/SQDG membrane showed molecular motions as high as 70-80% of the motional rates observed in straight-chained lipid membranes at 25°C. DPhPC membranes were found to be loosely packed compared to tetraether lipid ones by fluorescence measurements [80] and molecular dynamics calculations [81]. These findings support the view that molecular motions in DPhPC membranes are higher than those in MPL ones. However, the difference in motions between DPhPC and Mal$_3$(Phyt)$_2$ can not be ascribed solely to interchain packing because the chains of both lipids are similar as mentioned above. The bulkier saccharide moiety of Mal$_3$(Phyt)$_2$ may reduce motions and this is probably due to the formation of hydrogen-bond network among headgroups.

3.3. Correlation between permeation property and intramembrane environments

As mentioned above, the hydration in membranes observed on the time scale of nanoseconds can not fully account for the order of proton permeation rates, because the degree of hydration in DPhPC membranes, which show the lower proton permeability, are unexpectedly high. This implies that the coupling between proton and water is weak and that the lower proton permeability of branch-chained lipid membranes can not be ascribed solely to the lower hydration in membranes. Based on the proton wire model, however, it is likely that the arrangement, translocation rate, and reorientation rate of water molecules responsible for proton charge transport are much more important to govern proton permeation rates than the distribution itself of water molecules in membranes. For example, if one assumes that water penetration into membranes is governed by lateral diffusive motions of lipids [63] and a certain fraction of water permeability P_{H2O} is involved in a "proton wire" formation, the relative proton permeability for various membranes at 25°C can be roughly estimated from the relation: $P_{H/OH} \propto P_{H2O} \propto D_L/L^4$ [63] as 100 for EPC/SQDG, 105 for POPC/SQDG, 100 for DGDG/SQDG, 70 for DPhPC /SQDG, and 20 for Mal$_3$(Phyt)$_2$/SQDG, respectively. Obviously, the value for Mal$_3$(Phyt)$_2$ is reasonable to explain the lower proton permeability compared to that for EPC by a factor of about 4, although the value for DPhPC is too large because of its high motion in membranes. In addition to this dynamic view, it can be speculated that bulky methyl branches showing "lateral interdigitation" [82] may prevent water molecules from forming "proton wires" spanning a membrane, resulting in a reduction of an efficient proton transport. According to molecular dynamics calculations [61], the energy barrier of the formation of a "proton wire" spanning a membrane was estimated to be > 100 kJ/mol, indicating that a complete "proton wire" formation is a rare phenomenon. The presence of bulky branches may lower the formation frequency of such molecular assemblies, leading to the reduction of proton charge transport. Further molecular dynamics studies would provide more useful information on the behavior of water assemblies in branch-chained lipid membranes. At the present stage, however, it can be concluded that the restricted motion of chain segments rather than lower hydration accounts for the lower proton permeability of branch-chained lipid membranes including archaeal lipid membranes.

4. CONCLUSION

Membrane conductance and pH-sensitive fluorescent probe experiments revealed that the proton permeability of the novel synthetic phytanyl-chained glyceroglycolipid membrane is lower than that of conventional straight-chained lipid membranes and comparable to that of a branch-chained lipid DPhPC membrane. This supports the view that the rate-limiting step for an ion permeation is the translocation of ion across a hydrophobic chain region rather than a headgroup region. Nanosecond time-resolved fluorescence spectroscopy revealed that the lower proton permeability of branch-chained lipid membranes can be accounted for in terms of the restricted motion of chain segments rather than lower hydration in membranes. On the basis of the characteristic permeation property as well as the high stability, novel synthetic phytanyl-chained glyceroglycolipid membranes are considered to be promising materials for the reconstitution matrices for energy conversion membrane proteins which generate and/or consume of proton motive force. Actually, our preliminary investigations revealed that ion channels [31] and oxygen-evolving photosystem II complexes [83] can be functionally and stably incorporated into the proposed glyceroglycolipid membranes.

ACKNOWLEDGMENTS

We thank Professor Dr. Naoki Kamo (Graduate School of Pharmaceutical Sciences, Hokkaido University), Mr. Yoshiyuki Toshima (National Institute of Agrobiological Sciences (NIAS)), and Mr. Kyosuke Suzuki (Daiichi Pharmaceutical Co.), for their collaboration and helpful discussions on BLM measurements. We also thank Dr. Hiroyuki Saito (NIHS Osaka) for his help and advice with time-resolved fluorescence measurements, and Dr. Hiroaki Komatsu (Thomas Jefferson University, U. S. A.) for his valuable comments.

REFERENCES

[1] H. Wiegandt (ed.), Glycolipids, Elsevier, Amsterdam, 1985.
[2] W. Curatolo, Biochim. Biophys. Acta, 906 (1987) 111.
[3] W. Curatolo, Biochim. Biophys. Acta, 906 (1987) 137.
[4] B.L. Slomiany, V.L.N. Murty, Y.H. Liau and A. Slomiany, Prog. Lipid Res., 26 (1987) 29.
[5] S. Hakomori, Pure Appl. Chem., 63 (1991) 473.
[6] M.S. Webb and B.R. Green, Biochim. Biophys. Acta, 1060 (1991) 133.
[7] N. Kojima, Trends Glycosci. Glycotechnol., 4 (1992) 491.
[8] S. Hakomori, Biochem. Soc. Trans., 21 (1993) 583.
[9] K. Simons and E. Ikonen, Nature, 387 (1997) 569.
[10] D.D. Lasic, Liposomes: From Physics to Applications, Elsevier, Amsterdam, 1993.
[11] M. Rosoff (ed.), Vesicles, Marcel Dekker, New York, 1996.
[12] H.T. Tien and A.L. Ottova, Electorochim. Acta, 43 (1998) 3587.
[13] M. Hara and J. Miyake, Curr. Top. Biophys., 24 (2000) 47.
[14] O. Lockhoff, Angew. Chem., Int. Ed. Engl., 30 (1991) 1611.
[15] A. Gambacorta, A. Gliozzi and M. de Rosa, World J. Microbiol. Biotechnol., 11 (1995) 115.
[16] J.L.C.M. van de Vossenberg, A.J.M. Driessen and W.N. Konings, Extremophiles, 2 (1998) 163.
[17] L.C. Stewart, M. Kates, I. Ekiel and I.C.P. Smith, Chem. Phys. Lipids, 54 (1990) 115.

[18] M.G.L. Elferink, J.G. de Wit, R. Demel, A.J.M. Driessen and W.N. Konings, J. Biol. Chem., 267 (1992) 1375.
[19] K. Yamauchi, K. Doi, M. Kinoshita, F. Kii and H. Fukuda, Biochim. Biophys. Acta, 1110 (1992) 171.
[20] K. Yamauchi, K. Doi, Y. Yoshida and M. Kinoshita, Biochim. Biophys. Acta, 1146 (1993) 178.
[21] M.G.L. Elferink, J.G. de Wit, A.J.M. Driessen and W.N. Konings, Biochim. Biophys. Acta, 1193 (1994) 247.
[22] C.G. Choquet, G.B. Patel, T.J. Beveridge and G.D. Sprott, Appl. Microbiol. Biotechnol., 42 (1994) 375.
[23] H.-J. Freisleben, K. Zwicker, P. Jezek, G. John, A. Bettin-Bogutzki, K. Ring and T. Nawroth, Chem. Phys. Lipids, 78 (1995) 137.
[24] H. Komatsu and P.L.-G. Chong, Biochemistry, 37 (1998) 107.
[25] G.B. Patel and G.D. Sprott, Crit. Rev. Biotechnol., 19 (1999) 317.
[26] H. Minamikawa and M. Hato, Langmuir, 13 (1997) 2564.
[27] H. Minamikawa and M. Hato, Langmuir, 14 (1998) 4503.
[28] M. Hato, H. Minamikawa, K. Tamada, T. Baba and Y. Tanabe, Adv. Colloid Interface Sci., 80 (1999) 233.
[29] D.J. Mitchell and B.W. Ninham, J. Chem. Soc., Faraday Trans. II, 77 (1981) 601.
[30] S.M. Gruner, J. Phys. Chem., 93 (1989) 7562.
[31] T. Baba, Y. Toshima, H. Minamikawa, M. Hato, K. Suzuki and N. Kamo, Biochim. Biophys. Acta, 1421 (1999) 91.
[32] M. Takagi, K. Azuma and U. Kishimoto, Ann. Rept. Biol. Works, Fac. Sci., Osaka Univ., 13 (1965) 107.
[33] M. Montal and P. Mueller, Proc. Natl. Acad. Sci. USA, 69 (1972) 3561.
[34] R. Fettiplace, D.M. Andrews and D.A. Haydon, J. Membr. Biol., 5 (1971) 277.
[35] R. Benz, F. Beckers and U. Zimmermann, J. Membr. Biol., 48 (1979) 181.
[36] I.G. Abidor, V. B. Arakelyan, L.V. Chernomordik, Yu.A. Chizmadzhev, V.F. Pastushenko and M.R. Tarasevich, Bioelectrochem. Bioenerg., 6 (1979) 37.
[37] Yu.A. Shchipunov and G.Yu. Drachev, Biochim. Biophys. Acta, 691 (1982) 353.
[38] F. Gambale, M. Robello, C. Usai and C. Marchetti, Biochim. Biophys. Acta, 693 (1982) 165.
[39] H.T. Tien and A.L. Diana, Chem. Phys. Lipids, 2 (1968) 55.
[40] M. Robello and A. Gliozzi, Biochim. Biophys. Acta, 982 (1989) 173.
[41] L.V. Chernomordik, M.M. Kozlov, G.B. Melikyan, I.G. Abidor, V.S. Markin and Yu.A. Chizmadzhev, Biochim. Biophys. Acta, 812 (1989) 643.
[42] M. Winterhalter and W. Helfrich, Phys. Rev. A, 36 (1987) 5874.
[43] E.L.Chang, Biochem. Biophys. Res. Commun., 202 (1994) 673.
[44] C.G. Choquet, G.B. Patel and G.D. Sprott, Can. J. Microbiol., 42 (1996) 183.
[45] J. Gutknecht, Biochim. Biophys. Acta, 898 (1987) 97.
[46] S.B. Hladky, Biochim. Biophys. Acta, 352 (1974) 71.
[47] M. Inabayashi, S. Miyauchi, N. Kamo and T. Jin, Biochemistry 34 (1995) 3455.
[48] A. Ono, S. Miyauchi, M. Demura, T. Asakura and N. Kamo, Biochemistry 33 (1994) 4312.
[49] E.A. Disalvo and S.A. Simon (eds.), Permeability and Stability of Lipid Bilayers, CRC Press, Boca Raton, Florida, 1995, pp. 197-240.
[50] B. Ketterer, B. Neumacke and P. Läuger, J. Membr. Biol., 5 (1971) 225.
[51] R.F. Flewelling and W.L. Hubbell, Biophys. J., 49 (1986) 541.

[52] T. Baba, H. Minamikawa, M. Hato and T. Handa, Biophys. J., 81 (2001) 3377.
[53] T. Baba, L.-Q. Zheng, H. Minamikawa and M. Hato, J. Colloid Interface Sci., 223 (2000) 235.
[54] M. Swaine, J.-R. Brisson, G.D. Sprott, F.P. Cooper and G.B. Patel, Biochim. Biophys. Acta, 1345 (1997) 56.
[55] S.-L. Lo and E. L. Chang, Biochem. Biophys. Res. Commun., 167 (1990) 238.
[56] S. Grzesiek and N.A. Dencher, Biophys. J., 50 (1986) 265.
[57] A.A. Foley, A.P.R. Brain, P.J. Quinn and J.L. Harwood, Biochim. Biophys. Acta, 939 (1988) 430.
[58] S. Paula, G. Volkov, N. van Hoek, T.H. Haines and D.W. Deamer, Biophys. J., 70 (1996) 339.
[59] D.W. Deamer and J.W. Nichols, J. Membr. Biol., 107 (1989) 91.
[60] J.F. Nagle, J. Bioenerg. Biomembr., 19 (1987) 413.
[61] S.J. Marrink, F. Jähnig and H.C.J. Berendsen, Biophys. J., 71 (1996) 632.
[62] R. Pomès and B. Roux, Biophys. J., 75 (1998) 33.
[63] T.H. Haines, FEBS Lett., 346 (1994) 115.
[64] F.S. Abrams, A. Chattopadhyay and E. London, Biochemistry, 31 (1992) 5322.
[65] A.L. Maçanita, F.P. Costa, S.M.B. Costa, E.C. Melo and H. Santos, J. Phys. Chem., 93 (1989) 336.
[66] E.C.C. Melo, S.M.B. Costa, A.L. Maçanita and H. Santos, J. Colloid Interface Sci., 141 (1991) 439.
[67] E. Asuncion–Punzalan, K. Kachel and E. London, Biochemistry, 37 (1998) 4603.
[68] L. Stryer, J. Am. Chem. Soc., 88 (1966) 5708.
[69] C. Ho, S.J. Slater and C.D. Stubbs, Biochemistry, 34 (1995) 6188.
[70] R.V. McDaniel, Biochim. Biophys. Acta, 940 (1988) 158.
[71] C.-H. Hsieh, S.-C. Sue, P.-C. Lyu and W.-G. Wu, Biophys. J., 73 (1997) 870.
[72] S. Kawato, K. Kinoshita, Jr. and A. Ikegami, Biochemistry, 16 (1977) 2319.
[73] H.-J. Galla and E. Sackmann, Biochim. Biophys. Acta, 339 (1974) 103.
[74] H.-J. Galla, W. Hartmann, U. Theilen and E. Sackmann, J. Membr. Biol., 48 (1979) 215.
[75] H.C. Jarrel, K.A. Zukotynski and G.D. Sprott, Biochim. Biophys. Acta, 1369 (1998) 259.
[76] J.-F. Tocanne, L. Dupou–Cézanne and A. Lopez, Prog. Lipid Res., 33 (1994) 203.
[77] Y.L. Kao, E.L. Chang and P.L.-G. Chong, Biochem. Biophys. Res. Commun., 188 (1992) 1241.
[78] H. Degani, A. Danon and S.R. Caplan, Biochemistry, 19 (1980) 1626.
[79] L.C. Stewart, M. Kates, I.H. Ekiel and I.C.P. Smith, Chem. Phys. Lipids, 54 (1990) 115.
[80] T.K. Khan and P.L.-G. Chong, Biophys. J., 78 (2000) 1390.
[81] J.L. Gabriel and P.L.G. Chong, Chem. Phys. Lipids, 105 (2000) 193.
[82] O. Dannenmuller, K. Arakawa, T. Eguchi, K. Kakinuma, S. Blanc, A.M. Albrecht, M. Schumtz, Y. Nakatani and G. Ourisson, Chem. Eur. J., 6 (2000) 645.
[83] T. Baba, H. Minamikawa, M. Hato, A. Motoki, M. Hirano, D. Zhou and K. Kawasaki, Biochem. Biophys. Res. Commun., 265 (1999) 734.

Planar Lipid Bilayers (BLMs) and their Applications
H.T. Tien and A. Ottova-Leitmannova (Editors)

Chapter 23

Modulation of planar bilayer permeability by electric fields and exogenous peptides

A. Gliozzi[a*], F. Gambale[b] and G. Menestrina[c]

[a]Dipartimento di Fisica, Università di Genova, I-16146 Genova, Italy

[b]Istituto di Cibernetica e Biofisica CNR, I-16149 Genova, Italy

[c]CNR-ITC Centro Fisica Stati Aggregati, I-38050 Povo, Trento, Italy

1. ELECTRIC FIELD-DEPENDENT FORMATION OF PORES IN PURE LIPID MEMBRANES

1.1. Electropermeabilization of cells, liposomes and planar bilayers

Cell electropermeabilization is well known since many years [1-3]: it consists of an increase in membrane permeability to ions and molecules induced after exposure to an external electric field. In recent times, electropermeabilization is becoming more and more important due to its potential applications to biotechnology [2-4]. In fact, although numerous high-resolution techniques exist to detect the inner structure of single cells, few methods are able to control and manipulate biochemically their content. Electroporation is commonly used for gene transfection [5, 6], membrane protein insertion [7] and injection of membrane-impermeant molecules into cells [8]. Moreover, electrofusion of cells is more efficient after membrane electroporation [9-11].

Experimental data on electroporation come from cell suspensions, individual cells and bilayer lipid membranes (BLMs). In all these systems, the membrane plays a role in amplifying the applied electric field. For example, the transmembrane voltage V_m at different locations of a spherical cell in a homogeneous electric field E_c, is described by the Laplace equation. The solution, with appropriate boundary conditions, is:

$$V_m = 1.5 \ r_{cell} \ E_c \ \cos\theta \qquad (1)$$

where r_{cell} is the radius of the cell and θ is the angle with respect to the direction of the electric field. At the poles (θ=0, π) the applied electric field, E_c, is small compared to the transmembrane electric field and the potential drop across the membrane, V_m, is 75% of the potential drop in the region near the cell.

When the transmembrane potential reaches a value in the range 0.2-1.5V for mammalian cells, electropermeabilization occurs. During this process, impermeant solutes added to the extracellular medium can penetrate into the cell. Experiments performed using fluorescent dyes sensitive to the electric field have demonstrated that Eq. (1) applies as far as significant electroporation does not occur [3, 12]. Studies have been performed to image the time course of cell electropermeabilization in the millisecond time scale [13, 14]. The duration, strength and wave function of the applied electric field determine the extent of electropermeabilization. Moreover, combining patch clamp and fluorescence microscopy techniques, the kinetics of pore opening and closing in single cells during electroporation has been investigated. Using 1 ms rectangular waveform pulses in NG 108-15 cells, the transmembrane potential inducing electroporation was around 250 mV [15].

When electropermeabilization is a transient phenomenon, it is known as *reversible breakdown*. By contrast, if the electric field induces irreversible modifications in membrane permeability, the phenomenon named *irreversible breakdown* causes membrane rupture. As electropermeabilization is related to the bilayer structure of the cell membrane, a quantitative study under well-controlled conditions was made possible by experiments on BLMs.

In the following we shall describe some key experimental observations on electropermeabilization in BLMs, that appear central to understand basic aspects of the phenomenon. Results will be discussed in terms of current theoretical models, with particular emphasis on the transient aqueous pore model [16-21].

1.2. Phenomenological description and theoretical approaches

When a voltage step change is applied to a BLM, the current response is initially that of an RC circuit; at a certain time current fluctuations appear, followed by membrane rupture. However, if the potential step is applied for a short time (typically ranging from seconds to milliseconds, as transmembrane voltage varies from 0.2 to 0.4V), the membrane conductance temporarily increases; as the potential is set to zero again, the membrane recovers its initial conductance. Irreversible breakdown is a stochastic rather than deterministic process in which the mean membrane lifetime is strongly dependent on V_m.

There have been several models to describe BLMs instability due to high intensity electric fields. Earlier models do not explicitly deal with the formation of pores. One of the first theories suggested that electromechanical compression on the entire membrane may cause rupture when a critical potential value is exceeded

[22]. However, the absence of a marked change in C_m during electro-permeabilization [23] argues strongly against such interpretation. Moreover, this theory appears unable to account for other experimental data, for instance the stochastic nature of membrane rupture or the strong lifetime dependence on V_m. Other models include the involvement of electrocompressive forces [24], phase transitions in membranes treated by statistical physics [25] or the onset of an instability in a viscoelastic film [26].

We shall describe here in some detail the electroporation model based on the formation of aqueous pathways across the membrane [16-21]. This model leads to specific predictions which are in agreement with several experimental observations: the approximate value of transmembrane voltage associated to rupture or reversible breakdown, the stochastic membrane lifetime, the order of magnitude of transported charged species per single pulse. More recently, the dynamics of pore growth has been treated as a non-Markovian stochastic process [27] or investigated by Monte Carlo simulations [28]. Theoretical simulations also include pore disappearance in a cell after electroporation [29].

Following the theory of homogeneous nucleation, a BLM can be considered as a system in a metastable state with a fluctuating population of defects, or pores, viewed as the nuclei of a new phase. The balance between the energy necessary to form the edge of a pore and the energy released by the pore surface gives the work to form a new pore W(r)

$$W (r)=2\pi\gamma r - \pi\Gamma r^2 \qquad (2)$$

where γ is the line tension and Γ is the surface energy of the bilayer. Eq. (2) has only one stable minimum at r=0. There is a critical radius, $r=\gamma/\Gamma$, above which the pore is unstable. For typical estimates of line tension (10^{-6}dyn), thermally driven rupture requires a surface tension on the order of 1 dyn/cm [30]. For larger tension any transient pore will lead to rupture, while for lower tension it will close again.

The electroporation theory differs from the usual nucleation theory since the work to form a pore also depends on the applied potential V_m, and for that reason the surface energy is defined as an effective surface energy Γ_{eff}, given by

$$\Gamma_{eff} = \Gamma_+ aV_m^2 \qquad (3)$$

where a is a parameter defined as

$$a = \frac{\varepsilon_0}{2d}(\varepsilon_w - \varepsilon_1) \qquad (4)$$

and d is the bilayer thickness, ε_0 is the vacuum dielectric permittivity and ε_w, ε_1 are the relative permittivities, of water and lipid, respectively. When the radius of the pores overcomes a critical value dependent on V_m, membrane rupture occurs. So it is convenient to introduce a parameter, $\bar{\tau}$, accessible to experimental determination, that can be interpreted as the average lifetime of a metastable state. In the theory of homogeneous nucleation $\bar{\tau}$ is defined as

$$\bar{\tau} = A \ exp \ (W_c / k_B T) \tag{5}$$

where k_B is the Boltzman constant, T is the absolute temperature and W_c the energy barrier for pore formation, defined as

$$W_c = \frac{\pi \gamma^2}{\Gamma_{eff}} \tag{6}$$

The prefactor A is given by [16]

$$A = \frac{(k_B T)^{3/2}}{4\pi c_p SD\gamma \Gamma_{eff}^{1/2}} \tag{7}$$

where c_p is the pore concentration, S is the membrane area and D is the pore diffusion coefficient (in the space of radii).

The average lifetime $\bar{\tau}$ can be written explicitly, combining Eq. (6) and (7) and replacing in Eq. (5).

$$\bar{\tau} = \frac{(k_B T)^{3/2}}{4\pi c_p SD\gamma \Gamma_{eff}^{1/2}} \exp\left(\frac{\pi \gamma^2}{\Gamma_{eff} k_B T}\right) \tag{8}$$

Eq. (8) indicates that the higher is V_m, the smaller is $\bar{\tau}$. For example, the average lifetime of a dipalmitoyl phosphatidylcholine membrane decreases by more than six orders of magnitude, starting from 10s as V_m varies from 0.2 to 1.4V [21].

1.3. Experimental approaches

Studies on membrane electroporation have followed three different experimental protocols, namely voltage-clamp, charge pulse and current clamp. We shall briefly describe the first two techniques and shall treat in greater detail the third one, which can provide new information on the time evolution of pores [23, 31-33].

1.3.1 Voltage clamp

This technique is based on the application of a step change in the applied transmembrane voltage, V_m, and subsequent determination of the transmembrane current I(t) and average membrane lifetime $\bar{\tau}$ [16-21]. Unfortunately, membrane lifetime is so short that it is difficult to obtain information on pore dynamics. In order to increase BLM lifetime, membranes were treated with uranyl ions [34, 35].

The time behavior of the pores was obtained from the kinetics of membrane conductance after application of multistep voltage pulses. The results suggest that there are three processes with different time constants determining changes in membrane conductivity. One process is very rapid (<2µs), and it is interpreted as due to a non-ohmic behavior of the single pore. The second is within 1 to 10ms and it is related to changes in number and distribution of pores. The third, in the range of seconds or minutes, is related to the resealing process. The drawback of this approach is that the presence of uranyl ions might introduce uncontrolled effects in the membrane behavior.

In order to evaluate the fraction of membrane occupied by pores, computer simulation studies have been performed [36]. This fraction is very small at low V_m, but rises rapidly to a maximum value of 10^{-3}, which is consistent with the transport of water-soluble molecules.

1.3.2. Charge pulse relaxation studies

In charge pulse experiments [37] the membrane capacitance is connected to a pulse generator. As a result of an electric pulse, electropermeabilization or membrane rupture can occur, depending on its amplitude and duration. Analysis of the voltage decay gives information on the electrical parameters of the membrane and their time evolution. Typical values of applied voltages and pulse duration are on the order of hundred millivolts and from 0.1 to 100 µs, respectively. It is possible to record the discharge process of the pore with a time resolution below 1 µs, by applying single voltage pulses across the membrane [38]. On the basis of this technique, membranes made of neutral or charged lipids were analyzed under various conditions and the kinetics of pore formation and membrane rupture was determined [39, 40]. While the time at which electroporation occurred was randomly distributed, in all the membranes the conductance increased with time linearly and with a constant slope, independently of the lipid or bathing electrolyte. The authors propose that under their experimental conditions a single pore is formed (or occasionally two) leading to membrane rupture. All the membranes showed a constant radial pore-widening velocity, suggesting a rupture determined by inertia, as during the bursting of soap films. In contrast, polylysine-decorated BLMs [41] or dextran-supported BLMs [42] showed an exponential increase of conductance, suggesting a viscosity-determined rupture process. Also S-layer supported BLMs [43] showed such viscosity-determined widening of the pore.

1.3.3. Current clamp

In voltage clamp and charge pulse experiments the mean lifetime of the membrane rapidly drops with increasing voltage. In particular, in the case of BLMs, a steady potential difference in the order of 200-300 mV causes the membrane to break. Therefore, following this protocol only the action of very short electric pulses (lasting less than 1ms) can be investigated. The novelty of the approach, firstly introduced by our group [23], is to study electroporation under current clamp rather than the usual voltage-clamp conditions. A sketch of the experimental apparatus is shown in Fig.1. The membrane was in a feedback network of a high-impedance operational amplifier, which acts as a current-voltage converter. The current-voltage characteristic of the membrane shows a conductance transition to a higher value above an average potential V_c (Fig.2). Since the process is stochastic, this value (which represents the voltage above which electroporation occurs) is randomly distributed around a mean potential, named "critical potential" V_c. V_c depends on the length of lipid chains, on the number and position of double bonds, and on the chemical nature and charge of the polar heads [23, 44, 45].

It is worth noting that this phenomenon cannot be observed under voltage-clamp conditions owing to the very short lifetime of the membrane in the high-conductance state. In contrast, when the membrane is in a current-clamp configuration, the rise of conductance due to the opening of the pores causes a sudden decrease of the transmembrane potential, which in turn determines an increase of the membrane average lifetime. Fig. 2 shows that decreasing the membrane potential does not cause an immediate change in conductance, i.e. a sudden decrease of the pore size. In fact, pore resealing requires hundreds of seconds under current clamp conditions [23, 33, 44, 45].

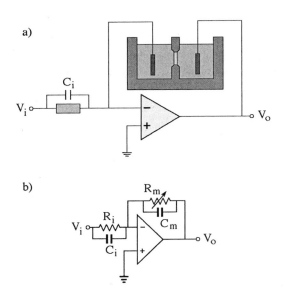

Fig. 1. Sketch of the experimental set-up. a) The membrane is in the feedback branch of a high impedance operational amplifier. b) The model configuration is shown with the membrane replaced by a time-dependent RC circuit (from Ref. [32]).

When the membrane is in the electroporated state, and the current is kept constant, it is possible to record long living voltage fluctuations (Fig. 3). The experimental protocol is the following: the membrane is stimulated by an input signal consisting of a current ramp (used to induce membrane electroporation) followed by a constant current. Performing noise analysis of the responses to the constant stimulus, we expected to find Lorentzian spectra, suggesting the presence of characteristic times for pore dynamics. In fact, although fast time constants (less than one millisecond) could be masked by the circuit cut-off, the closure process was slow enough (hundreds of seconds) to be described by noise analysis. It was quite surprising indeed to find a 1/f spectrum [31, 32] over at least four decades of frequency (Fig. 4). This suggests that electroporation forms pores which are quite different from ionic channels in biological membranes; their behavior is similar instead to a great number of physical systems (e.g. film resistors or solid state devices) displaying a power spectrum lacking of any characteristic time constant. The physical meaning of 1/f spectra is still under consideration [46].

On the other hand, noise analysis has also shown a non-stationary behavior of the voltage-voltage autocorrelation function [32], suggesting the presence of slow processes masking the time constant associated to the pore closure. Slow non-stationary fluctuations could be due to pore-pore interactions or mechanical forces

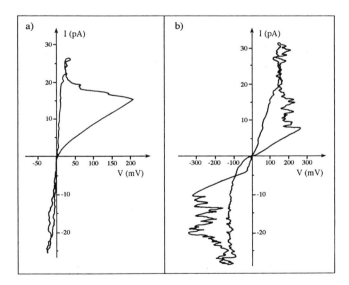

Fig. 2. Current-voltage characteristics of membranes made of: (a) egg PC/cholesterol (molar ratio 3:1) and (b) egg PC. Notice that the presence of cholesterol stabilizes the membrane in the state of higher conductance. In case (b) the phenomenon is almost completely reversible. In both cases the ionic concentration was 10^{-1} M KCl at pH 7 (modified from Ref. [23]).

generated by the electric field, causing membrane deformation and changes in rheological parameters [47].

An intriguing aspect of electroporation is the still open problem of number of pores and their diameter. Using simplified models, Benz et al. calculated values ranging from 1 to 10 nm for conductance changes of 2.8 and 280nS respectively [37]. Subsequent work predicted mean pore radius dependent on the waveform applied to the membrane: 5-10 nm for square pulses, 1.5-10 nm for exponential pulses, 1-3 nm for bipolar square pulses [48]. In the same vein, using the simplified model of an aqueous pore, Kalinowski et al. calculated a mean value of 5 nm in the case of current clamp configuration [33], the higher the current, the larger being the pore diameter. The authors propose that a single oscillating pore is being formed since currents passing through the pore cause structural changes in the membrane.

However, the number of pores formed during electroporation in artificial and natural membranes is still matter of debate and it is possible that it is related to the geometry of the system and the experimental protocol chosen to induce electroporation. Indirect evidence comes from electrofusion experiments. In fact, it is well known that the processes of electropermeabilization and electrofusion are related to each other, as shown by monitoring the fluorescence of voltage sensitive dyes in pulsed cells [12, 49]. Teissié and Ramos proposed that fusogenicity is due to a large number of small defects, created randomly during the pulse in a well defined part of the cell surface [10]. These defects, on the order of 3 to 5 $10^5/\mu m^2$, can be described as fluctuations in the packing density of the lipid matrix, and depend upon pulse duration.

The time scale for recovery has also been investigated in studies on cells [16, 29, 50, 51]. Recovery of the membrane occurs within 0.5-1s. Some cells were not fully recovered even after 5 min, especially those that had experienced higher fields.

We have seen above that during electroporation a feedback mechanism in current clamp conditions causes BLM lifetime to increase [23]. It is also possible to increase the pore lifetime in artificial giant vesicles immobilized on the tip of a pipette [52]. The pores are large (~ 1 µm in diameter) and quasi stable as they remain open for up to several seconds before snapping shut very quickly. It has been proposed that their dynamics is due to a peculiar mechanism involving the way such vesicles are formed and held in place [53]. More precisely, the key of the feedback is a dynamic interplay between the outflow of solution through the pore and the velocity of the vesicle edge as it is aspirated into the pipette, causing the surface tension to decrease.

1.4.   Effects of non-phospholipid membrane constituents on electroporation

1.4.1 Addition of surfactants

The high-intensity fields needed for electroporation can damage cells and tissues by Joule heating, or decrease the cell survival rates [54]. There are several examples of substances able to reduce intensity and duration of the electrical stimulus needed for electroporation. For instance, in an azolectin membrane, addition of the block copolymer poloxamer 188 to the bathing solution increased the thresholds by 5-15% [55], while addition of lysophosphatidylcholine decreases electroporation threshold in egg phosphatidylcholine membranes [18]. In terms of the electroporation model, such effect is due to the surfactant molecular shape, acting to change the spontaneous curvature of the membrane and therefore the linear tension of the pore edge. Troiano et al. [56] performed a quantitative study on the effect of a nonionic surfactant ($C_{12}E_8$), a molecule which is conically shaped like lysophosphatidylcholine, on BLMs made of 1-palmitoyl-2-oleoyl phosphatidylcholine. The authors found a threshold decrease of 30-40% with $C_{12}E_8$ 10 µM . Moreover, in the absence of surfactant the strength-duration relationship showed that threshold decreased linearly with the logarithm of time with a slope of ~ 45 mV/decade, while upon addition of $C_{12}E_8$ the slope decreased to 10

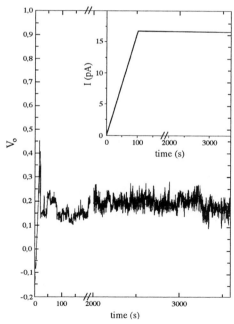

Fig. 3. Time-dependent voltage fluctuations of a membrane stimulated by the input current represented in the inset. The experiment is performed under current-clamp; the initial ramp is used to induce electroporation, which is shown to occur by the sharp potential decrease. The analysis of voltage fluctuations is performed when the current is kept constant and the membrane is in the electroporated state (from Ref. [32]).

mV/decade. All these experiments demonstrate that a proper use of surfactants could be useful for biotechnological applications of electroporation *in vivo*.

1.4.2 Addition of gramicidin or proteins

The effect of membrane proteins on electroporation is an important task, since a full understanding of this phenomenon in cell membranes must take into account that membrane proteins are a major component. Several studies [2, 57, 58] have shown that addition of proteins increases electroporation threshold. According to the pore theory, proteins or peptides pack along the pore edges, thus making membrane resealing more difficult [58]. A different explanation is provided by the electromechanical model of Needham and Hochmuth [59], which predicts that increasing the area expansivity modulus (and therefore mechanical stability), produces an increase in the electroporation threshold.

Transmembrane peptides are often studied to model membrane proteins. In particular, Troiano et al. [60] incorporated gramicidin D (gD) (a mixture of pentadecapeptides isolated from *Bacillus brevis*) into BLMs and giant unilamellar vesicles to record possible changes in electroporation. Both electroporation threshold and transmembrane conductance were found to be function of gD concentration. Moreover, when gD was incorporated into the vesicles, the membrane area expansivity modulus increased monotonically as the peptide/lipid ratio increased. The data have been interpreted in terms of an electromechanical model, which correlates increases in electroporation threshold with an increase in the mechanical stability of the bilayer.

1.5 Electroporation and mechanisms of transport in bolaform lipid membranes

In earlier studies we formed BLMs of bolaform lipids [61] using the conventional technique introduced by Mueller, Rudin, Tien and Wescott [62]. Bolaform lipids extracted from the Archeon *Sulfolobus solfataricus* consist of a polar headgroup at each end of a double C_{40} hydrocarbon chain, with variable degree of cyclization [63-65]. BLMs are formed from the hydrolytic fraction glycerol-dialkyl-nonitol tetraether (GDNT). Several studies have shown that both *in vivo* and in artificial systems the membrane is organized in a simple monolayer lacking middle plane region [63-68], this being the main reason of the unusual mechanical stability of artificial and natural bolaform membranes. It was shown that reversible electroporation occurs even in such rigid system, although with a voltage threshold much higher than in usual lipids [66-68]. This result, somehow expected, has been interpreted in terms of the pore model suggesting that bolaform molecules bend to form aqueous pores under the electric field; therefore, the polar headgroup of such U-shaped molecules would cover the pore interior [68]. On the other hand, the continuum model of electromechanical instability [59] can also

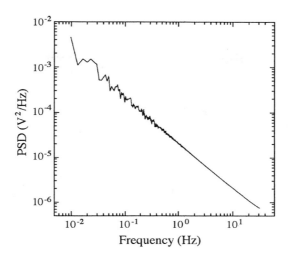

Fig. 4. The power spectrum density (PSD) of the voltage fluctuations of a membrane held at a constant current of 9 pA during electroporation. The initial ramp to induce electroporation had a slope of 0.09 pA/s and lasted 100 s. The slope of the PSD on a log-log scale is 0.99 ± 0.01 (from Ref. [32]).

explain the threshold increase, since the area expansivity coefficient is much higher in bolaform membranes. This can be obtained from the average interfacial tension, which is 3±0.8 mNm^{-1} [67], a value ten times higher than that of monopolar lipid BLMs.

Membranes formed by bolaform lipids are characterized by high stability against external stresses (temperature or pressure jumps), as well as unusually low hydrophobic thickness (from 3 to ≅ 2.4 nm, depending on the solvent). Potential technological applications are in areas such as device technology, drug delivery, biosensors and simulations of cell surfaces [69-71].

The relevance of model systems in nanoscopic applications led us to reconsider the clear-cut division between the two classical transport mechanisms, namely carriers and channels [72]. The first type of transport involves molecules which are embedded in the membrane and shuttle specific ions from one side to the other, while the other one consists of ionic channels which are fixed with respect to the hydrophobic core of the membrane during the transport. We have shown that the linear dependence on valinomycin concentration of the zero current conductance is not observed anymore when the thickness of bolaform membranes is reduced to the approximate length of two valinomycin molecules. Instead, a quadratic behavior was recorded, suggesting that a dimer permeation pore was formed; whereby the conduction mechanism changes from carrier to channel [72].

2   VOLTAGE DEPENDENCE IN PEPTIDE INDUCED CHANNELS

2.1. Indirect voltage modulation of the gramicidin channel

The mechanisms of ionic transport through biological membranes have been intensively investigated through the incorporation into phospholipid bilayers of polypeptides that are able to form ionic channels. The pentadecapeptide gramicidin A (gA) is one of the best known; having been first characterised in BLM in 1972 [73-76]. It is formed by an alternating sequence of D- and L-amino acids [77] whose monomers associate head to head to form a channel spanning the hydrophobic phase of the bilayer. The amino terminal association of two gA monomers is established by six intermolecular H-bonds and the resulting dimer forms a selective channel permeable to monovalent cations only [78].

Once the gramicidin channel is formed, the access of permeating ions is not regulated by any voltage-dependent gating mechanism; on the contrary, the formation rate of gramicidin dimers increases with transmembrane voltage (a factor of 10 between 0 and 200 mV). This depends on the probability of dimer formation from two encountering monomers which, in turn, is regulated by mechanisms like the voltage-dependent binding of permeating ions to gramicidin monomers and the polarization of monomers induced by the applied electric field [79]. As gA channel formation produces bilayer deformation, a decrease in membrane thickness, induced by an increase in the applied membrane potential, may also increase the probability of dimer formation [73, 79].

gA kinetics and lifetime also depend on membrane chemico-physical parameters and boundary conditions at the membrane-water interface [80]. For example, phospholipid composition and the membrane's elastic deformation capability clearly affect both gA single channel lifetime [73, 81-84] and intra-channel flickering [85]. The dependence of channel lifetime on membrane deformation (determined by the incorporation of micelle-forming molecules) has even been used to measure the membrane stiffness in planar lipid bilayers [86]. Moreover, a recent paper [87], on gramicidin monomers covalently linked with diolaxane rings, suggests that lipid composition also affects gA conductance. Specifically, the differences in gA proton conductance measured in membranes of different composition cannot be explained by surface charge effects or interfacial dipole potentials only since also the bilayer fluidity may substantially contribute to the modulation of channel conductance by acting directly on proton transfer within the permeation pore.

This modulation of the gA channel is of increasing interest for other channels as well, since biological membranes comprise several different types of lipids and local membrane composition, together with local elasticity, may well affect the properties of channels. The finding that the gA ionophore shows flickering and lifetime dependence on membrane composition may be valuable to the interpretation and understanding of similar processes that occur in proteins/channels, irrespective of the fact that gA channel gating involves dimerization. In fact, functional homomeric and heteromeric association of two or more subunits participate to the formation of several native channels such as the potassium [88] and chloride channels [89, 90].

The important role played by chemico-physical parameters in the modulation of the gA channel was demonstrated in experiments performed on membranes comprising both phospholipids and gangliosides i.e. glycosphingolipids carrying oligosaccharide chains characterised by sialic acid residues. Sialic acids are negatively charged sugars responsible for a large part of the electrical charges present at the plasma membrane of several cells.

Incorporation of glycolipids into the lipid bilayer affects the characteristics of the gA channel as demonstrated by the addition of ganglioside micelles to the aqueous solution bathing planar phospholipid membranes (BLM). Gramicidin was incorporated into monosialoganglioside (GM1) micelles, which were successively fused with bilayers formed by one of the following phospholipids: dioleoylphosphatidylcholine (DOPC), or phosphatidylserine (PS) or phosphatidylethanolamine (PE) [91]. The experimental protocol was designed to favor the incorporation into the lipid bilayer of gramicidin molecules entrapped in ganglioside micelles, preventing the incorporation into the bilayer of free gramicidin molecules. In DOPC and PS membranes, GM1 micelles containing gramicidin generate channels whose lifetimes and conductance are not significantly different from those observed in pure DOPC and PS bilayers. On the contrary, in PE membranes the presence of GM1 significantly modifies the characteristics of gA, increasing the mean lifetime of the single channel and decreasing its amplitude (Fig. 5) [91].

The membrane surface charge presumably plays a major role in regulating the interaction between PE and GM1 micelles as no single channel opening was observed in the absence of calcium at relatively low ionic strength of the bathing solution (100 mM KCl) and low gramicidin concentration. The addition of 5 mM $CaCl_2$ greatly increases the probability of incorporating ganglioside micelles into PE bilayers. This role played by calcium is not surprising, as different studies have shown that divalent ions affect fusion of phospholipid vesicles both by neutralizing (ganglioside) negative charges [92] and by conferring fusogenic properties to PE [93]. It has been suggested that ganglioside micelles incorporated into PE membranes may survive for relatively long periods and give rise to the formation of ganglioside microdomains [91]. Possibly, the existence of hydrogen bonds between the nitrogen and the non-esterified phosphate oxygens of adjacent PE phospholipids may confer to ganglioside domains a peculiarly long lifetime in PE bilayers [91].

646

Chapter 23

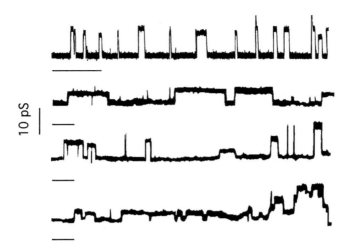

10 pS

Fig.5 Gramicidin single channels recorded in a pure PE bilayer (upper trace) and in a PE bilayer after gramicidin incorporation into ganglioside GM1 micelles (three lower traces). The segment below each trace represents a time calibration bar of 10 s. Upper trace, ionic solutions: KCl 0.1 M, pH≈6, applied potential V = +100 mV; three lower traces: ionic solution KCl 0.1 M, CaCl$_2$1 mM, pH≈6, applied potential V = +60 mV. The mean single channel conductance was 12 pS (upper trace) in PE bilayers while it ranged from a few pS up to 8 pS (three lower traces) when gramicidin was incorporated into GM1 vesicles fused with PE membranes.

Gramicidin presumably remains entrapped in these domains and experiences chemico-physical boundary conditions which are largely determined either by the hydrophobic chains of gangliosides and/or their negatively charged polar heads.

Therefore, in accordance with other papers which have described the dependence of gramicidin conductance [87] and lifetimes [73, 81-86] on membrane composition and bilayer tension the peculiar characteristics of GM1 micelles incorporated into PE bilayers also affect both channel conductance and lifetime.

Notably, free gramicidin molecules incorporated into bilayers made of PE:GM1 mixtures do not show current transitions significantly different from those observed in pure PE membranes. Possibly, in bilayers made of PE:GM1 mixtures, gangliosides are homogeneously distributed through the phospholipids milieu and therefore do not determine crucial modifications in the local chemico-physical characteristics of the membrane as they do in the presence of clustered ganglioside microdomains.

This evidence supports the possibility that the composition and properties of the membrane may modulate the activity of channels with a relatively simple structure. This will be further illustrated in the following sections where the incorporation of channel peptides into native membranes and BLM is reported.

2.2. Voltage-dependent peptide pore formers

A wealth of peptides able to increase membrane permeability is known and has been studied to date [94]. They range from natural products having antimicrobial or, more in general, antibiotic function to toxin components of animal or bacterial origin, to entirely synthetic peptides that either reproduce elements of natural channels or elaborate the concept of amphipathic membrane-active compounds. The majority of these peptides form channels that are not permanently open, but rather fluctuate between an open and a closed state, respectively, permitting or not the passage of ions through the membrane. The mechanism by which a channel switches between the closed and the open state, under the control of an external variable, is called 'gating'. For what concerns peptide pore-formers, the main parameter controlling channel gating is the applied voltage, however, other variables could exert the same effect, such as, for example, the binding of ligands to chemically activated channels (e.g. Ca^{++}-activated K^+ channels or receptor channels].

Gating of a single ion channel is a stochastic process. Usually an equilibrium probability P_o is introduced to define the fraction of time the pore is in the open state, or, in a population of identical channels, the fraction of pores that are in the open state at a given moment of time. If gating is described as a simple chemical equilibrium between two conformations, C (closed) and O (open), of one macromolecule (the pore) we can write

$$C \rightleftharpoons O \tag{i}$$

$$P_o = k_{co} / (k_{co} + k_{oc}) \tag{9}$$

where k_{co} and k_{oc} are the opening and closing rate respectively.

Different models have been introduced to describe the voltage dependence of the opening and closing rates depending on the peptide. Three representative cases will be discussed in the following.

2.2.1 barrel stave pores by natural and synthetic peptides

A central paradigm of pore formation by peptide and protein structures is that the secondary structure should be such that hydrogen bonding to surrounding molecules is avoided. The very first, most basic structure complying with this assumption was considered to be the α-helix bundle. This model originated from two considerations: 1) secondary structures with all-internal hydrogen bonding are only α-helix and β-sheet; 2) theoretical analysis of the primary structure of natural channels, as they were becoming available (e.g. acetylcholine receptor channel [95], Na^+ [96], K^+ [97] and Ca^{2+} [98] channels), suggested the presence of multiple hydrophobic stretches with α-helix propensity and average length of 20 residues.

They were just long enough to form a transmembrane helix when inserted perpendicular to the membrane surface. Though, the uniqueness of this model has been challenged by the existence of natural channels with transmembrane β-barrel conformation (the porins [99-101]) and by the later discovery of protein toxins and peptides that also adopt beta configuration (e.g. *Staphylococcus aureus* α-toxin [102] and protegrin [103]), the α-helix bundle has remained the model of choice for most of membrane active-peptides.

In the α-helix bundle pore model, a central pore, punched in an otherwise unperturbed bilayer, is surrounded by amphipathic α-helices long enough to span the bilayer (i.e. at least 20 residues for a bilayer thickness of 3 nm). The helices are disposed parallel to each other and perpendicular to the plane of the membrane. Their hydrophobic face is oriented towards the apolar lipid interior whereas the hydrophilic residues face the channel lumen where they can establish favorable interactions with solvent and permeant molecules. The bundle may be further stabilized by ion interactions between adjacent helices.

Residues on the hydrophilic face may be either charged or not. Very often, in antibacterial peptides, positive charges prevail to confer a basic character, which is advantageous for the interaction with bacterial outer membrane. Electrically charged residues may concur to generate a voltage-gating mechanism, whereby resting α-helical rods, lying flat on the surface of the membrane, are inserted to form the bundle pore by an imparted electric field.

Independently of the presence of electrical charges, α-helices have an intrinsic property, the helix dipole, which causes them to couple to the electric field [104, 105]. This originates from the fact that the intrinsic dipole moment of the peptide bond of each residue aligns, in an α-helix, to generate a macroscopic dipole moment approximately oriented as the helix axis. For an helix of 20 residues, this dipole amounts to 63 Debye, which corresponds to having approximately a charge of +1/2 e.u. at the N-terminus and one of -1/2 e.u. at the C-terminus.

The presence of this dipole can have important consequences. In fact, in the most general case, an amphipathic helical peptide would lie, in the absence of an electric field, parallel to the membrane surface (see Fig. 6C model a). However, upon application of a transmembrane voltage it will tend to orient with the helix axis parallel to the electric field, normal to the bilayer. The fraction of inserted peptide, $f_\perp$, is given by [94]

$$f_\perp = 1/(1+\exp(\Delta W + \mu V/d)/k_B T) \qquad (10)$$

where ΔW is the change in conformational energy between the two orientations, V the applied voltage, d the membrane thickness and μ the helix dipole. Assuming d = 3 nm, it can be calculated that a $V_{1/2}$ (voltage for half time at perpendicular position) requires a μ corresponding to 80 to 100 residues, which is not the practical case. However, simultaneous orientation or 4 or 5 helices of 20 residues, as it would occur for example in a bundle pore, can attain the same result, making the model significant. In the case of opening of an oligomeric bundle pore, a power dependence of the opening rate on the peptide concentration, C^δ, where δ is the number of helices in the bundle, is expected.

An alternative antiparallel-to-parallel model was introduced by Boheim for alamethicin [106], a natural peptaibol (i.e. a peptide containing the unusual aminoacid α-amino isobutyric acid, Aib, and an α-amino alcohol at the C-terminus). In this case, at the resting potential, the oligomeric seed of peptide molecules is already inserted into the membrane, forming a compact bundle of antiparallel helices in favorable self-interaction. Upon application of a sufficiently large electric field, odd-oriented dipoles are forced to reorient along the field, thus forming a bundle of parallel helices which, owing to self-repulsion, will lead to an open pore (Fig. 6). Because the transition is now due to a complete 180° turn of the dipoles (instead of 90° as in the previous case), a reasonable coupling with the electric field is obtained with just two helices reorienting in pre-existing a bundle of a minimum of 3 or 4 helices.

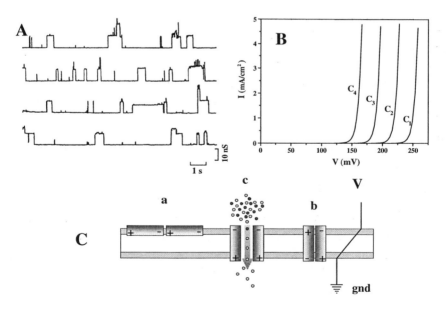

Fig. 6. A. Current fluctuations induced by the peptide Boc-(Ala-Aib-Ala-Aib-Ala)$_n$-Ome (with n = 4) in a BLM at an applied voltage of 220 mV and a concentration of 10 μg/ml (opening = upward deflections). B. Simulation of the current voltage curve induced by the same peptide according to Eq. 13. C1 to C4 correspond to 10, 20, 40 and 80 μg/ml respectively. α and δ (taken

from Ref. [108]) are 4.8 and 8.5 respectively. C. Model of generation of the helix-bundle pore. The bilayer is represented as a white slice enclosed by light-gray layers corresponding to the headgroups. The helical rods are amphipathic (clear where hydrophobic) and report opposite charges at the two ends generated by their dipole moment (charge is 0.5 e.u.). They can either lay flat on the membrane surface (a) or insert with antiparallel orientation (b). Upon application of a voltage they either insert in the bilayer or re-orient to generate an inserted parallel bundle (c). Mutual dipole repulsion, counterbalanced by helix-helix side interactions, leads to opening of a pore for the passage of cations (clear dots) or, to a lesser extent, anions (dark dots).

Since alamethicin has a net negative charge near the C-terminus, to test this model, a series of completely uncharged artificial peptides were generated reproducing the amphipathic character of the molecule. These peptides were constructed around a repeating unit: Boc-(Ala-Aib-Ala-Aib-Ala)$_n$-Ome with n assuming the values 1 to 4, to provide a variable length. All peptides assumed α-helical configuration in apolar solvents and were able to generate in a bilayer a strongly voltage-dependent conductance very similar to that of alamethicin [107, 108]. The characteristics of this can be summarized in: 1) a conductance-voltage curve that grows exponentially (e-fold every 5.2 mV) after a switch voltage Vc; 2) a concentration dependence of Vc which is decreased by 30 mV for each two-fold increase of the peptide concentration.

As for alamethicin, with these synthetic peptides the voltage and concentration dependence of the conductance G has the following form

$$G = Go \cdot \exp\left(\alpha\, eV/k_B T\right) \cdot C^\delta \tag{11}$$

From the observed voltage dependence of G (at constant C) one derives $\alpha = 4.8$. Since V_c is the value of V at which G first becomes $\neq 0$ we have

$$\exp\left(\alpha\, eV_c'/k_B T\right) \cdot C'^\delta = \exp\left(\alpha\, eV_c''/k_B T\right) \cdot C''^\delta \tag{12}$$

where C' and C" are two different peptide concentrations and V_c' and V_c'' the corresponding switch voltages. Now, in view of the mentioned dependence of V_c on C, for the case C" = 2 C' we can write

$$\exp\left(\alpha\, eV_c'/k_B T\right) \cdot C'^\delta = \exp\left(\alpha\, e(V_c' - 30\text{mV})/k_B T\right) \cdot (2C')^\delta \tag{13}$$

Solving Eq. (13) leads to $\delta = 8.5$, which represents the molecularity of the bundle. Therefore, the average peptide pore comprises 8 to 9 helices, and the voltage dependence indicates the movement of 4.8 charges through the electric field upon opening, i.e. roughly 0.5 charges moving for each helix. The same is true for alamethicin, for which it was found that $\alpha = 5.5$ and $\delta = 9.8$ [108, 109]. This is exactly the situation expected for the antiparallel-to-parallel helix dipole model, since in that case half of the helices of the bundle should undergo an 180° rotation, causing the net movement through the membrane of one charge: +1/2 of it located at the N-terminus and -1/2 at the C-terminus (which moves in the opposite direction) [108, 110]. On the other hand, also a situation in which, at rest, the helices lie flat on the membrane, and are all inserted simultaneously by the electric field (i.e. a 90° rotation with the equivalent displacement of only 1/2 charge/helix through the membrane), would lead to the same result. Actually, many different models have been provided and distinguishing between them only on the basis of electrical measurements seems to be difficult [94]. The fact that even short peptides were able to produce pores with similar characteristics was rationalized in terms of the formation of head to tail dimers inside the membrane. This was justified by the fact that the concentration of smaller peptide required to observe the same amount of pores was much larger than that of longer peptides (e.g. 100 times larger in the case of peptides with 5 and 20 residues respectively).

2.2.2. Lipocyclopeptide pores (syringopeptins and related peptides)

Syringopeptin 25 A ($SP_{25}A$) is a lipodepsipeptide (LDP) produced by the plant pathogen *Pseudomonas syringae* pv. *syringae* [111, 112] a phytopathogenic organism that affects several crop plants. The $SP_{25}A$ molecule is composed of three parts: a long unbranched 3-hydroxy fatty acid chain, a peptide moiety of 25 amino acids and a cyclic ring of 7 amino acids at the C-terminus. The cyclic ring contains two positively charged residues (2,4-diaminobutanoic acid, Dab) which impart $SP_{25}A$ its basic character [111]. The marked phytotoxic activity of $SP_{25}A$ indicates that it may significantly contribute to the virulence of *P. syringae*, and therefore be involved, with other LDPs, in a variety of plant diseases [113] and in antifungal activity [114].

Similarly to the case of other bacterial lipocyclopeptides, the peculiar amphipathic characteristic of $SP_{25}A$ suggests that its main site of action is the cell plasma membrane. Indeed we could show that $SP_{25}A$ incorporates into lipid monolayers and increases the permeability of unilamellar lipid vesicles, releasing their internal content [115]. Most informative was the analysis of its interaction with BLM [116]. The opening of $SP_{25}A$ channels was indicated by the appearance of uniform current increases when negative voltages were applied, which were quickly reverted when the voltage was switched positive. Gating was dependent also on lipid composition. In fact, rather large negative voltages were required to open pores in BLM comprised of mixtures of purified phospholipids, but much smaller in BLM comprised of asolectin, a natural plant lipid mixture containing approximately 20% of acidic lipids [116, 117]. Channel conductance, measured by the height of the current step, was about 40 pS at negative voltages and 30 pS at positive voltages (in

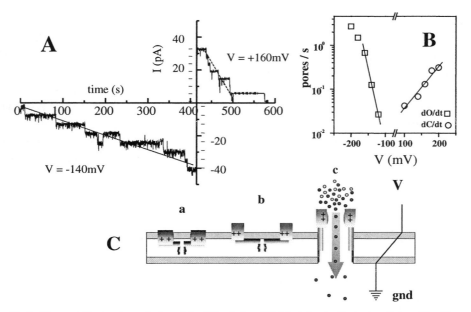

Fig. 7. A. Current fluctuations induced by $SP_{25}A$ in a BLM at the indicated voltage. The linear rate of opening and closing of the channels at negative and positive voltages are indicated respectively by a dotted and a dashed line. B. Voltage-dependence of the initial rates of opening (dO/dt) and closing (dC/dt) of the channels in half logarithmic scale (taken from Ref. [116]). C. Model of generation of the $SP_{25}A$ pore (the bilayer is represented as in Fig. 6). (a) Monomers absorb in a folded configuration, with the cyclic peptide moiety at the level of the glycerol backbone and the acyl chain inserted between the lipid tails. (b) Application of a negative voltage induces $SP_{25}A$ unfolding by attracting the positive charge of each lipopeptide (+2 e.u.) out of the membrane while the hydrophobic part is anchored to the apolar core. (c) A barrel-stave pore is formed. Due to the positive charge, the pore is selective for the passage of anions (dark dots).

0.1 M NaCl). Even in the presence of many channels, the I/V curve was markedly non-linear, in the sense that the conductance (slope of the I/V curve) was larger at negative than at positive potentials. The ratio of the two limiting conductances was around 2. The origin of this non-linearity is conceivably an asymmetric distribution of the charges on the pore (see the model in Fig. 7). In fact, the presence of the positive charges of the lactone ring near the pore entrance would accumulate locally the anions and generate a larger anion current when negative voltages, that push anions through the pore, are applied. $SP_{25}A$ channels were anion selective, independently of membrane composition, the permeability of Cl$^-$ being almost 3 times that of Na$^+$ [116]. Also such selectivity is conceivably due to the presence, at the pore entrance, of the uncompensated positive charges of the cyclic peptide moiety that would attract anions and repel cations.

Voltage gating. The initial rate of opening of the channels at negative voltages, and the rate at which the channels decayed to the closed state with large positive voltages, were both exponentially regulated by the applied voltage. This voltage-dependent gating can be explained by the two-state model previously introduced, Scheme (i), assuming that the transition between the open and the closed state corresponds to a major conformational change of the peptide, triggered by the movement of a charge through the electric field [118-120] (see Fig. 7).

In the case of N channels present in the membrane, O and C of Scheme (i) assume the value of the number of open and closed channels respectively (with C + O = N). The two rate constants k_{co} and k_{oc} of Eq. (9) are given by [121]:

$$k_{co} = f \exp(-(E^* - Ec)/k_B T) \tag{14a}$$

$$k_{oc} = f \exp(-(E^* - Eo)/k_B T) \tag{14b}$$

in which E^*, Eo, and Ec are the activation energy and the energies of the open and closed state respectively, and f is a frequency factor proportional to $k_B T/h$ (h, Planck constant).

By inspection of Fig.7, it is apparent that the applied electric field will contribute a term to the energies of the different states, which is QV for the open state, $(1-\delta)$ QV for the closed state (δ being the fraction of field felt by the charge at the equilibrium position of the closed state), and an intermediate value for the active state. Hence we may write:

$$Eo = Eo(0) + QV \tag{15a}$$

$$Ec = Ec(0) + (1-\delta) QV \qquad 0 < \delta < 1 \tag{15b}$$

$$E^* = E^*(0) + (1-\varepsilon\delta) QV \qquad 0 < \varepsilon < 1 \tag{15c}$$

where the terms Ei(0) (with i=o,c,*) denote conformational energies.

For the initial rate of opening of the channels, dO/dt, using the starting condition O = 0 and combining Eq. (14) and (15), we get:

$$dO/dt = N\, k_{co}(0)\ \exp -(1-\varepsilon)\, \delta\, QV / k_B T \tag{16a}$$

and for the initial rate of closing, dC/dt, and the starting condition C = 0

$$dC/dt = N\, k_{oc}(0)\ \exp(\varepsilon\, \delta\, QV / k_B T) \tag{16b}$$

Here $k_{co}(0)$ and $k_{oc}(0)$ are given by $f \cdot \exp(Ec(0) - E^*(0)/k_B T)$ and $f \cdot \exp(Eo(0) - E^*(0)/k_B T)$ respectively. As shown in Fig. 7, the two observed rates have indeed exponential voltage dependence with slopes (m and n) of opposite sign:

$$m = -(1-\varepsilon)\ \delta\ Q/\ k_B T \tag{17a}$$

$$n = \varepsilon\ \delta\ Q\ /\ k_B T) \tag{17b}$$

The rate of opening decreased e-fold every 12 mV, while that of closing increased e-fold every 45 mV. Inspecting Eq. (17), it is evident that the slopes are related to the gating charge by

$$\delta Q = (n-m) \cdot k_B T \tag{18}$$

In the case of $SP_{25}A$ we obtained $\delta Q = 2.6$ elementary charges. Considering that each $SP_{25}A$ carries two positive charges, provided by the two Dab residues located at the cyclic peptide moiety, and that pores may be pentameric [115], the gating should be due to the displacement of the ring through approximately one-quarter of the applied electric field ($\delta = 0.26$). Consequently we can derive $\varepsilon = 0.21$ for the fraction of voltage change in the activated state. Because a negative voltage on the side of toxin addition opens the pore, the involved events, outlined in Fig. 7, should be as follows. 1) Partitioning of the folded LDP monomer into the membrane, with partial embedding of the cyclic peptide moiety into the lipid at the level of the glycerol backbone and insertion of the acyl chain between those of the lipid. 2) Aggregation of the embedded molecules into the lipid film. 3) Extrusion of the positive charges from the membrane in response to a negative voltage. Since the hydrophobic part of the peptide is anchored to the apolar core of the membrane, a complete unfolding of the adsorbed LDP ensues, with the alignment of the unfolded hydrophobic part of the peptide to the lipid tails. 4) Formation of a barrel-stave pore. This implies that the cyclic peptide moiety remains always on the side of addition without crossing the bilayer even when the channels open and close [116].

A qualitatively similar behavior was reported also for SRE, however in that case an e-fold change every 13 mV for the opening and every 24-27 mV for the closing were observed [122].

2.2.3. Mixed lipid-peptide channels (toroidal pore model)

A new model for the formation of peptide-pores was recently proposed. It was suggested that the opening of the channel might be accompanied by the formation in the bilayer of a non-lamellar structure, i.e. a toroidal lipid pore surrounding the toxin structure [123]. In such toroidal arrangement, exemplified in Fig. 8, the lipid has a rather peculiar distribution. A positive curvature is observed perpendicular to the plane of the membrane, whereas a negative curvature is present in the membrane plane surrounding the pore. In such a configuration the outer monolayer joins the inner one at the pore site. Non lamellar lipid structures are formed which share features of both the cubic phase (negative curvature) and the micelle (positive curvature). The toroidal pore has been proposed for the action of the cationic anti-microbial peptide magainin [124, 125] and has recently found compelling experimental evidences [126]. Magainin pore is lined by 7 copies of α-helical arranged molecules, and has a diameter of around 3.5 nm.

A careful comparison of two other α-helical peptide pore-formers, alamethicin and melittin [127], has shown that while the former never makes a toroidal pore (but always an helix-bundle) the second may form the toroidal pore, in some conditions of lipid type and peptide concentration.

Very recently we have found that also sticholysin I and II (St I and St II), two pore-forming cytolysins from the sea anemone *Stichodactyla helianthus*, may work by forming a toroidal pore in the lipid membrane [128]. We observed, in fact, that the presence of minute amounts of lipids favoring a non-lamellar phase, strongly promoted the formation of toxin pores. Among the lipids tried, phosphatidic acid (PA) and cardiolipin (CL), were the most efficient pore promoters. Interestingly, they are also the strongest inducers of negative curvature in the bilayer [129-132]. It is possible that their presence favors the non-lamellar organization required by the toroidal pore. Besides inducers of negative curvature, like the negatively charged lipids above [131, 132], stearylamine (SA) had also a strengthening effect [128]. SA, being a single chain lipid, is an inducer of positive curvature [131, 132] and could help in stabilising the profile of the toroidal pore perpendicular to the membrane plane.

The diameter of the pore formed by sticholysins is around 2.2 nm [133, 134] and thus could involve only four or five transmembrane elements, in agreement with the fact that the average stoichiometry of actinoporin channels is believed to be a tetramer [135-137]. The diameter of the lipid pore surrounding the toxin would thus be around 3-4 nm, exactly in the range of that observed in the inverse hexagonal lipid phase (H_{II}), which varies between 3 and 5 nm, depending on the lipid present [131].

A putative region of the actinoporins, which could be involved in crossing the membrane, is an amphipathic α-helix located at the N-terminus which was predicted, on the basis of the primary structures, as a conserved element of all actinoporins [138-140]. Recently, the existence of an α-helix at the N-term of equinatoxin II (Eqt II), a parent actinoporin from the sea anemone *Actinia equina*, has been directly established by X-ray crystallography [141]. Interestingly, we anticipated that this helix has striking homology with bee venom melittin [138], another peptide capable of inducing the toroidal pore [127]. The helix carries, at its the C-terminus, a cluster of three positively charged residues (K^+-S-V-R^+-K^+) that could interact with acidic lipids in the negative curvature region, compensating the electrostatic repulsion that would arise by the close packing of the negatively charged head groups Fig. 8. Consistently, the interaction of the N-term region of Eqt II with the lipid membrane was demonstrated by cysteine scanning mutagenesis [142].

While the formation of channels by actinoporins is virtually voltage independent [133, 136], for magainin a voltage-dependence somehow similar to that in Fig. (6) was reported [143]. From those data, and using the treatment described by Eq. (11-13), we obtain $\alpha = 1.3 - 1.7$ and $\delta = 4.8 - 6.3$. These values are considerably smaller than with the peptaibols, and suggest that the size of the pore is penta- to hexameric and that for each monomer there is an average displacement of 0.27 charges (given by the ratio α/δ). In the case of larger proteins, like the actinoporins, local monomer-monomer interactions may stabilize the pore allowing channels to be formed even in the absence of applied voltage.

One implication of the toroidal model is that each pore provides a point of fusion of the inner lipid monolayer to the outer. Transbilayer movement of lipid molecules, which is normally severely restricted, would thus be favored. We have indeed observed that St I induces mobilization of inner layer lipids in a dose dependent way [128] thus lending further credit to the toroidal hypothesis.

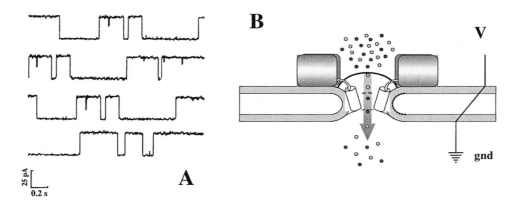

Fig. 8. A. Current fluctuations induced by the sea anemone toxin St II in planar lipid bilayers at an applied voltage of -40 mV (opening = downward deflection). B. Model of the toroidal-pore generation. Four or five absorbed toxin monomers insert a melittin-like helix into the membrane, stabilising a lipidic pore. This is favored by the presence of particular lipids. Those inducing positive curvature (e.g. SA) are favored in the section of the lipid torus perpendicular to the plane of the membrane (dotted line), whereas those inducing negative curvature (e.g. PA and CL) are favored in the central section parallel to the plane of the membrane (continuos line).

A structure with some similarities, called the 'carpet model', has been proposed as a general mechanism of action of amphipathic α-helical antimicrobial peptides [144]. Also this model implies the formation of non-lamellar lipid phases, however, the overall structure is much less ordered and involves the presence of a large amount of peptide.

3. VOLTAGE-GATING IN PROTEIN CHANNELS

3.1. The plant tonoplast: a native membrane model

The plant vacuole is an intracellular storage organelle that in mature plant cells occupies up to 90% of cytoplasmic volume. The membrane of the plant vacuole (tonoplast) can be used as a biological model system for studying the action of pore-forming proteins and peptides. While artificial membranes are typically constructed with pure lipids, the tonoplast contains a variety of phospholipids, glycolipids and neutral lipids [145, 146] together with channel regulators. Channel properties in this native membrane may differ significantly from those observed in artificial pure lipid bilayers. The plant tonoplast is a spherical stable and clean structure that well lends itself to investigation through the patch-clamp technique, while the presence of endogenous (vacuolar) channels represents a useful tool when monitoring membrane functionality during prolonged patch-clamp experiments [147, 148].

3.2. Incorporation of an exogenous peptide into the tonoplast is modulated by the lipid composition of the membrane.

As we have shown before, the plant pathogenic lipodepsipeptide $SP_{25}A$ produced by *P. syringae* pv. *syringae* [111, 112] forms ion channels in planar lipid bilayers [149]. To investigate the properties of $SP_{25}A$ in biological membranes, and to promote plant tonoplast as a more complete model system than BLM into characterization of exogenous channel formers, $SP_{25}A$ was incorporated into tonoplasts obtained from sugar beet and maize roots [147]. In sugar beet tonoplast, $SP_{25}A$ determines a rapid increase in membrane permeability, giving rise to the formation of discrete current transitions lasting several seconds (Fig. 9) which resemble ion channels formed by $SP_{25}A$ in planar lipid bilayers [147]. In plant tonoplasts the $SP_{25}A$ channel displays anion selectivity (with a Cl^-/K^+ permeability ratio of ≈ 7) and a selectivity sequence ($NO_3^- \approx Cl^- > F > gluconate^-$) similar to those observed in BLM.

As is the case in artificial membranes, also in vacuolar patches permeabilization induced by $SP_{25}A$ decays within a few minutes as soon as $SP_{25}A$ is removed from the bathing solution. This may be due either to the inactivation of the channel or to the desorption of the peptide from the membrane phase. As for BLMs, in patched tonoplasts the channel shows an intrinsic rectifying characteristic as a result of different single channel conductances at negative and positive voltages. This asymmetric conductance presumably depends on an asymmetric distribution of fixed positive charges at the cyclic ring which participate in the formation of channel's mouth [149].

Despite all the similarities in the behavior of SP$_{25}$A in BLM and in tonoplasts, experiments performed on plant vacuoles definitely confirmed that some of the characteristics of the SP$_{25}$A channel are significantly affected by the membrane's composition. In contrast with results obtained with BLM made of pure phospholipid mixtures (PC:PE:PS 2:2:1 or PC:PS 4:1), where the application of large negative voltages ($\approx$-140 mV, see Fig. 9 for the convention on voltage sign) is necessary to gate SP$_{25}$A channel activation [149], no voltage-dependent gating was observed in plant vacuoles, where the Sp$_{25}$A does not show any threshold of activation and channels open at very low transmembrane potentials.

On the other hand, as in the case with vacuolar membrane, in asolectin BLMs smaller membrane voltages are sufficient to trigger channel opening. It is worth noting that the phospholipid composition of the asolectin mixture should be similar to that of the vacuolar membrane [145, 146]. This lipid dependence of channel activity, also observed with other LDPs [151, 152], suggests that lipids participate in the SP$_{25}$A gating mechanism.

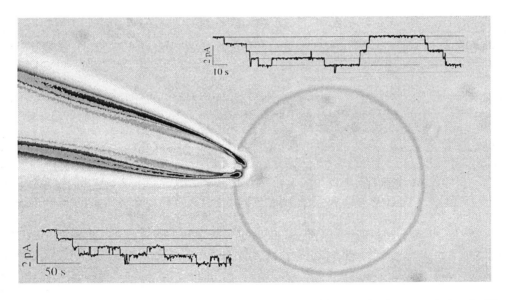

Fig. 9 Typical configuration showing a sugar beet vacuole with a patch pipette sealed to it. The current trace on the upper right corner of the figure represents single channel transitions induced by the phytotoxic peptide SP$_{25}$A in an excised tonoplast patch. Applied potential V=-50 mV (vacuolar side assumed as the reference potential). The sign convention for endomembranes was

adopted [150]. Symmetrical solutions (in mM): KCl 150, EDTA 1, MES 10, pH 6.0. For prompt comparison the trace in the lower left corner represents single channels recorded in an asolectin BLM. Applied potential V=-40 mV (trans side was assumed as the reference potential). Symmetrical ionic solution (in mM): NaCl 100, MES 10, EDTA 1, pH 6.0. In both cases SP25A was added on the cis side only.

With respect to the classical use of BLMs, the investigation of channel formers via a more complete and complex system, such as the plant tonoplast, clearly confirms that lipid composition plays a role in membrane-channel interaction and provides information which may be of great importance for understanding physiological mechanisms mediated by channels in vivo. Specifically, the presence of a lipid-dependent mechanism of incorporation of $SP_{25}A$ into natural membranes may be of primary importance to understand the processes affecting toxin penetration and its effects in plant and fungal membranes.

3.3. Intrinsic voltage dependence in ion channels

In the last decade several ion channels have been cloned from Bacteria and Eukarya and domains important for specific channel activity have been identified and characterised using primary structure site-directed mutagenesis and functional expression of mutated channels in heterologous cell lines.

Voltage-gated ion channels exhibit different ion selectivities for different monovalent and divalent cations (Na^+, K^+, and Ca^{2+}) but a similar mechanism of activation and a well conserved common structure (for a recent review see [153]). These proteins form the family of voltage-gated ion channels, which in animal cells are involved in major physiological roles like the generation of electrical signals. Their common structure (based on thousands of amino acids) is based on four homologous domains (for Na^+ and Ca^{2+} channels) or four alpha-subunits (for K^+ channels), each one comprising six putative helices spanning the hydrophobic core of the membrane. A highly conserved segment between helices S5 and S6 contributes to the formation of the channel pore and is responsible for the selectivity filter.

In these channels, gating is an intrinsic property of the protein as voltage-dependent activation relies on the presence of an electrically charged structure, identified with the S4 domain, which senses the transmembrane electrical field. The S4 segment comprises ≈ 21 amino acids with several positively charged basic amino acids, which participate in voltage-dependent gating. The application of a depolarizing membrane potential determines conformational changes in S4 and in contiguous channel regions, forcing the voltage sensor to move through the lipid bilayer.

3.4. Comparison of intrinsic voltage dependence in plant and animal channels

The potassium channels AKT1 and KAT1 cloned from *Arabidopsis thaliana* were the very first voltage-dependent channels identified and cloned in higher plants [154, 155]. KAT1, as well its homologous KST1 (cloned from potato [156]), are mainly expressed in leaves while AKT1 and its homologous channels (like AKT2 [157]) appear to be predominantly expressed in roots (for recent reviews on plant channels see [158-160]).

Although their topology closely resembles that of the outward rectifying *Shaker* channel, KAT1 and AKT1 and their homologous have actually been identified as inward rectifying channels from expression in *Xenopus laevis* oocytes or complementation of yeast mutants deficient in K$^+$ influx. Like in *Shaker* channels, six transmembrane domains, a highly conserved pore segment (connecting the S5 to the S6 domains, responsible for K$^+$ selectivity), and a S4 domain (which is the voltage sensor of the channel) have also been identified in the KAT1 and AKT1 channels [154, 155, 159].

Notably the primary structure of plant inward K$^+$ channels differs significantly from that of inward-rectifying channels cloned from animal systems. These last channels only have two main hydrophobic domains (surrounding a consensus sequence implicated in K$^+$ selectivity) while their rectification characteristic is not an intrinsic property of the protein but depends on a blockage of the pore by intracellular magnesium [161].

These observations have posed interesting questions on the regulation of voltage-dependent channels and have also contributed significantly to clarifying the nature of the molecular mechanisms at the basis of the gating processes in voltage-dependent channels.

The construction of chimeras containing the first four membrane-spanning segments of KAT1 and the S5 and S6 segments of the animal outward rectifying channel, Xsha2, demonstrated that the region comprised between the N-terminus and the S4-S5 linker determines hyperpolarisation-induced activation in plant channels [162]. Accordingly, there is evidence that negatively charged residues in other segments act as counter ions for S4 charges and affect the voltage sensing process. It has been demonstrated that S4 interacts with the negatively charged residues in the S2 and S3 segments [163, 164] and even with the S5 segment in the hERG channel [165]. Also the interaction of S4 with the N-terminus (mostly involved in channel inactivation and in subunit assembly [166]) may affect voltage dependence and voltage sensitivity in plant potassium channels [167]. The fact that S4 is just one of the components of the voltage sensor is also confirmed by studies performed on chimeras constructed by inserting S4 from several divergent potassium channels into a *Shaker* channel [168]. None of these chimeras displayed the gating properties of their correspondent S4 donor channel. Other properties like sialidation [169] and cooperative transitions contribute to modify the gating charge of the channel.

In conclusion, gating in potassium-selective voltage-dependent channels is a complex process which is regulated by the interaction of the voltage sensor S4 with several other protein amino acids. It is reasonable that in these proteins comprising several domains and thousands of amino acids the role played by the phospholipid medium would not be as crucial as in smaller polypeptide channel formers.

Nevertheless, minor modulation of channel properties due to the composition of the membrane have even been observed in relatively complex channels such as the voltage-dependent N-type calcium channels [86] and (VRAC) volume regulated anion channels [170].

In the former paper [86], it has been shown that the alteration of membrane stiffness affects the properties of both elementary channels (like gA) incorporated into lipid bilayers and complex channels (like the N-type calcium channel) in their native membrane. Specifically, membrane patches (from IMR32 cells) and planar bilayers were exposed for a few minutes to beta-octylglucoside and Triton X-100 (TX100), cone shaped and micelle-forming molecules which are able to change membrane stiffness. These molecules determine a reversible decrease in the current amplitude of N-type calcium channels and an hyperpolarizing shift in the inactivation curve. Instead no shift or modification in the current-voltage relationship was observed. In the same paper the authors demonstrate that the same molecules (TX100 and beta-octylglucoside) induced an increase in gA channel lifetime in DOPC bilayers. Notably, these modifications are definitely similar to those described in section 2.1 dealing with gramicidin incorporated into GM1 micelles and fused with PE bilayers [91].

In the latter case (VRAChannel), it was demonstrated that the cholesterol content of bovine aortic-endothelial cell membranes affects the equilibrium between the open and closed states of the VRAChannel. The enrichment of cell membranes with cholesterol determined a suppression of VRAC activation on the application of a mild osmotic gradient but had no effect on VRAC activation when the cells were stimulated with a strong osmotic gradient. The authors suggest that suppression of VRAC currents by an increase in cholesterol is due either to an increase in membrane deformation energy or to a specific sterol-protein interaction which possibly modifies the transition probability between the closed and open state of the channel [170]. These modifications may contribute to the impairment of endothelial functions in the development of atherosclerosis.

It should finally be noted that several other examples are reported in the literature where non-specific interactions (mediated by bilayer elasticity) and specific interactions (which depend on the chemistry of both lipids and proteins) affect the activity of channels, either incorporated into a model system or expressed in heterologous cells, or even in their native membranes [171-177].

REFERENCES

[1] E. Neumann, N. Sowers and C. Jordan (eds.), Electroporation and Electrofusion, Plenum, New York, 1989.
[2] T.Y. Tsong; Biophys. J. 60 (1991) 297.
[3] D.C. Chang, B.M. Chassy, J.A. Saunders and A.E. Sowers (eds.), Guide to Electroporation and Electrofusion, Academic Press, New York, 1992.
[4] M.R. Prausnitz; Adv. Drug Deliv. Rev. 35 (1999) 61.
[5] H. Wolf, M.P. Rols, E. Boldt, E. Neumann and J. Teissié; Biophys. J. 66 (1994) 524.
[6] M. Golzio, M.P. Mora, C. Raynaud, C. Delteil, J. Teissié and M.P. Rols; Biophys. J. 74 (1998) 3015.
[7] Y. Mouneimne, P.F. Tosi, Y. Gazitt and C. Nicolau; Biochem. Biophys. Res. Commun. 159 (1989) 34.
[8] M.P. Rols and J. Teissié; Biophys. J. 75 (1998) 1415.
[9] S. Hui, N. Stoicheva and Y. Zhao; Biophys. J. 71 (1996) 1123.
[10] J. Teissié and C. Ramos; Biophys. J. 74 (1998) 1889.
[11] C. Ramos and J. Teissié; FEBS Lett. 465 (2000) 141.
[12] K.J. Kinosita, I. Ashikawa, N. Saita, H. Yoshimura, H. Itoh, H. Nagayama and A. Ikegami; Biophys. J. 53 (1988) 1015.
[13] B. Gabriel and J. Teissié; Biophys. J. 73 (1997) 2630.
[14] B. Gabriel and J. Teissié; Biophys. J. 76 (1999) 2158.
[15] F. Ryttsén, C. Farre, C. Brennan, G.S. Weber, K. Nolkrantz, K. Jardemark, D.T. Chiu and O. Orwar; Biophys. J. 79 (2000) 1993.
[16] I.G. Abidor, V.B. Arakelyan, L.V. Chernomordik, Y.A. Chizmadzhev, V.F. Pastushenko and M.R. Tarasevich; Bioelectrochem. Bioenerg. 6 (1979) 37.
[17] L.V. Chernomordik and I.G. Abidor; Bioelectrochem. Bioenerg. 7 (1980) 617.
[18] L.V. Chernomordik, M.M. Kozlov, G.B. Melikyan, I.G. Abidor, V.S. Markin and Y.A. Chizmadzhev; Biochim. Biophys. Acta 812 (1985) 643.
[19] J.C. Weaver, R.A. Mintzer, H. Ling and S.R. Sloan; Bioelectrochem. Bioenerg. 15 (1986) 229.
[20] L.V. Chernomordik, S.I. Sukharev, S.V. Popov, V.F. Pastushenko, A.V. Sokirko, I.G. Abidor and Y.A. Chizmadzhev; Biochim. Biophys. Acta 902 (1987) 360.
[21] J.C. Weaver and Y.A. Chizmadzhev; Bioelectrochem. Bioenerg. 41 (1996) 135.
[22] J.M. Crowley; Biophys. J. 13 (1973) 711.
[23] M. Robello and A. Gliozzi; Biochim. Biophys. Acta 982 (1989) 173.
[24] U. Zimmermann, F. Beckers and H.G.L. Coster; Biochim. Biophys. Acta 464 (1977) 399.
[25] I.P. Sugar; Biochim. Biophys. Acta 556 (1979) 72.
[26] D.S. Dimitrov; J. Membrane Biol. 78 (1984) 53.
[27] W. Sung and P.J. Park; Biophys. J. 73 (1997) 1797.
[28] J.C. Shillcock and U. Seifert; Biophys. J. 74 (1998) 1754.
[29] G. Saulis; Biophys. J. 73 (1997) 1299.
[30] E. Evans and W. Rawics; Phys. Rev. Lett. 64 (1990) 2094.
[31] E. Scalas, A. Ridi, M. Robello and A. Gliozzi; Europhys. Lett. 43 (1998) 101.
[32] A. Ridi, E. Scalas and A. Gliozzi; Eur. Phys. J. 2 (2000) 161.
[33] S. Kalinowski, G. Ibron, K. Bryl and Z. Figaszewski; Biochim. Biophys. Acta 1369 (1998) 204.

[34] I.G. Abidor, L.V. Chernomordik, S.I. Sukharev and Y.A. Chizmadzhev; Bioelectrochem. Bioenerg. 9 (1982) 141.

[35] R.W. Glaser, S.L. Leikin, L.V. Chernomordik, V.F. Pastushenko and A.I. Sokiro; Biochim. Biophys. Acta 940 (1988) 275.

[36] S.A. Freeman, M.A. Wang and J.C. Weaver; Biophys. J. 67 (1994) 42.

[37] R. Benz, F. Beckers and U. Zimmermann; J. Membrane Biol. 48 (1979) 181.

[38] C. Wilhelm, M. Winterhalter, U. Zimmermann and R. Benz; Biophys. J. 64 (1993) 121.

[39] K.H. Klotz, M. Winterhalter and R. Benz; Biochim. Biophys. Acta 1147 (1993) 161.

[40] A. Diederich, G. Baehr and M. Winterhalter; Phys. Rev. 58E (1998) 4483.

[41] A. Diederich, G. Baehr and M. Winterhalter; Langmuir 14 (1998) 4597.

[42] A. Diederich, M. Strobel, W. Meier and M. Winterhalter; J. Phys. Chem. 103 (1999) 1402.

[43] B. Schuster, U.B. Sleytr, A. Diedrich, G. Baehr and M. Winterhalter; Eur. Biophys. J. 28 (1999) 583.

[44] M. Gambaro, A. Gliozzi and M. Robello; Progr. Colloid Polym. Sci. 84 (1991) 189.

[45] I. Genco, A. Gliozzi, A. Relini, M. Robello and E. Scalas; Biochim. Biophys. Acta 1149 (1993) 10.

[46] M.B. Weissman; Rev. Mod. Phys. 60 (1988) 537.

[47] P. Pawlowski, S.A. Gallo, P.G. Johnson and S.W. Hui; Biophys. J. 75 (1998) 721.

[48] S.A. Freeman, M.A. Wang and J.C. Weaver; Biophys. J. 67 (1994) 42.

[49] D. Gross, L.M. Loew and W.W. Webb; Biophys. J. 50 (1998) 339.

[50] M.N. Teruel and T. Meyer; Biophys. J. 73 (1997) 1785.

[51] E. Neumann, K. Toensing, S. Kakorin, P. Budde and J. Frey; Biophys. J. 74 (1998) 98.

[52] D. Zhelev and D. Needham; Biochim. Biophys. Acta 1147 (1993) 89.

[53] J.D. Moroz and P. Nelson; Biophys. J. 72 (1997) 2211.

[54] B. Gabriel and J. Teissié; Biochim. Biophys. Acta 1266 (1995) 171.

[55] V. Sharma, K. Stebe, J.C. Murphy and L. Tung; Biophys. J. 71 (1996) 3229.

[56] G.C. Troiano, L. Tung, V. Sharma and K.J. Stebe; Biophys. J. 75 (1998) 880.

[57] Y. Rosemberg, M. Rottemberg and R. Korenstein; Biophys. J. 67 (1994) 1060.

[58] L.V. Chernomordik, S.I. Sukharev, S.V. Popov, V.F. Pastushenko, A.V. Sokirko, I.G. Abidor and Y.A. Chizmadzhev; Biochim. Biophys. Acta 902 (1987) 360.

[59] D. Needham and R.M. Hochmuth; Biophys. J. 55 (1989) 1001.

[60] C.G. Troiano, K.J. Stebe, R.M. Raphael and L. Tung; Biophys. J. 76 (1999) 3150.

[61] A. Gliozzi, R. Rolandi, M. De Rosa and A. Gambacorta; J. Membrane Biol. 75 (1983) 45.

[62] P. Mueller, D.O. Rudin, H. Ti Tien and W.C. Wescott; Nature 194 (1962) 979.

[63] M. De Rosa, A. Gambacorta and A. Gliozzi; Microbiol. Rev. 50 (1986) 70.

[64] A. Gambacorta, A. Gliozzi and M. De Rosa; World J. Microbiol. Biotechnol. 11 (1995) 115.

[65] A. Gliozzi, A. Relini and P.L.G. Chong; J. Membrane Sci. 5191 (2002) 1.

[66] A. Gliozzi, M. Robello, A. Relini and G. Accardo; Biochim. Biophys. Acta 1189 (1994) 96.

[67] L. Fittabile, M. Robello, A. Relini, M. De Rosa and A. Gliozzi; Thin Solid Films 284 (1996) 735.

[68] G.B. Melikian, N.S. Matinyan, S.L. Kochanov, V.B. Arkelian, D.A. Prangishvili and K.G. Nadareishvili; Biochim. Biophys. Acta 1068 (1991) 227.

[69] A. Ottova-Leitmannova and H. Ti Tien; Progr. Surf. Sci. 41 (1993) 337.

[70] M. Montal; Annu. Rev. Biophys. Biomol. Struct. 24 (1995) 31.

[71] H. Haas, G. Lamura and A. Gliozzi; Bioelectrochemistry 54 (2001) 1.

[72] A. Gliozzi, M. Robello, L. Fittabile, A. Relini and A. Gambacorta; Biochim. Biophys. Acta 1283 (1996) 1.

[73] E. Bamberg and P. Laüger; J. Membrane Biol. 11 (1973) 177.

[74]   S.B. Hladky and D.A. Haydon; Biochim. Biophys. Acta 274 (1972) 294.
[75]   H. Kolb and E. Bamberg; Biochim. Biophys. Acta 464 (1977) 127.
[76]   V. Myers and D.A. Haydon; Biochim. Biophys. Acta 274 (1972) 313.
[77]   R. Sarges and S.B. Witkop; J. Amer. Chem. Soc. 87 (1965) 2011.
[78]   D.D. Busath; Annu. Rev. Physiol. 55 (1993) 473.
[79]   J. Sandblom, J. Galvanovskis and B. Jilderos; Biophys. J. 79 (2001) 2526.
[80]   A. Ring; Biophys. J. 61 (1992) 1306.
[81]   O.S. Andersen, H.J. Apell, E. Bamberg, D.D. Busath, R.E. Koeppe, F.J. Sigworth, G. Szabo, D.W. Urry and A. Woolley; Nature Struct. Biol. 6 (1999) 609.
[82]   M. Goulian, O.N. Mesquita, D.K. Fygenson, C. Nielsen, O.S. Andersen and A. Libchaber; Biophys. J. 74 (1998) 328.
[83]   E. Neher and H. Eibl; Biochim. Biophys. Acta 464 (1977) 37.
[84]   S.M. Saparov, Y.N. Antonenko, R.E. Koeppe II and P. Pohl; Biophys. J. 79 (2000) 2526.
[85]   A. Ring; Biochim. Biophys. Acta 856 (1986) 646.
[86]   J.A. Lundbaek, P. Birn, A.J. Hansen and O.S. Andersen; Biochemistry 35 (1996) 3825.
[87]   C.M.G. de Godoy and S. Cukierman; Biophys. J. 81 (2001) 1430.
[88]   D. Naranjo, in, R. Latorre and J.C. Saez (eds.), Plenum Press, New York and London (1997).
[89]   R. Dutzler, E.B. Campbell, M. Cadene, B.T. Chalt and R. MacKinnon; Nature 415 (2002) 287.
[90]   C. Miller and M.M. White; Proc. Natl. Acad. Sci. U. S. A. 81 (1984) 2772.
[91]   F. Gambale, C. Marchetti, C. Usai, M. Robello and A. Gorio; J. Neurosci. Res. 12 (1984) 355.
[92]   N. Ohki and N. Düzgünes; Biochim. Biophys. Acta 552 (1979) 438.
[93]   N. Düzgünes, J. Wilschut, R. Fraley and D. Papahadjopoulos; Biochim. Biophys. Acta 642 (1981) 182.
[94]   M.S.P. Sansom; Prog. Biophys. Molec. Biol. 55 (1991) 139.
[95]   M. Noda, H. Takahashi, T. Tanabe, M. Toyosato, S. Kikyotani, Y. Furutani, T. Hirose, H. Takashima, S. Inayama, T. Miyata and S. Numa; Nature 302 (1983) 528.
[96]   M. Noda, S. Shimizu, T. Tanabe, T. Takai, T. Kayano, T. Ikeda, H. Takahashi, H. Nakayama, Y. Kanaoka, N. Minamino, *et al.*; Nature 312 (1984) 121.
[97]   A. Kamb, L.E. Iverson and M.A. Tanouye; Cell 50 (1987) 405.
[98]   T. Tanabe, H. Takeshima, A. Mikami, V. Flockerzi, H. Takahashi, K. Kangawa, M. Kojima, H. Matsuo, T. Hirose and S. Numa; Nature 328 (1987) 313.
[99]   F. Jähnig; Trends Biochem. Sci. 15 (1990) 93.
[100] R. Chen, C. Kramer, W. Schmidmayr, U. Chen-Schmeisser and U. Henning; Biochem. J. 203 (1982) 33.
[101] R. Chen, W. Schmidmayr, C. Kramer, U. Chen-Schmeisser and U. Henning; Proc. Natl. Acad. Sci. U. S. A. 77 (1980) 4592.
[102] L. Song, M.R. Hobaugh, C. Shustak, S. Cheley, H. Bayley and J.E. Gouaux; Science 274 (1996) 1859.
[103] W.T. Heller, A.J. Waring, R.I. Lehrer and H.W. Huang; Biochemistry 37 (1998) 17331.
[104] A. Wada; Adv. Biophys. 9 (1976) 1.
[105] W.G. Hol, P.T. van Duijnen and H.J. Berendsen; Nature 273 (1978) 443.
[106] G. Boheim, W. Hanke and G. Jung; Biophys. Struct. Mech. 9 (1983) 181.
[107] G. Jung, E. Katz, H. Schmitt, K.P. Voges, G. Menestrina and G. Boheim, in Conformational requirements for the potential dependent pore formation of the peptide antibiotics alamethicin, suzukacillin and trichotoxin, G.Spach (ed.), Elsevier, North Holland (1983) 349.
[108] G. Menestrina, K.P. Voges, G. Jung and G. Boheim; J. Membrane Biol. 93 (1986) 111.

[109] G. Jung, K.-P. Voges, G. Becker, W.H. Sawyer, V. Rizzo, G. Schwarz, G. Menestrina and G. Boheim; Chem. Pept. Proteins 3 (1986) 371.

[110] G. Boheim, S. Gelfert, G. Jung and G. Menestrina, in alpha-helical ion channels reconstituted into planar bilayers, K.Yagi and B.Pullmann (ed.), Academic Press, New York (1987) 131.

[111] A. Ballio, D. Barra, F. Bossa, A. Collina, I. Grgurina, G. Marino, G. Moneti, M. Paci, P. Pucci, A. Segre and M. Simmaco; FEBS Lett. 291 (1991) 109.

[112] J.F. Bradbury, Guide to plant pathogenic bacteria, CAB International Mycological, Farnham Royal, Slough, England, 1986.

[113] V. Fogliano, M. Gallo, F. Vinale, A. Ritieni, G. Randazzo, M. Greco, R. Lops and A. Graniti; Physiol. Mol. Plant Pathol. 55 (1999) 255.

[114] N. Iacobellis, P. Lavermicocca, I. Grgurina, M. Simmaco and A. Ballio; Physiol. Mol. Plant Pathol. 40 (1992) 107.

[115] M. Dalla Serra, G. Fagiuoli, P. Nordera, I. Bernhart, C. Della Volpe, D. Di Giorgio, A. Ballio and G. Menestrina; Mol. Plant - Microbe Interact. 12 (1999) 391.

[116] M. Dalla Serra, P. Nordera, I. Bernhart, D. Di Giorgio, A. Ballio and G. Menestrina; Mol. Plant - Microbe Interact. 12 (1999) 401.

[117] A. Carpaneto, M. Dalla Serra, G. Menestrina and F. Gambale; Biophys. J. 76 (1999) A442.

[118] B. Hille, Ionic channels of excitable membranes, Sinauer Associates Publishers, Sunderland Massachussets, 1984.

[119] G. Schwarz; J. Membrane Biol. 43 (1978) 127.

[120] G. Ehrenstein and H. Lecar; Q. Rev. Biophys. 10 (1977) 1.

[121] G. Menestrina and M. Ropele; Biosci. Rep. 9 (1989) 465.

[122] A.M. Feigin, J.Y. Takemoto, R. Wangspa, J.H. Teeter and J.G. Brand; J. Membrane Biol. 149 (1996) 41.

[123] R.M. Epand; Biochim. Biophys. Acta 1376 (1998) 353.

[124] K. Matsuzaki; Biochim. Biophys. Acta 1376 (1998) 391.

[125] K. Matsuzaki; Biochim. Biophys. Acta 1462 (1999) 1.

[126] L. Yang, T.M. Weiss, R.I. Lehrer and H.W. Huang; Biophys. J. 79 (2000) 2002.

[127] L. Yang, T.A. Harroun, T.M. Weiss, L. Ding and H.W. Huang; Biophys. J. 81 (2001) 1475.

[128] C. Alvarez, M. Dalla Serra, C. Potrich, I. Bernhart, M. Tejuca, D. Martinez, I.F. Pazos, M.E. Lanio and G. Menestrina; Biophys. J. 80 (2001) 2761.

[129] J.M. Seddon, R.D. Kaye and D. Marsh; Biochim. Biophys. Acta 734 (1983) 347.

[130] S.B. Farren, M.J. Hope and P.R. Cullis; Biochem. Biophys. Res. Commun. 111 (1983) 675.

[131] J.M. Seddon; Biochim. Biophys. Acta 1031 (1990) 1.

[132] P.R. Cullis and B. de Kruijff; Biochim. Biophys. Acta 559 (1979) 399.

[133] M. Tejuca, M. Dalla Serra, M. Ferreras, M.E. Lanio and G. Menestrina; Biochemistry 35 (1996) 14947.

[134] V. De Los Rios, J.M. Mancheno, M.E. Lanio, M. Onaderra and J.G. Gavilanes; Eur. J. Biochem. 252 (1998) 284.

[135] A. Varanda and A. Finkelstein; J. Membrane Biol. 55 (1980) 203.

[136] G. Belmonte, C. Pederzolli, P. Macek and G. Menestrina; J. Membrane Biol. 131 (1993) 11.

[137] V. De Los Rios, J.M. Mancheno, A. Martinez del Pozo, C. Alfonso, G. Rivas, M. Onaderra and J.G. Gavilanes; FEBS Lett. 455 (1999) 27.

[138] G. Belmonte, G. Menestrina, C. Pederzolli, I. Krizaj, F. Gubensek, T. Turk and P. Macek; Biochim. Biophys. Acta 1192 (1994) 197.

[139] P. Macek, G. Belmonte, C. Pederzolli and G. Menestrina; Toxicology 87 (1994) 205.

[140] G. Anderluh, I. Krizaj, B. Strukelj, F. Gubensek, P. Macek and J. Pungercar; Toxicon 37 (1999) 1391.

[141]   A. Athanasiadis, G. Anderluh, P. Macek and D. Turk; Structure 9 (2001) 341.
[142]   G. Anderluh, A. Barlic, Z. Podlesek, P. Macek, J. Pungercar, F. Gubensek, M. Zecchini, M. Dalla Serra and G. Menestrina; Eur. J. Biochem. 263 (1999) 128.
[143]   H. Duclohier, G. Molle and G. Spach; Biophys. J. 56 (1989) 1017.
[144]   Z. Oren and Y. Shai; Biopolymers 47 (1998) 451.
[145]   M. Behzadipour, R. Ratajczak, K. Faist, P. Pawlitschek, A. Trémolières and M. Kluge; J. Membrane Biol. 166 (1998) 61.
[146]   E. Tavernier, D. Le Quoc and K. Le Quoc; Biochim. Biophys. Acta 167 (1993) 242.
[147]   A. Carpaneto, M. Dalla Serra, G. Menestrina, V. Fogliano and F. Gambale; J. Membrane Biol. (2002) .
[148]   F. Gambale, M. Bregante, F. Stragapede and A.M. Cantù; J. Membrane Biol. 154 (1996) 69.
[149]   M. Dalla Serra, P. Nordera, I. Bernha rt, D. Di Giorgio, A. Ballio and G. Menestrina; Mol. Plant-Microbe Interact. 12 (1999) 401.
[150]   A. Bertl, E. Blumwald, R. Coronado, R. Eisenben, G. Firlay, D. Gradmann, B. Hille, K. Kohler, H.A. Koll, E. MacRobbie, G. Meissner, C. Miller, E. Neher and P. Palade; Science 258 (1992) 873.
[151]   C. Julmanop, Y. Takano, J. Takemoto and T. Miyakawa; J. Gen. Microbiol. 139 (1993) 2323.
[152]   V. Malev, L. Schagina, J. Takemoto, E. Nestorovich and S. Bezrukov; Biophys. J. 80 (2001) 132a.
[153]   H. Terlau and W. Stuehmer; Naturwissenschaften 85 (1998) 437.
[154]   H. Sentenac, N. Bonneaud, M. Minet, F. Lacroute, J.M. Salmon, F. Gaymard and C. Grignon; Science 256 (1992) 663.
[155]   J.A. Anderson, S.S. Huprikar, L.V. Kochian, W.J. Lucas and R.F. Gaber; Proc. Natl. Acad. Sci. U. S. A. 89 (1992) 3736.
[156]   B. Mueller-Roeber, J. Ellenberg, N. Provart, L. Willmitzer, H. Busch, D. Becker, P. Dietrich, S. Hoth and R. Hedrich; EMBO J. 14 (1995) 2409.
[157]   Y. Cao, J.M. Ward, W.B. Kelly, A.M. Ichida, R.F. Gaber, J.A. Anderson, N. Uozumi, J.I. Schroeder and N.M. Crawford; Plant Physiol. 109 (1995) 1093.
[158]   K. Czempinski, N. Gaedeke, S. Zimmermann and B. Mueller-Roeber; J. Exp. Botany 50 (1999) 955.
[159]   D.P. Schachtman; Biochim. Biophys. Acta 1465 (2000) 127.
[160]   S. Zimmermann, T. Ehrhardt, G. Plesch and B. Mueller-Roeber; Cell. Mol. Life Sci. 55 (1999) 183.
[161]   C.A. Vandenberg; Proc. Natl. Acad. Sci. U. S. A. 84 (1987) 2560.
[162]   Y. Cao, N.M. Crawford and J.I. Schroeder; J. Biol. Chem. 270 (1995) 17697.
[163]   D.M. Papazian, X.M. Shao, S.A. Seoh, A.F. Mock, Y. Huang and D.H. Wainstock; Neuron 14 (1995) 1293.
[164]   R. Plannels-Cases, A.V. Ferrer-Montiel, C.D. Patten and M. Montal; Proc. Natl. Acad. Sci. U. S. A. 92 (1995) 9422.
[165]   W. Dun, M. Jiang and G.N. Tseng; Eur. J. Physiol. 439 (1999) 141.
[166]   L.Y. Jan and Y.N. Jan; Annu. Rev. Neurosci. 20 (1997) 91.
[167]   I. Marten and T. Hoshi; Biophys. J. 74 (1998) 2953.
[168]   C.J. Smith-Maxwell, J. Ledwell and R.W. Aldrich; J. Gen. Physiol. 111 (1998) 399.
[169]   C. Castillo, M.E. Diaz, D. Balbi, W.B. Thornhill and E. Recio-Pinto; Dev. Brain Res. 104 (1997) 119.
[170]   I. Levitan, A.E. Christian, T.N. Tulenko and G.H. Rothblat; J. Gen. Physiol. 115 (2000) 405.
[171]   T. Baukrowitz and B. Fakler; Biochem. Pharmacol. 60 (2000) 735.

[172] N. Dan and N.A. Safran; Biophys. J. 75 (1998) 1410.
[173] A.E. El-Husseini, J.R. Topinka, J.E. Lehrer-Graiwer, B.L. Firestein, S.E. Craven, C. Aoki and D.S. Bredt; J. Biol. Chem. 275 (2000) 23904.
[174] F. Gambale and M. Montal; Biophys. J. 53 (1988) 771.
[175] J.M. Gonzalez-Ros, A.M. Fernandez, J.A. Encinar and J.A. Poveda; J. Physiol. Paris 92 (1998) 432.
[176] G. Menestrina, C. Pederzolli, S. Forti and F. Gambale; Biophys. J. 60 (1991) 1388.
[177] S.I. Sukharev, P. Blount, B. Martinac, F.R. Blattner and C. Kung; Nature 368 (1994) 265.

Planar Lipid Bilayers (BLMs) and their Applications
H.T. Tien and A. Ottova-Leitmannova (Editors)

Chapter 24

Gravitational impact on ion channels incorporated into planar lipid bilayers

[a]M. Wiedemann, [b]H. Rahmann and [a]W. Hanke

[a]University of Hohenheim, Institute of Physiology, Stuttgart, Germany

[b]University of Hohenheim, Institute of Zoology, Stuttgart, Germany

Content

1. Introduction

Gravity is a small but permanently present physical stimulus on earth, becoming of high interest in times of live sciences under space conditions (Rahmann and Hirsch, 2001). During evolution about all biological systems thus have developed sensory systems to detect gravity. These mechanisms in higher biological systems very often are specialized organs as for example the vestibular apparatus of vertebrates and especially man (i.e. Greger and Windhorst, 1996, Krasnov, 1994); plants and also a variety of single cellular organisms on the other hand often use heavy particles for this purpose (Braun, 1997; Häder, 1999). Nevertheless especially some single cells systems are not known to have such specialized mechanisms and nevertheless are able to detect gravity (Machemer, 1997). The question of how these cells can detect gravity is still open, but during the last decade ideas have evolved about possible underlying processes. Fundamental to all these ideas is the cellular membrane being part of all cells, which is known to be, with all its components and interactions, involved in all sensory processes. Ion channels as integral membrane proteins which are partitioning in all sensory processes are integral parts of these membranes, and according to the question of gravity sensing became of high interest based on two possible aspects. First, it might be possible that gravity directly interacts with single membrane based proteins, including ion channels; second, gravity might interact with the two-dimensional thermodynamical system membrane, changing its parameters and thus effecting the properties of ion channels incorporated in the membrane. Changing other physical parameters than gravity in a variety of different experiments, for example temperature or pressure, both mechanisms have been shown to be possible using a variety of electrophysiological and other techniques (Hanke and Schlue, 1993). Especially the investigation of mechano-sensitive ion channels has contributed a lot to understanding of how membranes can interact with mechanical and other weak external forces (i.e. Garcia-Anoveras and Corey, 1997; Sukharev, 1999).

Although a variety of experimental approaches have been used to study the interaction of gravity with single cells, all these studies are limited, as far as the interpretation of the results is concerned, by the problems of understanding the complex interactions being present in single cells, this problem is even more obvious in higher biological organisms. Thus it is tempting to investigate the interaction of gravity with membranes directly on the molecular level. According to the investigation of ion channels mainly two electrophysiological techniques have been developed to investigate the properties of especially single ion channels, these are the bilayer reconstitution technique (Hanke and Schlue, 1993) and the patch-clamp technique (Hamill et al., 1981). Comparing the two techniques, it is immediately obvious that the bilayer approach allows the use of a more simplified model system finally consisting of one lipid species and, in the limiting situation, of one isolated protein molecule. By this, a highly simplified model system is given, which would allow to directly study the interaction of forces like gravity with either isolated proteins, including ion channels, or the simplified two dimensional thermodynamical membrane system, in an environment with unequivocally defined parameters like temperature, biochemical composition, membrane curvature and others. Finally, instead of using natural ion channels in a further step of simplification model pore forming polypeptides can be used in a first biophysical approach to understand the complex nature of gravity interaction with living matter.

When studying the interaction of gravity with ion channels, a second, technical aspect becomes important. As gravity is permanently present on earth, special experimental platforms must be used to be able to perform experiments under micro-gravity as well as under increased gravity. Both of these situations and of course the controls under 1 g are necessary to get acceptable information for the understanding of gravity interacting with living matter. Whereas it is comparably easy to get macro-gravity in a lab by just using a centrifuge, micro-gravity platforms are technically complicated and extremely expensive. Additionally, in all micro- and macro-gravity experiments a high degree of remote control systems is necessary to run the experiments successful. Furthermore, especially on micro-gravity platforms the number of possible experiments usually is limited due to technical and financial limitations. Consequently, it is useful to run a higher number of identical experiments in parallel on one platform to get an acceptable statistics for later data evaluation, as, according to the above statements micro-gravity experiments usually cannot be repeated to often. Both aspects, automatic experiments and parallel experiments, much easier can be realized in the bilayer technique compared to the patch-clamp technique.

After having decided to do bilayer experiments under micro-gravity investigating the dependence of single ion channels on gravity, the time scale of the experiments must be considered. The kinetics of single ion channels usually is found to be in the millisecond to second range, thus, investigating the conductances and lifetimes of such single ion channels, as well as the open state probabilities, it is sufficient to do the experiments on a second to minute time scale, at least at the beginning. According to the above statements this allows the use of so called small micro-gravity platforms, giving micro-gravity periods from seconds to minutes (see next paragraph for more details). These platforms are easier accessible, better reproducible, and cheaper than the long lasting missions, as for example the international space station (ISS) is.

1 g controls and macro-gravity experiments in a modified laboratory centrifuge are at least basically not due to such limitations and can be performed to any temporal excess in the laboratory.

In the studies presented in this paper, we have started to investigate the interaction of gravity with single ion channels following the above outlined strategies. First, we have developed the necessary technologies and then we investigated alamethicin (Boheim, 1974) and a purified porin (i.e. Berrier et al., 1992) reconstituted into planar lipid bilayers.

Construction plans of the set-ups and results from such experiments are given in the following chapters.

2. Platforms for micro-g and macro-g research

A variety of platforms has been developed to do experiments under micro-and macro-gravity. For macro-gravity experiments a modified laboratory centrifuge can be used without significant temporal limitations. g-values up some ten g are usually sufficient for biological studies and can be easily obtained in such a centrifuge. A centrifuge developed in our lab is described at the end of this chapter.

For micro-gravity experiments the situation is much more complicated. Nevertheless, due to research, technical, militaries, and other needs, a variety of platforms has been developed giving micro-gravity periods from seconds to month or years. In figure 1 an overview of the presently available systems is given together with the length of the delivered micro-gravity period.

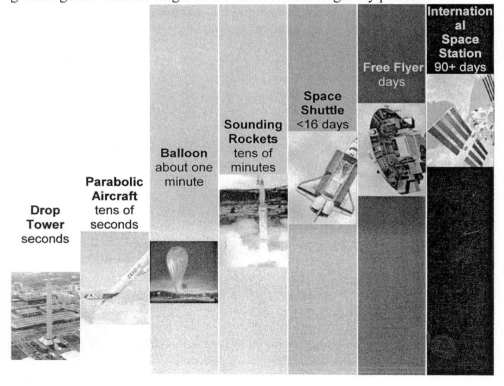

Figure 1 Presently available platforms for micro-gravity research. The mean duration of the micro-gravity period is given in the figure for each system separately. This figure is modified from the Space-Hub internet pages.

The quality of micro-gravity given by the different systems has a span from 10^{-6} g in the drop tower to 10^{-3} g in the parabolic flights, all of these values are sufficient for basic biological experiments.

In the following, the so called smaller opportunities which have been used for our planar lipid bilayer experiments with reconstituted ion channels are described in some more detail.

Experiments with planar lipid bilayers on bigger platforms seem to be possible based on our experience, an outlook about possible future experiments is given later in this article.

2.1. Drop-tower

The most simple way to produce micro-gravity at least for a limited period of up to some seconds is to lift the set-up to a certain level above ground and then let it fall down freely. This free fall principle is the concept of the drop tower in Bremen, Germany, which delivers about 4.5 s of micro-gravity. Other comparable systems exist around the world with micro-gravity periods from 1 s to about 9 s.

To reduce the friction of the falling set-up with air, which would significantly decelerate the free fall and thus duration and quality of the experiments, an evacuated tube of about 100 m height is used in which a drop capsule can fall really freely, delivering a period of about 4.5 s of micro-gravity with a quality of about 10^{-6} g. Under normal operating conditions two drops can be performed each day and a typical experimental mission lasts two weeks with about 15 drops.

Due to the evacuation of the tube, the set-up must be fully remote controlled, as no manual access is given to the drop capsule during operation. The evacuation periods itself needs about two hours, thus the last access to the experiments is about two hours before drop. After drop, the capsule is stopped in a container filled with small styrofoam pellets, a negative acceleration up to 50 g is given during this breaking period, consequently the set-up must be mechanically quite stable to survive this.

The air-tight sealed drop capsule into which the experimental set-up has to be fitted, is round with a diameter of about 40 cm, the length is up to 2 –3 m. Atmospheric pressure can be kept throughout the experiments in the capsule. DC power supply is given as well as data storage capabilities. During the pumping period, a video line is established to observe the experiments, during the free fall this line is disconnected. Remote control of any part of the set-up can be used during the pumping period by command lines, and the moment of release of the capsule can be determined, after the tube has been evacuated, by the experimenter.

A sketch of the drop tower system with all its support systems and a photo of the drop capsule are given in figure 2.

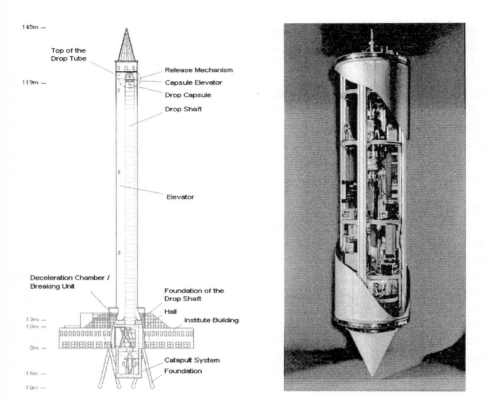

Figure 2 At the left a cartoon of the drop tower in Bremen is given with its main support systems. At the right a photo shows a drop capsule filled with an experimental set-up. The figure is taken in modified form from the internet pages of the drop tower (http.//www.zarm.uni-bremen.de/), here also more detailed technical and scientific information about the drop tower can be found.

2.2. Parabolic flights

A possible method to produce micro-gravity is to throw a desired object into a free parabolic curve. Throughout the parabolic flight period all forces acting onto the object are levelled off and thús it is free of any gravity. Parabolic flight with airplanes are utilising this principle for producing micro-gravity periods of up to about 25 seconds. The principle of such flights is given in figure 3. The European A-300 airplane is, controlled by the pilots, doing the parabolic flights at high altitude to reduce friction and allow a micro-gravity quality of up to 10^{-4} g under optimal conditions. Each micro-gravity period of about 25 seconds is accompanied by macro-gravity periods with about 2 g and a duration of some 20 seconds. Between two parabolic events, a 1 g phase of any desired length can be done, usually the 1 g phase lasts about 3 minutes. Within one flight typically 30 parable periods are done, and a complete mission consists of three flights, this means 90 parabolic periods.

Parabolic flights are offered from the European, the USA and the Russian space organizations with comparable conditions.

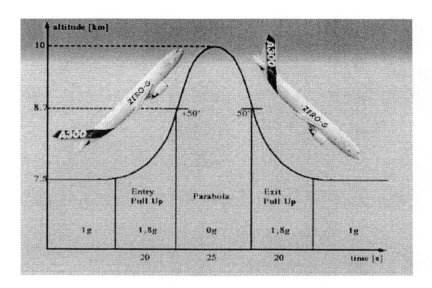

Figure 3 Principles of parabolic flights. The figure is taken in modified form from the Novespace internet pages (http.//www.novespace.fr/) where also more detailed information about the European parabolic flights can be found.

An advantage of the parabolic flights is that within the plane experiments can be handled in a way comparable at least partially to laboratory conditions. Free access is given to the set-up and thus no remote control of the complete system is required. The set-up can be constructed to a volume filling a significant part of the plane, thus enough space is given for typical biological experiments. Of course safety conditions are reasonably strict and thus a specific set-up has to be used. Also, for people working under conditions of permanently changing gravity, some limitations in their capabilities to handle experiments must be taken into account.

2.3. Stratospheric balloons (MIKROBA)

An enhanced version of the drop tower principle is to pull up a set-up with a stratospheric balloon to about 40 km above ground and then let it fall down freely. Such a system has been developed and tested successfully with a bilayer set-up. To overcome air-friction in the atmosphere the drop capsule has an additional rocket motor compensating friction by downward directed acceleration at lower altitude. A free-fall distance of about 20 km can be thus used delivering a micro-gravity period of about 1 minute. After the free-fall period the drop capsule has a velocity of about mach 2 and must be stopped by parachutes for save recovery. The ascent of the hydrogen filled balloon (diameter at higher altitude over 100 m) lasts for about 3 hours, and due to the low pressure and low temperature at higher altitudes the capsule is constructed air-tight and the experiments must be temperature controlled. The release of the drop capsule can be remotely controlled. The advantage of the system is that before the micro-gravity period begins, only 1 g is applied to the system, thus ideal conditions are given for sensitive biological experiments.

Figure 4 Take-off of a stratospheric hydrogen filled balloon, as has been used among others for a bilayer experiment. The right part shows the MIKROBA drop capsule. More information about the MIKROBA project can be obtained at the companies Kyser-Trade (Muenchen, Germany) and OHB (Bremen, Germany).

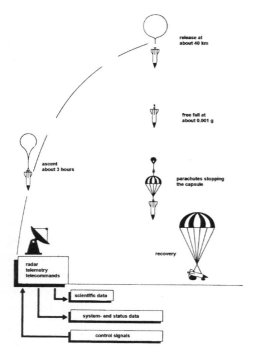

Figure 5 Profile of a MIKROBA mission Modified from the MIKROBA handbook, OHB, Bremen, Germany

A typical experimental set-up can have the following dimensions: round, with about 40 cm diameter and a length of about 50 cm. The experiments must be completely remote controlled, as after take-off no more access to the experiment is given. Late access to the experiment before take-off typically is limited to 4 hours, thus the experimental set-up must be stable over a reasonably long period of up to 10 hours. In figure 4 a photo shows the take-off of a balloon together with the drop capsule, Figure 5 shows the basic profile of a MIKROBA mission.

2.4. Centrifuge

In the laboratory macro-gravity experiments can be done with acceptable effort using a centrifuge. For biological applications macro-gravity values up to 10 g are sufficient for most experiments. The demands for biological macro-gravity experiments in our laboratory are fulfilled by a home-build centrifuge as shown in figure 6. This centrifuge has a rotor plate with a diameter of 50 cm on which a complete set-up can be fixed. A video transmitter can be installed for observation of the experiments during operation, power is delivered by rechargeable batteries for up to 10 hours, and a data transmission line with a bandwidth of 10 kHz is installed. The centrifuge can deliver horizontal accelerations up to about 15 g, and it has been successfully used in our lab for a variety of experiments up to several hours. The bilayer chamber itself is mounted on the centrifuge rotor in a swing-out holder, which can be seen in the figure at the upper left side of the rotor plate.
Detailed information about the construction of the centrifuge can be requested at the authors.

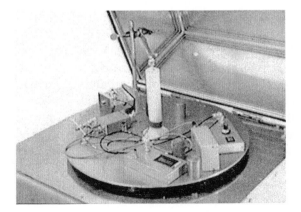

Figure 6 Photo of a home build centrifuge for biological macro-gravity with planar lipid bilayers experiments in the laboratory. The rotor plate with the complete bilayer set-up mounted can be seen on the photo.

3. Bilayer set-ups for micro-g and macro-g research

In macro-gravity experiments it has been proven better to use the classical Montal technique of painted bilayers (Montal and Mueller, 1972), because it can be easier adapted to automated systems and delivers somewhat less sensitive bilayers according to mechanical disturbances which, up to a certain degree, are always present in the applications described in this paper. The lipid used in the macro- and micro-gravity experiments, of course also in the 1 g laboratory controls, was either Asolectin, a soy-bean lipid extract frequently used in standard bilayer applications, or di-oleoyl-phosphatidyl-choline (DOPC). The lipid was dissolved in decane at 30 mg/ml. As aqueous solution in all experiments KCl solutions of concentrations as given in the results section, adjusted to pH=7.4 with 10 mm Tris and HCl, were used. Within the experiments especially the temperature was kept constant, at 20 °C.

More details about the bilayer technology itself can be found in the literature (i.e. Hanke and Schlue, 1993).

3.1. Eight chamber block module for the drop-tower and MIKROBA

According to limitations in the number of experiments in micro-gravity studies and to the need of a remote controlled system we developed an bilayer module having eight chamber for eight separate bilayer to be used at the same time. Additionally this module had to be sealed, to prevent solution loss during micro-gravity periods.

The module is shown in figure 7 as a complete block and in its main parts. It has as a basic component a system of eight small chambers being connected to a bigger central chamber. The bilayer carrying holes are in the walls to this central chamber. These walls are less than 1 mm thick and the holes have a diameter of about 200 μm. Into the eight smaller chambers and also into the central chamber silver-chloride electrodes are inserted, which are connected to BNC-connectors. The chamber is closed and sealed by a cover together with an o-ring. This cover holds a motor-driven wiper, which distributes small amounts of lipid dissolved in decane on the holes in the chamber walls. The motor can be remotely controlled. Also in the cover there are two valves for each chamber to allow to fill it with aqueous solution without air-bubbles. Below the chamber block, an aluminium case is mounted, which holds eight amplifiers for the eight bilayers. An additional electronics delivers the membrane potential for the bilayers and a signal for capacity control. The electronics is connected to the outer world by BNC-connectors. More details about this block module haven been published elsewhere (Klinke et al., 1998).

In operation, the chambers are filled with aqueous solution, lipid is added to the wiper, which distributes it over the bilayer carrying holes on external commands. Substances or drugs or vesicles with proteins integrated into their membranes (Hanke and Schlue 1993) are added to the central chamber and are thus accessing all eight bilayers at once. According to the given electronic, all bilayers are apposed to the same command voltage, and the current flowing across the bilayers is measured. The current trace are stored and in parallel are visualized on an oscilloscope.

Figure 7 Eight chamber block module, the left side shows the module mounted, the right side shows the separate parts, cover with motor, chamber block, eight bilayer chambers plus central chamber, and electronics.

The above described module was used in the drop-tower as well as in the MIKROBA set-up.

In the drop-tower as shown in figure 8, one of the modules was mounted together with the supporting electronics in the capsule. The supporting electronics mainly consists of power supply, data-transmission line, data storage facilities, and the systems to control the motor of the block module via remote control commands. In the experiments, the module was properly prepared, inserted into the drop-capsule, and then this was mounted into the drop-tower. During the evacuation period, lasting about 2 hours, the lipid was distributed over the bilayer carrying holes by use of the wiper and the formation of bilayers was monitored electrically by capacity measurements. After formation of bilayers and onset of channel- or pore current fluctuations, control measurements under 1 g were performed. After proper evacuation of the tower, the capsule was released in case proper bilayer activity could be observed. The procedure of bilayer formation could be repeated any time before release. Upon stopping the drop capsule in a pellet filled container, a negative acceleration of up to 50 g is given. This always destroys the membranes.

By the described procedure, up to eight functional bilayers could be observed in one run during a drop tower experiment, thus delivering sufficient data within one complete mission on the basis of the limited number of drops being possible in one experimental session.

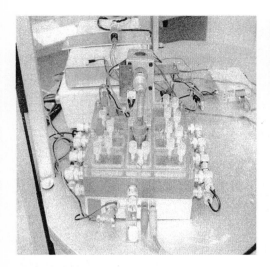

Figure 8 The left side shows one block as shown in figure 7, mounted in the drop tower capsule together with part of the supporting electronics. The right side shows the complete set-up mounted in the drop tower capsule.

The set-up for MIKROBA was constructed similar to the drop-tower system, however, due to the fact that usually only one mission is possible, the complete set-up had four of the eight chamber block modules, thus giving 32 bilayers in one experiment. The complete mounted set-up is shown in figure 9. This set-up was inserted into an air-tight chamber and the chamber was mounted into the MIKROBA drop capsule shown in figure 4.

Figure 9 Four of the blocks as shown in figure 7 mounted in the complete MIKROBA set-up together with the complete supporting electronics.

Details about the complete structure of a MIKROBA mission can be found in the literature. In short, the lift to the final altitude of about 40 km needs 2-3 hours, during this time, similar to then evacuation period of the drop tower, control experiments at 1 g can be done. At the final height temperature is quite low, thus the system has a build in temperature control. At given time the drop capsule is released and undergoes a free-fall period of about 1 min. After this its velocity is about 2 Mach and it must be stopped by parachutes, giving up to 12 g negative acceleration, thus typically terminating the experiments. A sketch of a mission profile was already given in figure 5.

3.2. Set-up for parabolic flights

Different to the fully remote controlled set-up for MIKROBA or the drop tower, in parabolic flights experiments can be done under conditions similar to laboratory conditions. Consequently, the set-up used is based on a standard bilayer set-up (Hanke and Schlue, 1993) with only one bilayer. A tightly sealed chamber is used to avoid solution loss during micro-gravity periods. A cartoon of such a chamber is shown in figure 10.

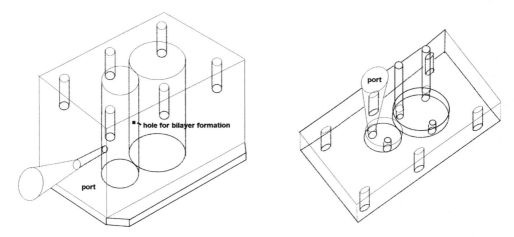

Figure 10 Cartoon of a sealed chamber for parabolic flight planar lipid bilayer experiments. The lower part of the chamber (left part of this figure) has the two compartments for aqueous solutions, divided by a 1 mm thick wall. Into this wall a hole with about 200 μm diameter is made for bilayer formation. Also in the lower part a port is integrated, being connected to a syringe (not shown). By use of this syringe the aqueous level in the front camber can be changed and thus bilayers can be formed by a modified Montal-Mueller technique (Mueller et al., 1962; Montal and Mueller, 1972). A small amount of lipid dissolved in decane (or hexane) is added to the chamber before sealing it for this purpose, together with the aqueous solutions. The cover of the chamber (right part of this figure) has an additional port for equilibration and two connectors for the silver chloride electrodes being inserted into both compartments of the chamber.

The complete set-up is shown in figure 11, being mounted on the floor of the A300 plane used for the parabolic flights.

Figure 11 Bilayer set-up for use in parabolic flights. The right part holds the "standard" electronics, the left part consists of the Faraday cage with a sealed bilayer chamber and a shock absorbing system.

According to safety limitations, the set-up had to be packed at least partially in styrofoam, to reduce the risk of violation of the experimenters under rough flight conditions outside the micro-gravity phases.
During the acceleration phases and due to vibration during the flight, frequently bilayers broke throughout the missions and thus new bilayers had to be formed, however, due to the about 90 parabolas of each mission, sufficient data could be collected without problems.

3.3. Concepts of set-ups for sounding rockets and long term flights
Although the duration of the discussed flight opportunities is sufficient for single pore experiments, as stated above, it could be useful to do some long time-scale experiments. For this purpose one choice could be sounding rockets, delivering micro-gravity periods of up to 15 min. For technical details about sounding rockets see for example the ESA and Esrange internet pages. A set-up for a sounding rocket could be constructed following the same principles as being given for the MIKROBA set-up. The only additional point which has to be taken into account is the starting phase of the rocket, giving accelerations of up to 12 g a rotation of up to 4 g and significant mechanical vibrations. Thus a possible set-up would have to be constructed for these environmental conditions. Most probably bilayers would have to be made after the starting phase. It is presently open, how well bilayer formation would take place in case of the method of painted bilayers at missing gravity. Besides these two questions, a sounding rocket bilayer set-up on our interpretation would not have further technical problems. As, however, a sounding rocket set-up typically is build by

an external company (i.e. Astrium, Germany) among others for safety reasons, different to the other set-ups, which are build in the workshops of the laboratory and the university, such a system would be reasonably expensive.

The situation becomes much more complicated in case of a set-up for long lasting flights on board a space-shuttle or the ISS. Here a complete new design would be necessary, fulfilling all the limitations of these missions. Principally, however, based on our experience such a set-up would be reasonable.

3.4. Centrifuge set-up
In the centrifuge we used a bilayer chamber similar to laboratory chambers (Hanke and Schlue, 1993). The chamber was mounted in a swing-out holder on the rotor plate. Bilayers were made with the centrifuge at rest, and after proper formation of the bilayer, the centrifuge run was started. The bilayer was observed online and the current fluctuation traces were recorded on a computer with proper hard- and software for later data evaluation. The bilayer could be mounted with its plain perpendicular to the acceleration vector, or in line with it. Accelerations up to 6 g were used in our experiments, and the run time of the centrifuge could be hours, changing rotating speed throughout the experiment at any desired time.

3.5. Set-up to change bilayer orientation relative to earth gravity vector
A first quite trivial approach to investigate the gravity dependence of ion-channels incorporated into planar lipid bilayers is to change the orientation of the bilayer relative to the earth gravitation vector (Hanke, 1995). For this purpose we have constructed a bilayer chamber, allowing to measure a bilayer under any desired orientation relative to the earth gravitational vector. The chamber is made of a basic compartment of about 3 x 3 x 5 cm dimensions. Through this compartment a horizontal tube (polycarbonate) is mounted tightly sealed relative to the basic compartments being with its interior the second compartment. A part of the wall of this tube is made thin, about 1 mm, and the bilayer carrying hole, diameter again about 200 μm, is drilled into this part of the wall. The bilayer can be formed on the hole using the method of painted bilayers, and after formation of the bilayer the tube can be rotated to any angle. Besides this rotating tube, the set-up with all its other parts is a normal set-up for painted bilayers. However, the system can also be changed to be used with folded bilayer in case this is necessary.

4. Gravity influence on alamethicin incorporated into bilayers

For a first set of experiments investigating the influence of gravity on pore and channel forming polypeptides and proteins, alamethicin (Boheim, 1974) was chosen as model system on a variety of reasons. First, it is easy to handle and to incorporate into planar lipid bilayers. However second, and even more important, due to its very specific pore forming mechanism it can be expected to be most sensitive to the action of week external forces on the membrane with its incorporated transport mechanisms.

The model of a flip-flop and aggregation of alpha-helices (i.e. Boheim et al., 1983), is still valid in its basics, including a turn of the complete helix through the hydrophobic core of the membrane, and allows a strong interaction of membrane thermodynamics with the pore forming polypeptides.

Just for reminding, in figure 12 a trace of the typical multi conductance current fluctuations induced in planar lipid bilayers by alamethicin is shown.

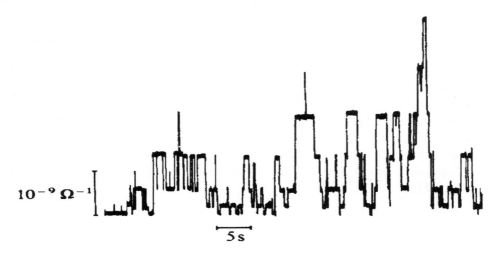

Figure 12 Typical alamethicin induced current fluctuation in a planar lipid bilayer made from asolectin according to the Mueller technique (Mueller et al., 1962). The multi conductance state of alamethicin fluctuations can be seen easily. The lowest conductance state is not resolved. Most probably the complete fluctuation trace belongs to one pore event (see also figure 15 for comparison).

Slightly different to other pores and channels, the behaviour of this type of pore fluctuations is defined mainly by the state conductances, the state lifetimes, the pore lifetime, the pore frequency, and the mean open pore state acquired. Accordingly the dependence of all these parameters on changing gravity must be considered.

4.1. Dependence of alamethicin activity on bilayer orientation relative to earth gravity

The simplest way to look in a ground based laboratory for the interaction of gravity with any system is to study its dependence on the orientation relative to the earth gravity vector. This was done with alamethicin pore fluctuations. It was found that the pore state conductances do not change significantly in dependence on the angle of the bilayer relative to the earth gravity vector, however, the lowest conductance state (< 20 pS in 1 M KCl) of the alamethicin pore could not be investigated in this study due to technical limitations, here possibly changes could be expected. The pore state lifetimes also were not dramatically effected, they were about 10 % longer in the perpendicular position of the bilayers compared to the horizontal position. A much more pronounced effect was, however, found in the pore kinetics within a preformed pore, being reflected by the mean pore state (number of pore states acquire on an average). The results are shown in figure 13. It is obvious that the mean pore state strongly decreases from the horizontal (0 degree) to the perpendicular (90 degree) position of the planar lipid bilayer as also has been shown previously in the literature (Hanke, 1996).

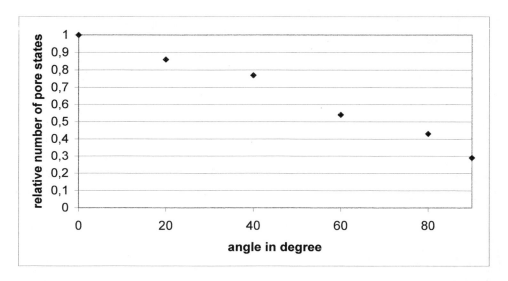

Figure 13 Dependence of the number of pore states on the positioning of the bilayer relative to earth gravity vector (0 degree is horizontal) in an alamethicin induced current fluctuation. Condition are similar to those given in figure 12. Data are normalized to the value of the horizontal position. The figure is modified from Hanke, 1996.

In figure 13 the data were fitted by a straight line, due to the scatter of the data this is a reasonable fit.

4.2. General gravity dependence of alamethicin-pore parameters

In another series set of experiments the dependence of alamethicin pore parameters on the amplitude of the gravitational vector was investigated. In the data presented here, results from centrifuge experiments (macro-gravity) drop tower experiments (micro-gravity, about 0 g) and parabolic flights (0 g, 1 g, 1.8 g) are pooled, without mentioning the specific set-up (Goldermann et al., 2000; Klinke et al., 1998, 1999, 2000). In macro-gravity and 1 g data the orientation is considered, as far as differences in the results are given, at 0 g this it not given by definition.

To find out whether the bilayer formation process or changes in the area of the bilayer itself effect the response of the system to gravity changes, we first measured the capacity of the bilayer as function of gravity amplitude, only a weak increase towards higher g values was found, as shown in figure 14.

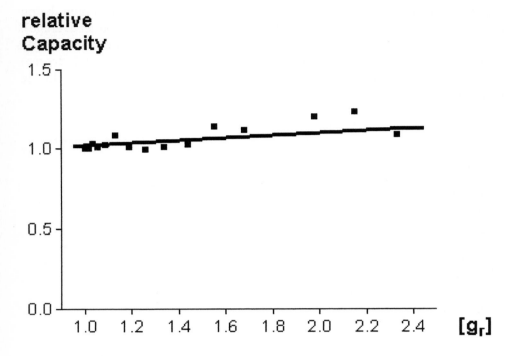

Figure 14 Dependence of bilayer capacity on gravity, the results were found to be independent on the orientation of the bilayers. Only a weak statistical not significant increase of bilayer capacity towards an increase in gravity amplitude was found. More details can be found in Klinke, 1999.

As we have confirmed in a variety of detailed experiments, the amplitude of gravity, the direction of the vector, and even more, also the rate of change of gravity, are effecting the alamethicin induced pore state current fluctuations (i.e. Klinke, 1999), making a detailed data evaluation quite complicated.

A first result again was the independence of pore state conductances on any change in gravity, whereas the pore kinetics were significantly effected. This can be easily seen from a number of original traces under different macro-gravity values as presented in figure 15. It is obvious that even at changes from 1 g to less than 2 g the number of fluctuation events (pore frequency, here reflecting the pore formation mechanism) increases significantly. The data were obtained with stepwise changes in gravity, after relaxation to a new steady-state value. Effects of continuously changed gravity and of changes in the direction of the gravity vector are different and much more complex (Hanke, 1996; Goldermann et al., 2000). The main part of the changes as can be seen in a

detailed statistical analysis was found to be in the pore frequency, which in this case mainly reflects the pore forming mechanism.
A trace of alamethicin induced current fluctuations measured in the drop tower later in this paper shows the effect of micro-gravity.

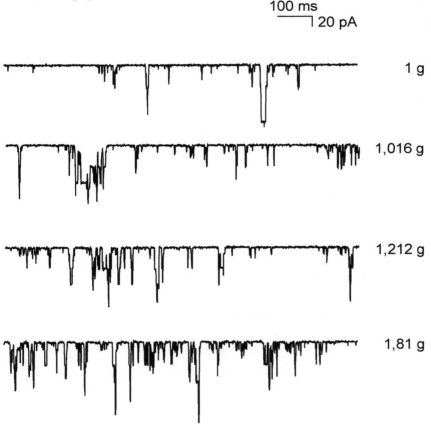

Figure 15 Traces of alamethicin induced current fluctuation in a bilayers under different gravitation values at –30 mV potential. The gravity was changed stepwise.

In figure 16 the data are summarized in dependence on gravity amplitude. As can be seen, the pore frequency increases significantly with increasing gravity amplitude; the results were comparable for horizontal and perpendicular orientation of the bilayer, thus mainly the gravity amplitude is relevant here.
Amazingly, at continuous changes in gravity an opposite behaviour was found (Hanke, 1996; Goldermann et al., 2000). Questions about the effects of vectorial changing gravity on alamethicin induced fluctuations are also complex and their investigation is still in progress.
Possibly, the distribution coefficient of alamethicin between the water phase and the lipid is effected by the way of changing gravity (amplitude, direction and dg/dt). This includes the association of alamethicin to the membrane surface and

the insertion of the monomeric molecules into the membrane core. The later pore formation is then due to aggregation and a flip-flop of the molecules under the action of electrostatic forces. Our present assumption is that incorporation and pore formation are in a different manner depending on gravity and changes of it, mainly, however, by gravity amplitude, whereas fluctuation within an existing pore are effected by gravity direction relative to the membrane surface. Possibly also the hysteresis found in the data when changing gravity in opposite directions can be explained by this approach.

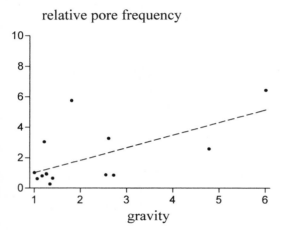

relative pore frequency

gravity

Figure 16 Mean relative pore frequency of alamethicin plotted as function of the gravity amplitude.

In a detailed evaluation of the pore state lifetimes we found only a week dependence on gravity, however, the direction of change was dependent on the direction and changing rate of gravity again.

To summarize the results shown up to now, it can be clearly said that alamethicin pore fluctuations are depending on gravity. The capacity measurements verify that we measured true changes in alamethicin activity and not in bilayer area.
The main part of the changes in alamethicin is due to changes in the alamethicin pore frequency, which reflects preferentially the pore formation process and seems to depend mainly on changes in gravity amplitude. Fluctuations within a preformed pore are obviously not that much depending on gravity amplitude, but somewhat more on changes in the direction of the gravity vector, however, this effect is smaller than the amplitude dependence on gravity amplitude. The pore state conductances are not effected by any gravity changes. Also the pore decay seems to be not that much effected by gravity. The last statement is verified for example by multi pore results as shown in figure 17 (see also Klinke et al., 2000; Goldermann et al., 2000).

In a drop tower experiment, after release of the capsule, this release delivers the transition from 1 g to 0 g, the alamethicin induced activity in a bilayer remains unchanged for some hundred milli-seconds and then abruptly decreases to a very low activity level. However, the activity is not zero, as can be seen in the insets, thus a functional bilayer is still given. The only explanation is that the decay of the pores is less dependent on gravity changes than the pore formation process.

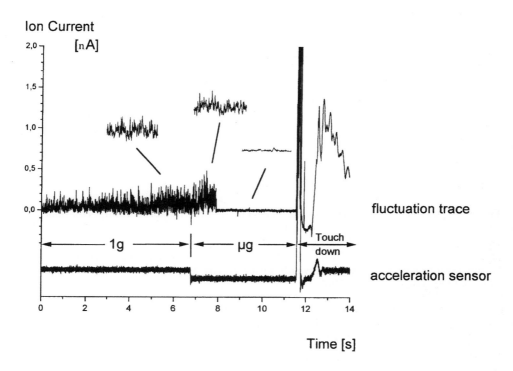

Figure 17 Multi pore alamethicin fluctuations in a drop tower experiment. After release to 0 g the fluctuations are about constant for some hundred milli-seconds and then a sudden decay of activity occurs. The low (but not zero) activity remains constant to the end of the experiment, which is given by the touch down. (Goldermann and Hanke, 2001)

5. Gravity influence on *e-coli* porins incorporated into bilayers

In a second set of experiments, we investigated a more physiological channel forming protein, a porin from *e-coli* in its dependence on gravity. This type of ion channel is known to be mechanosensitive and the channel is also formed as a homooligomeric structure of bundles of alpha-helices, thus a reasonable gravity dependence could be expected (Goldermann and Hanke, 2001).
In figure 18 the result from a drop tower experiment is shown. Obviously the channel activity decreases from 1g to micro-gravity. A statistical analysis

demonstrated this to be due to changes in the mean open state probability of the channel.

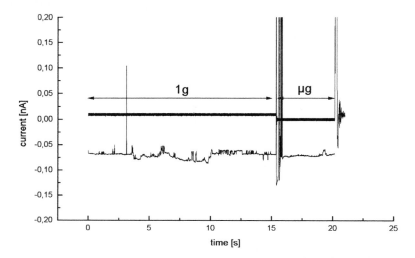

Figure 18 Current fluctuations (3 active channels) induced by porins from *e-coli* under 1 g and micro-gravity measured in the drop tower. The aqueous solution of the experiment had 200 mM KCl, 10 mM Tris, 1 mM $CaCl_2$ and was adjusted to pH=7.4 at 20 °C, The holding potential of the experiment shown was –40 mV. Channel activity (channel opening is given by downward changes) decreases towards lower gravity amplitude. The figure is modified from Goldermann and Hanke, 2001.

In centrifuge experiments the findings from micro-gravity experiments were confirmed toward higher gravity. The mean open state probability of porin channel fluctuations increased with increasing gravity amplitude as shown in figure 19 in a set of current fluctuation traces from one bilayer at different gravity values.

Similar to the results for alamethicin the pore conductance was not effected by gravity, however, the kinetics was. The changes in kinetics were exclusively found in the open state probabilities, which are theoretically related to the lifetime of the open and the closed state of the pore. The number of channels incorporated into the bilayers by fusion with protein containing liposomes (Hanke and Schlue, 1993) in our experiments was not effected by gravity.
Accordingly the situation as far as related to the data analysis was much simpler than with alamethicin, and in addition porins are more biologically relevant channel forming proteins. In figure 20, finally the mean open state probability of porins is given as a function of gravity amplitude; the data from drop tower experiments and from centrifuge experiments are pooled. A sigmoid increase of

mean open state probability with increasing gravity was found, the 50 % value is in the range of 1.5 g.

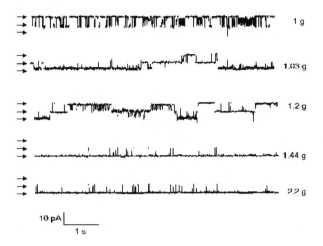

Figure 19 Porin channel fluctuations at different gravity values in a centrifuge. The experimental conditions were as given in figure 18. The open states are marked by arrows at the right side, opening of the channels is reflected by downward deviations of current. The mean open state probability increases with increasing gravity. The figure is modified from Goldermann and Hanke, 2001.

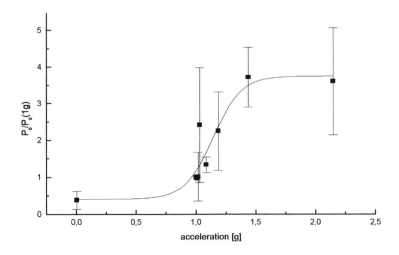

Figure 20 Dependency of the relative mean open state probability of porin channels on the gravity amplitude. The figure is modified from Goldermann and Hanke, 2001.

6. Discussion

The above presented experiments about the gravity dependence of ion channels incorporated into planar lipid bilayers are partially initiated by the ideas presented in earlier papers from some other groups according to gravity interaction with ion channels. Thus, the few significant studies out of the field are shortly summarized in the following, before the ideas and consequences of our results are discussed.

6.1. Short historical overview about studying the influence of gravity on ion channels

It has been shown that the conductance of gramicidin pores incorporated into planar lipid bilayers is depending on the orientation of the bilayer relative to the earth gravitational vector. Although a real convincing theoretical interpretation of these results is still missing, they demonstrate very clearly that not only channel kinetics but also channel conductances directly can be gravity dependent (Schatz, et al., 1996).

When comparing the gramicidin results with those from other channels and pores investigated under the influence of gravity, one should have in mind that all these channels and pores are basically formed by bundles of alpha-helices, whereas the gramicidin pore is build by a conducting pathway through the interior of a beta type helix. Possibly this is the reason for the high conductance effect of gravity.

Independently, other papers have demonstrated that the kinetics of gramicidin pores can be effected by membrane tension (i.e. Goulian et al., 1998). As it has also been demonstrated that alamethicin kinetics is tension dependent (Opsahl and Webb, 1994, Hanke and Schlue, 1993), and that also a variety of other channel forming proteins are interacting with membrane tension, i.e. porins (Sukharev, 1999), possibly a link can be constructed between the thermodynamical parameters of the membrane, in this specific case tension, and the interaction of membrane incorporated proteins with gravity. Our guess is that gravity at least partially interacts with the two-dimensional thermodynamical system membrane, possibly shifting the position of the membrane in the phase diagram. This could explain the effects of gravity on ion channels instead of a direct molecular interaction; gravity would act on the membrane tension, thereby effecting the kinetics and possibly the conductance of ion channels.

At least one study has qualitatively investigated gravity interaction with liposome reconstituted gap-junction channels and did not find any significant response (Claassen and Sponsor, 1989).

Although the other literature about the direct interaction of gravity with ion channels is not too much extended up to now, an increasing amount of studies of the interaction of gravity with living systems has shown the importance of these experiments, and also the number of experiments studying membrane transport under conditions of changing gravity directly, is increasing

6.2. Gravity dependence of ion channels and membranes

In this study we have shown that ion channels incorporated into planar lipid bilayers can be directly effected by the interaction with gravity. Two example have been chosen, the well characterised pore-forming polypeptide alamethicin (Boheim, 1974) and a channel forming porin from *e-coli*.

Alamethicin has a somewhat specific mechanism of pore formation by the assembly of monomeric alpha-helical structures, this mechanism most probably includes the flip-flop (Boheim et al., 1983) of the whole alamethicin molecule through the membrane, making it very sensitive to the properties of the lipid core of the membrane (i.e. Boheim et al., 1980).

Indeed we have found that the kinetic pore parameters of alamethicin are highly gravity dependent, in the range of 0 g up to 6 g. Significant effects can even be found in the range of 0 g to 2 g, values which are by sure physiologically relevant. The pore state conductances were not significantly effected by gravity. The lowest known conductance state (ca. 20 pS in 1 M KCl) could not be resolved in our studies due to technical limitations, possibly here some effects of gravity on the conductance can be expected; this question should be investigated in a future study.

Both, alamethicin state lifetimes and pore frequencies, depend on gravity in a complex manner. However, globally the activity of alamethicin induced currents in planar lipid bilayers increases with increasing gravity under steady state conditions. Deviations in this behaviour, partially found in centrifuge experiments, can be explained by the fact that not only the amplitude, but also the direction of the gravity vector is changing the alamethicin kinetics. To make the situation even more complicated, we found that the vectorial changing rate of gravity, dg/dt, is also effecting the system. Finally, some first not published results indicate a dependency of the measured effects on membrane area.

All the results about alamethicin could be best explained when assuming that gravity does not alone directly interact with the single molecules, but also has effects on the thermodynamics of the complex membrane system and modifies the distribution coefficient of alamethicin between water- and lipid phase. A theoretical description of this view is in progress.

The second ion channel investigated is a porin from *e-coli*, the pore forming mechanism of this ion channel is also well known and also is based on the assembly of alpha-helical structures. This channel is additionally known to be mechano-sensitive in its native membrane. In our experiments it was found that ion channels made by porins are clearly gravity dependent. Again, mainly the channel kinetics was effected, giving an increasing open state probability with increasing gravity. There was no significant effect on the channel conductance.

From our studies it can be concluded, that ion channels, the core of which are formed by bundles of alpha-helices, are depending on gravity in their kinetics. As found for gramicidin (Schatz et al., 1996) there is a possible difference to pores using another molecular mechanism, however, most when not all biological significant ion channels are formed according to the conducting core by the assembly of alpha-helical structure.

The physiological consequences of our findings manning fold, especially as the amplitude of the gravity vector used clearly is in the physiological range. The question, why humans exposed for example to increased gravity for a while, do not show such significant effects, must be investigated in more complex structures, from tissue preparations to global human physiology. It can be only speculated presently, that additional regulative mechanisms are involved in the gravity response of intact living systems, including humans.

6.3. Related studies
Of course the above discussed studies about the gravity influence on single ion channels are not standing only on themselves. It is obvious for example that changes in the kinetics of the potential dependent sodium channel would have severe consequences on action potential properties, thus effecting the complete biological signal processing. Indeed, at least basically such an effect has already been proven (Rüegg et al., 2000)
Additionally, an increasing number of studies is investigating the medical consequences of micro- and macro-gravity exposure. According to the fact that membrane transport processes, including ion channels, are involved in the function of about all living systems, the relevance of the above studies for other fields in life sciences is obvious.
Finally, in the recent years, the study of excitable media became an increasing field in natural sciences. According to the physical definitions, all biological systems are at least partially such excitable media (i.e. Keener and Sneyd, 1998; Paresi et al., 1998), a statement which also holds for biological membranes themselves. It has been shown in a variety of physico-chemical but also other experiments that gravity, as a weak but permanently present external force, interacts directly with such excitable media. Examples are the retinal spreading depression (Fernandes de Lima et al., 1999; Hanke et al., 2001) and the Belousov-Zhabotinsky reaction (Zaitkin and Zhabotinsky, 1970; Wiedemann et al., 2002). Interestingly both cited examples at least in theoretical aspects are very similar to the above cited action potentials. Consequently, also from such studies a direct effect of gravity on action potential properties can be postulated, however, possibly based on other mechanisms than direct interaction with ion channels in the membranes under investigation.

7. Outlook

Up to now it has been clearly demonstrated that ion channels and pores can be directly influenced by gravity and thus be involved in gravity perception or at least allow an interaction of gravity with any living system. However, the true molecular mechanism of this interaction is still open. A variety of studies thus are tempting to be done, for example it is necessary to investigate the dependence of the gravity effect on ion channels on membrane area. This is a study which could be possibly done in planar lipid bilayers.

An extended study of ion channels incorporated in liposomes under the action of varying gravity would be also useful (Claassen and Sponsor, 1989), especially when optical methods are used, as these allow to upscale the number of experiments done in parallel, quite easy.

Although planar lipid bilayers are convincing model systems, a study of single ion-channels (especially the potential dependent sodium channel and the nicotinic acetylcholine receptor might be of interest here) in true biological membranes by the patch-clamp technique would be also a desirable approach. A prerequisite of this would be the construction of a fully automated patch-clamp set-up additionally being stable to the rough mechanical environment which is given in most micro-gravity platforms at lift-off, in between the micro-gravity phases or at touch-down. The high demand of such an approach is obvious, also having in mind that up scaling of the number of parallel experiments in the patch-clamp-technique most probably is either not possible or far too expensive.

The presented studies in addition to the obtained scientific results have shown that it is technically possible to scale up bilayer experiments to 32 parallel fully automated experiments, but it would be easily possible to even proceed to 100 parallel experiments or more.

A very similar approach has been made with retinal spreading depression experiments, here also the automation and the up scaling (up to 8 parallel experiments) has been shown in micro-gravity experiments. Both techniques are of interest for example for high throughput screening systems for receptor ligand interaction for basic pharmacological research. In case of bilayers, purified receptors can be reconstituted into the bilayers and their interaction with drugs can be studied, in case of the retinal spreading depression (Fernandes de Lima et al., 1999) comparable studies can be done with intact central nervous tissue.

Obviously the technical needs of micro-gravity experiments sometimes result in highly interesting technical applications in other fields.

References

Berrier, C. et al. (1992) Fast and slow kinetics of porin from *e-coli* reconstituted into giant liposomes and studied by patch-clamp. FEBS, **306**, 251-256

Boheim, G. (1974) Statistical analysis of alamethicin channels in black lipid membranes. J. Membr. Biol., **19**, 277-303

Boheim, G., Hanke, W. und Eibl, H. (1980) Lipid phase transition in planar lipid bilayer membrane and its effect on carrier- and pore- mediated ion transport. Proc. Natl. Acad. Sci. USA **77**, 3403-3407

Boheim, G., Hanke, W. und Eibl, H. (1980) Lipid phase transition in planar lipid bilayer membrane and its effect on carrier- and pore- mediated ion transport. Proc. Natl. Acad. Sci. USA **77**, 403-3407

Boheim, G., Hanke, W. und Jung, G. (1983) Alamethicin pore formation, voltage dependent flip-flop of alpha-helix dipoles. Biophys. Struct. Mech. **9**,188-197

Braun, M. (1997) Gravitropoism in tip-growing cells. Planta, **203**, 11-19

Claassen, D.E. and Sponsor, B.S. (1989) Effects of gravity on liposome-reconstituted cardiac gap-junction channelling activity. Biochem. Biophys. Res. Comm., **161**, 358-362

Fernandes de Lima, V.M., Goldermann, M. and Hanke, W. (1999) The retinal spreading depression. Shaker Verlag, Aachen, Germany, (1999)

Garcia-Anoveros, J. And Corey, D.P. (1997) The molecules of mechanosensation. Ann. Rev. Neurosci., **20**, 567-594

Goldermann, M., Klinke, N. and Hanke, W. (2000) Der Einfluß der Gravitation auf künstliche Porenbildner. In: Bilanzsymposium Forschung unter Weltraumbedingungen, Keller, M.H. and Sahm, P.R., eds., RWTH-Aachen, pp. 509-515, (2000)

Goldermann, M. and Hanke, W. (2001) Ion channels are sensitive to gravity changes. J. Microgravity Sci. Technol., **XIII/1**, 35-38

Goulian, M., Mesquita, O.N., Feygenson, D.K., Nielson, C., Anderson, O.S. and Libachaber, A. (1998) Gramicidin channel kinetics under tension. Biophys. J., **74**, 328-3387

Greger, R. and Windhorst, U. (1996) Comprehensive human physiology. Springer Verlag, New York, USA

Häder, D.P. (1999) Gravitaxis in unicellular microorganisms. Adv. Space Res., **24**, 851-860

Hamill, O.P., Marty, A., Neher, E., Sackmann, B. and Sigworth, F.J. (1981) Improved patch-clamp-technique for high-resolution recording from cells and cell-free membrane patches. Pflügers Arch., **391**, 85-100

Hanke, W. and Schlue, W.-R. (1993) Planar lipid bilayer experiments: Techniques and application. Academic Press, Oxford, UK

Hanke, W. (1995) Studies of the interaction of gravity with biological membranes using alamethicin doped planar lipid bilayers as a model system. Adv. Space Rev., **17**, 143-150

Hanke, W., Wiedemann, M., Rupp, J. and Fernandes de Lima, V.M. (2001) Control of the retinal spreading depression by weak external forces. Faraday Discussions, **120**, 237-248

Keener, J. and Sneyd, J. (1998) Mathematical physiology. Springer, New York, USA

Klinke, N. (1999) Alamethicin als Sensor für Membraneigenschaften. Dissertation, University of Hohenheim, Stuttgart, Germany

Klinke, N., Goldermann, M., Rahmann, H. and Hanke, W. (1998) The bilayer block module: A system for automated measurement and remote controlled measurements of ion current fluctuations. Space Forum, **2**, 203-212

Klinke, N., Goldermann, M. and Hanke, W. (1999) Planar lipid bilayers doped with alamethicin as a sensor for gravity. III Workshop on Cybernetic Vision. da Fontoura et al., eds., IFCS-USP, Brazil, pp. 25-30

Klinke, N, Goldermann, M, and Hanke, W. (2000) The properties of alamethicin incorporated into planar lipid bilayers under the influence of micro-gravity. Acta Astronautica, **47**, 771-773

Krasnov, I.B. (1994) Gravitational Neuromorphology. Adv. Space Biol. Med., **4**, 85-110

Machemer, H. (1997) Unicellular responses to gravity transitions. Space Forum, **3**, 3-44

Montal, M. and Mueller, P. Formation of bimolecular membranes from lipid monolayers and a study of their electrical properties. Proc. Natl. Acad. Sci. USA, **69**, 3561-3566

Mueller, P., Rudin, D.O., TiTien, H.T. and Wescott, W.L. (1962) Reconstitution of cell membrane structure in vitro and its transformation into an excitable system. Nature, **194**, 979-980

Paresi, J., Müller, S.C. and Zimmermann, W. eds. (1998) A perspective look at nonlinear media. Springer Verlag, Berlin, Germany

Rahmann, H. and Hirsch, K.A. (2001) Mensch-Leben-Schwerkraft-Kosmos. Verlag G. Heimbach, Stuttgart, Germany

Rüegg, D.G., Kakebeeke, T.H. and Studer, L.M. (2000) Einfluss der Schwerkraft auf die Fortleitungsgeschwindigkeit von Muskel-Aktionspotentialen. In: Bilanzsymposium Forschung unter Weltraumbedingungen 1998, Kelle and Sahm, eds. WPÜF, Aachen, Germany, pp. 752-759

Schatz, A., Linke-Hommes, A. and Neubert, J. (1996) Gravity dependence of the gramicidin a channel conductivity: a model for gravity perception on the cellular level. Eur. J. Biophys., **25**, 37-41

Sukharev, S. (1999) Mechanosensitive channels in bacteria as a membrane tension reporter. FASEB, **13**, 55-61

Wiedemann, M., Fernandes de Lima, V.M. and Hanke, W. (2002) Gravity dependence of waves in the retinal spreading depression and in gel-type Belousov-Zhabotinsky systems, PCCP, in press

Zaitkin, A.N. and Zhabotinsky, A.M. (1970) Concentration wave propagation in two-dimensional liquid-phase self-organizing system. Nature, **225**, 535-537

Acknowledgement: This work was kindly supported by the German Space Agency, DLR, by grant No. 50 WB 9716.

Planar Lipid Bilayers (BLMs) and their Applications
H.T. Tien and A. Ottova-Leitmannova (Editors)

Chapter 25

Advantages and disadvantages of patch clamping versus using BLM

M. L. Kelly and Dixon J. Woodbury

Departments of Physiology and Pharmacology, Wayne State University School of Medicine, 540 E. Canfield, Detroit, MI 48201

1. INTRODUCTION

Biological membranes are interspersed with proteins that allow for the conduction of ions across the membrane. The desire to understand the mechanisms that allow for conduction and selectivity of these channel-forming proteins lead to the development of an array of electrophysiological techniques. One of the first inventions in the investigation of current, the movement of ions, in a biological system was the galvanometer. Created in 1820 by Schweigger, this simple devise allowed for the measurement of currents in a frog's leg (Galvani's twitching frog leg experiment).

Over a hundred years later, one of the greatest contributions to understanding the physiology of ion channels was the invention of the voltage clamp. Though Hodgkin and Huxley performed their famous voltage clamp experiments on giant squid axons in 1952 [1-4], the voltage-clamp was invented in 1940 by Kenneth Cole [5]. The invention made it possible to maintain or clamp the potential across the cell membrane. The process of voltage clamping introduces a current that is equal in amplitude but opposite in sign to that which flows across the membrane of the cell. This maintains the membrane potential since the net flow of current is equal to zero. By measuring the amount of current required to sustain the membrane potential, one is measuring the flow of current across the membrane. Electrophysiology has come a long way through the years. Techniques have advanced to the point in which the current passing through a single channel can be measured. This came with the inventions of the planar lipid bilayer technique by Mueller, Rudin, Tien, and Wescott, and the development of the patch clamp by Neher and Sakmann. Patch clamping has remained a powerful technique for the study of channels in intact membranes. It provides a way not only to study a cell as a whole, but in individual pieces. However, the patch clamp method has some limitations, as it is designed for the study of channels in cell membranes and not for purified proteins in a lipid environment. In order to study these proteins in isolation from other cellular constituents, a planar lipid bilayer system is more suited.

The first artificial bilayer experiment was performed by Mueller, Rudin, Tien and Wescott in 1962 [6]. The system that became known as the MRTW bilayer provided a way to create phospholipid bilayers that mimicked the asymmetric distribution of lipids found in biological membranes. These two scientists developed a system that allowed fully functional purified membrane proteins to be reconstituted into artificial membranes. There are many overlaps that exist between the bilayer and patch clamping systems. The primary point of intersection is single channel recordings. There are benefits and drawbacks to both techniques. Lets begin here with the patch clamp technique.

2. THE PATCH CLAMP TECHNIQUE

A major advantage of the patch clamping technique is the ability to examine the electrical properties of an ion channel in situ within the confines of the whole cell environment. There are several types of patches that can be formed and many of them will be discussed.

In voltage clamp mode in a patch clamping system, the experimenter sets the voltage to be maintained across the cell membrane and measures the current required to maintain that voltage. The amount of current at different voltages can be measured and plotted. The slope of the resulting current-voltage curve gives the conductance of the channel or channels in the membrane. A second mode, known as current clamp allows for the reverse to be measured. A known amount of current is injected across the cell membrane and changes in membrane voltage can be recorded.

Whereas a planar lipid bilayer is formed in a hole in a partition, patch clamping seals a membrane bilayer to the end of a pipette. The pipette is usually formed by pulling a heated piece of capillary glass until it breaks apart.

Typically, this will form pipettes that can then be heat-polished down to form an ~1 μm hole at the tip. The exterior pipette surface close to the tip can be coated to reduce pipette capacitance. Once the patch electrode is placed in the bath, the resistance of the pipette can be measured. If resistance is <1 MΩ, then the pipette tip is probably broken. A typical pipette used for patching cells will be between 3-6 MΩ. Larger resistances maybe needed for smaller cells and for sub-micron patching, since resistance is inversely related to the diameter of pipette tip. As the patch electrode makes contact with the cell surface, the resistance of the tip will increase as a seal is formed between the pipette and the cell. This type of patch is referred to as a "cell attached" patch. The seal is referred to as a "tight" seal if the resistance of the pipette becomes greater than 1 GΩ after the formation of the seal. Formation of a tight seal is aided by applying negative pressure or suction to the patch electrode. In this configuration, one is able to study channels under the tip of the electrode. The ability to observe the activity of these channels is dependent on the presence of open channels in the membrane that is not patched.

The best way to observe the total activity of channels in the cell is to break through the membrane of the cell that was sealed in the "cell attached" patch. Applying a large voltage (>500 mV) for a brief amount of time or increasing the amount of negative pressure/suction can do this. Once inside the cell, the pipette resistance will decrease. This configuration is called the "whole cell" patch.

Instead of breaking through the membrane to form a whole cell patch, one can simply pull the pipette away from the cell after the tight seal has been formed. This excises the patch away from the cell to form an "inside-out" patch since now the inside of the cell is exposed to the bath.

If one wishes the configuration of the patch to be the same as the cell, there is a way to achieve this. This type of patch is known as an "outside-out" patch and is formed by moving the electrode side to side after formation of a whole cell patch so that the cell membrane loops back around forming a seal. Then the pipette is backed away from the cell as is done with an inside out patch.

3. USES AND DRAWBACKS OF PATCH TECHNIQUES

This section will focus on the advantages and drawbacks of using the different patch configurations mentioned in the previous section and addresses some variations on these techniques. A paper giving an example for each technique will be presented in each subsection. For further explanation of techniques and reference to drawbacks see Sakmann and Neher, "Single-Channel Recordings" [7].

3.1. Cell-attached patch:

As shown in Fig. 1A, when the pipette first makes a tight seal with the cell, this configuration is known as a cell-attached patch. Cell attached recordings are used for observing a channel where the mechanism for gating is unknown. Because the cell remains intact during the recording, the physiological environment of the channel inside the cell is preserved. This allows one to distinguish whether the properties of the channel change after patch excision.

This type of patch can also be used to assess whether a second messenger is involved in activation. If a second messenger activates the channel, then addition of the agonist to the bath will result in activation of the channel even though the channel patch is secluded from the bath by the electrode. An example of this is the activation of potassium channels by clonidine.

Clonidine, used to decrease symptoms associated with opiate withdrawal, has been shown to act at the α_2-adrenoreceptor [8] and to activate potassium channels as well [9]. Murphy and Freedman used the cell-attached configuration on rat amygdalohippocampal dissociated neurons to examine K^+ channels that were activated by application of clonidine and compare the channels to those that were activated by morphine application. It was originally thought that the two chemicals worked through the same mechanism. However, they observed that a difference in the frequency of channel type either a 95 pS or 130 pS K^+ channel was related to whether clonidine or morphine respectively was applied. They concluded that clonidine acted through activation of a separate channel than morphine.

For this experiment, out of 59 patches only 30 percent contained the potassium channels in question. Since only channels directly under the tip can be observed, and are dependent on the opening of channels in the cell membrane in order to be observed, it is difficult to study channels that are expressed at low levels in the membrane. Another negative aspect of this technique is the lack of knowledge of cell resting membrane potential. The resting membrane potential adds to the applied pipette potential. There is also an inherent inability to control and change the ionic composition of the solutions on both sides of the patch during measurement.

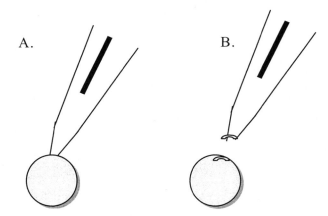

Figure 1 A. The "cell-attached patch". The pipette forms a seal between the glass of the electrode and the cell membrane. B. Inside-out patch. The pipette is pulled away from the docked cell and the seal remains between the glass of the pipette and the membrane that was beneath it.

3.2. Inside-out recording:

The inside-out recording (see Fig. 1B) allows one to readily change the cytosolic side of the patch. Thus, one can easily study the effects of drugs on ion channel activity at the single channel level. This type of patch is used for single channel analysis of voltage gated as well as second-messenger-activated channels. An example of this is looking at Ca^+ activated K^+ channels.

Neonatal rat hippocampal neurons contain large conducting Ca^{2+}-activated K^+ channels (maxi-K channels). In the intracellular fluid of these neurons there is glutathione, which is involved in reducing oxidative stress. Soh et al. [10] sought to determine if the concentration of glutathione in the neuron had an effect on the activity of the maxi-K channels. They cultured rat hippocampal neurons and formed inside-out patches on the neurons. Glutathione was then applied to the bath at various concentrations and found to increase channel activity with a half activation of 710 μM. They also find that the Maxi-K channel is inhibited by oxidized glutathione at a half maximal concentration of 500 μM. They conclude that the redox state of glutathione may play a pivotal role in the activation of the maxi-K channel.

The major disadvantage of the inside-out patch is the loss of key cytosolic factors controlling the behavior of the ion channels. For example, if activation of the channel is via a second messenger system and some of the key components are missing the channel will not activate.

3.3. Whole-cell recording

The whole-cell patch (Fig. 2A) enables recording of the ion currents of the whole cell. These can be from ion channels endogenous to the cell or exogenous channels expressed in the cell membrane. The technique is often employed in measuring exocytotic activity of secretory cells by observing cell membrane capacitance as well.

Sun et al. in their 2001 [28] paper use the whole cell patch method to help understand how ischemia can lead to epilepsy. They utilized hippocampal primary cultures from rat. The cells were then treated with 5 μM glutamate and the effects were monitored using current clamp before, during and after treatment for 30 minutes. During treatment, membrane depolarization and spiking occurred. However, continuous spiking did not occur. This is also observed during ischemic or anoxic brain injury. Other affects such as somatic swelling and a decrease in membrane input resistance were noted. Neuronal cell death, following exposure, became significant after 30 minutes of glutamate exposure. In this group, spontaneous recurrent epileptiform discharges (SREDs) were observed in ~86% of the surviving neurons.

Advantages of this technique are the ability to observe the ion channel activity of a whole cell, maintaining the natural environment of the channel in question, and the ability to take capacitance measurements. However, cytosolic factors dilute with the pipette solution over time.

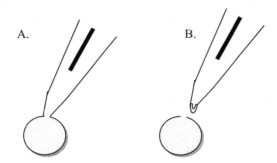

Figure 2   A.   The whole cell patch is formed by breaking through the cell membrane. The solution in the pipette becomes continuous with the cytosol. B. After formation of the whole cell patch, an outside-out patch can be formed by rocking the pipette back and forth until a seal is made and then backing the pipette away from the cell.

3.4. Outside-out recording

The outside-out patch (Fig. 2B) is similar to the inside-out patch in that it allows for single channel recording. This type of patch has the advantage of allowing the fluid on the extracellular side of the patch to be readily changed. The main purpose of this type of patch is studying receptor activated ion channels.   The technique has been used to study electrogenic membrane exchangers [11], transporter currents [12-15], channel currents [16-20], transporter and channel charge movements [21-23], and single-channel recording of low-density channels [24, 25].

Rat hippocampal neurons contain AMPA receptors. These receptors undergo rapid desensitization that is thought to contribute to the decay of synaptic excitatory currents and possibly in the modulation of synaptic plasticity [26]. AMPA receptors are a type of glutamate receptor like NMDA receptors. NMDA receptors are inhibited by protons within physiological pH. Lei et al. [27], wanted to know if protons also inhibited AMPA receptors. They isolated hippocampal CA1 neurons from 2-3 week old Wistar rats. Outside-out patches were formed and glutamate was applied.   They examined the effect of acidification on the desensitization of the AMPA receptor and found that protons reduced AMPA receptor desensitization. Also, protons decrease the maximum open probability of the receptors. They further noted that positive modulators of AMPA receptors reduce the proton sensitivity of the AMPA receptor to desensitization. They concluded that extracellular acidification decreases glutamate toxicity.

There are three technical limitations to this technique. First, cytosolic factors are again lost. Second, more steps are required to reach outside-out configuration (requires the formation of a whole-cell patch first). Finally, this requires that the cell be well adhered to the bottom of the chamber. If the cell is not adhered properly, the cell will pull up when the pipette is pulled away to form the outside out patch.

3.5. Perforated patch technique

One of the major problems of patch clamping is the loss of cytosolic constituents following patch excision or during long recordings using the whole cell configuration [33]. A way of maintaining the interior components of the cell is through the use of a perforated patch. This is done by placing a channel forming substance in the patch pipette solution. Compounds that form small channels such as gramicidin [34-41], ATP [42], nystatin [43], beta-escin [44], and amphotericin B [45-52], allow for the passage of small particles while preventing the washout of cytosolic factors.

An example of this technique is given by Yamashita et al., 2001 [53]. The authors examine the effects of a volatile anesthetic, isoflurane, on glycinergic miniature inhibitory postsynaptic currents. Glycine is a major inhibitory neurotransmitter in the central nervous system. The major role of glycine is suppression of electrical activity in the spinal cord and brain stem [54]. Dissociated neurons from the trigeminal region of Wistar rats' brains were patched using a solution containing the antifungal agent nystatin. (A discussion of the uses of nystatin in bilayers will be addressed also). Nystatin forms slightly cationic channels when incorporated into membranes. The channels are small and allow for the movement of ions while leaving major cytosolic components inside the cell. Glycine was applied to the neurons at various concentrations. A current induced by the addition of glycine was increased by the presence of isoflurane. An increase in miniature IPSC frequency was observed as well. They conclude that isoflurane increases glycine-induced chloride currents.

There are some drawbacks to this technique. An increase in noise associated with the recordings is often observed. The amount of time required to allow access to the interior of the cell varies and results in delays before measurements can be taken.

Detector patch

The general purpose of a detector patch is to observe the release of neurotransmitter by using a receptor that would normally detect release. This is done by taking an outside-out patch that contains the receptor and placing it close to a synaptic structure [55, 56]. An example would be to use acetylcholine receptor containing patches in close proximity to a cholinergic neuron in order to detect acetylcholine release.

This technique can be replaced by using ionophore-detecting pipettes. Detector pipettes are commonly used to detect the release of catecholamines from neurons. Adrenal chromaffin cells release catecholamines as well as ATP. Kasai et al, 1999, [57], measured the release of both catecholamines and ATP from these cells.

3.6. Double patch technique

The double patch technique allows measurement of current through two cells that are directly interacting. Two cells are patched using the whole cell configuration. Measurements are then taken simultaneously in the two cells.

The double patch can also be used with bilayers. In this case, a protein is incorporated into a bilayer formed as a seal on the tip of one or both pipettes. An example of this technique is given in Antonento and Pohl's 1998 paper. The investigators form gigaohm seal patches with patch pipettes placed on opposite sides of the of a planar lipid membrane. The membrane contains a channel forming protein known as gramicidin. Gramicidin naturally incorporates into lipid bilayers from solution. Gramicidin was added to both sides of the planar lipid membrane. Each patch pipette observed the activity of a gramicidin channel. One channel was used to operate as a source for proton conduction. The other was used as a current sink. The results of the measurements suggested that the proton enters the cell moves across the membrane and then directly into the second channel, supporting the theory of local proton coupling.

3.7. Pipette perfusion

Capillary tubes can be inserted into the patch pipette and placed close to the tip. This allows for exchange of the pipette solution within seconds to minutes [58- 61]. This makes the patch system more comparable to the bilayer system. However, the bilayer system holds the advantage in the rate and ease of perfusion.

4. BILAYER TECHNIQUES

One of the greatest advantages of the bilayer system is the ability to alter the composition of the membrane into which the protein is being introduced. Proteins may change configuration or orient differently when the composition of the membrane that surrounds them is altered. For example, nystatin requires the presence of ergosterol in order to form a stable channel. The M2 protein of influenza requires cholesterol in order to form a stable proton channel [62]. The movement of SNARE proteins in a membrane is dependent on the negatively charged lipids such as phosphatidylserine and phosphatidylinositol. All of these facts were determined by incorporating proteins into lipid bilayers, altering the composition of the bilayer and observing changes in activity.

Just as the patch clamping system has several techniques for looking at ion channel activity, so does the planar lipid bilayer system. Variations in this method may take the form of which lipids are used, the way in which a protein is introduced into the bilayer membrane, and the type of chamber system used.

4.1. Formation of the BLM technique

The technique used primarily today was contributed by Abramson et al. in 1965. The bilayer apparatus consisted of a milled out piece of plastic or Teflon in which a cup could be inserted. The cup was inserted snuggly into the apparatus so that a compartment in front of the cup formed one chamber and the inside of the cup formed a second chamber. These two chambers referred to as the "cis" and "trans" chamber were separated by the formation of a bilayer across a small whole in the cup. Originally, a fine sable haired brush trimmed to a few hairs was utilized to apply lipid to the hole. Thus, the technique was referred to as "brushing". Later, the brush was replaced by a syringe with a short length of tubing [63]. Today, 'brushing" a membrane can be done by simply applying lipid from the tip of a pipette to the hole of the cup.

4.2. The brush technique

The quickest way to introduce a protein into a planar lipid bilayer is to "brush" the protein into the membrane. This can be done in several ways. The purified protein can be applied at the end of a pipette tip to the already existing membrane, added to the lipids that are going to be used to form the membrane, or pieces of cell membranes that contain the protein can be brushed into the bilayer. The result is the same, direct incorporation of the protein into the bilayer.

One example of the Brush technique is given by Berdiev et al. (2001), [64]. The experimenters desired to examine the characteristics of Brain Na^+ Channel 2 (BNaC-2) incorporated into planar lipid bilayers. They express BNaC-2 in Xenopus oocytes by microinjection with BNaC-2 cRNA. The oocyte membranes are then prepared following the Perez et al. method [65]. Planar lipid bilayers using diphytanoylphosphatidylethanolamine and diphytanoylphosphatidylserine were painted onto a 200 μm hole in polystyrene cup with membrane capacitances between 250-350 picofarads with a 100 mM NaCl, 10 mM MOPS/TRIS buffer (pH 7.4). The membrane vesicles were introduced into the lipid membrane using a glass rod. Single channel conductances were recorded at pH 6.2 and at pH 7.4. They observed that the channel was open 90% of the time at pH 6.2 and only 8% of the time at pH 7.4. A flickery amiloride block verified the presence of the wild type BNaC-2 channel.

Previously described in C. elegans, mutations that cause neurodegeneration were discovered in a BNaC-2 homologous protein. These same mutations, result in BNaC-2 activation. By replacing BNaC-2 glycine 433 with phenylalanine, a constitutive activation of the channel was observed in the lipid bilayer. This same mutant was not affected by pH change.

The authors conclude that BNaC-2 forms functional channels in planar lipid bilayers, though there are some slight differences between the endogenous and the bilayer channels. They also conclude that the mutant channel is functional and has channel properties that differ from the wild type. Thus, the mutated amino acid may be involved in the formation of the pore and/or the selectivity filter. It has been suggested for the mutation in a similar position in BNaC-1 that the residue is responsible for inhibition of the channel.

The BNaC-2 channel could have been observed using the patch clamp technique by patching either neurons containing the protein or by patching the membranes of Xenopus oocytes in which the protein was expressed. Both Xenopus oocytes and the neurons in which the channel are found contain endogenous channels that would have to be blocked in order to study the BNaC-2 channel. By incorporating the channel into a bilayer membrane, these endogenous channels are no longer a complication. Also, the mutant form of the channel is lethal to cells and thus patching is not a possibility.

4.3. The nystatin/ergosterol technique

Directly brushing a protein into a bilayer can result in the incorporation of too much protein and decrease the chance of single channel conductance observation. It can also expose proteins to organic solvents or air, which may denature them. One way around these problems is through the use of the nystatin/ergosterol technique. Liposomes are prepared containing phospholipids, ergosterol and the antifungal agent nystatin. In the presence of an osmotic gradient, the channels promote fusion of the liposome into the bilayer membrane. The nystatin channel collapses due to the absence of ergosterol in the bilayer membrane. The result is incorporation of the protein or piece of cell/vesicular membrane into the bilayer.

Previously, an example of the nystatin/ergosterol technique using pieces of vesicular membrane has been reported [66]. A suspension of phosphatidylethanolamine, phosphatidylcholine, and phosphatidylserine, with ergosterol and nystatin was dried under a stream of nitrogen and resuspended in a buffered salt solution forming lipid vesicles. In this study, the ion channel activity of cholinergic synaptic vesicles was observed. Purified vesicles from

Torpedo californica and were added to the lipid vesicles. Freezing and thawing the mixture incorporated synaptic vesicle membrane into the lipid vesicles. The "joined" vesicles were subsequently added to the cis side of a bilayer chamber. By increasing the salt gradient across the membrane and holding the membrane at a negative potential, fusion of the joined vesicles with the planar lipid bilayer was promoted, allowing characterization of at least 4 different channel activities. It was concluded that these activities are the results of channels or transporters that exist in the native vesicle membrane.

The advantages of this technique are that once the protein or membrane is incorporated into the bilayer membrane, single channel conductances can be monitored. The amount of incorporation is measured electrically as a sudden increase in current through the membrane via the nystatin channel and followed by a decay in current as the nystatin channels collapse due to the diffusion of ergosterol in the bilayer membrane. Since it is a bilayer system, there is the advantage over patch clamping systems in the ability to readily change solutions on both sides of the bilayer membrane.

There are disadvantages of this technique as well. Cytosolic factors are lost. Membrane components necessary for channel activation may not be present in every bilayer. The amount of incorporation is visible and dependent upon delivery of the vesicles to the bilayers membrane. However, once incorporation begins, it does not always readily stop. Incorporation may continue for a period of time after stirring has been discontinued due to the amount of vesicles already present at the membrane. Thus, over-incorporation may occur.

The technique can also be used for the incorporation of purified proteins as well. An example of this is incorporation of a viral protein. The replication of human rhinovirus is a complex process that requires the interaction of several viral proteins. The viral mRNA encodes a long polypeptide chain. The polypeptide is cleaved by three virus-encoded proteases (including 2A and 3C) to produce 11 protein products from three polypeptide regions. The third region, P3, encodes four proteins: 3A, 3B, 3C and $3D^{pol}$. The viral protein 3C is a protease that cleaves the viral polypeptide into mature viral proteins and acts on transcription factors required for RNA polymerases II and III [67]. The viral protein $3D^{pol}$ is a primer–dependent RNA polymerase required for elongation of the nascent RNA chains from an ssRNA template. It is also responsible for adding the initial uridines to the RNA primer, VPg. It may also play a role in RNA duplex unwinding. The precursor protein 3AB serves as a donor of VPg (3B) to the replication complexes [67]. The viral protein 3A contains a conserved hydrophobic region of 22 amino acids. This conserved region associates with membranes [68]. The viral protein 3A is involved in sequestering the viral proteins needed for formation and function of the replication complexes [69]. Both 3A and 3AB proteins can be glycosylated. Inhibition of glycosylation results in the inhibition of RNA synthesis [70]. The viral protein 3B by itself is a highly basic [68], polar [69] protein. It is covalently linked to all newly synthesized viral positive and negative strand RNAs and to nascent strands of the replication intermediate RNA.

In poliovirus, 3A [71-73] and 3AB [68] proteins have been predicted to contain a membrane-spanning region. It has been shown that expression of 3AB protein in *Escherichia coli* causes the bacterial cell to become more permeable to several small molecules [71], and cells infected with the poliovirus have been shown to be more permeable than uninfected cells to monovalent cations and small molecules [74-76]. Thus, it was hypothesized that the 3AB protein of rhinovirus forms a channel. Purified 3AB protein was incorporated into vesicles in the same manner that synaptic vesicle membranes were incorporated into lipid vesicles (by directly adding purified protein to already formed vesicles and then freezing, thawing and sonicating the vesicles). The vesicles were then added to the cis side of the bilayer chamber and fused into a planar lipid bilayer membrane using a salt gradient. 3AB incorporated into planar lipid bilayers formed a fairly non-selective channel. The channel displayed three main conductance states. One of these was characterized by a fast flickering, medium amplitude (80 pS; ± 1 pS SEM; N=6) conductance (Fig. 3 +140 mV trace). The second state was a slow gating subconductance (29 pS ± 5 pS SEM; N=6) state

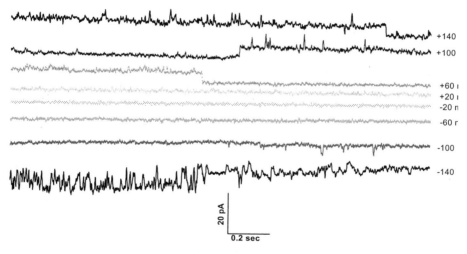

Figure 3. Ion channel activity of the 3AB protein incorporated into a 7:3 PE:PC bilayer membrane using the nystatin/ergosterol technique. Time courses of the observed currents while the bilayer membrane was held at voltages from −140 mV (bottom) to +140 mV (top) are depicted above. Solutions were 428 mM NaCl, 8 mM HEPES, cis and 150 mM NaCl, 8 mM HEPES, trans (pH 7.2). Trans chamber is virtual ground.

(Fig. 3 +60, +100 and +140mV traces) and a full conductance (103 pS ± 8 pS SEM; N=6) state characterized by rapid gating (Fig. 3, -140 mV trace, t<800 msec). The subconductance state was on most of the time for the negative voltages shown in Figure 3. The conductance of the channel was voltage dependent when studied in asymmetric solutions. This was seen with 450 mM NaCl in the cis chamber and 150mM NaCl in the trans chamber. The medium and full conductance states of the channel were prominently observed at voltages more positive than +40 mV.

The 3AB protein is found in the replication complex in virus-infected cells [77]. Therefore, the replication complex as a whole was examined for ion channel activity. Crude replication complexes were incorporated into planar lipid bilayers. The reconstituted replication complex also demonstrated ion channel activity (Fig. 4). The ion channel activity of the replication complex was similar to that of 3AB protein in two respects. First, both channels were fairly nonselective to various ions with a preference of Na>K. Second, both demonstrated mild pH sensitivity, with both showing a decrease in conductance as pH was decreased. The channels also differed in several ways. The ion channel activity of the replication complex had a greater open time with a

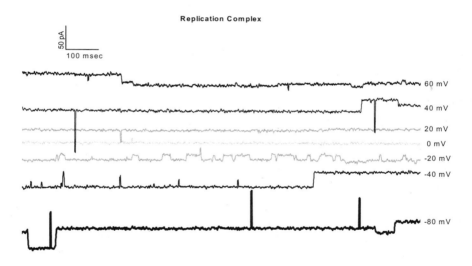

Figure 4. Ion channel activity associated with HRV14 replication complex when incorporated into a planar lipid bilayer held at voltages from -80 mV to +60 mV. Note that fewer conducting states are present than for the channel activity associated with the 3AB protein. Solutions were 428 mM NaCl in DEPC-treated dH$_2$0 cis and 150 mM NaCl in DEPC-treated dH$_2$O trans.

conductance of 313 pS $\pm$ 10 pS (SEM; N=3) than did that of 3AB protein. The ion channel activity produced by 3AB protein had more conductance states than did that of the replication complex.

4.4. The dipping technique

Bilayers can be formed on a metal loop or frame by dipping the loop in lipids before placing it into the bath. The "dipping technique" allows one to form a bilayer in a continuous bath or in one that is separated by two chambers. The continuous chamber eliminates problems with hydrostatic pressure, osmotic pressure, chemical composition or electrical potential.   The technique is primarily used for visual observation, photography, electron microscopy, and for testing the efficacy of membrane-forming solutions.

For separation in two compartments, a technique by van den Berg can be used. A bilayer supporting apparatus fits diagonally into a cross section of a cuvette.   Lowering the bilayer forming apparatus into the upper phase of membrane forming solution forms a bilayer. Thus, it is similar to the tip-dip technique in patch clamping. The apparatus is further lowered into aqueous phase and the bilayer is formed. Large bilayers up to 50 mm^2 can be formed using this method. The main problem with this technique is contamination occurring at the interfaces of the aqueous phase and the membrane forming solution. It is also not well suited for electrophysiological measurements.

4.5. Optical techniques

An advantage of the bilayer system over patch clamping is optical imaging of proteins in lipid systems. One technique referred to as total internal reflection fluorescence microscopy (TIRFM) allows one to observe proteins reconstituted into planar lipid bilayers. The interaction of membrane proteins and external proteins can be observed as well as the interaction of proteins with lipids. An example of the uses of this technique is demonstrated by Wagner and Tamm, [78].

Wagner and Tamm reconstituted proteins that are involved in synaptic transmission into planar lipid bilayers to look at the interaction between synaptic proteins and also between these proteins and membrane lipids.   They fluorescently labeled synaptic proteins and reconstituted them into lipid vesicles. A lipid monolayer was formed on a quartz slide. The vesicles containing plasma membrane proteins involved in synaptic transmission were allowed to sit on the monolayer until a bilayer was formed and the excess was washed away. The analogous protein found in synaptic vesicles was added to the bath and the interaction between the proteins was measured as a change in fluorescence.

They also investigated the interaction of the plasma membrane synaptic proteins with lipids. The mobility of proteins in different lipids can be observed in this system. Lipids that decrease the mobility of a protein in a bilayer most likely have an interaction with the protein in question. They found that the negatively charged lipids DOPS and PIP_2 decreased the mobility of synaptic plasma membrane proteins.

Ide and Yanagida, [79] used TIRMF in conjunction with the planar lipid bilayers to observe gramicidin and alamethicin channels in the bilayer. The horizontal bilayer used for this experiment consisted of an upper and lower chamber. The bottom of the upper chamber consisted of a thin polypropylene film with a small pore in the center. Bilayer formation took place by passing the upper chamber through the lipid layer of the lower chamber into the aqueous layer. The bottom of the lower chamber consisted of a coverslip coated with agarose. The upper chamber was lowered through the aqueous until the membrane came into contact with the agarose-coated glass. Cy3-labeled alamethicin peptides were added to the upper chamber. Within minutes of addition, ionic currents were detected and bright spots could were observed in the membrane.

Other techniques used to study proteins in bilayer membranes are confocal microscopy, nearfield microscopy, coupled plasmon-waveguide resonance spectroscopy (CPWR) and atomic force microscopy (AFM). CPWR combines the high sensitivity, spectral resolution and ability to measure anisotropies in refractive index and optical absorption coefficient in a sensing layer of waveguide spectroscopy with the simple and convenient optical coupling arrangement and isolation of the optical probe provided by SPR spectroscopy [80]. This technique has been used to study the interaction of cytochrome b6f in a phosphatidylcholine bilayer membrane [81].

Perhaps one of the greatest advancements in optics is the ability to view proteins in a membrane at the nanometer scale. AFM provides such a minute viewing scale. The technique was invented in 1986 by Binnig, Quate and Gerber. A probe which is simply a tip on the end of a cantilever which bends in response to the force between the tip and the sample is moved over the surface of a sample in a raster scan. Over the years, AFM has been used in conjunction with bilayers to study DNA/lipid interactions, bacteriophages [82], and proteins incorporated into artificial bilayers [83-86]. Recently, it was shown that when the amyloid β protein was incorporated into planar lipid bilayers, a channel with multiple conductance states was observed. Amyloid β proteins when observed through AFM form globular structures that have been determined to be tetramers and hexamers through biochemical analysis [87, 88].

5. MIXED BILAYER/PATCH CLAMPING TECHNIQUES

Often a researcher wishes to examine a specific protein believed to possess channel activity or to examine components of a known ion channel in order to determine the units that are necessary for proper channel function. This is often accomplished by expressing the suspected channel forming protein in a cell line that has little ion channel activity or where the ion channel activity is largely characterized. The basic requirement to begin such a task is to possess the RNA sequence from which the protein can be made. Once the RNA sequence for a protein is known, the protein can be synthesized by incorporating the sequence into a vector and expressing it in a cell line. Examples of cell lines for expression are HEK cells, Xenopus oocytes, and E. coli. Many other options exist. The cell line one chooses depends on the goal of the experimenter. If the goal is to examine the activity of the channel in a cell membrane using patch clamping, HEK cells and oocytes are ideal. Expression in Xenopus oocytes also allows for examination of the channel using a third technique referred to as two-electrode voltage clamp as well. (In this system two electrodes penetrate the oocyte membrane and a third electrode is used as a reference in the bath). However, if the goal is to mass-produce the protein, expressing the protein in E. coli provides the largest yield of protein. The protein, once purified, can be placed either into liposomes, brushed or incorporated into a bilayer or placed into lipid containing solution. The bilayer, tip-dip technique, or patch clamping of liposomes techniques can be used.

5.1. Tip-dip bilayer recording

This technique is used to study channels reconstituted into a lipid membrane at the end of a patch pipette tip. The pipette tip is repeatedly dipped into the lipid containing the channel forming protein mixture [89, 90] and single channel measurements are recorded. Slight variations on this technique exist. An example is Hara et al., [91] where the protein in located inside the pipette and the lipid is applied to the surface of the bath. The pipette is lowered through the lipid layer to form a membrane at the tip of the pipette. The protein then inserts into the lipid membrane. The advantage of this technique is that it decreases capacitance artifacts found in bilayer recordings. However, you sacrifice the ability to quickly and completely perfuse both sides of the membrane as in the bilayer system.

5.2. Submicron vesicle patch clamping

Liposomes with a diameter between 200 nm-1 μm can be made using phospholipids as mentioned in section 4.3 with only slight modifications can be manufactured for patch clamping. Briefly, a phospholipid mixture   containing phosphatidylethanolamine (PE) and phosphatidylcholine (PC) in a ratio of 7:3 (10 mg/ml in chloroform) can be dried down under a steady stream of either nitrogen or argon. The dried lipids once resuspended in a buffered salt solution, vortexed, and sonicated produce vesicles suitable for patching. A protein of choice can be added to the suspension (vesicles should be frozen and thawed 3 times before use to ensure proper distribution of the protein).

These vesicles can be plated onto poly-d-lysine coated cover slips. The vesicles are visible under an inverted light microscope. Fine submicron tipped pipettes can be pulled and used to examine the small liposomes using the vesicle attached, inside-out or whole vesicle technique.

This technique can also be used to look at pieces of cell or vesicular membrane. In this case vesicles are prepared as mentioned above and the desired membrane is added instead of protein. Channels in the membrane can be examined using the configurations described above.

The advantage of the technique is that it allows one to separate a cell or vesicular membrane into pieces. This aids in separating channels in a membrane that normally possesses many, thus, enabling one to classify the channels present individually. At the same time, the environment of the channel is somewhat maintained.

As with all techniques, the submicron patch has its disadvantages. When incorporating pieces of cell or vesicle membrane, the sidedness of the patch is always in question. You cannot guarantee that the channel you are looking for will be present in the patch and cytosolic or vesicular components may be lost.

5.3. Giant liposome patch clamping

Submicron vesicles may be hard to image and thus a technique that allows for the formation of larger liposomes can be used. Membrane from a cell or cell organelle can be incorporated into liposomes using the same or slightly varied procedure as that described for submicron vesicular patching. Giant liposomes can then be formed through a process of dehydration/rehydration.

Guihard et al., [92] use both the giant liposome clamp technique and the planar lipid bilayer technique to study liver nuclear ionic channels. The investigators purified liver nuclei from both dogs and rats. The purified nuclei were sonicated and treated with both DNase I and RNase. The remaining nuclear envelope (NE) vesicles were sedimented and resuspended in potassium buffered solution and stored in liquid nitrogen until use.

NE vesicles were then mixed with asolectin and frozen and thawed twice. These vesicles were dehydrated under a vacuum and rehydrated with a potassium, magnesium buffered solution. The giant proteoliposomes formed were patched using patch pipettes with a resistance of 5-10 MΩ. They form GΩ seals and then they exposed the patch to air to ensure a single bilayer membrane.

Both the planar lipid bilayer system and the patch clamp technique provide ways to examine single channel activity. The bilayer system provides the ability to remove a channel-forming protein from the constraints of the natural environment allowing for examination of individual characteristics of the protein. Conversely, the patch clamping system usually preserves the natural milieu and promotes an understanding of the role of the protein in the life cycle of the cell. Combinations of mixed bilayer patch clamp techniques open a new horizon in the evolution for techniques available to the electrophysiologist.

I would like to acknowledge the following for their contributions to this chapter. Funding for the 3AB project was provided by Lilly Research Laboratories of Eli Lilly and Company. 3AB protein and replication complex were purified by Patricia Brown-Augsburger, James Cook, Beverly Heinz, and Michele Smith at Eli Lilly. Lawrence Pinto, Northwestern University, provided laboratory equipment for 3AB studies and participated in data review. Dennis Przywara assisted with the vesicular submicron patch technique. Dixon Woodbury and David Giovannucci provided commentary and guidance in the editing of this chapter.

REFERENCES

[1] A.L. Hodgkin and A.F. Huxley, Currents carried by sodium and potassium ions through the membrane of the giant axon of Loligo, J. Physiol (London) 116, (1952), pp. 449-472.

[2] A.L. Hodgkin and A.F. Huxley, The components of the membrane conductance in the giant axon of Loligo. J. Physiol (London) 116, (1952), pp. 473-496.

[3] A.L. Hodgkin and A.F. Huxley, The dual effect of membrane potential on sodium conductance in the giant axon of Loligo. J. Physiol. (London) 116, (1952), pp. 497-506.

[4] A.L. Hodgkin and A.F. Huxley, A quantitative description of membrane current and its application to conduction and excitation in nerve. J. Physiol. (London) 117, (1952), pp. 500-544.

[5] K.S. Cole, Membranes, Ions and Impulses. University of California Press, Los Angeles,1972.

[6] P. Mueller, D. Rudin, H.T. Tien, and W.C. Wescott, Reconstitution of excitable cell membrane structure in vitro. Nature. 194, (1962), p. 979.

[7] B. Sakmann and E. Neher, Single-Channel Recordings, Plenum Press, New York and London, 1995.

[8] MS. Gold, DE Redmond, Jr., HD. Kleber, Clonidine blocks acute opiate-withdrawal symptoms. Lancet. 2(8090), (1978), pp. 599-602.

[9] R. Murphy and JE. Freedman, Morphine and clonidine activate different K+ channels on rat amygdala neurons. European Journal of Pharmacology. 415(1), (2001), pp R1-3.

[10] Soh H. Jung W. Uhm DY. Chung S. Modulation of large conductance calcium-activated potassium channels from rat hippocampal neurons by glutathione. Neuroscience Letters. 298(2), (2001), pp. 115-8.

[11] Hilgemann DW. Regulation and deregulation of cardiac Na(+)-Ca2+ exchange in giant excised sarcolemmal membrane patches. Nature. 344(6263), (1990), pp. 242-5.

[12] S. Matsuoka and DW. Hilgemann, Steady-state and dynamic properties of cardiac sodium-calcium exchange. Ion and voltage dependencies of the transport cycle. Journal of General Physiology. 100(6), (1992), pp. 963-1001.

[13] AE. Doering and WJ. Lederer, The mechanism by which cytoplasmic protons inhibit the sodium-calcium exchanger in guinea-pig heart cells. Journal of Physiology. 466, (1993), pp. 481-99.

[14] BA. Clark, M. Farrant, and SG. Cull-Candy, A direct comparison of the single-channel properties of synaptic and extrasynaptic NMDA receptors. Journal of Neuroscience. 17(1), (1997) pp. 107-16.

[15] LC. Liets and LM. Chalupa, Glutamate-mediated responses in developing retinal ganglion cells. Progress in Brain Research. 134, (2001) pp. 1-16.

[16] DW. Hilgemann and CC. Lu, Giant membrane patches: improvements and applications. Methods in Enzymology. 293, (1998), pp. 267-80.

[17] S. Cai, L. Garneau, R. Sauve, Single-channel characterization of the pharmacological properties of the K(Ca2+) channel of intermediate conductance in bovine aortic endothelial cells. Journal of Membrane Biology. 163(2), (1998), pp. 147-58.

[18] O. Zegarra-Moran and Galietta LJ. Biophysical characteristics of swelling-activated Cl-channels in human tracheal 9HTEo-cells. [Journal Article] Journal of Membrane Biology. 165(3), (1998), pp. 255-64.

[19] A. Bloc, T. Cens, H. Cruz, Y. Dunant, Zinc-induced changes in ionic currents of clonal rat pancreatic -cells: activation of ATP-sensitive K+ channels. Journal of Physiology. 529 Pt 3, (2000), pp. 723-34.

[20] HZ. Wang, SW. Lee and GJ. Christ, Comparative studies of the maxi-K (K(Ca)) channel in freshly isolated myocytes of human and rat corpora. International Journal of Impotence Research. 12(1), (2000), pp. 9-18.

[21] DW. Hilgemann, A. Collins, DP. Cash and GA. Nagel, Cardiac Na(+)-Ca2+ exchange system in giant membrane patches. Annals of the New York Academy of Sciences. 639, (1991), pp. 126-39.

[22] DW. Hilgemann, DA. Nicoll and KD. Philipson, Charge movement during Na+ translocation by native and cloned cardiac Na+/Ca2+ exchanger. Nature. 352(6337), (1991), pp. 715-8.

[23] DW. Hilgemann, Channel-like function of the Na,K pump probed at microsecond resolution in giant membrane patches. Science. 263(5152), (1994), pp. 1429-32.

[24] TC. Hwang, G. Nagel, AC. Nairn and DC. Gadsby, Regulation of the gating of cystic fibrosis transmembrane conductance regulator Cl channels by phosphorylation and ATP hydrolysis. Proceedings of the National Academy of Sciences of the United States of America. 91(11), (1994), pp. 4698-702.

[25] N. Gamper, SM. Huber, K. Badawi, F. Lang, Cell volume-sensitive sodium channels upregulated by glucocorticoids in U937 macrophages. Pflugers Archiv - European Journal of Physiology. 441(2-3), (2000), pp. 281-6.

[26] MV. Jones and GL. Westbrook, The impact of receptor desensitization on fast synaptic transmission. Trends in Neurosciences. 19(3), (1996), pp. 96-101.

[27] S. Lei, BA. Orser, GR. Thatcher, JN. Reynolds and JF. MacDonald, Positive allosteric modulators of AMPA receptors reduce proton-induced receptor desensitization in rat hippocampal neurons. Journal of Neurophysiology. 85(5), (2001), pp. 2030-8.

[28] DA. Sun, S. Sombati and RJ. DeLorenzo, Glutamate injury-induced epileptogenesis in hippocampal neurons: an in vitro model of stroke-induced "epilepsy". Stroke. 32(10), (2001), pp. 2344-50.

[29] DW. Hilgemann, Giant excised cardiac sarcolemmal membrane patches: sodium and sodium-calcium exchange currents. Pflugers Archiv - European Journal of Physiology. 415(2), (1989), pp. 247-9.

[30] CC. Lu and DW. Hilgemann, GAT1 (GABA:Na+:Cl-) cotransport function. Steady state studies in giant Xenopus oocyte membrane patches. Journal of General Physiology. 114(3), (1999), pp. 429-44.

[31] J. Rettinger, Novel properties of the depolarization-induced endogenous sodium conductance in the Xenopus laevis oocyte. Pflugers Archiv - European Journal of Physiology. 437(6), (1999), pp. 917-24.

[32] A. Diakov, JP. Koch, O. Ducoudret, S. Muller-Berger and E. Fromter, The disulfonic stilbene DIDS and the marine poison maitotoxin activate the same two types of endogenous cation conductance in the cell membrane of Xenopus laevis oocytes. Pflugers Archiv - European Journal of Physiology. 442(5), (2001) pp. 700-8.

[33] M. Pusch and E. Neher, Rates of diffusional exchange between small cells and a measuring patch pipette. Pflugers Archiv - European Journal of Physiology. 411(2), (1988), pp. 204-11.

[34] J. Brockhaus and K. Ballanyi, Synaptic inhibition in the isolated respiratory network of neonatal rats. European Journal of Neuroscience. 10(12), (1998), pp. 3823-39.

[35] F. Le Foll, H. Castel, O. Soriani, H. Vaudry and L. Cazin, Gramicidin-perforated patch revealed depolarizing effect of GABA in cultured frog melanotrophs. Journal of Physiology. 507 (Pt 1), (1998), pp. 55-69.

[36] N. Akaike, [Gramicidin perforated patch recording technique]. [Japanese] Nippon Yakurigaku Zasshi - Folia Pharmacologica Japonica. 113(6), (1999), pp. 339-47.

[37] JL. Kenyon, The reversal potential of Ca(2+)-activated Cl(-) currents indicates that chick sensory neurons accumulate intracellular Cl(-). Neuroscience Letters. 296(1), (2000) pp. 9-12.

[38] K. Nakanishi and F. Kukita, Intracellular [Cl(-)] modulates synchronous electrical activity in rat neocortical neurons in culture by way of GABAergic inputs. Brain Research. 863(1-2), (2000), pp. 192-204.

[39] C. Vale and DH. Sanes, Afferent regulation of inhibitory synaptic transmission in the developing auditory midbrain. Journal of Neuroscience. 20(5), (2000), pp. 1912-21.

[40] J. Eilers, TD. Plant, N. Marandi, A. Konnerth, GABA-mediated Ca2+ signalling in developing rat cerebellar Purkinje neurons. [Journal Article] Journal of Physiology. 536(Pt 2), (2001), pp. 429-37.

[41] C. Hirono, T. Nakamoto, M. Sugita, Y. Iwasa, Y. Akagawa and Y. Shiba, Gramicidin-perforated patch analysis on HCO3-secretion through a forskolin-activated anion channel in rat parotid intralobular duct cells. Journal of Membrane Biology. 180(1), (2001), pp. 11-9.

[42] M. Lindau and JM. Fernandez, A patch-clamp study of histamine-secreting cells. Journal of General Physiology. 88(3), (1986), pp. 349-68.

[43] R. Horn and A. Marty, Muscarinic activation of ionic currents measured by a new whole-cell recording method. Journal of General Physiology. 92(2), (1988), pp.145-59.

[44] JS. Fan and P. Palade, Perforated patch recording with beta-escin. Pflugers Archiv - European Journal of Physiology. 436(6), (1998), pp. 1021-3.

[45] J. Rae, K. Cooper, P. Gates and M. Watsky. Low access resistance perforated patch recordings using amphotericin B. Journal of Neuroscience Methods. 37(1), (1991), pp. 15-26.

[46] HK. Lee and KM. Sanders, Comparison of ionic currents from interstitial cells and smooth muscle cells of canine colon. Journal of Physiology. 460, (1993), pp. 135-52.

[47] C. Dart and NB. Standen, Activation of ATP-dependent K+ channels by hypoxia in smooth muscle cells isolated from the pig coronary artery. Journal of Physiology. 483 (Pt 1), (1995), pp. 29-39.

[48] PB. Hill, RJ. Martin and HR. Miller, Characterization of whole-cell currents in mucosal and connective tissue rat mast cells using amphotericin-B-perforated patches and temperature control. Pflugers Archiv - European Journal of Physiology. 432(6), (1996), pp. 986-94.

[49] CS. Bockman, M. Griffith, MA. Watsky, Properties of whole-cell ionic currents in cultured human corneal epithelial cells. Investigative Ophthalmology & Visual Science. 39(7), (1998), pp. 1143-51.

[50] RA. Nicholson, G. Lees, J. Zheng and B. Verdon, Inhibition of GABA-gated chloride channels by12,14-dichlorodehydroabietic acid in mammalian brain. British Journal of Pharmacology. 126(5), (1999), pp. 1123-32.

[51] SM. Ward and JL. Kenyon, The spatial relationship between Ca2+ channels and Ca2+-activated channels and the function of Ca2+-buffering in avian sensory neurons. Cell Calcium. 28(4), (2000), pp. 233-46.

[52] SM. Bryant, CE. Sears, L. Rigg, DA. Terrar, B. Casadei, Nitric oxide does not modulate the hyperpolarization-activated current, I(f), in ventricular myocytes from spontaneously hypertensive rats. Cardiovascular Research. 51(1), (2001) pp. 51-8.

[53] M. Yamashita, T. Ueno, N. Akaike and Y. Ikemoto, Modulation of miniature inhibitory postsynaptic currents by isoflurane in rat dissociated neurons with glycinergic synaptic boutons. European Journal of Pharmacology. 431(3), (2001), pp. 269-76.

[54] H. Betz, Glycine receptors: heterogeneous and widespread in the mammalian brain. Trends in Neurosciences. 14(10), (1991), pp. 458-61.

[55] S.H. Young and M.M. Poo, Spontaneous release of transmitter from growth cones of embryonic neurones. Nature. 305(5935), (1983), pp. 634-7.

[56] AD. Grinnell, CB. Gundersen, SD. Meriney and SH. Young, Direct measurement of ACh release from exposed frog nerve terminals: constraints on interpretation of non-quantal release. Journal of Physiology. 419, (1989), pp. 225-5.

[57] Y. Kasai, S. Ito, N. Kitamura, T. Ohta and Y. Nakazato, On-line measurement of adenosine triphosphate and catecholamine released from adrenal chromaffin cells. Comparative Biochemistry & Physiology. Part A, Molecular & Integrative Physiology. 122(3), (1999), pp. 363-8.

[58] M. Soejima, A. Noma, Mode of regulation of the ACh-sensitive K-channel by the muscarinic receptor in rabbit atrial cells. Pflugers Archiv - European Journal of Physiology. 400(4), (1984), pp. 424-31.

[59] J. Y. Lapointe and G. Szabo, A novel holder allowing internal perfusion of patch-clamp pipettes. Pflugers Archiv - European Journal of Physiology. 410(1-2), (1987), pp. 212-6.

[60] C. Oliva, IS. Cohen and RT. Mathias, Calculation of time constants for intracellular diffusion in whole cell patch clamp configuration. Biophysical Journal. 54(5), (1988), pp. 791-9.

[61] JM. Maathuis, AR. Taylor, SM. Assmann and D. Sanders, Seal-promoting solutions and pipette perfusion for patch clamping plant cells. Plant Journal. 11(4), (1997), pp. 891-6.

[62] DZ. Cleverley, HM. Geller, J. Lenard, Characterization of cholesterol-free insect cells infectible by baculoviruses: effects of cholesterol on VSV fusion and infectivity and on cytotoxicity induced by influenza M2 protein. Experimental Cell Research. 233(2), (1997), pp. 288-96.

[63] H. Ti Tien, Bilayer Lipid Membranes (BLM), Marcel Dekker, Inc, New York, 1974.

[64] BK. Berdiev, TB. Mapstone, JM. Markert, GY. Gillespie, J. Lockhart, CM. Fuller and DJ. Benos, pH alterations "reset" Ca2+ sensitivity of brain Na+ channel 2, a degenerin/epithelial Na+ ion channel, in planar lipid bilayers. Journal of Biological Chemistry. 276(42), (2001), pp. 38755-61.

[65] G. Perez, A. Lagrutta, JP. Adelman and L. Toro, Reconstitution of expressed KCa channels from Xenopus oocytes to lipid bilayers. Biophysical Journal. 66(4), (1994), pp. 1022-7.

[66] M.L. Kelly and D.J. Woodbury, Ion channels from synaptic vesicle membrane fragments reconstituted into lipid bilayers. Biophysical Journal. 70(6), (1996), pp. 2593-9.

[67] A. G. Porter, Picornavirus nonstructural proteins: emerging roles in virus replication and inhibition of host cell functions. J. Virol. 67, (1993), pp. 6917-21.

[68] J. S. Towner, T. V. Ho and B. L. Selmer, Determinants of membrane association for poliovirus protein 3AB. J. Biol. Chem., 271, (1996), pp. 26810-8.

[69] B. A. Heinz and L. M. Lance, The antiviral compound enviroxime targets the 3A coding region of Rhinovirus and poliovirus. J. Virol. 69, (1995), pp. 4189-97.

[70] J. Lama and L. Carrasco, Expression of poliovirus nonstructural proteins in Escherichia coli cells. Modifications of membrane permeability induced by 2B and 3A. J. Biol Chem., 267, (1992), pp. 15932-7.

[71b] U. Datta and A. Dasgupta, Expression and subcellular localization of poliovirus VPg-precursor protein 3AB in eukaryotic cells: evidence for glycosolation in vitro. J. Virol. 68, (1994), pp. 4468-77.

[72] J. Lama and L. Carrasco, Mutations in the hydrophobic domain of poliovirus protein 3AB abrogate its permeabilizing activity. FEBS Lett. 367, (1995), pp. 5-11.

[73] J. Lama, M. A. Sanz and P. L. Rodriguez, A role for 3AB protein in poliovirus genome replication. Journal of Biological Chemistry. 270(24), (1995), pp. 14430-8.

[74] J. Lama, L. Carrasco, Screening for membrane-permeabilizing mutants of the polivirus protein 3AB. J. Biol. Chem. 270, (1995), pp. 14430-8.

[75] A. Lopez-Rivas, J. L. Castillo and L. Carrasco, Cation content in poliovirus infected HeLa cells. J. Gen. Virol. 68, (1987), pp. 335-42.

[76] A. Munoz and I. Carrasco, Effect of interferon treatment on blockade of protein synthesis induced by poliovirus infection. Eur. J. Biochem. 137, (1983), pp 623-9.

[77] Doedens JR. Kirkegaard K. Inhibition of cellular protein secretion by poliovirus proteins 2B and 3A. EMBO Journal. 14(5), (1995), pp. 894-907.

[78] Wagner ML. Tamm LK. Reconstituted syntaxin1a/SNAP25 interacts with negatively charged lipids as measured by lateral diffusion in planar supported bilayers. Biophysical Journal. 81(1), (2001), pp. 266-75.

[79] Ide T. Yanagida T. An artificial lipid bilayer formed on an agarose-coated glass for simultaneous electrical and optical measurement of single ion channels. Biochemical & Biophysical Research Communications. 265(2), (1999), pp. 595-9.

[80] Z. Salamon and G. Tollin, Interaction of horse heart cytochrome c with lipid bilayer membranes: effects on redox potentials. Journal of Bioenergetics & Biomembranes. 29(3), (1997), pp. 211-21.

[81] Z. Salamon, D. Huang, WA. Cramer and G. Tollin, Coupled plasmon-waveguide resonance spectroscopy studies of the cytochrome b6f/plastocyanin system in supported lipid bilayer membranes. Biophysical Journal. 75(4), (1998), pp. 1874-85.

[82] N. Matsko, D. Klinov, A. Manykin, V. Demin and S. Klimenko, Atomic force microscopy analysis of bacteriophages phiKZ and T4. Journal of Electron Microscopy. 50(5), (2001), pp. 417-22.

[83] PS. Ajmani, W. Wang, F. Tang, MA. King, EM. Meyer and JA. Hughes. Transgene delivery with a cationic lipid in the presence of amyloid beta (betaAP) peptide. Neurochemical Research. 26(3), (2001), pp. 195-202.

[84] RF. Epand, CM.Yip, LV. Chernomordik, DL. LeDuc, YK. Shin and RM. Epand, Self-assembly of influenza hemagglutinin: studies of ectodomain aggregation by in situ atomic force microscopy. Biochimica et Biophysica Acta. 1513(2), (2001), pp. 167-75.

[85] FA. Schabert, C. Henn and A. Engel. Native Escherichia coli OmpF porin surfaces probed by atomic force microscopy. Science. 268(5207), (1995), pp. 92-4.

[86] G. Zuber and E. Barklis, Atomic force microscopy and electron microscopy analysis of retrovirus Gag proteins assembled in vitro on lipid bilayers. Biophysical Journal. 78(1), (2000), pp. 373-84.

[87] H. Lin, R. Bhatia and R. Lal, Amyloid beta protein forms ion channels: implications for Alzheimer's disease pathophysiology. FASEB Journal. 15(13), (2001), pp. 2433-44.

[88] H. Lin, YJ. Zhu and R. Lal, Amyloid beta protein (1-40) forms calcium-permeable, Zn2+-sensitive channel in reconstituted lipid vesicles. Biochemistry. 38(34), (1999), pp. 11189-96.

[89] R. Coronado and R. Latorre, Phospholipid bilayers made from monolayers on patch-clamp pipettes. Biophysical Journal. 43(2), (1983), pp. 231-6.

[90] BA. Suarez-Isla, K. Wan, J. Lindstrom and M. Montal, Single-channel recordings from purified acetylcholine receptors reconstituted in bilayers formed at the tip of patch pipets. Biochemistry. 22(10), (1983), pp. 2319-23.

[91] T. Hara, H. Kodama, Y. Higashimoto, H. Yamaguchi, M. Jelokhani-Niaraki, T. Ehara and M. Kondo. Side chain effect on ion channel characters of aib rich peptides. J Biochem (Tokyo). 130(6), (2001), pp. 749-55.

[92] G. Guihard, S. Proteau, MD. Payet, D. Escande and Rousseau E. Patch-clamp study of liver nuclear ionic channels reconstituted into giant proteoliposomes. FEBS Letters. 476(3), (2000), pp. 234-9.

Planar Lipid Bilayers (BLMs) and their Applications
H.T. Tien and A. Ottova-Leitmannova (Editors)

Chapter 26

Using bilayer lipid membranes to investigate the pharmacology of intracellular calcium channels

P. Koulen

Department of Pharmacology and Neuroscience, University of North Texas Health Science Center at Fort Worth, 3500 Camp Bowie Boulevard, Fort Worth, Texas 76107-2699, USA

1. ADVANTAGES OF BILAYER LIPID MEMBRANES VERSUS OTHER TECHNIQUES FOR THE INVESTIGATION OF THE BIOPHYSICAL PROPERTIES OF INTRACELLULAR CALCIUM CHANNELS

Intracellular calcium channels are expressed in membranes of intracellular organelles, such as the sarcoplasmic reticulum, endoplasmic reticulum, Golgi apparatus and nuclear envelope [1-2]. Therefore, standard electrophysiological techniques have been proven difficult for the analysis of intracellular calcium channels. Analogous to the measurement of whole cell currents produced by ion channels on the plasma membrane, population responses of intracellular calcium channels on isolated intracellular vesicles are often measured using calcium release assays [3-4]. Similarly, bilayer lipid membranes are used to study single channel properties of intracellular calcium channels analogues to single channel patch clamp electrophysiology [5].

1.1. Accessibility of calcium channels on intracellular membranes to patch clamping techniques

A number of studies describe the use of single channel patch-clamp electrophysiology, the standard method to investigate biophysical properties of ion channels, to investigate intracellular calcium channels [6-11]. However, these studies utilize the properties and especially the size of specialized intracellular organelles, such as the vacuoles of plant cells [6] and the nuclear envelope of animal cells [7-8] and oocytes [9-11]. Because of the technical difficulties involved in introducing patch clamp electrodes into the inside of cells without compromising neither the intracellular milieu nor the functionality of the patch clamp electrode a different technique is typically chosen to analyze the properties of intracellular channels: bilayer lipid membranes [5]. Vesicles or microsomes are prepared from intracellular membranes and incorporated into bilayer lipid membranes. Alternatively, purified channel proteins are reconstituted into micelles and subsequently inserted into bilayers lipid membranes. Using adequate protein concentrations, bilayer lipid membrane composition, incorporation and recording conditions as well as analysis tools allows one to analyze single channel properties [5, 12-14].

1.2 Biophysical properties of intracellular calcium channels that make bilayer lipid membranes the method of choice for their analysis

Intracellular calcium channels unlike their analogues of the plasma membrane, such as voltage and ligand gated calcium channels typically show a low selectivity for their current carrier calcium and conduct other divalent cations as well as monovalent cations [5, 14]. This is due to the fact that of the metal cations only calcium ions form a physiologically relevant electrochemical gradient across the membranes of sarcoplasmic reticulum, endoplasmic reticulum, Golgi apparatus and nuclear envelope, the expression sites of intracellular calcium channels [1, 2, 15-17]. Therefore, even if other intracellular channels such as potassium and chloride channels are present in membrane preparations of intracellular calcium channels, their contribution to ion channel currents measured in bilayer lipid membranes can be excluded by choosing the appropriate composition of buffers on either side of the bilayer lipid membrane. Typically a divalent current carrier in the several millimolar concentration range is chosen for the buffer on the luminal side of the channel protein (termed the trans side of the bilayer lipid membrane). A buffer containing a low concentration of the divalent current carrier is present at the cytosolic side of the channel protein (termed the cis side of the bilayer lipid membrane) [5, 12-14].

The low ion selectivity of intracellular calcium channels also makes the use of current carriers possible that do not interfere with ion channel function. For example, all intracellular calcium channels are modulated by the luminal and cytosolic calcium concentration and therefore barium is often chosen as a current carrier not affecting channel function [5, 12-14]. Because the function of intracellular calcium channels is modulated by interaction with regulatory molecules and proteins on both the luminal and the cytoplasmic side of the protein, the bilayer lipid membrane system offers unique opportunities to investigate the function of these channels [5].

1.3. Properties of the bilayer lipid membrane system that make it the system of choice for the analysis of intracellular calcium channels

The use of the bilayer lipid membrane system for the electrophysiological study of ion channel proteins inserted into the planar lipid bilayer offers several advantages that involve the control of the channel environment, of the phosphorylation status and of the binding partners of the channel [5]. A bilayer lipid membrane is formed in a hole with a defined diameter that represents the only electrical connection between two compartments. Because the composition of media on either side of the planar lipid bilayer electrophysiology system can be fully controlled, changes in the physiological conditions do not occur over time or if required can be introduced at given time points during the experiment [12-14].

Similarly, interactions of intracellular calcium channels with other cellular components can be controlled by removal or introduction of such components [18]. The absence of changing 'run down' parameters, of changes in the protein and buffer composition and the accessibility to both sides of the channel protein represent major advantages of the bilayer lipid membrane system over conventional patch clamp electrophysiology methods in addition to the properties discussed in 1.1.

2. CLASSES OF INTRACELLULAR CALCIUM CHANNELS

After the initial identification and description of intracellular calcium channels, their importance for cellular function and intracellular signaling has been recognized [19-24]. Several cellular and physiological processes such as secretion, gene expression, metabolism, contraction, cell death, cell proliferation, neuronal excitability, differentiation, learning and memory formation, and apoptosis are regulated by intracellular calcium channels [1, 5, 25-26]. Three intracellular calcium channels have been identified so far. The inositol 1, 4, 5-trisphosphate receptor and the ryanodine receptor are exclusively expressed in intracellular membranes, particularly the sarcoplasmic and endoplasmic reticulum membrane. Each of these two proteins forms tetrameric complexes and shares substantial sequence homology with the other protein in its functional domains [24]. The number of known cellular physiological agents that modulate intracellular calcium channels present in the cytosol or the lumen of the endoplasmic reticulum is small. Endogenous ligands of both the ryanodine receptor and the inositol 1, 4, 5-trisphosphate receptor include adenosine 5'-trisphosphate, calcium, cyclic adenosine diphosphate-ribose, inositol 1, 4, 5-trisphosphate, arachidonic acid and leukotriene B4 [5]. The activity of both receptors depends critically on the level of cytosolic free calcium, the presence of ATP and the concentration of calcium in the ER lumen (see Fig. 1). Splice variants or isoforms of both types that differ molecular and physiological properties are known. However, isoforms-specific agonists or antagonists have not been found despite the fact that biophysical and cell biological data for specific isoforms are available [25-26].

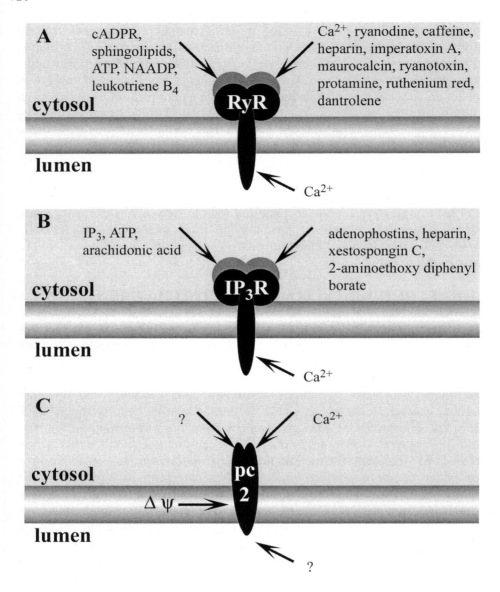

Fig. 1. Cellular (left) and pharmacological (right) modulators of intracellular calcium channels (A: RyR: ryanodine receptor; B: IP$_3$R: inositol 1, 4, 5-trisphosphate receptor; C: pc2: polycystin-2).

A novel, third class of intracellular calcium channels has been recently described [14]: The endoplasmic reticulum integral membrane protein polycystin-2 is a calcium channel of the TRP (transient receptor potential) channel superfamily and is distinct from any other cation channel described so far. Polycystin-2 has an ion-specificity and conductivity similar to inositol 1, 4, 5-trisphosphate receptors and the ryanodine receptors and is regulated by changes both in cytosolic free calcium and in the membrane potential [14]. Fig. 1 shows cellular and pharmacological modulators of the three types of intracellular calcium channels.

2.1. Ryanodine receptors

Ryanodine receptors (calcium-induced calcium release channels) play an essential role in most cell types including neurons, muscle cells, and epithelial cells. Mediating the release of calcium from the endoplasmic and sarcoplasmic reticulum into the cytosol ryanodine receptors convert different extracellular stimuli into intracellular calcium signals. Ryanodine receptors are distinct from the related inositol 1,4,5-trisphosphate-gated calcium channels of the endoplasmic and sarcoplasmic reticulum with respect to their biophysical and pharmacological properties [27-29].

Single channel electrophysiology experiments using bilayer lipid membranes showed that ryanodine receptors channels typically display a bell-shaped activity dependence on cytosolic calcium concentrations. Lack of activation by low, and inhibition by high cytosolic calcium concentrations tunes ryanodine receptor activity to a narrow, physiologically relevant range [12, 27]. The various ryanodine receptor isoforms differ in the width of this physiological range of calcium concentrations that already activate but do not yet inhibit the channel. This leads to tissue- and cell-specific distributions of isoforms most suited for particular physiological tasks [26]. Ryanodine receptor activity is independent of the membrane potential, a parameter that can be easily determined using voltage clamp electrophysisology in bilayer lipid membrane systems. The interactions of the ryanodine receptor with plasma membrane proteins and a number of intracellular proteins [18] as well as with signaling substances, such as cyclic adenosine diphosphate-ribose, nicotinic acid adenine dinucleotide phosphate, arachidonic acid and its derivatives, sphingolipids, and ATP [5] contributes to the modulation of ryanodine receptor channel function.

2.2. Inositol 1, 4, 5-trisphosphate receptors

Once activated by their specific ligands, G-protein coupled or tyrosine-kinase linked receptors mediate the hydrolysis of phosphatidyl inositol 4, 5-bisphosphate to inositol 1, 4, 5-trisphosphate and 1,2-diacylglycerol by phospholipase C. Inositol 1, 4, 5-trisphosphate binds to the inositol 1, 4, 5-trisphosphate receptor located on the sarcoplasmic and endoplasmic reticulum and induces the release of calcium from the intracellular calcium stores into the cytosol.

As outlined for the ryanodine receptor, such increases in intracellular calcium initiate and participate in a large number of cellular signaling processes [1, 25]. Besides binding inositol 1, 4, 5-trisphosphate, the inositol 1, 4, 5-trisphosphate receptor is regulated through binding proteins such as calmodulin and the immunophilin FK506 binding protein, phosphorylation, cytosolic adenosine 5'-trisphosphate and cytosolic as well as luminal calcium concentrations [5]. The three inositol 1, 4, 5-trisphosphate receptor isoforms known to date show 60-70% homology with one another and vary in their cell biological functions and biophysical properties [5, 25, 30].

2.3. Polycystin-2

Polycystin-2 is a transmembrane protein produced by the human gene PKD2, the gene mutated in type 2 autosomal dominant polycystic kidney disease and is a member of the polycystin family of the TRP (transient receptor potential) channel superfamily [31, 32]. The channel is expressed in the endoplasmic reticulum and pre-medial Golgi membranes. The protein functions as an intracellular calcium release channel increasing the calcium release from intracellular stores by other intracellular channels in a positive feedback mechanism [14, 33]. Pathogenic changes in the structure of polycystin-2 are accompanied with abnormal proliferation and apoptosis as well as with a de-differentiation of epithelial cells [32-34].

Polycystin-2 contains domains homologous to known voltage-activated calcium channels of the plasma membrane and therefore differs significantly from inositol 1, 4, 5-trisphosphate receptors and ryanodine receptors that are not dependent on the membrane potential [14, 32]. However, in its activation by low and inactivation by high cytosolic calcium concentrations polycystin-2 resembles properties of both the inositol 1, 4, 5-trisphosphate receptor and the ryanodine receptor [14].

3. BILAYER LIPID MEMBRANES AS AN ANALYSYS TOOL TO INVESTIGATE THE PHARMACOLOGY OF RYANODINE RECEPTORS

The use of bilayer lipid membranes for the electrophysiological measurement of single channel properties of ryanodine receptors is well established [12-13, 20-23, 27] and provides unique advantages over other electrophysiological and calcium release assay systems (see 1.). The identification of the interaction between ryanodine receptors and highly charged molecules has been one of the most challenging set of studies but at the same time it was very representative for a large number of investigations using bilayer lipid membranes to analyze ryanodine receptor function.

Negatively charged polyanions such as pentosan polysulfate, polyvinyl sulfate or heparin are potent activators of ryanodine receptor type 1 and several polyanion molecules need to bind to the channel in order to exert their full activity [35]. These results were obtained by addition of molecules to the cytoplasmic side of the ryanodine receptor in the bilayer lipid membrane using dose-response-curve analyses as well as established pharmacological agonists and antagonists of the ryanodine receptor in control experiments. Similarly, protamine was used to analyze the effects of a positively charged polycation binding to the ryanodine receptor [12]. The activity of single ryanodine receptor type 1 channels was measured by incorporating channels into planar lipid bilayers and using barium as the current carrier in the presence and absence of protamine sulfate at the cytoplasmic site of the channel. The electrophysiological measurement of ryanodine receptors in bilayer lipid membranes allowed testing the effect of protamine on the channel activity at different cytosolic calcium concentrations, the reversibility of the effect, and the interaction of protamine with other ryanodine receptor modulators as well as determining the Hill coefficient. The same assay system was employed to test the specificity of the molecular interaction of protamine with the ryanodine receptor that had been incorporated into the bilayer lipid membrane. After addition of protamine crosslinked with agarose to the cytoplasmic side of the ryanodine receptor channel activity did not change using comparable concentrations of crosslinked protamine and protamine. These control experiments tested for protamine-induced changes in the cytosolic calcium concentration, osmotic strength or medium composition. In addition, the binding of protamine to the agarose beads clearly excluded a direct interaction of protamine with the ryanodine receptors located in the bilayer lipid membrane. Impact of the agarose beads on the membrane, the only way a direct molecule interaction could take place, would lead to a de-stabilization of the bilayer lipid membrane [12]. This indicated a direct molecular interaction of protamine with the ryanodine receptor. The results suggested that protamine sulfate exerts a charge effect on the ryanodine receptor in a similar manner but with reversed sign as polyanions such as heparin, introducing cation-repelling positive charges to the channel. As shown in Fig. 2, binding of the ryanodine receptor to highly charged molecules such as heparin and protamine might increase the charge of the ryanodine receptor channel complex. This might lead to a changed affinity of calcium to the activating calcium-binding regulatory site of the ryanodine receptor and ultimately a resulting change in channel open probability. This contributes to the importance of a negative surface charge activated by localized increases in calcium concentration near the ryanodine receptor channel for the potentiation of conduction and selectivity of the channel [12, 27, 35-36].

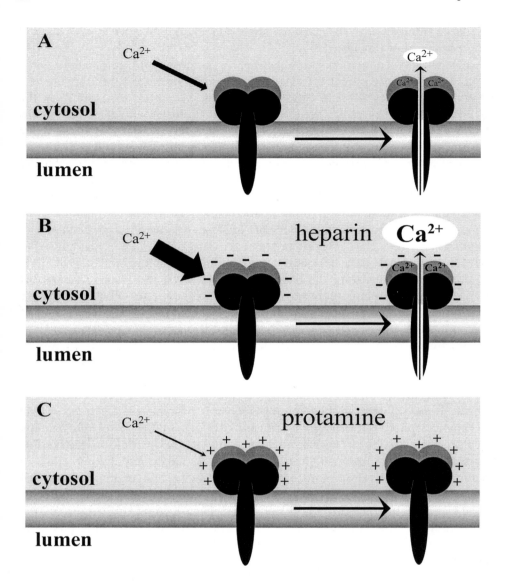

Fig. 2. Effect of highly charged ligands of ryanodine receptors on channel activity. Calcium binds (left, strength indicated by thickness of the arrow) to the ryanodine receptor channel and the resulting release of calcium into the cytosol depends on the degree of calcium binding to the calcium-binding regulatory site of the ryanodine receptor channel.

4. BILAYER LIPID MEMBRANES AS AN ANALYSYS TOOL TO INVESTIGATE THE PHARMACOLOGY OF POLYCYSTIN-2

Bilayer lipid membranes were used to determine the biophysical properties of a previously uncharacterized intracellular calcium channel, polycystin-2, at the single channel level [14]. Vesicles prepared from endoplasmic reticulum membranes of cells overexpressing polycystin-2 were fused to planar lipid bilayers in the presence of potassium chloride on the cytosolic side of the bilayer lipid membrane. Subsequently, large potassium and chloride ion channel currents could be recorded indicating fusion of the endoplasmic reticulum vesicles containing potassium and chloride ion channels. By removing potassium chloride from the cytosolic side of the bilayer lipid membrane the activity of intracellular calcium channels could be monitored using magnesium, calcium or barium ions on the endoplasmic reticulum luminal side of the bilayer lipid membrane as the current carriers. In the absence of the inositol 1, 4, 5-trisphosphate receptor ligand inositol 1, 4, 5-trisphosphate and in the presence of the ryanodine receptor inhibitor ruthenium red, calcium channel activity indicated a novel type of intracellular calcium channel, polycystin-2 [14].

The slope conductances of the polycystin-2 channel were 114, 95 and 90 pS when measured between 0 and -10 mV with barium, calcium and magnesium as current carriers respectively (barium, 2 pA; calcium, 1.7 pA; magnesium, 1.5 pA; measured at -10 mV). These values are similar to the ones observed for inositol 1, 4, 5-trisphosphate receptors and ryanodine receptors and increase by applying increasing negative holding potentials to the bilayer lipid membrane [14]. Like inositol 1, 4, 5-trisphosphate receptors and ryanodine receptors, the activity of polycystin-2 is regulated by the concentration of free calcium ions in the cytosol. Whereas submicromolar and micromolar concentrations of calcium activate the channel, millimolar concentrations of calcium decrease channel open probability, thus producing a bell-shaped calcium-dependence curve [14].

However, in contrast to both the inositol 1, 4, 5-trisphosphate receptor and the ryanodine receptor, polycystin-2 could be activated by an increasing negative holding potential applied to the bilayer lipid membrane. This voltage dependence of the channel activity parallels properties of other members of the of the TRP (transient receptor potential) channel superfamily but is a novel feature for an intracellular calcium channel. As depicted in Fig. 3, both the dependence of polycystin-2 channel activity on the cytosolic calcium concentration and on the membrane potential of the bilayer lipid membrane might serve as a positive feedback mechanism for calcium release from intracellular stores. This proposed mechanism of action for polycystin-2 function might explain the role of polycystin-2 in intracellular calcium signaling and the discrepancy between inositol 1, 4, 5-trisphosphate receptor and the ryanodine receptor activity measured in cells and in bilayer lipid membrane electrophysiology experiments [1, 2, 5, 16, 17, 25, 26].

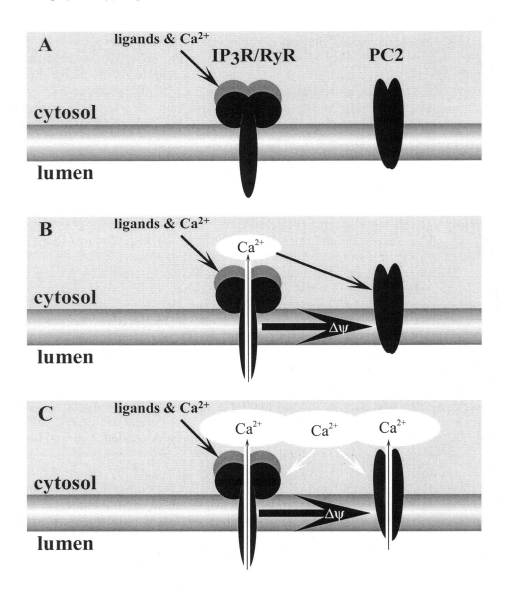

Fig. 3. Proposed mechanism of action for polycystin-2 functioning as an intracellular calcium release channel. Both the cytosolic calcium concentration and the membrane potential ($\Delta\psi$) of the endoplasmic reticulum membrane generated by inositol 1, 4, 5-trisphosphate receptors (IP$_3$R) or ryanodine receptor (RyR) activity trigger polycystin-2 channel activity.

REFERENCES

[1] M.J. Berridge, P. Lipp, and M.D. Bootman, Nat. Rev. Mol. Cell. Biol., 1 (2000) 11.
[2] R. Rizzuto, Curr. Opin. Neurobiol., 11 (2001) 306.
[3] A. Galione, H.C. Lee, and W.B. Busa, Science, 253 (1991) 1143.
[4] A. Galione, and G.C. Churchill, Sci. STKE. 41 (2000) PE1.
[5] P. Koulen, and E.C. Thrower, Mol. Neurobiol., 24 (2001) 65.
[6] G.J. Allen, S.R. Muir, and D. Sanders, Science, 268 (1995) 735.
[7] D. Boehning, S.K. Joseph, D.O. Mak, and J.K. Foskett, Biophys. J., 81 (2001) 117.
[8] D. Boehning, D.O. Mak, J.K. Foskett, and S.K. Joseph, J. Biol. Chem., 276 (2001) 13509.
[9] D.O. Mak, and J.K. Foskett, Am. J. Physiol., 275 (1998) C179.
[10] D.O. Mak, and J.K. Foskett, J. Gen. Physiol., 109 (1997) 571.
[11] D.O. Mak, S. McBride, V. Raghuram, Y. Yue, S.K. Joseph, and J.K. Foskett, J. Gen. Physiol., 115 (2000) 241.
[12] P. Koulen, and B.E. Ehrlich, Mol. Biol. Cell, 11 (2000) 2213.
[13] P. Koulen, T. Janowitz, F.W. Johenning, and B.E. Ehrlich, J. Membr. Biol., 183 (2001) 155.
[14] P. Koulen, Y. Cai, L. Geng, Y. Maeda, S. Nishimura, R. Witzgall, B.E. Ehrlich, and S. Somlo, Nat. Cell Biol., (2002) Feb 18 (epub ahead of print)
[15] T. Clausen, Can. J. Physiol. Pharmacol., 70 (1992) Suppl:S219.
[16] M. Iino, Jpn. J. Pharmacol., 54 (1990) 345.
[17] K.M. Sanders, J. Appl. Physiol., 91 (2001) 1438.
[18] J.J. MacKrill, Biochem. J., 337 (1999) 345.
[19] I. N. Pessah, A. O. Francini, D. J. Scales, A. L. Waterhouse, and J. E. Casida, J. Biol. Chem., 261 (1986) 8643.
[20] T. Imagawa, J. S. Smith, R. Coronado, and K. P. Campbell, J. Biol. Chem., 262 (1987) 16636.
[21] F. A. Lai, K. Anderson, E. Rousseau, Q. Y. Liu, and G. Meissner, Biochem. Biophys. Res. Commun., 151 (1988) 441.
[22] L. Hymel, M. Inui, S. Fleischer, and H. Schindler, Proc. Natl. Acad. Sci. U S A, 85 (1988) 441.
[23] L. Hymel, H. Schindler, M. Inui, and S. Fleischer, Biochem. Biophys. Res. Commun., 152 (1988) 308.
[24] G.A. Mignery, T.C. Südhof, K. Takei, and P. De Camilli, Nature, 342 (1989) 192.
[25] Patel S., Joseph S.K., and Thomas A.P., Cell Calcium, 25 (1999) 247.
[26] V. Sorrentino, V. Barone, and D. Rossi, Curr. Opin. Genet. Dev., 10 (2000) 662.
[27] J. Smith, R. Coronado, and G. Meissner, J. Gen. Physiol., 88 (1986) 573.
[28] S. Supattapone, P.F. Worley, J.M. Baraban, and S.H. Snyder, J. Biol. Chem., 263 (1988) 1530.
[29] P. Palade, C. Dettbarn, B. Alderson, and P. Volpe, Mol. Pharmacol., 36 (1989) 673.
[30] P. Koulen, T. Janowitz, L.D. Johnston, and B.E. Ehrlich, J. Neurosci. Res., 61 (2000) 493.
[31] J. T. Littleton, and B. Ganetzky, Neuron, 26 (2000) 35.
[32] T. Mochizuki, G. Wu, T. Hayashi, S.L. Xenophontos, B. Veldhuisen, J.J. Saris, D.M. Reynolds, Y. Cai, P.A. Gabow, A. Pierides, W.J. Kimberling, M.H. Breuning, C.C. Deltas, D.J. Peters, and S. Somlo, Science, 272 (1996) 1339.

[33] Y. Cai, Y. Maeda, A. Cedzich, V.E. Torres, G. Wu, T. Hayashi, T. Mochizuki, J.H. Park, R. Witzgall, and S. Somlo, J. Biol. Chem., 274 (1999) 28557.

[34] F. Qian, T. J. Watnick, L. F. Onuchic, and G. G. Germino, Cell, 87 (1996) 979.

[35] I.B. Bezprozvanny, K. Ondrias, E. Kaftan, D.A. Stoyanovsky, and B.E. Ehrlich, Molec. Biol. Cell, 4 (1993) 347.

[36] T.K. Ghosh, P.S. Eis, J.M. Mullaney, C.L. Ebert, and D.L. Gill, J. Biol. Chem., 263 (1988) 11075.

ACKNOWLEDGEMENTS

I thank Margaret R. Koulen and Sara E. Koulen for generous support and encouragement.

Planar Lipid Bilayers (BLMs) and their Applications
H.T. Tien and A. Ottova-Leitmannova (Editors)

Chapter 27

Systems Aspects of Supported Membrane Biosensors

I.R. Peterson and J.A. Beddow
Centre for Molecular and Biomolecular Electronics,
Coventry University SE, Priory Street, Coventry, CV1 5FB, UK

1. INTRODUCTION

By far the most sensitive and efficient biosensor systems known are those employed by biological organisms for neurotransmission and olfaction. Both examples involve gated channel proteins bound to a membrane. The conductivity of an open channel is many orders of magnitude greater than that of an equivalent area of virgin membrane, and the electrical change on opening signals the presence of the target substance. This system readily achieves the ultimate sensitivity of response to a single molecular event.

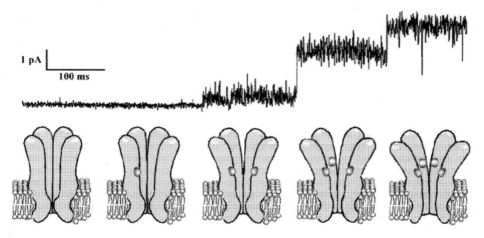

Figure 1. Typical patch clamp recordings showing different levels of current through a single channel in different states

It has long been proposed to construct high-sensitivity biosensors by mimicking these natural systems [1,2], and many groups are currently attempting implementation [3-6]. The configuration of the resultant sensor resembles that of the patch-clamp technique [7]. Figure 1 illustrates a representative patch-clamp response obtained from a single glutamate receptor as it binds progressively to more molecules of its neurotransmitter ligand. It is easy to distinguish four different states of an individual receptor differing in the number of adsorbed glutamate molecules. No greater sensitivity is conceivable

Basically, a membrane biosensor consists of a packaged membrane element customized by the incorporation of specific gated channels. The two decades of research effort into this principle have demonstrated the feasibility of its fundamental elements at the molecular level [8]. In an even simpler variant, packaged membrane elements with no gated channels could monitor pathological conditions involving attack on membranes [9]. Sensors based on packaged membranes can be expected to appear shortly in the marketplace, with applications foreseen in medicine, in the food-processing industry, and in farming. It is timely to consider the operation of these sensors at a higher level, and how the basic mechanism can be incorporated into a practical system that is helpful and easy to use in a clinical, industrial or agricultural context.

2. BIOSENSOR REQUIREMENTS

The presence and concentration of biological analytes in medicine and other areas is at the moment determined mostly using bioassays. However there is a real motivation to replace these with biosensors, which are expected to be both faster and more quantitative.

The speed possible in principle with bilayer-based sensors is illustrated by the current-time traces of Figure 1, in which different conductivity states of a single neuroreceptor can be distinguished within tens of milliseconds. However it must be kept clearly in mind that in spite of the scientific potential of the patch-clamp technique, and its use in investigating the mechanisms of the transmission of nervous impulses between neurons, a biosensor has requirements that are quite distinct from both.

The purpose of the junction between nerve cells is to transmit impulses with high speed and low error rate. The neurotransmitter is not particularly expensive to synthesize, so that sensitivity is not a major consideration, and in fact the rate of false positive responses at low levels must be minimized. Most neuroreceptors have a state-transition response leading them to respond preferentially to transients and much less strongly to maintained concentrations. In contrast, the biosensor must give an accurate indication of the current concentration of analyte, with a frequency response as featureless as possible.

The patch-clamp technique is used to investigate the behavior of individual channel-forming proteins in membranes. The behavior of interest covers not just the response to added substances, but also the molecular dynamics of the channel. The membranes, as well as the channels, are usually sampled by a glass micropipette directly from the walls of biological cells [10,11]. Since the focus of interest is the channel dynamics, efforts are often made to ensure that the membrane contains only one channel-forming protein because this simplifies the analysis.

In a biosensor, the behavior of the channel is of little interest in itself; the focus of interest is the analyte. A biosensor is required to give a useful indication of the concentration of an analyte or analytes in the adjacent fluid phase in as great a variety of different situations as possible. It is in general not required to give an indication of rates of change. Its output is required to be analog, distinguishing a whole range of different analyte levels, rather than digital, distinguishing just presence or absence relative to a reference level. The requirements are much closer to those for olfaction. In the present chapter, the specific demands of the biosensor application will be analyzed and research directions identified.

2.1. Dynamic range

The dynamic range of a sensor is a systems concept defining the range of analyte concentrations for which the sensor provides a useful indication. All sensors have a minimum and maximum detectable analyte concentration. Below the minimum concentration, also known as the sensitivity of the sensor, the output signal cannot be distinguished from the random fluctuations that always occur, even in the absence of analyte. Above the maximum, the output signal saturates and ceases to provide useful information about the precise amount present.

The dynamic range of a sensor is often given as the ratio of maximum to minimum concentration. In this case its sensitivity and dynamic range specifies the performance. These parameters are true systems parameters, and are not uniquely defined by the molecular arrangement employed. At the expense of complicating the design, it is usually possible to extend the dynamic range. If the random fluctuations of the sensor output in the absence of analyte are uncorrelated, then the sensitivity can be increased by filtering, with the trade-off of increased response time. The saturation that sets the maximum usable concentration is not usually a hard clipping of the output signal, but rather a smooth nonlinearity. Hence appropriate signal conditioning techniques can extend the range over which the output signal represents the analyte concentration, with the trade-off of increased noise. And so on. With the appropriate, more expensive, circuitry to calibrate and correct for the dominant sources of error, the range of useful response can be extended.

The ideal sensor gives an output related in a simple way to the analyte concentration. The simplest possible relationship is proportionality. In a membrane-based biosensor, the overall electrode current is just the sum of the currents from the different channels. As long as the voltage across the membrane remains constant, the overall current is directly proportional to the number of open channels and thus directly related to the molecular event signaling the analyte presence.

High sensitivity in a biosensor is of no use unless it is coupled with a large dynamic range. A sensor for, say, testosterone, with a sensitivity of 1 pM and a dynamic range of 10, would not be of great use since the clinically-interesting distinction between men and women occurs at levels of around 10 nM. However the membrane-based approach promises high dynamic ranges. The inherent sensitivity of the principle results because of the excellent insulating properties of bilayer membranes. The conductivity of an open channel can be of the order of 100 pS, equal to the conductivity of a square patch of membrane of side 100 μm or more, an area accommodating up to 10^8 protein molecules. It follows that these sensors are potentially capable of eight orders of magnitude of dynamic range.

2.2. Time response

Perhaps one of the major advantages to be expected from the change from bioassay to biosensor is that a result is obtained much faster. Once a molecular event has caused a molecular channel to open, this event is transmitted at electronic speed to the external electrical circuit. In the patch clamp technique, the noise level is such that a reliable indication of channel opening is obtained within milliseconds. The major source of delay is then the time required for the analyte molecules to diffuse to the active channels.

In the synapse, the smearing out of the nerve pulse resulting from diffusion across the synaptic gap is largely compensated by the refractory state of the neuroreceptors. Immediately after the binding to the neurotransmitter, the channel opens, but after some 20 ms the channel closes again with the neurotransmitter still bound. This refractory state is clearly of value for the process of nerve impulse propagation. However it is effectively lost time as far as a biosensor is concerned. A time constant of 20 ms may be appropriate to compensate for the effects of diffusion across a submicron gap, but is far too fast to perform the same function for diffusion across macroscopic distances. In addition, it degrades the sensitivity by reducing the average current through the channel at low analyte concentrations.

2.3. Noise

At a system level, "noise" can be defined as the error in the output of a device or system. Even in the absence of analyte, there will always be fluctuations. Some of these will be caused by changes in other parameters, e.g. temperature. Others will be related to random quantum events occurring in the sensor.

Considerations of noise are relevant to the decision whether the membrane channels should open, or close, in response to the arrival of a molecule of analyte. These may be called "positive-going" and "negative-going", respectively. In the nature of things, a negative-going sensor is easier to achieve, because the inherent steric encumbrance of any molecule makes it relatively simple to arrange for blockage of an initially open channel on analyte arrival. However the level of noise of a closed channel, and the leakage current of the membrane which surrounds it, is quite small. In contrast, the current through an open channel is very noisy, most likely as a result of thermal fluctuations of the channel size and shape. These differences are even more marked than apparent from Figure 1, because most of the noise shown when the channel is closed comes from the amplifier. While for many purposes the choice of signal polarity does not matter, a positive-going sensor is seen to be inherently much more sensitive. It is of interest that this is the polarity found in both neurotransmission and olfactory systems.

When using ligand-gate channels sensitive only to the analyte of interest, there is no advantage in limiting their number. The use of more receptors in a given area of membrane reduces the level of statistical fluctuation, which can be traded off against the time of response to low concentrations of analyte. An area of membrane of diameter 100μm, scarcely visible to the human eye, can accommodate 100 million receptors. When this number of receptors is used, the level of current through the membrane is much higher, reducing electronic design constraints and statistical variations.

There is one exception to this categorical statement. In stochastic sensing, the noise itself is the signal. Winterhalter [12] demonstrated that the levels of electrical noise emanating from maltoporin channels as maltose molecules translocate could be used as a measure of the maltose concentration of the adjacent phase. Bayley [13,14] has suggested that the concentrations of a number of analytes can be measured simultaneously, by analyzing the statistical distribution of the channel current considered as a time series. Stochastic sensors of this sort will have a trade-off between dynamic range and speed of response, which may be compensated for by their versatility.

2.4. Specificity

Intermolecular interactions can show exquisite specificity, and many aspects of the operation of biological systems are based on it. Perhaps the most striking example is that displayed by the nucleic acids DNA and RNA. Each helical chain binds strongly to the strand with its complementary base sequence. In this case the condition for strong binding can be written down in very simple form. In the general case, however, there is no simple rule describing the relationship between the primary chemical structures of two molecules that bind strongly to each other.

Among all the chemical substances normally present in the cell, the neuroreceptors in the post-synaptic membrane respond only to the neurotransmitter. This specificity is not complete and absolute. For example, the most common variety of neuroreceptor, the nicotinic acetylcholine receptor, responds also to carbamyl choline, nicotine, and a few other agonists [15]. This explains why tobacco plants secrete nicotine (as an insecticide), and why nicotine is addictive (related to the fact that nAChR receptors are more common in the sympathetic nervous system). However this does not detract from their very great specificity of response, achieved by exactly the same mechanism used by antibodies to achieve specific recognition of potential pathogens. A membrane customized by the incorporation of a receptor will respond only to a small set of compounds.

2.5. Parasitics

'Parasitics' can be defined as small but inevitable physical effects that lead to departures from a theoretical ideal. Not all systems have such an ideal: there are many examples where the operation cannot be analyzed in detail, because it involves rather complicated chains of cause and effect between the initial molecular stimulus and the resultant output. However these systems are usually readily perturbed by extraneous influences, and the interpretation of their output is often difficult. In contrast, robust systems are usually ones in which the sensor operation approximates to a very simple mechanism, with a minimum of complexity in the causal chain. Membrane-based sensors fit into the latter category. A molecule arrives and opens a hole in the membrane. The number of holes is determined by measuring the electrical conductivity of the membrane.

The real world is never this clear-cut. Parasitic effects, which mean that the output signal never measures the exact number of open channels, include leakage of the membrane matrix, variations in the conductivity of each channel, and fluctuations of the transmembrane potential.

The major parasitic influence on the transmembrane potential is the impedance in series with the membrane. The membrane current must flow through the electrolyte, across the electrolyte-electrode junction, and thence through a wire to measurement electronics. The total resistive drop changes the voltage across the channels. By far the most significant of these three voltage drops occurs across the electrode. Since the electrode characteristic is nonlinear, it is not readily compensated for. In the practical situation of analyzing biofluids, the resistive drop increases when the electrode is fouled by the nonspecific adsorption of proteins.

2. 6. Lifetime

The duration of a patch clamp experiment is limited by the lifetime of the patch of membrane sampled by the glass micropipette, which is typically on the order of an hour. This is totally inadequate for a commercially viable biosensor, which must remain functional for many months after fabrication, and survive handling, storage and transportation from the factory to the point of use. In spite of its apparently mundane nature, this problem is the reason that the great potential of membrane-based biosensors has as yet remained unfulfilled.

3. MEMBRANES

3.1. Phase Behavior

The bilayer membrane in a tethered-membrane sensor is not manufactured by the sort of process used to fabricate e.g. a polymer film. Normal extrusion-type processes produce films with statistical variations in thickness, as well as the occasional hole. Instead, the membrane self-assembles, in a process which ensures uniform thickness. Defects have a tendency to anneal. This spontaneous process of lipid self-assembly into a bilayer membrane occurs because under the particular conditions, the membrane structure minimizes the system free energy. The driving force for this self-assembly is the attempt by the molecules to minimize the area of high-energy interface between their hydrophobic parts and water, and the end-state of the process is the thermodynamic ground state.

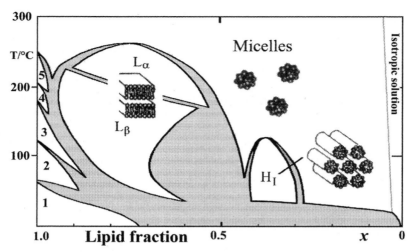

Figure 2. The (x,T) phase diagram of the sodium oleate/water system

Unfortunately the ground state is not always the desired membrane configuration. Under different conditions of temperature and concentration, the lipid molecules can aggregate in many different states of order, ranging from none at all through 0-, 1-, 2- and 3-dimensional structures. The conditions under which each state is obtained are shown most clearly in a phase diagram versus composition and temperature. Many of these possibilities are shown in the phase diagram of Figure 2 for mixtures of water and sodium oleate, a common component of soaps. For the regions shown white on the diagram and labeled 'Micelles', H_I , L_α , L_β , etc., the thermodynamic ground state of the system consists of one uniform phase. This is not the case in the gray regions of the diagram, and for these combinations (x, T) a uniform phase with composition is not possible. Instead, demixing occurs into phases that coexist in equilibrium, one with a lipid fraction greater than x, and the other less.

The case of no order at all is encountered at very low amphiphile concentrations. Under these conditions the mixture is a true solution, with isolated molecules completely surrounded by water. At higher concentrations the molecules start to aggregate together in spherical micelles with the hydrophilic headgroups pointing outwards in contact with the water, and the hydrophobic tails hidden away from it in the interior. These are zero-dimensional aggregates consisting of only a few molecules. They are mobile, and their ready incorporation of foreign molecules is responsible for the cleaning action of detergents. However they are useless as far as biosensing is concerned.

One-dimensional structures are encountered in the region of the phase diagram marked H_I. In this region the micelles have stretched into linear threads or columns, which pack in a hexagonal array. Like the micellar phase, the columnar phase H_I is unable to present a low conductance background against which the presence of open channels may be detected in an external electrical circuit. In order to provide a useful electrical insulation function, the micelles have to flatten into two-dimensional bilayer sheets, or lamellae. This occurs at higher lipid fractions x, in the regions L_α and L_β of the phase diagram, also described as liquid-crystalline and gel in fairly obvious correspondence to their mechanical and optical properties.

In the two lamellar phases L_α and L_β, the molecules form two-dimensional sheets and so from a purely geometric viewpoint appear equally suitable for biosensor application. However of these, the L_α phase is far better. The phase transition between the two regions signals a difference in the molecular order between the two phases. While in the higher-temperature L_α phase the molecules show no order in the plane of the membrane, which behaves like a two-dimensional liquid, in the lower-temperature L_β phase the molecules are arranged in a type of order called hexatic [16,17], showing pronounced local crystallinity. There are some functional proteins, like bacteriorhodopsin, which are capable of functioning correctly when inserted in a hexatic membrane, but the majority of ligand-gated channel proteins require a fluid membrane environment.

Even though the L_α phase region of sodium oleate shown in Figure 2 gives the correct fluidity, it is of course ruled out for biosensing application because it only occurs at temperatures (and pressures) well above ambient. Another reason for disqualification is apparent. The region of phase coexistence between L_α and the adjacent micellar phase is quite narrow, so that is feasible and easy for L_α membranes of this particular amphiphile to disintegrate into micelles. This means that sodium oleate membranes rapidly lose their integrity when completely surrounded by water.

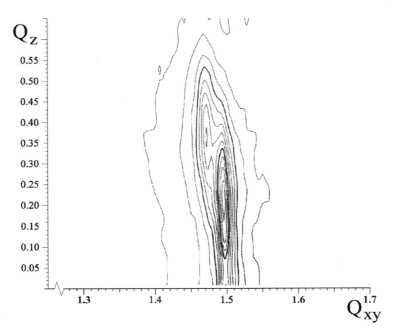

Figure 3. The grazing-incidence X-ray diffraction pattern from a stearic acid monolayer at room temperature and at pressures just below the tilt transition.

It has only fairly recently become possible to determine the molecular arrangement within the layers of amphiphilic molecules. Figure 3 shows the X-ray diffraction pattern obtained from a monolayer of stearic acid at temperatures slightly above ambient and at surface pressures just below the tilt transition [18]. The well-defined diffraction peaks are characteristic for one of the L_β phases, which for bulk phases has been called $L_{\beta F}$ [19] and for monolayers Ov [20]. L_α phases give no in-plane diffraction peaks.

It is mainly from studies like this that lamellar packings of amphiphiles are now known to have many states of order, illustrated by the generic phase diagram of Figure 4 [21]. The major subdivision is into packings with different degrees of ordering of the chain packing: crystalline, herringbone hexatic, rotator hexatic, and amorphous, apparent as the shaded bands in the Figure. The system of phases is far too rich to fit into the traditional system of nomenclature. L_β describes all the phases in the rotator hexatic band, and L_α corresponds to amorphous ordering. These bands are further subdivided into phases differing in the direction and magnitude of the molecular tilt. Under accessible conditions, a given substance will in general only show a few of the phases of Figure 4, but the positions of their regions of stability in the (x, T) plot are always found to lie in the same relative positions as in the Figure.

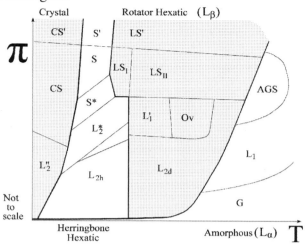

Figure 4. Generic phase diagram showing all the long-chain monolayer phases in their correct relative positions in respect of surface pressure and temperature.

Let us now look at the phase diagram of an amphiphile suited for membrane formation. Figure 5 shows that of a lecithin, dipalmitoyl phosphatidyl choline (DPPC), in mixtures with water. Similar lecithins, with perhaps different fatty acid 'tails', are a common component of biological cell membranes. There are a number of points worth making on comparison with sodium oleate. Firstly, the diagram shows only lamellar phases, so that DPPC only wants to form membranes. Secondly, there is a very wide two-phase region extending from the lamellar phase to essentially pure water. Membranes made of DPPC are stable for a long time in the presence of pure water, and have no tendency to disintegrate into less ordered phases.

Two disadvantages of DPPC can also be seen from the diagram. The phase suitable as a matrix for a neuroreceptor is the fluid L_α phase, which is stable only above 40°C. Lower transition temperatures can however be achieved by using lipids with shorter or unsaturated chains. More seriously, the phase boundaries of Figure 5 are drawn stylized, showing less detail than Figure 2. This is due to the difficulty of obtaining the pure lipid, a result of its chemical instability. In all phospholipids, the ester linkage connecting the fatty acid residues to the central carbon of the glycerol residue is slowly hydrolyzed by water [22]. Although phospholipids are otherwise almost perfectly suited to form membrane-based biosensors, this is a major practical shortcoming.

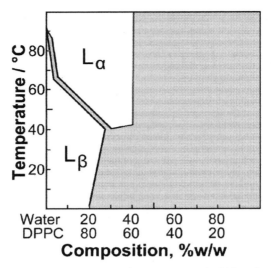

Figure 5. The phase diagram for mixtures of water with a lecithin, dipalmitoyl phosphatidyl choline.

From these considerations it is clear that a phase diagram contains a considerable amount of information about the ability of a particular lipid to form membranes, and its suitability in a biosensor.

3.2. Formation

The transformation of amphiphilic starting materials and water from some initial physical state into a functional membrane is a phase change, and in amphiphiles these may differ in important respects from those of more usual materials. Many phase transitions of amphiphile-water mixtures may be described as cooperative or topological. The energy of the hydrogen bonds in water is greater than kT, and many of these bonds are perturbed by the presence of an aliphatic chain. In a cooperative transformation, events in which individual amphiphiles leave the protection of the ordered environment are much less likely than events those in which the assembly as a whole undergoes origami-like transformations into the final equilibrium ordered state. This compares with 'normal' phase transitions e.g. the freezing of water to ice, in which ice crystals with random orientation nucleate and grow by the accretion of molecules from the other phase.

The difference is seen particularly clearly in the lamellar to columnar transition of layered systems, for example, the lamellar-to-columnar transition of amphiphilic mesophases [23,24]. Thin films retain their integrity on very slow heating.   Instead, near the transition temperature defects in the lamellae nucleate and propagate until the bilayers are "ripped to shreds". The fibrous shreds then rearrange to form hexagonally packed columns that are parallel to the original layers from which they came.

In many studies of artificial membranes, the bilayer is produced by the unrolling of vesicles in an aqueous phase [25]. This method is capable of producing membranes of high quality. However since the lipid is already present in the vesicles in the form of a fully hydrated bilayer, the driving force for a membrane to move to the desired location can only be small, and must involve surfaces with cohesive properties rather different from water, for example metals [26].

The driving force for the assembly of the bilayer in the desired location can be increased by using specific interactions not usually encountered in a biological context. The strong interaction between sulfur and the coinage metals was first used by Nuzzo and Allara to prepare self-assembled but tethered monolayers [27], and the technique has now been used widely to tether monolayers for biological purposes [28].

A significant driving force for self-assembly in the desired location can be achieved without the use of such non-biological materials, using variants of the Pockels spreading technique, in which the lipid is dissolved in a hydrophobic solvent. Pockels' technique [29,30] is widely used for the assembly of monolayers, and it is the standard method for assembling bilayer lipid membranes [31]. The present authors have used it to good effect to assemble membranes at the interface between two layers of hydrogel [32].

3.3. Maintenance of physical integrity

The practical application of developments in gated channel technology has been restricted by the difficult trade-off between membrane fluidity and durability. Bilayer lipid membranes have been known for forty years, but they are far too fragile to be commercialized, with lifetimes rarely exceeding an hour. The failure mechanism appears in this case to be mechanical: random turbulence and pressure variations in the surrounding fluid apply a fluctuating force to the membrane that eventually exceeds its yield strength (although this has not been fully documented).

There have been a number of promising developments in the direction of much longer bilayer lifetime. The first approach leading to significant improvements in membrane lifetime was to tether it adjacent to the surface of one of the electrodes used to measure its conductance [33]. In subsequent work, the membrane was linked chemically to the metal surface by a thiol group [34]. The strength of the thiol-metal bond is strong enough to prevent the lipids rearranging when the membrane dries out.

Unfortunately the thiol approach is not a complete answer to the stability problem, as there are a number of pathways of chemical instability that result in the breaking of the metal-sulfur bond [35]. Possible variations on the thiol theme that would not be as sensitive to electrochemical oxidation are organochlorine compounds [36]. Another problem with this approach is that the exchange current density of the tethering metal electrode is quite insufficient to allow direct measurement of the membrane conductivity. This point is discussed further in Section 5.2.

As a tethering surface for the bilayer membrane, hydrogels provide more gentle support. These are typically made from the polysaccharide agarose [37]. In initial work, the gel support was only on one side of the bilayer, leading to membrane lifetimes measured in days. More recently, Pockels-spread bilayer membranes between two layers of gel have been shown to give even longer lifetimes [32,38]. The gel does influence the bilayer phase but does not compromise fluidity or the normal operation of membrane-bound channel proteins, and allows the use of nonblocking electrodes separated from the membrane by many Debye lengths.

3.4. Chemical stability

Phospholipids are a major constituent of biological cell membranes, and the phase diagram of Figure 5 suggests this is because these materials overwhelmingly form lamellar phases stable in an excess of water. Cholesterol is another major constituent of natural biomembranes. By itself, cholesterol does not form a bilayer, but it enhances the barrier properties of the lipid bilayer and improves the function of ligand-gated channel proteins [39].

Unfortunately, neither phospholipids nor cholesterol are indefinitely chemically stable. Most natural phospholipids have unsaturated chains, and the double bonds are attacked by oxygen. Phospholipids have an appreciable rate of lipid hydrolysis, at the ester linkage in the 2-position of the glycerol residue [40]. Both mechanisms limit the lifetime of natural membranes to about a week [41]. This is not a problem for the cell with its constant metabolic activity capable of recycling degraded materials, but is difficult to implement in a sensor element which must be stored for a period of up to several months before use. The requirement of a commercially viable sensor is for a chemically stable lipid, which forms stable bilayers with properties similar to those of biological membranes.

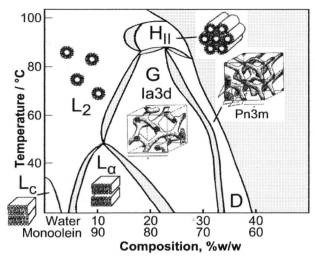

Figure 6. Phase diagram of 1-monoolein

The choice of an alternative, chemically stable lipid is complicated by the lack of understanding, in spite of some attempts [42], of the principles relating amphiphilic structure and phase behavior. One approach is to use the lipids of organisms living in extreme environments, the extremophiles. Bacteria belonging to the order *Archaea* are known which can live and reproduce at temperatures above 100°C. They solve the problem of the hydrolysis of the ester linkage by replacing it with an ether linkage [43]. Chain unsaturation appears to favor monolayer fluidity (i.e. stabilization of the L_α phase at the expense of hexatic L_β phases). Instead, archaebacterial lipid chains often have a turpenoid-like branched phytanol structure [44,45]. Synthetic lipids with ether linkages [46] and phytanyl chains [47] have been used for biosensor purposes.

There is also the approach of serendipity. It has been found by accident that dehydrocholesterol [48,49] and 1-monoolein [50,51] are much more stable than natural lipids and form suitable membranes. In the case of 1-monoolein the explanation for this is unclear, as in the phase diagram, shown in Figure 6, the room-temperature phase in excess of water is non-lamellar. Nevertheless this material works well.

4. RECEPTORS

The first proposal for a membrane-based biosensor involved the use of a natural receptor that responds to a particular ligand by opening a channel. Since then, the field has become broader, with other principles under investigation. New types of gated channel proteins have been synthesized [13], and known ones derivatized [52], to adapt a membrane to respond with high sensitivity and specificity to a range of analytes.

Before looking at these different approaches, it is of interest first to discuss an aspect common to all of them, namely, the process by which an analyte molecule finds its way from the bulk of the solution to the binding site on the receptor.

4.1. Molecular Arrival and Departure

The probability that the binding site is occupied by analyte at any given time is an important parameter of biosensor operation. However this is not the only parameter of interest. It is also important to know the rate at which binding and dissociation occurs. The problem of determining the stochastic dynamics of the response of receptors to analytes arriving by a diffusion process was solved by Smoluchowski [53], who modified the diffusion equation for a spherical sink of radius a in an infinite medium of molecules at concentration c. The desired quantity is the flux J of molecules into the sink. Starting with an equal distribution of molecules, the time - dependent solution is:

$$J = 4\pi a D c \left[1 + a(\pi D t)^{-0.5} \right] \qquad \ldots \ldots \ldots \ldots \ldots \ldots \text{(Equation 1)}$$

A large initial flux of molecules falls asymptotically to the steady state value:

$$J = 4\pi a D c \qquad \ldots \ldots \ldots \ldots \ldots \ldots \ldots \ldots \text{(Equation 2)}$$

. where D is the diffusion coefficient in $m^2.s^{-1}$. This is clearly not proportional to the volume of the sphere, which is advantageous for a biosensor, because it means that small objects are encountered much more often than would be expected on the basis of their volume or even their surface area.

This result is similar in many ways to the calculation in electrical theory of the spreading resistance of a sharp contact, except that a gradient of concentration rather than voltage is the dominant driving force. The "diffusion resistance" of a small parallelopiped of widths a and b and length c is taken to be c/abD. Appropriate combination of these resistances in series and in parallel gives the overall diffusion resistance. In the Smoluchowski case, the resistance due to a thin spherical shell of average radius r and thickness δr is given by $\delta r / 4\pi D r^2$, whose integral converges to the Smoluchowski value.

The integral to infinity does not converge for other sink dimensionalities. A sink arranged in a line gives an infinite resistance, because contributions from successive cylindrical shells decay only as r^{-1}. For a finite outer radius b and length l the diffusion resistance is $\ln(b/a)/4\pi D l$. For a sink covering a large area, the divergence is even faster. Of course no system is in fact infinite, so that the diffusion resistance is in all cases finite. Since ultimately the sensing action of the membrane biosensor is located in compact receptor molecules, the finite Smoluchowski diffusion resistance $1/4\pi D a$ is an inescapable microscopic contribution to the macroscopic diffusion resistance of the system.

The rate of arrival of an analyte at the binding site does depend on its diffusion constant D, but this is only a rather weak function of molecular weight and not at all specific. Specificity is provided by the rate of departure from the site. This occurs when thermal fluctuations exceed the binding energy ΔG, at a rate R given by the Eyring theory [54] as:

$$R = \frac{kT}{h} \cdot \exp\left(-\frac{\Delta G}{kT}\right) \qquad \ldots\ldots\ldots\ldots \text{(Equation 3)}$$

where ΔG is the Gibbs free energy of binding of the analyte. Since the rate of analyte departure is totally independent of its concentration in the solution, the analyte remains bound on average for a time $\tau = R^{-1}$ before departing.

It is the rate of departure which can be analyte-specific and which determines both the sensitivity and the response time of a sensor. Values of ΔG less than 0.7 eV give response times less than a second, and up to that value increases in ΔG are advantageous, leading to increased sensitivity. If however ΔG is greater than 1.1 eV the response time at room temperature lengthens to a year.

4.2. Constrained Dimer Channels

Although the first proposal for a membrane-based biosensor involved natural ligand-gated channels, the first commercial prototype involved a series of synthetic molecules with a significantly different principle [52]. The channels are based on gramicidin A, a bacterial antibiotic consisting of a single α helix. α helices built from natural L-amino acids as usually encountered have a pitch of 4.5 amino acid residues, and the cavity down the middle of the helix is far too small to allow ionic translocation. The normal strategy for constructing a pore, for example in nAChR and other neuroreceptors, involves 5 or 6 α helices held together as a unit only by tethers at the top and bottom. This construction is not particularly robust, and such channels are readily inactivated by heat at temperatures not far above normal physiological ambient. In contrast gramicidin is constructed from a mixture of L- and D- amino acids, giving a much looser helix. The pore along its axis is large enough to allow the passage of ions, and it is held together by several radial polypeptide strands.

The gramicidin pore is passive. It is always open, and no analytes are known which occlude it in the manner of maltoporin. However a single gramicidin pore is not long enough to span a complete bilayer. The membrane is only porated by dimers. A novel feature of the Cornell proposal is the way in which the probability of dimer formation is controlled by chemically attached constraints. The probability of dimer formation can in fact be modulated by constraints on the membrane [55].

A further novel feature of the constrained-dimer channel approach is the coupling of the gramicidin channel to an antibody. The neuroreceptors that are readily available respond only to neurotransmitters. In a clinical context, the most pressing need is to determine the level of hormones. Antibodies to these, and to a whole range of other biologically active molecules, are readily available and can be produced in large quantities by monoclonal technology.

The constrained-dimer approach has a number of less desirable features. The most easily implemented configuration is negative going, i.e. the membrane conductivity decreases with increasing analyte concentration, limiting the achievable dynamic range as previously discussed. Another factor limiting the dynamic range is the complex coupling of the equilibria between gramicidin monomers and dimers, and between bound and free antigen [56]. These features may yield to further research.

4.3. Occluding Channels

Ligand-gated channels, for example neuroreceptors, tend to open on ligand binding. However it is much easier to arrange for the blockage of the channel when, for example, the ligand binds to a site located inside the channel itself. Because of the vastly simpler synthetic task involved, there have been many attempts to achieve this, e.g. [46]. As explained in Section 2.3, this principle is not expected to lead to high dynamic range, because as is illustrated in Figure 1, the open state of a channel always has a much higher level of noise than the closed state.

However there is an alternative mode of operation of an occluding channel, demonstrated by both Bayley and Winterhalter. Rather than looking at the average membrane current, its statistics as a function of time are analyzed. There are advantages in observing just a single channel. In principle, such a stochastic sensor has the exciting potential to provide information about a wide range of analyte molecules present in the test liquid [57], in the form of a 'spectrum' in much the same way as chromatography. Stochastic sensing has been proposed as a powerful technique for DNA and RNA sequencing [58]. Bayley and coworkers use mainly genetically engineered variants of α-hemolysin, the toxin excreted by staphyloccus. Because of the β-barrel construction of its channel, α-hemolysin is even more robust than gramicidin.

In spite of the inherent reliance of stochastic sensing on the ability to detect and analyze individual molecular events, the diffusion resistance imposes a design constraint on the response time at low analyte concentrations. A typical diffusion constant for a biomolecule in water is $1E\text{-}10$ $m^2.s^{-1}$ for albumins of molecular weight ~20 kDa [59]. Putting $a=1$ nm and $J=1$ s^{-1} in Equation 2 gives $c=1E18$ m^{-3} , i.e. a concentration of 1.6 nM. At this concentration, a single receptor will be visited by analyte molecules once a second. At picomolar concentrations, the receptor is visited once an hour. Hence single-channel stochastic sensing is unlikely to compete for the ultrahigh sensitivity possible with ligand-gated channels. Of course, in a biosensor with a large number N of receptors, the first indication of analyte presence will be obtained N times sooner. The sensitivity limitations of the use of a single channel could be reduced by the use of large multiplexed arrays of nanofabricated occluding channel cells, monitored in real time by a computer, and this would be a feasible and attractive way of exploiting the capabilities of modern nanofabrication techniques.

4.4. Neuroreceptors

The function performed by the neuroreceptors in the post-synaptic membranes of nerve cells is not optimized for biosensor application. As shown in Figure 1 for a glutamate receptor, the current due to the binding of a single molecule cannot be detected against the background noise level. Two molecules of neurotransmitter molecule must normally bind to a receptor for an appreciable current to flow. The binding of additional neurotransmitter molecules causes the current to increase further, and faster than linearly.

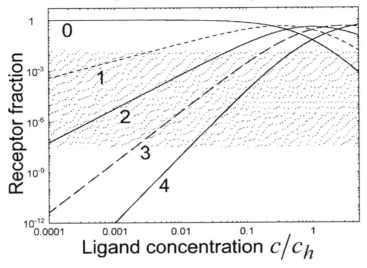

Figure 7. Fractions of receptor population with 0, 1, 2, . . bound molecules of ligand as a function of the ligand concentration in the bulk phase

This has implications for the dynamic range of a biosensor fabricated using neuroreceptors. It may be assumed that to a first approximation, the analyte binding sites on the different chains are independent. For a population of receptors in equilibrium with a concentration c of neurotransmitter, the expressions for the fraction f_0 , f_1 , . . . , f_r ,.. of receptors occupied by 0, 1, . . . , r , . . . neurotransmitter molecules, respectively, have the form:

$$f_0 = 1/(1 + c/c_h)^4$$
$$f_1 = 4(c/c_h)f_0 \qquad \ldots\ldots\ldots\ldots\ldots\ldots \text{(Equation 4)}$$
$$f_r = {}^4C_r(c/c_h)^r f_0$$

where c_h is the bulk concentration at which half the sites are occupied. These are plotted in Figure 7 on a logarithmic scale.

A logarithmic scale is appropriate because for a wide range of applicability, a biosensor should have a large dynamic range. It can be seen from the Figure that in any implementation the binding-site half-occupancy concentration c_h will be close to the upper limit at which the sensor output saturates.

As already noted, this upper limit is a system parameter, which can be extended by appropriate engineering effort at the expense of greater system complexity and cost. However the half-occupancy concentration is a convenient benchmark figure. Likewise, the lower limit of detection is likely to be somewhat less than the concentration at which the channel current is equal to the leakage through virgin membrane, but again, the latter concentration is useful as a benchmark. In both cases, extending the range beyond these points would require increasing effort and expense.

For the purposes of discussion, Figure 7 shows a stippled region of active receptor fraction (or equivalently, conductivity), which is intended to represent the range of fractional active channels that can be measured. In Figure 7 it covers six orders of magnitude, which is readily achievable for the ratio of electrolyte to virgin membrane conductivities, providing that the conductance of real electrodes has been compensated electronically. The stippled range shown does not extend to 100% channel activity. Clearly, there is no advantage in being able to measure beyond this point, and compensation for response nonlinearity becomes increasingly difficult as 100% is approached.

The problem with measuring high conductivities does not lie with the measurement electronics. The high-end limitation is the conductivity of the electrolyte in series with that of the membrane, and the range shown implies that the membrane conductivity with all channels open is greater than that of the electrolyte. Since the length of a pore is typically 4-5 orders of magnitude less than the distance between the electrodes, this is realistically achievable with receptors fairly densely packed in the membrane, even if the cross-section of an open pore is less than 1% that of the overall receptor.

Figure 1 however suggests that the conductivity of singly occupied receptors is negligible. It can be seen from Figure 7 that response is then square-law, giving a sensor dynamic range of three orders of magnitude. However if the conductivity of a receptor occupied by a single analyte molecule is more than 10% of that for a doubly-occupied receptor, then the response is linear over the whole usable range, giving a sensor dynamic range of six orders of magnitude. Actual measurements of nicotinic receptors are consistent with a conductivity ratio of approximately 5% [60]. Clearly this is an important parameter that must be established.

Since most of the useful range of the sensor lies at concentrations below c_h , it can be seen that tight binding of the analyte to the neuroreceptors is not desirable, although of course to provide specificity of response binding to the analyte must be stronger by an adequate margin than that to other commonly encountered substances. For nAChR, the half-occupancy concentration for acetylcholine is approximately 100 μM [60]. With eight orders of magnitude dynamic range the sensitivity would be 1 pM, at which concentration the rate at which analyte molecules visit individual receptors is once an hour. If however there are at least 10^8 receptors in the membrane, the response only needs averaging over less than a millisecond to establish the current concentration in the adjacent bulk phase.

The refractory state is another peculiarity of neuroreceptors, ideal for their function of nerve impulse transmission but not ideal for a biosensor. On initial binding of neurotransmitter molecules, the channel opens, but very shortly thereafter (on the experimental timescale) the channel closes again, in spite of the fact that the neurotransmitter is still bound. Much later, the neurotransmitter departs and the molecule resets to its initial closed but responsive state. As far as biosensor operation is concerned, the existence of the refractory state is purely deleterious. However it is not a necessary state, and mutant molecules are known with much reduced refractory durations [61].

4.2. Olfactory receptors

Nervous impulses are essentially digital, on or off. Just as the logic gates in a computer rely on nonlinearity to eliminate the effects of component variation and restore degraded signals to their ideal values of '0' and '1', so the neuroreceptor proteins of the postsynaptic membrane respond less than linearly to small concentrations of neurotransmitter. However the other biological biosensor system, olfaction, has the opposite requirement to detect even extremely small concentrations of the odor molecule. Olfaction therefore appears possibly to be a more appropriate model.

High sensitivity to smell is particularly marked in insects in general and in Lepidoptera (butterflies and moths) in particular. Their olfactory systems for the detection of sex pheromones have been demonstrated to respond to a single molecule, for example in the antennae of the silkworm moth, *Bombyx mori* [62,63]. The olfactory systems of beetles have been used as biosensors, albeit so far not very practical ones. By mounting a freshly severed antenna of a California potato beetle on the gate of a field-effect transistor, response in the low ppb range has been obtained to the characteristic potato-odor compound *cis*-3-hexen-1-ol [64].

The exquisite sensitivity of insect olfaction is achieved by molecular amplification. Like a neuroreceptor, an odor receptor has a binding site specific for a particular odor molecule. However it has no channel. Channels are eventually opened, leading to the flow of current across the membrane, but these are located in membrane-bound protein molecules remote from the original binding event. The initial event initiates a cascade, involving trimeric G-proteins [65], which gives rise to many molecules of cAMP, a neurotransmitter that opens the channels in adjacent neuroreceptors in the usual way.

While the great *in vivo* sensitivity of the olfactory system suggests its great promise as a model for biosensors, successfully mimicking it *in vitro* involves the simultaneous incorporation of not just one but three different proteins into an artificial membrane, together with the provision of controlled concentrations of various high-energy phosphates and enzymes. Whilst the chemical gain of the G-protein cascade allows the organism to detect individual molecules of important chemicals, biosensors have the alternative option of using electronic gain, and it is clear from patch-clamp results that the latter is fully sensitive enough to detect a single molecular event.

5. PERIPHERALS

By the "peripherals of the membrane" is meant those essential system components that establish the environment, both physical and chemical, for correct membrane function.

The current flowing through the membrane is composed of ions, while it is electrons that flow on to the measurement circuit. The two types of carrier exchange their charge at the junction between electrode and electrolyte. In systems of free-swimming black lipid membranes, this does not present a problem. However the membrane lifetime in such systems is at most a few minutes, totally unsuited to commercial applications. As discussed above, two apparently different solutions have been developed for achieving workable membrane lifetimes. In fact, they turn out to be essentially the same. In each case, the bilayer is supported some distance from an electrode by a system of hydrophilic organic chains surrounded by electrolyte. This system of chains is soft, so that it does not itself damage the membrane. Nevertheless it is solid. By transmitting small forces it protects the membrane from damage by convection currents, etc., in the test liquid.

In the case of thiol-tethered membrane systems, the membrane is approximately a Debye length from a noble-metal electrode [66]. No redox system is available at the electrode to allow the passage of direct current. Instead, this thickness of electrolyte provides a reservoir in which the ions that flow through the active channels can accumulate. The chains maintaining the separation between the membrane and the electrode are typically polyethylene glycol (PEG), which is well known to dissolve in water and is often used as the headgroup for nonionic detergents.

The alternative to tethering the membrane directly to the electrode surface is to assemble it on the surface of a gel, in one of two different ways. In the approach of Uto et al. [37], the membrane is assembled at the interface between a gel and liquid electrolyte. In the approach of the present authors, the membrane is assembled between two slabs of gel [41]. The chains are not PEG but polysaccharide; the function of the gel as a soft solid supporting and protecting the membrane while allowing the free passage of ions is identical.

There is a trade-off associated with the use of the gel layer on the front surface. Although it protects the bilayer from damaging agents, it also hinders access to it by the analyte species. Agarose gels are well-known to allow the passage of large biomolecules, and it has been shown in previous work that not only gramicidin [41] but also the membrane attack complex of human blood complement [38] can diffuse through to the bilayer.

5.1. Mass Transport through a Gel

The gel layers on the front and back surfaces of the sensor have the function in common of allowing the field-driven transport of ions to the membrane while protecting the latter from turbulent forces. The front gel is more critical and must additionally allow the diffusion of analyte. Diffusion, described by the diffusion coefficient D, and field-driven transport, described by the mobility μ, are intimately connected by the Einstein relationship:

$$\mu = \frac{Dq}{kT} \quad\quad \dots\dots\dots\dots\dots\dots\dots\dots\dots\dots\dots\dots \text{(Equation 5)}$$

The one-dimensional diffusion of material through a homogeneous medium is described by Fick's second equation [69]:

$$\frac{\partial c}{\partial t} = D\frac{\partial^2 c}{\partial x^2} \quad\quad \dots\dots\dots\dots\dots\dots\dots\dots\dots\dots\dots \text{(Equation 6)}$$

where $c(x,t)$ is the analyte concentration at the position x and time t. The well-known solution to this equation [69]:

$$c(x,t) = \frac{1}{\sqrt{\pi Dt}}\exp\left(-\frac{x^2}{4Dt}\right), \quad t > 0 \qquad \dots\dots\dots\text{(Equation 7)}$$

corresponds to the application of unit amount of the material to the interface of a semi-infinite diffusion medium. With the boundary conditions for the present sensor configuration, the numerically computed response is shown in Figure 9. It cannot be expressed exactly in terms of standard analytic functions, but for a gel of thickness L is well approximated by the following combination, which has the correct short-term and long-term behavior [67]:

$$f(t) = \exp\left[-1.09*\sinh\left(\left(Dt/0.608L^2\right)^{0.867}\right)^{-1.07}\right]$$

$$\dots\dots\text{(Equation 8)}$$

At a systems level, the important role of the gel is to protect the membrane, and so far, no minimum thickness has been found at which the protective function ceases. According to Equations 7 and 8, the degradation of response time varies as the square of the gel thickness. For individually solvated hydrophilic analytes, the diffusion time of a molecule of hydrodynamic radius 2 nm through 50 µm of gel is less than 30 s [68].

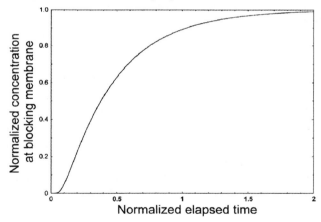

Figure 9. Calculated step response for a blocking membrane

5.2. Electrode-electrolyte charge transfer

Physiological saline has a bulk conductivity of approximately 3 S.m^{-1}. In gel-protected membrane elements, a distance through the electrolyte from the membrane to the electrode of 100 µm gives an electrolyte conductance per unit area of 30 kS.m^{-2}. This is typically six orders of magnitude greater than that of an intact membrane. Microlithographic techniques allow even higher conductances. Hence the series resistance of the electrolyte will not significantly degrade the sensor dynamic range.

The relationship between the current density through and the voltage applied across an electrode-electrolyte junction is the well-known Butler-Volmer equation [69]:

$$J = J_0\left\{\exp\left(\frac{eV}{kT}\right)-1\right\}\exp\left(-\frac{\beta eV}{kT}\right)$$

. . . . (Equation 9)

The parameter J_0 is known as the exchange current density and characterizes the balanced but spontaneous rate of the redox reaction under quiescent, equilibrium conditions. At low current densities, the conductance per unit area σ is given by $\sigma = \dfrac{eJ_0}{kT}$.

Electrode-redox systems are known with exchange current densities of the order of 100 A.m^{-2}, but these are all rather reactive and toxic. The best that can be achieved with mild electrode systems is approximately 1 A.m^{-2} using the well-known Ag/AgCl electrode. The corresponding low-current conductivity is 40 S.m^{-2}, less than two orders of magnitude higher than the values reported for some intact membranes [32,70]. Electrode contamination by the nonspecific adsorption of proteins from a test fluid reduces the room for maneuver even further.

5.3. Control of the Transmembrane Potential

The potential for high dynamic range of membrane-based sensors results from the conservation of charge. The macroscopic current measured in the external circuit is exactly equal to the sum of the currents through each individual part of the membrane. Providing that voltage across the membrane is kept constant, the membrane current is a precise measure of the number of open channels.

Unfortunately, the voltage across the membrane cannot always be kept constant. Electrically, the membrane is in series with a thickness of electrolyte, an electrode, and the wires connecting to electronic circuitry. Their series resistance limits the achievable dynamic range. Of these three, the electrode impedance represents the most significant design constraint.

In electrochemistry, the non-negligible impedance of available electrode systems is recognized. It is nevertheless possible to make precise measurements of potentials within an electrolyte by means of the potentiostat circuit. This is a simple feedback circuit in which the voltage on a reference electrode is used to control the voltage on a second, working, electrode so as to maintain the first at a given voltage.

For a membrane-based biosensor, two potentiostats are required, one on each side of the membrane [71,72]. In the present case, since the desired output signal is proportional to the number of open channels, the voltage across the membrane is kept constant. The desired output signal is the membrane current, which is readily monitored [67]. The four electrodes required by the use of a bipotentiostat can be arranged compactly and conveniently as shown in Figure 9.

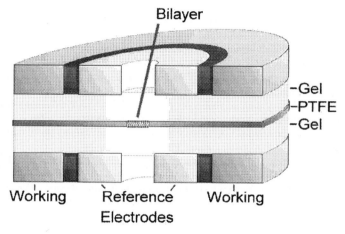

Figure 9. Biplanar conformation of electrodes giving a compact implementation of the bipotentiostat configuration of membrane operation

6. PERSPECTIVES

Biosensor technology based on supported bilayer membranes has the potential for application to a vast range of analytes present in biofluids in concentrations from micromolar to picomolar. There are no remaining scientific difficulties with this concept, but merely systems difficulties in implementing it so as to serve a useful purpose in medicine, industry and agriculture.

The major remaining systems problem is to extend the lifetime of packaged membrane elements sufficiently to allow transportation, handling and storage prior to use. A number of promising solutions to this problem are now being investigated.

REFERENCES

[1] M. Thompson, U.J. Krull and M.A. Venis, A chemoreceptive bilayer lipid membrane based on an auxin receptor. *Biochem. Biophys. Res. Commun.* 110 (1983) 300-304.
[2] P. Yager, A.W. Dalziel, J. Georger, R.R. Price and A. Singh, Acetylcholine receptor in planar polymerised bilayers toward a receptor-based biosensor. *Biophys. J.* **51** (1988) A143-A143.
[3] H.T. Tien and A. Ottova-Leitmannova, *Membrane Biophysics*, Elsevier: Amsterdam (2000).
[4] B.A. Cornell, L.E. Weir and F. Separovic, The effect of gramicidin A on phospholipid bilayers. *Eur. Biophys. J.* **16** (1988) 113-119.
[5] E.K. Sinner and W. Knoll,Functional tethered membranes. *Curr. Opin. Chem. Biol.* **5** (2001): 705-711.
[6] A. Palmeri, I.M. Pepe and R. Rolandi. Channel-formation activity of the antibiotic nisin on bilayer lipid membranes. *Thin Solid Films* **285** (1996) 822-824.
[7] F.J. Barrantes, The acetylcholine receptor ligand-gated channel as a molecular target of disease and therapeutic agents. *Neurochem. Res.*, **22** (1997) 391-400.
[8] M. Trojanowicz, Miniaturized biochemical sensing devices based on planar bilayer lipid membranes. *Fresenius J. Anal. Chem.* **371** (2001) 246-260.
[9] S.W. Evans, S.D. Evans and I.R. Peterson, UK Patent Application No GB 9817014.5, Tethered Bilayer Biosensor for Complement Activity.
[10] H.G. Breitinger. Fast kinetic analysis of ligand-gated ion channels. *Neuroscientist* **7** (2001) 95-103.
[11] L.K. Liem, J.M. Simard, Y. Song and K. Tewari, The patch clamp technique. *Neurosurgery* **36** (1995) 382-92.
[12] M. Winterhalter, Sugar transport through channels reconstituted in planar lipid membranes. *Coll. Surf. A* **149** (1999): 547-551.
[13] O. Braha, B. Walker, S. Cheley, J.J. Kasianowicz, L.Z. Song, J.E. Gouaux and H. Bayley, Designed protein pores as components for biosensors. *Chem. Biol.* **4** (1997) 497- 505.
[14] H. Bayley and C.R. Martin, Resistive-pulse sensing - From microbes to molecules. *Chem. Rev.* **100** (2000) 2575-2594.
[15] R. Anholt, J. Lindstrom and M. Montal, Stabilization of acetylcholine receptor channels by lipids in cholate solution and during reconstitution in vesicles. *J. Biol. Chem.* **258** (1981) 4377-4387.
[16] D.R. Nelson and B.I. Halperin, Dislocation-mediated melting in two dimensions. *Phys. Rev. B* **19** (1979) 2457.
[17] J.V. Selinger and D. R. Nelson, Theory of hexatic-to-hexatic transitions. *Phys. Rev. Lett.* **61** (1988) 416.
[18] I.R. Peterson, G. Brezesinski, B. Struths and E. Scalas, A Grazing-Incidence X-Ray Diffraction Study of Octadecanoic Acid Monolayers. *J. Phys. Chem. B* **102** (1998) 9437-9442.
[19] G.S. Smith, E.B. Sirota, C.R. Safinya, R.J. Plano and N.A. Clark, Structure of the L_β phases in a hydrated phosphatidylcholine. *J. Chem. Phys.* **92** (1990) 4519
[20] G.A. Overbeck and D. Möbius, A new phase in the generalized phase diagram of monolayer films of long chain fatty acids. *J. Phys. Chem.* **97** (1993) 7999.

[21] I.R. Peterson and R.M. Kenn, Equivalence between Two-dimensional and Three-dimensional Phases of Aliphatic Chain Derivatives, *Langmuir* **10** (1994) 4645-4650.

[22] R. Moog, M. Brandl, R. Schubert, C. Unger and U. Massing, Effect of nucleoside analogues and oligonucleotides on hydrolysis of liposomal phospholipids. *Int. J. Pharm.* **206** (2000) 43-53.

[23] P. Laggner and M. Kriechbaum, Phospholipid phase transitions – kinetics and structural mechanisms. *Chem. Phys. Lipids* **57** (1991) 121

[24] H.J. Merle, R. Steitz, U. Pietsch and I.R. Peterson, The Lamellar-Columnar Transition in Langmuir-Blodgett Multilayers of Cadmium Soaps, *Thin Solid Films* **237** (1994) 236.-243

[25] S.D. Ogier, R.J. Bushby, Y.L. Cheng, S.D. Evans, S.W. Evans, A.T.A. Jenkins, P.F. Knowles and R.E. Miles, Suspended planar phospholipid bilayers on micromachined supports. *Langmuir* **16** (2000) 5696-5701.

[26] M. Thompson and U.J. Krull, Biosensors and the transduction of molecular recognition. *Anal. Chem.* **63** (1991) A393.

[27] R.G. Nuzzo and D.L. Allara, Adsorption of bifunctional organic disulfides on gold surfaces. *J. Am. Chem. Soc.* **105** (1983) 4481

[28] J.B. Hubbard, V. Silin and A.L. Plant, Self assembly driven by hydrophobic interactions at alkanethiol monolayers: mechanisms of formation of hybrid bilayer membranes. *Biophys Chem* **75** (1998) 163-76.

[29] A. Pockels, *Nature* **43** (1891) 437.

[30] A. Pockels, *Science* **64** (1926) 304.

[31] D.P. Nikolelis and U.J. Krull, Bilayer lipid membranes for electrochemical sensing, *Electroanalysis* **5** (1993) 539-545.

[32] R.F. Costello, I.R. Peterson, J. Heptinstall, N.G. Byrne and L.S. Miller, A Robust Gel-Bilayer Channel Biosensor. *Advan. Mater. Opt. Electron.* **8**, (1998). 47-52.

[33] Y.E. He, M.G. Xie, A. Ottova and H.T. Tien, Crown-ether-modified bilayer lipid membranes on solid support as ion sensors. *Electroanalysis* **5** (1993) 691.

[34] A.L. Plant, M. Gueguetchkeri and W. Yap . Supported phospholipid/alkanethiol biomimetic membranes: insulating properties. .*Biophys. J.* **67** (1994) 1126.

[35] C.A. Widrig, C. Chung and M.D. Porter, The electrochemical desorption of *n*-alkanethiol monolayers from polycrystalline Au and Ag electrodes. *J Electroanal. Chem.* **310** (1991)335-359.

[36] I.R. Peterson, G. Veale and C.M. Montgomery, The Preparation of Oleophilic Surfaces for Langmuir-Blodgett Deposition. *J. Coll. Interface Sci.* , **109** (1986) 527-30.

[37] M. Uto, M. Araki, T. Taniguchi, S. Hoshi, and S. Inoue, Stability of an agar-supported bilayer lipid membrane and its application to a chemical sensor. *Anal. Sci.* **10** (1994) 943-946.

[38] R.F. Costello, S.W. Evans, S.D. Evans, I.R. Peterson and J. Heptinstall, Detection of Complement Activity using a Polysaccharide-Protected Membrane, *Enzyme Microbial Technol.* **26** (2000) 301-303.

[39] G. Gimpl, K. Burger and F. Fahrenholz. Cholesterol as modulator of receptor function. *Biochemistry* **36** (1997) 10959-10974.

[40] M. Grit and D.J.A. Crommelin, The effect of aging on the physical stability of liposome dispersions. *Chem. Phys. Lipids* **64** (1993) 3-18.

[41] R.F. Costello, I.R. Peterson, J.Heptinstall and D.J. Walton, Improved Gel-Protected Bilayers, *Biosens. Bioelectron.* **14** (1999) 265-271.

[42] I.R. Peterson, V. Brzezinski, R.M. Kenn and R. Steitz, Equivalent States of Amphiphilic Lamellae. *Langmuir* **8** (1992) 2995-3002.

[43] M. Kates, Biology of halophilic bacteria, Part II. Membrane lipids of extreme halophiles: biosynthesis, function and evolutionary significance. *Experientia* **49** (1993) 1027-1036.

[44] M. De Rosa and A. Gambacorta, The lipids of archaebacteria. *Prog Lipid Res.* **27** (1988) 153-175.

[45] H. Morii, M. Nishihara, M. Ohga and Y. Koga, A diphytanyl ether analog of phosphatidylserine from a methanogenic bacterium, *Methanobrevibacter arboriphilus. J Lipid Res* **27** (1986) 724-730.

[46] M. Smetazko, C. Weiss-Wichert, M. Saba, Th. Schalkhammer, New synthetic, bolaamphiphilic, macrocyclic lipids forming artificial membranes. *J. Supramol. Sci.* **4** (1997) 495-502.

[47] B. Raguse, P. Culshaw, J. Prashar and K. Raval, The Synthesis of Archaebacterial Lipid Analogues. *Tetrahedron Lett.* **41** (2000) 2971.

[48] H.T. Tien, Formation of Black Lipid Membranes by Oxidation Products of Cholesterol. *Nature* **212** (1966) 718-719.

[49] R.L. Robinson and A. Strickholm, Oxidized cholesterol bilayers. Dependence of electrical properties on degree of oxidation and aging. *Biochim. Biophys.* **509** (1978) 9-20.

[50] H.T. Tien, *J. Coll. Interface Sci.* **22** (1966) 438.

[51] S. Amemiya and A.J. Bard, Voltammetric ion-selective micropipet electrodes for probing ion transfer at bilayer lipid membranes. *Anal. Chem.* **72** (2000) 4940-4948.

[52] B.A. Cornell, V.L.B. BraachMaksvytis, L.G. King, P.D.J. Osman, B. Raguse, L.Wieczorek and R.J. Pace, A biosensor that uses ion-channel switches. *Nature* **387** (1997) 580-583.

[53] M.von Smoluchowski, Drei Vorträge über Diffusion, Brown'sche Bewegung und Koagulation von Kolloidteilchen. *Phys. Z.* **17** (1916) 557-571, 585-599.

[54] S. Glasstone, K.J. Laidler and H. Eyring, *The Theory of Rate Processes* McGraw-Hill: New York (1941).

[55] A. Hirano, M. Wakabayashi, Y. Matsuno and M. Sugawara, *Biosens. Bioelectron.* in press

[56] G. Woodhouse, L. King, L. Wieczorek and B. Cornell, Kinetics of the competitive response of receptors immobilised to ion channels which have been incorporated into a tethered bilayer. *Faraday Discuss.* **111** (1998) 247.

[57] H. Bayley and P.S. Cremer. Stochastic sensors inspired by biology. Nature 413 (2001) 226-230.

[58] S. Howorka, S. Cheley and H. Bayley, Sequence-specific detection of individual DNA strands using engineered nanopores. *Nat. Biotechnol.* **19** (2001) 636-639.

[59] L. Tao and C. Nicholson, 1996. Diffusion of albumins in rat cortical slices and relevance to volume transmission. *Neuroscience* **75**: 839-847.

[60] S.A. Forman and K. W. Miller. High acetylcholine concentrations cause rapid inactivation before fast desensitization in nicotinic acetylcholine receptors from Torpedo.*.Biophys. J.* **54** (1988) 149-158..

[61] C. Léna and J.P. Changeux, Allosteric nicotinic receptors, human pathologies. *J Physiol. Paris* **92** (1998) 63-74.

[62] K. E. Kaissling and E. Priesner, Die Riechschwelle des Seidenspinners. *Naturwiss.* **57** (1970) 23-28.

[63] B.H. Sandler, L. Nikonova, W.S. Leal, and J. Clardy, Sexual attraction in the silkworm moth: structure of the pheromone-binding-protein-bombykol complex. *J. Chem. Biol.* 7 (2000) 143-51.

[64] P. Schroth, M.J. Schöning, H. Luth, B. Weissbecker, H.E. Hummel and S. Schütz, Extending the capabilities of an antenna/chip biosensor by employing various insect species. *Sens. Actuat. B* **78** (2001) 1-5.

[65] H.R. Bourne, How receptors talk to trimeric G proteins. *Curr. Opin. Cell Biol.* **9** (1997) 134-142.

[66] G. Krishna, J. Schulte, B.A. Cornell, R. Pace, L. Wiecorek and P.D. Osman, Tethered bilayer membranes containing ionic reservoirs: the interfacial capacitance. *Langmuir* **17** (2001) 4858-4866.

[67] J.A. Beddow, I.R. Peterson, J. Heptinstall and D.J. Walton, Electrical monitoring of gel-protected bilayer lipid membranes using a bipotentiostat, *Proc. SPIE* **4414** (2001) 62-69.

[68] A. Pluen, P.A. Netti, R.K. Jain and D.A. Berk,. Diffusion of Macromolecules in Agarose Gels: Comparison of Linear and Globular Configurations. *Biophys. J.* **77** (1999) 542-552.

[69] P.W. Atkins, *Physical Chemistry*, Oxford UP (1998)

[70] D.R. Laver, J.R. Smith and H.G.L Coster, The Thickness of the Hydrophobic and Polar-regions of Glycerol Monooleate bilayers determined from the Frequency-dependence of Bilayer Capacitance. *Biochim. Biophys. Acta.* **772** (1984) 1-9.

[71] H. Hieda, T. Ishino, K. Tanaka and N. Gemma, Analysis on Electrostatic Tip Sample Interaction in Aqueous Solution under Potentiostatic Condition. *Jpn. J. Appl. Phys.* **34** (1995) 595-599.

[72] S.A. McClintock and W.C. Purdy, A Bipotentiostat for 4-Electrode Electrochemical Detectors. *Anal. Lett.* **14** (1981) 791-798.

Planar Lipid Bilayers (BLMs) and their Applications
H.T. Tien and A. Ottova-Leitmannova (Editors)
© 2003 Elsevier Science B.V. All rights reserved.

Chapter 28

Biosensors from interactions of DNA with lipid membranes

U. J. Krull[a], D. P. Nikolelis[b] and J. Zeng[a]

[a]Chemical Sensors Group, Department of Chemistry, University of Toronto at Mississauga, 3359 Mississauga Rd. N., Mississauga, Ontario, L5L 1C6, Canada

[b]Laboratory of Analytical Chemistry, Chemistry Department, University of Athens, Panepistimiopolis-Kouponia, Athens 15771, Greece

1. INTRODUCTION

Monitoring of pathogens, and of compounds such as insecticides, pesticides and herbicides, in environmental samples and foods is becoming increasingly important in the area of government regulations and quality control. Environmental monitoring requires the determination of very low concentrations of pollutants. For example, the maximum admissible concentrations of pesticides allowed in the European Community countries are 0.1 ng/mL [1], and the limits for water-borne markers for pathogens even in waste water is often on the order of a few hundred organisms per deciliter. Biosensor technology has been exploited to fill niche applications in the area of monitoring. Many sensitive biosensors that approach the detection limits of 0.1 ng/mL are based on the high affinity recognition of antigen by antibodies. A wide range of immunosensors based on immunochemical reactions that combine the classic ELISA format with amperometric, spectrophotometric, chromatographic and other detection systems have been developed (see [2,3] and references therein). Lipid films have been proposed as the basis for development of electrochemical biosensors that use immunodetection, and have been investigated for the rapid detection and monitoring of a wide range of compounds [4]. More recent methods have focused on the use of nucleic acid oligomers as reagents to develop biosensors for both chemicals and pathogens.

The concept of primers as used for polymerase chain reaction has been extended to the use of these short single-stranded oligonucleotides as immobilized probes to develop sensors that can detect the presence of pathogens [5]. Electrochemical redox, acoustic wave, and optical plasmon methods can directly detect the formation of double-stranded DNA (dsDNA), or the association of chemicals with dsDNA, without further amplification or use of markers [5]. The sensitivity of biosensors that are based on these methods can be improved by reliance on hybridization to promote a cascade of chemical interactions that cause large signal changes. For example, a redox experiment might rely on the transport of many electrons by one hybrid. Acoustic wave devices rely on extensive structural changes of water and ions at an interface caused by formation of hybrids. Plasmon methods rely on the disturbance of many electrons in oscillation by the formation of hybrids.

Lipid membranes operate as biosensors on the basis that a single binding event can alter the movement of a large number of ions on or across a membrane. This can provide a substantial intrinsic amplification of the initial binding event, and the amplification can be achieved without use of specialized labels. DNA can be immobilized on lipid films so that selective interactions with an analyte (e.g. another strand of DNA, or a molecule such as a mutagen) can be transduced by the lipid film as a measurable signal [6]. This approach to biosensor development offers advantages in the area of real time monitoring, and direct transduction of nucleic acid-analyte interactions. The use of stabilized metal and filter-supported lipid films that are allowed to associate with DNA as either a selective binding or a sensitising reagent has allowed progress to be made towards development of BLMs for detection of targets in real samples [4,7].

2. DETECTION OF DNA HYBRIDIZATION USING SELF-ASSEMBLED BILAYER LIPID MEMBRANES (s-BLMs)

We have investigated the use of self-assembled BLMs (s-BLMS) for the development of a simple, practical and low cost approach for monitoring of DNA hybridization [6]. The lipids used were phosphatidylcholine (PC), dipalmitoyl- and dimyristoyl-phosphatidylcholine (DPPC and DMPC, respectively). s-BLMs were supported on metal wire (silver wire). s-BLMs were prepared according to established techniques [8-10]. Hexane was used as a solvent to prepare "solventless" or "solvent-free" BLMs (these membranes are nominally solvent-free since a low level of the residual solvent may be retained in the BLM torus) [11]. The sensing electrode was connected to the power supply and the reference electrode to the electrometer. The same electrometer was used as a current-to-voltage converter.

The probe oligonucleotide was single stranded deoxyribonucleic acid (ssDNA) thymidylic acid icosanucleotide that was terminated at 5' end with a C16 alkyl chain to assist incorporation into s-BLMs (dT_{20}-C16). The complementary oligonucleotide was deoxyadenylic acid icosanucleotide (dA_{20}). The oligonucleotides were prepared by use of an automated solid-phase synthesizer using phosphoramidite synthons, and products were characterized for sequence integrity and purity by anion exchange HPLC. The electrochemical ion current across s-BLMs was found to change due to the presence of ssDNA and during the formation of dsDNA. Single-stranded oligonucleotides were found to interact with s-BLMs and caused the ion current to increase. The s-BLMs exhibited reproducible response caused by oligonucleotide adsorption, with response times on the order of minutes. Fig. 1 shows recordings of responses of s-BLMs to dT_{20}-C16 in 0.1 M KCl electrolyte solution using a silver wire of 0.5 mm diameter. These data indicate that for all concentrations of oligonucleotide, it took about the same delay time (8-10 s), for the current increases to be observed. The short and relatively invariable delay times did not appear to be related to the concentration of the modified (C16) oligonucleotide, and this was consistent with results observed for unmodified oligonucleotides [12]. Preformed dsDNA from the interaction of dA_{20} with dT_{20}-C16 did not induce any changes in ion permeability through BLMs, suggesting that dsDNA did not significantly partition onto or into s-BLMs.

The results suggest that the conductance enhancements of s-BLMs that are observed following the adsorption of dT_{20}-C16 and/or dA_{20} are due to the increase of surface charge density of the membrane caused by the negative charge of the phosphate groups of ssDNA. A consequence of adsorption may be the appearance of structural changes (i.e. defect sites) as is known to be induced by polyelectrolytes in other experiments [13-15].

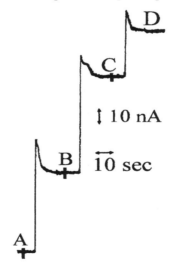

Fig. 1. Typical recordings obtained with s-BLMs prepared from PC upon stepwise additions of dT_{20}-C16 in 0.1 M KCl electrolyte solution (pH 6.8). The concentration of dT_{20}-C16 in the bulk electrolyte solution (ng/mL) was (A) 0.0; (B) 8.3; (C) 14.0 and (D) 15.2.

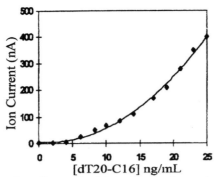

Fig. 2. Calibration curve for oligonucleotide concentration using 0.1 M KCl electrolyte solution. The s-BLMs were prepared from PC.

The ion current changes were related to the oligonucleotide concentration (Fig. 2). The BLM response towards dA_{20} provided similar results (i.e. rapid increases of ion conductivity), with a concentration range suitable for target detection between 1 and 14 ng/mL. The sharp (nearly quadratic) function of membrane current versus concentration of target may suggest the presence of cooperative processes leading to formation of aggregates. Reproducibility of response was found to be on the order of 4 to 8% (N=5, 95% confidence limit).

Differential scanning calorimetry studies using vesicles composed of DPPC or DMPC showed that dT_{20} and dA_{20} did not significantly alter the Tm of the lipids, which were determined to be 40 °C and 23 °C, respectively, [16]. Similar experiments using the C16 derivatized oligonucleotide have shown that the dT_{20}-C16 partitions onto BLMs and caused structural changes, as suggested by an increase of Tm from 22.4 °C to 29.0 °C when dT_{20}-C16 was mixed with DMPC. The incorporation of dT_{20}-C16 resulted in the appearance of more ordered regions of DMPC bilayer, which was reflected by the increased Tm value of these liposomes [17]. Preliminary results from deuterium NMR experiments also indicated an interaction of the dT_{20}-C16 with vesicular BLMs, which was not evident using dT_{20}. The results were consistent with adsorption of dT_{20}-C16 onto BLMs.

s-BLMs were used to electrochemically monitor DNA hybridisation (Fig. 3). The dT_{20}-C16 was used as probe, and dA_{20} as complementary DNA (cDNA). The probe ssDNA was injected in the electrolyte solution after stabilization of the s-BLMs. The ion current increased and stabilized. The injection of an equimolar quantity of cDNA (presuming that these oligonucleotides form a hybrid complex at a ratio 1:1) into the electrolyte solution resulted in a further increase of the ion current that gradually dropped to the initial background ion current values (Fig. 3). The ion current changes were completed within 30 min at 22 °C. At higher temperatures, less time was required for the ion current to drop to the initial background current level.

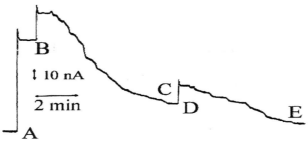

Fig. 3. Typical recording obtained using s-BLMs for the electrochemical monitoring of DNA hybridization using dT_{20}-C16 as probe DNA and dA_{20} as cDNA. Experiment was performed at 28 ± 1 ^{0}C. (A) Injection of 12.4 ng/mL probe DNA; (B) Injection of 6.23 ng/mL cDNA; (C) Formation of hybrid; (D) Injection of a further 6.23 ng/mL cDNA; (E) Recovery of background signal.

The return of the ion current to the background current level was indicative of the formation of the hybrid, which may subsequently have been excluded from the membrane. This was consistent with previous observations that showed no association of pre-formed dsDNA with BLMs in terms of generation of an electrochemical signal [12]. Our experiments also have investigated pressure-area curves, and fluorescence from lipid monolayers containing fluorescein-labelled dT_{20}-C16 to determine the presence of the oligonucleotides at monolayer surfaces [18]. The results of this work showed that upon hybridization, the double stranded material physically dissociated from the surface of lipid membranes even though a C16 'anchor' was present. This suggested that there is an interesting balance of energetics at play, which might be tuned for selective release of DNA for electrochemical signaling as well as sensor regeneration [18].

Injection of a smaller quantity of dA_{20}, as compared to the molar concentration of dT_{20}-C16 in bulk solution, resulted in a drop of the ion current to the level corresponding to the amount of dT_{20}-C16 that should have remained unhybridized. The injection of a dA_{20} in a quantity that was equimolar to the quantity of dT_{20}-C16 that was left unhybridized lead to complete recovery of the background ion current.

The time dependence of current change strongly depended upon the concentration of both probe DNA and cDNA (Table 1). The time required for current to reduce to the background ion current level increased with an increase of the concentration of the oligonucleotides. The results of hybridization could also be observed using dA_{20} as the probe ssDNA and dT_{20}-C16 as cDNA. However the time dependence of current change was much longer. For example, the use of 6.23 ng/mL dA_{20} as probe ssDNA first resulted in an increase of the ion current to 85 nA. The injection of dT_{20}-C16 resulted in a further increase of the ion current that eventually dropped to the background levels after a period of time of 135 min at

Table 1
Effect of Oligonucleotide Concentration on Response Time
(Experiments Performed at $28 \pm 1\ ^0C$)

[dT$_{20}$-C16] (ng/mL)	[dA$_{20}$] (ng/mL)	Response Time[a] (min)
6.2	6.3	5.5 ± 0.3
12.0	6.3	6.2 ± 0.3
12.0	12.5	12.8 ± 0.8
23.0	12.5	21.8 ± 1.2

[a]Average of five measurements.

22 ^0C. The increase of ion current on addition of complementary ssDNA was likely the result of additional ssDNA adsorbing to the s-BLM, as was seen in control experiments using non-complementary ssDNA. Hybridization was expected to reach completion over a period of minutes, providing ample time for ssDNA to partition and to cause an initial ion current increase.

Generally, an increase of the temperature resulted in a decrease of the time required for the completion of the electrochemical signal changes. For example, the hybridisation of 6.23 ng/mL dT$_{20}$-C16 and 6.23 ng/mL dA$_{20}$ took 30 min to complete at 22 $^{\circ}C$, whereas this time decreased to 5.5 min at 28 $^{\circ}C$ (see Table 1), and 4.5 min at 35 ^0C. This temperature sensitivity was likely due to structural changes of the lipid membrane and the mobility of ssDNA in/on the membrane. Hybridization of these short oligonucleotides in bulk solution should be substantially complete in 5 minutes at the temperatures that were used. The reduction of current suggests that dsDNA produced at the s-BLM was excluded from the membrane, leaving the s-BLM available for another cycle of experimentation. The maximum number of cycles that were attempted is 20. Drift of response during these experiments was negligible.

Further experimental work showed that BLMs could be used as electrochemical transducers of hybridization of mixed base sequences, and of sequences that were only partially complementary [7]. These experiments used synthetic sequences of up to 25 bases for both the immobilized probe and the target oligonucleotides. The trends in analytical results were similar to the experiments that used 20mer homopolymers of dT$_{20}$-C16 and dA$_{20}$. Base pair discrimination to a level of 2 mismatched base pairs was shown to be possible at room temperature for 20mers, and the use of ionic strength and higher temperatures may improve this selectivity.

This work has demonstrated that ssDNA could be incorporated into self-assembled metal supported bilayer lipid membranes. The ssDNA could undergo hybridization with complementary DNA introduced into the supporting electrolyte solution. The change of ion current signal caused by removal of DNA due to hybridisation may be transduced by s-BLMs, and therefore, hybridisation may be monitored by ion conductivity changes. Detection limits of the s-BLM electrochemical sensor were in the nM range. The price paid for not relying on labels or tags for detection of hybridization does leave the s-BLM system less sensitive than the best electrochemical [19] and fluorometric biosensors [20]. The time required for complete signal generation was found to depend upon concentration of the probe ssDNA and cDNA, and temperature. Ionic strength, supporting electrolyte ions, and pH should also be factors that must be controlled to adjust response levels.

3. APPLICATIONS OF BLMS THAT ARE MODIFIED WITH DNA FOR DETECTION OF SMALL ORGANIC MOLECULES

3.1. Control of biosensor sensitivity by incorporation of DNA onto s-BLMs

3.1.1. A DNA biosensor based on self-assembled s-BLMs on silver electrodes for the detection of hydrazines

A sensitive biosensor based on s-BLMs for the rapid detection of hydrazines was developed. Hydrazine compounds are widely used as fuels (in rocket propulsion systems), corrosion inhibitors, catalysts, emulsifiers and dyes [21]. A biosensor using a s-BLM that was modified by adsorption of ssDNA was developed, and detection was based on monitoring of changes in the ion current that were induced by hydrazine [22]. s-BLMs that were modified by adsorption of ssDNA displayed an analytically useful response for part-per-billion levels of hydrazine [22]. The biosensor offered a highly sensitive, rapid, and portable tool for monitoring of this environmentally and toxicologically significant compound.

Fig. 4 shows recordings of responses from s-BLMs without ssDNA (at pH 6.8) for different concentrations of hydrazine. These data indicate that for all concentrations of the analyte it took about the same time (10 s) for the current increases to be observed. The relatively short and invariable delay time did not appear to be related to the concentration of hydrazine. The ion current values could be used to quantify the concentration of hydrazine. Ion current values were linearly related to the concentration of hydrazine in bulk solution between 515 to 3000 ppm [I(nA) = 0.0571 C(ppm) - 21.77, r^2 = 0.989]. The detection limit of hydrazine (for S/N = 3) was found to be 515 ppm. The high detection limit and the non-selective nature of hydrazine adsorption to s-BLMs suggested that it would be advantageous to sensitize the membrane to the electrochemical effects induced by hydrazine interactions.

ssDNA incorporated in s-BLMs was used as a sensitizer for the detection of hydrazine. Injections of various concentrations of hydrazine into the bulk electrolyte solution were done after 6.22 ng/mL dT_{20}-C16 had been equilibrated with a lipid membrane. The addition of the oligomer resulted in an increase of the

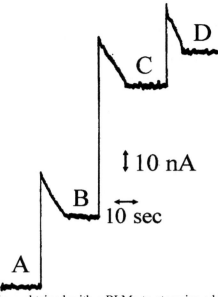

Fig. 4. Typical recordings obtained with s-BLMs to stepwise additions of hydrazine in 0.1 M KCl electrolyte solution (pH 6.8). The concentration of hydrazine in the bulk electrolyte solution (ppm) was (A) 515; (B) 1030; (C) 2056; and (D) 2600.

ion current, as was seen for the other experiments described in Section 2. The subsequent addition of hydrazine standard solution resulted in further increases of the ion current. The response time was 18-20 s and the increases of the ion current were linearly related to hydrazine concentration. The delay time for signal evolution was greater that that observed in the absence of ssDNA.

Optimization of the analytical signal was achieved by the use of different oligonucleotide concentrations (e.g. 19 and 25 ng mL^{-1}). Fig. 5 shows results with the two different concentrations of dT$_{20}$-C16. Higher concentrations of oligonucleotide (more than 24.9 ng mL^{-1}) resulted in substantially increased background ion current from the electrolyte, and limited the useful range of the analytical signals. The results shown in Fig. 5 indicate that the analytically useful concentration range for hydrazine determination was between 1.2 to 12 ppb (when using 19 ng mL^{-1} ssDNA) and 0.26 to 2.6 ppb (when using 24.9 ng mL^{-1} ssDNA), and that the ion current values were linearly related to hydrazine concentration (I(nA) = 64.64 C(ppb) + 189.06, r^2 = 0.985, for ssDNA concentration of 19 ng mL^{-1} and I(nA) = 281.01 C(ppb) + 330.73, r^2 =0.9686, for ssDNA concentration of 24.9 ng mL^{-1}). The detection limit was 10-fold lower than that observed in the absence of ssDNA, suggesting that sensitization of the membrane to surface charge and/or dipolar potential was achieved. Similar results were observed when using dA$_{20}$, although the response time was 80-100 s. This

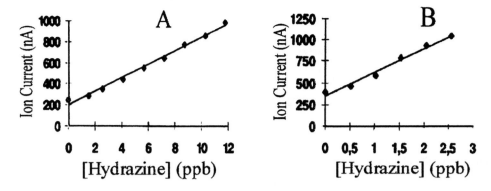

Fig. 5. Response curves for hydrazine in 0.1 M KCl electrolyte solution in the presence of (A) 19 ng mL^{-1} and (B) 24.9 ng mL^{-1} dT$_{20}$-C16.

was probably due to the absence of the alkyl chain that seemed to assist immobilization of the probe and would provide a higher partition coefficient for probes with the hydrocarbon tail.

The results suggest that s-BLMs that are modified with oligonucleotides can be used for rapid and sensitive detection of organic compounds and environmental pollutants. The detection limit for hydrazine was one of the lowest reported for any biosensor.

3.1.2. Stopped-flow analysis of mixtures of hydrazine compounds using filter-supported BLMs that were modified with DNA

Filter-supported lipid membranes were used as stable and reversible detectors for flow injection analysis (FIA) for hydrazine, methylhydrazine, dimethylhydrazine and phenylhydrazine [23]. The modification of the lipid membranes involved incorporation of dT$_{20}$-C16. The ssDNA did not act as a selective receptor for hydrazines. Rather, the presence of the ssDNA altered the electrochemical sensitivity of the BLM so that perturbations of the charge and structure of the membrane were amplified to provide greater changes in signal (ion current). The oligomer dT$_{20}$-C16 was incorporated in these BLMs by injection of solutions of the oligonucleotide (in a stopped-flow mode) prior to hydrazine monitoring.

A transient current signal as a single event was obtained by the interactions of hydrazine with the filter-supported BLMs that had been modified with the ssDNA (Fig. 6). A constant time delay for the appearance of the transient currents of (6.8 ± 1.3) s was observed. The magnitude of these transient responses was in direct proportion to the hydrazine concentration in the carrier electrolyte solution [I(pA) = 3.05 C (ppb) + 3.37, r^2 = 0.994). The

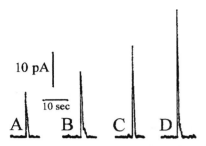

Fig. 6. Results obtained for hydrazine at pH 7.3 (0.1 M KCl) with BLMs composed of PC and 19.0 ng mL^{-1} dT$_{20}$-C16. The BLMs were supported in glass microfiber filters. Hydrazine concentrations of the solutions injected (75 uL) in the carrier electrolyte solution (ppb): (A) 2.57; (B) 5.15; (C) 7.40; (D) 10.3. The injection of each sample was made at the beginning of each recording. The flow rate of the carrier electrolyte solution was 0.7 mL min^{-1}.

variability of response of the BLMs to repetitive hydrazine injections expressed by the relative standard deviation was between 3.5 to 7.5% (N=5). Repetitive cycles of injection of hydrazines provided transient signals with no signal degradation over 30 sequential injections. The detection limit of hydrazine expressed as three times the signal noise was 2.5 ppb.

The presence of ssDNA substantially decreased the time required for signal generation in comparison to that observed for s-BLMs. Membranes that were modified with ssDNA either adsorbed hydrazines more quickly, or provided a different kinetics for association, or were more sensitive and reached the threshold concentration required for the production of the transient ion current peak more quickly. Similar transient signals to those of hydrazine were obtained with injections of methylhydrazine, dimethylhydrazine and phenylhydrazine in continuous flowing streams of a carrier electrolyte solution of pH 7.3 (0.1 M KCl). However, the time delays for the appearance of these transients were 35.8 ± 1.7 s for methylhydrazine, 36.9 ± 2.0 s for dimethylhydrazine and 58.8 ± 2.3 s for phenylhydrazine. The range of time delay for the appearance of the transient signals was between 5 to 8 s for hydrazine (N = 11), 34 to 38 s for methylhydrazine (N = 11), 34 to 39 s for dimethylhydrazine (N = 11) and 56 to 62 s for phenylhydrazine (N = 11). The difference in time for appearance of signals from the different hydrazines was likely due to their relative hydrophobicities. Fluorescence microscopy [18] of lipid membranes that were exposed to fluorescent dT$_{20}$-C16 indicated strong partitioning of the oligonucleotides with a dense coverage at the solution-lipid interface. This suggested that the planar and filter-supported membranes that were exposed to ssDNA would have a polar negatively charged surface, in contrast to membranes prepared from PC lipid alone, which would have an electrically neutral surface in terms of formal charge. The adsorption of molecules to a membrane surface would therefore be driven by the relative hydrophilicity, and this correlates with the properties of the hydrazines that produce electrochemical signals.

The results suggested that each transient signal caused a temporary change in the membrane. These current transients appeared as singular events. Further transients were not observed over periods of 10 min. The magnitude of ion current transients was found to increase with an increase of the concentration of methylhydrazine, dimethylhydrazine and phenylhydrazine. The magnitude of the transient signals could be used to quantify the concentrations of methylhydrazine, dimethylhydrazine and phenylhydrazine in the carrier electrolyte solution. Statistical treatment of results gave regression equations: I (pA) = 35.8 C (ppb) + 1.67, r^2 = 0.990, for methylhydrazine, I (pA) = 36.9 C (ppb) + 2.01, r^2 = 0.980, for dimethylhydrazine and I (pA) = 20.4 C (ppb) + 0.65, r^2 = 0.999, for phenylhydrazine. Replicate analyses of hydrazine samples indicated that the reproducibility was better than 6%. The calibration graph for the determination of hydrazines has shown linearity in the concentration range between 2.5 to 12.5 ppb for hydrazine, 0.2 to 1.2 ppb for methylhydrazine, 0.2 to 1.2 ppb for dimethylhydrazine, and 0.3 to 1.5 ppb for phenylhydrazine. The detection limits were on the order of 2.5, 0.1, 0.2 and 0.25 ppb for hydrazine, methylhydrazine, dimethylhydrazine and phenylhydrazine, respectively (for S/N = 3). The differences observed in the delay time for the appearance of signal for the hydrazines allowed the investigation of a simultaneous determination of a mixture of hydrazine, methylhydrazine or dimethylhydrazine and phenylhydrazine. Fig. 7 shows recordings obtained for such mixtures containing variable amounts of the different hydrazines. This figure shows that a discrete signal was obtained for each hydrazine in mixture and that a single transient corresponding to the effect of the hydrazines was observed only for methylhydrazine and dimethylhydrazine. The resolution of the signals of each hydrazine obtained in such mixtures was sufficient to permit reliable quantitative monitoring of these in mixture.

The present method offers a simultaneous and repetitive mode of detection of hydrazines that is faster and has lower cost than methods based on fluorescence or chromatography [24,25].

Analysis of these environmental pollutants can be done in protein-free water samples, however proteins or small peptides should be eliminated from other types of samples prior to analysis. These macromolecules may cause non-selective interference. Our results indicated that concentrations of albumin larger than 10 ug mL^{-1} provided ion current transients from BLM-based sensors when experiments used 12 ng mL^{-1} of ssDNA in solution. Such interference from albumin was achieved at a threshold concentration of 8 and 6 ug mL^{-1}, when the concentration of ssDNA was increased to 19 and 21 ng mL^{-1}, respectively. A number of inorganic ions and compounds were tested as potential interferents, and

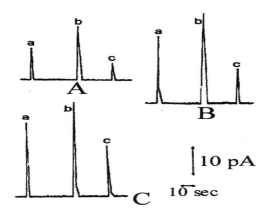

Fig. 7. Results obtained for mixtures of: hydrazine (a); methylhydrazine and dimethylhydrazine (b); and phenylhydrazine (c); at pH 7.3 (0.1 M KCl) with BLMs composed of PC and19.0 ng/mL dT$_{20}$-C16. The BLMs were supported in glass microfiber filters. The solution that was injected (75 uL) into the carrier electrolyte (flow rate of 0.7 mL min^{-1}) contained:
(A) hydrazine 2.57 ppb; methylhydrazine 0.22 ppb; dimethylhydrazine 0.20 ppb; phenylhydrazine 0.28 ppb.
(B) hydrazine 7.73 ppb; methylhydrazine 0.22 ppb; dimethylhydrazine 0.60 ppb; phenylhydrazine 0.55 ppb.
(C) hydrazine 7.73 ppb; methylhydrazine 0.22 ppb; dimethylhydrazine 0.60 ppb; phenylhydrazine 0.85 ppb.

validation of the technique for monitoring hydrazine compounds in reference water samples was reported.

The method was applied to the determination of hydrazines in water standards. A number of reference Water Standards (AlliedSignal, Riedel-deHaen, Seelze, Germany) were artificially contaminated with hydrazines at various concentrations and the hydrazine content was then assayed using the above method and a calibration equation. Recoveries within the range of ca. 94-103 % were obtained.

3.2. Detection by inhibition of hybridization

3.2.1. Detection of Aflatoxin M$_1$ based on electrochemical transduction by s-BLMs modified with DNA

A sensitive assay for the rapid screening of aflatoxin M$_1$ (AFM$_1$) has been developed [26]. AFM$_1$ is a hydroxylated metabolite of aflatoxin B$_1$ which has well established carcinogenic and mutagenic potentiality [27-29]. Aflatoxins are produced by certain fungi and occur naturally in a wide variety of foods. The most abundant of the group, aflatoxin B$_1$ (AFB$_1$), is a recognized carcinogen. When aflatoxin B_1 is ingested by cows, it is secreted as its hydroxylated metabolite, AFM_1. Due to the potential carcinogenicity of AFM_1, the detection and determination of AFM_1 in dairy products is of increasing interest [30]. An action level of 0.5 ppb (i.e., 1.5 nM) for AFM_1 in dairy products has been established by the U.S. Food and Drug Administration, and this level is even lower in the legislation of some European countries (e.g., 0.15 nM) [31]. To date, most of the procedures for AFM_1 determination are based on liquid chromatography, thin layer chromatography, immunoassays and fluorescence emission. The methods usually require cleanup steps using immunoaffinity chromatography [30,32-34].

A strategy to rapidly detect AFM_1 relies on the use of s-BLM biosensors for detection of hybridization of DNA. The approach detects changes in the time dependence of ion current resulting from DNA hybridization in the presence of the toxin. The probe oligomer dT_{20} -C16 was incorporated into s-BLMs and complementary dA_{20} was injected into the bulk electrolyte solution. The electrochemical ion current was found to increase due to the presence of ssDNA and decrease due to the formation of dsDNA. The rate of change of ion current and the time required for the completion of the signal change due to hybridization was found to depend on the concentration of aflatoxin. Measurements of the rate of signal generation provided a method for the rapid and sensitive detection of this toxin with a detection limit of 0.5 nM.

The initial rate of hybridization, $[dI/dt]_0$, was found to be 1.75 nA/min when using conditions of equimolar concentration of ssDNA and cDNA as described in Section 2. An increase of temperature resulted in a decrease of the time required for the completion of the electrochemical signal changes, as well as an increase in the initial rate of signal change. For example, the signal change resulting from interaction of 6.22 ng/mL dT_{20} -C16 and 6.24 ng/ml dA_{20} required 12.6 min to complete at 24^0C ($[dI/dt]_0$ =2.21 nA/min), whereas the time decreased to 5.5 min at 28^0C ([dI/dt]0 =3.12 nA/min) and 4.5 min at 35^0C ([dI/dt]0 =5.30 nA/min).

The effect of the toxin on the time dependence of the signal due to hybridization was assessed by additions of 4.19 nM of the toxin at various times following the initiation of hybridization (i.e. after association of 6.22 ng/mL dT_{20}-C16 with the membrane and the subsequent addition of 6.24 ng/mL dA_{20}). These experiments were repeated at various temperatures ranging from 22 to 35 °C. The results showed that the toxin prolonged the time that was required for signal to develop. The effect was strongly dependent upon the time of addition of the toxin, and decreased the initial rate of signal generation caused by hybridization. These effects were observed for all temperatures examined. The concurrent addition of the toxin and cDNA (t = 0 min) provided the largest effect at all temperatures. The effect was to increase, but not totally inhibit, the time required for the return of the current to background values, as well as to decrease the initial rate of current change. This may occur because the toxin impedes the process of hybridization, or because of changes in electrostatic fields at the membrane surface that alter interfacial diffusion.

The inhibitory effect was progressively lessened when the toxin was added after the hybridization reaction was initiated. For instance, when the addition of the toxin occurred 2 min after the addition of cDNA, the time required for the completion of signal generation was 45.5 min (at 22^{0}C), whereas when the toxin was added 7 min after the addition of cDNA, the time decreased to 30 min. This is possibly due to the fact that by the time the toxin is added, an amount of the duplex is already formed on the BLM surface and that the primarily acts on the unhybridized DNA. The addition of the toxin beyond a certain time (which varies with temperature) does not provide any effect upon the time course to reach completion of the current change. For example, the addition of the toxin beyond 8 min after the addition of cDNA (at 22^{0}C) provided results that were the same as if the toxin was absent. This may suggest that hybridization may be effectively complete within 8 min, and suggests that the time course of signal change is dependent at longer times on desorption of dsDNA rather than on hybridization. The time required for the return of the ion current to initial background levels suggests that there is a slow electrostatic and structural relaxation process based on reorganization at the s-BLM surface.

The effect of AFM$_1$ concentration on the initial rate of hybridization was studied at various temperatures ranging from 22 to 35^{0}C. The effects of the toxin became more prominent when increasing amounts of the toxin (up to 6.1 nM) were added at the same time as cDNA. The decrease of the observed initial rate of hybridization (at 35^{0}C) due to the presence of aflatoxin M$_1$ was used to quantify the concentration of the toxin in bulk solution. The reliable concentration range for toxin determination was between 0.5 and 6.1 nM, and the rate of hybridization was linearly related to AFM$_1$ concentration [I (nA) = 0.7286[aflatoxin M$_1$](nM) + 5.10, r^2=0.990]. This approach has allowed the detection of the toxin at low concentrations, and the detection limit (calculated as 3 SD) was found to be 0.5 nM. The addition of larger amounts of the toxin (more that 6.1 nM) increased the time required for complete signal generation by 10 fold, hence, the measurement of the initial rate was impractical.

This work has demonstrated that AFM$_1$ affected the time course of signal development caused by DNA hybridization at s-BLMs and that this phenomenon could be used to develop a detection scheme for this toxin. This approach can be rapid for the purposes of assay if it is based on measuring the decreases of the initial rate of hybridization, which depend on temperature and the concentration of the toxin. Lower concentrations of AFM$_1$ can be determined using the method than can be determined by using the observation of direct effects of the toxin on the electrochemical signals from s-BLMs.

3.2.2. Stopped-flow monitoring of Aflatoxin M_1 in cheese samples using filter-supported BLMs modified with DNA

A technique for the rapid and sensitive electrochemical flow injection monitoring of AFM_1 in cheese samples was developed [7]. Stabilized filter-supported BLMs modified with dT_{20}-C16 were used as detectors. The incorporation of dT_{20}-C16 in BLMs lowered the detection limit for this toxin by one to four orders of magnitude as compared with the detection limit obtained in the absence of ssDNA [35,36]. The detection limit was decreased to 0.05 nM, and was within the levels of interest for practical monitoring. The sensor system allowed a rapid direct determination of the toxin in cheese samples without sample pretreatment at levels equal or lower than those set by both U.S. Food and Drug Administration and most European countries [31]. The method significantly reduced the time required for determination of AFM_1 as compared to other analytical techniques for the determination of this toxin [30,32-34].

The preparation of cheese samples spiked with aflatoxin was done as follows: 1.0 g of cheese sample and a suitable quantity of toxin was dissolved in 10.0 mL of *n*-hexane by sonication for 1 h. The solvent was evaporated by passage of nitrogen gas. 100 mL of 0.1 M KCl buffered with 10 mM HEPES (pH 7.4) was added to the dry extract to dissolve the toxins.

A transient ion current signal as a single event (Fig. 8) was obtained upon injection of samples of AFM_1 in a continuous flowing stream (2.0 mL min^{-1}) of a carrier electrolyte solution (0.1 M KCl, 10 mM HEPES, pH adjusted to 7.4) with BLMs composed from egg PC that were modified with dT_{20}-C16. The magnitude of the transient response was directly proportional to the concentration of AFM_1 in the carrier electrolyte solution: ΔI (pA) = 39.2 $[AFM_1]$ (nM) – 0.127, r^2 = 0.9998. A detection limit of 0.05 nM was observed as determined by a noise level of 0.6 pA and a S/N ratio equal to 3. The variability of response of the BLMs to repetitive sample injections as expressed by the relative

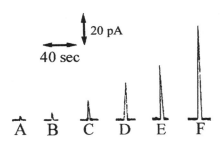

Fig. 8. Results obtained for AFM_1 using a carrier electrolyte of 0.1 M KCl with 10 mM HEPES at a pH of 7.4. BLMs were composed of phosphatidylcholine with adsorbed ssDNA, and were supported in glass microfiber filters. The flow rate was 2.0 mL min^{-1}. AFM_1 concentrations of the injected samples (75 uL) in the carrier electrolyte (nM) were: (A) 0.060; (B) 0.120; (C) 0.300; (D) 0.600; (E) 0.900; (F) 1.50.

standard deviation was on the order of ± 4-8%.

Reports in the literature have appeared that provide evidence that AFB_1 binds at the N^7-position of guanine in DNA, and adducts of the AFB_1-N^7-guanine are formed [37,38]. Various analytical methods have been developed for the detection and determination of the guanine-AFB_1 adduct [38]. There has been no evidence of interaction between ssDNA and AFM_1 [26]. The ssDNA may not act as a selective receptor for AFM_1. Rather, the presence of the ssDNA may alter the electrochemical sensitivity of the BLM so that perturbations of the charge and structure of the membrane are amplified to provide greater changes in signal (ion current).

Matrix effects due to cheese constituents were investigated to adapt the detection strategy to the direct determination of AFM_1 toxin in real samples. Lactose (a milk and cheese disaccharide) was examined as a potential interferent, and this compound was found not to cause any transient signals even at concentrations of 5 mM. Various cheese sample suspensions containing a variable concentration of cheese without AFM_1 were injected into the flowing carrier electrolyte solution to investigate the matrix effects. In the sample preparation process, toxins and proteins were dissolved in the aqueous phase [39] and most of the lipids were not dissolved. The mixture was allowed to stand a few minutes so that the lipids that were present precipitated, but proteoliposomes could still have been in the aqueous phase [40]. The effect of the flow rate of the carrier electrolyte solution on the transient ion current signal was examined. The results have shown that no interference is observed when using flow rates larger than 1.2, 1.0 and 1.7 mL min^{-1} for Edam, soft Greek (Anthotiro) and Feta cheese, respectively, for a cheese concentration of 0.01 g mL^{-1}. It appears that at higher flow rates, the interferences are eliminated, perhaps due to insufficient time of interaction with the lipid membrane and/or shear effects at the interface.

The effect of lactoglobulin (a common protein present in milk and cheese products) has been evaluated by injecting solutions containing variable concentrations of this protein into the flowing carrier electrolyte solution. Our results indicated that no ion current transients from BLM-based sensors were obtained for concentrations of lactoglobulin larger than 0.15% w/v (5 times larger than the cheese samples) at any flow rate. (used 19 ng mL^{-1} dT_{20}-C16).

As shown in Table 2, a number of commercial cheese products have been deliberately contaminated with AFM_1 at various concentrations, and the AFM_1 content has been assayed. Recoveries in the range of ca. 90-110 % have been obtained.

The detection system has shown good signal reversibility and reproducibility for repetitive determinations of AFM_1, with no sample carryover or membrane memory effects (the relative standard deviation was ca. 4-8%). The results have shown that there is an immediate return to the baseline after each measurement. This has permitted repetitive AFM_1 determinations with a rate of at

Table 2
Results of recovery studies of AFM_1 added to commercial cheese preparations (the numbers in parentheses are the added amounts of aflatoxin M_1). Results presented are the average of 5 determinations ± 1sd

Sample ID	Aflatoxin M_1 (nM)
FAGE Dairy Products, S.A., Edam cheese	(0.70) 0.66 ± 0.05
Delta Dairy Products S.A., Anthotiro (soft Greek cheese)	(2.10) 2.00 ± 0.11
Delta Dairy Products S.A., Feta cheese	(1.45) 1.50 ± 0.09

least 4 samples/min for cheese samples. There was no reduction of the signal magnitude during each subsequent cycle of addition. The number of injections performed was ca. 30, and was only limited by the syringe capacity (2.5 mL) of the Hamilton repeating dispenser that was kept on-line.

The toxin AFM_1 was determined in real cheese samples by using stabilized BLM-based sensors that were modified with dT_{20}-C16. The flow system provided very fast response times, on the order of 15 s or less. The detection limits were approximately the same as those obtained by immunoassays or chromatographic methods [30,32–34], and were within the action limits set by the FDA [30,31]. The approach has significant advantages over immunoassays or chromatographic procedures in terms of analysis times, sample volumes and preparation, and size and cost of instrumentation. The BLM biosensor system allowed rapid and direct determination of this toxin in cheese samples without sample pretreatment, significantly reducing the time required for determination of AFM_1.

3.3. Rapid Screening of Atenolol in Pharmaceutical Preparations Based on s-BLMs Modified with DNA

Atenolol [4-[2-hydroxy-3-isopropyl-aminopropoxy]–phenyl-acetamide] belongs to the category of β-blockers and, more specifically, it is a hydrophilic β1-receptor blocking agent. This drug is of therapeutic value in the treatment of various cardiovascular disorders, such as angina pectoris, cardiac arrhythmia and hypertention [41]. β-blockers are exceptionally toxic and most of them have a narrow therapeutic range [42]. Gas chromatography (GC) with mass spectrometry or electron capture detector has been used extensively for the determination of atenolol [42,43]. High performance liquid chromatography (HPLC) has also been extensively used for the determination of atenolol [44,45]. Immunoassays have also been used for the determination of this β-adrenoreceptor blocking agent [46].

s-BLMs that were modified with dT_{20}-C16 were used to construct a sensor of high response speed and sensitivity for detection of atenolol [47]. Such studies of interactions of atenolol with lipid membranes have not been reported in the literature despite the fact that it is well known that a wide range of drugs interact with lipid membranes (for example, see [4]). The interactions of atenolol with s-BLMs provided increases of ion current that reproducibly appeared within a few seconds after exposure of the lipid membrane to atenolol. The magnitude of the current increase was related to the concentration of atenolol in bulk solution in the micromolar range (20-200 uM). The sensors were found to be stable for periods of time greater than 48 hours.

The response times to establish 99% of steady-state current for this sensing device were on the order of a few seconds. The current values were linearly related to atenolol concentration within the range of 20 to 200 uM [ΔI (nA) = 11.1 C (uM) – 108.3, r^2 = 0.992]. The detection limit (for S/N = 3 and for noise levels of 7 nA) was 15 uM. A number of BLM-based sensors (independently prepared according to procedure) were used to determine the reproducibility of fabrication. These experiments indicated that the reproducibility was about ±4 to 8% (N=5, 95% confidence limits).

The pH effect on the signal magnitude was examined in the pH range of 3.0 to 9.0 [48]. The signal was found to be constant within experimental uncertainty for the pH range examined. PC is an amphiphilic molecule above pH 3.0 [49] and atenolol at any value above pH 3.0 is a neutral compound [50].

ssDNA that was allowed to associate with s-BLMs was used to sensitize s-BLMs for the detection of atenolol. Experiments were performed where injections of various concentrations of atenolol in the bulk electrolyte solution were done after incubation and stabilization of s-BLMs with 6.35 ng/ml dT_{20}-C16 in saline solution. Subsequent addition of atenolol standard solution resulted in increases of the ion current when the ssDNA concentration was kept constant. The response time was 18-20 s and the increases of the ion current were linearly related to atenolol concentration [ΔI (nA) = 15.9 C (uM) + 100, r^2 = 0.999] for 6.35 ng/mL of dT_{20}-C16. The detection limit that was established (for S/N = 3 and for noise levels of 7 nA) was 1.8 uM. The detection limit was 10 fold lower than that observed in the absence of ssDNA. Optimization of the analytical signal was achieved by the use of different oligonucleotide concentrations that were used to treat s-BLMs (e.g. 6.35 and 19.0 ng/mL). Larger concentrations (i.e. more than 19.0 ng/mL) resulted in substantially increased background ion current, and limited the useful analytical range. The results indicated that the ion current values were linearly related to atenolol concentration [ΔI (nA) = 15.9 C (uM) + 100, r^2 = 0.999 and ΔI (nA) = 0.958 C (uM) + 17.2, r^2 = 0.990 for 6.35 and 19.0 ng/mL dT_{20}-C16, respectively].

Competitive interference studies were done using both atenolol and interferent together in solution in a ratio of 1:1. The interferents included the most commonly found compounds in pharmaceutical preparations containing atenolol such as magnesium stearate, macrocrystalline cellulose, povidone and

Table 3
Quantification of atenolol in Tenormin tablets (The results presented herein are random from a large number of experiments).

Atenolol given by the manufacturer (mg)	Atenolol Found (mg)
25 mg	25.4
50 mg	51.2
100 mg	99.2

sodium starch glycolate. No significant interferences were noticed from the presence of these compounds.

The quantification of atenolol in Ternomin ("Cana" Ltd., Athens, Greece) tablets containing 25, 50 and 100 mg of atenolol provided the results given in Table 3. The results have shown good agreement with the atenolol composition stated by the manufacturer (Table 3). The recovery studies were done by spiking a known amount of atenolol in solutions of Tenormin. The recovery ranged between ca. 97 to 103%.

In conclusion, the results indicated that sensors based on stabilized BLMs have extremely fast response times (few seconds) for the rapid screening of atenolol, and can be reproducibly fabricated with simplicity and low cost. Furthermore, the results indicate that s-BLMs that are modified by the adsorption of ssDNA oligonucleotides can provide at least a 10-fold improvement of the detection limit of atenolol. The technique may find application in the determination of atenolol in pharmaceutical preparations and for the continuous monitoring and analysis of this compound.

REFERENCES

[1] C. Wittmann and B. Hock, J. Agric. Food Chem., 39 (1991) 1194.
[2] T. Kalab and P. Skladal, Electroanalysis, 9 (1997) 293.
[3] J. Rishpon and D. Ivnitski, Biosens. Bioelectron., 12 (1997) 195.
[4] D. P. Nikolelis, T. Hianik and U. J. Krull, Electroanalysis, 11 (1999) 7.
[5] P.A.E. Piunno, D. Hanafi-Bagby, L. Henke and U.J. Krull, in O. Sadik and A. Mulchandani, eds., Recent Advances in Environmental Chemical Sensors and Biosensors, American Chemical Society Symposium Series, 762 (2000) 257.
[6] C. G. Siontorou, D. P. Nikolelis, P.A.E. Piunno and U. J. Krull, Electroanalysis 9 (1997) 1067.
[7] C. G. Siontorou, D. P. Nikolelis and U.J. Krull, Electroanalysis, 12 (2000) 747.
[8] H. T. Tien and Z. Salamon, Bioelectrochem. Bioenerg., 22 (1989) 211.
[9] M. Zviman and H. T. Tien, Biosens. Bioelectron., 6 (1991) 37.
[10] M. Otto, M. Snejdarkova and M. Rehak, Anal. Lett., 25 (1992) 653.
[11] D. P. Nikolelis, J. D. Brennan, R. S. Brown, G. McGibbon and U. J. Krull, Analyst, 116 (1990) 1221.
[12] C. G. Siontorou, A. M. O. Brett and D. P. Nikolelis, Talanta, 43 (1996) 1137.

[13] D. P. Nikolelis, C. G. Siontorou, V. G. Andreou and U. J. Krull, Electroanalysis, 7 (1995) 531.
[14] D. F. Sargent and T. Hianik, Bioelectrochem. Bioenerg., 33 (1994) 11.
[15] T. Hianik, V. I. Passechnik, D. F. Sargent, J. Dlugopolsky and L. Sokolikova, Bioelectrochem. Bioenerg., 37 (1995) 61.
[16] T. Yoshida, K. Taga, H. Okabayashi, H. Kamaya and I. Ueda, Biochim. Biophys. Acta, 984 (1989) 253.
[17] O. G. Mouritsen and M. Bloom, Biophys. J., 46 (1984) 141.
[18] J. Zeng, D. P. Nikolelis and U. J. Krull, Electroanalysis, 11 (1999) 770.
[19] F. Azek, C. Grossiord, M. Joannes, B. Limoges and P. Brossier, Anal. Biochem., 284 (2000) 107.
[20] F. Kleinjung, F.F. Bier, A. Warsinke and F.W. Scheller, Anal. Chim. Acta, 350 (1997) 51.
[21] A. J. R. Teixeira, J. C. Klein, H. A. Heus and A. P. J. M. de Jong, Anal. Chem., 67 (1995) 399.
[22] C. G. Siontorou, D. P. Nikolelis, B. Tarus, I.Dumbrava and U. J. Krull, Electroanalysis, 10 (1998) 691.
[23] C. G. Siontorou, D. P. Nikolelis and U. J. Krull, Anal. Chem., 72 (2000) 180.
[24] G. E. Collins and S. L. Rose-Pehrsson, Analyst, 119 (1994) 1907.
[25] A. Safavi and M. R. Baezzat, Anal. Chim. Acta, 358 (1998) 121..
[26] C. G. Siontorou, D. P. Nikolelis, A. Miernik and U. J. Krull, Electrochim. Acta, 43 (1998) 3611.
[27] G. N. Wogan and S. Paglialunga, Food Cosmet. Toxicol., 12 (1974) 381.
[28] C. H. Schlatter, in Symposium on Health Risks by Aflatoxins, Institute of Toxicology, Schwerzenbach, Switzerland, 1979, p. 51.
[29] W. Muecke, Ernaehr.-Umsch., 25 (1978) 9.
[30] C. De Boevere and C. Van Peteghem, Anal. Chim. Acta, 275 (1993) 341.
[31] G.-S Qian, P. Yasei and G.C. Yang, Anal. Chem, 56 (1984) 2079.
[32] E. Ioannou-Kakouri, M. Christodoulidou, E. Christou and E. Constandinidou, Food Agric. Immun., 7 (1995) 131.
[33] T.J. Hansen, J. Food Prot., 53 (1990) 75.
[34] S. Diaz, M.A. Moreno, L. Dominguez, G. Suarez and J.L. Blanco, J. Dairy Sci., 76 (1993) 1845.
[35] V.G. Andreou, D.P. Nikolelis and B. Tarus, Anal. Chim. Acta, 350 (1997) 121.
[36] V.G. Andreou and D.P. Nikolelis, Anal. Chem., 70 (1998) 2366.
[37] D. Wong, Ed., Mechanism and Theory in Food Chemistry, Van Nostrad Reinhold, New York, 1989, p. 300.
[38] T. Vidyasagar, N. Sujatha ansd R.B. Sashidhar, Analyst, 122 (1997) 609.
[39] S. Williams, ed., Official Methods of Analysis of the Association of Official Analytical Chemists, Association of Official Analytical Chemists, Inc., 14th Edition, 1984, Arlington, Virginia, USA, p. 491.
[40] Alan Townshend, ed., Encyclopedia of Analytical Chemistry, Academic Press Limited, London, 4 (1995) 2520.
[41] H. Winkler, W. Ried and B. Lemmer, J. Chromatogr. Biomed Applic., 228 (1982) 223.
[42] H.Siren, M. Saarinen, S. Hainari and M.L. Riekkola, J. Chromatogr. Biomed. Applic., 632 (1993) 215.
[43] M. Ervik, K. Kylberg-Hanssen and P. Lagerstrom, J. Chromatogr. Biomed. Applic., 182 (1980) 341.
[44] M.S. Leloux and F. Dost, Chromatographia, 32 (1991) 429.

[45] M.T. Rosseel, A.N. Vermeulen and F.M. Belpaire, J. Chromatogr. Biomed. Applic., 568 (1991) 239.
[46] T. Kaila and E.J. Lisalo, Clin. Pharmacol., 33 (1993) 959.
[47] D. P. Nikolelis, M. V. Mitrokotsa and S.-S. E. Petropoulou, Bioelectrochem., in press.
[48] D.P. Nikolelis, J.D. Brennan, R.S. Brown and U.J. Krull, Anal. Chim. Acta, 257 (1992) 49.
[49] The Merck Index, 12th Edition, Merck & Co., Inc., NJ, USA,(1996) 145-146.

Planar Lipid Bilayers (BLMs) and their Applications
H.T. Tien and A. Ottova-Leitmannova (Editors)
© 2003 Elsevier Science B.V. All rights reserved.

Chapter 29

Structure and electrochemistry of fullerene lipid-hybrid and composite materials

Naotoshi Nakashima

Department of Materials Science, Graduate School of Science and Technology, Nagasaki University, Bunkyo, Nagasaki 852-8521 Japan

1. INTRODUCTION

In this review article, the structure and aqueous electrochemistry for films of fullerene lipids (fullerene-lipid hybrid) and of fullerene C_{60}/lipid composites are described.

Increasing attention is being given to the chemistry and physics of fullerenes including higher fullerenes and metallofullerenes[1-6]. Synthetic lipid bilayer membranes possess fundamental physico-chemical properties similar to those of biomembranes and can be immobilized onto a variety of substrates as multibilayer films by several means[7-9]. We[10-21] and others[22-35] have been interested in combining fullerene chemistry and the chemistry of lipid bilayer membranes, which may create a novel field in the chemistry and biochemistry of fullerenes. The formation of thin fullerene films including fullerene-lipid hybrid and fullerene-lipid composite bilayer membranes is of interest both from a fundamental and practical application point of view[10-41]. We have reported the design, synthesis and characterization of C60-bearing triple-chain lipids **1-3** (Fig. 1), where **1-3** possess a triple alkyl chain of C_{16}, C_{14} or C_{12}, respectively[10,14,20]. In this article, i) the structure and the phase transition behavior for thin films of **1-3**, ii) the formation of monolayers and Langmuir-Blodgett (LB) films of a fullerene lipid and their properties, and iii) the construction of electrode devices modified with fullerene lipid films and of fullerene/lipid composite films and their unique electrochemical behavior are briefly reviewed. We have earlier described that synthetic lipid bilayer membrane-modified electrodes provide unique electrochemical behavior based on the properties of the lipid bilayer membranes[42-53].

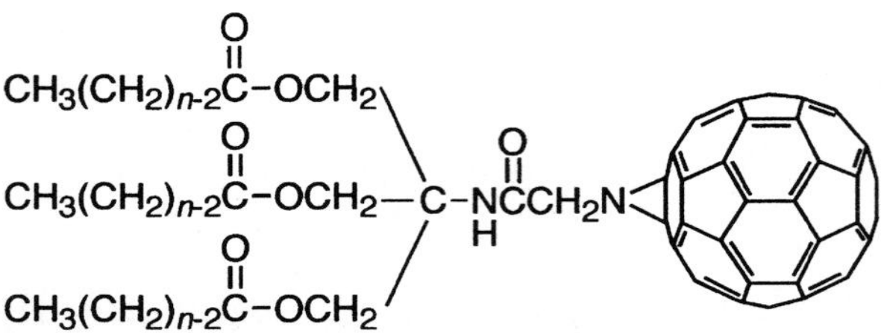

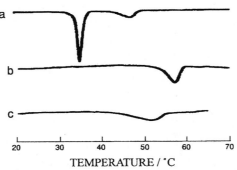

1 (*n*=16)
2 (*n*=14)
3 (*n*=12)

Fig. 1. Chemical structure of fullerene lipids **1-3**.

2. STRUCTURE OF FULLERENE LIPID BILAYERS

Fullerene lipids **1-3** are not soluble in water, whereas they are soluble in a variety of organic solvents such as DMF, DMSO, chloroform, benzene, and hexane. We can prepare cast films of **1-3** readily from organic solutions. The obtained films were characterized by means of differential scanning calorimetry (DSC), UV-visible and FTIR absorption spectroscopies and X-ray diffraction study.

The phase transition of lipid bilayers between crystalline and liquid crystalline phases is one of the most fundamental characteristics of lipid bilayer membranes and the lipid bilayer properties depend on the fluidity change of bilayers[54]. A typical DSC thermogram for a cast film of **1** in air is shown in Fig. 2 (trace a)[10]. Similar thermograms were obtained for a **1**-film in the

a

b

c

20   30   40   50   60   70

TEMPERATURE / °C

Fig. 2. DSC thermograms of **1**(a), **2**(b) and **3** (c).

presence of water, water containing an electrolyte or acetonitrile[10,20]. The major endothermic peaks arise from a transition between crystalline and liquid crystalline states similar to what has been observed for aqueous bilayer membranes[54,55], and the minor peaks observed near 48 °C is due to the structural change of the C_{60} moieties in the film. It is interesting to note that the phase transition behavior of **1** in various media was not much different. This means that the lipid possesses similar molecular orientation in these media. We have earlier reported that cast films of artificial lipids could retain an organized lipid bilayer structure and phase transition behavior in organic solvents[56]. DSC thermograms of cast films of **2** and **3** exhibit endothermic peaks at 57.0 and 51.5 °C, respectively (Fig. 2, traces b and c), which would be ascribable to the phase transitions of the fullerene moieties as is discussed later.

For aqueous liposomal[54] and synthetic lipid bilayer[55] membranes, the phase transition temperature decreases with a decrease in the alkyl chain length by ca. 13-18 °C/two methylene groups. The phase transition behavior of **2** and **3** differs from liposomal and synthetic lipid bilayer membranes since we could not detect a main transition peak like a **1**- film even when the DSC measurements were carried out in the range of –20 - 70°C.

Temperature dependent FT-IR and UV-visible spectra for cast films of **1-3** in air explain the DSC thermograms of the lipids[20]. The wavenumber of the asymmetric and symmetric methylene stretching vibrations in the FT-IR spectra of the **1** film changed drastically near 35 °C. The shifts of $v_{as}(CH_2)$ from 2917.9 to 2922.0 cm^{-1} and $v_s(CH_2)$ from 2849.2 to 2851.0 cm^{-1} are ascribable to the *trans-gauche* conformational change of the long alkyl chain[57], which leads to the phase transition of the thin film. The FT-IR result indicates that the main peak in the DSC thermogram of the **1** film is attributable to the bilayer phase transition typically observed for liposomal and synthetic lipid bilayer membranes. Frequencies $v_{as}(CH_2)$ and $v_s(CH_2)$ appears at 2922.7 and 2851.3 - 2851.8 cm^{-1} for a **2**-film and at 2921.7 and 2851.3 - 2851.8 cm^{-1} for a **3**-film and the spectra for both films showed no temperature dependence. The result shows the formation of *gauche* conformation in the alkyl chains in these films.

The UV-visible spectrum of a **1**-film in air at temperatures below 35 °C showed three bands with absorption peaks at 215.5, 262.5, and 329.5 nm, respectively; the peak maxima shifted to longer wavelength upon increasing the temperature above the subphase transition, thus indicating the existence of an electronic interaction between the C_{60} moieties in the **1** film. Similar temperature dependent spectral behavior was observed for films of both **2** and **3**. The breaks observed near 55 °C for the **2**-film and 50 °C for the **3**-film are close to the phase transition temperatures of the corresponding films determined by DSC. It is evident that the phase transition of the film regulates an electronic interaction between the C_{60} moieties on **1-3**.

Cast films of **1** form multi-bilayer structure with the molecular layer tilting by 42.8 degree from the basal plane[10]. X-ray diffraction diagrams for cast films of **2** and **3** gave diffraction peaks at $2\theta = 1.97$ and 2.12 degree, respectively, which can be assigned to the (001) plane of these films. By assuming the molecular length of **2**(3.28 nm) and **3** (3.04 nm) estimated from CPK space filling models, it is clear that the cast films of both **2** and **3** form a biomembrane-mimetic multi-bilayer structure[20] as is the case for **1**.

The X-ray diagrams of a **1**-film showed temperature dependence. The diffraction peak of the **1**-film was maintained at temperatures below 35 °C (which is the main transition of the film), but upon increasing temperature to 50 °C, a temperature higher than the subphase transition temperature, the diffraction peak almost disappeared. The structure of the **1**-film at the higher temperatures was found to be rather disordered. The observed temperature dependence was reversible.

The results of DSC when combined with the spectral and X-ray data suggest a possible schematic model for the structure and phase transition behavior for cast films of **1-3** (Fig. 3)[20]. Phase A consists of the lipids with a rigid crystalline state, in which all alkyl-chains form *trans* conformation. Lipid **1** only forms this state. Phase B is a fluid state in which the alkyl-chains contain

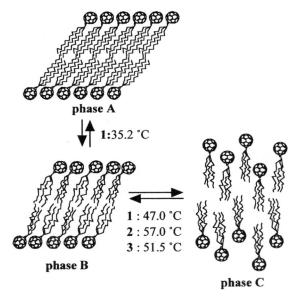

Fig. 3. A schematic model for the phase transition of **1-3**.

a *gauche* conformation while the orientation of the fullerene moieties remains constant. Phase C is a fluid state with a less ordered structure than phases A and B.

3. MONPLAYER AND LANGMUIR-BLODGETT FILMS OF A FULLERENE LIPID

The construction of fullerene ultrathin films with an ordered structure is quite interesting and important for the application and utilization of fullerenes[36-41,58-70]. The derivatives of fullerenes C_{60} and C_{70} form monolayers at the air-water interface and the formed monolayers can be transferred onto solid substrates[58-70]. Single component of C_{60} and C_{70} form monolayers on water under limited experimental conditions[66-68]. Here, stable monolayer formation from the fullerene lipid **1** at the air-water interface and the deposition of the monolayer onto a solid substrate as an LB film which possesses a phase transition are presented.

The surface pressure-area isotherms of **1** from the spreading solution of 1.0×10^{-4} M gave a limiting area of 0.27 nm²/molecule (Fig. 4). This suggests the formation of a multilayer. In contrast, the spreading from 1×10^{-5} M solution produces a monolayer on water. The limiting areas obtained from the higher and the lower pressure regions are 0.78 and 0.98 nm²/molecule, respectively, suggesting the formation of the monolayers with the hexagonal or the simple square packing, respectively. In general, spreading from diluted solutions ($<10^{-5}$ M) is required to form monolayers of fullerenes on water; instead, multilayers

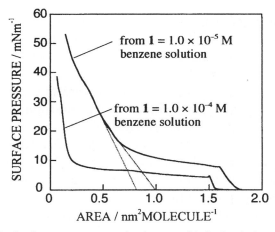

Fig. 4. Surface pressure-area isotherms of **1** in the dark condition.

are formed from the spreading of higher concentrations (>10^{-4} M).

The monolayer of **1** can be transferred onto solid substrates such as a quartz plate. The UV-visible absorption spectra of the LB films of **1** prepared in the dark (trace a) show a temperature dependence (Fig. 5). The shift to longer wavelengths at the lower temperatures compared to those in hexane solution suggests the existence of an electronic interaction between the fullerene moieties in the LB film. The peak maxima drastically changed around 47 °C, which is identical with the subphase transition temperature of the cast films of **1** (Fig. 6). The temperature dependence was reversible. In contrast, the LB film prepared in the light showed structureless electronic spectra with almost no temperature dependence. This may be due to the light-driven structural change in the fullerene moiety on **1**. Since cast films of **1** did not show such behavior, higher ordered molecular orientation in the ultrathin films is found to be sensitive to the weak light.

The molecular orientations of the fullerene moieties of the LB film and cast film of **1** were not significantly different. The fundamental property of the self-assembled bilayer membrane film was maintained in the LB film. The introduction of bilayer properties to the LB films of fullerenes may be useful for many applications.

4. ELECTROCHEMISTRY OF FULLERENE LIPID FILMS

Fullerenes form multiply-charged anions because of their high degrees of degeneracy of the LUMO, which lead to a variety of unique functions[1-6];however, the electrochemistry of fullerene thin films is complicated[71,72]. Here, the stable and unique electron transfer reactions of fullerene lipid-coated electrodes are presented.

The cyclic voltammograms (CVs) of a cast film of **1** on a basal plane graphite (BPG) electrode exhibit well-defined redox waves leading to the generation of fullerene dianion in water containing tetra-n-butylammonium chloride (TBAC) as an electrolyte (Fig. 7)[20]. The first redox process is stable on the potential cycling but the second reduction current is not so stable. The **2**- and **3**- modified electrodes gave similar cyclic voltammetric behavior. The formal potential (E^0) for the modified electrodes are not much different from each other. In contrast, electroactive amounts of the fullerene moiety on **1-3** increased with the decrease in the alkyl-chain length. The electroactive amounts C$_{60}$ moiety for **1-3** electrodes at 60 °C are 6.7, 21.3 and 48.0%, respectively. The fullerene moieties in the films were found to be disposed in a way to undergo facile electrochemical communication with the underlying electrodes[20].

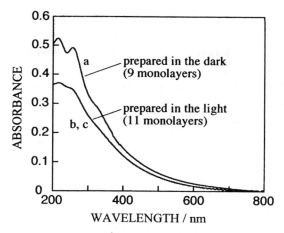

Fig. 5. UV-visible absorption spectra of LB films of **1** on quartz plates. Measured temperature: 25 °C for a and b, 50 °C for c.

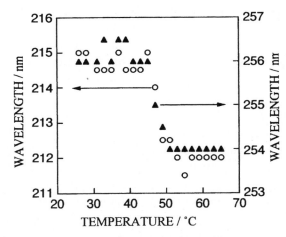

Fig. 6. Absorption peaks in the UV-visible spectra of a 1–LB film prepared in the dark condition *vs* temperature.

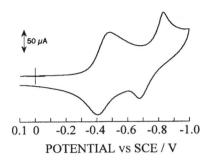

POTENTIAL vs SCE / V

Fig. 7. Cyclic voltammogram for a **1** film-coated electrode in water
containing 0.5 M TBAC at 55 °C. Scan rate, 100 mV/s.

For **1**-, **2**- and **3**-modified electrodes, both the cathodic and anodic peak
current were proportional to the potential sweep rates, υ, in the range of 10 - 30
mV/s, as expected for thin-layer electrochemical behavior. In contrast, the
currents were proportional to the square root of υ at 100 - 3,000 mV/s, as
expected for a diffusion-controlled process.

Osteryoung square-wave voltammetry was used to evaluate the effect of the
phase transition of the fullerene films. The temperature dependence of the
reduction current in the Osteryoung square-wave voltammograms (OSWVs) of
a **1**-modified electrode in water containing TBAC as a function of temperature
is shown in Fig. 8. The breaks near 35 and 48 °C are close to the respective
main and subphase transitions of **1**-film. This indicates that the phase transition
of the **1**-film regulates the electrochemistry of the fullerene modified
electrode[20].

Both **2**- and **3**-modified electrodes show an unexpected temperature
dependent OSWVs. At temperatures below 25 °C, the modified electrodes
showed almost no faradaic current and the breaks were seen near 30 °C, which
is not consistent with the transition temperatures of **2** and **3** obtained by DSC.
This temperature dependence might be derived from some structural changes of
the films that are undetectable by DSC. The observed temperature dependence
for the modified-electrodes was reversible. The construction of redox-
switchable molecular devices based on fullerene films would be of interest for
the application and utilization of fullerenes.

Thermodynamical treatment for the binding between the electrogenerated
fullerene anions in the films and electrolyte cations was possible[15].
Theoretical treatment for the binding was made with the following eqn(1) and
eqn(2),

$$E_{1/2,1} = E_1^{0'} + \frac{RT}{F} \ln K_1 + \frac{pRT}{F} \ln c \qquad (1)$$

$$E_{1/2,2} = E_2^{0'} + \frac{RT}{F} \ln \frac{K_2}{K_1} + \frac{(q-p)RT}{F} \ln c \qquad (2)$$

where $E_{1/2,1}$ and $E_{1/2,2}$ are the half-wave potential for the first and second redox processes, E_1^0 and E_2^0 are the standard redox potentials for the two elemental redox processes, p and q are numbers of cations bound to fullerene anions in the first and second reduction processes, respectively, K_1 and K_2 are the binding constants between the fullerene radical monoanions and the electrolyte cations, respectively, and c is concentrations of electrolytes.

The electrolytes used were: tetramethylammonium chloride (TMAC), trimethybenzylammonium chloride (TMBAC), TEAC, tetraethylphosphonium chloride (TEPC), triethylbenzylammonium chloride (TEBAC), tetra-*n*-propylammonium chloride (TPAC), tri-*n*-butylbenzylammonium chloride (TBBAC), TBAC and tetra-*n*-butylphosphonium chloride (TBPC). The electrochemistry for the modified electrodes was examined using differential pulse voltammetry. The results are summarized in Table 1. Similar electrolyte dependence was observed for **2**- and **3**-modified electrodes. The interesting features are as follows.

i) For all cations used, p is almost 1 and q is almost 2.

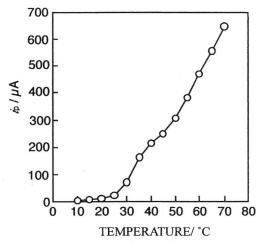

Fig. 8. Temperature dependence of cathodic current in the DPVs of a **1**-coated electrode device.

Table 1. Parameters for the binding of **1** anions and alkylammonium and phosophonium ions at the reduction processes of **1** films on BPG at 328K.

Electrolyte cation	p	q	K_1 / M^{-1}	$\Delta G_{1,328K}$ / kJ mol^{-1}
TMAC	1.03	—	3.80×10	-9.9
TMBAC	1.08	—	5.24×10^3	-23.4
TEAC	1.05	—	7.93×10^3	-25.7
TEPC	1.16	—	5.00×10^4	-29.5
TEBAC	1.04	—	1.79×10^5	-34.6
TPAC	1.10	2.15	0.96×10^6	-37.6
TBBAC	0.97	1.99	1.01×10^8	-50.3
TBAC	0.97	2.11	1.71×10^8	-50.6
TBPC	1.06	1.98	1.70×10^9	-58.0

ii) K_1 shows a strong alkyl chain length dependence. This is in sharp contrast with the binding of electrochemically generated fullerene anions in organic solvents with tetraalkylammonium cations, where no alkyl chain length dependence has been reported[73,74]. K_1 values for tetraalkylphosphonium cations are ca. one-order of magnitude greater than those for tetraalkylammonium cations with the same alkyl chain length. The softness of tetraalkylphosphonium cations would explain this stronger complex formation, since fullerene anions are soft anions.

iii) For the binding with fullerene radical anions, introduction of a benzyl moiety to the electrolyte shows stronger effect than for methyl or ethyl groups but a weaker effect than for butyl group.

5. STRUCTURE AND ELECTROCHEMISTRY OF FULLERENE/LIPID COMPOSITE FILMS

Composite films of C$_{60}$ and artificial lipids possess multi- bilayer structures. For example, the X-ray diffraction diagram for a cast film of C$_{60}$/dihexadecyldimethylammmonium poly(styrene sulfonate) (**4**) (molar ratio, 1/19) showed a diffraction peak at $2\theta = 1.8$ degree, which is identical with that for a cast film of a single component of the lipid[12]. The d-spacing calculated to be 4.8 nm from the Bragg s equation that is shorter than two-fold of the molecular length of **4**, suggesting that the film forms multi-bilayer structure with the molecular layers tilting by 47 degree from the basal plane.

We have reported that C_{60} embedded in cast films of cationic lipids on electrodes show evident electron transfer reactions[12-14, 16, 18, 19]. Here the results are summarized.

The electrogenerated C_{60} radical monoanion at the electrode modified with C_{60} alone in the aqueous system has been reported to be unstable[69,70]. CVs for a C_{60} embedded in a cast film of an artificial poly(ion-complexed) lipid, ditetradecyldimethylammonium poly(styrene sulfonate)(**5**) on an electrode show two reversible electron transfer in an aqueous solution and that the generated radical monoanion and the dianion are very stable (Fig. 9a). In contrast, the electrogenerated C_{60} radical monoanion at the electrode modified with C_{60} alone in the aqueous system has been reported to be unstable. The voltammograms for a cast film of $C_{60}/5$ on an electrode gave two couples of reversible redox waves

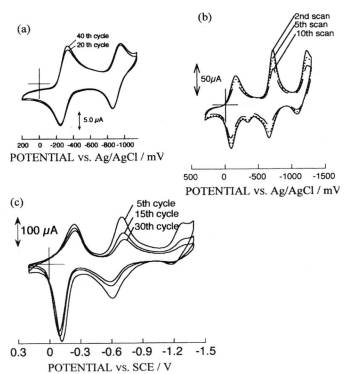

Fig. 9. Cyclic voltammograms for electrodes modified with films of $C_{60}/5$ (a),$C_{60}/6$ (b) and $C_{60}/7$ (c) in 0.5 M tetraethylammonium chloride aqueous solution. Scan rate, 0.1 V/s.

with E_1^0 =-290 and E_2^0 =-910 mV. The CVs were very stable; both the first and second redox couples did not change even after 50 cycles.

Since the electrochemical reduction of cast films of C_{60} involves incorporation of cations, supporting electrolytes are known to influence the voltammograms[67,68]. In our system, however, the CVs for cast films of C_{60}/**5** on BPG were essentially the same in 0.5 M NaCl, KCl and $CaCl_2$ aqueous solutions. The results suggest that these electrolyte cations are not major counter cations of the radical monoanion and the dianion of the fullerene, instead, ditetradecyldimethylammonium in **5** may act as the counter cation during the reduction of the fullerenes.

The cyclic voltammograms for a C_{60}/tridodecylmethylammonium bromide (**6**)-coated electrode gave three couples of almost reversible redox waves (Fig. 9b)[11]. The observed formal potentials are E_1^0 =-0.12, E_2^0 =-0.74 and E_3^0 =-1.21 V and the obtained E_1^0 and E_2^0 shifted positive by ca. 0.17 V relative to those of the C_{60}/**5**-coted electrode described above. This shift would be due to stronger binding between the fullerene anions and tridodecylmethylammonium cations and make possible to detect three reduction peaks within the potential window in the aqueous system. The radical monoanion of C_{60} generated in this system was unusually stable, that is, virtually no change in the CVs for the first redox wave was observed even after a 30 min hold at -0.5 V. The hold at -1.0 V, where the dianions of the fullerenes are generated caused a gradual decrease in the current, suggesting the formation of electro-inactive films on the electrodes. The radical trianion generated in this system were not so stable.

The chemistry and physics of organic gels that have potential applications in the construction of intelligent material systems are also of an exciting area. Organic gels are known to provide suitable microenvironments for the electrochemistry and regulated electrochemistry of some redox active molecules. The CVs for a gel-like membrane of C_{60}/tetraoctylphosphonium bromide (**7**) on a BPG electrode showed three redox couples corresponding to C_{60}/C_{60}^-, C_{60}^-/C_{60}^{2-} and C_{60}^{2-} / C_{60}^{3-} in the potential window in the aqueous medium (Fig. 9c)[16]. This result shows that the electrochemistry of fullerene C_{60} incorporated in a gel-like membrane on an electrode is possible. The formal potentials for electrochemistry of C_{60} in the membrane were E_1^0 =-0.17, E_2^0 =-0.66 and E_3^0 =-1.23 V, which are close to those for the C_{60}/**6**-coated electrode. The voltammograms were stable for potential cycling over the first two redox waves leading to the generation of C_{60}^{2-}. The amount of reacted C_{60} calculated from the cathodic current at the first reduction in the CV at a scan rate of 2 mV/s was 1.3×10^{-8} mol/cm^2, indicating that 63% of C_{60} in the film was electroactive. The finding affords the opportunity to undergo electrochemical communication of fullerenes and related materials at gel-modified electrode systems that possess many possible applications in a variety of areas.

Reports[75,76] describing the electrochemical properties of higher fullerenes have been limited due to their low abundance compared with the more readily available C_{60} and C_{70}. From the isolated pentagon rule, twenty-four isomers of C_{84} are predicted to exist and major two isomers of C_{84} with molecular symmetry D_2 and D_{2d} have been isolated and used for experiments. In organic solution, C_{84} has been reported to show four or five of reduction processes[77-80], of which the first three reduction processes are stable and the last two being unstable.

The CVs for a cast film of single component of C_{84} on an electrode gave no well-defined redox current in aqueous solution. In contrast, well-defined multiple electron transfer reactions for a composite film of C_{84}/tetaoctylammonium bromide (**8**) cast on an electrode occur (Fig. 10). The electrode gives multiple cathodic peaks together with their corresponding anodic peaks. The formal potentials for the six redox couples estimated from their CVs at 5 mV/s are: +477, +204, -145, -254, -636, -900 and -1,110 mV. Since C_{84} exists as two isomers, the assignment of the multiple redox process is complex. The large positive shift of the formal potentials of C_{84} at the modified electrode compared to those in organic solution[80] suggests the strong binding of C_{84} anions with tetraoctylammonium cation in the film[15]. Hydrophobic microenvironments afforded by the tetraoctylammonium matrix contribute to the generation and stability of C_{84} anions.

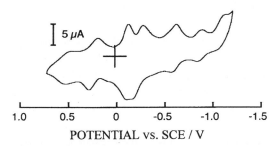

POTENTIAL vs. SCE / V

Fig. 10. Cyclic voltammogram for a C$_{84}$/**8** film-coated electrode. Scan rate, 5 mV/s.

6. CONCLUSION AND OUTLOOK

Combination of fullerene and lipids have been focused. The fullerene lipids formed multibilayer structure films that possess phase transitions attributable to the lipid bilayer phase transition typically observed for liposomal and synthetic lipid bilayer membranes and/or to the change in orientation of the C_{60}-moieties. We could schematically show the phase transitions of the fullerene lipid films. Regulated electrochemistry at the fullerene lipid-modified electrodes was possible. The electrogenerated radical monoanions and dianions of the fullerene lipids were found to bind strongly with soft electrolyte cations with large size tetra-n-butylammonium and tetra-n-butylphosphonium ions. We have pointed out the factors that play important role for the complex formation in aqueous systems.

We have discovered that in an aqueous system, both C_{60} and the higher fullerene C_{84} embedded in cast films of cationic lipids and C_{60} embedded in an organic gel-like membrane film on electrodes undergo multi-step electron transfer reactions with the electrodes. We have pointed a factor that plays important role for the generation of multiply-charged fullerene anions at the electrode devices. Very recently we have found that a metallofullerene La@C_{82}/cationic lipid-coated electrode shows evident electron transfer reactions in aqueous solution[81].

Although we have not described here, the discovery that a simple C_{60}-carrying ammonium amphiphile forms both fibrous and disk-like superstructures with 10–12 nm of thickness is of interest since the superstructure is formed through self-organization of the amphiphile in aqueous solution[82,83]. Fullerene-based biomembrane-like carbon-nanosuperstructures created in aqueous solution may be adapted for biochemical and biological use.

Finally, we would like to emphasis that the studies presented here guide us toward the construction of fullerene lipid bilayer devices, which might be useful in the field of nanoscience and nanotechnology of fullerenes.

REFERENCES

[1] W. E. Billups and M. A. Ciufolini (eds.), Buckminsterfullerenes , VCH , NY, 1993.
[2] K. Prassides(ed.), Physics and Chemistry of the Fullerenes, Kluwer Academic Publishers, Boston, 1994.
[3] H. W. Kroto, The Fullerenes; New Horizons for the Chemistry, Physics and Astrophysics of Carbon, Cambridge University Press, Cambridge, 1997.

[4] H. S. Nalwa (ed.), J. Chlistunoff, D. Cliffel, A. J. Bard, Handbook of Organic Conductive Molecules and Polymers, Vol. 1, John Wiley & Sons, Chichester, 1997, pp 333-412.

[5] A. Hirsch (ed.), Fullerenes and Related Structure, Springer, Berlin, 1999.

[6] K. M. Kadish and R. S. Ruoff (eds), Fullerenes: Chemistry, Physics and Technology, Wiley-Interscience, NY, 2000.

[7] N. Nakashima, R. Ando and T. Kunitake, Chem. Lett., 1983, 1577.

[8] T. Kunitake, A. Tsuge and N. Nakashima, Chem. Lett., 1984, 1783.

[9] N. Nakashima, M. Kunitake, T. Kunitake, S. Tone, T. Kajiyama, Macromolecules, 18 (1985) 485 and references cited therein.

[10] H. Murakami, Y. Watanabe and N. Nakashima, J. Am. Chem. Soc., 118 (1996) 4484.

[11] N. Nakashima, T. Kuriyama, T. Tokunaga, H. Murakami and T. Sagara, Chem. Lett. 1998, 663.

[12] N. Nakashima, Y. Nonaka, T. Nakanishi, T. Sagara and H. Murakami, J. Phys. Chem., B 102 (1998) 7328.

[13] N. Nakashima, T. Tokunaga, T. Nakanishi, H. Murakami and T. Sagara, Angew. Chem. Int. Ed., 37 (1998) 2671.

[14] T. Nakanishi, H. Murakami and N. Nakashima, Chem. Lett., 1998, 1219.

[15] T. Nakanishi, H. Murakami, T. Sagara and N. Nakashima, J. Phys. Chem. B, 103 (1999) 304.

[16] T. Nakanishi, H. Murakami, T. Sagara and N. Nakashima, Chem. Lett., 2000, 340.

[17] N. Nakashima, T. Ishii, M. Shirakusa, T. Nakanishi, H. Murakami and T. Sagara, Chem. Eur. J., 7 (2001) 1766.

[18] H. Ohwaki and N. Nakashima, Trans. Mater. Res. Soc. Jpn, 26 (2001) 933.

[19] N. Nakashima, N. Wahida, M. Mori, H. Murakami and T. Sagara, Chem. Lett., 2001, 748.

[20] M. Morita, H. Murakami,T. Sagara and N. Nakashima, Chem. Eur. J., 2002, in press.

[21] N. Nakashima, M. Sakai, H. Murakami, T. Sagara, T. Wakahara and T. Akasaka, J. Phys. Chem. B, 2002, in press.

[22] H. Hungerb hler, D. M. Guldi and K.-D. Asmus, J. Am. Chem. Soc., 115 (1993) 3386.

[23] R. V. Bensasson, J. -L. Garaud, S. Leach, G. Miquel and P. Seta, Chem. Phys. Lett., 141 (1993) 141.

[24] J. L. Garaud, J. M. Janot, G. Miquel and P. Seta, J. Membr. Sci., 259 (1994) 91.

[25] S. Niu and D. Mauzerall, J. Am. Chem. Soc., 118 (1996) 5791.

[26] J. M. Janot, P. Seta, R. V. Bensasson and S. Leach, Synthetic Metals, 77 (1996) 103.

[27] H. T. Tien, L.-G. Wang, X. Wang and A. L. Ottova, Bioelectochem. Bioenerg. , 42 (1997) 161.

[28] M. Hetzer, S. Bayerl, X. Camps, O. Vostrowsky, A. Hirch and T. M. Bayerl, Adv. Mater., 9 (1997) 913.

[29] M. Cassell, C. L. Asplund and J. M. Tour, Angew. Chem. Int. Ed., 38 (1999) 2403.

[30] M. S. F. Lie Ken Jie and S. -H. W. Cheung, Lipids, 34 (1999) 1223.

[31] M. Sano, K. Oishi, T. Ishii and S. Shinkai, Langmuir, 16 (2000) 3773.

[32] M. Brettreich, S. Burghardt, C. B ttcher, T. Bayerl, S. Bayerl and A. Hirsch, Angew. Chem. Int. Ed., 39 (2000) 1845.

[33] D. Jiang, J. Li, P. Diao, Z. Jia, R. Tong and H. T. Tien, J. Photochem. Photobiol., A, 132 (2000) 219.

[34] M. Braun, X. Camps, O. Vostrowsky, A. Hirch, E. Endress, T. Bayerl, M. Thomas, O. Birkert, G. Gauglits, Eur. J. Org. Chem., 2000, 1173.

[35] H. Gao, G. Luo, J. Feng, A. L. Ottova and H. T. Hien, J. Electroanal. Chem., 496 (2001) 158.

[36] J. Chlistunoff, D. Cliffel and A. J. Bard, Thin Solid Films, 257(1995) 166.

[37] S. E. Campbell, G. Luengo, V. I. Srdanov, F. Wudl and J. N. Israelachvili, Nature, 382, (1996) 520.

[38] P. Wang, Y. Maruyama and R. M. Metzger, Langmuir, 12 (1996) 3932.

[39] V. V. Tsukruk, M. P. Everson, L. M. Lander and W. J. Brittain, Langmuir, 12(1996) 3905.

[40] M. I. Sluch, I. D. W. Samuel, A. Beeby and M. C. Petty, Langmuir, 14, 3343, (1998).

[41] H. Imahori, H. Yamada, Y. Nishimura, I. Yamazaki and Y. Sakata, J. Phys. Chem., 104, 2000, 2099.

[42] N. Nakashima, K. Yamashita, T. Jorobata, K. Tanaka, K. Nakano and M. Takagi, Anal. Sci., 2 (1986) 589.

[43] N. Nakashima, K. Nakano, K. Yamashita and M. Takagi, J. Chem. Soc., Chem. Commun., 1989, 1441.

[44] N. Nakashima, H. Eda, M. Kunitake, O. Manabe and K. Nakano, J. Chem. Soc.,Chem. Commun.,1990, 443 and 1992, 455.

[45] N. Nakashima, K. Masuyama, M. Mochida, M. Kunitake and O. Manabe, J. Electroanal. Chem., 319 (1991) 355.

[46] T. Kunitake, A. Tsuge and N. Nakashima, Chem. Lett., 1984, 1783.

[47] N. Nakashima, S. Wake, T. Nishino, M. Kunitake and O. Manabe, J. Electroanal.Chem., 333 (1992) 345.

[48] N. Nakashima, Y. Yamaguchi, S. Kawahara, M. Kunitake and O. Manabe, Denki Kagaku (Electrochemistry), 60 (1992) 1156.

[49] F. A. Schultz and I. Taniguchi (eds.), N. Nakashima, K. Masuyama, Proc. Fifth Int. Symposium on Redox Mechanisms and Interfacial Properties of Molecules of Biological Importance, The Electrochemical Socity, Inc. Pennington, NJ, 1993, pp. 280-289.

[50] N. Nakashima and T. Taguchi, Interfacial Design and Chemical Sensing, ACS Symp. Ser. 561, Amer. Chem. Soc., Washington DC, 1994, pp.145-154.

[51] N. Nakashima and T. Taguchi, Colloids Surfaces A, 103 (1995) 159.

[52] N. Nakashima and Y. Yamaguchi, J. Electroanal. Chem., 384 (1995) 187.

[53] N. Nakashima, Y. Yamaguchi, H. Eda, M. Kunitake and O. Manabe, J. Phys. Chem. B., 101(1997) 215.

[54] D. Chapman, Biomembrane and Functions, Verlag Chemie, Weinheim, 1984.

[55] Y. Okahata, R. Ando and T. Kunitake, Ber. Bunsenges. Phys. Chem., 85 (1981) 789.

[56] N. Nakashima, K. Nakayama, M. Kunitake and O. Manabe, J. Chem. Soc., Chem. Commun., 1990, 887.

[57] N. Nakashima, N.Yamada, T. Kunitake, J. Umemura and T. Takenaka, J. Phys. Chem., 90 (1986) 3374.

[58] J. Chlistunoff, D. Cliffel and A. J. Bard, Thin Solid Films, 257(1995) 166.

[59] M. Matsumoto, H. Tachibana, R. Azumi, M. Tanaka, T. Nakamura, G. Yunome, M. Abe, S. Yamago and E. Nakamura, Langmuir, 11 (1995) 660.

[60] D. M. Guldi, Y. Tian, J. H. Fendler, H. Hungerb hler and K.-D. Asmus, J. Phys. Chem., 99 (1995) 17673.

[61] S. E. Campbell, G. Luengo, V. I. Srdanov, F. Wudl and J. N. Israelachvili, Nature, 382 (1996) 520.

[62] P. Wang, Y. Maruyama and R. M. Metzger, Langmuir, 12 (1996) 3932.

[63] V. V. Tsukruk, M. P. Everson, L. M. Lander and W. J. Brittain, Langmuir, 12 (1996) 3905.

[64] M. I. Sluch, I. D. W. Samuel, A. Beeby and M. C. Petty, Langmuir, 14 (1998) 3343.

[65] F. Cardullo, F. Diederich, L. Echyegoyen, T. Habicher, N. Jayaraman, R. M. Leblanc, J. F. Stoddart and S. Wang, Langmuir, 14, 1955 (1998).

[66] C. Jehoulet, Y. S. Obeng, Y.-T. Kim, F. Zhou and A. J. Bard, J. Am. Chem. Soc., 114 (1992) 4237.

[67] Y. Tomioka, M. Ishibashi, H. Kajiyama and Y. Taniguchi, Langmuir, 9 (1993) 32.

[68] M. Yanagida, A. Takahara and T. Kajiyama, Bull. Chem. Soc. Jpn., 73 (2000) 1429.

[69] P. Tundo, A. Perosa, M. Selva, L. Valli and C. Giannini, Collids Surf., A, 190 (2001)295.

[70] H. Tachubana, R. Azumi, A. Ouchi, M. Matsumoto, J. Phys. Chem, B, 105 (2001) 42.

[71] A. Szucs, A. Loix, J. B. Nagy and L. Lamberts, J. Electroanal. Chem., 397 (1995) 191.

[72] J. Davis, H. A. O. Hill, A. Kurz, A. D. Leighton and A. Y. Safronov, J. Electroanal. Chem., 429 (1997) 7.

[73] W. R. Fawcett, M. Opallo, M. Fedurco and J. W. Lee, J. Am. Chem. Soc., 115 (1993) 196.

[74] K. M. Kadish and R. S. Ruoff (eds.), B. S. -Guillous, W. Kutner, M. T. Jones and K. M. Kadish, Fullerenes, Recent Advances in the Chemistry and Physics of Fullerenes and Related Materials, The Electrochemical Society, Pennington, 1994, pp. 1020-1029.

[75] K. M.Kadish and R. S. Ruoff (eds.), L. Echegoyen, F. Diederich, L. E. Echegoyen, Fullerenes: Chemistry, Physics and Technology, Wiley-Interscience, New York, 2000, chap.1.

[76] C. L.Reed and R. D. Bolskar, Chem. Rev., 100 (2000) 1075.

[77] M. S. Meier, J. F. Guarr, J. P. Selegue and V. K. Vance, J. Chem. Soc. Chem. Commun., 1993, 63.

[78] P. L. Boulas, M. T. Jones, K. M. Kadish, R. S. Ruoff, D. C. Lorents and D. S. Tse, J. Am. Chem. Soc., 116 (1994) 9393.

[79] K. M. Kadish and R. S. Ruoff (eds.), J. P. Selegue, J. P. Show, T. F. Guarr, M. S. Meier, Recent Advances in the Chemistry and Physics of Fullerenes and Related Materials, The Electrochemical Society, Proceeding Series, Electrochemical Society; Pennington, NJ. 1994, 1274.

[80] P. L. Boulas, M. T. Jones, R. S. Ruoff, D. C. Lorents, R. Malhotra, D. S. Tse and K. M. Kadish, J. Phys. Chem., 100 (1996) 7573.

[81] N. Nakashima, M. Sakai, H. Murakami, T. Sagara, T. Wakahara and T. Akasaka, J. Phys. Chem. B, 2002, in press.

[82] H. Murakami, M. Shirakusa, T. Sagara and N. Nakashima, Chem. Lett., 1999, 815.

[83] N. Nakashima, T. Ishii, M. Shirakusa, T. Nakanishi, H. Murakami and T. Sagara, Chem. Eur. J., 7 (2001) 1766.

Planar Lipid Bilayers (BLMs) and their Applications
H.T. Tien and A. Ottova-Leitmannova (Editors)

Chapter 30

Analytical applications of planar bilayer lipid membranes

M. Trojanowicz

Department of Chemistry, Warsaw University, Pasteura 1, 02-093 Warsaw, Poland

1. INTRODUCTION

One of the most pronounced trends in recent 30 years in development of analytical instrumentation is miniaturization of chemical and biochemical sensors. The fundamental feature of analytical sensor is its possibility of easy use in different places, and such operation mode that sensor is introduced (*e.g.* immersed) into the sample and not sample introduced into the detection device. This reflects numerous important trends of contemporary analytical chemistry. One of them is a need to design analytical measuring devices that can be used by direct end-user out of the dedicated analytical laboratory, that means necessity of replacement of stationary laboratory equipment with smaller, portable devices. Another one is a need for a possibly large integration of various stages of analytical procedure and elimination of operations which are not necessary or those, which might be a source of errors, such as transport and transfer of sample to be analyzed. Third one - it is authentic need of miniaturization of measuring devices for implantation in living organisms for *in vivo* use or for incorporation into other devices, of which functioning requires continuous chemical monitoring (*e.g.* various process reactors).

These tendencies come across increasing fields of rapidly growing electronics and microelectronics, optoelectronics and material science, micromachining and computer science. All these branches of science and technology create favorable background for development of sensor technologies for chemical analysis.

Under general term of sensors of potential analytical application that can be employed for qualitative and quantitative determination of content of analyzed material one should understand a very different devices of typical physical principles of functioning (various semiconductor detectors), sensors based on recognition of analyte through defined chemical or biochemical interactions. In each of these cases an integrated sensors contain the element that recognizes a given property of the analyzed object and transducers, transforming this property into easy to process and transfer electric signal.

As the beginning of development of sensor technology in analytical chemistry a design of a potentiometric glass electrode for pH measurements in first years of previous century should be assumed [1,2]. A further continuous and fast development of chemical and biochemical sensors takes place since end of fifties, with such a milestones as development of amperometric oxygen probe by Clark [3], invention of first enzymatic biosensor also by Clark [4], and then numerous potentiometric membrane ion-selective electrodes and first optodes based of optical fiber technology [5]. At present scientific and technical literature on chemical and biochemical sensors include about 20.000 papers [6] and many of them have found wide application in routine chemical analysis.

A basic principle of construction of integrated chemical and biochemical sensors is the placement of a phase containing molecules that recognize the analyte in a close proximity of transducer. In case of electrochemical sensors almost always it is most favorable to incorporate such a recognizing element in the phase of a working or indicating electrode or on the surface of electrode. Hence, the often exploited advantages of voltammetric biosensors made of carbon paste, sensors with electro-active signal-generating materials adsorbed directly on the electrode surface or incorporated in various ways within a self-organized monolayers on electrode surfaces. Of great importance for such a designs is proper reproducibility of formation of such a systems, especially in miniaturized systems.

The artificial formation of planar BLMs, especially those produced on solid supports, is a potentially attractive technique for use in sensor technology. A quite extensive literature that deals with this subject, which is reviewed in this Chapter indicates that analytical applications of BLMs can be based both on direct interaction of various species with an artificially formed BLM structure as well as on various interactions of analytes with ionophores, biomolecules or other species incorporated for this purpose into BLM.

The properties, artificial formation and applications of BLMs are the subject of a very broad literature, including several specialized monographs [7-10]. Analytical applications of artificially formed BLMs are focused on their use for design of various analytical devices, which was pioneered by work on lipid films separating two aqueous saline solutions operating as transducers for detection of antigen-antibody and enzyme-substrate reactions [11]. Such applications, intensively developed especially in recent decade, are based on direct interaction of various species with BLM structure or on interaction of analytes with biomolecules incorporated for this purpose into BLM structure. This is already widely discussed in numerous original research papers and also in several reviews [12-20].

2. METHODS OF PREPARATION OF PLANAR BLMs

Since pioneering works by Mueller *et al.* [21,22] several methods of artificial formation of planar free-suspended BLMs as well as solid or gel supported once have been developed and reported for biophysical investigations and analytical applications.

2.1. Free-suspended planar BLMs

Several methods of artificial formation of planar BLMs across a small window (diameter 1 mm or less) between two aqueous phases made in polyethylene foil or other hydrophobic septum have been developed. In earliest works lipid membrane was formed by spreading a lipid solution with a thin brush across the window immersed in aqueous solutions [11,21,22]. When solvent diffuses into the bulk of aqueous phase a lipid bilayer is formed, which initially exhibits interference colors and then black area. As it was mentioned in one of works on BLM immunosensors [23], this procedure can be carried out with commercially available devices. A lipid solution can be also introduced into the hole with the aid of a small syringe or pipette with plastic tips. A comparably simple is also a "tip-dip" technique, where a capillary tip of glass pipette of 1-2 μm diameter is immersed into aqueous solution with a lipid monolayer on the surface, then is taken out and immersed again [24]. In the course of these operations on the tip of capillary filled with aqueous solution a planar BLM is formed. This method has been successfully used, for instance, for formation of ionophore incorporated BLMs [25]. The formation of lipid bilayer can be followed by microscopic observation of the black lipid membrane, by measurement of the capacitive current passing the BLM or resistance or also by electrochemical characterization using gramicidin [26], based on transport of monovalent cations through the membrane.

This method of BLM folding from monolayers can be significantly improved by appropriate use of Langmuir-Blodgett casting technique. In the earliest version described by Takagi *et al.* [26], a Teflon foil 25-500 μm thick with aperture 0.25-1 mm was placed in the middle of vessel with a hole above the level of aqueous solution. After introducing of a lipid layer on the surface of aqueous solution and evaporation of solvent a Teflon partition was lowered, which resulted in joining of two monolayers with hydrophobic parts of lipid molecules. In various later experiments it was shown, for instance, that surface of such formed membranes should not be larger than 0.1 mm^2, in order to obtain satisfactory stability [27]. The details of design of vessel used for this method of preparation of free-suspended planar BLMs used can be found in works by Umezawa *et al.* [25,28-32]. Such a cell allows also a simultaneous measurement of the trans-membrane current. The BLMs were formed across a small, smooth, circular aperture produced in a 7.5 μm polyimide film by passing an electrical spark. The aperture diameters used were of 120 and 20 μm for multi-channel and single-channel type sensing BLM, respectively [28]. In another version of this method, lipid solution was placed on the surface of aqueous solution from one side of the partition, and then the solution level was lowered by pumping out below the aperture, and then again it was elevated above the aperture [33]. At 0.32 mm aperture diameter the obtained BLM system was stable above 4 h with lifetime up to 24 h. BLMs prepared from solutions of oxidized cholesterol exhibit stability even after one month storage [34]. Lipids commonly used for formation of artificial BLMs are shown in Table 1. They include components extracted from various tissues of living organisms as well as synthetic lipids, and also other cationic, anionic and non-ionic amphiphilic compounds, porphyrins and pigments.

Free-suspended BLMs produced with methods described above, although employed for several analytical purposes, are extremely fragile, and when unmodified they exhibit also high resistivity, which makes them excellent insulators. A certain improvement of mechanical stability of BLM brings the use of various chemical additives in the solutions which are used for formation of membrane. A stabilizing effect of cholesterol was found both for free-suspended BLMs [35,36] and especially for described below in details a metal supported membranes (s-BLM) [37,38]. This effect has been attributed to the modification by cholesterol of the short-range repulsive interactions between phosphatidylcholine membranes [39]. A pronounced stabilizing effect on the s-BLM in a negative range of potentials has been found for stearylamine [37]. A destruction of BLM as result of dehydration can be prevented by the use of some saccharides in the bilayers, which is attributed to replacing water bound to polar heads in membrane phospholipids. Such effects were observed for polysaccharide derivatives bearing hydrophobic anchor groups [40], for sucrose [41] and trehalose [42]. It was also shown recently that planar BLMs formed by folding of monolayers of phosphatidylcholine containing bolaamphiphilic steroid dimer are more stable than those without dimer [43]. Due to decrease of electrocompressibility by approximately 20 times, the membrane formation is shorter and they can be used for construction of capacitance sensors.

Other methods of improvement of stability of free-suspended BLMs is a blending with hydrophobic polymers [44], use of polymerizable [458] or photoactivable [46] lipids or the use of the phospholipids with highly branched hydrophobic chains [47]. A large stability is exhibited by BLMs with *in situ* generated micro-crystalline inorganic semiconductors and possibility of preparing of arrays of BLM-supported semiconductors may potentially find application in design of analytical micro-devices [48].

Inconvenient high resistivity of planar, free suspended BLMs can be omitted by incorporation into membrane a conducting electron mediators such as 7,7,8,8-tetracyanoquinodimethane or tetrathiafulvalene [49-51], or also conducting polymers e.g. polypyrrole formed in BLM by chemical oxidation of monomer [52]. The polypyrrole-lectithin BLM exhibits electron conductivity and a good mechanical stability.

In spite of generally rather limited stability of free-suspended BLM systems, numerous reports on their analytical applications have been published. They were employed as transducers of immunochemical and enzymatic reactions [11,23], in ion-channel based sensors for glutamic acid [28], adenosine 5'-triphosphate [29], D-glucose and for evaluation of chemical selectivity of agonists towards receptor ion-channel proteins [31]. Potentiometric sensing with ionophore-incorporated BLMs was reported [32] and potential analytical use of BLMs with selected insecticides [53], herbicide atrazine [547] and metabolite aflatoxin [55] and artificial sweeteners [56]. The use of free-suspended BLMs in flow-injection system [57] and flow detection of ammonia in gas phase [58] have been also reported.

2.2. BLMs supported with additional membrane

A particular importance for application of BLMs in analytical determinations have methods of their preparation on various supports, which substantially improve long-term stability of BLMs. This is practically indispensable condition for their real analytical applications. Regarding a large progress that has been made in recent two decades in various technologies of manufacturing of durable and reliable chemical and biochemical sensors based on numerous novel materials, microelectronics and micro-mechanics, only novel concepts of sensors exhibiting especially competitive features for sensing or practical use may have a chance for mass production and routine use. Systems stable for few hours barely can have a chance in this competition. An important step in this direction, as far as BLM sensors are concerned, was discovery of the possibility of artificial formation of filter supported BLMs already in seventies, followed by metal supported systems (s-BLM), and in recent years also salt-bridge supported BLMs (sb-BLM).

Supporting a fragile structure of planar BLM with porous membranes allows to obtain sensing devices with a longer lifetime and a larger surface area [59-61]. Introduction of lipid solution into the pores of polycarbonate membrane (microporous Nucleopore filters) and then equilibrating membrane with aqueous solution on its both sides provides a matrix of micro-BLMs generated simultaneously [62]. In the simplest setup a polycarbonate membrane can be glued at the end of polycarbonate tube. In a series of papers [63-67] a polycarbonate membrane or glass micro-fiber supported BLMs have been employed for flow-injection determination of various analytes. The measuring systems used were not that far miniaturized as it can be achieved with s-BLMs, however, several interesting applications have been developed. BLMs on porous support have been formed in similar system as it is used for preparation of free-suspended BLMs using the monolayer folding technique [26], where a microporous filter was placed between the two plastic foil layers (about 10 μm thickness) and centered on the 0.32 mm diameter orifice. Membranes such prepared exhibited enhanced mechanical stability up to a flow rate about 20 ml/min. They were employed for determination of acetylcholine, urea and penicillin in the systems with appropriate enzymes immobilized in BLM [63], aflatoxin M_1 [65,66] and hydrazines [67] and sweetener sucralose [68] based on electrostatic perturbation of lipid membranes. In determination of aflatoxin it was found that filter supported BLMs provide larger signals than free-suspended ones [65]. Adding a single stranded DNA to the membrane resulted in an even more sensitive and robust detection system [66].

The bilayer membranes have been also formed on a smooth layer of crystalline cell surface (S-layer) protein lattices, deposited on a microfiltration Nylon membrane [141]. The obtained system was compared with folded bilayer and lipid bilayer generated on planar Nylon membrane in terms of conductance, capacitance and breakdown voltage. The S-layer produced a stabilizing effect as determined by the breakdown voltage and the life time. It was shown that obtained system provides a biomimetic environment for transmembrane proteins by the comparative reconstitution of α-hemolysis pores.

2.3. Solid supported BLMs

Even larger progress in improvement of stability of BLM detectors and possibility of their practical applications in miniaturized analytical devices has been made by inventing the formation of BLM structure on solid supports. The formation of such a structures permits a favorable biofunctionalization of inorganic solid surfaces (metallic electrodes, oxide semiconductors, and optoelectronic devices) and organic polymers, providing a natural environment for immobilization of proteins in non-denaturing conditions [69,70]. This method enables preparation of ultra-thin layers of large electric resistance on surface of conducting materials. Incorporation of ionophores or receptors in these layers leads to preparation of sensors and biosensors which can be easily miniaturized. The formation of BLM structure on a solid support allows to employ in measurements, besides-commonly used electric or electrochemical measurements, also several-surface measuring techniques such as microinter-ferometry, surface plasmon spectroscopy, reflectance FTIR and X-ray surface reflectivity.

The first obtaining of supported phospholipid bilayers was described in the middle of eighties [71]. Solid supported BLMs (s-BLMs) can be successfully produced directly on the surface of metallic electrodes and other conducting materials, or on surfaces functionalized with the inner layer fixed to the substrate, covered with ultrathin water layer or ultrathin hydrated polymer films. Formation of lipid layer on the surface of various metals is based on the interactions of the amphipathic molecules with the nascent metallic surface, which is highly hydrophilic and microscopically rough [72,73]. BLM is supposed to be stabilized on the metal surface by hydrogen bonds formed between the hydrophilic groups of the monolayer and the electronegative metal surface. The most often used technique in this case is formation of s-BLM on the surface of micro-disk obtained by cutting of metal wire insulated with hydrophobic polymer (Fig.1). In order to avoid oxidation of freshly exposed metal surface, the cutting with scalpel or miniature guillotine is performed in the drop of lipid solution of which BLM is formed [74] In the lipid solution on exposed metal surface a lipid monolayer is formed with polar heads directed towards the metal surface. This process takes about 10 min and after this period of time a metal disk should be immersed in aqueous solution, where during next few minutes a bilayer is self-assembled.

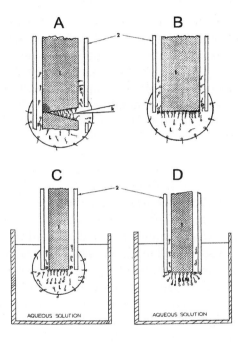

Fig. 1. Schematic diagram showing the process of self-assembly for a lipid bilayer on a freshly cleaved metal surface [74]. 1-metal wire, 2-polymer coating, k-scalpel, p-Plateau-Gibbs border. A: a metal wire cut when immersed in a lipid solution , B: a monolayer of lipid molecules adsorbed on the nascent metal surface, C: upon immersion of the lipid coated surface into aqueous solution a self-assembled lipid bilayer is formed, D: s-BLM is formed.

The formation of BLM can be monitored by measurements of increase of the capacitance or decrease of the conductance, until they reach a stable values. Good results were reported also when lipid solution was placed on silver pads (3.5 to 7.5 mm diameter), vacuum deposited on polyvinylidene fluoride polymer and then immersed into electrolyte solution [75]. A similar procedure was found satisfactory in the formation of s-BLMs also on thin-film platinum or gold supports [76]. In case of platinum disc electrode initial chemical cleaning and sonication was applied [77].

Miniaturized s-BLM biosensors have been prepared using various metals as support. It was shown, that platinum is better for this purpose than nickel or aluminum, although good results can be obtained for stainless steel support [78]. In the design of a glucose biosensor with glucose oxidase and s-BLM it was found, that the use of Pt instead of stainless steel provides in amperometric measurements 10 times larger signals [79]. The most commonly used metallic supports in this technique are stainless steel and silver. In another work [38], by comparison of normalized transmembrane resistance value for s-BLMs formed on stainless steel, silver and iridium, it was found that the steel was most convenient providing the most stable response. The worst stability was observed for iridium support. The diameter of a supporting metal disk did not affect significantly the stability of s-BLM sensors. Stainless steel supported BLMs were employed for development of an ionophore based potassium sensor [80], for enzymatic biosensors [81-87], immunosensors [88-90] and for examination of interaction of nitric oxide with s-BLM without and with fullerene C_{60} [91]. In studies of s-BLM immunosensors with stainless steel it was found, however, that more stable and reproducible than s-BLM is a system with lipid monolayer assembled on gold support modified with octadecanethiol [89,90].

For silver supported BLMs formed on the silver pads of diameter 3.5 to 7.5 mm ellipsometric studies have been carried out on the thickness and time of BLM formation [75]. Different results were obtained depending on the way of BLM formation, as both the thinning of dried lipid films and formation in the contact with aqueous electrolyte have been used. Depending on the time of the lipid layer formation, a monolayer, bilayer or trilayers were formed. No effect of the size of support on time of stabilization of s-BLM was reported.

Several different analytical applications of silver supported BLM were reported by Nikolelis, Krull *et al.* (see Table 1), including detection of aflatoxin M_1 [92], triazine herbicides [93] and several artificial sweeteners [94], based on direct interaction with analytes, for carbon dioxide [95] and cyanide [96] with chemically modified s-BLMs, and for ammonium ion [97] using s-BLM with gramicidin D as a channel-forming ionophore. Silver supported s-BLMs were proposed for electrochemical detection of hybridization of DNA [98,99], and s-BLMs with incorporated DNA were used for the detection of hydrazines [100].

Results of studies on formation of s-BLMs on a metallic supports have been utilized for preparation of miniaturized interdigitated sensors, creating a next step towards practical analytical devices based on artificial BLMs [101,102].

Stable BLMs can be also successfully formed on other solid supports. A metallic copper electrode, besides a stainless steel one, has been used in s-BLM modified with C_{60} fullerene as iodide sensor [103]. Glassy carbon disc electrode was employed as support of s-BLM with incorporated DNA [104] and for ion-channel sensor of ferricyanide based on s-BLM prepared from didodecyldime-thylammonium bromide [105]. Glassy carbon has been also used as support for preparation of s-BLM biosensors with immobilized horseradish peroxidase [106] and glucose oxidase [107]. The use of pyrolytic graphite as support for s-BLM was widely exploited by Nakashima *et al.* for examination BLM with immobilized chlorella ferredoxin [108 where it was indicated that the lipid bilayers provide suitable microenvironments for a stable immobilization of protein. A basal plane pyrolytic graphite was also used for s-BLM with immobilized glucose oxidase and mediating ferrocene [109], and for examination of multi-bilayer structures formed with C_{60} embedded in a cast film of an artificial lipid [110].

Various silicon based materials have been also successfully employed as solid support for formation of s-BLMs. A quartz substrate was reported for s-BLM with immobilized ferredoxin [108]. Lipid bilayer was deposited by the fusion of vesicles onto highly doped p-type Si/SiO_2 electrodes [111]. The resistance of s-BLM on Si/SiO2 electrode was observed to be approximately one order of magnitude larger than the reported for another semiconducting support indium-tin-oxide [112]. It has been concluded that membranes deposited on semiconductor surfaces exhibit an increased stability compared to suspended membranes, and can be stored for several weeks without a significant decrease of the membrane impedance [112]. Phospholipid solid supported bilayer was coated on the surface of poly(dimethylsiloxane) microchannels for fabrication of microfluidic chips for immunoassays with fluorimetric detection [113].

Another solid materials used successfully as support for artificial formation of BLMs are organic conducting polymers. Their application for s-BLM formation is especially convenient as they can be deposited on the surface of solid electrodes electrochemically with full control of the chemical form of deposited polymer and the thickness of its layer. Ion-exchanging properties of conducting polymers are advantageous for stabilization of BLM formed. Also, because of chemical sensitivity of these materials they may serve in some cases as internal transducer for s-BLM biosensor. The mentioned ion-exchange properties may additionally serve as diffusion barrier for discrimination of charged electroactive interferences in some biosensors. These properties have been exploited in design of urea [82, 114] and glucose [79] BLM biosensors. In glucose biosensor as solid support a platinum micro-disc modified with Nafion layer containing ferrocene was also employed [79].

The self-assembling of BLM directly on a solid support is not ideally suited, however, for measurements with membrane-spanning proteins because their mobility is limited due to friction with substrate [115], that disturbs the conformation and hence the function of immobilized protein [116]. Due to this reason, and also for the electrochemical monitoring of transmembrane current, the bilayer has to be fixed by a spacer to keep a certain distance to the solid surface [117]. Hence, from the middle of nineties several different concepts of formation of tethered lipid membrane (t-BLMs) have been reported [118-121]. They are solid-supported lipid films with hydrophilic spacer groups tethered covalently to the support via Au-S-, Si-O- or Si-O-Si- groups. Such systems function as fluid lipid membranes with aqueous layer separating BLM and support. In case of using gold as a solid support, as spacer can be used for instance thiopepetide lipids [117, 120, 121]. t-BLMs are formed by spreading lipid vesicles on a thiopeptide tethered lipid monolayer attached to a gold support (Fig.2a). It was shown for H⁺ - ATPase [122] and acetylcholine receptor [117, 123] that these biomolecules can be incorporated in a functionally active form using liposomes with reconstituted protein. Cytochrome c oxidase was incorporated by dilution of the detergent in which protein is solubilized below the critical micellar concentration in the solution above a preformed t-BLM [121] (Fig.2b). Biosensors with t-BLMs were also developed for ion-channel based sandwich and competitive immunoassays [121, 124, 125] and for detection of DNA via an ion-channel switch [126]. Using formation of self-assembled monolayers a solid-supported lipid layers were formed where fluorescent liposomes were attached to monolayers of mercaptoundecanoic acid on gold surface [127].

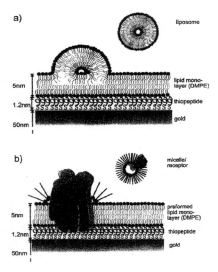

Fig. 2. Schematic model of vesicle fusion onto a peptide-supported lipid monolayer (a) and the incorporation of cytochrome c oxidase in a preformed lipid bilayer (b) [121]. Lipid thickness indicated are measured by surface plasmon resonance spectroscopy.

Three techniques of the liposome fusion method were compared to produce s-BLM or t-BLMs, containing GM1 receptor on silicon nitride surfaces for biosensing of cholera toxin using a resonant mirror method [128]. t-BLMs were formed on a streptavidin coated surface using liposomes containing biotinylated lipid. For biosensor application the most suitable was assumed method where liposomes were fused directly onto a bare silicon nitride surface, producing a bilayer in direct contact with the biosensor surface stable for several months.

The example of bilayer membrane resting on ultrathin, soft hydrated polymer film was reported with a polymer covalently linked to glass substrate upon which the lipid bilayer was deposited [116]. Glass slides were first functionalized with silane and covered by chemisorbed reactive polymer. Further reaction with diamines produced a polymer layer containing hydrophilic amino gropus. The Langmuir-Blodgett technique was used to transfer the first lipid monolayer onto this polymer, and then the second outer layer was attached by dipping the substrate horizontally through a prespread monolayer of lipid on the film balance by Langmuir-Schaefer technique (Fig.3). The fluorescence micro-scopy study has proved stability of this system over several days.

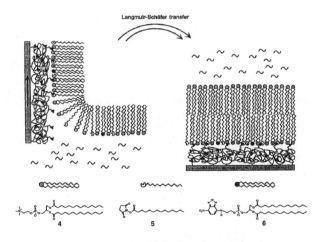

Fig. 3. Representation of the buildup of the bilayer by application of the Langmuir-Blotgett and the Langmuir-Schäfer technique [116]: NBD-DMPE (6. lipids with black headgroups), DMPC (4. lipids with shaded headgropus), anchor lipid (%. with white headgroups). Black triangles symbolize amino groups.
NBD-DMPE: 7-nitrobenzo-2-oxa-1,3-diazol-dimyristoylphosphatidylethanolamine
DMPC: dimyristoylphosphatidylcholine.

Another way of stabilization of BLM structure, favorable for biosensor design, is the use of attachment to solid supports with associated crystalline bacterial cell surface layers (S-layers) [129-131]. Initially, it was shown that S-layer formed on the top of s-BLM produced by chemisorption of thiolipids on gold and attachment of several monolayers by the Langmuir-Schaefer technique, stabilized and protects the underlying s-BLM [130]. In progress are studies on additional forming of S-layer between solid support and BLM covered at the top with S-layer [131].

2.4. Gel supported BLMs

A significant advantage of a solid supported BLMs is a better long-term stability compared to free-suspended systems. Their certain disadvantage in case of metallic support is a direct contact of BLM with electronic conductor, that prevents transport of ions from the membrane. This drawback can be omitted by the above mentioned t-BLMs systems or the use of underlaying polymer cushion layer, and potentially also organic conducting polymers, possessing both electronic and ionic conductivity. Much simpler these drawbacks can be solved by forming the bilayer membrane gel support such as used, for instance, in salt bridges in electrochemistry (hence sb-BLM). Such a system can be easily fabricated using "tip-dip" technique used for preparation of free-suspended BLMs on a tip of glass capillary [24], where instead of a solution capillary is filled with the appropriate gel [132,133]. A principle of this technique is schematically shown in Fig.4A. When capillary with gel is taken out of the electrolyte solution with lipid solution on the surface, a lipid monolayer is formed on the tip. BLM is formed during next immersion into the electrolyte solution. In this way of formation, sb-BLM achieves constant values of conductance and capacitance after about 30 min [132]. These systems can also be made by cutting the tip of plastic capillary filled with gel in the drop of lipid solution, and then by immersion of capillary to aqueous electrolyte [133,134].

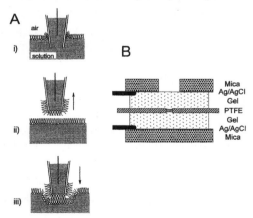

Fig. 4. A. The main stages of a BLM formation using "tip-dip" technique [25]. i) immersion of the capillary filled with aqueous solution or hydrogel into aqueous solution covered with the layer of lipid solution, ii) pulling up the capillary to form a lipid layer, iii) moving down the capillary into aqueous solution to form bilayer.
B. Scheme showing the major components of the bilayer cell with a gel-protected bilayer lipid membrane [140].

In another method reported in first works on gel based sb-BLM, lipid membranes made of lecithin and cholesterol were formed by self-assembling in a small aperture on an agar gel support [135]. sb-BLM based systems were developed for potentiometric alkali cation sensors [134,135], for biosensor for hydrogen peroxide with immobilized horseradish peroxidase [136], and for detection of DNA based on hemin embedded into the BLM [137].

An improved gel-protected lipid bilayers have been fabricated by sandwiching between two slabs of agarose gel [138-140]. On chloridated silver-covered mica 480 µm agarose-electrolyte gel was formed, then 10 µm PTFE sheet was laid on top with 250-500 µm diameter hole where BLM was formed by spreading. Bilayer cell was completed by putting 100 µm protective gel layer with Ag/AgCl electrode on mica with perforation serving as a sample cell (Fig.4B). Such a system responded to potassium when valinomycin-containing bilayer was used [138], to gramicidin D [139], to acetylcholine with nicotinic acetylcholine receptor in BLM [139]. It was also used for detection of the complement activity of human blood [140]. It was suggested that the greater robustness of this sb-BLM is due to better mechanical isolation of BLM. The median time to cell failure was 11 days, which is commensurate with the hydrolysis time of ester-linked phospholipids in an aqueous ambient [140].

3. ANALYTICAL APPLICATIONS OF PLANAR BLM

The possibility of artificial preparation of BLMs has opened not only new ways of biophysical investigation of membranes but has also has resulted in a wide examination of these self-assembled devices for their use in construction of chemical sensors and biosensors. This is caused by several reasons. One of them is sensitivity of BLMs to various chemical interactions and a possibility of instrumental control of changes of different electrical and optical parameters. Second one is the possibility of the use of BLMs as a matrix for ionophores, enzymes, antibodies, engineered receptor proteins or ion-channel switches, which with appropriate transducer may create different types of chemical or biochemical sensors. Of special importance is the fact, that BLMs create the natural environment for the immobilization of proteins in non-dentauring conditions and in well defined orientation. Supported BLMs, especially these separated from the support solid surface by ultrathin layer of electrolyte of hydrophilic polymer, with improved stability maintain also structural properties of free-suspended BLMs.

3.1. Applications of planar, unmodified BLMs

The physico-chemical property of bilayer lipid membranes prepared artificially *in vitro* from different components are often affected as result of contact of unmodified chemically BLMs with various kinds of chemical compounds. These in many cases can be used as analytical information. It is naturally more obvious that in order to get special response, BLM is modified with appropriate chemicals to generate a change of a given signal while contacting the analyte.

The complexicity of interactions of unmodified BLMs can be illustrated, for instance, by changes of properties of free-suspended membranes with phosphatidylcholine and cholesterol in the presence of antibiotic valinomycin [57] or local anesthetic tetracaine [142]. In case of valinomycin the current response of BLM has been shown to be biphasic with the first maximum of signal caused by perturbation of membrane surface dipoles. A semi-logarithmic dependence has been found between measured current and amount of injected valinomycin. The interaction with tetracaine causes a change of membrane surface potential and membrane capacitance and provides additional relaxation component. The main contribution to the change of surface potential was attributed to the dipole potential of membrane surface.

Several attempts have been reported to utilize such effects for analytical purposes with different configurations of BLMs. For free-suspended BLM made of phosphatidyl-choline and dipalmitoylphosphatidic acid a linear relationship between a transient current signal and concentration of the organophosphate and carbamate insecticides monocrotofos and carbofuran has been observed and attributed to the absorption of the lipophilic insecticide molecules and rapid reorganization of the membrane electrostatics [53]. Similar interactions with BLM of the same composition free-suspended [54], filter [64] and silver [93] supported ones have been reported for atrazine and other triazine herbicides. The transient signals observed for BLM interactions with these analytes have a duration of seconds and are indicative of alterations of the electrostatic fields of BLMs. The mechanism of signal generation was related to a rapid adsorption of analytes followed by a slow aggregation process of the herbicide at the surface of BLMs with a consequent rapid reorganization of the membrane electrostatics [64]. Additionally, the time of appearance of the transient response was different for different analytes, which allowed their simultaneous flow-injection determination with resolution similar to a high performance separation techniques [64] (Fig.5). Silver supported BLMs exhibited the shortest response times of examined BLM systems [91]. A similar type of transient response with several peaks and reversible return to base-line signal level was observed in study of the complement activity of human blood by monitoring the electrical impedance changes of a gel-protected BLM in contact with blood serum samples [140].

A sensitivity of Ag-supported BLMs of phosphatidylcholine has been also observed for interactions with acesulfame-K, cyclamate and saccharin, and the mechanism of detection current was attributed to electrochemical double layer and phase structure changes [94]. The same artificial sweeteners have been determined with freely-suspended BLMs, where the mechanism of signal current generation was investigated by differential scanning calorimetric studies and monolayer compression techniques [56]. The alteration of lipid density caused by the adsorption of sweeteners changes the effective area of the headgroup and the association of water, and hence the electrostatic fields at the surface of the BLM. Similar studies were carried out with surface-stabilized BLM for another sweetener sucralose [68].

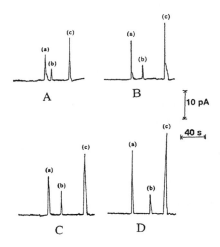

Fig. 5. Examples of recordings obtained for simultaneous flow injection determination of mixture of triazine herbicides: simazine (a), atrazine (b) and propazine (c) with BLM composed of 35% (w/w) DPPA and supported in glass microfiber filters [54]. The solutions injected (75 μl) in the carrier electrolyte contained: A: 30 ppb simazine, 70 ppb atrazine and 66.7 ppb propazine, B: 44.4 ppb simazine, 91.6 ppb atrazine and 100 propazine, C: 44.4 ppb simazine, 183 ppb atrazine amd 100 ppb propazine, D: 75 ppb simazine, 133 ppb atrazine and 150 ppb propazine. The injection of each sample was made at the beginning of each recording.

Freely-suspended BLMs have been also employed for examination of interactions of organotin(IV) and organolead(IV) compounds with unmodified [143] and cationic and anionic surfactants [144]. Planar BLMs were formed from azolecithin solution in *n*-decane and measurements of transmemebrane potentials were carried out. The obtained results showed that increase of length of alkyl chains, both for tin(IV) and lead(IV) organometallic compound results in a higher degree of membrane depolarization. The largest effect was found for phenyl derivatives at 10 nM level. The general conclusion from these studies was that interaction of examined analytes with BLMs was governed mainly by lipophilicity of analytes, while electrostatic interactions were secondary.

The silver supported BLM was employed for electrochemical detection of hybridization of DNA oligomers [98, 99]. This was based on observation that hybridization of single stranded DNA immobilized on the BLM surface significantly alters the ion current through a membrane. The stainless steel supported BLM was used for investigation of interactions of nitric oxide at biologically relevant micromolar concentrations with plain and C_{60} fullerene doped bilayer membranes. It was found that NO accumulated inside lipid layer increases the membrane capacitance, but presence of fullerene changes the dielectric constant of lipid bilayer, thus reducing effect of NO.

3.2. BLMs modified with redox mediators

It was already mentioned, that redox mediators can be easily incorporated into bilayer lipid membranes. Such modification of BLM offers convenient method for a modification of electron transfer at working electrode in electro-analytical measurements. The main aim of such modification is enhancement of sensitivity of redox processes occurring on modified electrodes. It was showed, for instance, that incorporation of tetrathiafulvane (TTF) in BLM significantly enhances sensitivity of hydrogen peroxide detection, while incorporation of TCNQ lowers it markedly [14].

Detailed studies have been carried on redox behavior of free-suspended BLM with incorporation of TCNQ mediator during membrane formation from solution of lipids in *n*-decane [148]. The TCNQ-doped BLM was first examined in the presence of a variety of redox couples using cyclic voltammetry and redox potentiometric titration. The obtained results indicate that mediator modified BLM can function as an electronic conductor in aqueous media taking part in redox reactions at the membrane/solution interfaces. The mechanism for electron translocation across the mediated BLM system was assumed to be tunneling of electrons through the potential barrier separating the TCNQ molecules embedded in the BLM. The same properties of conducting redox electrode were observed for BLMs doped with ferrocene and chloroplast extract. TCNQ-doped BLM in measurements of open-circuit membrane potential has shown almost Nernstian response to ascorbic acid down to concentration level 10 μM [148]. The comparison of electron transfer across BLM with various mediators (TTF, TCNQ, 2,2'-dipyridine, ferrocene, C_{60}) was carried out also for stainless-steel supported BLM [149]. The lipid layer is excellent environment for hydrophilic fullerenes, and it was shown aleady earlier [150], that the presence of C_{60} greatly enhances stability of s-BLM and significantly alters their electrical properties. It was found also [149] that a combination of C_{60} and other mediators is the most effective way in enhancing the s-BLM sensitivity to different redox species [149]. For illustration data for H_2O_2 and ascorbic acid are shown in Fig.6.

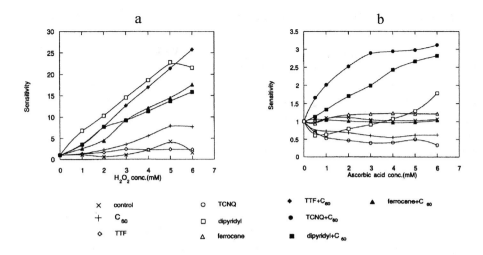

Fig. 6. The amperometric sensitivity (nA) of supported BLMs containing different electron transfer mediators to hydrogen peroxide (a) and ascorbic acid (b) [149].

The incorporation of benzoylferrocene into Pt-supported BLM essentially improves the sensitivity of detection for the system $Fe(CN)_6^{3-}/Fe(CN)_6^{4-}$ down to the sub-micromolar level. Properties of such immobilized mediators have been successfully utilized in construction of BLM based enzymatic biosensors, which will be discussed below. Pt-supported BLM formed of 4,4'-*n*-octyl-cyanobiphenyl in squalene with addition of TCNQ has showed to exhibit unique sensitivity to lead(II) ions [145]. The linear relationship between the oxidation peak current and Pb(II) concentration was reported from 0.1 to 0.6 mM. The incorporation of vinyl ferrocene into Pt supported BLM provides enhancement of electrochemical activity of cytochrome c at Pt electrode due tomore efficient electron transfer between the oxidized form of cytochrome c and BLM modified Pt electrode [146].

Incorporation of some quinoid compounds into BLM structure results in obtaining almost Nernstian sensitivity to pH changes [147]. This was observed for tetrachloro-*o*-benzoquinone and tetrachloro-*p*-benzoquinone with the use of stainless steel supported BLM [101].

3.3. Ionophore based BLMs

Ionophores are widely used compounds in various areas of analytical chemistry for improvement of selectivity for determination of analytes, which are bound by them in heterogeneous systems. Their widest use in electrochemistry can be dated back to sixties, since then they are commonly applied as potential-forming compounds in ion-selective electrodes, mostly with liquid and plasticised membranes. Their lipophilicity, together with that of complexes formed with analytes, provides a possibility of incorporation them into BLMs. Numerous such systems have been reported in the literature for various BLM systems.

In the pioneering work in this field, a free-suspended BLM formed of phosphatidyl-choline and cholesterol was loaded with nonactin, a widely used ionophore for ammonium ions [58]. It was applied in a flow cell for determination of ammonia in gas phase, where a linear relationship of logarithm of transmembrane current *vs.* ammonia concentration was observed within 5 orders of magnitude.

Various configurations of BLM systems containing valinomycin with measurements of various electrical quantities have been reported for BLM biosensors sensitive to potassium ions. A comparative study of potentiometric response of free suspended BLMs with conventional solvent polymer ion-selective electrode provided similar results, however, with a marked difference in respect of the effect of the membrane charge [151]. In a later work of the same research group it was shown, that simultaneous incorporation of an ionophore and appropriate anionic site drastically improves the potentiometric response of the ionophore-based BLMs, generating almost Nernstian response for potassium, sodium and calcium sensitive systems [32]. In the study of membrane conductance changes with the use of free-suspended BLMs it was shown, that the selectivity changes in transmembrane ion permeability were observed upon complexation of primary ions with membrane ionophores associated with their counter anions [25]. It was observed, that ion permeability of the micro BLMs formed by "tip-dip" method is much larger than for the macro BLMs formed by folding method that was attributed to less tight molecular packing of mico BLMs. This leads directly to an amplified signal transduction. Agar supported BLMs containing valinomycin also responded to changes of potassium ion concentration in both the membrane conductance and the membrane potential measurements [135]. Specific conductance changes in the presence of potassium ion were also reported for BLM sensor with agarose gel protecting both membrane surfaces when BLM contained valinomycin. In spite of the absence of a reservoir of ions in one side of membrane of the valinomycin based BLM sensor on stainless steel support, changes of potassium ion have been followed as a function of conductivity of BLM in the range of concentration of two orders of magnitude [80]. Also potentiometric Nernstian response to potassium ion was reported for steel supported BLM containing crown ether formed of cholesterol [78]. The outstanding sensitivity for potassium ion sensing has been reported in capacitance measurements with valinomycin-incorporated bilayer supported on a gold electrode [152]. Bilayer was formed by self-assembling layer of methyl sulfide on gold disc electrode and then by painting this layer with solution of N,N-dimethyl-N,N-dioctadecylammonium bromide in decane and incubation in valinomycin solutions. The formation of alkanethiol-lipid bilayer was confirmed by capacitance change. The obtained sensor has shown linear response of capacitance *vs.* log [K⁺] in the range from 0.5 µM to 5 mM (Fig.7).

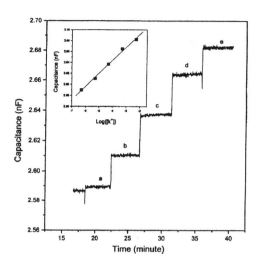

Fig. 7. Capacitance change of the valinomycin-incorporated bilayer supported on a gold electrode and made of *N,N*-dimethyl-*N,N*-dioctadecylammonium bromide versus potassium ion concentration [142]. a) 0.5 µM, b) 5.0 µM, c) 50 µM, d) 0.5 mM, e) 5.0 mM. The measurements were performed in 10 mM Tris-HCl buffer solution of pH 8.28.

Potentiometric measurements were also described for sb-BLM containing Calix[n]arene derivatives [134], however, response observed for various derivatives for potassium and cesium were much lower than monovalent Nernstian response.

3.4. Application of BLM ion-channel sensors

Facilitated diffusion especially of hydrophilic species through a bilayer lipid membrane requires the presence of some specific components of the membrane. One of the possibilities is a mentioned incorporation of ionophores as selective carriers into BLM that allows transport of certain ions in both directions according to the concentration gradient. Another way to activate such transport, even against a concentration gradient, is implementing ion-channel activity in BLM membranes. Substances responsible for this effect are most often membrane proteins incorporated in natural BLMs, which in most cases span the entire lipid bilayer. The most commonly known ion-channel forming pentadecapeptide gramicidin A added in sufficient amount to BLM may increase the membrane conductance 3 to 4 orders of magnitude higher than by the use of the carrier mechanism [10]. Channels formed by incorporated proteins operate through gates activated by different factors, hence they are classified as voltage-gated channels, ligand-gated channels and store-operated channels. All these mechanisms can be utilized for design of BLM-based electrochemical biosensors.

The principle of functioning of sodium/potassium cell pump is functioning in natural BLMs of the enzyme Na^+,K^+-ATPase, which provides transport of both ions through BLM, utilizing the chemical energy arising from the hydrolysis of adenosine 5'-triphosphate (ATP) in the presence of magnesium ions. A similar system was also employed in artificial BLMs for design of biosensors. The auxin-receptor H^+,K^+-ATPase incorporated into a BLM has generated transmembrane current responding to auxin [153]. Also, based on the fact that the rate of Na^+/K^+ exchange with Na^+,K^+-ATPase is dependent on the concentration of ATP, a free suspended BLM potentiometric biodetector for ATP was constructed, showing detection limit 1.0 μM [29]. With the use of gramicidin D in the membrane the amperometric s-BLM ammonium sensor was obtained [97]. Additional modification of BLM with a semi-synthetic platelet-activating factor allowed to reduce potassium interference and linear current response was found in the range from 0.001 to 5 mM.

Incorporation of gramicidin D into s-BLMs was examined for membranes deposited onto silicon-silicon dioxide electrodes [111], on indium-tin-oxide electrodes [112] and for free-suspended BLM protectively sandwiched between two slabs of agarose gel [139]. On the first case gramicidin was inverted into s-BLM from trifluoroethanol solution and obtained membrane exhibited high conductivity and reasonable selectivity to Na^+ and K^+ ions against chloride [111]. For ITO supported BLM gramicidin (and also purins) were transferred into the s-BLM by incubation of the electrode cell with dispersion of vesicles containing polypeptide [112].

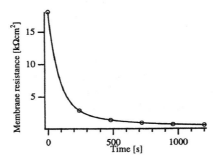

Fig. 8. Illustration of the transfer of gramicidin in supported BLMs by time dependence of the membrane resistance after addition of dimyristoylphosphatidylcholine/cholesterol vesicles containing 0.97 mol% gramicidin [112]: (o) measured membrane resistance; (-) single-exponential fit with a relaxation time of 118 s.

The ratio of gramicidin conductivity for K^+ and Na^+ for this s-BLM was 2.6, and a very low conductivity channel was found for chloride. Fig.8 shows effect of gramicidin on the s-BLM resistance.

    Gramicidin A ion-channels were employed for much more sophisticated detection systems with tethered BLMs [121,124,125]. Using gramicidin additionally functionalized with the water-soluble hapten digoxigenin, it was possible to cross-link gramicidin to fragments of antibody tethered at the membrane/electrolyte interface. The measured ionic conductivity was proportional to kinetics of hapten-protein interactions at the membrane surface [125]. Biosensor that use ion-channel switches can be employed for two-side sandwich assays and competitive assays [121]. In the first case, in the absence of analyte, the mobile ion channels diffuse within our monolayer forming conductive channels. The addition of analyte crosslinks antibody attached to lipids and prevents of formation ion-channels.

    Incorporation of nicotine acetylcholine receptor (AcChoR) in free-suspended BLM sandwiched between two slabs of agarose gel [139] and solid supported tethered membrane with the thiopeptide lipid monolayer coupled to gold elecytrode [117] (Fig.9) provides ion-channel biosensor for acetylcholine.

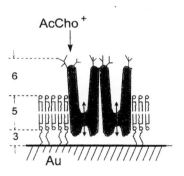

Fig. 9. Principle of the cholinergic sensory system based on solid-supported lipid bilayers containing the sensing acetylcholine receptor dimmer species [117]. The lipid membrane is covalently coupled to a gold surface via Au-S bonds of thiopeptide lipids in the monolayer facing the solid support. The approximated microdimensions of the spacings are 3 nm, 5 nm bilayer thickness and about 6 nm receptor vestibules.

    Immobilized protein converts the binding of an extra-cellular ligand molecule into electrical cross-membrane currents of alkali and alkaline earth metal ions. It is commonly emphasized that approximately 20 ms after the acetylcholine-binding event which opens AcChoR channel, the receptor switches to a refractory state, however, in examined sb-BLM it was observed that response lasts for five orders of magnitude longer [139]. For the practical use of t-BLM sensor with AcChoR it was stressed to be important to remove the bound acetylcholine from the receptor sites [117]. This can be achieved by the presence of acetylcholinesterase and the sensing device should contain a capillary for the supply of calcium or magnesium solution and a mobile esterase for regeneration of receptor. The reported covered concentration of acetylcholine ranged from 0.1 µM up to 1 mM [117].

In a free-suspended BLM system the closure of nicotinic acetocholine receptors was investigated, which can be employed for design of biosensors allowing extremely sensitive detection of even individual bindings antibody-antigen [23]. Receptors incorporated into BLM are inactivated (ion passage stopped) when two bispecific antibodies attached to the same receptor bind to a single antigen molecule.

The glutamate receptor ion-channel protein (GluR) has been used for development of a BLM biosensor for glutamic acid [28] and for determination of selectivity of agonists towards receptor ion-channel proteins [31]. For this ligand-gated ion channel receptor the binding of glutamate opens the channels for large number of ions. This was the principle for the design of a coulometric biodetector with filter supported BLM containing GluR. Based on integrated channel current a detection limit about 30 nM was obtained, together with a high selectivity for L-glutamate compared with D-glutamate [28]. In the system with free-suspended BLM for the detection of D-glucose a cotransport protein, Na^+/D-glucose cotransporter, was employed. [30]. The measured current signal, proportional to an ionic current through the BLM, is selectively induced by D-glucose.

Evaluation of selectivity of typical agonists to activate the N-methyl-D-aspartate (NMDA) receptor was reported with free-suspended BLM prepared by "tip-dip" method and containing NMDA receptor [155]. The method was based on the measurement of an agonist-induced integrated single-channel current corresponding to the number of total ions passed through the open channel. The comparison of selectivity of agonists was carried out at 50 µM agonist concentration.

Detection of DNA in nanomolar level has been reported with the use of ion channel switch t-BLM biosensor [126]. Two different biotinylated probes were bound via streptavidin either to the outer region of a gramicidin ion channel dimmer or to an immobilized membrane component. The ion channels are switched off upon detection of DNA containing complementary epitopes to the used probes, which is schematically showed in Fig.10. Addition of DNase switches the ion channels on. The measurement is completed in 10 min, without sample amplification or multiple washing steps. The two probe assay has mor potential benefit than one probe assay due to the greater specificity achievable.

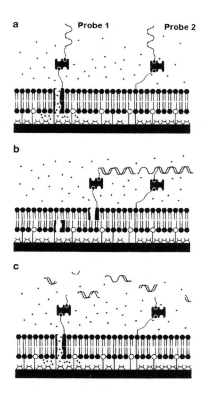

Fig. 10. Schematic diagram of ion channel switch biosensor [126]. (a) Immobilization of biotinylated oligonucleotide probes 1 and 2: ion channels are switched on. (b) Addition of target sequence cross-links the two probes and switches the ion channels off. (c) Addition of DNase I destroys the target DNA and switches the channels on.

As channel forming proteins has acted antibodies to *Campylobacter* species embedded into a stainless-steel supported BLM [156]. This was reported as a convenient way of fabrication of electrochemical biosensor for fast detection of pathogenic bacteria.

Using Pt support, the properties of BLM system with incorporated voltage-gated anion channel were examined [154]. From comparison with free-suspended BLM it was concluded that also solid supported BLMs might be useful for the fabrication of durable biosensors containing ion-channel proteins with reproducible conductivity changes caused by channel gating.

Ferricyanide, which can be used as a marker of ion-channel sensors has been found to act as a stimulus to regulate permeability of a glassy carbon-supported BLM prepared from didodecyldimethylammonium bromide [105]. Such an ion-channel sensor can detect 5 μM $Fe(CN)_6^{3-}$.

3.5. Enzymatic BLM biosensors

As it was mentioned above, the possibility of application of reconstituted *in vitro* BLMs for monitoring of the progress of biocatalytic enzymatic reactions has been indicated in a work with free-suspended BLM by Del Castillo *et al.* in 1966 [11], only few years since a stable "black" lipid membrane was artificially formed by Mueller *et al.* [21,22]. This pioneering work [11] was inspired by earlier observations that reactions involving coating proteins often result in changes of membrane permeability. Del Castillo *et al.* have observed that BLM exposed to enzymes such as trypsin, chymotrypsin or lactate dehydrogenase show marked changes of impedance when they are brought to contact with appropriate substrates. It was observed also, for instance, that micromolar amounts of Cu(II) or Hg(II) inhibit response of chymotrypsin treated membranes. Predictions of possibility of construction or novel biosensors by immobilization of enzymes in artificially formed BLMs can be also found in works on polypyrrole-lecithin BLMs [157,158].

Since then numerous different enzymatic biosensors have been already developed with the use of metal supported BLMs. Particularly, when oxidases are employed, such configuration allows the direct Faradaic amperometric detection of hydrogen peroxide formed during enzymatic reaction. With the use of stainless steel supported BLMs composed of phosphatidylcholine and cholesterol it was showed, that such a system is sensitive to changes of hydrogen peroxide diffusing through the BLM to the working electrode surface, although the sensitivity of response significantly depends on the composition of lipid membrane [37]. In several works on glucose BLM biosensors, glucose oxidase was attached to BLM using biotinylated phospholipid for BLM formation and conjugate of glucose oxidase with streptavidin [81] or avidin [79,82-84]. It was also immobilized directly in BLM formed on pyrolytic graphite together with electron transfer mediator ferrocene substituted with double long chains [109], and also in hydrogel covered on a glassy carbon-supported BLM, containing mediator TTF [107]. In biosensor with ferrocene mediator it was shown that the tunning of the lipid bilayer fluidity between crystal-to-liquid crystalline phase regulates the catalytic transfer reactor.

The modification of BLM by addition of the TCNQ mediator to the lipid solution during formation of BLM allows a decrease of the polarizing potential for the detection, and then changes of the membrane current for glucose detection were observed already at +0.15 V [84]. As mechanism of detection of biosensor with immobilized glucose oxidase was also proposed the tunneling of electrons through enzyme and lipid molecules towards the anode [83,84] instead of anodic oxidation of hydrogen peroxide through BLM to stainless steel support. Several different possibilities of reduction of the effect of electroactive interferences present in physiological fluids have been investigated for Pt supported BLM biosensor with glucose oxidase immobilized using avidin-biotin interaction [79]. For this purpose a Pt disc of 0.6 mm diameter was covered by electro-deposition with various organic polymers as well as with a layer of evaporated Nafion with incorporated ferrocene. The use of Pt support modified with conducting polypyrrole provide stable biosensor functioning for about 2 weeks with substantial elimination of interferences. A similar effect with simultaneous reduction of the detection potential has been found for modification of Pt support with Nafion layer containing ferrocene prior to the formation of BLM with glucose oxidase (Fig.11).

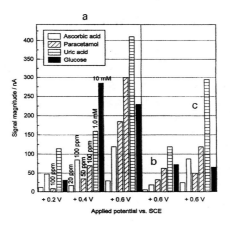

Fig. 11. Comparison of amperometric responses at different potentials obtained for Pt-BLM glucose microbiosensors for glucose and potential interferents in clinical samples [79]. (a) Pt support covered with Nafion with ferrocene; response at three different polarizing potentials. (b) Pt support covered with Nafion without ferrocene. (c) Base Pt support. Concentrations of examined compounds: 10 mM glucose, 20 and 100 mg/L ascorbic acid, 50 and 100 paracetamol mg/L and 1.0 mM uric acid.

Immobilization of enzyme into BLM by the use of the conjugate with avidin or streptavidin was employed also for xanthine oxidase [86], acetylcholinesterase and choline oxidase [87] and urease [82,114]. In all these cases stainless steel support was used for fabrication of micro-disc biosensors. More details are given in Table 1.Bienzymatic biosensor has been employed for the inhibition based indirect determination of the organophosphorus insecticide trichlorphon in the nanomolar range of concentrations and for the natural alkaloid eserine in a micromolar concentration range [87]. An urea BLM biosensor with attached urease-avidin conjugate was prepared on steel support modified with electrodeposited polypyrrole (Fig. 12A) [82,114]. This design allows the use of polypyrrole for amperometric detection of ammonia, formed during enzymatic hydrolysis of urea [159]. Fig. 12B shows the effect of conditions of formation of chemically modified BLM with urease on sensitivity of amperometric detection of urea.

BLM biosensors for detection of acetylcholine, urea and penicillin have been also fabricated by incorporating the enzyme into the lipid matrix at the air/electrolyte interface before the filter supported BLM formation [63]. Hydronium ions produced in enzymatic reaction of substrates caused in flow-injection measuring systems a dynamic alteration of the electrostatic field and phase structure of BLM, which were measured as transient ion current. Flow measurements were carried out at a flow rate of 3.2 ml/min and the measured current in the picoampère range was linearly dependent on the substrate concentration.

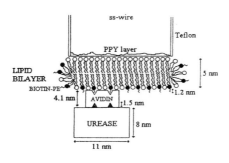

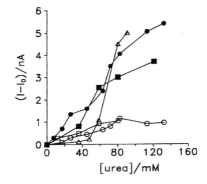

Fig. 12. Schematic diagram (A) of stainless steel supported BLM microbiosensor for amperometric detection of urea with supporting sensing layer of electrodeposited polypyrrole (PPY) and urease immobilized to BLM using avidin-urease conjugate with biotinylated phospholipids for BLM formation and calibration plots (B) obtained for different procedures of biosensor preparation [114].

In B: 1- stainless steel wire with PPY layer was first immersed into the lipid solution for 30 min (o) or 180 min (●) and then for 1 min in solution of urease-avidin conjugate, 2 – steel wire with PPY layer was immersed into the phospholipid solution containing urease-avidin conjugate for 2 min (□), 51 min (■) or 80 min
(△). Measurements at + 0.3 V in 25 mM borate buffer pH 8.44.

Amperometric biosensors for hydrogen peroxide have been developed by immobilization of horseradish peroxidase in glassy carbon-supported BLM [106], and in bilayer lipid membrane modified with lauric acid, which was formed on a salt bridge [136]. In both cases linear response was observed in micromolar range of hydrogen peroxide concentration.

The peptide-tethered membrane incorporating cytochrome c oxidase was reported for biosensing of cytochrome c [121]. The direct electron transfer between enzyme and supporting gold electrode was observed using square wave voltammetry and chronoamperometry. The total resistance of developed t-BLM biosensor decreased as a function of cytochrome c concentration at -200 mV or increased by addition of cyanide, that inhibits enzyme. Impedance spectra were shown to be correlated quantitatively with concentrations of substrate and inhibitor. Square wave voltammogram shown increase of peak current for cytochrome c in concentration range from 15 to 300 μM.

3.6. Immunochemical BLM probes

Del Castillo *et al.* [11] observed that addition of proteins to the solution in contact with BLM after the corresponding specific antigen or antibody treatment of membrane, results in sudden reduction in the electrical impedance of the bilayer membrane. This was observed for bovine serum albumin and antibody treated BLM. Since then several studies of immuno-chemical reactions were carried out on micro BLMs, *e.g.* [160,161], where it was found that changes of electrical properties of BLMs are sufficiently sensitive to observe immu- nological reactions occurring upon BLM matrix.

For analytical purposes both immobilization of antibody and antigen in BLM structure can be utilized. In stainless steel supported BLM of 50 μm diameter the antigen hepatitis B was immobilized, and significant changes of electrical resistance or the membrane capacitance were observed when this s-BLM containing antigen was put in contact with solution containing monoclonal antibody to this antigen [88]. Electrical properties of s-BLM were measured by cyclic voltammetry and obtained detection sensitivity was comparable to that of radioimmunoassay method. Antigen was dissolved in a lipid solution prior to the forming of s-BLM.

Similarly to the enzymatic BLM biosensors discussed above, the antibody can be immobilized in metal supported BLMs by attachment of its conjugate with avidin to s-BLM contained biotinylated phospholipids. This method was used for antibodies to herbicide 2,4-D (2,4-dichlorophenoxy acid) [89,90] and to human immunoglobulin E (IgE) [90]. The immunochemical reaction can be followed by changes of membrane conductance, elasticity modulus or mem- brane capacitance. More reproducible results than with s-BLM for immuno- chemical determination of 2,4-D were obtained, however, with lipid films supported on a gold via octadecanethiol binding with detection limit of about 1 μM, whereas for s-BLM about 50 μM [89]. Using addition of antibody directly on the surface of lipid layer, formed on gold modified with octanethiol, the detection limit for 2,4-D was improved by one order of magnitude (~ 5 μM). It was also reported that better detection limits can be achieved in simple conductance measurements for larger molecules of antigen, *e.g.* immuno-globulins.

Microfluidic on-chip immunoassays were also developed with the use of solid supported BLMs [113]. BLM was formed on the surface of poly-(dimethylsiloxane) microchannels and contained dinitrophenyl (DNP)-conjugated lipids for binding with bivalent anti-DNP antibodies. Twelve independent data points of surface coverage versus bulk protein concentration could be made simultaneously by forming a linear array of channels and flowing fluorescently labeled antibodies into them. This enabled an entire binding curve to be obtained in a single experiment.

3.7. DNA modified BLMs

This application is based on exploiting either selective molecular interaction between oligomers and analytes and detection of effects associated with hybridization. Current signals observed for silver supported BLMs in the presence of hydrazine are interpreted as the conductance enhancements of s-BLM resulting from adsorption of the analyte, that causes changes of the dipolar potential of the membrane [100]. Modification of BLM with a single stranded DNA (ssDNA) terminated with C_{16} alkyl chain to allow incorporation into the membrane, allows about 20 fold increase of signal magnitude for hydrazine. This suggested the presence of larger concentration of cations on the BLM surface with ssDNA, as no selective binding of hydrazine with ssDNA was found. Similar effects were observed for methylhydrazine, dimethylhydrazine and phenylhydrazine. For the same composition of BLM formed on a glass fiber differences in time of response for various substituted hydrazines were reported [67], as observed earlier for triazine herbicides [64]. A comparison of response with that of free-suspended BLM proves, that the material of filter support used is not responsible for a chromatographic artifact of the results.

Various configurations of BLM sensors without [55, 65] and with [66, 92] incorporated ssDNA were employed also for detection of aflatoxin M_1. The mechanism of signal current generation was attributed to alterations of the electrostatic fields at the surface of membranes, while incorporation of ssDNA into lipid membranes provides more sensitive and robust detection system. It was reported that flow-injection measurements were successfully employed for aflatoxin determination in milk preparations [65] and cheese [66].

3.8. Applications of other BLM-based sensors

A silver supported BLMs modified with hemoglobin [95] or methemoglobin [96] can be used for the detection of carbon dioxide and hydrogen cyanide, respectively. Determinations of carbon dioxide were carried out in the system where to buffer solution of pH 5.5 simultaneously solutions of $NaHCO_3$ and hemoglobin were added. The fast response of BLM sensor is interpreted by a rapid alteration of ion transport through s-BLM, which is attributed to the alterations of the surface charge density of membrane. The change of charge of hemoglobin in the reaction with carbon dioxide results in a significant change of the surface charge of a s-BLM. An additional 6 fold increase of sensitivity of detection was observed when into BLM structure a semi-synthetic platelet-activating factor AGEPC was incorporated. In the same work it was suggested that in silver supported BLM systems the silver surface is covered with a silver chloride layer which gives a symmetry to the membrane system [95]. A very similar system (PC in BLM instead of DPPA) with hemoglobin is sensitive to cyanide in the solution of pH 8.0 (present mostly in protonated HCN form at this pH value) [96]. The lack of interference from carbon dioxide was explained by a decrease of affinity of hemoglobin to CO_2 in alkaline solution. A decrease of the detection limit value to 4.9 nM cyanide was obtained by the use of met-hemoglobin in solution instead of hemoglobin.

The BLM sensor for iodide was developed by incorporation a fullerene C_{60} and elemental iodine into copper supported lipid membrane [103]. Reported studies were based on earlier work describing the sensitivity of a s-BLM sensor formed of oxidized cholesterol to iodide [162]. The incorporation of fullerene into the lipid membrane significantly increases the membrane capacitance and decreases a membrane resistance [163], while presence of iodine enhances these changes. The presence of these both components provides a BLM sensor for iodide responding in the concentration range from 10 nM to 10 mM. The mechanism of its functioning was explained by formation in the BLM phase a $(C_{60})I_3^-$ complex, which is discharged at the metal surface.

A salt bridge supported BLM with incorporated lauric acid and hemin embedded via electrostatic interaction has been employed for the study on the interaction of hemin with DNA [137]. It was shown that double-stranded DNA (dsDNA) formed a complex with hemin which was attributed to axial ligation of hemin porphyrin ring, and binding in a groove mode. It was also observed that single-stranded DNA did not interact with hemin when the double strand of DNA disappeared. For analytical purposes a linear relationship between cathodic peak current on cyclic voltammogram and concentration of dsDNA in the range from 84 to 510 μM can be used.

The immobilization of bilayer lipid membranes incorporating carbo-hydrate receptor GM1 on silicon nitride surfaces of a resonant mirror was reported for the binding of cholera toxin B – subunit [128]. The use of biological receptors has advantage over antibodies, as they are sites for numerous pathogenic organisms and toxins. Depending on method of s-BLM formation the long-term stability of fabricated resonant mirror ranged from < 2 days up to several months, hence, it can be considered for continuous monitoring of samples.

Table 1
Analytical applications of planar BLM-based biosensors

Analyte	Biosensor type	Lipid solution used	Chemical BLM modifier	Detection range	Ref.
Acesulfame-K	Ag	PC in *n*-hexane	None	1-15 µM	94
Acesulfame-K	FS	PC in *n*-hexane	None	1 –10 µM	56
Acetylcholine	filter s.	PC and DPPA in *n*-hexane:ethanol (4:1)	Acetylcholinesterase	1-30 µM	63
Acetylcholine	Au	PC	Nicotinic acetyl-choline receptor	$10^{-7} - 10^{-3}$ M	117
Aflatoxin M_1	Ag	PC in *n*-hexane	dT_{20}-C_{16}	0.5-21 nM	92
Aflatoxin M_1	filter s.	PC and DPPA in *n*-hexane and ethanol	None	0.585-21 nM	65
Antibody to Hepatitis B	ss	PC in *n*-decane or glycerol dioleate in squalene	Hepatitis B	1-50 ng/mL	88
Ascorbic acid	FS	PE and PS in chloroform	TCNQ	0.6 – 36 mM	148
ATP	FS	PC and cholesterol in *n*-hexane	Na^+, K^+-ATPase	10^{-6}-10^{-3} M	29
Atrazine	Ag	PC and DPPA in *n*-hexane and ethanol	None	15-400 ng/mL	93
Carbofuran	FS	PC and DPPA in *n*-hexane and ethanol	None	0.48-9 µM	53
Carbon dioxide	Ag	PC and DPPA in *n*-hexane and etha-nol (4:1)	AGEPC in BLM and hemoglobin in solution	38-566 nM	95
Cyanide	Ag	PC in n-hexane and Ethanol (4:1)	AGEPC in BLM and methemoglobin in solution	4.9-350 nM	96
Cyclamate	Ag	PC in n-hexane	None	10-160 µM	94
Cyclamate	FS	PC in n-hexane	None	3 – 30 µM	56

Cytochrome c	Au	PC	Cytochrome c oxidase	15 – 300 µM	121
2,4-Dichlorophe-noxyacetic acid (2,4-D)	ss	Biotiylated phospholipids in *n*-decane:butanol (8:1)	Conjugate of avidin with antibody	0.05-0.2 mM	89
Dimethylhy-drazine	filter s.	PC in *n*-hexane	dT_{20}-C_{16}	0.2-1.2 ng/mL	67
DNA	Au			0.1 – 10 nM	126
DNA	gel s.	PC, cholesterol and lauric acid in *n*-octane	Hemin	84 – 510 µM	137
Ferricyanide	GC	DDMA in chloroform	None	5 – 67 µM	105
Eserine	ss	Biotinylated phospholipid in *n*-decane:*n*-butanol (8:1)	Avidin conjugates with acetylcholinesterase and choline oxidase	0.368-12.5 mM	87
Glucose	GC	DMPG with TTF	Glucose oxidase	0.1 – 10 mM	107
Glucose	ss	Biotinylated phospholipid in *n*-decane	Streptavidin-glucose oxidase conjugate	0.1-15 mM	81
Glucose	ss	Biotinylated phospholipid in *n*-decane:butanol (8:1)	Avidin-glucose oxidase conjugate	0.1-15 mM	82-84
Glucose	Pt	Biotin DHPE and cholesterol in *n*-decane:*n*-butanol (8:1)	Avidin-glucose conjugate. Pt modif. with Nafion and ferrocene	0.1-20 mM	79
D-Glucose	FS	PC and cholesterol in *n*-hexane	Na^+/D-glucose co-transporter	10^{-9}-10^{-6} M	30
L-Glutamate	filter s.	PC and cholesterol in *n*-hexane	Glutamate receptor ion channel protein	0.03-5 µM	28
H_2O_2	GC	DMPG	Horseradish peroxidase	2 – 30 mM	106
H_2O_2	gel s.	PC, cholesterol and lauric acid in *n*-octane	Horseradish peroxidase	9.5 µM – 2 mM	136

Hydrazine	Ag	PC in *n*-hexane	dT_{20}-C_{16}	0.26-12 ng/mL	100
Hydrazine	filter s.	PC in *n*-hexane	dT_{20}-C_{16}	2.5-12.5 ng/mL	67
Iodide	Cu	Glycerol dioleate in squalene or PC in *n*-decane	Fullerene C_{60} and iodine	$10^{-8} - 10^{-2}$ M	103
K^+	FS	DOPC and cholesterol in hexane and chloroform	Valinomycin, NaTFBP	$10^{-5} - 10^{-2}$ M	32
K^+	ss	PC and cholesterol	Valinomycin	2.5-130 mM	80
K^+	Au	DODAB in decane	Valinomycin	$5 \times 10^{-8} - 5 \times 10^{-3}$ M	152
Methylhydrazine	filter s.	PC in *n*-hexane	dT_{20}-C_{16}	0.1-1.2 ng/mL	67
Monocrotofos	FS	PC and DPPA in *n*-hexane and ethanol	None	0.045-0.5 µM	53
Na^+	FS	PC and cholesterol in hexane and chloroform	Ionophore ETH 4120 potassium tetrakis(4-chloropheyl)borate	$4.36 \times 10^{-4} - 0.1$ M	32
NH_4^+ (NH_3 in gas phase)	FS	PC and cholesterol in *n*-hexane	Nonactin	$10^{-6} - 0.1$ M	58
NH_4^+	Ag	PC in *n*-hexane	Gramicidin D, AGEPC	0.001-5 mM	97
Nitric oxide	ss	PC and cholesterol in *n*-decane	None	$1 - 70$ µM	91
Pb(II)	Pt	4,4-*n*-octyl-cyano-biphenyl in squalene	TCNQ	$0.1 - 0.6$ mM	145
Penicillins	filter s.	PC and DPPA in *n*-hexane:ethanol (8:1)	Penicillinase	0.1-1 mM	63
Phenylhydrazine	filter s.	PC in *n*-hexane	dT_{20}-C_{16}	0.25-1.5 ng/mL	67
Propazine	Ag	PC and DPPA in *n*-hexane and ethanol	None	20-200 ng/mL	93
Saccharin	Ag	PC in *n*-hexane	None	0.4-7 µM	94
Sachcarin	FS	PC in *n*-hexane	None	$0.35 - 3.5$ µM	56

Simazine	Ag	PC and DPPA in *n*-hexane and ethanol	None	1-10 ng/mL	69
Sucraloze	filter s.	PC	None	5 – 50 μM	68
Trichlorphon	ss	Biotinylated phospholipid in *n*-decane:*n*-butanol (8:1)	Avidin conjugates with acetylcholinesterase and choline oxidase	3.9-116 nM	87
Urea	filter s.	PC and DPPA in *n*-hexane:ethanol (4:1)	Urease	10-100 μM	63
Urea	ss	Biotinylated phospholipid in n-decane	Avidin-urease conjugate. Pt/PPY	5-130 mM	114
Xanthine	ss	Biotinylated phospholipid in *n*-decane	Avidin-xanthine oxidase conjugate	0.02-1 mM	86

Abbreviations used:

Biosensors type:
Ag   - silver supported BLM
Au   - gold supported BLM
Cu   - copper supported BLM
filter s. – filter supported BLM
FS   – free suspended BLM
gel s.   - gel supported BLM
GC   - glassy carbon supported BLM
Pt   - platinum supported BLM
ss   - stainless steel supported BLM

Names of compounds:
AGEPC   – 1-*O*-alkyl-2-acetyl-*sn*-glyceryl-3-phosphorylcholine
ATP   – adenosine 5'-triphosphate
Biotin DHPE – N-(biotinoyl)-1,2-dihexadecanoyl-*sn*-glycero-3-phosphoethanoloamine, triethylammonium salt,
DDMA   - didodecyldimethylammonium bromide
DMPG   - dimytristoylphosphatidylcholine
DPPA   – dipalmitoylphosphatidic acid
DOPC   – dioleoyl L-α-phosphatidylcholine
DODAB   - *N,N*-dimethyl-*N,N*-dioctadecylammonium bromide
dT$_{20}$-C$_{16}$ – 5'-hexadecyl-deoxythymidylic acid icosanucleotide
NaTFBP   – sodium tetrakis [3,5-bis(trifluoromethyl)phenyl] borate
PC   – phosphatidylcholine
PE   - phosphatidylethanolamine
PS   - phosphatidylserine
TCNQ   – 7,7,8,8-tetracyanoquinodimethane

4. CONCLUSIONS

Biosensors based on artificial bilayer lipid membranes are nowadays a well recognized class of biochemical detection devices for chemical analysis. Generally, depending on BLM composition and configuration of sensing system, such devices can operate from few hours up to even few weeks. Measured analytical signals, mostly current response, also essentially depend on setup configuration and are in the range of pA up to fractions of μA. A diameters of artificially formed BLMs for analytical purposes are from several millimeters down to tens of micrometers.

Among advantages of BLM biosensors one can easily indicate the possibility of different modification of their sensitivity with various chemical and biochemical species and hence exploit different kinds of molecular interactions for detection. Important practical aspect is simplicity of fabrication of these devices, specially solid-supported ones.

REFERENCES

[1]   M. Cremer, Z. Biol., 47 (1906) 562
[2]   F. Haber and Z. Klemensiewicz, Z. Phys. Chem., 67 (1909) 385
[3]   L. C. Clark Jr., Trans. Am. Soc. Artif. Intern. Organs, 2 (1956) 41.
[4]   L. C. Clark Jr. and C. Lyons, Am. N. Y. Acad. Sci., 102 (1962) 29
[5]   W. R. Seitz, Anal. Chem., 56 (1984) 16A.
[6]   J. Janata, Anal. Chem., 73 (2001) 151 A
[7]   H. T. Tien, Bilayer Lipid Membranes (BLM): Theory and Practice, Marcel Dekker, New York 1974
[8]   W. Hanke and W.-R. Schule, Planar Lipid Bilayers, Academic Press, London 1993
[9]   T. Hianik and V. I. Passechnik, Bilayer Lipid Membranes: Structure and Mechanical Properties, Kluwer Academic Publishers, Dordrecht 1995
[10]   H. T. Tien and A. Ottova-Leitmannova, Membrane Biophysics as viewed from Experimental Bilayer Lipid Membranes, Elsevier, Amsterdam 2000
[11]   J. Del Castillo, A. Rodriguez, C. A. Romero and V. Sanchez, Science, 153 (1966) 185
[12]   D. P. Nikolelis, C. G. Siontorou and V. G. Andreou, Lab. Rob. Autom., 9 (1997) 285
[13]   D. P. Nikolelis and C. G. Siontorou, J. Autom. Chem., 19 (1997) 1
[14]   H. T. Tien, R. H. Barish, L.-Q. Gu and A. L. Ottova, Anal. Sci., 14 (1998) 3
[15]   M. Sugawara and A. Hirano, Bunseki Kagaku, 47 (1998) 903
[16]   H. T. Tien and A. L. Ottova, Electrochim. Acta, 43 (1998) 3587
[17]   D. P. Nikolelis, T. Hianik and U. J. Krull, Electroanalysis; 11 (1999) 7
[18]   P. Krysinski, H. T. Tien and A. Ottova, Biotechnol. Prog., 15 (1999) 974
[19]   L. Q. Luo and X. R. Yang, Chinese J. Anal. Chem., 28 (2000) 1165
[20]   M. Trojanowicz, Fresenius J. Anal. Chem., 371 (2001) 246.
[21]   D. Mueller, O. Rudin, H. T. Tien and W. C. Wescott, Nature, 194 (1962) 979
[22]   D. Mueller, O. Rudin, H. T. Tien and W. C. Wescott, J. Phys. Chem., 67 (1963) 534
[23]   S. R. B. J. Van Wie, H. Sutsina, D. F. Moffett, A. R. Koch; M. Silber, W. C. Davies, Biosensors Bioelectron., 11 (1996) 91
[24]   R. Coronado and R. Lattore, Biophys. J., 43 (1983) 231

[25]   H. Sato, H. Hakamada, Y. Yamazaki, M. Uto, M. Sugawara and Y. Umezawa, Biosensors Bioelectron., 13 (1998) 1035
[26]   M. Takagi, Chapter in T. Ohnishi (Ed.), Experimental Techniques in Biomembrane Research, Nankondo, Tokyo 1967
[27]   V. Vodyanoy and R. B. Murphy, Biochim. Biophys. Acta, 687 (1982) 189
[28]   H. Minami, M. Sugawara, K. Odashima, Y. Umezawa, M. Uto, E. K. Michaelis and T. Kuwana, Anal. Chem., 63 (1991) 2787
[29]   Y. Adachi, M. Sugawara, K. Taniguchi and Y. Umazawa, Anal. Chim. Acta 281 (1993) 577
[30]   N. Sugao, M. Sugawara, H. Minami and Y. Umezawa, Anal. Chem., 65 (1993) 363
[31]   M. Sugawara, A. Hirano, M. Rehak, J. Nakanishi, K. Kawai, H. Sato and Y. Umezawa, Biosensors Bioelectron., 12 (1997) 425
[32]   H. Sato, M. Wakabayashi, T. Ito, M. Sugawara and Y. Umezawa, Anal. Sci., 13 (1997) 437
[33]   D. P. Nikolelis and U. J. Krull, Talanta, 39 (1992) 1045
[34]   R. L. Robinson and A. Strickholm, Biochim. Biophys., 509 (1978) 9
[35]   T. Hanai, S. Morita, N. Koizumi and M. Kajiyama, Bull. Inst. Chem. Res., Kyoto Univ., 48 (1970) 147
[36]   D. Papahadjopoulos, M. Cowden and H. K. Kimelberg, Biochim. Biophys. Acta, (1973) 330
[37]   M. Otto, M. Snejdarkova and M Rehak, Anal. Lett., 25 (1992) 653
[38]   M. Snejdarkova, M. Rehak and M. Otto, Biosensors Bioelectron., 12 (1997) 145
[39]   T. J. McIntosh, A. D. Magid and S. A. Simon, Biochemistry, 28 (1989) 17
[40]   J. Moellerfeld, W. Prass, H. Ringsdorf, M. Hamazaki and J. Sunamoto, Biochim. Biophys. ACTA, 857 (1986) 265
[41]   J. Kotowski and H. T. Tien, Bioelectrochem. Bioenerg., 22 (1989) 69
[42]   T. Hianik, J. Dlugopolsky, M. Gyeppessova, B. Sivak, H. T. Tien, A. Ottova-Leitmanova, Bioelectrochem. Bioenerg., 39 (1996) 299
[43]   S. Kalinowski, Z. Lotowski and J. W. Morzycki, Cell. Mol. Biol.Lett., 5 (2000) 107
[44]   P. S. Ash, A. S. Bunce, C. R. Dawson and R. C. Hider, Biochim. Biophys. Acta, 510 (1978) 216
[45]   R. Benz, W. Prass and H. Ringsdorf, Angew. Chem. Suppl., 21 (1982) 869
[46]   F. Borle, M. Sanger and H. Sigrist, Biochim. Biophys. Acta, 1066 (1991) 144
[47]   W. R. Redwood, F. R. Pfeiffer, J. A. Weisbach and T. E. Thompson, Biochim. Biophys. Acta, 233 (1971) 1
[48]   X. K. Zhao, S. Baral, R. Rolandi and J. H. Fendler, J. Am. Chem. Soc., 110 (1988) 1012
[49]   H. T. Tien, J. Phys. Chem., 88 (1984) 3172
[50]   P. Krysinski and H. T. Tien, Bioelectrochem. Bioenerg., 19 (1988) 227
[51]   H. Yamada, H. Shiku, T. Matsue and I. Uchida, J. Phys. Chem., 97 (19983) 9547
[52]   J. Kotowski, T. Janas and H. T. Tien, Bioelectrochem. Bioenerg., 19 (1988) 283.
[53]   D. P. Niokolelis and U. J. Krull, Anal. Chim. Acta, 288 (1994) 187
[54]   D. P. Nikolelis and V. G. Andreou, Electroanalysis, 7 (1996) 643
[55]   V. G. Andreou, D. P. Nikolelis and B. Tarus, Anal. Chim. Acta, 350 (1997) 121
[56]   D. P. Nikolelis, S. Pantoulias, V. J. Krull and J. Zeng, Electrochim. Acta, 46 (2001) 1025
[57]   M. Thompson and U. J. Krull, Anal. Chim. Acta, 142 (1982) 207
[58]   M. Thompson, U. J. Krull and L. I. Bendell-Young, Talanta, 30 (1983) 919
[59]   M. Mountz and H. T. Tien, Photochem. Photobiol., 28 (1978) 395
[60]   Thompson, R. B. Lennox and R. A. McClelland, Anal. Chem., 54 (1982) 76

[61] H. Yoshikawa, T. Hayashi, H. Shimooka, H. Terada and T. Ishii, Biochem. Bipohys. Res. Commun., 145 (1987) 1092

[62] H. Nakanishi, Prog. Surf. Sci., 49 (1995) 197

[63] D. P. Nikolelis and C. G. Siontorou, Anal. Chem., 67 (1995) 936

[64] D. P. Nikolelis and C. G. Siontorou, Electroanalysis, 8 (1996) 907

[65] V. G. Andreou and D. P. Nikolelis, Anal. Chem., 70 (1998) 2366

[66] C. G. Siontorou, V. G. Andreou, D. P. Nikolelis and U. J. Krull, Electroanalysis, 12 (2000) 747

[67] C. G. Siontorou, D. P. Nikolelis and U. J. Krull, Anal. Chem., 72 (2000) 180

[68] D. P. Nikolelis and S. Pantovkas, Biosensors Bioelectron., 15 (2000) 439

[69] W. Müller, H. Ringsdorf, E. Rump, G. Wildburg, X. Zhang, L. Angermaier, W. Knoll, M. Litey and J. Spinke, Science, 262 (1992) 1706

[70] E. Sackmann, Science, 271 (1996) 43

[71] L. K. Tamm and H. M. McConnel, Biophys. J., 47 (1985) 105

[72] K. Miyano, Jap. J. Appl. Phys., 24 (1985) 1379

[73] H. Leidheiser and P. D. Deck, Science, 241 (1988) 1176

[74] H. T. Tien and Z. Salamon, Bioelectrochem. Bioenerg., 22 (1989) 211

[75] K. L. Chiang, U. J. Krull and D. P. Nikolelis, Anal. Chim. Acta, 357 (1997) 73

[76] V. Rehacek, I. Novotny, V. Tvarozek, F. Mika, W. Ziegler, A. Ottova-Leitmannova and H. T. Tien, Sens. Materials, 10 (1998) 229

[77] P. Diao, D. Jiang, X. Cui, D. Gu, T. Tong and B. Zhang, Bioelectrochem. Bioenerg. 45 (1998) 173

[78] Y. E. He, M. G. Xie, A. Ottova and H. T. Tien, Anal. Lett., 28 (1995) 443

[79] M. Trojanowicz and A. Miernik, Electrochim. Acta, 46 (2001) 1053

[80] M. Rehak, M. Snejdarkova and M. Otto, Electroanalysis, 5 (1993) 691

[81] M. Snejdarkova, M. Rehak and M. Otto, Anal. Chem., 65 (1993) 665

[82] T. Hianik, M. Snejdarkova, Z. Cervenanska, A. Miernik, T. Krawczynski vel Krawczyk and M. Trojanowicz, Chem. Anal. (Warsaw), 42 (1997) 901

[83] T. Hianik, M. Snejdarkova, V. I. Passechnik, M. Rehak and M. Babincova, 41 (1996) 221

[84] M. Snejdarkova, M. Rehak, M. Babincova, D. F. Sargent, T. Hianik, Bioelectrochem. Bioenerg., 42 (1997) 35

[85] M. Trojanowicz, T. Krawczyński vel Krawczyk, A. Miernik and B. Sivak, J. Flow Injection Anal., 15 (1998) 210

[86] M. Rehak, M. Snejdarkova and M. Otto, Biosensors Bioelectron., 9 (1994) 337

[87] M. Rehak, M. Snejdarkova and T. Hianik, Electroanalysis, 9 (1997) 1072

[88] L. G. Wang, Y. H. Li and H. T. Tien, Bioelectrochem. Bioenerg., 36 (1995) 145

[89] T. Hianik, V. I. Passechnik, L. Sokolikova, M. Snejdarkova, B. Sivak, M. Fajkus, S. A. Iwanov and M. Franek, Bioelectrochem. Bioenerg., 47 (1998) 47

[90] T. Hianik, M. Snejdarkova, L. Sokolikova, E. Meszar, R. Krivanek, V. Twarozek, I. Novotny and J. Wang, Sens. Actuators B, 57 (1999) 201

[91] L. Gu, L. Wang, J. Xun, A. Ottova-Leitmannova and H. T. Tien, Bioelectrochem. Bioenerg., 39 (1996) 275

[92] C. G. Siontorou, D. P. Nikolelis, A. Miernik and U. J. Krull, Electrochim. Acta, 43 (1998) 3611

[93] C. G. Siontorou, D. P. Nikolelis, U. J. Krull and K. L. Chiang, Anal. Chem., 69 (1997) 3109

[94] D. P. Nikolelis and S. Pantoulias, Electroanalysis, 12 (2000) 786

[95] C. G. Sionorou, D. P. Nikolelis and U. J. Krull, Electroanalysis, 9 (1997) 1043

[96] C. G. Siontorou and D. P. Nikolelis, Anal. Chim. Acta, 355 (1997) 227

[97] D.P. Nikolelis, C.G. Siontorou, U.J. Krull and P.L. Katrivanos, Anal. Chem., 68 (1996) 1735
[98] U. J. Krull, D. P. Nikolelis, S. C. Jantzi and J. Zeng, Electroanalysis, 12 (2000) 921
[99] C. G. Siontorou, D. P. Nikolelis, P. A. E. Piunno and U. J. Krull, Electroanalysis, 9 (1997) 1067
[100] C. G. Siontorou, D. P. Nikolelis, B. Tarus, J. Dumbrava and U. J. Krull, Electroanalysis, 10 (1998) 691
[101] V. Tvarozek, H. T. Tien, I. Novotny, T. Hianik, J. Dlugopolsky, W. Ziegler, A. Ottova, J. Jakabovic, V. Rehacek and M. Uhlar, Sens. Actuators B, 19 (1994) 597
[102] A. Ottova, V. Tvarozek, J. Racek, J. Sabo, W. Ziegler, T. Hianik and H. T. Tien, Supramolec. Sci., 4 (1997) 110
[103] L. G. Wang, X. Wang, A. L. Ottova and H. T. Tien, Electroanalysis, 8 (1996) 1020
[104] C. G. Siontorou, A. M. Oliveira Brett and D. P. Nikolelis, Talanta, 43 (1996) 1137
[105] X. Han and E. Wang, Anal. Sci., 17 (2001) 1171
[106] Z. Wu, B. Wang, Z. Cheng, X. Yang, S. Dong and E. Wang, Biosensors Bioelectron., 16 (2001) 47
[107] Z. Wu, B. Wang, S. Dong and E. Wang, Biosensors Bioelectron., 15 (2000) 143
[108] A. E. F. Nassar, J. F. Rusling, M. Tominaga, J. Yanagimoto and N. Nakashima, J. Electroanal. Chem., 416 (1996) 183
[109] M. Tominaga, S. Kusano and N. Nukashima, Bioelectrochem. Bioener., 42 (1997) 59
[110] N. Nakashima, Y. Nonaka, T. Nakaniski, T. Sagara and H. Murakami, J. Phys. Chem. B, 102 (1998) 7328
[111] O. Parrucker, H. Hillebrandt, K. Adlkofer and M. Tanaka, Electrochim. Acta, 47 (2001) 791
[112] S. Gritsch, P. Nollert, F. Jähnig and E. Sackmann, Langmuir, 14 (1998) 3118
[113] T. Yang, S. Jung, H. Mao and P. S. Cremer, Anal. Chem., 73 (2001) 165
[114] 114. T. Hianik, Z. Cervenenaska, T. Krawczynski vel Krawczyk and M. Snejdarkova, Mat. Sci. Eng. C, 5 (1998) 301
[115] M. Kühner, R. Tampe and E. Sackamnn, Biophys. J., 67 (1994) 217
[116] D. Beyer, G. Elender, W. Knoll, M. Kühner, S. Maus, H. Ringsdorf and E. Sackmann, Angew. Chem. Int. Ed. Engl., 35 (1996) 1682
[117] E. Neumann, K. Tönsing and P. Siemens, Bioelectrochem., 51 (2000) 125
[118] H. Lang, C. Duschl and H. Vogel, Langmuir 10 (1994) 197
[119] R. Neumann, A. Jonczyk, R. Kopp, J. von Esch, H. Ringsdorf, W. Knoll, Angew. Chem., 34 (1995) 2056
[120] N. Bunjes, E. K. Schmidt, A. Jonczyk, F. Rippmann, D. Beyer, H. Ringsdorf, P. Graber, W. Knoll and R. Naumann, Langmuir, 13 (1997) 6188
[121] R. Neumann, E. K. Schmidt, A. Jonczyk, K. Fendler, B. Kadenbach, T. Liebermann, A. Offenhäusser and W. Knoll, Biosensors Bioelectron., 14 (1999) 651
[122] R. Neumann, A. Jonczyk, C. Hampel, H. Ringsdorf, W. Knoll and N. Bunjes, Bioelectrochem. Bioenerg., 42 (1997) 241
[123] E. K. Schmidt, T. Liebermann, M. Kreiter, A. Jonczyk, R. Neumann and A. Offenhäuser, Biosensors Bioelectron., 13 (1998) 585
[124] B. Raguse, V. Braach-Maksvytis, B. A. Cornell, L. G. King, P. D. J. Osman, R. J. Pace and L. Wieczorek, Langmuir, 3 (1998) 648
[125] G. E. Woodhouse, L. G. King, L. Wieczorek and B. A. Cornell, Faraday Discuss. 11 1 (1998) 247
[126] S. W. Lucas and M. M. Hareling, Anal. Biochem., 282 (2000) 70
[127] J. Ufheil, F. M. Boldt, M. Börsch, K. Borgwarth and J. Heinze, Bioelectrchem., 52 (2000) 103

[128] M. I. Fisher and T. Tjärnhage, Biosensors Bioelectron., 15 (2000) 463
[129] U. B. Sleytr, P. Messner, D. Pum and M. Sara (Eds.), Crystalline Bacterial Cell Surface Proteins, R. G. Landes, Academic Press, Austin 1996
[130] B. Wetzer, D. Pum and U. B. Sleytr, J. Struct. Biol., 119 (1997) 123
[131] D. Pum and U. B. Sleytr, Trends Biotechn., 17 (1999) 8
[132] W. Ziegler, J. Gaburjakova, M. Gaburjakova, B. Sivak, V. Rehacek, V. Tvarozek and T. Hianik, Coll. Surf. A: Physicochem. Eng. Aspects, 140 (1998) 357
[133] H. P. Yuan, A. L. Ottova and H. T. Tien, Mat. Sci. Eng. C, 4 (1996) 35
[134] Y. L. Zhang, H. X. Shen, Y. Liu, C. X. Zhang and L. X. Chen, Anal. Lett., 33 (2000) 831
[135] M. Uto, M. Araki, T. Taniguchi, S. Hoshi and S. Inoue, Anal. Sci., 10 (1994) 943
[136] Y. L. Zhang, S. Z. Jin, C. X. Zhang and H. X. Shen, Electroanalysis, 13 (2001) 137
[137] Y. L. Zhang, H. X. Shen, C. X. Zhang, A. Ottova and H. T. Tien, Electrochim. Acta, 46 (2001) 1251
[138] R. F. Costello, I. R. Peterson, J. Heptinstall, N. G. Byrne and L. S. Miller, Adv. Meter. Opt. Electron., 8 (1998) 47
[139] R. F. Costello, I. R. Peterson, J. Heptinstall and D. J. Walton, Biosensors Bioelectron., 14 (1999) 265
[140] R. F. Costello, S. W. Evans, S. D. Evans, I. R. Peterson and J. Heptinstall, Enz. Microb. Technol., 26 (2000) 301
[141] B. Schuster, D. Pum, m. Sara, O. Braha, H. Bayley and W. B. Sleytr, Langmuir, 17 (2001) 499
[142] T. Hianik, M. Fajkus, B. Tarus, P. T. Frangopol, V. S. Markin and D. F. Landers, Bioelectrochem. Bioenerg., 46 (1998) 1
[143] H. Radecka, D. Zielińska and J. Radecki, Sci. Total Env., 234 (1999) 147
[144] D. Zielińska, H. Radecka and J. Radecki, Chemosphere, 40 (2000) 327
[145] Z. Salamon and H. T. Tien, Electronalysis, 3 (1991) 707
[146] Z. Salamon and G. Tollin, Bioelectrochem. Bioenerg., 25 (1991) 447
[147] H. T. Tien, S. H. Wurster and A. L. Ottova, Bioelectrochem. Bioenerg., 42 (1997) 77
[148] H. T. Tien, Bioelectrochem. Bioenerg., 15 (1986) 19
[149] K. Asaka, A. Ottova and H. T. Tien, Thin Solid Films, 354 (1999) 201
[150] H. T. Tien, L. G. Wang, X. Wang and A. L. Ottova, Bioelectrochem. Bioenerg. 42 (1997) 161
[151] H. Minami, N. Sato, M. Sugawara and Y. Umezawa, Anal. Sci., 7 (1991) 853
[152] Z. Cheng, L. Luo, Z. Wu, E. Wang and X. Yang, Electroanalysis, 13 (2001) 68
[153] M. Thompson, U. J. Krull and M. A. Venis, Biochem. Biophys. Res. Commun., 110 (1983) 300
[154] D. A. Stenger, T. L. Fare, D. H. Cribbs and K. M. Rusin, Biosensors Bioelectron., 7 (1992) 11
[155] A. Hirano, M. Sugawara, Y. Umezawa, S. Uchino and S. Nakajima-Iyima, Biosensors Bioelectron., 15 (2000) 173
[156] D. Ivnitski, E. Wilkins, H. T. Tien and A. Ottova, Electrochem. Commun., 2 (2000) 457
[157] J. Kotowski, T. Janas and H. T. Tien, J. Electroanal. Chem., 253 (1988) 283
[158] T. Janas, J. Kotowski and H. T. Tien, Bioelectrochem. Bioenerg., 19 (1988) 405
[159] M. Trojanowicz, A. Lewenstam, T. Krawczynski vel Krawczyk, I. Lahdesmaki and W. Szczepek, Electroanalysis, 8 (1996) 233
[160] J. D. Mountz and H. T. Tien, J. Bioenerg. Biomembr., 10 (1978) 139
[161] K. O'Boyle, F. Q. Siddiqui and H. T. Tien, Immunol. Commun., 13 (1984) 85
[162] D. Xiao, J. Li and R. Yu, Chemical Sensors, 2 (1994) 100

[163]   H. Gao, J. Feng, G. A. Luo, A. L. Ottova and H. T. Tien, Electroanalysis, 13 (2001) 49

Planar Lipid Bilayers (BLMs) and their Applications
H.T. Tien and A. Ottova-Leitmannova (Editors)

Chapter 31

Transmembrane voltage sensor

J. A. Cohen[a], B. Gabriel[b], J. Teissié[b], M. Winterhalter[b]

[a]Laboratory of Physiology and Biophysics, University of the Pacific,
School of Dentistry, 2155 Webster St., San Francisco, CA 94115, USA

[b]Institut Pharmacologie et Biologie Structurale UMR 5089,
205, rte de Narbonne, F-31 077 Toulouse, France

1. ABSTRACT

The biological membrane has many functions. Firstly, it is a barrier for most aqueous substances. Life requires compartmentalization: the cell needs an "inside" and an "outside", which are provided by the cell-wall membrane. However, specific substances must pass through this wall, in a controlled manner. Thus a major purpose of the cell membrane is to form a stable barrier with machinery for selective transport. A second membrane function is to provide a locale for chemical reactions having interfacial and surface kinetics. These features are all influenced by electric fields and electrical potentials which, when asymmetric with respect to the membrane, produce a transmembrane voltage. Here we describe briefly where such fields and potentials originate, how to measure them, and their biological implications.

2. INTRODUCTION

The principal component of biological membranes is the lipid bilayer, which has astonishing material properties. For example, the red blood cell membrane is a few nanometers thick and highly flexible, yet is stable enough to survive drastic changes in shape again and again during its passage through narrow blood vessels. Electrically, the lipid bilayer is an insulator better than porcelain. These features permit the cell to maintain non-equilibrium ion concentration gradients and transmembrane potentials for long times. According to the model by Singer and Nicolson, the lipid bilayer forms a fluid matrix which provides

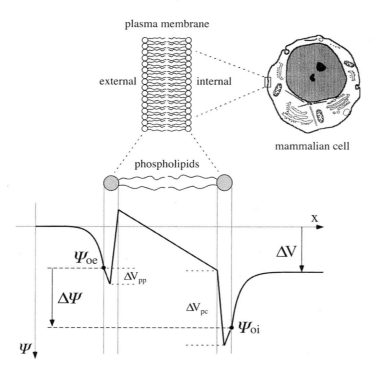

Fig. 1. Different contributions to the boundary potential across a plasma membrane. The major membrane component is the lipids, which contain two prominent dipoles: one from the carbonyl groups in the aliphatic-chain linkages, and one due to the negative phosphate and positive nitrogen in the headgroup. The orientation of the headgroup is almost parallel to the surface and results in a rather small negative surface-polarization potential (ΔV_{pp}). The dominant dipole contribution from the carbonyl groups creates a positive potential inside the membrane (ΔV_{pc}). Negatively-charged lipids in the outer monolayer produce a negative surface potential at the exterior surface (ψ_{oe}) which decays superexponentially into the aqueous phase. A higher surface-charge density inside the cell produces a more negative surface potential at the inner membrane surface (ψ_{oi}). The difference between the inner and outer surface potentials is the transmembrane potential ($\Delta\psi$). Electrodes in the bulk phase sense the macroscopic voltage difference ΔV, which results from ion gradients and selective ion permeability, electrogenic transport, or an externally applied voltage. For a mammalian cell ΔV is negative inside as shown. A change of ΔV also changes $\Delta\psi$. However, a change of $\Delta\psi$ due to an alteration of surface-charge density is rapidly screened and does not affect ΔV, hence cannot be sensed by the bulk electrodes.

a host environment for membrane proteins [1,2]. The proteins, in turn, regulate the translocation of substrates and signals across the lipid membrane barrier [2-4]. Cells can attach hydrophilic proteins in a reversible manner via phosphorylation (modifying the charge state) or via addition of hydrophobic anchors (e.g. myristoylation, the attachment of a myristoyl chain) [2,5,6]. Membrane-bound proteins trigger many important cellular reactions [2,7]. The lateral movement of proteins in the membrane depends on a characteristic diffusion constant that determines their reaction rates, which can be very different from those in bulk solution [8]. Due to electric fields, charged-substrate and ion concentrations, including [H$^+$], at the membrane surface can be one or two orders of magnitude different from those in the bulk [9], with consequent effects on interfacial reaction rates. In all these processes, surface potentials and transmembrane potentials are involved.

Depending on their chemical structure, on the ionic composition of the aqueous environment, and on their macroscopic alignment in a lamellar phase, the lipids create a characteristic electric-field distribution. A typical example is shown schematically in Fig. 1. The best-understood contributions are those due to charges located in the lipid headgroup region. They originate from deprotonated acidic or protonated basic groups, or from the binding of charged substances to the membrane surface.

A reasonable description of the behavior of ions in solution near a charged surface is based on the early Gouy-Chapman theory: ions in the aqueous phase are treated as point charges interacting with the mean electric field in a homogeneous dielectric, while charges fixed at the lipid-water interface are treated as a homogeneous surface-charge density σ_0 [10]. The mobile counterions in solution feel an electrostatic attraction to the charged surface, thus becoming more concentrated near the surface, which produces an increased osmotic force driving them back to the bulk. The balance between these two forces yields a Boltzmann distribution of the aqueous ions. Poisson's equation (i.e., Coulomb's law) in the electrolyte leads to the Poisson-Boltzmann equation, which relates the aqueous charge distribution to the electrostatic potential profile in the aqueous phase. The requirement of global electroneutrality then yields a relation between the surface-charge density and the electrostatic potential at the surface, called the Gouy equation. This approach nicely describes the main electrostatic properties of charged interfaces and also provides a characteristic length over which the electrostatic interaction is transmitted. The Debye screening length is, for the case of 1:1 electrolytes:

$$\lambda_{Debye} = \left(\frac{\varepsilon_w \varepsilon_0 kT}{2n_\infty e^2} \right)^{1/2}, \tag{1}$$

where ε_{w} is the dielectric constant of the aqueous solution, ε_0 the dielectric permittivity of the vacuum, k Boltzmann's constant, T the absolute temperature, e the electron charge, and n_∞ the bulk salt concentration. (For the case of multivalent ions the factor 2 is replaced by a sum, see textbooks.) For example, at biologically-relevant temperatures and ionic strengths (e.g. for ~100 mM salt concentration) this length is ~10 Å. For 10 mM salt it is ~30 Å and for 1 M salt it is ~3 Å. As the Poisson-Boltzmann equation is derived in detail in many textbooks, we add only a few comments. There is no general solution for this nonlinear equation and thus approximations are required. Often numerical solutions are given for the electrostatic potential as a function of distance for specific values of the surface-charge density σ_0 and ionic strength. For convenience the charge density may be expressed in dimensionless units: $p = \dfrac{\sigma_0 e}{2\varepsilon_{\mathrm{w}}\varepsilon_0 kT}\lambda_{Debye}$. For the simple case of a planar geometry and a 1:1 electrolyte, the Poisson-Boltzmann equation has an analytical solution. In this case the surface potential ψ_0 at the lipid/membrane interface is given in terms of σ_0 and λ_{Debye} by:

$$\psi_0 = \frac{2kT}{e}\ln(p+q) \quad \text{with} \quad q = \sqrt{1+p^2}. \tag{2}$$

At sufficiently low surface-charge density, known as the Debye-Hückel regime, the solution of the Poisson-Boltzmann equation for a planar geometry can be further simplified. For example, a membrane of neutral lipids containing less than 1-2 mol% of charged lipid in 100 mM KCl results in $p \ll 1$, which allows q to be approximated by 1, and $\ln(p+q) \approx \ln(1+p) \approx p$. Under such conditions, Eq. (2) shows the surface potential ψ_0 to be approximately linear in the surface-charge density σ_0. In the Debye-Hückel approximation the electrostatic potential in the aqueous phase decays exponentially with the simple form:

$$\psi(x) = \psi_0 \exp\left(-x/\lambda_{Debye}\right). \tag{3}$$

The Gouy-Chapman theory has been used quite successfully. The limit at very low surface-charge density was discussed by Nelson and McQuarrie, but it seems the effects for lipid membranes are less pronounced than predicted, probably due to lipid lateral mobility [10]. However, in the opposite limit of high surface-charge density even down to relatively modest densities, the theory overestimates the effect. Unfortunately, various corrections to the Gouy-Chapman theory do not produce convergent results, and higher-order theories do not improve agreement with the data [11]. Currently, discrepancies between experimental observations and predictions based on the Poisson-Boltzmann theory are reconciled by introducing an additional parameter which is intended to account for counterion binding or an immobile surface layer called the Stern layer. In many cases this procedure is satisfactory but may not be generalized. Recently a series of such discrepancies was discussed, and more fundamental approaches including ion-ion correlations [12] and Onsager and Lifshitz theories were proposed [13]. However, such treatments are rather complex. Further experimental evidence is still necessary to elucidate the nature of ion interactions with lipid membranes. In summary, the Gouy-Chapman theory describes the main results nicely. However, special care must be taken to avoid inaccuracies due to specific-ion effects.

Less known and understood are the contributions from the orientation of the lipid dipoles in the lamellar phase. The measured potential $V_{dipole} = V_{pp} + V_{pc}$ is related to the total perpendicular dipole moment $\mu_\perp$:

$$V_{dipole} = \mu_\perp / (\varepsilon_w \varepsilon_0 A_{lipid}) , \qquad (4)$$

where A_{lipid} is the area per lipid molecule [6,14,15]. The magnitude and variation of V_{dipole} for different lipids have been nicely characterized by measuring their monolayer boundary potentials at the air-water interface [14,15]. Most reported values are in qualitative agreement with those obtained for lipid bilayers by other techniques [14-18]. However, the boundary potentials of lipid monolayers are generally found to be about twice as large as those of lipid bilayers [17,18]. Many explanations have been given for this difference, but none is fully convincing [14,15]. Closely associated with the dipole potential is the role of oriented water bound to the membrane surface. For an introduction to this subject see Ref. [19].

The behavior of charges and potentials at the interface of a membrane/aqueous ionic solution is still a widely-investigated subject. Here we focus mainly on our own results and restrict our discussion to a few selected points.

3. TECHNIQUES TO MEASURE MEMBRANE POTENTIALS

Below we describe several different techniques using mainly polylysine and the protocols of our laboratories as examples of measuring membrane potentials in direct or indirect ways. Our intent is not to present a comprehensive treatise with conclusive results, but rather to describe the methods, daily difficulties, and unresolved questions in this field. For example, polylysine has been used as a model for the electrostatic interaction of a protein with a membrane [20].

Positively-charged polylysine is known to interact *only* electrostatically with negatively-charged lipid membranes. Incorporation of the peptide into the bilayer is not expected *a priori*. Nevertheless the peptide may cause pronounced structural rearrangements inside the lipid membrane, which in the case of vesicles may lead to the formation of multilamellar lipid-peptide complexes [20]. Neutralization of negative charges at the membrane surface can affect lipid packing and the hydrocarbon-chain phase-transition temperature [21]. Pentalysine has also been used to induce fusion of lipid vesicles.

3.1. Monolayer potentials

A very old method providing structural and functional information is the monolayer technique. Lipids are spread at the air-water interface and the surface pressure is recorded as a function of lipid density [14,22]. Later this technique was combined with a simultaneous recording of the surface potential [6,14-17,23,24]. For determination of the latter, two techniques are available: the first measures the reversal potential, i.e., the voltage required to produce zero current across the interface. To measure the current, one electrode is placed in the subphase and the other is positioned in the air very close to the top of the lipid layer. To render the small air gap conductive, an alpha-emitter is placed nearby to ionize the air.

In our laboratory we use a second method, the vibrating plate technique based on a commercial instrument (KSV Instruments Ltd., Helsinki, Finland). The apparatus is shown schematically in Fig. 2. A lipid monolayer is prepared by spreading lipids dissolved in chloroform (1 mg/ml) [23]. After spreading, the chloroform is allowed to evaporate for ~15 min, and the film is compressed to an empirical "bilayer equivalence pressure" of ~32 mN/m [23-25]. We should note that the equivalence pressure is defined as the pressure at which a monolayer behaves similarly to a bilayer with respect to a particular property, e.g., phase-transition temperature, binding constant for an amphiphilic substance, etc. Nowadays it is widely accepted that there is no unique equivalence pressure -- it depends on the investigated property [26].

An Ag/AgCl electrode is connected to the subphase via an agar salt bridge. This conductive bridge prevents additives from adsorbing on the electrode and improves stability of the signal. A metallic conducting plate (~2 cm diameter) is placed less than 1 mm above the air-lipid interface. The plate and the subphase electrode are separated by an adjustable potential. The air space forms the gap of a plate condensor. The plate is now vibrated up and down at ~60 Hz. As the plate vibrates, its distance from the lipid/water interface continually changes. This produces changes of capacitance and, depending on the steady-state voltage across the capacitor, to charge flow to and from

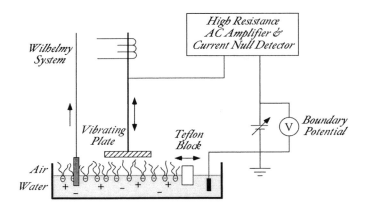

Fig. 2. Schema of the vibrating plate technique (VPT) combined with a Wilhelmy monolayer film balance. On the right is a teflon block that compresses the lipid surface. The change in surface pressure due to the presence of surface-active substances at the air-water interface is recorded by measuring the force needed to pull a filter platelet up through the surface (see textbooks [14]). The electrical setup is explained in the text.

the capacitor plate ($I = V \, dC/dt$). The current null detector determines the applied voltage between the plate and subphase electrode that yields zero current. From the above equation, it is seen that the total voltage across the air gap must then be zero. Under this condition the applied voltage is equal to the lipid-layer boundary potential.

The pure water-air interface is intrinsically polarized, and measurement of its absolute value is difficult [14]. However, as we are interested only in differential effects, the absolute value is not required. In our experiments we use the surface potential at the pure water-air interface as a criterion for cleanliness. If a large deviation (>50 mV) from the expected value occurs, the cleaning procedure is repeated. The lipids are spread as described and compressed to ~32 mN/m. As shown in Fig. 1, there is typically a positive potential between the bulk phase and the hydrophobic region of a lipid bilayer. For our monolayers, we observe potentials of ~400-600 mV, depending on the lipid. The resolution of the potential measurement is ~3 mV. Next we titrate the molecules of interest by additions to the subphase. For repeated experiments under identical conditions, the plateau values of a titrated adsorbent can vary by as much as 20 mV. The standard error of our measurements is <10 mV.

In Fig. 3A we show the results at low ionic strength for the titration of pentalysine with a lipid monolayer made of negatively-charged palmitoyloleoyl

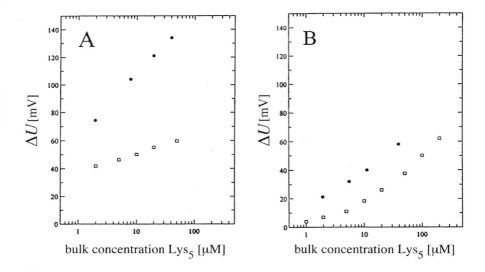

Fig. 3. Comparison of changes in the boundary potential upon adsorption of pentalysine to monolayers of POPS measured by the vibrating plate technique (VPT, filled circles), and also to bilayers of POPS measured by the inner-field-compensation technique (IFC, open squares). (See the next section for an explanation of IFC.) pH = 7.0 and T = 23° C. A: ionic strength 10 mM KCl. B: ionic strength 100 mM KCl.

phosphatidylserine (POPS) at pH 7, held at a pressure of 32 mN/m. Increasing the bulk concentration of pentalysine increases the adsorption of the positively-charged peptide onto the monolayer, producing positive shifts of the surface potential. In Fig. 3B we show the same experiment at a higher ionic strength. Raising the ionic strength reduces the response in 2 ways: (1) the negative POPS surface potential is diminished because of increased screening, which decreases its affinity for the positively-charged peptide, and (2) a given density of adsorbed peptide yields a smaller shift of surface potential, also due to increased screening. The Gouy equation (Eq. (2)) may be used to calculate charge-density increments, and thus the adsorption isotherms, from the measured shifts of surface potential.

3.2. Inner-field-compensation technique (IFC)

A second type of model membrane is the planar bilayer (or "black") lipid membrane (BLM). Such membranes generally are formed either by the "painting" technique of Mueller, Rudin, Tien and Wescott [27] or by the "folding" technique of Montal and Mueller [28]. In either case a thin bi-molecular film of ~30-80 Å thickness is formed, composed of specified phospholipids and varying amounts of organic solvent, spanning a small aperture in a hydrophobic partition separating two aqueous compartments. Bulk electrodes in the aqueous phases permit the measurement of currents and voltages across the film. The BLM has very high electrical resistance (10^8–10^{10} Ω-cm^2) and capacitance (0.3–1.3 μF/cm^2) [29]. BLMs have been shown to constitute an excellent model of the phospholipid phase of biological membranes [2,29,30]. The BLM setup has the advantage of allowing access to both sides of the membrane and permitting controlled modification of the membrane and its environment after formation.

IFC is a method well-suited to the study of transmembrane potentials across planar lipid membranes [6,31-39]. A nonvanishing transmembrane potential requires different surface-charge densities and/or different ionic strengths on the two sides of the BLM. Bilayers with asymmetric lipid compositions can be formed reproducibly by the so-called solvent-free "Montal-Mueller" technique [28,40]. However, this method is restricted to membranes of small diameter (~100 μm) with correspondingly small capacitances. Since the IFC method is based on capacitance measurements, we find these membranes too small to permit adequate resolution for our purposes. For our experiments, good resolution requires large membrane areas, thus we can use only solvent-containing "Mueller-Rudin" membranes [6,24,25,27,41]. These membranes are essentially symmetric, as there is nothing to produce asymmetry during formation. However, asymmetry can be achieved by forming the symmetric membrane first, then modifying one side only, e.g. by changing the pH or ionic strength or adding charged molecules that interact with the membrane. IFC then measures the difference between the two boundary potentials, i.e. the transmembrane potential $\Delta\psi$ (cf. Fig. 1).

It is thus possible to study the effect of peptide binding on the transmembrane potential. Here we describe the protocol for titrating small positively-charged peptides with a negatively-charged membrane [24]. In our laboratory we use homemade teflon cells containing two 5 ml chambers separated by a 1 mm thick teflon wall. A small section of the wall is drilled out to produce a 100-200 μm thick partition with an aperture of 0.5-2 mm diameter. Each side of the aperture is prepainted with about 1 μl of 1:99 (v:v) lipid-chloroform solution. After drying for 20 min, the chambers are filled with buffer, and 1 μl of 1:99 (v:v) lipid dissolved in n-decane is placed on a teflon loop and smeared across the aperture, closing the hole. Thinning of the film can be observed via a microscope ($\times$60 magnification) through a glass window in the cell wall. It is well-known that thinning of the membrane depends strongly on the details of prepainting the aperture [29,30,41]. Optical observation by reflected light 15 minutes after forming the membrane reveals in most cases a nearly black surface with silvery reflecting areas at the periphery. Prolonged waiting times or application of a series of voltage pulses does not then change the capacitance of the membranes significantly. The bulk solutions in both compartments are stirred gently with small magnetic stirrers. The front compartment is called *cis* and the rear compartment is *trans*.

From an electrical point of view a lipid bilayer can be regarded, to a good approximation, as a non-conducting condensor. The application of a transmembrane voltage induces mechanical forces that compress and thin the hydrophobic layer. These "Maxwell stresses" result from the attraction of opposite charges on the two sides of the bilayer as the membrane capacitor charges up. The thinning leads to a voltage-dependent increase of the BLM capacitance [29-39]. A transmembrane voltage can originate either from an externally applied voltage or from intrinsic membrane asymmetries. If a BLM has unequal intrinsic boundary potentials, the resulting intrinsic transmembrane potential thins the membrane and raises its capacitance. Application of a reverse external voltage reduces the Maxwell forces and decreases the capacitance, minimizing it when the net transmembrane potential is zero. This occurs when the applied voltage ΔV is equal and opposite to the intrinsic transmembrane potential $\Delta \psi$ (cf. Fig. 1). In this circumstance ΔV is known as the "compensation voltage".

In our experiments, shown in Fig. 4, we apply a sinusoidal ac voltage from the frequency generator of the lock-in amplifier (SR830 DSP, Stanford Research Systems, Sunnyvale, CA) at frequency 1061.9 Hz and amplitude 30 mV (all ac amplitudes given in rms). The ac voltage is applied to the membrane via a summator and causes a small oscillating compression of the membrane. The compressions produce changes of capacitance, which give rise to an

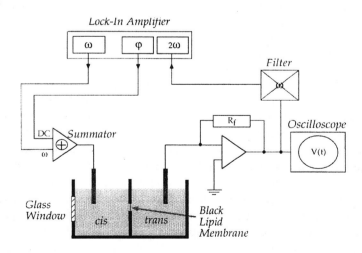

Fig. 4. Setup of the Inner-Field-Compensation technique (IFC), which measures changes in the boundary potential of a planar lipid membrane. See text for details.

oscillating capacitive current containing small components of higher harmonics [32,33]. This current is converted to voltage via a current amplifier, and the amplitude of the second harmonic is detected by the lock-in amplifier. As shown in the chapter by Ermakov, the amplitude of the second-harmonic signal is proportional to the voltage drop in the hydrophobic core of the bilayer, i.e., to the transmembrane potential [32-39]. The lock-in amplifier puts out a dc voltage proportional to the amplitude of the second harmonic signal. This voltage is then fed back to compensate the intrinsic voltage drop across the membrane. The higher the sensitivity of the lock-in amplifier, the better this compensation voltage corresponds to the negative of the intrinsic transmembrane voltage. However, the increase in sensitivity also increases noise and instability of the circuit. We thus perform measurements at a sensitivity for which a further sensitivity increase hardly changes the compensation voltage. The compensation voltage is finally converted to a digital signal by a DigiData 1200 Interface (Axon Instruments, Foster City, CA) and recorded on a PC computer at an acquisition rate of one point per second using the program pCLAMP6 (Axon Instruments). The data are smoothed with the program EasyPlot (Spiral Software, Brookline, MA) using the sliding data window option. The resolution of the IFC is ~1 mV. The standard error of our measurements is < 5 mV.

In Fig. 5 we show a typical titration curve of pentalysine interacting with a planar lipid membrane of diphytanoyl phosphatidylserine (DPhPS) lipids. Potentials in Fig. 5 are measured on the *cis* side relative to *trans*. The plot shows the compensation voltages necessary to hold the membrane at zero transmembrane potential, thus negative shifts of ΔU_{comp} correspond to positive shifts of $\Delta \psi_{cis}$. Within 2 minutes after pentalysine addition to the *cis* compartment, the IFC detected a positive shift of the transmembrane potential, indicating the adsorption of pentalysine. After 10 min the transmembrane potential reached a plateau value of ~8 mV in the presence of 100 mM KCl with 2 µM pentalysine in the bulk phase. Further increase of the bulk concentration of pentalysine on one side of the BLM increases is adsorption to the oppositely-charged membrane surface. In some cases we added pentalysine to both sides and waited until the transmembrane potential vanished, indicating symmetric adsorption. An additional control measurement was done with uncharged membranes made of diphytanoyl phosphatidylcholine (DPhPC) or palmitoyl-oleoyl phosphatidylcholine (POPC), and no binding could be detected. Here we assume that pentalysine does not translocate across the membrane, which seems reasonable since at this pH pentalysine has about 5 positive charges. However, in general, measurements must be performed to test for permeation of the additive. If translocation occurs within range of our time resolution, it can be

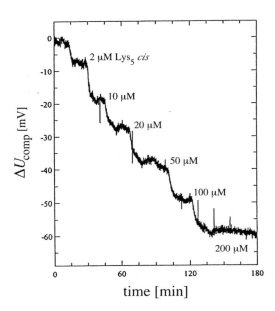

time [min]

Fig. 5. Experimental IFC trace for the titration of pentalysine with a DPhPS BLM in 100 mM KCl. The indicated concentrations are total pentalysine concentrations in the bulk phase. Negative shifts of the compensation voltage ΔU_{comp} correspond to positive shifts of the transmembrane potential $\Delta \psi$ on the *cis* side relative to *trans*.

detected by IFC as a time-dependent decrease of the initial voltage shift, and conclusions regarding the permeation can made.

Earlier in Fig. 3 we compared the changes of surface potential produced by additions of increasing amounts of pentalysine for two different model membranes: the lipid bilayer measured by IFC and the lipid monolayer measured by VPT. In both cases the titrations were performed with the same buffer solutions, same pH, and the same negatively-charged lipids (POPS). The boundary-potential shifts measured for bilayers were generally smaller than those measured for monolayers. For both systems pentalysine binding produced smaller changes of surface potential at high ionic strength, Fig. 3B, compared to those at low ionic strength, Fig. 3A, demonstrating the screening effect of the background ions. In Fig. 3A a small pentalysine concentration gave a large IFC signal, which increased only slightly with increasing concentration. These data were reproduced repeatedly, and the discrepancy between the monolayer and bilayer data was greater than experimental uncertainty. Currently we do not have an explanation for this observation, nor should this result be generalized. The discrepancy might be specific to pentalysine.

3.3. Conductance probes

As mentioned earlier, the electrostatic surface potential arises from the presence of uncompensated charges on the membrane, usually associated with deprotonated acidic (such as carboxyl) groups or protonated basic (such as amino) groups that are integral to the membrane lipids or proteins. Electrostatic surface potentials may be of order 100 mV, but they are undetectable by bulk electrodes. The reason is that in biological media the electrostatic surface potential is screened by aqueous counterions within ~10 Å from the membrane surface. Bulk electrodes are much too big to probe the membrane surface to within such microscopic distances. Nevertheless, any ion or charged macromolecule traversing the membrane or binding to its surface encounters these potentials, with profound biological effects.

Several methods for detecting and measuring these surface potentials have been discussed. Monolayer techniques measure a boundary potential that includes the electrostatic surface potential plus dipolar contributions (section 3.1). The IFC method detects asymmetric boundary potentials across planar bilayer membranes, which include contributions from electrostatic surface potentials as well as from dipole potentials (section 3.2). Here we describe a third method based on the incorporation of an extrinsic molecular probe, i.e., a molecule that senses the surface potential and in doing so generates a detectable signal. Whereas the monolayer and IFC methods are based on capacitance measurements, in this case the signal is a change of conductance. It is useful that some conductance probes are sensitive to electrostatic surface potentials but are relatively *in*sensitive to dipole potentials. Thus comparison with monolayer and IFC data permits separation of the various contributions.

BLMs, whether formed by the "painting" technique of Mueller, Rudin, Tien and Wescott [27] or by the "folding" technique of Montal and Mueller [28], form thin films with remarkably low electrical conductance. However bare BLMs can be doped with various molecules that significantly modify their transport characteristics [42]. Of interest here are ionophoric carriers and ion-specific channels which provide conductances that are sensitive to BLM electrostatic surface potentials and internal electric fields. Several examples are the electrogenic carriers valinomycin and nonactin, the voltage-gated channel-former monazomycin, and the cation-selective channel gramicidin A. These probes may be used with either the painted or folded BLMs.

3.3.1. Valinomycin and nonactin

Valinomycin is a neutral cyclic dodecadepsipeptide antibiotic having a 36-member ring, and nonactin is a neutral macrotetralide antibiotic with a 32-member ring [2,43-45]. Both are hydrophobic and partition strongly into the hydrocarbon region of the membrane. They also bind K^+ with some specificity. When present in a BLM, these molecules undergo association-dissociation reactions with aqueous K^+ at the membrane-water interface by substituting a polar molecular cage for the K^+ hydration shell.   The cationic complex, however, is still hydrophobic enough to remain dissolved in the membrane, thus can "carry" K^+ through the membrane interior when driven by an applied electric field.   The interfacial association-dissociation reactions follow mass-action laws, thus the concentration of complex formed at each interface is proportional to the aqueous K^+ concentration at the interface.

If K^+ is present equally on both sides of the membrane, then the membrane conductance G is proportional to $[K^+]$. However, it is the *interfacial* concentration $[K^+]_0$ that is relevant: $G \propto [K^+]_0$. If the membrane is charged and has an electrostatic surface potential ψ_0, then $[K^+]_0$ is given by the Boltzmann relation: $[K^+]_0 = [K^+]_{bulk} \exp(-e\psi_0/kT)$. Therefore measurement of G yields information about ψ_0. In order to conduct, the charged complex must jump over hydrophobic and dipole energy barriers inside the bilayer. These effects enter the proportionality constant above. One type of experiment is done under symmetric conditions, where the ionophore concentration and $[K^+]_{bulk}$ are held constant and a charged adsorbent is added symmetrically to both aqueous chambers.   If the adsorbent causes a shift of the surface potential $\Delta\psi_0$, it follows that:

$$\Delta\psi_0 = -\frac{kT}{e}\ln(G_2/G_1),\tag{5}$$

where G_1 and G_2 are the measured conductances before and after the addition, respectively [45]. In Eq. (5) it is assumed that the additive itself does not conduct, and that its adsorption does not affect dipole or hydrophobic potentials. The nonactin method has been used by McLaughlin et al. [45] to detect the adsorption of divalent cations to negatively-charged membranes.   The uncertainty of these measurements is about ±5 mV.

Asymmetric experiments are also possible.   The addition of a charged adsorbent to one side (the *cis* side) of a BLM under symmetric electrolyte conditions shifts ψ_0 on the *cis* side. ψ_0 on the *trans* side will shift only if the adsorbent permeates the membrane and delivers charge to the *trans* surface. In addition to a conductance change, the current-voltage curve now becomes asymmetric. The method thus can be used to detect not only adsorption, but also charge permeation.   However, quantitative analysis of the asymmetry of the current-voltage curve is problematic, as it depends on the detailed shape of the membrane energy barrier [46], which usually is not known. We have used this technique to monitor the interaction of charged liposomes with neutral BLMs, reasoning that adsorption or hemifusion should impart charge to the *cis* side only, while full fusion should add charge to both sides. Under our conditions, the charge appeared preferentially on the *cis* side of the BLM [47]. A difficulty in the interpretation of such experiments is that unambiguous assignment of observed charge alterations to the *cis* vs. *trans* sides of the membrane is not possible. In all experiments, but especially those involving liposomes, controls must be done to ascertain whether additives to the *cis* aqueous phase cause repartitioning of the conductance probe out of the BLM [47].

3.3.2. Monazomycin and LL-A491

A very different class of conductance probes is the voltage-gated channel-formers such as monazomycin or its close relative LL-A491. Monazomycin is a polyene-like antibiotic of molecular weight 1365 Da containing a 48-member lactone ring, a primary amino charge, and a mannose side-group [48]. The 3-dimensional structure has not been determined, but it has been proposed that the ring collapses to a rod-like shape ~40 Å long and 5-7 Å wide with the amino charge at one end and the polar mannose near the other. The intermediate region approximates a flat rod, one side exposing mainly polar hydroxyl groups and the other mainly nonpolar methyl groups [49,50]. Thus the molecule is a surfactant. It is soluble in water, methanol and ethanol, but adsorbs at lipid/water interfaces.

Its proposed mechanism of action [51] is as follows: When added to the aqueous phase on one side of a BLM (*cis* side), it adsorbs at the interface. A *cis*-positive voltage then drives the positively-charged amino end of the molecule through the bilayer to the *trans* side, while the neutral polar end remains anchored on the *cis* side. Molecules spanning the bilayer then aggregate into oligomeric channels of average size ~5, with their polar hydroxyls facing inward toward the lumen and nonpolar groups facing outward. The channels are selective for monovalent cations. Since monazomycin is inserted into the bilayer by positive electric fields, the current-voltage curve is highly asymmetric, behaving like a diode. Because of the oligomerization, the forward-biased I-V curve is very steep, changing *e*-fold in ~5 mV. It must be appreciated that this rectifying behavior is not due to the single-channel conductances, which are essentially ohmic [52], but follows from the field-dependent injection of channels into the bilayer. The utility of monazomycin for our purposes arises from the sensitivity of its macroscopic conductance to the intramembrane electric field, which is related to the transmembrane voltage and the electrostatic surface potentials.

Monazomycin is most useful when used under asymmetric conditions. In a typical experiment, monazomycin dissolved in water or ethanol is added to the *cis* chamber of a pre-formed BLM, with stirring, to give ~0.5-10 µg/ml final concentration. The monazomycin concentration required is somewhat variable depending on the BLM lipids, surface-charge density, and ionic strength. The criterion is that the steady-state monazomycin-mediated conductance be ~2 orders of magnitude higher than the bare BLM conductance. After the monazomycin addition, 10-20 mV positive bias is applied. The insertion appears to require a nucleating event, after which the conductance rises sigmoidally with time. We typically wait ~20 minutes to ensure the attainment of steady state, then switch the circuit to current-clamp, as shown in Fig. 6. We operate at a current of 0.5-2 nA, and the monitored voltage across the bulk electrodes is 10-100 mV.

The steady-state behavior of monazomycin in BLMs has been described in Refs. [51]. The monazomycin-mediated conductance is exponential in the applied voltage:

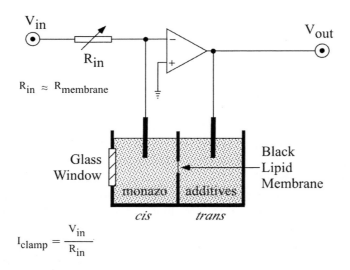

Fig. 6. Schema of current-clamp setup for measuring changes in the *trans* electrostatic surface potential of a BLM by the monazomycin technique as charged adsorbents such as pentalysine are added to the *trans* chamber. The high-impedance operational amplifier holds the *cis* side at virtual ground. The clamped current is V_{in}/R_{in}, and V_{out} is the voltage in the *trans* bulk compartment relative to ground. However the circuit inverts the output voltage, so $V_{out} = V_{cis} - V_{trans}$. Thus an applied current producing positive V_{out} forward-biases the membrane, injecting the monazomycin. See text for details.

$$G_{mon} = I/V \propto \left[mon^+ \right]_0^s \exp(neV/kT), \tag{6}$$

where V means $V_{cis} - V_{trans}$, $[mon^+]_0$ is the aqueous monazomycin concentration at the *cis* interface, *s* is a parameter related to the mean oligomer size of conducting channels, and *n* is a constant related both to the mean channel size and to the fraction of the transmembrane voltage sensed by monazomycin monomers during injection. The test for a monazomycin-dominated conductance is that ln G be linear in V. Proper behavior is confirmed by stepping the current clamp, typically by factors of 10, to several different values, then using Eq. (6) to evaluate *n*, which generally falls in the range 4-6. The value of *n* must be determined for each experiment and should also be checked after the addition of adsorbents. The parameter *s* is evaluated from titrations of G vs. $[mon^+]$ at constant V, and is typically ~5.

If electrostatic surface potentials are present, then the transmembrane potential driving monazomycin injection is $V + \psi_0^{cis} - \psi_0^{trans}$. In this case the interfacial value of $[mon^+]$ is given by the Boltzmann expression:

$$\left[mon^+ \right]_0^{cis} = \left[mon^+ \right]_\infty^{cis} \exp\left(-e\psi_0^{cis}/kT \right), \tag{7}$$

and Eq. (6) becomes:

$$G_{mon} \propto \left[mon^+ \right]_\infty^s \exp\left\{ \frac{(n-s)e}{kT} \psi_0^{cis} \right\} \exp\left\{ \frac{ne}{kT}\left(V - \psi_0^{trans} \right) \right\}. \tag{8}$$

Since $n \approx s$, the first exp factor ~1 and the dependence of G_{mon} on ψ_0^{cis} is very weak or absent. Any increase of injection due to a positive ψ_0^{cis} forward-biasing the internal electric field is nearly offset by a decrease of $[mon^+]_0$ due to electrostatic repulsion of monazomycin from the *cis* surface. If $n = s$, then the cancellation is exact. In the second exp factor, since $n \sim 5$, the dependence of G_{mon} on ψ_0^{trans} is strong.

Therefore when adsorbents are added to the system, the monazomycin-mediated conductance responds sensitively to surface-potential shifts $\Delta\psi_0$ on the *trans* side but not on the *cis* side. This situation has several advantages: (1) A *trans* additive does not interact with *cis* aqueous monazomycin, thus the experiment is relatively clean. (2) *Trans* additions of charged molecules permit straightforward measurement of adsorption isotherms on the *trans* surface. Alternatively, if a charged additive is added on the *cis* side, $\Delta\psi_0$ will be seen only if the charge permeates and appears on the *trans* side. (3) The asymmetric assignment of surface-charge alterations is relatively clear. Due to the high power of n, this method is considerably more sensitive than the carrier method for the detection of asymmetric shifts of surface potentials. The sensitivity to $\Delta\psi_0^{trans}$ is about ± 1 mV.

There are several advantages to performing these measurements in current-clamp. First, due to the very steep I-V curve, voltage-clamp yields noisy and difficult-to-control currents. However, in current-clamp the voltages are very stable. Second, because of its amphiphilic nature, monazomycin acts as a detergent and destabilizes the membrane. In current-clamp, as opposed to voltage-clamp, the amount of injected monazomycin remains nearly constant during shifts of ψ_0^{trans}, thus improving membrane stability. Third, it can be shown [53] that when adsorbents are added in current-clamp, the monitored voltage shift ΔV is related to $\Delta\psi_0^{trans}$ via:

$$\Delta\psi_0^{trans} = \Delta V + \frac{kT}{ne}\ln\frac{V_2}{V_1}, \tag{9}$$

where V_1 and V_2 are the voltages measured by the bulk electrodes before and after the addition, respectively, and $\Delta V \equiv V_2 - V_1$. Under normal conditions the second term in Eq. (9) is small. It is assumed in Eq. (9) that $s = n$; if not, a correction can be made. Thus, monitoring $V(t)$ yields $\Delta\psi_0^{trans}(t)$ in real time. As long as the BLM conductance is monazomycin-dominated, Eq. (9) is independent of the monazomycin concentration.

We have used this method to detect the adsorption of liposomes, polylysine, cytochrome c, Ca^{++} and Mg^{++} to planar BLMs [53,54], and have also used it to test the Gouy-Chapman-Stern theory by symmetric titrations of monovalent cations [55]. In addition, we have shown that the monazomycin technique can be used to monitor the electroneutral transport of Ca^{++} and Mg^{++} (each with countertransport of $2H^+$) by the ionophores X-537A and A23187 through highly-charged phosphatidylserine BLMs [56]. In this experiment, both monazomycin and the ionophore are present in the membrane simultaneously. The divalent cations are added *cis*, and upon being translocated to the *trans* surface by the ionophore, they are trapped by the *trans* unstirred layer where they screen and bind to the *trans* interface, thus producing a shift $\Delta\psi_0^{trans}$ which is sensed by the monazomycin. In this manner transmembrane ion fluxes, even if electroneutral, can be measured very sensitively by use of the monazomycin probe [56].

In Fig. 7 we show a *trans* titration of pentalysine adsorbing to a negatively-charged DPhPS BLM. In such experiments monazomycin (here LL-A491) and other additives are introduced in small concentrated aliquots to minimize dilution of the electrolyte. The positive voltage shifts observed

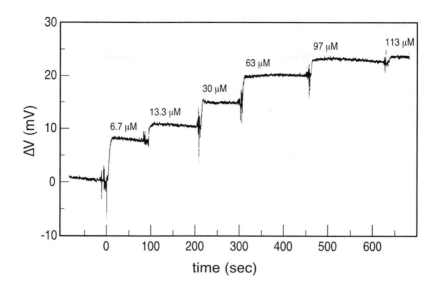

Fig. 7. Experimental trace by the monazomycin technique for the titration of pentalysine with DPhPS BLM in 100 mM KCl. Antibiotic LL-A491 has been added to the *cis* chamber at 0.6 μg/ml. The membrane is current-clamped at 2 nA. Indicated values correspond to total bulk pentalysine concentrations in the *trans* chamber. Positive values of ΔV correspond to positive shifts of the *trans* electrostatic surface potential $\Delta\psi_0^{trans}$.

following each addition correspond to $\Delta\psi_0^{trans}$, as given by Eq. (9). It is remarkable that voltage shifts ΔV measured by bulk electrodes "infinitely" far from the membrane can sense $\Delta\psi_0^{trans}$, which is fully screened within 10-100 Å from the membrane-electrolyte interface. In this sense, monazomycin may be said to function as a "molecular electrode".

In Fig. 8A we show the effects of pentalysine titrations on the boundary potentials of POPS membranes measured by three different techniques under identical conditions. As previously noted, the VPT monolayer measurements show larger shifts than the IFC bilayer measurements. The monazomycin method gives consistently smaller shifts than those measured by IFC.

The reasons for the smaller effect seen by the monazomycin probe relative to IFC are unknown, but there a number of possibilities. IFC determines the applied voltage that causes minimal compression of the BLM, which occurs when the transmembrane potential drop and internal electric field are zero. This null condition includes contributions from asymmetric dipole potentials as well as from asymmetric electrostatic surface potentials. Monazomycin, on the other hand, is quite *in*sensitive to changes of dipole potentials. We and others [57] have found that monazomycin responds very weakly to additions of phloretin,

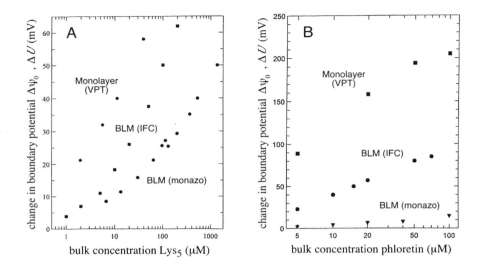

Fig. 8. Comparison of changes in the boundary potentials of POPS membranes measured by 3 different techniques: monazomycin probe with BLMs, IFC with BLMs, and VPT with monolayers. The conditions are the same in all 3 cases: 100 mM KCl, pH 7.0, T = 23°C. A: Pentalysine titration. (The IFC and VPT data are the same as in Fig. 3B.) B: Phloretin titration. See text for details.

an uncharged molecule that decreases membrane dipole potentials [58]. This feature is illustrated in Fig. 8B. Thus, one explanation for Fig. 8A could be that pentalysine alters the BLM electrostatic and dipole potentials, both of which are sensed by IFC. The monazomycin curve would then represent only the electrostatic component of the IFC curve.

It is difficult to reconcile monazomycin's lack of dipole-potential sensitivity with the fact that the ratio n/s is commonly found to be ~0.9 [51,57], suggesting that the monazomycin charge moves through only ~90% of the transmembrane-potential drop during insertion. If this were the case, however, the monazomycin charge should sense the dipole-potential jump that lies very near the interface, thus should be sensitive to dipole-potential alterations. This dilemma has not been resolved.

A further unknown factor is the extent to which monazomycin itself alters the membrane surface-charge density when it adsorbs on the *cis* surface or when it translocates to the *trans* surface as channels collapse [49]. The amount of monazomycin on the *cis* surface associated with conducting channels can be estimated from the conductance, but this enumeration does not include adsorbed non-conducting monomers or dimers, which may be present in significant amounts. Fortunately the monazomycin conductance is insensitive to *cis* surface charge, as previously discussed. In preliminary studies, we have begun to measure monazomycin adsorption isotherms by two methods: first by *trans* titrations of monazomycin, similar to experiments with other charged adsorbents (thus using *cis* monazomycin as a probe to monitor its *trans* adsorption), and second by use of the IFC technique. It is possible that some of the reduced monazomycin response observed experimentally is due to self-inactivation caused by charge translocated to the *trans* surface when the channels turn off and eject on the *trans* side [49]. A full characterization of the monazomycin technique awaits the resolution of these questions.

3.3.3. Gramicidin A

Finally, mention should be made of the technique of Rostovtseva et al. [59] for measuring electrostatic surface potentials with gramicidin-A single channels. These channels are uncharged and are highly cation-selective. Their conductance is related to the interfacial concentrations of the cations, which depend on the electrostatic surface potential via Boltzmann factors, similarly to the case of nonactin and valinomycin. An advantage to using single channels is that the measured conductances are microscopic, not macroscopic, thus do not depend on the amount of probe that partitions into the membrane. Hence this variable is eliminated. A single channel is either present or not, and the channel conductances are measured individually. For details of this elegant technique, see Ref. [59].

3.4. Membrane rupture

External electric fields have been used for a long time to permeabilize cells [60]. Highly-charged molecules such as polylysine are known to reduce the barrier for membrane destabilization by a similar mechanism [61]. This fact can be used to probe transmembrane potentials in an indirect way and to study the effects of these potentials on the mechanical properties of lipid membranes.

In our experiments we apply a short voltage pulse across a planar BLM, which induces surface charges of opposite sign at the two lipid/water interfaces [24,62]. These charged layers attract, producing so-called Maxwell stresses that depend quadratically on the transmembrane electric field. These are the same stresses that cause BLM thinning in the IFC technique (section 3.2). The membrane thus undergoes mechanical strain, and its ability to bear such strain is limited. When the limit is reached, the film ruptures. As before, the field driving the membrane response is the externally-applied field plus the intrinsic field resulting from asymmetric boundary potentials. Analysis of the applied-field strength needed for breakdown yields information on the pre-existing intrinsic field, i.e., on the intrinsic transmembrane potential.

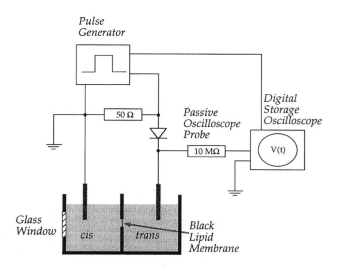

Fig. 9. Schema of the charge-pulse instrumentation used for measurement of the electric-field-induced irreversible breakdown of planar lipid membranes.

The rupture experiments were initiated using the experimental setup shown in Fig. 9. The Ag/AgCl electrode in the *trans* compartment is connected to a fast pulse generator (Tektronix PG 507) through a diode (reverse resistance $\gg 10^{11}\Omega$). The 50 Ω resistor guarantees a fast disconnect of the power supply at the end of the pulse. The *cis* electrode is grounded. The voltage between the two electrodes is recorded on a digital storage oscilloscope (LeCroy 9354A, input resistance 1 MΩ). Prior to rupture, the membrane capacitance is measured by charging the bilayer with a +100 mV rectangular pulse of 1 µs duration. The membrane capacitance is then determined from the RC time constant of the transmembrane voltage decay through the membrane capacitance and the parallel 10 MΩ resistance of the oscilloscope probe. Irreversible breakdown may be initiated by charging the membrane with a 300-800 mV rectangular pulse of 10 µs duration. Application of a high-voltage pulse causes many immediate defects and fast rupture of the membrane. To make quantitative measurements of membrane stability, we instead increase the applied voltage in small steps, which allows for the creation of single pores rather than multipores. Thus in our protocol the critical voltage for membrane rupture is approached carefully.

Typical traces of the transmembrane voltage vs. time are shown in Fig. 10. First the voltage pulse charges up the membrane. At the end of the charging pulse (here 20 µs), the voltage decays slowly due to RC discharge through the membrane capacitance and the parallel 10 MΩ resistor. Above

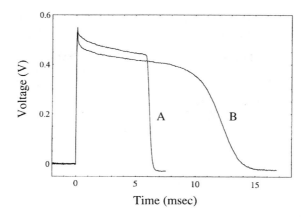

Fig. 10. Typical time course of the transmembrane voltage during electric-field-induced irreversible rupture. A: DPhPS membrane in 100 mM KCl in the presence of short polylysine ($N = 38$). B: The same membrane composition and electrolyte but in the presence of long polylysine ($N = 2007$). The rupture voltage in B is about 50 mV less than in A. The rupture velocity is strongly reduced by adsorption of the long polymer, and the membrane appears viscous. See text for details.

a critical voltage, the mechanical strain is large enough to nucleate an irreversible pore, which occurs after a delay (here 5-10 msec). The membrane then discharges rapidly through the highly-conductive pore, producing a fast drop of the transmembrane voltage. We define the "breakdown voltage" as the voltage on the discharge curve located at the transition between the slow and fast kinetic segments.

We should note that a breakdown voltage has analytical relevance only if the critical rupture voltage is approached carefully [62]. The application of higher voltages causes rupture as well, but the influence of the intrinsic transmembrane potential is masked. Here we analyze only voltages close to the critical voltage which lead to the creation of single -- or only a few -- pores. It is not obvious that membrane rupture follows from the widening of a single pore. The kinetics of pore growth is obtained from the time-dependence of the voltage discharge across the membrane. Pore appearance is characterized by a sudden superexponential decay. Statistical analysis of repeated measurements under identical conditions suggests that this decay is indeed caused by the widening of a single pore [62]. After carefully approaching the critical voltage leading to rupture, we usually find that the rupture velocity has a discrete minimal value, indicative of single-pore growth. In some cases integer multiples of this minimal value are seen, indicating multipores.

In a series of experiments we formed DPhPS membranes as described previously. After formation, polylysine was added to the bulk phase on one side of the BLM. Positively-charged polylysine adsorbs to the negatively-charged surface and induces a transmembrane electric field. The adsorption was monitored by use of the IFC technique as previously described. For additions of pentalysine (degree of polymerization $N = 5$) in 100 mM KCl, we observed at ~20 µg/ml (~30 µM) a transmembrane voltage of ~35 mV [24]. Polylysine of approximate size $N = 38$ gave 67 mV at similar monomer concentrations, and larger polylysines gave about 77 mV. After the polylysine binding reached equilibrium, we disconnected the IFC circuit and induced membrane rupture.

The effects of polylysine adsorption on the breakdown voltages of DPhPS membranes are summarized in Table 1. In 100 mM KCl with no polylysine present, DPhPS membranes rupture at ~530 mV. Symmetric addition of pentalysine ($N = 5$) show almost the same mean breakdown voltage as bare DPhPS membranes. Addition of pentalysine on one side of the BLM induces a transmembrane potential of 35 mV, which adds to (or subtracts from) the externally-applied voltage for *trans* (or *cis*) additions. Here the external voltages are applied on the *trans* side relative to the grounded *cis* side. As shown in Table 1, the *trans*-side adsorption of pentalysine reduces the critical applied voltage, whereas the *cis*-side addition increases it. Qualitatively similar effects are seen for the larger polylysines as well.

Table 1
Average breakdown voltages for DPhPS membranes in the presence of various polylysines

Membrane and Additives (100 mM KCl)	Breakdown Voltage (mV) for Symmetric Adsorption	Transmembrane Voltage (mV) for Asymmetric Adsorption	Breakdown Voltage (mV) for Asymmetric Adsorption	
			(trans)	(cis)
DPhPS, no PL	530 ± 20	---	---	---
DPhPS, PL (N=5)	525 ± 15	35 ± 2	482 ± 25	552 ± 22
DPhPS, PL (N=38)	540 ± 24	67 ± 2	485 ± 25	619 ± 25
DPhPS, PL (N=251)	500 ± 20	78 ± 3	425 ± 25	590 ± 27
DPhPS, PL (N=2007)	426 ± 20	76 ± 3	334 ± 25	486 ± 25

Polylysines (*PL*; $N =$ polymer index) are present at about 20 µg/ml as described in the text. Transmembrane voltages induced by asymmetric adsorption were measured by the IFC technique. External voltages are applied on the *trans* side relative to *cis*. Uncertainties are standard errors of the mean for at least 10 independent measurements. DPhPS, diphytanoyl phosphatidylserine.

In summary, we find that membrane rupture experiments can provide useful insight into the effects of both intrinsic and extrinsic transmembrane electric fields on the mechanical properties of bilayer lipid membranes.

3.5. Electrophoretic mobility

Another widely-used model lipid membrane system is liposomes. There is an elegant way to obtain information about the sign and magnitude of the liposome surface charge. The liposomes are placed in an electric field, and their steady-state velocity is measured. The velocity is related to their charge. An excellent treatment of this subject is given in the classic book by Hunter [63]. Here we summarize the main points and add some comments.

Various techniques permit the formulation of liposomes in a fairly reproducible manner. Our preferred method [64] is to dissolve ~10 mg lipids in ~1 ml chloroform, evaporate in a glass flask with a stream of N_2, then dry in vacuum for 6 h. The dried film can be stored under argon at $-20°C$ for a few days. The desired aqueous buffer is then added to the flask, and the film is suspended by gentle shaking followed by heavy vortexing for several min, which produces a polydisperse suspension of multilamellar vesicles. In these vesicles, each lamella is a phospholipid bilayer separated by a small water space from neighboring lamellae. The suspension is now frozen by immersion in a liquid-nitrogen bath for a few seconds, then warmed in a water bath. Repeating this freeze-thaw procedure about 10 times produces unilamellar vesicles. The vesicles can then be sized by extrusion through polycarbonate filters of 100 or 200 nm pore diameter. Dynamic light scattering reveals a rather homogeneous size distribution. These liposomes serve as a simple model of the phospholipid phase of biological cell membranes.

The force that drives the particle is qE, where q is the particle charge, and E is the magnitude of the externally-applied electric field. Three retarding forces oppose the particle's motion. One is the "viscous resistance force" due to Stokes friction between the particle and solvent, $f_v = -6\pi a\eta v$ for spheres, where a is the particle radius, η is the solvent viscosity, and v is the particle's velocity. The second is the "electrophoretic retardation force," due to viscous shearing of the solvent. The layer of water in contact with the particle surface sticks to the surface and moves with it ("no-slip" boundary condition). However, the bulk water remains stationary. Therefore the water phase must shear over a region extending outward from the surface toward the bulk. By Newton's third law, since the particle imparts a shearing force on the solvent, the solvent exerts a counterforce on the particle, which slows it down. The total shearing force acting on the particle may be calculated by analysis of the complementary electroosmosis problem in which field-driven electrolyte flows relative to a stationary charged surface. For flat surfaces, the one-dimensional

Navier-Stokes equation in combination with Poisson's equation yields an analytical solution for the bulk electroosmotic velocity, which leads to the famous Smoluchowski formula relating electrophoretic mobility to zeta potential (see below).

The third retarding force results from the "electrophoretic relaxation effect." The applied electric field acts also on counterions clustered around the charged particle, pushing them in a direction opposite to that of the particle's motion. The counterion cloud remains associated with the particle; however, the external field is strong enough to distort the cloud, elongating it "behind" the particle as it moves. The extra counterions behind the particle, relative to those in front of it, electrostatically pull back on the particle and slow it down.

Several points deserve mention: First, when the electric field is applied, the particle accelerates to its terminal velocity very quickly. The acceleration phase is not observed. Second, for experimentally-relevant applied fields, the particle's velocity is linear in the applied-field strength. Therefore it is customary to express the particle's motion in terms of velocity per unit field, $\mu_e \equiv v/E$, where μ_e is the "electrophoretic mobility". Third, the referred-to particle is a "hydrodynamic particle", and its total charge includes all charges located within its outer boundary known as its "shear surface", which sits further out than its physical surface. The net particle charge therefore includes all intrinsic charge, adsorbed charge, and counterion screening charge located within the shear surface. Thus, even in the absence of ion adsorption, the hydrodynamic particle's charge depends on the electrolyte ionic strength. Fourth, the electrostatic potential at the shear surface is defined as the "zeta potential". A major problem in the interpretation of electrophoretic data is uncertainty about the exact nature and location of the shear surface, and hence about the exact meaning of the zeta potential.

There is no general analytic solution for the particle charge in terms of the particle's velocity in an electric field. However, analytic solutions do exist for limiting cases, and analytic approximate solutions exist in certain regimes. The key parameter defining the various regimes is κa, where κ is the Debye-Hückel constant and a is the particle radius. Since the Debye screening length $\lambda_{Debye} \equiv \kappa^{-1}$, κa may be understood as the ratio of the particle radius to the Debye screening length. λ_{Debye} sets the length-scale for electrostatic effects. When $\kappa a \gg 1$ (i.e. $\lambda_{Debye} \ll a$), the screening layer is said to be "thin", and the screening charge sees a surface of relatively large radius of curvature. For this situation, the dominant retarding force is viscous shearing of the water. In the limit $\kappa a \to \infty$, known as the Smoluchowski limit, since κ is finite, $a \to \infty$ and the surface is flat. No Stokes friction and no relaxation effect are possible in this limit. The mobility is determined, as mentioned, by calculating the electroosmotic velocity, v_{EO}, of field-driven bulk electrolyte flowing past a stationary, charged, infinite plane. Using $\mu_e = -v_{EO}/E$, this solution leads directly to the Smoluchowski equation: $\zeta = (\eta/\varepsilon)\mu_e$ [63,64]. For water at 25°C, the formula in practical units is $\zeta(mV) = 12.80\,\mu_e(\mu m\cdot cm/V\cdot sec)$ [64]. For $\kappa a > 500$ and not-too-large zeta potentials, the error associated with the use of this formula is $< 5\%$ [63].

We shall not discuss "thick" screening layers $\kappa a \ll 1$ (i.e. $\lambda_{Debye} \gg a$), since they are generally not relevant for liposomes, except to note that in this regime the dominant retarding force is Stokes friction. The limit $\kappa a \to 0$ is known as the Hückel limit, and an analytic expression for ζ in terms of μ_e also exists. Consult [63] for details.

The crossover region between the Smoluchowski and Hückel regimes is complex. Henry calculated a smooth function joining the two limits, but did not include relaxation effects which are important. Excellent numerical results covering the whole range of κa values, including relaxation effects, are given in a now-classic work by O'Brien and White [65]. Very unusual behavior was found in the range $\kappa a \sim 2\text{-}10$ for zeta potentials >100 mV, where the dependence of ζ on μ_e becomes indeterminate

For liposomes at experimentally relevant ionic strengths, the electrokinetic regime depends on the liposome size. In our experiments with mutilamellar liposomes (average diameter $\sim 3.5\,\mu m$), our most extreme condition was $\zeta \sim -100$ mV at 0.5 mM ionic strength, thus $\kappa a \sim 135$. The analytic approximate formula of O'Brien and Hunter [66], which is accurate to several percent or better in this range, was used to compute the zeta potential from the mobility. This calculated value was about 20% larger than the Smoluchowski value. The situation is much more severe when mobilities are measured for unilamellar liposomes or DNA-lipid complexes of typical diameter 100 nm. Here 0.5 mM ionic strength gives $\kappa a \sim 4$, which lies in the above-mentioned anomalous regime where only zeta potentials less than ~ 100 mV can be determined with acceptable accuracy [65]. To obtain well-defined zeta potentials, it is beneficial to use large liposomes, especially with low ionic-strength electrolytes.

A useful comment is that commercial zeta-potential instruments often provide the Smoluchowski zeta potential as a programmed readout. It should be appreciated that this number is calculated from the experimentally-measured mobility by simple use of the Smoluchowski formula and *a priori* has no further significance. The experimenter must determine κa from size measurements and decide whether the Smoluchowski value is applicable. For given values of η and ε, the Smoluchowski zeta potential may be viewed simply as a mobility expressed in mV units. Its relevance to the actual zeta potential depends on κa.

For a given mobility, the Smoluchowski value always underestimates the true zeta potential.

Once the zeta potential has been determined, the particle charge can be found. In the Smoluchowski regime the net surface-charge density of a particle in an electrolyte is related to the electrostatic potential at the particle's surface by the Gouy equation. The zeta potential thus yields the surface-charge density of the hydrodynamic particle. For a planar geometry and a monovalent salt, the equation is [67]:

$$\sigma^* = \left(8n_\infty \varepsilon_w \varepsilon_0 kT\right)^{1/2} \sinh\left(e\zeta/2kT\right), \tag{10}$$

where σ^* is the surface-charge density of the hydrodynamic particle and n_∞ is the bulk concentration (number of molecules/volume) of the salt. For aqueous solvent at 25°C, and in useful units, Eq. (10) is:

$$\sigma^* = 7.331\times10^{-3}\, C^{1/2} \sinh\left(\zeta/51.39\right), \tag{11}$$

where σ^* is in number of charges $/\text{Å}^2$, C is the bulk salt concentration in mol/liter, and ζ is in mV. If the distance Δx of the shear surface from the particle's physical surface is known (or assumed), then the electrostatic potential and charge density at the particle's physical surface may also be calculated. Assuming the Poisson-Boltzmann potential profile $\psi(x)$ [64], the particle's surface potential ψ_0 in terms of ζ and Δx is:

$$\psi_0 = \frac{2kT}{e}\ln\left[\frac{1+\alpha}{1-\alpha}\right], \quad \text{where} \quad \alpha = \left[\frac{\exp\left(e\zeta/2kT\right)-1}{\exp\left(e\zeta/2kT\right)+1}\right]\exp\left(\kappa\Delta x\right). \tag{12}$$

For $e\zeta/2kT \ll 1$, i.e. $\zeta \ll 50$ mV at 25°C, Eq. (12) reduces to the Debye-Hückel expression:

$$\psi_0 = \zeta \exp\left(\kappa\Delta x\right). \tag{13}$$

Once ψ_0 is obtained, the surface-charge density σ_0 of the physical particle is calculated using Eq. (10) with ψ_0 substituted for ζ. Here we note that Eq. (10) can be inverted, and using the identity $\text{arc}\sinh(x) = \ln[x + (x^2+1)^{1/2}]$ with the bulk salt concentration expressed in terms of the Debye length via Eq. (1), we get a relation between surface potential and surface-charge density in the form of Eq. (2).

The fact that the zeta potential yields a surface-charge density does not necessarily mean the particle's charge is located on its surface. σ_0 is an *effective* surface-charge density, i.e. the total particle charge / $4\pi a^2$. However, for liposomes the interior charge generally is compensated by included counterions, so the uncompensated charge does indeed sit at or near the physical surface. Incomplete screening of the liposome surface charge within the hydrodynamic particle occurs when the distance of the shear surface from the liposome charge surface (i.e. the surface of phosphate, headgroup, and adsorbed-ion charges) is less than several Debye screening lengths. When the shear-surface distance is greater than several Debye lengths, the liposome charge is completely screened within the hydrodynamic particle, and the electrophoretic mobility is zero. This concept explains why, for a fixed surface-charge density on the physical particle, the electrophoretic mobility decreases with increasing ionic strength, even when no counterion adsorption is occurring. For multilamellar liposomes, complete electrostatic screening between the lamellae is generally assumed even at low ionic strength, but this fact has not been demonstrated.

In addition to the well-known ionic-strength dependence of the Debye screening length, there is circumstantial evidence for an ionic-strength dependence of the location of the shear surface. McDaniel et al. [68], measuring zeta potentials of negatively-charged liposomes at various NaCl concentrations, found their data consistent with Gouy-Chapman-Stern theory if they assumed the distance of the shear surface from the membrane charge surface to be 2 Å in 100 mM salt, 4 Å in 10 mM salt, and 10 Å in 1 mM salt.

Deviations from Gouy-Chapman screening behavior for zeta potentials measured as a function of ionic strength have been attributed to ion adsorption [69,70]. The practice is to assume an adsorption isotherm such as a Langmuir isotherm, then to fit the binding constants, using interfacial ion concentrations calculated from the bulk concentrations and the Boltzmann factors involving ψ_0 [69-71]. For multivalent ions this procedure appears to work well. Some multivalent ions can reverse the sign of the mobility, indicating sign reversal of the net surface charge. Since charge reversal cannot result from screening alone, this result is considered incontrovertible evidence for binding [70]. For monovalent ions the situation is less satisfactory. The apparent binding constants for alkali-metal cations interacting with acidic lipids are small and decrease in order of increasing ionic size [69]. It is not clear whether perturbation of interfacial water structure, or some other interfacial effect, could also account for the observations.

Experimental methods require a device to apply the field across a hydrodynamically well-defined cell and to measure the particle velocities. For giant vesicles, their movement can be followed through a microscope with appropriate illumination and a calibrated eyepiece grid. We routinely use the Rank Brothers Mark II apparatus (Bottisham, Cambridge, U.K.) for multilamellar liposomes. However, smaller particles are difficult or impossible to observe, and fluorescent-labeled liposomes have not yet been used for this purpose. Several commercial instruments are available to characterize the electrophoretic mobilities of nanometer to micrometer-sized particles. In our laboratories we use the Zetasizer 4 (Malvern Ltd., England) and the Delsa 440-SX (Beckman Coulter Inc., Hialeah, FL). These instruments employ laser-Doppler methods to determine the velocity distribution of a particle suspension in an electric field. They require fairly dilute liposome samples to avoid multiple scattering; our concentration is typically about 0.25 mg/ml. It is important to note that the cells have fairly large glass surfaces, and many peptides, especially positively-charged ones, can bind significantly to the glass, which lowers their bulk concentration. Control measurements at different peptide concentrations must be done.

In Fig. 11 we show a comparison of pentalysine titrations with POPS membranes by two very different methods: electrophoretic mobilities with liposomes and the IFC technique with planar lipid bilayers. Surface potentials are calculated from the zeta potentials using a shear-surface distance of 2Å (cf. Eqs. (12) and (13)).

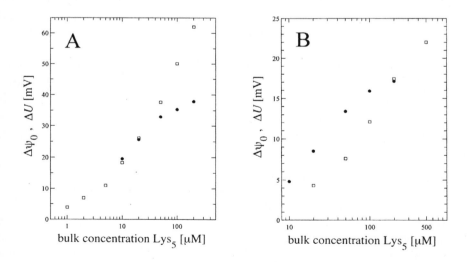

Fig. 11. Comparison of changes in surface potential calculated from the zeta potential (filled circles) with changes in the boundary potential measured by IFC (open squares) upon adsorption of pentalysine. The bulk solution is 100 mM KCl at pH 7. A: Results for pure POPS membranes. (The IFC data are the same as in Figs. 3B and 8A.) B: Results for membranes of 2:1 (mol:mol) POPC:POPS mixtures. For the zeta potential calculation, the shear plane is assumed to be 2 Å from the vesicle surface.

The discrepancies in Fig. 11 between the zeta-potential and IFC results have not yet been explained.

An interesting feature of particle electrophoresis is its sensitivity to hydrodynamic drag. This effect can be used to study the properties of "hairy" surfaces. In Fig. 12 we show a series of mobilities measured for multilamellar liposomes modified with end-grafted neutral polymers of poly(ethylene glycol) (PEG) of various lengths at different ionic strengths [64]. The PEG-grafted lipids (PEG-PE) are negatively-charged. In these experiments the PEG-lipid mole fraction and the surface-charge density are held constant. The reduction of mobility seen with increasing PEG size (at each ionic strength) results from polymer-induced hydrodynamic drag. These data have been analyzed for the hydrodynamic thicknesses of the polymer coats as a function of polymer size. The inferred thicknesses range from about 25 to 130 Å [64]. In this experiment the membrane charge is not itself the subject of investigation, but is used as a "handle" to drive the liposomes so that the hydrodynamic properties of their surface coats may be studied.

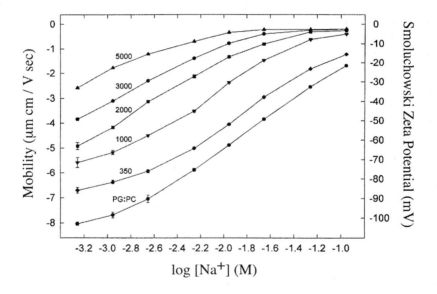

Fig. 12. Electrophoretic mobility vs. log [Na$^+$] for 1:9 (mol:mol) PEG-PE:PC and PG:PC multilamellar liposomes, with PEG molecular weights as shown. PG has the same charge as PEG-PE but lacks the PEG polymer, thus serves as a control for PEG-mediated effects. The Smoluchowski zeta potentials are also indicated. Error bars for each data point are standard errors of the mean for at least two independent sets of three or more runs. PEG, poly(ethylene glycol); PE, phosphatidylethanolamine; PC, phosphatidylcholine; PG, phosphatidylglycerol. From Ref. [64], with permission.

3.6. Potentiometric optical dyes

Fluorescence traditionally has been an important tool for studying the biophysical properties of biological membranes. Since the early 1970s and the pioneering works of Cohen and co-workers [72,73], considerable effort has been aimed at developing fluorescent indicators for sensing the plasma membrane potential (potentiometric dyes). Nowadays, by the use of cytometry and/or digital fluorescence imaging techniques, potentiometric dyes are used to map membrane-potential changes among cell populations or at the single-cell level with spatial resolution and sampling-time frequency that are not approachable using microelectrodes. Furthermore, these dyes enable membrane potential measurements to be performed in intracellular organelles for which the use of microelectrodes is not possible.

Potentiometric dyes are currently divided into two classes: slow-response (with response time of seconds) and fast-response (with response time of milliseconds). Slow-response dyes are charged compounds and include cationic carbocyanines with short alkyl tails, cationic rhodamines, anionic oxonols, and merocyanine [74] (see Fig. 13). Each of these chromophore types has specific properties that make it applicable to various experimental analyses. The basic process that underlies the response of these dyes is their potential-dependent partitioning between the extracellular medium and either the plasma membrane, the cytoplasm, or organelles of the cell.

The carbocyanine dyes are symmetrical compounds with varying alkyl tail lengths and heterocyclic nuclei with delocalized positive charges. These cationic dyes accumulate on hyperpolarized membranes and insert into the phospholipid bilayer. The magnitude and direction of the fluorescence change depend on the structure and concentration of the dye. Dye aggregation occurs at high dye concentrations and results in a large self-quenching of fluorescence. Another limit in the use of these dyes is their relative toxicity when they interact with mitochondria.

The methyl and ethyl ester cationic tetramethylrhodamine dyes (TMRM and TMRE) were developed in order to provide quantitative assays of membrane potentials in living cells [75]. The charge and ability of these dyes to permeate cell membranes make them good "Nernstian dyes", because they accumulate in the cytoplasm of cells and come to electrochemical equilibrium with the cell's transmembrane potential. The dye's Nernst potential can be determined from the ratio of fluorescence intensities measured inside (F_{in}) and outside (F_{out}) of the cell, since these intensities are proportional to the dye concentrations in those compartments. The cell's transmembrane potential, which equals the dye's Nernst potential, is then:

$$\Delta\psi = -\frac{RT}{zF}\ln\left(F_{in}/F_{out}\right) \tag{14}$$

where R is the ideal gas constant, F is Faraday's constant, and z is the valence of the dye.

Anionic bis-barbituric acid oxonol dyes (DiBac) are also "Nernstian dyes". These dyes enter depolarized cells and exhibit increased intracellular fluorescence. On the other hand, hyperpolarization results in extrusion of the anionic dyes and thus a decreased fluorescence. The maximum excitation wavelength of DiBac$_4$(3) (490 nm) allows its use for determining plasma-membrane potentials by flow cytometry [76]. While oxonol dyes have less tendency to aggregate than the cyanine dyes, they are also less sensitive.

Fast-response dyes are usually membrane stains whose spectral responses are perturbed by the transmembrane voltage difference. They include cationic and zwitterionic styrylpiridiniums and impermeant anionic oxonols [74].

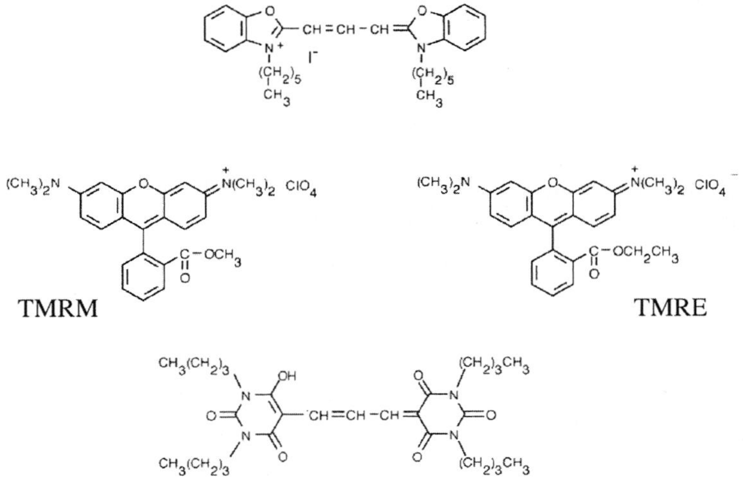

Fig. 13. DiOC$_6$(3) is a cell-permeant, green-fluorescent, lipophilic carbocyanine dye that is selective for the mitochondria of live cells when used at low concentrations. TMRM and TMRE are tetramethylrhodamine 'Nernstian' dyes, which are currently used for determining membrane potentials by quantitative imaging. DiBac$_4$(3) is a bis-barbituric acid oxonol 'Nernstian' dye which enters depolarized cells. (Obtained from Molecular Probes website, www.probes.com)

The mechanisms underlying the fast response of these dyes is based on potential-dependent intramolecular rearrangements such as internal charge migration (electrochromism) [77], which induces rapid but small changes in the fluorescent response of the dye (2-10% fluorescence change per 100 mV).

The aminonaphthylethenylpyridinium (ANEP) dyes were developed in Loew's laboratory (Di-4-ANEPPS and Di-8-ANEPPS) (Fig. 14). Like other styryl dyes, the ANEP dyes are essentially nonfluorescent in aqueous solution, while the membrane-bound dye displays fluorescence. ANEP dyes have good photostability and low toxicity. The optical response of these dyes is a decrease of the 440/530 fluorescence excitation ratio upon membrane hyperpolarization. Such a spectral shift permits the use of a ratiometric method to correlate the changes in fluorescence with variations in membrane potential.

RH dyes include a large series of dialkylaminophenylpolyphenyl-pyridinium dyes [74] (Fig. 15). As with ANEP dyes, they undergo significant

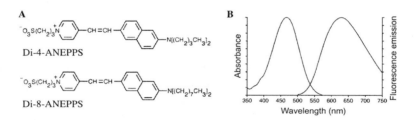

Fig. 14. A: Molecular structures of Di-4-ANEPPS and Di-8-ANEPPS. B: Absorption and fluorescence emission spectra of Di-8-ANEPPS bound to phospholipid bilayer membranes. (Obtained from Molecular Probes website, www.probes.com)

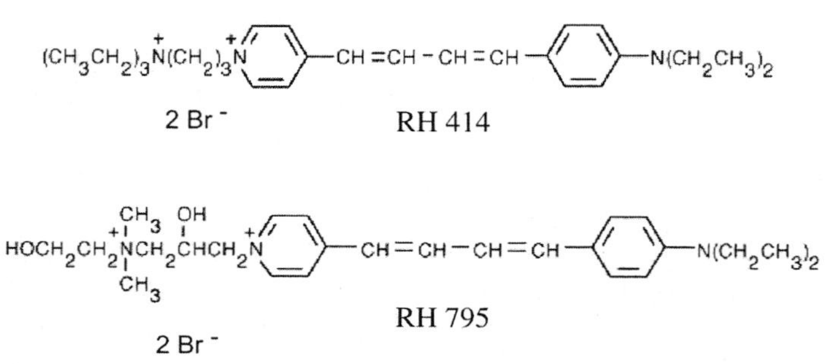

Fig. 15. RH 414 and RH 795 belong to the dialkylaminophenylpolyenylpyridinium dye series originally synthesized by Rina Hildesheim and are the most widely-used RH dyes in cell biology. (Obtained from Molecular Probes website, www.probes.com)

fluorescence spectral shifts in response to membrane potential changes, and a ratiometric method can be used. However, physiological effects of the different RH dyes are not equivalent and must be analyzed systematically.

Impermeant oxonols are oxonol dyes with phenylsulfonate substituents. While faster, their potentiometric responses are smaller than their slow-response dye equivalents. Adsorption of modified oxonols to the cell membrane increases upon depolarization. Their potentiometric properties are based on both changes in absorption (red shift) and in fluorescence (increase with depolarization).

Selection of the best potentiometric dye for a given experiment or cell type is essential and must take into account the time-response, optical properties, localization, and potent side-effects of the various dyes.

4. DISCUSSION

Inspection of the results stemming from different techniques clearly demonstrates inconsistencies among them. Although the techniques have been available for many years, and although this problem is well-known, a systematic comparison among the various methods is sorely needed. Prior to generalizing results obtained by one technique, other techniques should be tested to determine whether the results are mutually consistent.

A second issue requiring systematic evaluation is the discrepancy between the Gouy-Chapman theory and experimental observations. The theory does not explain the effects of different electrolyte compositions or of ion specificity. An introduction to this question may be found in the work of Ninham [13]. A better understanding of these issues will be of great benefit to biological studies.

As stated in the introduction, surface potentials play a very important role in the signal cascade of cells. How the cell uses this property to translocate signals locally across the lipid membrane has not yet been investigated. Intrinsic electric fields play a major role in biological processes. For example, electrostatic interactions often regulate the binding of peripheral proteins to lipid membranes. An example of such regulation that occurs intracellularly is the so-called myristoyl switch [5]. A secondary effect of intrinsic fields on cells is their influence on shape transitions. For example, a pH variation is able to deliver enough energy to trigger a conformational change of protein shape. With regard to artificial model membranes, intrinsic electric fields influence the stability and size distribution of liposomes.

Externally applied electric fields constitute a unique tool for the characterization and manipulation of cells or liposomes. For instance, a common technique for transfection is to permeabilize cells by short electric-field pulses, which allows the uptake of macromolecules into the cell interior.

Electric fields also enhance drug delivery through the skin, a method known as iontophoresis. Although these techniques are in widespread use, little is known about the underlying mechanisms. Recently the notion that local changes in membrane electric fields might strongly control fundamental membrane processes was introduced by the concept of "*molecular electroporation*" [78]. In this technique a local membrane permeability change is induced by exposure of living cells to brief and intense external electric-field pulses (electropermeabilization or electroporation) [62,79]. The driving force for electroporation is a large and rapid change in the transmembrane electrical potential which induces a change in organization of the membrane interface. Local membrane electroporation triggers a whole-cell response. One major cell response is the production of oxygen-reactive species (oxidative jump) [80] which generally are generated during cell stress. By changing the dynamics and local organization of the membrane, electroporation stresses the cell. The capacity of the whole cell to counter this stress must govern the reversibility of the local process and the long-term behavior of the cell. Cellular responses associated with local changes in transmembrane electric fields must be elucidated in order to refine the "molecular electroporation" concept and to clarify its reverse regulation by the cell.

It is hoped that well-defined physical studies of model membrane systems such as those discussed here, and resolution of the questions raised, will provide valuable insight into the interfacial and transmembrane electrical properties of living cell membranes.

ACKNOWLEDGMENTS

We thank Dr. Yu. A. Ermakov for his valuable contribution to our chapter. We gratefully acknowledge the help and support of Professor G. Schwarz, in whose laboratory at the Biozentrum Basel many of the results reported here were obtained. Much of the work was done with the help of our collaborators Dr. A. Diederich and Dr. G. Bähr. We also thank Neal Johnson and F. Viala for excellent graphics support, and Drs. Rudi Podgornik and Adrian Parsegian for valuable discussions. Financial support of the Research Training Network - "Nanocapsules with functionalized surfaces and walls" (HPRN-CT-2000-00159) is gratefully acknowledged.

REFERENCES

[1]   S.J. Singer and G.L. Nicolson, Science 175 (1972) 720.
[2]   See the classic textbook: B. Alberts, D. Bray, J. Lewis, M. Raff, K. Roberts and J.D. Watson, The Molecular Biology of the Cell, Garland Publishing Inc., New York (1989).

[3] L. Kullman, M. Winterhalter and S.M. Bezrukov, Biophys. J. 82 (2001) 803.
[4] C. Bieri, O.P. Ernst, S. Heyse, K.P. Hofmann and H. Vogel, Nat. Biotechnol. 17 (1999) 1105.
[5] S. McLaughlin and A. Aderem, Trends Biochem. Sci. 20 (1995) 272; D. Murray, N. Ben-Tal, B. Honig and S. McLaughlin, Structure 5 (1997) 985.
[6] G. Bähr, A. Diederich, G. Vergères and M. Winterhalter, Biochemistry 37 (1998) 16252.
[7] See A. Arbuzova, A.A.P. Schmitz and G. Vergères, Biochem. J. 362 (2002) 1.
[8] P.G. Saffman and M. Delbruck, Proc. Natl. Acad. Sci. USA 72 (1975) 3111.
[9] F.C. Tsui, D.M. Ojcius and W.L. Hubbell, Biophys. J. 49 (1986) 459; S. Lund-Katz, M. Zaiou, S. Wehrli, P. Dhanasekaran, F. Baldwin, K.H. Weisgraber and M.C. Phillips, J. Biol. Chem. 275 (2000) 34459.
[10] Concerning the effects at a lipid membrane see: S. McLaughlin, Curr. Topics Membr. Transp. 9 (1977) 71; S. McLaughlin, Annu. Rev. Biophys. Chem. 18 (1989) 113; G. Cevc, Biochim. Biophys. Acta 1031 (1990) 311. See also classic textbooks such as: P.C. Hiemenz, Principle of Colloids and Surface Chemistry, Marcel Dekker, New York (1986). Other sources that develop the physical background nicely are: E.J.W. Vervey and J.T.G. Overbeek, Theory of the Stability of Lyophobic Colloids, Elsevier, Amsterdam (1948); A.P. Nelson and D.A. McQuarrie, J. Theor. Biol. 55 (1975) 13; D. Andelman, in Handbook of Biological Physics 1B, R. Lipowski and E. Sackmann (eds.), Elsevier, Amsterdam (1995) 603.
[11] For a recent work see: S. Lamperski and C.W. Outwaite, Langmuir 18 (2002) 3423.
[12] P. Attard, J. Phys. Chem. 99 (1995) 14174.
[13] B.W. Ninham and V. Yaminsky, Langmuir 13 (1997) 2097.
[14] G.L. Gains, Insoluble Monolayers at the Liquid-Gas Interface, Wiley, New York (1966).
[15] For an overview see: H. Brockman, Chem. Phys. Lipids 73 (1994) 57.
[16] A. Diederich, G. Bähr and M. Winterhalter, Phys. Rev. 48E (1998) 4883.
[17] R. Cseh and R. Benz, Biophys. J. 77 (1999) 1477; R. Cseh, M. Hetzer, K. Wolf, J. Kraus, G. Bringmann and R. Benz, Eur. Biophys. J. 29 (2000) 172.
[18] D. Pickar and R. Benz, J. Membr. Biol. 44 (1978) 353.
[19] K. Gawrisch, D. Ruston, J. Zimmerberg, V.A. Parsegian, R.P. Rand and N. Fuller, Biophys. J. 61 (1992) 1213.
[20] B. Roux, J.M. Neumann, M. Bloom and P.F. Devaux, Eur. Biophys. J. 16 (1988) 267; N. Ben-Tal, B. Honig, R.M. Peitzsch, G. Denisov and S. McLaughlin, Biophys. J. 71 (1996) 561; B. De Kruijff, A.B. Rietveld, N. Telders and A. Vaandrager, Biochim. Biophys. Acta 820 (1985) 295.
[21] G. Laroche, D. Carrier and M. Pezolet, Biochemistry 27 (1988), 6220.
[22] I. Weis, P.B. Welzel, G. Bähr and G. Schwarz, Chem. Phys. Lipids 105 (2000) 1.
[23] M. Winterhalter, H. Bürner, S. Marcinka, R. Benz and J.J. Kasianowicz, Biophys. J. 69 (1995) 1372.
[24] A. Diederich, G. Bähr and M. Winterhalter, Langmuir 14 (1998) 4597.
[25] A. Diederich, M. Strobel, W. Meier and M. Winterhalter, J. Phys. Chem. B 103 (1999) 1402.
[26] A. Giehl, T. Lemm, O. Bartelsen, K. Sandhoff and A. Blume, Eur. J. Biochem. 261(1999) 650; A. Seelig, Biochim. Biophys. Acta, 899 (1987) 196; R.C. MacDonald, S.A. Simon, Proc. Natl. Acad. Sci. USA 84 (1987) 4089.

[27] P. Mueller, D.O. Rudin, H.T. Tien and W.C. Wescott, J. Phys. Chem. 67 (1963) 534.

[28] M. Montal and P. Mueller, Proc. Nat. Acad. Sci. USA 69 (1972) 3561.

[29] H. Ti Tien, Bilayer Lipid Membranes (BLM): Theory and Practice, Marcel Dekker, New York (1974).

[30] Other textbooks on BLM: M.K. Jain, The Bimolecular Lipid Membrane, Van Nostrand Reinhold, New York (1972); W. Hanke and W.R. Schlue, Planar Lipid Bilayers: Methods and Applications, Academic Press, New York (1993).

[31] O. Alvarez and R. Latorre, Biophys. J. 21 (1978) 1.

[32] A.V. Babakov, L.N. Ermishkin and E.A. Liberman, Nature 210 (1966) 953.

[33] W.J. Carius, J. Colloid Interface Sci. 57 (1976) 301.

[34] Yu.A. Ermakov, A.Z. Averbakh, A.I. Yusipovich and S.I. Sukharev, Biophys. J. 80 (2001) 1851.

[35] Yu.A. Ermakov, I.S. Fevraleva and R.I. Ataulakhanov, Biol. Membr. 2 (1985) 1094.

[36] T. Hianik and V.I. Passechnik, Bilayer Lipid Membranes: Structure and Mechanical Properties, Ister Science, Bratislava (1995).

[37] D.F. Sargent and T. Hianik, Bioelectrochem. Bioenerg. 33 (1994) 11.

[38] P. Schoch and D.F. Sargent, Experimentia 32 (1976) 811.

[39] V.S. Sokolov and S.G. Kuzmin, Biofizika 25 (1980) 170 [in Russian].

[40] A. Wiese, M. Munstermann, T. Gutsmann, B. Lindner, K. Kawahara, U. Zahringer and U.J. Seydel, Membr. Biol. 162 (1998) 127; their technique is based on M. Montal, Biochim. Biophys. Acta 298 (1973) 750.

[41] S.H. White, Biophys. J. 12 (1972) 432; H. Ti Tien, J. Gen. Physiol. 52 (1968) 125s.

[42] S. McLaughlin and M. Eisenberg, Ann. Rev. Biophys. Bioenerg. 4 (1975) 335.

[43] R. Benz, O. Frohlich and P. Läuger, Biochim. Biophys. Acta 464 (1977) 465.

[44] M. Dobler, Ionophores and their Structures, John Wiley & Sons, New York (1981).

[45] S.G.A. McLaughlin, G. Szabo, G. Eisenman and S.M. Ciani, Proc. Nat. Acad. Sci. USA 67 (1970) 1268; S.G.A. McLaughlin, G. Szabo and G. Eisenman, J. Gen. Physiol. 58 (1971) 667; O. Alvarez, M. Brodwick, R. Latorre, A. McLaughlin, S. McLaughlin and G. Szabo, Biophys J. 44 (1983) 333.

[46] J.E. Hall and R. Latorre, Biophys. J. 15 (1976) 99; J.E. Hall, C.A. Mead, and G. Szabo, J. Memb. Biol. 11 (1973), 75.

[47] J.A. Cohen and M.M. Moronne, J. Supramolec. Struct. 5 (1976) 409. Reprinted in: Prog. Clin. Biol. Res. 15: Z. Hall, R. Kelly and C.F. Fox (eds.), Cellular Neurobiology, Alan R. Liss, New York (1977) 13.

[48] H. Nakayama, K. Furihata, H. Seto and N. Ōtake, Tetrahedron Lett. 22 (1981) 5217.

[49] E.J. Heyer, R.U. Muller and A. Finkelstein, J. Gen. Physiol. 67 (1976) 731; R.U. Muller and C.S. Peskin, J. Gen. Physiol. 78 (1981) 201.

[50] R. Dutzler and J.A. Cohen, unpublished calculations.

[51] R.U. Muller and A. Finkelstein, J. Gen. Physiol. 60 (1972) 263; ibid., 285.

[52] O.S. Andersen and R.U. Muller, J. Gen. Physiol. 80 (1982) 403; R.U. Muller and O.S. Andersen, loc. cit., 427.

[53] J.A. Cohen and M.M. Moronne, Biochem. Biophys. Res. Commun. 83 (1978) 1275.

[54] J.A. Cohen and M.M. Moronne, Biophys. J. 25 (1979) 12a; J.A. Cohen and D.M. Bradford, Biophys. J. 59 (1991) 632a.

[55] J.A. Cohen, Biophys. J. 61 (1992) A239.

[56] M.M. Moronne and J.A. Cohen, Biochim. Biophys. Acta 688 (1982) 793.

[57] R.U. Muller, G. Orin and C.S. Peskin, J. Gen. Physiol. 78 (1981) 171.

[58] O.S. Andersen, A. Finkelstein, I. Katz and A. Cass, J. Gen. Physiol. 67 (1976) 749; P. Pohl, T.I. Rokitskaya, E.E. Pohl and S.M. Saparov, Biochim. Biophys. Acta 1323 (1997) 163.

[59] T.K. Rostovtseva, V.M. Aguilella, I. Vodyanoy, S.M. Bezrukov and V.A. Parsegian, Biophys J. 75 (1998) 1783.

[60] E. Neumann, A. Sowers and C.A. Jordan (eds.), Electroporation and Electrofusion in Cell Biology, Plenum Press, New York (1989); D.C. Chang, B.M. Chassy, J.A. Saunders and A. Sowers (eds.), Guide to Electroporation and Electrofusion, Academic Press, New York (1992).

[61] W. Zauner, A. Kichler, W. Schmidt, K. Mechtler and E. Wagne, Exp. Cell Res. 232 (1997) 137; A. Karshikov, R. Berendes, C. Burger, H.D. Kux and R. Huber, Eur. Biophys. J. 20 (1992) 337; J.M. Ruysschaert, E. Goormaghtigh, F. Homble, M. Andersson, E. Liepinsh and G. Otting, FEBS Lett. 425 (1998) 341.

[62] M. Winterhalter, in Nonmedical Application of Liposomes, D.D. Lasic and Y. Barenholz (eds.), CRC Press, Boca Raton (1996) 285; C. Wilhelm, M. Winterhalter, U. Zimmermann and R. Benz, Biophys. J. 64 (1993) 121.

[63] R.J. Hunter, Zeta Potential in Colloid Science: Principles and Applications, Academic Press, London (1981).

[64] J.A. Cohen and V.A. Khorosheva, Coll. Surf. A: Physicochem. Eng. Aspects 195 (2001) 113.

[65] R.W. O'Brien and L.R. White, J. Chem. Soc. Faraday II 74 (1978) 1607. A Fortran program for this calculation is available from Prof. D.Y.C. Chan, Department of Mathematics and Statistics, The University of Melbourne, Parkville, Victoria 3010, Australia.

[66] R.W. O'Brien and R.J. Hunter, Can. J. Chem. 59 (1981) 1878.

[67] Modified from: S. McLaughlin, Curr. Topics Membr. Transp. 9 (1977) 71.

[68] R.V. McDaniel, A. McLaughlin, A.P. Winiski, M. Eisenberg and S. McLaughlin, Biochemistry 23 (1984) 4618. See also A. McLaughlin, W.-K. Eng, G. Vaio, T. Wilson and S. McLaughlin, J. Memb. Biol. 76 (1983) 183.

[69] M. Eisenberg, T. Gresalfi, T. Riccio and S. McLaughlin, Biochemistry 18 (1979) 5213.

[70] S. McLaughlin, N. Mulrine, T. Gresalfi, G. Vaio and A. McLaughlin, J. Gen. Physiol. 77 (1981) 445.

[71] J.A. Cohen and M. Cohen, Biophys. J. 36 (1981) 623.

[72] L.B. Cohen, R.D. Keynes and B. Hille, Nature 218 (1968) 438.

[73] H.V. Davila, B.M. Salzberg, L.B. Cohen and A.S. Waggoner, Nature New Biol. 241 (1973) 159.

[74] R.P. Haugland, Handbook of Fluorescent Probes and Research Chemicals, Sixth Ed. (1996) 585.

[75] B. Ehrenberg, V. Montana, M-D. Wei, J.P. Wuskell and L.M. Loew, Biophys. J. 53 (1988) 785.

[76] Z. Krasznai, T. Marian, L. Balkay, M. Emri and L. Tron, J. Photochem. Photobiol. 28 (1994) 93.

[77] L.M. Loew, Fluorescent and Luminescent Probes for Biological Activities, W.T. Mason (ed.), Academic Press, New York (1993) 151.

[78] K. Rosenheck, Biophys. J. 75 (1998) 1237; R.B. Sutton, D. Fasshauer, R., Jahn, and
     A.T. Brunger, Nature 395 (1998) 347; M. Montal, FEBS Lett. 447 (1999) 129;
     C. Ramos and J. Teissié, FEBS Lett. 465 (2000) 141.
[79] B. Gabriel and J. Teissié, Biophys. J. 73 (1997) 2630; loc. cit., 76 (1999) 2158.
[80] B. Gabriel and J. Teissié, Eur. J. Biochem. 223 (1994) 25; loc. cit., 228 (1995) 710;
     N. Sabri, B. Pelissier and J. Teissié, Eur. J. Biochem. 238 (1996) 737; B. Galutzov and
     V. Ganeva, Bioelectrochem. Bioenerg. 44 (1997) 77; P. Bonnafous, M.C. Vernhes,
     J. Teissié and B. Gabriel, Biochim. Biophys. Acta 1461 (1999) 123.

Planar Lipid Bilayers (BLMs) and their Applications
H.T. Tien and A. Ottova-Leitmannova (Editors)
© 2003 Elsevier Science B.V. All rights reserved.

Chapter 32

Domains, cushioning, and patterning of bilayers by surface interactions with solid substrates and their sensing properties

Ruxandra Vidu, Timothy V. Ratto, Marjorie L. Longo and Pieter Stroeve*

Chemical Engineering and Materials Science, University of California at Davis, 1 Shields Avenue, Davis, CA, 95616

1. INTRODUCTION

Supported lipid bilayers are unique, self-assembling, two-dimensional fluid systems that can be formed by spontaneous fusion of lipid bilayer vesicles with a very few number of hydrophilic surfaces such as mica, glass, or certain hydrophilic self-assembled monolayers [1, 2]. The planar geometry is ideal for imaging and quantification using microscopy and a variety of surface-sensitive techniques. Additionally, certain applications (e. g. separations) favor the use of bilayers supported on curved surfaces.

We focus in section 2 on the use of supported lipid bilayers to study the morphology and growth of two-dimensional lipid bilayer domains. As with Langmuir monolayers, fluorescent probes are conveniently employed to control domain morphology and to allow for imaging of the domains. An advantage of the supported lipid bilayer format in comparison to Langmuir monolayers is that atomic force microscopy (AFM) is used to obtain high-resolution images of domain features of the scale that would be expected in a biological membrane, i.e. nanometer scale domains and defects. In addition, further partitioning can be engineered into these domain-containing supported bilayers through surface patterning as discussed in section 5.

For sensor formats that employ supported bilayers, patterning of lipid bilayers and/or targeting of vesicle fusion to specific locations is necessary. In section 3 we present work aimed at identifying surface functionalization (through self-assembled monolayers (SAM) on gold) which attract or repel cationic and zwitterionic vesicle fusion. In addition, the lipid bilayer/SAM/gold system can be probed by sensitive optical and electrochemical techniques such as Surface Plasmon Resonance (SPR) and impedance spectroscopy, and cyclic voltammetry (CV). In section 5, we develop the vesicle targeting further by employing a patterned SAM as support for two dissimilar lipid bilayers contained within specified regions (but adjacent to each other) by virtue of differences in the nature of the supporting thiol and of the bilayer.

The sensing of, and screening for, the insertion of molecules into lipid bilayers is one of the most promising applications of supported lipid bilayers. Successful developments of such systems could impact such areas as the design and study of viral fusion peptides, antimicrobial peptides, and surfactants. Therefore, in section 4 we present the development of a supported lipid bilayer system for detecting and studying peptide insertion. Lipid bilayers interacting with SAM surfaces can have diminished lateral mobility. In addition it has been demonstrated that the close proximity of the support to the lipid bilayer can interfere with movement of inserted proteins that have substantial segments exterior to the lipid bilayer [3]. In the case of antimicrobial peptides those exterior segments exist but can be quite short (e.g. several amino acids). Therefore, our supported lipid bilayer system is slightly lifted from the SAM surface by a hydrated adsorbed polymer layer (~1nm) or multilayers (~3 nm) formed by charge interaction. Bilayers supported on this polymer layer have a mobility that is only slightly diminished in comparison to free bilayers.

2. DOMAINS IN LIPID BILAYERS

Supported lipid bilayers serve as an alternative to monolayers for studying basic aspects of two-dimensional domain formation and phase co-existence. These systems have the benefit of being flat over long expanses as in Langmuir monolayers; however, they possess the biologically relevant lipid bilayer structure. Since these systems are deposited on a rigid support, such as mica or silicon oxide, they may be studied by high resolution scanning probe surface analysis techniques such as AFM) and Near Field Scanning Optical Microscopy (NSOM) [4]. If hydrophilic surfaces such as clean glass or freshly cleaved mica are used, the solid support provides a local charge density that acts in concert with surrounding ions to control the density of adsorbed charged species[5]. This mimics the ionic screening effects that would be encountered by cell membranes in ionic environments. A very relevant example can be found in the formation of simple lipid bilayer structures on charged mineral surfaces which has been hypothesized as one of the essential steps in molecular evolution (for a review see Ref. [6]).

Here, we discuss domain growth and topology in supported lipid bilayers. We used either cationic (positively charged) lipids or zwitterionic (neutral) lipids to form our supported bilayers and included a fluorescent NBD-PC lipid probe ((1-palmitoyl-2-(-6({7-nitro-2-1,3-benzoxadiazol-4-yl}amino)caproyl)-sn-glycero-3-phosphatidylcholine)) that partitions into the least-ordered phase when two lipid phases are present (for example, see Ref. [7]). Probes allow us to use fluorescence microscopy to distinguish lipid phases, and therefore characterize domain shape and size, and also use Fluorescence Recovery after Photobleaching (FRAP) to assess lipid mobility. We used AFM to characterize bilayer topology and identify domains that were too small to detect with optical microscopy. We identified two classes of heterogeneities that exist in lipid bilayers. One type, which can exist in cationic bilayers that only contain one lipid component, consists of lipids that are in two distinct phase states (for example, one phase less ordered than the other phase.) Another class of heterogeneity is caused by phase separation of two or more lipids that display different physical characteristics, such as acyl chain length. Both classes of heterogeneities can be created and controlled by variations in heating and cooling, and are very sensitive to acyl chain length. Additionally, the supported cationic lipid bilayers are also affected by ionic conditions and interactions between the charged lipids and charged substrates.

2.1. Experimental procedures for fluorescence microscopy of cationic bilayers

A 25 mm diameter freshly cleaved mica disk (Muscavite, Pelco®, Ted Pella, Inc., Redding, CA) or a 25 mm diameter pre-cleaned glass slide (Fisher Scientific), a PTFE spacer, a rectangular glass cover slip, and an aluminum platform were used to form an assembly suitable for fluorescence microscopy viewing (assembly and bilayer formation procedure is described in detail in[8]). The assembly was filled with a sonicated cationic lipid vesicle solution and incubated at room temperature for 30 min in small aluminum and glass chamber humidified by drops of water (see [8], for detailed description). The assembly was then heated to 70 °C for 30 min in a laboratory oven. Cooling was accomplished either by turning off the oven (the oven was allowed to cool to room temperature, this required several hours) or by moving the sample to a room temperature environment (required ~30 min). To eliminate unfused lipid vesicles, the internal contents of the assembly were then rinsed with approximately ten internal volumes of the solution used to suspend the vesicles. The mica or glass supported membrane remained hydrated at all times during this preparation and subsequent fluorescence microscopy.

2.2. Experimental procedures for atomic force microscopy of cationic bilayers

For all AFM work, an 80 µl drop of lipid vesicle solution in water or salt solution was placed on the surface of a freshly cleaved mica disk and allowed to incubate at room temperature in a petri dish containing drops of water for 30 minutes. If a heating and cooling cycle was applied, the sample was heated to 70 °C for 30 minutes in an oven or a water bath. The sample was then cooled by moving it to a room temperature environment and, after cooling, rinsed ten times with 20 mM NaCl. When no heating was applied the mica was simply rinsed with 20 mM NaCl after 90 minutes of room temperature incubation with lipid vesicles. Finally, samples were usually stored overnight in the refrigerator before AFM imaging (we noted no difference in sample topology or appearance with or without refrigeration). Samples were imaged the following day with a Digital Instruments NanoScope® IIIa (Santa Barbara, CA) in Force Mode with either a 'J' or 'E' scan head. Sharp, coated AFM microlevers, model MSCT-AUHW (Park Scientific Instruments) were used for all scans. Hydration of the samples during scanning was maintained using Digital Instruments' AFM TappingMode Fluid Cell, model MMTFC. All scans were done at room temperature and in contact mode. Set points ranged between 0.25 V to 1.0 V with scan rates between 1.5 Hz to 8 Hz.

2.3. Experimental procedure for mixed lipid zwitterionic bilayer imaging

Similarly to the sample preparation for the AFM studies, a 150µl drop of a heated mixture of DLPC/DSPC (1,2 Dilaural-sn-Glycero-3-phosphocholine/ 1,2 Disteroyl-sn-Glycero-3-phoshocholine) vesicle solution containing 2 mol % NBD-PC was placed on the surface of a freshly cleaved cooled mica disk. After cooling the sample was allowed to incubate at room temperature for 30 minutes prior to rinsing away the unbound vesicles with 20 mM NaCl solution. Samples were imaged with a Digital Instruments NanoScope® IIIa (Santa Barbara, CA) in force mode with either a 'J' or 'E' scan head. Sharp, coated AFM microlevers, model MSCT-AUHW (Park Scientific Instruments) were used for all scans. Hydration of the samples during scanning was maintained using Digital Instruments' AFM TappingMode Fluid Cell, model MMTFC. All scans were done at room temperature and in contact mode or tapping mode. Set points ranged between 0.0 V to 1.0 V with scan rates between 4 Hz to 10 Hz.

2.4. Fluorescence microscopy results

Deposition of 1,2 Dipalmitoyl-trimethyl-ammonium-propane (DPTAP) and 1,2 Dimyristoyl-trimethyl-ammonium-propane (DMTAP) vesicles (containing 3% NBD-PC) on mica followed by heating to 70°C and cooling to 25°C resulted in domain formation visible with the fluorescence microscope. The domain shapes seen in these cationic bilayers at room temperature depended on the lipid acyl chain length (C16 for DPTAP and C14 for DMTAP), solution salt concentration, and cooling rate. Fractal-like domains were seen for DPTAP (Fig. 1 A, B, F) compared to elongated triangular domains for DMTAP (Fig. 2 A, B).

Note that domains in which tip-splitting is clearly evident (as observed in the case of DPTAP) are usually referred to as fractal, whereas, dendritic domains contain thin straight branches with stable tips [7, 9]. Ramified morphologies such as the fractal and feathery textures that we observed for cationic bilayers on mica are characteristic of instability in the two-dimensional domain growth process. It has been demonstrated previously that very similar domain shapes are seen in Langmuir monolayers when the LE (liquid expanded) phase is compressed or cooled rapidly into the LE-LC (liquid condensed) phase co-existence regime (for example, see Ref. [9]). These same fractal or dendritic shapes in LE-LC monolayers can also originate because of the presence of a probe that controls the domain growth process[7]. In both cases, slow diffusion rates of lipid in comparison to domain formation time results in unstable growth of the domain wall. Analogously, both the rapid cooling rates and the presence of the NBD-PC probe contribute to the ramified appearance of domains in cationic bilayers.

The presence of fractal patterns in DPTAP (for example, vs. dendritic patterns) indicates that domain growth is not strongly correlated with crystallographic directions in the gel phase of DPTAP and thus takes place isotropically [7, 9]. We did not observe any relaxation to rounded structures, even after several days because our lipids are relatively immobile at room temperature (diffusion rates in the gel phase are approximately four orders of magnitude less than in the liquid phase). In contrast, the molecules in fractal or dendritic domains in Langmuir monolayers are more mobile and will relax into more rounded equilibrium structures after several hours[7, 9]. We did observe rounding in DPTAP domains if slower cooling rates were used. Similarly, in monolayers, slower compression rates give more rounded LC domains (see for example [10]).

Surprisingly, we observed very different behavior for bilayers of DMTAP on mica in comparison to DPTAP. DMTAP domains resemble "flying ducks" (note that the dark pattern is approximately mirrored in the light pattern due to the symmetry of the pattern.) The domains were relatively symmetric about their long axis but unsymmetric from "nose" to "feathery tail". Additionally, straight, or slightly curved, edges existed over relatively long distances (10-100 microns). The anisotropic nature of the DMTAP domains may indicate that domain growth is correlated along crystallographic directions of the DMTAP two-dimensional lattice.

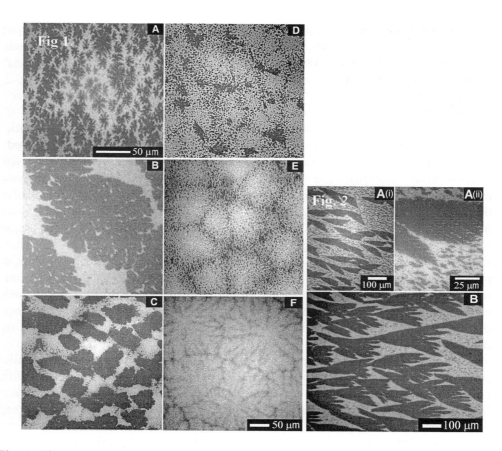

Fig. 1. Fluorescence microscopy images of mica supported DPTAP + NBD-PC bilayers showing effect of cooling rate (compare A and B), probe concentration (compare A and C), and NaCl concentration (compare A, D, E, and F). Magnification is the same in A, B, C, D, and E. Supported bilayers were formed by vesicle fusion and then subjected to heating (to 70 °C) and cooling (to room temperature.) **A.** Cooling rate ~ 2 °C/min. 3% NBD-PC probe. Buffer solution 20 mM NaCl. Note that the fractal domains are dark in this image. In comparison to the sample in image A, the samples in the following images were made with: **B.** Cooling rate ~0.2 °C/min ; **C.** 2% NBD-PC; Lower NaCl concentration, **D.** 15 mM NaCl, **E.** 5 mM NaCl, and **F.** 0 mM NaCl. Note that the dark domains have percolated (all connected) between 0 and 5 mM NaCl. (Reproduced with permission from Biophys. J., 2001, 79, 2605-2615. Copyright 2000 by the Biophysical Society)

Fig. 2. Fluorescence microscopy images of mica supported DMTAP + 3% NBD-PC bilayers in 20 mM NaCl that were subjected to a heating and cooling cycle. Domains are elongated and triangular shaped with a feathery texture. A high density of small, elongated, domains can be seen in the region between the large domains. **A.** Cooling rate ~ 2 °C/min. (i) triangular shaped domains, (ii) higher magnification image shows that domains have feathery texture. **B.** Cooling rate ~0.2 °C/min. (Reproduced with permission from Biophys. J., 2001, 79, 2605-2615. Copyright 2000 by the Biophysical Society)

2.5. Atomic force microscopy results

AFM can be used to determine height differences between lipid domains and surrounding continuous lipids and layer height (if deep defects exist in the layer). Additionally, subtle features, such as sub-micron sized domains that can not be seen in an optical microscope can be observed[4]. We scanned DPTAP + 3% NBD-PC mica supported membranes in 20 mM NaCl (cooled at ~2 °C/min) and observed fractal-like domains (Fig. 3A) that were higher by ~1.4 nm (Fig. 3B) than their surroundings. These domains were of similar size and shape to the dark domains seen in our fluorescence images of samples prepared at the same cooling rate (Fig. 1A). Thus, the dark regions in the fluorescence images are thicker by ~1.4 nm than the bright regions. Occasionally, holes of ~4 nm, in depth in comparison to the level of the domains, were found at domain edges. Therefore, the behavior observed here occurred in bilayers of lipids rather than monolayers. We measured a consistent height difference of about 1.4 nm at the domain edges in DPTAP bilayers on mica. These abrupt steps at domain edges occur because of a difference in acyl chain tilt between neighboring phases as has been previously demonstrated using scanning probe microscopy of supported monolayers [4, 11]. Therefore, bright regions in our fluorescence images are composed of a lipid phase with "more tilt" with respect to the surface normal, in comparison to the dark regions, which contain a lipid phase that is "less tilted" (see Fig. 4).

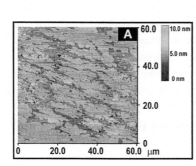
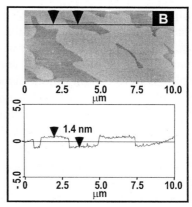

Fig. 3. AFM images of mica supported DPTAP + 3% NBD-PC bilayer in 20 mM NaCl that was subjected to heating and cooling cycle (cooled at rate ~ 2 °C/min). A. 60 *μm* X 60 *μm* scan showing that the fractal domains seen in Fig. 1A are higher than the surrounding continuous region. B. Section demonstrating that domains are ~ 1.4 nm thicker than surrounding regions. Occasional deep holes adjacent to the fractal domains are ~ 4 nm in depth in comparison to domain height (not shown in this section). (Reproduced with permission from Biophys. J., 2001, 79, 2605-2615. Copyright 2000 by the Biophysical Society)

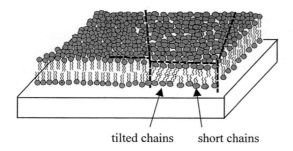

tilted chains   short chains

Fig. 4. This cartoon demonstrates the two different supported bilayer systems discussed in this section.

Previous observations show that NBD-PC fluorescent probes prefer to partition into the more disordered phase when two lipid phases co-exist. Therefore, the "more tilted" phase is more disordered in comparison to the "less tilted" phase. We saw no significant fluorescence in the "less tilted" phase in comparison to the "more tilted" phase. Thus, it is likely that there is superposition of the phase behavior of opposing monolayers in the "less tilted" regions. Similar coupling of monolayer phases has been demonstrated in giant vesicles containing phase-separated lipid bilayers [12].

Deposition of heated DLPC/DSPC vesicle solutions containing 2 mol % NBD-PC onto cooled mica resulted in the formation of solid domains that were 1.8 nm higher than the surrounding area (see Fig. 5). As with the cationic bilayer results discussed above, this height difference indicates that the monolayer phases in this system are also coupled. The domains were circular, were generally evenly dispersed across the bilayer, and, at low DSPC concentrations, were roughly monodisperse in terms of their size (see Fig. 6A). At higher DSPC concentrations the domains overlap and become continuous (see Fig. 6B.) The diameter of the domains was controlled by how quickly the vesicle solution was cooled by contact with the substrate. When the 70 ^{0}C DLPC/DSPC vesicle solution was added to a mica substrate cooled to approximately 10 ^{0}C (rapid quenching) the domains were between 90- 140 nm in diameter.

Less rapid cooling resulted in the formation of larger domains. These domains are circular because the probe lipid is being expelled from the growing solid-phase domains into a fluid environment. The diffusion of the probe away from the domain is therefore much faster than the rate of formation of the solid domain and thus the growth of the domain is two dimensionally stable.

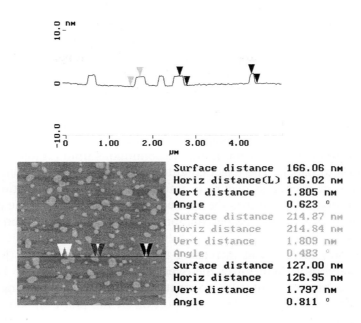

Fig. 5. AFM section analysis of a gel to fluid bilayer showing a solid area fraction of 0.17 (1/6). The height difference between phases is in keeping with the predicted value of 1.826 nm for a DSPC bilayer domain embedded in a DLPC bilayer due to the difference in acyl chain length [13].

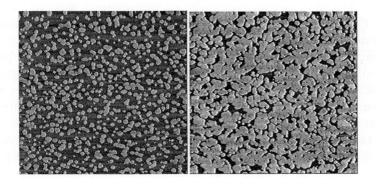

Fig. 6. Two AFM height/deflection images showing the generation of solid phase DSPC domains surrounded by a fluid DLPC bilayer. The lighter colors denote higher areas, and thus the solid domains. The second image was prepared with a higher concentration of DSPC than the first image, note the progress toward continuity of the solid phase. Scale bar ~ 1000 nm.

2.6. I

[handwritten note: 5NK → TZ 3:50 PM 4/26/2002]

V   that ionic screening plays in the observed
doma   + 3% NBD-PC supported bilayers (cooled
at a ra   hown), 5 and 0 mM NaCl (Fig. 1 D, E, and
F). A   d, the dark domains decreased in size and
surfac   into a network (see for example, Fig. 1 D
and E). Between 5 mM and 0 mM NaCl, the dark domains join to form a
continuous network that surrounds large bright fractal domains (Fig. 1 F). Thus,
by decreasing the salt concentration, the continuous lipid regions are exchanged
(from bright to dark). The exact condition (in this case NaCl concentration) in
which such an exchange occurs is referred to as the "percolation threshold"
where "percolation" describes the connectivity of one phase in the presence of a
co-existing phase. Thus, the percolation threshold occurs between 0 mM and 5
mM of NaCl.

It can be expected that increasing the headgroup repulsion (by decreasing
NaCl concentration) will lower the overall density of the layer. It can also be
expected, in the case of two co-existing phases, that lowering the overall
density, will increase the relative proportion of the phase of lower density[7,
14]. We observed that increasing headgroup repulsion (by decreasing salt
concentration) resulted in an increase in the relative proportion of "more tilted"
phase. This result also indicates that the "more tilted" phase is of lower density
than the "less tilted" phase. In our system, if the lower density "more tilted"
phase is in larger abundance, it will nucleate and grow into fractal domains
surrounded by a continuous network of "less tilted" lipids. In comparison to
Langmuir monolayers, if the lower density LE phase is in larger abundance than
the LC phase, it does not form domains of well-defined geometry since it is a
relatively disordered non-crystalline phase. This difference demonstrates that
the less dense phase in our study is crystalline in nature.

Within the two-lipid DLPC/DSPC system the amount of one lipid phase
present in the bilayer relative to the other can be controlled by regulating the
ratio of DLPC to DSPC in the vesicles (see Fig. 6.) By increasing the amount of
solid phase DSPC compared to the fluid phase DLPC it was possible to tailor
our system to display whatever percolating characteristic was required, e.g. gel-
phase lipid "obstacles" surrounded by a percolating fluid bilayer, or the
opposite, fluid "pools" bounded by an immobile gel bilayer. The percolation
threshold occurred at a solid lipid area fraction (the percentage of sample
covered by the solid phase lipid) of approximately 0.65 percent.

2.7. Comments on interactions with the substrate

Deposition of DPTAP vesicles (containing 3% NBD-PC) on glass followed
by heating to 70°C and cooling to room temperature resulted in the formation of
domains with similar fractal-like shapes in comparison to domains formed on

mica (data not shown). However, under the same ionic conditions, more of the glass surface was visibly covered with dark domains in comparison to mica surfaces. The differences in domain surface coverage between glass and mica can be qualitatively explained in terms of surface charge. Mica readily undergoes ion exchange in which negative surface sites are occupied by solute ions including H^+. As a result of the differences in size of ions, the negative surface charge on mica in water is very small, approximately 0.005 C/m^2, while it is approximately an order of magnitude higher in 20 mM NaCl. The negative surface charge of glass, which does not bind sodium ions stronglyand therefore should not be significantly affected by NaCl concentration, can be estimated to be approximately 0.1 C/m^2 in weak NaCl solutions or in pure water. This estimate is based on the surface charge of silica which has been shown to closely approximate the surface charge on glass [5] at pH ~5.5 in which approximately 19% of the OH groups are charged. Thus, cationic DPTAP bilayers are more condensed on glass in comparison to mica under the same NaCl concentrations because of the relatively larger surface charge of glass. (Parts of this text were reproduced with permission from Biophys. J., 2001, 79, 2605-2615. Copyright 2000 by the Biophysical Society)

3. ADSORPTION OF LIPID BILAYERS ON SUPPORTED SELF-ASSEMBLED MONOLAYERS

Adsorption of lipid bilayers on surfaces has attracted much interest because of the possibility of constructing biomembrane-based sensors and for investigating fundamental biomembrane properties. Self-assembled monolayers (SAM) are advantageous for adsorption of lipid bilayers since the surface chemistry can be easily varied and patterned.

In addition, in the case of SAMs on metal films, sensitive optical and electrochemical techniques such as Surface Plasmon Resonance (SPR) and impedance spectroscopy, and cyclic voltammetry can be applied to study adsorption phenomena and pore and defect formation. In the past, attachment of lipid bilayers to SAMs has received some attention[15] and therefore served as a backdrop for initiating our work in this area.

Here, we present Surface Plasmon Resonance (SPR) and fluorescence results in which we have tested several surface functionalizations of SAMs in order to identify functional groups on which cationic bilayers from vesicles readily deposit, and functional groups which repel the deposition of cationic bilayers. We studied the layer formation of a self-assembled system consisting of a base layer of a negatively charged SAM chemisorbed on gold and a top layer of an electrostatically adsorbed cationic lipid bilayer (see Fig. 7). The formation of the lipid and alkylthiol layers was monitored by SPR.

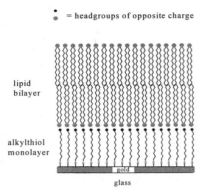

Fig. 7. A schematic of a lipid bilayer electrostatically adsorbed to a self-assembled monolayer chemisorbed to gold on a glass substrate. (Reproduced with permission from Langmuir, 1999, 15, 8133-8139. Copyright 1999 Am. Chem. Soc.)

Cationic lipids readily formed layers with thickness between 32-33 Å on self-assembled alkylthiols possessing terminal carboxylic acid groups within 24 h and at pH>2. Fluorescence bleaching experiments indicated that these layers were homogeneous and relatively immobile. For comparison, we found that cationic lipids do not form layers on alkylthiols possessing a terminal alcohol group, whilst zwitterionic lipids formed bilayers on these surfaces and on the carboxylated surfaces with a thickness of approximately 38-44 Å. The use of self-assembled alkylthiols with diethyleneglycol groups prohibited the formation of both cationic and zwitterionic lipid layers.

3.1. Experimental materials and methods

Self-assembly of the alkylthiol layer was achieved by immersing the gold plated glass slides in a 5 mM solution of the desired alkylthiol in ethanol for at least 18 h at room temperature. Just before using the gold-plated glass slide in the SPR or fluorescence experiments, the slide was removed from the alkylthiol solution and thoroughly rinsed with ethanol and dried under a stream of N_2. For the vesicle solutions, 100 μl of a 10 mg/ml chloroform solution of the desired lipid was transferred to a 3 ml glass vial and the chloroform was evaporated using a stream of N_2. The lipid was suspended again by adding 2 ml of a 5 mM buffer solution (either Tris or MES, adjusted to the desired pH by addition of 0.1 N HCl) and swirling in a water bath at approximately 60 °C until all of the lipid was dispersed. The milky solution was then ultrasonicated with an ultrasonic tip (Branson 250, 20 W output, continuous) for 1-2 min or until the solution became clear. The lipid solution was immediately used in the case of the SPR experiments. For the fluorescence experiments, 2 mol % of 4-Di-10-ASP was added to the glass vial prior to resuspending in the buffer solution and the vesicle solution was stored at 4 °C and used within 4 days.

The Kretschmann configuration was used for the SPR setup [16]. A Teflon™ cell of approximately 0.8 ml volume was used on which the gold-coated glass slide was mounted. On top of the glass slide a LaSFN9 glass prism was mounted and monobromonaphtalene (Bellingham + Stanley Ltd., England) was used as the matching fluid between the prism and the glass slide. A laser beam ($\lambda = 633$nm) was incident on the base of the prism and the reflected beam was detected using a charge-coupled device. The intensity of the reflected beam is a function of the angle of incidence and a minimum is observed at the angle where there is optimal resonance with the plasmon surface polaritons. The minimum angle is a function of the refractive indices of the glass, gold, the organic layers, the medium in the Teflon cell (usually the buffer solution) and the thickness of the gold layer and the organic layers. Since most of these variables are either known or can be determined, the thickness of the organic layers can be calculated using the Fresnel equations (using SPR software: WASPLAS version 2.1, Max-Planck-Institute for Polymer Research, Mainz, Germany). For kinetic profiles, the angle was kept constant at approximately 0.5 degrees lower than the minimum angle and the intensity of the reflected beam was monitored. After 10 min a vesicle solution was injected into the Teflon cell. Once the intensity of the reflected beam did not increase by more than 10 mV/h, the cell was rinsed with a buffer solution.

3.2. Experimental results

3.2.1 Deposition of lipids on 11-mercaptoundecanoic acid (MUA)

Deposition of vesicle solutions of cationic lipids (DHDAB, DOTAP and DPTAP) prepared by ultrasonication (Tris-buffer at pH 8.5) resulted in all instances in the formation of a layer on top of the self-assembled MUA-layer (Fig. 8). In all cases, the average thicknesses were in a narrow range between 32-33 Å (Table 1). These measurements are slightly lower than previous measurements of hydrated cationic bilayers (between 37-40 Å, Table 1) but considerably higher than expected for a monolayer (between 18-20 Å, Table 1). Our measured thicknesses also fall bellow the cationic bilayer thicknesses estimated from molecular dimensions (Table 1), i.e. twice the length of a fully stretched DHDAB molecule. However, it is known from supported bilayer and monolayer studies [17] of a chemically similar compound, hexadecyl-trimethyl-ammonium bromide, that the measured bilayer or monolayer thickness is smaller than expected from its molecular dimensions (by at least a few Angstroms), probably due to intermixing of the alkyl chains and/or chain tilt.

The adsorption of the cationic lipids was reasonably fast at first (up to 40% of the layer is formed within the first 30 minutes) and then slowed down until it reached a more or less stable value (increasing less than 10 mV/h) within 24 h.

The slight increase in the signal still occurring at that time may have been due to the weak adsorption of vesicles to the formed lipid layer. Indeed, in most instances rinsing with buffer resulted in a slight drop in reflectance intensity, indicating that some weakly absorbed lipids were flushed from the surface.

Interestingly, the zwitterionic lipid POPC also formed a layer on MUA (Fig. 8; Table 1), and more rapidly than the cationic lipids, i.e. within less than an hour the final thickness was reached. The absence of any significant increase in the thickness of the layer after one hour provides supporting evidence of bilayer formation. The layer thickness of approximately 38 Å is only slightly lower than of an unsupported POPC bilayer (40 Å, Table 1). Slight differences in lipid density due to surface interactions may account for the 2 Å disparity. Like the cationic lipids, POPC also possesses a positively charged quaternary ammonium atom that may be attracted to the negatively charged surface. Additionally, forming flat bilayers from curved vesicles will decrease the bending energy of the lipid.

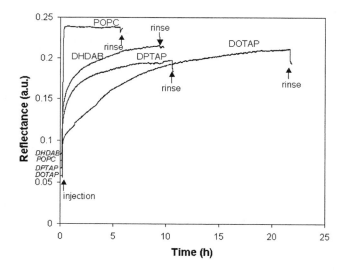

Fig. 8. Kinetic profile of adsorption of cationic lipids DHDAB, DOTAP and DPTAP and the zwitterionic lipid POPC on gold surfaces covered with a monolayer of MUA. First arrow pointing at a line is time of injection of lipid vesicle solution. Second arrow pointing at the same line indicates time when the Teflon cell was rinsed with Tris buffer (pH 8.5). Lipid names in italic indicate the initial intensity of the reflected beam before injection of the vesicle solution. (Reproduced with permission from Langmuir, 1999, 15, 8133-8139. Copyright 1999 Am. Chem. Soc.)

Table 1
Thickness of lipid bilayers on negatively charged MUA. Numbers between parentheses indicate number of experiments. (Reproduced with permission from Langmuir, 1999, 15, 8133-8139. Copyright 1999 Am. Chem. Soc.)

Lipid	Bilayer thickness (Å)	SPR thickness (Å)
DHDAB	42[a]	32±3 (5)
DOTAP	37[b]; 50[a]	32±2 (3)
DPTAP	40[c]; 50[a]	33 (2)
POPC	40[d]; 44[a]	38 (2)

[a] Based on molecular mechanics calculations
[b] Based on X-ray diffraction measurements of liposome mixture [18]
[c] Based on AFM measurements of holes in bilayers [19]
[d] Based on X-ray diffraction measurements of liposome mixture [20]

3.2.2. Deposition of lipids on 11-mercaptoundecanol/6-mercaptohexanol (MUO/MHO)

To investigate the importance of a negative surface charge for the adsorption of cationic lipids, we used self-assembled alkylthiols with a terminal alcohol group, i.e. MUO or MHO yielding a hydrophilic yet uncharged surface. None of the cationic lipid vesicles (in Tris-buffer at pH 8.5) formed any layer on top of the MUO/MHO layers (Fig. 9 and Table 2). This lack of adsorption must be an effect of the deposited alkylthiols since cationic lipids are partially adsorbed on bare gold [21]. The lack of any electrostatic attraction may prohibit the attachment of vesicles to these layers. Additionally, if any cationic lipid does adsorb, the surface will become positively charged restricting any further adsorption.

In contrast, POPC (in Tris-buffer at pH 8.5) did rapidly form a layer on MUO/MHO. The average measured thickness of this layer (45±5 Å) is slightly higher than the layer formed on MUA and slightly higher than a free POPC bilayer (38 Å). These small height discrepancies may be due to differences in bilayer packing densities as a result of surface interactions. Furthermore, the kinetics profile indicates that no increase in thickness is observed after a certain time, consistent with the formation of a bilayer. Bilayer formation may be favorable in the case of POPC because the surface should maintain approximately the same charge during lipid adsorption and the bending energy of the lipids will be decreased as the lipids adsorb to the surface.

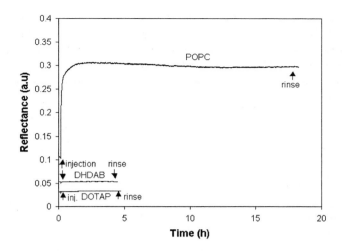

Fig. 9. Kinetic profile of adsorption of cationic lipids DHDAB, DOTAP and DPTAP and the zwitterionic lipid POPC on gold surfaces covered with a monolayer of MUA. First arrow pointing at a line is time of injection of lipid vesicle solution. Second arrow pointing at the same line indicates time when the Teflon cell was rinsed with Tris buffer (pH 8.5). Lipid names in italic indicate the initial intensity of the reflected beam before injection of the vesicle solution. (Reproduced with permission from Langmuir, 1999, 15, 8133-8139. Copyright 1999 Am. Chem. Soc.)

Table 2
Thickness of lipid bilayers on alkylthiols possessing terminal alcohol groups (MUO/MHO) and alkylthiols possessing terminal diethyleneglycol groups (MUDEG). Numbers between brackets indicate number of experiments. (Reproduced with permission from Langmuir, 1999, 15, 8133-8139. Copyright 1999 Am. Chem. Soc.)

Lipid	Thickness (Å) on MUO/MHO	Thickness (Å) on MUDEG
DHDAB	0.5 (2)	0 (2)
DOTAP	0 (2)	0 (1)
DPTAP	0 (1)	0 (1)
POPC	45±5 (3)	0 (1)

3.2.3. Deposition of lipids on mercaptoundecanediethyleneglycol (MUDEG)
     To investigate the potential of micropatterning cationic lipid bilayers on chemically modified gold surfaces, it is necessary to identify alkylthiols that coat the surface, but prohibit the adsorption of any lipid. Based on previous studies [22] alkylthiols possessing a terminal oligo(ethylene)glycol group are

good candidates for blocking adsorption of cells and proteins. Our experiments showed that deposition at pH 8.5 of cationic lipids and the zwitterionic lipid on alkylthiols possessing ethyleneglycerol moieties, i.e. MUDEG, did not result in any layer formation (Table 2). For the cationic lipids this is not too surprising since MUDEG has alcohol functional groups and, as has been shown above, cationic lipids do not form bilayers on alcohol groups. However, it is surprising that the zwitterionic lipid does not adsorb to MUDEG, since our earlier results showed that POPC does form a bilayer on alcohol groups. (Parts of this text were reproduced with permission from Langmuir, 1999, 15, 8133-8139. Copyright 1999 Am. Chem. Soc.)

4. SUPPORTED BILAYERS ON A POLYION/ALKYLTHIOL LAYER PAIR

Phospholipid bilayers supported on flat solid substrates are both of practical and scientific interests as model systems to study the structure and function of natural membranes, as well as those of membrane-bound biomolecules [2, 23, 24]. Most biomolecules require a fluid membrane environment to retain their bioactivities, which makes it important for the supported biomembrane to be laterally mobile. A fluid membrane environment requires the minimization of the influence of the solid support on the membrane since the solid substrate has been found to restrict the motion of supported organic layers [2]. Hydrated polymer layers have been used as "cushions" to lift biomembranes from solid surfaces [2]. The hydrated space created by the addition of a polymer layer is not only advantageous for decreasing the substrate effect on the membrane itself, but is also desirable for maintaining the activities of incorporated biomolecules. For example, transmembrane proteins can protrude from the membrane and become immobilized by direct interaction with the solid surface [3]. Moreover, a polymer cushion is similar to the cytoskeleton structure supporting mammalian plasma cell membrane.

4.1. Phosphatidylserine/phosphatidylcholine bilayers on a polyion/alkylthiol layer pair

Here we describe the use of a polyelectrolyte/alkylthiol layer pair that lifts the model membranes away from the solid substrates, creating aqueous environments on both sides of the membranes and retaining the mobility of the membranes. We focused on a model membrane system of lipid mixtures including an anionic phospholipid (i.e. 1-stearoyl-2-oleoylphosphatidylserine (SOPS)) and a zwitterionic phospholipid (i.e. 1-palmitoyl-2-oleoylphosphatidylcholine (POPC)). Negatively charged and zwitterionic phospholipids are major components of natural biomembranes. Poly(diallyldimethylammonium chloride) (PDDA) was used as the polymer to support the lipid bilayer. Extensive research has been done on PDDA adsorption on solid substrates to form a stable layer and it has been shown that the PDDA layer thickness depends on the ionic strength of the buffer solution that dissolves the polycation [25]. Additionally, self-assembled monolayers (SAMs) of alkylthiols are able to form on gold by the chemisorption of the thiol group to gold [26] and can be used to control the adsorption of the following layers. For this system, an alkylthiol with hydrophilic acidic headgroups, 11-mercaptoundecanoic acid (MUA), has been used to form a homogeneous negatively charged surface for the adsorption of a PDDA layer. The PDDA/MUA layer pair lifts the biomembrane away from the solid gold surface. It should be noted here that both sides of the lipid membrane are in hydrated environments. A schematic drawing of this system is shown in Fig. 10.

4.2. Experimental materials and methods

4.2.1. Experimental procedures for adsorption experiments

The adsorption experiments were monitored with SPR as described in Section 3. A desired buffer was injected into the Teflon cell and was allowed to incubate for approximately 15 minutes to equilibrate with the surrounding room temperature before SPR measurements were taken. Then a 0.2 M PDDA solution dissolved in the same buffer was exchanged into the cell for at least 20 minutes before it was rinsed with buffer. The cell was rinsed until there was no further decrease of reflectivity. The freshly prepared vesicle solutions were then injected into the cell. The growth of the bilayer, including a rinsing step, was again monitored by a SPR kinetic measurement (Fig. 11).

4.2.2. Experimental procedures for fluorescence recovery after photobleaching (FRAP)

The mobility of the supported bilayers was tested at room temperature by FRAP experiments on a Nikon Diaphot 300 fluorescence microscope. Two half-circular Teflon spacers with a thickness of 60 μm were placed on a cleaned microscope cover slide. About 60 μl of a 0.2 M PDDA solution in Tris 8.5 buffer was added in the middle of the two spacers.

A gold-coated glass slide covered by a self-assembled MUA monolayer was then mounted on the spacers. The setup was allowed to incubate in a humid environment for about 30 minutes. The polymer solution captured between the slides was exchanged with a buffer solution and subsequently a lipid vesicle solution was introduced to the space by exchanging solutions. The setup was then allowed to incubate for at least three hours. The vesicle solution was then exchanged with Tris 8.5 buffer solution for at least 10 times.

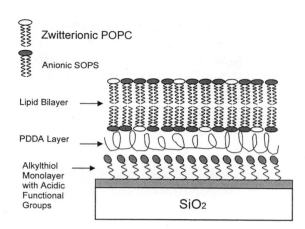

Fig. 10. Schematic representation of the model membrane system. The alkylthiol (MUA) layer self-assembled on a gold surface. The negatively charged headgroups of MUA adsorb a cationic polymer (PDDA) layer. A lipid bilayer with negative charges is then deposited on the PDDA/MUA layer pair. (Reproduced with permission from Langmuir, 2000, 16, 5093-5099. Copyright 2000 Am. Chem. Soc.)

The NBD fluorescent probe was excited by blue light filtered from a Mercury lamp and emitted bright green light. Using a 20X objective lens on the microscope, a spot of 60 μm in diameter was bleached for 3 minutes with the diaphragm closed and with both neutral density filters out. The bleached spot was then viewed using the same objective lens with the diaphragm open. Images were captured by a digital camera every 15 minutes until the spot was totally diffused and could not be observed on the screen.

On-line software, *Scion Imaging* (www.scioncorp.com), was used to analyze the images. For each image, an optical intensity profile was generated. The data were smoothed and the differences of maximum and minimum values of the curves were normalized and plotted versus time. The lateral diffusion coefficient (D) was estimated from D (cm^2/s) = 0.224 ω^2(cm)/$t_{1/2}$(s), where ω is the radius of the bleached spot and $t_{1/2}$ is the half-life of the fluorescence recovery.

4.3. Experimental results
4.3.1. PDDA on self-assembled MUA on gold

PDDA, a positively charged polymer, can be deposited onto a negatively charged surface formed by MUA/gold only at a concentration of 0.2 M (based on its monomer MW = 161.7). At PDDA concentrations below 0.2 M, the adsorbed PDDA layer was not sufficiently thick to overcome the negative charges of the MUA surface and thus prevented vesicle fusion.

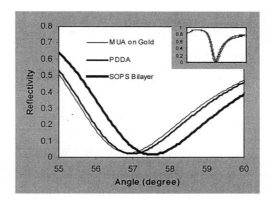

Fig. 11. Part of the SPR curves near the minimum angles, where the reflectivity is at its lowest value. The curves shift to the right with successive deposition of PDDA and a lipid bilayer. A SPR curve measured after the deposition of a SOPS bilayer is shown as an example. Inset: Complete SPR curves measured from 45° to 70°. The parts near the minimum angles are enlarged for better resolution. (Reproduced with permission from Langmuir, 2000, 16, 5093-5099. Copyright 2000 Am. Chem. Soc.)

For PDDA concentration of 0.2 M, simulation gave a thickness increase of 10 ± 1 Å assuming a refractive index of 1.41 for the adsorbed PDDA layer. No significant difference was observed under the three different pH conditions (i.e. 8.5, 7.2 and 5.0). The PDDA layer provides an aqueous cushion for the deposition of a lipid bilayer. Thickness of aqueous polyelectrolyte layers adsorbed from low ionic strength solutions has been reported to be in a similar range as measured here. Small angle x-ray reflectivity study of a polyethyleneimine (PEI) layer (adsorbed from a PEI solution of 0.03 mol/L in water) between a quartz substrate and a lipid bilayer revealed a thickness of 11-12 Å, including the lipid headgroup thickness [24].

The relatively small thickness of the adsorbed PDDA layer can be explained by the low ion concentration in the buffer solutions. In the presence of few counter ions, most of the charges of the PDDA are not screened, and thus the charged groups on the polymer chains repel each other to maximize inter-chain distance, leading to a relatively flat conformation onto the solid surface [25, 27].

It should be noted here that thicker aqueous polymer layers can be achieved by employing the layer-by-layer polyion adsorption technique [25]. Additional layers of PDDA can be adsorbed by interleaving with a polyanion such as PSS and will be presented in Section 5.

4.3.2. Lipid bilayers on PDDA/self-assembled MUA/gold

Lipid layers were deposited on the PDDA covered substrates by vesicle fusion at pH 8.5. The pH 8.5 is used here because it ensures that the acidic headgroups of the alkylthiol render negative charges at this pH, and this ensures the strong adsorption of the PDDA layer to the substrate. The depositions of the negatively charged lipid, SOPS, and zwitterionic lipid, POPC, as well as lipid mixtures made out of the combination of these two lipids are shown in Fig. 12. The reflectivity changes were minor before and after rinsing.

As shown by SPR kinetic profiles, pure SOPS followed a sharp and fast deposition, and the reflectivity increased to reach an equilibration within an hour. In Fig. 13, the thickness values are obtained from the simulation of the SPR curves, before and after adsorption. A thickness of 32 ± 2 Å of the adsorbed lipid membrane was calculated using a refractive index of 1.49 for the lipid layer. If a refractive index of 1.45 is used, the average simulated thickness is 43 Å.

Vesicle fusion of pure POPC showed a different kinetic response on PDDA than SOPS. The reflectivity continued to increase and was still changing slowly when the experiments were stopped at about 10 hours after the vesicle solution was injected. Thickness calculation gave a result of 140 ± 15 Å of the POPC layers. When mixtures with different compositions of SOPS and POPC were used for deposition, the final thickness of lipid adsorbed varied (Fig. 13). The thickness increased from a single bilayer thickness to that of a multilayer when the ratio of POPC was beyond 25%. When the POPC composition in the lipid mixtures increases to 50%, a "kink" in the kinetic curve was often observed when the reflectivity approached 0.2, and the final thickness was more than that of a single bilayer.

Fluorescence recovery after bleaching was performed to determine the lateral mobility of single lipid layers adsorbed on PDDA. In all cases, the fluorescence intensity of the bleached spot recovered about 90% at 90 minutes after bleaching. The bilayer formed with 25% POPC/75% SOPS had a faster recovery than that formed by pure SOPS (Fig. 14). The calculations for the lateral diffusion coefficients (D) showed that for the mixture with 75% SOPS and 25% POPC on PDDA $D{\sim}2 \times 10^{-9} cm^2/s$, whereas for the bilayer formed from pure SOPS bilayer $D{\sim}1 \times 10^{-9} cm^2/s$.

4.4. Comments on relationship between charge and bilayer thickness

Vesicle fusion of the negatively charged lipid SOPS versus the zwitterionic lipid POPC showed very different results. Negatively charged SOPS vesicles were attracted to the PDDA layer by strong coulombic interaction and were disrupted to form a single bilayer. Possibly, the negatively charged lipid surface repelled further adsorption of other negatively charged lipid molecules, and thus the lipid system equilibrated as a single bilayer structure. The fluorescence experiments showed a homogeneous distribution of the fluorescent probe in the bilayer.

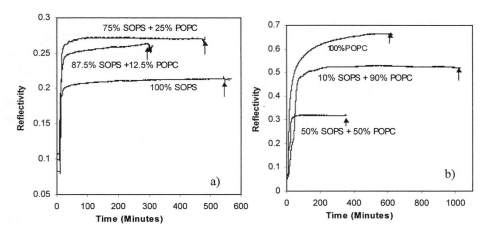

Fig. 12. SPR kinetic profiles of lipid adsorption with different compositions of SOPS and POPC (a: 100% SOPS; 87.5% SOPS/12.5% POPC; 75% SOPS/25% POPC; b: 50% SOPS/50% POPC; 100% POPC) on PDDA/MUA covered gold surfaces at pH 8.5. Note that the reflectivity scales are different in Fig. 12a and Fig. 12b. For clarity purposes, the kinetic profiles shown in the figures connect the bilayer adsorption kinetics and the rinsing kinetics. The perpendicular arrows indicate rinses with buffer to eliminate unbound vesicles. The reflectivity changes were negligible before and after rinsing. (Reproduced with permission from Langmuir, 2000, 16, 5093-5099. Copyright 2000 Am. Chem. Soc.)

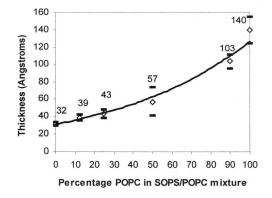

Fig. 13. Lipid layer thickness changes versus percentage of POPC in SOPS/POPC vesicles used for vesicular deposition on PDDA/MUA covered gold surfaces at pH 8.5. Each data point represents the average of at least three measurements. Similar thickness data were obtained before and after rinsing. The data presented here are those after rinsing. The error bars represent the standard deviations of the experimental data. The thickness of the lipid layers increases from a single bilayer thickness to multilayers at about 50% POPC in the mixture. (Reproduced with permission from Langmuir, 2000, 16, 5093-5099. Copyright 2000 Am. Chem. Soc.)

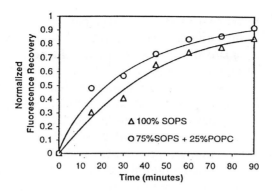

Fig. 14. FRAP data curves obtained for bilayers (100% SOPS; 75% SOPS/25% POPC) formed on PDDA/MUA covered gold surfaces at pH 8.5 under room temperature. The fluorescence intensities of the two bilayers were normalized independently. Data were collected every 15 minutes. The time interval for taking images was restricted within 3 seconds. The shutter was closed in between each datum collection to prevent excessive bleaching. (Reproduced with permission from Langmuir, 2000, 16, 5093-5099. Copyright 2000 Am. Chem. Soc.)

It should be noted that the thickness of the SOPS bilayer is 32 Å, lower than the usual bilayer thickness ~40 Å for long chain phospholipids in the fluid phase [28]. This is possibly due to two reasons. The first possibility is that there was interdigitation between the two lipid monolayers. Secondly, it is possible that there were very small defects in the bilayer that are beyond the resolution ability of the fluorescence microscope. In the case of the zwitterionic lipid POPC, vesicle fusion led to multilayer thicknesses on PDDA. Possibly the interaction of the POPC lipid layers and the PDDA layer was not strong enough so that the van der Waals interaction between the lipid layers dominated, leading to the formation of multilayer aggregates or the presence of undisrupted vesicles on the surface. When mixtures of the two lipids were used (POPC composition > 25%), the electrostatic interaction and van der Waals interaction were in competition and resulted in some intermediate thickness.

The lipid mixtures with 12.5% POPC and 25% POPC followed similar kinetics in comparison to pure SOPS, i.e., a quick steep growth period and an almost flat equilibrium stage. The thickness increased slightly with an increase in POPC content. In these systems, it appears that the electrostatic attraction from the polymer layer was the driving force for bilayer adsorption, and then the electrostatic repulsion between the lipid bilayer and the vesicles prevented further deposition of lipid molecules. The somewhat larger bilayer thickness is possibly due to the fact that the addition of zwitterionic lipids diluted the charge density of the lipid membrane. The weaker electrostatic attraction due to the charge dilution effect by zwitterionic POPC molecules may result in less interdigitation between the two lipid monolayer pairs. It is also possible that zwitterionic lipid molecules filled in the defects and made a more complete bilayer. The increasing diffusion coefficient of the membrane with increasing percentage of POPC supports this hypothesis.

The kinetics and thickness for the lipid mixture with 50% POPC indicated an intermediate stage for adsorption. The "kink" in the kinetic profiles observed at the reflectivity of ~0.2 is possibly due to the fact that a bilayer was formed before other lipid molecules were deposited as multilayers or as undisrupted vesicles on the surface. Another possibility is SPR line broadening and sharpening at the fixed angle due to a highly inhomogeneous layer formation. It is interesting that the kinks only show up when POPC composition is at or above 50%. Zwitterionic POPC molecules reduced the charge density of the membrane so that there was not sufficient repulsion to prevent further vesicle deposition.

The lateral diffusion coefficients of lipid bilayers measured for both SOPS and 75% SOPC/25%POPC mixture in this system are within the range of those of fluid bilayers. It should also be pointed out here that the lateral diffusivity values are well above the values characteristic for lipid membranes in the rigid L_β phase (partly ordered, untilted, $D<10^{-11}$ cm^2/s). The mobility of the solid-supported bilayer plays a significant role in the mechanism of membrane – bound molecules.

4.5. Interactions with macromolecules

A study of macromolecule interaction with a supported lipid bilayer using surface plasmon resonance (SPR), electrochemical cell for cyclic voltammetry (CV) and atomic force microscopy (AFM) has been recently published [29-31]. The results have shown that a pore forming β-sheet peptide, protegrin-1, has increased membrane permeability significantly (as measured by CV) in POPC and cholesterol containing membranes, therefore mimicking its natural antimicrobial activity in the negatively charged membranes, but the effect was less dramatic in 100% SOPS membranes. Supporting AFM experiments showed that 20-30 nm holes were formed by protegrin in POPC + SOPC membranes but not in the case of 100% SOPS membranes even though similar amount of peptide adsorbed to the membrane as shown by SPR. A more hydrophilic mutant protegrin, (Leu$_3$, Thr$_{14}$)-PG-1, actually decreased the permeability of the lipid membranes and no pores were observed by AFM even though peptide did associate with the membrane as observed with SPR. These results suggest that a transmembrane orientation of the native peptide and the formation of pores is essential for its normal activity. (Parts of this text were reproduced with permission from Langmuir, 2000, 16, 5093-5099, Langmuir, 2002, 18, 1318-1331, and J. Colloid Interf. Sci., 2000, 228, 82-89. Copyright 2000 Am. Chem. Soc. and 2000 Academic Press)

5. PATTERNING SUPPORTED BILAYERS

Adsorption of lipid bilayers on surfaces has attracted much interest because of the possibility of constructing biomembrane-based sensors and for investigating fundamental biomembrane properties. In order to retain the fluid nature of the lipid that is important for cellular functions, various materials have been used to support the lipid bilayer and thereby minimize the effect of the solid substrate. Self-assembled monolayers (SAM) and SAM-supported polycationic layers have served as supports for lipid bilayers [21, 29-32]. In addition to retention of fluidity, it is advantageous to confine the bilayer to pre-specified areas on the surface in order to study multiple analyte-specific interactions.

Partitioning of the fluid bilayer, so as to retain spatial information and control spatial geometry, has been accomplished by employing barriers to lipid diffusion such as metal lines and proteins that are deposited as patterns on the solid surface. Solid patterned lines and selective immobilization of lipids have also been employed as barriers to diffusion. Recently, a blotting and stamping method that exploits the self-limiting lateral expansion of the lipid bilayer and avoids the use of a second material on the solid surface has been developed.

5.1. Patterned self assembled monolayers

In this work, we have employed patterned SAMs as supports for two dissimilar lipid bilayers with the objective of investigating the feasibility of containing a bilayer within specified regions (but adjacent to a dissimilar lipid) by virtue of differences in the nature of the supporting thiol and of the bilayer. The pattern was created in the form of squares separated by grid lines.

We show by fluorescence microscopy and fluorescence recovery after photobleaching (FRAP) that mobility for the lipids is restricted within the patterned region due to the confinement of the lipids to their respective thiol SAM supports. The boundaries to mobility are sustained when a second, different lipid bilayer is deposited on a second thiol (different from the stamped thiol) that is in turn chemisorbed on the remaining bare gold surface of the stamped substrate.

5.1.1. Experimental methods and results

The thiols 2-mercaptoethylamine, or cysteamine (CA, Sigma, St. Louis, MO) and mercaptoundecanoic acid (MUA, Sigma) were used as the materials for self-assembly on the gold surface. POPC and SOPS were used as the lipids (both from Avanti Polar Lipids, Alabaster, AL). 1-palmitoyl-2-(6-((7-nitro-2-1,3-benzoxadiazol-4-yl)amino)caproyl)-sn-glycero-3-phosphoserine (NBD-PS), 1-palmitoyl-2-(6-((7-nitro-2-1,3-benzoxadiazol-4-yl)amino)caproyl)-sn-glycero-3-phosphocholine (NBD-PC) (both from Avanti), (Texas Red)-1,2-dihexadecanoyl-sn-glycero-3-phosphoethanolamine triethylammonium salt

(Texas Red – DHPE from Molecular Probes, Eugene OR) were used as the fluorescent probes. Tris-(hydroxymethyl) aminomethane (tris, 99.9%, Sigma) at pH = 8.5 was used as the buffer.

Microcontact printing of thiols was carried out as described in the literature [33]. Typically, CA was stamped on the gold surface to form the discrete squares and MUA was chemisorbed on to the continuous grid lines. Lipid vesicle solutions were prepared tip sonication as described in earlier sections.

Figure 15 shows the sequence of FRAP images obtained for the case of SOPS containing NBD-PS deposited on CA. The CA was stamped on the gold slide as discrete squares and SOPS was subsequently deposited on the gold slide [34]. Since the CA monolayer has a positive terminal charge, the anionic lipid bilayer fused preferentially on the CA-stamped surfaces, thus defining the pattern. Figure 15a shows the observed pattern prior to bleaching. One of the squares is completely dark at the end of the bleaching period (Fig. 15b) while the peripheral squares are partially bleached. Figure 15c shows the pattern 45 minutes after the end of the bleaching period. The figure shows complete mixing in the case of the peripheral squares as evidenced by the even brightness in those squares and no recovery in the case of the completely bleached square. This provides evidence that the lipids within each square is isolated from its neighboring squares and demonstrates the possibility of confining the lipid to a pre-determined area in the absence of extrinsic barriers. The boundaries of the supporting SAM serve as self-limiting barriers to the diffusion of the lipid and therefore expansion of the bilayer does not occur as observed on glass supports, [5] this ensures the integrity of the pattern.

Deposition of a second lipid on to a surface patterned with both CA and MUA was accomplished by deposition of SOPS followed by POPC on the pattern. When the two lipids were deposited with different fluorescent probes (Texas Red for SOPS and NBD for POPC) and the slide imaged, it was possible to observe the green POPC regions (grid-line region in Fig. 16a) and the red SOPS regions (squares in Fig. 16b).

The figure shows that in general, the boundaries between CA/SOPS regions and MUA/POPC regions are intact when both lipid bilayers are present. These boundaries do not degrade for at least 24 hours (longer times were not studied). In this case, we believe that mixing of the two lipids is mainly prevented by the repulsion between SOPS and MUA that retains SOPS on the CA monolayer. We see evidence of transport of the thiol from the stamp to the surface in the noncontact regions through the vapor phase. This results in some sporadic degradation of the pattern where CA has deposited either diffusely or in spots in the grid-lines.

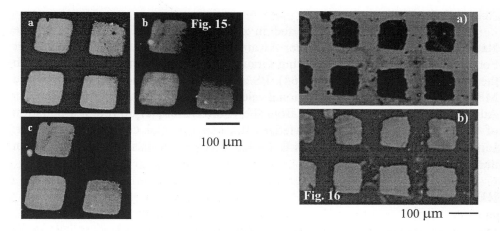

Fig. 15. FRAP images of SOPS (2% NBD-PS) deposited on CA-stamped pattern. 1a: fluorescence image before bleaching; 1b: image after bleaching for 10 min; 1c: image after recovery period of 60 min. (Reproduced with permission from Langmuir, 2001, 17, 7951-7954. Copyright 2001 Am. Chem. Soc.)

Fig. 16. Fluorescent images showing the POPC (channel regions, 3a) and SOPS (squares, 3b) areas on the gold slide. NBD-PC and Texas Red were used as fluorescent probes for POPC and SOPS, respectively. Fig. 16a was obtained using blue light and Fig. 16b using green light. Light spots in the grid lines (b) can be explained by vapor phase deposition of CA during the stamping procedure as explained in the text. (Reproduced with permission from Langmuir, 2001, 17, 7951-7954. Copyright 2001 Am. Chem. Soc.)

Subsequently, when SOPS and POPC are deposited, we observe sporadic regions of diffuse red fluorescence or spots (see Fig. 16b) associated with SOPS within the continuous grid-lines. These spots of red fluorescence are maintained for at least 24 hours, again indicating that mixing does not occur between SOPS (confined by CA) and POPC. This observed sporadic pattern degradation could probably be almost completely avoided by replacing hand stamping, as performed here, by a more mechanical method in which the contact time and pressure could be more carefully controlled.

Diffusion coefficients obtained from recovery curves were found to be 2.2 x 10^{-8} cm^2/s and 7.6 x 10^{-9} cm^2/s, for SOPS and POPC, respectively. These values are consistent with measured diffusivities of bilayers on glass. (Parts of this text were reproduced with permission from Langmuir, 2001, 17, 7951-7954. Copyright 2001 Am. Chem. Soc.)

Acknowledgement

MLL and TVR were supported in part by the MRSEC Program of the National Science Foundation under Award Number DMR-9808677, a Whitaker Foundation Biomedical Engineering Grant (RG-98-0276) and the NSF through the Career Program (BES-9733764). PS and RV were supported in part by the MRSEC Program of the National Science Foundation (DMR-9808677). Additional support was obtained from the Clorox Company through a donation of a SPR. PS and MLL acknowledges the experimental contributions by Dr. Liqin Zhang, Dr. Chen Ma, Dr. A.E. McKiernan and Professors M.P. Srinivasan and S. Schouten.

REFERENCES

[1]   M. Seitz, E. Ter-Ovanesyan, M. Hausch, C.K. Park, J.A. Zasadzinski, R. Zentel, J.N. Israelachvili, Langmuir, 16 (2000) 6067

[2]   E. Sackmann, Science, 271 (1996) 43.

[3]   J. Salafsky, J.T. Groves, and S.G. Boxer, Biochem., 35 (1996) 14773.

[4]   C.W. Hollars and R.C. Dunn, Biophys. J., 75 (1998) 342.

[5]   P.S. Cremer and S.G. Boxer, J. Phys. Chem. B, 103 (1999) 2554.

[6]   J.T. Trevors, Antonie Van Leeuwenhoek Intl. J. Gen. Molec. Microbiol., 71 (1997) 363.

[7]   C.M. Knobler, Science, 249 (1990) 870.

[8]   A.E. McKiernan, R. I. MacDonald, R.C. MacDonald, D. Axelrod, Biophys. J., 73 (1997) 1987.

[9]   G. Weidemann and D. Vollhardt, Biophys. J., 70 (1996) 2758.

[10]   G. Weidemann and D. Vollhardt, Thin Solid Films, 264 (1995) 94.

[11]   J. Masai, T. Shibataseki, K. Sasaki, H. Murayama, K. Sano, Thin Solid Films, 273 (1996) 289.

[12]   J. Korlach, P. Schwille, W.W. Webb G.W. Feigenson, Proc. Natl. Acad. Sci. U. S. A., 96 (1999) 9966.

[13]   J.N. Israelachvili, Intermolecular and surface forces, 2nd ed., Academic Press London, London, 1992.

[14]   H. Mohwald, Ann. Rev. Phys. Chem., 41 (1990) 441.

[15]   M. Liley, J. Bouvier, H. Vogel, J. Colloid Interface Sci., 194 (1997) 53.

[16]   E.F. Aust, M. Sawodny, S. Ito and W. Knoll, Scanning, 16 (1994) 353.

[17]   W.A. Hayes and D.K. Schwartz, Langmuir, 14 (1998) 5913.

[18]   J.O. Radler, I. Koltover, A. Jamieson, T. Salditt and C.R. Safinya, Langmuir, 14 (1998) 4272.

[19]   J.X. Mou, D.M. Czajkowsky, Y.Y. Zhang, Z.F. Shao, Febs Lett., 371 (1995) 279.

[20]   K. Kinoshita, S. Furuike and M. Yamazaki, Biophys. Chem., 74 (1998) 237.

[21] S. Schouten, P. Stroeve and M.L. Longo, Langmuir, 15 (1999) 8133.
[22] T.B. Dubrovsky, Z.Z. Hou, P. Stroeve and N.L. Abbott, et al., Anal. Chem., 71 (1999) 327.
[23] J. Majewski, J.Y. Wong, C.K. Park, M. Seitz, J.N. Israelachvili and G.S. Smith, Biophys. J., 75 (1998) 2363.
[24] U. Sohling and A.J. Schouten, Langmuir, 12 (1996) 3912.
[25] G. Decher, Comprehensive Supramolecular Chemistry, Pegamon Press, New York, 1996.
[26] A. Ulman, An Introduction to Ultrathin Organic Films, from Langmuir-Blogett to Self-Assembly, Academic Press, San Diego, 1996.
[27] G. Decher, B. Lehr, K. Lowack, Y. Lvov and J. Schmitt, Biosensors & Bioelectronics, 9 (1994) 677.
[28] D. Marsh, CRC Handbook of Lipid Bilayers, CRC Press, Inc., Boca Raton, Florida, 1990.
[29] L.Q. Zhang, M.L. Longo and P. Stroeve, Langmuir, 16 (2000) 5093.
[30] L.Q. Zhang, R. Vidu, A.J. Waring, R.I. Lehrer, M.L. Longo and P. Stroeve, Langmuir, 18 (2002) 1318.
[31] R. Vidu, L. Zang, A.J. Waring, M.L. Longo and P. Stroeve, *Mat. Sci. Eng. B,* (in press, 2002).
[32] L.Q. Zhang, C.A. Booth and P. Stroeve, J. Colloid Interface Sci., 228 (2000) 82.
[33] A. Kumar and G.M. Whitesides, Science, 263 (1994) 60.
[34] M.P. Srinivasan, T.V. Ratto, P. Stroeve and M.L. Longo, Langmuir, 17 (2001) 7951.

Planar Lipid Bilayers (BLMs) and their Applications
H.T. Tien and A. Ottova-Leitmannova (Editors)

Chapter 33

Supported Planar Lipid Bilayers (s-BLMs, sb-BLMs, etc.)

A. Ottova[a,b], V. Tvarozek[b] and H.T. Tien[a,b]

[a]Membrane Biophysics Laboratory, Biomedical and Physical Sciences Building Physiology, Michigan State University, East Lansing, Michigan 48824 (USA)

[b]Center for Interface Sciences, Microelectronics Department, Faculty of Electrical Engineering and Information Technology Slovak Technical University, Bratislava, Slovak Republic

1. INTRODUCTION

The interaction of interfacial chemistry with cell physiology and molecular biology has given rise to an exciting area of research termed membrane biophysics, which integrates up-to-date findings on molecules and processes involved in inter- and intra-cellular recognition and communication. Knowledge of the ideas and findings resulting from such interdisciplinary research are now being used for practical applications in analytical chemistry, immunology, photobiology, chemical/bio- sensors, and in molecular electronics [1]. The initial discovery of planar BLMs, and later supported BLMs have made it possible for the first time to study, directly, the electrical properties and transport phenomena across a 5 nm lipid bilayer separating two interfaces (i.e., *aqueous solution | BLM | aqueous solution; aqueous gel | BLM | aqueous solution; metal | BLM | aqueous solution*). In this connection other systems such as liquid-liquid and liquid-membrane interfaces are also of interest, when amphiphilic compounds are present. A modified BLM (or a supported BLM) is viewed as a self-sealing and self-assembling entity that changes as a function of time in response to environmental stimuli. A functional biomembrane system, based on a self-assembled lipid bilayer and its associated proteins, carbohydrates and their complexes, is in a liquid-crystalline and dynamic state. In molecular and electronic terms; a functional membrane system can facilitate both ion and electron transport, and is the site of cellular activities in that it functions as a 'device' for either energy conversion or signal transduction. Such a system, as we know intuitively, must act as some sort of a transducer capable of gathering information, processing it, and then delivering a response based on this received information. With the availability of supported s-BLMs and sb-BLMs, a host of compounds have been and may be embedded in the ultrathin lipid bilayer for detecting their counterparts present in the environment. Owing to its long-term stability, ease of formation, and low cost in its construction, a supported BLM offers an approach especially advantageous in the research and development of lipid bilayer-based sensors and devices. BLMs can be modified with many different constituents to perform diverse functions. The membrane can be modified with molecules, which make the membrane more sensitive to substances in the bathing solution. Other modifiers act to increase the stability of the BLM.

Planar BLMs (lipid bilayers) are now classified into two major types: conventional and supported. A conventional BLM (c-BLM) is considered as a free standing ultrathin film separating two aqueous solutions, and as such it is not very stable for long-term investigations, whereas supported planar lipid bilayers (s-BLMs) are systems of choice for practical applications, owing to their durability. The cogent reason that self-assembled BLMs are of sustained scientific and practical interest is owing to the fact that most physiological activities involve some kind of lipid bilayer-based ligand-receptor contact interactions. For instance, by embedding a receptor or a host of specific entities into a supported BLM, it is possible to create a sensor that will interact with its selected environmental counterpart. In this connection, the c-BLM along with the liposome has been extensively used to elucidate the molecular mechanisms of biomembrane function such as ion sensing, material transport, electric excitability, gated channels, antigen-antibody binding, signal transduction, and energy conversion, to name a few [1-3]. In this paper, we will summarize some of the highlights of research of the past decade on supported s-BLMs as practical sensors and devices, albeit c-BLMs will be described in certain selected cases.

2. EXPERIMENTAL

2.1. Methods of formation and study

2.1.1. BLM-forming solutions
Some of the typical lipid solutions used to form BLMs are as follows: (a) 2-5% phosphatidylcholine (PC or lecithin) in n-decane, (b) 1:1 mol % PC and cholesterol in n-decane, (c) 1% glycerol monooleate in squalene, (d) oxidized cholesterol in n-octane, (e) 2% PC in n-decane and squalene (1:1 v/v), and many others [1, 2]. In this connection, a wide variety of molecules, providing various chemical and physical properties, have been incorporated into the lipid bilayer. These molecules are termed *modifiers*. Adding a modifier to either the bathing

solution or the BLM-forming solution modifies the lipid bilayer. To make a significant change in the membrane composition, only trace amounts need to be embedded. Lipid solutions for supported bilayer lipid membranes are essentially the same. For example, 1 % phosphatidylcholine (PC) and 1 % cholesterol (CH) are dissolved in n-decane, or in 1:1 by vol. n-decane and squalene. Alternatively, for BLM-forming solutions 1% egg PC by wt. in squalene:decane (1:1 v/v) is often used. Teflon-coated stainless steel (ss) wire was cut and immersed in a BLM-forming solution for 5-10 min. The hydrophilic polar groups of lipid molecules form a monolayer on the freshly cut wire tip, which serves as a base for the second monolayer to assemble. Upon immersing the lipid-coated wire tip into an aqueous solution, excess lipid solution is drained away leading to the formation of a self-assembled, supported BLM.

2.1.2. BLM chambers and electrodes

The actual hardware of the system can also be varied, such as the chamber components, aperture supports, thermostat provisions, electrodes, metal wires, and insulation. Shielding intended to eliminate extraneous light and electrical disturbance may be accomplished, respectively, by either painting the apparatus black or by using black Lucite, and by enclosing the setup in a Faraday cage made of copper wire mesh. Platinum, silver, gold, or stainless steel wire can be used for supported BLMs ranging from 0.02 to 1 mm or more in diameter. Teflon coating, 0.075 mm, is usually present on the wire. The reference electrodes in the cell chamber may be either a saturated calomel electrode (SCE) or a chlorided Ag wire electrode (see Fig. 1).

2.2　Conventional BLMs (c-BLMs or planar lipid bilayers)

The conventional bilayer lipid membranes (c-BLMs), also known as planar lipid bilayers, since their inception in 1962, have been extensively investigated, in particular, as models of biomembranes [1,2]. Experimentally, a conventional BLM or c-BLM can be formed by supporting on a small hole (diameter 1.0 mm) in the wall of a Teflon cup. The c-BLM separates an electrolyte solution into inner and outer section, and two SCEs inserted in the two sections, respectively, were linked to the external circuits. Although the conventional BLM has proved to be very useful, it has one major deficiency, namely stability. Due to the nature of the thin membrane, the conventional BLM setup provided very unstable support. The lipid layer could unexpectedly rupture at any time. This was very frustrating for researchers to deal with until this problem was finally resolved. Nevertheless, c-BLMs are still extensively employed in membrane reconstitution experiments. For practical applications, supported BLMs are the systems of choice, as described in Fig. 1 below.

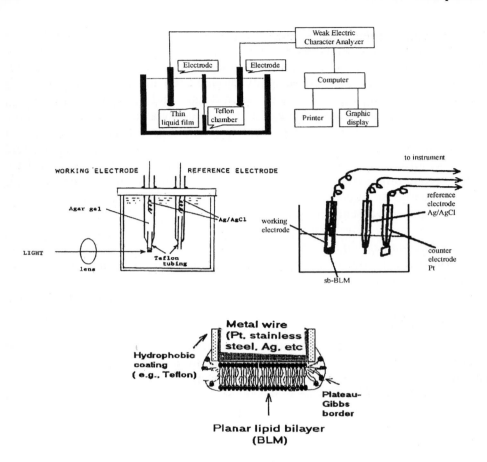

Fig. 1. Experimental arrangements for investigating conventional Bilayer Lipid Membranes (planar lipid bilayers) (c-BLMs), Hydrogel- or salt-bridge-supported sb-BLMs, and metal-supported s-BLMs. *Top:* c-BLMs;   *Middle:* sb-BLMs;   *Bottom:* metal-supported s-BLMs [4-12].

2.3   The patterning of ultrathin film structure

Several planar microelectrochemical chips with thin-film electrodes of different shapes and arrangements on Si or glass substrates have been developed and utilized in voltametric, conductometric and potentiometric sensors [7,44,45]. Selected planar forms of thin-film microelectrodes and arrays fabricating in our lab are shown in Fig. 2. Thin film electrodes were fabricated by techniques of silicon technology on silicon wafers (3 inches in diameter) with crystallographic orientation of (111). The wafers were covered by SiO$_2$ with thickness of 400 nm. Before sputtering, the lift-off mask was patterned by image reversal technique with photoresist AZ 5214 E (Hoechst), photomasks were fabricated in the Lab. of E-Beam Lithography & Microtechnology ICS-SAS. Deposition of thin films was done by r.f. sputtering; adhesion and barrier layers had a thickness of Ti / 50 nm and Pd / 50 nm (Fig. 2). Top layer of Au or Pt of 150 nm in thickness displayed polycrystalline structure with preferred orientation of their grains in the (111) direction. The Si-substrates were sliced to chips in dimensions of 10 mm x 3 mm [50,54].

Fig. 2:   *The patterning of thin-film structure*

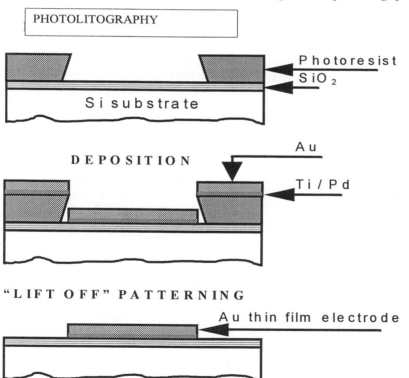

2.3.1. Microelectrode arrays

We have designed and realized the 4-electrode array (disc-shaped electrodes of 700 μm in diameter) supported on Si-chip of dimension of 11 x 8 mm (Fig. 3, left). The simple analog multiplexer was designed for signal switching of individual sensor electrodes. That electronic switch is applicable for potentiometric and cyclic voltametry methods, particularly by using very high resistive sensor elements based on BLM, but can be adapted also for

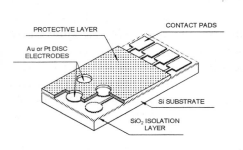

PROTECTIVE LAYER   CONTACT PADS

Au or Pt DISC
ELECTRODES

Si SUBSTRATE

SiO₂ ISOLATION
LAYER

Fig. 3. Thin film 4-electrode array chip.

conductometric / impedimetric measurements.

The arrays of working and reference electrodes realized in one chip offer some advantages as a multicomponent sensing, and increase of sensor reliability and reproducibility. We designed a multielectrode chip for biosensors (Fig. 4) consisting of array of 8 thin film metal electrodes (Au, Pt) and a reference Ag/AgCl electrode. .The reference electrode is placed around working electrodes and its area is larger by a

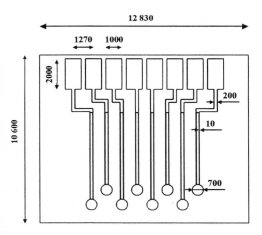

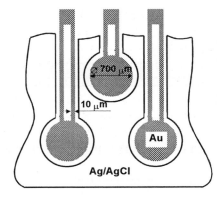

factor 1.5 as a total area of working electrodes. Disc working electrodes of 700 μm in diameter are connected by double 10 μm lines with contact pads for an improving reliability [44,45,50,54].   Fig. 4.

Of special interest to note here is the article by Neumann, Tonsing and Siemens [64]. They discussed perspectives for microelectrode arrays for biosensing and electroporation (*see* Chapter 1). Two representative cases in biomedical research are selected, aiming at new diagnostic and therapeutic clinical applications. One example is from the field of biosensing cholinergic neurotransmitter substances by the nicotinic acetylcholine receptor (AChoR) in solid-supported lipid bilayer membrane and the other one refers to new developments of electrode systems for the electrochemical delivery of drugs and genes to biological cell aggregates and tissue by the powerful method of membrane electroporation. In both cases, the new developments include the use of electrical feedback control of electrode arrays for biosensing processes as well as for the extent and duration of tissue electroporation. In line with the impressive advances in medical microsurgery, where increasingly smaller organ targets become accessible, microelectrode systems have become a continuous technical challenge for bioanalytical purposes and for the new field of the electroporative delivery of effector substances like drugs and genes, using miniaturized electrochemical electrode arrays, according to the authors [64].

2.4. Supported lipid bilayers (s-BLMs, sb-BLMs, and t-BLMs)

Recently, many attempts have been made to improve the mechanical stability of BLMs by assembling them onto a solid or to a filter support. Many reports have been published on self-assemblies of molecules as 'advanced materials' or 'smart sensors' [1,4-8]. Supported bilayer lipid membranes (s-BLMs) can now be employed for embedding a variety of compounds such as peptides, enzymes, antibodies, receptors, ionophores and redox species in detecting their respective counterparts, such as substrates, antigens, hormones, ions and electron donors or acceptors. The s-BLM is formed on the tip of a solid support, usually a Teflon-coated Pt or stainless steel (ss) wire, and proved to be much more stable and longer lasting than the conventional BLM. A s-BLM is usually formed on the tip of a freshly cut, Teflon-coated stainless steel wire (0.25 mm in diameter with a cross-section area of 4.9×10^{-4} cm^2) according to the method described in references [4-6]. The procedure used in forming a s-BLM is quite different from that used for the c-BLM. This new procedure, as the name implies, provides greater support for the stability of the BLM enabling the membrane to last much longer. The formation of a s-BLM is done in two basic steps (see Ref. 1). In the first step, a metal wire, e.g. platinum or stainless steel, is immersed in a lipid droplet. The wire is usually coated with some type of insulator, usually Teflon. It should be noted that the wire normally has sufficient time to react with the oxygen in the air causing the metal to form oxide on its surface. The usual procedure by flaming a wire, e.g., Pt, produces an ultrathin oxide film very well, hence should be avoided in preparing s-BLM probes. To prepare a fresh metal surface, the wire is submerged in the lipid solution, the tip is cut off using a sharp knife or a miniature guillotine [4,7].   The cutting of the wire, while immersed it in lipid solution was found out later to be unnecessary, providing the freshly cut tip is immersed in the BLM-forming solution as soon as possible. The nascent surface of the freshly cut metal tip is highly hydrophilic and microscopically rough. The rough surface has many pits and crevices in it, which makes it difficult for the polar heads of the lipid molecules to bind to the support. However, the 'fluid' and dynamic nature of the lipid bilayer obviates this problem.

In the second step, the freshly cut metal wire with the lipid droplet is submerged in an aqueous solution. After a few minutes, a stable lipid bilayer will be formed on the tip of the wire. Electrical properties are utilized to observe the thinning process of the membrane. When the increasing slope of the capacitance and the decreasing slope of the resistance begin to level off, this indicates the presence of a self-assembled planar lipid bilayer (s-BLM). The hydrocarbon interior of the bilayer membrane is responsible for these electrical properties [4]. It is important to note that the BLM formed here is liquid-crystalline in structure that enables it to be modified for future basic studies [1].

After immersing the cut end of the wire in the lipid solution for about 10 minutes, it is then transferred into an aqueous solution in the measuring cup. Several minutes are needed for the BLM to self-assemble on the tip of the stainless steel wire. The construction of the s-BLM made it very easy to be tested for many electrical parameters. Already having an electrical contact on one side of the membrane, essentially all that is needed to test, for example, for the potential difference across the BLM is to have a reference electrode (e.g., SCE) in the bathing solution. Care must be given, however, to insulate all electrical devices to ensure that the only potential being measured is that one across the planar lipid bilayer.

2.4.1. Preparing a solid supported BLM

Conventional BLMs which separate two aqueous solutions, are very fragile, therefore, preclude their use in practical applications. This problem of stability of planar lipid bilayers was finally solved by a novel technique of membrane formation. BLMs can now be easily formed on metallic substrates and hydrogel supports with much longer stability [4-6]. These supported bilayer lipid membranes (s-BLMs and sb-BLMs) greatly promote the application of BLMs in practical devices such as biosensors. Recently, many s-BLM-based sensors have been reported for the detection of ions and biomolecules of diverse interests. Past investigations have shown that BLMs possess both high capacitance and resistance like biomembranes [1,2].

The addition of certain modifiers can also be used to produce dramatic stabilization effects on the membrane [2,7]. Introduction of certain saccharides, for example, onto the bilayer may provide the possibility of membrane storage in a dehydrated form without causing damage to the structure [8]. Researchers are currently working towards devising ways in which s-BLMs may be of practical use as a biosensor to clinicians and laboratory technicians. The two key criteria which must be satisfied when attempting to apply a s-BLM as a practical biosensor are stability and cost. The biosensor membrane must obviously possess excessive stability in order to provide any useful applications.

2.4.2. BLMs on metallic substrates (s-BLMs)

Unlike conventional BLMs, the structural state of s-BLMs and therefore many of the mechanical and electrical parameters can be modified by applying a d.c. voltage. Further, a s-BLM formed in the manner described is remarkably stable; it can not be removed by simple washing or mechanical agitation. A s-BLM may be detached from its metallic substrate, however, by sonication, electrochemically or by drastic chemical treatments. Although usually depicted, the lipids are oriented perpendicular to the metal surface, they are however most likely tilted in some angle from the normal. To cover any surface by a layer of lipid molecules at the molecular dimension is an extremely difficult task, since the morphology of the substrate is not likely to be 'smooth'. As an experimental fact, monolayers and multilayers of lipids prepared by the L-B technique are often full of pinholes. These defects are hard to avoid owing to the nature of the substrate at the atomic level. Thus, a freshly cleaved metal surface is not smooth at the atomic level; it is most likely to be very rough with grain and edge boundaries. However, the lipid solution used, being a fluid, is able to interact with the 'bumpy' terrain of the newly cut metal surface and to form an intimate attachment within its indentations, pits, and crevices. The lipid monolayer adjacent to the solid metal support is presumed to be stabilized by hydrogen bonds arising between the hydrophilic groups of the monolayer and the electronegative metal surface. The hydrophobic alkyl tails of the amphipathic lipid molecules are arranged in such a way, which allows the polar head, groups to pack more closely. The final self-assembling lipid bilayer is stabilized because of intermolecular forces. The breakdown voltage of s-BLMs under these conditions is several folds higher than conventional BLMs (up to 1.5 V or more). The most important factor in the process of the s-BLM preparation seems to be the time the cut end of the wire is allowed to remain in the membrane forming solution (~ 10 min) prior to its transfer into the aqueous solution for a supported BLM to self-assemble. It should be mentioned that s-BLMs may be formed from a lipid droplet deposited on an air-water interface by the method of Wardak et al. (for details see Refs. 4, 9-11).

2.4.3. S-BLMs stabilized by trehalose

From an experimental viewpoint, the stability of BLMs has always been a common concern. Recently, we have investigated the effect of trehalose and glucose on the elasticity modulus perpendicular to the membrane plane and the electrical capacitance C_m of s-BLMs on Teflon-coated Ag wire [8]. Addition of the saccharide (trehalose) into the electrolyte resulted in decrease of elasticity modulus of the s-BLM formed from soybean phosphatidylcholine in n-hexadecane, while C_m did not change. Further, trehalose had a considerable stabilizing effect on the other parameters of the s-BLM, consistent with our earlier findings [1,2]. Treatment of the s-BLM in electrolyte containing 300 mM trehalose allowed s-BLMs to be kept in dry conditions and stored for several days in a refrigerator with subsequent recovery of membrane parameters after dipping the wire into the electrolyte. The addition of trehalose to the lipid membrane precluded the usual damage that the membrane endures from exposure to air, enabling it to be stored in a dehydrated form for up to many days. When the dehydrated membrane is dipped back into the electrolyte solution, the electrical and physical properties of the BLM are amazingly recovered [8]. This feat is accomplished when the trehalose is removed from the membrane due to the hydrogen bonding forces between the water molecules and the phosphate heads and also by the reordering of the hydrophobic region of the membrane. Therefore, trehalose modified membranes provide many potential possibilities for practical applications of s-BLMs as biosensors in the future.

2.4.4. Preparation of s-BLM on conducting glass

Briefly, two consecutive steps formed the ITO-supported BLM: (a) The lipid solution (1:1mol % phosphatidylcholine and cholesterol were dissolved in n-decane) was placed in contact with the ITO surface. (b) The ITO conducting glass, having become coated with lipid solution, was immersed in a buffer of 0.1 M KCl + 0.1 M tris/HCl (1:1 v/v), at pH 7.0. The preparation of ITO/s-BLM/MCF-7 nucleoli is described in details in Refs. 12, 13. A super thin glass cell was devised with the ITO glass mounted on one side. After the lipid bilayer was formed spontaneously on the ITO glass, MCF-7 nucleoli with cytosol S-150 in tris buffer were deposited on the side of BLM supported by ITO in the super thin glass cell to form the self-assembled MCF-7 'cell' (see Fig. 2 *Middle*).

2.4.5. Supported planar lipid bilayers on hydrogels (sb-BLMs)

Solid supported planar bilayer membranes proved to be very useful and easy to work with in the field of membrane research and solved many of the shortcomings of the conventional variety. However, the nature of s-BLMs poses a new problem when considering the passage of ions across the membrane [5-7]. The solid support that provides such great stability for the membrane precludes any passage of materials across it. Therefore, until a few years ago the pursuit of a simple method for obtaining long-lived, planar BLMs separating two aqueous media has been an elusive one. The procedure for forming a planar BLM on agar or agarose gel has been published [5-8]. Briefly, a small diameter (~ 0.5 mm) Teflon tubing is filled with a hot hydrogel solution (e.g., 0.3 g agar in 15 ml 3 M KCl). For electrical connection as well as serving as a reference electrode, an Ag/AgCl wire is inserted at one end. The other end of the agar-filled Teflon tubing is cut in air and then the cut end is immediately immersed into the lipid solution for about 5-10 minutes. The next step is to put the lipid-coated tip into an aqueous solution for a sb-BLM to self-assemble. In Fig. 4 *Middle* is illustrated a planar BLM on a hydrogel substrate. The sb-BLM relied on the idea of finding a substance which can act as a support and still possesses fluid properties allowing for the translocation of particles [1]. The substance used was a hydrogel (agar or agarose) commonly used as a salt bridge in electrochemistry. This gel not only proved to be viscous enough to provide valuable support for the thin membrane, but also fluid and polar enough to act as an aqueous interface. This new bilayer lipid membrane system allows for the passage of materials across a stable ultrathin membrane as well as possessing the electrical properties of its evolutionary BLM ancestors. Therefore, the sb-BLM model sheds new light on the mechanisms involved in membrane transport, ion selectivity, voltage and ligand gated channels, electron transfer, and light transduction.

Alternatively, the sb–BLM may be prepared as follows, in the procedure consisting of three steps. In the first step, the Teflon tube is cleaned. The steel (or Ag) wire is polished using ultrafine Emery paper and then it is cleaned. A section of the wire (about 3 cm long) is inserted into the Teflon tube. The tube is filled with a hot smelting agar solution (0.3 g agar dissolved in 12 ml of 1 M KCl solution). In this way, a steel (or Ag) wire–Teflon tubing salt bridge (sb) is constructed. In the second step, the tip of the other end of the Teflon salt bridge is cut *in situ* with a sharp knife while immersed in a BLM forming solution. In the last step, the Ag–Teflon salt bridge with a freshly coated lipid solution is transferred to the supporting electrolyte to allow the formation of the sb-BLM.

2.4.6. Syringotoxin-induced channels in a sb-BLM

Of special interest is that cation channels in BLMs can be induced by a number of toxins such as α-endotoxins [7,11,14]. With the availability of highly stable sb-BLMs, we decided to embed certain phytotoxic substances such as syringotoxin (S-toxin) into the lipid bilayer, since the dimensions of this S-toxin are almost in the same range as the lipid bilayer thickness [7].

2.4.7. sb-BLMs and DNA

A novel method for investigating the interactions between double-stranded DNA (dsDNA) and hemin using the sb–BLM modified with lauric acid and hemin was developed. As a biosensor, this kind of sb–BLM could be easily prepared at low cost and with a good mechanical stability. A biological compatibility, simplicity and reliability characterize the method. Further studies

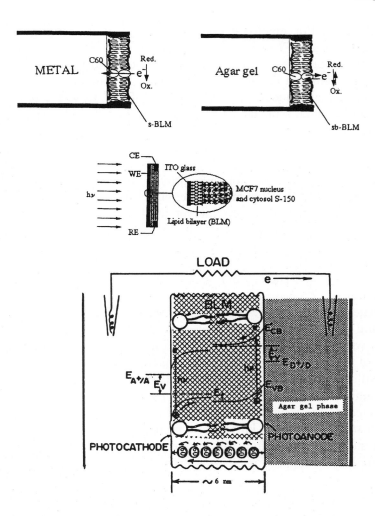

Fig. 5. Fullerene C$_{60}$-modified supported s-BLMs and sb-BLMs and their interaction with light.

Top: S-BLM on metal wire and sb-BLM on hydrogel [9-12,37];
Middle: s-BLM on conducting ITO glass [13,40,51,56];
Bottom: Mechanism of C$_{60}$-doped supported BLM [7,34].

are necessary for obtaining the interacting mechanism and kinetic parameters of the detailed electron transfer. The other result of this work is considering a possibility of a direct detection of DNA in the environment close to the biological condition. Also studied was the electrochemical behavior of horseradish peroxidase (HRP) incorporated in sb-BLM), using electrostatic interaction between HRP and lauric acid, which resulted in the enhancement of the electron exchange between the protein and the sb-BLM surface. A quasi-reversible electron transfer of HRP was observed even in the absence of mediators. The electrocatalytically kinetic behavior of HRP and the electrode-kinetic process were investigated with this biosensor [55,56]. In this regard is the paper by Nikolelis and Petropoulou [63] who reported the interactions of a resorcin[4]arene receptor with BLMs that can be used for the electrochemical detection of dopamine and ephedrine. BLMs were composed of egg PC and 35% (w/w) dipalmitoyl PA in which the receptor was incorporated. These BLMs modified with the resorcin[4]arene receptor can be used as one-shot disposable sensors for the direct electrochemical sensing of these energizing-stimulating substances. The interactions of these compounds with the lipid membranes were found to be electrochemically transduced in the form of a transient current signal with a duration of seconds, which reproducibly appeared within 8 and 20 s after exposure of the membranes to dopamine and ephedrine, respectively, The magnitude of the transient current signal was related to the concentration of the stimulating agent in the µM range.

2.4.8. Tethered planar lipid bilayers (t-BLMs)

Upon modifying the terminal -CH_3 group of the alkanethiols by a suitable functional group, bearing the negative (carboxyl) or positive (amino) charge, it is possible to attach pure phospholipid bilayers by electrostatic interactions [9-11]. It was shown that the function of many proteins (e.g., toxin, mellitin, Na,K-ATPase and gramicidin) [1] tested in such supported bilayers maintain their characteristic function. However, it was recognized, that for a membrane protein to function correctly, it is fundamental to have water on both sides of the lipid bilayer into which such a protein is reconstituted. Moreover both layers have to be in the liquid crystalline state. The presence of long chain alkanethiol molecules, forming the solid state film, regardless whether they were obtained by the self-assembly or LB technique, was a major drawback in such structures. Therefore a hydrophilic "spacer" molecules were introduced, such as polymers with polyoxyethylenoxy chains, functionalized on one side with thiol to bind to the gold electrode and with cholesteryl residue on the opposite side, to interact with the membrane, peptides or polyelectrolytes. The role of such spacer was to form bilayer lipid membranes bound to, but structurally decoupled from the solid support by a flexible network of molecules. In such a way the functions of bacteriorhodopsin, H,K-ATPase, Ca-ATPase, ATP-ase CF_0F_1 and other membrane-bound proteins were demonstrated to be perfectly unaffected. In this connection, highly stable supported membranes were formed, ion channel were functionally integrated into these membranes and molecules interactions of analytes with embedded ion channels were monitored [11]. It is also important to note that along this line of study a new class of lipids was synthesized, named "thiolipids" which allow for the direct attachment of the first thiolipidic layer to the surface of gold with the subsequent transfer of the "normal" lipid layer on such pre-treated electrode (t-BLMs - tethered bilayer lipid membranes [11,15]), this group also includes the BLMs with other spacer molecules - molecular tethers for fixing the BLM to the electrode surface. The purpose of using such novel synthetic lipids was to develop a stable, covalently bound to the surface of electrode system, which mimics the native biomembranes with both hydrophilic sides in contact with aqueous "ionic reservoirs". This approach provides the required hydrophilic/hydrophobic/hydrophilic environment for the intrinsic membrane protein to retain their functions, either for the development of highly selective bioelectronic devices or for the immobilization of macromolecules for structural studies. The properties of such t-BLMs [11,15-17] were studied by various techniques, including spectroscopic and electrochemical techniques. These studies reveal that such a system has comparable capacitance to those of conventional BLMs, indicating the formation of bilayer lipid membranes similar to BLM [9-11,52]. For such systems the measurements of electron transfer rates show that the lipid monolayer improves the blocking properties of assembled first thiolipid monolayer and the ionic sealing ability was similar to that of conventional BLMs.

2.4.9. Alkanethiol/lipid hybrid bilayer lipid membranes (h-BLMs)

The use of LB techniques in fabrication of ordered layered structures on solid supports has grown rapidly in recent years [9-11]. However, the LB technique does not allow one to mimic those biophysical properties and processes that require a fluid membrane, e.g., the functioning of membrane integral proteins, as this technique produces mono-, or multilayer assemblies that are polycrystalline or amorphous in nature. This two-step technique, in which a monolayer of exactly known composition is initially spread and compressed on the water surface and then transferred onto an electrode substrate, offers substantial flexibility in selection of molecules to be immobilized on the surface of the electrode. This is also the only approach that allows one a quantitative control of the surface density of all species of a multi-component monolayer assembly in a wide range of concentrations extending down to 10^{-17} mol/cm^2. The ability to operate at this low level of gate molecules is limited by the defect density in the passivating LB monolayer constituting a source of background current. It was shown that when the surface concentration of the gate sites is less than 10^{-15} mol/cm^2, they begin to behave as an array of independent molecular size submicro-electrodes. In another example, a novel model bilayer lipid membrane is prepared by the addition of phospholipid vesicles to alkanethiol monolayers on gold [11,18]. This supported BLM is said to be rugged, easily and reproducibly prepared in the absence of organic solvent, and is stable for very long periods of time. This type of s-BLMs formed from phospholipids and alkanethiols are pinhole-free, as in conventional BLMs. Capacitance values suggest a relative dielectric constant of 2.7 for phospholipid BLMs, which is about 30% higher. These researchers also tested the protein toxin, melittin (a pore-forming peptide, [18]), on this type of s-BLMs and found that it destroys the insulating capability of the phospholipid layer without significantly altering the bilayer structure.

2.5   Techniques of investigation
2.5.1. Measurements of Electrical Properties
The electrical properties of a planar BLM separating two aqueous solutions (i.e., aqueous solution | BLM | aqueous solution) or a metal/gel substrate and an aqueous solution (i.e., metal | BLM | aqueous solution or gel | BLM | aqueous solution) can be easily determined [2,19]. Generally, the BLM resistance R_m is several orders of magnitude higher than those of the combined resistances of the contacting electrodes and aqueous solutions. A good electrometer with a 10^9 input impedance together with a picoammeter should be adequate. For BLM experiments involving cyclic voltammetry [19], a block containing two adjacent 2 cm diameter chambers (one of which holds a 10 ml Teflon cup), was commonly used. The Teflon cup was referred to as the inside, and the other chamber as the outside. A three-electrode system for obtaining voltammograms has been used in the following configuration: one reference electrode (SCE or Ag/AgCl) is placed in the Teflon cup and two other reference electrodes are on the outside. The voltammograms of the BLM are obtained using an X-Y recorder fed by a picoammeter and the voltage generator (e.g., Princeton Applied Research, Universal Programmer, Model 175). The voltage from the programmer is applied through the potentiometer to the reference electrode immersed in the inside solution. Another reference electrode immersed in the outside solution is connected to the picoammeter. The important feature of the setup is a very weak dependence of its input voltage on the current being measured. This means that the current is measured under "voltage clamp" with accuracy $\pm$ 1 mV. In voltammetry the potential of the cell is varied and the corresponding current is monitored. The graph with the current plotted on the vertical axis vs. the potential on the horizontal axis is called the voltammogram, which is characterized by several parameters. The working electrode (WE) may be made of Pt, Au, or carbon paste, glassy carbon, semiconductor SnO_2 or in our case one side of a BLM. An Ag/AgCl or a saturated calomel electrode (SCE) is

used as the reference electrode (RE). The potential scan or sweep is carried out between two potential values of interest (e.g., from about 1.2 to -0.8 V vs. SCE). The scan rates can be anywhere from 0.1 mV to 100 V or more per second but values between 10 mV to 400 mV s^{-1} are frequently used. The current response of the processes at a metal electrode is indicative of the nature of the redox reaction at the interface.

There are a few basic methods used to modify the bilayer membrane. The membrane constituent(s) to be incorporated into the bilayer may be added to the lipid solution prior to the assembly of the membrane. If modification is desired after a BLM assembly, then the modifier can be added to the bathing solution. The bilayer membrane can be sufficiently modified by the addition of only trace amounts of the modifier. The reason for this is a result of the large ratio between the surface area and thickness of the membrane.

Measurements with a modified membrane begin once the electrical parameters such as membrane resistance or capacitance reach a steady state value and a membrane of the bilayer thickness has been reached. Changes regarding membrane properties in response to a stimulus from the bathing solution may change the electrical potential across the membrane which will propagate a signal to be sensed by the electrodes. This signal can then be quantified using electrical measuring devices. The extreme fragility of the conventional BLM seriously limits its utility as a practical tool since it cannot be easily fabricated and will not endure rugged handling. This problem of fragility was finally overcome by forming BLMs on either freshly created metallic surfaces or on smooth substrates (e.g., gels), as already mentioned.

3. RESULTS OF SOME RECENT STUDIES

As already mentioned above, and described previously [1-3], the development of electrochemical biosensors is growing at a rapid pace since the early 1980s. In the biomimetic approach, a lipid bilayer is used. The functions of biomembranes are mediated by specific modifiers, which assume their active conformations only in the lipid bilayer environment. Further, the presence of the lipid bilayer greatly reduces the interference and effectively excludes hydrophilic electroactive compounds from reaching the detecting surface, which may cause undesired reactions. Thus, from the specificity, selectivity, and design points of view, the BLM is a natural environment for embedding a host of materials of interest for the biosensor development. Supported BLMs can be employed for the incorporation of a number of compounds such as enzymes, antibodies, protein complexes (receptors, membrane fragments or whole cells), ionophores and redox species for the detection of their counterparts, respectively, such as substrates, antigens, hormones (or other ligands), ions, and electron donors or acceptors. Hence the rationale behind the development of BLM-based biosensors is remarkably simple: to function in a biological environment the sensing element must be biocompatible. The lipid bilayer (BLM) meets this criterion and is, therefore, an ideal choice upon which to develop a new class of electrochemical biosensors. The essential aspect of biosensors consists of a biosensing element for detection and a transducer for converting the detected signal for an appropriate display. Biosensing elements include membrane-bound antigens, antibodies, DNA, carriers, enzymes, receptors, tissues, cell organelles, or whole cells. The transducing components can be electrodes, optical fibers, piezoelectric crystals, etc. responsible for the observed transmembrane interactions.

Recent success in interdisciplinary research in biology combine with electronics has led to exciting new developments based on enzymology and transducer techniques. They are known as enzyme electrodes, enzyme thermistor, CHEMFET/ENFET devices, and immunosensor or enzyme transistor. Collectively, they are called "biosensors or biochips". A common feature of all these devices is a close connection between the enzyme and a transducing system, which is used to follow the enzymatic reaction. The essential principle of the devices, broadly speaking, is predicated upon the ligand-receptor contact interactions. Application of such developments in the fields of medicine, pharmaceuticals, biochemistry, environmental settings, robotics, and the food industry are obvious. For example, enzyme thermistors make use of the heat, which is liberated during an enzymatic reaction. Their usual sensitivity is around $10^{-2}\,C^0$. A recent modification of enzyme thermistor is the "TELISA" electrode, which achieves a sensitivity of about 10^{-13} M using an immunoabsorbent. It is expected that this measuring technique will find a broad application in continuous measurements of the release of hormones and/or antigens-antibodies in blood circulation. The quick achievement of a new steady state in the reaction occurring at an enzyme electrode after a random perturbation makes the latter ideally suited to monitor an industrial process, e.g., the production of antibodies. Classic calorimetric methods require much more time than an enzyme thermistor assay to perform a quantitative analysis. Two other interesting developments are ellipsometry and piezoelectric crystals. In ellipsometry, a close connection between the enzyme and the transducing device is not required. The method relies on the change in the angle of polarization of incident light that is reflected by a layer of biomolecules bound to a solid surface. A change in the thickness and conformation of this layer, under the influence of other macromolecules interaction with the layer, can be easier monitored. This principle is now used in the fermentation industry. Piezoelectric crystals can be used in the analysis of the traces of certain compounds, mainly anesthetics. The frequency of the crystal depends strongly on the absence or presence of adsorbed molecules on the surface of the crystal. A coating process may increase the selectivity of crystals toward a given compound, e.g., with hydrophobic substances such as oils and fats [20-23].

Another exciting new research area is the combination of semiconductor technology with enzymes and other biological macromolecules. Here, mostly field-effect transistors (FETs) are used. If sensitivity of a FET toward certain chemicals or ions can be achieved, the prototype of an 'ISFET" is born. A common feature of all these devices is the use of MOS (metal oxide semiconductor) structure. In combination with a thin layer of palladium, a high sensitivity toward gaseous hydrogen can be achieved. In this case, the membrane separates the gaseous from the liquid phase. Addition of traces of certain metals (e.g., Ir) to the Pd-MOS device also increased its sensitivity toward ammonia. It has been shown that such a device is capable of monitoring reliably the production of hydrogen by microorganisms, e.g., *Clostridium acetobutylicium* [1,23].

3.1   Molecular electronics and planar lipid bilayer-based biosensors

Molecular electronics uses molecular materials in which the molecules retain the separate identities. As a result, the properties of such materials depend on the molecular arrangement, properties, and interactions. Theory seeks to guide the design and synthesis of effective molecular materials. It does so by analysis, interpretation, and prediction, leading to the development and evaluation of concepts, models, and techniques. The role of theory in treating molecular properties (mainly by molecular orbital methods), arrangement (by electromagnetic or quantum-mechanical approaches) is of importance. When these factors are combined, the material properties can be treated more successfully in cases where the interactions are not essential for the existence, e.g., in nonlinear optics as opposed to electronic transport properties.

The major advantage of molecular electronics with a lower limit of μm is the further development of lithographic techniques. The changed physical properties in the submicroscopic region are the major obstacles to further miniaturization in the semiconductor technologies. The physical border for the silicon technologies is about 100 nm, because one cannot overcome the characteristic lengths such as diffusion, Debye, and tunnel lengths. With still smaller dimensions, we enter the realm of biological and molecular systems. Human brain (as well as our sensor organs), is also, without the silicon, enormously capable. Although biotransducers function much slower than silicon-based devices and are not very reliable, they are extremely efficient. Also, despite their disadvantages, nature's molecular device functions more generally and is superior to technical computers or sensors. In contrast to macromolecular biological systems, the main advantage of molecular devices, purportedly, is their relatively simple construction. In this sense, molecular devices may be readily synthesized and are always easily accessible experimentally from a quantitative point of view.

The main elements of molecular electronics are molecular wire, conducting material, molecular-specific transducers of signals similar to the particles, and molecular switches, memories, emitters, detectors, etc. The flux of information between the molecules can be released in many ways. One of the most important is the transfer of individual charges in terms of electrons, holes, or hydrogen ions, or of other shapes similar to the elements, like solitons, soliton waves, or excitons. Molecular switches may be optical, electrical, magnetic, or thermally reversible systems. Storage of information in a molecular system can be realized through a change in the electronic as well as geometric structures of the molecules in reversible thermal reactions, e.g. conformational or configurational changes upon replacement of hydrogen or protons.

The key advantage of molecular and biomolecular computing is specificity. The large number of variations that are possible with organic polymers allows for fine-tuning of electronic motions to a much greater extent than is possible in organic materials. In biological molecules, unclear configurational motions are comparable in significance to electronic motions. This is certainly the case in all conformation-based recognition processes. Enzymatic recognition is itself a basic form of information processing. When proteins and other macromolecules are combined into highly integrated complexes, it becomes possible for conformational switching processes to propagate over significant distances. The cytoskeleton is a good candidate for such a signal-processing network.

One of the most delicate 'molecular wires' is the so-called hydrocarbon chain, which is the best represented through the chain of carbon atoms in polymers. Most organic polymers are well-known insulators. However, polyacetylene $(CH)_x$, polydiacetylene, and polysulfinitrid $(SN)_x$, with their conjugated double bonds, are semiconductors or superconductors. Such conjugate systems form the group of organic conductors and semiconductors. The most important organic electroactive polymer is polyacetylene. The foundation of the electron-hole pairs and the positive and negative charges are quite well known. In the outer electric field, the electron and the hole are accelerated in opposite directions. These properties can be used in optical switches for switching on and off the flux of information. Combining molecular 'wire' and switchable molecules could lead to the construction of electronic systems based on molecules. Present research is oriented toward discovering peptides/proteins that can transduce electrical current or exist in two-electrical stages. This would lead to a future 'biochip'. Research on biochips could lead not only to a better understanding of higher nerve function, but also to the foundation of qualitative computer systems that could provide many of the activities currently performed only by the human brain.

Biochips can be considered as highly sophisticated biosensors. The unique properties of biochips are their analog and digital computing potential, self-perpetuating and potentially self-repairing. Biochips hold promise in a variety of applications such as bionic implants, memory-intensive systems, image processing and storage, artificial intelligence, language processing and molecular computers. For instance, the analog capability of biochips could enable the creation of "artificial intelligence". As such, biochips are at a very early stage of research and development. As with biosensors, the current problem is our inability to produce uniform high-activity, stable biomolecular layers and their associated transduction systems. These problems notwithstanding, it seems likely that the initial application of biochips as advanced biosensors, based on ligand-receptor contact interaction, may be in the clinical setting, where they could serve as automated control devices for drug delivery. It also appears probable that in order to extend the capabilities of present silicon-only systems, hybrid biochips and silicon-chip devices would be first produced for computing and memory-intensive systems. The key to the successful application of biochips will be to fill places that are not well served by current silicon chip technology. Thus, the future development of biochips requires the successful technologies of stable biomolecule immobilization, biotransduction, and molecular lithography. Urgent problems to be solved are biologically based amplification, molecular switching, electron transport, and memory function. In the coming decade, the answers to some of these problems will undoubtedly be found. In this connection, the development of lipid-bilayer-based sensors and biological electronic devices seems to be a logical first step. With the BLM system, especially the s-BLM, we now have an experimental approach for testing new ideas in the development of sensors for practical applications [1,4-10,22-24].

4.2   Advances of self-assembled lipid bilayer based biosensors

In the last decade or so, there have been a number of reports on self-assembled molecules or structures described as advanced materials or smart materials. Without question, the inspiration for this exciting work comes from the biological world, where the lipid bilayer of cell membranes plays a pivotal role. Past and recent achievement on self-assembled lipid bilayers as biosensors will now be described below.

4.2.1. pH sensors

The pH glass electrode is routinely used for the detection of hydronium ions (H_3O^+). However, the large size and fragility of pH glass electrodes preclude their use in many situations such as *in vivo* cell studies and in monitoring membrane boundary potentials. For example, the hydrolysis of membrane lipids by phospholipid enzymes (lipases A and C) could alter the boundary potential of BLM because of a local pH change. These findings suggest that s-BLMs can be used as a pH probe in membrane biophysical research and in biomedical fields where the conventional glass electrode presents many difficulties. To test our concept, we incorporated a number of quinonoid compounds (chloranils) into s-BLMs and found that, indeed, s-BLMs containing either TCoBQ (tetrachloro-o-benzoquinone) or TCpBQ (tetrachloro-p-benzoquinone) responded to pH changes with a nearly theoretical slope (55 ± 3) mV [7,24]. This pH-sensitive s-BLM offers prospects for ligand-selective probe development using microelectronics technologies (see below).

4.2.2. IDA/compact electrodes vertically arranged

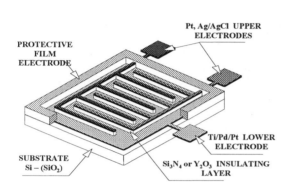

We have developed the thin-film cell consisted of the vertically arranged interdigitated array (IDA) of electrodes with a continuos Pt- basis and an insulation layer in the middle separating the upper Pt- or Ag/AgCl-IDA electrodes (Fig. 6). The closely spaced array of electrodes (width of gaps are in μm-/subμm- range) has found a very favourable application in voltammetric sensors [54].

When both reduction and oxidation potential is applied at an IDA pair, the effect of redox recycling and the high collection efficiency are observed, that are causing a high current amplification, i.e. a significant lowering the detection limit of the sensor. The redox cycling method with using microelectrodes was proven for Fe-trace determination in ultrapure spectral C in the range of Fe content 1×10^{-5} to 1.5×10^{-4} wt. %. The gap between bottom and upper microelectrodes is determined by the thickness of the insulating layer (from 0.2 to 1 μm). Therefore the separating insulative layer should fulfil several important requirements: high electric resistivity and strength, low leakage current and dielectric losses, good mechanical properties as high strength and stability, low intrisic stress and plastic deformation.

4.2.3 IDS-based sensor for urea

Several investigators have reported conductometric urea sensors using interdigitated electrodes. Using interdigitated electrodes (IDS) of our own design, urease was embedded on the surface of an IDS with Pt 'fingers' of 25 on each lead in a measuring window 0.3 x 0.3 mm. We have also experimented with IDS of a circular design (6 mm diameter) with 7 and 8 fingers for each electrode, with similar results. The response of this type of sensors to urea was linear up to 15 mM with good time resolution [54]. Both the time resolution and stability of the sensor may be improved. For example, by incorporating an electron mediator such as fullerene C_{60} with the enzyme, the speed of the response should increase [28,34].

4.2.4 Hydrogen peroxide

Considered next is the investigation of hydrogen peroxide sensitivity to membranes modified with TCNQ and DP-TTF [4,7,9-11]. When a s-BLM was modified with DP-TTF, the range of sensitivity to H_2O_2 in the bathing solution and the stability of the membrane both increased. The red/ox current increased across the membrane as the H_2O_2 concentration in the bathing solution was increased. The current was found to originate from a red/ox reaction with the s-BLM and/or the stainless steel electrode surface. DP-TTF molecules proved to be good modifiers in improving the membranes response to hydrogen peroxide and stability of both the s-BLM and the sb-BLM. However, when the membrane was modified with TCNQ, the s-BLM showed to be less sensitive to hydrogen peroxide than the unmodified s-BLM and also less stable. Therefore, TCNQ provides no potential benefits towards the formation of a H_2O_2 biosensor. This was not entirely unexpected since TCNQ should behave as an electron acceptor [19]. If highly conjugated compounds such as TCNQ are incorporated in the s-BLM forming solution, the resulting s-BLM was able to detect the presence of ascorbic acid, which is consistent with the findings obtained with conventional BLMs [3,18,25-29,30-33,37].

4.2.4. Ion sensors

S-BLMs containing six different kinds of crown ethers were synthesized and investigated using CV [19-21]. In particular, s-BLMs formed from a liquid crystalline *aza*-18-crown-6 ether and cholesterol-saturated n-heptane solution was found to be sensitive to K^+ in the concentration range of 10^{-4} to 10^1 M having a Nernstian slope. The specificity for three alkali metal cations and NH_4^+ of five different kinds of bis-crown ethers in BLMs were investigated. The order of specificity for most of these bis-crown ethers was found to follow hydrated radii of cations, i. e., $NH_4^+ > K^+ > Na^+ > Li^+$. The results obtained with these s-BLMs compare favorably with conventional BLMs containing related compounds such as valinomycin. Umezawa and his colleagues, for example, earlier reported in an elegant comparative study on the potentiometric responses between a valinomycin-based BLM and a solvent polymeric membrane [10]. Minami et al. found that the detection limit, dynamic range and optimum valinomycin concentration for the K^+ ion-induced Nernstian slope for the BLM were comparable to those of the polymeric membranes. On the basis of their findings, Umezawa and co-workers concluded that they have made the first successful planar lipid bilayers using totally synthetic lipids [10]. In this connection, Kunitake and colleagues reported the use of calixarenes, which are one of the most sophisticated synthetic host (receptor) molecules as selective transporters in BLM [10,18,21]. Kimizuka et al. concluded that calix(4)arene displayed a mediated transport of alkali metal ions selectively across BLMs formed from double-chained amphiphiles with cationic, anionic and non-ionic polar groups. They further stated that synthetic BLMs not only provide highly organized molecular matrices for carrier translocation but also affect the ion-complexation and release processes at the membrane-solution interface (see Refs. 9-11 for detailed citations). In this connection, potentiometric behaviors of a salt-bridge supported sb-BLM modified with calix[n]arene (n=4, 6, 8) derivatives were studied for some alkali metal ions [55,56]. The modified sb-BLM was used as an alkali cation sensor. The membrane potentials were observed to generate Nernstian responses to the concentration of alkali metal ions in electrolyte. The sb-BLM modified with the calix[n]arenes show high selectivity for individual alkali metal ions (e.g. K^+ and Cs^+).

4.2.5. Sensitive to Pb(II)

Interesting results were obtained regarding the TCNQ modified s-BLM biosensor designed to detect lead ions in the bathing solution. When TCNQ was incorporated into a cyanobiphenyl bilayer membrane, the lead ion sensitivity of the platinum electrode increased three fold from the unmodified membrane [7,21,24]. This specific s-BLM shows high efficiency of electron transfer across the membrane. The high efficiency may be due to the charge transfer complex ($8CB^+TCNQ^-$) which is created in the membrane that cooperates in red/ox reaction enabling the transfer of the charge. Therefore, it is evident that the electron exchange between the ($8CB^+TCNQ^-$) complex and the lead ion has a lower energy barrier than the transfer of an electron between the lead ion and the bare Pt electrode. It is of interest to note that TCNQ-containing BLMs have been investigated by Schriffin and colleagues. Cheng et al. [26] reported a potential dependence of transmembrane electron transfer across PC- BLMs mediated by ubiquinone, whereas Cunnane and Schiffrin studied the kinetics of ionic transfer across adsorbed PC planar lipid bilayers (see Refs. 7, 9-11 for detailed citations). More recently, Wang et al. [34], using CV, have examined the electrical behavior of TCNQ-modified s-BLMs, displaying different peaks as a function of scan rates. Also using the CV technique, Sabo et al. reported an investigation of s-BLMs on Pt substrate with a number of electron mediators (see Refs. 18, also 9,10 for detailed citations).

4.2.6. Molecular sensors

From a technical point of view, transducers for use in biosensors can be divided into four categories: electrochemical, semiconductive, optical and others such as piezoelectric. We shall be concerned only with the electrochemical category here, which consists of potentiometric and amperometric approaches. Many researchers have reported sensors for the detection of molecular species besides glucose such as antigens and antibodies. For example, we have reported in a feasibility study of an antigen-antibody reaction using s-BLMs as biosensors with electrical detection [29]. The antigen (HBs-Ag -- Hepatitis B surface antigen) was incorporated into a s-BLM, which was then interacted with its corresponding antibody (HBs-Ab -- Monoclonal antibody) in the bathing solution. This Ag-Ab interaction resulted in some remarkable changes in the electrical parameters of s- BLMs. The magnitude of these changes were directly related to the concentrations of the antibody in the bathing solution. The linear response was very good, ranging from 1 to 50 ng/ml of antibody, demonstrating the potential use of such an Ag-Ab interaction via the s-BLM as a transducing device.

4.2.7. Glucose sensors

Earlier, we have reported the embedding of glucose oxidase in a polypyrrole-lecithin BLM with good results [9,10,24]. In the case of the glucose sensor, the potentiometric approach has been less successful than the amperometric one. The steady-state current for amperometric glucose sensors is largely determined by the effective membrane thickness and the concentration of the embedded enzyme. A glucose sensor has been tested by embedding glucose oxidase (GOD) in avidin on s-BLM formed from biotinylated phospholipids [11]. Essentially, after biotinylated s-BLM formation, the coated wire tip was immersed in a 2.7 µM avidin-GOD solution to allow the coupling between avidin-GOD and s-BLM to establish. This was evidenced by a current reduction of about one order of magnitude. When the glucose was added to the cell, an increase in redox current was observed, which was a function of the applied voltage having a maximum at +670 mV. This avidin-GOD complexed biotinylated s-BLM sensor was capable of detecting glucose with a linear response up to 9 mM. Since the lipid bilayer is 'liquid crystalline' with self-sealing property, the presence of pinholes (defects) seems unlikely. Hence, a credible explanation for this glucose sensor would be based on electron transfer from the enzyme GOD embedded in the lipid bilayer to the metallic substrate, where the major barrier for the electron pathway is most likely to exist. One way to test this hypothesis is by incorporating an electron carrier (mediator) such as fullerene C_{60} in the lipid bilayer phase [1,7,24].

Additional examples of biosensors constructed using s-BLMs are as follows: Xanthine can be detected when the membrane is modified with xanthine oxidase, an enzyme present in purine metabolism, whereas glucose can be detected when the membrane is modified with glucose oxidase or electron mediators [39]. Carbon dioxide can be detected in a bathing solution when the s-BLM is modified with hemoglobin [9-11,21,32].

4.2.8. Experiments with fullerenes

Within the last decade fullerenes (e.g., C_{60}) have been of great interest in materials science, since it has been shown that C_{60} is a good electron mediator and behaves as a n-type semiconductor (bandgap = 1.6 eV). Our interest in fullerenes as a BLM modifier is owing to their most unique properties. Unmodified C_{60} is water-insoluble and highly hydrophobic. Hence, a BLM is an ideal environment for the compound. Recently we have investigated electron transfer across a BLM containing C_{60}, which can act both as a mediator and a photosensitizer [7,9-11,25-28,30-37].

4.2.9. S-BLM-based sensor for iodide

The outstanding properties of geodesic fullerenes C_{60}, C_{70}, etc. appear to be excellent modifiers for investigating electronic processes in BLMs [40]. Thus, to test fullerenes' efficacy as an electron mediator in a lipid bilayer environment, we have chosen an iodine-modified BLM system, since it has been previously shown that such a system was sensitive to iodide [5,29,34]. The results of our new findings are presented as follows. First, the presence of C_{60} greatly strengthens the stability of the s-BLM and dramatically alters its electrical properties [41]. The cyclic voltammograms contain distinct redox peaks, which are not symmetrical. We have found that C_{60} in the lipid bilayer had the opposite effects on membrane resistance R_m and capacitance C_m, respectively; it caused the R_m to decrease and C_m to increase. The presence of iodine in conjunction with C_{60} in the s-BLMs further accentuated the effects. It is apparent that both C_{60} and iodine, when embedded in BLMs, facilitate electrical conduction, thereby lowering the R_m. Since the relative dielectric constant ε_r of a typical, unmodified BLM is about 2, the presence of C_{60} and iodine should exert a great influence on ε_r. Since C_m depends on a number of factors such as the surface charge of the BLM, the nature of hydrocarbon chains, and embedded modifiers, an increase in ε_r, and consequently in C_m, is therefore expected. By incorporating C_{60} into our s-BLMs, we have found that the presence of C_{60} in the lipid bilayer increased the detection limit for I^- by 100 times (to 10^{-8} M). Thus, C_{60} greatly facilitates the discharge of I_3^- at the metal surface, which demonstrates clearly that the embedded C_{60} is indeed an excellent electron mediator [34].

4.2.10. Photoelectric responses of supported BLMs modified with fullerenes

Electron transfer and redox reactions across biomembranes is central to vital processes such as green plant photosynthesis and mitochondrial respiration as well as in the visual process, and have been the focal point of numerous BLM studies [1,25]. For example, Yonezawa, Fujiwara and Sato have reported transient photocurrents in pigmented BLMs in response to alternate illumination with the visible and the ultraviolet light. In an earlier report, Toda and his colleagues reported photo-switched current through BLMs containing spirobenzopyran (see Refs. 9-11 for detailed citations). In this connection Yamaguchi and Nakanishi have investigated the photoresponses of azobenzene(AZ)-containing BLMs in detail and found that the electrical changes induced by exposure to light of 360 and 450 nm, alternatively, resulted in reversible changes in the BLM structure initiated by the photoisomerization of AZ. Nakanishi and Yamaguchi concluded that the structural change of AZ seems to be transmitted directly and swiftly to the neighboring molecules because of the *fluid* and *ultrathin* structure of the BLM. Their findings are in dramatic contrast to the results of other types of artificial membranes such as L-B films and polymers, as pointed out by the authors (see Ref. 1 for detailed citations).

The above referenced work is consistent with the fact that BLMs and s-BLMs represent the simplest self-assembled liquid-crystalline structure that separate two phases (liquid/liquid or liquid/solid). We have recently shown that certain electron mediators can indeed modify the electrical parameters of supported BLMs [19,40]. Further, in an preliminary voltammetric study [42], we have found that a s-BLM modified with C_{60} can detect FMN. Thus, the insights gained in these findings may facilitate the future research in the use of s-BLMs for practical applications. For instance, in our recent experiments with the sb-BLM containing fullerenes (C_{60}) we have found that C_{60} function both as a mediator and a photosensitizer. In this connection, Ding and Wang [43] investigated methylene-blue modified s-BLMs for examining the behavior of dyes in the lipid bilayer. They suggested that the system should be further investigated by combining the CV method with other surface-sensitive techniques to determine the exact nature of dye-BLM interactions.

4.2.11. Interdigitated structure-based sensors

By forming s-BLMs on IDS (InterDigitated Structure) made on platinum with a window of 0.5 mm x 0.5 mm, we obtained the following interesting results [44,45]. First, when an IDS coated with a BLM formed from asolectin, it responded to pH changes with (15 ± 2) mV/decade slope. The conductance of s-BLMs on IDS was about 50 times higher than that of the usual s-BLMs. Second, when an IDS coated BLM formed from asolectin plus TCoBQ (or TCpBQ), the pH response was linear with a (50 ± 1) mV/decade slope. This very interesting finding suggests that (i) the lipid bilayer, the fundamental structure of all biomembranes, could be attached to an IDS whose responses were not unlike those found in s-BLM [7,44,45], (ii) this type of structure (i.e., s-BLM on interdigitated electrodes) can be used to investigate ligand-receptor contact interactions, and (iii) s-BLMs on an IDS can be manufactured using microelectronics technologies which already exist without the explicit need of special modification. In this connection it should be mentioned that the experiment on IDS-chips modified with a BLM is based on a common basic aspiration. That is to self-assemble a lipid bilayer containing receptors so that a host of physiological activities, such as ion/molecular recognition can be investigated. The structures to be reconstituted are inherently dynamic. Receptors and ligands in such close contact normally will vary as a function of time, frequently resulting in non-linear behavior. With IDS-chip modified with BLMs, we now have at last a most unique system for extensive experimentation which will be limited only by our imagination.

4.1.12 Ultrathin film microelectrode arrays applicable in the study of red blood cell sedimentation

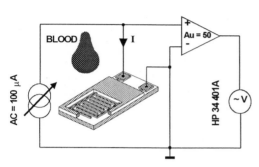

Fig. 7 *Layout of blood impedanc measurement*

Ultrathin-film interdigitated array of electrodes (IDAE) was used for study of the red blood cell sedimentation by impedance method. For investigation of the blood sedimentation process we utilized Si chips with thin-film Au or Pt IDAE of different finger/gap widths of values from 5/5 μm to 400/400 μm. The red blood cell sedimentation rate was monitored by measurement of impedance |Z| after applying a drop of blood to planar IDAE (Fig. 7). An improvement of that method is presented by the determination of a change of impedance rate with time. That dynamic parameter is different for the blood of healthy (or inflammatory) and cancer patients (Fig. 8).

The bioelectrical impedance method by means of planar Au IDAE looks very perspective to analyse the electrophysical forces among blood elements and plasma components during the aggregation phase of red blood cell sedimentation in the first 10 minutes as well as for fast differential diagnosis between cancer and inflammatory diseases [44,50,54].

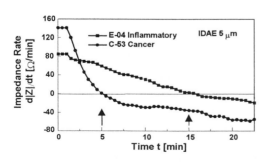

Fig. 8 *Typical dependence of impedance rate on time during the sedimentation of inflammatory and cancer blood*

4.1.13   Au microelectrode support for affinity biosensors

A novel immunosensor was constructed and based on self-assembled Au supported systems of biotinylated bilayer lipid membranes (s-BLM) (Fig. 9). Avidin modified monoclonal antibody originating from E2/G2 clone (A-Ab) was used as the antibody and 2,4-dichlorophenoxyacetic acid (2,4-D) herbicide as antigen (Ag) to provide an immunological interaction that could be transduced by the presented s-BLM sensor. An increase of elasticity modulus in the direction perpendicular to the membrane plane, $E_\perp$ and decrease of membrane capacitance, C, were experimentally proved by the addition of Ag to the s-BLM and used as a sensor response [54].

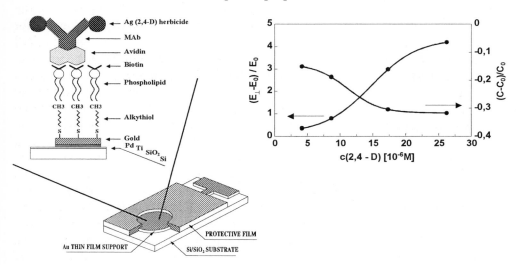

Fig. 9 *Immunosensor for (2,4-D) herbicide based on supported lipid membranes modified by monoclonal antibody* [54]

3.1.   Supported lipid bilayer-based sensors for ligand-receptor interaction

Biomembranes play an important role in signal transduction and information processing. This is due to the fact that most physiological activities involve some kind of lipid bilayer-based ligand-receptor contact interactions. Planar BLMs provide a natural environment for embedding a host of compounds such as proteins, receptors, membrane/tissue fragments, and even whole cells under nondenaturing conditions and in a well-defined orientation [1]. The lipid bilayer acts as an ultrathin electric insulator, as a framework for antigen-antibody binding, as a bipolar electrode for redox reactions, as a reactor for energy conversion (e.g. light to electric and/or to chemical energy). Further, a modified lipid bilayer performs as a transducer for signal transduction (i.e.

sensing), but it carries out other functions as well. BLMs (planar lipid bilayers) have been used in a number of investigations ranging from basic membrane biophysics studies to the conversion of solar energy via water photolysis, and to biosensor development [4-7,9-11]. Certain compounds incorporated into the BLM are used as channel formers. The channels can be classified into several categories: (a) voltage-gated channels, (b) ligand-gated channels, which are activated by contacting with their specific ligands, and (c) mechanosensitive channels [46]. When channels are open, ions flow through the membrane according to the difference in the voltage and the ion concentration on both sides of the BLM. This principle was applied as a strategy for the ion-channel biosensor design. The operation of the biosensors is based on the transmembrane ion-current modulation caused by a selective interaction of analyte with a membrane-bound receptor. Detection of a channel opening or closing can either be obtained directly by monitoring the membrane conductivity or by monitoring a transient change in the ion concentration. The concept that used the complexation of functional molecular interactions at the surface of a BLM with the transmembrane ion current modulation has the following potential advantages. This approach provides the direct transformation of the biochemical reaction into an electric signal for different biosystems. It may be a general approach for designing sensors for biologically active substances, including non-electroactive species. Simply changing the kind of the biological element in the BLM on the electrode surface can alter the selectivity of the BLM for one substance over others. As a specific example, Wang, Hou et al. studied receptor-ligand interactions using a c-BLM [47]. In this work, after reconstituting, the receptor into a c-BLM, it still retains its ligand activity. Furthermore, the relationship between receptor-ligand interactions and electrical properties of reconstituted BLM such as the membrane capacitance C_m and membrane resistance R_m was studied. When glycophorin in erythrocyte and asialoglycoprotein in hepatocyte were taken as examples, it was found that the resistance of reconstituted BLM decreased when adding blood type monoclonal antibody or the solutions of galactose, respectively, and the decrease is ligand-concentration dependent, however, the membrane capacitance was not influenced. This provides a simple, practical approach to determine the interactions between the receptor and its ligand.

Concerning ligand-receptor interactions, Wang, Hou et al. studied receptor-ligand interactions using a c-BLM [48]. The method reported is to reconstituting membrane receptors into c-BLMs. After reconstituting, the receptor still retains its ligand activity. Furthermore, the relationship between receptor-ligand interactions and electrical properties of reconstituted BLM such as the membrane capacitance C_m and the membrane resistance R_m was studied. When glycophorin in erythrocyte and a sialoglycoprotein in hepatocyte were taken as examples, it was found that the resistance of reconstituted BLM decreased when adding blood type monoclonal antibody or the solutions of galactose, respectively, and the decrease is ligand-concentration dependent, however, the membrane capacitance was not influenced. This provides a simple, practical approach for determining the interactions between the receptor and its ligand. Experimentally, erythrocyte (RBC) membrane and hepatocyte membrane were prepared as described as described [48]. Formation and thinning of the BLM were monitored electrically. Only when membranes had attained a capacitance C_m of 10^{-9} F and a resistance R_m of 10^9 Ω, the below experiments were then performed. In the blood type-B erythrocyte receptors-reconstituting BLM system, 50, 100, 200 μl anti-B monoclonal antibody was added to the outer chamber, respectively. C_m and R_m were recorded. As a control, a bathing solution or anti-A monoclonal antibody was added, and C_m and R_m were observed, too. Similarly, in the hepatocyte receptors-reconstituting BLM system, 100, 200 μl 0.5 M or 200 μl 1 M galactose solution was added to outer chamber, and C_m and R_m were recorded also. Plain BLM (EPC : Chol = 5:3, mol/mol) served as a control. In the blood type-B erythrocyte receptors-reconstituting BLM system, when adding anti-B monoclonal antibody, C_m was not influenced, but R_m decreased. In the hepatocyte receptors-reconstituting BLM system, adding galactose solution caused R_m to decrease, but C_m remained unchanged. Similar to the RBC receptors-reconstituting BLM system, the decrease of R_m is positively related to the amount of galactose. But different from that the erythrocyte receptors-reconstituting BLM system, after adding galactose solution, the hepatocyte receptors-reconstituting BLM did not rupture during the experiment. As to plain BLM only consisting of EPC-Chol, galactose did not effect C_m, R_m and the membrane stability. It should be mentioned, however, there is one shortcoming of the conventional BLM system, i.e. its stability which rarely lasts more than 8 hours. This shortcoming has been, however, overcome by forming the BLM on either metallic substrate or hydrogel support (s-BLM or sb-BLM) with the long-term stability. The above results showed that certain membrane receptors could be reconstituted into BLM [48].

3.4. Cancer

As a drug carrier and in other applications, liposomes have been studied for many years [1,16]. The main problem of liposomes is that they will be mainly uptaken by reticuloendothelial system when administrated *in vivo* and cannot get to the target sites, which greatly reduces the therapeutic index of medicines, especially for anticancer drugs. One way to overcome the difficulty is by means of immunoliposomes that is the liposomes linked with the concerned antibody or its fragments. In recent years several types of immunoliposomes have been developed and showed promise in this field [4-7]. The difference between conventional liposomes and immunoliposomes lies on the ability whether they can be recognized by the receptors of target cells. The molecular recognition will result in a series of physicochemical changes on lipid membrane, especially the change of electrical properties. Concerning this, it is well-established that the bilayer lipid membrane (the planar BLM) possesses the similar structure as the liposomal membrane, e.g., double phospholipid bilayer. A modified BLM can be a simple model for investigating the molecular recognition which occurred on liposomes *in vitro*. In another aspect, the electrical change of membrane is sensitive to the specific interactions such as antigen-antibody, receptor-ligand, enzyme-substrate and so on [1]. Hu, Hou et al. [49] recently carried out research on the design of an immunoliposome system for molecular recognition using reconstituted, hydrogel-supported bilayer lipid membranes (sb-BLMs). To study the influence of anticancer drug Adriamycin (ADM), 1 mg/ml solution was added drop-wise and the change of the BLM was monitored from the screen in the reconstituted LOVO-sb-BLM system, a conventional sb-BLM was used as a control in this experiment. The test of ADM and BLM shows that the molecular recognition must have taken place on the reconstituted BLM, but the accurate functional sites are not clear at present. As an effective anticancer drug, ADM generally is deemed to act on DNA of tumor cells [49]. This experiment revealed another potential mechanism of ADM, namely the function on a BLM may play part role in the anti-tumor process. In this connection, it has been reported that standard anti-cancer drugs, Adriamycin (Doxorubicin) and Taxol have been assessed by apoptosis; DNA synthesis; growth rate by MIT assay; uptake of amino acid, and by morphological changes [49]. The BLM is a very good model and the determination of its electrical parameters is direct and straight-forward, and the sensitivity of the method is excellent (the concentration of Ag or Ab usually less than 10^{-9}M). At the same time the method can be coupled with the computer technology, so the detection online may be realized, as will be reported in due course. In this connection, Uma Maheswari et al. [50] reported that cisplatin, an anti-tumor drug, exhibits a non-specific action on the BLM which might be responsible for its neurotoxic side effects.

3.5. Apoptosis

Cell death in a multi-cellular organism can occur by two distinct mechanisms: apoptosis and necrosis. The former can be distinguished from the latter by a number of characteristics, such as nuclear chromatin condensation, plasma membrane blabbing, cell shrinkage, nuclear fragmentation into apoptotic bodies, and the degradation of the nuclear DNA into oligonucleosome chains [1,12,51]. So, the internucleosomal cleavage has been shown to accompany apoptosis occurring in a wide variety of cell types and the DNA electrophoresis is used extensively for identifying the process. Cell-free nucleus apoptosis is a new way for evaluating apoptotic effects. Recently, photoelectric behavior of mammalian cells was found having bioanalytical significance. Concerning this, photoelectric effects of bilayer lipid membranes (BLMs) and electron mediator modified BLMs have been extensively studied, on account of their potential applications in understanding the mechanism of natural photosynthesis, and in developing photoelectric devices [1-3,52-54]. Mimicking the functionalities in the natural photosynthesis, which are represented by photoactive groups, electron donors and acceptors [4], various attempts have been made to realize the artificial photosynthesis and solar-energy conversion system under laboratory conditions. For example, synthetic dyes have been used to dope BLMs and corresponding photoresponses investigated [2,5-6]. Experimental findings indicate that electron mediator-doped BLMs can accelerate the photoinduced electron transfer across membranes, and enhance the photoelectric conversion efficiency [2,4]. Fullerenes, in particular C_{60} as a modifier of BLM, have attracted much interest in the study of photoelectric conversion because of the affinity of these molecules for electrons and also, because of their highly hydrophobic properties for doping BLMs [6-9]. However, past experiments on the photoelectric property of C_{60} modified BLMs were mostly conducted on the conventional planar BLMs [6,8], the defect of which is the fragility, thus precludes protracted investigations and practical applications. In contrast, BLMs self-assembled on the solid support (dubbed as s-BLM) showed much more stability, and exhibited electrochemical and photoelectric conversion properties. This kind of s-BLMs has many applications in the area of membrane biophysics and in the development of biosensors [2,8,10-12]. In the present work, a simpler method for forming s-BLMs for photoelectric conversion studies is reported. S-BLMs are easily self-assembled on ITO (indium-tin oxide) conducting glass, and the photoelectric properties of the lipid bilayer, as well as C_{60} modified BLM are systematically studied. The mechanism of the facilitation effect by C_{60} on the photoinduced electron transfer across the BLM, as well as the potential application of s-BLMs in photodynamic therapy is discussed [12,15,18,19].

Here, we introduce the photoelectric method used for analyzing the apoptosis of the nucleoli of human breast cancer cells (MCF-7 line) induced by Taxol (paclitaxel, an anticancer drug). The cell-free MCF-7 nucleoli are deposited on self-assembled bilayer lipid membranes (BLMs) on ITO conducting glass (ITO = indium-tin oxide). The photoelectric behavior of the "cell" and the nucleolus-related biological behavior, apoptosis, were investigated. Compared with the traditional techniques used to estimate apoptosis, such as the morphological observation and the agarose gel electrophoresis, the photoelectric analytical method of apoptotic system may provide a rapid and sensitive way to evaluate the nucleus apoptosis in earlier time.

The photoelectric current measurements were performed using a model 600 voltage analyzer (CH instruments Inc., USA). The light source was a Xe light (USHID Inc., Japan) with the light intensity of 121.4 mW cm^{-2}. A super thin cell made of glass was used as the photoelectric cell in which there have been three electrodes, as shown in Fig. 2. The working electrode was made of ITO-coated glass with the area of 2.0 cm^2. After the ITO glass was mounted, the width of the cell was 0.5 mm. The counter electrode was platinum metal, and the reference electrode was an Ag/AgCl electrode. In experiments, ITO conducting glass electrode was mounted in the light path, and the entire ITO window was shined in the light path. The voltage of between the reference electrode and the working electrode was set to zero, which was a consideration from the previous study [2,13]. The dark current (light off) was first measured, and then the light current was measured with the light on. The light-induced current was determined as the difference between the two measured values.

Taxol is the agent that can cause apoptosis and during the apoptosis, nuclear DNA will degrade into oligonucleosome chains. In this experiment, the degradation of the nuclear DNA was verified by the examination of the gel electrophoresis and fluorescence microscopy study of the total DNA. After the nucleoli were incubated with Taxol for 20 min, the chromatin began to condense around the nuclear periphery. The peripheral chromatin ring began to condense into discrete masses after 40 min, and then faint DNA ladder emerged. After incubating for 1 hour, the chromatin masses blabbed of from the nuclear surface and became apoptotic bodies. During this time, the DNA ladder became distinct. These results indicated that the decreasing tendency of the photoelectric current had close relationship with the cleavage of the chromatin into oligonucleosome chains. The specificity of Taxol on nuclear DNA also displayed that the photoelectric current of the ITO/s-BLM/MCF-7 nucleus assemblage is mainly dictated by the nucleoli. Nucleoli skeleton is known to be essential not only in maintaining nucleoli structure, but also in energy transfer. The apoptosis of nucleoli is accompanied by the disassembly of the nuclear lamina, which leads to the damage of nuclear skeleton, and thereby a decreasing of the photoelectric current as well. The decreasing tendency of the photoelectric current of the apoptotic nucleoli is related to the cleavage of the chromatin. In the interpretation of the photoelectric response of nucleoli, the possibility has been considered that the DNA double helix, which contains a stacked array of heterocyclic base pairs, could be a suitable medium for electron transfer over long distance [12,51]. So the nuclear DNA can serve as an "electric wire" for photo-induced electron transfer by "hopping" from base to base. One widely observed property of apoptotic cells is the cleavage of the DNA into fragments at sites separated by the internucleosomal spacing. The cleavage of nuclear DNA resulted in the damage of DNA molecules as photo-induced electron-transferring bridge. So, the photoelectric current decreasing was in accordance with the cleavage of the nucleoli. In this connection, Gao, Luo and their associates [40,51] have carried out experiments of s-BLMs without or with fullerene C_{60} that have been self-assembled on indium-tin oxide (ITO) glass.

The photoelectric properties of the ITO supported planar lipid bilayers were studied. The light intensity of irradiation, bias voltage, and the concentration of donors, have been found to be limiting factors of the transmembrane photocurrent. Additionally, the facilitation effect of C_{60} doped BLMs on the photoinduced electron transfer across the BLM has been considered.

The s-BLM/cytosol nucleoli assemblage responded to white light (200-800 nm). Electron transfer along the DNA double helix and along nuclear skeleton is invoked in our interpretation. This novel photoelectric analytical method may be useful and could provide a rapid and sensitive technique in evaluating apoptosis by photodynamic therapy (PDT). The apoptotic response appears to be a function of both the photosensitizer and the cell line. One widely observed property of apoptotic cells is the cleavage of the DNA into fragments at sites separated by the internucleosomal spacing. The cleavage of nuclear DNA resulted in the damage of DNA molecules as photo-induced electron-transferring bridge. If so, a decreasing of the photoelectric current would be detected in accordance with the cleavage of the nucleoli, as evidence of apoptosis. In the present paper, our findings using supported planar lipid bilayers (s-BLMs), based on combined methods of cyclic voltammetry and photochemistry [8,18] are described below.

The photoconductance of C_{60}-containing BLM is higher than that of undoped BLM, as calculated from the slope of the I/V characteristics [9]. The above data indicate that C_{60} doped in the BLM accelerates the photoinduced electron transfer process across membrane self-assembled on the ITO support. As has been described above, the bilayer lipid membrane formed on the ITO substrate prevents the transmembrane electron transfer, thus reduces the intensity of the generated photocurrent. The comparatively higher photoconductivity of bilayer lipid membranes doped with fullerenes can be interpreted as the electron transporting effect of fullerenes. Once photoexcited, the fullerene in its long lived triplet state is reduced by the electron donors in the solution and forms the radical anion C_{60}^{-}. Since fullerenes are free to move about within the lipid bilayer environment, due to their geodesic structure, molecular dimensions and highly hydrophobic properties, a subsequent electron transfer from a photoproduced anion to another fullerene in its photoexcited state occurs. This electron transporting effect of C_{60} propagates the electron flux from the donors in the solution through the membrane towards the ITO electrode, which acts as an electron acceptor. Thus, experimental findings have shown that a BLM self-assembled on metal electrode blocked the electron transfer across the electrode and solution. The present photoelectric conversion experiment indicates that photoinduced electron could transfer across BLMs self-assembled on ITO conducting glass and the mediators doped in BLMs could facilitate this transmembrane photoinduced electron transfer. Since the ITO/BLM probes possess biological compatibility, therefore biomaterials could be embedded and their photoresponse properties investigated. It seems evident that this novel self-assembled ITO/BLM probes will be a promising tool for the study of light-induced properties of biomembranes (e.g., photodynamic therapy) and the development of biomimetic photoelectric devices (see Fig. 2 *Middle*).

3.6 DNA detection using sb-BLMs

There has been an increasing research interest in studying the electrochemistry of nucleic acids (DNA), focusing on the detection of DNA sequence, on the reduction of DNA, on the electrochemical probes, the DNA electrochemical biosensors for environmental monitoring, and on the interaction of DNA with drugs and other substances. Recently Zhang et al. have reported [55,56] a novel method for investigating the interaction of DNA using sb-BLMs. The interaction of DNA (calf thymus) with hemin (a naturally occurring iron-porphyrin) was studied by means of cyclic voltammetry (CV) in salt bridge supported bilayer lipid membrane (sb-BLM) system. Sb-BLM was modified with lauric acid (LA) dissolved in the membrane forming solution, then hemin molecules in electrolyte could also be embedded into the BLM by electrostatic interaction between hemin and LA. Hemin showed a well–defined CV behavior. The cathodic peak current (I_{Pc},) of hemin decreased in the presence of DNA, which is consistent with the results of decreasing hemin concentration and illustrates the existence of interaction between hemin and DNA. The interacting mechanism was studied. The effect of pH on the interaction was also considered. The I_{Pc} of hemin decreases in proportion to the DNA concentration from 8.44×10^{-5} mol l^{-1} to 6.75×10^{-4} mol l^{-1}, which provided a possibility of detecting DNA with a natural biological substance in a biomimetic membrane system. As reported, a typical voltammogram of hemin in the sb-BLM was seen. No peak current signals were observed when the BLM was composed only of PC and CH. However, the cathodic peak of hemin was presented (about -0.35 V) when the BLM was formed by a mixture of LA with PC and CH. The result showed that anionic sites promoted the adsorption and incorporation of hemin into the BLM by electrostatic interactions between the positively charged hemin molecules and the negatively charged BLM, which is favorable for the electron transfer of hemin on the surface of electrode. Furthermore, adding anionic sites decreased the membrane resistance, which resulted in improvement of weak current measurements. The cathodic peak current I_{Pc} increased with the increase of hemin concentration. The cathodic peak current of hemin was reduced in the presence of double-stranded DNA (dsDNA). This can result from two factors: (1) The competitive adsorption of dsDNA on the sb-BLM; (2) The complex without electrochemical activity was formed between dsDNA and hemin. However, the former factor was precluded by further investigation, as follows: (i) the results obtained from UV-visible spectrometry showed that the absorbance of hemin decreased distinctly in the presence of dsDNA, while there was no change in its absorption wavelength. It illustrates the existence of interaction between dsDNA and hemin; (ii) the phosphate group of bases in the nucleic acid bases of DNA is negatively charged, what is unfavourable to the adsorption of dsDNA on the negatively charged sb-BLM; (iii) the peak current of hemin did not disappear completely with the increase of dsDNA concentration, what is not the character of competitive adsorption. According to these results we think that dsDNA interacted with hemin and formed a complex. In the presence of dsDNA, the equilibrium concentration of hemin decreased, which resulted in a decrease of the peak current. This suggests the different nature of interactions with the change of ionic strength. In the low ionic strength range (μ=0.96 mmol l^{-1} to μ=13.5 mmol l^{-1}), the interaction between dsDNA and hemin was in the electrostatic mode. With increasing ionic strength (μ>13.5 mmol l^{-1}), the nature of interaction changed from the electrostatic interaction to the other mode. Generally, those porphyrins with axial ligands families bind nucleic acids in a groove mode [42]. The interaction between hemin with dsDNA belongs to this category at a high ionic strength. Contrary to dsDNA, the peak current and peak potential of hemin didn't change in the presence of single-stranded DNA (ssDNA) at a high ionic strength. It could be seen that ssDNA did not interact with hemin when the double strand of DNA disappeared. It also illustrated the existence of the groove binding mode at a high ionic strength. This indicates that this type of sb-BLM biosensor can provide information not only about the extent of interaction between DNA and other molecules but also the difference in the nature of interactions. The results obtained from UV-visible spectrometry coincide with the result mentioned above. The absorption curve of ssDNA+hemin completely overlapped with the curve of hemin within a mimetic membrane system, using only natural chemical compounds in the detection procedure. Thus, the distinctive merit of this biosensor reported here is its biological compatibility.

3.7. Drug testing

Before describing practical applications using supported BLMs as biosensors for drug testing, a brief mention is made concerning the results of some recent studies of conventional planar lipid bilayers. One of the interesting findings is related to the chirality of biomolecules. Tsai et al. [57] reported in liposomes (i.e., BLMs of spherical configuration) that the phosphate head group of phospholipids is chiral. This is very exciting since chirality is 'yes' or 'no', or '0' or '1' in the computer language. Further, the chirality of drugs is very important in their biofunction. In this connection, Li, Hou and colleagues [48] found certain selectivity of permeating chiral complexes in BLMs. For example, frequently only one of antimers is efficient. The difference in the transmembrane permeability rates of these optical antimers could be one of the important reasons for their different bioactivities. Thus, the BLM along with s-BLMs described below, could be developed into a significant part of a new method for screening chiral drugs.

Electrical properties such as membrane potential E_m of planar BLMs are readily measured. Planar BLMs have been extensively used as models of biomembranes. In this paper we report BLMs formed in the solutions containing chiral complexes -- d-K[Co(EDTA)], l-K[Co(EDTA)]; d-[Co(C$_2$O$_4$)(en)$_2$]I, and l-[Co(C$_2$O$_4$)(en)$_2$]I, whose E_m display great differences, implying strong chiral selectivity. The permeability ratios of different chiral complexes calculated from E_m are the same as those obtained from human erythrocyte experiments. Electrical properties are important characteristics of the biomembranes. Both transport of charged species through membrane and the interaction of electro-active species with membrane cause changes of electric parameters. Some of these changes influence many important physiological functions [2-5]. In the present work, we report permeating courses of two kinds of chiral complexes with different charges, and measured their permeability ratios by monitoring membrane potential E_m. The results of E_m are compared with those obtained with human erythrocyte experiments. These results showed that chirality selectivity of cell uptake was mainly caused by the chirality of the membrane phospholipid itself. An explanation for the observed chirality selectivity is discussed. Further, we suggest that the BLM system may be developed into a useful tool for drug screening.

Specifically, the solution of chiral complexes was prepared by PIB buffer (NaCl 145 mmol dm^{-3}, KCl 5 mmol dm^{-3}, glucose 10 mmol dm^{-3}, Tris 10 mmol dm^{-3}, pH 7.4). 1 x 10^{-3} mmol dm^{-3} d-K[Co(EDTA)] and l-K[Co(EDTA)] solutions 1.5 ml were mixed with 2 ml erythrocytes suspension, respectively. After 0, 30, 60, 120, 180 min at 37 °C, cobalt content in cells was measured by atomic absorption spectrophotometry. The uptake quantities of chiral complexes and the first-order rate constant were calculated. 5.00 mmol.dm^{-3} d-[Co(C$_2$O$_4$)(en)$_2$]I and l-[Co(C$_2$O$_4$)(en)$_2$]I 1.5 ml were mixed with 0.5 ml erythrocytes suspension, respectively. After 3 hrs at 37 °C, cobalt content was measured using the same method above. These procedures were repeated three times and the results were calculated as a mean value.

The permeability ratio calculated from E_m was in good agreement with the results from human erythrocyte (RBC) experiments. The composition of the human RBC membrane is complex, containing many different proteins and glycolipids, but the experimental BLMs used in our study contain phosphatidylcholine (PC) and cholesterol only, and only phosphatidylcholine has the chirality. These two kinds of membranes have the same chirality selectivity and permeability ratio, which demonstrates that the chirality selectivity of cell uptake is mainly due to the chirality of membrane lipids, in other words, due to the selective recognition of natural phospholipid's chiral carbon in the chiral molecule. The selectivity can only come from a direct interaction between the chiral carbon of lipid and its antimers.

These above-mentioned advantages make the BLM system a useful tool for studying membrane transport, and drug-membrane interaction, as well as their mechanisms. The BLM electrical measurement is more simple, rapid and sensitive than the measurements of RBCs. The composition of BLMs can be controlled accurately, which eliminates the disturbance of other components, leading to reliable results. By studying the permeating behavior of other chiral complexes and the electrical properties of BLMs with different compositions, details of this selective interaction could be further investigated.

3.8. Sensors for bacteria

The bilayer lipid membrane (BLM) system is extensively used as an experimental model for a variety of biomembranes. One of great advantages of the BLM is its electrical properties that may be easily measured. The electrical parameters of the BLM provide a sensitive and rapid tool for monitoring the interactions of membrane lipids with other entities in the environment, such as bacteria. Recently, Ivnitski and collegues [46] reported a system for direct detecting *Campylobacter* species by an electrochemical ion-channel biosensor using antibodies (Abs) to *Campylobacter* embedded in the BLM. *Campylobacter* species are now recognized as the most important causes of acute diarrheal disease in humans throughout the world. The two predominant species responsible for most human gastroenteritis are *C. jejuni* and *C. coli*. In developed countries, *C. jejuni* has been demonstrated in the stools of 4 to 14% of all patients with a diarrheal disease and it is estimated to be responsible for as many as two million cases of infection in the U.S. each year.

In the report of Ivnitski and collegues [46], the BLMs on the electrode surface were formed from phosphatidylcholine (PC; Walgreen Laboratories, Inc.) in squalene (Estman Kodak Co.). Mortalized (by heat treatment) lyophilized pathogenic *Campylobacter* species, and affinity purified antibody to *Campylobacter* were obtained from Kirkegaard and Perry Laboratories (KPL, Inc., Gaithersburg, MD). The sensing element of the biosensor is composed of a stainless steel working electrode, which is covered by a planar BLM. Antibodies to bacteria embedded into the BLM are used as channel forming proteins. The biosensor has a strong signal amplification effect, which is defined as the total number of ions transported across the BLM. The total number of (univalent) ions flowing through the channels are 10^{10} ions/sec. The biosensor showed a very good sensitivity and selectivity to *Campylobacter* species. The sensing element of the biosensor is shown in Fig. 3. It is composed of a stainless steel wire (ss-wire), whose tip is covered with an artificial bilayer lipid membrane. A BLM is used simultaneously: (1) as a very thin electrical insulator, (2) as a matrix for the incorporation of antibodies, and (3) for the suppression of nonspecific ligand binding. Antibodies to bacteria embedded in the BLM are used as channel forming-proteins. The antibody channel pore (Fig. 3) is formed from two identically heavy and two light protein chains, which are held together by non-covalent forces and covalent disulfide bonds. The operation of the biosensor is based on the transmembrane ion-current modulation caused by a selective interaction of bacteria with the membrane-bound Abs. The bacteria-antibody binding on the BLM surface is accompanied by conformation perturbations of antibody molecules. This perturbation disturbs the ion permeability of the BLM, whereby ions are transferred through the channel and the current is detected amperometrically. To demonstrate this principle, we used the heat-killed *Campylobacter* species (*jejuni* cells) and the affinity purified antibody to *Campylobacter* species. Mortalized bacteria were used for safety reasons and for ensuring a static (non-proliferating) sample population for quantitative purposes. The results of Ivnitski and collegues [46] demonstrate the potentiality of the electrochemical ion-channel biosensor based on antibody-modified supported bilayer lipid membranes on stainless steel electrodes for direct detection of *Campylobacter* species in water samples. The biosensor has a strong signal amplification effect, which is defined as the total number of ions transported across the BLM. The principle of signal transduction that is described herein is simple and the complete assay is carried out in 10 min. The biosensor may be used for detection of different microorganisms in water, food, and air without pre-enrichment. Work now in progress is aimed towards optimizing such parameters like the sensitive and reproducible assays of bacteria.

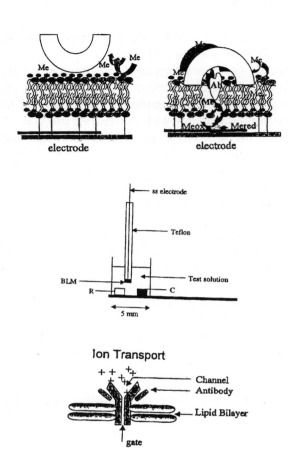

electrode   electrode

Fig. 3. Ion-channel sensor for bacteria using supported BLMs [15,46].
Top: a proposed scheme for ion-channel sensor;
Middle: experimental cell arrangement for ion-channel sensor studies;
Bottom: an ion channel in the lipid bilayer (BLM) formed from an antibody.

3.9. Cyclic voltammetry of s-BLMs modified with redox species

In one application of the self-assembled BLM on a solid support, the redox reactions of embedded pigment-protein complexes are of interest, where the transfer of electrons is assumed from a pigment to a substrate. To model these processes we have studied the electron transfer from cytochrome c to a Pt electrode covered with a modified s-BLM [31,47]. The water soluble cytochrome c is often denatured on the clean metallic surfaces. In the case of s-BLMs modified with AQS (anthraquinone-2-sulfonic acid), the cytochrome c in the solution gives no observable current responses. When 1 mM diaminodurene (DAD) was added to the cytochrome c solution, the cyclic voltammogram with corresponding oxidation and reduction peaks was recorded. The redox potentials did not change with the different scan rates. In another experiment, the heights of the cyclic voltammogram peaks increased, when only diaminodurene has been present in the solution. The current was changed independently from the concentration of the cytochrome c, when the s-BLM has been modified by AQS and with DAD present in the solution. It is reasonable to assume that cytochrome c is physically adsorbed onto the BLM and thereby effectively excludes the DAD molecule from the s-BLM surface [18]. The analogous experiments were also carried out with s-BLMs modified by 4,4'-bipyridyl and with ferricyanide present in the solution.

3.10. Antigen-antibody sensitivity

The next study conducted an investigation of membranes modified with the hepatitis B antigen utilizing both s-BLM and a sb-BLM systems [9,10]. This type of BLM set up allowed an immune reaction to occur between the antigen and antibody. The antigen/antibody interaction was observed when the resistance across the membrane decreased and the capacitance across the membrane increased. The antigen modified s-BLM was shown to be active, stable, and suitable for practical applications as a biosensor.

In this connection, Hong Gao et al. [40,51] investigated certain electrochemical features of s-BLMs using impedance spectroscopy. In their case, s-BLMs used were modified by C_{60}, a well known electron mediator. The membrane resistance and capacitance are calculated from the whole impedance spectrum, which is said to be more informative for the characterization of lipid membranes compared to capacitance and resistance calculated at certain frequencies [12]. The time course of membrane capacitance of these two kinds of s-BLMs under different AC frequencies is also investigated with the attempt to achieve a better clarification of the structure of the s-BLM. Further, to demonstrate the electron transfer function of fullerenes, the cyclic voltammetry (CV) and the electron impedance spectroscopy (EIS) of fullerene-modified s-

BLMs with $Fe(CN)_6^{3-}$ in the bathing solution were investigated. Electrons can transfer across the BLM and the oxidation-reduction (redox) reaction can take place. Modifiers such as electron mediators or carriers (e.g., TCNQ = tetracyanoquino-dimethane) can facilitate such transfer and redox process [18,40,41,51]. For example, fullerenes doped BLMs containing I_2 enhanced the stability of BLMs and increased the sensitivity toward I^- in the bathing solution [21].

Their results are in accord with the value obtained on red blood cell membranes (RBC), estimated by Fricke nearly 6 decades ago [2]. The impedance technique, nowadays also known as electrochemical impedance spectroscopy (EIS), has been subsequently used by many others [13,16,39,42,48,49,57,59]. The BLM-Gramicidin A interaction can be characterized by the changes observed in the low frequency range of the impedance spectra. The sequence of conductivity of alkali cations, as reported earlier by Steinem et al., follows the hydrated radii (i.e., $Cs^+ > K^+ > Na^+ > Li^+$).

5   CONCLUDING REMARKS

Cells are the basic unit of life. The cell membrane plays a critical role in the proper functioning of the cell by providing a barrier between the inside and outside of the cell. This barrier is essential for the driving forces of life. The dynamic and ingenious properties of the cell membrane have been duplicated to some extent in the lab allowing for a much more profound understanding of lipid membrane functions and also the practical applications of biosensors.

Research with planar lipid bilayers began with the formation of the conventional BLM. A great deal was learned about the nature of lipid membranes from this model but its poor stability created enormous limitations for researchers. The metal and gel supported BLMs described here solved many of the shortcomings of the conventional BLM and opened a door to numerous possibilities for the membrane and biosensor research. It was learned that when a membrane is formed on the tip of a freshly cut solid support (e.g., stainless steel wire) the membrane can last indefinitely. The solid support enabled the membrane to possess great stability, become modified, and be tested for electrical properties. However, the solid support precluded the translocation of ions across the membrane. This presents in some cases a limitation until the solid support was replaced with an agar gel. Sb-BLMs overcame the shortcomings of both the conventional BLM and the s-BLM by allowing the passage of ions across the lipid bilayer.

In the past decades, our work has been benefited by cross fertilization of ideas among various branches of sciences, including colloid and interface chemistry, solid-state physics, molecular biology and microelectronics. It seems likely that the sensors and devices may be constructed in the form of a hybrid structure, for example, utilizing both inorganic semiconducting nanoparticles and synthetic lipid bilayers (i.e., supported BLMs). The biomimetic approach to sensor development is unique and full of exciting possibilities. We have been mimicking Nature's approach to 'smart' materials science or life, as we understand it, which may be summarized by one word *'trial-and-error'*. This approach was fine for Nature but it is not a viable one for us now, since we do not have unlimited time and resources at our disposal. Nevertheless, we can glean the design principles from Nature's successful sensing receptor cells such as in vision, taste buds, and olfactory bulbs, apply them to our search for better materials from which advanced devices ultimately depend. Indeed, the single ion channel activities of this S-toxin-containing sb-BLM were observed, which appear as a square-shaped current fluctuation during the transitions of the channels between different conducting states. From plots of frequency vs. single-channel conductance histogram, and the current-voltage curve, a linear relationship was obtained. In the latter case, the slope conductance was found to be 125 pS for the S-toxin-modified sb-BLM, which is in accord with known values [7]. It is informative to reiterate that the lipid bilayer is the principal component of all biomembranes, and the signal transduction begins at the receptor embedded in it [1]. Researchers are currently working towards ways to further increase the stability and decrease the cost of BLM-based biosensors, allowing for practical use in lab and clinical settings. Once this is accomplished, the possibilities that BLM biosensors will be limited are determined only by one's imagination.

REFERENCES

[1]   H.T. Tien and A.L. Ottova, Membrane Biophysics: As viewed from experimental bilayer lipid membranes (planar lipid bilayers and spherical liposomes), Elsevier, Amsterdam & New York, 2000

[2]   A.L. Ottova and H.T. Tien, Prog. Surf. Sci., 41 (1992) 337.

[3]   H.T. Tien, J. Clinical Lab. Analysis, 2, (1988) 256.

[4]   R. Birge (ed.), Molecular and Biomolecular Electronics, ACS, Washington, D.C., 1994, Chap. 17, pp. 439-454.

[5]   H.P. Yuan, A. Ottova and H.T. Tien, Mater. Sci. Eng. C, 4 (1996) 35.

[6]   X.-D. Lu, A. Leitmannova-Ottova and H.T. Tien, Bioelectrochem. Bioenerg., 39, (1996) 285.

[7]   Ottova, V. Tvarozek, J. Racek, J. Sabo, W. Ziegler, T. Hianik and H.T. Tien, Supramolecular Science, 4 (1997) 110.

[8]   T. Hianik, J. Dlugopolsky, M. Gyeppessova, B. Sivak, H.T. Tien and A. Ottova-
        Leitmannova, Bioelectrochem. Bioenerg., 39 (1996) 299

[9]   J.-M. Kauffmann (ed.), Special issue devoted to electrochemical biosensors,
        Bioelectrochem. Bioenerg., 42, (1997) 1-104

[10]   Y. Umezawa, S. Kihara, K. Suzuki, N. Teramae and H. Watarai (eds.), Molecular
        recognition at liquid-liquid interfaces: fundamentals and analytical applications: A
        Special Issue, Analytical Sciences, 14 (1998) 1-245

[11]   P. Krysinski, H.T. Tien and A. Ottova, Biotechnology Progress, 15 (1999) 974.

[12]   J. Feng, C.Y. Zhang, A.L. Ottova et al., Bioelectrochem., 51 (2000) 187

[13]   M. Zviman and H.T. Tien, Biosensors & Bioelectronics, 6 (1991) 37.

[14]   O. Rossetto, M. Seveso, P. Caccin, G. Schiavo and C. Montecucco, Toxicon, 39
        (2001) 27.

[15]   B. Raguse, V. Braach-Maksvytis, B.A. Cornell, L.G. King, P.D.J. Osman, R.J. Pace
        and L. Wieczorek, Langmuir, 14 (1998) 648

[16]   M. Seitz, E. Ter-Ovanesyan, M. Hausch, C.K. Park, J.A. Zasadzinski, R. Zentel, J.N.
        Israelachvili, Langmuir, 16 (2000) 6067

[17]   J.W. Bunger and N.C. Li (eds.), Chemistry of Asphaltenes, Advances in Chemistry
        Series, Washington, D.C., 1979, Vol. 195, pp. 39-52

[18]   J. Sabo, A. Ottova, G. Laputkova, M. Legin, L. Vojcikova and H.T. Tien, Thin Solid
        Films, 306 (1997) 112.

[19]   H.T. Tien, J. Phys. Chem., 88 (1984) 3172; J. Electroanal. Chem., 174 (1984) 299;
        211 (1986) 19.

[20]   K.L. Mittal, D.O. Shah (eds.), Surfactants in Solution, Plenum Press, New York, 1992,
        Vol. 11, pp. 61-88.

[21]   R.F. Taylor, J.S. Schultz (eds.), Handbook of Chemical and Biological Sensors,
        Institute of Physics Publishing, Philadelphia, 1996, pp.221-256.

[22]   H.T. Tien, Z. Salamon, P. Kutnik, J. Krysiñski, J. Kotowski, D. Lederman, T. Janas, J.
        Mol. Electronics, 4 (1988) 1.

[23]   F.L. Carter, R.E. Siatkowski, H. Wohltjen (eds.), Molecular Electronics Devices,
        Elsevier, Amsterdam, 1988

[24]   C.M.A. Brett, A.M. Oliveira-Brett (eds.), Electroanalytical Chemistry: A Special
        Issue, Electrochimica Acta, 43 (1998) 3587-3610.

[25]   H. Yamada, H. Shiku, T. Matsue and I. Uchida, J. Phys. Chem., 97 (1993) 9547.

[26]   Y.F. Cheng and D.J. Schiffrin, J. Chem. Soc. Faraday Trans., 90 (1994) 2517.

[27]   K. Asaka, A. Ottova and H.T. Tien, Thin Solid Films, 354 (1999) 201.

[28]   P. Diao, D. Jiang, X.L. Cui, D.P. Gu, R. Tong and B. Zhong, Bioelectrochem.
        Bioenerg., 45 (1998) 173.

[29]   L. Gu, L. Wang, J. Xun, A.L. Ottova and H.T. Tien, Bioelectrochem. Bioenerg., 39
        (1996) 275.

[30]   W.S. Xia, C.H. Huang, L.B. Gan, H. Li, X.S. Zhao, J. Chem. Soc., Faraday Trans., 92
        (1996) 769.

[31]   K. Uosaki, T. Kondo , X.Q. Zhang and M. Yanagida, J. Am. Chem. Soc., 119 (1997)
        8367.

[32]   H. Fujiwara and Y. Yonezawa, Nature, 351 (1991) 724.

[33]   E. Tronel-Peyroz, G. Miquel-Mercier, P. Vanel and P. Seta, Chemical Physics Letters,
        285 (1998) 294

[34]   H.T. Tien, L.G. Wang and X. Wang, Bioelectrochem. Bioenerg., 42, (1997) 1.

[35]   J.L. Garaud, J.M. Janot and G.J. Miquel, J. Membrane Science, 91 (1994) 259

[36]   I.Novotny, V.Breternitz, R.Ivanic, Ch.Knedlik, V.Tvarozek, L.Spiess,

V.Rehacek,44rdInternational Scientific Colloquium Technical University of Ilmenau (1999) 20-23

[37]   K. Asaka, H.T. Tien and A. Ottova , J. Biochem. Biophys. Methods, 40 (1999) 27

[38]   K.J. Donoan, R.V. Sqdiwala and E.G. Wilson, Mol. Cryst. Liq. Cryst., 194 (1991) 337

[39]   P.L. Katrivanos, A.J. Purnell, A.A. Aleksandridis, C.G. Siontorou and C. White, Laboratory Robotics and Automation, 10 (1998) 239.

[40]   H. Gao, J. Feng, G.A. Luo, A.L. Ottova and H.T. Tien, Electroanalysis, 13 (2001) 49.

[41]   J.G. Webster (ed.), Wiley Encyclopedia of Electrical and Electronics Engineering, John Wiley & Sons, Inc., 1999, pp. 479-493; Addendum - an updated version was published in 2001

[42]   C.G. Siontorou, D.P. Nikolelis and U.J. Krull, Electroanalysis, 9 (1997) 1043.

[43]   L. Ding and E. Wang, Bioelectrochem. Bioenerg., 43 (1997) 173.

[44]   V. Tvarozek, A. Ottova-Leitmannova, I. Novotny, V. Rehacek, F. Mika and H.T. Tien, 21st International Conference On Microelectronics - Proceedings, IEEE, New York, Vol. 2, 1997, pp. 543-546.

[45]   V. Rehacek, I. Novotny V. Tvarozek, F. Mika, W. Ziegler and A.L. Ottova, Sensors & Materials, 10 (1998) 229.

[46]   D. Ivnitski, E. Wilkins, H.T. Tien and A. Ottova, Electrochemistry Communications, 2 (2000) 457.

[47]   L. Wang, X..P. Hou, A. Ottova et al., Electrochem. Commun., 2 (2000) 287.

[48]   W. Li, X.P. Hou, A. Ottova and H.T. Tien, Electrochem. Commun., 2 (2000).

[49]   X. Hu, S.H. Wang, Y.F. Zhang, X.-D. Lu, X.P. Hou, A. Ottova and H.T. Tien, J. Pharmaceutical and Biomed. Analysis, 26 (2001) 219.

[50]   Tvarozek, V., Novotny, I., Rehacek, V., Ivanic, R., Mika, F.: Thin film electrode chips for microelectrochemical sensors, NEXUS Research News, 1, (1998), pp. 15-17

[51]   H. Gao, G.A. Luo, J. Feng, A. L. Ottova and H.T. Tien, J. Photochem. Photobiol. B: Biology, 59 (2001) 87.

[52]   D.L. Jiang, J.X. Li, P. Diao, Z.B. Jia, R.T. Tong, H.T. Tien and A.L. Ottova, J. Photochem. Photobiol. A-Chemistry, 132 (2000) 219.

[53]   H.T. Tien and A.L. Ottova, Research Trends, 9 (2000).

[54]   Tvarozek, V., R.Ivanic, A. Jakubec, Novotny, I., Rehacek, V., Proc. SPIE, 4414 (2001) 183-188

[55]   Y.L. Zhang, Z.X. Guo, H.X. Shen, Chinese J. Anal. Chem., 27 (1999) 1096.

[56]   Y.L. Zhang, H.X. Shen, C.X. Zhang, A. Ottova and H.T. Tien, Electrochim. Acta, 46 (2001) 1251.

[57]   T.C. Tsai, R.T. Jiang and M.D. Tsai, Biochem., 23 (1984) 5564.

[58]   H.B. Fang, L. Zhang, L. Luo, S. Zhao, J. An, Z.C. Xu, J.Y. Yu, A. Ottova and H.T. Tien, J. Colloids Interf. Sci. (2001) 177.

[59]   P. Diao, D.L. Jiang, X.L. Cui, D.P. Gu, R.T. Tong and B. Zhong, Bioelectrochem. Bioenerg., 48 (1999) 469

[60]   K. Uma Maheswari, T. Ramachandran and D. Rajaji, Biochim. Biophys. Acta, 1463 (2000) 230.

[61]   G. Biczo (ed.), MEBC, 7th Newsletter (1993) 23-40; 8th Newsletter (1997) 7-27.

[62]   D.G. Wu, C.H. Huang and L.B. Gan, Langmuir, 14 (1998) 3783.

[63]   D.P Nikolelis, Petropoulou SSE, Biochimica Et Biophysica Acta-Biomembranes 1558 (2002): 238-245

[64]   E. Neumann, Tonsing K, Siemens Bioelectrochemistry, 51 (2000): 125-132

Planar Lipid Bilayers (BLMs) and their Applications
H.T. Tien and A. Ottova-Leitmannova (Editors)
© 2003 Elsevier Science B.V. All rights reserved.

Chapter 34

Photoinduced Charge Separation in Lipid Bilayers

D. Mauzerall* and K. Sun

The Rockefeller University, 1230 York Avenue, New York, New York 10021

.1. INTRODUCTION

The lipid bilayer is the basic structure of membranes in living cells. It separates the inside and outside and so defines a cell. It also defines various cell organelles such as the mitochondria and the chloroplasts. Electron transfer and ion transport across these lipid membranes are processes fundamental to cellular functions. The discovery of the planar lipid bilayer by Mueller and co-workers [1] some four decades ago was the break-through that allowed direct experimental measure of the electrical properties of these membranes. It is the archetype of the self-organizing nanosystem discovered well before these buzz-words were invented. Ion transport and photochemical electron transfer reactions could now be conveniently studied in a realistic model.

Over the last three decades, many efforts have been made to assemble biomimetic photodevices, but poor efficiency of the transmembrane charge transport in artificial systems has limited progress. A study of charge transport mechanisms is not only important for the development of biomimetic photodevices, but also to the understanding of many biological processes. Some antibiotics function by relaxing ion gradients across the cell membrane via channel formation or by acting as ionophores, and some anesthetics are lipophilic ions. Finding methods to increase and control ion transport across lipid bilayers will be helpful in developing efficient biomimetic systems.

Time-resolved photoelectric measurements are a powerful method to probe transmembrane charge transport. In this article we review progress in the photoelectrochemical studies of lipid bilayer systems in our laboratory. We have described rapid interfacial charge separation and have distinguished between electronic and ionic transmembrane conductivity. We have found that the rate of transport of borate anions is greatly increased by the presence of photo-formed porphyrin cations. The transport rate of the porphyrin cations can be similarly enhanced by two orders of magnitude in the presence of lipophilic borate anions. Thus the system is an organic analogue of a photo transistor. When porphyrin cations are formed at a single interface, the photoformed electric field and the nonlinear enhancement of ionic conductivity cause ion pumping, i.e. formation of an ion gradient across the lipid bilayer. This asymmetric photoformation of porphyrin cations and their resulting local electrostatic fields in the lipid bilayer can also cause dissociation of lipophilic weak acids by shifting their pK_a, and thus result in proton pumping across the lipid bilayer. The underlying concept is the importance of the local electric fields brought out by small dimensions and low dielectric coefficient inside the bilayers. These subjects will be discussed in the following sections.

.2.   1) Interfacial Charge Separation

A major advantage of the planar bilayer system is the ability to use fast electrochemical measurements for study of photo-induced electron transfer reactions. This advantage is enhanced by the simple separation of donors and acceptors across the lipid-water interface. By use of a lipophilic pigment and ionic reactants this separation is achieved automatically. The experimental set-up is shown in Fig. 1. The lipid bilayer is formed in a ~1 mm hole in a Teflon partition separating aqueous solutions. The small amount of lipid-hydrocarbon solution placed in the opening spontaneously thins to the bilayer structure. It is proven to be a bilayer by its capacitance (~5 nF mm^{-2}) and high resistance ($>10^8$ ohms mm^{-2}). Voltages or currents are measured with suitable amplifiers [2-7].

Photo induced currents or voltages across a lipid bilayer containing pigments and separating electron donors or acceptors were first observed three decades ago [8-10]. At that time these photoinduced signals were claimed to be evidence for electron conduction across the lipid bilayer [8, 11-13], but were proven later by time-resolved measurements [9, 14-15] to be mainly interfacial charge separation, i.e. displacement currents. The triplet state of porphyrins was suggested to be the active excited state [2,15]. The presence of oxygen had negligible effect on the photoinduced interfacial electron transfer from photo-excited closed-shell metalloporphyrins in the lipid bilayer to stable electron acceptors, such as ferricyanide [16]. These reactants were chosen purposely for the lack of reaction of their products with oxygen on single electron transfer. The small effect of oxygen showed that the initial electron transfer was faster than the quenching rate of triplet state of porphyrins by oxygen. The measured

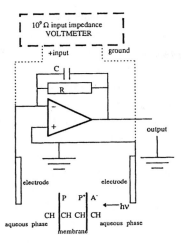

Fig. 1. Schematic diagram of membrane systems and probing methods. The top circuits with solid and dashed lines represent the conditions for current (short circuit) and voltage (open circuit) measurements, respectively. Two calomel electrodes are immersed into the aqueous phases on both sides of the lipid bilayer. The gain and time constant for current measurements are set by adjusting the variable resistor R and capacitor C. A variable DC voltage source is connected on the ground side in series when needed. The lipid bilayer membrane is formed across a 1.5 mm diameter hole in a thin Teflon sheet. P and P^+ respectively represent neutral and positively-charged porphyrins. A^- represents the reduced electron acceptor. See the publications 5-7 for other details.
(Reprinted from ref #7 with permission from PNAS

apparent second-order rate constant for the system with MgOEP (magnesium octaethylporphyrin) and ferricyanide was $10^9 \, M^{-1} \, s^{-1}$ i.e. close to the encounter limit [3]. The rate constant increased to $10^{11} \, M^{-1} \, s^{-1}$ when AQS (anthraquinone-2-sulfonate) was used as electron acceptor [3], faster than the encounter limit, indicating binding of the AQS to the interface. Similar fast interfacial electron transfer was observed when MgOEP was replaced by chlorophyll, but the product cations were less stable.

The reverse interfacial electron transfer from aqueous electron donors, such as ferrocyanide and ascorbate, to the photoformed porphyrin cation was not sensitive to oxygen either, confirming the known very slow reaction of porphyrin radical cations with oxygen [17]. The interfacial reaction of MgOEP cation and ferrocyanide has an apparent second-order rate constant of $10^7 \, M^{-1} \, s^{-1}$. The transit time of the $MgOEP^+$ cation across the lipid bilayer was found to be very slow, 100-500 ms [2]. The photoformed cations of chlorophyll have a yet lower mobility in lipid bilayers than $MgOEP^+$ cations [5].

In contrast, the direct photoinduced electron transfer from the aqueous phase to the lipid bilayer was observed only under anaerobic conditions because of the known sensitivity of porphyrin anions to oxygen. The electron transfer from ascorbate to triplet MgOEP occurred with a ~10 μs time constant [18]. The interfacial electron transfer in this direction was still slow, 6 μs, even when a strong excited electron acceptor, C_{70} or C_{60} was incorporated into the lipid bilayer and saturating concentrations of ascorbate was the donor [19]. Faster electron transfer, <15 ns, in this direction was observed only when the stronger electron donor dithionite was used with the fullerene system [20].

Our studies showed that there was no appreciable barrier to interfacial electron transfer from the lipid to water phases beyond the 1 ns time scale [3]. The apparent rate constants were similar to those observed in solution. Electrostatic interactions indicated that the ionically charged acceptor molecules penetrate to the level of the polar headgroups of lipids with differing orientations. Such a photoinduced charge separation had a high quantum yield (>10%) and long life time (seconds), because of enforced separation of reaction products. However, there is a possible barrier to the interfacial electron transfer in the reverse direction, i.e. from the water to lipid phases as indicated by the slower rate constants. The local electric fields at the water side of the lipid-water interface might be favorable to the former and unfavorable to the latter electron transfer. Thus the interface itself may have a function in the electron transfer reactions. The photosynthetic systems use membrane-spanning proteins to optimally separate and orient similar pigments and reactants to achieve efficient charge separation across the membrane.

.2.1. 2) Electronic and Ionic Conductance

Photoinduced charge separation across membranes and interfaces is the basis of photosynthesis and of artificial solar energy conversion. It was soon found after the discovery of the planar bilayer that photocurrents could occur in bilayers containing photosensitive pigments [8]. There were two explanations of the "photoconduction". One proposed that the photoinduced current was electronic i.e. it occurs by electron hopping between pigment molecules inside the lipid bilayer [8, 11-13]. The other suggested that photoformed mobile ions in the lipid bilayer are the current carriers [9, 14, 15, 21] Ion transport across lipid bilayers is slow. It takes >100 ms for free lipophilic porphyrin cations to cross the lipid bilayer [2] although they can be photochemically formed on the ns time scale by interfacial electron transfer [3]. The transport of ions directly across the lipid bilayer is limited by the large electrostatic energy of the ion in the hydrocarbon core, and in the case of relatively high concentrations of lipophilic ions, by space charge [4] in the bilayer. Biological systems use specialized ion channel proteins to transport small ions across the lipid bilayer. Transmembrane ion movements through biological ion channels can take as short a time as 100 ns [22], and are controllable. Thus evolution selected this mechanism for biological ion transport.

There are two fairly direct ways to distinguish electronic and diffusive ionic conductance in a lipid bilayer. The first is by cooling the lipid below the phase transition temperature, which inhibits diffusion of ions. The second method is to show that the charge transit time is shorter than that expected for ion diffusion across the lipid bilayer. The former method was unsuccessful in our hands but the latter produced useful results.

Our experiments with porphyrins as carriers produced only ion limited conductances, but fullerenes were found to produce a large photoelectric conductance when incorporated into lipid bilayers [20, 23]. Under most conditions, photoinduced electron transfer across the lipid bilayers was limited by interfacial transfer from aqueous electron donor to the excited fullerene [19]. Photovoltage measurements showed that the interfacial charge separation occurred on the >1 μs time scale when relative stable electron donors ascorbate and amphoteric zinc deuteropoyphyrin were used as electron donor. Since this was the time range in which ionic diffusive and electronic conduction could not be well separated, a faster interfacial reaction was needed. When C_{70} was incorporated into the lipid bilayer and dithionite was added to one aqueous phase, the photovotage (Fig. 2A), with a positive sign on the dithionite side, occurred with a <15 ns rise time on pulsed light excitation of 7 ns [23] On adding ferricyanide to the trans side as acceptor, the amplitude of photovoltage was increased by a factor of 15, and the rise time remained the same as before. Photocurrent measurements on the separate addition of dithionite and ferricyanide to the aqueous phases showed similar changes in charge transfer, but with less time resolution (Fig. 2B). The nanosecond charge transit time is proof of electron conduction, since the transmembrane diffusion of lipophilic anions with a similar size to the fullerene anions occurs on the many μs to ms time range. Even the fastest transmembrane ion transport which proceeds through ion channels, requires ~100 ns. The C_{70} molecule has a ~1 nm diameter, while lipid bilayers have a 6-8 nm thickness, so it cannot bridge the lipid bilayer. The electron conduction most likely occurs through photoactive fullerene aggregates.

Experimental evidence for fullerene aggregates came from photovoltage measurements at different C_{70} concentrations in the lipid bilayer [23]. The nonlinearity suggested that fullerene aggregates became more fully connected in the lipid bilayer when the C_{70} concentration in the lipid forming solution was >0.5 mM. There is spectral evidence for aggregates of C_{70} in organic sovents [24] and the action spectrum of the photoconductivity [23b] followed that of the aggregates, not that of the monomer. The photoactive aggregates in the lipid

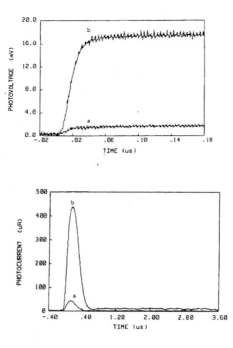

Fig. 2. Photovoltages (top) and photocurrents (bottom) across the lipid bilayer containing C_{70} induced by a 7 ns light pulse. The membrane-forming solution contains 0.6 mM C_{70}. The aqueous solution is deoxygenated with glucose oxidase and glucose. Sodium dithionite at 100 mM is present on the positive electrode side as electron donor. The excitation is at 532 nm and of 100 μJ energy. (a) No electron acceptor. (b) Ferricyanide at 10 mM is present on the trans side of the donor as electron acceptor.
(Reprinted from ref #23b with permission from ACS).

bilayer would be very efficient in transporting charge from donor interface to acceptor interface by electron hopping between the molecules in the aggregates.

The only proven instances of electron hopping across the membrane involve photosynthetic reaction centers incorporated across a lipid bilayer [25], some long chemically integrated molecules which can bridge the lipid bilayer [26, 27] and the fullerene system discussed above. For lipid bilayers containing other photosensitive pigments, however, transmembrane electron hopping has been immeasurable. In most systems, it has been demonstrated that the movements of photoformed ions across the lipid bilayer and the lipid-water interface cause the photo conduction.

.3.   3) Non-linear effects in the bilayer

.3.1. 3a) Photogating of Ion Transport

When the initial charge separation is caused by interfacial electron transfer, the photoformed ions in the lipid bilayer are near the interface due to limited electron tunneling distance and minimization of electrostatic energy. The metallo-porphyrin cations have a low mobility as discussed above. The movements of water soluble hydrophobic ions across the lipid bilayer is limited by the electrostatic energy barrier in the hydrocarbon region, and also by space charge at high charge density [4, 28]. Thus it was expected that the current of such hydrophobic ions would be enhanced by photo-formation of oppositely charged ions on interfacial electron transfer. The ideal test case was the system with mobile $TPhB^-$ (tetraphenylborate) anions and relatively immobile photo-formed $MgOEP^+$ cations. The experimental results showed that in the linear range of current to $TPhB^-$ concentration, the enhancing effect was 30-50% on symmetric photoformation of $MgOEP^+$ cations near the two interfaces [4, 29]. However, the enhancing effect reached ~1000%, 10 fold, at high concentrations of $TPhB^-$ where its current progressively saturates with increase in concentration, and 100 fold with halo substituted tetraphenylborates (Fig. 3).

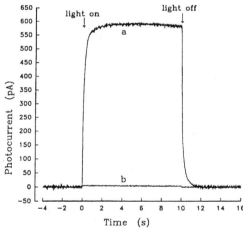

Fig. 3.   The steady-state photodriven currents caused by 130 mW cm^{-2} of white continuous illumination of the asymmetric lipid membrane are plotted versus time. MgOEP (3.6 mM) is present in the DPPC bilayer, and saturating concentrations of acceptor (0.5 mM AQS) and donor (4mM ferrocyanide) are present on opposite sides, curve a, 4 μM p-TCPhB$^-$ is added to both sides, curve b, No borates are present in the system. The presence of p-TCPhB$^-$ increases the low-frequency (40-60-Hz) noise by ~10 times under both dark and photo conditions (*curve a*) because of a 10-fold increase in the membrane capacitance.
(Reprinted from ref.#5 with permission from BS).

This enhancement of the transmembrane ionic current was named the photogating effect [29]. It is an organic analogue of a photo-field-effect transistor [30]. The effect in the linear range was expected from our estimate of the electrostatic fields inside the bilayer. That in the nonlinear range could be partly explained as electrostatic cancellation of the accumulated space charge in the lipid bilayer, which itself quantitatively accounts for the peculiar progressive saturation (the result of a simple implicit equation for the internal potential [4]) of the borate current with concentration [28]. An ion chain mechanism was proposed to explain the full 10 fold enhancing effect [31].

The activation energies (E_a) of borate translocation across the lipid bilayer were obtained by measuring the temperature dependence of the ionic currents over the 6-30 °C range [5]. The E_a values of the photogating currents were 2-3 times those in the absence of porphyrin cations. The average value for all the borates in the presence of $MgOEP^+$ was 13 kcal mol^{-1}, while that in the dark was 5 kcal mol^{-1}. The differences in the E_a values between different borates were small (<2 kcal mol^{-1}). These results indicated that any change in the Born electrostatic energy caused by different effective ion radii of borates had little effect on the activation energy. If the photogating effect was caused simply by reducing ionic electrostatic energy in the lipid bilayer, one would expect the E_a of borate movements to decrease compared to that in the dark. The 2-3 fold larger E_a of the enhanced ion translocation suggests that the enhancement arose from intermolecular interactions other than electric field effects.

The enhanced transport of porphyrin cations and borate anions may be explained by a dynamic ion chain mechanism [5, 31]. In this model, chains of alternatively charged ions across the lipid bilayer are formed by transient binding between photoformed cations and borate anions (Fig. 4). This ion chain mechanism is similar to the mechanism for the conductance of inorganic cations or anions through biological ion channels, in that these ions are thought to hop along the polar sites of the channels [22]. It is also analogous to the Grotthus mechanism for anomalous proton and hydroxide ion conductivity in water. The electrostatic energy of the ion chain of alternate charges is significantly less than that of the individual ions. Both the porphyrin cation and the borate anion can hop along the ion chain. This mechanism can explain the larger activation energy of the enhanced ion transport. Rotation of the porphyrin cation-borate anion ion pairs to reorder the conducting ion-chain requires an activation energy estimated to be ~10 kcal mol^{-1}, which rationalizes the observed increase of 8 kcal mol^{-1} over the dark activation energy. Our kinetic analysis of the time-resolved photoinduced currents strongly supports this dynamic ion chain mechanism [5]. The calculated current time courses from this model fit the observed photodriven currents very well. The different rate constants of

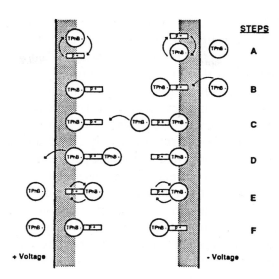

Fig. 4. Schematic of the ion-chain mechanism for the photogating of ionic currents across the lipid bilayer. The width of the membrane and the size of the MgOEP$^+$ (*rectangles*), and the TPhB$^-$ (*circles*), are to scale. The lipid head groups are represented by the shaded areas. Individual ions are represented as rectangles or circles, whereas ion pairs and aggregates are shown as connecting circles and rectangles.
(Reprinted from ref #31 with permission from BS).

formation, dissociation, and translocation also rationalize the structural sensitivity of the enhancing effect to different borate and porphyrins.

.3.2. 3b) Photoformed Fields and Ion Pumping 1—Donor Only System

An ion gradient across the cell membrane is characteristic of all living cells. These ion gradients are created and maintained by biological ion pumps, which are membrane proteins. The energy for maintaining the gradients is supplied by ATP. Conversely, the formation of ATP is driven by proton or electrochemical potential gradients across the membrane which, in turn, are driven photochemically in photosynthesis or by oxidation of foods (from photosynthesis) by respiration. Although all analysis of these processes have been in terms of transmembrane potentials, we have found that ion pumping across a membrane can be effected by photodriven electron transfer across a single lipid water interface [6].

The experimental system was composed of a MgOEP-containing lipid bilayer with p-halo borate anions present symmetrically on both sides and AQS$^-$

present on only one side, i.e. with no donor on the trans side. On excitation of MgOEP by a 1 µs pulse of light, a striking negative voltage developed on the time scale of seconds, following decay of the expected initial fast positive voltage formed by interfacial electron transfer from excited MgOEP to AQS⁻ (Fig. 5). The buildup of a consistent negative voltage was also shown by similar experiments using continuous light. Current measurements on continuous excitation (Fig. 6b) further confirmed the larger opposite ion flow than the steady-state electron transfer current. Since only the borate anions and the photoformed porphyrin cations were membrane permeable in the system, the striking negative voltage and current can be caused only by the transmembrane movements of borate anions from the side with AQS⁻ to the other (trans) side. The ~20 s decay time of the negative voltage (Fig. 5) was longer than the 5 s RC time constant of the membrane circuit. This was conclusive evidence that a concentration gradient of borate anions was created across the lipid bilayer. The negative voltage on the long time scale must be caused by the Nernstian potential of the concentration gradient of borate anions between the two aqueous phases across the lipid bilayer. The efficiency of the ion pumping was estimated, using the ratio of the negative borate ion current to the maximum electron transfer current, to be 0.28.

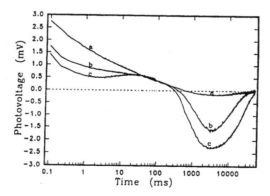

Fig. 5. Voltage signals of the system with only electron donor caused by a light pulse of 1 µs are plotted versus log time. AQS⁻ is present in the aqueous phase of the positive electrode side as electron acceptor. (a) No borate is present. (b) TFPhB⁻ anions are present in the aqueous phase on both sides. (c) TCPhB⁻ is present instead of TFPhB⁻.
(Reprinted from ref #6 with permission from BS).

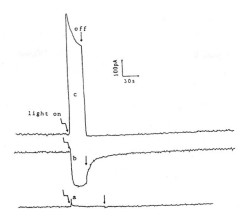

Fig. 6. The currents caused by continuous illumination ($\sim$130 mW cm^{-2}) of the membrane system with 0.5 mM AQS⁻ as acceptor on one side are plotted versus time. No voltage is applied across the membrane. (a) Neither borates nor electron donor is present. (b) 4 μM p-TCPhB⁻ is added in the absence of donor. The half-decay time of the reversed current is within 2-3 min under illumination. (c) 4 mM ferrocyanide is added on the opposite side to the acceptor in the presence of p-TCPhB⁻.
(Reprinted from ref # 6 with permission from BS).

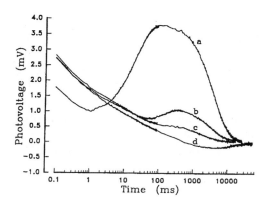

Fig. 7. Photovoltages across the MgOEP-containing lipid bilayer caused by the light pulse are plotted versus log time. AQS⁻ is present on the positive electrode side as electron acceptor. Except for measuring curve (d), ferrocyanide as electron donor is on the trans side of AQS⁻. (a) TCPhB⁻ anions are present in the aqueous on both sides. (b) TPhB⁻ is present instead of TCPhB⁻. (c) No borate is present. (d) Neither borate nor electron donor is present.
(Reprinted from ref #5 with permission from BS).

.3.3. 3b) 2—Donor-Acceptor System

The presence of p-TCPhB⁻ or TFPhB⁻ in the system with aqueous electron acceptor and donor on opposite sides, increased dramatically (100 fold, Fig. 3) the photodriven charge flow across the MgOEP-containing lipid bilayer [5]. The results of time-resolved measurements confirmed that the transmembrane diffusion rate of porphyrin cations was increased ~100 fold in the presence of p-TCPhB⁻ or p-TFPhB⁻. On pulsed excitation of 1 μs, the photoformed voltage (Fig. 7) showed that the MgOEP$^+$ cations took an average of only ~10 ms to cross the lipid bilayer in the presence of p-halo borate anions. The striking resurgence of the positive photovoltage on the millisecond time scale and its decay time longer than the RC time of the lipid bilayer (Fig. 7), are conclusive evidence that negative charges were transported from the aqueous phase containing ferrocyanide to the aqueous phase containing AQS⁻. In other words the rate of reduction of the MgOEP$^+$ cation by ferrocyanide was much increased by the increased mobility of the cation by the lipophilic borate ion.

As shown in Fig. 6c on continuous illumination a large positive transmembrane current and voltage occurred when the electron donor, ferrocyanide was in the opposite side to AQS⁻. By analyzing the relationship of total current amplitude to the concentration of ferrocyanide, it was shown that the presence of the trans electron donor greatly reduced the negative current of borate anion pumping by reducing the population of porphyrin cations inside the lipid bilayer, even though the charge separation between AQS^{2-} and Fe(CN)$_6^{3-}$ increased the total external driving voltage across the lipid bilayer. This suggested that the external voltage was not the main driving force for the ion pumping. In fact the photoinduced voltage across the lipid bilayer was zero on the >100 μs time scale in measurements of the ion pumping current because of the voltage-clamp circuit. The transmembrane pumping of borate anions must arise mainly from the internal electric field caused by the photoformed porphyrin cations and their gating effect on the internal transport of borate anions inside the lipid bilayer. As the porphyrin cation concentration diminished with the reverse electron transfer from AQS^{2-} to MgOEP$^+$ and the enhanced ionic conductance vanished, the borate anion gradient across the lipid bilayer was temporarily trapped and became visible as a negative voltage (Fig. 5).

The mechanism of this ion pump is very different from the known biological ion pumps. Because this system does not require proteins, but only lipophilic ions, simple photosensitive molecules and ionic donors or acceptors, it is an excellent candidate for a prebiological ion pump.

.3.4. 3c) Proton Pumping by Single Interface Charge Transfer

Proton gradients created by light-driven proton pumps in many biological systems play a crucial role in supplying energy in the form of ATP. As the complexities of the integral membrane proteins constructing these proton pumps have so far prevented a complete description of their mechanism, a light-driven proton pump composed of the lipid bilayer and simple molecules is of some interest. When a lipophilic weak acid, CCCP (carbonylcyanide m-chlorophenyl-hydrazone) or FCCP (carbonylcyanide p-trifluoromethoxyphenylhydrazone), was used to replace the p-halo borate in the above described ion-pumping system, composed of the MgOEP-containing lipid bilayer with AQS⁻ present on one side, became an efficient photodriven proton pump [7].

In the presence of CCCP or FCCP, the positive photoinduced current and voltage were gradually replaced by increasingly negative currents and voltages, as the pH of the aqueous phase on both sides of the lipid bilayer was decreased from 10 to 6. This negative charge transport was proven to require the porphyrin cation in the bilayer by adding electron donors to the acceptor side to shorten its lifetime: the negative currents and voltages disappeared. The sensitivity of the negative current and voltage to pH suggested that protons were moved in the same direction as the interfacial electron transfer. This was in accord with the direction of the photogenerated local field: $MgOEP^+$ (lipid)/ AQS^{2-} (aqueous). The larger amplitude of the negative voltage than that of the positive photovoltage showed that the ionic charges were separated across the lipid bilayer, i.e. much further apart than the photoformed ions. The proton movement was confirmed by a striking isotope effect on substituting D_2O for H_2O in the system (Fig. 8). The negative ionic current was <u>increased</u> by 2-3 fold on D_2O substitution at pH 7-8 in both the steady-state and pulsed measurements, while the positive photo current and voltage in the absence of the proton carriers were not affected. Thus the negative current must be a proton pumping current across the lipid bilayer.

The 2-3 fold larger pumping current of the D_2O system suggested a specific proton pumping mechanism. The usual kinetic D_2O effect is a decrease of rate. The membrane with D_2O and proton carriers showed a 20% smaller conductivity than that with H_2O in the dark. The D_2O substitution most likely increased the proton pumping current by increasing the concentration of neutral proton carriers in the lipid bilayer. A 3-6 fold decrease of ionization of weak acids was known to be caused by using D_2O instead of H_2O as solvent [32], which was partially explained in terms of zero-point energy difference of the O-H and O-D bonds. The calculated pK_D - pK_H difference of CCCP and FCCP (both pK_H 6.0-6.2) on isotope substitution was 0.5. The UV-absorption measurements of CCCP

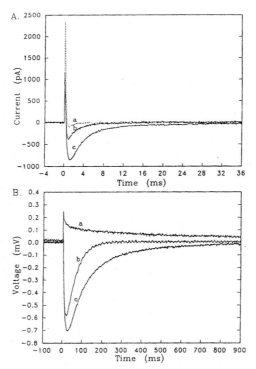

Fig. 8. Currents (A) and Voltages (B) induced by a 10-ns pulse of light for the D_2O and H_2O systems are plotted versus time. The time constant of the current amplifier in A is 0.1 ms. The bathing solutions have a pH of 7.0 or a pD of 7.5. Curve a, no proton carrier in D_2O or H_2O. In A, this system shows a displacement current arising from the forward and the rapid component of the reverse interfacial electron transfer. The integral of the total current is zero. Curve b, 10 µM FCCP in H_2O. Curve c, 10 µM FCCP in D_2O. The integral in A is not zero. (Reprinted from ref #7 with permission from PNAS).

and FCCP showed that their partition coefficients between benzene and water were enhanced 2-4 fold by replacement of H_2O with D_2O [7, 33]. The partition coefficients between the lipid bilayer and the water phase might be similarly increased by D_2O substitution. This could cause the observed increase of the negative ionic current, since there are more neutral CCCP molecules releasing protons on photoformation of $MgOEP^+$ cations in the lipid bilayer of the D_2O system.

Since the photovoltage across the lipid bilayer was zero in the "short circuit" current measurements, the electrostatic driving force of the proton pumping must be local electric fields of photoformed porphyrin cations. The striking D_2O effect suggested that the proton pumping was initiated by interfacial deprotonation of neutral CCCP at the interface containing the acceptor. The photoformed local field or direct effect of $MgOEP^+$ cations by binding $CCCP^-$ anions is favorable to deprotonation of the neutral CCCP from the interface to the aqueous phase, i.e. the interfacial pK of the weak acid is lowered. A decrease of 0.1 in pK represents a free energy difference equivalent to 6 mV, which is a ~10 fold stronger driving force than the maximum photovoltage observed in the "open circuit" measurements. Thus the proton pumping was initiated by the interfacial electrostatic deprotonation. The transmembrane diffusion of released $CCCP^-$ or $FCCP^-$ anions which are known to be mobile in lipid bilayers [34], carried the ionic current across the hydrocarbon region. The transported $CCCP^-$ was protonated at the opposite interface, and then diffused back the interface with the acceptor. This constituted the loop of the transmembrane proton pumping. The estimated efficiency from the pumped proton charges divided by the photoformed charges was 15-30% [7].

The suggestion in [35] that O_2^- formed by photooxidation of CCCP is responsible for the observed currents is untenable. Their system is completely different: an interfacial ionic pigment sensitizer, long term continuous illumination and much longer time scales. Their suggestion cannot even explain the pH dependence of our observed currents.

Our simple transmembrane proton pump involves an electrostatic pK shift mechanism, which is similar to some known biological proton pumps [36, 37]. This system also pumps protons across the membrane on the millisecond time scale, but it is inherently much more "leaky" than biological pumps because it uses a mobile proton carrier, not an integral protein containing a one-way proton chain.

.4. Conclusion

The nanometer size and the complex dielectric properties of the lipid bilayer combine to make it a rich barrier which defines a living cell. Cellular evolution has developed proteins which function as gateable ion channels and as ion pumps. Our work has shown that the small size and dielectric properties of the lipid bilayer can be used to form gateable ion currents and pumps without specialized proteins. Calculations and experiments show that the potentials and the resulting electric fields generated at one bilayer-water interface extend and influence ions at the second interface. Such reactions may have occurred in pre-biotic times as precursors to the present complex biological systems.

Acknowledgements

The NIH supported this research for many years with grant GM-25693. We thank our many coworkers whose names appear in the references. Their patient work with the fragile planar bilayers achieved the results quoted in this paper.

References

[1] P. Mueller, D. O. Rudin, H. T. Tien and W. C. Wescott. Nature, 194 (1962) 979.
[2] M. Woodle and D. Mauzerall, Biophys. J., 50 (1986) 431.
[3] M. Woodle, J. W. Zhang and D. Mauzerall, Biophys. J. 52 (1987) 577.
[4] D. Mauzerall and C. M. Drain, Biophys. J. 63 (1992) 1544.
[5] K. Sun and D. Mauzerall, Biophys. J. 71 (1996) 295.
[6] K. Sun and D. Mauzerall, Biophys. J. 71 (1996) 309.
[7] K. Sun and D. Mauzerall, Proc. Natl. Acad. Sci. USA. 93 (1996) 10758.
[8] (a) H. T. Tien, Nature, 219 (1968) 272. (b) H. T. Tien, J. Phys. Chem., 72 (1968) 4512. (c) H. T. Tien and N. Kobamoto, Nature, 224 (1969) 1107. (d) H. T. Tien and S. P. Verma, Nature, 227 (1970) 1232.
[9] (a) D. Mauzerall and A. Finkelstein, Nature, 224 (1969) 690. (b) F. T. Hong and D. Mauzerall, Nature New Biol., 240 (1972) 154.
[10] T. R. Hesketh Nature, 224 (1969) 1026.
[11] T. L. Jahn, J. Theoret. Biol., 2 (1962) 129.
[12] S. W. Feldberg, G. H. Armen, J. A. Bell, C. K Chang and C. B. Wang, Biophys. J., 34 (1981) 149.
[13] W. E. Ford and G. Tollin, Photochem. Photobiol., 38 (1983) 441.
[14] H. U. Lutz, H. W. Trissl and R. Benz, Biochim. Biophys. Acta, 345 (1974) 257.
[15] F. T. Hong and D. Mauzerall, J. Electrochem. Soc., 123 (1976) 1317.
[16] A. Ilani, T. M. Liu and D. Mauzerall, Biophys. J. 47 (1985) 679.
[17] J. Furhop and D. Mauzerall, J. Am. Chem. Soc., 91 (1969) 4174.
[18] A. Ilani, M. Woodle, D. Mauzerall, Photochem. Photobiol. 49 (1989) 673.
[19] K. C. Hwang and D. Mauzerall, J. Am. Chem. Soc. 114 (1992) 9705.
[20] K. C. Hwang and D. Mauzerall, Nature, 361 (1993) 138.
[21] B. R. Masters and D. Mauzerall, J. Membr. Biol., 21 (1978) 377.
[22] B. Hille, Ion Channels of Excitable Membranes, Sinauer, Sunderland, MA, (1984) 426.
[23] (a) K. C. Hwang, S. Niu, D. Mauzerall, ECS Fullerene Symp. Proc., 94-24 (1994) 845. (b) S. Niu and D. Mauzerall, J. Am. Chem. Soc., 118 (1996) 5791.
[24] Y. P. Sun and C. E. Bunker, Nature 365 (1993) 398.
[25] M. Schonfeld, M. Montal and G. Fehor, Proc. Natl. Acad. Sci. USA, 76 (1979) 6351.
[26] P. Seta, E. Bienvenue, A. L. Moore, P. Mathis, R. V. Bensasson, P. Liddell, P. J. Passiki, A. Joy, T. A. Moore and D. Gust, Nature, 316 (1985) 651.
[27] G. Steinberg-Yfrach, P. A. Liddell, S. C. Hung, A. L. Moore, D. Gust and T. A. Moore, Nature 385 (1997) 239.
[28] R. F. H Flewelling,.W. L. Hubbell. Biophys. J. 49 (1986) 531.
[29] C. M. Drain, B.Christensen and D. Mauzerall, Proc. Natl. Acad. Sci. USA, 86 (1989) 6959.
[30] C. M. Drain and D. Mauzerall, Bioelectrochem. Bioenerg. 24 (1990) 263.

[31] C. M. Drain and D. Mauzerall, Biophys. J. 63 (1992) 1556.
[32] P. M. Laughton and R. E. Robertson, Solute-Solvent Interactions, J. F. Coetzee, and C. D. Ritche (eds), Dekker, New York, (1969) 399.
[33] P. Gross and A. Wischin, Trans. Faraday Soc. 32 (1936) 879.
[34] H. Tarada, Biochim. Biophys. Acta, 639 (1981) 225. (b) J. Kasianiowicz, R. Benz, S. and McLaughlin, J. Membr. Biol. 82 (1984) 179.
[35] V. S. Sokolov, M.Block, I. N. Stozhkova, and P. Pohl, Biophys. J. 79 (2000) 2121.
[36] E. Bamberge, H. J. Butt, A. Eisenrauch and K. Fendler, Q. Rev. Biophys. 26 (1993) 1.
[37] L. S. Brown, J. K. Lanyi, Proc. Natl. Acad. Sci. USA 93 (1996) 1731.

Planar Lipid Bilayers (BLMs) and their Applications
H.T. Tien and A. Ottova-Leitmannova (Editors)

Chapter 35

Photosynthetic pigment-protein complexes in planar lipid membranes

W. I. Gruszecki[a] and A. Wardak[b]

[a]Department of Biophysics, Institute of Physics, Maria Curie-Sklodowska University, 20-031 Lublin, Poland

[b]Institute of Physics, Technical University of Lublin, Lublin, Poland

1.WHY STUDY PHOTOSYNTHETIC PIGMENT-PROTEIN COMPLEXES IN PLANAR LIPID MEMBRANES?

Life on Earth is driven by energy of the Sun and photosynthesis is a process that "translates" light into forms of energy that can be utilized by living organisms. The light processes of photosynthesis responsible for such a "translation" take place in a highly sophisticated photosynthetic apparatus comprising numerous functional proteins embedded in the lipid membranes called thylakoids. Fig. 1 presents a model of the photosynthetic apparatus of higher plants. The protein complexes, elements of the photosynthetic machinery, are presented in the model in the order (from left hand side to right hand side) corresponding to the direction of electron transfer within the framework of the so-called linear electron flow. This electron transfer begins in the Photosystem II (PS II) reaction center in consequence of the absorption of a light quantum by the special pair of chlorophyll a molecules (P680). Singlet excited state of P680 promotes an instant electron transfer to pheophytin (Pheo), the first acceptor, followed by further electron transfer to plastoquinone A (Q_A) and plastoquinone B (Q_B) along with the direction of a decrease in the oxidation-reduction (redox) potential. Doubly reduced Q_B binds two protons at the outer surface of the thylakoid membrane and leaves its binding pocket in PS II migrating towards the cytochrome b_6-cytochrome f (cyt b_6-f) complex where further electron/proton exchange processes take place. Oxidized P680 is again reduced at the expense of an electron originating from the water splitting enzyme.

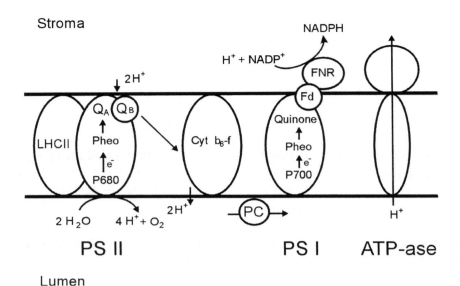

Fig. 1. Simplified model of the thylakoid membrane of higher plants with indicated selected pigment-protein complexes and the pathways of photosynthetic linear electron transfer chain and proton flow. Explanation of abbreviations are in the text.

It means that the process of enzymatic oxidation of water is a source of electrons that are transferred along the linear electron transfer chain, and protons in the inner space of the thylakoid membranes. Molecular oxygen, another product of water splitting reaction, diffuses out of chloroplasts and is a major source of atmospheric oxygen. It has to be noted that hundreds of photosynthetic accessory pigments absorb light quanta and transfer electron excitation energy to P680 in order to maintain high rate and efficiency of photosynthesis. According to the current knowledge photosynthetic pigments (both chlorophylls and carotenoids) are bound in vivo to the functional pigment-protein complexes. Several light-harvesting pigment-protein complexes serve PS II by supplying the reaction center with excitation energy. Among the antenna complexes of PS II is the major one, LHCII, comprising about half of chlorophyll pool on Earth. Electron exchange among Q_B pool and cyt b_6-f complex is associated with net proton transfer from the outer towards the inner thylakoid space, which is a second source of proton gradient (apart of PS II linked functionally to the water splitting enzyme). This proton gradient across the thylakoid membrane drives ATP synthesis in accordance with the chemiosmotic mechanism of activation of ATP-synthase (ATP-ase). The next element of the linear photosynthetic electron transfer chain in higher plants is plastocyanin (PC), a water-soluble protein that carries an electron from cyt b_6-f to Photosystem I (PS I). PS I is another system of the light-harvesting pigment-protein complexes absorbing light energy and transferring it to the reaction center in order to drive electron flow from the primary electron donor P700 to the reaction center-bound electron acceptor, ferredoxin (Fd), via the intermediate electron carriers, and eventually NADP-reductase (FNR) that catalyses the synthesis of NADPH. NADPH is a "reducing power" that is utilized not only for the photosynthetic assimilation of CO_2 but also in other metabolic processes in a living cell. Owing to dense packing of all functional protein complexes in the thylakoid membrane and extremely high rate and efficiency of both excitation energy flow and electric charge flow, the transfer of a single electron from H_2O to NADPH is completed in the photosynthetic apparatus of higher plants in a relatively short time of 20 ms (see [1] for a detailed description of the photosynthetic charge transfer processes).

As may be seen from the brief description above, the photosynthetic activity relies on electric charge transport and exchange among protein complexes embedded in a lipid membrane. Lipid membranes, prepared in a laboratory, containing incorporated protein complexes isolated from thylakoids seem to provide an excellent model to study physical processes that take place in a single functional protein or several different protein constituents assembled in the same system. Such studies have been carried out with application of liposomes [2,3], monomolecular layers [4-6] and also planar lipid bilayers (as discussed below). The latter system seems to be particularly suited for experimentation on electric charge transfer owing to the fact that redox conditions may be easily manipulated at the opposite sides of a planar lipid membrane. The planar lipid membrane system is also suitable for precise optical experiments with application of polarized light, primarily owing to well defined plane of the membrane and localization and orientation of photosynthetic pigment-protein complexes strictly determined by physical properties of the lipid matrix. This system has already been applied several times at early stages of photosynthesis research, at which researchers believed that majority of photosynthetic pigments were not protein-bound but present directly within the lipid phase of the thylakoid membrane (see [7] and [8] for a review). The same system appears also very useful in model studies of photosynthetic pigment-protein complexes.

2. INCORPORATION OF PHOTOSYNTHETIC PIGMENT-PROTEIN COMPLEXES INTO PLANAR LIPID BILAYERS

There have been several reports describing the specific techniques employed to incorporate reaction centers (RC) into planar bilayer lipid membranes. This section summarizes methods concerning membranes separating two aqueous solutions assembled by the "painting" ("brush") technique [9-13] or by the monolayer technique that results in formation of "solvent free" membranes [14-15].

Bacterial Reaction Centers (BRC) were reconstituted into BLM of both types by using the solution of RC/lipid complex in an organic solvent (hexane, octane) to form a membrane. The method to produce such a complex is based on techniques developed for other membrane proteins. It has been shown that BRC preserve their characteristic spectral and photochemical properties in organic solvents [16]. BRC in membranes formed according to this method are distributed symmetrically between two possible orientations. To generate electrical signals upon illumination one BRC population must be preferentially modified to obtain "net functional orientation". This was achieved by adding potassium ferricyanide to one aqueous phase which causes the oxidation of the primary electron donor complement of one BRC population [9,10], or ferrocytochrome c to the other side, to enable multiple electron turnovers of one pool of the reaction center population [9,10,14]. The agreement between the action spectrum of the photoresponses and the absorption spectrum of BRC in the membrane-forming solution used to be applied as a test of functional reconstitution of BRC into a membrane [9,10,14].

Instead of using hexane or decane extracts of BRC to assembly "solvent free" membrane, vesicles supplemented with BRC may be directly transferred into monolayers at the air-water interface [15]. Theoretically, if liposomes with BRC had been introduced into one compartment and only lipids into other one, asymmetry would have been achieved, but it was not and cytochrome c^{2+} had to be added to one compartment [15].

Functional reconstitution of the reaction centers from the higher plant thylakoid membranes, PS I or PS II was pursued using the fusion [11,12] or partial-fusion [11] approach. Addition of liposomes containing PS I to one side of the planar bilayer formed from negatively-charged lipids, in the presence of Ca^{2+} or Mg^{2+}, resulted in partial-fusion of liposomes and the planar membrane [11]. If phenazinemethosulfate (PMS), the artificial carrier of reducing equivalent, was present at the same side as liposomes were added or vitamin K_3 was present in the liposome membranes, photoeffects were observed.

Osmotic gradients across the liposome bilayers and across the planar membrane appeared to be an additional requirement for effective fusion of PS I- or PS II-containing liposomes with planar lipid membranes, which results in a functional reconstitution of the complexes [11,12]. Addition of PMS to the same side of the protein complex-modified planar lipid membrane as liposomes, enabled the generation of photopotentials. If PMS was present on the opposite side, very small photovoltage of opposite polarity to that one observed before was recorded.

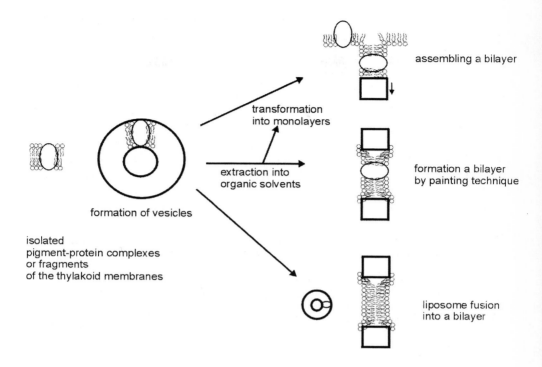

Fig. 2. Schematic representation of the methods of incorporation of photosynthetic pigment-protein complexes into planar lipid membranes showing intermediate steps.

This finding may suggest that PS I incorporated into planar lipid membrane by this method was distributed asymmetrically between two possible orientations.

Incorporation of PS II and LHCII into "painted" planar lipid bilayers was achieved also by fusion method as in the case of PS I [12,13].

The methods of incorporation of photosynthetic pigment-protein complexes into planar lipid membranes are summarized in Fig. 2.

3. EXAMPLES OF STUDYING THE PHOTOSYNTHETIC PIGMENT-PROTEIN COMPLEXES IN PLANAR LIPID MEMBRANES

Basically, the photosynthetic pigment-protein complexes may be classified into two separated and functionally distinctive classes: reaction centers (PS II, PS I and BRC) and light harvesting complexes (such as the largest light-harvesting complex of PS II, LHCII). Nevertheless, very recent findings reveal that also proteins believed to play functions solely in the electric charge transfer in the photosynthetic apparatus, such as cyt b$_6$-f complex, bind photosynthetic pigments, chlorophyll *a* and β-carotene in the conformation 9-*cis* [17,18]. The physiological function of these particular pigment molecules is still not known.

3.1. BRC

Major part of research on photosynthetic pigment-proteins carried out with the application of model planar lipid membranes concerns the electron flow across the reaction centers, in particular BRC. As early as in 1979 Schönfeld, Montal and Feher [14] succeeded to reconstitute photosynthetic reaction center from *Rhodopseudomonas sphaeroides* R-26 in planar lipid bilayers by apposing two reaction center-containing lipid monolayers. The fact that the membrane separated two water bulk phases containing cytochrome c and ubiquinone, the secondary electron donors and acceptors to BRC, respectively, enabled the observation of light-induced electric current and voltage across the protein-containing membrane. The wavelength dependence of the photoresponse matched the absorption spectrum of reaction centers, which is a proof of functional reconstitution of the protein complex. In 1982 Packham, Dutton and Mueller [10] demonstrated light-induced electric current and potential responses across planar lipid membranes formed with the mixture of phospholipids, containing incorporated reaction centers isolated also from the photosynthetic bacterium *Rhodopseudomonas sphaeroides*. The authors applied experimental conditions (the single short flashes of strong light and manipulation of concentration of extra ubiquinone or ferrocytochrome c) that made it possible to detect and analyze the single electron turnover in the BRC. One year latter Apell, Snozzi and Bachofen [9] were able to perform detailed kinetic analysis, based on electric measurements, of the electron transfer across the photosynthetic RC isolated from the same bacterium organism and reconstituted in bimolecular lipid membranes. In particular, the interaction between the electron-donating ferrocytochromes was analyzed in a quantitative manner and the mathematical model has been presented to describe the observed phenomena. The experimental setup for membrane formation was mounted on an optical bench and placed on a vibration-insulating table, which made in addition possible to perform precise optical measurements of the system. The linear dichroism measurements carried out enabled to determine the orientation angles between the axis normal to the plane of the membrane and the dipole transition moments of chromophores giving rise to three main absorption bands in the long-wavelength spectral region (at 757 nm, 801 nm and 860 nm). This kind of measurements would not be possible in a liposome model. Gopher and coworkers applied the model of BRC reconstituted in planar lipid bilayers to examine the effect of an applied electric field on the charge recombination kinetics [15]. The authors interpreted the photocurrents observed in a function of applied voltage, in terms of charge recombination process, between the reduced primary quinone (Q_A) and the oxidized bacteriochlorophyll donor. From a quantitative analysis of the voltage dependence on the recombination rate it was concluded that the component of the distance between the intermediate electron acceptor I (most probably pheophytin $Pheo_A$?) and Q_A along the axis normal to the membrane is about one-seventh of the thickness of the membrane. More recently, Salafsky, Groves and Boxer applied the modified planar lipid membrane model called supported lipid bilayer to study photosynthetic BRC [19]. The system was prepared by fusion of small unilamellar liposomes containing incorporated reaction centers with a glass surface. Interestingly such a procedure yields unidirectional orientation of protein complexes with the H-subunit facing inside and with the cytochrome c binding surface exposed to the bulk solution. The quality of the homogeneous supported lipid bilayers was characterized by epifluorescence microscopy.

3.2. PS I and PS II

Photosynthetic reaction centers isolated from higher plants were also a subject of studies carried out with application of planar lipid membrane model. Lopez and Tien [11] reported successful reconstitution of the PS I reaction center isolated from spinach chloroplasts into planar bilayer lipid membranes. The presence of exogenous carriers of reducing equivalents such as phenazinemethosulfate or vitamin K_3 in the bathing solution or in the lipid phase, respectively, as well as the presence of divalent cations (Ca^{+2} or Mg^{+2}) enabled the generation of light-induced voltage changes on the membrane and short-circuit photocurrent across protein-containing lipid bilayer. It was shown that the photocurrent action spectrum obtained in this work matched very well the action spectrum of the PS I reaction center. The results have also clearly supported the concept according to which the charge separation in the reaction center of PS I results in the generation of a potential difference across the thylakoid membrane. Gruszecki and coworkers [12] applied a planar lipid membrane model to study photoelectric phenomena associated with illumination of PS II with red light, absorbed exclusively by chlorophyll pigments, and blue light absorbed both by chlorophylls and carotenoids bound to the PS II pigment-proteins. PS II particles were incorporated to bimolecular lipid membranes from phospholipid liposomes containing fragments of the thylakoid membranes enriched with PS II, called BBY particles (Berthold-Babcock-Yokum particles [20]). The light-driven electron transport across PS II-containing membrane was observed upon the presence of ferricyanide in the bathing solution at one side of the membrane. This charge flow was interpreted as a vectorial electron transfer within PS II: between water and ferricyanide. The photocurrent across the membrane, generated by short-wavelength light was distinctly higher than the photocurrent generated by long-wavelength light, despite comparable number of light quanta absorbed by the system in these two spectral regions. Such a difference was discussed in terms of the cyclic electron transfer around PS II, regulated by excitations of accessory carotenoid pigments. These findings corroborate with the rate of photosynthetic oxygen evolution induced by red light and blue light, monitored by bare platinum electrode oxymetry of liposome-bound PS II particles [21].

3.3. LHCII

The main physiological function of photosynthetic antenna complexes, such as LHCII, is harvesting light quanta and transferring excitation energy towards the reaction centers. According to the current knowledge these pigment-proteins are not directly involved in the linear or cyclic photosynthetic electron transfer in the thylakoid membranes. Due to this fact, the examination of electric properties of planar lipid membranes containing incorporated antenna pigment-protein complexes was not a popular subject of studies. Also optical measurements of antenna complexes were rather carried out in other oriented systems such as monomolecular layers [5] or stretched polyacrylamide gel films [22]. Nevertheless, according to the findings of Jahns and coworkers [23,24], the largest light-harvesting pigment-protein complex of PS II is directly involved in the transfer of protons from the lumenal to the outer space within the chloroplasts, thus providing a framework to short-circuiting protons across the thylakoid membrane. Such a finding inspired Wardak and coworkers [13] to examine the effect of LHCII on ion transport across lipid membranes. The researchers applied the liposome model but also the model of planar lipid membranes to study this phenomenon. It appeared that specific conductivity of the LHCII-containing membrane was highly-dependent on the ionic strength of a bathing solution (see Fig. 3). Such a dependence was not observed in the control phospholipid membranes and indicated directly on the effect of ions on structural properties of LHCII in the lipid phase that influence distinctly the membrane conductivity. According to the 77 K chlorophyll *a* fluorescence emission measurements the increased concentration of ions was associated with aggregation of LHCII. The possible physiological importance of aggregation of LHCII in the thylakoid membranes was discussed in terms of providing a "safety valve" protecting the integrity of the thylakoid membranes under conditions of overexcitation and excessive accumulation of protons in the lumenal space of the thylakoid membranes. Currently, the photo-isomerization

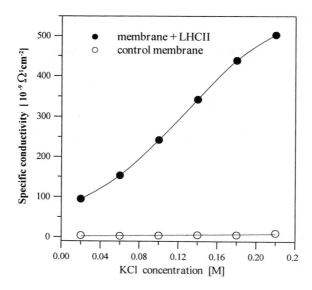

Fig. 3. Dependence of specific conductivity of planar lipid membrane containing incorporated LHCII or a control membrane (indicated) on concentration of KCl in the bathing solution. Figure based on the results reported in [13].

of violaxanthin, an accessory xanthophyll pigment in LHCII, is investigated as a mechanism that is most probably directly involved in a structural rearrangement of the antenna protein resulting in the alterations in the membrane permeability to ions.

ACKNOWLEDGEMENTS

Research on photosynthetic pigment-protein complexes in planar lipid membranes in the author' laboratory is currently financed by the State Committee of Scientific Research of Poland under the project 6P04A00219.

REFERENCES

[1] Govindjee and W.J. Coleman (Y.P. Abrol, P. Mohanty and Govindjee eds.), Photosynthesis. Photoreactions to Plant Productivity, Oxford & IBH Publishing Co. Pvt. Ltd., New Delhi, 1993, pp.83-108.
[2] W.I. Gruszecki, P. Kernen, Z. Krupa and R.J. Strasser, Biochim. Biophys. Acta, 1188 (1994) 235.
[3] G. Steinberg-Yfrach, P.A. Liddell, S.-Ch. Hung, A.L. Moore, D. Gust and T. A. Moore, Nature, 385 (1997) 239.
[4] P. Kernen, W.I. Gruszecki, M. Matula, P. Wagner, U. Ziegler and Z. Krupa, Biochim. Biophys. Acta, 1373 (1998) 289.

[5] W.I. Gruszecki, W. Grudzinski, A. Banaszek-Glos, M. Matula, P. Kernen, Z. Krupa and J. Sielewiesiuk, Biochim. Biophys. Acta, 1412 (1999) 173.

[6] S. Yu. Zaitsev, N. A. Kalabina, V.P. Zubov, G. Chumanov, D. Gaul and T.M. Cotton, Colloids and Surfaces A, 78 91993) 211.

[7] H. Ti Tien and A. Ottova-Leitmannova (eds.), Membrane Biophysics, Elsevier, Amsterdam, 2000.

[8] D.S. Berns, Photochem. Photobiol., 24 (1976) 117.

[9] H.J. Apell, M. Snozzi and R. Bachofen, Biochim. Biophys. Acta, 724 (1983) 258.

[10] N.K. Packham, P.L. Dutton, and P. Mueller, Biophys. J., 37 (1982) 465.

[11] J.R. Lopez and H. Ti Tien, Biochem. Biophys., 7 (1984) 25.

[12] W. I. Gruszecki, A. Wardak, W. Maksymiec, J. Photochem. Photobiol. B: Biol., 39 (1997) 265.

[13] A. Wardak, R. Brodowski, Z. Krupa, W.I. Gruszecki, J. Photochem. Photobiol. B: Biol., 56 (2000) 12.

[14] M. Schönfeld, M. Montal, and G. Feher, Proc. Natl. Acad. Sci. USA, 76 (1979) 6351.

[15] A. Gopher, Y. Blatt, M. Schönfeld, M.Y. Okamura, G. Feher, and M. Montal, Biophys. J., 48 (1985) 311.

[16] M. Schönfeld, M. Montal, and G. Feher, Biochemistry, 19 (1980) 1536.

[17] U. Bronowsky, S.-O. Wenk, D. Schneider, C. Jäger and M. Rögner, Biochim. Biophys. Acta, 1506 (2001) 55.

[18] J. Yan, Y. Liu, D. Mao, L. Li and T. Kuang, Biochim. Biophys. Acta, 1506 (2001) 182.

[19] J. Salafsky, J.T. Groves, and S.G. Boxer, Biochemistry, 35 (1996) 14773.

[20] D.A. Berthold, G.T. Babcock and C.F. Yocum, FEBS Lett., 134 (1981) 231.

[21] W.I. Gruszecki, K. Strzalka, A. Radunz, J. Kruk and G. H. Schmid, Z. Naturforsch., 50c (1995) 61.

[22] G. Cinque, R. Croce, A. Holzwarth and R. Bassi, Biophys. J., 79 (2000) 1706.

[23] P. Jahns, A. Polle and W. Junge, EMBO J., 7 (1988) 589.

[24] P. Jahns and W. Junge, Eur. J. Biochem., 193 (1990) 731.

Planar Lipid Bilayers (BLMs) and their Applications
H.T. Tien and A. Ottova-Leitmannova (Editors)

Chapter 36

Biochemical Applications of Solid Supported Membranes on Gold Surfaces: Quartz Crystal Microbalance and Impedance Analysis

Andreas Janshoff,[a] Hans-Joachim Galla,[b] and Claudia Steinem[c*]

[a]Institut für Physikalische Chemie, Johannes-Gutenberg Universität, Jakob-Welder-Weg 11, 55128 Mainz, Germany

[b]Institut für Biochemie, Westfälische Wilhelms-Universität, Wilhelm-Klemm-Str. 2, 48149 Münster, Germany

[c]Institut für Analytische Chemie, Chemo- und Biosensorik, Universität Regensburg, 93040 Regensburg, Germany

1. INTRODUCTION: SOLID SUPPORTED MEMBRANES

Since their inception in 1985 by Tamm and McConnell [1], solid supported lipid bilayers have been widely used as model systems for cellular membranes [2]. They have been applied in fundamental and applied studies of lipid assemblies on surfaces, to study the structure of membranes and membrane dynamics, lipid-receptor-interactions and electrochemical properties of membranes [3-5]. Several attempts have been made to apply solid supported membranes (SSM) in biosensor devices [6]. Planar lipid membranes can be formed on various surfaces, *i.e.* glass, silicon, mica or metal surfaces such as platinum or gold. Surface attachment of the lipids is typically achieved following two different strategies, the deposition of Langmuir-monolayers or more easily by self-assembly techniques. The major advantage of this membrane type is their attachment to a solid support, resulting in long-term and high mechanical stability. They can be combined with all kinds of surface sensitive techniques and electrochemical methods provided that the support is a conducting surface such as metals, inorganic materials (indium-tin oxide) or conducting polymers.

2. SOLID SUPPORTED MEMBRANES ON GOLD SURFACES

2.1. Functionalization of gold surfaces

If one is interested in performing electrochemical measurements on supported lipid membranes it is most straightforward to use noble metal surfaces and in particular gold surfaces as they are inert and can be easily modified by sulfur-bearing reagents. The best characterized and most widely studied self-assembly systems of sulfur-bearing compounds are long-chain thiols, $HS(CH_2)_nX$ [7, 8]. These compounds adsorb spontaneously from solution onto gold and for $n \geq 10$, they form densely packed defect-free monolayers that are stable against acids or bases. The choice of the functional group X determines the properties of the self-assembled organic surface. Hydrophobic alkanethiol monolayers ($X = CH_3$) are the basis for the formation of a second lipid monolayer atop it. pH-dependent groups ($X = COO^-$ or $X = NH_3^+$) confer hydrophilicity and charge to the surface allowing membrane attachment via electrostatic interactions. In this case, even short chain thiols ($n = 2$-3) have been successfully used [9]. A different approach is to use phospholipids or cholesterol, whose headgroups are modified by a hydrophilic thiol-terminated spacer [10, 11]. Based on these surface modifications lipid bilayers can be formed on gold surfaces. In the last couple of years more complex surfaces were created by coadsorbing two or more thiols with different X-functionalities or different chain lengths [12]. For patterning two or more different thiols are applied on a gold surface in a spatially defined manner using techniques such as microcontact printing and microwriting. The first thiol compound is spatially delivered on the surface while the second one fills up the residual areas. Micromachining and photolithography selectively remove thiols from a preformed monolayer and the resulting pattern is then filled by a second thiol species.

2.2. Formation of lipid membranes on gold surfaces

Supported membranes might be classified according to their different architectures on the gold surface: (A) hybrid bilayers composed of a lipid monolayer on a hydrophobic support (B) lipid bilayers on a hydrophilic or charged surface, and (C) covalently anchored lipid bilayers on a hydrophilic surface.

(A) For the formation of a hybrid bilayer composed of a lipid monolayer on a hydrophobic support, the gold surface is first functionalized by an alkanethiol, or a hydrophobic hairy-rod polymer [13]. Various techniques are available to deposit a second lipid monolayer atop the first hydrophobic layer. Apart from the Langmuir-Schäfer dip, where a lipid monolayer is transferred from the air-water interface to the hydrophobic solid support all techniques are based on the physisorption of lipids. Techniques comprise the deposition of a lipid monolayer from organic [14] or detergent solution [10], or from vesicle suspension [15], where the latter one leads to solvent and detergent free bilayers. These supported lipid layers are well suited for the investigation of reactions at membrane surfaces, *i.e.* the interaction of membrane-confined receptors anchored in one leaflet only with their ligands in solution, and association of peptides and peripheral proteins.

(B) Functionalization of gold surfaces with charged molecules enables one to adsorb oppositely charged lipid membranes onto the surface. It is also possible to attach negatively charged lipid membranes onto negatively charged functionalized gold surfaces through calcium ion bridges. However, electrostatic attachment of lipid membranes makes them susceptible to increasing ion strength resulting in a detachment of the membrane from the surface [9].

(C) Membranes can be covalently attached to solid supports by using lipids with hydrophilic spacer groups terminated by a thiol or disulfide functionality. These tethered lipid membranes have been shown to act as fluid lipid bilayers possessing an aqueous phase separating the membrane and the support. This is advantageous with respect to a functional reconstitution of large membrane spanning proteins. The spacer acts as an elastic buffer, decoupling the lipid layer from the solid surface and generating a water layer in between [16, 17].

3. PROTEIN MEDIATED ION TRANSPORT IN SOLID SUPPORTED MEMBRANES

The insertion of fully functional membrane spanning proteins is a major goal in the field of solid supported membranes with respect to biosensor applications. However, not all membrane systems are suited to host membrane proteins. Hybrid lipid bilayers on a hydrophobic support are in general not reasonable for the insertion of transmembrane proteins in a functionally active form. This is due to the rigid and often crystalline structure of those membranes preventing a proper insertion and folding of the protein component. Moreover, they are directly attached to the solid support thus the proper folding of the extramembraneous parts of membrane proteins is also abolished.

However, tethered membranes acting as fluid lipid bilayers with a hydrophilic aqueous phase separating them from the support enable one to incorporate ionophores such as valinomycin, alamethicin and gramicidin [6, 18-21]. Some examples are also given for a successful incorporation of complex proteins. For example, Salamon et al. [22] managed to insert rhodopsin and cytochrome c oxidase. Naumann et al. [23-25] incorporated the F_0F_1 ATPase from chloroplasts and cytochrome c oxidase in peptide-tethered lipid bilayers and Steinem et al. [26] inserted fully functional bacteriorhodopsin into lipid bilayers on gold surfaces.

Our primary goal was to incorporate ion-transporting peptides and proteins in lipid membranes attached to gold surfaces and investigate the specific ion transport mediated by those components by impedance spectroscopy.

3.1. The technique: Impedance spectroscopy

Impedance spectroscopy (IS) is a versatile technique for investigating the electrical properties of a variety of different materials, which may be ionic, semiconducting or even insulating [27]. It provides information about both the materials' bulk phase (*e.g.* conductivity, dielectric constant) and their inner and outer interfaces (*e.g.* capacitance of the interfacial region and derived quantities). The method is based on measuring the frequency dependent impedance of the electrochemical system of interest followed by its analysis applying equivalent circuits modeling the electrical properties of the system. There are basically two different approaches to acquire impedance data that differ with respect to the excitation signal. The first and most common technique (continuous wave IS) is to measure the impedance in the frequency domain by applying a single sinusoidal voltage of small amplitude with a defined frequency to the system and recording the corresponding current. Applying a discrete set of different frequencies provides a frequency spectrum of the system's impedance. The second approach (Fourier Transform IS) makes use of a transient excitation signal that is applied to the electrochemical system. The system's response is monitored in the time domain and subsequently transformed to the frequency domain by Fourier Transformation providing the frequency dependent impedance of the system.

A   B

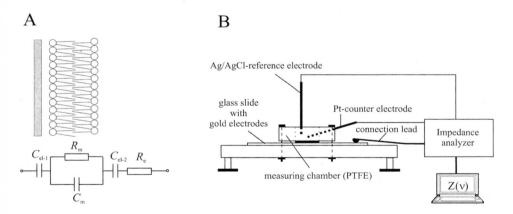

Fig. 1. A Equivalent circuit representing a solid supported lipid bilayer adjacent to an electrolyte solution. **B** Typical setup used for impedance analysis of solid supported lipid bilayers immobilized on gold surfaces.

One interesting and important application of impedance spectroscopy in biochemistry and biophysics is the characterization of solid supported lipid bilayers on a conductive surface. Impedance analysis allows studying the formation process of solid supported bilayers and investigating their long-term-stability and their stability towards experimental conditions such as temperature, pH, etc. [9]. By applying appropriate equivalent circuits membrane specific parameters can be readily extracted from impedance spectra (Fig. 1A). Moreover ion transport through these bilayers mediated by ion carriers or channel proteins is readily accessible by IS.

A typical setup used for the investigation of solid supported membranes on gold surfaces is depicted in Fig. 1B. It basically consists of a gold electrode serving as the working electrode and a counter electrode, which might be either a second gold electrode equally functionalized or a platinized platinum wire. Gold electrodes may be easily prepared by evaporating gold on a glass substrate through a mask. The glass substrate is mounted in a cell made of PTFE sealed with an O-ring. The electrodes are connected to an impedance analyzer connected to a personal computer. To apply an external d.c. potential a reference electrode is included such as an Ag/AgCl electrode.

Impedance analysis is ideally suited to determine the coverage of electrodes and the thickness of dielectric layers. Since alkanethiols form almost defectfree and insulating monolayers the resulting impedance spectra can be represented by a simple capacitance C_m in series to an Ohmic resistance R_e representing the resistance of the electrolyte and the wire connections (Fig. 2).

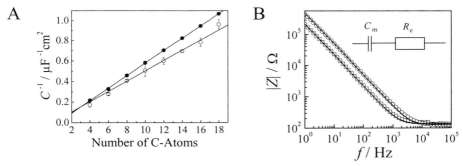

Fig. 2. A (O) Inverse capacitance as a function of alkanethiol chain length as determined from impedance analysis of chemisorbed thiols. (●) Calculated inverse capacitance taking the chain length (*all-trans* conformation) and a dielectric constant of 2 into account. The deviation is probably due to a small number of defects. **B** Impedance spectra together with the fitting results using the displayed equivalent circuit of an octanethiol monolayer (□, C_m = (2.4 ± 0.2) µF/cm^2) and a lipid bilayer composed of octanethiol and POPC (O, C_m = (1.0 ± 0.2) µF/cm^2).

Fitting the parameters of the equivalent circuit to the data results in characteristic values for the capacitance of the alkanethiol monolayer indicative of the formation of a complete insulating lipid layer. As the thickness of an alkanethiol monolayer increases its capacitance decreases assuming a simple plate condenser (Fig. 2A). These results indicate that the representation of an alkanethiol monolayer immobilized on a gold surface by a plate condenser is a good first approximation. Impedance analysis also allows one to follow the formation of a second phospholipid monolayer atop the first hydrophobic alkanethiol monolayer (Fig. 2B).

3.2. Impedance analysis of ion transport mediated by peptides and proteins

3.2.1. Gramicidin D mediated ion transport in solid supported membranes

To garner information about the selective transport of ions through solid supported membranes we focused on the development of a solid supported membrane system on gold surfaces containing the channel-forming peptide gramicidin D [19]. Gramicidin D, an antibiotic synthesized by *Bacillus brevis*, is a linear pentadecapeptide forming a channel in lipid bilayers consisting of two antiparallel oriented monomers bound to each other by six hydrogen bonds. The resulting dimer has a length of 26 Å sufficient to span the hydrophobic part of a membrane. The van-der-Waals size of the pore is 4 Å in diameter and strongly selective for monovalent cations. The sequence of conductivity is determined to be $H^+ > NH_4^+ > Cs^+ > Rb^+ > K^+ > Na^+ > Li^+$.

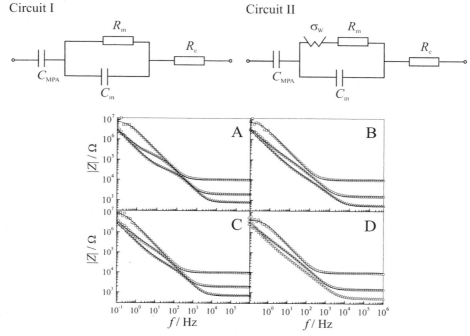

Fig. 3. Impedance spectra ($|Z|$) of a gramicidin D (1 mol%) doped DODAB bilayer electrostatically immobilized on a MPA-monolayer in the absence (□) and presence of 7.4 mM (○) and 21.8 mM (△) of the corresponding cation. **A)** LiCl, **B)** NaCl, **C)** KCl, **D)** CsCl. The continuous lines are results of the fitting procedure with equivalent circuit I (without alkali cations) and II (with alkali cations), respectively, shown in the figure.

The peptide was reconstituted into large unilamellar vesicles (LUV) of dimethyldioctadecylammoniumbromide (DODAB), which were fused onto a negatively charged monolayer of 3-mercaptopropionic acid (MPA). Formation of solid supported bilayers was monitored by impedance spectroscopy. First, the specific capacitance of the MPA-monolayer was determined by fitting a series connection of a capacitance C_{MPA} and a resistance R_e to the data. Typical values for MPA-monolayers lie in the range of C_{MPA} = 9-10 μF/cm^2. Second, large unilamellar DODAB vesicles were added resulting in bilayer formation within an hour (Fig. 3 and 4).

Selective cation transport through the incorporated gramicidin channels was followed by means of impedance spectroscopy. Altogether, the impact of four different monovalent cations in various concentrations on the bilayer was investigated. In Fig. 3 impedance spectra of a DODAB bilayer with 1 mol% gramicidin D in the absence and presence of LiCl, NaCl, KCl, and CsCl in the bulk phase are presented.

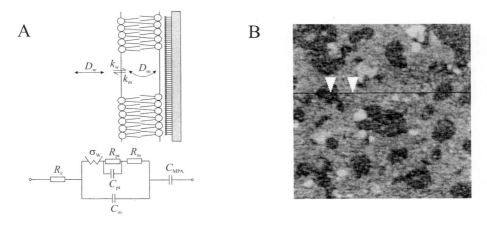

A   B

Fig. 4. A Electrochemical model of ion transport facilitated by gramicidin in a DODAB bilayer electrostatically adsorbed on a MPA monolayer on gold. The scheme shows processes that are involved in passive ion transport through a solid supported lipid bilayer taking into account aqueous diffusion and first order interfacial kinetics. Subscript w indicates the water and m the membrane phase. k_m and k_w denote the rate constants, D_m and D_w the diffusion constants of the ions in the membrane and water phase. The corresponding equivalent circuit consists of the electrolyte resistance R_e, the membrane capacitance C_m in parallel to the Warburg impedance σ_w and the membrane resistance R_m together with the phase transfer resistance R_{pt} and capacitance C_{pt}. The capacitance C_{MPA} represents the membrane/solid interface, which is the capacitance of the preformed self-assembled monolayer of MPA. The phase transfer resistance R_{pt} may be merged with the membrane resistance R_m, since C_{pt} can be neglected. **B** Atomic force microscopy image of a DODAB bilayer electrostatically attached to a MPA-monolayer chemisorbed on gold. The height of the bilayer is 5.4 nm. Image size: 8 × 8 μm^2.

It is obvious that the presence of alkali cations alters the impedance of the electrochemical system considerably. For a quantitative analysis an equivalent circuit adapted from a work of de Levie was applied (Fig. 4A) [28, 29]. In a comprehensive theoretical study de Levie derived an equivalent circuit for the ion transport of membrane soluble ions through planar lipid membranes with five parameters that are related to ion transport, C_m, R_m, and C_{pt}, R_{pt} accounting for the phase transfer of the ions as well as σ_w, the Warburg impedance representing the mass transport from the bulk solution to the electrode.

This model had to be modified to account for the solid support, while neglecting the phase transfer of the ions. Moreover, a complete description of the electrochemical system requires an additional capacitance C_{MPA} accounting for the self-assembled monolayer and the electrolyte resistance R_e (Fig. 3, circuit II)). Impedance spectra were fitted according to this equivalent circuit keeping C_m and C_{MPA} fixed during the fitting routine. Notably, it turned out that the values for σ_w are orders of magnitude larger than expected from theory. One has to take into account the dramatically reduced area as ion transport takes place merely through a gramicidin dimer. Plotting R_m vs. the inverse cations' bulk concentration results in a linear relation as supposed from theory (Fig. 5). The slope decreases in the following sequence: $Li^+ > Na^+ > K^+ > Cs^+$ corresponding to the known gramicidin sequence of conductivity. The increasing conductivity of the larger alkali-cations arises from the decreasing hydration enthalpy.

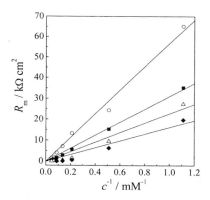

Fig. 5. Dependence of the resistance R_m on the reciprocal alkali cation concentration in the bulk phase. The plot shows the nearly linear relation between R_m and c^{-1} dependent on the alkali cation; (○) LiCl, (■) NaCl, (△) KCl, (◆) CsCl.

Monovalent ions exhibiting ion radii larger than 347 pm are excluded from the gramicidin channel, which was demonstrated by addition of $N(CH_3)_4Cl$ in a concentration range of 1-22 mM. No decrease in R_m was observed upon addition of tetramethylammonium ions being 347 pm large.

Even though the evaluation of impedance spectra with equivalent circuits is discussed controversially, the application of the modified semiempirical network derived by de Levie leads to cation concentration dependent R_m values confirming the selectivity sequence of alkali cations published for gramicidin as well as the exclusion size of the channel rationalizing the procedure.

3.2.2. Insertion of an ion carrier in solid supported membranes

In contrast to gramicidin as a small membrane channel, which can be incorporated into gel-like lipid membranes without loosing channel activity, the functionality of an ion carrier requires high membrane fluidity to allow for diffusion of the carrier in the membrane phase. Several attempts have been made to improve the flexibility of the first chemisorbed monolayer. Lang et al. synthesized phospholipids based on phosphatidic acid linked to one, two or three ethoxy groups with a terminal thiol anchor allowing chemisorption of those lipids [10]. We synthesized a phospholipid based on phosphatidylethanolamine linked via succinic acid to four ethoxy groups (Scheme 1, compound **1**) [20].

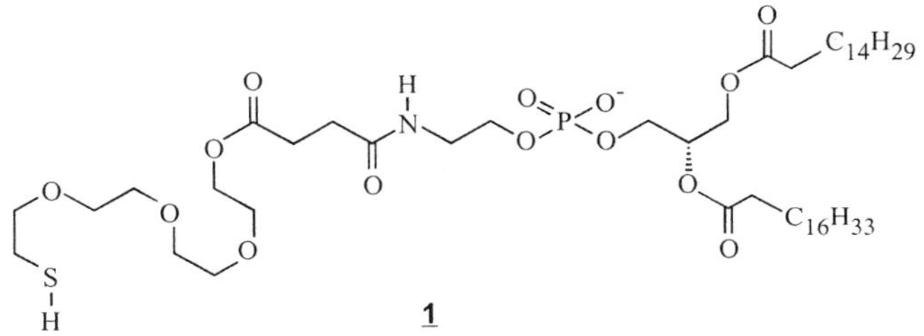

Scheme 1. Structure of compound **1**.

This very long hydrophilic spacer is terminated by a thiol group allowing the formation of self-assembled monolayers on gold, which were characterized by XPS and impedance spectroscopy. Starting with these monolayers, solid supported lipid bilayers with a second POPC monolayer were produced and valinomycin was inserted by adding it to the preformed bilayer [20]. Valinomycin is a cyclic depsipeptide consisting of three identical units. It mediates the transport of alkali and alkaline earth cations in organic films or solvents of low polarity and acts as an ion carrier in lipid membranes. It is known to selectively facilitate the transport of K^+ and Rb^+ over Li^+, Na^+ or alkaline earth cations by forming a three-dimensional complex with the cations.

Impedance analysis of monolayers formed from compound **1** reveals that the resistance of such monolayers is too low to be detected. However, fusion of POPC vesicles onto the hydrophobic functionalized monolayer results in an impedance spectrum that can be analyzed by the equivalent circuit shown in Fig. 5. The mean value obtained by six independent measurements of a lipid bilayer composed of compound **1** and POPC in 10 mM Tris, 50 mM $N(CH_3)_4Cl$, pH 7.0 amounts to $C_m = (1.0 \pm 0.2)$ $\mu F/cm^2$ and $R_m = (11000 \pm 1000)$ Ω cm^2. Compared to SSM composed of alkanethiols and POPC the membrane resistance is considerably smaller. This might be explained in terms of the larger flexibility of the self-assembled monolayer and bilayer resulting in less ordered films. Incorporation of the carrier into the preformed lipid bilayers was achieved by adding valinomycin, dissolved in dimethyl sulfoxide to the aqueous phase, which decreased the magnitude of impedance in a frequency range of 50-500 Hz, in which the membrane resistance is predominately determined.

A   B

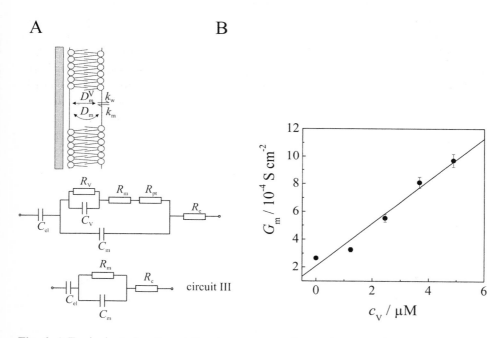

Fig. 6. A Equivalent circuit modeling ion transport via membrane carriers. $\leftrightarrow$ symbolizes the diffusion of the free uncharged carrier in the membrane with the diffusion constant D_m^V and $\curvearrowright$ the diffusion and migration of the ion-carrier-complex in the a.c. field with the diffusion constant D_m. $\rightleftharpoons$ represents the interfacial reaction of the cation and the carrier which is assumed to be of first order with the rate constants k_w and k_m. The corresponding equivalent circuit consists of the capacitance of the gold electrode C_{el}, the membrane capacitance C_m, the membrane resistance R_m, the resistance due to the interfacial kinetics R_{pt}, the capacitance and resistance C_V and R_V due to the diffusion of the unloaded carrier in the membrane and the electrolyte resistance R_e. The equivalent circuit III is the simplified model used in this study to analyze the obtained impedance spectra. **B** Dependence of the membrane conductance G_m on the valinomycin concentration in the bulk phase in the presence of 25 mM KCl.

In order to evaluate the impedance spectra we applied the equivalent circuit shown in Fig. 6A (circuit III) to the experimental data, which was derived from adapting an equivalent circuit, based on the theory of de Levie (Fig. 6A) [28, 30]. The resulting conductance G_m increases almost linearly with increasing valinomycin concentrations in the bulk as expected from theory (Fig. 6B). However, the membrane capacitance C_m also increases upon addition of valinomycin. The insertion of the peptide presumably causes an increase in the dielectric constant of the lipid bilayer.

To investigate the cation selectivity of valinomycin embedded in the solid supported membrane the influence of different K^+- and Na^+-concentrations in solution on R_m in the absence and in the presence of valinomycin was investigated. In order to determine the increase in conductance solely caused by valinomycin we assumed that the membrane resistance R_m^0, which is elicited by defects in the membrane, is parallel to the resistance R_m^V representing the transport of cations in the membrane as mediated by valinomycin. Hence, the conductance due to the transport of cations facilitated by valinomycin G_m^V for each alkali cation concentration in solution can be evaluated using the following expression:

$$G_m^V(c_w^+) = \left(G_m^{0,V}(c_w^+) - G_m^{0,V}(0)\right) - \left(G_m^0(c_w^+) - G_m^0(0)\right) \tag{1}$$

$G_m^{V,0}$ is the conductance obtained from the impedance spectra in the presence of valinomycin and G_m^0 in the absence of valinomycin. Fig. 7 shows the conductance G_m^V versus different concentrations of potassium and sodium cations in solution.

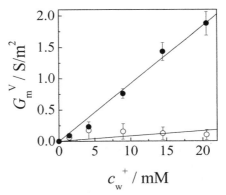

Fig. 7. Plot of the conductance dependent on sodium (O) and potassium ions (●) in the aqueous phase. In agreement with theory, a linear relationship between the potassium ion concentration and the specific conductance was observed.

The membrane conductance increases significantly in the presence of potassium ions and only slightly in the presence of sodium ions. A linear relation between the cation concentration and the conductance G_m^V is expected from theory and confirmed by the experiments. The slope for potassium ions is about twelve times larger than that for sodium ions. This is in accordance to reported observations that valinomycin is highly selective for potassium.

3.2.3. Reconstitution of an anion channel of Clavibacter michiganense ssp. nebranskense in solid supported membranes

Besides small antibiotic peptides it is even more interesting to insert proteins into SSM to study their channel activity. In this context, it is also conceivable to use SSM to sense the toxicity of a protein due to the protein's channel activity. Such a phytotoxin is produced by the *Corynebacterium Clavibacter michiganense ssp. nebraskense* causing Goss' wilt and blight in *Zea mais* [31, 32]. In general, members of the genus *Clavibacter* have been described to produce phytotoxins, which were classified as high molecular mass polysaccharides (*Clavibacter michiganense ssp. michiganense*) [33] or glycolipids (*Clavibacter rahtayi*) [34]. However, the presence of polysaccharides did not explain the discrete action on chloroplasts and their membranes. Schürholz et al. [35, 36] were the first, who systematically searched for toxic activities excreted into the culture medium of *Clavibacter michiganense ssp. nebraskense* and identified a membrane-active component that forms anion selective channels in planar lipid bilayers. The relative molecular mass of the *Clavibacter* anion channel was determined to be 25 kDa by functional reconstitution of the channel protein from SDS gels. Voltage clamp measurements demonstrated the specific anion selectivity ($Cl^- > F^- > SCN^- > I^- > C_2O_4^- > SO_4^{2-}$) and that channel activity can be abolished by protease treatment and anion specific channel inhibitors such as indanyloxyacetic acid (IAA-94) and by OH^--ions (pH >10). Voltage clamp experiments also revealed that single channel conductance increases exponentially with voltages up to 200 mV saturating at 250 mV and that the channels are closed at negative potential relative to the side of insertion. The average number of open channels also increases with the applied potential.

According to Schürholz et al. [36] the *Clavibacter* anion channel (CAC) inserts spontaneously into planar lipid membranes when culture fluid of this species is added to the aqueous phase. Our major objective was to study the CAC by incorporating the channel into solid supported lipid bilayers immobilized on gold surfaces and thoroughly characterize the electrical parameters of the systems by means of impedance spectroscopy. Two different membrane systems were investigated, one based on the electrostatic immobilization of DODAB membranes on a negatively charged MPA-monolayer and one based on a chemisorbed thiol lipid with a hydrophilic spacer and a second physisorbed phospholipid monolayer [37, 38]. Both systems led to similar results and as an example, the results obtained from the electrostatically immobilized DODAB membranes will be summarized here.

Similar to the experiments performed by Schürholz et al. [35, 36] preformed DODAB bilayers were incubated with the culture fluid containing CAC while applying a d.c. potential of 50 mV and impedance spectra were taken after 15 min. For data reduction, equivalent circuit II shown in Fig. 3 was used. From the obtained parameters we concluded that the anion channel is inserted into the SSM resulting in an increased conductance of the membrane. To confirm this hypothesis, an experiment was performed in which the protein channel was digested by adding a protease before adding the culture fluid to the lipid bilayer. Indeed, no significant increase in conductance was observed after treatment of the membrane with the culture fluid. For a more detailed analysis of channel activity, voltage dependent measurements were performed. The most characteristic feature of the CAC is the exponential increase in single channel conductance with a rising applied d.c. voltage [36]. We investigated changes in the electrical membrane parameters (R_m and C_m) with an increasing d.c. potential. It turned out that in the absence of CAC, C_m and $G_m = R_m^{-1}$ do not alter with increasing d.c. potentials. However, after adding CAC to the lipid bilayer the membrane resistance R_m and the Warburg impedance σ_W decrease exponentially with increasing potential indicative of the formation of voltage dependent ion channels.

In Fig. 8A the conductance G_m is plotted vs. the applied d.c. potential. To solely account for the change in conductance induced by CAC we defined the conductance of CAC (G_{CAC}) as:

$$G_{CAC}(V) = \left(G_{m-1}(V) - G_{m-1,0}\right) - \left(G_{m-2}(V) - G_{m-2,0}\right) \qquad (2)$$

G_{m-1} and G_{m-2} are the conductances of DODAB bilayers with and without CAC dependent on the applied d.c. potential. $G_{m-1,0}$ and $G_{m-2,0}$ are the conductances of the DODAB bilayers before addition of CAC to account for variations in R_m of different bilayer preparations. Each conductance was determined by fitting R_m to the corresponding impedance spectrum. Calculating G_{CAC} according to Eq. (2) leads to the plot shown in Fig. 8B.

To further support the idea that indeed CAC is incorporated into solid supported DODAB bilayers we investigated the influence of increasing anion concentrations, *i.e.* chloride, on the conductance of the lipid bilayer and increasing CAC concentration in the bulk. In both cases, an increase in chloride concentration and CAC concentration in the bulk phase, respectively leads to an increase in conductance. The conductance G_{CAC} depends on the chloride-concentration and the CAC-concentration, respectively in a linear fashion in the observed concentration range. Schürholz et al. [35] demonstrated that the *Clavibacter* anion channel can be inactivated by chloride selective channel inhibitors. We decided to use diphenylamine-2-carboxylic acid (DPC) previously shown by single channel analysis to act on the *cis*-side that corresponds to the aqueous phase facing side. G_{CAC} was monitored with and without DPC dependent on the applied d.c. potential (Fig. 9).

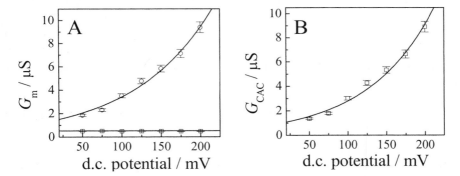

Fig. 8. A Conductance G_m dependent on the applied d.c. potential (□) in the absence and (○) presence of CAC. **B** Conductance G_{CAC} obtained according to Eq. (2) vs. the applied d.c. potential. The solid lines are the results of fitting an exponential function to the data.

The addition of DPC results in a considerable decrease in G_{CAC} indicative of a selective inhibition of channel activity. A remaining channel activity was detected due to non-blocked channels. This result rules out that the increase in membrane conductance is caused by non-specific defects, which are formed due to the incubation of the solid supported membrane in the culture fluid of *Clavibacter michiganense ssp. nebraskense*.

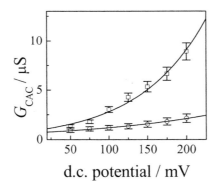

Fig. 9. Conductance G_{CAC} dependent on the applied d.c. potential. (□) represents the data obtained in the presence of CAC and (○) those in the presence of CAC together with 0.13 mM DPC. The overall chloride concentration was 4.6 mM. The solid lines are the results of fitting an exponential function to the data.

The experiments outlined above clearly reveal that the presence and activity of the *Clavibacter* anion channel in the culture fluid of *Clavibacter michiganense ssp. nebraskense* can be monitored and also thoroughly characterized by using solid supported lipid bilayers electrostatically immobilized on planar gold electrodes. Besides biophysical characterization on a long-term stable membrane system the detection of toxic channel activity in bacterial fluids by means of solid supported membranes combined with impedance spectroscopy might also be a promising step for the development of new biosensor devices scanning a solution with unknown composition for toxic ingredients.

4. SPECIFIC PROTEIN BINDING TO SOLID SUPPORTED MEMBRANES

Lipid membranes immobilized on a solid support are highly ordered and thus well suited to orient membrane confined receptor molecules such as receptor lipids or proteins on surface. The surface density of receptor molecules can be readily adjusted so that steric crowdence on the surface is minimized. Moreover, it is well established that a fully membrane covered solid substrate prevents non-specific adsorption of proteins very efficiently. Hence, these membranes are frequently used to study membrane confined adsorption phenomena. In recent years, the quartz crystal microbalance technique evolved as a labelfree method to study the interaction of proteins with lipid membranes in a time-resolved manner [39].

4.1. The technique: the quartz crystal microbalance

The quartz crystal microbalance (QCM) technique in non-biological applications has been well known for many decades. The core component of the device is a thin quartz disk, which is sandwiched between two evaporated metal electrodes and commonly referred to as a thickness shear mode (TSM) resonator. As this quartz crystal is piezoelectric in nature, an oscillating potential difference between the surface electrodes leads to corresponding shear displacements of the quartz disk. This mechanical oscillation responds very sensitively to any changes that occur at the crystal surfaces. It was Sauerbrey [40], who first established in 1959 that the resonance frequency of such a quartz resonator alters linearly when a foreign mass is deposited on the quartz surface in air or vacuum (Eq. (3)):

$$\Delta f = \frac{-2 f_0^2 \Delta m}{A \sqrt{\rho_q \mu_q}} = S_f \Delta m ,$$
(3)

where f_0 denotes the fundamental resonant frequency, A the electrode area, ρ_q the density of the quartz ($\rho_q = 2.648$ g/cm^3) and μ_q the piezoelectric stiffened shear modulus of the quartz ($\mu_q = 2.947 \cdot 10^{11}$ dyne/cm^2). S_f denotes the integral mass sensitivity, which amounts to 0.036 Hz·cm^2/ng. From these resonance frequency readings it was possible to detect mass deposition on the quartz surface in the sub-nanogram regime and accordingly the device was named a *microbalance*.

For the majority of bioanalytical applications it was, however, necessary to follow adsorption processes under physiological conditions most notably in a liquid environment. The development of high-gain oscillator circuits that can overcome the viscous damping of the shear oscillation under liquid loading eventually paved the way to apply the QCM technique to many biological problems. Nowadays QCM measurements are frequently used to follow a multitude of biomolecular recognition processes like, for instance, antigen-antibody binding or ligand-receptor interactions [39, 41]. In these applications either the ligands or the receptors are immobilized on the quartz surface and the corresponding counterpart is offered from solution. From readings of the characteristic shear wave parameters like the resonance frequency, it is then possible to extract binding constants and also binding kinetics with outstanding time resolution. Recently, it was also demonstrated that the QCM provides invaluable data about the interaction of adherent cells with solid substrates. [39, 42-44].

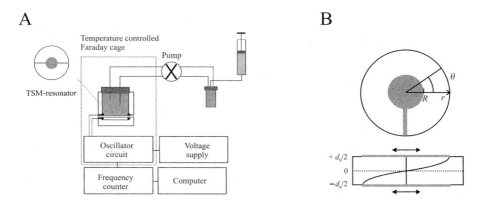

Fig. 10. A Schematic diagram of the experimental setup used for QCM measurements. The quartz resonator is mounted in a crystal holder, which itself is placed in a temperature controlled Faraday cage. The quartz resonator is connected as the frequency-controlling element to an electronic oscillator circuit that compensates damping losses and thereby drives the shear oscillation. Addition of analyte is performed using a syringe, while buffer flows continuously over the quartz plate. **B** Schematic representation of the shear wave in the quartz plate.

Since the electrodes of the quartz plates are commonly made of gold, solid supported bilayers can be readily prepared on quartz plates. A typical setup used to monitor adsorption of proteins on solid supported membranes by means of the quartz crystal microbalance technique is depicted in Fig. 10. The core component is an AT-cut quartz disk with a fundamental frequency of 5 MHz (d = 14 mm). Gold electrodes are evaporated on both sides of the quartz crystal to allow excitation of the quartz oscillation and to serve as substrate for membrane immobilization (Fig. 11A). The quartz plate is mounted in a holder made of PTFE. The oscillator circuit capable of exciting the quartz crystal to its resonance frequency in a liquid environment has been developed in our laboratory and supports the resonance frequency of minimum impedance.

The crystal holder is designed to minimize parasitic damping due to mounting losses and to prevent the occurrence of longitudinal waves. Crystal holder and oscillator circuit are placed in a temperature controlled Faraday cage, while the resonance frequency is recorded with a frequency counter outside the cage. A personal computer is used to control the measurement and record the data.

Using this active oscillator mode without amplitude control only one parameter, the shift of the resonance frequency can be monitored. A time resolution of practically 0.1-1 s can be readily achieved with a frequency resolution of 0.2-0.5 Hz for a 5 MHz resonator. In Fig. 11B a typical time course of protein binding to a solid supported membrane obtained from frequency recording is shown. Upon specific binding of the protein to the surface the resonance frequency of the quartz plate decreases and levels off at the point of binding equilibrium. From the obtained data, the resonance frequency shift Δf can be extracted.

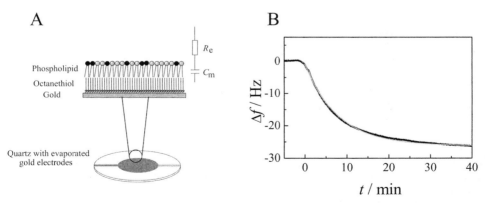

Fig. 11. A Schematic drawing of a functionalized quartz plate. **B** Time course of the frequency shift of a 5 MHz quartz plate upon addition of a protein to a functionalized lipid membrane and specific adsorption of the protein to the membrane. The solid line is the result of fitting the parameters of Eq. (7) to the data.

4.2. Specific adsorption of proteins to solid supported membranes

Protein-receptor interactions at lipid membranes, for example ganglioside-toxin interactions play an essential role in biological processes. The first contact of a protein, virus or bacterium with its receptor at a biological membrane initiates a variety of reactions within and at the cell membrane. Solid supported membranes immobilized in a highly ordered fashion on gold surfaces are well suited for studying adsorption processes by means of the quartz crystal microbalance technique. The quartz crystal microbalance in combination with solid supported membranes composed of an alkanethiol monolayer and a second lipid monolayer obtained by vesicle fusion allows the determination of thermodynamic and kinetic parameters of protein-receptor couples without labeling the protein. The success of bilayer preparation can be easily followed using impedance spectroscopy. Direct control of bilayer preparation guarantees high reproducibility of the adsorption processes and minimizes the amount of non-specific adsorption.

4.2.1. Specific binding of a lectin to solid supported membranes

To provide an example for the suitability of this approach the interaction of peanut agglutinin (PNA) with gangliosides monitored by quartz crystal microbalance technique is described. The lectin PNA from *Arachis hypogaea* (M = 110 kDa) is composed of four identical subunits each with a molecular weight of 27-28 kDa exhibiting a specific binding site for carbohydrates of the composition β-Gal$p$-(1→3)-GalNAc.

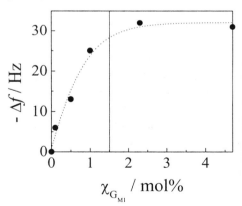

Fig. 12. Dependence of the frequency shift on the G_{M1}-content in the POPC monolayer. A PNA concentration of 2 µM was added and the resonance frequency recorded. The solid line indicates the minimal value that is necessary for a complete PNA-coverage on surface, while the dotted line represents a spline.

To investigate the interaction of PNA with a specific receptor on surface a lipid monolayer physisorbed on an octanethiol monolayer is doped with different concentrations of the receptor lipid G_{M1} (Scheme 2) and the frequency shift Δf of the quartz crystal is monitored upon addition of PNA [45].

Fig. 12 demonstrates that a dopant concentration of 1.5 mol % of the receptor lipid G_{M1} is sufficient to achieve maximum protein surface coverage. A calculation of the theoretical value of the minimum number of necessary G_{M1} molecules within the lipid matrix assuming a homogenous distribution of the receptor lipids and correct values for the geometry of the protein leads to a value of 1.5 mol %. A comparison of the theoretical value with the experimentally obtained one implies that the utmost monomeric protein coverage on surface has to be close to one. Similar maximum protein coverage using a lipid matrix doped with 2 mol % of the receptor lipid was corroborated by Ebato et al. [46] who investigated the streptavidin-biotin couple with the quartz crystal microbalance.

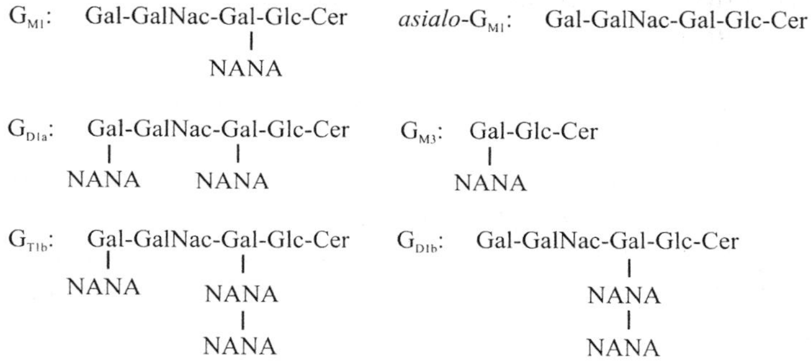

G_{M1}:   Gal-GalNac-Gal-Glc-Cer   *asialo*-G_{M1}:   Gal-GalNac-Gal-Glc-Cer
                          |
                      NANA

G_{D1a}:   Gal-GalNac-Gal-Glc-Cer G_{M3}:   Gal-Glc-Cer
                    | | |
                  NANA   NANA   NANA

G_{T1b}:   Gal-GalNac-Gal-Glc-Cer G_{D1b}:   Gal-GalNac-Gal-Glc-Cer
                    | | |
                  NANA   NANA   NANA
                                  | |
                               NANA   NANA

Scheme 2. Schematic representation of various gangliosides.

In order to determine the binding constant of a protein-receptor couple the frequency shift Δf has to be monitored dependent on the protein concentration c_0 in solution. Assuming that the binding sites on the surface are energetically equivalent and that there is a homogeneous distribution of the receptor lipids, the binding constant K_a can be obtained by fitting the parameters of a Langmuir adsorption isotherm (Eq. (4)) to the data:

$$\Theta(t) = \frac{K_a c_0}{1 + K_a c_0} \tag{4}$$

The established binding constants present information about the required chemical structure of the receptor essential for an appropriate binding, as demonstrated by the adsorption of PNA to G_{M1} and *asialo*-G_{M1}. While the binding constant of PNA to G_{M1} is $K_a = (0.83 \pm 0.04) \cdot 10^6$ M^{-1}, it is determined to be $K_a = (6.5 \pm 0.3) \cdot 10^6$ M^{-1} for *asialo*-G_{M1}, almost a factor of 10 larger. This difference is attributed to the fact that *N*-acetylneuramic acid of G_{M1} sterically hinders PNA binding.

This is an example of how the molecular structure of a receptor molecule can be illuminated by varying the receptor molecules embedded in the lipid membrane using the quartz crystal microbalance. Upon quantifying the inhibition of binding in solution this method is capable of clarifying carbohydrate structures that play a pivotal role for receptor function. Monitoring the frequency shift upon binding of PNA to G_{M1} in the presence of an inhibitor allows determining the binding constant K_I of the inhibitor in solution [45]. A prerequisite for the determination of K_I is an appropriate ratio between K_I and K_a. If the binding constants have similar orders of magnitude, an exact determination of the binding constant K_I is practicable since the frequency changes continuously with the inhibitor concentration in solution. If there are several magnitudes between K_I and K_a, the protein binds either almost unaffectedly to the surface or not at all.

4.2.2. Interaction of bacterial toxins with membrane-embedded receptor molecules

Cholera toxin, the enterotoxin of *Cholera vibrio*, is an 87 kDa protein composed of six subunits (AB_5) in which the five identical B subunits form a pentagonal ring surrounding the A subunit. The B subunits harbor the binding sites for the cell surface receptor. Using solid supported membranes, binding of cholera toxin to G_{M1} containing POPC-membranes was quantified [47]. Adsorption isotherms of cholera toxin binding to POPC monolayers doped with either 10 mol% G_{M1} or 10 mol% *asialo*-G_{M1} were recorded and association constants were obtained by fitting Eq. (4) to the data. For G_{M1} doped lipid layers, a binding constant of $K_a = (1.8 \pm 0.1) \cdot 10^8$ M^{-1} with a maximum frequency decrease of $\Delta f_{max} = (111 \pm 2)$ Hz was obtained, while for *asialo*-G_{M1} doped bilayers a significant lower binding constant of $K_a = (1.0 \pm 0.1) \cdot 10^7$ M^{-1} with a maximum frequency decrease of $\Delta f_{max} = (34 \pm 2)$ Hz was obtained (Table 1) [48]. The higher affinity of cholera toxin to G_{M1} clearly demonstrates the importance of the sialinic acid in the receptor structure. Simple electrostatic interaction driven by the negative charge of the sialinic acid could be excluded since no adsorption was detected on G_{M3} doped lipid layers.

In a similar approach the receptor structure for tetanus toxin binding was investigated [47, 48]. Different gangliosides as receptor lipids serving as binding sites for tetanus toxin were analyzed. The exotoxin of *Chlostridiuum tetanii* is known as one of the most effective toxins from bacteria. It consists of two subunits, fragment B ($M = 99$ kDa) and C (52 kDa). Fragment C harbors the specific receptor binding site. Four different gangliosides G_{M3}, G_{D1a}, G_{D1b} and G_{T1b} (Scheme 2) were incorporated into the outer leaflet of a solid supported lipid bilayer by fusing POPC vesicles doped with 10 mol% of the corresponding ganglioside on a hydrophobic octanethiol monolayer chemisorbed on the gold electrode of the quartz plate. Concentration dependent measurements allow the determination of binding constants applying Eq. (4). The results are summarized in Table 1.

Table 1
Bindings constants and maximum frequency shifts of the interaction of cholera toxin and the C-fragment of tetanus toxin to various gangliosides.

Toxin	Ganglioside	K_a / M^{-1}	- Δf_{max} / Hz
Cholera toxin	G_{M3} / 10 mol%	—	—
	G_{M1} / 10 mol%	$1.8 \cdot 10^8$	111
	asialo-G_{M1} / 10 mol%	$1.0 \cdot 10^7$	34
C-fragment	G_{M3} / 10 mol%	—	—
	G_{D1a} / 10 mol%	$2.4 \cdot 10^6$	28
	G_{D1b} / 10 mol%	$3.0 \cdot 10^6$	99
	G_{T1b} / 10 mol%	$1.7 \cdot 10^6$	66

4.2.3. Adsorption of Raf to solid supported membranes

The Ras/Raf/MEK/ERK cascade plays a pivotal role in the regulation of cell growth and –differentiation. One major constituent of this cascade is Raf, a member of the serine/threonine protein kinase family that mediates signals from the cell surface to the nucleus via activation of the mitogen activated protein kinase [49-51].

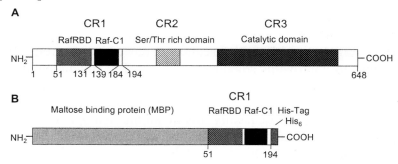

Fig. 13. **A** Schematic drawing of the structure of full-length c-Raf-1. The protein is composed of three conserved regions CR1, CR2 and CR3. CR1 comprises the Raf-Ras-binding domain (RafRBD) and the cysteine-rich domain termed Raf-C1. CR2 is a serine-threonine rich domain, which is a phosphorylation site. CR3 is the catalytic domain located near the C-terminus. **B** Schematic representation of the protein construct used in this study. CR1 including amino acids 51-193 comprising RafRBD and RafC1 is fused to maltose binding protein. A His-tag composed of six histidine residues is added at the C-terminus for a facilitated purification procedure of the protein construct.

A critical step in the activation of Raf is its interaction with membrane-anchored Ras, a small GTPase, via its Raf-Ras binding domain (RafRBD). In its active GTP-bound state, Ras recruits Raf to the plasma membrane in vivo, which is the first step in Raf-activation [52, 53]. However, Ras-interaction alone is not sufficient to activate the Raf-kinase, other events such as Raf-phosphorylation may also be required [51].

Though there is little known about the mechanism of Raf-activation, the structure of Raf is well resolved. Three isoforms of Raf can be distinguished in mammals; A-Raf, B-Raf and C-Raf-1, the latter being the best studied one [54]. C-Raf-1 consists of a N-terminal non-catalytic region and a C-terminal kinase domain (Fig. 13A). If the N-terminal region is missing (v-Raf oncoprotein) the kinase is constitutively active indicating that the N-terminal part locks the kinase in an inactive conformation and is thus responsible for its regulation [55]. The non-catalytic N-terminus of Raf is composed of two regions (CR1 and CR2) that are highly conserved between different members of the Raf family. The first conserved region (CR1) consists of two modules that are referred to as RafRBD (amino acids 51-131) and a C1-type cysteine rich domain (Raf-C1, amino acids 139-184). While Ras binding to RafRBD is well understood, the role of the Raf-C1 cysteine-rich domain has remained elusive.

The objective of our study was to develop a quantitative in vitro assay for the determination of thermodynamic and kinetic data of the Raf-C1 interaction with lipid bilayers employing solid supported bilayers in conjunction with the quartz crystal microbalance technique. With this assay it was possible to investigate the influence of variations in the lipid bilayer composition on the binding behavior of Raf-C1 [56].

For all experiments, a protein construct composed of amino acids 51-194 including the RafRBD and Raf-C1 fused to a maltose binding protein (MBP) at the N-terminus to improve its solubility and a His-Tag containing six histidine residues at the C-terminus (Fig. 13B) was used and termed MBP-Raf-C1. Lipid bilayers composed of a first chemisorbed octanethiol monolayer and a second phospholipid monolayer subsequently fused onto the first one were prepared on the gold surface of a quartz plate for binding experiments. Time resolved frequency shifts were monitored upon addition of the protein. For a quantitative analysis of the kinetics of protein binding we assumed that the rate-limiting step is the adsorption of the protein on the surface, while diffusion-limiting steps are neglected and that all individual protein binding sites are independent of each other, *i.e.* no cooperativity occurs. The binding kinetics can then be ascribed by Eq. (5):

$$\Theta(t) = \frac{c_0}{K_d + c_0}\left(1 - \exp\left(-\frac{t}{\tau}\right)\right), \tag{5}$$

where $\Theta(t)$ is the surface coverage at any given time, K_d the dissociation constant of the monomolecular reaction and c_0 the protein concentration of the bulk. τ is defined as the lifetime (Eq. (6)):

$$\tau(c_0) = \frac{1}{k_{on}c_0 + k_{off}}.$$

(6)

with k_{on}, the rate constant of association and k_{off}, the rate constant of dissociation. Since the resonance frequency shift Δf is proportional to the amount of adsorbed material, Eq. (5) can be rewritten as (Eq. (7)):

$$\Delta f(t) = \Delta f_e \left(1 - \exp\left(-\frac{t}{\tau}\right)\right),$$

(7)

where Δf_e is the equilibrium frequency shift for a given bulk protein concentration c_0. By fitting the parameters of Eq. (7) to the data the equilibrium frequency shift and the lifetime τ can be obtained.

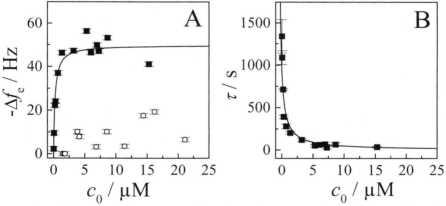

Fig. 14. A Adsorption isotherm of MBP-Raf-C1 (■) and MBP (O). A lipid bilayer immobilized on a 5 MHz quartz plate composed of DMPC/DMPS (7:3) was used for each experiment. Δf_e and τ were obtained from fitting those parameters to the time course of the resonance frequency shift after addition of the corresponding amount of protein using Eq. (7). By assuming a Langmuir adsorption isotherm the dissociation constant K_d and the maximum frequency shift Δf_{max} were extracted. **B** τ vs. c_0 plot. The rate constants of association and dissociation of MBP-Raf-C1 binding were determined by fitting Eq. (6) to the data. The values are summarized in Table 2.

To obtain dissociation constants K_d and rate constants of association k_{on} and dissociation k_{off}, protein concentration dependent measurements were performed on lipid bilayers composed of octanethiol and DMPC/DMPS (7:3). The concentration of MBP-Raf-C1 was varied between 0-16 µM and the equilibrium resonance frequency Δf_e and τ were extracted. The results are shown in Fig. 14 as Δf_e vs. c_0 (A) and τ vs. c_0 plots (B).

By varying the phosphatidylserine content we addressed the question, whether the DMPS content and hence the effective negative surface charge density affects the thermodynamic and kinetic parameters of Raf-C1 binding to the bilayer. We monitored binding isotherms and concentration dependent lifetimes for bilayers containing 10 mol%, 30 mol% and 100 mol% DMPS. The thermodynamic and kinetic data are summarized in Table 2.

Table 2
Thermodynamic and kinetic data of adsorption of MBP-Raf-C1 to various lipid bilayers immobilized on gold surfaces of 5 MHz quartz plates.

χ_{DMPS} / mol%	$-\Delta f_{max}$ / Hz	K_d / 10^{-7} M	k_{on} / 10^3 (M s)$^{-1}$	k_{off} / 10^{-4} s^{-1}	K_d^* / 10^{-7} M
10	19 ± 3	1.5 ± 0.2	1.4 ± 0.1	5.5 ± 0.9	3.9 ± 0.7
30	50 ± 3	2.4 ± 0.1	2.0 ± 0.1	10.5 ± 0.3	5.4 ± 0.3
100	96 ± 6	8.3 ± 0.3	1.8 ± 0.2	10 ± 1	5.8 ± 0.8

* K_d is obtained from the kinetic constants of adsorption and desorption.

The most significant difference between the three bilayer systems under investigation is the increase in Δf_{max} with increasing DMPS content. The dissociation constants only slightly increase with increasing DMPS content and no considerable change in the rate constants of association and dissociation were detected. In summary, solid supported membranes allow one to quantify thermodynamics and kinetics of lipid-protein interactions.

5. CONCLUSIONS

Solid supported membranes on gold surfaces have been shown to be well suited to investigate protein mediated ion transport through membranes, and ligand-receptor interactions at the membrane surface delivering thermodynamic and kinetic data for different types of lipid-protein couples. Impedance spectroscopy mainly monitors conductance variations, whereas the quartz crystal microbalance allows to follow biomolecular recognition events at the membrane surface.

Together with new developments in the design of nanoscaled biofunctionalized lipid arrays [57, 58] solid supported membranes will open up a new avenue within the fields of bioanalytics and nanobiotechnology including chip technology. Moreover, the possible applications will expand to more complex systems using whole cells and their mimics as sensor devices [59]. Indeed, both techniques already allow studying cellular processes like cell-surface adhesion, cell-cell-interaction and even signal transduction processes if morphological cell changes are involved [43, 44].

REFERENCES

[1]   L.K. Tamm and H.M. McConnell, Biophys. J., (1985) 105.
[2]   E. Sackmann, Science, 271 (1996) 43.
[3]   H.T. Tien and A.L. Ottova, Electrochimica Acta, 43 (1998) 3587.
[4]   S. Heyse, T. Stora, E. Schmid, J.H. Lakey and H. Vogel, Biochim. Biophys. Acta, 85507 (1998) 319.
[5]   M.L. Wagner and L.K. Tamm, Biophys. J., 79 (2000) 1400.
[6]   B.A. Cornell, V.L.B. BrachMaksvytis, L.G. King, P.D.J. Osman, B. Raguse, L. Wieczorek and R.J. Pace, Nature, 387 (1997) 580.
[7]   J.I. Siepmann and I.R. Mcdonald, Thin Films, 24 (1998) 205.
[8]   D.S. Karpovich, H.M. Schessler and G.J. Blanchard, Thin Films, 24 (1998) 43.
[9]   C. Steinem, A. Janshoff, W.-P. Ulrich, M. Sieber and H.-J. Galla, Biochim. Biophys. Acta, 1279 (1996) 169.
[10]   H. Lang, C. Duschl and H. Vogel, Langmuir, 10 (1994) 197.
[11]   B. Raguse, V. Braach-Maksvytis, B.A. Cornell, L.G. King, P.D.J. Osman, R.J. Pace and L. Wieczorek, Langmuir, 14 (1998) 648.
[12]   Y. Cheng, S.D. Ogier, R.J. Bushby and S.D. Evans, Rev. Mol. Biotechnol., 74 (2000) 159.
[13]   H. Sigl, G. Brink, M. Seufert, M. Schulz, G. Wegner and E. Sackmann, Eur. Biophys. J., 25 (1997) 249.
[14]   E.-L. Florin and H.E. Gaub, Biophys. J., 64 (1993) 375.
[15]   A.L. Plant, M. Gueguetchkeri and W. Yap, Biophys. J., 67 (1994) 1126.
[16]   W. Knoll, C.W. Frank, C. Heibel, R. Naumann, A. Offenhäusser, J. Rühe, E.K. Schmidt, W.W. Shen and A. Sinner, Rev. Mol. Biotechnol., 74 (2000) 137.
[17]   E. Sackmann and M. Tanaka, Trends Biotechnol., 18 (2000) 58.
[18]   A.T.A. Jenkins, R.J. Bushby, N. Boden, S.D. Evans, P.F. Knowles, Q.Y. Liu, R.E. Miles and S.D. Ogier, Langmuir, 14 (1998) 4675.
[19]   C. Steinem, A. Janshoff, H.-J. Galla and M. Sieber, Bioelectrochem. Bioenerg., 42 (1997) 213.
[20]   C. Steinem, A. Janshoff, K. von dem Bruch, K. Reihs, J. Goossens and H.-J. Galla, Bioelectrochem. Bioenerg., 45 (1998) 17.
[21]   T. Stora, J.H. Lakey and H. Vogel, Angew. Chem. Int. Ed., 38 (1999) 389.
[22]   Z. Salamon and G. Tollin, Biomed. Health Res., 20 (1998) 186.
[23]   R. Naumann, A. Jonczyk, R. Kopp, J. van Esch, H. Ringsdorf, W. Knoll and P. Gräber, Angew. Chem. Int. Ed., 34 (1995) 2056.
[24]   R. Naumann, A. Jonczyk, C. Hampel, H. Ringsdorf, W. Knoll, N. Bunjes and P. Gräber, Bioelectrochem. Bioenerg., 42 (1997) 241.
[25]   R. Naumann, E.K. Schmidt, A. Jonczyk, K. Fendler, B. Kadenbach, T. Liebermann, A. Offenhäusser and W. Knoll, Biosens. Bioelectronics, 14 (1999) 651.
[26]   C. Steinem, A. Janshoff, F. Höhn, M. Sieber and H.-J. Galla, Chem. Phys. Lipids, 89 (1997) 141.
[27]   J.R. Mcdonald (1987) Impedance spectroscopy, John Wiley & Sons, New York.

[28]   R. de Levie, Adv. Chem. Phys., 37 (1978) 99.

[29]   R. de Levie, N.G. Seidah and H. Moreira, J. Membr. Biol., (1974) 17.

[30]   R. de Levie, Chem. Inter. Electrochem., 58 (1975) 203.

[31]   D.S. Wysong, A. Vidaver, K., H. Stevens and D. Sternberg, Plant. Dis. Rep., 57 (1973) 291.

[32]   M.C. Metzler, M.J. Laine and S.H. De Boer, FEMS Microbiologie Letters, 150 (1997) 1.

[33]   P.V. Rai and G.A. Strobel, Phytopathology, 59 (1968) 47.

[34]   P. Vogel, B.A. Stynes, W. Coackley, G.T. Yeoh and D.S. Petterson, Biochem. Biophys. Res. Commun., 105 (1982) 835.

[35]   T. Schürholz, M. Wilimzig, E. Katsiou and R. Eichenlaub, J. Membrane Biol., 123 (1991) 1.

[36]   T. Schürholz, L. Dloczik and E. Neumann, Biophys. J., 64 (1993) 58.

[37]   A. Michalke, H.-J. Galla and C. Steinem, Eur. Biophys. J., 30. (2001) 421.

[38]   A. Michalke, T. Schürholz, H.-J. Galla and C. Steinem, Langmuir, 17 (2001) 2251.

[39]   A. Janshoff, H.-J. Galla and C. Steinem, Angew. Chem. Int. Ed., 39 (2000) 4004.

[40]   Sauerbrey, Z. Phys., 155 (1959) 206.

[41]   A. Janshoff and C. Steinem, Sensors Update, 9 (2001) 313.

[42]   J. Wegener, A. Janshoff and H.-J. Galla, Eur. Biophys. J., 28 (1998) 26.

[43]   J. Wegener, J. Seebach, A. Janshoff and H.-J. Galla, Biophys. J., 78 (2000) 2821.

[44]   J. Wegener, A. Janshoff and C. Steinem, Cell Biochem. Biophys., 34 (2000) 121.

[45]   A. Janshoff, C. Steinem, M. Sieber and H.-J. Galla, Eur. Biophys. J., 25 (1996) 105.

[46]   H. Ebato, J.N. Herron, W. Müller, Y. Okahata, H. Ringsdorf and P. Suci, Angew. Chem. Int. Ed., 31 (1992) 1087.

[47]   A. Janshoff, C. Steinem, M. Sieber, A. el Bâya, M.A. Schmidt and H.-J. Galla, Eur. Biophys. J., 26 (1997) 261.

[48]   C. Steinem, A. Janshoff, J. Wegener, W.-P. Ulrich, W. Willenbrink, M. Sieber and H.-J. Galla, Bionsens. Bioelectronics, 43 (1997) 339.

[49]   G. Daum, I. Eisenmann-Trappe, H.-W. Fries, J. Troppmair and U.R. Rapp, Trends in Biochem. Sci., 19 (1994) 474.

[50]   J. Avruch, X. Zhang and K. J.M., Trends Biochem. Sci., 19 (1994) 279.

[51]   D.K. Morrison and R.E. Cutler, Cur. Op. Cell Biol., 9 (1997) 174.

[52]   D. Stokoe, S.G. Macdonald, K. Cadwallader, M. Symons and J.F. Hancock, Science, 264 (1994) 1463.

[53]   S.J. Leevers, H.F. Paterson and C.J. Marshall, Nature, 369 (1994) 411.

[54]   W. Kolch, Biochem. J., 351 (2000) 289.

[55]   G. Heidecker, M. Huleihel, J.L. Cleveland, W. Kolch, T.W. Beck, P. Lloyd, T. Pawson and U.R. Rapp, Mol. Cell. Biochem., 10 (1990) 2503.

[56]   A. Eing, A. Janshoff, C. Block, H.-J. Galla and C. Steinem, ChemBioChem, (2002) (in press).

[57]   S. Künneke and A. Janshoff, Angew. Chem. Int. Ed., 41 (2002) 314.

[58]   A. Janshoff and S. Künneke, Eur. Biophys. J., 29 (2000) 549.

[59]   B. Pignataro, C. Steinem, H.-J. Galla, H. Fuchs and A. Janshoff, Biophys. J., 78 (2000) 487.

Planar Lipid Bilayers (BLMs) and their Applications
H.T. Tien and A. Ottova-Leitmannova (Editors)
© 2003 Elsevier Science B.V. All rights reserved.

Chapter 37

Simultaneous Measurement of Spectroscopic and Physiological Signals from a Planar Bilayer System

Yoshiro Hanyu

Institute of Molecular and Cell Biology
National Institute of Advanced Industrial Science and Technology
1-1-1 Higashi, Tsukuba, 305-8566 Japan

1. Introduction

Voltage-gated ion channels such as sodium and potassium channels of nerve represent a particularly important and interesting class of membrane proteins [1,2]. Characterizing the molecular dynamics of these membrane proteins is one of the greatest challenges in the field of structural biology [3]. The channel conformations, closed or open, are regulated by the membrane potential, which implies that a voltage sensor must link the voltage to these conformational changes. Site-directed mutagenesis has provided evidence for the involvement of charged segments of the channel in the gating event [4]. However, dynamical changes in structure with gating event have not been resolved. New approaches to resolve the structural changes of this class of protein are needed. We think that it is necessary to study the structure and function of channel proteins simultaneously, to elucidate the structural changes that are directly related to the function of channel proteins. Technical problems make it very difficult to measure the conformations of channel proteins while precisely controlling the functional state. Since the experiments considered the structure and function of channels separately; structure was studied in a liposome system and function was studied in a planar lipid bilayer system. In the liposome system, the functional state of the channels is not properly controlled. On the other hand, the planar lipid bilayer system is good for studying the function of ion channels [5], such as their activation curve and ionic selectivity, but it yields little information on their structure. It is necessary to develop the new experimental system for studying the conformational changes of channel proteins. That system should enable the simultaneous measurement of the structure and function of channel proteins, to elucidate the structural changes with gating. To detect the structural changes involved in gating, precise spatio-temporal control of the functional state is necessary while studying the changes in structure. Thus, both the function and the structure of an ion channel in a planar bilayer membrane must be measured simultaneously, while the functional state of the channel is properly controlled. We have developed the system designed to optically record ion channel gating in a planar lipid bilayer [6]. We can measure the ionic current (function) and fluorescence emission (structural change) of an ion channel in an artificial planar lipid bilayer, while controlling the membrane potential. In this study, a 22-mer peptide with a sequence identical to that of the S4 segment of the electric eel sodium channel domain IV [7] was used as a model amphiphilic helix that can form ion channels. A fluorescence-labeled S4 peptide was synthesized and incorporated into the bilayer. Structural changes in the S4 peptide were observed while precisely controlling its functional state with a voltage clamp method. This study is the first to present results on the movement of functioning channel-forming peptides in a planar lipid bilayer.

2. Experimental Methods

2.1. Chamber for the Optical Measurement

A specially designed chamber was used for the simultaneous electrical and optical measurements [6]. The chamber consists of two blocks of poly-tetrafluoroethylene (PTFE) and PTFE film with a cover glass for the optical windows. A piece of PTFE film (Nichiryo, Japan) 25μm thick, with a hole 120μm in diameter (made by an electric arc), was clamped in the chamber so that the hole was in the center. The chambers on each side of the hole were pretreated with 0.5μL of 1% n-hexadecane in hexane. To form a folded bilayer [8], buffer was added to each side of the chamber so that the level was below the hole in the PTFE film. A solution of lipid in hexane (5 mg/mL) was spread on the top of the buffer solution and left for a few minutes to allow the hexane to evaporate. The buffer level in both compartments was raised to form a planar lipid bilayer. The formation of the bilayer was monitored by capacitance measurements. Peptide was added to the *cis* side of the chamber, with stirring. When fluorescence-labeled lipids and spin-labeled lipids were used, they were first mixed with an asolectin solution in hexane before being applied to the top of the buffer solution.

2.2. Experimental Set-up for Simultaneous Measurement

The fluorescent emissions from the planar lipid bilayer were measured in the newly developed experimental system using the chamber described above. A schematic diagram of the experimental system is shown in Figure 1. The excitation light was focused on the planar lipid bilayer with an objective lens (Olympus UPlanFl/4 N/A = 0.13, Japan). The optical system was adjusted so that only an area of the planar bilayer with a diameter of 80μm was irradiated. This was achieved by monitoring the focused beam on the planar bilayer directly with a microscope. The fluorescent emissions were collected through the objective lens and sent to the photomultiplier (R643S, Hamamatsu Photonics, Japan). Any fluorescence from other areas was blocked by pinholes placed after the objective lens. Scattered light was eliminated by a dichroic mirror and two long-pass glass filters (RG550). The intensity of fluorescence was measured by photon-counting methods (PC-545AS, NF Electrics Inc., Japan). The emission spectrum was measured with a multichannel analyzer (IRY-700, Princeton Instruments, NJ).

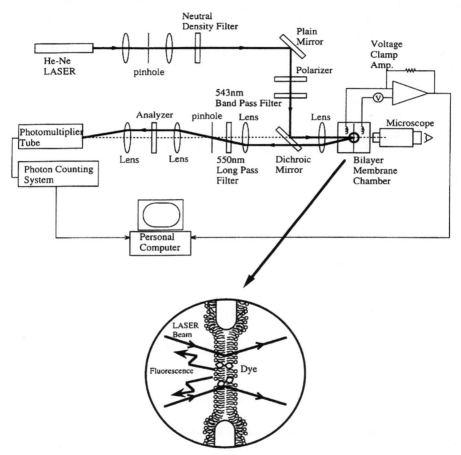

Figure 1   A schematic diagram of the experimental system. Excitation light from a He-Ne laser was focused with an objective lens. The focal point was monitored with a microscope and positioned in the restricted area of the folded membrane. The backward emitted florescence was collected and measured by photon-counting method.

The data were stored in and analyzed with a PC computer. Electrical measurements were made with Axopatch200A (Axon Instruments Inc., CA) using the chamber described above. The voltage was applied through an agar bridge and a Ag/AgCl electrode on the *cis* side of the cell, the side to which the peptides were added. The *trans* side was virtually grounded through an agar bridge and a Ag/AgCl electrode.

2.3. Measuring Fluorescence from a Planar Bilayer

The fluorescent emissions from a N-6(6-tetramethylrhodaminethio carbamoyl) 1,2-dihexadecananoyl-sn-glycero-3-phosphoethanolamine (TR-PE) containing planar lipid bilayer were measured. The intensity of fluorescence and spectral shape from a planar bilayer were measured successfully, with a high signal-to-noise ratio (Figure 2). The system can measure the intensity of fluorescence from a restricted area of the planar bilayer, with a diameter of 70μm and a focal depth of 15μm. The fluorescent emissions were observed around 590 nm using a single 550-nm long-pass filter in the emission light-collecting path (Figure 2A). Scattered light from the excitation laser beam appeared as a sharp line at 543 nm. Two 550-nm long-pass filters eliminated almost all of the scattered light, while conserving the shape of the fluorescence spectrum. Therefore, this system was able to measure fluorescent emission exclusively.

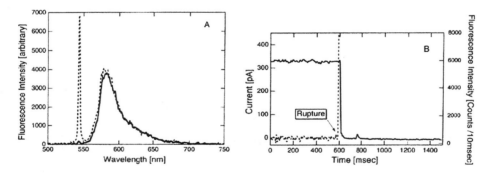

Figure 2   Fluorescent emission from a planar lipid bilayer containing TR-PE. Lipids:Asolectin was added to the *cis* side, and a mixture of asolectin- TR-PE 100:1 was added to the *trans* side. The buffer solution on both sides was 100 mM sodium chloride and 50 mM Tris buffer (pH = 7.5). (A) Emission spectra through one and two 550-nm long-pass glass filters are shown in dashed and solid lines, respectively. (B) Change in the intensity of fluorescence after rupturing the membrane. The electric current and intensity of fluorescence are shown with dashed and solid lines, respectively. Reproduced with permission from *Biochemistry* 1998, 37, 15376-15382. Copyright 1998 Am.Chem.Soc.

The background level of the signals was checked in the following manner: The membrane was ruptured by applying 1.3 V for 1 ms. When the membrane ruptured, the conductance became very large and therefore the current was overloaded. When this happened, the intensity of fluorescence decreased quickly to less than 5% of the original value, as shown in Figure 2B. This shows that more than 95% of the measured fluorescence comes from the planar lipid bilayer. The low background signal was achieved by optimizing the optical system. The diameters of the pinholes in the excitation light pathway were adjusted so that the diameter of the area of the planar membrane irradiated was 70μm. The diameters of the pinholes in the emission light pathway were also adjusted so that only emission light from the irradiated area was detected. Thus, the optical components were adjusted to work confocally. No photobleaching was observed with the intensity of excitation used in the experiment. The power of the He-Ne laser (0.75 mW) used as a light source for excitation was attenuated by 10^{-3}-10^{-4} with spatial and neutral density filters. At these excitation intensities, nonspecific conductance leaks due to the photodynamic effect [9] were not observed. We used an objective lens with a relatively small numerical aperture (N/A = 0.13), so that any optical aberration produced by the thick layer of water in the chamber would be reduced. The small aperture reduces the optical aberration, but measurement of the fluorescent emissions is poor. If the appropriate compensation for an objective lens with a large aperture was used, the system would be able to measure a larger amount of the fluorescent emission with a higher signal-to-noise ratio, due to the smaller focal depth.

3. Detection of Movement of Ion Channel with Gating

3.1. Formation of Voltage-Dependent Ion Channel by R-S4

A 22-mer peptide with a sequence identical to that of the S4 segment of the electric eel sodium channel domain IV [7] was synthesized by a solid-phase method. This peptide has the following structure: Arg-Val-Ile-Arg-Leu-Ala-Arg-Ile-Ala-Arg-Val-Leu-Arg-Leu-Ile-Arg-Ala-Ala-Lys-Gly-Ile-Arg. Its N-terminal was labeled with tetramethyl rhodamine isothiocyanate. The rhodamine-labeled S4 peptide of the sodium channel was denoted R-S4. The current-voltage relationship (I-V curve) shown in Figure 3 was obtained with a ramp voltage clamp after adding R-S4 and perfusing the solution. The relationship was independent of the ramp speed applied in the range from 0.5 mV/s to 10 mV/s. As shown in Figure 3, R-S4 formed a voltage-dependent ion channel. Ionic current was only generated with depolarization. The dependence of current on voltage above 100 mV could be fitted by an exponential relationship very well. The current increased *e*-fold for every 19.6 mV increase in voltage.

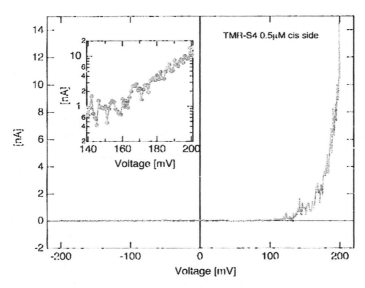

Figure 3   Current-voltage (I-V) relationship for R-S4. The buffer solution on the both sides of chamber was 1 M sodium chloride and 50 mM Tris buffer (pH = 7.5). The voltage-dependent current (I) could be fitted with the following exponential relationship: $I = I_0 \exp(V/Ve)$, where Ve is 19.6 mV.

The I-V curve for R-S4 was quantitatively different from that for S4 (Tosteson et al. [10]). They reported that the conductance for positive potentials (above 40 mV) increased e-fold for every 10 mV increase in potential. The number of charged residues in R-S4 is less than that of S4 because N-terminus of R-S4 is labeled with tetramethyl rhodamine isothiocyanate. This may cause the differences in I-V curve. These differences also may be due to the different lipids used; they used a negatively charged phospholipid, while we used a neutral phospholipid. Brullemans et al. showed that the Ve of the S4-S45 peptide for the negatively charged phospholipid is 9 mV, while that for the neutral phospholipid is 19 mV [11].

The current-voltage relationship was monitored with ramps until a stable relationship was obtained. It took about 10 min to reach a steady current-voltage state. The conductance was increased until the steady state was reached. This increase in conductance was not due to the binding of aqueous R-S4 to the membrane, because the unbound soluble R-S4 was quickly removed after the addition of R-S4 by perfusion. We measured the intensity of fluorescence from R-S4 during this period using the experimental system in Figure 1. The intensity of fluorescence also increased with time. The increase in the intensity of fluorescence means that the quantum yield of rhodamine of membrane-bound

The background level of the signals was checked in the following manner: The membrane was ruptured by applying 1.3 V for 1 ms. When the membrane ruptured, the conductance became very large and therefore the current was overloaded. When this happened, the intensity of fluorescence decreased quickly to less than 5% of the original value, as shown in Figure 2B. This shows that more than 95% of the measured fluorescence comes from the planar lipid bilayer. The low background signal was achieved by optimizing the optical system. The diameters of the pinholes in the excitation light pathway were adjusted so that the diameter of the area of the planar membrane irradiated was 70μm. The diameters of the pinholes in the emission light pathway were also adjusted so that only emission light from the irradiated area was detected. Thus, the optical components were adjusted to work confocally. No photobleaching was observed with the intensity of excitation used in the experiment. The power of the He-Ne laser (0.75 mW) used as a light source for excitation was attenuated by 10^{-3}-10^{-4} with spatial and neutral density filters. At these excitation intensities, nonspecific conductance leaks due to the photodynamic effect [9] were not observed. We used an objective lens with a relatively small numerical aperture (N/A = 0.13), so that any optical aberration produced by the thick layer of water in the chamber would be reduced. The small aperture reduces the optical aberration, but measurement of the fluorescent emissions is poor. If the appropriate compensation for an objective lens with a large aperture was used, the system compensation for an objective lens with a large aperture was used, the system would be able to measure a larger amount of the fluorescent emission with a higher signal-to-noise ratio, due to the smaller focal depth.

3. Detection of Movement of Ion Channel with Gating

3.1. Formation of Voltage-Dependent Ion Channel by R-S4

A 22-mer peptide with a sequence identical to that of the S4 segment of the electric eel sodium channel domain IV [7] was synthesized by a solid-phase method. This peptide has the following structure: Arg-Val-Ile-Arg-Leu-Ala-Arg-Ile-Ala-Arg-Val-Leu-Arg-Leu-Ile-Arg-Ala-Ala-Lys-Gly-Ile-Arg. Its N-terminal was labeled with tetramethyl rhodamine isothiocyanate. The rhodamine-labeled S4 peptide of the sodium channel was denoted R-S4. The current-voltage relationship (I-V curve) shown in Figure 3 was obtained with a ramp voltage clamp after adding R-S4 and perfusing the solution. The relationship was independent of the ramp speed applied in the range from 0.5 mV/s to 10 mV/s. As shown in Figure 3, R-S4 formed a voltage-dependent ion channel. Ionic current was only generated with depolarization. The dependence of current on voltage above 100 mV could be fitted by an exponential relationship very well. The current increased *e*-fold for every 19.6 mV increase in voltage.

R-S4 becomes larger. It suggests that the rhodamine is moving from the surface to the inside of the membrane since the quantum yield of rhodamine is higher inside the membrane. When the intensity of the fluorescence did not change, the I-V curve remained unchanged. These results show that R-S4 penetrates the membrane after binding to the membrane. An increase of ionic current over a similar length of time also occurs with the channel-forming peptide melittin [12], although the adsorption of melittin was very quick, on the order of milliseconds [13]. It has been suggested that this period is necessary for the membrane -adsorbed peptide to rearrange its configuration in the membrane, to form the channel. One of the rearrangement of membrane-bound R-S4 is the penetration into the membrane.

3.2. Movement of R-S4 with Voltage Gating

Changes in the intensity of the fluorescence of R-S4 in a planar membrane due to voltage gating were measured. When a steady current-voltage state had been reached, a rectangular depolarizing or hyperpolarizing command voltage with a duration of 1.75 s was applied to the planar bilayer (Figure 4). Changes in the intensity of fluorescence of R-S4 in a planar membrane when the command voltage lasted between 100msec and 10s were measured. With a shorter duration, the statistical deviation of the relative changes in the intensity of fluorescence became larger. With a longer duration, we could measure the smaller changes with better resolution, but lost the temporal resolution. With a duration of 1.75 s, the statistical deviation of the relative changes in the intensity of fluorescence was as small as 0.3%, which was small enough to detect the changes in the intensity of fluorescence due to the voltage gating. The background fluorescence was measured after each experiment by rupturing the membrane. It was smaller than 5% each time, so the observed fluorescence was almost all from the membrane-incorporated R-S4. The drift of the intensity of fluorescence was less than 20 counts/10 ms per minute with each experiment. The I-V characteristics of the R-S4 did not change according to the presence or absence of excitation irradiation. The mean values of the intensity of fluorescence and the current during the command were calculated and compared with those from the resting state (V = 0 mV) are shown in Table 1. As a control experiment, 1 M NaCl was added to both sides of the chamber (Figure 4A). In this case, the relative changes in the intensity of fluorescence following hyperpolarizing and depolarizing voltages were smaller than 0.2%, which is within the statistical deviation. To study the changes in the environment surrounding R-S4 with voltage gating, fluorescence-quenching agents were added to the chamber. When the quenching agent was added, a significant change in the intensity of fluorescence was detected with the application of a depolarizing voltage that generated an ionic current. When the aqueous quencher potassium iodide was added, the intensity

Figure 4 Simultaneous measurements of ionic current and the intensity of fluorescence with voltage clamp. A rectangular depolarizing or hyperpolarizing command voltage, ±150 (mV) with a duration of 1.75 s was applied to the planar bilayer. (A) 1 M sodium chloride on both sides of the chamber; (B) 0.5 M potassium iodine and 50 mM sodium chloride on both sides of the chamber; and (C) 10 mol % 5-SLPC in the *trans* side of the membrane, and 1 M sodium chloride on both sides of the chamber. In all cases, 50 mM Tris buffer (pH = 7.5) was used; solid lines, ionic current (nA); dots, intensity of fluorescence (counts/10 ms); dashed lines, command voltage ±150 (mV). Reproduced with permission from *Biochemistry* 1998, 37, 15376-15382. Copyright 1998 Am.Chem.Soc.

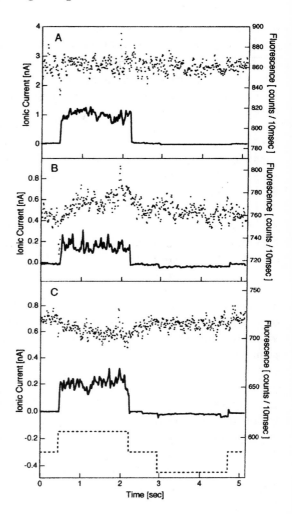

of fluorescence increased by 0.98% at the depolarizing voltage (Figure 4B). When the lipophilic quencher (1-palmitoyl-2-stearoyl(5-DOXYL)-sn-glycero-3-phosphocholine; 5-SLPC) was on the *trans* side of the bilayer membrane, the intensity of fluorescence decreased by 1.22% at the depolarizing voltage (Figure 4C). However, at the hyperpolarizing voltage, which does not generate an ionic current, no changes in the intensity of fluorescence were observed when the fluorescence quencher was used on either side of the membrane. These results show that the fluorophore of the N-terminal of S4 migrates into the membrane when the channel opens with depolarization and generates an ionic current.

Table 1   Changes of Ionic Current and Fluorescence Intensity with Voltage Gating

environment	command voltage (mV)	fluorescence (count/10 ms)	δF (%)	ionic current (pA)
NaCl	0	861.7 ± 1.6		0.0 ± 2.1
(control)	+150	863.0 ± 1.5	0.15 ± 0.25	982.6 ± 23.3
	-150	860.3 ± 1.4	-0.16 ± 0.25	-35.9 ± 0.5
KI	0	763.5 ± 1.4		0.0 ± 3.6
	+150	771.0 ± 1.4	0.98 ± 0.26	163.3 ± 9.9
	-150	763.4 ± 1.3	-0.02 ± 0.25	-30.3 ± 1.1
5-SLPC	0	717.1 ± 1.5		0.0 ± 2.8
	+150	708.3 ± 1.2	-1.22 ± 0.27	201.9 ± 7.4
	-150	714.9 ± 1.2	-0.31 ± 0.27	-18.9 ± 1.0

The data of the intensity of fluorescence and the current in Figure 4A-C were averaged during the command. δF is the relative change of the intensities of fluorescence at the command which was normalized with those from the resting state (V = 0 mV). Reproduced with permission from *Biochemistry* 1998, 37, 15376-15382. Copyright 1998 Am.Chem.Soc.

The degree of movement in the membrane can be estimated from the experiment involving depth-dependent fluorescence quenching of a tryptophan residue on the ColicinE1 channel [14]. In this paper, the authors studied the quenching of tryptophan fluorescence from single-tryptophan mutants of the ColicinE1 channel, using a lipophilic quencher (5-SLPC and 12-SLPC). When 10 mol % SLPC was present in the liposomes, the same amount as in our experiment, the intensity of fluorescence with 5-SLPC was compared with that with 12-SLPC. The intensities of fluorescence were reduced by 4% and 15% for tryptophan 355 and 460, respectively. The difference in the depth of the quenchers between 5-SLPC and 12-SLPC was about 6.5 Å. The intensity of the fluorescence was reduced by approximately 1-2%/Å. The changes that we detected were about 1% at 150 mV. If all the peptides in the membrane formed the voltage-dependent channel, the movement was estimated to be only 1 Å in depth. This value seems to be too small for gating. Most of the fluorescence might come from the membrane-incorporated peptides that do not form voltage-dependent channels. John and Jähnig [15] studied the aggregation states of melittin in lipid vesicle membranes with fluorescence energy transfer and revealed that the percentage of aggregated melittin was on the order of 10%. If this is also true in our case, the movement with voltage gating is larger than 1 Å. On the other hand, when the quencher was in the aqueous phase, the intensity of fluorescence increased by 0.98% with current generation. This change can be interpreted as the movement of fluorophores into the interior of the membrane when voltage gating opens a channel. It is not possible to estimate the movements of the peptide quantitatively in this case. However, the opposite sign of the changes with aqueous and lipophilic quenchers revealed that the N-terminal of the S4 peptide moves into the interior of the membrane with voltage gating. Larger changes of fluorescence with the movement of the fluorophore are expected when the quencher is located near the fluorophore. In our experiment, the lipophilic quencher was located on the trans side of the membrane while the peptide was added to the *cis* side of the chamber. If the fluorophore is located in the *cis* membrane, the lipophilic quencher should also be on the cis side of the membrane to obtain the larger signals. We could not measure the signals, since the planar lipid bilayer was unstable under these conditions.

4. Conclusion

We developed an experimental system that can measure spectroscopic and physiological signals simultaneously from ion channels reconstituted in a planar lipid bilayer, to study the relationship between the structure and function of the ion channels. With this system, structural changes of ion channels can be detected with precise spatio-temporal control of their functional state. While the membrane potential was clamped, fluorescent emission and ionic currents were measured simultaneously. The fluorescent emissions from a planar bilayer constructed in a specially designed chamber were monitored exclusively, and the signal intensity was measured with a photon-counting system. The intensity of fluorescence and spectral shape were measured successfully from the planar bilayer, with a high signal-to-noise ratio. The system can measure the intensity of fluorescence from a restricted area of the planar bilayer, with a diameter of 70μm and a focal depth of 15μm. The low background signal was achieved by optimizing the optical system. More than 95% of the measured fluorescence comes from the planar lipid bilayer. A 22-mer peptide with a sequence identical to that of the S4 segment of the electric eel sodium channel domain IV was synthesized and fluorescence-labeled. This peptide formed a voltage-dependent ion channel in a planar bilayer. The changes in the intensity of the fluorescence accompanying ionic currents generated by a voltage clamp suggest that voltage gating involves the insertion of the N-terminal of the peptide into the membrane. The electrical and optical signals were measured with a gate time of 10 ms. This measurement enabled the detection of movement of the membrane-incorporated peptides with channel opening. This method should prove to be a useful tool for elucidating structure-function relationships for other membrane proteins.

Confocal design of optical system enables the measurement of fluorescence exclusively from a planar bilayer. But the signal/noise ratio needs to be improved to detect structural changes more precisely. There are some points to be improved in the system. It is necessary to collect a larger amount of fluorescent emission. One way is using the objective lens with a large aperture. This makes not only the signal larger but also the signal-to-noise ratio much higher, due to the smaller focal depth. But it is necessary to develop the special lens with an appropriate compensation for the thick layer of water in the chamber. Another way is reducing the distance between the objective lens and the membrane. It makes also possible to collect a larger amount of fluorescent emission. But it would be very difficult to form a folded membrane in this type of chamber.

Other spectroscopic techniques should be applied for further improvement. By using an environment-sensitive [16] or conformation-sensitive [17] dye instead of rhodamine, changes in the polarity around the fluorophore could be determined more precisely. Also, information on the degree of insertion could be obtained quantitatively using parallax methods [18-20]. Furthermore, to gain insight into the ion channel properties of an intact ion channel, techniques used to introduce a probe into an active channel and to reconstitute the labeled channel in planar bilayers with a proper orientation should be developed [21-23]. Further improvement of the system will make it possible to measure fluorescent and electrical signal from the single channel in a planar bilayer. This would elucidate the molecular dynamics of channel gating. A detailed study of the structure-function relationship of voltage-gated channels will not only lead to valuable insights regarding voltage gating mechanism of ion channels, but should also broaden the knowledge of dynamics of membrane proteins.

REFERENCES

[1] B.A. Yi, D.L. Minor Jr, Y.F. Lin, Y.N. Jan and L.Y. Jan, Proc. Natl. Acad. Sci. U.S.A. 98 (2001) 11016.
[2] A.O. Grant, Am. J. Med. 110(2001) 296.
[3] D.A. Doyle, J.Morais Cabral, R.A. Pfuetzner, A .Kuo, J.M. Gulbis, S.L. Cohen, B.T. Chait and R. MacKinnon, Science 280 (1998) 69.
[4] E. Perozo, L .Santacruz-Toloza, E. Stefani, F. Bezanilla and D.M. Papazian, Biophys. J. 66 (1994) 345.
[5] C. Miller (eds.), Ion Channel Reconstitution, Plenum Press, New York, 1986
[6] Y. Hanyu, T. Yamada and G. Matsumoto, Biochemistry 37 (1998) 15376.
[7] M. Noda, S. Shimizu, T. Tanabe, T. Takai, T. Kayano, T. Ikeda, H. Takahasi, H. Nakayama, Y. Kanaoka, N. Minamino, K. Kanagawa, H. Matsuo, M.A. Raftery, T. Hirose, S. Inayama, H.

Hayasida, T. Miyata and S. Numa, Nature 312 (1984) 121.

[8] J.S.,Huebner, Photochem. Photobiol., 30 (1979) 233

[9] L. Kunz and G. Stark, Biochem. Biophys. Acta 1327 (1997) 1.

[10] M.T. Tosteson, D.S. Auld and D.C. Tosteson, Proc. Natl. Acad. Sci. U.S.A. 86 (1989) 707.

[11] M. Brullemans, O. Helluin, J.-Y. Dugast, G. Molle and H. Duclohier, Eur. Biophys. J. 23 (1994) 39.

[12] M. Pawlak, S. Stankowski and G. Schwarz, Biochem. Biophys. Acta 1062 (1991) 94.

[13] G. Schwarz and G. Beschiaschvili, Biochem. Biophys. Acta 979 (1989) 82.

[14] L.R. Palmer and A.R Merril, J.Biol.Chem. 269 (1994) 4187.

[15] E. John and F. Jähnig, Biophys. J. 60 (1991) 319.

[16] D. Rapaport, M. Danin, E. Gazit and Y. Shai Biochemistry 31 (1992) 8868.

[17] T. Yamamoto and N. Ikemoto, Biochemistry 41 (2002) 1492.

[18] S.E. Malenbaum, A.R. Merrill and E.London, J Nat Toxins 7 (1998) 269.

[19] X. Chen, D.E. Wolfgang and N.S. Sampson, Biochemistry 39 (2000) 13383.

[20] F.J. Barrantes, S.S. Antollini, M.P. Blanton and M. Prieto, J. Biol. Chem. 275 (2000) 3733.

[21] A. Cha and F. Bezanilla, Neuron 5 (1997) 1127.

[22] K.S. Glauner, L.M. Mannuzzu, C.S. Gandhi and E.Y. Isacoff, Nature 402 (1999) 813.

[23] C.S. Gandhi, E. Loots and E.Y. Isacoff, Neuron 3 (2000) 585.

INDEX